M. Schidlowski, S. Golubic, M.M. Kimberley,
D.M. McKirdy, P.A. Trudinger (Eds.)

Early Organic Evolution

Implications for Mineral and Energy Resources

With 256 Figures and 50 Tables

Springer-Verlag
Berlin Heidelberg New York
London Paris Tokyo
Hong Kong Barcelona Budapest

Managing Editor

Prof. Dr. Manfred Schidlowski
Chairman, IGCP Project 157

Max-Planck-Institut für Chemie (Otto-Hahn-Institut),
Saarstr. 23, W-6500 Mainz, FRG

Editors

Prof. Dr. Stjepko Golubic
Biological Science Center
Boston University
Boston, MA 02215, USA

Prof. Dr. Michael M. Kimberley
Dept. of Marine, Earth and
Atmospheric Sciences
North Carolina State University
Raleigh, NC 27695-8208, USA

Dr. David M. McKirdy, Sr. Lecturer
Dept. of Geology & Geophysics
University of Adelaide
Adelaide, S.A. 5001, Australia

Dr. Philip A. Trudinger
Bureau of Mineral Resources
Division of Continental Geology
GPO Box 378
Canberra, A.C.T. 2601, Australia

Proceedings of the 9th Alfred Wegener Conference being the Final Meeting of IGCP-Project 157 (Early Organic and Mineral & Energy Resources), Maria Laach, FRG, September 19–23, 1988.

Library of Congress Cataloging-in-Publication Data
Alfred Wegener-Conference (9th : 1988 : Abtei Maria Laach)
Early organic evolution : implications for mineral and energy resources : proceedings of the 9th Alfred Wegener Conference, being the final meeting of IGCP-Project 157 (early organic evolution and minerals & energy resources), Maria Laach, FRG, September 19–23, 1988 / M. Schidlowski ... [et al.] (eds).
Includes index.
ISBN-13: 978-3-642-76886-6 e-ISBN-13: 978-3-642-76884-2
DOI: 10.1007/ 978-3-642-76884-2

1. Organic geochemistry-Congress. 2. Geology, Stratigraphic-Precambrian-Congresses. 3. Geology, Stratigraphic-Paleozoic-Congresses. 4. Stromatolites-Congresses. 5. Paleopedology-Congresses. 6. IGCP Project 157-Congresses. I. Schidlowski, M. II. IGCP Project 157. III. Title. QE516.5.A44 1988 553'.1-dc20

Typesetting: Fotosatz-Service Köhler, Würzburg

32/3020-543210 – Printed on acid-free paper

Preface

This volume is the final outcome of a conference designed to wrap up IGCP Project 157 ("Early Organic Evolution and Mineral and Energy Resources") after a decade of prolific activity. The picturesque solitude of Maria Laach Abbey in the Eifel Mountains (FRG) provided the appropriate setting for a conclave of some 80 specialists from the various walks of the field who, during the week of Sept. 19–23, 1988, strived hard to define the state of the art in the principal segments of this Earth Science frontier.

The following pages contain the essence of the conference transactions, giving a vivid cross-section of the activities pursued by IGCP Project 157 during its final years. The coverage of topics is not necessarily complete, but rather eclectic in part. With regard to single papers dealing with modern analogues of ancient processes, the book title might even be considered a grave misnomer. Nevertheless, all contributions relate to the subject in the widest sense, and the reader should be reminded that much of the heterogeneity reflected by the volume derives from the fact that it is primarily a research report from a highly interdisciplinary field rather than a textbook. To secure at least some balance between these opposite goals, each of the four subprojects is introduced by one or more review papers that summarize the field as a whole (or major facets thereof) for the benefit of those readers in need of introductory guidance. Considering the principal character of the book as a progress report from an actively evolving field, editorial efforts to eliminate occasional inconsistencies among individual contributions were kept to a minimum, the rationale being that there must be room for different viewpoints and a continuing dialogue.

As is nowadays almost common for conference proceedings, this volume also had an extended gestation period. However, the potentially adverse effects of such delay are largely offset by the fact that most papers do not concern themselves with ephemeral issues, and that the authors were given ample opportunity to apply a final touch to their texts during the copy-editing procedure. Accordingly, we may safely claim that the gist of the concepts set out on the subsequent pages is contemporaneous with the year given on the copyright page. By summarizing our current knowledge of the field, the book as a whole will hopefully serve as a springboard for shifting frontiers further into the unknown. I am convinced that future generations of students interested in early organic evolution and biologically mediated geochemical transformations will detect in this volume several ideas that have stood the test of time.

As organizer of the conference that has materialized in this volume, it remains for me to acknowledge the generous support given by the UNESCO-IGCP Organisation (Paris), the Alfred Wegener Foundation (Bonn), the Deutsche Forschungsgemeinschaft (Bonn) and the Max Planck Society (Munich). Moreover, special thanks are due to my subproject leaders for their unfailing commitment, notably as coeditors of this volume (which, in the case of Dave McKirdy, included the virtual rewriting of five papers of his section!). Hubert

Wiggering volunteered to compile the index, and a legion of reviewers (see separate list) helped to secure a high scientific standard. Last, but not least, I should like to mention the monasterial hospitality offered by the resident Benedictine community, and a donation by Mr. Harald Eckes, Nieder-Olm, which substantially added to both the flavour and the spirits of the meeting.

Mainz, October 1991 Manfred Schidlowski

The Editors would like to greatfully acknowledge the help of the following reviewers:

P. Aharon
R. Alexander
S. M. Awramik
B. D. Batts
R. A. Berner
E. N. Cameron
A. R. Chivas
J. Connan
A. C. Cook
D. J. Des Marais
T. H. Donnelly
S. Eldridge
M. H. Engel
C. B. Foster
H. Friedrichsen
A. P. Gize
S. Golubic
J. D. Gorter
D. E. Grandstaff
J. Hahn
T. C. Hoering
C. F. Hoffmann
H. D. Holland
B. Horsfield
H. Jannasch
R. J. F. Jenkins
R. I. Kagi
M. M. Kimberley
A. H. Knoll
A. Krause-Mirus
G. W. M. Lijmbach
S. A. Macko
D. M. McKirdy
D. A. Mossman
S. Pickering
T. G. Powell
R. Raiswell
G. Retallack
A. E. Ringwood
J. D. Saxby
M. Schidlowski
Shen Su Sun
G. W. Skyring
M. Smyth
B. Spiro
E. F. Stumpfl
R. E. Summons
G. H. Taylor
W. R. Taylor
P. A. Trudinger
H. G. Trüper
J. Veizer
J. K. Volkmann
M. R. Walter
M. A. Wilson

List of Contributors

Aharon, P., Prof. Dr.
Department of Geology and Geophysics, Louisiana State University,
Baton Rouge, LA 70803-4104, USA

Al-Biatty, H. J., Dr.
Department of Earth Sciences, University of Leeds, Leeds, LS2 9JT, UK

Awramik, S. M., Prof. Dr.
Department of Geological Sciences, University of Santa Barbara,
Santa Barbara, CA 93106, USA

Banerjee, D. M., Dr.
Department of Geology, University of Delhi, Delhi 110007, India

Bertrand-Sarfati, J., Dr.
CNRS, Centre Geologique et Geophysique de l'Université des Sciences et
Techniques, Place Eugène Bataillon, 34060 Montpellier Cedex, France

Bharati, S., Dr.
Geology Department, University of Oslo, Postboks 1047, 0316 Blindern,
Oslo 3, Norway

Brongersma-Sanders, M., Dr.
National Naturhistorisch Museum, P.O. Box 9517, 2300 RA Leiden,
The Netherlands

Bustin, M., Dr.
Department of Geological Sciences, University of British Columbia,
Vancouver, V6T 2B4, Canada

Carl, N., M. Sc.
Environmental Science Division, School of Applied Science and Technology,
The Hebrew University, Jerusalem 91904, Israel

Crick, I. H., Dr.
Bureau of Mineral Resources, Division of Continental Geology, GPO Box 378,
Canberra, ACT 2601, Australia

Deb, M., Dr.
Department of Geology, University of Delhi, Delhi 110007, India

Deines, P., Prof. Dr.
Department of Geosciences, The Pennsylvania State University,
University Park, PA 16802, USA

Des Marais, D. J., Dr.
NASA Ames Research, Center, Moffett Field, California 94035, USA

Donnelly, T. H., Dr.
CSIRO Division of Water Resources, P.O. Box 1666, Canberra, ACT 2601,
Australia

Dor, I., Dr.
Environmental Sciences Division, School of Applied Science and Technology,
The Hebrew University, Jerusalem 91904, Israel

Douglas, A. G., Dr.
NRG in Fossil Fuels of Environmental Geochemistry,
University Newcastle-upon-Tyne,
Drummond Building, Newcastle upon Tyne, NE1 7RU, UK

Dypvik, H., Dr.
Geology Department, University of Oslo, Postboks 1047, 0316 Blindern,
Oslo 3, Norway

Eglinton, T. I., Dr.
Organic Geochemistry Unit, Technical University of Delft, 2628 RZ Delft,
The Netherlands

Fan, P., Prof. Dr.
Lanzhou Institute of Geology, Academia Sinica, Lanzhou 730000, PR China

Farrow, C. E. G., Dr.
Department of Geology, Carleton University, Ottawa, Ontario, K1S 5B6,
Canada

Foster, C. B., Dr.
Western Mining Corp. Ltd., Petroleum Division, 28 Ventnor Avenue,
West Perth W.A. 6005, Australia

Fowler, M. G., Dr.
Institute of Sedimentary and Petroleum Geology, Geological Survey of Canada,
3303-33rd St. N.W., Calgary, Alberta T2L 2A7, Canada

von Gehlen, K., Prof. Dr.
Institut für Geochemie, Petrologie und Lagerstättenkunde,
Senckenberg-Anlage 28, W-6000 Frankfurt a.M. 1, FRG

Glaessner, M. F., Prof. Dr. (deceased)
Department of Geology and Geophysics, The University of Adelaide,
Adelaide, SA 5001, Australia

Goodarzi, F., Dr.
Institute of Sedimentary and Petroleum Geology, Geological Survey of Canada,
3303-33rd St. N.W., Calgary, Alberta T2L 2A7, Canada

Golubic, S., Prof. Dr.
Department of Biology, Boston University, Boston, MA 02215, USA

Gorzawski, H., Dr.
Max-Planck-Institut für Chemie, Saarstr. 23, W-6500 Mainz, FRG

Hayes, J. M., Prof. Dr.
Biogeochemical Laboratories, Department of Chemistry and Geology, Indiana University, Bloomington, IN 47405, USA

Hirner, A. V., Prof. Dr.
Mineralogisch-Petrographisches Institut der Universität, Theresienstr. 41, W-8000 München 2, FRG

Holland, H. D., Prof. Dr.
Harvard University, Department of Earth and Planetary Sciences, Cambridge, MA 02138, USA

Horsfield, B., Dr.
Institut für Erdöl und Organische Geochemie, Forschungszentrum Jülich GmbH, Postfach 1913, W-5170 Jülich, FRG

Hutchinson, R. W., Prof. Dr.
Department of Geology and Geological Engineering, Colorado School of Mines, Golden, CO 80401, USA

Imbus, S. W., Dr.
School of Geology and Geophysics, University of Oklahoma, Norman, OK 73019, USA

Kimberley, M. M., Prof. Dr.
Department of Marine, Earth and Atmospheric Sciences, North Carolina State University, Raleigh, NC 27695-8208, USA

Knoll, A. H., Prof. Dr.
Botanical Museum, Harvard University, Cambridge, MA 02138, USA

Lambert, I. B., Dr.
Elisian Resources, 2 Bonwick Place, Garran, Canberra, ACT 2605, Australia

Larter, S. R., Dr.
Geology Department, University of Oslo, Postboks 1047, 0316 Blindern, Oslo 3, Norway

de Leeuw, J. W., Dr.
Organic Geochemistry Unit, Technical University of Delft, 2628 RZ Delft, The Netherlands

Leistner, F., Dr.
Institut für Erdöl und Organische Chemie, Forschungszentrum Jülich GmbH, Postfach 1913, W-5170 Jülich, FRG

Li, J., Dr.
Lanzhou Institute of Geology, Academia Sinica, Lanzhou 730000, PR China

Li, Y., Dr.
Laboratory of Biogeochemistry and Gas Geochemistry, Academia Sinica, Lanzhou 730000, PR China

Li, Z., Dr.
Lanzhou Institute of Geology, Academia Sinica, Lanzhou 730000, PR China

Liew, T. C., Dr.
Max-Planck-Institut für Chemie, Saarstr. 23, W-6500 Mainz, FRG

Littke, R., Dr.
Institut für Erdöl und Geochemie, Forschungszentrum Jülich GmbH, Postfach 1913, 5170 Jülich, FRG

Makarikhin, V., Dr.
Geological Institute, Karelian Branch, USSR Academy of Sciences, Petrozavodsk, 385 610, USSR

Marmo, J. S., Dr.
Geological Survey of Finland, 02150 Espoo, Finland

Marzi, R., Dr.
Institut für Erdöl und Organische Chemie, Forschungszentrum Jülich GmbH, Postfach 1913, W-5170 Jülich, FRG

McKirdy, D. M., Dr.
Department of Geology and Geophysics, University of Adelaide, Adelaide, SA 5001, Australia

Meng, Q., Dr.
Lanzhou Institute of Geology, Academia Sinica, Lanzhou 730000, PR China

Meyers, P. A., Prof. Dr.
Department of Geological Sciences, The University of Michigan, Ann Arbor, MI 48109-1063, USA

Miller, A. R., Dr.
Geological Survey of Canada, 601 Booth Street, Ottawa, Ontario, K1A 0E8, Canada

Mossman, D. J., Prof. Dr.
Department of Geology, Mount Allison University, Sackville, New Brunswick, E0A 3C0, Canada

Nagy, B., Prof. Dr.
Laboratories of Organic Geochemistry, Department of Geosciences, The University of Arizona, Tucson, AZ 85721, USA

Needham, R. S., Dr.
Bureau of Mineral Resources, Geology and Geophysics, GPO Box 378, Canberra, ACT 2601, Australia

Neumann-Mahlkau, P., Prof. Dr.
FB 9, Geologie, Universität, Gesamthochschule Essen, Postfach 103764, W-4300 Essen 1, FRG

Ohmoto, H., Prof. Dr.
Department of Geosciences, Pennsylvania State University, University Park, PA 16802, USA

Pentecost, A., Dr.
Division of Biosphere Sciences, King's College, Campden Hill Road, London W8 7AH, UK

Pflug, H. D., Prof. Dr.
Geologisches Institut der Justus-Liebig-Universität, W-6300 Gießen, FRG

Powell, T. G., Dr.
Bureau of Mineral Resources, GPO Box 378, Canberra, ACT 2601, Australia

Premović, P. I., Prof. Dr.
Laboratory for Geochemistry, Department of Chemistry, Faculty of Sciences, University of Niš, 18000 Niš, Yugoslavia

Raiswell, R., Dr.
Department of Earth Sciences, University of Leeds, Leeds, LS2 9JT, UK

Reimer, T. O., Dr.
Bernhard-May-Str. 43, W-6200 Wiesbaden, FRG

Reitz, E., Dr.
Geologisches Institut der Justus-Liebig-Universität, W-6300 Gießen, FRG

Retallack, G. J., Prof. Dr.
Department of Geology, University of Oregon, Eugene, OR 97403-1272, USA

Robinson, B. W., Dr.
Institute of Nuclear Sciences, DSIR, P.O. Box 31312, Lower Hutt, New Zealand

Ronov, A. B., Prof. Dr.
Vernadsky Institute of Geochemistry and Analytical Chemistry, USSR Academy of Sciences, 117975 Moscow V-334, USSR

Rullkötter, J., Dr.
Institut für Erdöl und Organische Geochemie, Forschungszentrum Jülich GmbH, Postfach 1913, W-5170 Jülich, FRG

Schenk, H. J., Dr.
Institut für Erdöl und Organische Geochemie, Forschungszentrum Jülich GmbH, Postfach 1913, W-5170 Jülich, FRG

Schidlowski, M., Prof. Dr.
Max-Planck-Institut für Chemie, Saarstr. 23, W-6500 Mainz, FRG

Selbach, H.-J., Dipl.-Geol.
FB 9, Geologie, Universität, Gesamthochschule Essen, Postfach 10 37 64,
W-4300 Essen 1, FRG

Shen, P., Dr.
Laboratory of Biogeochemistry and Gas Geochemistry, Academia Sinica,
Lanzhou 730000, PR China

Sinninghe Damsté, J. S., Dr.
Organic Geochemistry Unit, Technical University of Delft, 2628 RZ Delft,
The Netherlands

Strauss, H., Dr.
Ruhr-Universität Bochum, Institut für Geologie, W-4630 Bochum 1, FRG

Stuart-Smith, P. G., Dr.
Bureau of Mineral Resources, Geology and Geophysics, GPO Box 378, Canberra,
ACT 2601, Australia

Summons, R. E., Dr.
Bureau of Mineral Resources, Geology and Geophysics, GPO Box 378, Canberra,
ACT 2601, Australia

Trudinger, P. A., Dr.
Bureau of Mineral Resources, Division of Continental Geology, GPO Box 378,
Canberra, ACT 2601, Australia

Wiggering, H., Dr.
FB 9, Geologie, Universität, Gesamthochschule Essen, Postfach 10 37 64,
W-4300 Essen 1, FRG

Xu, Y., Prof.
Laboratory of Biogeochemistry and Gas Geochemistry, Academia Sinica,
Lanzhou 730000, PR China

Yu, X., Dr.
Lanzhou Institute of Geology, Academia Sinica, Lanzhou 730000, PR China

Table of Contents

3. Antiquity and Metallogenic Potential of Bacterial Sulfate Reduction

4. Fossil Microbial Ecosystems of the "Stromatolitic" Type and Their Modern Analogs

Introduction

Early Organic Evolution and Mineral and Energy Resources: A Farewell Address to IGCP Project 157

Manfred Schidlowski

Chairman, IGCP Project 157, Max-Planck-Institut für Chemie, Saarstr. 23, W-6500 Mainz, FRG

The Final Meeting of IGCP Project 157 in the form of the 9th Alfred Wegener Conference (Maria Laach, Sept. 19–23, 1988) was pre-eminently an occasion to draw a balance sheet of the work performed in a scientific venture that, in the terminal assessment of the IGCP Board (1989), has been rated "a highly successful project producing a number of major advances in the field and resulting in a number of important publications." Since the inauguration of the project in the late 1970s, our project community had understood itself as an interdisciplinary task force bent on exploring those blurred areas in which economic geology, organic geochemistry, and evolutionary biology seem to blend and overlap. At this stage, a brief look back at a decade's activities would seem in order, to retrace recent developments at one of the most fascinating frontiers in Earth Sciences.

IGCP Project 157 had been launched in 1977 by an Australian steering committee composed of D. M. McKirdy, P. A. Trudinger and M. R. Walter, with M. F. Glaessner as adviser (see Walter et al. 1979). On a first meeting of the project community at the occasion of the 10th International Sedimentology Congress (Jerusalem 1978), chairmanship was entrusted to M. Schidlowski, who immediately proceeded with the organization of an international conference in the form of a Dahlem Workshop. Apart from the release of a series of newsletters edited by J. H. Oehler and A. Button, progress reports were published at regular intervals (Oehler and Schidlowski 1980; Schidlowski 1983, 1985, 1987a). Detailed accounts of the principal activities have been given in a fair number of project-related monographs and book publications (inter alia Holland and Schidlowski 1982; Nagy et al. 1983; Retallack 1986; Johns 1986). Moreover, aspects of the work of Project 157 have figured prominently in other recent publications (e.g., Schopf 1983; Cohen et al. 1984; Friedman and Krumbein 1985). Since 1979, the project had organized a series of regional and international conferences some of which have proved seminal in the field [notably the Dahlem Workshop "Mineral Deposits and the Evolution of the Biosphere" (Berlin, 1980) and a conference "Developments and Interactions of the Precambrian Atmosphere, Lithosphere, and Biosphere" (Mexico City, 1982)].

In accordance with the "Vernadskyan concept" that views life as a powerful geochemical agent responsible for a number of large-scale transformations in the exogenic cycle (Vernadsky 1930), the project community had concerned itself with the geochemical impact of the Earth's biota on processes of exogenic mineral and hydrocarbon formation. By virtue of their accumulation of negative entropy, living systems are apt to impose on their surroundings a marked thermodynamic gradient which is the driving force for a number of quantitatively important and environmentally relevant geochemical reactions. Most conspicuous among these are oxidation processes driven by the redox imbalance imposed on terrestrial near-surface environments by photosynthetic oxygen. Another biologically mediated reaction is the dissimilatory breakdown of organic matter by sulfate, which releases large quantities of hydrogen sulfide to the aquatic (specifically marine) realm.

Within the broad frame of biologically powered geochemical processes, the work of IGCP Project 157 has particularly addressed the timing of major quantum steps in early bioenergetic evolution such as the emergence of oxygen-evolving photosynthesis and of dissimilatory sulfate reduction. It is self-evident that specifically these two biological innovations were bound to induce profound changes in the chemical regime at the Earth's surface by the release to the environment of molecular oxygen and hydrogen sulfide in geochemically relevant quantities. This, in turn, was apt to significantly affect surface processes including those responsible for the formation of economic mineral and fossil fuel deposits.

Since the initiation of the project, it had been practical to pursue relevant investigations in four working groups or subprojects, each of which focusing on a key topic of the field. In the following, a brief

M. Schidlowski et al. (Eds.)
Early Organic Evolution: Implications for Mineral and Energy Resources

characterization will be given of these subprojects along with their principal achievements.

1 Precambrian Weathering Horizons

(co-ordinator: M. M. Kimberley)

Although this subject had figured very prominently in the original working program (cf. Walter et al. 1979), corresponding activities remained, unfortunately, marginal to the mainstream of the work of the project until the mid-1980s. The significance for paleoenvironmental research of ancient weathering horizons derives from their character as *reaction rims* formed at the interface between crust and atmosphere plus hydrosphere in which the liquid and gaseous reactants involved in these processes (H_2O, CO_2, O_2, etc.) have entered, and been fixed in, solid mineral phases. Accordingly, fossil weathering profiles are storing (albeit selectively) information on the composition of the Earth's volatile envelope. Consequently, Precambrian weathering horizons and notably fossil soils ("paleosols") may hold crucial clues to atmospheric evolution during the Earth's early history. Unfortunately, the primary chemistry of weathering processes tends to be largely clouded or even obliterated during lithification processes since diagenetic reconstitution of sediments usually takes place in a reducing environment. As a result, the role played by oxygen in the original weathering process may be thoroughly masked, with the Fe^{3+} component of the primary detritus often completely reduced to bivalent iron.

With the aim of a first systematic review of the field, Project 157 had sponsored an "*International Symposium on Precambrian Paleosols*" organized by M. M. Kimberley at North Carolina State University, Raleigh, N. C. (1985). Being the first of its kind, this conference has clearly set a milestone in paleosol research, its proceedings having appeared as a Special Issue of *Precambrian Research* (Retallack 1986). The papers assembled in this volume, as well as those appearing in Section 1 of the present book, reflect the current status of the topic and will serve to both stimulate and direct further work in this field.

2 Organic Constituents of Precambrian and Early Paleozoic Sediments with Special Reference to Pre-Devonian Oils

(co-ordinator: D. M. McKirdy)

A principal achievement of this subproject over the last decade was the definitive establishment of a continuous organic record in the form of kerogen and derivative graphite over 3.8 Ga of geological history, indicating that microbial (prokaryotic) life had been extant on Earth not long after its formation. Moreover, the bias in favor of the light carbon isotope (^{12}C) inherent in the principal carbon-fixing enzymatic reaction of the photosynthetic pathway could be traced back in sedimentary organic matter over practically the same time span (Hayes et al. 1983; Schidlowski et al. 1983; Schidlowski 1987b; see also chapters by Schidlowski and Aharon, and McKirdy and Imbus, this Vol.). These findings give eloquent testimony to the continuity on this planet of photoautotrophy as both a biochemical process and a geochemical agent. A project-sponsored workshop addressing these and related questions (*Organic Matter in Sediments: The Molecular and Isotopic Record of Life over the Last 3.8 Ga*) had been organized by P. Albrecht, G. Eglinton, and M. Schidlowski at the occasion of the 2nd EUG Meeting in Strasbourg, 1983 [see meeting report in *Episodes, 1983* (3), p. 44]. At the 27th International Geological Congress in Moscow (1984), M. Schidlowski and L. M. Mukhin had convened an Intersectional Symposium entitled *Origin and Evolution of Life on Earth* [cf. *Episodes, 1984* (4), p. 39].

On the applied side, several members of the community (D. M. McKirdy, R. E. Summons, T. G. Powell, M. G. Fowler, and others) had devoted themselves to a thorough survey of the biomarker geochemistry of ancient kerogens and oils as a potential exploration tool in the search for pre-Devonian (including mid- to late Proterozoic) hydrocarbon deposits. Biological markers or "chemofossils", respectively, are molecules or molecule fragments of the biological source materials that are capable of withstanding humification process and subsequent diagenetic alterations in the sediment with little or no change. Prior to the appearance of widespread Devonian continental floras, sedimentary organic matter was almost solely derived from bacteria, algae, and marine invertebrates, all of which have left specific sets of biomarker compounds in the record.

As a result of the work performed by our project community, it is safe to say that the Earth's early microbial biomass has contributed substantially to the hydrocarbon-generating potential of the sedimentary shell and is specifically to be credited for the formation of the oil and gas occurrences of late Proterozoic and early Paleozoic rocks. Inter alia, it has been possible to identify individual microbial species and genera as key contributors to oils and oil-prone kerogens of this age. It was, moreover, shown that Proterozoic oils and sediment extracts commonly contain only low abundances of the "classical" biomarker molecules (such as steranes and hopanes), but are rich in iso- und anteiso-alkanes, alkylcyclohexanes and mid-chain mono-methyl alkanes all pointing to bacterial and cyanobacterial progenitor materials. This work has resulted in a wealth of publications a major part of which have been listed in the *IGCP Catalogue*, Vol. II (1983). A project-related monograph on biological markers in general has been recently edited by Johns (1986).

The formidable collection of papers presented in Section 2 of this Volume gives a vivid and updated summary of this fascinating field, underlining the geochemical role and economic potential of those organic materials that were destined to end up as constituents of ancient sedimentary rocks.

3 Antiquity and Metallogenetic Potential of Bacterial Sulfate Reduction

(co-ordinator: P. A. Trudinger)

The quest for the antiquity, and the evaluation of the metallogenetic potential, of bacterial (dissimilatory) sulfate reduction had been in the focus of IGCP Project 157 since its very beginnings. Dissimilatory sulfate reduction is an energy-yielding process that couples the reduction of sulfate to hydrogen sulfide with the oxidation of organic substances (e.g., $2CH_2O + SO_4^{2-} \rightarrow H_2S + 2HCO_3^-$), which may be regarded as a form of anaerobic respiration since sulfate instead of oxygen serves as oxidant in this reaction. Although utilized by only few bacterial genera that are mostly restricted to the anaerobic layer of the littoral sea floor, this process is of paramount importance from the geochemical point of view as it catalyzes a large-scale conversion of sulfate to sulfide under the low-temperature regime of terrestrial near-surface environments (cf. Trudinger 1979). This conversion is a pivotal link in the global sulfur cycle but, for thermodynamic reasons, depends on biological mediation at temperatures below 150 °C.

It is well established that the large quantities of hydrogen sulfide released by this process display $\delta^{34}S$ values typically between 30 and 60‰ more negative than those of the parent seawater sulfate. The isotopically light (^{32}S-enriched) H_2S thus generated is, in turn, instrumental in precipitating iron and various base metals in the form of sulfide minerals (mostly pyrite) which basically inherit the isotopic composition of the bacteriogenic H_2S-precursor. Accordingly, sedimentary sulfides commonly exhibit the isotopic signature of bacterial sulfate reduction.

Being responsible for a large-scale scavenging of dissolved iron and base metals from seawater, the economic potential of this process is quite formidable. Its role in the formation of sediment-hosted (stratiform) sulfide deposits is necessarily conditional on the previous emergence of the underlying bioenergetic mechanism. Therefore, efforts of Project 157 have preferentially centered on pinning down the advent of dissimilatory sulfate reduction in the general context of the evolution of bioenergetic processes. Presently available isotopic evidence (Fig. 1) would suggest that bacteriogenic sulfur isotope patterns in sedimentary sulfides made their appearance about 2.8 Ga ago which is astoundingly late in view of the fact that photosynthesis almost certainly arose 1 billion years earlier (dissimilatory sulfate reduction is commonly regarded as an adaptive reversal of bacterial photosynthesis that utilizes H_2S as a reductant instead of H_2O.)

It is, moreover, well known that sulfate was available in the ancient seas as from at least 3.5 Ga ago. Therefore, it is most likely that early tendencies to bring about a microbially mediated isotopic differentiation of marine sulfur were either suppressed or camouflaged by processes or environmental conditions on the Archaean Earth that are as yet not fully understood. It has been argued elsewhere (Schidlowski 1987c) that a large-scale isotopic reequilibration between marine sulfur and the primordial sulfur reservoir of the early oceanic crust could have possibly tethered the $\delta^{34}S$ values of both sulfide and sulfate closely to a mean of about zero permil (Fig. 1, Nos. 1–5). With vast areas of newly formed oceanic crust exposed to the Archaean oceans, and a substantially higher heat flow, basalt-seawater interactions should have been at a pitch in those times, resulting in a largely mantle-dominated chemistry of the ancient seas. Alternatively, and perhaps even more plausibly, Ohmoto and Felder (1987) have invoked a combination of increased SO_4^{2-} reduction rates in the surficial sedimentary layer of a warmer Archaean ocean, and a system closed or semi-closed to sulfate (due to the absence of a burrowing fauna), to account for the markedly reduced isotopic fractionation between sulfide and sulfate observed in the oldest sediments (cf. Fig. 1). In any case, it seems well established that bacterial sulfate reduction did not develop its full metallogenetic potential prior to Proterozoic times, since sedimentary sulfide deposits of commercial importance have invariably proved to be younger than 2 Ga.

Work of the project community during the last decade was primarily concerned with the antiquity of dissimilatory sulfate reduction, the economic assessment of the geologically oldest stratiform base metal sulfides, and questions related to the microbial control

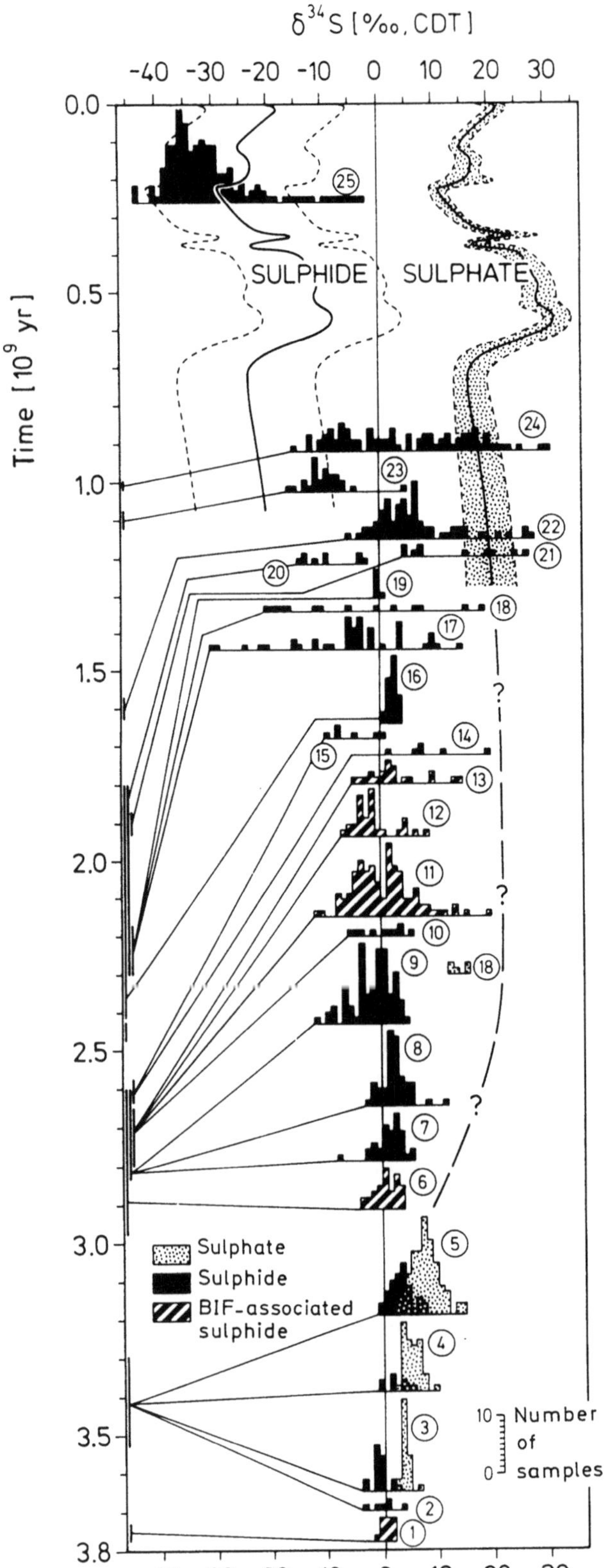

of the global sulfur cycle. Part of these investigations were summarized on a project-sponsored symposium *Sulfide Mineralization in Sediments: Current Status of Syngenetic Theory* (Canberra, 1981) and published as Special Issue of the BMR Journal of Australian Geology and Geophysics (Trudinger and Walter 1981). The seven contributions assembled in Section 3 of the present Volume follow up selected aspects of the field in the widest sense.

4 Fossil Microbial Ecosystems of the "Stromatolitic" Type and Their Modern Analogues

(co-ordinators: S. Golubic and W. E. Krumbein)

The investigation of fossil laminated ("stromatolite-type") microbial ecosystems is of immediate relevance for the assessment of the oil-bearing potential of pre-Devonian rocks since prokaryotic and eukaryotic microorganisms constituted the principal source material of the geologically oldest hydrocarbon assemblages. As part of the work of this subproject, a concerted effort had been made to systematically explore the paleoecology and organic geochemistry of benthic microbial (bacterial and algal) communities from both ancient and contemporary sedimentary environments to elucidate their roles as potential petroleum generators and metal accumulators. This work has yielded a large body of data that serve to vividly illustrate the impact of the microbial world on sediment-forming processes.

It could be demonstrated that lithified microbial mats and related biosedimentary structures ("stromatolites"), apart from representing the oldest un-

Fig. 1. Isotopic evolution of sedimentary sulfide and sulfate over 3.8 Ga of Earth history. Note that the microbially induced isotopic differentiation between the two sulfur species tends to fade as from about t > 1.3 Ga (mainly as a result of the low preservation potential of sulfate evaporites). Therefore, the oldest isotopic evidence of bacterial sulfate reduction primarily rests on the $\delta^{34}S$ distribution patterns of bacteriogenic sulfides that are characterized by extended spreads coupled with a "Rayleigh tail" at the positive end reflecting a reservoir effect [such as exemplarily displayed by the Permian Kupferschiefer of Central Europe (No. 25)]. Applying these criteria to the early record, the presumably oldest bacteriogenic sulfide patterns appear around 2.7–2.8 Ga ago [Deer Lake Greenstone Belt, Minnesota (No. 8); Michipicoten, Woman River and Lumby-Finlayson Lakes banded iron formations, Canada (Nos. 11–13)]. *BIF* = banded iron formation. (Schidlowski 1987c) equivocal morphological evidence of life, figure among the most conservative and persistent biogenic features of the record, covering a time span of some 3.5 Ga from their first appearance in the Archaean to their manifestations in the present world (Walter 1983). It has been shown, furthermore, that the ecology and physiology of extant mats serve as excellent model systems, helping to better understand their fossil progenitors. Corresponding investigations have principally addressed cyanobacterial ecosystems from coastal hypersaline habitats of the Sinai Peninsula (Solar Lake, Gavish Sabkha), the Persian Gulf (Abu Dhabi), Lower California (Laguna Mormona), Western Australia (Shark Bay), as well as from artificial hypersaline ponds. These studies have precipitated a large number of publications including a monograph on the Gavish Sabkha, a small-scale microbial ecosystem situated on the Sinai coast (Friedman and Krumbein 1985). Another seminal book publication dealing with the stromatolite subject in the widest sense has been edited by Cohen et al. (1984). As a result of this and related work, our perception of the nature and the geochemical role of benthic microbial ecosystems has considerably improved during the last decade, furnishing a wealth of new information on the subject from microbial taxonomy to methanogenesis. This has greatly enhanced our understanding of the geochemical significance of microbially mediated geochemical transformations in the exogenic cycle during the first 3 Ga of recorded Earth history when microbiota (notably prokaryotes) held dominion over the terrestrial biosphere.

At the occasion of the 12th International Sedimentological Congress (Canberra, 1986), IGCP Project 157 had co-sponsored a symposium *Microorganisms and the Sedimentary Record* (conveners C. Monty and M R. Walter) that was devoted to a review of the totality of interactions between the microbial world and sedimentary processes. The stromatolite topic was also in the focus of another project-sponsored workshop (*Paleomicrobiology and Paleoenvironments*) convened by M. Schidlowski as part of the 8th International Symposium on Environmental Biogeochemistry (Nancy, 1987). The papers appearing in Section 4 of this Volume stand in the continuation of previous work, either summarizing major parts of the investigations conducted during the last decade or addressing single important problems posed by extant and fossil microbial ecosystems and the corresponding paleontological record.

Summary and Outlook

As this brief outline of the activities of IGCP Project 157 has shown, important advances have been made since investigations commenced in 1978. A graphic synopsis of the current state of our knowledge is given in Fig. 2, where the records of microbial life and the concomitant biogeochemical evidence are traced back over 3.8 Ga of recorded Earth history.

Summing up, there is no doubt that the original project proposal drafted by the Australian founding team had exerted a considerable catalytic impact on the work pursued at this frontier. Altogether, the working program had constituted an ideal blend of basic and applied research, one line probing into the very nature and evolution of life and its interactions with geochemical cycles, the other aiming at an

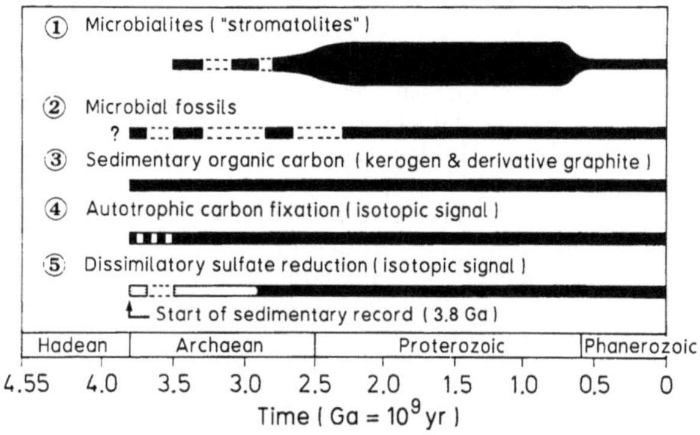

Fig. 2. Record of fossil microbial life (*1, 2*) and selected categories of biogeochemical evidence (*3–5*) within the temporal framework of 4.55 Ga of Earth history. The Hadean, Archaean and Proterozoic eons are conventionally summarized as "Precambrian", with the Hadean (> 3.8 Ga) lacking a sedimentary record and thus being virtually devoid of geological documents pertaining to the Earth's surficial environment. The stromatolite diagram (*1*) was designed to reflect the exuberant proliferation of microbial ecosystems after the establishment of extended stable marine shelves during the Proterozoic as well as their sudden decline following the rise of metazoan grazers at the dawn of the Phanerozoic. The extension to 3.8 Ga ago of the record of microbial fossils (*2*) is contingent on the biogenicity of putative microfossils from the Isua Suite (Pflug and Jaeschke-Boyer, 1979). The isotopic signature of autotrophic carbon fixation (*4*) as revealed by sedimentary organic carbon quite uniformly through Earth history bears a metamorphic overprint as from t > 3.5 Ga (*broken line*) that is well understood today (see Schidlowski and Aharon, this Vol.). A corresponding isotopic index line of dissimilatory sulfate reduction (*5*) apparently fades between 2.7 and 2.8 Ga ago (see also Fig. 1), but it seems most likely that the underlying biologically mediated isotope effect has been camouflaged (*stippled signature*) by either geochemical or environmental controls that were selectively operative during the Earth's earliest history

improvement of the human condition by concerning itself with the increasing need of modern industrial societies for energy and raw materials. Moreover, the project was highly successful in promoting international communication and cooperation among scientists working in the field. Hence, the statement seems warranted that the present state of the field as a whole owes much to the co-ordinating efforts and stimulating effects that have emanated from our project community. It may be reasonably expected that the impetus thus imparted to the topic will radiate into various directions and spark off further efforts commensurate with the intellectual challenge of the field.

References

Cohen Y, Castenholz RW, Halvorson HO (eds) (1984) Microbial mats: stromatolites. MBL Lectures in Biology 3. Liss, New York, xviii + 498 pp

Friedman GM, Krumbein WE (eds) (1985) Hypersaline ecosystems: the Gavish Sabkha. Ecological Studies 53. Springer, Berlin Heidelberg New York, 484 pp

Hayes JM, Kaplan IR, Wedeking KW (1983) Precambrian organic geochemistry: preservation of the record. In: Schopf JW (ed) Earth's earliest biosphere: its origin and evolution. Univ Press, Princeton, NJ, pp 93–134

Holland HD, Schidlowski M (eds) (1982) Mineral deposits and the evolution of the biosphere. Springer, Berlin Heidelberg New York, 333 pp

Johns RB (ed) (1986) Biological markers in the sedimentary record. Methods in Geochemistry and Geophysics 24. Elsevier, Amsterdam, 364 pp

Nagy B, Weber R, Guerrero JC, Schidlowski M (eds) (1983) Developments and interactions of the Precambrian atmosphere, lithosphere and biosphere. Developments in Precambrian Geology 7. Elsevier, Amsterdam, xii + 475 pp

Oehler JH, Schidlowski M (1980) Early biological evolution in relation to mineral and energy resources: IGCP Project 157. Rev Inst Fr Petrol 35:319–324

Ohmoto H, Felder RP (1987) Bacterial activity in the warmer, sulphate-bearing Archaean oceans. Nature (London) 328:244–246

Pflug HD, Jaeschke-Boyer H (1979) Combined structural and chemical analysis of 3.800-Myr-old microfossils. Nature (London) 280:483–486

Retallack GJ (ed) (1986) Precambrian paleopedology. Precambrian Res 32(2–3) Spec Iss:93–259

Schidlowski M (1983) Early organic evolution and mineral and energy resources. Geological Correlation 1983. UNESCO, Paris, Spec Iss:120–122

Schidlowski M (1985) Early life and mineral resources. Nat Res 21(2):11–17

Schidlowski M (1987a) Project 157: Early organic evolution in relation to mineral and energy resources. Geological Correlation 1987. UNESCO, Paris, Spec Iss:81–83

Schidlowski M (1987b) Application of stable carbon isotopes to early biochemical evolution on Earth. Annu Rev Earth Planet Sci 15:47–72

Schidlowski M (1987c) Evolution of the early sulphur cycle. In: Rodriguez-Clemente R, Tardy Y (eds) Geochemistry and mineral formation in the earth surface. Consejo Superior Invest Cient, Madrid, pp 29–49

Schidlowski M, Hayes JM, Kaplan IR (1983) Isotopic inferences of ancient biochemistrie: carbon, sulfur, hydrogen and nitrogen. In: Schopf JW (ed) Earth's earliest biosphere: its origin and evolution. Univ Press, Princeton, NJ, pp 149–186

Schopf JW (ed) (1983) Earth's earliest biosphere: its origin and evolution. Univ Press, Princeton, NJ, xxv + 543 pp

Trudinger PA (1979) The biological sulfur cycle. In: Swaine DJ, Trudinger PA (eds) Biogeochemical cycling of mineral-forming elements. Elsevier, Amsterdam, pp 293–313

Trudinger PA, Walter MR (eds) (1981) Sulphide mineralization in sediments: current status of syngenetic theory. BMR J Aust Geol Geophys 6(4) Spec Iss:279–337

Vernadsky VI (1930) Geochemie in ausgewählten Kapiteln. Akademische Verlagsgesellschaft, Leipzig, 370 pp

Walter MR (1983) Archean stromatolites: evidence of the earth's earliest benthos. In: Schopf JW (ed) Earth's earliest biosphere: its origin and evolution. Univ Press, Princeton, NJ, pp 187–213

Walter MR, McKirdy DM, Trudinger PA (1979) Early organic evolution and mineral and energy resources (project proposal). IGCP Proj 157 Newslett 3:3–5

1. Precambrian Weathering and Paleosols

Introduction to Precambrian Weathering and Paleosols

M. M. Kimberley [1] and H. D. Holland [2]

[1] Dept. of Marine, Earth, and Atmospheric Sciences North Carolina State University Raleigh NC 27695-8208, USA
[2] Dept. of Earth and Planetary Sciences Harvard University Cambridge MA 02138, USA

Abstract

The influx of energy from the Sun probably has exceeded the outflow of heat from the interior of the Earth since shortly after accretion. The present influx of solar energy is about four orders of magnitude greater than the outflow of internal energy. Although the outflow of internal energy has been limited, the apparently continuous outflow of volatiles, including carbon dioxide, has resulted in intensive weathering throughout most of Earth history. The Precambrian represents 88% of Earth history; weathering during this long era played a major role in shaping the Earth and in creating its present condition.

There is ample indirect evidence for Precambrian weathering in Precambrian sedimentary rocks. However, there is little direct evidence in the form of preserved Precambrian paleosols. The few well-preserved Precambrian paleosols that have been studied to date can provide a record of the atmosphere at the time of their formation and perhaps of the evolution of terrestrial life. This record has not yet been deciphered clearly, since the study of Precambrian paleosols is still in its infancy.

1 Introduction

1.1 Definitions

A paleosol is defined in the American Geological Institute (AGI) glossary as "a buried soil horizon of the geologic past" (Gary et al. 1974). Unfortunately, the word, soil, is one of the least well defined among common scientific terms. The definitions in some textbooks would not exclude a graded sedimentary bed, e.g., that of Brewer (1976, p. 7): "Soil is the collection of natural bodies formed by alteration of sedimentary and/or igneous bodies due to exposure at the earth's surface and having an anisotropic arrangement of properties along an axis normal to the earth's surface." Several textbooks define scores of soil-related terms without defining "soil" itself, whereas others refrain from a serious attempt at a definition, e.g., FitzPatrick (1980, p. 341), "Soil: (1) the natural time-space continuum occurring at the surface of the earth and supporting plant life (2) anything so called by a competent authority".

A central issue in the study of Precambrian paleosols is the record of ambient microbial activity. The Precambrian record of continental plant life is not yet clear, but usage of the term "soil" is restricted by some definitions to require organic activity, e.g., the AGI Glossary of Geology which notes that "the lower limit (of soil) is normally the lower limit of biological activity" (Gary et al. 1974). It is unlikely that the lower limit of biologic activity ever will be determined precisely in any Precambrian paleosol and so it is impractical to distinguish between organically influenced "paleosol" and inorganic "paleoregolith" (cf. Robb et al. 1986). For the purpose of this chapter, the terms "soil" and "paleosol" are used without any biologic connotation.

In an accompanying chapter within this Volume, G. J. Retallack defines soil formation as any alteration in place downward from a land surface. Since we know relatively little about terrestrial Precambrian environments, this generalized definition may be preferable to a more elaborate one, e.g., "all material at the Earth's surface which was not permanently covered with water and which became altered in place by surface processes, such as interaction with rain water, atmospheric gases and/or plant life". Both definitions imply that the weathering material did not experience any lateral movement other than soil creep, and so exclude sedimentary processes. The latter definition excludes glacial effects and may require further specification concerning the origin and composition of the reactive gases. Atmospheric gases presumably always have included at least one species which formed an acid upon dissolution in water but they may not always have included a significant concentration of oxidizing species. The most important acid-forming atmospheric gas during the weathering of known

M. Schidlowski et al. (Eds.)
Early Organic Evolution: Implications for Mineral and Energy Resources

Precambrian paleosols almost certainly was carbon dioxide (Holland 1984).

Following burial, most Precambrian paleosols experienced substantial physical, compositional, and mineralogical modification. Some Precambrian paleosols may have been preferentially altered after burial because of their location along an unconformity or between beds of differing degress of cementation. The effects of multiple alteration events are particularly difficult to deconvolve, even in Cenozoic paleosols (Pavich and Obermeier 1985).

Recognition of Precambrian Paleosols

Criteria for the recognition of Precambrian paleosols have been discussed by several authors (e.g., Retallack 1981, 1986a, 1988; Grandstaff et al. 1986; Holland and Zbinden 1988). Methods of paleosol discovery and criteria for their recognition are discussed in an accompanying chapter by G. J. Retallack.

2.1 Precambrian Einvironments

It is well established that metazoans did not appear on Earth until toward the end of the Precambrian Era (Glaessner 1984), perhaps because earlier surficial environments may not have been conducive to the evolution of metazoan life. A better understanding of Precambrian atmospheric and surficial environments would place constraints on theories of biologic evolution, and clearly has a bearing on our understanding of the origin and age distribution of some metalliferous Precambrian strata, e.g., cherty iron formations.

Paleosols are just one of several potential indicators of Precambrian environmental conditions. All sedimentary rocks record something of the ambient environment in which they were deposited, but paleosols can provide a good deal more information, because their evolution from parent rocks often can be traced unequivocably. However, the total volume of well-documented Precambrian paleosols is so small at present that they may represent (collectively) a biased sample of Precambrian terrestrial environments. The record of Precambrian sedimentary rocks also may be biased, because sediments which accumulated on subducting or rapidly sinking (metamorphosing) crust have been destroyed preferentially. Nonetheless, the great volume and variety of preserved Precambrian sedimentary rocks probably ensures that they are a reasonably representative sample of the spectrum of Precambrian environments.

2.2 Weathering and Continental Growth

Weathering has been driven by a combination of solar energy and degassing of volatiles from the Earth. Gravity at the Earth's surface has been sufficiently strong that there has been little loss of degassed volatiles to outer space other than hydrogen and helium, except during rare impact events (Holland 1984; Chapman 1989). Most reactive volatiles (primarily water and various compounds of carbon, sulfur, and chlorine) therefore have remained and have been cycled through the weathering-metamorphic-igneous rock cycle of our dynamic planet. Compaction has forced most water trapped in sediments to return to the Earth's surface from quite shallow depths, but horizontal and vertical temperature gradients have forced the deep convection of seawater through the oceanic crust.

Some volatile-rich continental and oceanic weathering products have survived deep burial and have released their volatiles during metamorphism. The deep release of these volatiles has affected the accumulation of granitic continents and continental mineral deposits. The average composition of the bulk of the upper continental crust lies within the range of intermediate granites (formerly called quartz monzonites or adamellites). The release of volatiles into such rocks may induce melting, because water can depress the melting point of granitic rocks by as much as 400 °C. Whitney (1988) has suggested that the metamorphism of hydrous sediments and of hydrated oceanic crust has been an important factor in the accumulation of large continents.

3 Precambrian Paleosol Models:

Atmospheric-Abiologic Model for Precambrian Paleosols

A semi-quantitative model for Precambrian soil formation has been proposed by H. D. Holland and coauthors (Holland and Zbinden 1988; Pinto and Holland 1988). One assumption in this model is that the presence of biota on or within Precambrian soils modified the O_2 and CO_2 content of soil water only slightly. This condition is not true of any known modern soil. Pinto and Holland (1988) note that the O_2 flux required for oxidizing dissolved organic matter in many modern soils is on the order of 20 times greater than the flux required to oxidize the Fe^{2+} dissolved during the weathering of average rock.

It is difficult to estimate the concentration of dissolved organic carbon under a land cover of cyanobacteria and fungi. Pinto and Holland (1988) have suggested that an experimental "Precambrian garden" devoid of vascular plants be cultivated to determine the potential flux of dissolved organic carbon to soil water. Isotopic studies of Precambrian paleosols may record the presence of organic activity, but the results of the most thorough study of this type remain controversial (Beeunas and Knauth 1985; Vahrenkamp and Rossinsky 1987).

If experimentation or isotopic studies eventually indicate that a nonvascular land cover can contribute substantial quantities of dissolved organic carbon to soil water, the Pinto-Holland model would require reevaluation. No alternative semi-quantitative model has been proposed for the biologic case and no simple modification of the Pinto-Holland model would be appropriate. Models that include the effects of a land biota are complicated due to the potential fluctuation in the production of organic compounds due to climatically-induced fluctuations in plant growth. Campbell (1979) cogently argues that Precambrian paleosols may well have been coated with a mat of nonvascular plants. If so, the rate of carbon fixation in these cyanobacteria and fungi probably was correlated more closely with rainfall than is the rate of carbon fixation by vascular plants today. Dormant nonvascular plants initiate fixation of nitrogen within half an hour of becoming wet (Whitton and Potts 1982).

4 Archean Paleosols

4.1 Introduction

Fluvial sedimentary rocks of Archean age generally are about as chemically mature as their first-cycle Phanerozoic equivalents (Hyde 1980; Watchorn 1980). The composition of Archean shallow-marine sedimentary rocks has also been interpreted as recording subaerial weathering and erosion (Holland 1984, p. 142). Paleosols probably will offer more direct evidence of atmospheric conditions than these sediments, but the number of reported Archean paleosols is so small, and the chemical and textural evidence in them is so meager, that inferences based on available paleosol information must be viewed with caution.

4.2 Lack of Paleosols in Generally Rugged Archean Environments

Kimberley and Grandstaff (1986) described an Archean paleosol directly beneath the 2.7 Ga Timiskaming Group. This group consists largely of sedimentary rocks with a thickness of up to 3500 m. Their depositional environment ranges from fluvial to marine; no paleosol has been recognized within this sequence. This lack of paleosols is consistent with the observation by Hyde (1980) that Timiskaming fluvial sediments graded laterally into turbidites without any intermediate shallow-water facies, presumably due to steep depositional slopes. It is possible that relief generally was greater during the Archean than during subsequent portions of Earth history, and that this mitigated against paleosol preservation (e.g., Ojakangas 1985). However, Archean shallow-water sediments are reported from both Australia and South Africa (e.g., Barley et al. 1979; Watchorn 1980; Lowe 1983; Lowe et al. 1985).

4.3 Peculiarities of Archean Paleosols

Holland and Zbinden (1988) illustrate the time dependence through the Precambrian of the relationship between the proportion of iron lost during weathering to the ratio of oxidant/acid required for complete weathering of the parent rock (Fig. 1). Holland and Zbinden (1988) divide the Precambrian data into two paleosol groups, one which spans the Archean and the Proterozoic to ca. 2.0 Ga and a younger group (Fig. 1). In the older group, the only paleosols which seem to have retained iron during weathering are those which developed on iron-poor rocks. The youngest illus-

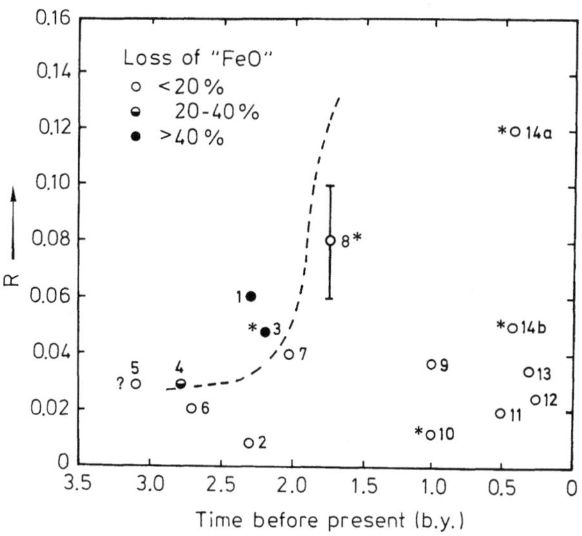

Fig. 1. Summary plot of the behavior of iron in paleosols, reproduced from Holland et al. (1989, Fig. 8). *R* is the ratio of O_2 demand to CO_2 demand during weathering. Locations of paleosols are listed in Holland et al. (Table 9). *Starred* paleosols are the best documented.

trated paleosol in the older group is the ca. 2.2 Ga Hekpoort (Button 1979; Retallack 1986b) and the oldest paleosol in the younger group is the ca. 1.8 Ga Flin Flon paleosol (Holland et al. 1989).

A 2.2 to 2.5 Ga age independently has been suggested to have evolutionary significance for marine sulfur species, based on isotopic data. Pyrite correlates closely with organic carbon in Archean sedimentary rocks, and this association is suggestive of a bacterial influence on precipitation (Dimroth and Kimberley 1976); but Archean pyrite is not enriched in ^{32}S as is pyrite in younger sediments (Holland 1984). One group of researchers interprets this evolutionary change to indicate that marine sulfate was essentially absent in the Archean oceans (Cameron 1982; Veizer et al. 1982; Hattori et al. 1983; Walker and Brimblecombe 1985; Cameron and Hattori 1987). In contrast, Ohmoto and Felder (1987) propose that the Archean oceans contained at least 1 mM sulfate (compared to 28 mM today). Holland (1984) has noted that the concept of a sulfate-free Archean ocean is inconsistent with the evidence of oxidation in some Archean paleosols and with the scattered occurrences of gypsum casts in Archean sedimentary rocks.

5 Lower Proterozoic Paleosols

5.1 Paleosols Associated with Pyritic Conglomerate

Proterozoic paleosols are considerably more plentiful than Archean examples; this is consistent with the greater proportion of Proterozoic supracrustal rocks. One of the oldest Proterozoic examples, on basement beneath the Huronian Supergroup, also is one of the best documented, because fresh profiles are available in the Elliot Lake uranium mines in east-central Canada (Gay and Grandstaff 1980; Kimberley et al. 1980; Farrow and Mossman 1988). Sub-Huronian paleosols developed on a hilly landscape of both granitic and basaltic rocks, and were covered by fluvial clastic sediments (Roscoe 1969; Kimberley et al. 1984). A possibly contemporaneous paleosol formed on tonalite, gneiss, and amphibolite in the Mistassini Lake area, some 800 km northeast of Elliot Lake (Chown and Caty 1983).

The Mistassini Lake paleosol developed on Archean crystalline rocks. Where it underlies fluvial clastic sedimentary rocks, the paleosol resembles the sub-Huronian paleosol. However, where the paleosol is covered by dolostone, dolomite replaces Archean mineral grains within the paleosol (Chown and Caty 1983). A similarly calcareous Lower Proterozoic paleosol is developed on Archean crystalline rocks about 500 km west of Elliot Lake near Keweenaw Bay, Lake Superior (Kalliokoski and Lynott 1987). The Keweenaw Bay carbonate has been attributed to a marine transgression over the paleosol (Kalliokoski and Lynott 1987).

Pyritic uraniferous conglomerate in the Huronian Supergroup resembles pyritic uraniferous-auriferous conglomerate in the Witwatersrand Supergroup of South Africa (e.g., Kimberley 1978). Like the Huronian, the Witwatersrand Supergroup is underlain by thick sericitic paleosol (Robb et al. 1986; Robb and Meyer 1985, 1986). Paleosols in both regions are depleted in Na_2O and moderately enriched in K_2O. Both Al_2O_3 and TiO_2 are enriched in the uppermost portion of the sub-Witwatersrand paleosol (Robb et al. 1986). A regional association of paleosols with pyritic conglomerate recurs in the Lower Proterozoic of Karelia, a district which covers eastern Finland and the northwestern Soviet Union (Koryakin 1981; Marmo, this vol.).

5.2 Paleosols Developed on Lower Proterozoic Iron Formations

Chemically reduced ore deposits are particularly susceptible to weathering, and Precambrian paleosols may be expected on shallow-water ore deposits like the locally oolitic and sideritic iron formations which rim Lake Superior (Lougheed 1983). Iron formations which became weathered during the Precambrian may represent some of the most voluminous examples of paleosols (e.g., Sokolov and Heiskanen 1984; Cannon 1976). The Precambrian age of the alteration is well established for some iron formations which suffered subsequent Precambrian metamorphism, but interpretations of the alteration process remain controversial (Morris 1985).

The evidence for pervasive oxidation in the Lower Proterozoic paleosols which developed on iron formations is difficult to reconcile with the concept of supratidal or intertidal sedimentation of siderite mud (cf. Dimroth 1976; Lougheed 1983). Simonson (1985) concludes that sedimentation of sideritic Precambrian iron formations occurred subtidally although water depth may have been very shallow locally.

5.3 Upper Proterozoic Paleosols

Upper Proterozoic carbonate-rich paleosols have been reported from a thick sequence of fluvial clastic sedimentary rocks and continental basalt in the Keweenawan Peninsula of Michigan (Kalliokoski 1986). In a nearby area, Zbinden et al. (1988) described a 1.1 Ga paleosol developed on greenstone. A sub-Torridonian paleosol in northwestern Scotland resembles a pedalfer where developed on biotite gneiss and a pedocal where developed on amphibolite (Williams 1968). In the Soviet Union, bauxite of Riphean age has been interpreted to be paleosol (Sokolov and Heiskanen 1984).

5.4 Potassium Metasomatism in Precambrian Paleosols

The most consistent peculiarity of Precambrian paleosols is their high K/Na ratio (e.g., Koryakin 1977; Sidorenko et al. 1978; Button 1979; Gay and Grandstaff 1980; Kimberley et al. 1980). Kimberley et al. (1984) noted that similarly high ratios of K/Na occur to depths of over 5 m in modern soil-saprolite which is weathering under subtropical conditions in the southeastern U.S.A. (Calvert et al. 1980; Lechler et al. 1981). Potassium is about an order of magnitude more abundant than sodium in Piedmont soil-saprolite of the southern Appalachians, largely because of the survival of muscovite relative to plagioclase feldspar. Nesbitt et al. (1980) report a similar fractionation of sodium from potassium in modern soil-saprolite on granodiorite in eastern Australia. They attribute the retention of potassium to an inherent preference by clay minerals for large cations.

The relevance of modern potassic soil-saprolite to typical Precambrian paleosol is uncertain, because thick soil-saprolite in the southern Appalachians commonly contains only a few percent clay (Liddle 1984), whereas the sericite in most Precambrian paleosols appears to be regenerated from clay rather than preserved from unweathered muscovite (e.g., Retallack 1986 a, b). Potassium is as leached as sodium in clayey soil on Hawaiian basalt (Patterson and Robertson 1961). Potassium also is scarce in intensely weathered soil of the Amazon basin (Kronberg et al. 1979). However, the sub-tropical soil described by Nesbitt et al. (1980) from Australia is both clayey and potassic.

Given the removal of potassium from the most intensely weathered modern soils, its abundance in thick and originally clayey Precambrian paleosols represents either peculiar Precambrian soil-forming processes or post-depositional alteration. Basu (1981) proposed that potassium would have been retained more readily in soil prior to the advent of vascular plants. However, Graustein and Velbel (1981) considered this proposal to be inconsistent with the potassium-concentrating processes which characterize modern forests, and Holland (1984) considered it to be inconsistent with the age distribution of the potassium content of sedimentary rocks.

Holland and Zbinden (1988) propose that the abundance of potassium in Precambrian paleosols may be attributed to late diagenetic uptake of this element from descending groundwater. In the case of paleosol burial under thick sand, they envision potassium uptake by deeply descending groundwater which had permeated feldspar-bearing sand. The resulting potassium-rich groundwater could convert potassium-poor clay in the paleosol to sericite.

6 Conclusion

The study of Precambrian paleosols remains in its infancy. There is considerable variety among the few examples which have been well documented and this variety can only be expected to increase with additional work. Many Precambrian paleosols are associated with sedimentary ore deposits. The study of Precambrian paleosols may therefore prove to be useful for ore-deposit exploration as well as for reconstructing the history of the atmosphere and of near-surface environments.

References

Barley ME, Dunlop JSR, Glover JE, Groves DI (1979) Sedimentary evidence for an Archean shallow-water volcanic-sedimentary facies, eastern Pilbara Block, Western Australia. Earth Planet Sci Lett 43:74–84

Basu A (1981) Weathering before the advent of land plants: Evidence from unaltered detrital K-feldspars in Cambrian-Ordovician arenites. Geology 9:132–133

Beeunas M, Knauth LP (1985) Preserved stable isotopic signature of subaerial diagenesis in the 1.2-b.y. Mescal Limestone, central Arizona: implications for the timing and development of a terrestrial plant cover. Geol Soc Am Bull 96:737–745

Birkeland PW (1974) Pedology, weathering, and geomorphological research. Oxford Univ Press, New York, 285 pp

Brewer R (1976) Fabric and mineral analysis of soils. Krieger Publ, Huntington, NY, 482 pp

Button A (1979) Early Proterozoic weathering profile on the 2200-Myr-old Hekpoort Basalt, Pretoria Group, South Africa: preliminary results. Inf Circ Econ Geol Res Unit, Univ Witwatersrand, Johannesburg, 133:20 pp

Calvert CS, Buol SW, Weed SB (1980) Mineralogical characteristics and transformations of a vertical rock-saprolite-soil sequence in the North Carolina Piedmont. J Soil Sci Soc Am 44:1096–1112

Cameron EM (1982) Sulphate and sulphate reduction in early Precambrian oceans. Nature (London) 296:145–148

Cameron EM Hattori K (1987) Archean sulphur cycle: evidence from sulphate minerals and isotopically fractionated sulphides in Superior Province, Canada. Chem Geol 65:341–358

Campbell SE (1979) Soil stabilization by a prokaryotic desert crust: implications for Precambrian land biota. Orig Life 9:335–348

Cannon WF (1976) Hard iron ore of the Marquette range, Michigan. Econ Geol. 71:1012–1028

Chapman CR (1989) Snowbird II: global catastrophes. Eos Trans Am Geophys Un 70:217–218

Chauvel J-J, Dimroth E (1974) Facies types and depositional environments of the Sokoman Iron Formation, central Labrador Trough, Quebec, Canada. J Sediment Petrol 44:299–327

Chown EH, Caty, J-L (1983) Diagenesis of the Aphebian Mistassini regolith, Quebec, Canada. Precambrian Res 19:285–299

Cloud PE (1973) Paleoecological significance of the banded iron formations. Econ Geol 68:1135–1143

Dimroth E (1976) Aspects of the sedimentary petrology of cherty iron formation. In: Wolf KH (ed) Handbook of strata-bound and stratiform ore deposits, vol 7, Elsevier, Amsterdam, pp 203–254

Dimroth E, Kimberley MM (1976) Precambrian atmospheric oxygen: evidence in the sedimentary distributions of carbon, sulfur, uranium, and iron. Can J Earth Sci 13:1161–1185

Eagleman JE (1980) Meteorology, the atmosphere in action. Van Nostrand, New York, 384 pp

Farrow CEG, Mossman DJ (1988) Geology of Precambrian paleosols at the base of the Huronian Supergroup, Elliot Lake, Ontario, Canada. Precambrian Res 42:107–139

FitzPatrick EA (1980) Soils, their formation, classification and distribution. Longman, London, 353 pp

Gary M, Mc Afee R Jr, Wolf CL (eds) (1974) Glossary of geology. American Geological Inst, Washington, DC, 805 pp

Gay AL, Grandstaff DE (1980) Chemistry and mineralogy of Precambrian paleosols at Elliot Lake, Ontario, Canada. Precambrian Res 12:349–373

Glaessner MF (1984) The dawn of animal life, a biohistorical study. Univ Press, Cambridge, 244 pp

Gough DO (1981) Solar interior structure and luminosity variations. Solar Phys 74:21–34

Grandstaff DE, Edelman MJ, Foster RW, Zbinden E, Kimberley MM (1986) Chemistry and mineralogy of Precambrian paleosols at the base of the Dominion and Pongola Groups (Transvaal, South Africa). Precambrian Res 32:97–131

Graustein WC, Velbel MA (1981) Comment on 'Weathering before the advent of land plants: evidence from detrital K-feldspars in Cambrian-Ordovician arenites'. Geology 9:505

Gruner JW (1922) The origin of sedimentary iron formations: the Biwabik Formation of the Mesabi Range. Econ Geol 17:407–460

Hattori K, Krouse HR, Campbell FA (1983) The start of sulfur oxidation in continental environments about 2.2×10^9 years ago. Science 221:549–551

Holland HD (1978) The chemistry of the atmosphere and oceans. Wiley-Interscience, New York, 351 pp

Holland HD (1984) The chemical evolution of the atmosphere and oceans. Univ Press, Princeton, NJ, 582 pp

Holland HD, Zbinden EA (1988) Paleosols and the evolution of the atmosphere, pt 1. In: Lerman A, Meybeck M (eds) Physical and chemical weathering in geochemical cycles. Reidel, Dordrecht, pp 61–82

Holland HD, Feakes CR, Zbinden EA (1989) The Flin Flon paleosol and the composition of the atmosphere 1.8 b.y.b.p.. Am J Sci 289:362–389

Hunt CB (1972) Geology of soils, their evolution, classification, and uses. W. H. Freeman and Co., San Francisco, 344 pp

Hyde RS (1980) Sedimentary facies in the Archean Timiskaming Group and their tectonic implications, Abitibi greenstone belt, northeastern Ontario, Canada. Precambrian Res 12:161–195

Kalliokoski J (1986) Calcium carbonate cement (caliche) in Keweenawan sedimentary rocks (~ 1.1 Ga), Upper Peninsula of Michigan. Precambrian Res 32:243–259

Kalliokoski J, Lynott JS (1987) Huron River: Precambrian unconformities and alteration at and near Big Eric's Crossing, Michigan. In: Biggs DL (ed) Geol Soc Am, North Cent Sec, Centennial Field Guide 3:273–276

Kalliokoski J, Welch EJ (1986) Keweenawan-age caliche paleosol in the lower part of the Calumet and Hecla Conglomerate, Centennial Mine, Calumet, Michigan. Geol Soc Am Bull 96:1188–1193

Kasting JF (1987) Theoretical constraints on oxygen and carbon dioxide concentrations in the Precambrian atmosphere. Precambrian Res 34:205–229

Kimberley MM (1978) Origin of stratiform uranium deposits in sandstone, conglomerate, and pyroclastic rock. In: Kimberley MM (ed.) Uranium deposits, their mineralogy and origin. Mineral Assoc Can Short Course Handb 3:339–381

Kimberley MM, Tanaka RT, Farr MR (1980) Composition of Middle Precambrian uraniferous conglomerate in the Elliot Lake-Agnew Lake area of Canada. Precambrian Res 12:375–392

Kimberley MM, Grandstaff DE, Tanaka RT (1984) Topographic control on Precambrian weathering in the Elliot Lake uranium district, Canada. J Geol Soc London 141:229–233

Kimberley MM, Grandstaff DE (1986) Profiles of elemental concentrations in Precambrian paleosols on basaltic and granitic parent materials. Precambrian Res 32:133–154

Koryakin AS (1977) Examples of metamorphosed weathering crusts in Karelia. In: Timofeev PP, Kholodov VN, Kazanski YuP, Nikolaeva IV, Koporlin VI (eds) Current problems of lithology and sedimentary mineral products (Problemy Sovremennoy litologii i osadochnykh poleznykh iskopaemykh). Proc 9th Int Sedimentological Congr, Nice, Fr; Izdatelstvo, Novosibirsk, pp 72–80 (in Russian)

Koryakin AS (1981) Results of a study of Proterozoic weathering crusts in Karelia. Int Geo Rev 13:973–980

Kronberg BI, Fyfe WS, Leonardos OH, Santos AM (1979) The chemistry of some Brazilian soils: element mobility during intense weathering. Chem Geol 24:211–229

Lechler PJ, Ray WR, Leininger RK (1981) Major and trace element analysis of 12 reference soils by inductively coupled plasma-atomic emission spectrometry. Soil Sci 130:238–241

Liddle SK (1984) Trace-element analysis of the groundwater at a hazardous-waste landfill in the Piedmont of North Carolina. MS Thesis, NC State Univ, Raleigh, 83 pp

Lougheed MS (1983) Origin of Precambrian iron formations in the Lake Superior region. Geol Soc Am Bull 94:325–340

Lowe DR (1983) Restricted shallow-water sedimentation of Early Archean stromatolite and evaporitic strata of the Strelley pool chert, Pilbara block, Western Australia. Precambrian Res 19:239–283

Lowe DR, Byerly GR, Ransom BL, Nocita BW (1985) Stratigraphic and sedimentologic evidence bearing on structural repetition in Early Archean rocks of the Barberton greenstone belt, South Africa. Precambrian Res 21:165–186

Morris RC (1985) Genesis of iron ore in banded iron-formation by supergene and supergene-metamorphic processes – a conceptual model. In: Wolf KH (ed) Handbook of strata-bound and stratiform or deposits, vol 13. Elsevier, Amsterdam, pp 73–235

Nesbitt HW, Young GM (1982) Early Proterozoic climates and plate motions inferred from major element chemistry of lutites. Nature (London) 299:715–717

Nesbitt HW, Markovics D, Price RC (1980) Chemical processes affecting alkalis and alkaline earths during continental weathering. Geochim Cosmochim Acta 44:1659–1666

Newman MJ, Rood RT (1977) Implications of solar evolution for the Earth's early atmosphere. Science 198:1035–1037

Ohmoto H, Felder RP (1987) Bacterial activity in the warmer, sulphate-bearing, Archean oceans. Nature (London) 328:244–246

Ojakangas RW (1985) Review of Archaean clastic sedimentation, Canadian Shield: major felsic volcanic contributions to turbidite and alluvial fan-fluvial facies associations. In: Ayres LD, Thurston PC, Card KD, Weber W (eds) Geol Assoc Can Spec Pap 28:23–48

Patterson SH and Robertson CE (1961) Weathered basalt in the eastern part of Kauai, Hawaii. US Geol Surv Profess Pap 424-C:195–198

Pavich MJ, Obermeier SF (1985) Saprolite formation beneath Coastal Plain sediments near Washington, D.C. Geol Soc Am Bull 96:886–900

Pinto JP, Holland HD (1988) Paleosols and the evolution of the atmosphere, pt 2. In: Reinhardt J, Sigleo WR (eds) Paleosols and weathering through geologic time: principles and applications. Geol Soc Am Spec Pap 216:21–34

Reimer TO (1987) The Late-Archaean Dominion conglomerates (South Africa): new aspects of their derivation and their relationship with those of the Witwatersrand. N Jahrb Miner Abh 158:13–46

Retallack GJ (1981) Fossil soils: indicators of ancient terrestrial environments. In: Niklas KJ (ed) Paleobotany, paleoecology and evolution, vol 1. Praeger, New York, pp 55–102

Retallack GJ (1986a) The fossil record of soils. In: Wright VP (ed) Paleosols: their recognition and interpretation. Blackwell, Oxford, pp 1–57

Retallack GJ (1986b) Reappraisal of a 2200-Ma-old paleosol near Waterval Onder, South Africa. Precambrian Res 32:195–232

Retallack GJ (1988) Field recognition of paleosols. In: Reinhardt J, Sigleo WR (eds) Paleosols and weathering through geologic time: principles and applications. Geol Soc Am Spec Pap 216:1–20

Retallack GJ, Feakes CR (1987) Trace fossil evidence for Late Ordovician animals on land. Science 235:61–63

Robb LJ, Meyer M (1985) The nature of the Witwatersrand hinterland: Conjectures on the source-area problem. Univ Witwatersrand Econ Geol Res Unit, Inf Circ 178:25 pp

Robb LJ, Meyer M (1986) The nature of the Archaean basement in the hinterland of the Witwatersrand Basin: 1 – The Rand anticline between Randfontein and Rysmierbult. Univ Witwatersrand Econ Geol Res Unit Inf Circ 187:14 pp

Robb LJ, Meyer M, Franz MF (1986) Palaeoregolith profiles in the granitic basement of the Witwatersrand hinterland. In: Int Earth Sci Congr, Univ Witwatersrand, Johannesburg RSA, Extended Abstr, pp 193–196

Ronov AB (1988) Common evolutionary trends in the chemical composition of sedimentary and igneous rocks of the Earth's crust. Terra Cognita 8:212

Roscoe SM (1969) Huronian rocks and uraniferous conglomerates in the Canadian Shield. Can Geol Surv Pap 68–40, 193 pp

Schmitz B (1988) Origin of microlayering in worldwide distributed Ir-rich marine Cretaceous/Tertiary boundary clays. Geology 16:1068–1072

Sidorenko AV, Tenyakov VA, Sidorenko SvA, Sozinov, NA et al. (1978) Precambrian and problems of Earth crust formation. Nauka, Moscow, 311 pp (in Russian)

Simonson BM (1985) Sedimentological constraints on the origin of Precambrian iron formations. Geol Soc Am Bull 96:244–252

Sokolov VA, Heiskanen KI (1984) Developmental stages of Precambrian crusts of weathering. In: Proc 27th Int Geol Congr, Moscow, vol 5: Precambrian geology. VNU, Utrecht, Neth, pp 73–94

Turcotte DL (1980) On the thermal evolution of the Earth. Earth Planet Sci Lett 48:53–58

Vahrenkamp VC, Rossinsky V Jr (1987) Discussion of "Preserved stable isotopic signature of subaerial diagenesis in the Mescal Limestone, central Arizona: implications for the timing of a terrestrial plant cover". Geol Soc Am Bull 99:595–596

Veizer J, Compston W, Hoefs J, Nielson H (1982) Mantle buffering of the early oceans. Naturwissenschaften 69:173–180

Walker JCG, Brimblecombe P (1985) Iron and sulfur in the pre-biologic ocean. Precambrian Res 28:205–222

Watchorn MA (1980) Fluvial and tidal sedimentation in the 3000 Ma Mozaan basin, South Africa. Precambrian Res 13:27–42

Whitney JA (1988) The origin of granite: the role and source of water in the evolution of granitic magmas. Geol Soc Am Bull 100:1886–1897

Whitton BA, Potts M (1982) Marine littoral. In: Carr NG, Whitton BA (ed) The biology of cyanobacteria. Univ Cal Press, Berkeley, 688 pp

Williams GE (1968) Torridonian weathering, and its bearing on Torridonian palaeoclimate and source. Scott J Geol 4:164–184

Zbinden EA, Holland HD, Feakes CR (1988) The Sturgeon Falls paleosol and the composition of the atmosphere 1.1 Ga BP. Precambrian Res 42:141–163

How to Find a Precambrian Paleosol

Gregory J. Retallack

Dept. of Geology, University of Oregon, Eugene, OR 97403, USA

Abstract

Precambrian paleosols have been discovered in unprecedented numbers in recent years and it is likely that many remain unrecognized for what they are. This essay attempts to aid the search by outlining diagnostic versus common but not necessarily distinctive features of Precambrian paleosols. Three general classes of criteria can be used to recognize Precambrian paleosols: biological traces, soil horizons, and soil structures.

Precambrian paleosols do not contain the root traces that are so helpful in identifying Phanerozoic paleosols. Moreover, the microbial traces in paleosols, such as ministromatolites, rock varnish, and microbial borings, are not yet distinguishable from similar phenomena on the floors of shallow lakes or marine shelves. The most diagnostic microbial indicator of Precambrian paleosols yet reported is the upward lightening of isotopic ratios of ^{18}O vs. ^{16}O and ^{13}C vs. ^{12}C in weathered carbonate rocks.

Soil horizons of Precambrian paleosols, like those of the Phanerozoic, ideally exhibit gradational changes in color, texture, mineralogy, and chemical composition downward from a sharp upper contact. Subsurface horizons enriched in quartz, clay, sesquioxides, or humus have not yet been demonstrated in paleosols older than the Devonian advent of trees. There are, however, Precambrian examples of surface horizons enriched in clay and of subsurface horizons enriched in carbonate minerals, gypsum, amorphous silica, or manganese. There also are Precambrian examples of bauxitic, kaolinitic, and iron-rich deep weathering.

Some soil structures are diagnostic in that they demonstrate multiple episodes of breaking, infilling, and closure due to randomly oriented, small-scale extensional and compressional forces at atmospheric temperatures and pressures. Examples include subangular near-equant soil clods (blocky peds), veins of clay washed down soil cracks (clay skins or illuviation argillans), tepee or pseudoanticline (mukkara) structures of swelling-clay soils, sand wedges and other features of periglacial soils, and certain kinds of microfabric in which birefringent clays are arranged as random streaks in a less-oriented matrix (insepic, mosepic, and masepic plasmic fabric).

Most Precambrian paleosols are deformed, metamorphosed, and poorly exposed, so can be confused with upward-fining fluvial sequences, mudflow deposits, ash beds, marine hardgrounds, fault mylonites, or zones altered by either groundwater or hydrothermal fluids. Argillic hydrothermal alteration is particularly troublesome because its effects cannot yet be distinguished from the wide range of known soil-forming conditions on compositional criteria alone. To distinguish a paleosol from hydrothermally altered rock, evidence is best sought from horizonation, soil structure, and geologic context.

1 Introduction

Are Precambrian paleosols rare? Compared to Precambrian komatiites, placer uraninites, stromatolites, microfossils, or fluvial rocks, the number of known Precambrian paleosols remains meager (Sokolov and Heiskanen 1984; Pinto and Holland 1988). However, Precambrian paleosols have been discovered in unprecedented numbers in recent years and it is likely that many remain unrecognized for what they are. Few geologists have extensive knowledge of soil science, and few soil scientists have carefully studied Precambrian rocks. It is only in the past two decades that the expertise of soil science has been diligently applied to Phanerozoic fluvial rocks. As a result, literally thousands of paleosols have been discovered (Bown and Kraus 1981; Retallack 1983; Allen 1986).

Paleosol studies have not been as successful in the Precambrian as in the Phanerozoic for at least two reasons. Precambrian paleosols lack obvious and diagnostic biologic features, such as root traces (Retallack 1988 a) and many Precambrian paleosols have

M. Schidlowski et al. (Eds.)
Early Organic Evolution: Implications for Mineral and Energy Resources

been so obscured by metamorphic alteration that they are difficult to distinguish from hydrothermally altered rocks (Lowe et al. 1985; Duchac and Hanor 1987). Despite these problems, considerable progress has been made recently on where and how to look for Precambrian paleosols. This essay outlines diagnostic versus common but not necessarily distinctive features of Precambrian paleosols. In looking for Precambrian paleosols, as in all science, fortune favors the prepared mind.

Given that it is premature to judge how common Precambrian paleosols may be, it is difficult to predict their significance for reconstructing Precambrian life and landscapes. Precambrian, like Phanerozoic, paleosols potentially may provide evidence of past climates, organisms, topographic relief, parent materials, and duration of soil formation (Retallack 1986b). The history of Precambrian atmospheric oxygenation (Holland 1984; Pinto and Holland 1988) and the tectonic style of continental accretion and differentiation (Retallack 1990) are two perennial problems of Precambrian geology that are being profitably addressed through study of paleosols. A comparison of Precambrian paleosols with surficial alteration of equivalent or greater age on the Moon, Mars, Venus, and other planetary bodies may elucidate the early history of our Solar System (Retallack 1988b, 1990). Paleosols also provide evidence for the origin and early evolution of life on land (Retallack 1986a, 1990).

The discovery of additional Precambrian paleosols would enhance these studies. Three aspects of Precambrian paleopedology are emphasized herein to facilitate such discoveries: where to look for Precambrian paleosols, what features are diagnostic of them, and how they are distinct from other geologic phenomena.

2 Where to Look for Precambrian Paleosols

Perhaps the most direct way to find Precambrian paleosols is to advertise among geologists. Holland (1988) has tried this, with some success. A less direct, but perhaps more promising, approach involves "reading between the lines" of Precambrian studies. As in Phanerozoic sequences, Precambrian paleosols may be expected along major unconformities and within nonmarine sedimentary or volcanic strata.

Paleosol reconnaissance should start with a geological map. Unconformities overlain by fluvial rocks are especially promising candidates for paleosols (Gay and Grandstaff 1980; Farrow and Mossman 1988). Some rivers erode canyons deep into fresh bedrock but others bury weathered hillsides or plains under alluvium (Fig. 1). Colluvium also may bury soil. Irregularity of a mapped geological contact occasionally records paleotopography rather than tectonism (Williams 1969). Highly irregular contacts characterize some carbonate sequences and some of these irregularities have proven to be ancient karst (Button and Tyler 1981; Schau and Henderson 1983; James and Choquette 1987).

Fig. 1. Sheigra clay paleosol beneath Torridonian (1 Ga) alluvial fan deposits in sea cliffs near Sheigra, NW Scotland. (Retallack 1986a)

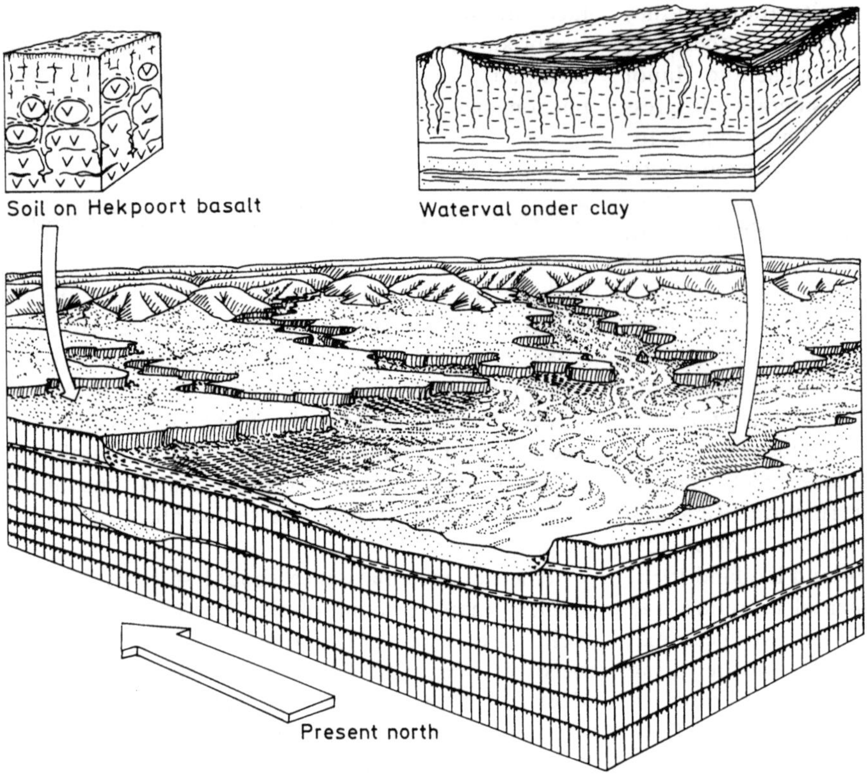

Fig. 2. Reconstructed sedimentary environment of the Waterval Onder clay (in basal Dwaal Heuvel Formation) and of undescribed paleosol on the 2.2 Ga Hekpoort Basalt in eastern Transvaal, South Africa. (Retallack 1986b)

Reconnaissance should also extend to mining records and publications. Some Precambrian aluminum, iron, manganese, and uranium ores formed as or near paleosols. This is not to say that they all are necessarily paleosols but rather that they deserve reexamination from this perspective. Crustal igneous rocks average about 15.6% Al_2O_3 but enrichments greater than 45% are found in pedogenic bauxite ores. Gibbsite, boehmite, and diaspore are characteristic of these paleosols. These three minerals become metamorphosed to kyanite, sillimanite, or corundum (Reimer 1986). Comparable degrees of iron enrichment are found in some Oxisols and paleosols (McFarlane 1976). Hematite in these soils and paleosols is remarkably stable up to greenschist-facies metamorphism. It alters to magnetite within the amphibolite and granulite facies (Thompson 1972). Manganese ores have been found in some paleokarst (Button and Tyler 1981; Schau and Henderson 1983) and some uranium ores are associated with major unconformities (Meyer 1985). Beware that Precambrian ores of sedimentary, volcanogenic or hydrothermal origin fortuitously also occur at major unconformities. In addition, the long periods of geological time represented at major uncormities means that they may preserve confusingly overlapping evidence of successive weathering episodes. Nevertheless, mining records for unconformity-associated ore deposits may well lead to the discovery of a Precambrian paleosol.

Studies of Precambrian sedimentary environments (Fig. 2) are also potential souces of paleosol information. Paleosols are abundant in Phanerozoic floodplain, eolian, alluvial-fan, and deltaic facies (Retallack 1986a). However, paleosols are difficult to distinguish in nonmarine Precambrian sequences because these rocks lack the fossil root traces which distinguish Phanerozoic paleosols. Useful criteria in Precambrian sequences mostly reflect climatic extremes. Ice wedges, sand wedges, pingos, and other periglacial features characterize soils which are classified as Cryorthents and Cryopsamments in the U.S. (Soil Survey Staff 1975). There is a long Precambrian fossil record of these soils of frigid climates (Williams 1986). The formation of salt crusts in deserts, playas, and coastal sabkhas is also a soil-forming process, resulting in soils identified as Salorthids in the U.S. taxonomy. Precambrian paleosols of this kind are among the most ancient known (Lowe 1983). Micritization, boring, and nodule formation in limestone and coastal beachrock also are soil-forming processes. The resulting soils are classified as Orthents and Calciorthids. Precambrian examples occur in shallow-marine and intertidal carbonates (Bertrand-Sarfati and Moussine-Pouchkine 1983; Grotzinger 1986). These distinctive alterations in place are the most obvious of a variety of subtle weathering effects which are detectable in Precambrian sedimentary rocks.

Studies of Precambrian volcanic rocks offer yet another potential source of information about Precambrian paleosols. Sequences of pillow lavas are not a promising place to look for paleosols because pillows form subaqueously. However, terrestrial sequences of lava flows may include paleosols, as well documented in the Irish Tertiary (Eyles 1952). Lahars, ignimbrites, and ash beds also are separated by paleosols in Phanerozoic volcaniclastic sequences (Retallack 1983) and a few Precambrian examples are known (Blades and Bickford 1976). Surficial weathering is not the only cause of stratiform alteration within volcanic sequences. Alternative methods of alteration include groundwater moving through porous pyroclastics after burial, hydrothermal alteration driven by volcanic heat, and downward baking by directly overlying lava (Fisher and Schminke 1984). None of these alternatives generally explains thick clayey horizons between Phanerozoic flows because the clays are less permeable than the columnar and tabular joints, because the mixing of hot lava and cool water is more likely to produce breccia, scoria, or lava spines than extensive interflow alteration, and because anhydrous volcanic rocks are such good insulators that baking generally is limited to just a few centimeters. Nevertheless, the foregoing nonpedogenic alternatives should be considered when evaluating possible paleosols in Precambrian volcanic sequences (Smith et al. 1982; Lowe et al. 1985; Duchac and Hanor 1987).

3 How to Look for Precambrian Paleosols

Three features of Phanerozoic paleosols are most useful for distinguishing them from enclosing rocks: root traces, soil horizons, and soil structure. Root traces are not found in Precambrian paleosols because they predate the advent of multicellular land plants. Nevertheless, other structures and effects of nonmarine microbes can be found. Similarly, the soil horizons and structures of Precambrian paleosols may not be as obvious or diverse as those in Phanerozoic paleosols, but comparable features have been found.

3.1 Microbial Traces

Even among soil scientists there is some disagreement about the exact definition of soil, but almost all definitions include material which supports the growth of vascular plants. The most diagnostic features of Phanerozoic paleosols are fossil roots or their clay or crystal-lined holes. The fossil record of roots is securely known only back to the Devonian (Retallack 1985, 1986a). Silurian paleosols contain some possible rhizome traces (Schopf et al. 1966; Retallack 1985). Paleosols as old as Late Ordovician contain burrows of animals so large that they probably fed on multicellular plants. They also contain erosional scours around reduction spots that may represent clumps of plants (Dewey, in Boucot et al. 1974; Retallack and Feakes 1987). The Late Ordovician is the time of the earliest fossil spores of land plants (Gray 1985). Prior to Late Ordovician, the land may have been covered with unicellular algae, bacteria, and other microbes (Wright 1985).

No indisputable microfossils have yet been found in a Precambrian paleosol despite several tantalizing possibilities (Retallack 1986a). Encouragement for the search comes from the discovery of traces of supratidal microbes in Cambrian phoscretes (Southgate 1986). These microbial traces can be distinctive but are very similar in soils and shallow marine or lacustrine deposits (Siegel 1977; Golubic and Campbell 1979). For example, some Precambrian paleosols contain structures like microbially produced rock varnish (Fig. 3). Modern iron-manganese varnish occurs on the upper but not lower sides of exposed pebbles within soils, streams, and lakes (Dorn and Oberlander 1982).

Modern carbonate crusts on cliffs and in caves (Reams 1989) and opaque rock varnishes on exposed rocks (Perry and Adams 1978) are microbially laminated and mounded (Fig. 4). These resemble the widespread Precambrian trace fossils called ministromatolites (Hofmann and Jackson 1987). Ministromatolites are not diagnostic of paleosols but neither should they be assumed to be exclusively aquatic. Possible Precambrian pedogenic ministromatolites include the laminated andalusite caps on quartz grains illustrated by Grandstaff et al. (1986).

Like ministromatolites, traces of endolithic microbes are similar in subaerial and aquatic environments. The bulbous or vase-shaped hollows created by endolithic algae in carbonate rocks occur in both marine (Knoll et al. 1986) and nonmarine settings (Folk et al. 1973). Subaerial endolithic algae produce spongy dark limestone which is known as phytokarst. Possible fossil examples are as old as Devonian (Folk and McBride 1976). Endolithic algae also are thought responsible for networks of surficial grooves on both silicate and carbonate grains in Phanerozoic marine and nonmarine sandstones (Ross and Fisher 1986). Endolithic microbes presently produce thin leached or ferruginized zones just below exposed rock surfaces

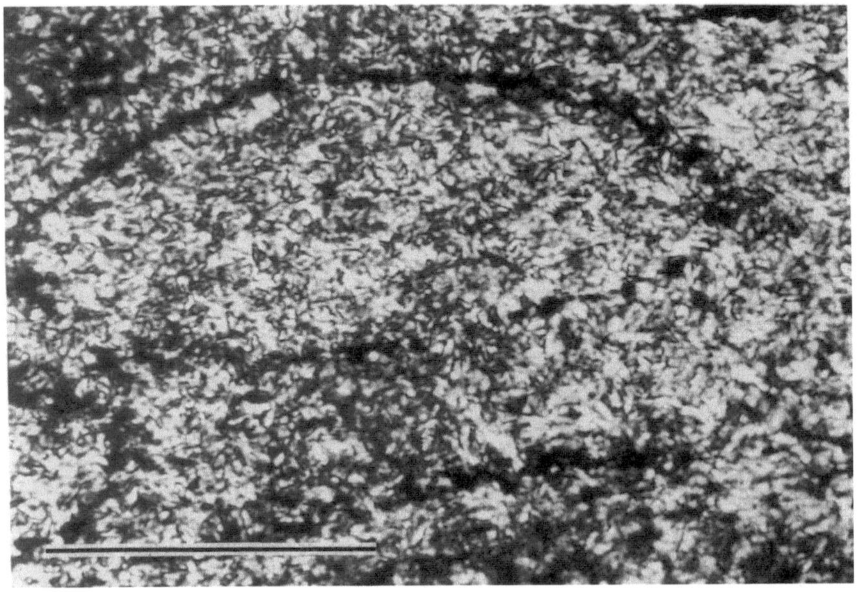

Fig. 3. Possible recrystallized rock varnish on clay granule in a petrographic thin section of surface (*A*) horizon of Waterval Onder clay paleosol of Fig. 2. *Bar* 0.1 mm. (Retallack 1986b)

Fig. 4. Ministromatolitic rock varnish in thin section from exposure near Phoenix, Arizona. Field is 0.15 mm wide. (Perry and Adams 1978)

(Friedman and Weed 1987) and similar submarine processes may be responsible for the discoloration of hardgrounds and boulders (Bathurst 1975; Bromley and Hanken 1981). The search for these kinds of microbial traces in Precambrian paleosols has just begun.

An isotopic signature may be diagnostic for pedogenic microbial colonization of subaerially exposed rock and sediment. An upward increase in lighter stable isotopes of both oxygen ($^{18}O/^{16}O$) and carbon ($^{13}C/^{12}C$) characterizes micritized limestone bedrock beneath modern Vertisols in Bermuda (Vernon and Carroll 1965; Allan and Matthews 1982) and paleosols which have developed on carbonate rocks ranging in age back to 1.2 Ga (Beeunas and Knauth 1985). Enhancement of the light isotopes is attributed to photosynthetic organisms but such isotopic fractionation may be compromised by burial alteration and metamorphism (Vahrenkamp et al. 1987). Precambrian occurrences are best evaluated by concurrent study of such indicators of alteration as illite crystallinity, kerogen maturation, and metamorphic petrography.

Paleosols may exhibit a surficial enrichment in either organic carbon or the trace elements which normally become complexed with organic matter, such as P, Cu, Ni and Y (Retallack 1986b). However, these elements also are concentrated in phoscrete and marine hardgrounds. Although minor-element distributions and microbial traces are not diagnostic of paleosols, they are important evidence of life in a paleosol which has been recognized by other criteria.

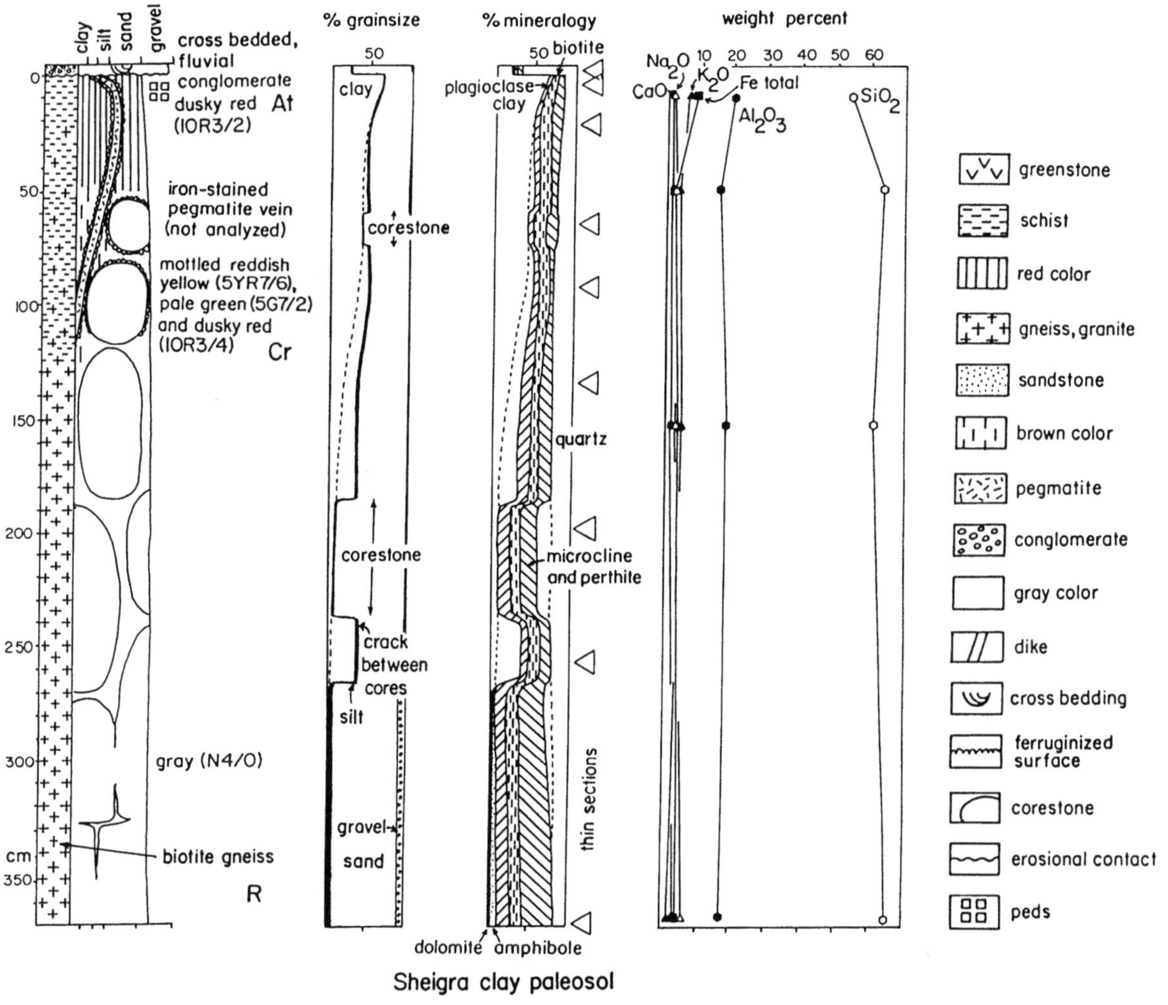

Fig. 5. Field section (by G. J. Retallack), grain-size distribution and modal analysis (by G. S. Smith) and chemical analysis (by G. E. Williams) of the Sheigra clay paleosol of Fig. 1. (Retallack 1990)

3.2 Soil Horizons

Soil formation essentially is alteration downward from a land surface. Ideally, it results in gradational changes downward through one or more distinct kinds of soil toward a little-altered parent material. Paleosols generally exhibit a sharp contact with overlying strata due to erosion of the uppermost soil. Boundaries between horizons within soils and paleosols are gradational over centimeters to meters (Figs. 1, 5). Vertical profiles of color, clay content, or elemental abundance through a typical paleosol exhibit only minor irregularities around nodules or less-weathered corestones. In contrast, sedimentary beds, hydrothermal vein networks, and igneous intrusions typically exhibit either sharp contacts or erratically zig-zagging compositional profiles.

One exception to the norm of a sharp upper contact and gradational boundaries is a floodplain soil subjected to increased frequency of flooding. For interpreting paleoenvironments, it is important to differentiate the paleosol from overlying transported soil material. For example, the NAN profile of the Elliot Lake paleosol (Mossman and Farrow, this Vol.) appears to be capped by transported soil (above 2.3 m).

Other exceptions are lithological differences in the parent material, which become inherited as sharp boundaries, color contrasts, compositional anomalies, or stone lines. One example is the Waterval Onder paleosol of the basal Dwaal Heuvel Formation in the Transvaal, South Africa. It was initially interpreted as a single thick paleosol developed on basalt (Button and Tyler 1981). However, the mineralogical and

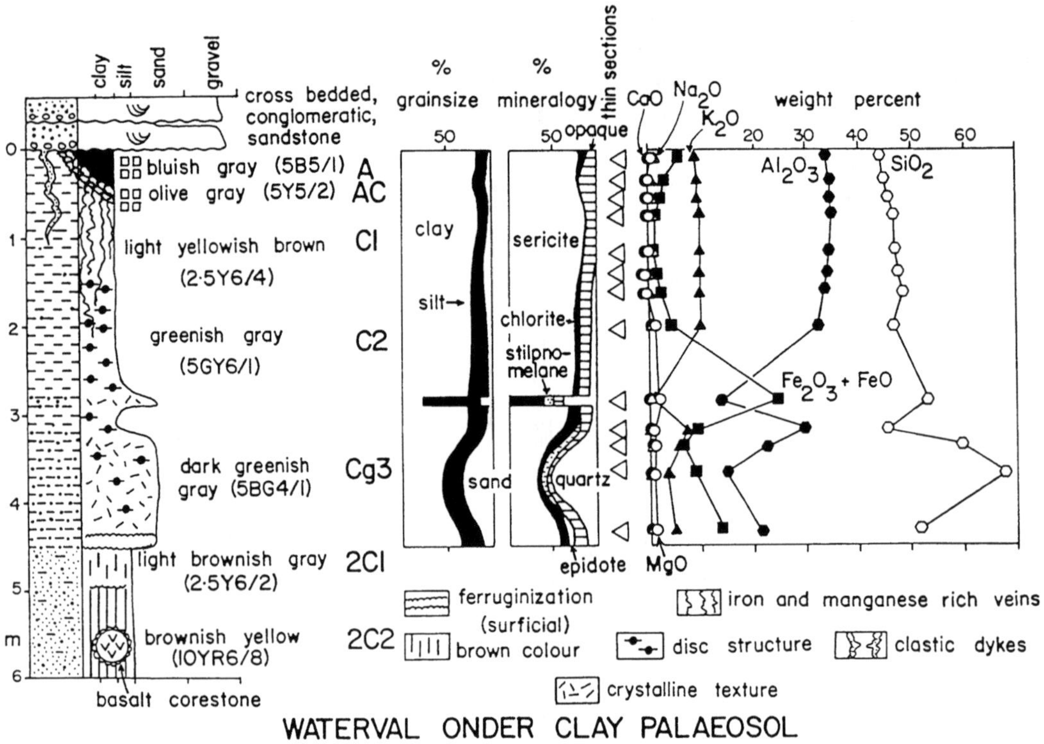

Fig. 6. Field section and grain-size distribution (by G. J. Retallack), petrographic composition and chemical analysis (by A. Button) of the Waterval Onder paleosol of Fig. 2. (Retallack 1986b)

chemical profile is erratic below 2.5 m (Fig. 6). Examination by the author has revealed relict beds and sedimentary structures of a fining- and thinning-upwards sequence of metamorphosed sandstone and shale overlying the basalt (Retallack 1986b). In contrast, the upper portion of this profile (above 2.5 m) shows the smooth chemical and petrographic variation expected in a paleosol.

Paleosol horizons can be specified using widely accepted abbreviations and concepts for soil horizons (Guthrie and Witty 1982). These include surface horizons of peat (organic or O) and near-surface horizons enriched in quartz (eluvial or E). Subsurface horizons may be enriched in clay (argillic or Bt), in humus and sesquioxides of iron and aluminum (spodic or Bs), or in carbonate (calcic or Bk).

Clayey soils of peaty swamps and marshes have become the rooted underclays of Phanerozoic coal seams (Gardner et al. 1988). Precambrian organic soils could potentially yield comparable evidence of ministromatolites, mudcracks, clay-flake breccias, and gradations of microbial mats into underlying shale.

Phanerozoic ganister paleosols with near-surface quartzose (E) horizons are found in Euramerican coal measures (Percival 1986). Similar effects are difficult to detect in cherty Precambrian paleosols because abundant silica has been supplied by the illitization and sericitization of clay minerals during late diagenesis (Curtis 1985) and hydrothermal alteration (Rinehart 1980). In thin section, near-surface quartzose (E) horizons of paleosols may exhibit evidence of hydrolytic weathering in a loose medium, e.g., abundant quartz and K-feldspar grains with subordinate etched mafic grains, all surrounded by thin clay skins (Dumanski and St. Arnaud 1966).

Pedogenic clay in subsurface clayey (Bt) horizons is characteristically laminated along former cracks in a paleosol (Brewer 1976). Pedogenic iron coatings around mineral grains in subsurface spodic (Bs) horizons exhibit internal lamination which is interrupted by deep radial cracks that extend inward from the surface of the coating (de Coninck et al. 1974). These petrographic criteria may prove useful in recognizing organic, eluvial, argillic, and spodic horizons in Precambrian paleosols, but they have not yet been found in paleosols older than Devonian (Retallack 1986a). Perhaps they do not occur in Precambrian paleosols because their formation is promoted by large plant roots which influence soil formation more than is possible under microbial ecosystems.

A common kind of surficial horizon in Precambrian paleosols is one enriched in clay (horizon At in Fig. 5). In some Early Precambrian paleosols, this horizon is a distinctive lime green color and is chemically reduced (high Fe^{2+}/Fe^{3+}), with a thick subsurface horizon (saprolite or C horizon) which exhibits the crystalline, metamorphic, or sedimentary structures of its parent material. These reduced but formerly well-drained paleosols are of great interest as evidence for a very weakly oxidizing primeval atmosphere (Holland 1984).

Some subsurface horizons in Precambrian paleosols are enriched in carbonate minerals (calcic or Bk). The carbonate may resemble shallow-marine carbonates in being nodular or pisolitic. A characteristic feature of pedogenic carbonate nodules is displacive fabric, e.g., spar-filled cracks or wedge-shaped cavities which surround the nodules or clasts due to their movement in loose soil. Replacive fabric is common both in soil and marine nodules. In this fabric, embayed margins of grains are filled by and float in a micritic cement. Almost all Precambrian carbonates are recrystallized to some extent, and this may obscure original fabrics (Kalliokoski 1986). Isotopic (Frank et al. 1982) and cathodoluminescence studies (Dickson and Coleman 1980) may be useful in unraveling their burial alteration history. Dolomite is a common carbonate in Precambrian paleosols (Retallack 1986a) whereas low-magnesium calcite is characteristic of calcareous Phanerozoic paleosols (Doner and Lynn 1977). It remains unclear whether this mineralogical difference is a product of burial alteration or records secular variation in the chemical conditions of soil formation. Similarly, the primary or diagenetic origin of marine dolostones remains disputed (Zenger et al. 1980; Given and Wilkinson 1987).

Precambrian paleosol horizons include near-surface enrichments in gypsum (gypsic or By: Lowe 1983), silica (duripan or Bq: Ross and Chiarenzelli 1985), and manganese (placic: Button and Tyler 1981). The classification of thick bauxitic, kaolinitic, and ferruginous horizons is controversial (Goudie 1973) but here they are interpreted to be deep weathering zones (Co, Cv and Ct, respectively). Bauxite, laterite, and china clay have all been reported in Precambrian rocks (Button and Tyler 1981; Bardossy 1982; Morris 1985).

3.3 Soil Structure

Soil structures progressively overwhelm the structures of the parent material, such as sedimentary bedding, metamorphic foliation, or igneous crystal outlines. These soil structures may prove critical for recognizing paleosols among their enclosing rocks. Nontechnical terms often used to describe soil structures in paleosols include "massive," "structureless," "jointy," "slickensided," "veined," "mottled," and "nodular". There is a much wider array of technical terms for soil structures (Brewer 1976) but this account will focus on peds, cutans, mukkara, periglacial patterned ground, and sepic plasmic fabric. These can be used to characterize paleosols even as ancient as Precambrian.

Peds are the natural aggregates (clods) of soil material that tumble loose when one digs in soil. Cutans (clod skins) are the naturally altered surfaces of peds, such as films of clay washed down cracks, or the rims of peds stained by iron oxides. Peds also are defined by cracks, pores, vugs, vesicles, and other open spaces, but these almost always collapse during burial of clayey paleosols. Former open spaces filled with later cements may be found in paleosols which developed on vesicular lava, volcanic ash, or limestone. In most cases, however, peds are recognized in paleosols by means of their bounding cutans. Peds are not necessarily the pieces of rock that come loose from a paleosol upon impact with a hammer. These fragments more often reflect jointing, foliation, and other nonpedogenic features.

The clearest known remnant of peds and cutans in a Precambrian paleosol occurs in the subsurface (AC in soil-science notation) horizon of the Waterval Onder clay paleosol, Transvaal, South Africa (Retallack 1986b). Here, blocky angular peds are defined by thick dark cutans of washed-in clay with opaque oxide grains of amorphous iron and manganese minerals (Fig. 7). On first inspection, one might be tempted to dismiss this structure as jointing, hydrothermal veining, secondary porosity, or tectonic breccia. However, the individual peds have healed interior faces of former cutans and boundaries that are either gradational or irregular with wispy flakes. The dark clayey cutans (clay skins or illuviation argillans) are internally banded, laterally continuous, and show local wedge-shaped cross-sections, as if the peds had been loose enough to rotate slightly. Furthermore, the cutan material is identical to the overlying surface (A) horizon from which it was washed downward. The ped material, on the other hand, is identical to the subsurface (C) horizon from which it was loosened. These observations indicate a complex history of multiple episodes of breaking, infilling, and closure under randomly oriented, extensional, and compressional stresses at moderate temperatures and negligible pressures. Such conditions are unlikely in any purely sedimentary, volcanic, igneous, or metamorphic environment. Other possible blocky peds in

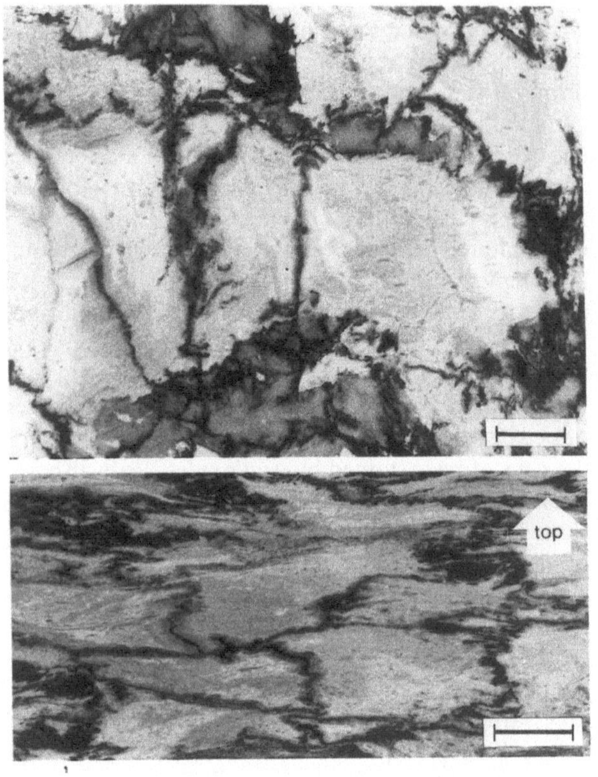

Fig. 7. Angular blocky peds in polished slabs cut horizontally (*above*) and vertically (*below*) in subsurface (AC) horizon of Waterval Onder clay paleosol of Fig. 2. *Scales* graduated in cm. (Retallack 1986b)

Precambrian paleosols worthy of further examination are the clay lined units of soil material illustrated by Gay and Grandstaff (1980) and the so-called "zebraic fracturing" of Lowe (1983), Lowe et al. (1985), and Duchac and Hanor (1987).

Even more diagnostic of paleosols than blocky peds are the pelletlike granular peds which characterize grassland soils (Retallack 1988a). However, other kinds of peds are not much more than mudcracks and may be indistinguishable from subaqueous synaeresis cracks or hydrothermal veins in deformed rocks. Among cutans, those formed by successive lining of cracks (illuviation cutans) are most diagnostic of paleosols (Fig. 7). Cutans which have formed by shear in soils (stress cutans) may show slickensides in a distinctively scattered orientation (Gray and Nickelsen 1989) but these are not markedly different from slickensides formed by by faulting, igneous injection, or compaction around coarse clasts. Cutans formed by chemical modification inward from a surface (diffusion cutans) also may not be appreciably different from features formed by development of secondary porosity during deep burial (Schmidt and McDonald 1979) or by hydrothermal alteration of porous rock (Edwards and Atkinson 1986). Putative peds and cutans therefore should be studied carefully and in context to determine their true nature.

Mukkara structure is a diagnostic feature of some paleosols (Paton 1974) and some modern swelling-clay soils (Vertisols of Soil Survey Staff 1975). Such features commonly are reported as "tepee structures" or "pseudo-anticlines" (Allen 1986). They may be differentiated from otherwise similar stratabound contorted laminations in that they have less conspicuous internal bedding and are more symmetrical and angular. The term mukkara is Australian aboriginal for "finger" and refers to the ridge of subsoil that is exposed between depressions floored by the disrupted surface horizon. This surface undulation is called gilgai microrelief. Gilgai is the aboriginal word for the small waterholes which abound on these undulating clayey soils after rainfall. Mukkara and gilgai are found in clayey soils, many of them rich in smectite, in subhumid to semi-arid climates with a strongly seasonal distribution of rainfall. The swelling of wet clay creates the large-scale structure as well as the radiating fans of intersecting shear planes that define slickensided lentil peds in some of these soils (Krishna and Perumal 1948). Numerous examples of these structures have been reported from Phanerozoic paleosols (Allen 1986) and the Waterval Onder clay is a well-known Precambrian example (Fig. 8).

Cryogenic features of paleosols include ice wedges, sand wedges, and patterned ground. Their record extends back to 2.3 Ga (Williams 1986). Sand wedges have vertically layered sandy fill and ice wedges are filled with horizontally bedded sediments. Both taper more strongly than cracks of mukkara structure. Periglacial patterned ground develops at or near the land surface during annual freeze-thaw cycles. Some of the patterns are restricted to specific climatic regimes (Washburn 1980).

Nodules and crystals are obvious features of many paleosols in outcrop but similar features can form during sedimentation, diagenesis, metamorphism, and cooling of lava. Like most nodules in sedimentary rocks, paleosol nodules are composed of minerals which form at low temperature and pressure. Diagnostic features of soil nodules and crystals include cement-filled voids which surround the nodules in a pattern which indicates that they could rotate slightly (displacive fabric). Paleosol nodules also exhibit good preservation of easily weathered minerals versus poor preservation (etching and pitting) of such minerals outside the nodules. Nodule-bearing paleosols may be overlain by fluvial paleochannels which contain similar nodules as pebbles. Paleosol nodules commonly congregate in horizons at a fixed depth below an

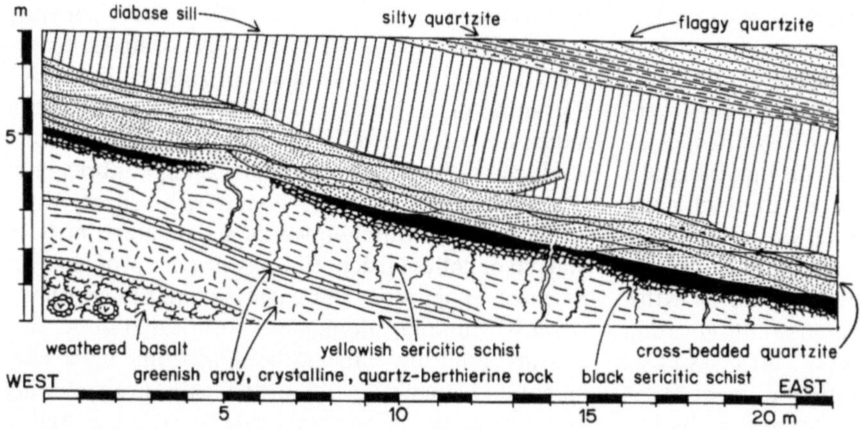

Fig. 8. Field sketch of mukkara structure in roadcut exposure of the Waterval Onder clay paleosol reconstructed in Fig. 2. (Retallack 1986b)

ancient erosional surface. In paleosols developed on sedimentary rocks, nodules may exhibit particularly well-preserved fossils, stromatolites, or sedimentary lamination inherited from the parent rock.

Among microscopic soil structures, the most clearly diagnostic for paleosols are certain kinds of bright-clay microfabric (sepic plasmic fabric). This is a pattern of randomly arranged streaks of highly birefringent clay in a less birefringent matrix (Brewer 1976). Such fabrics are observed in thin section at intermediate magnification (100 to 300 power) under cross-polarized light. The section should be cut perpendicular to the ancient land surface. Sepic plasmic fabric is a microscopic expression of peds and cutans and is therefore best expressed in strongly developed, well-drained, clayey soils and the corresponding paleosols (Brewer and Sleeman 1969).

Not all kinds of sepic fabric are diagnostic of paleosols. Grain-coating bright clay (skelsepic plasmic fabric of Brewer 1976) can be formed by the rolling of grains during sedimentation, by compaction of a clayey matrix during burial around hard mineral grains, and by grain-margin stress generated during shrinking and swelling of a soil. Sedimentary bright-clay rinds generally have smooth outer margins, and compactional clay rinds generally are thickest in the direction of principal strain. In contrast, bright clay around soil grains generally is wispy and randomly oriented. However, such distinctions are difficult to make in many cases. Cavity-lining bright clay (vosepic plasmic fabric) can be formed by infilling of secondary porosity generated by acidic groundwater during deep burial, as well as by water flow in soil cracks. Deep burial vosepic fabric may remain uncrushed or show highly discontinuous lamination unlike pedogenic examples, but these criteria also are difficult to use in practice.

In woven bright clay (omnisepic plasmic fabric), almost all the clay is bright and complexly interwoven. This can result from illitization and sericitization of clays during late diagenesis as well as through the accumulation of highly birefringent clay streaks over a long period of soil formation. Even so, when all the phyllosilicates are highly birefringent in metamorphosed paleosols, relative differences in brightness could represent original bright streaks of a less pervasively birefringent original fabric.

This leaves as diagnostic soil microfabrics incipient bright clay (insepic), streaky bright clay (mosepic), and intersecting bright clay (masepic). In these, the highly birefringent clay forms distinct wisps in a less birefringent matrix, indicative of a highly deviatoric, stratabound system of small-scale stresses that are difficult to envisage outside of soil environments. If the wisps within the highly birefringent matrix (omnisepic) of the thin section illustrated here (Fig. 9) can be considered to reflect an originally less pervasively sepic fabric, it would then be identified as clinobimasepic in the soil terminology of Brewer (1976).

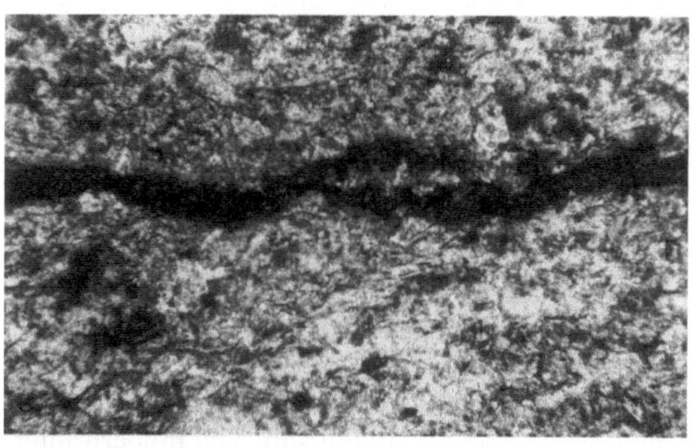

Fig. 9. Cross-polarized thin section of the subsurface (C) horizon of the Waterval Onder clay paleosol of Fig. 2. Note woven bright clay (omnisepic plasmic) microfabric due to sericitization, with grayish streaks (third order purple birefringence) reminiscent of clinobimasepic plasmic fabric. Field is 0.8 mm wide

Paleosol microfabrics may include all the igneous, metamorphic, hydrothermal and sedimentary fabrics of their parent materials. These parent microfabrics are especially well preserved in paleosols that were young, waterlogged, or salty, and so less affected by biological activity or wetting and drying. Massive, opaque, isotropic or nearly isotropic (inundulic and undulic) fabrics formed by cementation or clay flocculation in soils also can form during sedimentation. Colloform cavity fills, cavity-lining crystal growths, grain overgrowths, grain etching or dissolution within clay skins, and complex vein networks formed during soil formation are not much different from those formed in the subsurface by groundwater, hydrothermal alteration, or igneous instrusion. Microfabrics produced during shallow burial are difficult to distinguish from those produced during soil formation even in paleosols as young as Quaternary. A soil origin should be considered whenever such fabrics are observed but a distinction between pedogenetic and diagenetic origins may require a careful evaluation of the paragenetic sequence of mineral growth, together with isotopic, cathodoluminescence, and geochemical-modeling studies.

4 What Else Could it Be?

Paleosols are not always easy to recognize and it is therefore useful to consider what else could resemble a paleosol. There are several possibilities. Some beds superficially similar to paleosols form by chemical alteration or biologic activity on the floors of oceans or lakes. During long periods between sedimentation events, burrowing by marine organisms or crystallization of salt may destroy primary bedding within sediment (Reineck and Singh 1975). Shallow tropical seafloors become cemented with aragonite and high-magnesium calcite and then become further modified (micritized) by boring algae. Deep marine hardgrounds may exhibit a complex history of dissolution and multiple encrustation of iron and manganese (Bathurst 1975). These marine and lacustrine hardgrounds and burrowed beds can be differentiated from paleosols by their sedimentary structures or fossil content.

Fluctuation of shorelines produces material gradational between sediment and soil. The modern intertidal zone does not have soil in the conventional sense except where vegetated by reeds, salt marsh, or mangroves. Fluctuation of a lacustrine shoreline in cycles of tens to thousands of years produces more complex situations (Freytet and Plaziat 1982). Fossil root traces are the best evidence for such Phanerozoic paleosols, and distinctive assemblages of stromatolites and microfossils may eventually prove useful guides to Precambrian examples (Knoll 1985a, b). Other useful indicators of paleosols in such a context include mudcracks, bright-clay (sepic plasmic) fabric, and clayey, sesquioxidic- or carbonate-rich soil horizons.

Massive or fining-upwards sedimentary beds in alluvial or colluvial deposits may present problems for paleosol recognition. For example, mudflows usually occur by slumping along a mud-lubricated slope. In a volcanic mudflow (lahar), the lubricating water may be hot and chemically aggressive, extensively altering both the muddy matrix and contained clasts as well as underlying materials (Fisher and Schmincke 1984). Mudflows, however, do not exhibit subsurface horizons of mineral enrichment, soil structures, or microbial trace fossils.

Zones of liquification during mass wasting may superficially resemble paleosols where there is contorted lamination and water-escape structures. However, paleosols tend to be more massive than contorted laminae and paleosol root traces or burrows tend to be more complex than water-escape structures. The latter typically branch upward at a narrow angle (Cloud and Lajoie 1980).

Upward-fining sequences deposited from waning flood flow (Reineck and Singh 1975) also have some of the general appearance of paleosols. They lack biological traces, subsurface horizons of mineral enrichment and etched minerals within the wisps of their clayey and sesquioxidic weathering products.

Faulting produces fine-grained and massive mylonite (Suppe 1985). However, mylonite generally is readily distinguishable from paleosol, even where developed along a bedding-parallel thrust fault because mylonite generally has uniformly directed penetrative deformation (Gray and Nickelsen 1989).

Igneous rocks are generally distinguishable from paleosol given their distinctive chemical composition, crystalline textures, typically sharp contacts with surrounding rocks, and cross-cutting relationships. However, sills and tuff beds may be deceptive if extensively altered by metamorphism and hydrothermal fluids (Fisher and Schmincke 1984). Sharp lower and upper boundaries, lack of biological traces, and the persistence of relict shards or crystalline textures in thin section commonly betray their origin.

Rock altered by the passage of groundwater during deep burial may resemble paleosols (Pavich and Obermeier 1985). This is particularly true along a major unconformity where a sharp contrast in permeability commonly occurs. Precambrian unconformities may have experienced a long history of

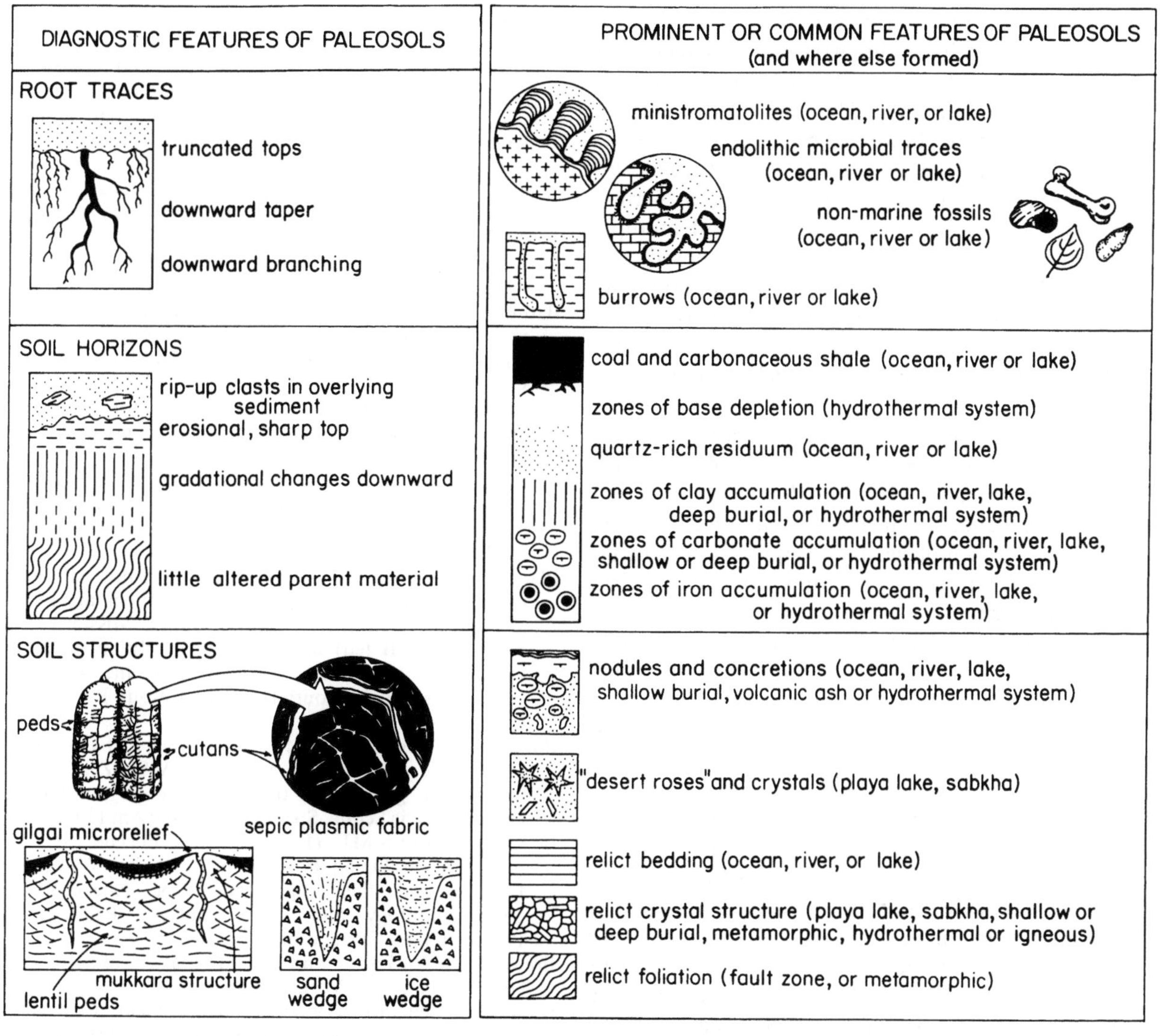

Fig. 10. Summary of diagnostic paleosol features and similar nonpedogenic features

repeated leaching and cementation. However, the resulting diagenetic structures generally are simpler than soil structures. Moreover, groundwater alteration rarely is oxidizing whereas weathering more commonly produces reddening from iron oxides.

Rock alteration by aggressive geothermal water is a special case of groundwater alteration (Rinehart 1980). Like weathering, this hydrothermal alteration may convert primary minerals to base-poor, alumina-rich clays (argillic alteration). In most cases, hydrothermal alteration is readily identifiable from the geologic context, such as pronounced lateral variation in alteration within a short distance, an association with volcanic rocks, and extension along a fault plane rather than along bedding.

Attempts have been made to distinguish paleosols from hydrothermal rocks on the basis of trace-element and alumina concentrations (Schreyer et al. 1981) and potassium enrichment (Duchac and Hanor 1987; Palmer et al. 1989). However, these criteria are not convincing, given the extreme alumina enrichment of bauxitic paleosols (Edwards and Atkinson 1986), the likelihood of even more aggressive weathering by Early Precambrian acid rain (Holland 1984), the alkali enrichment of many modern soils (Retallack 1986a), the diversity of trace-element abundances in soils (Kabata-Pendias and Pendias 1984), and the common late-diagenetic enrichment of potassium in paleosols (Curtis 1985). Chemical analyses commonly are insufficient to distinguish a paleosol from hydrothermally altered rocks.

Mineralogy is potentially more useful. Hydrothermal chlorite (propylitic alteration) and sericite (potassic alteration) are widespread whereas neither mineral typically results from weathering. The rupture of country rock by hydrothermal solutions may form a three-dimensional boxwork of veins that is superficially similar to soil structure. This may be distinguished from soil structure, however, because the minerals within the veins generally differ from typical soil minerals. Cracks in soil differ from hydrothermal boxwork by being filled more commonly with multiple generations of clay minerals. When exposed and weathered to a gossan, boxwork ores contain low-temperature minerals such as hydrous iron oxides but the primary ore minerals, such as sphalerite, form under conditions which are uncharacteristic of paleosols.

Zones of metamorphic alteration seldom are confused with paleosols because they contain distinctive minerals and cut across bedding (Kisch 1983). Superficial resemblance to paleosols may occur where metamorphic assemblages are juxtaposed due to metamorphism of contrasting protoliths or metamorphism along a fault zone. Schistosity and other metamorphic structures are quite dissimilar from biological traces and other soil structures. Furthermore, metamorphic alteration tends to be strongly reducing (Thompson 1972) whereas soil formation typically ranges from oxidizing to mildly reducing.

Many of the foregoing alternatives to interpreting a rock as a paleosol may seem so removed from soil formation as to be far-fetched. However, for small, tectonically complicated and metamorphosed outcrops, mistakes are on record in indentifying them as paleosols and in failing to identify paleosols that were present. Given that almost all Precambrian paleosols are metamorphically altered to some extent, the unraveling of metamorphic, hydrothermal, and soil components of alteration can prove challenging. In some cases, laboratory study is essential but the most tangible and direct evidence often is discovered by field observation. In searching for Precambrian paleosols, a clear concept of their diagnostic features is needed (Fig. 10). Paleopedology remains largely a field science, for which the most valuable instrument is a prepared mind.

Acknowledgments. This paper was written with fond memories of the splendid isolation, good company, and high science of the IGCP 157 meeting at Maria Laach, Germany, in 1988. I thank especially Manfred Schidlowski, who graciously facilitated my attendance. I also acknowledge more recent discussions with H. D. Holland and C. R. Feakes.

References

Allen JRL (1986) Pedogenic calcretes in the Old Red Sandstone facies (Late Silurian-Early Carboniferous) of the Anglo-Welsh area, Southern Britain. In: Wright VP (ed) Paleosols: their recognition and interpretation. Blackwells, Oxford, pp 58–86

Allen JR, Matthews RK (1982) Isotopic signatures associated with early meteoric diagenesis. Sedimentology 29:797–817

Bardossy G (1982) Karst bauxites. Elsevier, Amsterdam, 441 pp

Bathurst, RGC (1975) Carbonate sediments and their diagenesis, 2nd edn. Elsevier, Amsterdam, 628 pp

Beeunas MA, Knauth LP (1985) Preserved stable isotopic signature of subaerial diagenesis in the 1.2 b.y. Mescal Limestone, central Arizona: implications for the timing and development of a terrestrial plant cover. Bull Geol Soc Am 96:737–745

Bertrand-Sarfati J, Moussine-Pouchkine A (1983) Pedogenetic and diagenetic fabrics in upper Proterozoic Sarnyère Formation (Gourma, Mali). Precambrian Res 20:225–242

Blades EL, Bickford ME (1976) Rhyolitic ash-flow tuffs and intercalated volcaniclastic tuffaceous sedimentary rocks at Johnson Shut-ins, Reynolds County, Missouri. In: Kisvarsanyi EB (ed) Studies in Precambrian geology, with a guide to selected parts of the St Francois Mountains, Missouri. Missouri Dep Nat Resour Rep Invest 61:91–104

Boucot AJ, Dewey JF, Dineley DL, Fletcher R, Fyson WK, Griffin JG, Hickox CF, McKerrow WS, Zeigler AM (1974) The geology of the Arisaig area, Antigonish County, Novia Scotia. Spec Pap Geol Soc Am 139:191 pp

Bown TM, Kraus MJ (1981) Lower Eocene alluvial paleosols (Willwood Formation, northwest Wyoming, U.S.A.) and their significance for paleoecology, paleoclimatology and basin analysis. Palaeogeogr Palaeoclimatol Palaeoecol 34:1–30

Brewer R (1976) Fabric and mineral analysis of soils, 2nd edn. Krieger, New York, 481 pp

Brewer R, Sleeman JR (1969) The arrangement of constituents in Quaternary soils. Soil Sci 107:435–441

Bromley RG, Hanken NM (1981) Shallow marine bioerosion at Vardo, arctic Norway. Bull Geol Soc Denmark 29:103–109

Button A, Tyler N (1981) The character and economic significance of Precambrian paleoweathering and erosional surfaces in southern Africa. In: Skinner BF (ed) Economic geology 75th anniversary volume, 1905–1980. Economic Geology, El Paso, pp 676–699

Cloud PE, Lajoie KR (1980) Calcite-impregnated defluidization structures in littoral sands of Mono Lake, California. Science 210:1009–1012

Curtis CD (1985) Clay mineral precipitation and transformation during burial diagenesis. Philos Trans R Soc London Ser A315:91–105

de Coninck F, Righi D, Maucorps J, Robin AM (1974) Origin and micromorphological nomenclature of organic matter in sandy spodosols. In: Rutherford GK (ed) Soil microscopy. Limestone, Kingston, pp 263–280

Dickson JAD, Coleman ML (1980) Changes in carbon and oxygen composition during limestone diagenesis. Sedimentology 27:107–118

Doner HE, Lynn WC (1977) Carbonate, halide, sulfate and sulfide minerals. In: Dixon JB, Weed SB (eds) Minerals in soil environments. Soil Sci Soc Am, Madison, WI, pp 75–98

Dorn RI, Oberlander TM (1982) Rock varnish. Prog Phys Geogr 6:317–367

Duchac KC, Hanor JS (1987) Origin and timing of the metasomatic silicification of an Early Archaean komatite sequence, Barberton Mountain Land, South Africa. Precambrian Res 37:125–146

Dumanski J, St. Arnaud RJ (1966) A micropedological study of eluvial soil horizons. Can J Soil Sci 46:287–292

Edwards R, Atkinson K (1986) ore deposit geology. Chapman & Hall, London, 466 pp

Eyles VA (1952) The composition and origin of Antrim laterites and bauxites. Mem Geol Surv N Ireland, 90 pp

Farrow CEG, Mossman DJ (1988) Geology of Precambrian paleosols at the base of the Huronian Supergroup, Elliot Lake, Ontario, Canada. Precambrian Res 42:107–139

Fisher RV, Schmincke HU (1984) Pyroclastic rocks. Springer, Berlin Heidelberg New York, 472 pp

Folk RL, McBride EF (1976) The Caballos Novaculite revisited, part 1: origin of novaculite members. J Sediment Petrol 46:659–669

Folk RL, Roberts HH, Moore CH (1973) Black phytokarst from Hell, Cayman islands, British West Indies. Bull Geol Soc Am 84:2351–2360

Frank JR, Carpenter AB, Oglesby TW (1982) Cathodoluminescence and composition of calcite cement in the Taum Sauk Limestone (Upper Cambrian), southeast Missouri. J Sediment Petrol 52:631–638

Freytet P, Plaziat JC (1982) Continental carbonate sedimentation and pedogenesis – Late Cretaceous and Early Tertiary of southern France. Contributions in sedimentology 12. Schweitzerbart, Stuttgart, 213 pp

Friedmann EI, Weed R (1987) Microbial trace fossil formation, biogenous and abiotic weathering in the Antarctic cold desert. Science 236:703–705

Gardner TW, Williams EG, Holbrook PW (1988) Pedogenesis of some Pennsylvanian underclays: groundwater, topographic and tectonic controls. In: Reinhardt J, Sigleo WR (eds) Paleosols and weathering through geologic time: principles and application. Spec Pap Geol Soc Am 216:81–101

Gay AL, Grandstaff DE (1980) Chemistry and mineralogy of Precambrian paleosols at Elliot Lake, Ontario, Canada. Precambrian Res 12:349–373

Given RK, Wilkinson BH (1987) Dolomite abundance and stratigraphic age: constraints on rates and mechanisms of Phanerozoic dolostone formation. J Sediment Petrol 57:1068–1078

Golubic S, Campbell SE (1979) Analogous microbial forms in recent subaerial habitats and in Precambrian cherts, *Gloeothece caerulea* Geitler and *Eosynechococcus moorei* Hofmann. Precambrian Res 8:201–217

Goudie A (1973) Duricrust in tropical and subtropical landscapes. Clarendon, Oxford, 174 pp

Grandstaff DE, Edelman MJ, Foster RW, Zbinden E, Kimberley MM (1986) Chemistry and mineralogy of Precambrian paleosols at the base of the Dominion and Pongola Groups. Precambrian Res 32:97–131

Gray J (1985) The microfossil record of early land plants: advances in understanding of early terrestrialization, 1970–1984. Philos Trans R Soc London Ser B309:167–195

Gray MB, Nickelsen RP (1989) Pedogenic slickensides, indicators of strain and deformation processes in redbed sequences of the Appalachian foreland. Geology 17:72–75

Grotzinger JP (1986) Cyclicity and paleoenvironmental dynamics, Rocknest Platform, northwest Canada. Bull Geol Soc Am 97:1208–1231

Guthrie RL, Witty JE (1982) New designations for soil horizons and the new Soil Survey Manual. Soil Sci Soc Am J 46:443–444

Hofmann HJ, Jackson GD (1987) Proterozoic ministromatolites with radial-fibrous fabric. Sedimentology 34:963–971

Holland HD (1984) The chemical evolution of the atmosphere and oceans. Univ Press, Princeton, 582 pp

Holland HD (1988) Wanted! paleosols (paid advertisement). Geotimes 33:40

James NP, Choquette PW (eds) (1987) Paleokarst. Springer, Berlin Heidelberg New York, 576 pp

Kabata-Pendias A, Pendias H (1984) Trace elements in soils and plants. CRC, Boca Raton, 315 pp

Kalliokoski J (1986) Calcium carbonate cement (caliche) in Keweenawan sedimentary rocks (−1.1 Ga), Upper Peninsula, Michigan. Precambrian Res 32:243–259

Kisch HJ (1983) Mineralogy and petrology of burial diagenesis (burial metamorphism) and incipient metamorphism in clastic rocks. In: Larsen G, Chillingar GV (eds) Diagenesis in sediments and sedimentary rocks, vol 2. Elsevier, Amsterdam, pp 289–493

Knoll AH (1985a) The distribution and evolution of microbial life in the Late Proterozoic era. Annu Rev Microbiol 39:391–417

Knoll AH (1985b) A paleobiological perspective on sabkhas. In: Friedman GM, Krumbein WE (eds) Hypersaline ecosystems. Ecological Studies 53. Springer, Berlin Heidelberg New York, pp 407–425

Knoll AH, Golubic S, Green J, Swett K (1986) Organically preserved microbial endoliths from the Late Proterozoic of East Greenland. Nature (London) 321:856–857

Krishna PC, Perumal S (1948) Structure in black cotton soils of the Nizamsager Project area, Hyderabad state, India. Soil Sci 66:29–38

Lowe DR (1983) Restricted shallow water sedimentation of Early Archaean stromatolitic and evaporitic strata of the Strelley Pool Chert, Pilbara Block, Western Australia. Precambrian Res 19:239–283

Lowe DR, Byerly GR, Ransom BL, Nocita BW (1985) Stratigraphic and sedimentological evidence bearing on structural repetition in Early Archaean rocks of the Barberton Greenstone Belt, South Africa. Precambrian Res 27:165–186

McFarlane MJ (1976) Laterite and Landscape. Academic Press, New York London, 151 pp

Meyer C (1985) Ore metals through geological history. Science 227:1421–1428

Morris RC (1985) Genesis of iron ore in banded iron-formation by supergene and supergene-metamorphic processes – a conceptual model. In: Wolf KH (ed) Handbook of strata-bound and stratiform ore deposits, vol 13. Regional studies and specific deposits. Elsevier, Amsterdam, pp 73–235

Palmer JA, Phillips GN, McCarthy TS (1989) Paleosols and their relevance to Precambrian atmospheric composition. J Geol 97:77–92

Paton TR (1974) Origin and terminology for gilgai in Australia. Geoderma 11:221–242
Pavich MJ, Obermeier SF (1985) Saprolite formation beneath coastal plain sediments near Washington, D.C. Bull Geol Soc Am 76:886–900
Percival CJ (1986) Paleosols containing an albic horizon: examples from the upper Carboniferous of northern England. In: Wright VP (ed) Paleosols: their recognition and interpretation. Blackweels, Oxford, pp 87–111
Perry RS, Adams JB (1978) Desert varnish: evidence of cyclic deposition of manganese. Nature (London) 276:488–491
Pinto J, Holland HD (1988) Paleosols and the evolution of the atmosphere, pt 2. In: Reinhardt J, Sigleo WR (eds) Paleosols and weathering through geologic time: principles and application. Spec Pap Geol Soc Am 216:21–34
Reams MW (1989) Stromatolitic humid climate carbonates: a variety of calcrete? In: Douglas LA (ed) Int Work Meet Soil Micromorphology. Elsevier, Amsterdam, pp 381–386
Reimer TO (1986) Alumina-rich rocks from the Early Precambrian of the Kaapvaal craton as indicators of paleosols and as products of other decompositional reactions. Precambrian Res 32:155–179
Reineck HE, Singh IB (1975) Depositional sedimentary environments. Springer, Berlin Heidelberg New York, 439 pp
Retallack GJ (1983) Late Eocene and Oligocene paleosols from Badlands National Park, South Dakota. Spec Pap Geol Soc Am 193:82 pp
Retallack GJ (1985) Fossil soils as grounds for interpreting the advent of large plants and animals on land. Philos Trans R Soc London Ser B309:105–142
Retallack GJ (1986a) The fossil record of soils. In: Wright VP (ed) Paleosols: their recognition and interpretation. Blackwells, Oxford, pp 1–57
Retallack GJ (1986b) Reappraisal of a 2200-Ma-old paleosol near Waterval Onder, South Africa. Precambrian Res 32:195–252
Retallack GJ (1988a) Field recognition of paleosols. In: Reinhardt J, Sigleo WR (eds) Paleosols and weathering through geological time: techniques and applications. Spec Pap Geol Soc Am 216:1–20
Retallack GJ (1988b) Expected form of Precambrian paleosols by comparison with extraterrestrial paleosols. Abstr 9th Alfred Wegener Conf. Terra Cognita 8:214
Retallack GJ (1990) Soils of the past. Unwin Hyman, London, 520 pp
Retallack GJ, Feakes CR (1987) Trace fossil evidence for Late Ordovician animals on land. Science 235:61–63
Rinehart JS (1980) Geysers and geothermal energy. Springer, Berlin Heidelberg New York, 223 pp
Ross GM, Chiarenzelli JR (1985) Paleoclimatic significance of widespread Proterozoic silcretes in the Bear and Churchill Provinces of the northwestern Canadian Shield. J Sediment Petrol 55:196–204
Ross KA, Fisher RV (1986) Biogenic grooving on glass shards. Geology 14:571–573
Schau MK, Henderson JB (1983) Archaean weathering at three localities on the Canadian Shield. Precambrian Res 20:189–202
Schmidt V, McDonald DA (1979) The role of secondary porosity in the course of sandstone diagenesis. In: Scholle PA, Schluger RP (eds) Aspects of diagenesis. Soc Econ Paleontol Mineral, Tulsa, Spec Publ 26:175–207
Schopf JM, Mencher E, Boucot AJ, Andrews HN (1966) Erect plants in the early Silurian of Maine. Prof Pap US Geol Surv 550D:69–75
Schreyer W, Werding G, Abraham K (1981) Corundum-fuchsite rocks in greenstone belts of southern Africa: petrology, geochemistry, and possible origin. J Petrol 22:191–231
Siegel BZ (1977) *Kakabekia*, a review of its physiological and environmental features and their relation to its possible ancient affinities. In: Ponnamperuma C (ed) Chemical evolution of the early Precambrian. Academic Press, New York London, pp 143–154
Smith RE, Perdrix JL, Parks TC (1982) Burial metamorphism in the Hammersley Basin, Western Australia. J Petrol 23:75–102
Soil Survey Staff (1975) Soil taxonomy. USDA Handb 436:754 pp
Sokolov VA, Heiskanen KI (1984) Developmental stages of Precambrian crusts of weathering. In: Bodanov IA (ed) Precambrian geology. Proc 27th Int Geol Congr, Moscow. VNU, Utrecht 5:73–94
Southgate PN (1986) Cambrian phoscrete profiles, coated grains and microbial processes in phosphogenesis: Georgina Basin, Australia. J Sediment Petrol 56:429–441
Suppe J (1985) Principles of structural geology. Prentice-Hall, Englewood Cliffs, 537 pp
Thompson JB (1972) Oxides and sulfides in regional metamorphism of pelitic schists. In: Gill JE (ed) Geochemistry. Harpells, Gardenvale. Proc Int Geol Congr, Montreal 10:27–35
Vahrenkamp VC, Rossinsky V, Knauth LP, Beeunas MA (1987) Preserved stable isotope signature of subaerial diagenesis in the 1.2 b.y. Mescal limestone, central Arizona: implications for the timing of a terrestrial plant cover: discussion and reply. Bull Geol Soc Am 99:595–597
Vernon KC, Carroll DM (1965) Barbados. Regional Research Centre, University of the West Indies, Trinidad (Soil and Land Use Surveys) 18:38 pp
Washburn AL (1980) Geocryology. John Wiley & Sons, New York, 406 pp
Williams GE (1969) Characteristics and origin of a Precambrian pediment. J Geol 77:183–207
Williams GE (1986) Precambrian permafrost horizons as indicators of paleoclimate. Precambrian Res 32:233–242
Wright VP (1985) The precursor environment for vascular plant colonization. Philos Trans R Soc London Ser B309:143–145
Zenger DH, Dunham JB, Ethington RL (eds) (1980) Concepts and models of dolomitization. Spec Publ Soc Econ Paleontol Mineral 28:320 pp

Experimental Procedures to Simulate Weathering Under Atmospheres Which May Have Characterized the Early Archean

Hubert Wiggering, Peter Neumann-Mahlkau and Hans-Joachim Selbach [1]

[1] FB9, Geologie, Universität Gesamthochschule Essen Postfach 10 37 64, W-4300 Essen 1, FRG

Abstract

Weathering experiments with mafic and ultramafic rock material were carried out under different atmospheric conditions, such as a CO_2, a $N_2-CO_2-H_2-H_2S$, and a $N_2-O_2-CO_2$ atmosphere. Thus, a primordial, possible Early Archean, and a present-day atmosphere have been simulated. The experiments were carried out in an especially designed apparatus equipped with pH and redox electrodes to monitor pH and the oxidation potential over the whole period of the experiments.

The experimental results show that weathering under the CO_2-rich, H_2S-bearing, hypothetical Archean atmosphere was much more intense than under the modern atmosphere. Ferrous iron released by dissolution remained in solution together with Ca, Mg, Na, and K until the solubility products of various ferrous sulfides were exceeded. Given the experimental conditions ($Eh-pH-Fe^{2+}_{aq}$), the precipitated ferrous sulfides lie within the monosulfide stability field on appropriate phase diagrams.

Deduction of a rate law for the cation dissolution is difficult for our experiments because of the superimposition of reactions involving oxidation of Fe^{2+}, precipitation of FeS, and scavenging by precipitates of various solute species.

1 Introduction

Knowledge of Precambrian weathering is derived from the study of Precambrian paleosol, Precambrian shale (weathering products), and Precambrian sandstone (residue of weathering). Precambrian weathering depended upon the surface temperature and oxidation state of the early atmosphere and hydrosphere. Frictional heating due to accretion probably had a short-term effect on surface temperature (Walker 1977). The primary control soon became a balance between absorption of solar radiation and thermal radiation to space (Goody and Walker 1972). Estimates of average temperatures in the range of 25–40 °C are attributed to varying conditions, including a greenhouse effect due to water vapor and increasing CO_2 (Owen et al. 1979; Walker et al. 1981; Lovelock and Whitfield 1982).

Various models of the degassing of Earth infer an anoxic atmosphere during the early Archean (Rubey 1951, 1955; Turekian 1964; Fanale 1971; Hamano and Ozima 1978; Hart and Hogan 1978; Holland 1978, 1984a, b). Hypothetized atmospheric components include N_2, CO_2, H_2O, Ar, minor amounts of H_2, plus constituents such as H_2S, HCl, HF, CO, CH_4, NH_3, etc. (Schidlowski 1980, 1987; Kasting 1987; Schidlowski and Wiggering 1988).

Most models of the early atmosphere-hydrosphere system start with an immediately outgassed and condensed ocean system equilibrated with CO_2 (Rubey 1951, 1955; Abelson 1966; Lafon and Mackenzie 1974; Holland 1978, 1984a, b). It is assumed that weathering kept place with the injection of acidic volcanic gases into the atmosphere and hydrosphere. These initial reducing conditions changed with an increase of O_2 and simultaneous decrease of CO_2.

Weathering experiments under simulated Archean atmospheres permit evaluation of the exogenic processes which may have characterized the early Earth. This paper reports on experimental weathering to test the various models of Earth's early atmosphere-hydrosphere system and compares the experimental products with Precambrian paleosols and sediments.

2 Previous Work

The experiments reported herein differ from the numerous experiments which have been carried out on the weathering of rocks and specific minerals under such conditions as high temperature, high pressure, and low pH (e.g., Harker and Tuttle 1956; Correns

M. Schidlowski et al. (Eds.)
Early Organic Evolution: Implications for Mineral and Energy Resources

1961; Lagache 1965, 1976; Wollast 1967; Luce et al. 1972; Bischoff and Dickson 1975; Busenberg and Clemency 1976; Holdren and Berner 1979; Siever and Woodford 1979; Eggiman et al. 1980; Popp and Frantz 1980; Lin and Clemency 1981; Schott et al. 1981; Seyfried and Bischoff 1981; Berner and Schott 1982; Thornber et al. 1983; Schott and Berner 1985; Holdren and Speyer 1985; Furrer and Stumm 1986; Wiggering et al. 1986; Zinder et al. 1986).

Fewer dissolution experiments have been conducted at room temperature, atmospheric pressure, and intermediate pH. For example, Lin and Clemency (1981) placed varieties of muscovite in deionized water saturated with 1 atm partial pressure of carbon dioxide at room temperature and pressure for about 1000 h. The resultant carbonic acid buffered the bulk solution between pH 4.6 and 5.1 throughout the entire duration. At the end of the experiment, the concentration of dissolved silica had reached $10^{-4.1}$ mol/l.

Siever and Woodford (1979) studied the kinetics of the dissolution of obsidian, basalt, and various minerals at room temperature in pH-buffered solutions (4.5 to 5.5 pH) under air, N_2, and O_2. In each case, dissolution proceeded more rapidly initially but continued slowly for periods up to 6 months. Dissolution was favored by a lower pH and exclusion of air. For example, at one atmosphere of pO_2 and a pH of 4.5, basalt released silica up to 31×10^{-4} mol/l whereas this increased to 55×10^{-4} mol/l when the pO_2 was reduced to zero at the same pH.

3 Methods

Special vessels of Duran glass were designed for our experiments (Fig. 1 a). Large vessels were manufactured to produce a large reactive surface (Fig. 1 b). The gas supply was controlled electronically through an analog-to-digital converter attached to a Commodore C-64 microcomputer (Fig. 2). The Duran-glass vessels were equipped with two pH- and two redox-electrodes (Fig. 1 b). By interposition of switch registers, interferences were excluded. Samples were removed with a syringe from the vessels and placed in a nitrogen-box. Sampled solutions were placed on a glass slide. They were then dried by evaporation at room temperature in the N_2 atmosphere. The dry preparations were covered with a so-called liquid cover glass to protect them against atmospheric alteration. Precipitations on the mounts were analyzed by X-ray diffraction.

To facilitate observation of the crystallization of minerals from the sampled solutions, an electrochemical cell (Fig. 3) was fitted onto the stage of a reflecting microscope (Thornber et al. 1983). Sampled solutions were introduced with a syringe into the cell between the cover slip and a ring electrode, above a plate electrode (Fig. 3). The electrodes were gold-plated tungsten. Precipitates onto the plate electrode could be viewed through the cover slip.

To avoid the encroachment of air during analysis of the weathering residues, the sample holder was designed so that it could be purged continually by N_2 while the sample was being X-rayed (Fig. 4). To avoid effects of preferred orientation, the samples were

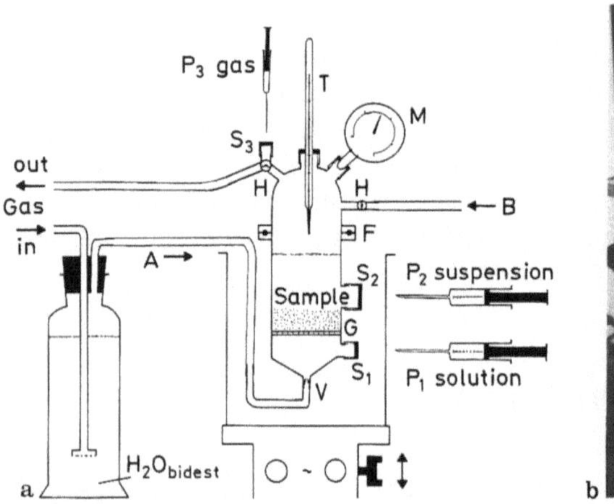

Fig. 1 a, b. a The apertures S1–3 are closed with silican stoppers. Through these stoppers samples (gases, solutions and suspensions) are removed with a special syringe (P1–3). Gas is supplied through the apertures *A* and *B*. The samples cannot pass through the glass frit (*G*), which is permeable to the gas. The sample is stirred by the gas-supply through aperture A which is secured by a nonreturn valve (*V*). Thermometers (*T*) and manometers (*M*) are fitted to check temperature and pressure. **b** In a later stage of the experiments, bigger vessels are equipped with pH- and redox-electrodes

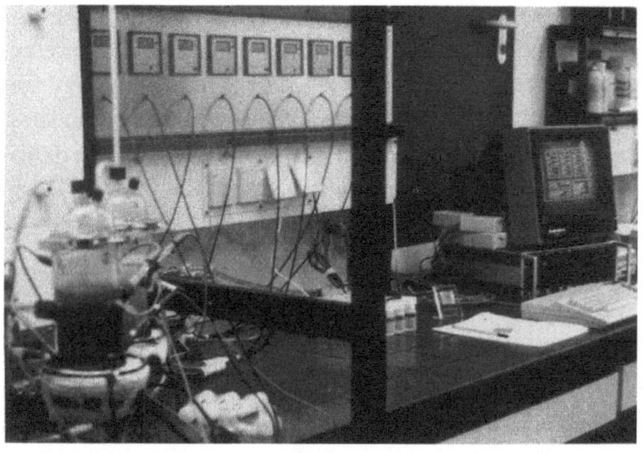

Fig. 2. The Commodore C64 microcomputer controls the gas supply and continuously records the measured data

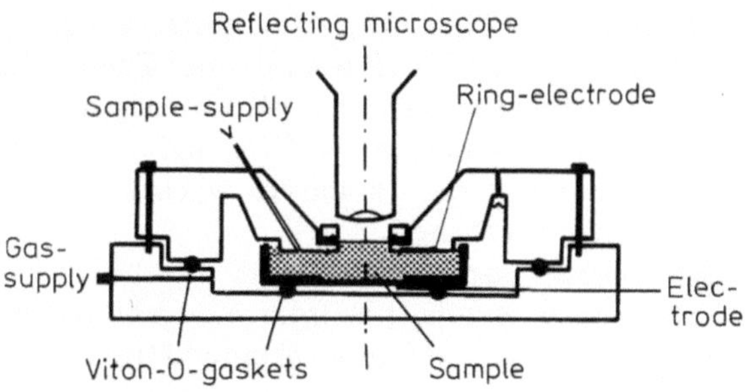

Fig. 3. Sketch of the electrochemical cell

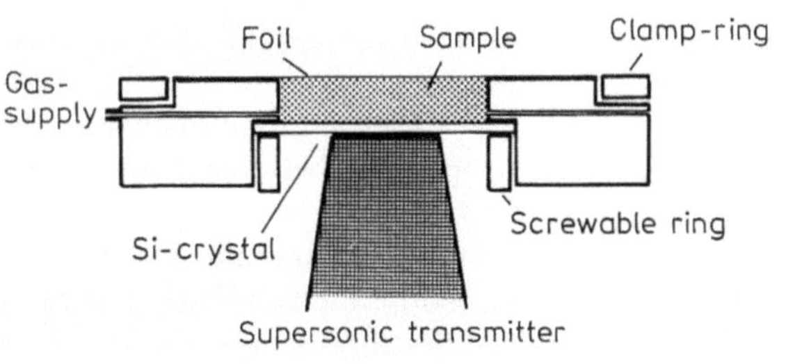

Fig. 4. Sketch of the sample holder for X-ray diffraction

vibrated supersonically during measurement. A supersonic dispenser was installed directly at an X-ray amorphous Si crystal to allow sample particles to be dispersed statistically.

4 Materials

Three types of mafic and ultramafic rock were subjected to weathering, i.e., alkali-olivine basalt, Paringa basalt, and peridotite/lherzolite (Table 1). Prior to crushing, the samples were carefully selected to exclude altered and weathered fragments. Dry sieving was used to separate the 63–180 μm size from samples of alkali-olivine basalt and peridotite/lherzolite. Preliminary tests showed that a rapid increase in the solute content of the weathering solution probably was caused by the initially rapid dissolution of ultra-fine, super-soluble particles that were produced during grinding and which adhered to larger grains (Schott et al. 1981; Wiggering et al. 1986). To avoid this effect, these samples were cleaned ultrasonically in distilled water to remove most adhering fine particles. Pulverized samples were used from the Paringa basalt. After grinding, all samples were maintained in a N_2-atmosphere.

To avoid the effect of microbiological encroachment (e.g., the catalytic effect of *Thiobacillus ferrooxidans* on the oxidation of Fe^{2+} to Fe^{3+}), both the vessels and samples were sterilized by a 30-min exposure in an autoclave at a temperature of 120 °C and a pressure of 2 bar. Electrodes were sterilized in 70% ethanol.

Table 1. Compositions of experimentally weathered rocks

1. Alkali-olivine basalt — Reykjanes ridge, NW Iceland
2. Peridotite/lherzolite — Dreiser Weiher, Eifel, FRG
3. Paringa basalt — Mt. Hunt, Kalgoorlie, W. Australia

	1	2	3
SiO_2	44.57	41.22	52.80
TiO_2	1.24	0.09	0.65
A^{l}O^{5}	13.95	2.00	14.36
F^{e}O^{5}	10.83	8.28	11.61
MnO	0.16	0.12	0.17
MgO	9.25	39.11	8.26
CaO	11.83	1.76	7.57
Na_2O	2.30	–	1.40
K_2O	0.11	0.03	0.78
PO	0.05	0.03	0.01
L.O.I.	3.04	3.12	2.53

Note: Chemical composition by X-ray fluorescence of the three mafic and ultramafic rocks used in the experiments. Units are wt.% (1 and 2: Wiggering et al. 1986; 3: Wiggering 1990).

5 Experimental Conditions

From the prepared samples, 100 g were subjected to both subaerial and subaqueous weathering under various types of atmospheres (Tables 2). The various gases were passed continuously through the sample-containing vessel at the same rate (2 l/h) whether the sample was submerged under water or exposed dry. A low excess pressure (0.1–0.2 bar) prevented atmospheric contamination. Variation in the composition

Table 2. Rationale for the atmospheric compositions used experimentally

1. Primordial atmosphere: pure CO_2.
   Rubey (1955) assumed an atmosphere in equilibrium with 91% CO_2 before the onset of rock weathering
2. Hypothetical Archean atmosphere: 65% N_2, 31% CO_2, 3% H_2S, 1% H_2.
   This composition is permissible according to models for the evolution of the atmosphere proposed by Rubey (1955), Walker (1977), Owen et al. (1983), Holland (1984a), Kasting (1987), and others
3. Present-day atmosphere: 78.04% N_2, 20.99% O_2, 0.03% CO_2.
   With the presumed increase of O_2 and decrease of CO_2 during evolution of the atmosphere, the weathering mechanisms changed simultaneously from predominantly acidic to oxidative weathering

Note: Substantial H_2S has been included in #2 to quantify the precipitation of H_2S within FeS (Wiggering 1988, 1990). Primordial volcanic gases in equilibrium with native iron would have contained more H_2S than SO_2 (Holland 1984a).

of the gases passing through the samples was checked continuously by gas chromatography. The temperature was kept at 30 +/− 1°C.

Experiments with pulverized Paringa basalt were carried out solely under submerged conditions whereas the crushed alkali-olivine basalt and the lherzolite-peridotite were exposed to both dry and submerged conditions. From the reactors with Paringa basalt, 10 ml of the covering solutions were removed and subsequently replaced by the same volume of N_2-saturated, distilled, sterilized water. This procedure facilitated normalization of Eh and pH measurements.

From subaqueous experiments with alkali-olivine basalt and lherzolite-peridotite, the reactant solutions were completely removed every second day and replaced by distilled water that had been brought to equilibrium with the respective gas. During the subaerial (dry) experiments, the sample material was irrigated with 100 ml of water that also had been brought to equilibrium with the gas. The water was forced through the sample material and thereafter drained. It took approximately 1 h for a single batch (100 ml) to pass by suction through the sample and the glass frit. Sampling by suction through the glass frit in both subaerial (dry) and subaqueous experiments avoided contamination by particulates in the sampled solution.

The removed samples were treated with HF and H_3BO_3 to dissolve probable precipitates and colloidal materials, and to preserve them before analysis by atomic-absorption spectrophotometry. Analysis of Na, K, Mg, Ca, and Fe was conducted by flame-technique atomic-absorption spectrophotometry (AAS), whereas Al and Si were analyzed by the graphite furnace technique. The weathering residues were dried by evaporation at room temperature in an inert N_2 atmosphere and prepared for X-ray diffraction.

6 Experimental Results

Three types of atmospheres were used in the experiments (Table 2), i.e., pure CO_2, a mixture of 65% N_2, 31% CO_2, 3% H_2S, and 1% He, and simulated modern atmosphere (74.04% N_2, 20.99% O_2, and 0.03% CO_2). The pH remained fairly constant in vessels which were filled only with distilled water and continually purged by each gas mixture. The results for CO_2 are shown in Fig. 5. The pH varied between 4 and 4.5 in vessels purged by CO_2 and $N_2-CO_2-H_2-H_2S$ and between 6 and 6.5 when a present-day atmosphere was used.

When a rock sample was added, dissolution immediately increased the pH from 4 to about 6 in the $N_2-CO_2-H_2-H_2S$ atmosphere and from about 5 to about 8 in the present-day atmosphere. In reactors purged by CO_2 and $N_2-O_2-CO_2$ (present-day atmosphere), the pH remained fairly constant throughout the experiments at about 5 and 8, respectively. In the reactors purged by N_2-C0_2 H_2 H_2S, the pH varied initially between 6.5 and 5.2, and then dropped to between 3 and 4 at a very late stage (Fig. 5).

Subaerial (dry) dissolution experiments for alkali-olivine basalt and peridotite/lherzolite are illustrated in Fig. 6a, b for a simulated present-day atmosphere ($N_2-O_2-CO_2$). Ca was the AAS-analyzed element which became most rapidly leached from the alkali-olivine basalt, followed by Mg, Na, Al, Fe, Si, and K in order of decreasing rates. After an initially roughly linear increase, the solution rates of Na, K, Mg, Ca, and Fe decreased slightly in the course of the experiments. Due to the high Mg content of the ultrabasic parent rock, Mg was the element most rapidly released from the peridotite/lherzolite under the present-day atmosphere, followed by Fe, Ca, Na, Si, Al, and K (Fig. 6a, b).

A change in the gas supply to $N_2-CO_2-H_2S$ drastically increased the dissolution rates under subaerial (dry) conditions (Fig. 7a, b). Mg release after 60 days was 250 times higher than in the simulated present-day atmosphere for both the alkali-olivine basalt and the peridotite/lherzolite. The increased

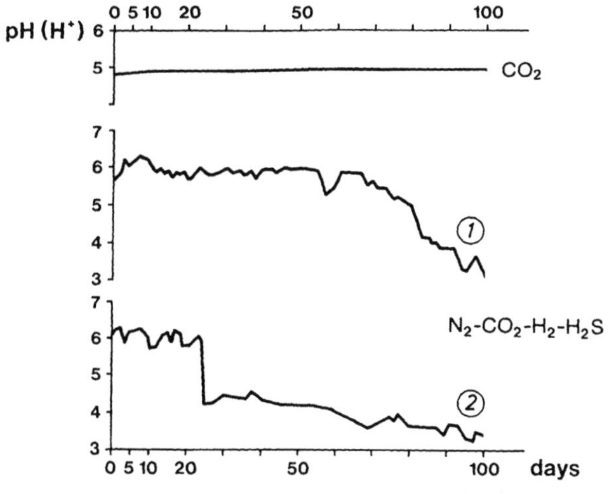

Fig. 5. Progressive change in pH during subaqueous weathering experiments. The pH remained fairly constant at about 5 in vessels filled with powedered Paringa basalt and purged with CO_2. In reactors purged by $N_2-CO_2-H_2-H_2S$ above the same basalt (*line 1*), the pH varied initially between 6.5 and 5.2 and then dropped between 3 and 4. With coarser-grained alkali-olivine basalt (*line 2*), the pH dropped between 3 and 4 more quickly

solution rates of the other elements were more moderate (Fe > Ca > Si > Na > K > Al).

The foregoing experiments were repeated under subaqueous conditions. An additional single run with pulverized Paringa basalt was conducted under a pure CO_2 atmosphere. The subaqueous solution rates for Ca and Mg from alkali-olivine basalt under the present-day atmosphere (Fig. 8a) were slightly higher then under dry conditions (Fig. 6a). The release rate decreased for Al. A rapid initial increase in cation dissolution was followed by a gradual decrease, as shown by a flattening of the curves (Fig. 8a).

Under a present-day atmosphere, dissolution of the peridotite/lherzolite occurred more rapidly subaqueously than under subaerial (dry) conditions (Figs. 6b, 8b). The solution rates increased initially and then rapidly diminished throughout the course of the experiments. This is particularly obvious for aluminum.

Under subaqueous conditions with $N_2-CO_2-H_2-H_2S$ gas supply, the rate of Mg solution from both alkali-olivine basalt and peridotite/

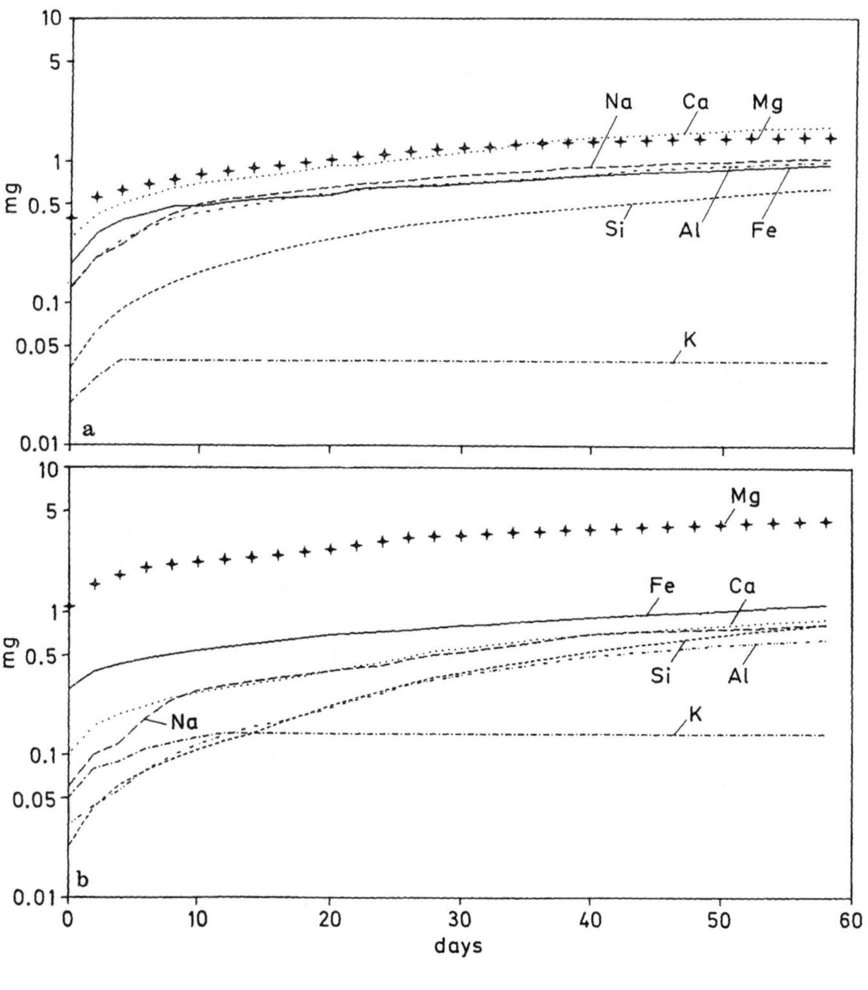

Fig. 6a, b. Release of Na, K, Mg, Ca, Fe, Al, and Si under simulated present-day atmospheric conditions (total mass of each element in solution at end of each 2-day period). **a** From alkali-olivine basalt. **b** From peridotite/lherzolite

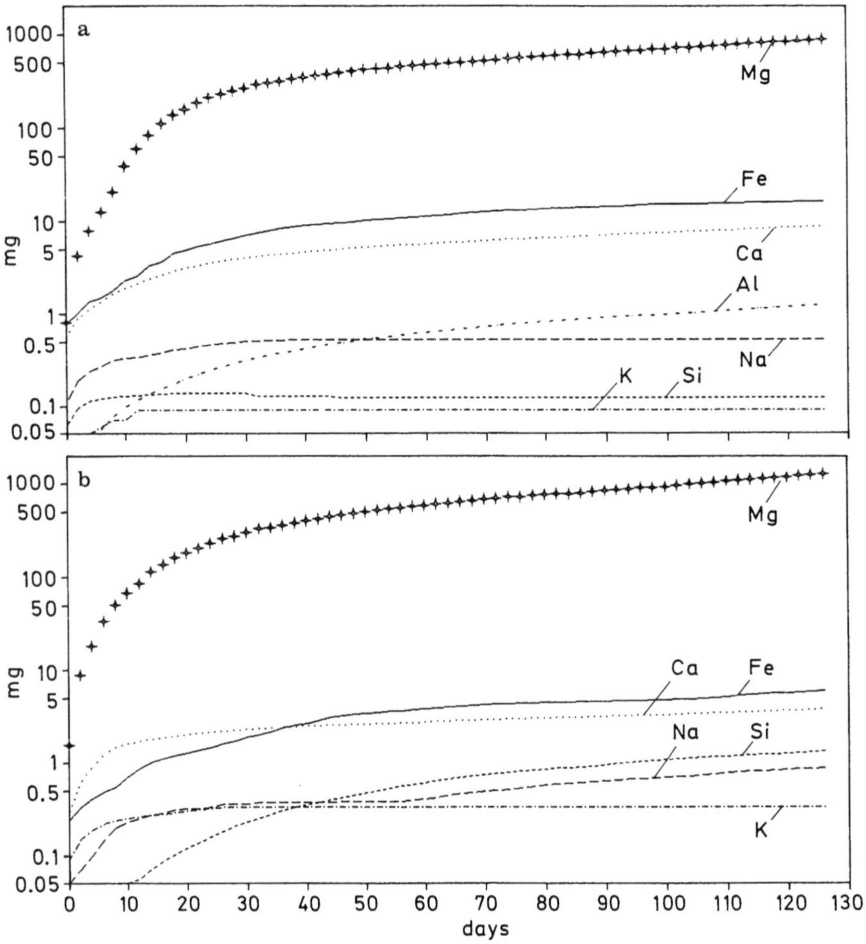

Fig. 7a, b. Release of Na, K, Mg, Ca, Fe, Al, and Si under $N_2-CO_2-H_2-H_2S$ atmospheric conditions (total mass of each element in solution at end of each 2-day period). **a** From alkali-olivine basalt. **b** From peridotite/lherzolite

lherzolite significantly diminished relative to subaerial dissolution with the same gas mixture (Fig. 9a, b). Si release also decreased subaqueously for both rock types. Alkali-olivine basalt released more Al under subaqueous conditions whereas peridotite/lherzolite released negligible Al subaqueously (Figs. 9a and b). The rate of release of Na, K, Ca, and Fe did not differ significantly between dry-versus-wet conditions. Alkali-olivine basalt continued to release Si and Mg at an increasing rate at the termination of the subaqueous experiment (after 60 days) but the release of other elements had stabilized (Fig. 9a).

Mg release also dominated dissolution of the pulverized Paringa basalt exposed to $N_2-CO_2-H_2-H_2S$ under subaqueous conditions. Mg was followed, in order of decreasing rates, by K, Ca, Na, Si, Fe, and Al (Fig. 10). The dominance of K over Ca dissolution is noteworthy.

7 Discussion

Our $N_2-CO_2-H_2-H_2S$ gas mixture (Table 2) is a possible recipe for the early Archean atmosphere. Our experiments show that weathering under such a CO_2-rich, H_2S-bearing atmosphere would have been much more intense than under the modern atmosphere. Ferrous iron released by dissolution would have remained in solution together with Ca^{2+}, Mg^{2+}, Na^+, and K^+ until the solubility products of various ferrous silicates, phosphates, or sulfides were exceeded. Of these, only ferrous sulfide could have precipitated rapidly enough (Wiggering 1990) to have formed an armour analogous to that of $Fe(OH)_3$ in modern soil (Siever and Woodford 1979; Wiggering et al. 1986).

Ferrous sulfide precipitated in three detectable ways during the weathering experiments, i.e., as an armour on the weathering grains, as a precipitate from solution (placed between a glass-slide and its cover-slip and evaporated under an N_2-atmosphere), and a coating on electrodes of the electrochemical cell. Given our experimental conditions (Eh – pH –

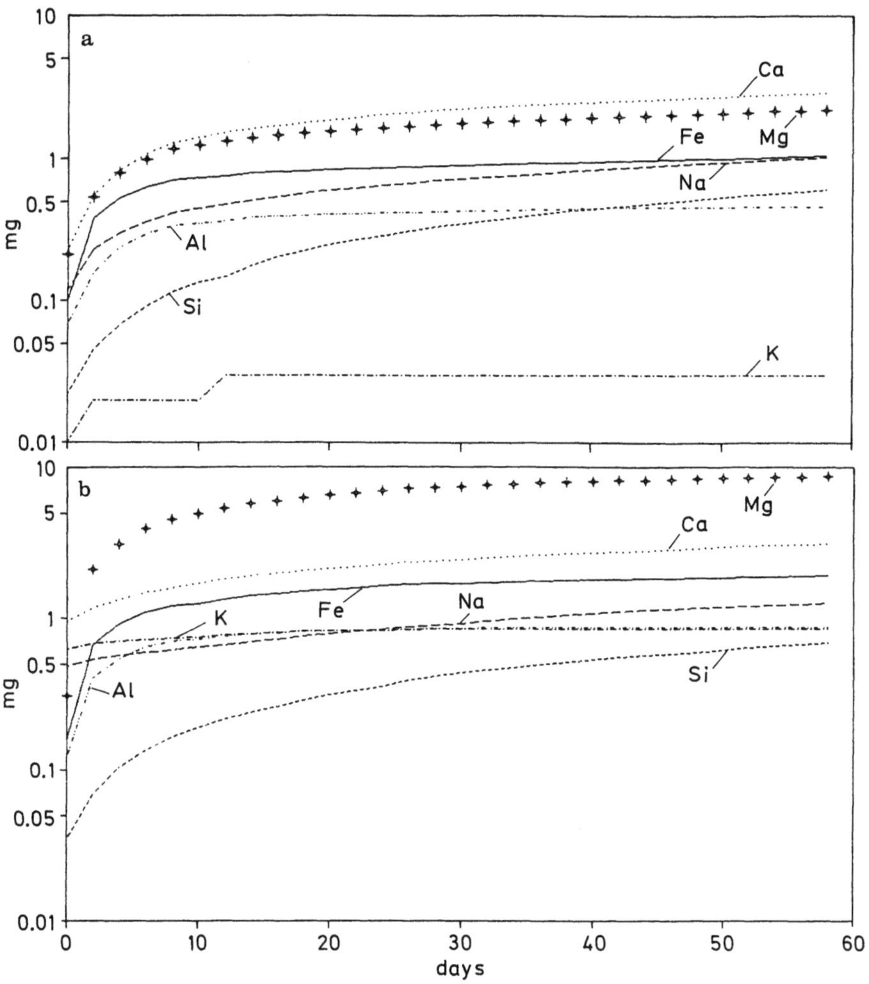

Fig. 8a, b. Release of Na, K, Mg, Ca, Fe, Al, and Si under simulated present-day subaqueous conditions (total mass of each element in solution at end of each 2-day period). **a** From alkali-olivine basalt. **b** From peridotite/lherzolite

Fe^{2+}_{aq}), this ferrous sulfide lies within the monosulfide stability field on the appropriate phase diagrams (e.g., Ewers 1983). No sulfide is identifiable in the X-ray diffractograms of sulfide-armoured weathered grains (Fig. 11). However, both smythite (Fe_9S_{11}) and mackinawite ($[Fe,Ni]_9S_8$) became identifiable after alteration of the armoured grains by agitation.

Inorganic oxidation of sulfide to sulfate generally is slow under conditions of low pO_2 (Holland 1962). Sulfide precipitation like that observed in our experiments therefore may have occurred in subaqueous environments even after free oxygen initially appeared in the Early Precambrian atmosphere. Simultaneous subaerial environments may have been characterized by the armouring of ferromagnesian minerals with $Fe(OH)_3$ which retarded dissolution.

The ferrous sulfide which formed in our anoxic experiments could have adsorbed substantial dissolved silica (Siever and Woodford 1979). Although abundant dissolved silica was released during our anoxic experiments (Figs. 7, 9), the rate of silica dissolution did decrease with increasing pO_2 and decreasing pCO_2 (Figs. 6, 8). Deduction of a rate law for silicate dissolution is difficult for our experiments because of the superimposition of reactions involving oxidation of Fe^{2+}, precipitation of FeS, and scavenging by that precipitate of various solute species, including silica.

Other investigators have found that weathering dissolution of pyroxene and amphibole is controlled by surface reactions rather than by bulk diffusion, neither diffusion in solution nor through a protective surface layer (Schott et al. 1981; Berner and Schott 1982). The products of dissolution, i.e., silica and other cations, precipitate as phases such as clay minerals which may or may not crystallize in a topotactic relationship with the underlying primary silicates (Eggleton and Boland 1982).

Our experiments show that it is not possible to characterize the release of silica and other cations as being unconditionally parabolic with time, consistent

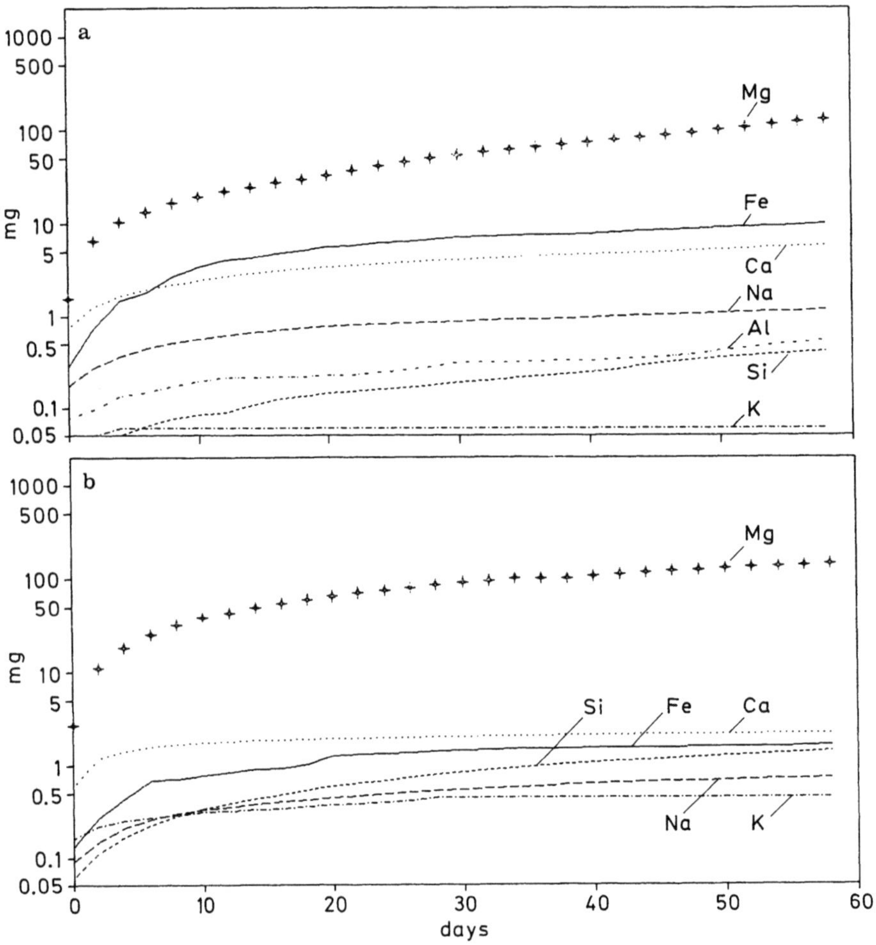

Fig. 9a, b. Release of Na, K, Mg, Ca, Fe, Al, and Si under subaqueous conditions with $N_2-CO_2-H_2-H_2S$ gas supply (total mass of each element in solution at end of each 2-day period). **a** From alkali-olivine basalt. **b** From peridotite/lherzolite

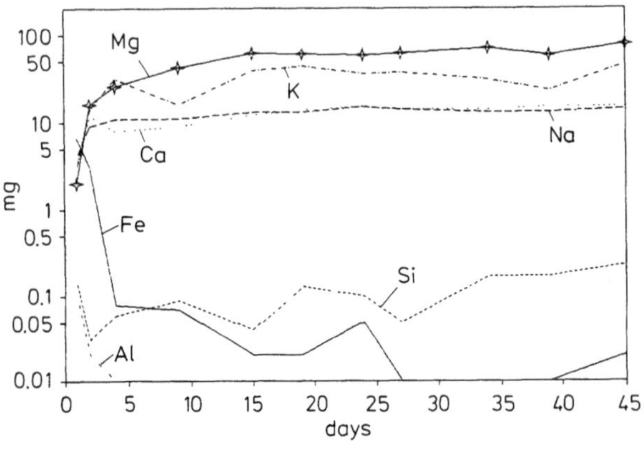

Fig. 10. Release of Na, K, Mg, Ca, Fe, Al, and Si (total mass) from Paringa basalt under subaqueous conditions with $N_2-CO_2-H_2-H_2S$ gas supply

with the results of Luce et al. (1972) and Grandstaff (1977). Like Holdren and Adaris (1982), we conclude that the parabolic dissolution behavior observed for many silicates partly results from simultaneous precipitation. If so, there is no discrepancy between parabolic behavior and the concept of dissolution control by surface reactions. Berner (1978) has shown that surface reactions are particularly important when the solubility is less than 10^{-4} to 10^{-3} mol/l. An alternative control due to solute transport may predominate at such low solubilites. Solute transport generally controls dissolution when the solubility exceeds $10^{-2.7}$ mol/l.

Syn-dissolution precipitation removes silica and aluminum from solution. Holdren and Adaris (1982) estimate that the resulting depression of solute concentrations may lower the apparent dissolution rate to half that of the true dissolution rate. Precipitation therefore must be considered when evaluating apparent dissolution rates like those presented herein. Prevention of precipitation by such means as depression of pH should enhance solute concentrations. For example, dissolved silica may approach saturation with respect to amorphous silica (Wollast et al. 1968).

Weathering in the Early Precambrian may have occurred under an H_2S-bearing, CO_2-rich atmosphere,

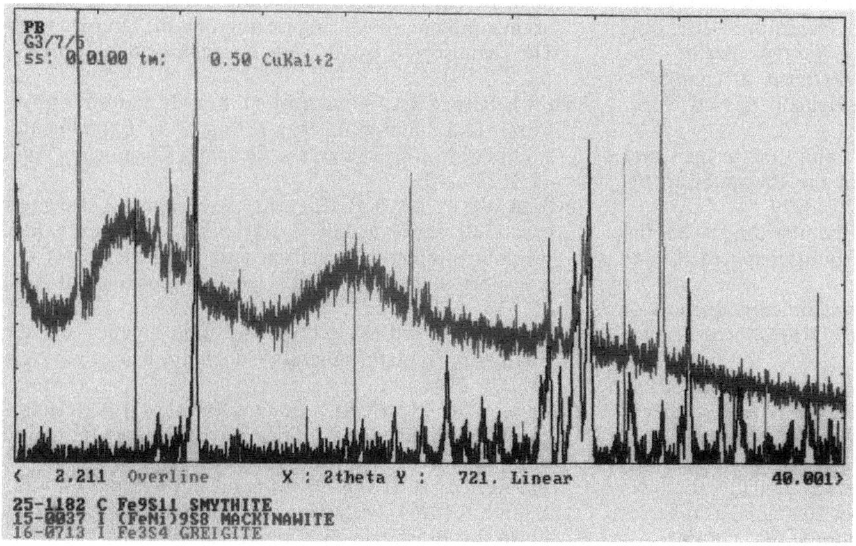

Fig. 11. Ill-defined monosulfides in the X-ray analyses of the post-experiment weathering products of Paringa basalt

like that employed in the weathering experiments reported herein. If so, such anoxic Early Precambrian weathering would have been considerably more rapid than weathering after the advent of substantial molecular oxygen in the atmosphere.

Acknowledgments. Financial support by the Deutsche Forschungsgesellschaft is gratefully acknowledged (Grants # Wi 822/1-2 and Ne 303/2-2). Alfred Lenk of Essen designed and constructed the analog-to-digital conversion system for the recording of measurements by a microcomputer.

References

Abelson PH (1966) Chemical events on the primitive Earth. Proc Natl Acad Sci USA 55:1365–1372

Berner RA (1978) Rate control of mineral dissolution under earth surface conditions. Am J Sci 278:1235–1252

Berner RA, Schott J (1982) Mechanism of pyroxene and amphibole weathering II. Observation of soil grains. Am J Sci 282:1214–1231

Bischoff J, Dickson FW (1975) Seawater-basalt interaction at 200 °C and 500 bars: Implications as to the origin of sea floor heavy metal deposits and regulation of seawater chemistry. Earth Planet Sci Lett 25:385–397

Busenberg E, Clemency CV (1976) The dissolution kinetics of feldspars at 25 °C and 1 atm. CO_2 partial pressure. Geochim Cosmochim Acta 40:41–49

Correns, CW (1961) The experimental chemical weathering of silicates. Clay Minerals Bull 26:249–265

Eggiman DW, Manheim FT, Betzer PR (1980) Dissolution and analysis of amorphous silica in marine sediments. J Sediment Petrol 50:215–225

Eggleton RA, Boland JN (1982) Weathering of enstatite to talc through a sequence of transitional phases. Clays Clay Minerals 30:11–20

Ewers WE (1983) Chemical factors in the deposition and diagenesis of banded iron-formation. In: Trendall AF, Morris RC (eds) Iron formation: facts and problems. Elsevier, Amsterdam, pp 491–512

Fanale F (1971) A case for catastrophic early degassing of the Earth. Chem Geol 8:79–105

Furrer G, Stumm W (1986) The coordination chemistry of weathering. I. Dissolution kinetics of μ-Al_2O_3 and BeO. Geochim Cosmochim Acta 50:1847–1860

Grandstaff DE (1977) Some kinetics of bronzite orthopyroxene dissolution. Geochim Cosmochim Acta 41:1097–1104

Goody RM, Walker JCG (1972) Atmospheres. Prentice Hall, Englewood Cliffs, NJ, 150 pp

Hamano Y, Ozima M (1978) Earth-atmosphere evolution model based on Ar isotopic data. In: Alexander EC Jr, Ozima M (eds) Terrestrial rare gases. Jpn Sci Soc Press, Tokyo, 173–183 pp

Harker RI, Tuttle OF (1956) Experimental data on the pCO_2-T curve for the reaction: calcite + quartz + wollastonite + carbon dioxide. Am J Sci 254:239–256

Hart R, Hogan L (1978) Earth degassing models and the heterogeneous vs. homogeneous mantle. In: Alexander FC Jr, Ozima M (eds) Terrestrial rare gases, Jpn Sci Soc Press, Tokyo, 193–206 pp

Holdren GR Jr, Adaris JE (1982) Parabolic dissolution kinetics of silicate minerals: an artifact of nonequilibrium precipitation processes. Geology 10:186–190

Holdren GR Jr, Berner RA (1979) Mechanism of feldspar weathering – I. Experimental studies. Geochim Cosmochim Acta 43:1161–1171

Holdren GR Jr, Speyer PM (1985) Reaction rate-surface area relationships during the early stages of weathering-I. Initial observation. Geochim Cosmochim Acta 49:675–881

Holland HD (1962) Model for the evolution of the Earth's atmosphere. In: Engel AEJ, James HL, Leonard BF (eds) Petrological studies. A volume in honor of A.F. Buddington. Geol Soc Am, New York, pp 447–478

Holland HD (1978) The chemistry of the atmosphere and oceans. John Wiley & Sons, New York, 351 pp

Holland HD (1984a) The chemical evolution of the atmosphere and oceans. Univ Press, Princeton, NJ, 582 pp

Holland HD (1984b) Degassing of the earth. In: Holland HD, Trendall AF (eds) Patterns of change in Earth evolution. Springer, Berlin Heidelberg New York, pp 303–311

Kasting JF (1987) Theoretical constraints on oxygen and carbon dioxide concentrations in the Precambrian atmosphere. Precambrian Res 34:205–229

Kasting JF (1988) Effect of heavy bombardment on the chemical composition of the early atmosphere. Lunar Planet Inst Contrib 681:43

Kasting JF, Ackerman TP (1986) Climatic consequences of very high CO_2 levels in Earth's early atmosphere. Science 234:1383–1385

Lafon MG, Mackenzie FT (1974) Early evolution of the ocean – a weathering model. In: Hay WW (ed) Studies in paleooceanography. Soc Econ Paleontol Mineral, Tulsa, Oklahoma, Spec Publ 20:205–218

Lagache M (1965) Contribution à l'étude de l'altération des feldspathe, dans l'eau entre 100 et 200 °C, sous diverses pressions de CO_2, et application à la synthèse des minéraux argileux. Soc Fr Min Cristall Bull 88:223–253

Lagache M (1976) New data on the kinetics of the dissolution of alkali felspars at 200 °C in CO_2 charged water. Geochim Cosmochim Acta 40:157–161

Lin FC, Clemency CV (1981) The kinetics of dissolution of muscovites at 25 °C and 1 atm. CO_2 partial pressure. Geochim Cosmochim Acta 45:571–576

Lovelock JE, Whitfield M (1982) Life span of the biosphere. Nature (London) 269:561–563

Luce RW, Bartlett RW, Parks GA (1972) Dissolution kinetics of magnesium silicates. Geochim Cosmochim Acta 36:35–50

Owen T, Cess RD, Ramanathan V (1979) Enhanced CO_2 greenhouse to compenstate for reduced solar luminosity on early Earth. Nature (London) 277:640–642

Popp RK, Frantz JD (1980) Mineral-solution equilibria-III. The system $Na_2O-Al_2O_3-SiO_2-H_2O-HCl$. Geochim Cosmochim Acta 44:1029–1037

Rubey WW (1951) Geological history of seawater. Bull Geol Soc Am 62:111–1148

Rubey WW (1955) Development of the hydrosphere and atmosphere with special reference to probable composition of the early atmosphere. In: Poldervaart A (ed) Crust of the Earth. Geol Soc Am Spec Pap 62:631–650

Schidlowski M (1980) The atmosphere. In: Hutzinger O (ed) The handbook of environmental chemistry, vol 1, pt A. Springer, Berlin Heidelberg New York, pp 1–16

Schidlowski M (1987) Photoautothrophie und Evolution des irdischen Sauerstoffbudgets. In: Jaenicke H (ed) Atmosphärische Spurenstoffe. VCH, Weinheim, pp 377–396

Schidlowski M, Wiggering H (1988) Die Erdatmosphäre im Präkambrium. Entwicklung des atmosphärischen Sauerstoffs. Geowissenschaften 6:197–226

Schott J, Berner RA (1985) Dissolution mechanisms of pyroxenes and olivines during weathering. In: Drever JI (ed) The chemistry of weathering. Reidel Dordrecht, pp 35–53

Schott J, Berner RA, Sjoberg EL (1981) Mechanism of pyroxene and amphibole weathering – I. Experimental studies of iron-free minerals. Geochim Cosmochim Acta 45:2123–2135

Seyfried WE Jr, Bischoff JL (1981) Experimental seawater-basalt interaction at 300 °C, 500 bars, chemical exchange, secondary mineral formation and implications for the transport of heavy metals. Geochim Cosmochim Acta 45:135–147

Siever R, Woodford N (1979) Dissolution kinetics and the weathering of mafic minerals. Geochim Cosmochim Acta 43:717–724

Thornber MR, Morris RC, Couper RG (1983) An electrochemical cell that allows optical microscopic examination of a mineral surface while it is reacting. Commonw Sci Ind Res Org (CSIRO) Div Mineral Res Rev, pp 223–225

Turekian KK (1964) Degassing of argon and helium from the Earth. In: Brancazio PJ, Cameron AGW (eds) The origin and evolution of atmospheres and oceans. John Wiley & Sons, New York, 74–82

Walker JCG (1977) Evolution of the atmosphere. Macmillan, New York, pp 318

Walker JCG (1986) Carbon dioxide on the early Earth. Origin Life 16:117–127

Walker JCG, Hays PB, Kasting JF (1981) A negative feedback mechanism for the long-term stabilization of Earth's surface temperature. J Geophys Res 86:9776–9782

Walker JCG, Klein C, Schidlowski M, Schopf JW, Stevenson DJ, Walter MR (1983) Environmental evolution of the Archean-early Proterozoic Earth. In: Schopf JW (ed), Earth's earliest biosphere: its origin and evolution. Univ Press, Princeton, NJ, pp 260–290

Wiggering H (1988) Weathering as a control for Precambrian conditions. Lunar Planet Inst Houston Contrib 681:108

Wiggering H (1990) Removal of hydrogen sulfide from simulated Archean atmospheres by iron sulfide precipitation. Chem Geol 85:311–320

Wiggering H, Selbach HJ, Neumann-Mahlkau P (1986) Experimental weathering of mafic and ultramafic rocks in a possible Archean atmosphere. Precambrian Res 32:181–193

Wollast R (1967) Kinetics of the alteration of K-feldspar in buffered solutions at low temperature. Geochim Cosmochim Acta 31:635–648

Wollast R, Mackenzie FT, Bricker OP (1968) Experimental precipitation and genesis of sepiolite at earth-surface conditions. Am Mineral 53:1645–1662

Zinder B, Furrer G, Stumm W (1986) The coordination chemistry of weathering-II. Dissolution of Fe(III) oxides. Geochim Cosmochim Acta 50:1861–1869

The Lower Proterozoic Hokkalampi Paleosol in North Karelia, Eastern Finland

Jukka S. Marmo

Geological Survey of Finland, SF-02150 Espoo, Finland

Abstract

An extensive Lower Proterozoic paleosol which has a minimum age of 2.2 Ga is reported from northern Karelia, eastern Finland. The paleosol suffered a greenschist facies regional metamorphism during the Svecofennian Orogeny 1.9 Ga ago. Interpretation of the Hokkalampi paleosol is based on variations in the chemical and mineral composition of its profiles and its association with overlying aluminous and quartz arenitic metasedimentary rocks. The paleosol consists of quartz-sericite schist with an increasing proportion of kyanite and andalusite toward the top. Paleosol thickness reaches 80 m but typically is much thinner. The original soil developed on both Archean granitoids and Lower Proterozoic glaciogenic rock.

The Hokkalampi paleosol clearly grades upward from its parent rocks through zones of increasing alteration. Relative to its parents, the paleosol is consistently depleted in ferrous iron, sodium, calcium, and magnesium. In the upper, most thoroughly weathered portions, ferric iron and potassium decrease, in some profiles considerably. Estimates of the extent of weathering (based on the ratio of alumina to more mobile components; the CIA index) are low near the bottom, moderate in the middle, and very high at the top. Silica and alumina constitute nearly all of the uppermost paleosol; alumina reaches 30%. The primary soil minerals probably were kaolinite and quartz.

The chemical maturity and great thickness of the paleosol and also associated metasediments are interpreted to record intense chemical weathering under a warm and humid climate, comparable to a modern tropical climate. Chemical reduction characterized the weathering until shortly before the soil became partially eroded and buried under clastic sediments. Soil erosion produced the overlying aluminous quartzite, conglomerate, and orthoquartzite.

The deep weathering which produced the Hokkalampi paleosol may have spanned an entire continent because other paleosols which are at least penecontemporaneous with the Hokkalampi are scattered across the eastern part of the Fennoscandian Shield. Lower Proterozoic paleosols have been attributed to "tropical" weathering on several other Precambrian cratons. A similar association of Lower Proterozoic aluminous rocks overlying glaciogenic deposits has been reported from the Laurentian Shield of Canada. Unfortunately, scarce paleomagnetic data and poor dating presently impedes an evaluation of either contemporaneity of Lower Proterozoic paleosols or paleolatitudinal control on climate.

1 Introduction

In eastern Finland, the Archean consists of orthogneiss, paragneiss, and remnants of greenstone belts. Collectively, these constitute the basement for the Lower Proterozoic Karelian sediments. U–Pb ages for the basement rocks plot between 3.1 and 2.5 Ga (Simonen 1980). The overlying sediments of the Karelia Supergroup (down to 2.0 Ga old) are traditionally divided into three groups, i.e., the Sariola, Jatuli and Kaleva (Pekkarinen 1979; Meriläinen 1980; Luukkonen and Lukkarinen 1986). A complete sequence of the Sariola Group consists of texturally and mineralogically immature or moderately mature arkose and conglomerate with an intermediate unit of mafic volcanics. The group is unconformably overlain by more mature quartzite and conglomerate of the Jatuli Group, which has dolomite, pelite and black schist in its upper part. These rocks were intruded and covered by mafic magma 2.2 to 2.1 Ga ago (Huhma 1986). At the top of the Karelian sequence, several thousand meters of marine mudstone and greywacke accumulated as the Kaleva Group.

The "kyanite-quartzite" of North Karelia has been known since the 1930s, when an English company quarried to evaluate its potential for the refractory industry. Reevaluation followed World War

M. Schidlowski et al. (Eds.)
Early Organic Evolution: Implications for Mineral and Energy Resources

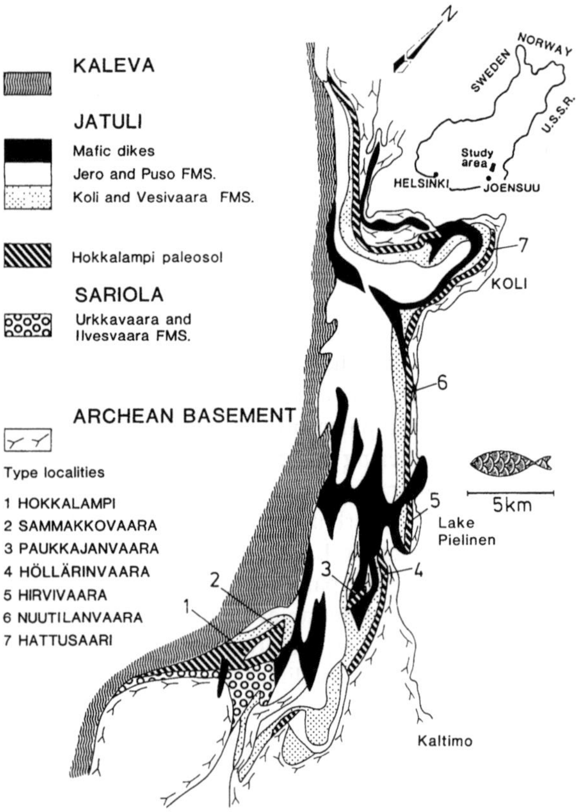

Fig. 1. Simplified lithostratigraphic map of the Koli-Kaltimo Area in eastern Finland, with type localities for the Hokkalampi paleosol. Paleosol thickness is exaggerated. In the north, the underlying Sariola sediments are included in the profile

II. New deposits were discovered and small-scale quarrying supplied material for tests (Aurola 1959). Aurola observed that the deposits are stratigraphically controlled and he was the first to suggest that they are either paleosols or their reworked equivalents. In 1979, the Geological Survey of Finland (GSF) again paid attention to aluminum silicates. In 1981–1984, the author was responsible for field studies, which reported almost 30 million tons of rock with 3 to 4 million tons of kyanite and andalusite in the Hokkalampi area (Fig. 1). Some 3000 samples were obtained by percussion drilling for placer gold deposits in the sandstone and conglomerate in the study area. Minor showings of gold were discovered in the conglomerate overlying the kyanite-andalusite deposits. Several epigenetic uranium mineralizations also occur in the study area (Piirainen 1968; Äikäs and Sarikkola 1987).

The author renamed the aluminous rock "quartz-sericite schist" because he interpreted field relationships to indicate an origin by weathering and wanted to avoid the implication of clastic sedimentation which is inherent to the term "quartzite". Using a name for a small lake from the type locality, the quartz-sericite schist is called the Hokkalampi paleosol. This informal usuage is similar to that of Senior and Mabbuth (1979), who discussed the problem of mapping paleosols. Given that the present study deals with the premetamorphic history of the study area, use of metamorphic rock names has been avoided in this paper.

This study is based on analysis of almost 3000 m of drillcore which intersects the Hokkalampi profile and associated sedimentary rocks. Over 500 chemical analyses have been performed. Silicate analyses by X-ray fluorescence were conducted in the laboratories of the GSF and Rautaruukki Company. Ferrous iron was determined by titration (at the GSF). Some trace elements and low contents of sodium were analyzed by neutron activation techniques in the Technical Research Centre of Finland. Whole rock analyses are all recalculated as waterfree to 100%, given that the loss on ignition was less than 5%. Mineral compositions of representative samples were determined by point counting of thin sections. The degree of chemical weathering was estimated with the chemical index of alteration (CIA of Nesbitt and Young 1982). Values for the CIA are given as a percentage for the ratio of molecular percentages of

$$[Al_2O_3/(Al_2O_3 + CaO + Na_2O + K_2O)] \times 100,$$

calculated from the whole rock silicate analyses, where CaO refers only to lime incorporated within silicates.

2 Regional Stratigraphic Setting

The Koli-Kaltimo area forms the northeastern part of the Schist Belt of North Karelia. All groups of the Karelia Supergroup occur in the region (Fig. 1). Although the rocks were subjected to deformation and metamorphism during the Svecofennian orogeny 2.0–1.8 Ga ago, the stratigraphic relations and primary features generally have been well preserved.

Sariola sediments were deposited directly on Archean gneiss (Marmo and Ojakangas 1984; Marmo 1986) and have been preserved only in the southern and northern parts of the study area (Figs. 1, 2). They have been interpreted as continental deposits; the Ilvesvaara and the Hattusaari Formations were deposited in alluvial environments during a rifting stage. The Urkkavaara Formation is glaciogenic. Its

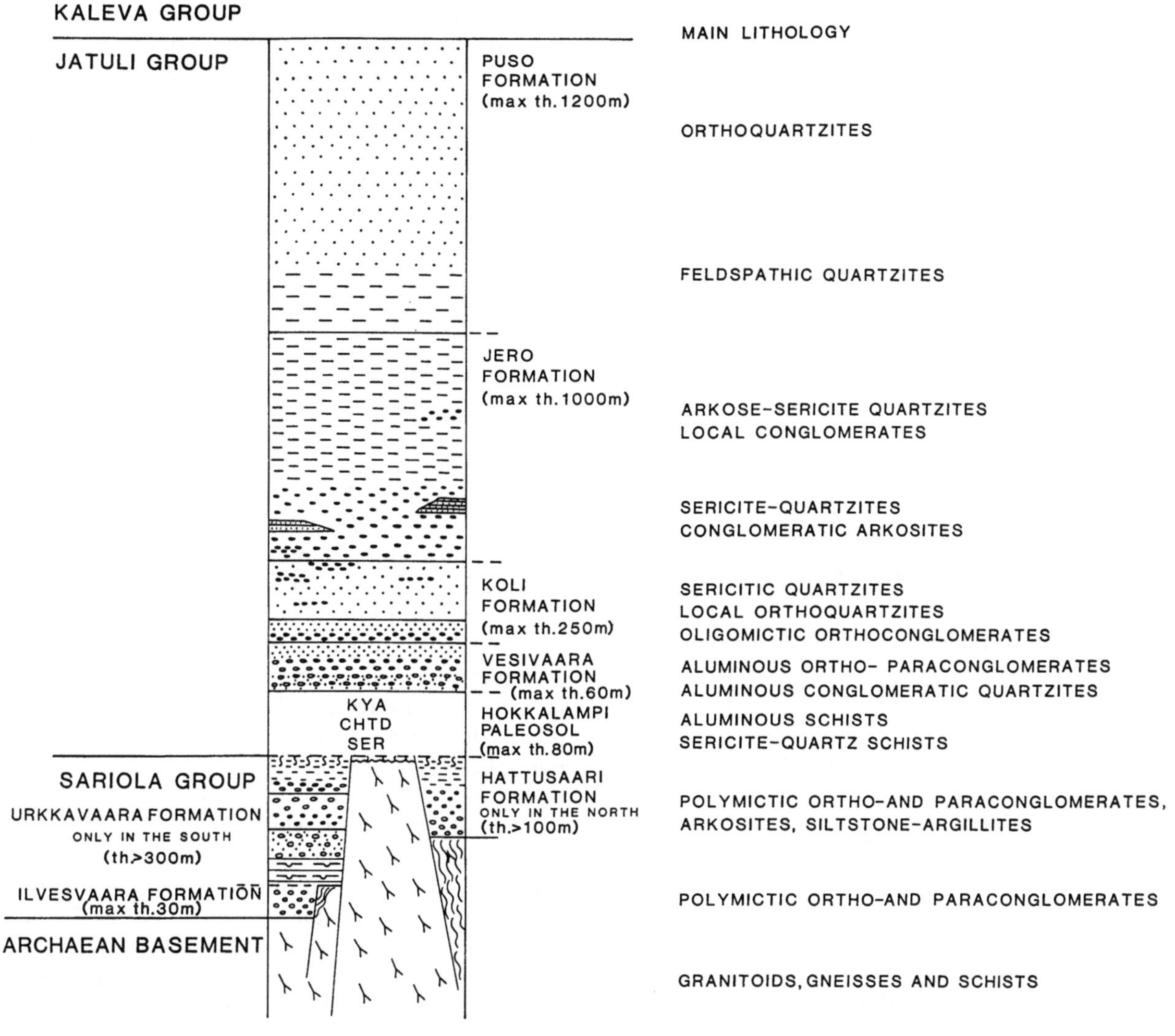

Fig. 2. Schematic stratigraphy of the Sariola and Jatuli Groups in the study area. *KYA* kyanite; *CHTD* chloritoid; *SER* sericite

lower members were deposited in a glaciomarine environment whereas the upper members formed as glacial outwash deposits during deglaciation. The Sariolan sediments are mineralogically immature and generally poorly sorted. The two lower formations had a local source, whereas the glaciogenic sediments were derived from an extensive source area which consisted primarily of felsic plutonic rocks that had undergone little if any chemical weathering prior to erosion (Marmo et al. 1988).

The Hokkalampi paleosol developed on the Sariola sediments in the southern and the northern extremities of the study area, but is generally associated with Archean basement elsewhere. Mafic rock which intrudes the paleosol is dated at 2.2 Ga (Vuollo 1991). In addition to quartz and sericite, the Hokkalampi paleosol contains appreciable quantities of aluminosilicates in its upper part. This contrasts sharply with the immaturity of the underlying rocks. The paleosol has been at least partly preserved throughout the study area and consistently underlies the lower Jatuli sedimentary sequence. The original lower contact of the paleosol is always gradational, whereas the upper contact generally is erosional.

Most of the Jatuli sequence has been interpreted as alluvial sediments (Marmo et al. 1988). Three main stages of deposition have been distinguished within the group. First, the Vesivaara Formation and the lower part of the Koli Formation were deposited as proximal braided river deposits, the upper part having formed as more distal deposits. These rocks are mineralogically mature but only poorly to moderately sorted. The corresponding source area was presumably covered by a soil which consisted of quartz and residual clays. The Jero Formation consists of sericitic quartzite, arkosic conglomerate and sericite-rich subarkose which were deposited in braided rivers during a rifting stage. The overlying rock is a well-sorted quartz arenite of the Puso Formation, which has been interpreted to represent a near-shore environment related to transgression. Paleocurrent measurements for the Jatuli sediments (Marmo et al. 1988) indicate a general westward transport of the detritus. Consequently, the source area was probably located to the east.

3 The Hokkalampi Paleosol and Associated Sedimentary Cover on the Archean Basement

3.1 Rock Descriptions

In the middle part of the study area (Fig. 1), the Hokkalampi paleosol is typically thin because the Lower Proterozoic soil was largely eroded and reworked to become aluminous quartzite and conglomerate. However, thicker profiles have been preserved at Hirvivaara, Höllärinvaara, and in the abandoned uranium mine of Paukkajanvaara. The maximum thickness of the paleosol on the Archean does not exceed 45 m and usually is less than 15 m. The paleosol is overlain by aluminous, hematite-rich conglomerate more than 30 meters thick. These sediments also contain paleoplacers rich in hematite and other heavy minerals. Only sparse analytical and petrographic data are available for these more complete sections. The following detailed description is mainly based on three drillholes at Nuutilanvaara, where the thickness of the profile varies from 11 to 16 m. The aluminous sediments overlying the Hokkalampi profile, the Versivaara Formation, were also intersected in the drillholes. Throughout this region the paleosol clearly exhibits the following, mutually gradational zones:

Zone 1 (top)	quartz-kyanite-andalusite schist	0–20 m
Zone 2	quartz-sericite schist	< 15 m
Zone 3 (bottom)	carbonate-bearing quartz-feldspar-sericite rock	< 6 m

The basal zone is characterized by a gradual vertical transition over a few metres of a relatively unaltered granitoid (Fig. 3a) into a quartz-sericite schist (Fig. 3b) which lacks internal features (Fig. 3c). In the lower part of Zone 3 a relict of the original plutonic texture is still visible. The rock is gray with a brownish or greenish hue. At the bottom of the zone, potassium feldspar and plagioclase have become partly disintegrated due to replacement by sericite and by carbonate or epidote. The disintegration of feldspars increases upward through the zone and the alteration products fill cracks within the primary grains. Chloritization of biotite also is typical of this basal zone. In addition, quartz grains locally exhibit corrosion.

In the intermediate zone, Zone 2, most of the feldspars have been destroyed, and the light gray or green rock consists of quartz and sericite (Fig. 3c). Relict textures cannot be discerned in thin section or in outcrop. The rock is now composed of scattered quartz grains of variable size and form within a matrix which consists of sericite and fine-grained quartz. Accessory minerals include titanium minerals, zircon, and tourmaline.

The uppermost paleosol zone is more intensively deformed than either the underlying or overlying more competent rocks. This zone differs from the intermediate zone mostly in the abundance of aluminosilicates (Fig. 3d). Sericite is increasingly replaced by kyanite and andalusite and the color pales to light gray or white. These alkali-free silicates locally constitute 25% of the rock. Textural characteristics, in addition to a matrix-supported framework, include the presence of pebble-size quartz and sparse, black tourmaline-quartz fragments. Chloritoid-rich quartz-sericite schist commonly occurs between the quartz-sericite schist and the overlying quartz-kyanite schist. The chloritoid content locally is sufficient to impart a dark green color to the rock (Fig. 4).

Figure 5 illustrates variation in the mineral composition in drillholes which intersect the two lower zones of the Hokkalampi paleosol and the overlying aluminous sedimentary rocks. The exact contact

Fig. 3a–d. Photomicrographs showing the gradual transition of Archean granitoid (**a**) into quartz-sericite schist of lower paleosol zone (**b**), middle paleosol zone (**c**) and finally into kyanite-schist of upper zone (**d**). Locality: Nimettömänkallio (Paukkajanvaara profile). In **a**, biotite is largely chloritized, and unaltered feldspar occurs with altered feldspar. In **b**, feldspar grains appear dusty and are intensively sericitized to form a pseudomatrix. Some quartz is finely crystalline. No feldspar remains in **c** and the rock consists entirely of sericite and quartz. The appearance of kyanite in **d** records greater loss of alkalies. *Q* quartz; *F* feldspar; *S* sericite; *K* kyanite; *B* chloritized biotite

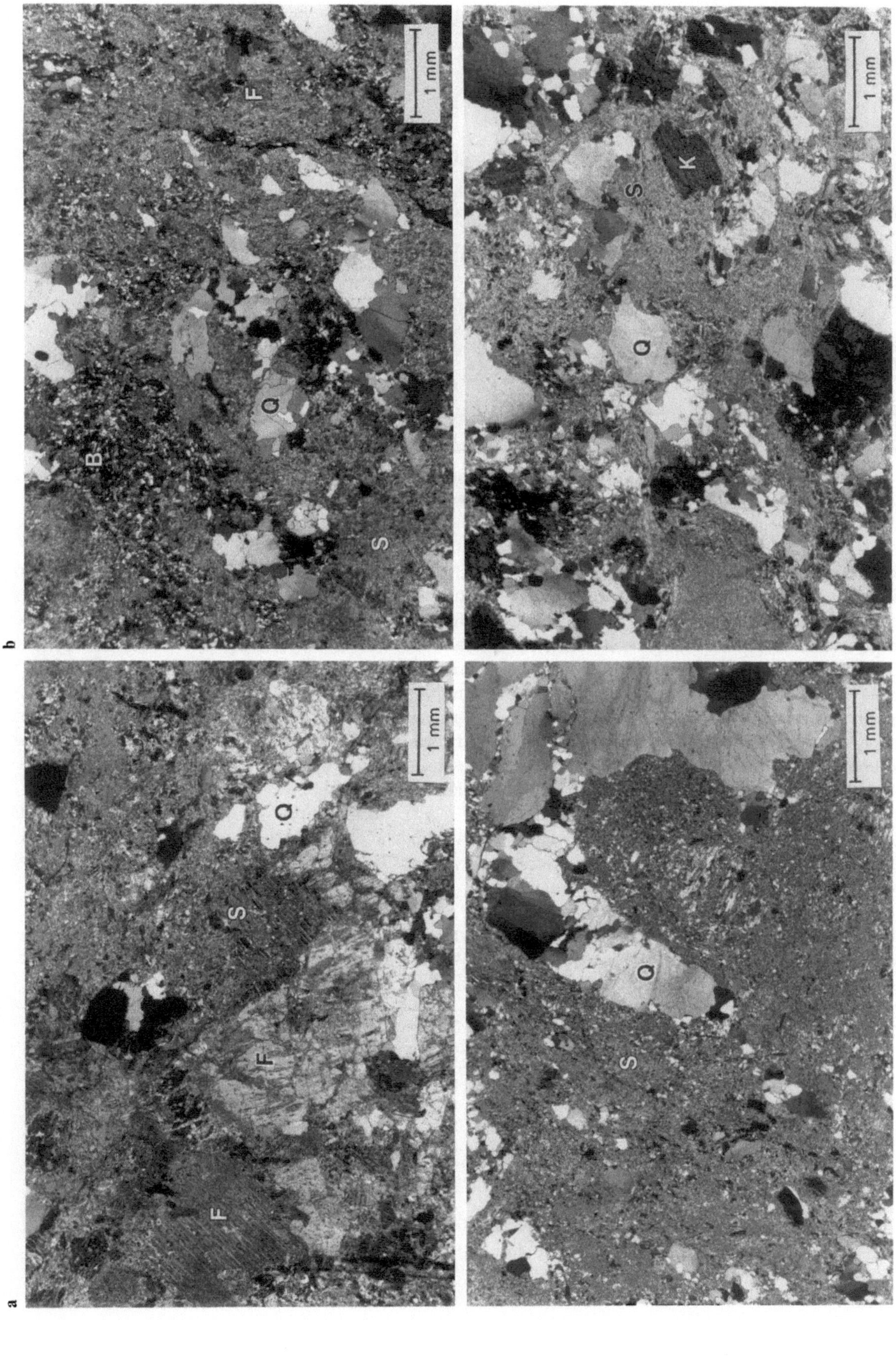

Fig. 3a–d

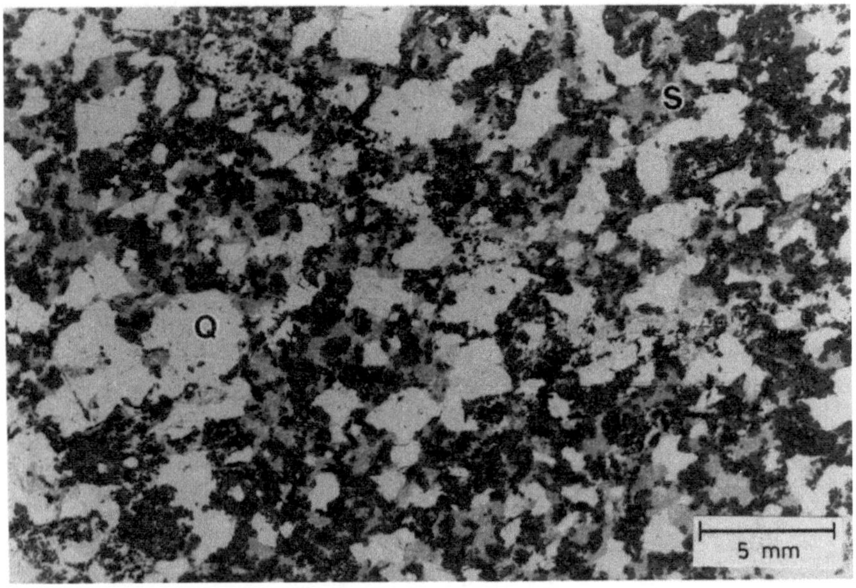

Fig. 4. Photomicrograph of a chloritoid-rich quartz-sericite schist. Locality: Höllärinvaara. Chloritoid porphyroblasts appear as dark crystal aggregates in ground-mass of quartz and sericite. *Q* quartz; *S* sericite

between the sediments and the underlying quartz-sericite schist is difficult to define texturally and has been placed at the lowermost appearance of a reddish hue. A thin layer of quartz-sericite schist, suspected of being an in situ alteration of arkosic sediments, has locally been observed in the upper part of the intermediate zone.

Poorly sorted aluminous conglomerate overlies the paleosol in the middle part of the study area. This conglomerate is, in turn, overlain by better-sorted quartz-pebble conglomerate (Fig. 6), which is rich in hematite that appears to be clastic. Excluding hematite, both units consist of quartz, sericite, and aluminosilicates. Most of the quartz fragments in the conglomerate of both Paukkajanvaara and Nuutilanvaara are sheared metamorphic quartz and common quartz. Also present are chert-like fragments (Fig. 7a), which include small amounts of sericite and titaniferous minerals (Marmo et al. 1988, Fig. 7b).

3.2 Chemical Composition

Table 1 lists the averages for eight representative granodiorite-tonalite samples and three aplitic samples from the Archean basement in this area. Note that most iron in these basement rocks is ferrous and that the values for the CIA index are low. Compositional profiles from Archean granitoids through the overlying paleosol are listed in Tables 2, 3, and 4. Table 4 also shows chemical composition of the overlying aluminous sediment. The CIA values for the lowermost paleosol (Zone 3) lie between 60 and 70 whereas they range from 70 to 80 in the intermediate zone (Zone 2) and from 80 to 90 in the upper zone (see also Table 5). The aluminous sedimentary rocks also exhibit moderate to high values. In the drillcore, a gradual vertical increase in the values for the CIA index characterizes the Hokkalampi paleosol.

The distributions of the individual elements through the Hokkalampi paleosol conform fairly well with the upward increase in the CIA values.

Table 1. Average chemical composition of granodioritic-tonalitic (1) and aplitic granitoids (2) from the study area

	1.	2.
SiO_2	69.19	76.04
Al_2O_3	15.60	14.61
TiO_2	0.39	0.06
Fe_2O_3(t)	3.78	0.70
Fe_2O_3	1.29	0.30
FeO	2.24	0.38
MgO	1.38	0.29
CaO	2.25	1.49
Na_2O	3.36	5.73
K_2O	3.80	1.02
MnO	0.05	0.01
P_2O_5	0.20	0.02
CIA	53.2	52.5
No. analyses	8	3

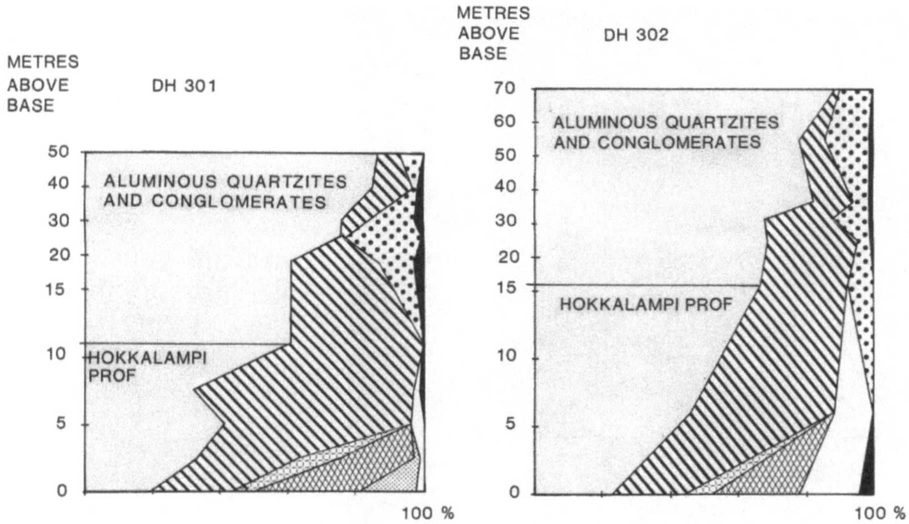

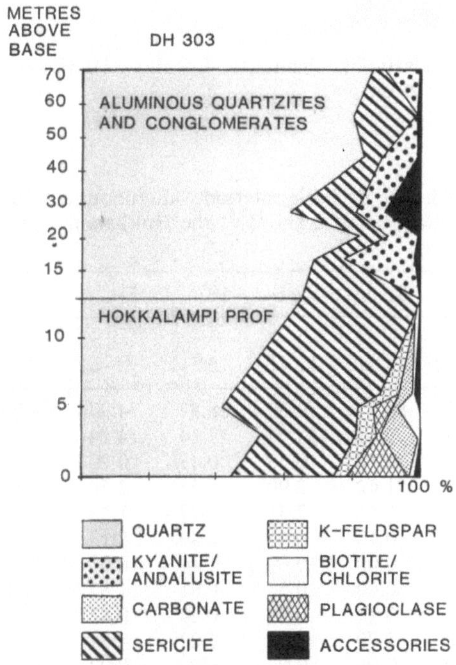

NORMATIVE MINERAL COMPOSITION OF BASEMENT SAMPLES FROM THE NUUTILANVAARA AREA

	1.	2.	3.
QUARTZ	28.80	30.65	36.11
K-FELDSPAR	25.50	19.03	12.23
PLAGIOCLASE	33.70	44.85	48.54
ENSTATITE+ FERROSILITE x	5.35	4.31	1.26
MAGNETITE+ ILMENITE	2.85	2.72	0.70
APATITE	0.53	0.19	0.05
CORUNDUM	3.02	0.15	0.72

1. AVERAGE OF 4 GRANODIORITES
2. MIXED DRILLCORE SAMPLE OF GRANODIORITE AND APLITE
3. APLITE

X CORRESPONDING TO BIOTITE + HORNBLENDE

Fig. 5. Vertical changes of mineral composition in three drillholes which intersect aluminous sediments of the Vesivaara and Koli Formations, the middle and lower zone of the Hokkalampi paleosol, and its parent rock (Archean granitoids). CIPW norms are shown for relatively unaltered Archean plutonic rocks in this area

Alumina silica and titania all tend to be concentrated upward, albeit somewhat irregularly. Total iron generally decreases through the paleosol profile, given a plutonic parent rock. However, some iron contents are actually distinctly higher than the content of iron in the parent rocks. The contents of FeO and Fe_2O_3 exhibit opposing trends, in that the former decreases rather smoothly upward whereas the latter increases. Total iron has been strongly depleted from the uppermost paleosol zone.

Lime and magnesia exhibit similar distributions. Both initially increase through the lower part of the paleosol and then markedly decrease the virtual absence. The contents of Na_2O and K_2O show opposite trends. The content of soda markedly decreases upward to near the limit of analytical detection whereas K_2O clearly increases upward within the basal and intermediate zones. Potassium strongly correlates with aluminum (Tables 2, 3, and 4). The vertically increasing values for the CIA index in the

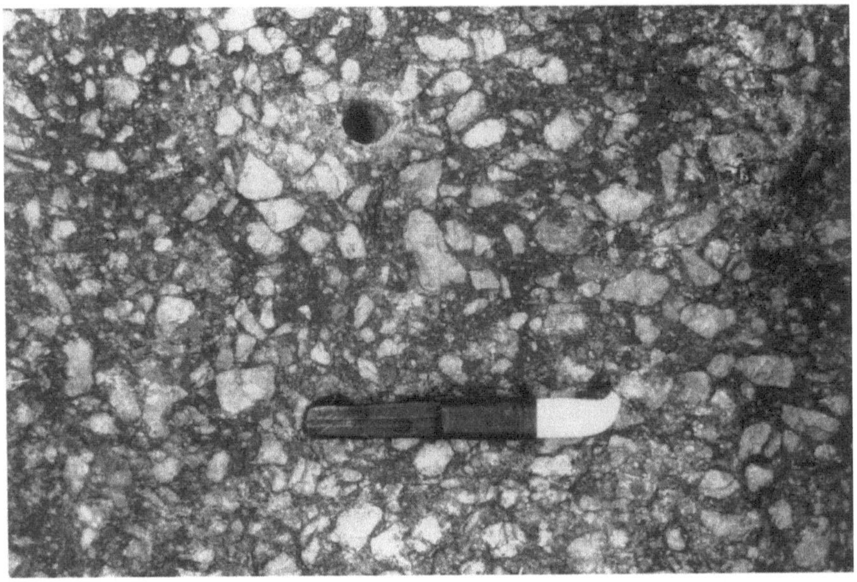

Fig. 6. Aluminous conglomerate of the Vesivaara Formation; moderately sorted quartz-pebble conglomerate rich in zircon and clastic-like hematite. Locality: Hirvivaara. Length of knife 20 cm

Table 2. Vertical variation of the chemical composition in DH 4313/84/R 303. The drillhole intersects aluminous quartzite of the Koli Formation, aluminous conglomerate of the Vesivaara Formation, zones 2 and 3 of the Hokkalampi profile and Archean granitoid. Locality: Nuutilanvaara

Analyses[a]	1	2	3	4	5	6	7	8	9	10	11	12
Meters above Base	57–69	48–57	39–48	31–39	28–31	22–28	19–22	13–19	11–13	6–11	0–6	–3
SiO_2	92.87	89.55	89.94	89.37	88.63	88.33	87.56	86.71	70.39	68.87	74.97	69.03
Al_2O_3	6.27	7.43	7.66	7.82	8.88	8.74	10.64	9.61	18.57	17.14	14.24	16.32
TiO_2	0.10	0.14	0.15	0.19	0.14	0.14	0.14	0.16	0.55	0.47	0.20	0.39
Fe_2O_3(t)	0.14	0.79	1.50	1.76	1.36	1.75	0.86	1.62	2.90	5.53	2.56	4.12
Fe_2O_3	–	–	1.33	1.58	1.19	1.57	0.72	1.51	2.41	2.77	1.32	1.42
FeO	n.d.	n.d.	0.15	0.15	0.15	0.16	0.12	0.10	0.43	2.49	1.13	2.42
MgO	0.00	0.00	0.00	0.00	0.00	0.00	0.00	0.00	0.69	1.41	0.91	1.12
CaO	0.00	0.00	0.00	0.00	0.00	0.00	0.00	0.00	0.55	0.81	1.83	2.20
Na_2O	0.04	0.15	0.06	0.09	0.03	0.13	0.11	0.19	0.39	0.24	0.83	2.48
K_2O	0.57	1.93	0.68	0.74	0.95	0.90	0.68	1.70	5.73	5.23	4.34	4.12
MnO	0.00	0.00	0.00	0.00	0.00	0.00	0.00	0.00	0.02	0.05	0.03	0.04
P_2O_5	0.01	0.01	0.01	0.03	0.01	0.01	0.01	0.01	0.21	0.25	0.09	0.18
Σ_{BR}[b]	98.91	98.63	99.38	98.57	98.31	98.89	99.39	98.90	99.37	98.78	99.20	98.48
CIA[c]	90.2	89.9	90.2	89.2	89.2	88.0	92.1	81.7	70.3	69.8	60.3	56.6

[a] Analysis 1, aluminous quartzites (distal sediments derived from the Hokkalampi profile); analyses 2–8, aluminous conglomerates (proximal sediments derived from the Hokkalampi profile); analyses 2–4, moderately sorted unit; analyses 5–8, poorly sorted unit; analyses 9–10, quartz-sericite schist of zone 2 of the Hokkalampi profile; analysis 11, slightly altered granitoid of zone 3 of the profile; analysis 12, Archean granitoid.

[b] Sum of analysis before recalculation.

[c] Value for the CIA index.

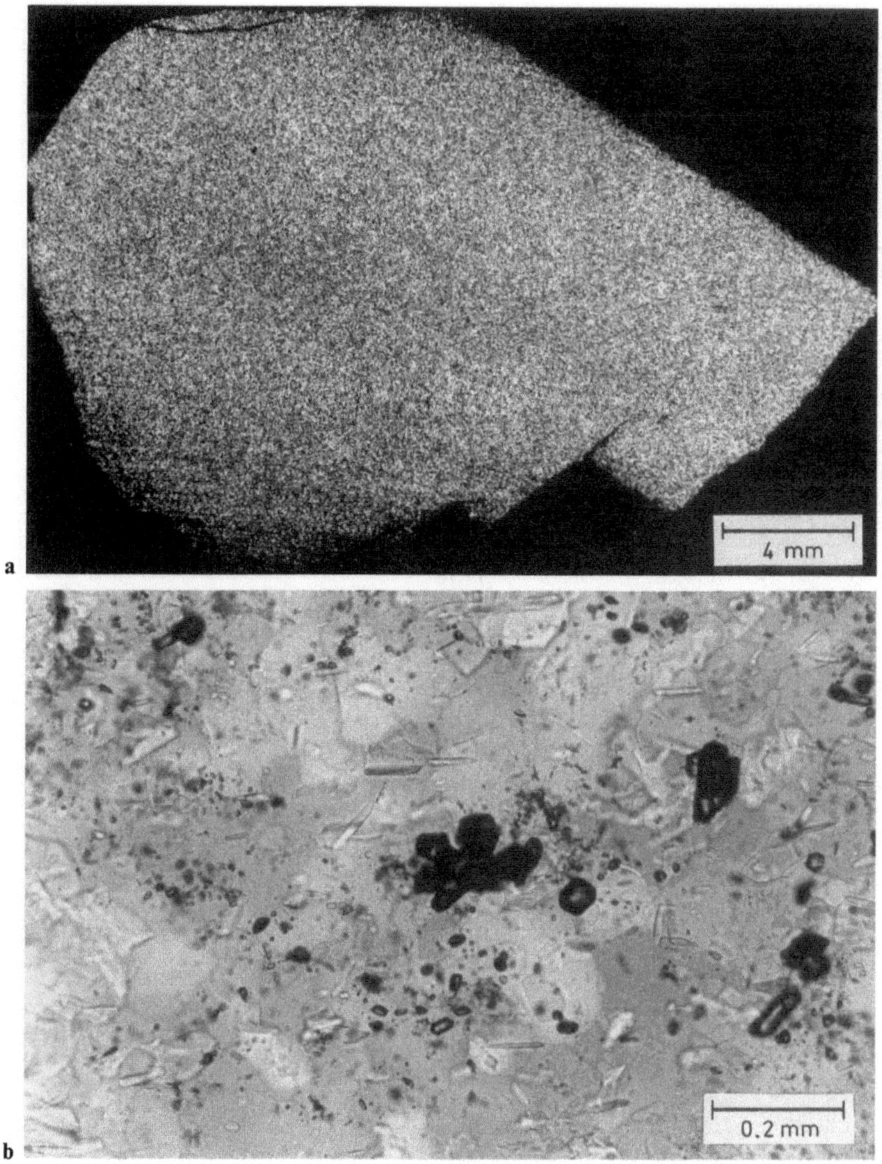

Fig. 7. Photomicrographs of clast of chert-like quartz which contains some titaniferous minerals. **a** is a fragment in aluminous conglomerate (5 ×). **b** shows detail of same clast (100 ×). Titaniferous minerals appear as dark, needle-like crystals or crystal aggregates. Locality: Paukkajanvaara

profile mainly reflect decreasing contents of sodium and alkaline earths.

In the Vesivaara Formation the content of potassium relative to the content of aluminum is much lower than in the underlying quartz-sericite schist, and the vertical increase in the CIA index here is dependent on relatively decreasing potassium. Silica smoothly increases upward through the Vesivaara Formation, whereas alumina decreases. A striking feature of these overlying aluminous sedimentary rocks is the absence of ferrous iron.

4 The Hokkalampi Paleosol Developed on the Sariola Glaciogenic Sequence

4.1 Rock Descriptions

In the southern part of the study area (Fig. 1) the Hokkalampi profiles developed on the Sariola glaciogenic rocks. These rocks have been preserved on the northern side of a dome-like structure of Archean gneiss. Sariola rocks also are exposed in the north in the Hattusaari area about 10 km north of

Table 3. Vertical variation of the chemical composition of the Hokkalampi profile and Archean granitoid in DH 4313/84/R 301. Upper contact of the profile is erosional and is overlain by 30-m-thick aluminous conglomerate derived from the Hokkalampi profile. Locality: Nuutilanvaara

Analyses[a]	1	2	3	4	5	6
Meters above Base	7–11	5–7	3–5	0–3	0	–10
SiO_2	84.28	76.52	69.38	74.78	72.03	69.85
Al_2O_3	11.41	13.85	20.50	14.60	15.15	15.26
TiO_2	0.19	9.50	0.60	0.21	0.30	0.36
Fe_2O_3(t)	1.11	4.71	2.52	3.31	3.35	3.52
Fe_2O_3	1.01	4.61	2.15	1.82	1.41	1.14
FeO	0.09	0.09	0.33	1.34	1.72	2.14
MgO	0.00	0.17	0.48	0.81	1.15	1.27
CaO	0.00	0.06	0.20	0.45	1.31	1.90
Na_2O	0.13	0.16	0.22	2.23	3.38	3.13
K_2O	2.85	3.15	6.02	3.39	3.22	4.50
MnO	0.01	0.01	0.01	0.03	0.03	0.03
P_2O_5	0.02	0.07	0.07	0.19	0.08	0.18
Σ_{BR}	98.73	99.35	95.95	98.82	98.75	99.16
CIA	77.8	75.0	74.0	64.2	57.0	53.1

[a] Analyses 1–2, quartz-sericite schist of zone 2 altered from sediments; analysis 3, quartz-sericite schist of zone 2 altered from granitoid; analysis 4, slightly altered granitoid of zone 3; analyses 5–6, Archean granitoid.

Table 4. Vertical variation of chemical composition of the Hokkalampi profile and Archean granitoid in DH 4313/84/R 302. The profile has erosional upper contact and is overlain by 40-m-thick aluminous conglomerate derived from the Hokkalampi profile

Analyses[a]	1	2	3	4	5
Meters above Base	8–14	5–8	3–5	0–3	–10
SiO_2	82.91	74.28	76.05	67.45	67.91
Al_2O_3	11.26	15.18	14.79	16.54	16.41
TiO_2	0.20	0.25	0.30	0.44	0.43
Fe_2O_3(t)	1.74	3.08	2.54	4.38	3.64
Fe_2O_3	1.53	1.81	1.57	1.80	1.25
FeO	0.18	1.14	0.88	2.32	2.17
MgO	0.16	2.11	0.94	1.48	1.00
CaO	0.03	0.17	0.19	3.44	2.49
Na_2O	0.22	0.26	0.21	1.57	3.55
K_2O	3.44	4.51	4.81	4.37	4.31
MnO	0.00	0.02	0.01	0.06	0.03
P_2O_5	0.04	0.14	0.16	0.29	0.24
Σ_{BR}	99.19	98.40	99.35	97.86	98.50
CIA	73.1	73.0	71.5	54.9	52.2

[a] Analysis 1, quartz-sericite of zone 2 altered from sediments; analyses 2–3, quartz sericite schist of zone 2 altered from granitoid; analysis 4, slightly altered granitoid of zone 3; analysis 5, Archean granitoid.

Table 5. Chemical composition of four representative samples of kyanite-quartz schist from zone 1 of the Hokkalampi profile on the Archean Basement. Localities: 1–2, Hirvivaara; 3–4, Paukkajanvaara

	1	2	3	4
SiO_2	63.28	74.92	78.93	78.07
Al_2O_3	29.92	21.53	18.46	19.34
TiO_2	0.60	0.33	0.19	0.25
Fe_2O_3(t)	0.62	0.31	0.17	0.25
Fe_2O_3	0.23	–	0.00	0.18
FeO	0.39	n.d.	0.16	0.06
MgO	0.10	0.00	0.03	0.03
CaO	0.03	0.00	0.00	0.01
Na_2O	0.39	0.99	0.08	0.08
K_2O	5.01	1.79	2.10	2.22
MnO	0.00	0.00	0.00	0.00
P_2O_5	0.04	0.00	0.04	0.00
Σ_{BR}	96.61	98.16	94.90	96.45
CIA	83.1	85.8	88.5	88.4

Koli (Fig. 1). Paleosol thickness in the south (near Hokkalampi Lake) is about 80 m but at Hattusaari the maximum thickness does not exceed 40 m due to erosion soon after it was formed. Less erosion in the south may be recorded in the smaller grain size of clastic sediment which covered the paleosol, i.e., white to gray sericite quartzite which formed as braided stream deposits. Chemical and mineralogical data are presented herein only from the type locality of the paleosol but the studies of the Hokkalampi profile in the north have revealed a close resemblance to the Hokkalampi area.

These three zones, developed on the Archean granitoids, also are present in the Hokkalampi area, where they are considerably thicker:

Zone 1	quartz-aluminous silicate schist	40 m
Zone 2	quartz-sericite schist	20 m
Zone 3	quartz-feldspar-sericite rock with carbonates	15 m

The mineral composition through the paleosol profile near Hokkalampi Lake is presented in Fig. 8a. The basal paleosol (Zone 3) is characterized by a gradual vertical transition, over 15 m, of a relatively unaltered conglomerate or arkose of the glaciogenic Urkkavaara Formation into a quartz-sericite schist (Fig. 9), which in the upper part of the basal zone lacks relict sedimentary features. Similar gradations have been observed on several outcrops throughout the southern area. As at Nuutilanvaara, there is a gradual upward decrease in the feldspar content and a simultaneous increase in sericite in both the matrix and the clasts. The sericit-

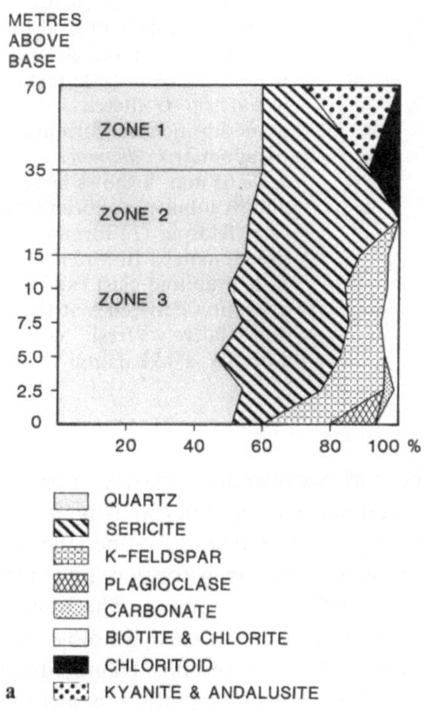

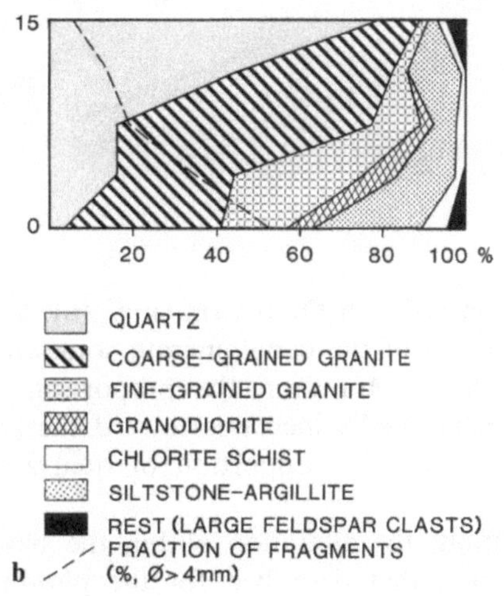

Fig. 8a, b. Vertical changes in the mineral composition through a Hokkalampi profile developed on glaciogenic sedimentary rocks (**a**). **b** shows vertical variation among the rock fragments through the lower paleosol zone

a

b

Fig. 9a, b. Textural changes in a conglomerate subjected to weathering. **a** shows ghost-like relict of conglomerate texture in outcrop. *G* altered granodioritic clast forming pseudomatrix: *diameter of scale* 65 mm. **b** shows large, almost totally sericitized clast of K-feldspar (*F*) forming pseudomatrix. In lower half of **b**, granitoid clast exhibits gradation between matrix and relatively "fresh" core. Locality: Hokkalampi

ization is first observed in the matrix (Fig. 9b), and then in the pebbles of the conglomerate somewhat higher in the zone. Moreover, the proportion of matrix (Fig. 8b) gradually increases toward the top of the zone, as does the proportion of quartz to other clast types.

In thin section, the alteration within the basal paleosol resembles that described for the plutonic parent rocks. Incipient alteration of rock fragments is characterized by indistinct boundaries between the fragments and surrounding matrix, apparent both in thin sections and in outcrop (Fig. 9). Toward the top of the basal paleosol zone, the rock fragments increasingly are transformed into pseudomatrix until they finally disappear. The intensity of alteration is correlative with clast lithology (Fig. 8b) in that granodiorite and tonalite fragments disappear before either argillite or mafic igneous fragments. True granite (both coarse- and fine-grained) is decomposed last. Much of the matrix quartz is so

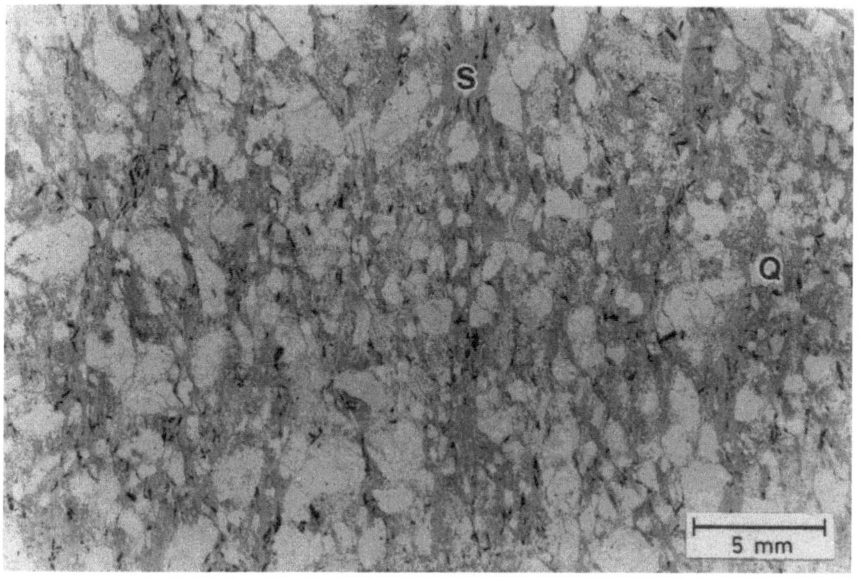

Fig. 10. Photomicrograph of typical quartz-sericite schist in middle paleosol zone. Note tiny disseminated porphyroblasts of chloritoid. *Q* quartz; *S* sericite. Locality: Sammakkovaara

fine-grained that it resembles chert; this may have precipitated from silica released during the conversion of feldspar to sericite.

Carbonate minerals are absent from the unaltered Sariola rocks but occur in the overlying basal paleosol. Carbonate first appears in the matrix, and then (somewhat higher) also in the plutonic fragments. This records a greater susceptibility of the porous matrix to alteration.

The intermediate zone of the paleosol consists of monotonous, gray to greenish quartz-sericite schist. Quartz grains of variable size float in a coarse-grained matrix of sericite and subordinate quartz (Fig. 10). Some of the fragments are chert. Also present are elongated, greenish sericitic lenses (2 to 10 cm thick) and black, angular to rounded (4–50 mm) clasts which consist of fine-grained tourmaline and quartz. Tourmaline also occurs as an accessory mineral along with zircon, sphene and rutile. In the upper part of this intermediate zone, chloritoid, kyanite and andalusite appear. Green chloritoid porphyroblasts range from 0.5 to 2.0 mm in length and are iron-rich with 29.6% FeO and 1.0% MgO (Marmo 1981).

In addition to quartz and sericite, the upper zone of the Hokkalampi paleosol is composed of kyanite and andalusite. The rock is typically white to light gray but locally reddish (Fig. 11). Texturally, the upper zone resembles the intermediate zone except for larger fragments of quartz and tourmaline-quartz. Kyanite and andalusite typically occur together as porphyroblasts within a matrix of white mica and fine-grained quartz (Fig. 11b). Kyanite prisms, commonly white or gray, are usually 2 to 3 mm in length whereas the pink andalusite crystals are larger, averaging 5 to 8 mm and locally exceeding 15 mm. Andalusite generally is largely kaolinized whereas kyanite is almost totally "fresh". Kyanite commonly forms separate bands 2 to 6 mm thick. These alternate with much thicker quartz bands, resulting in a gneissose texture (Fig. 11). Veins of intergrown kyanite and quartz are common whereas andalusite-quartz veins are rare. Both the kyanite and andalusite are essentially devoid of iron, but associated mica may contain up to 2.0% FeO based on microprobe data (Marmo 1981).

4.2 Chemistry

Given that the Hokkalampi paleosol here formed by alteration of the underlying glaciogenic sedimentary rocks, mainly of arkosic composition, analyses of these sedimentary rocks are presented in Table 6. As Archean granitoids were the sediment source, the sediments chemically resemble these basement rocks (Table 1). The prime differences in the chemical compositions are more potassium and less sodium and iron in the sedimentary rocks.

The Hokkalampi profile has been examined at several localities in the type area, but most of the chemical data are from continuous drillcore through the two upper zones. Systematic sampling of the basal zone was performed at only one locality. Elemental distributions through this basal zone and its underlying conglomerate are presented in Table 7.

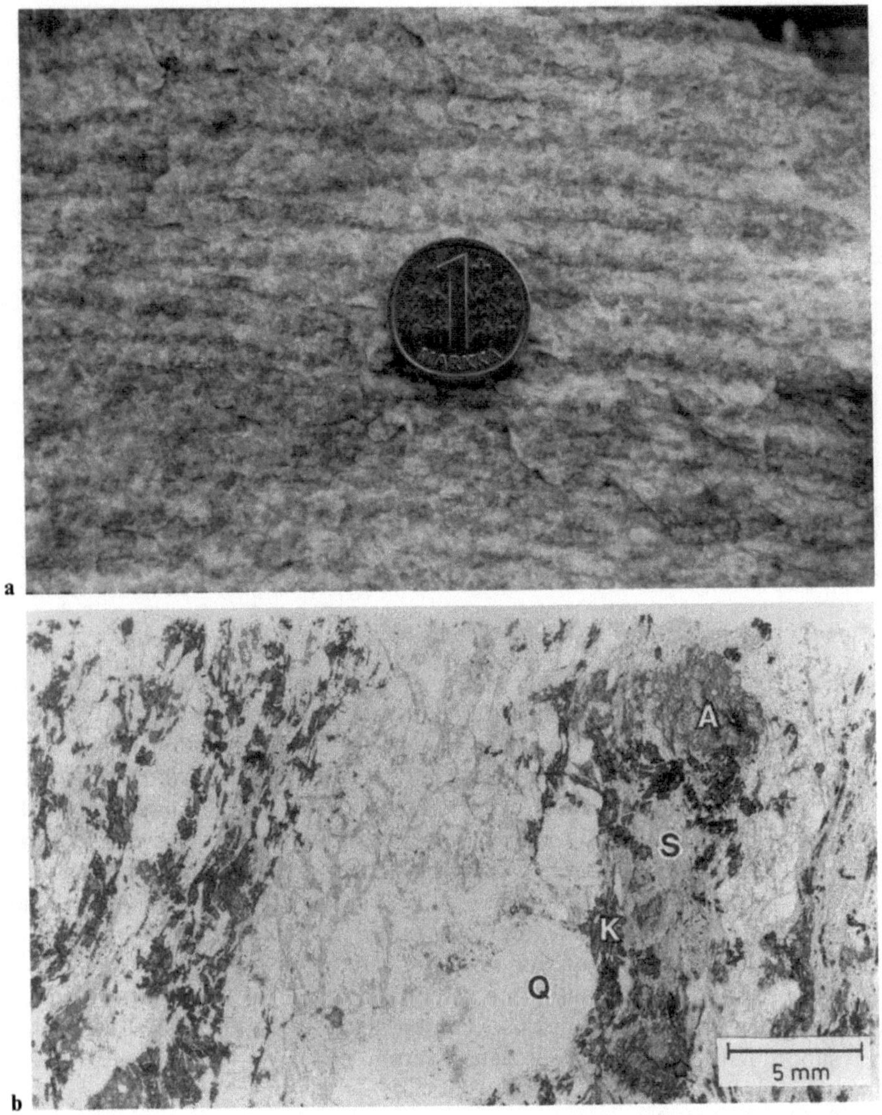

Fig. 11a, b. Kyanite-quartz schist of zone 1 upper paleosol zone. The rock exhibits compositional layering due to intense deformation. **a** Alternating gray bands are rich in andalusite and kyanite, whereas white bands are rich in quartz. *Diameter of coin* 25 mm. Locality: Hokkalampi. **b** is microphotograph; *A* andalusite; *K* kyanite; *S* sericite and pyrophyllite; *Q* quartz

Analyses of iron-poor and iron-rich quartz-sericite schists of the overlying intermediate zone are listed in Table 8 (see also Table 11). The vertical variation of chemical composition in the upper zone is presented in Tables 9 and 10. The corresponding variation through the quartz-kyanite schist and the underlying iron-rich quartz-sericite schist are shown in Table 11. The upward decrease of ferrous iron through this zone is most apparent in Table 10.

As with the granitoid-derived paleosol, the chemical index of alteration systematically increases upward through the paleosol developed on glaciogenic rock. CIA values for this rock average 60% whereas CIA rises to 65% in the basal Zone 3, 80% in the intermediate Zone 2, and as high as 96% in the upper Zone 1.

Al_2O_3 and TiO_2 are distinctly richer at the top than at the bottom of the paleosol. They are also higher than in the underlying arkosic rocks, alumina having increased by 50% and the titania by 100%. Silica increases slightly toward the top. Overall, iron decreases upward. Excluding the chloritoid-rich horizons, FeO is virtually absent from the uppermost zone but starts decreasing within the lower-

Table 6. Chemical composition of representative samples of the Urkkavaara Formation

	1	2	3	4
SiO_2	59.48	63.84	77.86	75.35
Al_2O_3	22.76	18.64	11.95	13.24
TiO_2	0.78	0.56	0.13	0.17
Fe_2O_3(t)	3.59	5.45	1.84	1.64
Fe_2O_3	n.d.	n.d.	0.50	0.42
FeO	n.d.	n.d.	1.19	1.10
MgO	2.78	4.12	0.65	0.71
CaO	0.88	0.55	1.31	1.45
Na_2O	3.68	4.61	2.94	2.64
K_2O	5.98	2.08	3.23	4.73
MnO	0.00	0.05	0.03	0.04
P_2O_5	0.05	0.09	0.06	0.08
Σ_{BR}	99.88	97.83	99.70	–
CIA	61.7	63.2	52.7	52.3

[a] 1, Argillite, lower siltstone-argillite member; 2, diamictite, diamictite member; 3, sandstone, cross-bedded conglomerate member; 4, average of sandstones.

Table 8. Average chemical composition of iron-poor (1) and iron-rich (2) quartz-sericite schists of zone 2 of the Hokkalampi profile. Analyses are averages recalculated to 100%. Locality: Hokkalampi

	1	2	SD
SiO_2	79.58	77.62	1.27
Al_2O_3	15.06	15.82	0.47
TiO_2	0.37	0.44	0.03
Fe_2O_3(t)	0.67	2.77	0.96
Fe_2O_3	–	0.78	–
FeO	n.d.	1.79	0.67
MgO	0.03	0.07	0.04
CaO	0.02	0.02	0.01
Na_2O	0.10	0.20	0.01
K_2O	4.10	3.01	0.23
MnO	0.01	0.02	0.01
P_2O_5	0.01	0.03	0.01
CIA	76.2	81.4	–
No. analyses	3	10	–

most zone, whereas Fe_2O_3 does not decrease until the two upper zones. The content of Fe_2O_3 in basal Zone 3 is actually higher than in the underlying sedimentary rocks.

The alkalies, lime, and magnesia all decrease upward. Their contents at the top of the profile are only small fractions of their contents at the bottom. As in the profiles developed on Archean granitoids, slight maxima in the contents of CaO and MgO occur in the lower part of the basal transition zone. Contents as high as 3.0% for CaO and 1.5% for MgO have been analyzed. The maximum for MgO is reached a little higher in the zone that for CaO. K_2O decreases smoothly and gradationally through the intermediate and upper zones, but is still somewhat more concentrated in the basal zone than in the underlying arkose. Phosphorus and manganese are so scarce in the paleosol and in the parent rocks that an interpretation of their distribution is difficult. However, an upward depletion of phosphorus by at least 50% is obvious.

There are some irregularities in the behavior of iron. Iron-rich horizons with up to 2% of Fe (presented as 3% Fe_2O_3 of total iron) locally occur

Table 7. Vertical variation of the chemical composition within zone 3 of the Hokkalampi profile (carbonate-bearing quartz-feldspar-sericite rock) on the conglomerate of the Urkkavaara Formation. Analyses are averages recalculated to 100%. Locality: Hokkalampi

	1	2	3	4	5	6	7
Meters above Base	0	2.5	5	7.5	10	12.5	15
SiO_2	73.31	75.82	74.56	76.03	76.15	77.13	78.40
Al_2O_3	15.37	13.79	14.51	14.11	13.46	13.64	13.13
TiO_2	0.29	0.23	0.26	0.20	0.24	0.19	0.16
Fe_2O_3(t)	2.28	2.35	2.42	1.86	2.34	1.80	1.54
Fe_2O_3	1.15	1.38	1.42	0.98	1.05	1.20	0.88
FeO	1.02	0.87	0.90	0.79	1.16	0.54	0.59
MgO	0.78	0.78	0.91	0.84	1.07	0.76	0.82
CaO	0.69	0.79	1.25	0.98	0.52	0.42	0.39
Na_2O	1.27	1.29	1.48	1.26	0.92	1.12	0.86
K_2O	6.03	4.87	4.48	4.70	5.25	4.87	4.56
MnO	0.00	0.00	0.00	0.00	0.00	0.00	0.00
P_2O_5	0.09	0.09	0.10	0.08	0.10	0.07	0.06
CIA	60.9	61.0	60.3	61.2	62.3	63.4	65.1
No. analyses	2	4	5	6	3	3	4

Table 9. Vertical variations in chemical composition in DH 4224/82/R 307. Drillhole intersects kyanite-quartz schist of zone 1 of the Hokkalampi profile. Estimated length of the section is 40 m. Locality: Hokkalampi

	Top						Bottom
SiO_2	77.31	79.54	82.85	82.85	81.69	82.30	81.86
Al_2O_3	20.90	18.61	14.38	15.17	16.07	15.25	15.23
TiO_2	0.50	0.47	0.34	0.31	0.33	0.29	0.30
Fe_2O_3(t)	0.31	0.37	0.80	0.33	0.38	0.46	0.54
MgO	0.19	0.19	0.26	0.21	0.21	0.22	0.27
CaO	0.09	0.09	0.10	0.09	0.08	0.09	0.10
Na_2O	0.09	0.09	0.14	0.13	0.14	0.12	0.16
K_2O	0.54	0.57	1.06	0.89	1.06	1.16	1.26
MnO	0.00	0.00	0.00	0.00	0.00	0.00	0.00
P_2O_5	0.06	0.05	0.05	0.05	0.05	0.04	0.04
Σ_{BR}	98.21	97.54	99.54	101.46	98.29	97.54	99.52
CIA	95.9	95.2	90.2	91.9	91.3	90.4	89.4

Table 10. Vertical variation of chemical composition in DH 4224/84/R 330. Drillhole intersects kyanite-quartz schist of zone 1 of the Hokkalampi profile. Note enrichment of ferrous iron toward the bottom of the section. Estimated length of the section is 25 m. Locality: Hokkalampi

	Top							Bottom
Analyses[a]	1	2	3	4	5	6	7	8
SiO_2	78.97	79.66	78.87	80.90	78.10	77.20	78.27	78.57
Al_2O_3	19.40	18.54	19.34	17.11	17.47	18.78	17.93	17.94
TiO_2	0.43	0.41	0.43	0.39	0.40	0.45	0.45	0.42
Fe_2O_3(t)	0.26	0.34	0.36	0.46	0.73	1.85	1.49	1.54
Fe_2O_3	0.14	0.21	0.21	0.26	0.25	0.25	0.32	0.12
FeO	0.10	0.12	0.13	0.18	0.43	1.44	1.05	1.28
MgO	0.21	0.17	0.20	0.19	0.23	0.27	0.28	0.28
CaO	0.10	0.09	0.10	0.11	0.11	0.12	0.10	0.10
Na_2O	0.10	0.12	0.23	0.07	0.12	0.14	0.14	0.12
K_2O	0.52	0.66	0.57	0.75	0.96	1.19	1.25	0.98
MnO	0.00	0.00	0.00	0.00	0.00	0.00	0.00	0.00
P_2O_5	0.00	0.00	0.00	0.00	0.00	0.00	0.05	0.04
Σ_{BR}	97.34	99.35	98.93	98.92	100.21	98.35	99.07	99.07
CIA	95.6	94.5	95.1	93.8	92.4	92.9	91.0	92.6

[a] Analyses 1–5, kyanite-quartz schists;
analyses 6–8, chloritoid-bearing kyanite-quartz schists.

within the two upper zones. These are chloritoid-bearing horizons with concentrations of ferrous iron.

5 Paleoenvironmental Interpretation

5.1 Evidence for a Weathering Origin of the Hokkalampi Paleosol

The Hokkalampi profile resulted from an intense chemical alteration of Sariola sedimentary rocks in the southern and in the northern portions of the study area, and of Archean plutonic rocks in the central portion. Recognition of alteration of the Sariola glaciogenic rock is facilitated by its mineralogical immaturity which contrasts sharply with the supermaturity of the quartzite overlying the Hokkalampi paleosol. Further evidence for a paleosol origin for the profile is its great extent over different types of parent rock throughout the study area.

The Hokkalampi profile possesses the principal attributes of a weathering mantle in that the parent rocks have gradually altered into $SiO_2-Al_2O_3$ material with an almost total loss of Na, K, Ca, and Mg. An upward decrease in the content of SiO_2 and a concomitant increase in the content of Al_2O_3 and TiO_2 are consistent with both generalized studies of chemical weathering (Birkeland 1974) and the

Table 11. Vertical variation in chemical composition in DH 4224/83/R 311. Drill-hole intersects kyanite-quartz schist of zone 1 and chloritoid-rich quartz-sericite schist of zone 2 of the Hokkalampi profile. Locality: Sammakkovaara

	1	2	3	4	5	6
Meters above Base	50–60	40–50	35–40	30–35	25–30	20–25
SiO_2	81.33	80.07	74.52	74.89	75.22	75.96
Al_2O_3	16.46	16.56	16.55	17.40	16.95	15.73
TiO_2	0.44	0.46	0.46	0.48	0.49	0.50
Fe_2O_3(t)	0.61	0.64	4.03	2.34	2.52	3.30
Fe_2O_3	n.d.	n.d.	1.51	0.73	0.54	0.76
FeO	n.d.	n.d.	2.27	1.45	1.78	2.29
MgO	0.27	0.25	0.51	0.28	0.44	0.62
CaO	0.10	0.10	0.03	0.02	0.02	0.02
Na_2O	0.20	0.33	0.29	0.31	0.33	0.29
K_2O	0.57	1.55	3.83	4.41	4.21	3.80
MnO	0.00	0.00	0.00	0.00	0.00	0.00
P_2O_5	0.00	0.04	0.04	0.03	0.02	0.03
Σ_{BR}	99.31	98.31	98.03	98.02	97.71	96.47
CIA	93.6	87.3	78.0	76.6	76.7	77.3

chemical zonation within kaolinitic soils and Phanerozoic paleosols (Kuzwart 1969, 1980; Störr et al. 1978; Wopfner 1983; Valeton 1983). Within the Hokkalampi profile, however, the separation of SiO_2 and Al_2O_3 generally is not very pronounced. Moreover, the concentration of potassium into the lower zones and the depletion of ferric iron toward the top of the paleosol are unlike Quaternary soils and Phanerozoic paleosols. The observed discrepancies may be attributed to original inhomogeneity within the weathering rocks, to peculiarities of Early Proterozoic weathering, or to diagenetic and metamorphic addition of potassium and silica.

5.2 Comparison with Other Precambrian Paleosols and Quaternary Kaolinitic Soil

The original soil which became the Hokkalampi paleosol apparently was reworked by partial erosion, as well as by diagenetic, metamorphic, and tectonic processes. These post-burial effects complicate modeling of the weathering environment. However, weathering effects become apparent in the compositional profiles of the paleosol where developed on both Archean granitoids and Lower Proterozoic glaciogenic rocks (Fig. 12). Only the potassium profile correlates with the parent lithology.

The Hokkalampi paleosol is compared to other Precambrian paleosols and to Quaternary soils in two types of triangular compositional diagrams, one with axes of $K_2O-SiO_2-Al_2O_3$ (Fig. 13) and the other with $K_2O-Fe_2O_3-Al_2O_3$ (Fig. 14). The parent rock in all cases is a felsic plutonic rock or an arkosic sandstone. Both the paleosol and soils exhibit enrichment in aluminum. The most obvious difference between the Precambrian paleosols and Quaternary soils is the depletion of potassium from the latter. However, potassium also is depleted from the upper two zones of the Hokkalampi paleosol and from a similar paleosol in the Hallavaara area.

In the $Fe_2O_3-K_2O-Al_2O_3$ diagram, potassium enrichment characterizes Precambrian paleosols with intermediate CIA values, whereas kaolinitic profiles are depleted in potassium. Most of the illustrated Precambrian paleosols are depleted in total iron. Kaolinitic soils (Fig. 14b) show depletion, where drainage conditions are good (Coultas 1980).

Potassium enrichment in the basal zone of the Hokkalampi paleosol is not readily attributable to compositional variation in the parent rock. Although the parent glaciogenic rock potentially exhibited compositional variation, plutonic rock which was the parent elsewhere presumably was homogenous. The compositional profile above the arkosic parent rock closely resembles the profiles on plutonic rock, particularly in the fluctuations in the contents of SiO_2, Al_2O_3, and TiO_2. Nonetheless, some of the discrepancies may well be attributed to original inhomogeneity.

Some of the abundant SiO_2 in the paleosol may have been precipitated from soil water. Under poor drainage conditions, silica transportation may become inhibited (Valeton 1983) and silica-rich soils may develop near the water table. However, the nearly complete removal of iron, alkalies, and al-

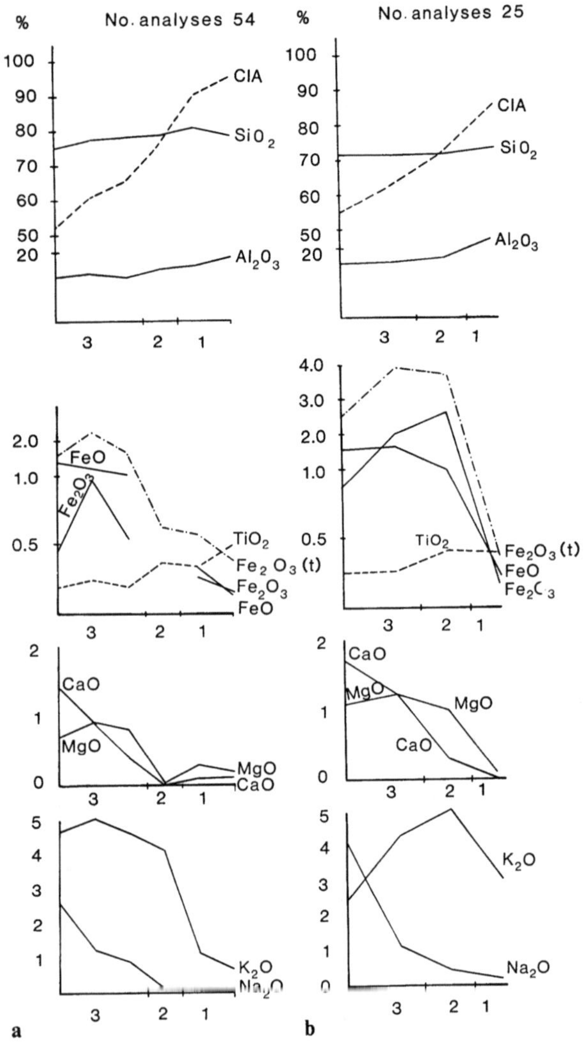

Fig. 12a, b. Composite diagrams of variations of chemical composition based on compositional averages within the three paleosol zones. **a** is for paleosol developed on sedimentary rocks and **b** is for paleosol developed on granitoids

kaline earths from the uppermost paleosol zone apparently records sufficiently good drainage for aqueous silica to have escaped, too.

The pH values of natural waters range between 4 and 9 (Mason and Moore 1982). Between pH 5 and 8, alkalies and alkaline earths released through chemical weathering are completely mobile (Loughnan 1969). Ferric iron, titania, alumina, and primary quartz are rendered practically immobile. However, at pH values less than about 4, alumina solubility becomes at least as great as that of silica; consequently, these two components cease to become separated during weathering. At a sufficiently low pH, $Ti(OH)_4$ also becomes released through the decomposition of titanium-bearing sili-

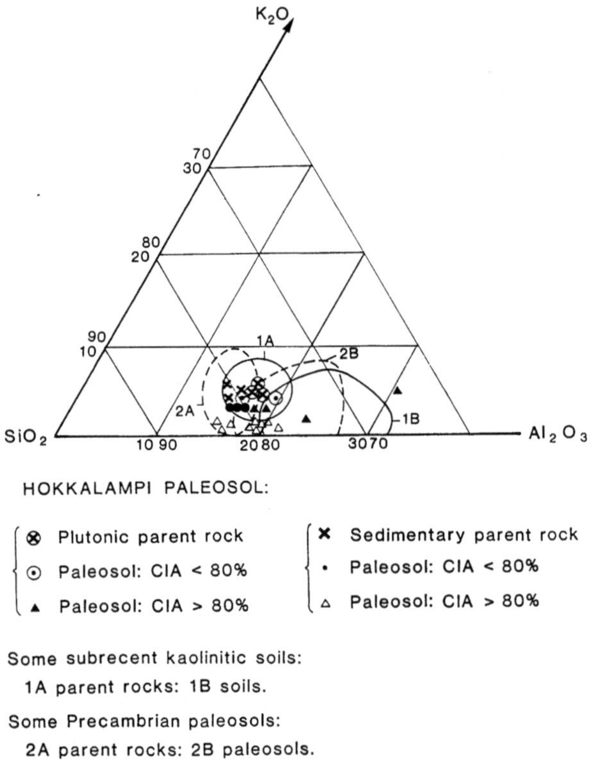

Fig. 13. Composition of the Hokkalampi paleosol and parent rocks plotted on $K_2O-SiO_2-Al_2O_3$ ternary diagram (same samples as Fig. 14). Also plotted are other Precambrian paleosols and Quaternary kaolinitic soils, along with their parents

cates. Titania clearly exhibits an upward increase within the profiles studied, reflecting its typical insolubility, but local slight concentration (Tables 8, 10 and 11) into chloritoid-rich horizons may record transient acidic soil water.

The addition of potassium into the lower paleosol zone, especially pronounced at Nuutilanvaara, is typical of Precambrian paleosols (Retallack 1986a). This enrichment may have been metasomatic but more likely occurred during weathering (leaching) of the uppermost, potassium-depleted zone.

Chert-like fragments present within the aluminous clastic rocks which overlie the Hokkalampi paleosol may have been derived from silcretes. Silcrete presently develops in deeply weathered, tropical to subtropical regions where surficial weathering of kaolinite is producing silica-rich groundwater which silicifies deep soil horizons (Millot 1970; Stephens 1971). The resulting silcretes characteristically contain about 1% TiO_2 (Summerfield 1983). However, silcrete-forming processes seem inadequate to explain the pervasive silica in the Hokkalampi paleosol. Much of the observed silica is at-

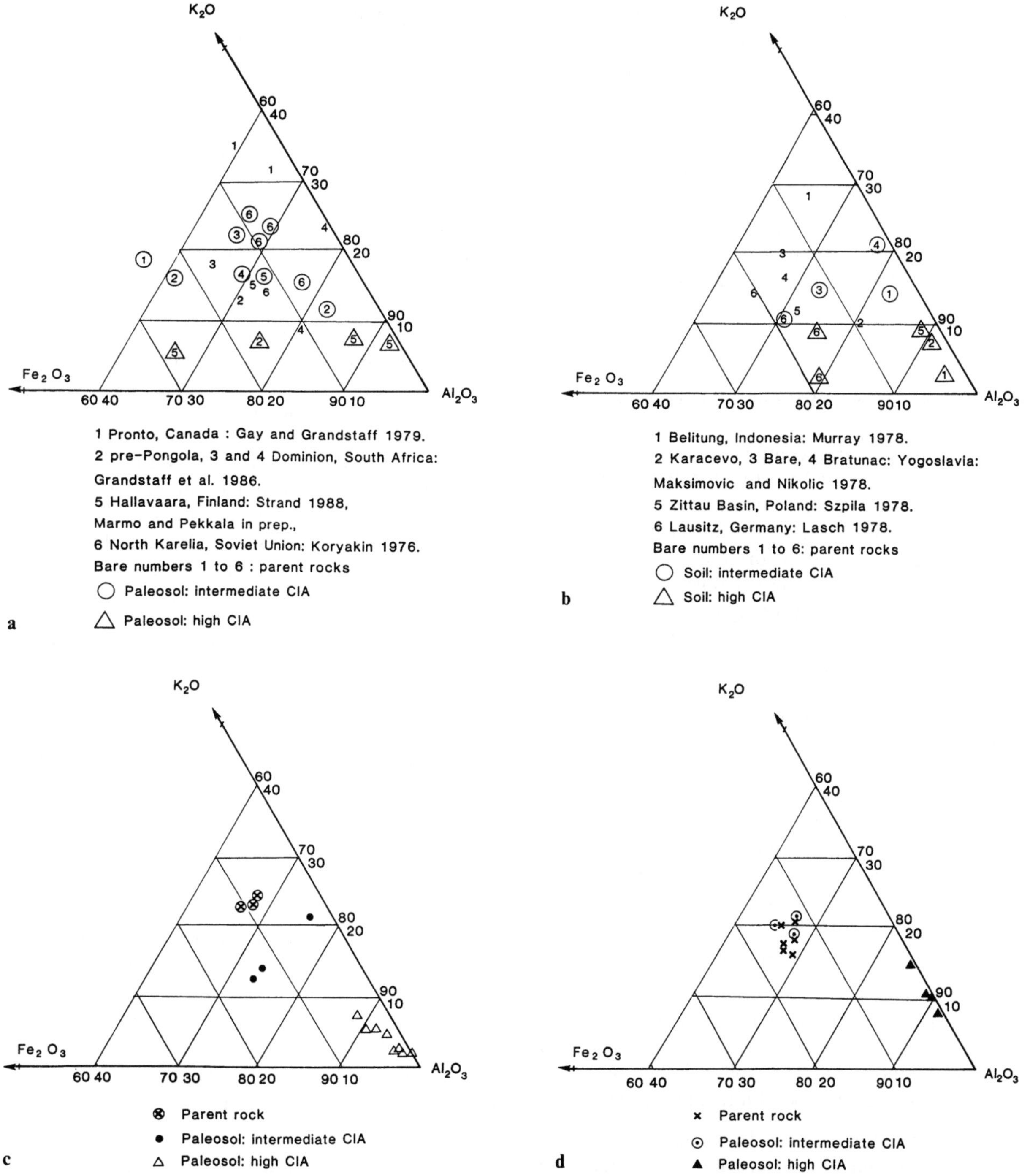

Fig. 14a–d. Composition of the Hokkalampi paleosol and parent rocks plotted on $K_2O-Fe_2O_3-Al_2O_3$ ternary diagram. Also plotted are other Precambrian paleosols and Quaternary kaolinitic soils, along with their parents. **a** shows Precambrian samples. **b** is Quaternary samples. **c** is Hokkalampi paleosol on the Sariola sequence. **d** is Hokkalampi paleosol on Archean granitoids

tributable to low-grade regional metamorphism, which featured reactions producing vast amounts of silica ionto the migrating metamorphic fluid (Marmo 1981).

The full extent of diagenetic and metamorphic alteration of the Hokkalampi paleosol remains uncertain. However, the gradually increasing values of the CIA index toward the top of the paleosol and the compositional resemblance of the upper part of paleosol to modern kaolinitic soil support the interpretation, which invokes only minor chemical changes following burial.

5.3 Metamorphism and Primary Mineralogy

The primary mineral composition of a metamorphosed paleosol may be estimated from metamorphic reactions of primary quartz and clay minerals. According to Winkler (1979), the very low grade is characterized by the transformation of kaolinite to pyrophyllite, by the gradual change of illite into sericite with gradually improving crystallinity and, by gradual improvement in the crystallinity of the chlorite. The subsequent low-grade reactions include the transformation of pyrophyllite to andalusite or kyanite, the complete substitution of illite by muscovite or sericite, and the appearance of chloritoid in rocks with high Al content relative to the contents of Na, K and Ca.

Following Winkler (1979), the mineral composition of the original soil may be estimated for each of the three zones of the Hokkalampi paleosol. Changes in the water table probably controlled the precipitation of both Fe^{2+} and Ti^{4+} and affected boundaries between the following original zones:

Upper Zone (Zone 1). This kyanite/andalusite-quartz schist corresponds to a zone of intensive kaolinization above the water table, given the removal of all soluble components. The soil weathering residue consisted essentially of kaolinite and quartz.

Intermediate Zone (Zone 2). This quartz-sericite schist corresponds to a mixture of kaolinite and quartz. The kaolinization of mica and illite did not occur below this zone.

Lower Zone (Zone 3). This quartz-feldspar-sericite rock corresponds to a zone of a partially disintegrated parent rock. The characteristic weathering reactions included the transformation of biotite into chlorite (vermiculite), alteration of plagioclase, and, higher in the zone, conversion of alkali feldspars into illite, kaolinite, and possibly smectite.

5.4 Semi-Quantitative Estimation of Weathering Rate

Weathering may be estimated quantitatively, by assuming that primary titaniferous minerals survive weathering (Birkeland 1974). This procedure involves normalizing abundances of the principal components and their sum to a constant titanium content (that of the parent rock), as explained by Pettijohn (1975, p. 488).

Figure 15 illustrates titania-normalized profiles for the Hokkalampi paleosol. A calculation is also shown separately for the upper zone. Assuming that these profiles are accurate, the percentage removed by leaching from the original parent rock varied between 60 and 40%. The maximum loss occurred between the lower and intermediate zones, as expected because of the total decomposition of the primary silicates. Figure 15 shows that separation of Al_2O_3 and SiO_2 did not occur below the upper zone (Zone 1). The other main components were relatively mobile, the least mobile being Fe_2O_3 and K_2O. Collectively, these trends are compatible with the theoretical scheme of Perel'man (1977) for the mobility of cations in the supergene zone. One peculiarity is that titania-normalized contents of total iron in the lower zone behave differently where the paleosol developed on granitoids than where it formed on glaciogenic rocks. Total iron increases upward above the plutonic rock whereas it decreases above the sedimentary rock.

5.5 Paleoclimate and Topography

Extensive kaolinization is presently best developed under tropical forests but also occurs locally in the subtropics. According to Strakhov (1967) and Störr et al. (1978), the optimal conditions for extensive kaolinization include a mean annual temperature of at least 16 to 18 °C, annual precipitation of 1.2 to 3 m, groundwater pH between 4 and 6.5, geomorphologic conditions which promote drainage, and

Fig. 15a–c. Compositional profiles of Hokkalampi paleosol normalized to average titania content of each parent rock. Profiles extend from lower paleosol zone (labeled *3*) through middle zone (*2*), the lower part of the upper zone (1_{LP}), and the top of the upper zone (1_{UP}). **a** is paleosol developed on glaciogenic rocks which average 0.17% TiO_2. **b** developed on granitoids with 0.24% TiO_2. Data of **a** and **b** correspond to that of Fig. 12. **c** is a drillhole which only intersected the upper paleosol zone. Compositions are normalized to the 0.30% TiO_2 of lowermost paleosol cored. Data of **c** come from Table 9

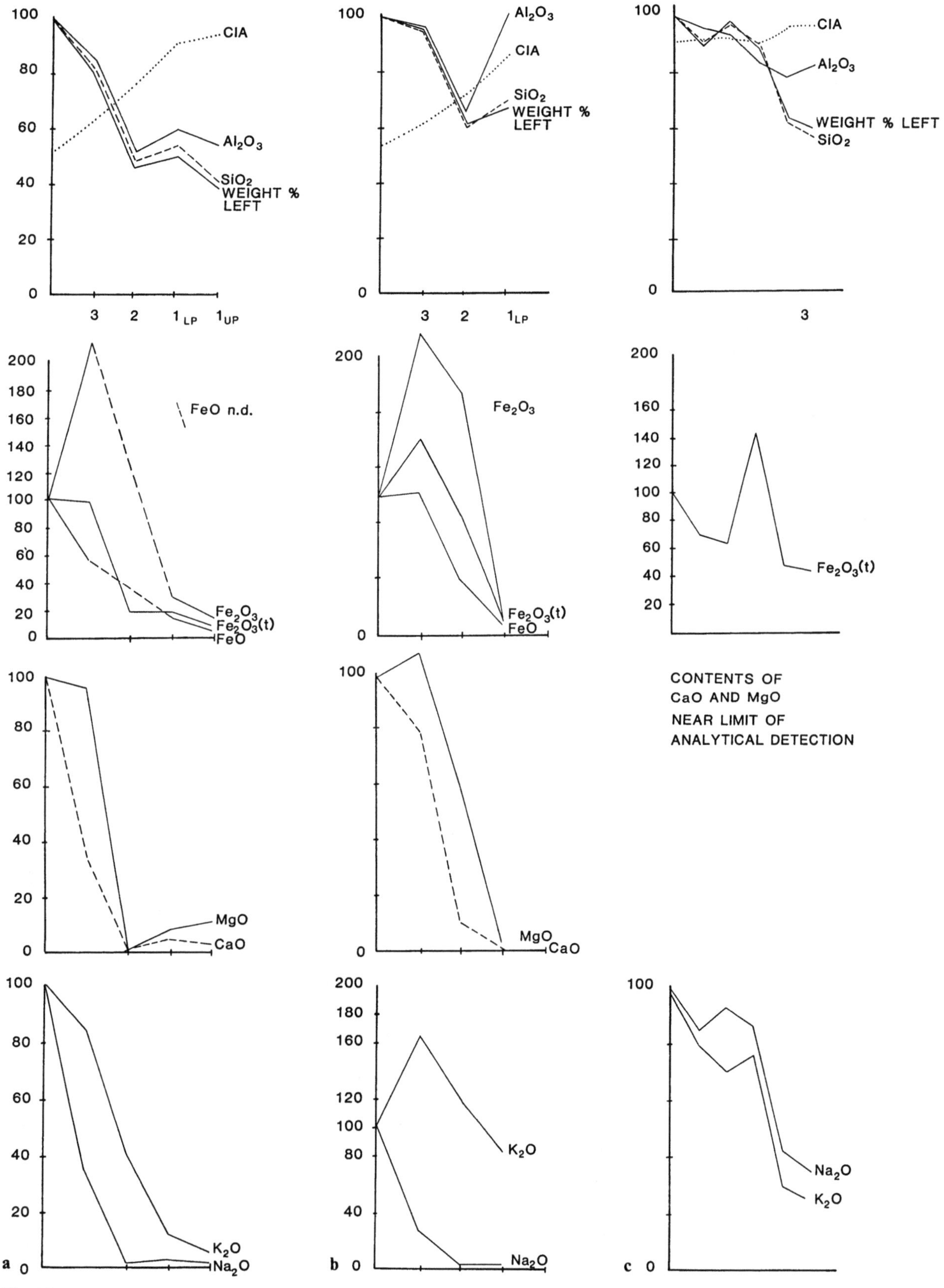
CIA
Al_2O_3
SiO_2
WEIGHT % LEFT
3 2 1_{LP} 1_{UP}
FeO n.d.
Fe_2O_3
$Fe_2O_3(t)$
FeO
CONTENTS OF CaO AND MgO NEAR LIMIT OF ANALYTICAL DETECTION
MgO
CaO
K_2O
Na_2O
a
b
c

Fig. 15a–c

stable tectonic conditions. The interpreted original mineral composition of the Hokkalampi paleosol is compatible with such conditions except for the presence of carbonate minerals and enrichment of potassium in the lower paleosol zone.

There is some evidence of Oxisol development, because the heavy mineral placers within overlying aluminous sedimentary rocks typically contain much hematite which appears to be clastic. If this iron was actually introduced by late fluids, other features of the sedimentary rocks and underlying paleosol may have resulted from nonpedogenic processes. The presence of carbonate-rich horizons (as well as Oxisols) could be attributed to a later drying of the climate. Caliche usually develops today in hot semi-arid regions as a result of evapotranspiration (Birkeland 1974). However, the characteristic soda enrichment has not been observed within the Hokkalampi paleosol.

The Hokkalampi paleosol is interpreted to have originated in a stable tectonic environment as a result of deep chemical weathering under the influence of climatic conditions that resembled those of the present humid tropics. Chemical weathering may have proceeded beyond the kaolinite stage to produce bauxite which subsequently reacted with silica-rich metamorphic fluids. Even if the weathering did not proceed to the bauxite stage, it would have required substantial time to form such a deep kaolinitic soil. Assuming a modern annual rate of kaolinization of 0.1 to 0.01 mm (Kuzwart 1980), soil formation lasted several million years. Paleoslopes must have been sufficiently high for the leachates to drain. The uneven distribution of iron-rich horizons probably records ancient water tables as they extended under the hilly topography. Much of the original soil apparently eroded to form the overlying aluminous alluvial sediment. Areas of subsidence and preferential paleosol preservation became the sites of braided river plains, as sketched in Fig. 16, which illustrates a hypothetical paleotopographic setting during partial erosion and burial.

5.6 Oxidation Reduction of the Paleosol

There is no direct knowledge of the composition of the Precambrian atmosphere. According to recent studies (Holland 1984; Pinto and Holland 1988), the partial pressure of O_2 in the Early Proterozoic atmosphere was considerably lower than at present and CO_2 was more abundant. However, any interpretation of the redox potential of the early Proterozoic atmosphere which is based on paleosol compositions is necessarily speculative, partly because of uncertainty about postburial alteration. Despite the highly oxidizing state of the modern atmosphere, kaolinitic soils are forming under reducing conditions due to organic decay within the soil. The ferrous iron which presently is leached from organic-rich surficial soil horizons becomes oxidized near the water table.

Perhaps the best evidence for generally reducing conditions during early Proterozoic chemical weathering is the existence of voluminous banded iron formations in all Precambrian shields (Walker et al. 1983; Reimer 1987). If continental weathering provided most of the iron in these iron formations, the iron must have been transported to and within the oceans while dissolved in the ferrous state. As the oxygen content of the atmosphere gradually increased, iron presumably stopped being dissolved from the continents and the iron dissolved in the oceans simultaneously was oxidized and precipitated.

There is no evidence that biogenic activity on land during the Early Proterozoic was sufficiently extensive to have contributed organic decay to chemical weathering. Moreover, the content of oxygen in the atmosphere may have been very low during Hokkalampi weathering. This is supported by the depletion of total iron toward the top of the paleosol, where atmospheric interaction was greatest, indicating that iron was dissolved in the ferrous state and mobilized. The presence of abundant ferrous iron in the chloritoid-rich parts of the paleosol may be regarded as further evidence for generally reducing conditions in that the chloritoid may record precipitation of ferrous iron near the water table (Fig. 16).

The lower paleosol zone might record some oxidation, given an increase in ferric iron relative to the parent rock (Figs. 12 and 15). In the south, this oxidation may be attributed to inhomogeneity in the parent rock. At Nuutilanvaara, however, excess ferric iron was associated with penecontemporaneous erosion of the paleosol (Fig. 16). A late-stage oxidation is further indicated by the almost complete absence of ferrous iron in the overlying sedimentary rocks, whose sediment was derived from the Hokkalampi soil. Paleoplacers within these overlying quartzites contain much clastic hematite and thorium-bearing minerals. If these hematite grains were produced within the contemporaneous soil, they record its oxidation. Moreover, variation in the ratio of uranium/thorium within the aluminous sediments, including the paleoplacers, may indicate oxidation associated with sedimentation. The ratio

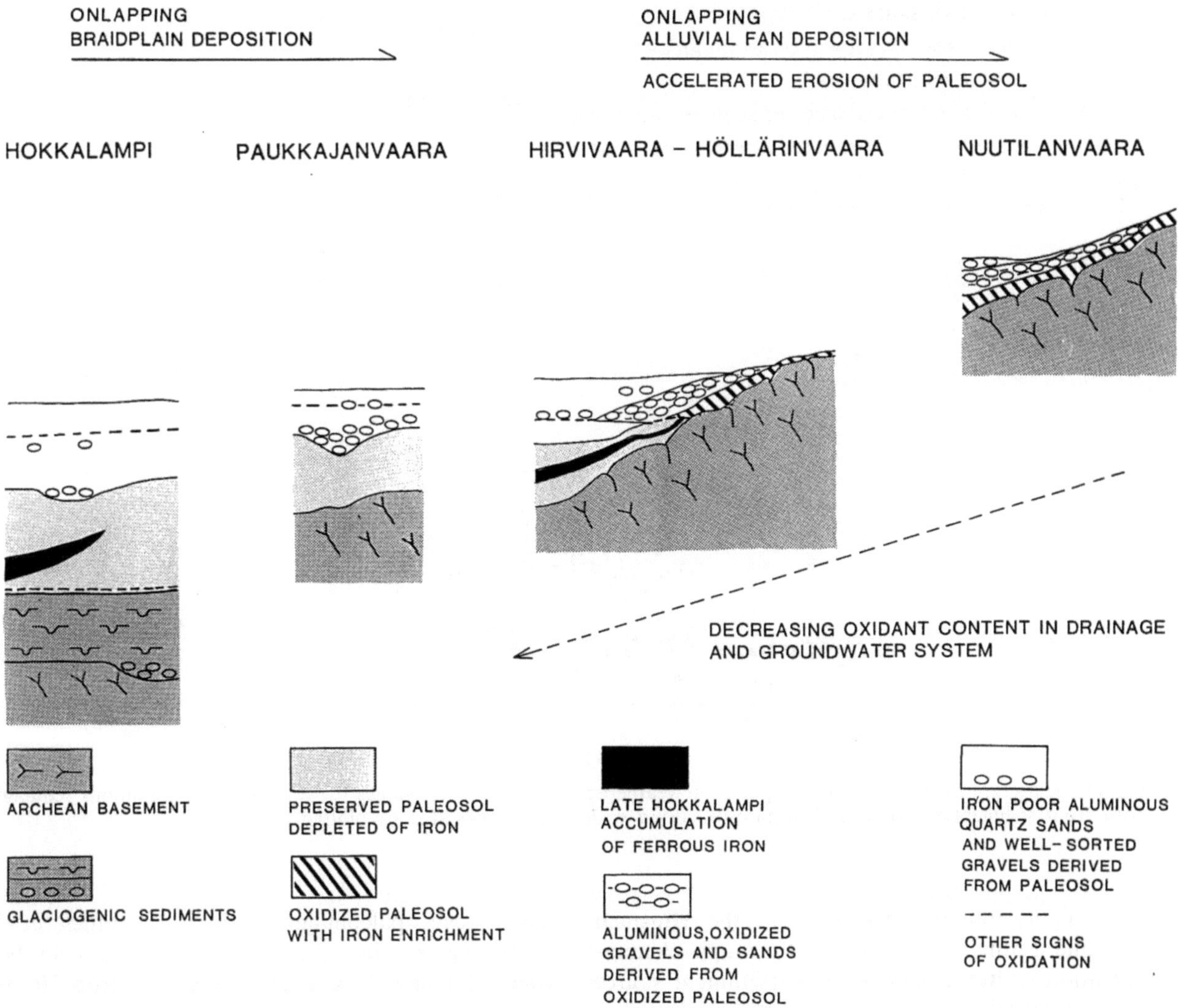

Fig. 16. Hypothetical paleogeographical reconstruction of the Hokkalampi area

generally is much higher both in the parent rocks and paleosol than within the overlying aluminous sedimentary rocks (Fig. 17). Uranium apparently became oxidized to soluble species which were removed by groundwater, whereas insoluble thorium became trapped in the heavy minerals of the sediments. Obviously, given that their contents are generally low, neither has been enriched.

5.7 Correlation with Other Precambrian Paleosols

Lower proterozoic paleosol underlies quartz-arenite in the Karelides of the Svekofennian Shield both in Finland (Meriläinen 1980; Strand 1988; Laajoki 1986; Idman and Eilu 1988; Laajoki et al. 1989) and to the east in Soviet Karelia (Koryakin 1971; Shokolov and Heiskanen 1984). Paleosol in Kainuu, 300 km north of the present study area, is several tens of meters thick and consists of quartz-sericite schist with aluminosilicates and chloritoid-rich units. It resembles the Hokkalampi paleosol in that it also developed on sedimentary rocks and is overlain by aluminous quartzite. In Lapland, a thin remnant of an eroded paleosol recently has been reported to underlie quartzite and consist of quartz-sericite schist. In Soviet Karelia, the paleosols are relatively thin but are similarly composed of quartz-sericite schist.

Lower Proterozoic paleosols have been reported from south-central Canada (Young 1973; Gay and Grandstaff 1979), Brazil (Cox 1967) and South Africa (Button and Tyler 1981; Kimberley and Grandstaff 1986; Retallack 1986b). Young (1973) inter-

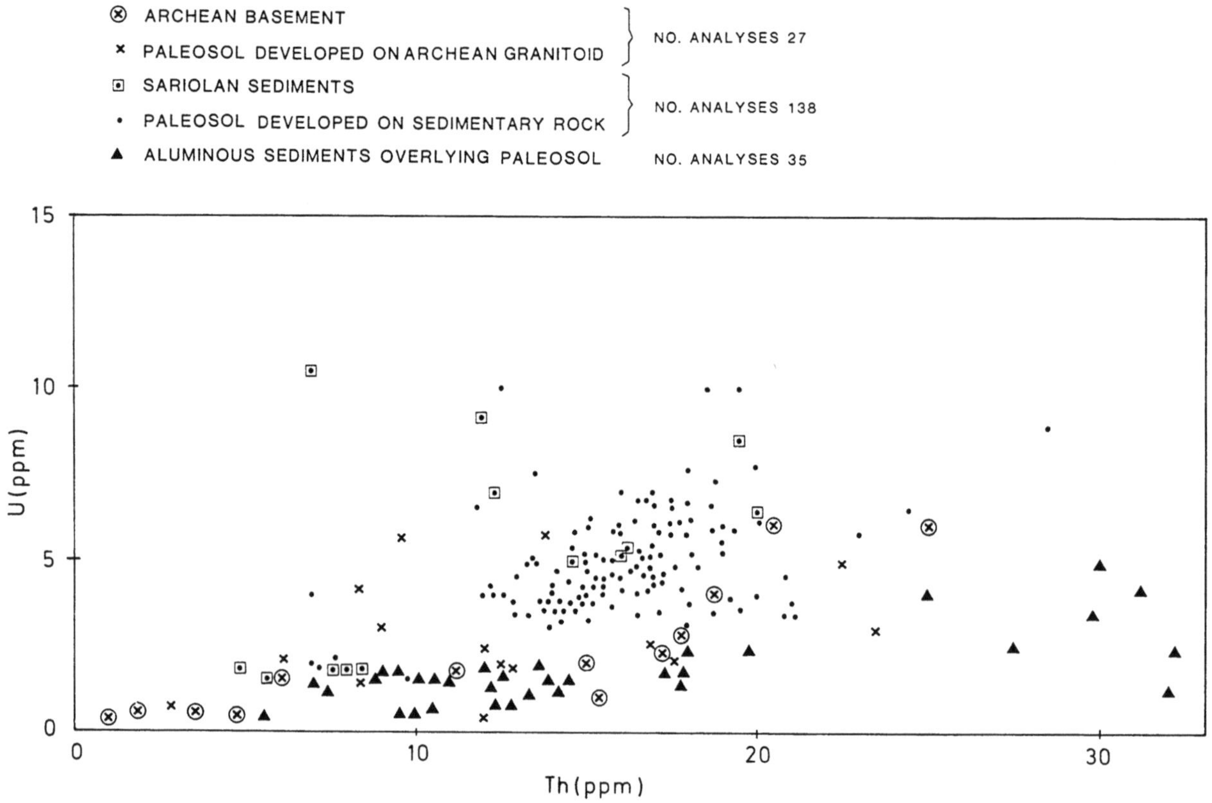

Fig. 17. Scattergram of U and Th in paleosol weathered both from sedimentary rocks and granitoids, along with their parents. Values for the overlying aluminous sedimentary rocks also are plotted

prets highly aluminous silicates in the Huronian Lorraine Formation to record thorough chemical alteration, either in situ or at the sediment source. These paleosols resemble the Hokkalampi paleosol in being composed of quartz-sericite schist but they are thinner. Most authors attribute their potassium enrichment to late-diagenetic metasomatism.

CIA index values for other Lower Proterozoic paleosols are generally lower than those of the Hokkalampi and Hallavaara paleosols. The most drab-colored of these paleosols have been informally called "green clays" by Retallack (1986a). However, all these paleosols are interpreted to have been formed during climatic conditions like those which induce modern tropical weathering.

It is unclear if climatic zoning on the Early Proterozoic Earth resembled that of the present. According to Pesonen and Neuvonen (1981), the paleolatitude for the Fennoscandian Shield 2.5 to 2.1 Ga ago was between 30° and 55° N and the paleolatitude of the Laurentian Shield ranged from 35° to 65° N. Both shields therefore were situated too far north for intensive chemical weathering, assuming the present climatic zonation. However, most recent paleomagnetic data (Mertanen et al. 1989) suggest that the Fennoscandian Shield between 2.6 and 2.2 Ga ago wandered from 70° to 30° N. The continent then rotated along this paleolatitude for several hundreds of million years. Thus, if there was climatic zoning during the Early Proterozoic, wandering of the Karelian Craton toward lower and warmer latitudes is in accordance with the observed development of the Hokkalampi paleosol on the older glaciogenic rocks. These is also some evidence that Precambrian surface temperatures were considerably higher than today (Karhu and Epstein 1986). Early Proterozoic soil formation therefore may have been characterized by an intense, deep chemical weathering. The nonuniformitarian accumulation of voluminous platformal quartzite in many Precambrian shield areas may also be attributed to prolonged chemical weathering.

6 Summary and Conclusions

The Lower Proterozoic Hokkalampi paleosol of eastern Finland is now composed of quartz-sericite schist with increasing proportions of kyanite and andalusite toward the top. The paleosol had a minimum original thickness of 70 m but is typically much thinner due to erosion prior to burial. The erosion resulted in the deposition of voluminous aluminous sand and gravel, now aluminous quartzite, metaconglomerate and orthoquartzite. The metamorphosed paleosol has a gradational base and exhibits a vertical zonation. The premetamorphic minerals are interpreted to have been kaolinite and residual quartz, given the present chemical and mineral composition of both the paleosol and overlying sedimentary rocks.

The paleosol is interpreted to have resulted from intense chemical weathering like that in the modern humid tropics. Its great thickness probably records prolonged tectonic stability in an area with some relief. The Hokkalampi paleosol is chemically reduced where thickest but slightly oxidized where it became eroded near topographic highs. The overlying clastic sedimentary rocks lack ferrous iron and contain abundant hematite.

Paleosols and associated quartzite units like those reported herein occur elsewhere in the Lower Proterozoic of the Fennoscandian Shield. If all these are contemporaneous, the entire area of the Archean Karelian Craton (400000 km^2) must have experienced a prolonged period of intense chemical weathering.

Acknowledgements. I thank the heads of the Exploration Department of the Geological Survey of Finland, especially Jouko Talvitie, Pentti Ervamaa and Yrjö Pekkala, for financial support. This work would lie unfinished without the support of Professor Ilmari Haapala of the University of Helsinki. Credit also is due to my colleagues at the Kyykkä base camp, which was built on the Hokkalampi paleosol. Väinö Hoffren made most of the XRF analyses and Risto Saikkonen provided ferrous iron determinations at the Geochemistry Department. Pirkko Kurki and her group drafted the figures. Aune Romberg typed the tables and Jari Väätäinen provided the photomicrographs. Gregory Retallack and Thomas Reimer reviewed and improved the manuscript.

References

Äikäs O, Sarikkola R (1987) Uranium in lower Proterozoic conglomerates of the Koli Area, eastern Finland. In: Uranium deposits in Proterozoic quartz-pebble conglomerates. Rep Work Group Uranium Geology, IAEA-TECDOC, pp 189–234

Aurola E (1959) Kyaniitti- ja pyrofylliittiesiintymät Pohjois-Karjalassa. Summary: The kyanite and pyrophyllite occurrences in North Karelia. Geol Tutkimusl Geoteknol Julk 63:36 pp

Birkeland PV (1974) Pedology, weathering, and geomorphologic research. Univ Press, Oxford, 285 pp

Button A, Tyler N (1981) The character and economic significance of Precambrian paleoweathering and erosion surfaces in southern Africa. Econ Geol 75:686–709

Coultas CL (1980) Soils of marshes in the Appalachicola, Florida, Estuary. Proc Soil Sci Soc Am 44:348–353

Cox DB (1967) Regional environment of the Jacobina auriferous conglomerate, Brazil. Econ Geol 62:773–780

Gay AL, Grandstaff DE (1979) Chemistry and mineralogy of Precambrian paleosols at Elliot Lake, Ontario, Canada. Precambrian Res 12:349–373

Holland HD (1984) The chemical evolution of the atmosphere and oceans. Univ Press, Princeton, 582 pp

Huhma H (1986) Sm–Nd, U–Pb and Pb–Pb isotopic evidence for the origin of the Early Proterozoic Svekokarelian crust in Finland. Geol Surv Finland Bull 337: 48 pp

Idman H, Eilu P (1988) The alteration of granitoids in the western part of an Early Proterozoic greenstone belt – an example of a chemical weathering (translated from Finnish). In: Lappalainen V, Papunen H (eds) Studies in the field of geology. Ann Univ Turkuensis Ser C Tom 68:39–52

Karhu J, Epstein S (1986) The implication of the oxygen isotope records in coexisting cherts and phosphates. Geochim Cosmochim Acta 50:1745–1756

Kimberley MM, Grandstaff DE (1986) Profiles of elemental concentrations in Precambrian paleosols on basaltic and granitic parent materials. In: Retallack GJ (ed) Precambrian paleopedology. Precambrian Res 32:133–154

Koryakin AS (1971) Results of a study of Proterozoic weathering crusts in Karelia. Int Geol Rev 13:973–980

Kuzwart M (1969) Kaolin deposits of Czechoslovakia. In: 23rd Int Geol Congr Prague, vol 15, pp 47–73

Kuzwart M (1980) IGCP Project Genesis of kaolins, results 1971–1979. Vetnik Ustredniho Ust Geol 55(3): 174–180

Laajoki K (1986) The Precambrian supracrustal rocks of Finland and their tectono-exogenic evolution. Precambrian Res 33:67–85

Laajoki K, Strand K, Härmä P (1989) Lithostratigraphy of the Early Proterozoic Kainuu Schist Belt in the Kurkikylä-Siikavaara area, northern Finland, with emphasis on the genetic approach. Bull Geol Soc Finland 61: 65–93

Lasch G (1978) Zu einigen Aspekten der Genese der Lausitzer Kaoline. Schriftenr Geol Wiss Berlin 11:165–178

Loughnan FC (1969) Chemical weathering of the silicate minerals. Elsevier, New York, 154 pp

Luukkonen E, Lukkarinen H (1986) Explanation to the stratigraphic map of Middle Finland. Geol Surv Finland Rep Inv 74:47 pp

Maksimovic Z, Nikolic D (1978) The primary kaolin deposits of Yugoslavia. Schriftenr Geol Wiss Berlin 11: 179–197

Marmo JS (1981) The Hokkalampi kyanite deposit at Kontiolahti. The geology of the environment and the origin of aluminium (translated from Finnish). Masters Thesis, Univ Helsinki, 81 pp

Marmo JS (1986) Sariolan stratigraphy and sedimentation in the Koli-Kaltimo Area, North Karelia, eastern Finland. In: Shokolov V (ed) Proc Symp Lower Proterozoic formations on the eastern part of the Baltic Shield, Petrozavodsk, 1985, pp 149–190
Marmo JS, Ojakangas RW (1984) Lower Proterozoic glaciogenic deposits, eastern Finland. Geol Soc Am Bull 95:1055–1062
Marmo JS, Kohonen JJ, Sarapää O, Äikäs O (1988) Sedimentology and stratigraphy of the lower Proterozoic Sariola and Jatuli Groups in the Koli-Kaltimo Area, eastern Finland. GSF Spec Pap 5:11–28
Mason B, Moore CB (1982) Principles of geochemistry, 4th edn. John Wiley & Sons, New York, 344 pp
Meriläinen K (1980) On the stratigraphy of the Karelian formations. In: Silvennoinen A (ed) Jatulian geology in the eastern part of the Baltic Shield. Proc Finn Sov Symp, Finland 1979, pp 97–112
Mertanen S, Pesonen LJ, Huhma H, Leino MAH (1989) Paleomagnetism of the Early Proterozoic layered intrusions, northern Finalnd. Geol Surv Finland Bull 347:40 pp
Millot G (1970) Geology of clays. Springer, Berlin Heidelberg New York, 429 pp
Murray HH (1978) Alteration of a granite to kaolin mineralogy and geochemistry. Schriftenr Geol Wiss Berlin 11:197–208
Nesbitt HW, Young GM (1982) Early Proterozoic climates and plate motions inferred from major element chemistry of lutites. Nature (London) 299:715–717
Pekkarinen LJ (1979) The Karelian formations and their depositional basement in the Kiihtelysvaara-Värtsilä Area, East Finland. Geol Surv Finland Bull 301:141 pp
Perel'man AI (1977) Geochemistry of elements in the supergene zone. Keter, Jarusalem, 266 pp
Pesonen LJ, Neuvonen KJ (1981) Paleomagnetism of the Baltic Shield – implications for Precambrian tectonics. In: Kröner A (ed) Precambrian plate tectonics. Elsevier, Amsterdam, pp 623–648
Pettijohn FJ (1975) Sedimentary rocks. Harper & Row, Singapore, 628 pp
Piirainen TA (1968) Die Petrologie und die Uranlagerstätten des Koli-Kaltimo-Gebiets im finnischen Nordkarelien. Bull Comm Geol Finlande 237:99 pp
Pinto JP, Holland HD (1988) Paleosols and the evolution of the atmosphere, pt 2. In: Reinhardt J, Sigleo WR (eds) Paleosols and weathering through geologic time. Geol Soc Am Spec Pap, pp 21–34
Reimer TO (1987) Weathering as a source of iron in iron formations: the significance of alumina-enriched paleosols from the Proterozoic of southern Africa. In: Appel P, LaBerge G (eds) Precambrian iron formations. Theophrastus, Athens, pp 601–619
Retallack GJ (1986a) The fossil record of soils. In: Wright VP (ed) Paleosols, their recognition and interpretation. Blackwells, London, pp 1–57
Retallack GJ (1986b) Reappraisal of 2200 Ma-old paleosol near Waterval, South Africa. In: Retallack GJ (ed) Precambrian paleopedology. Precambrian Res 32:195–232
Senior BR, Mabbuth JA (1979) A proposed method of defining deeply weathered rock units based on regional mapping in southwest Queensland. J Geol Soc Aust 26:237–254
Simonen A (1980) The Precambrian of Finland. Geol Surv Finland Bull 304:58 pp
Shokolov VA, Heiskanen KI (1984) Developmental stages of Precambrian crusts of weathering. In: Proc 27th Int Geol Congr, vol 5. VNU Science Press, Utrecht, pp 73–94
Szplila K (1978) Kaolinisierungsprozesse des kristallinen Untergrundes des Zittauer Beckens. Schriftenr Geol Wiss Berlin 11:293–304
Stephens CG (1971) Laterite and silcrete in Australia: A study of the genetic relationship of laterite and silcrete and their companion materials, and their collective significance in the formation of weathered mantle soils, relief and drainage of the Australian continent. Geoderma 5:5–52
Störr M, Kuzwart M, Neuzil J (1978) Age and genesis of weathering crusts of the Bohemian Massif. Schriftenr Geol Wiss Berlin 11:265–282
Strakhov NM (1967) Principles of lithogenesis, vol 1. Olivier & Boyd, Edinburg, 245 pp
Strand K (1988) Alluvial sedimentation and tectonic setting of the early Proterozoic Kurkikylä and Kainuu Groups in northern Finland. In: Laajoki K, Paakkola J (eds) Sedimentology of the Precambrian formations in eastern and northern Finland. Geol Surv Finland Spec Pap 5:75–90
Summerfield MA (1983) Geochemistry of weathering profile silcretes, southern Cape Province, South Africa. In: Wilson RCL (ed) Residual deposits. Geol Soc London Spec Publ 11:167–178
Valeton I (1983) Palaeoenvironment of lateritic bauxites with vertical and lateral differentiation. In: Wilson RCL (ed) Residual deposits. Geol Soc London Spec Publ 11:77–90
Vuollo J (1991) Proterozoic mafic rock associations in North Karelia. In: Piirainen T (ed) Archean and Proterozoic geologic evolution and related ore-forming processes in North Karelia. North Karelia Ore Project, University of Oulu. Report of investigation 31
Walker JCG, Klein C, Schidlowski M, Schopf JW, Stevenson DJ, Walter MR (1983) Environmental evolution of the Archean-Early Proterozoic Earth. In: Schopf JW (ed) Earth's earliest biosphere; its origin and evolution. Univ Press, Princeton, pp 260–289
Winkler H (1979) Petrogenesis of metamorphic rocks. Springer, Berlin Heidelberg New York, 301 pp
Wopfner H (1983) Environment of silcrete formation. In: Wilson RCL (ed) Residual deposits. Geol Soc London Spec Publ 11:151–158
Young GM (1973) Tillites and aluminous quartzites as possible time markers for middle Precambrian (Abhebian) rocks of North America. In: Young GM (ed) Huronian stratigraphy and sedimentation. Geol Assoc Can Spec Pap 12:97–128

Paleosol and Ore-Forming Processes in the Elliot Lake District of Canada

David J. Mossman[1] and Catharine E. G. Farrow[2]

[1] Department of Geology, Mount Allison University, Sackville, New Brunswick, Canada E0O 3C0
[2] Department of Geology, Carleton University, Ottawa, Ontario, Canada K1S 5B6

Abstract

The existence of Precambrian paleosols has been confirmed in the Elliot Lake district of Canada. However, the lithified weathering products exhibit substantial modification due to diagnesis and low-grade metamorphism. There is considerable chemical and mineralogical similarity among paleosols derived from different source rocks. The extent of oxidation in the paleosols is not simply related to paleotopographic conditions but to a variety of factors, including ancient climatic and atmospheric conditions. Calculation of the oxygen demand during weathering of a granodiorite parent is consistent with oxygen-deficient atmospheric conditions during the Early Proterozoic.

Uraniferous conglomerate at and near the base of the Huronian Supergroup was derived from paleosol which developed over uraniferous Archean granite. A modified-placer hypothesis explains the origin and distribution of the heavy minerals, i.e., brannerite, uraninite, monazite, and coffinite. Brannerite developed during diagnesis (the "Pronto reaction") within the conglomerate and perhaps also by weathering within uppermost soil horizons.

Mineralization was related to soil formation in several ways. These include the loss of thorium in eluvial soil horizons prior to their compaction, the concentration of rutile, leucoxene, and uranium-thorium minerals in uppermost soil horizons, minor concentrations of C_{org}, and modest enrichment of Au above crustal background. Sub-Huronian paleosols therefore record important ore-forming processes.

1 Introduction

The Huronian Supergroup (2.25–2.45 Ga) forms the southern boundary of the Superior province of the Canadian Shield. Uranium-bearing conglomerate was discovered in the base of the supergroup in the Elliot Lake district during the early 1950s. The supergroup consists of Lower Proterozoic metasedimentary and subordinate metavolcanic rocks which overlie dominantly granitic Archean rocks (e.g., Farrow and Mossman 1988). Separating the Archean rocks from the Huronian Supergroup is a profound unconformity upon which an extensive paleosol developed, with preserved thicknesses which commonly reach 15 m (Roscoe and Steacy 1958). This sub-Huronian paleosol is called the Elliot Lake paleosol throughout this paper.

Collins (1925) was the first to describe the Elliot Lake paleosol. Since then, various geologic and geochemical studies have addressed the environmental conditions recorded by the paleosol (Roscoe 1973). Farrow and Mossman (1988) review the studies of Roscoe (1957, 1969), Pienaar (1963), Gay and Grandstaff (1980) and Kimberley et al. (1984). The present study unequivocably establishes the identity and polygenetic character of the Elliot Lake paleosol and substantiates the concept of a prevalently oxygen-deficient atmosphere during the Early Proterozoic.

Evolution of the Elliot Lake paleosol involved several factors, i.e., duration of exposure, composition of the parent rock, climate, topography, microorganic activity, diagnesis, and metamorphism. The purpose of the present paper is to examine these factors in greater detail and discuss their relationship to ore-forming processes.

2 Paleosol Lithology

Paleosols in the Elliot Lake area commonly directly underlie basal uraniferous conglomerate of the Matinenda Formation, the major source of uranium in Canada. Where developed on greenstone, the paleosol is lighter green than the parent whereas it is yellow-green to light gray where overlying granit-

M. Schidlowski et al. (Eds.)
Early Organic Evolution: Implications for Mineral and Energy Resources

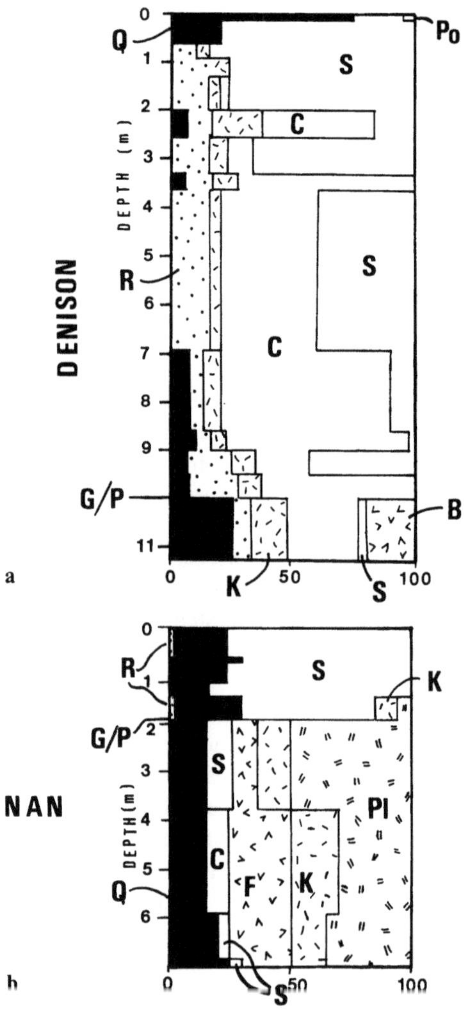

Fig. 1a, b. a Mineralogical profile through Elliot Lake paleosol developed on granodiorite. Core from North American Nuclear (NAN) property. Main minerals are sericite (*S*), chlorite (*C*), quartz (*Q*), K-feldspar (*K*), plagioclase (*Pl*). Minor minerals include ferromagnesians (*F*), rutile/leucoxene (*R*), pyrrhotite (*Po*), biotite (*B*). *G/P* base of paleosol. (After Farrow and Mossman 1988.) **b** Mineralogical profile through Elliot Lake paleosol developed on greenstone. Symbols as in **a**

oids. Primary structures in the paleosol and immediately overlying conglomerate include clay-filled desiccation cracks and rip-up clasts of paleosol in the conglomerate (Farrow and Mossman 1988). Particular paleosol horizons exhibit eluviation of clays and attendant loss of alumina. Gay and Grandstaff (1980) report ped structure (Soil Survey Staff 1975).

The Elliot Lake paleosol developed on Archean rocks, both granodiorite and greenstone. Mineralogical profiles through both paleosol types are shown in Fig. 1. The profile which developed on granodiorite comes from the North American Nuclear (NAN) drill site whereas that on greenstone is from the Denison mine. Sericite and quartz predominate in the groundmass of paleosol developed on both granodiorite and greenstone. Substantial rutile and leucoxene (5 to 8%) accompany these two minerals in the greenstone-derived paleosol. In the parent greenstone, rutile occurs as an accessory mineral and is partially derived from primary ilmenite.

The proportion of quartz characteristically increases upward through the profiles whereas the proportion of feldspar decreases until it disappears. Chlorite also decreases upward except for a horizon of Fe-rich chlorite within the upper portion of the greenstone-derived profile. As noted by Gay and Grandstaff (1980), there is an upward loss of iron in the chlorite of all profiles.

3 Paleosol Geochemistry

Vertical variations in elemental concentrations are reported herein for paleosol developed on both granodiorite and greenstone (Fig. 2). These generally resemble previously published compositional profiles through the Elliot Lake paleosol (Gay and Grandstaff 1980; Kimberley et al. 1980; Kimberley et al. 1984). The most pronounced changes in elemental concentration occur within the basal transition to bedrock. The granodiorite-derived paleosol particularly demonstrates the upward depletion in magnesium, calcium, sodium, and total iron (Fig. 2). There is a concomitant increase in potassium, perhaps because of late-diagnetic addition from descending groundwater (Holland 1984; Zbinden et al. 1988).

Graphical illustration of the progressive extent of weathering commonly is performed by normalizing concentrations against either alumina or titania to estimate gains and losses from a constant volume (Gay and Grandstaff 1980; Retallack et al. 1984). Such normalization is inappropriate, however, for the data in Fig. 2 because alumina apparently has been selectively eluviated from particular soil horizons (in the Denison profile) and titania unexpectedly decreases upward in some profiles.

There are similar trends in the vertical variation of major-element concentrations for both granodiorite and greenstone profiles despite differences in absolute abundances (Fig. 3). Total iron generally decreases upward even though there is a concomitant increase in the Fe^{3+}/Fe^{2+} ratio. This increase in Fe^{3+}/Fe^{2+} therefore does not necessarily record

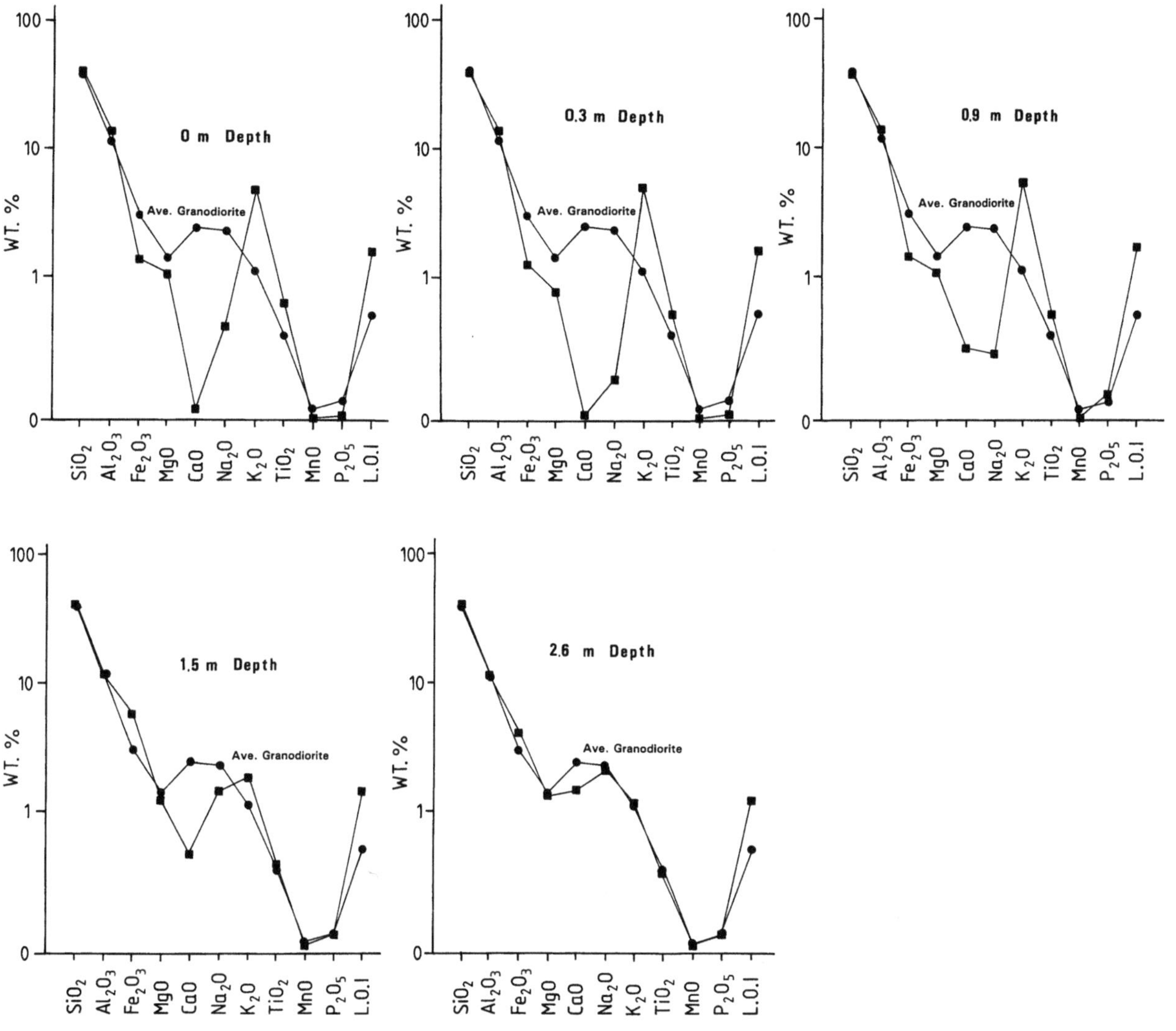

Fig. 2. Elemental composition of Elliot Lake paleosol developed on granodiorite, at depths of 0, 30, 90, 150, and 260 cm below unconformity surface. North American Nuclear core. (After Farrow and Mossman 1988)

oxidation, as noted by Roscoe (1969). The Denison and Stanleigh profiles (both developed on greenstone) exhibit a slight enrichment of TiO_2 and substantial enrichment of Al_2O_3 relative to the parent greenstone (Fig. 3) even though titania decreases upward through the Denison profile. Manganese also decreases upward in the Denison profile (Table 1).

Barium and rubidium are both concentrated toward the top of the Denison profile (1500 and 400 ppm respectively), possibly because of original concentration in clays (Farrow and Mossman 1988). Base metals are scarce and their trends equivocal. The heavy metals are more instructive. Two profiles in Fig. 4 illustrate an upward increase in Zr, Y, and Nb. The abundance of Zr and the corresponding Zr/Ti ratios near the top of the Denison

Table 1. Chemical analysis for MnO, P_2O_5, TiO_2 (wt. %) and for U and Th (ppm) in greenstone-derived Denison profile, Elliot Lake. (Analyst: K. Cameron)

Sample Name	MnO	P_2O	TiO_2	U	Th
D-8	0.05	0.28	2.18	14.0	23
D-10	0.06	0.15	3.56	2.4	11
D-12	0.07	0.12	3.56	2.2	11
D-14	0.15	0.14	3.13	1.8	–
D-16	0.09	0.06	3.90	2.2	9
D-20	0.12	0.58	3.78	2.1	10
D-25	0.16	0.65	3.32	1.7	9
D-30	0.21	0.48	2.88	0.9	5
D-35	0.26	0.47	3.05	1.0	–

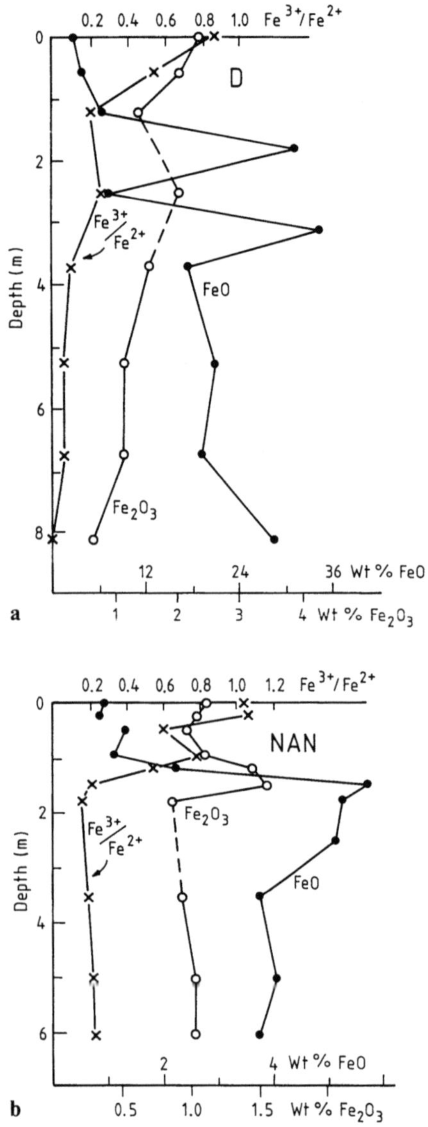

Fig. 3a, b. **a** Depth profile of iron (FeO and Fe_2O_3) in Elliot Lake paleosol developed on greenstone. Denison mine profile. *Dashed lines* indicate lack of data for Fe_2O_3. **b** Depth profile of iron (FeO and Fe_2O_3) in Elliot Lake paleosol developed on granodiorite. North American Nuclear profile. *Dashed lines* indicate lack of data for Fe_2O_3

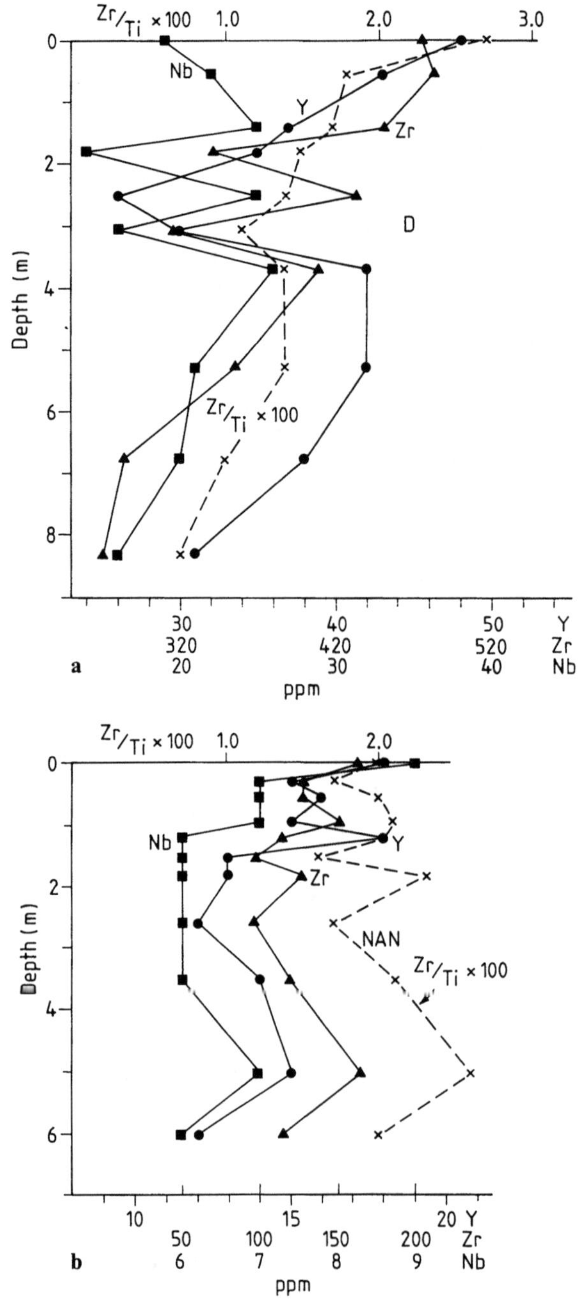

Fig. 4a, b. **a** Profiles of Nb, Y, Zr, and Zr/Ti through Elliot Lake paleosol developed on greenstone (Denison). **b** Profiles of Zr, Y, Nb, and Ti through Elliot Lake paleosol developed on granodiorite (North American Nuclear)

profile (Table 1) are attributable to detrital influx during or after development of the paleosol.

Gay and Grandstaff (1980) have attributed the apparent loss of TiO_2 at Denison to destruction of sphene by weathering. In contrast, Holland (1984) attributed the unusual behavior of titanium to the effect of fluids from nearby diabase dikes. An alternative suggestion by G. J. Retallack (quoted in Farrow and Mossman 1988) is that there is a second paleosol in the Denison profile located above an increment of sediment. According to this hypothesis, the decrease in TiO_2 in the upper portion of the Denison profile is attributable to sorting of heavy minerals within an overlying clayey horizon. Existence of such a sedimentary component elsewhere in the Elliot Lake area would explain the considerable thickness of the greenstone-derived paleosol profiles (e.g., Gay and Grandstaff 1980).

Table 2. Chemical analysis of North American Nuclear (NAN) granodiorite. (Sample # NAN 1273.) Analyst: K. Cameron

SiO_2	64.02
Al_2O_3	16.24
Fe_2O_3	1.06
FeO	3.78
MgO	2.30
CaO	4.17
Na_2O	4.25
K_2O	1.58
TiO_2	0.57
MnO	0.07
P_2O_5	0.13
L.O.I.	0.70
Total	99.29

Vertical profiles of U and Th appear to be inconsistent through the Elliot Lake paleosol. Roscoe and Steacy (1958) reported an upward decrease in U and a corresponding upward increase in the ratio, Th/U, in profiles developed on granodiorite. In contrast, Farrow and Mossman (1988) found a distinct upward increase in U and Th in both the Denison and NAN paleosols (Table 1) accompanied by erratic Th/U ratios (not shown here). The eluviated horizon within the Denison profile is surprisingly lacking in Th (Table 1).

4 Factors in Paleosol Development: The Early Proterozoic Atmosphere

A surge of interest in Earth's early atmosphere in the 1950s was triggered by the discovery of uraninite in Witwatersrand paleoplacers. Sedimentary preservation of the mineral was attributed to its virtual insolubility under reducing atmospheric conditions (Davidson 1953). Most researchers presently agree that some free oxygen has existed in the atmosphere since at least 3.0 Ga (Cloud 1983; Holland and Zbinden 1986). The question is, how much?

Various lines of evidence have been used to estimate the oxidation state of the Precambrian atmosphere, as reviewed by Mossman and Dyer (1985, p. 305–306). However, none of this evidence is compelling, including that from paleosols (Pinto and Holland 1988). This is certainly true for the comparatively well-studied Elliot Lake paleosol.

Paleosols have been aptly described to be reaction rims that have fixed various liquid and gaseous reactants during their development at the border between the Earth's crust and ancient atmosphere (Schidlowski 1988). As such, they hold crucial clues to early atmospheric evolution. Despite overprinting due to various processes operative during lithification, paleosols offer one of the best mediums for the study of ancient atmospheric conditions. In the case of the Elliot Lake paleosol, the prime observation is that the surficial oxidation of iron indicates that the Early Proterozoic atmosphere was not anoxic.

Holland (1984) and Holland and Zbinden (1986) have quantitatively assessed the relationship between pO_2 of the Archean atmosphere and iron loss in paleosols, using Gay and Grandstaff's (1980) analyses of Elliot Lake paleosol at the Denison and Pronto mines and Button's (1979) work on paleosol developed on the Hekpoort basalt of South Africa. The key equation is

$$R = D[O_2]/D[CO_2] = \frac{0.2\,M[FeO] + 0.5\,M[MnO] + 3.75\,M[FeS_2] + M[C]}{2(M[CaO] + M[MgO] + M[Na_2O] + M[K_2O]) - 4\,M[FeS_2]},$$

where R is the ratio of O_2 demand to the CO_2 demand during the weathering of parent rock. $D[O_2]$ = oxygen demand; $D[CO_2]$ = acid demand; $M[C]$ = moles of organic carbon, and $M[FeS_2]$ = moles of pyrite.

Following this scheme, rocks with high R values were likely to generate paleosols with reduced iron. This class is well represented by basaltic rocks. Conversely, rocks with low R values, such as granitoids, were likely to develop oxidized paleosols.

Calculation of an R value for either the Denison or Stanleigh profiles would be inappropriate because of abundant sulfides which may be secondary. Farrow and Mossman (1988) calculated an R value of 0.036 for the NAN paleosol which developed on granodiorite. This value is slightly higher than that for the comparable Pronto profile analyzed by Gay and Grandstaff (1980). The iron loss displayed in the chemical analyses of both of these granodiorite-derived paleosols substantiates the validity of Holland's (1984) procedure (Fig. 5).

Grandstaff et al. (1986) recognized that the coexistence of oxidized and reduced profiles along the same paleosol horizon may reflect differences in parent rocks and/or varying degrees of ventilation during weathering. Applying Holland's (1984) procedure, Grandstaff et al. (1986, Fig. 12) constructed a graph for estimating coexisting pO_2 and pCO_2

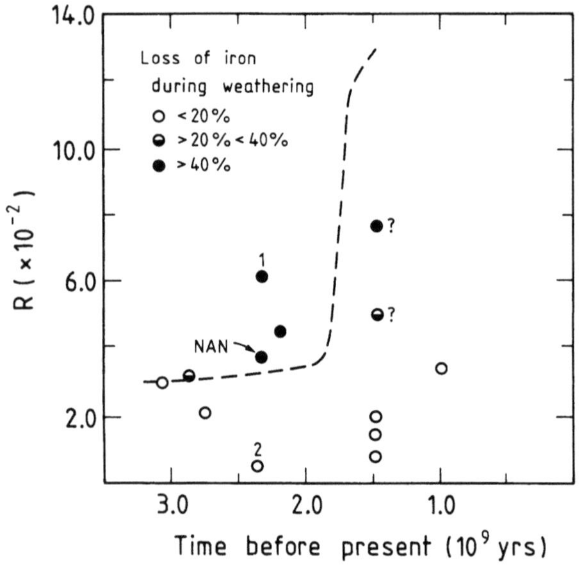

Fig. 5. Plot of Elliot Lake profiles on diagram of Pinto and Holland (1988, p. 32). Elliot Lake profiles developed on granodiorite are labeled *NAN* (North American Nuclear of this chapter) and *2* (Pronto profile of Gay and Grandstaff 1980). Profile developed on greenstone is labeled *1* (Denison mine). Other points are from Pinto and Holland (1988). The *R* value is the ratio of O_2 demand to CO_2 demand during weathering, as defined in the text. *Broken line* separates paleosols which have lost little or no iron during weathering from those which have lost appreciable iron

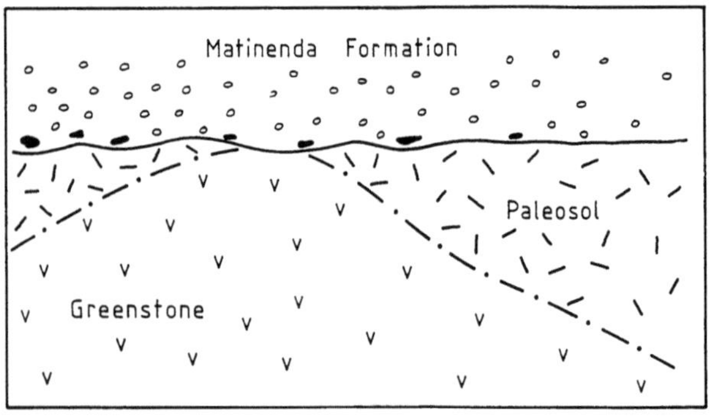

Fig. 6. Sketch of a paleotopographic high of greenstone showing erosion of paleosol prior to conglomeratic sedimentation. Rip-up clasts of paleosol (*solid black*) lie adjacent to the basal contact. View in Denison mine drift # 40988. (After Goddard 1987)

values. Application of this graph to their chemical analyses of parent rock and paleosol resulted in an estimate of pO_2 at 0.02% of the present atmospheric level (PAL) and pCO_2 at 5–30% PAL (cf. Grandstaff 1974, 1980). Application of the same graph to the composition of the NAN profile and parent rock, with its R value of 0.036, yields < 1% PAL of O_2.

Elliot Lake granitoids have a greater oxygen demand (R value) than the South African granitoids analyzed by Grandstaff et al. (1986), given a greater abundance of ferromagnesian minerals at Elliot Lake. The Elliot Lake paleosol therefore would tend to be more chemically reduced under the same partial pressures of CO_2 and O_2 than contemporaneous paleosols developed on the South African granitoids.

5 Paleotopography

According to Gay and Grandstaff (1980), paleotopography determined whether the Elliot Lake (sub-Matinenda) paleosol was oxidized or reduced. They hypothesized that granitic areas formed well-oxidized topographic highs whereas greenstone areas formed poorly drained lowlands. However, the granitic areas were not uniformly oxidized. Roscoe (1969, Table A) documents a thick, chemically reduced paleosol profile which developed on granitic rock down-slope from ore in the Panel Mine. Moreover, well-exposed greenstone-derived paleosol profiles in the Stanleigh and Denison mines are overlain by cross-bedded arkosic sandstone and quartz-pebble conglomerate of the Matinenda Formation. These coarse sediments almost certainly were deposited by fast-running water (Fig. 6) and the weathering surface under them must have been well drained. Topography presumably exerted some control on weathering but that control will not become clear until proprietary information is released on the thousands of drill holes through the paleosol.

6 Possible Role of Microorganisms

Stratiform kerogen seams in the basal Matinenda Formation are demonstrably syngenetic, just as in the Witwatersrand (Willingham et al. 1985; Nagy and Mossman, this Vol.). In both areas, the kerogen contains anomalous concentrations of U and Au and is understood to have originated, at least partly, from prokaryotic microorganisms. Prokaryotes were the dominant life from during the Early Precambrian. As primary photosynthetic producers, cyanobacteria presently generate some of the reduced carbon on which heterotrophs depend. In the Early Proterozoic, as now, they probably occupied many niches, including fluvial systems, nearshore settings, and almost certainly soils.

Organic carbon in the form of kerogen constitutes up to 0.25% of the upper 2 m of the paleosol profile at Denison. This may well represent the remains of organisms which grew upon or within soil. Unpublished isotopic analysis of this C_{org} indicates the expected depletion of ^{13}C. Strauss et al. (this Vol.) record several occurrences of ^{13}C-depleted C_{org} in slightly older (Late Archean) sedimentary rocks in Canada.

Life in the Elliot Lake paleosol may well have resembled that envisioned by Myers (1981) who proposes that microorganisms in Early Precambrian soils lived anaerobically by maintaining cycles of sulfate-to-sulfide and CO_2-to-C_{org} at roughly neutral pH. Such a concept of ubiquitous microbial activity is difficult to reconcile with observed sulfur isotopic ratios. Roscoe (1969) reports that pyrite in Elliot Lake conglomerate has a sulfur isotopic ratio ($^{34}S/^{32}S$) which is close to the meteoritic standard whereas all comparable Phanerozoic sedimentary pyrite is enriched in ^{32}S due to bacterial fractionation. The lack of enrichment of ^{32}S at Elliot Lake is characteristic of sedimentary pyrite of this age and older. Myers (1981) attributes this lack of ^{32}S enrichment in Early Precambrian pyrite to such low concentrations of sulfur solutes that isotopic fractionation was inefficient.

The ratios of Se/S and Co/Ni in Elliot Lake pyrite have been measured by Hawley and Nichol (1961), Roscoe (1969, Table V), and Kimberley et al. (1980). These data sets collectively indicate moderate concentrations of selenium and high Co/Ni ratios. The data previously have been interpreted to indicate a hydrothermal affinity (Hawley and Nichol 1961; Kimberley et al. 1980) but a microbially mediated origin of the pyrite is preferred herein.

7 Diagenetic and Metamorphic Alteration of the Paleosol

As documented elsewhere, the Lower Huronian metasedimentary rocks (and their kerogen) have a complex diagenetic and metamorphic history (Willingham et al. 1985; Nagy and Mossman this Vol). The immediately subjacent paleosol must have experienced a similar history. The prime alterations to the paleosol were compaction of initially porous clayey soils, variable to complete destruction of the primary fabric, and recrystallization of original clay mineral to sericite, with or without quartz cement. Quartz is widely distributed in the paleosol as mm-wide veinlets which are associated with rutile. Development of rutile and leucoxene may be a manifestation of one of the late-stage hydrothermal events recorded by Nagy and Mossman (this Vol.).

Pyrite is extremely abundant in uraniferous conglomate which overlies the Elliot Lake paleosol. Some investigators attribute the abundance of reduced sulfur in this conglomerate to sulfur-rich hydrothermal fluids generated by late-stage volcanic activity (e.g., Ruzicka and Steacy 1976; Ruzicka 1981). Whatever its origin, the pyrite was partially remobilized diagenetically.

Primary mineralogy, texture, and structures in the paleosol have been overprinted metamorphically as well as diagenetically. The lower greenschist facies predominates in the Elliot Lake area (Mossman 1987; Farrow and Mossman 1988). A key metamorphic assemblage of chlorite + Fe-rich chloritoid + quartz + sericite occurs toward the base of the Denison paleosol. Metamorphism of the paleosol generally is more obvious than that of the parent rock, presumably due to an original abundance of clay. The presence of kaolinite elsewhere in the Huronian Supergroup has been taken to indicate sub-greenschist-facies metamorphism (Card 1978; Willingham et al. 1985). Pyrophyllite has not been reported in the Huronian Supergroup but is associated with pyritic conglomerate in the Witwatersrand, South Africa (Phillips 1988).

8 Relationship of Elliot Lake Paleosol to Uranium Ore

There is a well-documented sorting of heavy minerals in the ore-grade paleoplacer which overlies the Elliot Lake paleosol (e.g., Roscoe 1969, 1981; Theis 1979). Erosion of soil presumably supplied most of the heavy minerals. The source of uranium at Elliot Lake was nearby uraniferous Archean granite (Ruzicka 1981; Robertson 1987). Not surprisingly, paleotopography strongly influenced ore disposition (Theis 1979; Farrow and Mossman 1988). Soil-forming processes had a subtler influence on ore-forming processes.

The U contents of both the Denison and NAN paleosol profiles are roughly proportional to the U concentrations in the respective parent rocks (ca. 1:10). As previously noted, the distribution of U is generally unlike that of Th through the paleosol profiles. Kimberley and Grandstaff (1986) attributed this fractionation of oxidative weathering. An alternative explanation is that U became concentrated near the top of the paleosol due to diagenetic processes which simultaneously affected overlying fluvial sediment.

Uranium enrichment near the top of the paleosol probably is related to the coexisting enrichment of rutile and leucoxene. Both enrichments may be attributed to Pronto-type reactions (Ramdohr 1957) in which the alteration of uraninite produced uraniferous pore fluids which reacted with titanium to form brannerite, i.e., $UO_2 + (2 \text{ to } 3)\, TiO_2 = UTi_{(2 \text{ to } 3)}O_{(6 \text{ to } 8)}$. Brannerite remains undetected in the Elliot Lake paleosol but is a major constituent of the overyling ore at the Denison mine, where heavy-mineral sorting played a key role in ore genesis. Theis (1979) argues against a detrital origin for brannerite but a detrital origin would be compatible with the bed-to-bed variation in the proportion of brannerite to uraninite (Robertson and Steenland 1960; Schidlowski 1981, p. 25). The alteration of uraninite to brannerite probably occurred in both the depositional and source (weathering) areas. Similar evidence of alteration (weathering) in both the source and depositional areas is reported by Minter et al. (1988) for some Witwatersrand paleoplacers.

The Elliot Lake paleosol is modestly enriched in gold (Farrow and Mossman 1988). This enrichment is attributable to minor concentrations of C_{org} within the upper portion of the paleosol. The carbon is attributable to microorganisms. The efficacy of microorganisms in ore-forming dissolution and precipitation of Au and/or U is well known (Mossman and Dyer 1985).

9 Conclusions

Although stratigraphically thin, the Elliot Lake paleosol yields substantial information about the complex interplay of geologic processes which were operative during an otherwise profound hiatus between Late Archean and Early Proterozoic time. The remaining problems of paleosol interpretation are largely attributable to diagenetic and metamorphic alteration. However, a consensus exists on several important issues:

1. The Elliot Lake paleosol is characterized by an upward depletion of alkalis and alkaline earths, a marked increase in potassium, a loss of iron, and an increase in the ratio of Fe^{3+}/Fe^{2+}. The similarity of mineralogical and chemical characteristics in paleosol profiles which developed at widespread locations on various parent rocks indicates a common response to weathering conditions. The oxidation of iron indicates that the Early Proterozoic atmosphere was not anoxic.

2. Calculation of the extent of oxidation of a granodiorite parent rock using Holland's (1984) procedure yields results compatible with the oxygen-deficient atmospheric conditions envisaged by Grandstaff et al. (1986). The extent of paleosol oxidation is not entirely attributable to topographically controlled drainage conditions.

3. Although diagenesis and lower greenschist metamorphism have substantially overprinted earlier-developed chemical and mineralogical features, it is clear that the paleosols are polygenetic, given preservation of such features as fossil eluvial horizons and desiccation structures.

4. Sulfur-cycling microbial communities probably played an important role in both soil formation and ore-forming processes at Elliot Lake. Evidence for this lies in certain soil structures, the moderate concentration of organic carbon, the elevated concentration of reduced sulfur, and the isotopic compositions of both carbon and sulfur.

5. The fractionation of uranium from thorium in the Elliot Lake paleosol is attributable to a combination of weathering, physical sorting, and diagenesis. Pronto-type diagenetic reactions also concentrated rutile and leucoxene toward the top of the paleosol. The alteration of uraninite to brannerite probably is related to the degree of weathering in both the source and depositional areas.

Acknowledgments. This contribution to IGCP 157 was facilitated by Prof. Dr. Manfred Schidlowski of the Max-Planck-Institute für Chemie, Mainz, FRG, who kindly provided office and laboratory space during a sabbatical leave (by DJM). Discussions with numerous colleagues and a grant from the National Sciences and Engineering Research Council of Canada (to DJM) are gratefully acknowledged.

References

Button A (1979) Early Proterozoic weathering profile on the 2200 m.y. old Hekpoort Basalt, Pretoria Group, South Africa: preliminary results. Univ Witwatersrand Johannesburg, Inf Circ 133:50 pp

Card KD (1978) Geology of the Sudbury-Manitoulin area, districts of Sudbury and Manitoulin. Ontario Geol Surv Rep 166:238 pp

Cloud P (1983) The biosphere. Sci Am 249:176–189

Collins WH (1925) North shore of Lake Huron. Geol Surv Can Mem 143:160 pp

Davidson CF (1953) The gold-uranium ores of the Witwatersrand. Min Mag (London) 88:73–85

Farrow CEG, Mossman DJ (1988) Geology of Precambrian paleosols at the base of the Huronian Supergroup, Elliot Lake, Ontario, Canada. Precambrian Res 42:107–139

Gay AL, Grandstaff DE (1980) Chemistry and mineralogy of Precambrian paleosols at Elliot Lake, Ontario, Canada. Precambrian Res 12:349–375

Goddard CE (1987) Geology of paleosols at the Lower Huronian-Archean unconformity, Elliot Lake, Ontario, Canada. BSc Thesis, Mount Allison Univ, Sackville, New Brunswick, Can, 128 pp

Grandstaff DE (1974) Uraninite oxidation and the Precambrian atmosphere (abstract). Trans Am Geophys Un 55:457

Grandstaff DE (1980) Origin of uraniferous conglomerates at Elliot Lake, Canada, and Witwatersrand, South Africa: implications for oxygen in the Precambrian atmosphere. Precambrian Res 13:1–26

Grandstaff DE, Endelman MJ, Foster BW, Zbinden E, Kimberley MM (1986) Chemistry and mineralogy of Precambrian paleosols at the base of the Dominion and Pongola Groups (Transvaal, South Africa). Precambrian Res 32:97–131

Hawley JE, Nichol I (1961) Iran elemente in pyrite, pyrrhotite and chalcopyrite of different ones. Econ Geol 56:467–487

Holland HD (1984) Chemical evolution of the atmosphere and oceans. University Press, Princeton, NJ, 656 pp

Holland HD, Zbinden EA (1986) Paleosols and the evolution of the atmosphere, pt. 1. In: Lehrman A, Meybeck M (eds) Physical and chemical weathering in geochemical cycles. NATO Adv Stud Inst. Reidel, Dordrecht, pp 61–82

Kimberley MM, Grandstaff DE (1986) Profiles of elemental concentration in the Precambrian paleosols on basaltic and granitic parent materials. Precambrian Res 32: 133–154

Kimberley MM, Tanaka RT, Farr MR (1980) Composition of Middle Precambrian uraniferous conglomerate in the Elliot Lake-Agnew Lake area of Canada. Precambrian Res 12:375–392

Kimberley MM, Grandstaff DE, Tanaka RT (1984) Topographic control on Precambrian weathering in the Elliot Lake uranium district, Canada. J Geol Soc London 141:229–233

Minter WEL, Feather CE, Glatthaar CW (1988) Sedimentological and mineralogical aspects of the newly discovered Witwatersrand placer deposit that reflects Proterozoic weathering, Welkom Gold Field, South Africa. Econ Geol 83:481–491

Mossman DJ (1987) Stratiform gold occurrences of the Witwatersrand type in the Huronian Supergroup, Ontario, Canada. S Afr J Geol 90:481–491

Mossman DJ, Dyer BD (1985) The geochemistry of Witwatersrand-type gold deposits and the possible influence of ancient prokaryotic communities on gold dissolution and precipitation. Precambrian Res 30:303–319

Myers WB (1981) Genesis of uranium-gold pyritic conglomerates. In: Armstrong FC (ed) Genesis of uranium and gold bearing Precambrian pebble conglomerate. US Geol Surv Prof Pap 1161:AA1–AA26

Phillips GJ (1988) Widespread fluid infiltration during metamorphism of the Witwatersrand goldfields: generation of chloritoid and pyrophyllite. J Metamorph Geol 6:311–332

Pienaar PJ (1963) Stratigraphy, petrology and genesis of the Elliot Lake Group, Blind River, Ontario, including the uraniferous conglomerate. Geol Surv Can Bull 83: 140 pp

Pinto JP, Holland HD (1988) Paleosols and the evolution of the atmosphere, pt 2. Geol Soc Am Spec Pap 216: 21–34

Ramdohr P (1957) Die „Pronto-Reaktion". N Jahrb Mineral Monatsh 1957:217–222

Retallack GJ, Grandstaff DE, Kimberley MM (1984) The promise and problems of Precambrian paleosols. Episodes 7:8–12

Robertson DS, Steenland NC (1960) On the Blind River uranium ores and their origin. Econ Geol 55:659–694

Robertson JA (1987) The Blind River (Elliot Lake) uranium deposits. In: Int Atom Energ Ag (IAEA) Symp Uranium resources and geology of North America, Aug. 31–Sept. 4, 1987, Saskatoon, Saskatchewan, Can

Roscoe SM (1957) Geology and uranium deposits, Quirke Lake-Elliot Lake, Blind River area, Ontario. Geol Surv Can Pap 56–7:100 pp

Roscoe SM (1969) The Huronian rocks and uraniferous conglomerates. Geol Surv Can Pap 68–40:205 pp

Roscoe SM (1973) The Huronian Supergroup, a Paleoaphebian succession showing evidence of atmospheric evolution. Geol Assoc Can Spec Pap 12:31–47

Roscoe SM (1981) Temporal and other factors affecting deposition of uraniferous conglomerates. In: Armstrong FC (ed) Genesis of uranium and gold-bearing Precambrian quartz pebble conglomerates. US Geol Surv Prof Pap 1161:W1–W17

Roscoe SM, Steacy HR (1958) On the geology and radioactive deposits of Blind River Region. In: United Nations survey of raw material resources. In: 2nd Int Conf Peaceful use of atomic energy, Geneva, Switz, 1–13 Sept 1958, Proc, vol 2, pp 473–483

Ruzicka V (1981) Some metallogenic features of the Huronian and post Huronian uraniferous conglomerates. In: Armstrong FC (ed) Genesis of uranium and gold-bearing Precambrian quartz pebble conglomerates. US Geol Surv Prof Pap 1161:V1–V8

Ruzicka V, Steacy HR (1976) Some sedimentary features of conglomeratic uranium ore from Elliot Lake, Ontario. Geol Surv Can Pap 76–1A:343–346

Schidlowski M (1981) Uraniferous constituents of the Witwatersrand conglomerates: Ore microscopic observations for the Witwatersrand metallogeny. In: Armstrong FC (ed) Genesis of uranium – and gold-bearing Precambrian quartz – pebble conglomerate. US Geol Surv Prof Pap 1161:N1–N29

Schidlowski M (1988) Early organic evolution: implications for mineral and energy resources – a brief portrait of IGCP Project 157. Terra Cognita 8:205–209

Soil Survey Staff (ed) (1975) Soil taxonomy. USDA Handb 436:754 pp

Theis NJ (1979) Uranium-bearing and associated minerals in their geochemical and sedimentological context, Elliot Lake, Ontario. Geol Surv Can Bull 304:50 pp

Willingham TO, Nagy B, Nagy LA, Krinsley DH, Mossman DJ (1985) Uranium-bearing stratiform organic matter in paleoplacers of the Lower Huronian Supergroup, Elliot Lake-Blind River regions, Canada. Can J Earth Sci 22:1930–1944

Zbinden EA, Holland HD, Feakes CR, Dobos SK (1988) The Sturgeon Falls paleosol and the composition of the atmosphere 1.1 Ga BP. Precambrian Res 42:141–163

Mineralogy and Geochemistry of the Pre-1.65 Ga paleosol under Kombolgie Formation sandstone of the Pine Creek Geosyncline, Northern Territory, Australia [1]

A. R. Miller [2], R. S. Needham [3] and P. G. Stuart-Smith [3]

[2] Geological Survey of Canada, 601 Booth Street, Ottawa, Ontario, K1A 0E8, Canada
[3] Bureau of Mineral Resources, Geology and Geophysics, GPO Box 378, Canberra, A.C.T. 2601, Australia

Abstract

Prior to deposition of the 1.69 to 1.65 Ga Kombolgie Formation in the Northern Territory of Australia, deep weathering occurred in the region around Katherine and Maningrida. Exposed Archean and Lower Proterozoic rocks are altered to saprolite in a clayey profile which reaches 50 m in thickness. Rock color grades downward from an upper hematitic zone through a transitional zone to a lower chloritic zone. The hematitic zone is divisible into an upper subunit of diaspore + hematite + kaolinite and a lower subunit of Fe-chlorite + hematite + illite. The transitional zone contains hematite + chlorite (both Fe- and Mg-rich varieties) + illite. The underlying chloritic zone contains chlorite (both Fe- and Mg-rich) + illite + relict feldspar. The chloritic zones grades downward into unaltered protolith.

Elemental concentrations and ferric/ferrous ratios vary downward through the profile. The hematitic zone is characterized by enrichment of Ni, Cr, V and depletion of Mg, Ba, Sr, Ca, Na, Li, U, Cu, Pb, and Zn. Beneath this zone, the concentrations of all these elements approach those of the underlying protolith. The ferric/ferrous ratio decrease downward. Diagenetic alteration of the paleosol included recrystallization of illite, redistribution of iron near the contact with the overlying Kombolgie Formation, and an introduction of phosphorus into the upper hematitic zone, presumably from fluids which had descended through Kombolgie sediment.

1 Introduction

Modern chemical weathering deeper than 100 m has been documented in hot humid regions (Ollier 1969). Thick profiles reflect tectonic stability, slow erosion, and potent leaching. Climate, topography and parent lithology control congruent and incongruent solution, hydration, and oxidation. Remnants of such weathering are scattered throughout the geologic column: in the Archean (Schau and Henderson 1983), in the Proterozoic (Williams 1968; Roscoe 1969; Button 1979; MacDonald 1980; Gay and Grandstaff 1980) and throughout the Phanerozoic (Sigleo and Reinhardt 1988). Interpretation of this ancient weathering is complicated by post-burial alteration of the original weathering products. Burial diagenesis within a paleosol may be elucidated by studying the diagenesis of overlying clastic sedimentary rocks, as in the present study of a Precambrian paleosol in the northern part of the Northern Territory of Australia. This paleosol developed on both Archean rocks and Early Proterozoic rocks of the Pine Creek Geosyncline.

[1] Geological Survey of Canada Contribution # 32989.

2 Previous Work

Needham et al. (1980) identified a regionally extensive paleosol underlying the Middle Proterozoic Kombolgie Formation of the Katherine River Group in the Northern Territory of Australia. They noted that this argillaceous profile locally exceeds 50 m in thickness and displays obvious variations in color. The study of Precambrian weathering in this area is complicated by Cenozoic weathering, particularly the intense weathering of the Early Tertiary. However, Eupene (1980) noted that the depth of recent weathering diminishes toward the edge of the Arnhem Land Plateau due to the ongoing retreat of the escarpment and the resulting exhumation of the sub-Kombolgie basement.

Numerous cliff-forming exposures of the sub-Kombolgie paleosol have been examined and sampled for this study (Fig. 1). None of these exposures reveal a complete profile through the paleosol, partly because of little lateral variation in the

M. Schidlowski et al. (Eds.)
Early Organic Evolution: Implications for Mineral and Energy Resources

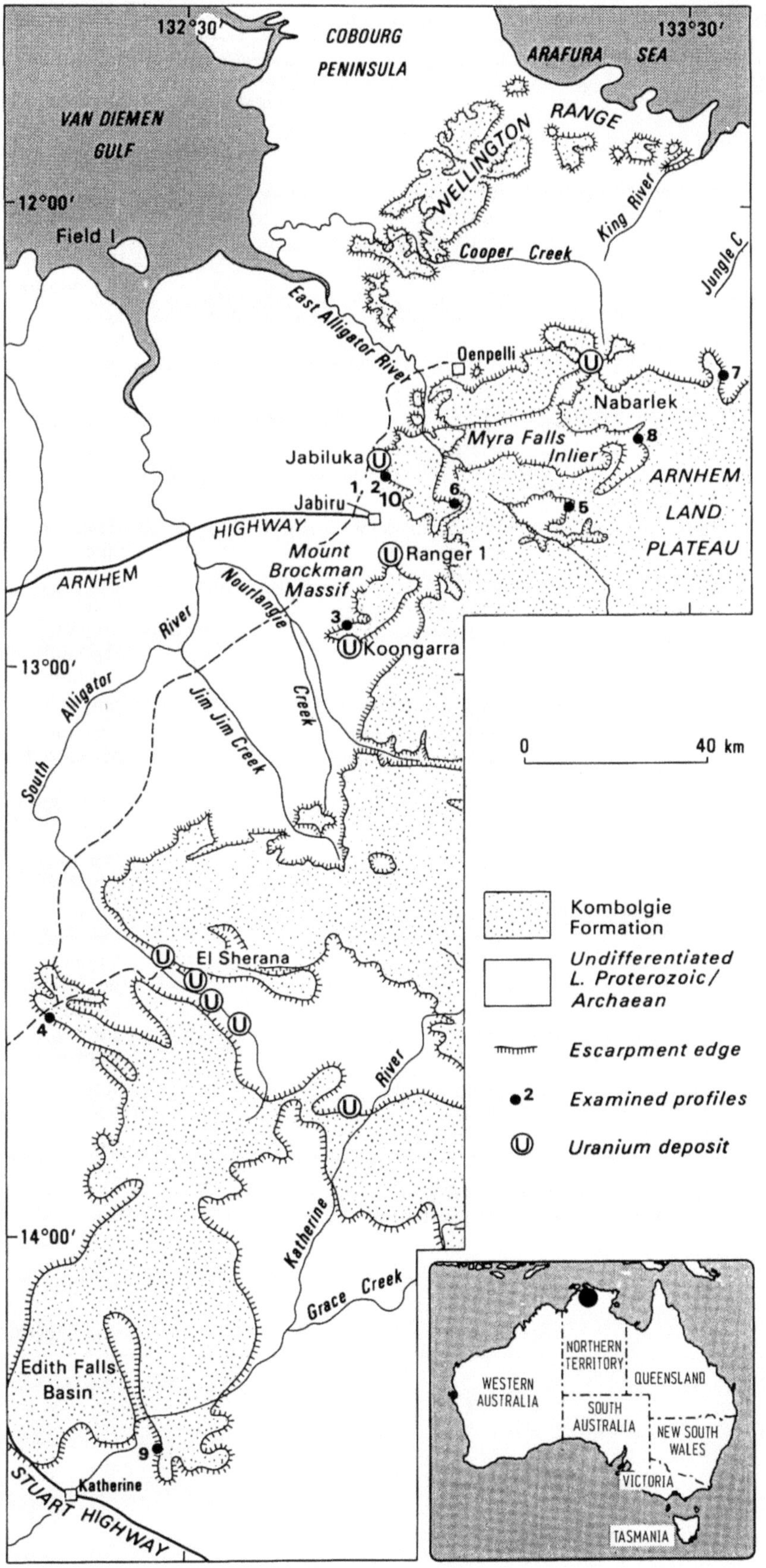

Fig. 1. Location of paleosol profiles. Locations *1* and *10* refer to profiles GT5 and GT6 respectively. Location *2* is the Granite Hill profile, *3* is the Baroalba Creek profile, and *4* is the Tent Creek profile. Locations *5* through *9* are not differentiated in the text

Table 1. Paleosol profiles, Pine Creek Geosyncline, N.T., Australia

No.	Location name	Grid reference (AMG)	1:100000 Map sheet	Formation	Protolith age	Rock type	Paleoweathered profile		
							Thickness (m)	No. of observed zones	Mineralogy
1.	GT5	716160 (1.2 km SSE of Jabiluka)	Cahill	Nanambu Complex	Arch./E. Prot.	Retrograded muscovite-biotite schist and gneiss (metasediments)	max. ≈ 60	3	Hematite, illite, trace kaolinite, Fe- and Mg-chlorite
2.	Granite Hill	720130 (4.1 km SSE of Jabiluka)	Cahill	Nanambu Complex	Arch.	Granite gneiss	max. ≈ 20	1	Kaolinite, hematite, illite, diaspore
3.	Baroalba Creek	655796 (3.4 km N of Koon-garra)	Cahill	Cahill Formation "upper member"	E. Prot.	Muscovite-feldspar-quartz schist	min. 10	1	Hematite, illite,
4.	Tent Hill	030972	Ranford Hill	Plum Tree Creek Volcanics	E. Prot.	Porphyritic feldspar rhyodacite ignimbrite	min. 8	1	Hematite, illite
5.	Beatrice Inlier	130073	Howship	Nimbuwah Complex	E. Prot.	Biotite-muscovite-feldspar-quartz metasediment	[a]	2 possibly 3	Illite, hematite, Mg- and Fe-chlorite
6.	Algodo	872065	Howship	Nourlangie Schist	E. Prot.	Muscovite-biotite feldspar-quartz schist, muscovite quartzite	[a]	[b]	Illite, hematite, Mg-chlorite, trace kaolinite
7.	Goomadeer	435363	Goomadeer	Nimbuwah Complex	E. Prot.	Equigranular biotite-hornblende granite	[a]	[b]	Illite, hematite, goyazite
8.	Tin Camp Creek	287226	Oenpelli	Tin Camp Granite	E. Prot.	Porphyritic quartz-feldspar-biotite granite	[a]	[b]	Illite, hematite, kaolinite, minor Mg-chlorite, trace goyazite
9.	Mount Shepherd	245082	Katherine	Plum Tree Creek Volcanics	E. Prot.	Porphyritic feldspar ignimbrite	[a]	[b]	Kaolinite, hematite, minor illite and goyazite
10.	GT6	720130	Cahill	Nanambu Complex	Arch./E. Prot.	Micaceous metasediment	min. 43	3	Hematite, kaolinite, illite, Mg- and Fe-chlorite, diaspore, trace apatite

[a] indeterminate. [b] insufficient samples.

level of erosion prior to burial under Kombolgie sandstone. However, complete profiles into the protolith were obtained in core drilled about 1.2 km south of the Jabiluka deposit (provided by Pancontinental Mining Ltd.). Table 1 lists the profiles examined for this study and additional samples collected by R. S. Needham during his regional mapping of the Pine Creek Geosyncline from 1971 to 1982.

3 Nomenclature and Methods

Nomenclature in this chapter follows modern usage by soil scientists; e.g., "Oxisol" is used instead of "laterite" (Buol et al. 1980). Oxisol is a product of intense weathering and is largely composed of secondary iron and aluminum sesquioxides. It is nearly devoid of bases and may contain variable amounts of quartz and kaolin-group clays. "Saprolite" is a thoroughly decomposed weathering product of igneous or metamorphic rock in which relict structures and textures are visible.

Outcrop samples were collected near and immediately below cliffs of the Kombolgie Formation. Drill-core samples were obtained from Pancontinental holes GT5 and GT6. These holes were drilled for an engineering study of a proposed decline into the Jabiluka II uranium-gold orebody. Drill-hole GT5 lies 1150 m south of Jabiluka II and has an inclination of 60° along an azimuth of 347°. The hole extends 125 m through sub-Kombolgie metasedimentary rocks. The upper 18 m were not cored but samples were collected through the underlying length to a depth of 81 m. Drill-hole GT6 was collared in Kombolgie Formation sandstone and cored through 149.5 m of sandstone before continuing through 47.5 m of basement metasedimentary rocks. Pegmatite appeared near the bottom of the hole.

Samples for whole-rock elemental analysis were crushed and then pulverised in a tungsten-carbide Siebtecknik mill. Major elements were determined on fusion discs using a Philips PW-1450 automated X-ray fluorescence spectrometer. Most trace elements were analyzed by XRF using powder pellets whereas Cr, Cu, Zn, Li, and Ni were determined by atomic-absorption spectrophotometry. The mineralogy of phyllosilicates and hydroxides was identified using a Philips PW-1010 diffractometer. A PW-1050-25 goniometer was used for diffractometer traces and a Dedije Scherrer camera for powder mounts.

Quantitative electron microprobe analyses of phyllosilicates and hydroxides were obtained with an energy-dispersive X-ray spectrometer at the Research School of Earth Sciences, Australian National University and at the Geological Survey of Canada, Ottawa. Autoradiographs of the more radioactive samples (GT series) were obtained with thermoset polymer CR-39, using the procedure of Basham (1981).

4 Regional Geology

Age relationships in the study area have been reviewed by Page et al. (1980). The regional geology and evolution of the Pine Creek Geosyncline are discussed by Needham et al. (1980), Stuart-Smith et al. (1980), Needham (1988), and Needham et al. (1988). The Nanambu Complex forms the basement to the supracrustal rocks of the Pine Creek Geosyncline between the East and South Alligator Rivers (Fig. 1). This complex is composed of metamorphosed granitoids (2.47 Ga) and a mantle of Early Proterozoic metasedimentary rocks. At Granite Hill, 4.1 km SSE of Jabiluka, both U/Pb and Rb/Sr dating of Nanambu granite gneiss indicate an age of about 2500 Ma, specifically 2470 (+ 47/− 40) Ma (U/Pb in zircon) and 2468 (+/− 26) Ma (Rb/Sr whole rock) (Page et al. 1980).

Erosion of the 2.47 Ga basement supplied part of the Early Proterozoic sediment which hosts uranium deposits in the Alligator Rivers area. Sedimentation occurred between 2.2 and 2.0 Ga, according to Rb/Sr data (Page et al. 1980). Regional metamorphism at about 1.8 Ga produced the metamorphic assemblages and fabrics of these rocks. Synmetamorphic igneous activity included postorogenic felsic extrusives of the Edith River Group (1863 +/− 5 Ma in Page et al. 1985), granitoid intrusives (1800 Ma in Needham et al. 1980), and mafic sills of the Oenpelli Dolerite (1688 +/− 13 Ma in Page et al. 1980).

Sandstone, pebbly sandstone, conglomerate, and minor siltstone of the Kombolgie Formation overlie older rocks with marked regional unconformity. The Nungbalgarri Volcanic Member of the Kombolgie establishes a minimum age for the lower beds at 1648 +/− 29 Ma. Needham et al. (1980) describe the pre-Kombolgie surface as subdued and gently undulating, having had a relief of about 20 m between scattered basement hills. Tectonic stability is implied by both the minimal relief and the regionally extensive zone of deep weathering along the unconformity. About 40 Ma elapsed between the intrusion of post-orogenic Oenpelli Dolerite and the first volcanic member in the Kombolgie Formation.

The basal Kombolgie contains cobbles of paleosol within high-energy fluvial sediments, a diagnostic feature of paleosols (Retallack et al. 1984).

5 Drill Core GT5

5.1 Introduction

Drill hole GT5 was collared about 200 m west of the Kombolgie Formation escarpment. It lies about 1150 m south of the Jabiluka II ore body and the nearest known intersection of ore-grade mineralization (drill hole M0996V) occurs 1050 m to the northwest. No cross cutting zones of brecciation or intense fracturing are apparent in the GT5 core. Muscovite-biotite schist occurs along the entire length of retrieved core, i.e., from 18 to 125 m. This schist is a psammitic component of the Nanambu Complex. The schist is color-stratified along the profile, with gradational contacts between zones. A red hematitic zone extends from 18 to 23.6 m, a greenish red transitional zone from 23.6 to 31 m, and a chloritic zone from 31 to 61 m. The chloritic zone contains an upper pinkish green subzone from 31 to 45 m and a lower green gray subzone from 45 to 61 m. The protolith extends below 61 m. Metamorphic textures and fabrics are well preserved throughout the profile, even though alteration progressively increases upward from 61 to 18 m.

5.2 Drill Core GT5: Hematitic Zone of Paleosol

The uppermost (hematitic) zone of the paleosol in GT5 is characterized by the assemblage: chlorite + hematite + illite + quartz. Quartz retains its equant to elongate form and this defines the metamorphic fabric. Hematite occurs as irregular wispy aggregates along quartz-grain contacts, as irregularly disseminated grains within a groundmass of illite, as a partial to complete replacement of muscovite grains, and within chlorite, which is replacing biotite (Fig. 2a).

Illite is the dominant phyllosilicate in this zone. It coexists with chlorite, relict metamorphic muscovite, hematite, and traces of kaolinite. Illite is present as extremely fine-grained mats which pseudomorphose feldspar. Locally, illite partially replaces the chlorite which occurs along cleavage planes in relict biotite. Chlorite occurs as platelets within the illite-rich matrix. These platelets superficially resemble those which replace biotite in the protolith, but here they frequently are partially to totally replaced by hematite. Chlorite plates in this hematitic zone are characterized by colorless to green pleochroism and a berlin-blue interference color. The well-preserved platy habit records a metamorphic lineage. Chlorite with these characteristics occurs throughout the profile and is designated "chlorite I".

Traces of anatase, epidote, apatite, tourmaline, and zircon are present within the illite matrix. Very fine anhedral anatase and epidote are associated with relict chlorite grains, suggesting that these are a product of greenschist retrogression. Traces of secondary pyrite with a string-like morphology occur within some knots of hematite.

5.3 Drill Core GT5: Transitional Zone of Paleosol

The transitional zone contains the same essential minerals as the hematitic zone but hematite is less abundant and there is an additional type of chlorite, here designated "chlorite II". Chlorite II is colorless to very pale green, with little or no pleochroism. Its birefringence is black to first-order gray and its interference colors are violet to khaki. It occurs as either fine-grained spherulitic to scaly grains which are intergrown with illite or as fine mats which are partially replacing feldspar (Fig. 2b). Plagioclase is completely replaced by illite in the transitional zone but minor relict cores of K-feldspar occur within illite-chlorite II mats. The matrix is composed primarily of clay and chlorite with finely disseminated

Fig. 2a–d. a Photomicrograph in plane light of hematitic zone, 19.1 m down GT5 profile of sub-Kombolgie paleosol. Equigranular quartz (*Q*) occurs in a matrix of illite (*I*) and finely crystalline hematite (*h*). Note replacement of metamorphic mica by coarser hematite (*H*). **b** Photomicrograph in plane light of transitional zone, 24.5 m down GT5 profile. Note metamorphic mica being replaced by chlorite I (Ch_1). Plagioclase is being replaced by mixture of chlorite II (Ch_2) and illite (*I*). Quartz (*Q*) and metamorphic muscovite (*Mu*) are stable. **c** Photomicrograph in cross-polarized light of chlorite zone, 38.8 m down GT5 profile. Metamorphic textures have been excellently preserved by paleosol minerals. Subhedral plagioclase which occurred within microcline (*Kf*) has been completely replaced by illite (*I*). Microcline is weakly altered to irregular dark areas of finely crystalline chlorite II (Ch_2). Granoblastic quartz (*Q*), muscovite (*Mu*) and chlorite I (Ch_1) occur throughout the section. **d** Photomicrograph in cross-polarized light of chloritic zone, 44.8 m down GT5 profile. Fine-grained illite (*I*) has replaced albite exsolution lamellae in perthite. *Dark areas* in the perthite are finely crystallized chlorite II (Ch_2). Illite (*light streaks*) has partially replaced chlorite I (Ch_1) along its cleavage planes

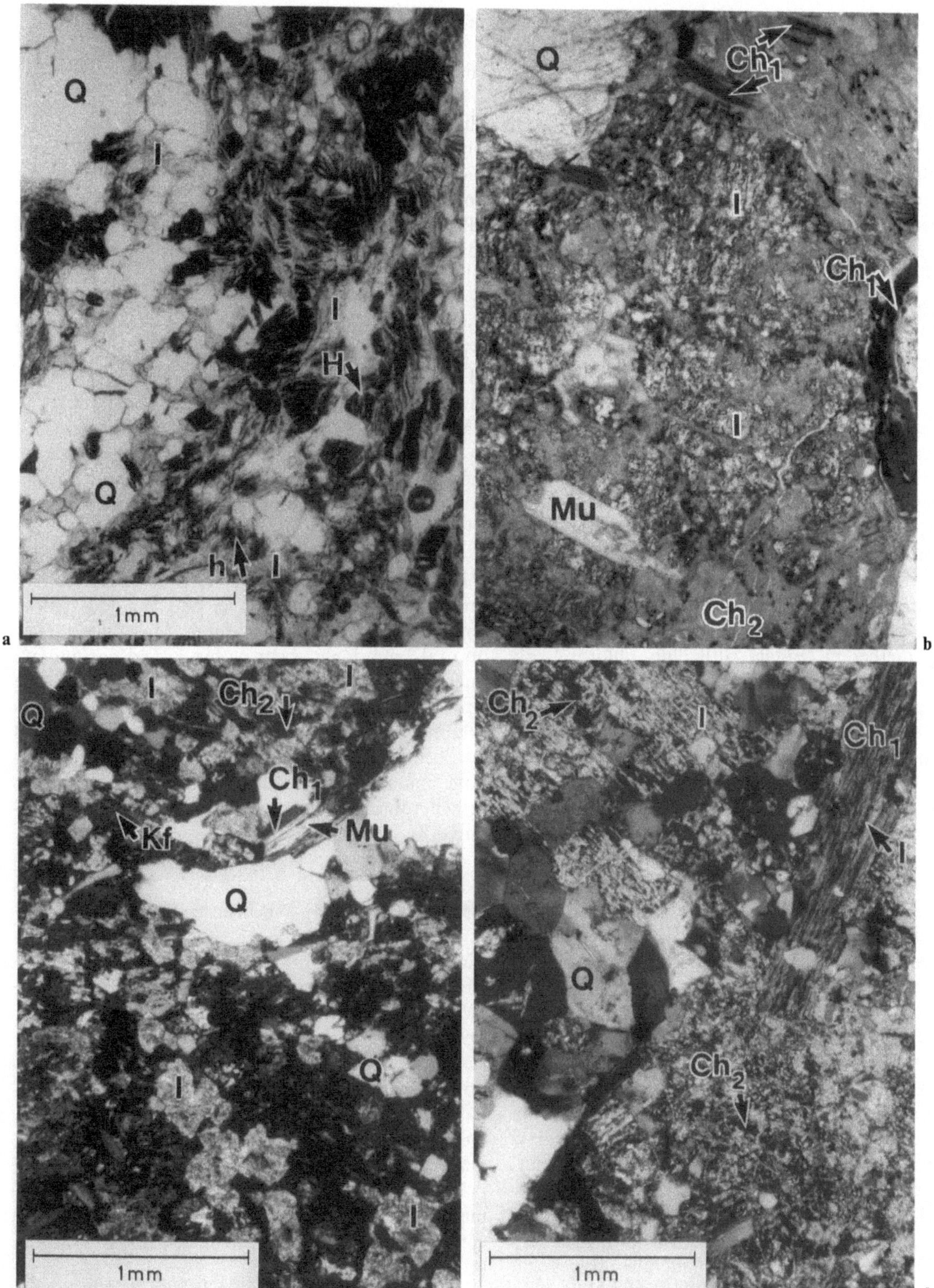

Fig. 2a–d

hematite. Very fine grains of equant pyrite are present, but rare, at the base of this zone.

5.4 Drill Core GT5: Chloritic Zone of Paleosol

The chloritic zone is distinguishable by its greenish hues and is further divisible into an overlying pinkish green subzone and an underlying grayish green subzone. Metamorphic textures are well preserved throughout the chloritic zone. The essential mineral assemblage of the pinkish green subzone is quartz + illite + chlorite I + microcline perthite + muscovite. The principal mineralogical variation with depth is a decreasing replacement of feldspars by clay and chlorite II. Chlorite II is more abundant than in the transitional zone. Here it replaces microcline perthite, either partially or completely. Illite largely pseudomorphs plagioclase (Fig. 2c) but locally is intergrown with chlorite II, both pseudomorphing K-feldspar (Fig. 2d). Alteration in this subzone apparently was isovolumetric, given the preservation of metamorphic textures and the replacement of albite and perthite twins by chlorite and illite. Metamorphic muscovite is unaltered and appears stable, whether in contact with illite, chlorite I, or chlorite II (Fig. 2a). Hematite is markedly less abundant than in the overlying zone and occurs as minute concentrations along the cleavage of muscovite and chlorite I plates. Fine-grained equant pyrite is sparsely disseminated throughout the rock.

The grayish green subzone of the chloritic zone differs from the overlying pinkish green subzone primarily in the greater preservation of feldspars. This rock is generally similar to the underlying protolith except for partial replacement of feldspar by either illite or chlorite II and the virtually complete replacement of biotite by chlorite I. The assemblage is quartz + microcline perthite + chlorite I + chlorite II + plagioclase + muscovite. At 54.4 m, cores of relict plagioclase occur within fine-grained aggregates of illite. Some microcline perthite locally exhibits variable replacement by mats of chlorite II and illite. Elsewhere, microcline perthite exhibits an ultra-fine dusting of opaque inclusions which may be hematite. Rare cores of remnant biotite occur within chlorite I plates. The modal abundance of plagioclase is less than that of K-feldspar and this proportionality is maintained downward into the protolith. Accessory minerals include traces of epidote and anatase within or adjacent to chlorite I, apatite, zircon, and tourmaline. Pyrite, chalcopyrite, and rare bornite are sparsely disseminated throughout this subzone.

5.5 Drill Core GT5: Paleosol Protolith

The protolith is an equigranular, medium-grained (2–4 mm), foliated quartzofeldspathic paragneiss which is mottled in pinkish gray-green hues. Preservation of feldspars, both megascopically and microscopically, clearly distinguishes the protolith from overlying paleosol zones. Clusters of white to pale pink feldspars occur in a gray-green matrix of phyllosilicate + quartz + feldspar, giving the protolith a pseudo-augen texture. Feldspar compositions of An_{25} are readily determined despite variable replacement by extremely fine-grained illite (Fig. 3b). Microcline perthite is essentially unaltered but locally displays traces of illite overgrowths. Moreover, the microcline perthite appears dusty in plane-polarized transmitted light, presumably due to micron-sized hematite. Metamorphic strain of the large quartz grains is recorded in undulatory extinction, sutured grain margins, and minor polygonization.

Accessory minerals in the protolith include chlorite I, muscovite, zircon, epidote, anatase, biotite, green-brown tourmaline, and opaques (probably magnetite and/or hematite). Muscovite plates, up to 0.2 mm long, occur within K-feldspar and are intergrown with chlorite (Table 2b). These chlorite grains are of type I (replacing biotite) because they exhibit berlin-blue to violet interference colors and are locally associated with trace quantities of epidote, anatase, and opaques. The few grains of green-brown tourmaline and red-brown biotite are armoured by quartz (Table 2b).

5.6 Drill Core GT5: Geochemistry of Paleosol

Figure 4 illustrates the compositional variation with stratigraphic depth (calculated from depth of core along the inclination of 60°). The chemical composition is presented as molecular ratios instead of oxide weight percents to emphasize the weathering reactions (Reiche 1943). Hydrolysis of feldspar and oxidation appear to have been the principal reactions. Qualitative assessment of major-element mobility may be obtained from their variation through the paleosol zones. A more quantitative assessment by mass-balance calculations was performed, given the composition of primary minerals in the protolith (Gresens 1967).

Silica and alumina display antipathetic trends through the paleosol profile. However, neither the mean concentration of SiO_2 nor its standard deviation vary much among the zones and there is only a

Fig. 3a, b. **a** Photomicrograph in cross-polarized light of chloritic zone, 47.7 m down GT5 profile. Plagioclase is partially replaced by an ultra-fine-grained intergrowth of illite (*I*) and chlorite II (Ch_2). Illite occurs as bright unoriented platelets whereas chlorite occurs as gray to black irregular patches within illite. Quartz (*Q*) and metamorphic muscovite (*Mu*) also are present. **b** Photomicrograph in cross-polarized light of the protolith, 70.05 m down GT5 profile. Albite-twinned plagioclase (*Pc*) is weakly sericitized. Microcline perthite (*Kf*), chlorite I (Ch_1), and metamorphic muscovite (*Mu*) are all unaltered

small decrease in Al_2O_3 through the lower portion (chloritic zone and protolith) compared to the hematitic and transitional zones. The preservation of feldspar in and below the chloritic zone accounts for this uniformity.

Iron and magnesium display significant variation and reflect the major mineralogical changes through the profile (Table 3). Total iron (FeO_T) is highest in the hematitic zone, averaging 2.59%, and decreases to a mean of 1.15% in the protolith. The decrease in total iron with depth is accompanied by a decreasing ratio of Fe^{III}/Fe_T, calculated as $Fe_2O_3/(FeO + Fe_2O_3)$, from a mean of 0.90 in the hematitic zone to 0.39 in the protolith. This variation is related to a decrease in hematite with depth, to an increase in ferrous-ferric phases like chlorite, and to an increase in trace amounts of sulfides ($\ll 1\%$). The range of the Fe^{III}/Fe_T ratio through one standard deviation is 0.87 to 0.93 within the hematitic zone and 0.63 to 0.71 in the transitional zone. The ratio varies little between the chloritic zone, with a range of 0.34–0.40, and the protolith, 0.34–0.44. Ratios of FeO_T/MgO and of $FeO_T/(FeO_T + MgO)$ similarly reflect zonation in the GT5 profile.

Magnesium is depleted in the hematitic zone but dramatically increases in the transitional zone, followed by a gradual decrease toward the base of the profile. This trend corresponds to the total degradation of Mg-bearing phases in the hematitic zone compared to partial preservation of chlorite I, formation of chlorite II, and the crystallization of magnesian illite within the transitional zone. This distribution is attributed to the original (Middle Proterozoic) weathering, given the high solubility of magnesium in neutral to weakly acidic groundwater.

Sodium, potassium, and calcium are depleted upward through the hematitic zone. The loss of sodium and potassium is attributed primarily to hydrolysis of the feldspars to illite. The distribution of calcium through the profile reflects the complete replace-

Table 2a. Electron microprobe analyses of diaspore, kaolinite, illite, relict biotite and muscovite from the kaolinite subzone of the hematite zone, Granite Hill profile, 0.3 m below unconformity

Analysis	Diaspore				Kaolinite				Illite	Red-brown biotite in quartz	
	1	2	3	4	1	2	3	4	1	1	2
SiO_2	0.63	1.90	0.73	1.02	43.17	48.14	46.48	46.79	47.01	35.61	35.72
TiO_2	ND	ND	ND	ND	ND	ND	ND	ND	ND	3.41	3.37
Al_2O_3	84.11	82.05	83.41	83.82	35.88	39.43	38.60	38.61	34.78	18.76	18.73
FeO_T	0.18	0.45	0.28	0.76	8.38	0.91	2.71	0.45	0.57	21.58	21.68
MgO	0.47	0.30	0.48	0.36	0.25	0.34	0.25	0.26	0.54	7.10	7.32
CaO	ND	ND	ND	ND	ND	ND	ND	ND	0.09	ND	ND
Na_2O	0.42	0.36	0.39	0.28	0.33	0.30	0.28	0.22	ND	ND	0.14
K_2O	ND	0.22	ND	ND	0.67	0.12	0.09	0.78	5.72	9.40	9.37
P_2O_5	ND	ND	ND	ND	0.11	ND	ND	0.11	ND	ND	ND
Cl	ND	ND	ND	ND	0.09	0.16	0.05	0.10	0.11	0.08	0.07
SO_3	ND	ND	ND	ND	0.13	0.12	ND	ND	ND	ND	ND
Total	85.81	85.28	85.30	86.25	89.03	89.51	88.41	87.32	88.84	95.94	96.41

Table 2b. Electron microprobe analyses of muscovite and relict biotite from the transition and chlorite zones of the GT5 profile; depths in m from drill collar

Analysis	Muscovite					Red-brown biotite in quartz
	1	2	3	4	5	1
Depth (m)	19.1	24.5	24.5	35.2	54.4	35.2
SiO_2	50.31	50.36	47.45	46.16	46.03	36.27
TiO_2	ND	ND	0.15	0.13	ND	1.95
Al_2O_3	30.42	27.51	34.74	33.87	33.93	19.01
FeO_T	1.29	2.54	1.33	1.02	1.38	16.46
MgO	2.01	3.00	1.09	1.16	0.93	10.93
CaO	0.08	ND	ND	ND	ND	ND
Na_2O	ND	ND	0.52	0.44	0.44	ND
K_2O	9.51	10.00	10.52	10.26	10.25	9.64
P_2O_5	ND	ND	ND	ND	ND	ND
Cl	ND	ND	ND	ND	ND	ND
SO_3	ND	ND	ND	ND	ND	ND
Total	93.62	93.42	95.70	93.05	92.96	94.26

ND = not detected.

ment of plagioclase by illite in the upper zones. Smectite is absent. The partial to total replacement of feldspar by illite records either Middle Proterozoic weathering or late-stage potassic metasomatism. If K-metasomatism, it selectively occurred along the sub-Kombolgie unconformity. Weathering therefore seems more likely.

Titanium is fairly evenly distributed through the profile, occurring in anatase, chlorite I, and chlorite II. There are minor to multifold increases in Li, Rb, Ba, and Sr which correspond to variation of the more abundant elements in the respective periodic groups (Table 3). In the protolith, these four elements occur in feldspars and so their depletion through the paleosol apparently reflects feldspar alteration. Rubidium is the least depleted, given an average concentration within both the transitional and hematitic zones which is 0.76 of that in the protolith. Strontium is the most depleted, with protolith-normalized concentration ratios of 0.17 in the transitional zone and just 0.14 in the hematitic zone. Protolith-normalized concentration ratios of barium and lithium are both 0.59 in the transitional zone but lithium is more depleted in the hematitic zone (0.33) than is barium (0.50).

Lead averages 27.3 ppm in the protolith, ranging from 17 to 33 ppm. Given the paucity of sulfides, this lead probably occurs in K-feldspar and mafic

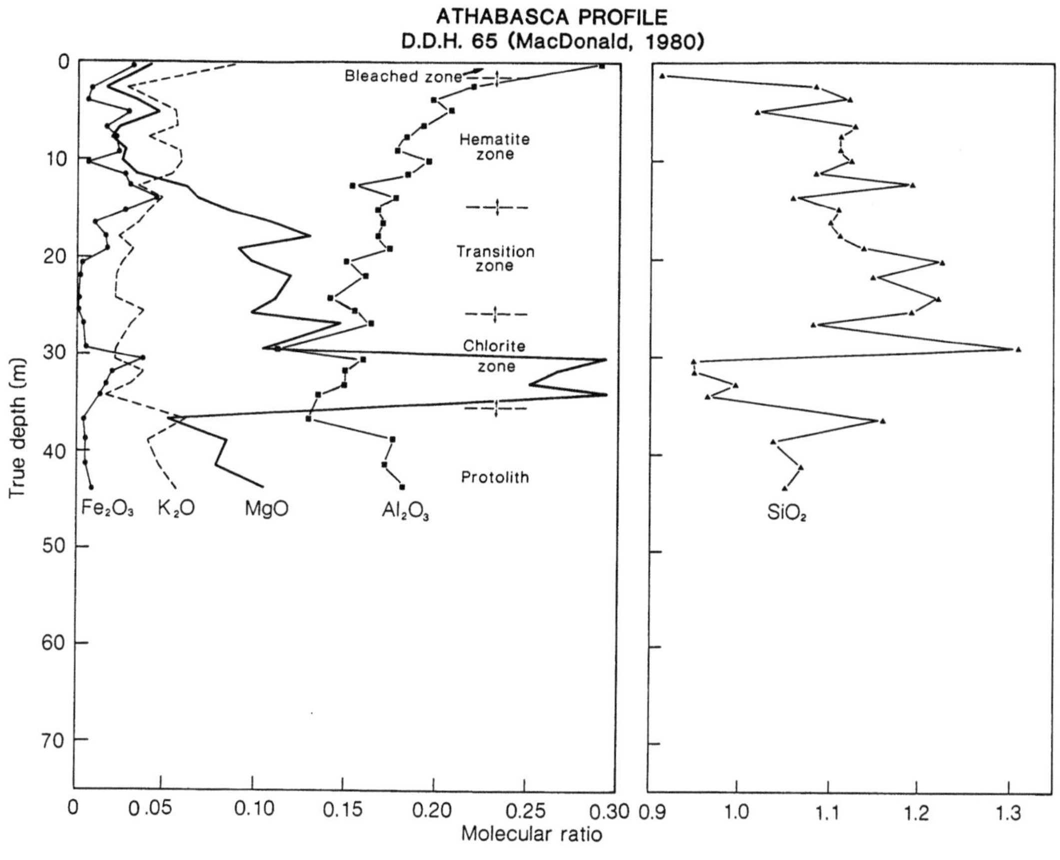

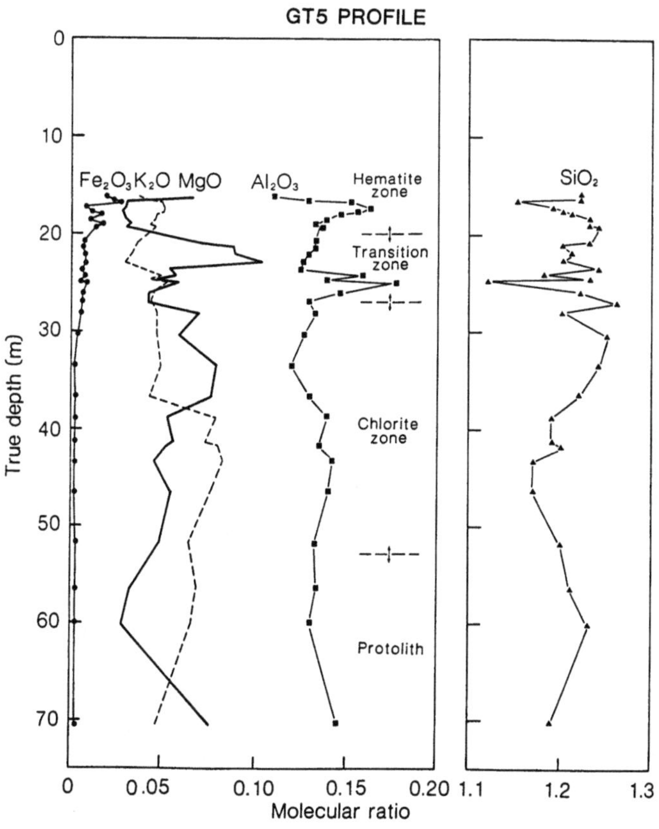

Fig. 4. Compositional profiles for Fe_2O_3, K_2O, MgO, Al_2O_3, and SiO_2 through the sub-Kombolgie paleosol (GT5 profile) and sub-Athabasca paleosol where developed on similar protoliths

Table 3. Geochemistry of the sub-Kombolgie paleosol (Pancontinental drillhole GT5) developed on the psammitic metasedimentary component of the Nanambu Complex. Depth in m from drill collar

Zone	Hematite			Transition		Chlorite								Protolith		
Depth (m)	19.1	20.3	22.3	24.5	28.1	31.2	32.5	35.2	38.8	42.2	47.7	48.2	54.4	65.05	70.05	81.1
SiO_2	73.28	72.47	74.75	72.39	71.24	76.28	72.43	75.49	74.50	73.30	71.41	72.41	70.48	73.15	73.91	71.82
TiO_2	0.12	0.14	0.06	0.11	0.15	0.06	0.07	0.08	0.11	0.09	0.09	0.08	0.08	0.12	0.13	0.05
Al_2O_3	13.37	16.05	13.83	13.56	16.20	13.26	13.56	12.95	12.20	13.25	13.91	13.77	14.40	13.72	13.35	14.77
Fe_2O_3	3.57	1.74	2.02	1.01	1.10	1.00	0.79	0.61	0.32	0.47	0.41	0.31	0.44	0.43	0.41	0.55
FeO	0.28	0.23	0.16	0.85	0.40	0.27	0.69	0.46	0.87	0.83	1.00	0.84	0.98	0.80	0.79	0.62
MnO	0.01	< 0.01	< 0.01	< 0.01	< 0.01	< 0.01	< 0.01	< 0.01	< 0.01	< 0.01	< 0.01	< 0.01	0.01	0.01	0.01	< 0.01
MgO	1.72	1.14	1.25	3.52	2.31	1.71	2.84	2.41	3.21	3.09	2.27	2.00	2.24	1.33	1.15	3.05
CaO	0.24	0.16	0.21	0.22	0.15	0.19	0.20	0.21	0.16	0.18	0.20	0.14	0.17	0.19	0.20	0.18
Na_2O	0.04	0.09	0.09	0.05	0.10	0.09	0.05	0.07	0.08	0.09	0.22	0.26	0.23	1.23	1.71	0.10
K_2O	4.08	4.82	4.32	3.34	4.80	4.04	4.42	4.51	4.62	4.18	6.98	7.58	7.28	6.55	6.23	4.49
P_2O_5	0.15	0.11	0.13	0.15	0.09	0.12	0.13	0.14	0.11	0.12	0.16	0.13	0.15	0.13	0.10	0.13
LOI	3.50	3.28	3.22	4.24	3.67	3.33	4.01	3.39	3.45	3.86	2.72	2.31	2.72	1.99	1.67	3.92
Total	100.36	100.22	100.04	99.45	100.20	100.33	99.19	100.30	99.61	99.46	99.37	99.84	99.18	99.66	99.65	99.68
Ba	25	156	183	24	278	154	82	98	133	113	261	332	289	320	295	183
Rb	102	104	104	97	134	114	114	112	93	108	175	175	180	163	159	116
Sr	4	9	7	9	7	7	12	11	24	18	49	52	48	47	52	38
Pb	2	5	4	3	6	3	5	6	8	8	28	33	28	32	33	17
Th	11	15	8	19	23	10	18	18	12	18	17	16	14	16	16	14
U	3	7	7	5	7	7	3	4	3	7	7	8	16	9.5	11	14
Zr	70	93	96	89	145	85	106	110	83	110	89	89	84	94	97	61
Nb	5	9	4	5	8	5	4	5	4	6	3	4	4	4	4	2
Y	7	9	7	7	9	7	6	6	3	6	5	4	8	11	13	10
La	7	18	12	19	18	10	16	15	11	14	16	14	13	17	17	14
Ce	16	33	22	39	38	20	36	29	25	33	33	28	25	36	37	31
Nd	7	14	8	17	16	9	14	12	10	14	13	11	10	14	15	11
Li	14	15	8	25	22	16	24	21	37	40	28	31	35	33	27	85
Sc	5	2	< 2	< 2	4	< 2	< 2	< 2	< 2	3	2	< 2	3	2	2	< 2
V	33	19	15	6	31	5	4	4	7	5	6	4	5	5	5	3
Ni	12	9	2	8	6	3	2	2	4	3	3	2	2	< 2	3	< 2
Cu	7	3	4	4	3	3	< 2	< 2	3	7	17	12	31	12	2	11
Zn	18	12	5	14	10	10	11	8	9	11	17	13	13	23	39	18
Cr	21	11	5	5	15	4	4	5	6	5	4	5	4	4	4	< 2
Gs	13	19	13	14	20	16	14	15	11	14	15	13	13	16	15	16
As	1	< 1	< 1	< 1	< 1	< 1	< 1	< 1	< 1	< 1	< 1	< 1	< 1	0.5	1	< 0.5
Mo	< 3	< 3	< 3	< 3	< 3	< 3	< 3	< 3	< 3	< 3	< 3	< 3	< 3	< 2	< 2	< 2
Sn	< 2	6	3	5	5	3	3	5	< 2	6	5	4	5	4	4	3
Bi	< 2	< 2	< 2	< 2	< 2	< 2	< 2	< 2	< 2	< 2	< 2	< 2	< 2	< 2	< 2	< 2

phyllosilicates. Upward alteration of these phases probably accounts for the depletion of lead. Lead averages just 3.6 ppm in the hematitic zone, ranging from 2 to 5 ppm.

Transition-series trace elements exhibit complicated patterns. There are local enrichments of Sc, Zr, V, Nb, Cr, Ni, Cu, the most noteworthy being V, Cr, and Ni. Uranium ranges from 14 to 16 ppm in the lower chloritic zone and protolith whereas it ranges from 2.5 to 11 ppm in the hematitic zone. No significant variation is observed in the distribution of Y, Ga, As, Bi, Sn, Mo, Ce, La, Nd, Th, or Zn.

6 Drill Core GT6

6.1 Introduction

This drill core provides the only intersection through a substantial thickness of Kombolgie Formation sandstone and the bulk of the underlying paleosol, well into its basal chloritic zone. The color-stratified paleosol is overlain by 149.5 m of medium- to coarse-grained sandstone. From the unconformity at 149.5 m to a depth of ca. 170 m, the profile consists of alternating bands of red, green, and mottled red-green micaceous rock. This interval is correlated with the hematitic and transitional zones of GT5. From ca. 170 m to the bottom of the hole at 197.0 m, the profile consists of green micaceous metasedimentary protolith with subordinate pegmatite. This interval is correlated with the chloritic zone of GT5.

6.2 Drill Core GT6: Kombolgie sandstone overlying Paleosol

Directly overlying the paleosol in drill core GT6, the basal unit of the Kombolgie Formation is ca. 9.5 m of coarse quartz arenite (sand- to granule-sized grains). Above this, the grain size diminishes to that of a fine- to medium-grained quartz arenite. This upward fining characterizes the basal Kombolgie in the region (Gustafson and Curtis 1983). Sandstone in the uppermost ca. 33 m of the drill core exhibits patchy colors, ranging from white to yellow-red, red, and cream. This contrasts with gray to green-gray hues in underlying Kombolgie sandstone. Alteration in the uppermost ca. 33 m is interpreted to be a product of Cenozoic weathering. Between this ca. 33 m depth and the unconformity at 149.5 m, the most notable features are an intersection of massive chlorite at 129.4 m, a thin intersection of chlorite-quartz breccia, and a red-gray mottling which extends from 122.0 to 149.5 m.

Below 33 m, the arenite consists of quartz-cemented quartz grains in a matrix with variable proportions of chlorite + illite +/− hematite (Fig. 5a). The framework quartz occurs either as strained singular or composite grains. The grains range from equant and rounded to elongate and subrounded. Original quartz-grain outlines may be readily differentiated from commonly occurring partial overgrowths of authigenic quartz because of an intervening thin dusting of opaques (hematite?). These overgrowths and opaques are characteristic of Kombolgie sandstone (Ojakangas 1979; Gustafson and Curtis 1983). Pressure solution of framework quartz grains also is common. Interstitial illite and chlorite occur either separately or intergrown, locally accompanied by subhedral to euhedral chlorine-bearing apatite (Fig. 5b). The illite is fine-grained and highly birefringent unlike the pale green chlorite. The apatite is interpreted to be a cement which postdates both the hematite and quartz overgrowths. There are trace quantities of detrital muscovite and some resistates, i.e., zircon, tourmaline and rare magnetite.

The maturity of Kombolgie sandstone is evident in the predominance of silica through the unweathered interval from 35.4 to 149.35 m (Table 4). Excluding the chloritic sample at 129.4 m, the weight percent of SiO_2 ranges from 76.41 to 90.72%. Potassium ranges from 0.03 to 1.08% K_2O due to minor illite and scarce potassic feldspar. Plagioclase is extremely scarce, given that Na_2O ranges from < 0.01 to 0.11% and CaO is < 0.01%. Magnesium ranges from 1.40 to 7.42% MgO.

The effect of Cenozoic weathering on the composition of the upper 33 m of the core provides evidence in support of a weathering (rather than hydrothermal) origin for the sub-Kombolgie paleosol. Recent weathering has removed magnesium and ferrous iron from the upper 33 m. The resulting difference between the recently weathered Kombolgie sandstone and underlying sandstone is similar to the difference between the hematitic and chloritic zones in the GT5 and GT6 profiles (Tables 3, 4; Fig. 4).

Metasomatism is invoked to explain the unit of massive, ultra-fine-grained chlorite within the Kombolgie Formation. Very finely serrated contacts occur between framework quartz grains in sandstone adjacent to the massive chlorite rock. Gustafson and Curtis (1983) have interpreted similar textures in altered sandstone to indicate quartz instability during chloritization. The massive chlorite rock contains disseminated anhedral apatite which also is attributed to metasomatism. The rock averages 0.39% P_2O_5 and 0.37% CaO. Anomalous concentrations of Sr, Th, Zr, Y, La, Ce, Nd, Li, V, Ni, and Ga also indicate post-Kombolgie mineralization, similar to the interpretation for Athabasca mineralization (Wallis et al. 1983). Phosphatic chloritic metasomatism is commonly associated with unconformity-type mineralization elsewhere (Binns et al. 1980; Nash and Frishman 1982). Diagenetic phosphate was not identified within the enclosing sandstone, possibly because authigenic quartz had occluded the porosity prior to the introduction of phosphatic fluids. Elevated contents of U, Zr, La, and Ce at the unconformity probably reflect an increase in refractory detritus.

Gustafson and Curtis (1983, Table 2) have proposed a compositional classification of alteration in the Kombolgie sandstone, based on average oxide content, FeO_T, MgO, the ratio of Fe^{2+}/Fe^{3+}, and selected trace elements. Using their criteria, the interval from 33 to 149.35 m lies within the range of sericitic-hematitic and weakly chloritized sandstone.

6.3 Drill Core GT6: Hematitic and Transitional Zones of the Paleosol

The hematitic and transitional zones of drill core GT6 (149.5 to ca. 170 m) resemble comparable

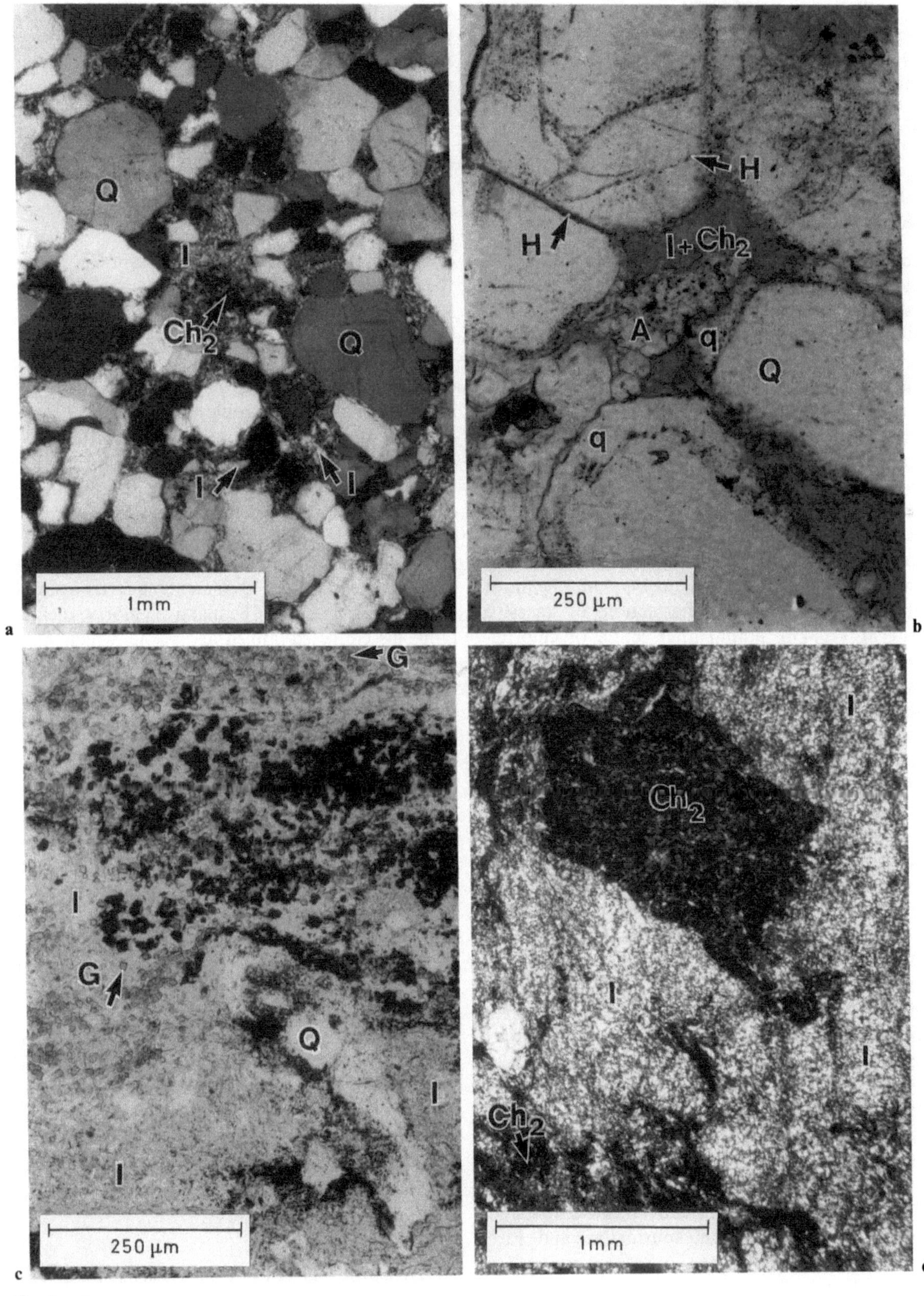
Q
I
Ch2
Q
I
I
1mm
a
H
H
I+Ch2
A
q
Q
q
250 µm
b
G
I
G
Q
I
I
250 µm
c
I
Ch2
I
I
Ch2
1mm
d

Fig. 5a–d

Table 4. Geochemistry of the Kombolgie Formation and sub-Kombolgie paleosol (Pancontinental drillhole GT6) which was developed on micaceous metasedimentary rocks. Depth in m from drill collar

Zone	Kombolgie Formation sandstone								Saprolite Altered hematite-transition					Chlorite	
Depth (m)	14.5	24.1	35.4	58.1	90.5	121.4	142.9	149.35	150.0	154.9	156.7	160.8	166.8	177.8	186.9
SiO_2	92.58	94.24	76.41	90.00	90.72	30.75	86.84	85.58	37.13	45.33	37.33	53.59	46.97	36.25	39.70
TiO_2	0.07	0.04	0.10	0.04	0.02	0.22	0.18	0.22	0.62	0.49	0.63	0.45	0.34	0.51	0.45
Al_2O_3	3.53	2.44	6.72	5.00	2.65	19.08	4.44	4.21	17.29	17.08	20.28	13.61	14.46	17.50	16.27
Fe_2O_3	1.74	1.47	2.32	0.37	0.50	2.21	0.64	0.75	1.49	13.14	2.71	5.89	4.67	2.88	1.76
FeO	0.10	0.09	1.22	0.45	0.98	8.78	1.04	1.37	1.98	2.26	3.04	1.69	2.24	2.04	2.07
MnO	<0.01	<0.01	<0.01	<0.01	<0.01	<0.01	<0.01	<0.01	<0.01	0.02	0.01	0.02	0.01	0.01	0.01
MgO	0.05	0.16	7.42	1.40	2.89	24.54	4.07	4.42	7.43	9.07	22.09	15.10	20.46	26.67	26.93
CaO	<0.01	<0.01	<0.01	<0.01	0.01	0.37	<0.01	<0.01	13.62	1.12	0.13	0.15	0.12	0.09	0.12
Na_2O	0.06	0.01	0.07	<0.01	0.11	0.07	<0.01	0.06	0.12	0.03	0.01	0.01	0.05	0.04	0.06
K_2O	0.19	0.02	0.01	1.08	0.03	<0.01	0.22	0.28	3.51	3.26	1.37	1.13	0.52	0.02	0.11
P_2O_5	0.07	0.06	0.02	0.01	0.02	0.39	0.01	0.02	9.84	0.59	0.08	0.08	0.06	0.02	0.04
S	<0.01	<0.01	<0.01	<0.01	<0.01	0.03	0.01	<0.01	<0.01	<0.01	<0.01	<0.01	<0.01	<0.01	<0.01
LOI	1.64	1.43	5.81	1.55	1.82	12.69	2.61	2.60	6.03	6.93	11.84	7.99	9.69	13.31	12.17
Total	100.03	99.95	100.08	99.88	99.73	99.15	100.15	99.51	99.07	99.33	99.52	99.72	99.59	99.33	99.69
Ba	17	14	12	20	13	<3	14	7	94	72	4	45	31	<3	<3
Rb	4	<1	<1	20	1	<1	4	5	50	48	22	21	13	1	5
Sr	23	33	19	12	10	135	13	29	62	15	19	14	22	5	5
Pb	4	4	2	3	2	3	2	2	12	4	11	5	3	14	4
Th	7	5	15	5	4	32	15	17	18	13	16	11	13	15	13
U	6.0	1.5	1.5	0.5	2.0	3.5	1.5	7.5	31	3.0	3.0	2.0	1.5	3.0	3.5
Zr	84	60	143	63	47	372	107	234	153	159	122	97	92	130	149
Nb	2	2	3	1	1	4	3	3	10	8	10	11	6	9	8
Y	5	7	2	3	2	9	2	3	68	14	6	7	11	5	3
La	36	65	47	20	26	362	35	74	138	36	40	30	60	16	<2
Ce	63	118	59	30	40	543	48	103	203	53	64	51	83	26	4
Nd	22	47	18	12	11	171	13	31	95	21	22	19	29	13	<2
Li	10	16	47	9	16	54	20	15	29	37	202	76	164	288	273
Sc	<2	<2	<2	<2	<2	<2	<2	<2	22	9	10	7	6	8	5
V	5	2	7	<2	<2	17	3	7	54	72	48	31	39	54	31
Ni	6	11	39	6	17	81	8	7	13	60	67	44	56	31	31
Cu	7	3	20	3	4	9	2	2	4	9	28	17	14	26	13
Zn	14	16	21	25	12	24	13	12	13	24	78	59	97	72	61
Cr	2	2	7	2	2	5	4	6	100	8	77	179	113	92	77
Ga	3	3	9	3	4	29	4	5	28	23	22	18	22	20	22
As	2.0	6.0	1.0	<0.5	<0.5	4.0	2.0	1.5	5.5	2.0	2.0	1.0	1.0	<0.5	<0.5
Mo	<2	<2	<2	<2	<2	<2	<2	<2	<2	3	<2	2	<2	<2	<2
Sn	<2	<2	2	<2	<2	12	<2	<2	5	<2	3	3	2	3	9
Bi	<2	<2	<2	<2	<2	<2	<2	<2	<2	<2	<2	<2	<2	<2	<2

Fig. 5a–d. a Photomicrograph in cross-polarized light of the Kombolgie sandstone at 58.1 m down GT6 drill hole. Quartz (*Q*) forms clast-supported framework. Matrix is finely crystalline illite (*I*) and chlorite II (Ch_2). Chlorite II occurs as *dark areas* within bright fiberlites of illite. **b** Photomicrograph in plane light of Kombolgie sandstone at 149.35 m down GT6 drill hole. Quartz clasts (*Q*) have overgrowths of quartz (*q*). Interstitial areas consist of euhedral apatite (*A*) and intergrown illite-chlorite II ($I + Ch_2$). Fine opaque grains are hematite (*H*). **c** Photomicrograph in plane light of phosphatic Goomadeer profile of sub-Kombolgie paleosol. Euhedral goyazite (*G*) is disseminated through paleosol which consists of illite (*I*), relict quartz (*Q*), and hematite (*black*). **d** Photomicrograph in cross-polarized light of the chloritic zone within the sub-Kombolgie paleosol, down 177.8 m in GT6 drill hole. Chlorite II (Ch_2) and illite (*I*) are both ultra-finely crystalline

zones of GT5 in the preservation of grain outlines by secondary minerals. Metamorphic textures are thereby preserved. Drill core GT6 differs, however, in exhibiting less regularity in the spatial distribution of alterion zones. Some portions of GT6 display the intense red hues which characterize the hematitic zone of GT5 whereas other sections resemble the transitional zone of GT5 in that green hues (variably light to dark) alternate with pink to red hues. Intensely red hematitic rocks exhibit green to white bands (veinlets) which parallel the foliation. This intimate association of reduced and oxidized minerals is attributed to a diagenetic introduction of reducing fluids.

Paleosol in the GT6 profile developed on compositionally heterogeneous metasedimentary rock. Metamorphic textures and fabrics are variably preserved in quartzose and mica-feldspar layers. As in the GT5 protolith and elsewhere (Granite Hill and Tent Hill profiles), there are pseudomorphs of hematite and clay after crenulated muscovite or biotite. Moreover, ultra fine-grained hematite is disseminated throughout mats of fine-grained quartz + kaolinite +/− illite. Two other features are shared with the subsequently described Granite Hill profile, i.e., sparse anhedral sieve-textured diaspore and pseudomorphs of kaolinite + hematite after metamorphic mica.

The contact between the hematitic and transitional zones is obscured by mats and veinlets of magnesian chlorite, illite, and phosphate minerals. These are interpreted to be diagenetic. The transitional zone grades downward into the chloritic zone with an increase in both types of chlorite, i.e., plates of fine-grained chlorite I (Fe > Mg) with berlin-blue birefringence, and very fine-grained felted mats of chlorite II (Mg > Fe) with low birefringence.

6.4 Drill Core GT6: Chloritic Zone of Paleosol

The interval from ca. 170 to 197 m in drill core GT6 is interpreted to be the chloritic zone based on a decreased abundance of hematite and the textures of the assemblage: chlorite I, chlorite II, quartz, illite, disseminated hematite, trace phosphate, and anatase (Fig. 5d). This chloritic zone differs from that of GT5 in that no feldspar has been preserved.

6.5 Drill Core GT6: Geochemistry of Paleosol

Alumina is abundant in this profile, ranging from 13.61 to 20.28% Al_2O_3, as in the GT5 and Granite Hill profiles (Table 4). This range of values is attributed both to primary variation in the protolith and to differential weathering (kaolinization). Short-range variation in the oxidation state of iron is attributed to diagenesis. A high and variable concentration of magnesium, 7.43–22.09% MgO, records late chlorite-apatite mineralization of the paleosol. The presence of apatite is reflected in the coupled variation of calcium and phosphorus, i.e., 13.62% CaO and 1.12% P_2O_5 to 9.84% CaO and 0.59% P_2O_5 through the upper 5 m of the paleosol.

Late mineralization of the paleosol is further indicated by particular elemental trends in the hematitic-transitional zone of GT6 which differ significantly from those of the comparable zones in the GT5 and Granite Hill profiles. The uppermost apatite-bearing zone exhibits anomalous Ba, Rb, Sr, U, Y, La, Ce, Nd, Sc, and possibly V contents compared to the lower portions of this and the other profiles (Table 4). These trace elements generally tend to be concentrated in phosphatic sediments (Altschuler 1980; Miller 1983). Moreover, Ferguson and Winer (1980) have demonstrated a common relationship among these elements in altered metasedimentary rocks of the Pine Creek Geosyncline.

The hematitic-transitional zone also contains abundant Cr, Ni, Zn, Cu, and possibly V. These elements similarly are concentrated in the hematitic zones of the GT5 and Granite Hill profiles relative to the respective protoliths. High concentrations of these immobile transition-group elements could be a cryptic relict of weathering. However, the correlation of nickel with Mg-metasomatism may well indicate nonweathering processes (Ferguson and Winer 1980).

7 Granite Hill Profile

7.1 Introduction

The sub-Kombolgie paleosol is exposed through a vertical distance of 21 m on the western side of Granite Hill, 4.1 km SSE of Jabiluka. This outcrop is mapped as part of the Archean component of the Nanambu Complex (Needham 1982; Page et al. 1980). Silicified conglomeratic sandstone of the Kombolgie Formation overlies a hematitic paleosol which retains a well-developed metamorphic fabric (Fig. 6a). The paleosol displays a vertical color variation which grades from blotchy reddish purple and cream at the unconformity surface (Fig. 6a) to an evenly colored light red to crimson red within a meter below the unconformity. The lower parts of the profile are covered by modern soil and boulders. Unaltered protolith could not be located in situ with confidence except at levels deeper than about 21 m beneath the unconformity.

7.2 Granite Hill Profile: Mineralogy and Petrography

The protolith is a foliated muscovite-biotite granite gneiss with K-feldspar augen. The gneiss is a fine- to medium-grained, equigranular to elongate mosaic of quartz, plagioclase (An_{8-15}) and microcline perthite. The minor minerals are red-brown biotite and muscovite, with accessory zircon, apatite, and tourmaline. Retrogression to the greenschist facies is marked by partial to complete replacement of the red-brown biotite by iron-rich chlorite, epidote, and needles of anhedral aggregates of anatase. Associated with chloritized biotite are very fine-grained mats of white mica and traces of carbonate replacing plagioclase. An ultra fine-grained dusting of opaques (hematite?) occurs through the K-feldspar.

The exposed hematitic zone of this saprolitic profile is composed of quartz + kaolinite + illite + hematite + diaspore with varying trace amounts of relict muscovite, accessory zircon, and apatite. There are relict grains of red-brown biotite and brown tourmaline, both armoured by quartz (Fig. 6c). Elongate to equant quartz grains with strained quartz overgrowths occur throughout the exposed saprolite. Hematite is texturally uniform through the profile interval which lies between 1.2 and 6.1 m below the unconformity. Hematite occurs here as disseminated, ultra-fine, irregularly opaque to translucent grains up to 50 microns across, as streaky discontinuous seams that mimic the metamorphic foliation, and as opaque clusters which consist of a mixture of hematite and clay. The hematite-clay clusters are particularly well developed in the uppermost interval of the profile (above 1.2 m) and account for the blotchy texture which is observed in outcrop and on freshly broken rock surfaces (Fig. 6b).

Subdivision of the Granite Hill profile has been based on the relative proportions of kaolinite and illite. The upper kaolinitic subzone extends from the unconformity down to 1.2 m whereas the lower illitic subzone extends between 1.2 and 6.1 m. In the kaolinitic subzone, kaolinite is present as ultra-fine-grained mats, with local areas of coarser spherulitic to plumose grains up to 0.3 mm across, and as discontinuous vermiform veinlets. Trace quantities of ultra-fine illite are distributed through the clayey matrix. Tabular subhedral to equant anhedral diaspore, 0.2–0.8 mm long, is present throughout this subzone whereas only trace amounts occur in the lower illitic subzone. Diaspore varies from colorless to cloudy depending on the ferrous iron content (Table 2a). It is considered to be authigenic because it exhibits a sieve texture (with quartz inclusions) and lacks any preferred orientation in the kaolinitic matrix. In summary, this aluminous zone consists essentially of clay + diaspore and is interpreted to have formed by intense weathering.

In the illitic subzone, very-fine-grained illite is much more abundant than kaolinite. Relict muscovite plates exhibit varying degrees of replacement by kaolinite and hematite. Degraded muscovite plates are splayed at the ends and display replacement by kaolinite and hematite along the basal cleavage.

Throughout the sampled saprolitic profile, the metamorphic foliation is retained to various degrees by the textural relationships between quartz and replaced feldspar. Within the kaolinitic subzone, metamorphic textures have been obliterated by the growth of clay to form a finely felted mat, whereas the illitic subzone is characterized by ghost outlines of feldspar grains. Observed textural modification may be due to recrystallization of the clay matrix, redistribution of the hematite, and/or compaction of the paleosol following burial under the overlying Kombolgie Formation.

7.3 Granite Hill Profile: Geochemistry

Even though a complete profile is not available at Granite Hill, analyses of the protolith at 21.3 m and three additional analyses of the Nanambu Complex in this area (Table 5) permit limited comparison between the protolith and the clayey paleosol.

Silica and alumina contents of the paleosol resemble those of the protolith. The preservation of metamorphic shapes of quartz grains suggests little or no desilication within the preserved hematitic zone. However, the presence of authigenic sieve-textured diaspore suggests that alumina may have moved downward from a higher unpreserved portion of the profile.

Iron mobility is not apparent among the few samples available from this profile. The distribution of ferric and ferrous iron is similar to that in the GT5 profile. In the highly oxidized hematitic zone, the ratio of $Fe_2O_3/(FeO + Fe_2O_3)$ ranges from 0.91 to 0.94. As in GT5, the content of magnesium in this zone is very low, 0.05 to 0.15% MgO, because of

Fig. 6a–d

Table 5. Geochemistry of paleoweathered granite gneiss. Nanambu Complex, Granite Hill. Samples marked PCG are unaltered Nanambu gneiss (Ferguson and Winer 1980). Depth in m below unconformity

Zone	Hematite					Protolith	PCG 322	PCG 322	PCG 327
Subzone	Kaolinite subzone			Illite subzone					
Depth (m)	0.03	1.2	3.1	4.6	6.1	21.3			
SiO_2	72.71	73.00	71.39	74.45	73.60	70.03	73.93	74.04	73.44
TiO_2	0.20	0.21	0.30	0.30	0.31	0.34	0.21	0.12	0.19
Al_2O_3	18.22	18.01	18.30	16.53	17.33	14.59	14.01	13.77	13.91
Fe_2O_3	1.05	1.19	1.42	1.43	1.14	0.26	0.19	0.21	0.32
FeO	0.08	0.08	0.10	0.09	0.11	1.61	1.11	0.60	0.88
MnO	<0.01	0.01	0.01	0.01	0.02	0.04	0.01	<0.01	0.01
MgO	0.05	0.06	0.15	0.09	0.08	1.00	0.56	0.28	0.43
CaO	0.03	0.02	0.03	0.04	0.03	0.67	1.00	0.61	0.92
Na_2O	0.05	0.02	0.08	0.02	0.05	3.05	3.39	3.85	4.05
K_2O	0.28	0.82	2.88	1.93	2.38	6.06	5.22	5.36	5.17
P_2O_5	0.04	0.03	0.04	0.09	0.08	0.07	0.16	0.06	0.07
LOI	6.40	5.81	4.50	4.58	4.47	1.42	0.88	1.44	1.12
Total	99.12	99.27	99.20	99.57	99.61	99.14	100.57	100.44	100.41
Ba	19	28	37	46	38	680	416	376	403
Rb	9	20	67	41	49	172	174	139	167
Sr	55	58	89	259	186	99	144	98	137
Pb	6	6	6	8	7	46	51	43	49
Th	51	48	63	69	75	62	46	30	56
U	1.5	2	2	1.5	1.5	7.5	7	5	8
Zr	209	183	242	279	279	290	175	111	185
Nb	10	10	14	14	13	16	10	3	10
Y	2	2	2	3	3	13	10	5	7
La	34	43	52	93	86	71	51	35	59
Ce	64	71	94	165	160	132	112	69	118
Nd	14	17	21	49	50	50	NA	NA	NA
Li	22	20	17	19	18	19	17	10	15
Sc	<2	<2	<2	<2	<2	2	4	2	2
V	5	7	11	12	11	18	42	9	15
Ni	4	3	3	<2	2	4	2	3	5
Cu	3	<2	<2	<2	<2	5	5	2	2
Zn	4	5	11	5	19	27	27	18	30
Cr	13	8	18	10	9	14	NA	NA	NA
Ga	16	17	22	20	19	18	18	16	17
As	1.5	1.5	1.5	2	2.5	<0.5	<2	<2	<2
Mo	<3	<3	<3	<3	<3	<3	4	<2	3
Sn	<2	2	2	3	4	<2	2	2	2
Bi	<2	<2	<2	<2	<2	<2	NA	NA	NA

NA = not analyzed.

Fig. 6a–d. a Unconformity at Granite Hill, northern Australia, between flat-lying Kombolgie Formation sandstone (*KFm*) and underlying paleosol (*GHs*) which developed on granite gneiss of the Nanambu Complex. Note well-preserved metamorphic foliation indicated by an *arrow* and the color gradation away from the unconformity. **b** Hand specimen of paleosol, 0.3 m below unconformity at Granite Hill. Mottled texture produced by remobilization of hematite. **c** Photomicrograph in plane light of kaolinitic subzone of hematitic zone 0.3 m below unconformity at Granite Hill. Diagnostic mineral assemblage is diaspore (*D*), kaolinite (*Ka*), disseminated hematite (*black*), and relict quartz (*Q*). **d** Photomicrograph in plane light of hematitic zone within Tent Hill profile of sub-Kombolgie paleosol. Paleosol developed here on the Plum Tree Creek Volcanics with excellent preservation of igneous textures. Embayed phenocrysts of quartz (*Q*) are mantled by illite after feldspar. Feldspar phenocrysts (outlines by *dashed line*) are replaced by illite (*I*) and disseminated hematite (*black*) the complete degradation of metamorphic chlorite. The remaining traces of magnesium occur in accessory clays. Magnesium metasomatism which is related to uranium mineralization elsewhere in this area has not modified the Granite Hill paleosol.

The hematitic zone and protolith differ markedly in their abundances of Na_2O, K_2O, and CaO. This difference is comparable to that in the GT5 profile, considering the difference in protolith between GT5 and Granite Hill. The potassium content of the illitic subzone, 1.93 to 2.88% K_2O, is similar to that of the hematitic zone of GT5 whereas the overlying kaolinitic subzone contains only 0.28 to 0.82% K_2O. The range of titanium in the hematitic zone, 0.20 to 0.31% TiO_2, is similar to that in the protolith and its vertical distribution is similar to that in GT5.

Significant variations of Rb and Ba within the hematitic zone are similar to those in GT5. Both rubidium and barium are depleted in the hematitic zone relative to the protolith (77 vs. 163 ppm for Rb; 33 vs. 469 ppm for Ba; Table 5). This depletion is attributed to the alteration of feldspar and the chemistry of illite. As in GT5, lead is depleted in the hematitic zone (6 to 8 ppm) relative to the protolith (43 to 51 ppm).

No trends are apparent for Sc, Zr, Nb, Cr, Ni, Cu, Ga, As, Bi, Sn, Mo, Ce, La, Nd, and Th. However, V and Y differ from GT5 in being depleted in the hematitic zone. Uranium is strongly depleted throughout the hematitic zone whereas Th has remained constant under the same oxidizing conditions.

8 Baroalba Creek Profile

A sub-Kombolgie exposure occurs near Baroalba Creek, 3.4 km north of Koongarra. This is hematitic throughout a vertical exposure of about 10 m. The protolith is a crenulated micaceous quartzose metasedimentary rock of the Early Proterozoic "upper member" of the Cahill Formation (Needham 1982). A well-preserved metamorphic fabric is defined by crenulated muscovite bands and granoblastic triple-point textured quartz grains. The essential mineral assemblage of the micaceous quartzite is quartz + muscovite + illite + hematite. Muscovite plates locally are partially replaced by hematite.

Intergranular areas between quartz and muscovite grains are occupied by finely crystalline illite and hematite. The habit of the illite-hematite aggregates, similar in size to the quartz grains, suggests that these may have been feldspar. Hematite is evenly disseminated as very fine grains throughout the aluminous rock. It occurs as discontinuous seams along the foliation and locally coalesces into irregular opaque aggregates. These textural features resemble those of the uppermost GT5 and Granite Hill profiles. Unaltered remnants of blue-green tourmaline, zircon, and apatite are distributed through the micaceous layers. Traces of relict brown biotite and brown tourmaline are observed only where armoured by quartz.

9 Tent Hill Profile

In this locality, the sub-Kombolgie paleosol is developed on the Plum Tree Creek Volcanics of the Edith River Group (Needham and Stuart-Smith 1988). The profile has a vertical exposure of about 8 m and is developed upon a porphyritic biotite-feldspar rhyolitic ignimbrite. A subhorizontal eutaxitic foliation is defined by aligned altered feldspar and altered recrystallized shards. Ignimbritic textures are perfectly preserved even though illite and hematite pervasively replace the feldspars. The replacement of plagioclase by illite and minor hematite perfectly mimics the complex magmatic zoning of these phenocrysts (Fig. 6d). Argillic alteration of the feldspar has produced very-fine-grained illitic aggregates which are equal in size to quartz in the equigranular groundmass. Hematite is pervasively disseminated through the finely crystalline groundmass and locally replaces feldspar (partially to completely), mafic phenocrysts, and euhedral opaques (originally oxides or sulfides).

10 Other Sub-Kombolgie Paleosol Profiles

Additional profiles (from 5 through 9 in Fig. 1) have been described from samples in the rock repository at Canberra. The mineralogical assemblages of each profile display replacement textures identical to those described in the aforementioned profiles (Table 1). However, up to 5% of a colorless to pale yellow crandellite-group mineral was recognized in the clayey hematitic paleosol of the Goomadeer, Tin Camp Creek, and Mount Shepherd profiles. This mineral has been identified as goyazite, a strontium aluminium phosphate, on the basis of energy-dispersive spectra. The spectra record major Sr, Al, P, minor Ca and S, and trace amounts of Ce, La, Ba, Nd, and Ga. Goyazite occurs as minute euhedral to subhedral cubes, up to 50 microns, which occur as clusters, disseminations, or sparsely distributed grains (Fig. 5c).

11 Phosphate in the Basal Kombolgie and Related Deposits

Phosphate in the basal Kombolgie is far more widespread than previously recognized. Apatite has long been known from the breccia-type ore of the Jabiluka uranium deposit (Binns et al. 1980) but it is here documented from the basal few meters of unmineralized Kombolgie sandstone (drill hole GT6) and the uppermost sub-Kombolgie paleosol.

Phosphatic concentrations similarly occur in the sub-Athabasca paleosol of Saskatchewan (MacDonald 1980), the sub-Thelon paleosol, Northwest Territories of Canada (Miller 1983; Miller et al. 1989) and in the basal Hornby Bay Group, Northwest Territories (Ross 1983). Like the sub-Kombolgie paleosol, the sub-Thelon is Middle Proterozoic (1720–1647 Ma) and underlies thick sandstone. Phosphatic cement in siliciclastics of the Thelon Formation is attributed to late diagenesis and the phosphate in underlying paleosol is attributed to the downward infiltration of those phosphatic solutions, especially along structures. As in the Pine Creek Geosyncline of Australia, the phosphate in Canada is associated with illitic and hematitic rocks. It is concentrated immediately below an unconformity which became covered by a thick sequence of quartzose fluvial sediments, i.e., the Athabasca Group (Wilson 1985a, b; MacDonald 1980, 1985; Cumming et al. 1987).

12 Phyllosilicates in the Paleosol

12.1 Chlorite

Two varieties of chlorite have been differentiated in the sub-Kombolgie paleosol on the basis of texture, composition, and optical properties. Electron-microprobe analyses reveal differences in FeO_T, MgO, and Al_2O_3 (Table 6). These two chlorite types are differentiated in Fig. 7 on the basis of Si versus $Fe_T/(Fe_T + Mg)$. Fe_T has been calculated from the structural formula, assuming that all iron is in the reduced state. Chlorite I is iron-rich and clusters in the ripidolite field but exhibits minor scattering through the pycnochlorite to diabanite fields with increasing magnesium. Chlorite II is magnesium-rich and clusters in the clinochlore-penninite fields with some variation into the pycnochlorite and penninite fields. Both types of chlorite are so intimately intergrown with fine-grained illite that variable K_2O contents were determined during electron-microprobe analyses. Given that little potassium, sodium, or calcium can substitute into the chlorite structure, analyses with greater than 1% combined $K_2O + CaO + Na_2O$ have not been plotted on the chlorite diagrams although they are listed in Table 6. A more comprehensive listing of the chlorite analyses is available from the first author (ARM) upon request.

In general, octahedral Al tends to be slightly less abundant than tetrahedral Al in common chlorite (Foster 1962). Most analyses of chlorite I plot within this field with some scatter to more abundant octahedral Al (Fig. 8). Both the textural relationships and composition of chlorite I support the interpretation that it has formed from biotite, either

Table 6. [a]Composition of representative chlorites from GT5 profile; depths in m from drill collar

	Chlorite I						Chlorite II				
Depth (m)	19.1	24.5	35.2	38.8	47.7	54.4	24.5	35.2	38.8	47.7	54.5
SiO_2	34.21	33.73	28.71	26.90	26.36	24.17	33.03	28.92	33.70	30.54	36.87
TiO_2	ND	ND	ND	ND	ND	ND	ND	ND	ND	ND	ND
Al_2O_3	23.63	21.59	20.08	21.89	21.80	20.45	27.05	21.09	26.15	22.71	27.18
FeO_T	12.56	16.94	24.64	24.10	28.19	29.45	6.32	15.49	8.77	13.27	5.63
MnO	ND	0.16	0.20	0.40	0.13	0.52	ND	ND	ND	ND	ND
MgO	17.57	16.70	17.33	17.72	12.99	11.45	24.18	19.69	21.27	22.00	15.43
CaO	ND	ND	ND	ND	0.08	ND	ND	ND	ND	ND	ND
Na_2O	ND	ND	ND	ND	ND	ND	ND	ND	ND	ND	ND
K_2O	0.38	0.67	0.08	ND	0.21	ND	ND	ND	0.37	0.32	2.49
Cl	ND	ND	ND	ND	0.05	ND	ND	0.06	ND	0.05	ND
Total	88.35	89.80	91.04	91.01	89.71	86.16	90.58	85.26	90.26	88.90	87.60

ND = not detected.
[a] Reader may write to the first author for the complete listing of chlorite analyses.

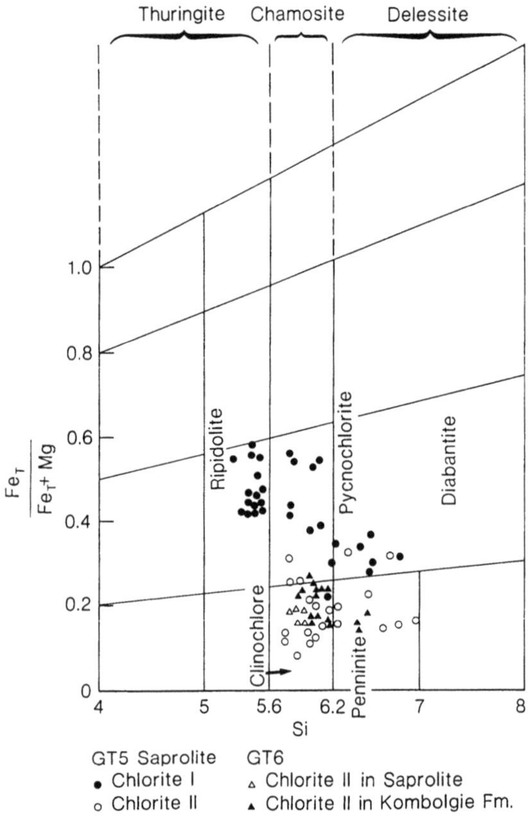

Fig. 7. Classification of chlorites in the GT5 and GT6 paleosol profiles after Hey (1954)

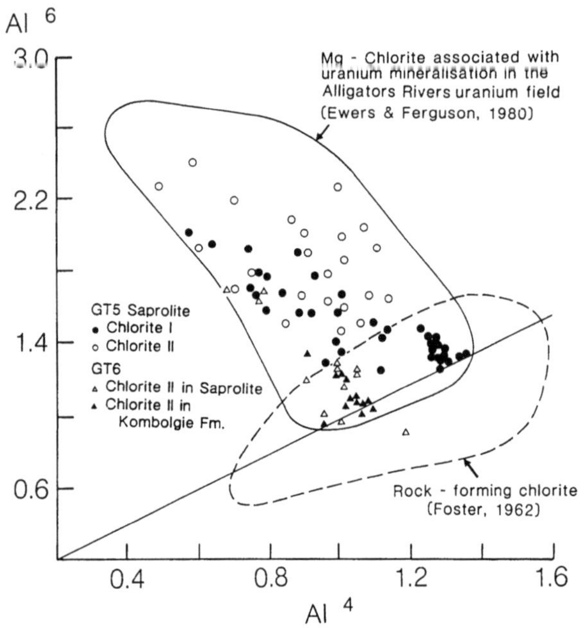

Fig. 8. Numbers of atoms of octahedrally coordinated aluminum (labeled Al^6) versus tetrahedrally coordinated aluminum (labeled Al^4) in Mg-poor chlorite (chlorite I) and Mg-rich chlorite (chlorite II) of the Kombolgie sandstone and sub-Kombolgie paleosol (profiles GT5 and GT6)

during prograde or retrograde metamorphism. Chlorite II has a greater proportion of octahedral Al (Fig. 8) and lies within the field of the magnesian chlorites which are associated with uranium deposits in the Alligator Rivers uranium field. The abundance of alumina in chlorite associated with unconformity-type uranium mineralization is interpreted by Ewers and Ferguson (1980) to result from gibbsitic layers in the chlorite structure. Nash and Frishman (1982) conclude that aluminum was mobile during uranium mineralization.

Magnesian chlorite is not necessarily associated with uranium mineralization. In a study of the sub-Athabasca Group paleosol distant from uranium mineralization in northern Saskatchewan, MacDonald (1980) identified magnesian chlorite intergrown with illite and replacing feldspar within the middle and lower sections of the paleosol profile. Semi-quantitative analysis of the chlorite revealed abundant octahedral Al and minimal iron (MacDonald pers. commun. 1984). This Mg-chlorite is more aluminous than the chlorite II of this study (Fig. 9) but occupies a similar position in the paleosol profile and displays mineralogical associations and textural features similar to those of chlorite II in GT5. Both the Athabasca chlorite and sub-Kombolgie chlorite II lie within the field of the Mg-Al chlorites which are associated with mineralization and alteration of the Cahill Formation and overlying sandstone at Jabiluka II (Gustafson and Curtis 1983; Fig. 9).

It is noteworthy that chlorite II was not identified in the hematitic zones of either the GT5 or Granite Hill profiles but was present in both the sandstone and underlying paleosol of GT6. If chlorite II were strictly related to the Mg-metasomatism which is associated with uranium mineralization, the entire profile should contain this chlorite type. Given its irregular distribution, it is concluded that Mg-Al chlorite may form by processes other than the Mg-metasomatism which characterizes unconformity-type uranium deposits (Nabarlek deposit: Ewers et al. 1983; Ewers and Ferguson 1980; Jabiluka deposit: Gustafson and Curtis 1983; Binns et al. 1980; Ranger deposit: Nash and Frishman 1982; Rabbit Lake deposit: Hoeve and Sibbald 1978; Kiggavik deposit – formerly called Lone Gull deposit: Miller and LeCheminant 1985).

12.2 Phyllosilicates in the Paleosol: Illite

Illite analyses exhibit cryptic variations in FeO_T, MgO, and CaO vertically through the GT5 and Granite Hill profiles (Tables 2a, 7). Total iron in illite

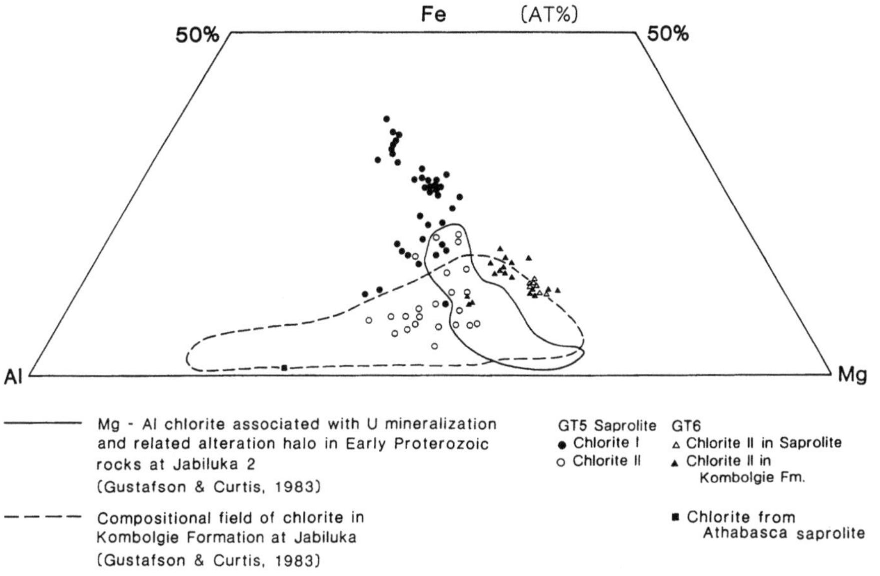

Fig. 9. Compositional differentiation between chlorite I (poorer in Al and Mg) and chlorite II (richer in Al and Mg) in the GT5 and GT6 profiles of the sub-Kombolgie paleosol (called saprolite). Also plotted are compositions of chlorite from Kombolgie sandstone, U-mineralized veins at Jabiluka 2, and sub-Athabasca paleosol in Canada

Table 7. Electron microprobe analyses of illite, GT5 profile; depths in m from drill collar

Depth (m)	19.1	19.1	19.1	19.1	19.1	19.1	
SiO_2	50.67	51.26	51.98	52.76	53.23	52.70	
Al_2O_3	32.33	31.98	31.79	31.41	28.17	27.43	
FeO_T	1.01	1.28	1.05	1.18	1.02	1.17	
MgO	1.73	1.70	1.92	2.24	2.68	2.84	
CaO	ND	0.08	ND	ND	0.26	0.19	
K_2O	9.69	9.38	9.54	9.59	8.39	8.34	
Total	95.43	95.68	96.29	97.19	93.75	92.67	
Depth (m)	24.5	24.4	24.5	24.5	35.2	35.2	35.2
SiO_2	51.09	50.87	52.33	50.80	51.72	53.53	53.04
Al_2O_3	32.28	30.94	28.40	31.74	27.21	27.36	27.56
FeO_T	1.03	1.76	1.19	1.38	1.75	1.38	1.43
MgO	2.71	2.86	2.56	2.29	4.18	3.27	2.93
CaO	0.28	ND	0.23	0.15	0.21	0.12	0.20
K_2O	8.08	9.73	8.65	9.27	8.33	8.68	8.93
Total	95.47	96.16	93.35	95.63	93.40	94.34	94.10
Depth (m)	38.8	38.8	38.8	38.8	47.7	54.4	
SiO_2	53.82	53.37	54.55	52.95	52.87	54.17	
Al_2O_3	27.36	26.03	26.75	27.37	26.05	25.85	
FeO_T	0.82	0.61	0.66	0.95	0.59	0.48	
MgO	3.59	3.53	3.61	4.15	4.21	3.32	
CaO	0.24	0.20	0.20	0.28	0.21	0.16	
K_2O	8.48	8.12	8.29	8.46	7.99	7.89	
Total	94.31	91.86	94.08	94.17	91.93	91.92	

ND = not detected.

increases with increasing modal hematite and with whole-rock total ferrous iron. Illite magnesium is lowest in the hematitic zone and increases into the underlying chloritic zone. Illite calcium generally correlates with both iron and magnesium but its distribution is more irregular. This elemental variation in illite is interpreted to record an original chemical stratification in the paleosol profile due to differential mobility of the elements.

Clay mineralogy in the original soil profile may have been stratified upward as in some modern soils, e.g., from vermiculite to overlying smectite, celadonite, and kaolinite. Diagenesis of the original clays presumably occurred during burial beneath the thick sediments of the McArthur Basin. Nonetheless, the original chemically stratified profile is apparent megascopically, and is cryptically preserved in clay-mineral compositions. Although volume reduction commonly accompanies the transformation of clays, the preservation of metamorphic textures in all but the kaolinitic subzone of the Granite Hill profile suggests that isochemical conditions prevailed during diagenesis. This is corroborated by the retention of compositional zonation through the profile.

The oxidized and leached nature of the paleosol provides qualitative insight into the groundwater chemistry which was responsible for profile development. Alkalis and alkaline earths were removed during feldspar hydrolysis, implying weakly acidic to neutral pH (Garrels and Christ 1965). The depletion of uranium implies oxidation and acidic to neutral conditions.

In a study of the modern hydrogeochemistry around uranium deposits in the Alligator Rivers field, Giblin and Snelling (1983) have documented high magnesium concentrations in oxidized, neutral to weakly acidic groundwater. Magnesium similarly may have been dissolved from primary metamorphic ferromagnesium minerals during the Middle Proterozoic weathering and thereby depleted from the sub-Kombolgie paleosol. Higher Mg values in the lower portion of the profile partly reflect the presence of chlorite II and suggest changing Eh-pH conditions with depth. However, the timing of chlorite II in the development and maturation of the profile must be considered because the textural relationships of chlorite II imply an authigenic paragenesis which is post-retrograde metamorphism. Chlorite with the composition of chlorite II has not been recognized in modern weathering profiles and may be strictly related to diagenetic reactions.

Mixed-layer clays may have been a precursor to chlorite II. Mixed-layer clays, e.g., illite/smectite, commonly form in the intermediate to lower sections of modern weathering profiles. Mixed-layer clays result from an interlayering of octahedral sheets, tetrahedral sheets, and hydrated or nonhydrated interlayer cations (Brindley 1981). Such clay minerals are readily transformed during maturation. In general, the sensitivity of clay minerals to cation exchange and restructing permit them to be used to track the diagenetic history of sediments (Dunoyer De Segonzac 1970). The occurrence of fine-grained intergrowths of illite and chlorite II within both the paleosol (Table 6) and overlying sandstone (Gustafson and Curtis 1983) are interpreted to record an advanced stage of diagenesis. The illite-chlorite II mixtures are attributed to diagenesis of a mixed-layer clay and the distribution of both illite and chlorite II are interpreted to reflect the original weathering profile.

13 Interpretation of the Sub-Kombolgie Paleosol Profile

A composite paleosol profile can be synthesized from the GT5, GT6, and Granite Hill profiles (Fig. 10). This composite profile is mineralogically and color stratified into hematitic, transitional, and chloritic zones. The hematitic zone is subdivided into an upper alumina-rich subzone which is characterized by diaspore and kaolinite and a lower subzone which is characterized by illite and Fe-chlorite. Illite and Fe-chlorite persist to depth but are associated with authigenic Mg-chlorite in the transitional and chloritic zones. The chemical expression of this mineralogical zonation is a depletion of both alkali and alkaline earths and an enrichment of certain transition-group metals in the uppermost zone of the profile. These features are comparable with those of modern soil profiles.

Intense tropical weathering initially involves kaolinization of the protolith, followed by desilicification and an increase in iron and/or aluminum (Valeton 1972; McFarlane 1976). The difference between iron-rich oxisol and bauxite does not lie in different weathering processes but simply differences in the relative amounts of iron, alumina, and silica. Bauxite is defined to contain 45 to 50% Al_2O_3, not more than 20% Fe_2O_3, and 3 to 5% SiO_2. Based on this definition, the hematitic zone in the GT5 and Granite Hill profiles clearly does not approach the composition of bauxite. Even the aluminous upper zone of the Granite Hill profile exhibits negligible desilicification.

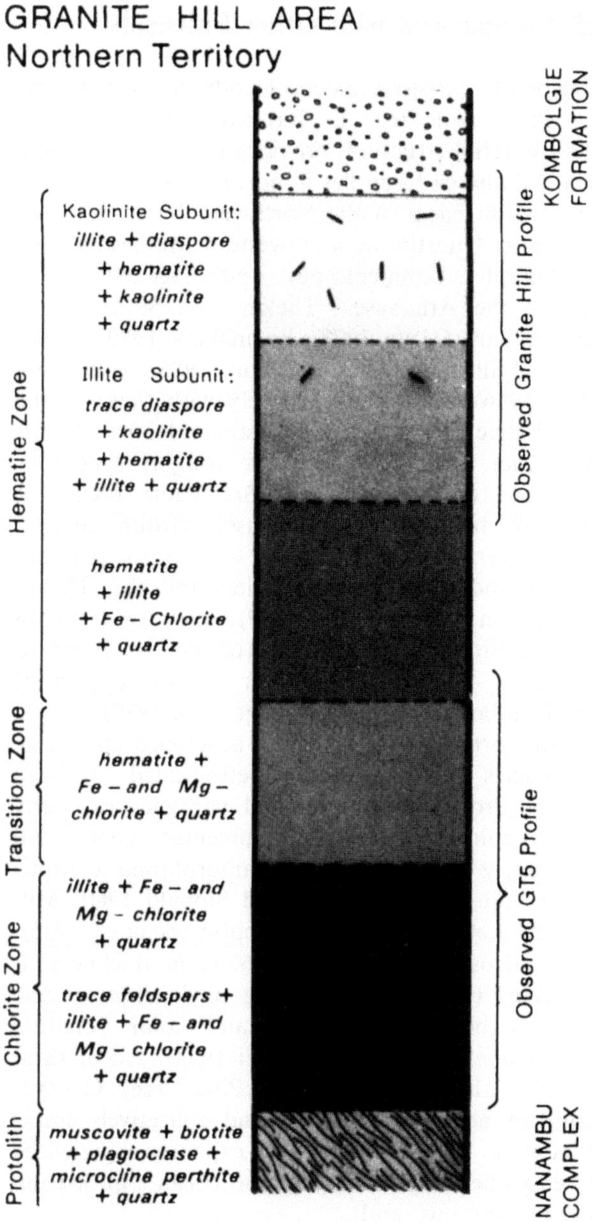

Fig. 10. Idealized composite profile for sub-Kombolgie paleosol, collectively based on several profile segments in the Pine Creek Geosyncline (Needham 1988)

The sub-Kombolgie paleosol is comparable to the lower (saprolite) portion of thick Cenozoic oxisols (including bauxite) in India, Australia, Africa, and the USA (Valeton 1972; Matheis 1983). Modern oxisol which has developed on acidic igneous and metamorphic sedimentary rocks exhibits an impoverishment in soluble elements similar to that which characterizes the sub-Kombolgie paleosol (Gordon et al. 1958; Tietz 1983). The enrichment of transition-group elements in modern oxisol relative to a variety of protoliths also is similar to that observed in the sub-Kombolgie profiles (Valeton 1972; Matheis 1983; de Oliveira 1983; Menegotto and Formoso 1983; Golightly 1981).

There is clear evidence of bauxitic weathering on the pre-Kombolgie land surface. The upper aluminous (kaolin-diaspore) subunit of the Granite Hill profile is interpreted to have been transitional between an overlying (eroded) bauxite and underlying saprolite. All of the examined sub-Kombolgie profiles are overlain by high-energy fluvial siliciclastics. The preservation potential of porous bauxite would be minimal in such a sedimentary environment. A search for remnant bauxite should be focused on areas where the basal Kombolgie beds represent a lower-energy facies (interbedded fine sandstone, siltstone, and mudstone). Other prospective areas include the tops of former plateaus and former exposures of the Zamu Dolerite. Dolerite may have weathered to a more indurated, iron-rich oxisol. The eventual discovery of paleosol which includes uppermost soil horizons may provide remnants of delicate soil structures such as cutans and pisolites.

14 Diagenetic Modification of the Paleosol and Overlying Sandstone

Burial diagenesis generally alters detrital clay, mica, and feldspar (Eberl 1978; Boles 1981; Hower 1981; Curtis et al. 1985). Diagenetic reactions depend upon the composition of interstitial solutions, porosity, permeability, and sediment mineralogy. Diagenetic maturity is measurable by the vertical variation in the ratio of illite-smectite interlayering, progressive illitization of smectite, conversion of montmorillonite to chlorite, and conversion of kaolinite to either illite or chlorite (Dunoyer de Segonzac 1969, 1970; Velde 1985; Wilson and Nadeau 1985).

Diagenetic features of the Kombolgie sandstone are similar at all sampled localities, i.e., 1 m above the unconformity at Granite Hill, at Baroalba Creek, and through 149.5 m of GT6 core. The rock is a texturally and compositionally mature quartz arenite with up to 5 modal percent quartz-rich lithic fragments. Detrital muscovite commonly is bent, indicating post-depositional compaction. Quartz overgrowths are distinguishable from underlying quartz grains by thin opaque rinds. These features also have been described from other Kombolgie sandstone distant from mineralization (Gustafson and Curtis 1983). Intergrain volumes which are not occupied

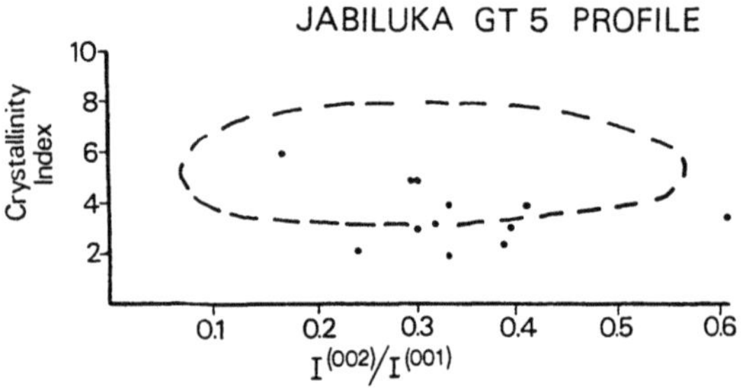

Fig. 11. Illite crystallinity in GT5 profile of sub-Kombolgie paleosol. Crystallinity is measured as a ratio of XRD intensities from two planes, [002]/[001]. *Dashed area* is Esquevin's (1969) "lower zone of late diagenesis" for clayey sediment

by secondary quartz contain aggregates of kaolinite and/or illite with chlorite II and minor disseminated hematite. Textural relationships between quartz overgrowths and diagenetic clay aggregates are similar to those depicted by Gustafson and Curtis (Fig. 8 of 1983). The diagenetic solutions which silicified the Kombolgie sandstone must have influenced the directly underlying paleosol.

Illite in the GT5 paleosol profile is so well crystallized (Fig. 11) that it is characteristic of Esquevin's (1969) "zone of late diagenesis". The crystallinity of illite has been quantified by measuring the width of the basal (001) diffraction peak at half the peak height (Dunoyer de Segonzac 1969). Illite crystallinity is influenced by compositional variation, particularly Al (Esquevin 1969), and the presence of Na-rich phyllosilicates (Kubler 1968). However, the Al content of illite is uniform within each zone of the profile (Table 7) and Na-rich phyllosilicates are absent. Whole-rock analyses record the scarcity of sodium (Table 3). The uniformly excellent crystallinity of illite in the paleosol (Fig. 11) indicates that recent weathering has had little effect.

Several features of the paleosol are attributable to diagenetic mobility of iron and phosphorus. At Granite Hill and elsewhere, blotches of hematite occur directly beneath the unconformity surface (Fig. 6b) whereas hematite is evenly distributed deeper in the profile. At Algodo, reduction spots occur in both Kombolgie sandstone and in the underlying hematitic paleosol. Trace quantities of secondary sulfides occur as fine-grained stringer-like pyrite within the hematitic zone of the GT5 and Granite Hill profiles. Epigenetic phosphate occurs in the GT6, Goomadeer, Tin Camp, and Mount Shepherd profiles.

15 Comparison with Other Paleosols

Regionally extensive paleosol underlies several major Proterozoic (ca. 1.7 Ga) sandstone basins in the Churchill structural province of North America. The Athabasca basin lies in northern Saskatchewan, the Thelon basin in the Northwest Territories, and the Sioux Quartite in southwestern Minnesota. The stratigraphy, sedimentology, and diagenetic histories of the Athabasca, Thelon, and Sioux basins are similar (Cecile 1973; Raemakers 1980; Miller 1983; Miller and LeCheminant 1985; Southwick et al. 1986) and compare closely with those of the Kombolgie Formation (Gustafson and Curtis 1983). Moreover, all these basins are roughly synchronous: 1632(+/− 32) Ma (Rb/Sr, whole rock) for paleosol beneath the Athabasca Group (Fahriq and Loveridge 1981), 1720 (+/− 6) to 1647 Ma (U/Pb and Pb/Pb, fluorapatite) for the Thelon Formation (Miller 1983, 1989), 1760 to 1630 Ma for the Sioux Quartzite and 1688(+/− 13) Ma to 1648(+/− 29) Ma (Rb/Sr) for the lower part of the Kombolgie Formation (Page et al. 1980).

In North America, paleosol developed on diverse lithologies which included retrograded Archean and/or Proterozoic ortho- and paragneisses, Early Proterozoic shelf-type metasedimentary rocks, and late Early Proterozoic unmetamorphosed volcanic rocks and redbeds (Lewry and Sibbald 1980; Miller and LeCheminant 1985; Southwick et al. 1986). The paleosol locally exceeds 60 m in thickness. It is clayey (especially K-bearing phyllosilicates) and exhibits both mineralogical and color zonation (MacDonald 1980; Chiarenzelli 1983). All of these features also characterize the Pine Creek Geosyncline of northern Australia and collectively imply that at ca. 1.7 Ga vast cratonic expanses experienced intense chemical weathering and became buried under voluminous sand.

MacDonald (1980) has documented the mineralogy and geochemistry of the sub-Athabasca paleosol where developed on a variety of Early Proterozoic metasedimentary lithologies distant from mineralization. Figure 4 illustrates compositional zonation through a sub-Athabasca paleosol which developed on metamorphosed arkose. The depth-dependent chemical trends are similar to those of the GT5 paleosol despite that fact that the average concentrations of the elements differ between the sub-Athabasca and sub-Kombolgie paleosols. Common features include decreasing oxidation with depth, slightly enhanced alumina concentrations in the middle to upper portions, an increase in magnesium in the middle to lower portions, and potassium-bearing clay minerals throughout (but with different trends). These common relationships also characterize the sub-Thelon paleosol (Miller unpublished data).

A widespread paleosol developed on the 2200 Ma Hekpoort Basalt in the Pretoria Group within the Transvaal basin of South Africa (Button 1979). Several mineralogical and geochemical features of this paleosol, especially elemental concentrations within the lower portions, resemble those of the sub-Kombolgie and sub-Athabasca paleosols. Cenozoic oxisols have developed on a variety of lithologies in the Darling Plateau of Western Australia (Davy 1979). As in the sub-Kombolgie paleosol, these exhibit virtually complete leaching of sodium and calcium, partial leaching of magnesium and potassium, depletion of barium and strontium, a gain of water, and a local increase in alumina, iron, and titanium.

16 Comparison of the Sub-Kombolgie Paleosol to Jabiluka Mineralization

Binns et al. (1980) have documented elemental concentrations associated with uranium mineralization in various rock types of the Cahill Formation at the Jabiluka deposit. Some of these rock types are similar to the apparent protolith of the GT5 paleosol near Jabulika, i.e., a homogeneous psammitic metasedimentary rock of the Nanambu Complex. Mineralization at Jabiluka is compared to paleosol alteration in Fig. 12. Both the mineralization and paleosol alteration are illustrated by a series of compositional ratios. Each ratio in the paleosol compares an average elemental concentration in the hematitic and transitional zone to that in the underlying chloritic zone. Each ratio at the Jabiluka deposit compares an average elemental concentration in rocks which have been affected by mineralization to the corresponding average concentration in comparable unmineralized rocks. Mineralized samples were collected in the immediate vicinity of uraninite-bearing veins and breccias (Group 2D of Binns et al. 1980) whereas samples from the more distant amphibolite-facies schist had not experienced retrograde metamorphism (Group 2A of Binns et al. 1980).

Binns et al. (1980) have noted enrichment of Li and Mg and depletion of Na, Ca, Sr, and Ba in ore-associated altered schist relative to comparable unaltered schist. A similar depletion of Na, Sr, and Ba is clear in the GT5 paleosol (Fig. 12). However, the ore-associated enrichment of Li and Mg at Jabiluka contrasts with a consistent moderate depletion of Li

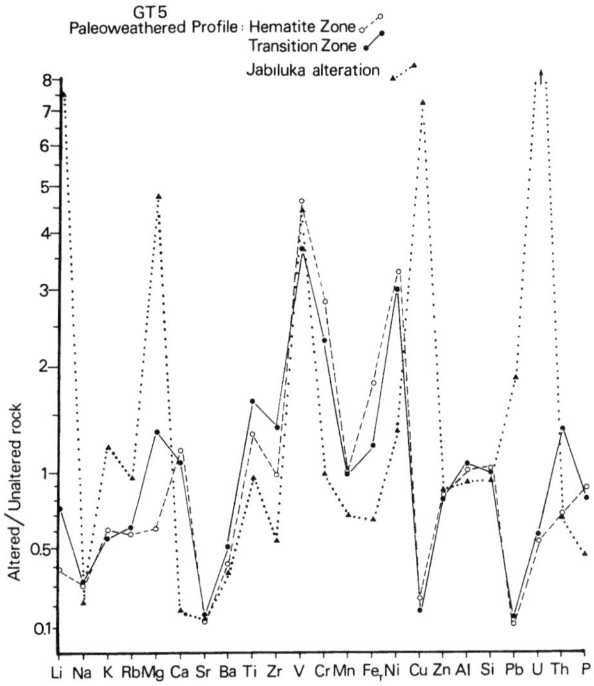

Fig. 12. Comparison of elemental concentrations in sub-Kombolgie paleosol (GT5 profile) with those of alteration halo at Jabiluka deposit. Vertical axis is a ratio of altered to unaltered rock, i.e., paleosol vs. protolith and mineralized vs. unmineralized rock

and variable enrichment or depletion of Mg in the GT5 paleosol. Moreover, iron is depleted near Jabiluka whereas Fe_T is variably enriched or depleted through the paleosol. The GT5 core did not extend completely through the paleosol to protolith and so the elemental ratios of Fig. 12 must be viewed with some caution, particularly the ratio for calcium given that calcium probably has been depleted relative to fresh protolith.

Chromium became enriched during both paleosol development and mineralization whereas uranium, copper, and lead were only enriched by mineralization. Mineralization slightly depleted zirconium and titanium but these two elements were stable to slightly enriched in the paleosol. In summary, the elemental signatures of these two processes differ significantly. It is concluded that the GT5 paleosol has not been subjected to the metasomatism which produced mineralization.

17 Economic Implications of Middle Proterozoic Paleosols

Unconformity-type uranium deposits are spatially associated with unconformities which overlie Early

Proterozoic metasedimentary rocks and/or Archean ortho- and paragneisses, and underlie or are adjacent to Middle Proterozoic, ca. 1.70 Ga, sandstone basins. The Early Proterozoic rocks sustained regional retrograde metamorphism prior to regional weathering and burial beneath extensive sand. The persistent association with sandstone suggests that clastic sedimentation was an integral part of mineralization, either active or passive.

Preliminary study of the Kombolgie sandstone indicates that its diagenetic history has resembled that documented for the Athabasca and Thelon sandstone basins, Canada. In Canada, a significant uranium fixation event at 1.72–1.65 Ga was associated with phosphorus enrichment of the siliciclastic sediments. In the Kombolgie, Rb/Sr metasomatic ages of 1610(+/−40) Ma to 1534(+/−90) Ma date altered sandstone which overlies ore at Jabiluka (Page et al. 1980; Riley et al. 1980). This dated alteration may or may not be associated with mineralization. Mineralization in the Alligator Rivers uranium province spans the period from 1.73 to 1.44 Ga (Ludwig et al. 1987).

Each Middle Proterozoic unconformity-type uranium deposit has some unique characteristics but there are several common features. The basement and/or overlying sandstone display varying degrees of enrichment in illite, chlorite, silica, phosphate, dravite (Mg-rich tourmaline), and either hematite or goethite (Binns et al. 1980; Nash and Frishman 1982; Gustafson and Curtis 1983; Sopuck et al. 1983; Wallis et al. 1983; Davidson and Gandhi 1989; Fuchs and Hilger 1989). Magnesian chlorite is commonly considered to be an indicator of mineralization but this study has shown that Mg-Al chlorite in the sub-Kombolgie paleosol is similar to that associated with ore. The occurrence of this type of chlorite in the paleosol is not attributed to the metasomatic fluids which produced uranium mineralization elsewhere in northern Australia. It is recommended that Mg-chlorite be used as an exploration guide for uranium only when integrated with geochemistry, coexisting mineral assemblage, lithology, and structure.

18 Summary

The Middle Proterozoic sub-Kombolgie paleosol has been documented in several profiles. An idealized composite profile consists of an upper hematitic and lower chloritic zone, separated by a transitional zone. An even higher, surficial bauxitic zone probably existed prior to the erosion which accompanied high-energy sedimentation of the overlying fluvial sand. The hematitic zone is divisible into an upper aluminous subzone, with diaspore and kaolinite, and a lower subzone characterized by illite and Fe-chlorite. The illite and Fe-chlorite persist to depth but become associated with authigenic Mg-rich chlorite in the transitional and chloritic zones. Diagenetic modification of the paleosol appears to have been restricted to the uppermost portions. Diagenetic changes include recrystallization of illite, minor redistribution of hematite, and phosphate impregnation.

Mineralogical and geochemical features of the sub-Kombolgie paleosol resemble those of modern oxisol and other Middle Proterozoic paleosols. The Middle Proterozoic weathering documented herein records a stable tectonic setting with little relief, a humid oxidizing environment, and a well-drained hydrologic system (Grubb 1982).

Acknowledgments. Mr. M. Hancock of the Pancontinental Mining Ltd. kindly provided access to core and permission to publish related data. Dr. N. G. Ware of the Australian National University permited use of the electron microprobe at the Research School of Earth Sciences. At the Bureau of Mineral Resources in Canberra, B. I. Cruikshank and P. G. Pyke performed the chemical analyses, W. Pappas computed geochemical plots, and J. L. Kamprad interpreted X-ray diffraction data. This research was conducted when the first author was on secondment at the Australian Bureau of Mineral Resources from 1983 to 1984. The secondment was part of an intergovernmental scientific exchange between Canada and Australia. This paper is published with the permission of the Director, Bureau of Mineral Resources, Geology and Geophysics, Canberra, A.C.T. The manuscript was improved through reviews by M. M. Kimberley, G. Retallack, and R. T. Bell.

References

Aldahan AA, Morad S (1986) Mineralogy and chemistry of diagenetic clay minerals in Proterozoic sandstones from Sweden. Am J Sci 236:29–80

Altschuler ZS (1980) The geochemistry of trace elements in marine phosphorites, pt 1. Characteristic abundances and enrichment. Soc Econ Paleontol Mineral Spec Publ 29:19–30

Basham IR (1981) Some applications of autoradiographs in textural analysis of uranium-bearing samples – a discussion. Econ Geol 76:974–982

Binns RA, Ayres DE, Wilmshurst JR, Ramsden AR (1980) Petrology and geochemistry of alteration associated with uranium mineralization at Jabiluka, Northern Territory, Australia. In: Ferguson J, Goleby AB (eds) Uranium in the Pine Creek Geosyncline. IAEA Proc Ser, pp 417–438

Boles JR (1981) Clay diagenesis and effects on sandstone cementation (case histories from the Gulf Coast Tertiary). In: Longstaffe FJ (ed) Clays and the resource geologist. Mineral Assoc Can Short Course Handb 7:148–168

Brindley GW (1981) Structure and chemical composition of clay minerals. In: Longstaffe FJ (ed) Clays and the resource geologist. Mineral Assoc Can Short Course Handb 7:1–21

Buol SW, Hole FD, McCracken RJ(1980) Soil genesis and classification, 2nd edn. Iowa State Univ Press, Ames, 406 pp

Button A (1979) Early Proterozoic weathering profile on the 2200-Ma-old Hekpoort basalt, Pretoria Group, South Africa. Univ Witwatersrand, Econ Geol Res Unit Inf Circ 133:20 pp

Cecile MP (1973) Lithofacies analysis of the Proterozoic Thelon Formation, Northwest Territories (including computer analysis of field data). MSc Thesis, Carleton Univ Ottawa, Can, 119 pp

Chiarenzelli JR (1983) Mid-Proterozoic chemical weathering, regolith and silcrete in the Thelon Basin, Northwest Territories. MSc Thesis, Carleton Univ, Ottawa, Can, 205 p

Cumming GL, Krstic D, Wilson JA (1987) Age of the Athabasca Group, Northern Alberta. Geol Assoc Can Annu Meet Abstr 12:35

Curtis CD, Hughes CR, Whiteman JA, Whittle CK (1985) Compositional variation within some sedimentary chlorites and some comments on their origin. Min Mag 49:375–386

Davidson GI, Gandhi SS (1989) Unconformity-related U-Au mineralization in the Middle Proterozoic Thelon sandstone, Boomerang Lake prospect, Northwest Territories, Canada. Econ Geol 84:143–157

Davy R (1979) A study of laterite profiles in relation to bedrock in the Darling Range near Perth, W.A. Geol Surv W Aust Rep 8:87 pp

de Oliveira JJ (1983) Lateritic soils profile in semiarid environment. Pedological, geochemical and mineralogical data. In: Melfi AJ, Carvalho A (eds) Lateritisation processes. Balkema, Rotterdam, Neth, pp 499–506

Dunoyer de Segonzac G (1969) Les minéraux dans la diagenèse; passage au metamorphism. Mem Serv Carte Geol Alsace Lorraine 29:320 pp

Dunoyer de Segonzac G (1970) The transformation of clay minerals during diagenesis and low-grade metamorphism; a review. Sedimentology 15:281–346

Eberl D (1978) Reaction series of dioctahedral smectites. Clay Clay Mineral 26:327–340

Esquevin J (1969) Influence de la composition chemique des illites sur leur crystallinite. Cent Rech Pau (SNPA) Bull 3:147–154

Eupene GS (1980) Stratigraphic, structural and temporal control of mineralization in the Alligator Rivers uranium province, N.T. Australia. In: Ridge JD (ed) Proc 5th Quadr Symp Int Assoc Genesis of ore deposits. Schweierbart, Stuttgart, pp 347–376

Ewers GR, Ferguson J (1980) Mineralogy of the Jabiluka, Ranger, Koongarra and Nabarlek uranium deposits. In: Ferguson J, Goleby AB (eds) Uranium in the Pine Creek Geosyncline. IAEA Proc Ser, pp 363–374

Ewers GR, Ferguson J, Donnelly TH (1983) The Nabarlek uranium deposit – some petrological and geochemical constraints on genesis. Econ Geol 78:823–837

Ferguson J, Winer P (1980) Pine Creek Geosyncline: statistical treatment of whole rock chemical data. In: Ferguson J, Goleby AB (eds) Uranium in the Pine Creek Geosyncline. IAEA Proc Ser, pp. 191–208

Foster MD (1962) Interpretation of the composition and a classification of the chlorites. US Geol Surv Prof Pap 414A:33 pp

Fuchs HD, Hilger W (1989) Kiggavik (Lone Gull): An unconformity related uranium deposit in the Thelon Basin, Northwest Territories, Canada. In: Uranium resources and geology of North America. IAEA Rep IAEA-TECDOC-500, pp 429–454

Garrels RM, Christ CL (1965) Solutions, minerals and equilibria. Harper and Row, Hagerstown, MD, 450 pp

Gay AL, Grandstaff DE (1980) Chemistry and mineralogy of Precambrian paleosols at Elliot Lake, Ontario, Canada. Precambrian Res 12:349–373

Giblin AM, Snelling AA (1983) Applications of hydrogeochemistry to uranium exploration in the Pine Creek Geosyncline, Northern Territory, Australia. J Geochem Explor 19:33–55

Golightly JP (1981) Nickeliferous laterite deposits. Econ Geol Publ, New Haven, CT, 75th Anniv Vol, pp 710–735

Gordon M, Tracey JI, Ellis MW (1958) Geology of the Arkansas bauxite region. US Geol Surv Prof Pap 299:268 pp

Gresens RL (1967) Composition-volume relationships of metasomatism. Chem Geol 2:47–65

Grubb PLC (1981) A review of current information and research on laterite and lateritic weathering products in South Africa. In: Melfi AJ, Carvalho A (eds) Lateritisation processes. Balkema, Rotterdam, Neth, pp 455–464

Gustafson LB, Curtis LW (1983) Post-Kombolgie metasomatism at Jabiluka, Northern Territory, Australia, and its significance in the formation of high-grade uranium mineralization in Lower Proterozoic rocks. Econ Geol 78:26–56

Hey MH (1954) A new review of the chlorites. Mineral Mag 30:277–292

Hoeve J, Sibbald TII (1978) On the genesis of Rabbit Lake and other unconformity-type uranium deposits in northern Saskatchewan, Canada. Econ Geol 73: 1450–1473

Hower J (1981) Shale diagenesis. In: Longstaffe FJ (ed) Clays and the resource geologist. Mineral Assoc Can Short Course Handb 7:60–80

Kubler BV (1968) Evalution quantitative du métamorphisme par la cristallinité de l'illite. Cent Rech Pau (SNPA) Bull 2:385–397

Lewry JF, Sibbald TII (1980) Thermotectonic evolution of the Churchill Province in northern Saskatchewan. Tectonophysics 68:45–82

Ludwig LR, Grauch RI, Nutt CJ, Nash JT, Frishman D, Simmons KR (1987) Age of uranium mineralization at the Jabiluka and Ranger deposits, Northern Territory,

Australia: New U-Pb isotopic evidence. Econ Geol 82:857–874
MacDonald C (1980) Mineralogy and geochemistry of a Precambrian regolith in the Athabasca Basin. MSc Thesis, Univ Saskatchewan, Can, 151 p
MacDonald C (1985) Mineralogy and geochemistry of the sub-Athabasca regolith near Wollaston Lake. In: Sibbald TII, Petruk W (eds) Geology of uranium deposits. Can Inst Min Metall Spec Vol 32:155–158
Matheis G (1983) Geochemical bedrock reflection in lateritic covers – case histories from Nigeria. In: Melfi AJ, Carvalho A (eds) Lateritisation processes. Balkema, Rotterdam, Neth, pp 309–316
McFarlane MJ (1976) Laterite and landscape. Academic Press, New York London, 151 pp
Menegotto E, Formoso MLL (1983) Geochemical study of the weathering of the ultrabasic Pedras Pretas Massif, Cao Sñepe (RS), Brazil. In: Melfi AJ, Carvalho A (eds) Lateritisation processes. Balkema, Rotterdam, Neth, pp 317–330
Miller AR (1983) A progress report: uranium-phosphorous association in the Helikian Thelon Formation and sub-Thelon saprolite, Central District of Keewatin, N.W.T. In: Current research, pt A. Geol Surv Can Pap 83-1A: 449–456
Miller AR, LeCheminant AN (1985) Geology and uranium metallogeny of Proterozoic supracrustal successions, Central District of Keewatin, N.W.T. with comparisons to Northern Saskatchewan. In: Sibbald TII, Petruk W (eds) Geology of uranium deposits. Can Inst Min Metall Spec Vol 32:167–185
Miller AR, Cumming GL, Krstic D (1989) U-Pb, Pb-Pb and K-Ar isotopic study of uraniferous phosphate-bearing rocks in the Thelon Formation, Dubwant Group, Northwest Territories, Canada. Can J Earth Sci 26: 867–880
Nadeau PH, Wilson MJ, McHardy WJ, Tait JM (1985) The conversion of smectite to illite during diagenesis: evidence for some illitic clays from bentonites and sandstones. Mineral Mag 49:393–400
Nash JT, Frishman D (1982) Progress report on geologic studies of the Ranger orebodies, Northern Territory, Australia. US Geol Surv Open File Rep 82/986:49 pp
Needham RS (1982) Cahill, Northern Territory, 1:100,000 Geological Map Commentary, Aust Bur Mineral Resourc Geol Geophys, 33 pp
Needham RS (1988) Geology of the Alligator River uranium field, Northern Territory. Aust Bur Mineral Resourc Geol Geophys Bull 224:96 pp
Needham RS, Stuart-Smith PG (1988) Stratigraphy and tectonics of the Early to Middle Proterozoic transition, Katherine-El Sherana area, Northern Territory. Aust J Earth Sci 32:219–230
Needham RS, Crick IH, Stuart-Smith PG (1980) Regional geology of the Pine Creek Geosyncline. In: Ferguson J, Goleby AB (eds) Uranium in the Pine Creek Geosyncline. IAEA Proc Ser, pp 1–22
Needham RS, Stuart-Smith PG, Page RW (1988) Tectonic evolution of the Pine Creek inlier, Northern Territory. Precambrian Res 40/41:543–564
Ojakangas RW (1979) Sedimentation of the basal Kombolgie Formation (upper Precambrian-Carpentarian) Northern Territory Australia: possible significance in the genesis of the underlying Alligator River unconformity-type uranium deposits. US Dep Energ Subcontr 77-054-E Natl Uranium Resources Evaluation, 35 pp
Ollier C (1969) Weathering. Oliver & Boyd, Edinburgh, 304 pp
Page RW, Compston W, Needham RS (1980) Geochronology and evolution of the late-Archean basement and Proterozoic rocks in the Alligator Rivers uranium field, northern Territory. In: Ferguson J, Goleby AB (eds) Uranium in the Pine Creek Geosyncline. IAEA Proc Ser, pp 39–68
Page RW, Williams IS, Compston W, McCulloch MT (1985) Age constraints for tectonic processes in Early to Middle Proterozoic terrains of Northern Australia. Aust Nat Univ Res School Earth Sci Annu Rep 1985:84–86
Raemakers P (1980) Stratigraphy and tectonic history of the Athabasca Group (Helikian) of Northern Saskatchewan; summary of investigations 1980. Saskatchewan Geol Surv Misc Rept 80-4, pp 99–106
Reiche P (1943) Graphic representation of chemical weathering. J Sediment Petrol 13:58–68
Retallack G, Grandstaff D, Kimberley M (1984) The promise and problems of Precambrian paleosols. Episodes 7:8–12
Riley GH, Binns RA, Craven SJ (1980) Rb-Sr chronology of micas at Jabiluka. In: Ferguson J, Goleby AB (eds) Uranium in the Pine Creek Geosyncline. IAEA Proc Ser, pp 457–468
Roscoe SM (1969) Huronian rocks and uraniferous conglomerates. Geol Surv Can Pap 68-40:205 pp
Ross GM (1983) Geology and depositional history of the Hornby Bay Group, Proterozoic, Northwest Territories, Canada. Thesis, Carleton Univ, Ottawa, Ontario, Can, 334 pp
Schau M, Henderson JB (1983) Archean weathering at three localities on the Canadian Shield. Precambrian Res 20:189–224
Sigleo W, Reinhardt J (1988) Paleosols from some Cretaceous environments in the southeastern United States. In: Reinhardt J, Sigleo W (eds) Paleosols and weathering through geological time; principles and applications. Geol Soc Am Spec Pap 216:123–142
Sopuck VJ, de Carle A, Wray EM, Cooper B (1983) Application of lithogeochemistry to the search for unconformity-type uranium deposits in the Athabasca Basin. In: Cameron E (ed) Uranium exploration in the Athabasca Basin, Saskatchewan, Canada. Geol Surv Can Pap 82-11:191–206
Southwick DL, Morey GB, Mossler JH (1986) Fluvial origin of the Lower Proterozoic Sioux quartzite, southwestern Minnesota. Geol Soc Am Bull 97:1432–1441
Stuart-Smith PG, Wills K, Crick IH, Needham RS (1980) Evolution of the Pine Creek Geosyncline. In: Ferguson J, Goleby AB, Uranium in the Pine Creek Geosyncline. IAEA Proc Ser, pp 23–38
Tietz GF (1983) Chemical and mineralogical alterations in the lateritic cover of Nigeria, West Africa. In: Melfi AJ, Carvalho A (eds) Lateritisation processes. Balkema, Rotterdam, Neth, pp 359–370
Valeton I (1972) Bauxite. Elsevier, Amsterdam, 226 pp
Velde B (1985) Possible chemical controls of illite/smectite composition during diagenesis. Mineral Mag 49:387–391
Wallis RH, Saracoglu N, Brummer JJ, Golightly JP (1983) Geology of the McClean uranium deposit. In: Cameron E (ed) Uranium exploration in the Athabasca Basin, Saskatchewan. Geol Surv Can Pap 82-11:71–110

Williams GE (1968) Torridonian weathering and its bearing on Torridian palaeoclimate and source. Scott J Geol 4:164–184

Wilson JA (1985a) Crandallite group minerals in the Helikian Athabasca Group in Alberta, Canada. Can J Earth Sci 22:67–72

Wilson JA (1985b) Geology of the Athabasca Group in Alberta. Alberta Geol Surv Bull 49:78 pp

Wilson MJ, Nadeau PH (1985) Interstratified clay minerals and weathering processes. In: Drever JI (ed) The chemistry of weathering. NATO ASI Series C: mathematical and physical sciences, vol 149. Reidel, Dordrecht, pp 97–11

Carbonaceous High-Alumina Shale in the Transvaal Supergroup: Evidence of Early Proterozoic Karstic Wheathering in a Marine Environment

Thomas O. Reimer

Bernhard-May-Str. 43, W-6200 Wiesbaden, FRG

Abstract

Extensive beds of carbonaceous shale occur within a thick sequence of Lower Proterozoic dolostone, the lower Transvaal Supergroup of South Africa. The shale contains such high concentrations of Al_2O_3 (21.9%) and K_2O (6.5%) that it is compositionally similar to Precambrian paleosols which developed on a variety of rock types. It is proposed that the alumina was concentrated from contemporaneously exposed dolostone through the removal of over 98% of the original mass by weathering. The shale contains angular fragments and blocks of chert which are attributed to an insoluble component of the dolostone. The abundant C_{org} also may have been inherited from the dolostone but the abundant potassium is attributed to the introduction of diagenetic pore fluids.

Stratigraphic relationships reveal that the carbonaceous shale was deposited extensively in the photic zone of a shallow epeiric sea. The shale occurs throughout virtually the entire preserved extent of the Transvaal Supergroup (250000 km^2), presumably indicating widespread emergence in adjacent areas and the weathering of voluminous dolostone. The dissolved ions may well have reprecipitated within the basin as dolostone which subsequently weathered, resulting in a chemical cycle. Chert breccia within the shale is attributed to locally high relief which induced debris flows.

1 Introduction

Precambrian paleosols are becoming recognized with increasing frequency, notably in Canada and South Africa (e.g., Schau and Henderson 1983; Button and Tyler 1981). Paleosol studies are addressing questions of atmospheric composition and sedimentary ore genesis. Indirect evidence of Precambrian weathering is interpreted from highly aluminous marine shale which is described herein.

Reimer (1986; Table 1) demonstrated that beds of shale with over 20% Al_2O_3 commonly occur within Archean and Early Proterozoic sequences within the Kaapvaal Craton of southern Africa. Aluminous shale beds are particularly widespread in the Transvaal Supergroup (Fig. 1), a 12-km-thick sequence dated at 2.25 to 2.05 Ga. Transvaal sediments were deposited under shallow-marine to continental conditions. The preserved extent (exposed and known from drill core) is 250000 km^2, divided between the Transvaal and Northern Cape basins (Fig. 1). The original extent probably exceeded twice the preserved extent but stratigraphic correlation across the intervening Lobatse Arch is not consistently feasible. Fine-grained rocks with more than 20% Al_2O_3 occur throughout the Transvaal Supergroup (Table 1). These have been interpreted to be products of wide-

Table 1. Composition of highly aluminous rock and paleosols

	1	2	3	4	5
SiO_2	56.25	54.78	46.80	46.50	61.14
TiO_2	1.02	0.87	1.12	1.68	1.14
Al_2O_3	27.63	23.80	34.35	33.39	21.47
Fe_2O_3	1.25	13.39	3.34	3.63	0.96
MgO	1.03	1.48	0.60	0.50	1.57
CaO	0.07	0.04	0.00	0.08	0.40
Na_2O	0.02	0.86	0.23	0.54	0.36
K_2O	8.53	1.74	9.05	9.16	6.19
MnO	0.03	0.11	0.03	0.05	0.03
P_2O_5	0.11	0.03	0.14	0.05	0.03
LOI	4.23	N.D.	4.71	5.35	7.36

1 Subaqueously weathered fine-grained tuff with quartz phenocrysts, Archean Fig Tree Group (Reimer 1983).
2 Paleosol developed on basaltic (?) tuff in Witwatersrand Supergroup. Green Bar bed in Western Deep Levels gold mine (average of 8, from Oberthür 1983).
3 Paleosol developed on basalt in Ventersdorp Supergroup (Tyler 1979).
4 Paleosol developed on Hekpoort basalt, Transvaal Supergroup (average of 9, Button 1979).
5 Carbonaceous shale from Malmani Dolomite, Transvaal Supergroup (average of 2, see also Table 2).

M. Schidlowski et al. (Eds.)
Early Organic Evolution: Implications for Mineral and Energy Resources

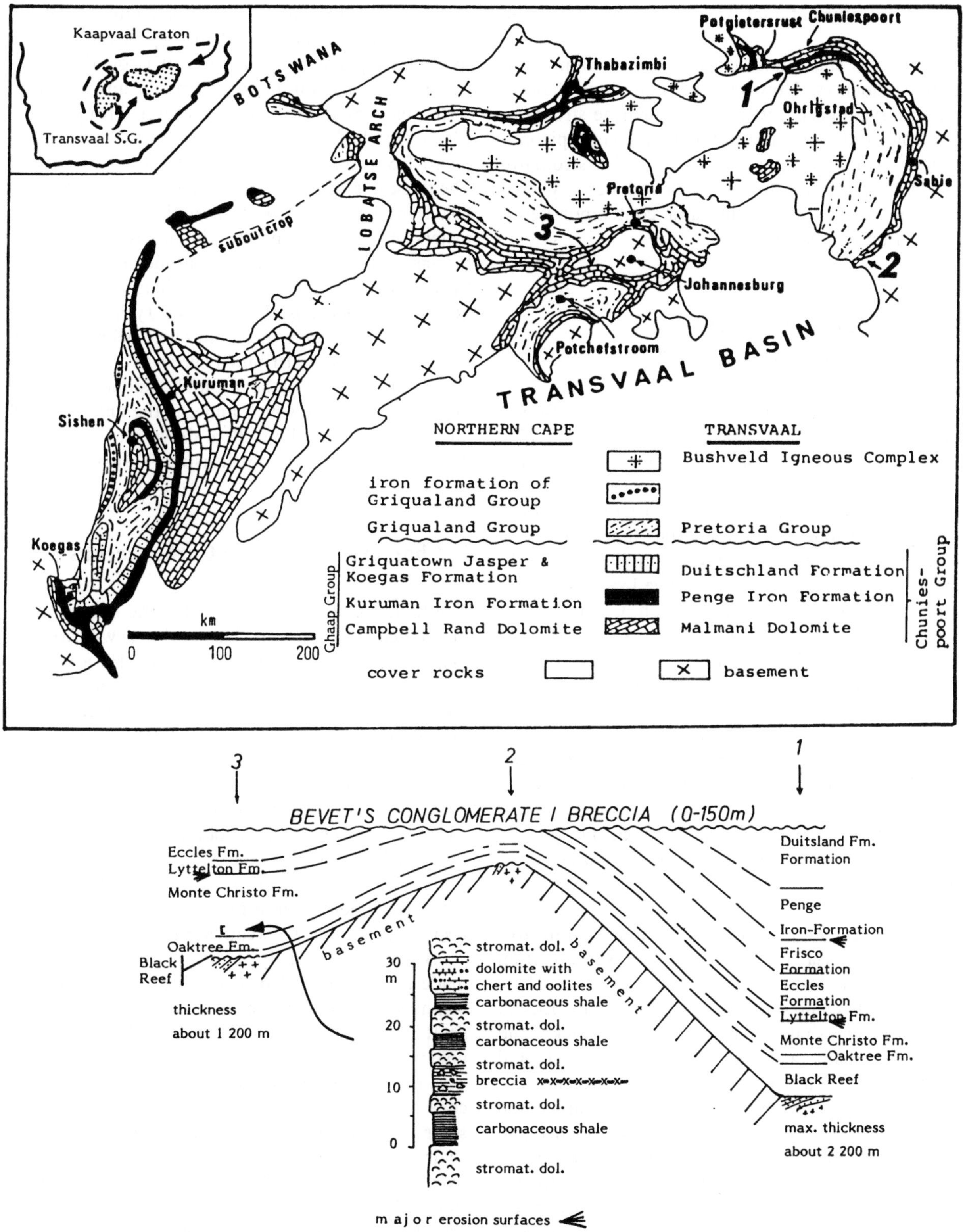

Fig. 1. Geological sketch map of Transvaal Basin and stratigraphic sections of Chuniespoort Group

spread intraformational weathering of shale and lava.

Paleosols characteristically develop within continental sequences. However, the aluminous shale of the Transvaal occurs with shallow-marine stromatolitic dolostone which precipitated on a wide epicontinental shelf. This paper interprets the stromatolite-shale association to record the sedimentary environment of the carbonates and contemporaneous weathering in adjacent areas.

2 Stratigraphic Context of the Carbonaceous Shale

The stratigraphic column of the Transvaal Supergroup in the Transvaal Basin is sketched in Fig. 1. The Black Reef Quartzite Formation is the lowest formation of more than local extent and is overlain conformably by the Chuniespoort Group which is dominantly dolostone and dolomitic limestone. The Chuniespoort starts with the Malmani Dolomite Subgroup. This is overlain by the Penge Iron Formation and finally the Duitsland Formation. Correlation of these three units with their equivalents in the Northern Cape area is shown in Fig. 1.

Malmani carbonates have been studied by Button (1973), Eriksson (1972), and Eriksson and Truswell (1974a, b). Beukes (1986) studied the correlative Ghaap Group in the Northern Cape. These studies correlate members and some individual beds (commonly stromatolitic) across an extensive platform which shallowed from subtidal through intertidal to supratidal environments. Evaporitic conditions are locally recorded by anhydrite casts and tepee structures (Clay 1986).

The foregoing authors record extensive intraformational breccia of chert and dolostone within a matrix of dolostone or carbonaceous shale. These breccia beds commonly are associated with beds of carbonaceous shale, many over 1 m thick, and probably share a related origin. Truswell and Eriksson (1975) attribute the recurring breccia-shale units to repeated minor intraformational uplift and erosion over vast areas, followed by subsidence and renewed sedimentation. The position of the major erosional surfaces is indicated in Fig. 1. In the Northern Cape area, extensive beds of carbonaceous shale, locally with chert or dolostone breccia, serve as marker horizons, as do stromatolitic beds. Beukes (1986) and Wheatley et al. (1986) attribute the carbonaceous shale to euxinic sedimentation and the breccia to submarine weathering.

The most prominent breccia-shale unit is "Bevet's conglomerate", which lies at the top of the Chuniespoort Group over most of its extent across the Transvaal Basin. This unit underlies a major unconformity at the base of the overlying Pretoria Group, along which the thickness of Chuniespoort Group is reduced from over 2000 m to locally less than 100 m. The thickness of the breccia-shale unit is highly variable, ranging from 0 to 150 m within a few hundred meters. The "conglomerate" is largely coarse chert breccia with blocks up to several meters long in the thicker portions. The matrix is argillaceous and commonly carbonaceous (Button and Tyler 1981). Locally, the unit occurs as a thin chert-pebble conglomerate. This unit is attributed herein to erosion of soil which accumulated as the insoluble residue of carbonate rocks like those which underlie the unit. Paleosol alteration extends about 30 m down into the underlying carbonate in the northeastern Transvaal.

3 Characteristics of the Carbonaceous Shale

The highly aluminous nature of some of the aforementioned carbonaceous shale (Table 1) was first recognized as unusual by Reimer (1986). He attributed the peculiar chemical composition to the accumulation of insoluble debris, locally including chert, during weathering of the carbonates. The mode of organic-carbon concentration was not deduced but a shallow subaqueous origin was suggested. Given the generally shallow-water setting of the carbonates which alternate with the carbonaceous shale, concentration of carbon by euxinic sedimentation seems unlikely. The organic carbon is attributed herein to detrital sedimentation of insoluble carbon which had been concentrated by weathering of the carbonates.

A case for the allogenic origin of carbon may be made from core drilled in the West Rand goldfield 40 km west of Johannesburg (Fig. 1; surface of no. 2 shaft in the Cooke section of the Randfontein Estates gold mine). The Transvaal succession at this site starts with the roughly 10-m-thick Black Reef Quartzite. It consists of conglomerate and coarse quartzite with some argillaceous interlayers and a thin carbonaceous shale bed at the top. The total cored thickness of the overlying dolostone is 240 m. The lower two-thirds of this is the Oaktree Formation which is largely dark gray to black homogeneous dolomicrite, mottled to variegated dolostone, minor chert, and domical stromatolites.

The upper third of the cored section is the Monte Christo Formation, which consists mainly of thinly

Table 2. Composition of carbonaceous shale and stromatolite

	1	2	3	4	5	6	
SiO_2 (%)	60.47	61.80	8.43	6.72	0.25	0.89	
TiO_2	1.03	1.24	0.10	0.03	1.06	*	
Al_2O_3	21.86	21.08	2.61	0.60	1.0	0	
Fe_2O_3	0.67	1.24	3.79	1.33	0.020	4.15	
MgO	1.43	1.71	18.22	19.52	0.002	20.56	
CaO	0.68	0.11	24.96	27.41	0.0004	28.48	
Na_2O	0.24	0.47	0.02	0.01	0.001	*	
K_2O	6.47	5.91	0.27	0.05	3.46	*	
MnO	0.05	0.02	0.67	1.04	0.83	0.76	
P_2O_5	0.03	0.03	0.01	0.03	0.028	0.01	
LOI	7.53	7.18	39.39	43.46	0.17	43.95	
S	0.036		0.52	0.008		0.59	
C	3.93	4.00					
As (ppm)	68		22	5.4		16	(1)
Be	4.48		0.52	0.11		*	
Cd	0.041		0.355	0.24		0.33	(0.0x)
Co	15	14	12	1.06		12	(0.01)
Cr	468	427	72	3.2		22	(20)
Cu	32		30	4.5		32	(4)
Hg	2.77		0.40	0.45		0.07	(0.04)
Ni	39	83	75	5.0		74	(20)
Pb	6.30	15	1.10	13		*	
Tl	1.70		16	0.011		18	(0.0X)
V	257	186	42	2.7		18	(11)
Zn	39		27	78		22	(20)
Rb		196					
Ba		317	424			439	(10)
Cs		20	0.68			*	
Sr		23					
Th		22	1.20			*	
U		4.3	0.38			*	
Sc		23	5.90			3.50	(1)
Y		33					
Zr		223					
Nb		21					
Hf		7.3	0.42			*	
Ta		1.8	0.10			*	
La		61	3.60			*	(11.5)
Ce		124	7.50			*	(1.1)
Sm		8.5	1.5			0.51	(1.3)
Eu		1.50	0.42			0.27	(0.2)
Tb		1.00	0.24			0.14	(0.2)
Yb		3.70	0.84			0.44	(0.5)
Lu		0.53	0.13			0.07	(0.2)

1 Carbonaceous shale from base of Monte Christo Formation (See Fig. 1) (analysis by Dyckerhoff AG, plant laboratory).
2 Breccia zone above stromatolite of column 3, base of Monte Christo Formation (analysis by K. C. Condie in Socorro, New Mexico).
3 Stromatolite below breccia of column 2, base of Monte Christo Formation (analysis by K. C. Condie in Socorro, New Mexico).
4 Dolostone directly below lowermost carbonaceous shale in Fig. 1, Monte Christo Formation (analysis by Dyckerhoff AG, plant laboratory).
5 Each component is the average of columns 1 and 2 (carbonaceous shales), divided by column 4 (dolostone) and multiplied by the ratio of (Al_2O_3 in column 4)/(Al_2O_3 in columns 1 and 2). Values less than 1.0 indicate Al-normalized depletion in the carbonaceous shale relative to dolostone.
6 Theoretical carbonate component of dolostone in column 4 after deduction of silicate component, assuming composition of silicate component to be average of columns 1 and 2. Recalculated to 100% after subtraction of the 12.4% of mass attributed to silicate component. Calculated negative concentrations are replaced by asterisk (*). Values in brackets are for average carbonate rocks (Turekian and Wedepohl 1969).

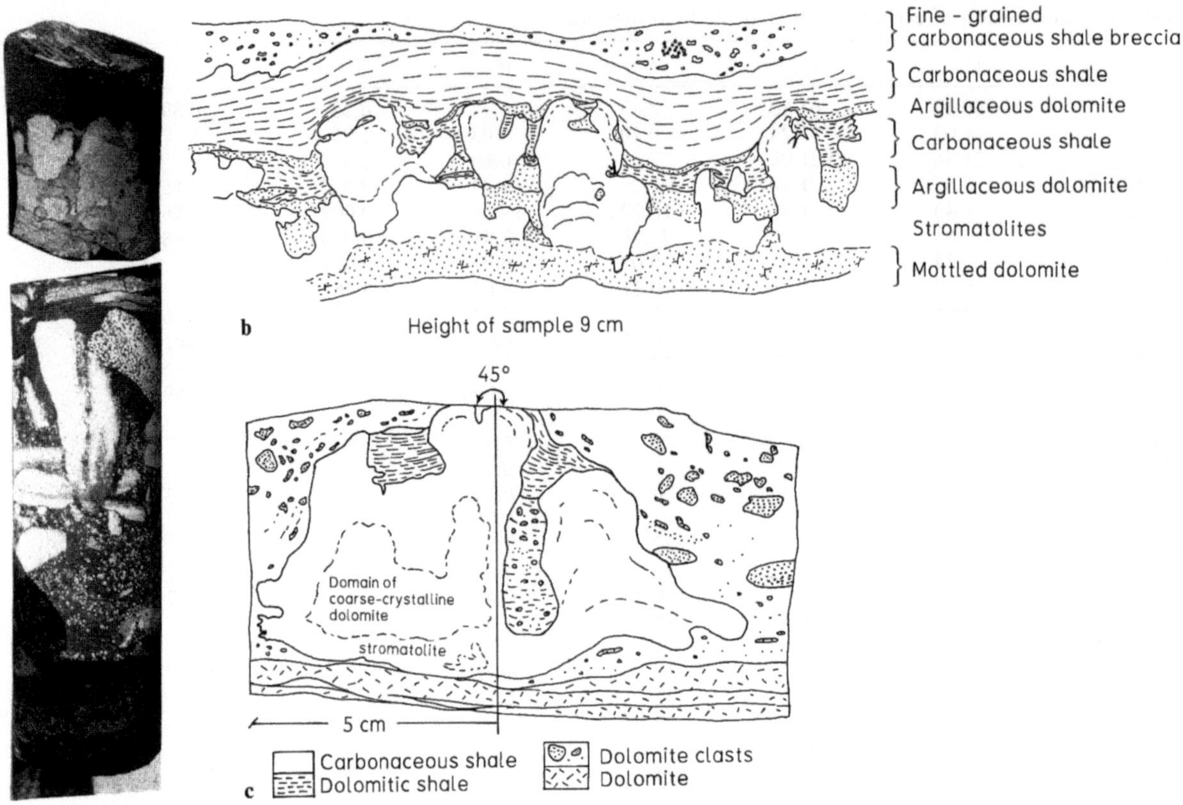

Fig. 2 a–c. Stromatolites associated with carbonaceous shale and chert breccia in Monte Christo Formation, Transvaal Basin. **a** Drill cores (7 cm diameter) with stromatolite above breccia layer. **b** "Rolled-off" view of stromatolite core from **a**. **c** Section through stromatolite, about 0.5 m above **a**

bedded, mottled to milky, cherty dolostone with abundant ripple marks. Large domical stromatolites and other microbially laminated sediments occur in close association with oolite. The stratigraphic column of Eriksson (1972) illustrates carbonaceous shale beds in this formation, mainly in the lower portion. A generalized column through the cored carbonaceous shale section is shown in Fig. 1. It was not possible to log the core in detail, however, because of disturbance during storage. Only a general drilling log was available.

The cored section consists of 25 m of alternating carbonaceous shale and laminated dolostone with poorly preserved domical stromatolites. The shale is mostly pitch black with subordinate lighter-colored shale that contains a higher proportion of carbonate. Laminae in the shale contain angular fragments of chert, up to 1 cm long, and subordinate light-colored dolostone. Isolated ooids up to 1.5 mm in diameter locally occur with these fragments. Organic carbon is concentrated on shear planes which parallel the bedding. Pyrite is rare along these planes and also rare within thin irregular veinlets. Moreover, chemical analyses show the shale to be virtually devoid of sulfides (Table 2). Mineralogically the shale consists of quartz and up to about 60% sericite, together with diabantite chlorite and rectorite ($K_6Al_2Si_4O_{10}[OH]_2 \cdot H_2O$), as well as traces of ankerite and dolomite (H. Wiggering pers. commun. 1988).

The breccia zone in which the stromatolites occur is about 4.9 m thick. It consists of beds with chert fragments up to 30 cm long in a matrix of carbonaceous shale. The proportion of fragments varies among the beds, as do fragment size and angularity. Dolostone fragments are rare but fragments of oolite in various stages of silicification locally are prominent. The silicification apparently occurred prior to erosion and sedimentation because platy fragments with less intensely silicified centers have shed isolated ooids into the shale wherever they became fractured after sedimentation.

The stromatolites shown in Fig. 2 occur near the base of the breccia zone, separated from an underlying layer of oolite fragments by 2 to 3 cm of dolostone. The layer of oolite fragments was emplaced by a mass flow, as illustrated by a large fragment standing on end (Fig. 2a). The poorly laminated columnar stromatolite which grew above this layer eventually may have been "drowned" by carbonaceous mud which contained chert and dolostone fragments as well as a few ooids. Although individual shale laminae cannot be traced into the stromatolite, an interruption of stromatolite growth near the top (arrow in Fig. 2b) may have resulted from the influx of silicate mud. The stromatolite itself contains finely disseminated pyrite.

Assuming that the stromatolites were formed by photoautrophic microorganisms, the carbonaceous mud associated with them must have been deposited in fairly shallow water. There is no evidence that the silicate mud came from the hinterland, 200 to 300 km to the north, because the proportion of shale to carbonate does not increase toward the basin margin. Both the mud and associated breccia are attributed to local sources. They probably represent the insoluble residue of dolostone which was being subjected to karstic weathering on nearby emergent islands. Emergence is attributed to a lowering of sea level. Silicate mud is envisaged to have been the usual product of erosion, punctured by mass flows of chert fragments when transgression created unstable slopes around the islands.

Abundant carbon is the shale probably is detrital, i.e., an insoluble weathering residue from the dolostone. The accumulation of detrital carbon could have caused anoxia within the sediment while oxygenated conditions prevailed in overlaying sea water. According to this model, the carbonaceous shale represents rapid sedimentation which occasionally drowned stromatolite, rather than background sedimentation which became apparent only when stromatolite growth had ceased for unknown reasons.

4 Composition

A high alumina content in sedimentary or metasedimentary rocks generally indicates the presence of continental weathering products (Reimer 1985a, 1987). However, submarine weathering related to volcanic exhalation can produce similar sediments (Reimer 1986). To illustrate this, the chemical composition of a subaqueously weathered Archean tuff is shown in Table 1.

The compositions of six paleosols and their parent rocks are shown in Fig. 3, a ternary diagram with corners of Al_2O_3, SiO_2, and K_2O. Figure 3 also illustrates highly aluminous Precambrian sedimentary rocks of South Africa, including the carbonaceous shale featured in this paper. All of these rocks clearly

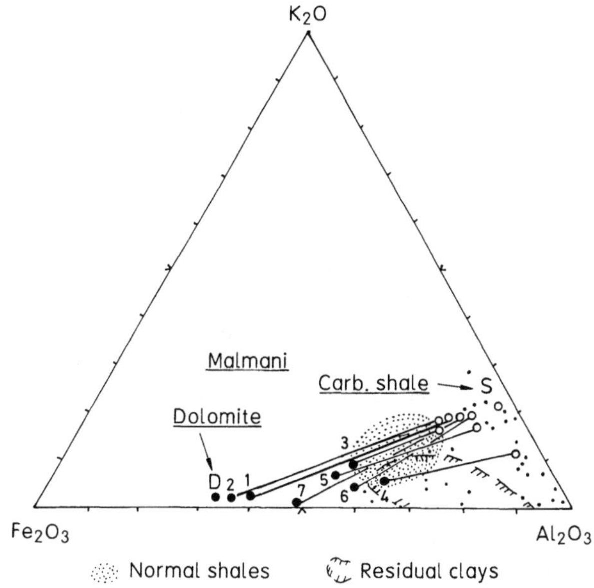

Fig. 3. Ternary diagram of SiO_2-Al_2O_3-K_2O for Transvaal carbonaceous shale, dolostone, and various pairs of paleosols and their parent rocks. Parent rock *1* is Bird amygdaloidal basalt of the Witwatersrand Supergroup in the East Rand. Parent rocks *2*, *3*, and *4* are Jeppestown shale of the Witwatersrand Supergroup. Parent rock *2* is in the Heidelberg area, *3* at the Marievale gold mine, and *4* in the East Rand. Parent rock 5 is Ventersdorp basalt (cf. Table 1, col. 3) whereas *6* is Hekpoort basalt (cf. Table 1, col. 4), and *7* is Archean greenstone (Denison Mine, Ontario, Canada)

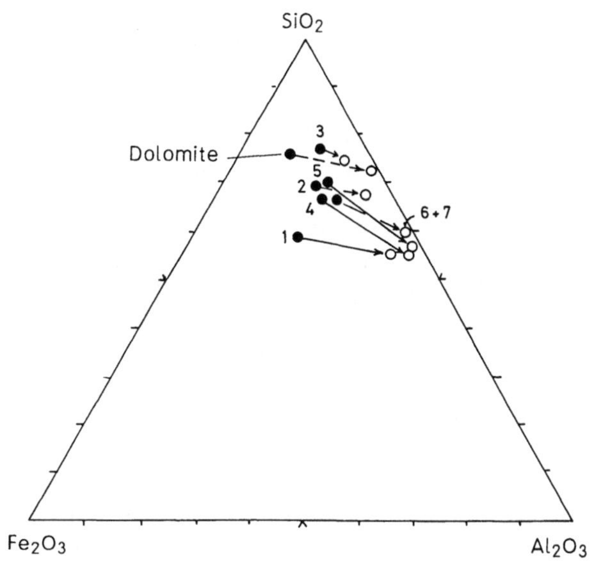

Fig. 4. Ternary diagram of SiO_2-Al_2O_3-Fe_2O_3 for carbonaceous shale, dolostone, and various paleosols and their parent rocks. See legend to Fig. 3 for paleosol locations

differ in composition from both typical shale and modern marine muds. In Fig. 3, carbonaceous shale from the Malmani Dolomite plots within the field of the highly aluminous rocks whereas the host dolostone which is the presumed source for the shale, plots well outside the field. A similar relationship is apparent in a comparable ternary diagram with Fe_2O_3 replacing K_2O as a corner (Fig. 4).

The carbonaceous shale averages 3.96% C_{org}. This lies well outside the range of 0.2 to 0.5% C_{org} which characterizes the Malmani dolostone and other carbonate rocks globally (e.g., Hall 1938). Given the abundant organic carbon, it is remarkable that sulfur is scarce (0.036% S in Table 2).

5 Weathering Reactions

During weathering at a normal pH, aluminum largely is retained within the clay minerals which form in the soil (Jackson et al. 1948). Titanium and zirconium also are typically immobile but inconsistencies occur, especially for zirconium because of its scarcity and its virtual restriction to a single heavy mineral, zircon. It is therefore more appropriate to assume conservation of Al than either Zr or Ti when normalizing the chemical composition of a weathering product against that of its parent rock. Figure 5 compares the Al-normalized compositions of several paleosols to their respective parents. As expected, most elements are depleted in the weathered residue (ratio for residue/parent less than 1). The residue/parent ratio for carbonaceous shale/dolostone similarly reveals that most elements, except potassium, are depleted (Fig. 5). Organic carbon also is depleted, given a residue/parent ratio which ranges from 0.2 to 0.5.

A measure of the mass lost during weathering may be obtained with the following parameter, [1-(Al_2O_3 in parent rock/Al_2O_3 in weathered residue)] 100%. A parametric value of 0% corresponds to no loss. Parametric values approximate 55% for decomposed lava and about 40% for shale (Reimer 1987). A parametric value of 98% is calculated for the carbonaceous shale, assuming Malmani dolostone to be the source rock. This calculation ignores diagenetic effects. If potassium or silica have been added to the shale, the proportional loss of dolostone mass would have been even greater than that calculated. Good drainage would have been required to remove such a large proportion of dissolved components. This rules out an arid weathering environment.

Potassium is enriched in the carbonaceous shale relative to average Malmani dolostone (Fig. 5). This enrichment may have been either diagenetic or inherited from the karstic soil. A diagenetic origin is more likely because the silicates within shale-associated stromatolite do not exhibit comparable enrichment in potassium. Within the presumed-source dolostone, the ratio of Al_2O_3/K_2O is 12. A comparable value of 9.7 characterizes the stromatolite (Table 2). This stromatolite presumably incorporated some of the same silicates which had washed into the basin to form the carbonaceous shale. In contrast, the ratio for the shale (3.4) indicates substantial potassium enrichment, presumably during diagenesis when the stromatolite already was impermeable to potassic solutions. Shale porosity originally approximated 30 to 40%, as calcu-

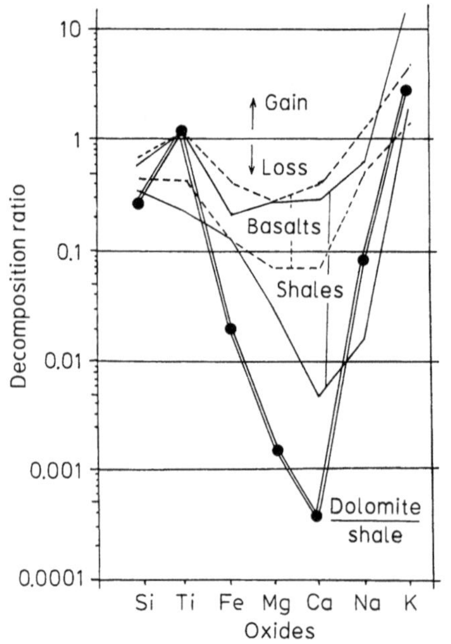

	Dolomite	Shale	Decomposition ratio (DR)
SiO_2	6.72%	61.09%	0.26
Al_2O_3	0.60	21.47	1.0
CaO	27.41	0.40	0.0004
K_2O	0.05	6.19	3.46

$$\frac{Al_2O_3 \text{ shale}}{Al_2O_3 \text{ dolomite}} = 35.6 \qquad \frac{SiO_2 \text{ shale}}{SiO_2 \text{ dolomite}} = 9.09$$

$$DR = \frac{9.09}{35.6} = 0.26$$

Fig. 5. Loss of material during weathering. *Double line* illustrates Transvaal carbonaceous shale vs. dolostone (presumed weathering source of the shale). A *pair of dotted lines* shows the range for paleosols developed on basalts whereas a pair of solid lines shows the range for paleosols on shales (from Figs. 3 and 4)

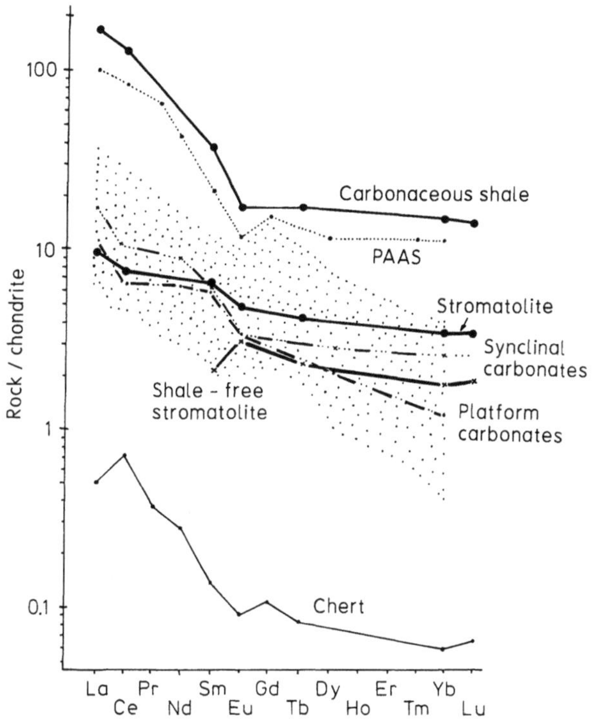

Fig. 6. Chondrite-normalized rare earth concentrations of Transvaal carbonaceous shale and dolostone. PAAS = post-Archean average shale of Nance and Taylor (1976). Synclinal and platform carbonates are from Ronov et al. (1974). Transvaal chert is from Fryer (1977)

lated from differential compaction around the stromatolite. This presumably permitted diagenetic percolation of potassium-bearing seawater.

Except for potassium, the detrital component of the stromatolite is assumed to be compositionally identical to the adjacent carbonaceous shale. The ratio of each element to aluminum should be the same in the detrital component as in the shale. Given the aluminum content of the stromatolite (2.6% Al_2O_3), one may subtract this detrital component to calculate the composition of the pure carbonate component (normalized to 100% in Column 6 of Table 2). This calculation reveals a depletion of the light rare-earth elements, La and Ce, relative to the heavy rare earths. Depletion of La and Ce appears to be a common feature of carbonate rocks, as illustrated in Fig. 6. Reimer (1985b) has attributed such rare-earth fractionation to sedimentary processes.

Normalized rare-earth distributions are illustrated in Fig. 6 for both the stromatolite and carbonaceous shale, along with a rare-earth distribution reported by Fryer (1977) for chert from the Malmani Dolomite. The rare-earth distribution of the carbonaceous shale is similar to that of average post-Archean shale ($La^N/Yb^N = 11.1$ vs. 9.2) and the shale-free stromatolite is similar to average carbonate rock (Nance and Taylor 1976).

6 Paleogeographic Significance

Widespread karstic weathering is invoked to explain the occurrence of carbonaceous shale, frequently associated with chert breccia, throughout almost the entire area of lower Transvaal carbonates. The emergence necessary for karst formation may have been caused either by eustatic fluctuation of sea level or crustal uplift. Crustal uplift probably is responsible for Bevet's conglomerate, the thick chert-shale breccia which marks the unconformity above the Chuniespoort Group. Uplift in the east and southeast reduced the thickness of the Transvaal Supergroup from over 2 km in the northeastern portion of the basin to almost zero toward the southeast.

There is no evidence that substantial uplift occurred within the area of Malmani sedimentation, given the lack of short-range thickness variation of its members. Moreover, its shallow-water stromatolites are developed almost undisturbed throughout the basin. The hypothetical karst weathering is therefore attributed to exposure of low-lying islands following a eustatic drop in sea level.

It is possible that the detrital component of the dolostone was largely supplied by aeolian transport from a distant hinterland. An aeolian depositional rate as high as 31 g/m^2 per year has been recorded on island off the coast of southern California (Muhs 1983). Moreover, an aeolian supply has been proposed for some bauxite in Western Australia (Brimhall et al. 1988). Aeolian supply seems unlikely, however, for the Malmani shale because of the aforementioned compositional difference between the shale and the shaley component of the adjacent stromatolite.

A rough estimate of the volume dissolved during the hypothetical karstic weathering can be made by assuming that the dolostone averages between 2–5% carbonaceous shale and chert breccia, and that these are the weathering residue of carbonate which contained 10% insoluble matter. It follows that between 20–50% of the original sedimentary mass was dissolved. The corresponding rock volume is between 160 and 400 $\times 10^{12}$ m^3. Although it is possible that a considerable portion of this reached the deep ocean beyond the Malmani shelf, it is more likely that the majority became recycled within the basin itself by e.g., precipitation onto stromatolites.

The weathering model presented in this paper probably is applicable also to other Proterozoic sedimentary sequences. Application to the Transvaal Supergroup has been facilitated by the conspicuous nature of both the carbonaceous shale and chert breccia and by the documentation of these units in the many drilling logs which have resulted from extensive mineral exploration.

Acknowledgements. Professors David Mossman (Sackville, NB, Canada) and Stanley Awramik (Santa Barbara, California) have kindly reviewed the manuscript. Dr. Hubert Wiggering (Essen, Germany) provided the XRD analyses and Prof. Kent Condie (Soccorro, New Mexico) provided some chemical analyses. Mrs. M. Schuch kindly typed the original manuscript.

References

Beukes NJ (1986) The Transvaal Sequence in Griqualand West. In: Anhaeusser CR, Maske S (eds) Mineral deposits of southern Africa. Geol Soc S Afr 1:819–828

Brimhall GH, Lewis CJ, Ague JJ, Dietrich WE, Hampel J, Teague T, Rix P (1988) Metal enrichment in bauxites by deposition of chemically mature aeolian dust. Nature (London) 333:819–824

Button A (1973) The stratigraphic history of the Malmani Dolomite in the eastern and northeastern Transvaal. Trans Geol Soc S Afr 77:229–248

Button A (1979) Early Proterozoic weathering profile on the 2,200 m.y.-old Hekpoort Basalt, Pretoria Group, South Africa. Univ Witwatersrand Econ Geol Res Unit Inf Circ 133:19 pp

Button A (1986) The Transvaal sub-basin of the Transvaal Sequence. In: Anhaeusser CR, Maske S (eds) Mineral deposits of southern Africa. Geol Soc S Afr 1:811–817

Button A, Tyler N (1981) The character and exonomic significance of Precambrian weathering and erosion surfaces in Southern Africa. Econ Geol 75th Anniv Vol, New Haven, pp 686–709

Clay AN (1986) The stratigraphy of the Malmani Subgroup in the Carletonville area, Transvaal: Genetic implications for lead-zinc mineralization. In: Anhaeusser CR, Maske S (eds) Mineral deposits of southern Africa. Geol Soc S Afr 1:853–860

Eriksson KA (1972) Cyclic sedimentation in the Malmani Dolomite, Potchefstroom Synclinorium. Trans Geol Soc S Afr 75:85–97

Eriksson KA, Truswell JF (1974a) Tidal flat association from a lower Proterozoic sequence in South Africa. Sedimentology 21:293–309

Eriksson DA, Truswell JF (1974b) Stratotypes of the Malmani Subgroup northwest of Johannesburg, South Africa. Trans Geol Soc S Afr 77:211–222

Fryer BJ (1977) Rare earth evidence in iron formations for changing Precambrian oxidation states. Geochim Cosmochim Acta 41:361–367

Hall AL (1938) Analyses of rocks, minerals, ores, coal, soils, and waters from Southern Africa. S Afr Geol Surv Mem 32:876 pp

Jackson ML, Tyler SA, Willis AL, Bourbeau GA, Pennington RP (1948) Weathering sequence in clay size minerals in soils and sediments. I. Fundamental generalizations. J Phys Coll Chem 52:1237–1260

Muhs DR (1983) Airborne dust fall on the California Channel islands, USA. Arid Environ 6:223–238

Nance WB, Taylor SR (1976) Rare earth element patterns and crustal evolution I. Australian post-Archean sedimentary rocks. Geochim Cosmochim Acta 40:1539–1551

Oberthür T (1983) Metallogenetische Überlegungen zur Bildung des Carbon Leader Reef, Carletonville Goldfield, Witwatersrand, Südafrika. Thesis, Cologne Univ, 214 pp

Reimer TO (1983) Pseudo-oolites in rocks of the Ulundi Formation, lower part of the Fig Tree Group, South Africa. Precambrian Res 20:375–390

Reimer TO (1985a) Volcanic rocks and weathering in the early Proterozoic Witwatersrand Supergroup, South Africa. Geol Surv Finland Bull 331:33–49

Reimer TO (1985b) Rare earth elements and the suitability of shales as indicators for the composition of the Archean continental crust. N Jahrb Min Abh 152:211–223

Reimer TO (1986) Al_2O_3-rich rocks from the early Precambrian of the Kaapvaal Craton as indicators for paleosols and other products of decompositional reactions, Precambrian Res 32:155–179

Reimer TO (1987) Weathering as a source of iron in iron formations: the significance of alumina-enriched paleosols from the Proterozoic of South Africa. In: Appel PW, La Berge G (eds) Precambrian iron formations. Theophrastus, Athens, pp 601–619

Ronov AB, Balashkov YU, Girin YP, Bratishks RK, Kazakov GA (1974) Regularities of rare earth element distribution in the sedimentary shell and in the crust of the earth. Sedimentology 21:171–193

Schau M, Henderson MB (1983) Archean chemical weathering at three localities on the Canadian Shield. Precambrian Res 20:189–224

Truswell J, Eriksson KA (1975) Paleoenvironmental interpretation of the early Proterozoic Malmani Dolomite from Zwartkops, South Africa. Precambrian Res 2:277–303

Tyler N (1979) Stratigraphy, geochemistry, and correlation of the Ventersdorp Supergroup in the Deerdeport area, Western Transvaal. Trans Geol Soc S Afr 82:133–147

Turekian KK, Wedepohl KJ (1969) Distribution of elements in the earth's crust. Geol Soc Am Bull 72:175–192

Wheatley CJ, Whitfield GG, Kenny KJ, Birch A (1986) The Pering carbonate-hosted Zn-Pb deposit, Griqualand West. In: Anhaeusser CR, Maske S (eds) Mineral deposit of southern Africa. Geol Soc S Afr 1:867–874

Significance of Precambrian Paleosols

Michael M. Kimberley

Dept. of Marine, Earth and Atmospheric Sciences, North Carolina State University, Raleigh, NC 27695-8208, USA

Abstract

The economic significance of Precambrian paleosols largely lies in their common association with strata-bound uranium ore. Their significance to resolution of Earth history depends upon their origin. There are four possible genetic models for known paleosols, i.e., the biologic, abiologic, exhalative, and bolide-impact models. Biologic soil formation is well understood from studies of modern soils. The abiologic model assumes that land plants have measurably affected soil formation only since the Ordovician. The exhalative model assumes that terrestrial exhalation of deeply-circulating fluids has contributed to surficial alteration. The bolide-impact model assumes that impacts have modified atmospheric chemistry long enough to enhance weathering detectably.

Insufficient data presently exist to evaluate the four alternative models clearly. The bolide-impact model is considered inappropriate for all known paleosols and none of the other three models is considered adequate to account for all Precambrian paleosols. The abiologic model has been developed in recent literature and the exhalative model is introduced in this paper. In all four models, the composition of the ambient atmosphere would have influenced paleosol development. However, deduction of that composition would be extremely difficult from biologic, exhalative, or bolide-impact paleosols.

1 Introduction

A paleosol is considered to be the product of any areally extensive chemical alteration downward from a former land surface. This definition excludes alteration which is essentially physical, e.g., compaction under an ice sheet, but is intentionally vague about the cause of chemical alteration. To many geologists, usage of "paleosol" implies that the reactants represented a long-term atmospheric composition (e.g., Palmer et al. 1989a, b; Retallack 1989). This restriction is avoided herein because it is impractical. Typical atmospheric reactants would have been involved in any areally extensive chemical alteration downward from a land surface, whether or not exhalation or bolide impact contributed additional reactants. The degree to which atypical (transient) reactants may have been involved potentially graded continuously, hence defying any attempt to define a practical alternative field term. A single term is needed and the proposed choice is "paleosol".

Usage of "paleosol" requires that at least some of the observed alteration occurred before burial under the overlying stratigraphic unit. Pervasive post-burial alteration may obscure all evidence of primary alteration (e.g., Pavich and Obermeier 1985). In such cases, usage of "paleosol" involves conjecture; usage of a neutral term like "altered horizon" may be preferable.

Most Phanerozoic paleosols are readily attributable to downward migration of ordinary atmospheric reactants, like virtually all modern soils. However, portions of the Precambrian probably were characterized by two alternative sources of reactants and these alternatives could have influenced paleosol production. One was bolide impact and the other was exhalation of volatile-rich fluids. Both of these could help to produce paleosols which were not fully representative of typical atmospheric conditions.

Bolide impact has been invoked to explain the apparent increase in cratonic weathering rate at the very beginning of the Tertiary, as recorded by elevated ratios of $^{87}Sr/^{86}Sr$ in marine precipitates (Macdougall 1988). Bolide impact could produce nitric acid, hence acid rain (Prinn and Fegley 1987). There is no known evidence that bolide impact contributed to any Precambrian paleosol but the possibility deserves consideration. Three other genetic models receive more attention herein, i.e., the biologic, abiologic, and exhalative models.

In both the biologic and abiologic models, all reactive volatiles are attributed to the worldwide Precambrian atmosphere. They differ in that the

M. Schidlowski et al. (Eds.)
Early Organic Evolution: Implications for Mineral and Energy Resources

biologic model assumes that a vegetative cover on land affected soil-gas composition, as in modern soil, whereas the abiologic model negates any measurable biologic effect. The exhalative model invokes the influence of regional fluid exhalation on surficial alteration. Theoretically, the exhalative fluids may be of igneous, sedimentary (compactive), or metamorphic-tectonic origin but only metamorphic-tectonic exhalation is a likely candidate for any well-documented Precambrian paleosol. Exhalative fluids may flow over the land or the rising volatiles may dissolve in surficial groundwater. In either case, the exhalative model differs from the abiologic model primarily in a greater lateral inhomogeneity in the intensity of alteration.

Among the three paleosol models, only the abiologic one has been thoroughly explored. This model is so clearly presented in the original literature that the reader should refer to these papers rather than any review of them (Holland and Zbinden 1988; Pinto and Holland 1988). A good appreciation of this or any other scientific model requires comparison with alternative models. This paper therefore provides preliminary development of the alternative biologic and exhalative models. In particular, the exhalative model has not previously received any known consideration (Holland and Feakes 1989). There is insufficient data to evaluate the relative merits of any paleosol model clearly. This paper is intended to provide incentive for the collection of additional relevant data.

Studies of Precambrian paleosols generally have addressed the contemporaneous climate and/or atmospheric composition. Inherent to most of these studies is the concept that the Precambrian atmosphere, on time scales of 10^8 years, either maintained a specific composition (Fig. 11-7 of Walker et al. 1983) or changed unidirectionally at a miniscule rate (e.g., Fig. 7 of Kasting 1987; Fig. 1 of Goodwin 1981). The author previously subscribed to this concept of long-term consistency (Dimroth and Kimberley 1976), but has found this concept to be incompatible with the geologic record of iron formations. Iron formations are attributable to exhalation of sufficient potency to affect atmospheric chemistry on a time scale of 10^6 years (Kimberley 1989). Even an effectively instantaneous (catastrophic) compositional excursion could have a major effect on a given paleosol, given that tropical soil formation may occur on a time scale of 10^2 years (Buringh 1979; Williams and McBirney 1979; Macdougall 1988).

The concept of long-term atmospheric consistency is difficult to reconcile with the growing evidence of rapid compositional variations during just the past 150000 years. Each transition from a glacial to interglacial stage has been marked by an increase in pCO_2 of over 25% on a time scale of 10^2 years (Lorius et al. 1988). There is little reason to suppose that compositional change during the Precambrian was more sluggish than during the Quaternary. If anything, the combination of a younger Sun, radioactively hotter (more exhalative) Earth, and less diverse biota would be consistent with more rapid compositional change.

Soil formation is highly dependent upon climate. Climate has varied tremendously on a time scale of 10^3 years throughout the Quaternary (Shackleton and Pisias 1985; Petit et al. 1990). This time scale is longer than that required to produce many tropical soils (Williams and McBirney 1979). If climate can change so quickly and the time lapse between observed Precambrian paleosols typically exceeds 10^7 years, these paleosols cannot provide a very reliable record of paleoclimatology.

Precambrian nonmarine sedimentary rocks are vastly more abundant than paleosols and provide a more representative record of climatic change, albeit more difficult to decipher than that of paleosols. Shallow-marine sedimentation also provides a useful record. The composition of shale has evolved through Earth history toward higher concentrations of K, Rb, and rare-earth elements, accompanied by lower concentrations of Mg, Ni, Co, and Cr. Ronov (1988) interprets this to record compositional evolution in the weathering rocks due to chemical differentiation of the Earth, rather than evolution in the weathering process due to atmospheric change.

Extensive paleosols are potentially as useful as sedimentary rocks for study of the tectonic evolution of Earth, although they have not yet been used for this purpose. However Precambrian paleosols may have formed, the preservation of thick and extensive paleosols probably required tectonic disruption of drainage. The normal development of a graded stream system effectively removes soil (Gerrard 1981). Some thick paleosols may be products of a tectonic regime in which a coastal area began to rise and thereby trap sediment within an interior basin, e.g., during incipient rifting. If the exhalative model is valid for any Precambrian paleosol, incipient rifting may have contributed to its development as well as its preservation.

2 Biologic Model of Precambrian Paleosols

2.1 Introduction

Soil scientists have thoroughly documented the biologic model of weathering under modern vascular-plant cover but little is known about the potential effect of primitive plants on soil-gas composition (e.g., Jensen et al. 1986). Since the Ordovician, vascular plants have invaded virtually all soil-forming environments on Earth (Retallack and Feakes 1987) but a few tiny areas remain where only cyanobacteria (blue-green algae) thrive. One of these is the southwestern tip (Punta Arenas) of Margarita island, Venezuela, where supratidal stromatolite covers a few hectares. Study of such a natural area would be more straightforward than fabrication of the artificial "Precambrian garden" proposed by Pinto and Holland (1988, p. 32).

It is probable that cyanobacteria have inhabited terrestrial environments for about as long as they have inhabited marine environments (Campbell 1979). However, the terrestrial environment is more prone to catastrophic change of atmospheric temperature and/or acidity (Budyko et al. 1988), due either to bolide impact (Bohor et al. 1987) or some type of exhalation (Crocket et al. 1988). Early Precambrian atmospheric catastrophes probably eliminated terrestrial cyanobacteria occasionally, necessitating recolonization from the better-buffered sea. Some atmospheric catastrophes may have had more effect on bacterial variation than biomass, given the capacity of some archaebacteria to proliferate in extreme environments (Brock 1978). For example, the population of *Thermoproteus* commonly exceeds 10^6 cells/ml in hot springs where the temperature ranges from 70 to 100 °C and pH ranges from 1.5 to 7 (Stetter 1986).

Except following rare catastrophes, the Precambrian terrestrial environment probably was more suitable for development of high-density plant life than the planktonic marine environment, as it is today, because the nutrient supply to terrestrial flora generally does not vary so unpredictably. Even if one accepts the notion of widespread cyanobacteria on Precambrian land, one can adopt an abiologic model for Precambrian paleosols if one assumes that that photosynthesizing cover had no significant effect on soil gases (Pinto and Holland 1988). Two seemingly unrelated questions bear on this assumption. One involves the extent to which the few known paleosols represent typical Precambrian conditions. The other involves the range of variation of Precambrian conditions, both spatially and temporally.

2.2 Biologic Enhancement of Optimal Conditions for Soil Formation

The few Precambrian paleosols may represent only optimal climatic conditions at a few times of optimal atmospheric composition. Optimization is difficult to demonstrate for any paleosol but unfavorable conditions are clearly demonstrable for the time-space of the Lower Proterozoic Gowganda Formation. Virtually no chemical weathering affected the sediments which accumulated to become the 1-km-thick Gowganda Formation in Ontario, Canada (Card et al. 1977; Nesbitt and Young 1982).

It is possible that the biologic model is appropriate for some Precambrian paleosols whereas the abiologic model is appropriate for others. If this is true, one may speculate whether optimization of conditions would have greater potential effect on organic than inorganic processes. Optimization clearly favors organic processes today and there is no reason to suspect the opposite in the Precambrian. One would therefore expect any biologic paleosols to be better developed than abiologic examples. The abiologic model presently is applied to some of the best-developed paleosols (Holland and Zbinden 1988).

The importance of optimization may be partially surmised from the chronostratigraphic record. Temporal variation in Precambrian paleosol development is addressed subsequently during discussion of the exhalative model. Here it is sufficient to note that Precambrian paleosols differ from Phanerozoic paleosols in that they apparently constitute a smaller proportion of the stratigraphic record but are more extensive on average, e.g., the sub-Huronian paleosol (Kimberley et al. 1980), the Hokkalampi paleosol (Marmo this Vol.), and the sub-Transvaal paleosol (Button 1979). It may be argued that this proportion-extent paradox is the biased result of metamorphic disguising of those Precambrian paleosols which were poorly developed and correspondingly less extensive. This argument clearly has some merit but does not seem to resolve the proportion-extent paradox completely. Extensive worldwide mapping of Precambrian rocks during the past two decades has not proportionally increased the number of known paleosols. If valid, the proportion-extent paradox presumably records enhancement of production and/or preservation at special times in Earth history.

If pCO_2 could suddenly increase by over 25% just 18000 years ago, comparably rapid variation must be considered likely occasionally during the 4 Ga of Precambrian time. However, most models of Precambrian atmospheres assume that pCO_2 was consistently very high (Newmann and Rood 1977; Walker et al. 1983; Holland 1984; Des Marais 1985; Kasting 1985, 1987). Within the Phanerozoic, anomalously high pCO_2 has been hypothesized for the Permo-Carboniferous, the time of greatest sedimentation of terrestrial plant matter (Berner and Raiswell 1983). By analogy, one may hypothesize that Precambrian times of particularly elevated pCO_2 would have enhanced accumulation of terrestrial photosynthesizers. This would have enhanced the biologic influence on soil-gas CO_2 and thus increased weathering rates.

Biologic enhancement of soil-gas pCO_2 over atmospheric pCO_2 is related to the local rate of carbon fixation from the atmosphere. Climatic variation presently induces great variation in the annual rate of photosynthesis, ranging from the high rate of a tropical rain forest to nil near the poles. Geographic variation in carbon fixation presumably also characterized the Precambrian.

Proper evaluation of the biologic model will require analysis of organic matter in the best-preserved Precambrian paleosols. At present, relevant data are sparse. This is most apparent from the number of citations of the four analyses (0.01 to 0.25% C) by Gay and Grandstaff (1980) in the sub-Huronian paleosol at Elliot Lake, Canada (e.g., Mossman and Farrow this Vol.; Holland 1984, p. 282). Indirect biologic information may come from studying both the enhancement of soil-gas CO_2 and the types of inorganic reactions which presently occur under terrestrial cyanobacterial mats.

3 Exhalative Model of Precambrian Paleosols

3.1 Introduction

For many geologists, the term "exhalation" is synonymous with "volcanic exhalation" but nonvolcanic processes collectively cause a greater volume of volatiles to exhale at the Earth's surface (Lasaga et al. 1985). Sediment loading in deltas presently induces exhalation of water and biogenic methane in mudlumps (Lynch 1954). However, sedimentary (compactive) exhalation has no apparent relevance to Precambrian paleosols. If anything, observed Precambrian paleosols are antithetic to deltaic environments.

Under surficial conditions, the solubility of methane is less than one-twentieth that of CO_2 (Stumm and Morgan 1981, p. 109). Methane generally reaches the atmosphere without causing much chemical alteration of the surficial material through which it passes. Spectacular brecciation by Quaternary methane exhalation abounds in northeastern Venezuela but chemical alteration has been minimal within most of these breccia pipes.

Metamorphism may produce soluble, acid-forming volatiles like CO_2 and SO_2. These may be released rapidly during the decompression which accompanies initiation of deep fractures. Permeable fractures may result from either pure crustal extension (Vann 1985) or wrench faulting (Sibson 1987; Sylvester 1988). Rifting and wrench faulting are regional processes which may induce regional exhalation, generally accompanied by seismic activity. A seismic event in Japan ($M = 6.3$) initiated a year-long exhalation of about 10^7 m^3 of warm Na–Ca–Cl brine (Sibson et al. 1975; Fyfe et al. 1978).

Exhalation of metamorphic fluids is a prime ore-forming process (Kerrich 1989). Surficial exhalation of deep fluids is a more viable explanation for most sedimentary ore deposits than is surficial weathering (Kimberley 1989). A large proportion of Precambrian paleosols occur close to sedimentary ore, particularly uranium ore (Button and Tyler 1981). Two-thirds of the paleosol profiles listed by Holland (1984, p. 297) occur near stratabound uranium ore. Like Precambrian paleosols, these uranium deposits constitute a vanishing small proportion of all Precambrian rocks. On a statistical basis alone, the exhalative model therefore deserves serious consideration.

3.2 Metamorphic-Tectonic Exhalation

Any mass-balance calculation of volatile recycling, e.g., that of CO_2, reveals the importance of metamorphic-tectonic exhalation (Lasaga et al. 1985). The average duration of sandstone and shale throughout the Phanerozoic has been on the order of roughly 600 Ma (million years) whereas limestone has lasted only about 300 Ma and evaporites about 200 Ma (Garrels and Mackenzie 1971, p. 272). The shorter life-span of chemical than clastic sedimentary rock partly reflects metamorphic-exhalative recycling. The contrast in average life-span presumably was even greater during early Earth history when the planet contained a higher proportion of heat-generating radioactive isotopes (Turcotte 1980).

Metamorphic-tectonic exhalation typically is an integral phenomenon but its integral nature is not apparent in many research papers, papers written either by a metamorphic geochemist (Kerrich 1989) or a structural geologist (McCaig 1988). Moreover, there is a continuum between sedimentary and metamorphic generation of rising volatiles. A conceptual distinction is that sedimentary processes involve compaction and biodegradation whereas metamorphism involves nonbiologic chemical production of volatiles. Low-temperature metamorphism starts at about 200 °C (Frey 1987).

Tectonism drives metamorphism (e.g.. Harte and Dempster 1987). It also controls the location of resulting exhalation both by passive opening of subvertical conduits and by active pumping along those factures. Diverse pumping models have been proposed for a wide variety of tectonic regimes (Sibson 1987; Sibson et al. 1988; McCaig 1988; Moore et al. 1988). However, the relative pumping potential of these various regimes is not yet well quantified.

3.3 Association of Precambrian Uranium Deposits with Paleosols

Interest in the exhalative model of Precambrian paleosols is inspired largely by their association with uranium deposits. The most voluminous Precambrian uranium deposits commonly are associated with paleosols. The oldest well-known examples occur in the Witwatersrand where the ore beds themselves locally became weathered during the Archean (Minter et al. 1988). Numerous paleosols and associated highly aluminous sedimentary beds (> 20% Al_2O_3) occur within the Witwatersrand Supergroup. Reimer (1985, p. 41) concludes that, "...an origin through decomposition by subaqueous volcanic exhalations appears to be excluded" whereas Palmer et al. (1989a) present evidence of hydrothermal influence.

Phillips (1988) argues that abundant chloritoid and pyrophyllite in the Witwatersrand represents exclusively intrastratal alteration, not paleosol. A partially surficial exhalative origin is preferred herein because the aluminum-titanium-niobium association reported by Phillips (1988) does not negate surficial alteration as he claims. These three elements generally are correlative in bauxite (Valeton 1972). Kaolinite deposits in the southeastern USA commonly also are titaniferous (Jepson 1988).

An extensive paleosol underlies uraniferous conglomerate in the Elliot Lake district of Canada. Sedimentation of uraniferous gravel locally coincided with partial erosion of that paleosol (Holmes 1957). This paleosol developed on various Archean rock types during the Early Proterozoic (Gay and Grandstaff 1980). The thickness varies from a few meters (Kimberley et al. 1980) to a few tens of meters (Farrow and Mossman 1988). Other paleosols have not yet been documented in the overlying Huronian Supergroup but highly aluminous sediments accumulated in the upper portion of the thick Huronian sequence (Young 1973).

The most impressive association of paleosol with uranium ore characterizes the Middle Proterozoic "unconformity-type" deposits. The paleosol near the Jabiluka deposit in north-central Australia is described by Miller et al. in this Volume. Elsewhere, Miller et al. (1989) have used uraniferous phosphate to date penecontemporaneous paleosol in the Thelon basin of north-central Canada. The Thelon Formation is underlain by a remarkably thick paleosol (> 50 m) and contains highly siliceous paleosols within it, reaching 30 m in thickness (Ross and Charenzelli 1985). If the arid climatic interpretation of Ross and Charenzelli (1985) is correct, the thickness of these paleosols presumably records either exhalation or a more reactive atmosphere than the modern one.

Uranium-gold mineralization is associated with the sub-Thelon paleosol (Davidson and Gandhi 1989). Some of the richest of all uranium deposits are associated with Middle Proterozoic paleosol beneath the adjacent Athabasca basin in northern Saskatchewan, Canada (Hoeve and Sibbald 1978; Hoeve and Quirt 1984). Chemical and mineralogical variations downward through the sub-Athabasca paleosol are described by Voultsidis et al. (1982) and Macdonald (1980, partially reproduced by Holland 1984). Initial paleosol development probably was followed by prolonged post-burial hydrothermal alteration. Mineralization associated with the Middle Proterozoic paleosol in northern Australia (Miller et al. this Vol.) apparently occurred through a time interval as long as 30 Ma (Ludwig et al. 1987).

Robertson et al. (1978, Fig. 5) appropriately describe uranium exploration near Middle Proterozoic paleosols as "good prospecting" versus "no hope" away from them. In general, paleosols are so intimately associated with stratabound Precambrian uranium deposits that the best advice one could offer on "how to find a Precambrian paleosol" is to start near a stratabound uranium deposit (cf. Retallack this Vol.). Candidate localities without known paleosols are briefly described by Robertson (1974) and D'Elboux (1984).

3.4 Genetic Models of Voluminous Precambrian Uranium Deposits

An understanding of the origin of voluminous uranium deposits clearly would enhance an understanding of associated Precambrian paleosols. Unfortunately, all of these voluminous deposits remain controversial. The prime issue is whether the concentration of uranium was driven by exhalation or a combination of weathering and gravity. In weathering-gravity models, the separation of uranium from other elements is achieved either by surficial weathering or by groundwater. Surficial weathering involves either dissolution of uraniferous minerals or dissolution of adjacent minerals. In the latter case, liberated uraniferous minerals become placer concentrates.

The groundwater-ore model assumes that uranium may be leached from a thick sedimentary-volcanic pile and deposited where the oxidation state of the groundwater becomes sufficiently low, given that chemical reduction precipitates uranyl solutes (Langmuir 1978). Although most thoroughly developed for Phanerozoic deposits (Samana 1986), this groundwater model also has been applied to major paleosol-related deposits which underlie Proterozoic sandstone sequences (Langford 1983).

The placer model has become firmly established in the literature for Middle Precambrian conglomeratic uraninite-brannerite-uranothorite deposits (Robertson 1962; Roscoe 1969; Pretorius 1975; Theis 1978; Robinson and Spooner 1984; Minter et al. 1988; Gabelman 1988). It is generally assumed that uraninite and accessory minerals, e.g., gold, became concentrated from widely disseminated veins and transported distances on the order of 10^2 km. In the words of Köppel and Saager (1974, p. 318), "These results, therefore, constitute strong evidence that the greenstone belts of the Swaziland Sequence and their numerous gold deposits contributed detritus to the Witwatersrand depository." Like the groundwater model, this distant-source placer model may be classified as a gravity-weathering model because the concentration of uranium and gold to ore grade is attributed more to fluvial separation of heavy minerals than any exhalation which may have produced the widely disseminated veins.

3.5 Relationship of Precambrian Stratabound Uranium Deposits to Paleosols

Detrital uraninite is attributable to erosion from chemically precipitated veins. Other types of uranium deposits are more direct products of chemical processes. To someone interested in paleosols, a prime question about all of these chemical processes is their timing relative to that of associated paleosol production. Soil formation and uraninite concentration within veins are distinct geochemical processes. If the time interval between these two geochemical processes is great, they probably have no direct genetic relationship, even where spatially associated. The reverse is not necessarily true. Demonstration of contemporaneity does not demonstrate linkage between the two chemical processes but does imply some common physical relationship to the regional tectonic setting. For example, incipient rifting could release exhalative uraniferous fluids and simultaneously preserve an extensive soil from erosion. The preserved soil need not have experienced any chemical alteration because of rift-related exhalation.

The relative timing of mineralization and associated paleosol alteration is not yet precisely known for any Precambrian stratabound uranium deposit, particularly deposits formed more than 2 Ga ago. The largest of these ancient deposits occur in the Witwatersrand of South Africa and the Elliot Lake district of Canada (Pretorius 1981). All ancient deposits have been affected by subsequent heating and assumptions must be made about the timing of that heating to interpret much of the available radiometric data (e.g., Meddaugh and Holland 1981).

Until recently, the radiometric dating of Witwatersrand uraniferous-auriferous conglomerate has been interpreted to prove that Archean uraninite had been eroded during the Early Proterozoic (Nicolaysen et al. 1962; Pretorius 1981). Now it is known that the Witwatersrand Supergroup itself is Archean and that extensive hydrothermal activity accompanied detrital uranium-gold mineralization (Pretorius 1989). Uraninite within the Elliot Lake conglomerate has been interpreted to be 200 Ma older than the conglomerate host (Robinson and Spooner 1984). This is similarly questionable.

Although the ages of uranium mineralization and paleosol development are not known precisely for the Elliot Lake district, there is general agreement on an Early Proterozoic age for both. The oldest estimate for the age of the conglomerate at Elliot Lake is 400 Ma younger than the upper limit of Witwatersrand ages and 750 Ma younger than the lower limit (Roscoe 1969; Pretorius 1989). This time gap of 400 to 750 Ma is inconsistent with the concept that the global occurrences of uraninite-bearing conglomerate are tightly "time-bound". These conglomeratic ores are nonetheless characteristic of the prolonged Archean-to-Proterozoic transition (Robertson et al. 1978). This transition was not globally synchronous. Following cratonization in each area, the new granitic crust typically developed rifts which may have channeled upward migration of deeply circulating fluids (e.g., Blake and Groves 1987; Kerrich 1989).

3.6 Uranium Mobilization by Deep Hydration of Granitic Rocks

The Late Archean (3.1 to 2.5 Ga) was the prime time of continental crustal growth, according to McLennan and Taylor (1982) and Veizer and Jansen (1979). This concept is favored herein relative to an alternative view that prime growth occurred prior to 3.8 Ga (Fyfe 1978; Armstrong 1981; Reymer and Schubert 1984). During continental growth, the mantle preferentially expels incompatible elements like uranium. These may become richly disseminated throughout a granitic rock body, e.g., the 10–20 ppm U in the Triassic Conway granite of New Hampshire (Richardson 1964). Subsequent hydration of such rock by through-flowing hydrothermal fluids could induce substantial transport of uranium. Such a hot fluid apparently is a more potent transport agent than is low-temperature groundwater (Rich et al. 1977; Kimberley 1978b; Langmuir 1978).

Gabelman (1988) attributes voluminous nonsedimentary uranium deposits to late-stage hydration of granitic rocks. He attributes the hydration either to "magmatic or metasomatic" fluids (Gabelman 1988, p. 57). A nonmagmatic source is preferred herein. All voluminous uranium deposits are attributed herein to deep hydration of granitic rocks, typically by descending-ascending water in an area of tectonic pumping (Sibson 1987; McCaig 1988). No voluminous uranium deposit is considered attributable to shallow continental weathering (cf. Barbier 1974; Langford 1983). In the deep-hydration model, the optimal time in Earth history for uranium mineralization would have followed the prime time of granitic production, particularly during rifting and wrench faulting of the voluminous granitic rocks.

Stratabound deposits are attributed herein to surficial exhalation of the same type of fluid which has produced the "replacement-dissemination" deposits described by Gabelman (1988). It is envisioned that uraniferous fluids preferentially rose through uplifting blocks and that near-surface precipitates became eroded shortly after precipitation. Most fluids presumably rose along basin-margin faults as initial uplift was beginning to induce clastic sedimentation. The fluvial-transport distance therefore would have been short, on the order of 10^1 instead of 10^2 km. Mass-balance calculations reveal that the concept of the longer-distance, more-disseminated source is inconsistent with Witwatersrand stratigraphy (Reimer 1984). The proposed short-range, exhalative-placer model differs from the distant-source placer model in the timing as well as the areal extent of primary mineralization.

A comprehensive description and explanation of the proposed exhalative-placer model would be as lengthy as the author's recent paper on iron formations (Kimberley 1989). The present discussion is limited to a few notes. Even this brief discussion would be inappropriate if irrelevant to Precambrian paleosols. However, many Precambrian paleosols are closely associated with uranium mineralization of probable exhalative origin. Even if exhalation has been contemporaneous with stratabound uranium mineralization, it is possible that exhalation has not been sufficiently potent to affect an associated extensive paleosol. The inherent potency of terrestrial exhalation therefore will be examined briefly prior to further consideration of uranium deposits.

3.7 Modern Metamorphic-Tectonic Exhalation

Metamorphic-tectonic exhalation presently is active in two areas which both lie near euxinic basins, i.e., the Black Sea and Cariaco Basin, but which occur on opposite sides of the globe. One exhalative area forms the northern rim of the Black Sea (Kerch and Taman peninsulas) whereas the other lies east of the Cariaco Basin, along the northeastern coast of Venezuela. Both areas also share the distinction of being the only places on Earth where iron-silicate ooids have recently accumulated in a saline water body, i.e., the voluminous Pliocene Kerch oolitic ironstone (Yakontova et al. 1985) and the Quaternary Mala Pascua deposit (Kimberley 1989). Both of these oolitic iron deposits are interpreted to be exhalative in origin (Kimberley 1989).

Drilling for hydrocarbons under the Kerch and Taman peninsulas has revealed that the source of exhalative fluids occurs at a depth of at least 8 km (Shnyukov et al. 1986). More than 80 exhalative vents are fairly evenly distributed over an area of about 50 km by 150 km. This area rivals the extent of most Precambrian paleosols. Ongoing eruptions are spreading mud for several kilometers beyond major vents, carrying high concentrations of both reactive volatiles and several metallic ions (Shnyukov et al. 1986). Exit temperatures are locally near boiling but quickly cool as the slurry flows over land.

East of the Cariaco Basin in Venezuela, similar but smaller mud volcanoes are exposed in sea cliffs of Quaternary fanglomerate on eastern Coche island (Kimberley and Llano in press). Continuous exhalation of boiling water occurs at a vent on the adjacent mainland, 6 km west of El Pilar, and diverse minerals are precipitating within this vent system. These precipitates may be sampled in ejected blocks which are scattered around the vent.

The mineral-dissolution processes in the deep source regions of both the Black Sea and Cariaco areas could be described as metamorphic if one considers metamorphism to start at 200 °C (Frey 1987). Rising metamorphic fluids are believed to be precipitating authigenic iron silicates in several areas along the northeastern coast of South America (Kimberley 1989). Over 10% of the seafloor grains contain authigenic iron silicates throughout one area which is comparable in extent to that of the Kerch-Taman peninsulas, i.e. 50 by 150 km (Kimberley 1989,

Fig. 21). Exhalation also characterizes the northern slope of the Gulf of Mexico (Aharon et al. 1989).

3.8 Origin of Precambrian Stratabound Uranium Deposits

Like many classical debates in geology, the origin of conglomeratic uranium-gold mineralization presently is advancing by incorporating concepts which were developed by opposing protagonists. The arguments about physical-vs.-chemical concentration are legendary for these ores (e.g., Pretorius 1975). All physical models advocate placer sedimentation but not all chemical models invoke hydrothermal activity (Derry 1960; Reimer 1984).

Drilling through the base of the Witwatersrand Supergroup has revealed an underlying paleosol and extensive hydrothermal alteration of subjacent granitic rock (Robb and Meyer 1985, 1986; Robb et al. 1986). It is concluded that these hydrothermally altered granitic rocks "...formed the principal source rocks to much of the adjacent placer gold-uranium deposits" (Robb and Meyer 1986). This has become the first corroboration of the argument by Kimberley (1978 a) that penecontemporaneous hydrothermal and detrital processes produced the ancient uranium-gold conglomeratic ores. However, the additional argument that uranium became concentrated into eroding veins by shallow groundwater no longer is tenable (cf. Kimberley 1978 a; Kimberley et al. 1980; Reimer 1984).

Deep-source exhalation is envisioned herein to have precipitated uranium within near-surface veins in an uplifting area, just as modern exhalation characterizes the rising Kerch-Taman peninsulas adjacent to the subsiding Black Sea (Shnyukov et al. 1986). Minor satellite exhalation would be expected within an alluvial basin adjacent to an uplifting area. Minor exhalation may account for the local reddening of quartzite at the Pronto mine, Elliot Lake district, as proposed by Heinrich (1958). The author has observed similar local reddening underground at another Elliot Lake mine, the Denison. Local exhalation probably supported the uranium-rich cyanobacterial mats which characterize Elliot Lake-Witwatersrand deposits, given their peculiar chemical composition (Hallbauer and Van Warmelo 1974; Willingham et al. 1985; Nagy and Mossman, this Vol.).

It has long been recognized that uraniferous conglomerate beds consistently overlie some type of unconformity, locally a disconformity (Roberts and Kransdorff 1938). Semantically, conglomeratic beds are as much of an "unconformity-type" deposit as the mineralogically distinct deposits which are associated with Middle Proterozoic unconformities (cf. Nash et al. 1981). Uranium oxide in the latter deposits typically is cryptocrystalline (pitchblende) whereas it typically occurs as crystalline uraninite within conglomerate. Gabelman (1988, p. 91) notes that this grain-size distinction is not readily attributable to differing temperatures of precipitation but may be related to the oxidation state of uraniferous fluids. He suggests that more oxidizing fluids transport uranyl ions (U^{6+}) along with similarly sized ions, including Fe^{3+}, whereas more reducing fluids transport uranous ions (U^{4+}) along with more Th^{4+}, REE, Ti^{2+}, and Fe^{2+}.

Genetic interpretation of uraniferous conglomerate is complicated by late-diagenetic and metamorphic alteration. Even well-preserved deposits like Elliot Lake and the Witwatersrand exhibit recrystallization and reaction of uraninite with titaniferous minerals to form brannerite (Ramdohr 1957; Theis 1976).

Phosphorus commonly is concentrated in uranium deposits (e.g., Little 1970, Table 1) and uranium generally is concentrated in phosphorite (De Voto and Stevens 1979). All gradations occur between phosphorite and uranium deposits (Castaño and Menicucci 1984; Angeiras 1988). The source of phosphorus in phosphorite generally is interpreted to be weathering of either the land or seafloor but exhalation is a more likely source of ore-forming marine phosphate, as noted by Kimberley (1989). Further work on an exhalative model for phosphorite would be relevant to stratabound uranium deposits and their associated paleosols.

3.9 Ratio of Potassium/Sodium in Uranium-Associated Paleosols

The most persistent peculiarity of Precambrian paleosols is a high ratio of K/Na, as initially emphasized by Kimberley (1978 a) and now widely recognized (e.g., Sokolov and Heiskanen 1984). Abundant potassium is particularly characteristic of uranium-associated paleosols and remains problematic for genetic interpretation of paleosol compositions (Palmer et al. 1989 b).

Within the exhalative paleosol model, there could be dissolution of exhalative pyrite from a highland and subsequent groundwater precipitation of potassic sulfosalts like alunite, $KAl_3(SO_4)_2(OH)_6$, within an adjacent paleosol-accumulating basin (Kimberley 1978 a). During late diagenesis, alunite could react to form sericite, the dominant potassic mineral in all

Precambrian paleosols (Kimberley 1978 a). The hypothetical, potassium-rich groundwater could not have been very acidic by the time it reached the basin because of the preservation of feldspar within basinal sediment, as noted by Robinson and Spooner (1984). An alternative exhalative explanation of potassium enrichment would be direct input from a deep-source fluid, as interpreted for modern glauconite in coastal Venezuela (Kimberley 1989).

3.10 Atmospheric Evidence in Stratabound Uranium Deposits

Purely academic interest in stratabound uranium deposits largely has involved their potential as an indicator of the oxidation state of the contemporaneous atmosphere (e.g., Holland 1984). Reinterpretation of uraninite-bearing conglomerate as an exhalative-placer deposit diminishes but does not eliminate its usefulness as an atmospheric indicator. Even if the distance of fluvial transport was on the order of 10^1 instead of 10^2 km, uraninite grains probably would have suffered oxidation if pO_2 had equalled the modern partial pressure (Grandstaff 1976, 1980).

The difference in pO_2 between modern and Late Archean-Early Proterozoic levels may have been as great as the couple of orders of magnitude recommended by Holland (1984) for basal Huronian sedimentation, but probably not a couple of tens of orders of magnitude (cf., Holland 1962). Sedimentological study has revealed that basal Huronian uraniferous conglomerate may have been virtually unweathered glacial outwash (Fralick and Miall 1987). The fact that similar but younger (1.9 Ga) conglomeratic deposits contain hematite and gold but no uraninite may well record greater oxidation during younger fluvial transport (Vogel 1987).

4 Paleosols, Karst, and Highly Aluminous Sedimentary Rocks

Aluminum and titanium generally are the least mobile of the major elements during weathering. Sediments with a great enrichment of aluminum, e.g., bauxite, usually are interpreted to record intensive weathering at or near the site of accumulation. The relationship of aluminum to weathering deserves some attention because many Precambrian rocks have been interpreted to record paleosol development largely on the basis of aluminium content (e.g., Vogel 1975; Schau and Henderson 1983; Reimer 1986, and this Vol.). Highly aluminous rocks constitute a remarkably large proportion of some of the oldest Archean sequences, e.g., the 3.5 Ga Holenarasipur schist belt of India. A description of these by Naqvi et al. (1983, p. 138) concludes "...it appears unlikely that high-Al sediments derived by the weathering of acid plutons would have been deposited in the absence of complementary quartzites of clastic origin."

The prime alternative to an atmospheric-weathering (biologic or abiologic) origin for highly aluminous paleosols is an exhalative-weathering origin, i.e., an intense augmentation of biologically induced weathering due to local exhalation of nutrients like phosphorus and/or an intense augmentation of acidic leaching due to exhalation of acid-forming volatiles which dissolve in groundwater. Exhalation almost certainly was a more important process during the Precambrian than it is today. Nonetheless, Paleozoic exhalation has produced aluminous hot-spring deposits in the southeastern USA, some with gold (Stuckey 1967; Feiss and Weslowski 1986).

Many geologists consider Phanerozoic bauxite to be the clearest example of biologic (nonexhalative) paleosol, including slightly transported biogenic soil (Valeton 1972). However, the evidence commonly is less clear than is popularly perceived. A summary of compositional data from numerous bauxite deposits reveals peculiarities which are difficult to reconcile without invoking an exhalative influence for some of these paleosols. On average, the element which is most concentrated in bauxite relative to the crustal mean is sulfur (5 times), followed by aluminum (4.05), vanadium (3.1), titanium (2.56), iron (2.38), copper (2.3), chromium (1.83), phosphorus (1.08), and zinc (1.05) (Beneslavsky 1963, quoted in Valeton 1972).

Subsequent work has added to the list of elements concentrated in bauxite, notably rare-earth elements (REE) and boron. The facies distribution of boron in extensive deposits reveals that boron cannot be attributed to post-depositional uptake from seawater (Petracheck 1989). Moreover, the stratigraphic profiles of REE abundance cannot always be attributed to physical inheritance from the underlying parent rock (Topp et al. 1984, 1986). Nonetheless, the great bulk of elemental concentrations in all bauxite deposits which are developed upon an observable igneous rock readily may be attributed to that rock (e.g., Schorin and Puchelt 1987). If exhalation has helped to produce bauxite, its role largely has involved the supply of reactants which have enhanced surficial weathering.

A major problem in some bauxite modeling is the common lack of a partially weathered parent, as for one of the youngest major deposits. About 1 Ma ago, voluminous bauxite accumulated in a sandy sequence along the northern edge of the Amazon rift (Aleva 1981). There is no known source underneath or adjacent to this bauxite. A related deposit (Paragominas) occurs in the southeastern corner of the Amazon rift (Grubb 1979). Bird et al. (1989, p. 1411) note "...The oxygen- and hydrogen-isotope composition of gibbsite from the Paragominas deposit in Brazil is markedly out of isotopic equilibrium with modern tradewind-derived regional rainfall...". Difficulty in isotopic interpretation of this Quaternary paleosol does not bode well for the attempted isotopic interpretation of Precambrian paleosols (Beeunas and Knauth 1985; Vahrenkamp and Rossinsky 1987).

A large proportion of bauxite fills karst caverns (Sinclair 1967; Comer 1974; Petrascheck 1989). Karst caverns commonly are attributed to exclusively nonexhalative weathering (e.g., Samana 1986). However, exhalation has contributed to such giant caverns as Carlsbad, New Mexico (Hill 1987, 1990). Like Carlsbad, the voluminous Guáchero cavern near Maturín, Venezuela lies updip from a major petroleum field which is leaking volatiles. Exhalative karst caverns abound in the southern USSR (Dublyanskiy 1980). The origin of the bauxite-filled karst caverns therefore warrants additional study, comparable to the ongoing study of Pb-Zn-filled karst caverns (Olson 1984).

5 Paleosols and Iron Formations

Virtually all models of Precambrian atmospheric conditions emphasize the existence of voluminous Precambrian iron formations. However, many of these papers misrepresent the characteristics and age distribution of iron formations. The most common misconceptions are that cherty iron formations represent a unique time in Earth history, the Early Proterozoic (Cloud 1973; Eriksson and Truswell 1978), that a warm tropical climate was essential for iron formations (Gruner 1922), and that oxidation induced precipitation of essentially all the iron (Kasting 1987). Cherty iron formations are actually widely scattered through Earth history, including several Late Proterozoic and a few Paleozoic examples (Kimberley 1989). Glaciation coincided with the Late Proterozoic cherty iron formations (Yeo 1984; Young 1988). Many cherty iron formations subsequently have become oxidized (Morris 1985) but all unweathered deposits contain sufficient ferrous iron to be out of chemical equilibrium with more than a few molecules of oxygen in the entire atmosphere (Kimberley 1989). Iron formations are not subaqueous "oxidized analogues" of red beds (cf. Towe 1983, p. 163) but may share some genetic processes with red beds (Kimberley 1989).

Although iron formations apparently do not require tropical or significantly oxidizing environments, iron formations provide some record of atmospheric conditions because a few were deposited in extremely shallow water (Chauvel and Dimroth 1974) and others became weathered in the Early Proterozoic shortly after sedimentation (Cannon 1981; Holland and Beukes 1989). It is possible that weathering and ferrous-iron transport occurred at different times and that these two processes record some temporal variability in atmospheric composition, on the scale of 10^7 years. However, even if the oxidation state varied somewhat, Holland (1984) appropriately concludes that it is difficult to reconcile the evidence of oxidative weathering with the concept of fluvial transport of ferrous iron to produce Proterozoic iron formations (cf. Lepp and Goldich 1964; Reimer 1987).

6 Conclusion

The pre-metazoan Precambrian record offers little evidence for the continuity of atmospheric composition. Precambrian life was dominated by prokaryotes which collectively display a phenomenal ability to survive in diverse environments. In contrast, the continuity of metazoan life from the Late Precambrian through the Phanerozoic has required a greater continuity of atmospheric composition.

Throughout Earth history, the ocean has provided a buffer for marine metazoans against rapid atmospheric change but the smaller mass of the atmosphere has limited is capacity to buffer exhalative or bolide-induced effects on terrestrial metazoans. The continuity of terrestrial metazoan life since the mid Paleozoic therefore has required an even more stringent continuity of atmospheric composition than during the Late Precambrian-Early Paleozoic time of exclusively marine metazoans. Nonetheless, the composition of the post-Paleozoic atmosphere has been far from constant, as evidenced by variation during just the past 10^5 years (e.g., Lorius et al. 1988).

With luck, Precambrian specialists eventually will learn that the foregoing suspicions about the abiologic paleosol model are unfounded and that semi-quantitative estimation of Precambrian atmospheric conditions really can be performed from paleosol compositions, as proposed by Holland (1984). Wherever the abiologic model is applicable, it is inappropriate to negate estimation of atmospheric conditions. Negative comments are common, however, e.g., Schau and Henderson (1983, p. 189), who conclude from a study of Archean paleosols, "Hence weathering profiles, old and new, are more a response to local environmental conditions within the rock than the composition of the atmosphere".

Comments like the foregoing are overly pessimistic if it is assumed that the Precambrian atmosphere was the sole source of both oxidants and acid-forming volatiles, and that biologic activity had no detectable effect on Precambrian soil formation. Additional work is needed to test both assumptions. The first is least likely to be valid for paleosols which are associated with stratabound ore deposits.

Acknowledgements. Like others who have had the privilege to study with Dick Holland, I derive much of my inspiration from him.

References

Aharon P, Roberts H, Sassen R, Wheeler C, Feng J (1989) An isotope study of cold vents deposits from the continental slope, Gulf of Mexico: progress report. Geol Soc Am Abstr (with Prog) 21, 6:A273

Aleva GJJ (1981) Essential differences between the bauxite deposits along the southern and northern edges of the Guiana Shield, South America. Econ Geol 76:1142–1152

Angeiras AG (1988) Geology and metallogeny of the northeastern Brazil uranium-phosphorus province emphasizing the Itataia deposit. Ore Geol Rev 3:211–225

Armstrong RL (1981) Radiogenic isotopes: the case for crustal recycling on a near-steady-state no-continent-growth Earth. Philos Trans R Soc London Ser A 301:443–472

Barbier J (1974) Continental weathering as a possible origin of vein type uranium deposits. Mineral Depos 9:271–288

Beeunas M, Knauth LP (1985) Preserved stable isotopic signature of subaerial diagenesis in the 1.2-b.y. Mescal Limestone, central Arizona: implications for the timing and development of a terrestrial plant cover. Geol Soc Am Bull 96:737–745

Beneslavsky SJ (1963) Minérologie des bauxites (critères d'évaluation de la quantité et des propriétés technologiques des minérais bauxitiques d'après leur composition). Gosgeolve Khizdat, Moscow, 170 pp (Translation #4706 of Bur Res Geol Min, Paris)

Berner RA, Raiswell R (1983) Burial of organic carbon and pyrite sulfur in sediments over Phanerozoic time: a new theory. Geochim Cosmochim Acta 47:855–862

Bird MI, Chivas AR, Andrew AS (1989) A stable-isotope study of lateritic bauxites. Geochim Cosmochim Acta 53:1411–1420

Birkeland PW (1974) Pedology, weathering, and geomorphological research. Oxford Univ Press, New York, 285 pp

Blake TS, Groves DI (1987) Continental rifting and the Archean-Proterozoic transition. Geology 15:229–232

Bohor BF, Triplehorn DM, Nichols DJ, Millard HT Jr (1987) Dinosaurs, spherules, and the "magic" layer: A new K-T boundary clay site in Wyoming. Geology 15:896–899

Brock TD (1978) Thermophilic microorganisms and life at high temperatures. Springer, Berlin Heidelberg New York, 465 pp

Budyko MI, Golitsyn GS, Izrael YA (1988) Global climatic catastrophes. Springer, Berlin Heidelberg New York, 99 pp

Buringh P (1979) Introduction to the study of soils in tropical and subtropical regions. 3rd edn. Centr Agric Publ Doc, Wageningen, Neth, 124 pp

Button A (1979) Early Proterozoic weathering profile on the 2200-Myr-old Hekpoort Basalt, Pretoria Group, South Africa: preliminary results. Econ Geol Res Unit, Univ Witwatersrand, Johannesburg, Inf Circ 133:20 pp

Button A, Tyler N (1981) The character and economic significance of Precambrian paleoweathering and erosion surfaces in southern Africa. In: Skinner BJ (ed) Econ Geol 75th Anniv Vol Econ Geol Publ, New Haven, pp 686–709

Campbell SE (1979) Soil stabilization by a prokaryotic desert crust: implications for Precambrian land biota. Orig Life 9:335–337

Cannon WF (1976) Hard iron ore of the Marquette range, Michigan. Econ Geol 71:1012–1028

Card KD, Innes DG, Debicki RL (1977) Stratigraphy, sedimentology, and petrology of the Huronian Supergroup in the Sudbury-Espanola area. Ontario Div Mines, Geosci Stud 16, 99 pp

Castaño R, Menicucci S (1984) Geologia y mineralización urano-fosfática en el sinclinal de Berlin, Caldas (Colombia). In: Anon (ed) Geología y metalogénesis de los depósitos y manifestaciones uraníferos de sudamérica. IAEA-STI-PUB-641:239–243

Chauvel J-J, Dimroth E (1974) Facies types and depositional environments of the Sokoman Iron Formation, central Labrador Trough, Quebec, Canada. J Sediment Petrol 44:299–327

Cloud PE (1973) Paleoecological significance of the banded iron-formations. Econ Geol 68:1135–1143

Comer JB (1974) Genesis of Jamaican bauxite. Econ Geol 69:1251–1264

Crocket JH, Officer CB, Wezel FC, Johnson GD (1988) Distribution of noble metals across the Cretaceous/Tertiary boundary at Gubbio, Italy: iridium variation as a contraint on the duration and nature of Cretaceous/Tertiary boundary events. Geology 16:77–80

Davidson GI, Gandhi SS (1989) Unconformity-related U-Au mineralization in the Middle Proterozoic Thelon Sandstone, Boomerang Lake prospect, Northwest Territories, Canada. Econ Geol 84:143–157

D'Elboux CV (1984) Principales modelos brasileños de mineralizaciones uraniferas. In: Anon (ed) Geología y metalogénesis de los depósitos y manifestaciones uraníferos de Sudamérica, IAEA-STI-PUB-641:143–176

Derry DR (1960) Evidence of the origin of the Blind River uranium deposits. Econ Geol 55:906–927

Des Marais DJ (1985) Carbon exchange between the mantle and the crust, and its effect upon the atmosphere: today compared to Archean time. In: Sundquist ET, Broecker WS (eds) The carbon cycle and atmospheric CO_2: natural variations Archean to present. Am Geophys Un Geophys Monogr 32:602–627

De Voto RH, Stevens DN (1979) Uraniferous phosphate resources, United States and Free World, Vol. 1 of Uraniferous phosphate resources and technology and economics of uranium recovery from phosphate resources, United States and Free World. US Dep Energ, Grand Junction, Col, Rep GJBX-110, 724 pp

Dimroth E, Kimberley MM (1976) Precambrian atmospheric oxygen: evidence in the sedimentary distribu-

tions of carbon, sulfur, uranium, and iron. Can J Earth Sci 13:1161–1185
Dublyanskiy VN (1980) Hydrothermal karst in Alpine folded belt of southern part of USSR. Univ Katowice, Pol, Kras Speleol 3, 12:18–36
Eriksson KA, Truswell JF (1978) Geologic processes and atmospheric evolution in the Precambrian. In: Tarling DH (ed) Evolution of the Earth's crust. Academic Press, New York London, pp 219–238
Farrow CEG, Mossman DJ (1988) Geology of Precambrian paleosols at the base of the Huronian Supergroup, Elliot Lake, Ontario, Canada. Precambrian Res 42:107–139
Feiss G, Weslowski D (1987) Oxygen isotopic evidence for the interaction of meteoric-hydrothermal waters with granites and subaerial volcanics in pyrophyllite/Au deposits of the Carolina Slate Belt. Geol Soc Am Abstr (with Prog) 18:600
Fralick PW, Miall AD (1987) Glacial outwash uranium placers? Evidence from the lower Huronian Supergroup, Ontario, Canada. In: Pretorius D (ed) Uranium deposits in Proterozoic quartz-pebble conglomerates. IAEA-TECDOC-427:133–154
Frey M (1987) Very low-grade metamorphism of clastic sedimentary rocks. In: Frey M (ed) Low temperature metamorphism. Chapman & Hall, New York, pp 9–58
Fyfe WS (1978) Evolution of the Earth's crust: modern plate tectonics to ancient hot spot tectonics? Chem Geol 23:89–114
Fyfe WS, Price NJ, Thompson AB (1978) Fluids in the Earth's crust. Elsevier, Amsterdam, 383 pp
Gabelman JW (1988) Representative uranium deposits and occurrence exclusive of sandstone impregnations and quartz conglomerates. Ore Geol Rev 3:31–159
Garrels RM, Mackenzie FT (1971) Evolution of sedimentary rocks. Norton, New York, 397 pp
Gay AL, Grandstaff DE (1980) Chemistry and mineralogy of Precambrian paleosols at Elliot Lake, Ontario, Canada. Precambrian Res 12:349–373
Gerrard AJ (1981) Soils and landforms. Allen & Unwin, London, 219 pp
Goodwin AM (1981) Precambrian perspectives. Science 213:55–61
Grandstaff DE (1976) A kinetic study of the dissolution of uraninite. Econ Geol 71:1493–1506
Grandstaff DE (1980) Origin of uraniferous conglomerates at Elliot Lake, Canada and Witwatersrand, South Africa: implications for oxygen in the Precambrian atmosphere. Precambrian Res 13:1–26
Grubb PLC (1979) Genesis of bauxite deposits in the Lower Amazon Basin and Guianas coastal plain. Econ Geol 74:735–750
Gruner JW (1922) The origin of sedimentary iron formations: the Biwabik Formation of the Mesabi Range. Econ Geol 17:407–460
Hallbauer DK, Van Warmelo KT (1974) Fossilized plants in thucholite from Precambrian rocks of the Witwatersrand, South Africa. Precambrian Res 1:199–212
Harte B, Dempster TJ (1987) Regional metamorphic zones: tectonic controls. Philos Trans R Soc London Ser A 321:105–127
Heinrich EW (1958) Mineralogy and geology of radioactive raw materials. McGraw-Hill, New York, 654 pp
Hill CA (1987) Geology of Carlsbad cavern and other caves in the Guadalupe Mountains, New Mexico and Texas. New Mex Bur Mines Mineral Resourc, Socorro, Bull 117:150 pp
Hill CA (1990) Sulfuric acid speleogenesis of Carlsbad cavern and its relationship to hydrocarbons, Delaware Basin, New Mexico and Texas. Am Assoc Petroleum Geol Bull 74:1685–1694
Hoeve J, Quirt D (1984) Mineralization and host rock alteration in relation to clay mineral diagenesis and evolution of the Middle Proterozoic Athabasca Basin, northern Saskatchewan, Canada. Saskatchewan Res Counc Publ R-855-2-A-84
Hoeve J, Sibbald TII (1978) On the genesis of Rabbit Lake and other unconformity-type uranium deposits. Econ Geol 73:1450–1473
Holland HD (1962) Model for the evolution of the Earth's atmosphere. In: Engel AEJ, James HL, Leonard BF (eds) Pretrologic studies: a volume to honor A. F. Buddington. Geol Soc Am, New York, pp 447–477
Holland HD (1984) The chemical evolution of the atmosphere and oceans. Univ Press, Princeton, NJ, 582 pp
Holland HD, Beukes NJ (1989) Evidence for a dramatic rise in atmospheric oxygen between 2.2 and 1.9 b.y.b.p. Geol Soc Am Abstr (with Prog) 21, 6:A 24
Holland HD, Feakes CR (1989) Paleosols and their relevance to Precambrian atmospheric composition: a discussion. J Geol 97:761–762
Holland HD, Zbinden EA (1988) Paleosols and the evolution of the atmosphere, pt 1. In: Lerman A, Meybeck M (eds) Physical and chemical weathering in geochemical cycles. Reidel, Dordrecht, pp 61–82
Holmes SW (1957) Pronto Mine. In: Structural geology of Canadian ore deposits, vol 2, 6th Commonw Min Metall Congr, pp 324–339
Jensen V, Kjoller A, Sorensen LH (eds) (1986) Microbial communities in soil. Elsevier, Amsterdam London, 447 pp
Jepson WB (1988) Structural iron in kaolinites and in associated ancillary minerals. In: Stucki JW, Goodman BA, Schwertmann U (eds) Iron in soils and clay minerals. Reidel, Dordrecht, pp 467–536
Kasting JF (1985) Photochemical consequences of enhanced CO_2 levels in Earth's early atmosphere. In: Sundquist ET, Broecker WS (eds) The carbon cycle and atmospheric CO_2: Natural variations Archean to present. Am Geophys Un Geophys Monogr 32:612–622
Kasting JF (1987) Theoretical constraints on oxygen and carbon dioxide concentrations in the Precambrian atmosphere. Precambrian Res 34:205–229
Kerrich R (1989) Archean gold: relation to granulite formation or felsic intrusion? Geology 17:1011–1015
Kimberley MM (1978a) Origin of stratiform uranium deposits in sandstone, conglomerate, and pyroclastic rock. In: Kimberley MM (ed) Uranium deposits: their mineralogy and origin. Mineral Assoc Can Short Course Handb 3:339–381
Kimberley MM (1978b) High-emperature uranium geochemistry. In: Kimberley MM (ed) Uranium deposits: their mineralogy and origin. Mineral Assoc Can Short Course Handb 3:101–104
Kimberley MM (1989) Exhalative origins of iron formations. Ore Geol Rev 5:13–145
Kimberley MM, Llano M (in press) Structural lineaments in the Margarita-Araya region of Venezuela: boundaries of ecologic environments. Soc Cienc Nat La Salle, Venezuela, Mem
Kimberley MM, Tanaka RT, Farr MR (1980) Composition of Middle Precambrian uraniferous conglomerate in the Elliot Lake-Agnew Lake area of Canada. Precambrian Res 12:375–392

Kimberley MM, Niederreither MS, Llano M (in press) Effect of marine currents, tides, wind, and Pleistocene sea-level changes on sedimentation in the Margarita-Araya region of Venezuela. Soc Cienc Nat La Salle, Venezuela, Mem

Köppel VH, Saager R (1974) Lead isotope evidence on the detrital origin of Witwatersrand pyrites and its bearing on the provenance of the Witwatersrand gold. Econ Geol 69:318–331

Langford FF (1983) Proterozoic uranium deposits and the Precambrian atmosphere. In: Medaris LG Jr, Byers CW, Mickelson DM, Shanks WC (eds) Proterozoic geology: selected papers from an international Proterozoic symposium. Geol Soc Am Mem 161:237–243

Langmuir D (1978) Uranium solution-mineral equilibria at low temperatures with applications to sedimentary ore deposits. Geochim Cosmochim Acta 42:547–569

Lasaga AC, Berner RA, Garrels RM (1985) An improved geochemical model of atmospheric CO_2 fluctuations over the past 100 million years. In: Sundquist ET, Broecker WS (eds) The carbon cycle and atmospheric CO_2: Natural variations Archean to present. Am Geophys Un Geophys Monogr 32:397–411

Lepp H, Goldich SS (1964) Origin of Precambrian iron formations. Econ Geol 59:1025–1060

Little HW (1970) Distribution of types of uranium deposits and favourable environments for uranium exploration. In: Anon (ed) Uranium exploration geology. IAEA-PL-391:35–48

Lorius C, Barkov NI, Jouzel J, Korotkevich YS, Kotlyakov VM, Raynaud D (1988) Antarctic ice core: CO_2 and climatic change over the last climate cycle. Eos Trans Am Geophys Un 69:681–684

Ludwig LR, Grauch RI, Nutt CJ, Nash JT, Freishman D, Simmons KR (1987) Age of uranium mineralization at the Jabiluka and Ranger deposits, Northern Territory, Australia: new U-Pb isotopic evidence. Econ Geol 82:857–874

Lynch SA (1954) Geology of the Gulf of Mexico. In: Galtsoff PS (ed) Gulf of Mexico: its origin, waters, and marine life. US Fish Wildlife Serv Fish Bull 89, 55:67–86

Macdonald CC (1980) Mineralogy and geochemistry of a Precambrian regolith in the Athabasca Basin. Master's Thesis, Univ Saskatchewan, Saskatoon, Can

Macdougall JD (1988) Seawater strontium isotopes, acid rain, and the Cretaceous-Tertiary boundary. Science 239:485–487

McCaig AM (1988) Deep fluid circulation in fault zones. Geology 16:867–870

McLennan SM, Taylor SR (1982) Geochemical constraints on the growth of the continental crust. J Geol 90:347–361

Meddaugh WS, Holland HD (1981) Age and origin of uraninite in the Elliot Lake, Ontario, uranium deposits. Geol Soc Am Abstr (with Prog) 13:509

Miller AR, Cumming GL, Krstic D (1989) U-Pb, Pb-Pb, and K-Ar isotopic study and petrography of uraniferous phosphate-bearing rocks in the Thelon Formation, Dubawnt Group. Northwest Territories, Canada. Can J Earth Sci 26:867–880

Minter WEL, Feather CE, Glatthaar CW (1988) Sedimentological and mineralogical aspects of the newly discovered Witwatersrand placer deposit that reflect Proterozoic weathering, Welkon Gold Field, South Africa. Econ Geol 83:481–491

Moore JC, Mascle A, Taylor E, Andreieff P, Alvarez F, et al. (1988) Tectonics and hydrology of the northern Barbados Ridge: results from Ocean Drilling Program Leg 110. Geol Soc Am Bull 100:1578–1593

Morris RC (1985) Genesis of iron ore in banded iron-formation by supergene and supergene-metamorphic processes – a conceptual model. In: Wolf KH (ed) Handbook of strata-bound and stratiform ore deposits, vol 13. Elsevier, Amsterdam, pp 73–235

Naqvi SM, Condie KC, Allen P (1983) Geochemistry of some unusual Early Archean sediments from Dharwar Craton, India. Precambrian Res 22:125–147

Nash JT, Granger HC, Adams SS (1981) Geology and concepts of genesis of important types of uranium deposits. In: Skinner BJ (ed) Econ Geol 75th Anniv Vol, New Haven, pp 63–116

Nesbitt HW, Young GM (1982) Early Proterozoic climates and plate motions inferred from major element chemistry of lutites. Nature (London) 299:715–717

Newman MJ, Rood RT (1977) Implications of solar evolution for the Earth's early atmosphere. Science 198:1035–1037

Nicolaysen LO, Burger AJ, Liebenberg WR (1962) Evidence for the extreme age of certain minerals from the Dominion Reef conglomerates and the underlying granite in the Western Transvaal. Geochim Cosmochim Acta 26:15–23

Olson RA (1984) Genesis of paleokarst and stratabound zinc-lead sulfide deposits in a Proterozoic dolostone, northern Baffin Island, Canada. Econ Geol 79:1056–1103

Palmer JA, Phillips GN, McCarthy TS (1989a) Paleosols and their relevance to Precambrian atmospheric composition. J Geol 97:77–92

Palmer JA, Phillips GN, McCarthy TS (1989b) Paleosols and their relevance to Precambrian atmospheric composition: a reply. J Geol 97:764–765

Pavich MJ, Obermeier SF (1985) Saprolite formation beneath coastal plain sediments near Washington, D.C. Geol Soc Am Bull 96:886–900

Petit JR, Mounier L, Jouzel J, Korotkevich YS, Kotlyakov VI, Lorius C (1990) Palaeoclimatological and chronological implications of the Vostok core dust record. Nature (London) 343:56–58

Petrascheck WE (1989) The genesis of allochthonous karst-type bauxite deposits of southern Europe. Mineral Depos 24:77–81

Phillips GN (1988) Widespread fluid infiltration during metamorphism of the Witwatersrand goldfields: generation of chloritoid and pyrophyllite. J Metamorph Geol 6:311–332

Pinto JP, Holland HD (1988) Paleosols and the evolution of the atmosphere, pt 2. In: Reinhardt J and Sigleo WR (eds) Paleosols and weathering through geologic time: principles and applications. Geol Soc Am Spec Pap 216:21–34

Pretorius DA (1975) The depositional environment of the Witwatersrand goldfields: a chronological review of speculations and observations. Mineral Sci Eng 7:18–46

Pretorius DA (1981) Gold and uranium in quartz-pebble conglomerates. In: Skinner BJ (ed) Econ Geol 75th Anniv Vol, New Haven, pp 117–138

Pretorius DA (1989) New concepts of the age, tectonic setting, and mineralization of the Witwatersrand Basin. Geol Soc Am Abstr (with Prog) 21, 6:A131

Prinn RG, Fegley B Jr (1987) Bolide impacts, acid rain, and biospheric traumas at the Cretaceous-Tertiary boundary. Earth Planet Sci Lett 83:1–15

Ramdohr P (1957) Die "Pronto-Reaktion". N Jahrb Mineral Monatsh 1957:217–222
Reimer TO (1984) Alternative model for the derivation of gold in the Witwatersrand Supergroup. J Geol Soc London 141:263–272
Reimer TO (1985) Volcanic rocks and weathering in the Early Proterozoic Witwatersrand Supergroup, South Africa. Finland Geol Surv Bull 331:33–49
Reimer TO (1986) Alumina-rich rocks from the Early Precambrian of the Kaapvaal Craton as indicators of paleosols and as products of other decompositional reactions. Precambrian Res 32:155–179
Reimer TO (1987) Weathering as a source of iron in iron formations: the significance of alumina-enriched paleosols from the Proterozoic of southern Africa. In: Appel PWU, La Berge GL (eds) Precambrian iron formations. Theophrastus, Athens, pp 601–619
Retallack GJ (1989) Paleosols and their relevance to Precambrian atmospheric composition: a discussion. J Geol 97:763–764
Retallack GJ, Feakes CR (1987) Trace fossil evidence for Late Ordovician animals on land. Science 235:61–63
Reymer A, Schubert G (1984) Phanerozoic addition rates to the continental crust and crustal growth. Tectonics 3:63–78
Rich RA, Holland HD, Petersen U (1977) Hydrothermal uranium deposits. Elsevier, Amsterdam, 264 pp
Richardson KA (1964) Thorium, uranium and potassium in the Conway granite, New Hampshire, U.S.A. In: Adams JAS, Lowder WM (eds) The natural radiation environment. Univ Press, Chicago, pp 39–50
Robb LJ, Meyer M (1985) The nature of the Witwatersrand hinterland: conjectures on the source-area problem. Univ Witwatersrand, Johannesburg, Econ Geol Res Unit Inf Circ 178, 25 p
Robb LJ, Meyer M (1986) The nature of the Archean basement in the hinterland of the Witwatersrand Basin: I – The Rand anticline between Randfontein and Rysmierbult. Univ Witwatersrand, Johannesburg, Econ Geol Res Unit Inf Circ 187:14 pp
Robb LJ, Meyer M, Ferraz MF (1986) Palaeoregolith profiles in the granitic basement of the Witwatersrand hinterland. In: Geocongress '86, Univ Witwatersrand, Johannesburg, Extended Abstr Vol, pp 193–196
Roberts ER, Kransdorff D (1938) The upper Witwatersrand system at Randfontein Estates. Trans Geol Soc S Afr 41:225–247
Robertson DS (1962) Thorium and uranium variations in the Blind River ores. Econ Geol 57:1175–1184
Robertson DS (1974) Basal Proterozoic units as fossil time markers and their use in uranium prospection. In: Anon (ed) Formation of uranium ore deposits. IAEA-SM-183:495–512
Robertson DS, Tilsley JE, Hoog GM (1978) The time-bound character of uranium deposits. Econ Geol 73:1409–1419
Robinson A, Spooner ETC (1984) Postdepositional modification of uraninite-bearing quartz-pebble conglomerates from the Quirke Ore Zone, Elliot Lake, Ontario. Econ Geol 79:297–321
Ronov AB (1988) Common evolutionary trends in the chemical composition of sedimentary and igneous rocks of the Earth's crust. Terra Cognita 8:212
Roscoe SM (1969) Huronian rocks and uraniferous conglomerates. Can Geol Surv Pap 68-40:205 pp
Ross GM, Charenzelli JR (1985) Paleoclimatic significance of widespread Proterozoic silcretes in the Bear and Churchill provinces of the northwestern Canadian Shield. J Sediment Petrol 55:196–204
Samana J-C (1986) Ore fields and continental weathering. Van Nostrand Reinhold, New York, 326 pp
Schau M, Henderson JB (1983) Archean chemical weathering at three localities on the Canadian Shield. Precambrian Res 20:189–224
Schorin H, Puchelt H (1987) Geochemistry of a ferruginous bauxite profile from southeastern Venezuela. Chem Geol 64:127–142
Shackleton NJ, Pisias NG (1985) Atmospheric carbon dioxide, orbital forcing, and climate. In: Sundquist ET, Broecker WS (eds) The carbon cycle and atmospheric CO_2: Natural variations Archean to present. Am Geophys Un Geophys Monogr 32:303–317
Shnyukov EF, Sobolevsku YuV, Gnamenko GI, Naumenko PI, Kumnu VA (1986) Mud volcanoes in the Kerch-Taman region. Kiev Nauk Dumka, 152 pp (in Russian)
Sibson RH (1987) Earthquake rupturing as a mineralizing agent in hydrothermal systems. Geology 15:701–704
Sibson RH, Moore JMcM, Rankin AH (1975) Seismic pumping – a hydrothermal fluid transport mechanism. J Geol Soc London 131:653–659
Sibson RH, Robert F, Poulsen KH (1988) High-angle reverse faults, fluid-pressure cycling, and mesothermal gold-quartz deposits. Geology 16:551–555
Sinclair IGL (1967) Bauxite genesis in Jamaica: new evidence from trace element distribution. Econ Geol 67:482–486
Sokolov VA, Heiskanen KI (1984) Developmental stages of Precambrian crusts of weathering. In: Proc 27th Int Geol Congr, Moscow, 1984, vol 5, Precambrian geology. VNU, Utrecht, pp 73–94
Stetter KO (1986) Diversity of extremely thermophilic archaebacteria. In: Brock TD (ed) Thermophiles – general, molecular, and applied microbiology. Wiley-Interscience, New York, pp 39–74
Stuckey JL (1967) Pyrophyllite deposits of North Carolina. N C Div Min Res Bull 80:38 pp
Stumm W, Morgan JJ (1981) Aquatic chemistry, 2nd edn. Wiley-Interscience, New York, 780 pp
Sylvester AG (1988) Strike-slip faults. Geol Soc Am Bull 100:1666–1703
Theis NJ (1976) Uranium-bearing and associated minerals in their geochemical and sedimentological context, Elliot Lake, Ontario. Ph D Thesis, Queen's Univ, Kingston, Can, 158 pp
Theis NJ (1978) Mineralogy and setting of Elliot Lake deposits. In: Kimberley MM (ed) Uranium deposits: their mineralogy and origin. Mineral Assoc Can, Short Course 3:331–338
Topp SE, Salbu B, Roaldset E, Jørgensen P (1984) Vertical distribution of trace elements in laterite soil (Suriname). Chem Geol 47:159–174
Topp SE, Salbu B, Roaldset E, Jørgensen P (1986) Vertical distribution of trace elements in laterite soil (Suriname) – reply. Chem Geol 56:161–163
Towe KM (1983) Precambrian atmospheric oxygen and banded iron formations: a delayed model. Precambrian Res 20:161–170
Turcotte DL (1980) On the thermal evolution of the Earth. Earth Planet Sci Lett 48:53–58
Vahrenkamp VC, Rossinsky V Jr (1987) Discussion of "Preserved stable isotopic signature of subaerial diagenesis in the Mescal Limestone, central Arizona: Implications for the timing of a terrestrial plant cover". Geol Soc Am Bull 99:595–596

Valeton I (1972) Bauxites. Elsevier, Amsterdam, 226 pp

Vann IR (1985) To stretch a continent. Nature (London) 316:293–294

Veizer J, Jansen SL (1979) Basement and sedimentary recycling and continental evolution. J Geol 87:341–370

Vogel DE (1975) Precambrian weathering in acid metavolcanic rocks from the Superior Province, Villebon Township, south-central Québec. Can J Earth Sci 12:2080–2085

Vogel W (1987) The mineralized quartz-pebble conglomerates of Ghana. In: Pretorius D (ed) Uranium deposits in Proterozoic quartz-pebble conglomerates. IAEA-TECDOC-427, pp 235–254

Voultsidis V, Von Pechmann E, Clasen D (1982) Petrography, mineralogy and genesis of the U-Ni deposits, Key Lake, Sask., Canada. In: Amstutz GC, Goresy AE, Frenzel G, Kluth C, Moh G, Wauschkuhn A, Zimmermann RA (eds) Ore genesis, the state of the art. Springer, Berlin Heidelberg New York, pp 469–490

Walker JCG, Klein C, Schidlowski M, Schopf JW, Stevenson DJ, Walter MR (1983) Environmental evolution of the Archean-Early Proterozoic Earth. In: Schopf JW (ed) Earth's earliest biosphere, its origin and evolution. Univ Press, Princeton, 543 pp

Walker N (ed) (1975) Soil microbiology. John Wiley & Sons, New York, 262 pp

Whitney JA (1988) The origin of granite: the role and source of water in the evolution of granitic magmas. Geol Soc Am Bull 100:1886–1897

Williams H, McBirney AR (1979) Volcanology. Freeman, San Francisco, 397 pp

Willingham TO, Nagy B, Nagy AL, Krinsley DH, Mossmann DJ (1985) Uranium-bearing stratiform organic matter in paleoplacers of the lower Huronian Supergroup, Elliot Lake-Blind River region, Canada. Can J Earth Sci 22:1930–1944

Yakontova LK, Andreeva NYa, Cipurskij SI, Naumenko PI (1985) New data on the mineralogy and formation conditions of the Kerch iron ore. Min Zh (Kiev Nauk Dumka) 7:29–43 (in Russian)

Yeo GM (1984) The Rapitan Group: relevance to the global association of Late Proterozoic glaciation and iron formation. Ph D Thesis, Univ Western Ontario, London, Can, 603 pp

Young GM (1973) Tillites and aluminous quartzites as possible time markers for Middle Precambrian (Aphebian) rocks of North America. In: Young GM (ed) Huronian stratigraphy and sedimentation, Geol Assoc Can Spec Pap 12:97–127

Young GM (1988) Proterozoic plate tectonics, glaciation and iron formations. Sediment Geol 58:127–144

2. Organic Matter in Precambrian and Early Paleozoic Sediments

Mantle Carbon: Concentration, Mode of Occurrence, and Isotopic Composition

Peter Deines

Department of Geosciences, The Pennsylvania State University, University Park, PA 16802, USA

Abstract

The chemical forms in which carbon occurs in the mantle are numerous and their relative abundances highly variable. Carbon concentrations measured on mantle xenoliths and estimated for the average mantle vary between a few parts per million and a percent. In mantle samples the element has been observed in fluid inclusions, commonly in the form of CO_2, but also as the minor species CO, CH_4, and COS. In addition, mantle carbon occurs as graphite, diamond, carbonates, moissanite, and carbonaceous films on cracks in minerals, and it has been proposed that the element is present in solid solution in silicates. The carbon isotopic composition of diamonds from kimberlites has a wide range from about 0 to -30‰ versus PDB which is comparable to that of the major reduced carbon phases in meteorites. The diamond $\delta^{13}C$ frequency distribution has a major mode at -5.5‰ and probably a minor one between -15 and -19‰. The weighted mean of the distribution is -7‰. The $\delta^{13}C$ frequency distribution of carbonatite carbonates, a second major mantle carbon sample, shows very limited variability, is Gaussian, has a mean of -5.4‰, and is distinct from that of diamonds. In comparison, the weighted mean $\delta^{13}C$ value of meteoritic bulk carbon lies probably below -10‰. There is evidence that carbon reservoirs of differing isotopic compositions exist in the mantle and that on the average $\delta^{13}C$ values may be lower at greater mantle depth. Geochemical models which assume that the mantle can be represented by a reservoir that is uniform with respect to chemical form, concentration, and isotopic composition of carbon are bound to be unrealistic. If a limited variability of these parameters is assumed in mantle models, it will be necessary to test how sensitive any derived conclusions are to deviations from the assumed values.

1 Introduction

The geochemistry and isotopic composition of carbon in the deeper parts of the Earth have attracted interest because the element may be determinative for mantle properties and processes, as well as for the formation of specific rock types and the stability of specific minerals. Knowledge about the ^{13}C-distribution in the mantle also provides crucial constraints on the carbon budget of the Earth and the global geochemical cycle of carbon.

Mantle properties which may in part depend on the geochemistry of carbon include the mantle's oxidation state and electrical conductivity. It has been suggested also that the oxygen fugacity in the mantle is spatially variable and that carbon species must be considered as potentially important redox-controlling reactants. In order to assess the magnitude of the mantle's oxygen fugacity and over what volume and what time span differences in redox state might persist, it is important to establish the chemical forms in which carbon resides.

Mantle processes such as melting, magma movement, subduction, and convection may be in part elucidated through a detailed understanding of the mantle carbon geochemistry. High pressure and temperature experiments have established the important influence of CO_2- or CH_4-containing fluids on the beginning of melting as well as on the composition of the partial melts formed. Because photosynthesized carbon compounds are depleted in ^{13}C, the isotope label constitutes a geochemical indicator through which the movement of some surface carbon into the mantle might be traced. The degree of isotopic heterogeneity of the mantle reflects the scale on which various forms of carbon have been homogenized and hence measures to some extent the intensity of mantle mixing processes.

The parameters which are significant for the complete understanding of the geochemistry of mantle carbon, i.e. its concentration, mode of occurrence, and isotopic composition are the subject of this chapter.

M. Schidlowski et al. (Eds.)
Early Organic Evolution: Implications for Mineral and Energy Resources

2 Carbon Concentration in the Mantle

The concentration and distribution of carbon in the mantle have been elusive properties of establish. There are numerous reasons for this, the most prominent of which is the difficulty of obtaining undisturbed, representative samples from the mantle. Xenoliths which are found in igneous and pyroclastic rocks, and for which depths of origin appropriate for the mantle can be rationalized on the basis of their mineralogy and composition, are considered to be samples of the mantle. Such xenoliths are most frequently found in alkali basalts and kimberlites. Because these volcanic rocks are by no means uniformly distributed over the surface of the Earth and because only a limited mantle depth range is sampled by volcanic processes, this type of mantle sample is biased.

The interpretation of the carbon content of xenoliths is complicated by additional factors. When or how the carbon concentration in a particular sample was established within the mantle, or how it might have been affected during transit to the Earth's surface are not known. One has to consider also that the carbon content of the mantle may have changed over time.

Finally, because carbon is a trace element, there are major sources of possible contamination both in the fiel and in the laboratory. Hence, considerable precautions are necessary to establish correct concentrations and isotopic compositions.

The most frequently observed type of mantle xenoliths are peridotite nodules, and to date, detailed studies of mantle carbon have been focussed mainly on them. While the early studies (Hoefs 1965; Freund et al. 1980) indicated that the total carbon content of these rocks might be in excess of 200 ppm, later work (Mathez et al. 1984a, b; Pineau and Mathez 1990; Nadeau et al. 1990) found significantly lower concentrations. Recent data for xenoliths from alkali basalts indicate that for many of them the total carbon content is below 100 ppm and that concentrations above 200 ppm are rare. A very significant fraction (up to 75%) of the total carbon can be combusted below 800 °C (Nadeau et al. 1990). The source, nature, and geochemical significance of this low-temperature-released carbon is a matter of active discussion. The data also suggest that websterites and wehrlites may have slightly higher carbon contents than other ultramafic rocks and that clinopyroxenes contain more carbon than coexisting olivine. Whether small differences in carbon content exist between xenoliths of the same rock type from different volcanoes remains to be established.

Lherzolite xenoliths from kimberlites studied by Mathez et al. (1984a) have slightly higher carbon contents (89–766 ppm) than do xenoliths from basalts. This could indicate that the deeper portions of the mantle, sampled by kimberlites, are richer in carbon. Mathez et al. (1984a) suggest, however, that some of the carbon analyzed in these xenoliths is introduced from the host and that the concentrations of carbon in xenoliths from kimberlites are not representative of mantle abundances.

While in the majority of the peridotites studied carbon occurs in dispersed form and as fluid inclusions, the element has also been observed as grains of graphite and diamond (Sobolev et al. 1969; Dawson and Smith 1975; Eggler and McCallum 1975; Pokhilenko et al. 1977; Shee et al. 1983; O'Neill et al. 1986; Jaques et al. 1990). The number of specimens described is, however, relatively small. Because harzburgite should be the most common host of diamonds in the mantle, the paucity of diamondiferous harzburgites among kimberlite xenoliths is considered remarkable (Gurney 1989), and has led to the suggestion that these rocks disaggregate as they are removed from their mantle source.

The diamond concentration in the Argyle peridotite xenoliths reaches 1.5 ppm C (O'Neill et al. 1986); Jaques et al. (1990) report a range of 1 to 77 ppm, with a mean of about 4.5 ppm. The garnet lherzolites from Finsch (Shee et al. 1982) have diamond abundances which correspond to 10 ppm and 60 ppm C, respectively. The diamond grade of these nodules is 50 to 300 times that of the kimberlite in which the nodules are found.

Generally, the study of ultramafic xenoliths from alkali basalts and kimberlites leads to the conclusion that the carbon content of mantle peridotites averages below 100 ppm. The chemical nature of these rocks indicates that they have undergone fractionation processes within the mantle so that the carbon contained in them must be considered residual. At what stage of the Earth's history this residuum was established for the individual xenoliths is unknown, as is the representativeness of these samples for the mantle as a whole.

Very much higher carbon contents, in the form of graphite, were established for ultramafic xenoliths (harzburgites and pyroxenites) from the alkali basalts of Tissemt in the Eggéré volcanic district of Algeria (Pineau et al. 1987). These xenoliths, which have graphite concentrations between 400 and 10000 ppm, are thought to be derived from a Precambrian layered intrusion, the parent magma of which was generated within the mantle and emplaced in the deep crust. The authors present arguments, based on the isotopic composition of the graphite in these samples, that the measured carbon content represents a minimum value, either for the original source of the magma (degassing has occurred) or for the original carbon content of the xenoliths (oxidation during residence in the alkali basalt magma). This line of reasoning suggests that portions of the undegassed mantle have carbon concentrations in excess of 10000 ppm.

Nodules of eclogite are another important type of mantle xenolith recovered from kimberlites. However, their carbon content has rarely been determined. Mathez et al. (1984a) analyzed four samples from the Roberts Victor kimberlite and measured carbon contents in the range from 35 to 258 ppm.

In contrast to the few diamond peridotite xenoliths found to date – in spite of extensive searches – many examples of graphite- and diamond-bearing mantle eclogites have been described (Robinson 1979; Robinson et al. 1984). Gurney (1989) has provided a summary of the currently documented occurrences of diamondiferous eclogites. Their relative abundance among eclogite xenoliths may be fairly high in a particular kimberlite. For example, in the Orapa diatreme, diamondiferous samples constitute about 0.7% of the total eclogite xenolith population (Gurney 1989).

The diamond content of these xenoliths is highly variable, although their diamond grade generally exceeds that of the host kimberlite by a factor of 3. In some xenoliths, which can be expected to represent larger mantle volumes, diamond concentrations as high as 3800 ppm have been determined. Even higher diamond abundances (a few percent of the specimen) occur in some small xenoliths; these, however, are thought to be less representative (Robinson 1979).

It is interesting to note that at Orapa, Botswana, only part of the diamonds in the kimberlite concentrate can be the product of a break-up of eclogite nodules. This deduction is based on a comparison of diamonds from the kimberlite concentrate with those observed in the eclogites and on the compositional differences between the eclogite-xenoliths' garnets and clinopyroxenes and the mineral inclusions in diamonds (Robinson et al. 1984). The same conclusion was reached independently by Deines (1984) through $\delta^{13}C$ measurements on diamonds from the eclogite xenoliths and diamonds occurring free in the kimberlite. Hence, in addition to the eclogitic material, there must be other diamond (and by implication carbon) sources beneath the Orapa kimberlite. The study of eclogite xenoliths leads to the conclusion that there are mantle regions in which the carbon concentrations are significantly greater than those indicated by peridotite xenoliths.

The observed relative abundances of diamondiferous peridotite and eclogite xenoliths has uncovered a paradox in mantle carbon geochemistry (Gurney 1989). While the peridotitic mineral association is the most frequently observed paragenesis among mantle xenoliths, and among mineral inclusions in diamonds, diamondiferous peridotite xenoliths (the expected diamond source rock) are rarely found. On the other hand, the relative abundances of eclogitic mantle nodules and diamonds with eclogitic inclusions are generally lower than those of xenoliths and diamonds of perioditic affinity, yet among mantle eclogites diamondiferous specimens are relatively common.

At this time the analytical data are far too limited to serve as a firm basis for accurate deductions about the past or present concentration and distribution of carbon in the mantle. It is clear, however, that the carbon content of the mantle is quite variable. In the absence of a consensus on the average mantle carbon concentration based on measurements of actual mantle samples, various geochemical models have been used to estimate it. Such estimates range from a few hundred to 10000 ppm (e.g., Javoy et al. 1982; Des Marais 1985; Marty and Jambon 1987).

3 Form of Carbon in Mantle Xenoliths

The interpretation of the carbon abundance and distribution in mantle xenoliths is complicated by the multitude of chemical and morphological forms in which the element occurs. Since Roedder (1965) pointed out that mantle minerals contain primary CO_2 inclusions, this form of mantle carbon has received considerable attention. The most common fluid composition reported from ultramafic xenoliths is pure CO_2 (e.g., Bilal and Touret 1976; Murck et al. 1978; Touret and Bilal 1978; Bergman 1981, 1982).

Other carbon-containing fluid species have occasionally been detected. These include CH_4 in diamonds (Melton et al. 1972; Melton and Giardini 1974), and COS in fluid inclusions (Murck et al. 1978). Carbon monoxide was first reported by Bergman and Dubessy (1984) in fluid inclusions in a veined amphibole-bearing wehrlite. Micro-thermometric and Raman spectroscopic studies revealed that the fluid inclusions in the host wehrlite are nearly pure CO_2 (> 99%), whereas those in the vein contain from 8.5 to 12 mol% CO in CO_2. Hydrogen gas, CH_4, H_2S, N_2, and SO_2 were searched for but not observed.

Some of the inclusions – which are generally thought to reflect the composition of fluid trapped in the mantle – have a lining of basaltic glass indicating that the trapped fluid was associated with a melt (Murck et al. 1978). The opinion that inclusions may represent samples of phases existing within the mantle during formation of their host minerals contrasts with the viewpoint expressed by Green and Radcliffe (1975). These authors suggest that the presently observed fluid inclusions are formed secondarily by precipitation (exsolution) from the solid. They base their conclusion on the observation that the smallest fluid inclusions, many of them below optical resolution, are attached to crystal defects – induced by deformation and exsolution – and to grain boundaries. The authors propose that formation of the smallest bubbles precedes incorporation of the xenoliths into the host magma and occurs by solid-state precipitation on deformation and exsolution features. The bubbles subsequently concentrated along migrating grain boundaries during syntectonic recrystallization.

Micro-inclusions have been reported for cubic diamonds and diamond coats (Navon et al. 1988; Turner et al. 1990). They appear to be rich in H_2O, CO_3^{2-}, SiO_2, K_2O, CaO, and FeO, with a bulk composition resembling that of a potassic magma. Navon et al. (1988) suggest that the inclusions represent a volatile-rich (40%) fluid or melt in the upper mantle from which the diamond grew. Estimates of the water content of the fluid range from 50 to 90 mol%; the remainder is CO_2. The high water content of these fluids is surprising in view of the generally low concentration of water in the fluid inclusions of mantle silicates. A high chlorine and a correlated excess ^{40}Ar content were reported by Turner et al. (1990) for inclusions of this type.

Carbonates have rarely been encountered in mantle xenoliths. McGetchin and Besançon (1973) observed carbonate inclusions in mantle-derived chrome-pyrope garnets from kimberlite diatremes in northern Colorado. The compositions of the carbonate inclusions are highly variable; some are Mg- and Fe-rich, while others are almost pure $CaCO_3$. The presence of the carbonate inclusions was taken as evidence that a free carbonate phase existed in the mantle. This remained a speculation until experiments (Wyllie 1978; Eggler 1978; Wyllie et al. 1983) showed that a depths greater than about 50 km dolomite and magnesite are stable minerals.

Brucite-calcite intergrowths [$Mg(OH)_2/CaCO_3$] have been described for mantle-derived lherzolites and wehrlite from the kimberlite pipes of Kimberley, South Africa (Berg 1986). The parageneses are interpreted as a product of a dolomite decarbonation reaction in response to decompression during the rise of the kimberlite from below 100 km:

$$CaMg(CO_3)_2 = CaCO_3 + MgO + CO_2 .$$

Subsequently the periclase hydrates to form brucite.

Nadeau et al. (1990) reported that although interstitial carbonate was rarely observed in thin sections of the xenoliths which they investigated, CO_2 (5 to 24 ppm carbon) could be liberated from the samples with concentrated phosphoric acid. When and how these secondary carbonates were formed remains to be established.

Of the reduced carbon phases present in the mantle, graphite and diamond have been mentioned already. Mathez and his co-workers have demonstrated that reduced carbon may also occur in finely dispersed form in mantle xenoliths as intercalation and "organic" compounds. Mathez and Delaney (1981) observed that in a coarse granular chromite-bearing dunite carbon occurs in intergrain cracks, along grain boundaries and on the walls of fluid inclusions. The minerals also contain abundant vapor inclusions (CO_2). In peridotite xenoliths from alkali basalts most of the carbon exists as inclusions of CO_2-rich vapor, whereas condensed carbonaceous matter occurs in the form of discrete platy grains and as films on natural surfaces (Mathez et al. 1984a). The particles contain other elements, in particular chlorine, silicon, potassium, and nitrogen. The carbon films have been characterized by photo-electron spectroscopy and found to consist of a mixture of "organic" (alcohols, organic acids) and graphitic compounds unstable at elevated temperatures (Mathez 1987). Similar observations were made by Tingle et al. (1990) for single crystals of olivine from San Carlos, Arizona, and Hualalai Volcano, Hawaii.

Most cracks appear to have originated during the cooling of the host magma and the carbonaceous films are thought to form on them by condensation from hot volcanic gases. Because the crack surfaces are chemically active, the low-temperature compounds could form from chemically simpler high-temperature condensates through continued reaction during cooling (Mathez 1987; Pineau and Mathez 1990). According to this hypothesis, the bulk of the carbon present on cracks in mantle xenoliths would actually represent contaminants from the host lava and would hence carry no direct information on mantle carbon. Pineau and Mathez (1990), however, also describe other cracks in websterites which appear to have been partially annealed before the xenoliths were incorporated into the host magma. It would follow that the solid carbon which occurs in these megascopic inclusion zones would have been introduced into the xenoliths prior to their incorporation into the host magma, i.e. within the mantle.

In serpentinized olivines of kimberlites, graphite occurs as flakes, blades, spherules and finely disseminated grains and can reach concentrations as high as 6 to 9 wt%. Pasteris (1981) considers it most likely that such carbon was introduced as CH_4 under reducing conditions as part of the serpentinization process; hence, any relationship to the mantle is very indirect.

The question of whether carbon can occur in solid solution in mantle silicates has been debated. The possibility was first suggested following the observation of submicroscopic inclusions in mantle olivines (Green 1972, 1985; Green and Guegen 1983). Near the melting point of the host minerals carbon atoms are thought to occupy cation vacancies in the silicate lattice (Freund et al. 1980; Freund 1981). With decreasing temperature the atoms are believed to transfer onto interstitial sites where they are highly mobile (Oberheuser et al. 1983). It is thought possible that the carbon atoms could react with the lattice oxygen to form CO_2, or with co-dissolved hydrogen to yield a variety of hydrocarbons (Freund 1981).

Mathez et al. (1984 b), Tsong and Knipping (1986), and Tingle et al. (1988) have questioned the validity of the earlier results (see also Freund 1986a, b) and Mathez et al. (1987) observed that the carbon solubility in mantle olivines was very low (below the detection limit of 65 ppm). Hence, the more recent work casts doubt on the importance of solid solution in olivine for the storage of mantle carbon, although its occurrence remains a possibility (Nadeau et al. 1990).

Carbon may also occur in the mantle in the form of moissanite (SiC). The mineral has been identified in diamonds as an inclusion, as well as in mineral concentrates from kimberlites and lamproites (e.g., Jaques et al. 1989). In diamonds, moissanite has been observed to coexist with peridotitic pyroxenes (Moore et al. 1986; Otter and Gurney 1989), with eclogitic garnets, clinopyroxenes, and rutile (Jaques et al. 1989), and with garnets which display the solid solution of pyroxene indicative of an unusually high-pressure origin (Moore et al. 1986). Hence, moissanite is present in a wide range of mantle regimes.

The relative abundance of the various chemical forms of carbon in the mantle cannot be established with the data at hand. It can be noted, however, that the most reduced forms have been reported less frequently than CO_2 and elemental carbon. Because carbon occurs in the mantle in its most oxidized, as well as its most reduced, form, large local differences in the mantle oxidation state must exist. The available data (Virgo et al. 1988; Haggerty 1989; Mattioli et al. 1989; Canil et al. 1990; Luth et al. 1990) indicate that the oxygen fugacity may vary by at least four orders of magnitude and be as high as the Ni–NiO buffer and lower than the iron-wüstite buffer. Whether the mantle is zoned with regard to this property, and hence with respect to the chemical form in which carbon occurs (Haggerty and Tompkins 1983; Haggerty 1989; Wood et al. 1990), or whether the differences in oxygen fugacity simply reflect local physico-chemical properties of the mantle, is not yet clear. In either case, however, the variable oxygen fugacity indicates a multifaceted mantle carbon geochemistry.

4 The Isotopic Composition of Mantle Carbon

No definitive characterization of the ^{13}C-content of the various carbon phases described above for peridotite xenoliths has been carried out as yet; however, the work of Pineau and Mathez (1990) and Nadeau et al. (1990) has started to outline the complexity of the problem.

A few analyses of graphite and diamond from mantle xenoliths are available. Jaques et al. (1990) report $\delta^{13}C$ values of -3.7 and -5.6 for diamonds from a peridotite, Deines et al. (1987) measured $\delta^{13}C$ of graphite and diamond in two graphite-diamond eclogite xenoliths from the Roberts Victor kimberlite. These isotopic compositions fall between -4 and -7‰ versus PDB. In both samples ^{13}C was enriched in the graphite with respect to diamond (1.2 and 0.6‰). The fractionation is opposite in sign to that expected on the basis of the computations of Bottinga (1969). From the Orapa kimberlite, Botswana, some 70 graphite- and diamond-bearing eclogites were investigated (Deines 1984; Deines et al. 1991). While the isotopic composition of the graphite varies between -4.6 and -7.8‰, that of diamonds falls in the range -4 to -23‰.

The isotopic composition of carbon observed on cracks in mantle xenoliths and fluid inclusions was studied by Pineau and Mathez (1990) and Nadeau et al. (1990). In both investigations carbon was extracted through stepwise heating permitting a differentiation of the isotopic composition of different carbon species. The carbon extracted below 800 °C is characterized by $\delta^{13}C$ values of about -25‰, and is thought to represent graphitic and organic material present on cracks in the xenoliths. The carbon fraction extracted at high temperatures commonly has compositions in the range from -2 to -10‰ and is thought to represent mainly carbon from fluid inclusions. However, in both studies xenoliths were found for which the high temperature carbon fraction is also ^{13}C-depleted and isotopically indistinguishable from the low-temperature fraction. The authors suggest that this carbon resides on cracks formed and partially annealed before the xenoliths were incorporated into the magma and, hence, they make a case for the existence of ^{13}C-depleted carbon in the mantle.

Leung and Friedman (1990) have reported a $\delta^{13}C$ value of −24‰ for silicon carbide from a diatreme near Fuxian in the Liaoning Province of China. Diamonds from the kimberlite have carbon isotopic compositions between −2.9 and −4.8‰ (average −4.1‰). Whether or not this moissanite has the same site of origin as the diamonds remains to be established. Hence, it is not clear whether the carbon in the moissanite and that in the diamonds are derived from the same reservoir, nor whether the observed isotopic differences between the two carbon species would thus reflect an isotope effect.

The most detailed record of the isotopic composition of mantle carbon has been established through the study of carbonates from carbonatites and kimberlites and of kimberlite diamonds. The available data have been reviewed in detail by Deines (1989a).

Carbonatite carbonates and diamonds were recognized early as samples of possible mantle carbon, and it has been generally assumed that their $\delta^{13}C$ sampling frequency distributions reflect that of mantle carbon (Craig 1953; Wickman 1956). Early studies of sulfur and oxygen isotope ratios of mantle-derived rocks and minerals suggested that $\delta^{18}O$ and $\delta^{34}S$ may be normally distributed in the mantle, with a mean close to that of meteorites and a relatively small standard deviation. Consequently, some workers may have tacitly assumed that this might also be the case for carbon isotopes. As we shall see below, such an assumption is not warranted.

Carbonatite carbonates which are most likely to have preserved the mantle ^{13}C-record have $\delta^{18}O$ values indicative of oxygen isotope equilibrium with mantle silicates. Such samples were selected from the literature (Deines 1989a) and used to compute mean $\delta^{13}C$ values for 30 carbonatites (Fig. 1). If the mean for each carbonatite is taken to be one sample of mantle carbon, the distribution of carbonatite $\delta^{13}C$ means can be regarded as a sampling distribution of the carbon isotopic composition in the source region of carbonatites. This distribution is Gaussian (mean = − 5.4‰, standard error of the mean = 0.2‰; skewness = 0.1, standard error = 0.4; kurtosis = − 0.6, standard error = 0.8; number of carbonatite means = 30). Individual carbonatite complexes may

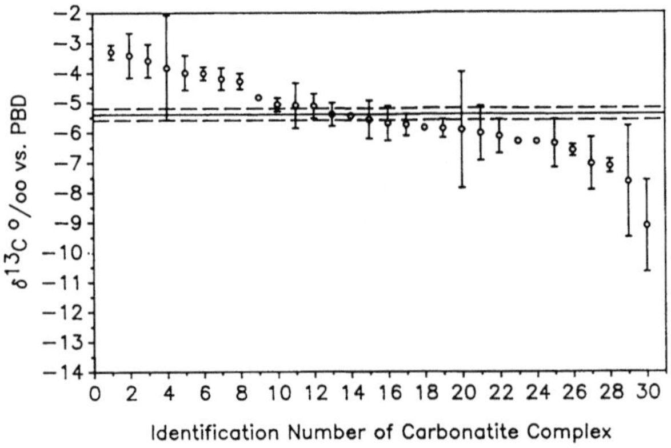

Fig. 1. Mean carbon isotopic composition of carbonatite carbonate and 95% confidence intervals. The ^{13}C-content is expressed as $\delta^{13}C = [(^{12}C/^{13}C)_{sample}/(^{12}C/^{13}C)_{reference} - 1] \times 10^3$; the PDB carbonate is used as reference. *1* Tororo; *2* Mbeya; *3* Sukulu; *4* Homa Mountain; *5* Phalaborwa; *6* Cargill; *7* Kovdors; *8* Mud Tank; *9* Wet Mountains; *10* St. Honoré; *11* Alnoe; *12* Tatarsky; *13* Soekli; *14* Magnet Cove; *15* Iron Hill; *16* Fen; *17* Oka; *18* Lueshe; *19* Kaiserstuhl; *20* Prairie Lake; *21* Chernigov; *22* Bailundo; *23* Maly Sayanski; *24* Bolshoy; *25* Monte Verde; *26* Nemegosenda Lake; *27* Ile de Fuerte Ventura; *28* Laacher See; *29* Ile de Fogo; *30* Oldoinyo Lengai

have mean $\delta^{13}C$ values which differ significantly from the grand mean isotopic composition.

Studies of diamonds also have led to the conclusion that different parts of the mantle may have a distinct carbon isotope record. Significant differences in the $\delta^{13}C$ distribution of mantle carbon related to location have been observed among diamond suites from nine different southern African kimberlites (Deines 1980; Deines et al. 1984, 1987, 1989). This was also noted by Galimov (1984), who investigated a comprehensive suite of Russian diamonds.

Study of $\delta^{13}C$ variations in individual diamonds (e.g., Swart et al. 1983; Javoy et al. 1984; Galimov 1984; Boyd et al. 1987) have revealed that in some cases the outer layers of diamonds may have slightly higher ^{13}C-contents (1 to 2‰) than the interiors. However, this is not a universal observation. Boyd et al. (1987) noted that the overgrowths of three octahedral diamonds had isotopic compositions in the range −6 to −8‰ which coincides with the $\delta^{13}C$ range of cubic diamonds from Zaire. Because cubic diamonds from Russia (Galimov 1984) have isotopic compositions which fall in this range as well, the authors suggested that diamonds have multiple growth histories in which part of the carbon is derived from a deeper reservoir and part from a shallower worldwide reservoir of an isotopic composition of −6 to −8‰.

Results for the Orapa (Deines 1984) and Roberts Victor (Deines et al. 1987) kimberlites indicate that diamonds with $\delta^{13}C$ values between −10 and −20‰ may come from restricted depths. Diamonds from Jagersfontein occluding garnetites for which a depth of origin of up to 450 km can be postulated (Deines and Harris 1988) are all depleted in ^{13}C by more than 10‰ compared to the carbonatite mean, suggesting that on the average $\delta^{13}C$ may be lower for mantle carbon at greater depth.

There are two continents – the U.S.S.R. and southern Africa – for which extensive diamond $\delta^{13}C$ data sets are available. This provides an opportunity to compare the mantle carbon isotopic composition on a continental scale. In order to assure a uniform base of comparison, only those samples were considered which can be genetically linked to the mantle most directly and through similar processes, i.e., diamonds from kimberlites. Thus data for placer diamonds, for which a variety of origins might be possible (see Galimov 1984; Sobolev and Shatsky 1990), were set aside.

Statistical comparison of the cumulative frequency distributions of Russian and southern African diamonds, in which the distribution for each diatreme is given equal weight, indicates no significant difference in $\delta^{13}C$ between them. Both distributions have the same wide $\delta^{13}C$ range, a major mode at −5.5‰, and a weighted mean of −7‰. This is a very remarkable result because it suggests that the large range in $\delta^{13}C$ and the relative high abundance of ^{13}C-depleted diamonds are fundamental mantle properties. Furthermore, if the wide range in $\delta^{13}C$ is a result of an isotope fractionation process which introduces the variability into a mantle of an initially uniform $\delta^{13}C$ value, this process must be reproducible on a continental scale.

In order to characterize the carbon isotope variation in the mantle and provide a $\delta^{13}C$ distribution against which various geochemical models may be tested, a summary distribution giving equal weight to the Russian and African samples (500 samples each) was constructed (Fig. 2). This distribution reflects, in a general way, the isotopic composition of mantle carbon represented by kimberlite diamonds, as well as the nature of the sampling process of mantle carbon by kimberlites. The distribution is highly skewed towards low $\delta^{13}C$ values (skewness = −2.1, standard error = 0.077), has a mode at −5.5‰ (kurtosis = 3.8, standard error = 0.15) and a weighted mean of −7‰. The summary distribution (Fig. 2) could conceivably be bimodal with a prominent mode at −5.5‰ and a minor mode between −15 and −19‰. Whether these two modes reflect worldwide distinct carbon reservoirs, and whether and how these might be related, remain major unanswered questions.

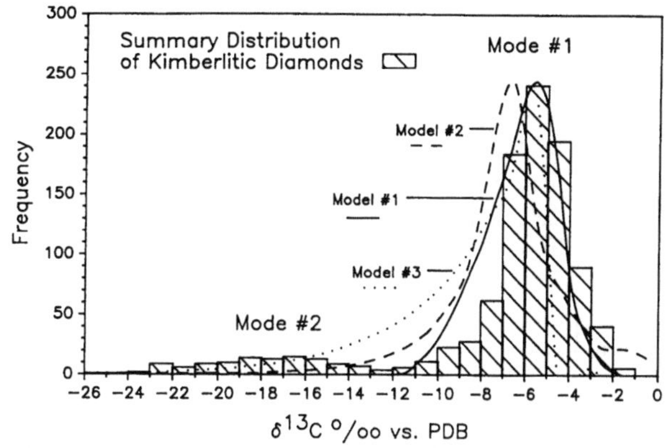

Fig. 2. Summary frequency distribution of $\delta^{13}C$ of kimberlite diamonds from Russia and South Africa, the sum of all frequencies equals 1000. *Model 1* details of the computation of the model curve (carbon precipitation through the reaction $CH_4 = C + 2\,H_2$) are given by Deines (1980), *Model 2* random sampling of subducted sedimentary carbon, end member $\delta^{13}C$ values assumed are 0 and −25‰, see text. *Model 3* Rayleigh fractionation process, $\alpha = 1.0038$, $\delta_0 = -5.5$‰, weighted mean $\delta^{13}C = -8.7$‰, see text. For all model distributions the total number of samples matches that of the diamond distribution

5 Carbon Content and Isotopic Composition of Basalts

Measurements of the concentration and isotopic composition of carbon in basalts may also provide insights into the geochemistry of mantle carbon. Any information gained is indirect, however, because the carbon content and isotopic composition of basaltic magmas are, in part, determined by element and isotope fractionation during partial melting of the mantle, and the transport and eruption of the magma. Much of the data on the carbon contents and carbon isotopic compositions of basalts have been summarized in recent reviews (Mattey 1987; Deines 1989a).

The total carbon contents and $\delta^{13}C$ values reported for the basalts vary greatly (Fig. 3) and the causes for the diversity in isotopic compositions and concentrations are not well understood. Part of the variability may be due to differences in the sample preparation techniques used by different investigators.

In the earlier studies samples were cleaned carefully before reaction in an atmosphere of O_2. More recently the samples have been combusted by stepwise heating under O_2 or, alternatively, a low-temperature

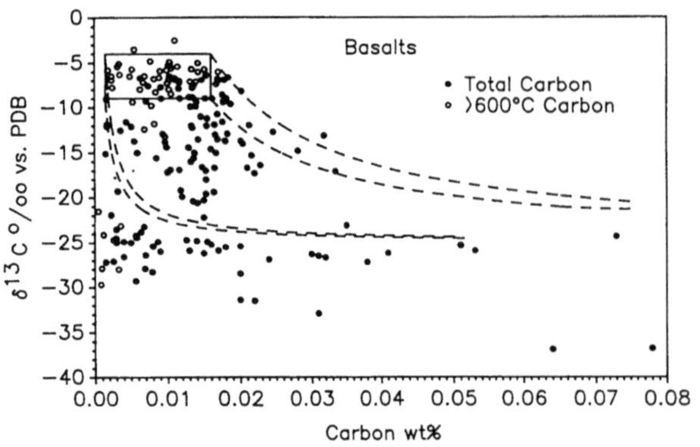

Fig. 3. Carbon isotopic composition and carbon content of basalts. For sources of data see Deines (1989 a). The rectangle includes most of the analytical data obtained in high-emperature combustion procedures; the hyperbolae indicate mixing lines of such carbon with carbon of $\delta^{13}C = -25$‰

combustion step was followed by heating in vacuum. The stepwise combustion procedures have shown that the CO_2 formed below 600 °C is generally depleted in ^{13}C compared to CO_2 formed above 600 °C. This observation has led to the suggestion that the low-temperature CO_2 results from surficial organic contamination, and that carbon contamination prior to and during analysis may be an important factor.

However, low $\delta^{13}C$ values have been reported in the high temperature fractions, and highly ^{13}C-depleted carbon has been observed in many samples subjected to careful low-temperature cleaning (e.g., Galimov and Gerasimovskiy 1978; see also the extended discussion by Nadeau et al. 1990). In Fig. 3 the $\delta^{13}C$ values of basalts have been plotted against their carbon contents; also indicated are the concentration and isotopic composition range into which most of the high-temperature combustion results fall. If this range were characteristic for indigenous carbon in basalts, and the variation in $\delta^{13}C$ and total carbon concentration were the result of the introduction of secondary contaminants of $\delta^{13}C = -25$‰, the mixtures should lie in the region between the mixing hyperbolae. Very high relative and absolute amounts of contamination would have to be postulated to explain a large portion of the data set. This seems very unlikely in view of the care taken by all investigators to minimize contamination of the samples analyzed. Secondary contamination, a possible explanation for some of the measured low $\delta^{13}C$ values, is clearly ruled out for many samples. Hence, there are basalts containing indigenous carbon with $\delta^{13}C$ values considerably lower than −10‰.

The question of how to interpret the isotopically light indigenous carbon has also been debated. One school attributes the low $\delta^{13}C$ carbon to organic material either subducted or caught up in the eruptive process (Mattey 1987, and references cited therein), while another (Javoy et al. 1986) explains the low ^{13}C-content of some of the carbon in basalts as the result of a degassing process. In such a process, CO_2 enriched in ^{13}C is removed from the magma (Javoy et al. 1978; Mattey et al. 1989, 1990) and can be trapped in vesicles from which it may be subsequently liberated by crushing of the samples. The isotopic composition of this CO_2 ranges from −4 to −10‰ with a peak between −5 and −6‰. The residual carbon, thought to be released during combustion, is depleted in ^{13}C. Arguments have been raised both in favor of, and in opposition to, the two hypotheses, and discussions concerning the measurement techniques and the interpretation of the results are continuing.

Although this uncertainty exists, mantle reservoirs of differing isotopic compositions have been postulated on the basis of $\delta^{13}C$ studies of basalts and basaltic glasses (see e.g. Exley et al. 1986 and Mattey et al. 1989 and references cited therein). The source of mid-ocean-ridge basalts has been assigned an average isotopic composition of −6.4‰ and that of carbon of hotspot sources a ^{13}C-content which is about 1‰ higher. Back-arc-basin basalts are thought to have up to 5‰ lower ^{13}C-concentrations than mid-ocean-ridge basalts due to input of surface organic carbon.

Carbon dioxide emanating from volcanic vents has also been considered as a source of information on the isotope composition of magmatic carbon. For such gasses $\delta^{13}C$ values between 0 and −31‰ have been observed, with a mode between −4 and −5‰. Emanations enriched in ^{13}C have been interpreted to be the product of thermal decomposition of limestones, while ^{13}C-depleted compositions have been attributed either to contamination by organic material or to isotope fractionation as a result of magma degassing. Welhan and Craig (1983) make a case that methane in hydrothermal fluids from the East Pacific Rise with a carbon isotopic composition in the range −15 to −17.6‰ cannot have an organic origin. The methane could be in isotopic equilibrium with CO_2 of −7‰ at 640 to 770 °C. In the absence of independent constraints which could serve to limit possible interpretations, data for volcanic CO_2 provide only very limited information on the isotopic composition of mantle carbon.

6 Discussion and Summary

A notable feature of mantle carbon is the high degree of variability in its concentration, chemical form, and isotopic composition. If one attempts to characterize the mantle's carbon content and isotopic composition, one has to consider that various types of mantle carbon samples may represent different mantle volumes and depth ranges, in finite geologic settings. Moreover, these samples may record the effects of chemical processes active during the extraction of carbon from the mantle and its transport to the Earth's surface. Models for the geodynamic cycle of carbon which ignore these complexities are bound to be oversimplified. This point can be amplified using as an example the best established geochemical parameter of mantle carbon, i.e., its isotopic composition.

Characteristic mantle $\delta^{13}C$ values and ranges that have been discussed in the literature can be summarized as follows:

$\delta^{13}C$‰ vs. PDB	Sample type, statistic
−5.5	Carbonatite, carbonate mean
	Kimberlite, carbonate mean
	Diamonds, mode # 1
	Hot spot basalts, mean
−6.4	Mid-ocean ridge basalts, mean
−7	Diamonds, weighted mean
−6 to −8	Some cubic diamonds
	Some diamond coats
−15 to −25‰	Diamonds, mode # 2
	Some asthenospheric diamonds
	High-temperature carbon from some mantle xenoliths
	Moissanite

Before the mantle carbon isotopic composition can be used to constrain computations of the global carbon cycle, the size and relationship of the various reservoirs of differing $\delta^{13}C$ must be understood.

Variations in the isotopic composition of mantle carbon can be the result of: (1) isotope effects in the precipitation of graphite or diamond as a result of changes in pressure, temperature, oxygen fugacity, and carbon reservoir depletion; (2) subduction of sedimentary carbon; (3) mantle degassing; and (4) variations remaining from accretion. It is also possible that isotope effects which have not yet been recognized may be important. These might occur in the formation of Si−C bonds, either in the dissolution of carbon in silicates or in the formation of moissanite.

Carbon precipitation models have been devised which incorporate changes in oxidation state, pressure, and temperature as well as reservoir depletion effects (Deines 1980). Calculations were carried out to test what effect these variables would have on graphite or diamond $\delta^{13}C$ if the minerals were formed from a carbon reservoir with a mean isotopic composition and variability similar to that of carbonatite magmas. An example of such model calculations is indicated in Fig. 2. The particular distribution was chosen from a large number of results – computed for a range of pressure, temperature and oxygen fugacity changes, and different degrees of carbon reservoir depletions – because it fitted best the diamond $\delta^{13}C$ distribution. One can conclude that the ^{13}C-depletion of diamonds attributable to changes in the tested variables is small compared to the total observed $\delta^{13}C$ range. However, while they cannot account for the total mantle carbon isotopic composition range, some of the smaller isotopic differences within individual diamonds can be readily explained as a result of changes in pressure, temperature, and oxygen fugacity, in addition to carbon reservoir depletion effects, during the growth of the mineral.

Several authors (Sobolev and Sobolev 1980; Milledge et al. 1983; Mattey 1987; Kesson and Ringwood 1989) have suggested that the isotopic variability of mantle carbon results from the recycling of sediments into the mantle through subduction of oceanic lithosphere. While this may be a ready explanation for low $\delta^{13}C$ values in the mantle (photosynthesized carbon has an isotopic composition of about −25‰), the hypothesis remains ad hoc as long as there is no independent geochemical evidence which would link the ^{13}C-depleted mantle carbon directly to a sedimentary source; I have searched for such evidence but to date I have found none.

If one assumes for argument's sake that sedimentary carbon can actually be subducted, the resulting $\delta^{13}C$ sampling frequency distribution of mantle carbon can be examined through the construction of a random sampling model. Because at this time no evidence has been presented for an isotopic composition difference between the crustal and mantle carbon reservoir, the mean $\delta^{13}C$ value of the subducted carbon is assumed to be that of the weighted diamond distribution (−7‰). The extent to which the $\delta^{13}C$ values of the sedimentary end members (organic carbon −25‰, limestones 0‰) are preserved, or altered through isotope exchange and homogenization during subduction, is unknown. In the absence of this knowledge, the retention of the end-member isotopic compositions and the establishment of any intermediate composition through isotope exchange are assumed to be equally likely. The only restriction imposed is that, on the average, the weighted mean $\delta^{13}C$ of the exchanging carbon compounds remains −7‰.

The volume over which the carbon isotopic composition is homogenized during subduction is also important for the resulting $\delta^{13}C$ sampling distribution. If the volume is very small the extreme sedimentary $\delta^{13}C$ values are more likely to be retained in the mantle; if the volume is very large the weighted mean of −7‰ must be attained. The sampling model is set up so that homogenization over a range of volumes can be simulated.

One example of the sampling frequency distributions generated by the random sampling algorithm is shown in Fig. 2 (Model 2). From over 100 examples which were examined a distribution was chosen which had the largest number of low $\delta^{13}C$ samples and a sample frequency at its mode which matched that of the diamond $\delta^{13}C$ Mode 1. Still, there remains a very significant discrepancy at low $\delta^{13}C$ values; low $\delta^{13}C$ diamonds occur *more frequently* than predicted by the model. A better match at low $\delta^{13}C$ values can be attained if the volume over which homogenization occurs is reduced. If model results are chosen for which the sample frequency is matched at low $\delta^{13}C$ values, neither the sample frequency of the Mode 1 (model sample frequency is only 50% of that observed for diamonds) nor that of $\delta^{13}C$ values between 0 and −3‰ (model sample frequency is 240% of that observed for diamonds) approaches that of the diamond distribution. With the constraints given above it was impossible to match more than one of the parameters of model and diamond $\delta^{13}C$ distribution at a time, or to generate a mode at −5.5‰. In all model distributions the mode occurred close to −7‰. Analysis of the model computation indicates that the relatively high abundance of low $\delta^{13}C$ diamonds is not balanced by ^{13}C-enriched mantle carbon as expected.

If the low $\delta^{13}C$ of diamonds were to represent subducted organic carbon, this would require either a process which introduces such carbon selectively into the mantle, or the existence of an as yet unrecognized carbon reservoir which stores the excess ^{13}C in the mantle. Although not impossible, it is difficult to envision a geologic process which would recycle selectively the organic sedimentary component into the mantle. A necessary consequence would be a concomitant increase in the relative ^{13}C-content of the crustal carbon reservoir. Evidence for this has not been presented by proponents of the subduction hypothesis. A mantle reservoir which stores excess ^{13}C in the mantle has not been recognized. Hence it is unlikely that the relatively high abundance of ^{13}C-depleted carbon in the mantle is the result of the subduction of sediments. This does not exclude, however, the possibility that some ^{13}C-depleted metamorphic diamonds (Sobolev and Shatsky 1990) might represent carbon fractionated at the Earth's surface.

An alternate hypothesis attributes the high degree of ^{13}C-depletion in some mantle samples to the loss of ^{13}C-enriched CO_2 during the degassing history of the Earth (Javoy et al. 1982) which can be described in a first-order approximation by a Rayleigh fractionation process. From the relative abundance of diamonds of varying degrees of ^{13}C-depletion an isotope fractionation factor of $\alpha = 1.0038$ can be estimated for the CO_2 loss. The value is of the same magnitude as that determined experimentally by Javoy et al. (1978) for the ^{13}C-fractionation between CO_2 and a tholeiitic melt. More recent experiments by Mattey et al. (1990) indicate, however, a significantly smaller value (1.002). A smaller fractionation is also expected on the basis of the known carbon species in silicate melts under oxygen fugacity conditions close to the quartz-fayalite-magnetite buffer and reasonable estimates of their isotope effects.

In Fig. 2 the sampling frequency distribution based on a Rayleigh fractionation degassing process is shown (Model 3). The curve was computed assuming a fractionation factor of 1.0038 and an initial reservoir isotopic composition such that a mode at −5.5‰ resulted. The sample frequency computed for the mode depends on the initial isotopic composition variability of the reservoir, which was adjusted by trial and error so that the mode of the model computations had about the same sample frequency as the diamond distribution. The weighted mean $\delta^{13}C$ value for the sample set generated by the model is −8.7‰, which is significantly lower than the weighted mean of diamonds (−7‰). In addition, compared to the diamond distribution, there is a very conspicuous over-representation of $\delta^{13}C$ values in the range from −7 to −15‰, and a significant under-representation in the range −1 to −4‰. Extensive trials showed that it was impossible to match simultaneously the skewness, the mode's $\delta^{13}C$ value and sample frequency, as well as the weighted mean $\delta^{13}C$ value of any model sample set with that of the diamond $\delta^{13}C$ sampling frequency distribution.

On the basis of the degassing hypothesis one would expect a positive correlation between ^{13}C- and nitrogen-content of diamonds. For none of the diamond suites studied by us has such a correlation been found (Deines et al. 1984, 1987; Deines 1989b).

The size of the required isotope effect, and the impossibility of matching simultaneously the major characteristics of the model frequency distribution with the observed diamond distribution, as well as the absence of a correlation between the ^{13}C- and nitrogen-content of diamonds, lead to the conclusion that it is unlikely that a Rayleigh-fractionation-type process is responsible for the large $\delta^{13}C$ range of mantle carbon.

The difficulty of explaining the observed diamond $\delta^{13}C$ distribution through chemical or physical isotope fractionation processes, or admixing of surface carbon to the mantle, leads to consideration of the possibility that $\delta^{13}C$ differences might remain from the accretion of the Earth. The large $\delta^{13}C$ range of bulk-reduced meteoritic carbon coincides with that of diamonds. It is also very interesting to observe that nodular carbon in iron meteorites commonly has $\delta^{13}C$ values close to -5‰ (diamond mode M1), but that carbon in taenite and cohenite which coexist with these nodules has $\delta^{13}C$ values between -18 and -22‰ (diamond Mode M2). This large, apparently systematic fractionation is an unexplained feature in the isotope geochemistry of meteoritic carbon.

Although these similarities between the meteorite and diamond $\delta^{13}C$ distributions exist, one finds that the mean isotopic compositions for carbon from stony and iron meteorites probably lie close to -14 and -10‰ respectively (Deines 1989a), i.e., at significantly lower $\delta^{13}C$ values than one deduces for mantle carbon from kimberlite or carbonatite carbonates (-5.4‰) or from diamonds -7‰). This raises the question whether ^{13}C-depleted carbon is under-represented in the mantle carbon samples considered to date. The preliminary data on asthenopsheric diamonds might suggest that the relative abundance of ^{13}C-depleted carbon is higher in the lower mantle.

The difference in the mean $\delta^{13}C$ of carbonatite carbonates and diamonds, and in their isotopic variability, can be rationalized on the basis of carbon isotope variations in the mantle and the manner in which the element is extracted from it and brought to the surface. The processes through which erupting kimberlites and accumulating carbonatite magmas sample mantle carbon must differ. Because carbon is a trace element in the mantle, the carbon present in a carbonatite must be extracted from a mantle volume more than a thousand times the size of the carbonatite intrusion, unless an origin via subducted limestone is proposed (see discussion by Wyllie 1987). In this process the ^{13}C-content is integrated over the volume from which the carbon is extracted, reducing variability. Diamond, because of its small size and rare occurrence, represents a local sample of mantle carbon, reflecting little or no averaging of mantle $\delta^{13}C$ variations. Thus, the difference in degree of $\delta^{13}C$ variability between diamonds and carbonatite carbonates can be understood.

The strong mode of diamond $\delta^{13}C$ at -5.5‰, which coincides with the mean value for carbonatite carbonates, indicates that carbon of this particular isotopic composition is widespread in those mantle regions sampled by both kimberlite and carbonatite magmas. However, the diamond data also demonstrate the existence of mantle carbon isotopic compositions which differ significantly from this value. If the relative abundance of ^{13}C-depleted diamonds belonging to Mode 2 (Fig. 2) compared to diamonds belonging to Mode 1 was higher at greater mantle depths, but was systematically under-represented in diamond suites due to the manner in which kimberlites form and incorporate diamonds, the bimodality of the distribution and the relative size of the two modes would be plausible. Also, if the ^{13}C-depletion occurred at depths that are sampled by kimberlite diamonds but not by carbonatite magmas, the difference in the weighted $\delta^{13}C$ means of carbonatites and diamonds can be rationalized. Finally, a higher relative abundance of ^{13}C-depleted carbon in lower parts of the mantle, which are under-represented among the mantle samples analyzed to date, could explain the difference in the mean carbon isotopic composition of meteoritic bulk carbon and the present estimates of the mean isotopic composition of mantle carbon. The proof of this concept would have far-reaching consequences for our understanding of the geochemical cycle of carbon.

The meteorite data may indicate also another possible cause for the large range in isotopic composition of mantle carbon. The systematic ^{13}C-fractionation between graphite on the one hand and cohenite and taenite on the other suggests the possibility that the isotope effects of Fe$-$C (and by inference Si$-$C) bonds might be significantly larger at higher temperatures than those of other carbon species. The reported low $\delta^{13}C$ value of silicon carbide would be consistent with such a proposal.

This discussion has shown that it is not appropriate to assume that the mantle carbon isotopic composition distribution is Gaussian and has a small standard deviation, i.e., is similar to the carbonatite carbonate $\delta^{13}C$ distribution. Realistic geodynamic carbon cycle models must consider the observed large variability of $\delta^{13}C$ in mantle carbon. The discussion has also pointed out arguments for and against the possible linkages between mantle $\delta^{13}C$ variability and potential components of the carbon cycle (subduction, degassing). The importance of these processes for the mantle carbon isotope record remains a topic of debate. The isotopic composition of mantle carbon can provide a powerful constraint in the formulation of carbon cycle models, provided the causes for the $\delta^{13}C$ variability in the mantle have been properly identified and accounted for.

Acknowledgments. Financial support for the preparation of this chapter was provided in part by National Science Foundation Grant EAR 8416386 to the author.

References

Berg GW (1986) Evidence for carbonate in the mantle. Nature (London) 324:50–51

Bergman SC (1981) Fluid inclusions in xenoliths: samples of the mantle metasomatic fluid? Geol Soc Am Abstr 13:379

Bergman SC (1982) Petrogenetic aspects of the alkali basaltic lavas and included megacrysts and nodules from the Lunar Crater Volcanic Field, Nevada, USA. Ph D Thesis, Princeton Univ

Bergman SC, Dubessy J (1984) CO_2-CO fluid inclusions in a composite peridotite xenolith: implications for upper mantle oxygen fugacity. Contrib Mineral Petrol 85:1–13

Bilal A, Touret J (1976) Les inclusions fluides des enclaves catazonales de Bournac. Bull Soc Fr Mineral Cristallogr 99:134–139

Bottinga Y (1969) Carbon isotope fractionation between graphite, diamond and carbon dioxide. Earth Planet Sci Lett 5:301–307

Boyd SR, Mattey DP, Pillinger CT, Milledge HJ, Mendelssohn M, Seal M (1987) Multiple growth events during diamond genesis: an integrated study of carbon and nitrogen isotopes and nitrogen aggregation state in coated stones. Earth Planet Sci Lett 86:341–353

Canil D, Virgo D, Scarfe CM (1990) Oxidation state of mantle xenoliths from British Columbia, Canada. Contrib Mineral Petrol 104:453–462

Craig H (1953) The geochemistry of the stable carbon isotopes. Geochim Cosmochim Acta 3:53–92

Dawson JB, Smith JV (1975) Occurrence of diamond in a mica-garnet lherzolite xenolith from kimberlite. Nature (London) 254:580–581

Deines P (1980) The carbon isotopic composition of diamonds: relationship to diamond shape, color, occurrence and vapor composition. Geochim Cosmochim Acta 44:943–961

Deines P (1984) Evidence for a systematic depletion in ^{13}C in parts of the mantle underlying the Orapa, Botswana, kimberlite. Geol Soc Am Abstr 16:485

Deines P (1989a) Stable sotope variations in carbonatites. In: Bell K (ed) Carbonatites – genesis and evolution. Unwin Hyman, Boston, pp 301–359

Deines P (1989b) Regularities in the ^{13}C and nitrogen content of the mantle revealed through studies of diamonds and the chemistry of their inclusions. In: Ext Abstr Worksh Diamonds, 28th Int Geol Congr, Washington DC, pp 18–20

Deines P, Harris JW (1988) Carbon isotope and nitrogen variations in deep diamonds. Geol Soc Am Abstr 20:A280

Deines P, Gurney JJ, Harris JW (1984) Associated chemical and carbon isotopic composition variations in diamonds from Finsch and Premier kimberlite, South Africa. Geochim Cosmochim Acta 48:325–342

Deines P, Harris JW, Gurney JJ (1987) Carbon isotopic composition, nitrogen content and inclusion composition of diamonds from the Roberts Victor kimberlite, South Africa: evidence for ^{13}C depletion in the mantle. Geochim Cosmochim Acta 51:1227–1243

Deines P, Harris JW, Spear PM, Gurney JJ (1989) Nitrogen and ^{13}C content of Finsch and Premier diamonds and their implications. Geochim Cosmochim Acta 53:1367–1378

Deines P, Harris JW, Robinson DN, Gurney JJ, Shee SR (1991) Carbon and oxygen isotope variations in diamond and graphite eclogites from Orapa, Botswana. Geochim Cosmochim Acta 55:515–524

Des Marias DJ (1985) Carbon exchange between the mantle and the crust, and its effect upon the atmosphere: today compared to Archean time. In: Sundquist ET, Broecker WS (eds) The carbon cycle and atmospheric CO_2: natural variations archean to present. Geophys Monogr 32:602–611

Eggler DH (1978) The effect of CO_2 upon partial melting of peridotite in the system $Na_2O-CaO-Al_2O_3-MgO-SiO_2-CO_2$ to 35 kb, with an analysis of melting in a peridotite-H_2O-CO_2 system. Am J Sci 278:305–343

Eggler DH, McCallum ME (1975) Diamond-bearing peridotite in a Wyoming kimberlite pipe. Geol Soc Am Abstr 7:1965

Exley RA, Mattey DP, Clague DA, Pillinger CT (1986) Carbon isotope systematics of a mantle "hotspot": a comparison of Loihi Seamount and MORB glasses. Earth Planet Sci Lett 78:189–199

Freund F (1981) Mechanism of the water and carbon dioxide solubility in oxides and silicates and the role of O^-. Contrib Mineral Petrol 76:474–482

Freund F (1986a) Solute carbon and carbon segregation in magnesium oxide single crystals – a secondary ion mass spectrometry study. Phys Chem Minerals 13:262–276

Freund F (1986b) Reply to "Comment on solute carbon and carbon segregation in magnesium oxide single crystals – a secondary ion mass spectrometry study". Phys Chem Minerals 13:280

Freund F, Kathrein H, Wengeler H, Knobel R, Heinen HJ (1980) Carbon in solid solution in forsterite – a key to the untractable nature of reduced carbon in terrestrial and cosmogenic rocks. Geochim Cosmochim Acta 44:1319–1333

Galimov EM (1984) Variations of isotopic composition of diamonds and their relation to conditions of diamond-formation. Geokhimiya 1984:1091–1118

Galimov EM, Gerasimovskiy VI (1978) Isotopic composition of carbon in Icelandic magmatic rocks. Geochem Int 1978:1–6

Green HW (1972) A CO_2-charged asthenosphere. Nat Phys Sci 238:2–5

Green HW (1985) Coupled exsolution of fluid and spinel from olivine: evidence for O^- in the mantle? In: Schock RN (ed) Point defects in minerals. Geophys Monogr Ser 31, AGU Washington, DC, pp 226–232

Green HW, Guegen Y (1983) Deformation of peridotite in the mantle and extraction by kimberlite: a case history documented by fluid and solid precipitates in olivine. Tectonophysics 92:71–92

Green HW, Radcliffe SV (1975) Fluid precipitates in rocks from the Earth's mantle. Geol Soc Am Bull 86:846–852

Gurney JJ (1989) Diamonds. In: Kimberlites and related rocks 2. Geol Soc Aust Spec Publ 14:935–965

Haggerty SE (1989) Upper mantle oxidation state and diamond genesis. In: Ext Abstr Worksh Diamonds, 28th Int Geol Congr, Washington, DC, pp 26–28

Haggerty SE, Tompkins LA (1983) Redox state of Earth's upper mantle from kimberlite ilmenites. Nature (London) 303:295–300

Hoefs J (1965) Ein Beitrag zur Geochemie des Kohlenstoffs in magmatischen und metamorphen Gesteinen. Geochim Cosmochim Acta 29:399–428

Jaques AL, Hall AE, Sheraton JW, Smith CB, Sun S-S, Drew RM, Foudoulis C, Ellingsen K (1989) Composition of crystalline inclusions and C-isotopic composition of Argyle and Ellendale diamonds. In: Kimberlites and related rocks. 2. Geol Soc Aust Spec Publ 14:966–989

Jaques AL, O'Neill HStC, Smith CB (1990) Diamondiferous peridotite xenoliths from the Argyle (AK1) lamproite pipe. Western Australia. Contrib Mineral Petrol 104:255–276

Javoy M, Pineau F, Iiyama I (1(78) Experimental determination of the isotopic fractionation between gaseous CO_2 and carbon dissolved in tholeiitic magma. Contrib Mineral Petrol 67:35–39

Javoy M, Pineau F, Allègre CJ (1982) Carbon geodynamic cycle. Nature (London) 300:171–173

Javoy M, Pineau F, Delorme H (1986) Carbon and nitrogen isotopes in the mantle. Chem Geol 57:41–62

Javoy M, Pineau F, Demaiffe D (1984) Nitrogen and carbon isotopic composition in the diamonds of Mbuji Mayi (Zaire). Earth Planet Sci Lett 68:399–412

Kessen SE, Ringwood AE (1989) Slab-mantle interactions, 2. The formation of diamonds. Chem Geol 78:97–118

Leung IS, Friedman I (1990) Evidence of silicon carbide-diamond paragenesis: implications for carbon isotopic composition of the mantle. EOS Trans Am Geophys Un 71:644

Luth RW, Virgo D, Boyd FR, Wood BJ (1990) Ferric iron in mantle-derived garnets. Implications for thermobarometry and for the oxidation state of the mantle. Contrib Mineral Petrol 104:56–72

Marty B, Jambon A (1987) $C/^3He$ in volatile fluxes from the solid Earth: implications for carbon geodynamics. Earth Planet Sci Lett 83:16–26

Mathez EA (1987) Carbonaceous matter in mantle xenoliths: composition and relevance to the isotopes. Geochim Cosmochim Acta 51:2339–2347

Mathez EA, Delaney JR (1981) The nature and distribution of carbon in submarine basalts and peridotite nodules. Earth Planet Sci Lett 56:217–232

Mathez EA, Dietrich VJ, Irving AJ (1984a) The geochemistry of carbon in mantle peridotites. Geochim Cosmochim Acta 48:1849–1859

Mathez EA, Blacic JD, Beery J, Maggiore C, Hollander M (1984b) Carbon abundances in mantle minerals determined by nuclear reaction analysis. Geophys Res Lett 11:947–950

Mathez EA, Blacic JD, Hollander JBM, Maggiore C (1987) Carbon in olivine: results from nuclear reaction analysis. J Geophys Res 92:3500–3506

Mattey DP (1987) Carbon isotopes in the mantle. Terra Cognita 7:31–37

Mattey DP, Exley RA, Pillinger CT (1989) Isotopic composition of CO_2 and dissolved carbon species in basalt glass. Geochim Cosmochim Acta 59:2377–2386

Mattey DP, Taylor WR, Green DH, Pillinger CT (1990) Carbon isotopic fractionation between CO_2 vapour, silicate and carbonate melts: an experimental study to 30 kbar. Contrib Mineral Petrol 104:492–505

Mattioli GS, Baker MB, Rutter MJ (1989) Upper mantle oxygen fugacity and its relationship to metasomatism. J Geol 97:521–536

McGetchin TR, Besançon JR (1973) Carbonate inclusions in mantle-derived pyropes. Earth Planet Sci Lett 18:408–410

Melton CE, Giardini AA (1974) The composition and significance of gas released from natural diamonds from Africa and Brazil. Am Mineral 59:775–782

Melton CE, Salotti CA, Giardini AA (1972) The observation of nitrogen, water, carbon dioxide, methane and argon as impurities in natural diamonds. Am Mineral 57:1518–1523

Milledge HJ, Mendelssohn MJ, Seal M, Rouse JE, Swart PK, Pillinger CT (1983) Carbon isotopic variation in spectral type II diamonds. Nature (London) 303:791–792

Moore RO, Otter ML, Rickard RS, Harris JW, Gurney JJ (1986) The occurrence of moissanite and ferro-plagioclase as inclusions in diamond. In: Ext Abstr 4th Int Kimberlite Conf, Perth, Geol Soc Aust 16:409–410

Murck BW, Burruss RC, Hollister LS (1978) Phase equilibria in fluid inclusions in ultramafic xenoliths. Am Mineral 63:40–46

Nadeau S, Pineau F, Javoy M, Francis D (1990) Carbon concentrations and isotopic ratios in fluid-inclusion-bearing upper-mantle xenoliths along the northwestern margin of North America. Chem Geol 81:271–297

Navon O, Hutcheon ID, Rossman GR, Wasserburg GJ (1988) Mantle-derived fluids in diamond micro-inclusions. Nature (London) 335:784–789

Oberheuser G, Kathrein H, Demortier G, Gonska H, Freund F (1983) Carbon in olivine single crystals analyzed by the $^{12}C(d,p)^{13}C$ method and by photoelectron spectroscopy. Geochim Cosmochim Acta 47:1117–1129

O'Neill HStC, Jacques AL, Smith CB, Moon J (1986) Diamond bearing peridotite xenoliths from Argyle AK1 pipe. In: Ext Abstr 4th Int Kimberlite Conf, Perth, Geol Soc Aust Abstr 16:300–302

Otter ML, Gurney JJ (1989) Mineral inclusions in diamonds from the Sloan diatremes, Colorado-Wyoming State Line kimberlite district, North America. In: Kimberlites and related rocks, vol 2. Geol Soc Aust Spec Publ 14:1042–1053

Pateris JD (1981) Occurrence of graphite in serpentinized olivines in kimberlite. Geology 9:356–359

Pineau F, Mathez EA (1990) Carbon isotopes in xenoliths from the Hualalai Volcano, Hawaii, and the generation of isotopic variability. Geochim Cosmochim Acta 54:217–227

Pineau F, Javoy M, Kornprobst J (1987) Primary graphite in ultramafic xenoliths: II. Isotopic composition of the carbonaceous phases in xenoliths and host lava at Tissemt (Eggéré, Algerian Sahara). J Petrol 28:313–322

Pokhilenko NP, Sobolev NV, Lavrent'ev YuG (1977) Xenoliths of diamondiferous ultramafic rocks from Yakutian kimberlites. In: Ext Abstr 2nd Int Kimberlite Conf, Santa Fe, New Mexico

Robinson DN (1979) Diamond and graphite in eclogite xenoliths from kimberlite. In: Boyd FR, Meyer HOA (eds) The mantle sample: Inclusions in Kimberlites and Other Volcanics. Proc 2nd Int Kimberlite Conf. Amer Geophys Un, Washington, DC, 2:50–58

Robinson DN, Gurney JJ, Shee SR (1984) Diamond and graphite eclogite xenoliths from Orapa, Botswana. In: Kornprobst J (ed) Developments in petrology 11 B, kimberlites II. The mantle and crust – mantle relationships. Proc 3rd Int Kimberlite Conf. Elsevier, Amderdam, pp 11–24

Roedder E (1965) Liquid CO_2 in olivine-bearing nodules and phenocrysts in basalts. Am Mineral 50:1746–1786

Shee SR, Gurney JJ, Robinson DN (1982) Two diamond-bearing peridotite xenoliths from the Finsch kimberlite, South Africa. Contrib Mineral Petrol 8:79–87

Sobolev NV, Shatsky VS (1990) Diamond inclusions in garnets from metamorphic rocks: a new environment for diamond formation. Nature (London) 343:742–745
Sobolev VS, Sobolev NV (1980) New proof on very deep subsidence of eclogised crustal rocks. Dokl Acad Nauk SSSR 250:88–90
Sobolev VS, Nai BS, Sobolev NV, Lavrentiev YuG, Poselova LN (1969) Xenoliths of diamond-bearing pyrope serpentinite from the Aykhal pipe, Yakutia. Dokl Acad Sci USSR Earth Sci Sec 188:168–170
Swart PK, Pillinger CT, Milledge HJ, Seal M (1983) Carbon isotopic variations within individual diamonds. Nature (London) 303:793–795
Tingle TN, Green HW, Finnerty AA (1988) Experiments and observations bearing on the solubility and diffusivity of carbon in olivine. J Geophys Res 93:15,289–15,304
Tingle TN, Hochella MF, Becker CH, Malhorta R (1990) Organic compounds on crack surfaces in olivine from San Carlos, Arizona, and Hualalai volcan, Hawaii. Geochim Cosmochim Acta 54:477–485
Touret J, Bilal A (1978) Inclusions fluides carbonique dans les roches du manteau supérieur. In: Abstr 6. Réunion Annu Sci Terre, Orsay, Fr, p 392
Tsong IST, Knipping U (1986) Comment on "Solute carbon and carbon segregation in magnesium oxide single crystals – a secondary ion mass spectrometry study" by F. Freund. Phys Chem Minerals 13:277–279
Turner G, Burgess R, Bannon M (1990) Volatile-rich mantle fluids inferred from inclusions in diamond and mantle xenoliths. Nature (London) 344:653–655
Virgo D, Luth RW, Ulmer GC (1988) Constraints on the oxidation of the mantle: an electrochemical and ^{57}Fe Mössbauer study of mantle-derived ilmenites. Geochim Cosmochim Acta 52:1781–1794
Welhan JA, Craig H (1983) Methane, hydrogen and helium in hydrothermal fluids at 21°N on the East Pacific Rise. In: Rona PA, Boström K, Laubier L, Smith KL (eds) Hydrothermal processes at the spreading centers. Plenum, New York, pp 391–410
Wickman FE (1956) The cycle of carbon and the stable carbon isotopes. Geochim Cosmochim Acta 9:136–153
Wood BJ, Bryndzia, Johnson KE (1990) Mantle oxidation state and its relationship to tectonic environment and fluid speciation. Science 248:337–345
Wyllie PJ (1978) Mantle fluid compositions buffered in peridotite-CO_2-H_2O by carbonates, amphibole, and phlogopite. J Geol 86:687–713
Wyllie PJ (1987) Transfer of subcratonic carbon into kimberlites and rare earth carbonatites. In: Mysen BO (ed) Magmatic processes: physicochemical principles. Geoch Soc Spec Publ 1:107–119
Wyllie PJ, Huang WL, Otto J, Byrnes AP (1983) Carbonation of peridotites and decarbonation of siliceous dolomite represented in the system $CaO-MgO-SiO_2-CO_2$ to 30 kbar. Tectonophysics 100:359–388

Carbon Cycle and Carbon Isotope Record: Geochemical Impact of Life over 3.8 Ga of Earth History

Manfred Schidlowski[1] and Paul Aharon[2]

[1] Max-Planck-Institut für Chemie (Otto-Hahn-Institut), W-6500 Mainz, FRG
[2] Dept. of Geology and Geophysics, Louisiana State University, Baton Rouge, LA 70803-4104, USA

Abstract

With sedimentary organic carbon (in the form of kerogen and its graphitic derivatives) dating back to 3.8 Ga ago, and the isotopic signature of autotrophic carbon fixation persisting over the same time span if allowance is made for a metamorphic overprint of the oldest (> 3.5 Ga) record, we may state with fair confidence that biological modulation of the terrestrial carbon cycle had commenced very early in geological history. Ever since then, carbon transformations in the Earth's exogenic system have proceeded in the form of a *biogeochemical* cycle comprising an organic (reduced) and an inorganic (oxidized) carbon branch. Since the partitioning of the element between the reduced (kerogen) and oxidized (carbonate) moiety is coupled with the isotopic compositions of both carbon species by an isotope mass balance, fluctuations in the $^{13}C/^{12}C$ record, notably of marine carbonates, reflect changing ratios C_{org}/C_{carb} in the total carbon flux from the surficial exchange reservoir to the sedimentary shell that has subsequently come to store the far bulk ($\approx 6 \times 10^{22}$ g) of carbon residing in and above the crust. Relying on the sensor function of $\delta^{13}C_{carb}$, it is possible to identify numerous imbalances in the past operation of the carbon cycle which, however, appear to be demonstrably tethered to a time-invariant average state. On time scales approaching the length of currently documented Earth history, these oscillations are largely smoothed out, with the sedimentary carbon isotope record as a whole constituting a long-term index line of autotrophic carbon fixation that primarily reflects the isotope-discriminating properties of ribulose-1,5-bisphosphate (RuBP) carboxylase, the principal CO_2-fixing enzyme of the Calvin cycle.

1 Introduction

Carbon exists in the Earth's sedimentary shell in two different forms, namely, as oxidized or carbonate carbon (mostly limestone and dolomite) and as reduced or organic carbon, the latter constituting the fossil residue of primary biogenic substances. There is no doubt that the affluence of modern industrial societies is based, for the most part, on the exploitation of solar energy that has come to be stored through the ages in the various forms of the organic moiety (such as coal, oil, and gaseous hydrocarbons). The antiquity and evolution of photosynthetic carbon fixation as the underlying energy-generating process is, therefore, a topic of paramount scientific and practical concern.

Of late, various disparate lines of evidence have merged to indicate that photoautotrophy as the quantitatively most important process of biological carbon fixation must be a very ancient achievement in the evolution of life on Earth. Extrapolating from currently available paleontological and biogeochemical data, a strong case can be made for the operation of a biologically modulated carbon cycle on the juvenile planet as from at least 3.8 Ga ago (Schidlowski 1988, 1990). Specifically, stromatolite-type microbial ecosystems commonly dominated by microbial photoautotrophs appear to be widespread already in sediments as old as 3.5 Ga; likewise, cellular morphologies attributable to prokaryotic microorganisms have been found in rocks of the same or even older age. Moreover, the isotopic signature of photoautotrophic carbon fixation dates back to 3.8 Ga ago if allowance is being made for a metamorphic overprint of the oldest record with a concomitant isotopic reequilibration between organic and carbonate carbon. For an overview of these questions, the reader is referred to a comprehensive summary given in a volume edited by Schopf (1983) as well as to other relevant publications (e.g., Pflug and Jaeschke-Boyer 1979; Schidlowski et al. 1979; Awramik et al. 1983; Buick 1984; Schopf and Packer 1987; see also the contributions by Awramik and by Pflug and Reitz, this Vol.).

Here, a brief summary will be given of the biogeochemical evidence bearing on the antiquity of

M. Schidlowski et al. (Eds.)
Early Organic Evolution: Implications for Mineral and Energy Resources

life on our planet, with special reference to the emergence of photoautotrophy as a second sink for surficial inorganic carbon and its long-standing impact on the global carbon cycle. Moreover, we shall review the role of the sedimentary carbon isotope age curve as a sensor of biospheric activity over 3.8 Ga of recorded Earth history.

2 Evolution of the Terrestrial Carbon Cycle: A Brief Outline

It is well established that carbon figures among the more ubiquitous element species in the Universe, occupying the fourth place in the list of cosmic element abundances subsequent to hydrogen, helium, and oxygen. More recently, radio astronomical molecular spectroscopy has furnished evidence of an active involvement of interstellar carbon in the synthesis of a suite of simple organic molecules including potential intermediates in the formation of high-molecular-weight organic substances that are important for life processes (cf. Irvine and Knacke 1989). Moreover, it has been demonstrated that the submicron-sized frozen dust particles that constitute the bulk of matter in interstellar molecular clouds are sites of a complex organic chemistry driven by photochemical reactions. Such photoprocessing by cosmic UV of the interstellar medium in both gas phase and solid state has given rise to a sizeable reservoir of intergalactic organic substances estimated to account for about one permil of the total mass of the Milky Way, most of which being concentrated in the refractory organic mantles of interstellar dust grains (Greenberg 1984).

With all interstellar dust clouds destined to be ultimately consumed by star formation, a substantial fraction of their carbonaceous component is bound to end up – after extensive high-temperature processing – in reconstituted form in the chondritic constituent of newly formed planets. There is, accordingly, little doubt that the totality of carbon processed, over the ages, in the Earth's exogenic cycle ultimately stems from the primary endowment of chondritic carbon bequeathed to the planet during the accretion of its upper mantle. Whether or not a late-stage cometary component has substantially added to the surficial carbon inventory (and that of other volatile elements) remains a matter of debate; for the time being it seems reasonable conjecture that the bulk of crustal carbon, and notably that stored in the sedimentary shell, has been degassed from the mantle during the Earth's post-accretionary history.

There is widespread agreement today that the chondritic material which went into the making of the mantle had been scavenged by the protoplanet in the latest stage of its accretion when temperatures in the ambient nebula had largely fallen below 750 K. In this temperature range, any remaining metallic iron was apt to be oxidized to Fe^{2+} and subsequently (at $t < 400$ K) to Fe^{3+} as the prevalent chemical equilibria were conducive to the formation of water as a potential oxidizing agent (Grossman and Larimer 1974). Consequently, the Fe-component of the late-formed (low-temperature) condensates of the residual nebula was already oxidized by the time it arrived on Earth. Since the oxidation state of the gases released from the juvenile planet was necessarily dependent on the overall oxidation state of its outer veneer assembled from such late-stage condensates in an inhomogeneous accretion process, the heterogeneous equilibria of mantle-derived (e.g., basaltic) magmas place clear-cut constraints on the composition of any degassed volatile fraction. Such emanations should have been just weakly reducing like contemporary volcanic gases, with specifically $CO_2 \gg \sum CO + CH_4$, $H_2O \gg H_2$, and $N_2 \gg NH_3$. The largely time-invariant Fe^{2+}/Fe^{3+} ratios of basaltic and ultramafic rocks observed since Archaean times give implicit testimony that the redox state of magmatic volatiles had never differed very much from that of modern volcanic gases. Specifically, the preponderance of oxidized iron in mantle-derived rocks was always high enough to exclude an equilibration of mantle material with metallic iron and, consequently, highly reduced gas species such as methane (CH_4), ammonia (NH_3), and hydrogen (H_2). Given this scenario, it can be argued convincingly (e.g., Walker 1977; Walker et al. 1983) that carbon dioxide (CO_2) was the prevalent volatile carbon species released during the whole of the Earth's degassing history. It is this primordial carbon dioxide that was subsequently processed in a sequence of inorganic and biological reactions that make up the terrestrial carbon cycle (Fig. 1).

Due to a rapid involvement of the degassed carbon constituents in the exogenic carbon cycle, the individual CO_2 molecule had only a transient existence in the ancestral atmosphere. The quantitatively most important pathway for the removal of primordial CO_2 from the atmospheric reservoir was the precipitation as a chemical sediment (carbonate). Since the surface conditions on the primitive Earth had ostensibly favored an early condensation of the bulk of water vapor discharged from the mantle, sizeable quantities of CO_2 were bound to

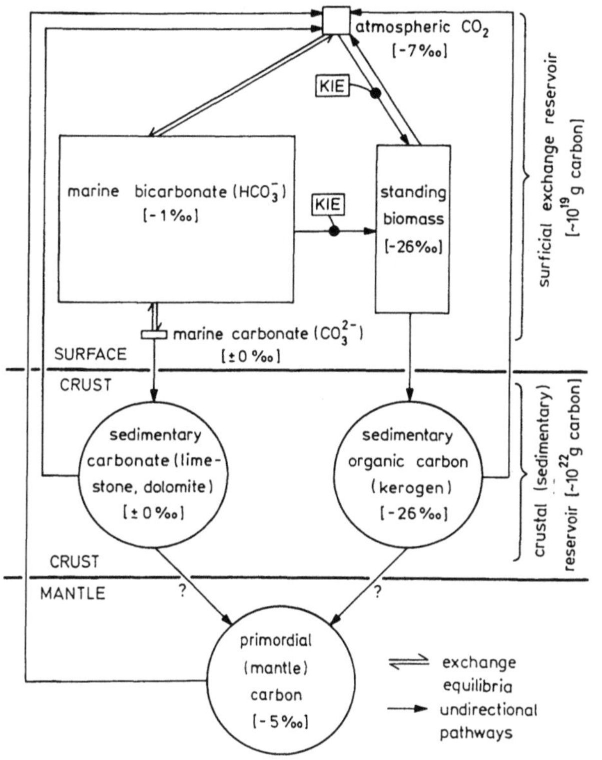

Fig. 1. Box model of the global carbon cycle showing characteristic bipartition into an inorganic (*left*) and organic branch (*right*). The relative proportions of the surficial reservoirs (*rectangular boxes*) are drawn approximately to scale; lithospheric reservoirs are represented as *circles* not indicative of any size hierarchy ($\delta^{13}C$ averages of individual reservoirs are given *in brackets*). Carbon fluxes into the biosphere are beset with kinetic isotope effects (*KIE*) responsible for a marked enrichment of ^{12}C in biogenic matter relative to the inorganic feeder pool, with the residual ^{13}C mostly accumulated in oceanic bicarbonate (the largest of the surficial carbon reservoirs). The isotopic disparity between the surficial carbon species is subsequently propagated into the rock section of the cycle where organic carbon and carbonate have piled up huge reservoirs within the sedimentary compartment (note that the amount of carbon stored in the crust exceeds that residing on the surface by three orders of magnitude)

go into solution in the newly formed oceans. With the subsequently formed bicarbonate and carbonate ions susceptible to precipitation as sedimentary carbonate, the equilibrium system

$$CO_{2\,(gas)} \rightleftharpoons CO_{2\,(aq)} \rightleftharpoons HCO_3^- \rightleftharpoons CO_3^{2-} \qquad (1)$$

provided a most effective channel for transferring volatile and dissolved carbon dioxide from the atmosphere and ocean to the solid crust, thereby sparing the planet any excessive accumulation in its gaseous envelope of a strong infrared absorber (and possibly the Venusian fate of a "thermal hell"). Incidentally, it can be shown that a reconversion to CO_2 of the totality of carbon currently stored in the Earth's sedimentary shell ($\sim 6 \times 10^{22}$ g) would yield a reservoir of the same order of magnitude as the CO_2-content of the present Venus atmosphere (5.3×10^{23} g).

A second sink for atmospheric carbon dioxide was established after the initiation of life processes on the ancient Earth, and notably the large-scale proliferation of the oldest prokaryotic microbial ecosystems as from early Archaean times. Ever since then, biological or "autotrophic" carbon fixation in the widest sense (involving the enzymatic reduction of CO_2) has effectively competed with the carbonate-forming process for the removal of carbon dioxide from the surficial exchange reservoir. This biological flux has ultimately resulted in the deposition within the crust of sizeable quantities of kerogen and related substances that constitute the bulk of fossil organic matter.

The dichotomy of the carbon flux from the atmosphere to the crust has given rise to a glaring bipartition of both the surficial and sedimentary carbon reservoirs as represented by the box model of Fig. 1. Continuous processing in the surficial compartment of the carbon dioxide released from the mantle has, in the fullness of time, not only brought about a large-scale partitioning of the planet's original endowment of primordial carbon between an oxidized (carbonate) and a reduced (kerogen) reservoir, but also a conspicuous isotopic disproportionation, with "light" carbon preferentially assigned to the organic moiety and the "heavy" species accumulating in the carbonate reservoir. Since both this partitioning and the isotopic differentiation of terrestrial carbon result from the influence of life processes, the basic configuration of the global carbon cycle as outlined in Fig. 1 clearly reflects the impact of the Earth's biosphere. Accordingly, the terrestrial carbon cycle may be regarded as the prototype of a *biogeochemical element cycle*.

We may reasonably infer that the equilibrium system carbon dioxide-bicarbonate-carbonate [Eq. (1)] had provided the first sink for atmospheric CO_2 on the primitive Earth, and that the second (biological) sink was established at a later stage as a result of the emergence of CO_2-fixing ("autotrophic") organisms, primarily prokaryotic photoautotrophs. It can, furthermore, be expected that the first major involvement of microbial photoautotrophs in the carbon cycle had been monitored by both the preservation of kerogenous materials in contemporaneous sediments and attendant carbon isotope evidence. Specifically, the isotopic manifestations of any large-scale activity of CO_2-fixing biota should have expressed themselves in the form of complementary geochemical signals in carbonates and kerogens of that particular time. With both carbon species commonly preserved in sediments, relevant isotopic evidence should have been encoded in the sedimentary record as from its start about 3.8 Ga ago.

The availability of a quasi-continuous carbon and carbon isotope record that covers almost 4 billion years of Earth history provides us with a unique store of data retrievable for the elucidation of the history of terrestrial biota as mirrored in the operation of the global carbon cycle. The following discourse will, after setting out some relevant physicochemical background information, retrace the basic features of this biogeochemical record and summarize their implications for the antiquity and evolution of life on Earth.

3 Isotope Fractionations in the Global Carbon Cycle

Carbon consists principally of a mixture of two stable isotopes, one light (^{12}C) and one heavy (^{13}C); a third, shortlived radioactive nuclide, ^{14}C, occurs only in trace amounts. All geochemical transformations of the element in the exogenic cycle are beset with thermodynamic and/or kinetic isotope effects that cause newly formed substances to be isotopically distinctive from their precursor materials. The largest isotope effects are brought about during biological processing of inorganic carbon in the various pathways of autotrophic (specifically photosynthetic) carbon fixation. It is these biologically mediated fractionations that are ultimately responsible for the isotopic disproportionation of the Earth's primordial carbon, with the subsequent establishment of two crustal carbon reservoirs of different isotopic composition.

Differences in the isotopic composition of the various forms of carbon are expressed in terms of the conventional δ-notation that gives the permil deviation in the $^{13}C/^{12}C$ ratio of a sample (sa) relative to that of a standard (st), i.e.,

$$\delta^{13}C = \left[\frac{(^{13}C/^{12}C)\ sa}{(^{13}C/^{12}C)\ st} - 1\right] \times 1000\,(‰,\ PDB).$$

The standard defining zero permil on the δ-scale is Peedee belemnite (PDB) with $^{12}C/^{13}C = 88.99$. Positive values of $\delta^{13}C$ indicate an enrichment of heavy carbon (^{13}C) in the sample, while negative values stand for a depletion in ^{13}C.

3.1 Isotope Fractionations in Biological (Autotrophic) Carbon Fixation

Autotrophic carbon fixation is primarily the assimilation of carbon dioxide (CO_2) and dissolved bicarbonate ion (HCO_3^-) by plants and microorganisms that proceeds by a limited number of biochemical pathways and carbon-fixing reactions (Table 1). Assimilation of carbon monoxide (CO) is unimportant from a quantitative point of view, whereas the rate of methane (CH_4) utilisation in methylotrophic pathways probably does not exceed a few percent of total biological carbon fixation amounting to some 10^{16} g organic carbon per year (cf. Woodwell et al. 1978). Hence, biological carbon fixation is largely synonymous with assimilation of carbon dioxide since the bulk of primary production in the present carbon cycle is clearly due to either chemically or solar-powered reduction of CO_2 to organic matter.

Since the pioneering work by Nier and Gulbransen (1939), Murphey and Nier (1941) and Rankama (1948) it is known that the conversion of inorganic carbon to living matter leads to a marked enrichment of the light carbon isotope (^{12}C) in the newly generated organic substances, while the heavy species (^{13}C) is preferentially retained in the residual substrate. Subsequent work performed by other investigators (cf. Craig 1953; Park and Epstein 1960) has confirmed that all common assimilatory pathways (and specifically the photosynthetic ones) discriminate against ^{13}C, mostly as a result of a kinetic isotope effect inherent in the first irreversible enzymatic CO_2-fixing reaction. With assimilatory reactions largely enzyme-controlled, and living systems constituting dynamic states undergoing rapid cycles of anabolism and catabolism, it has become generally accepted that most biological isotope fractionations are due to kinetic rather than equilibrium effects (Vogel 1980; O'Leary 1981; for dissent see Galimov 1985).

Kinetic isotope fractionations specifically reflect differences in either the reaction or translocation rates between the heavy and light carbon isotope and are primarily imposed on two steps in the primary metabolism of autotrophic organisms, namely, (1) the diffusion of external CO_2 to the assimilatory reaction centers, and (2) the first irreversible enzymatic fixation of CO_2 in the carboxyl (COOH) group of an organic (carboxylic) acid. On the other hand,

Table 1. Roster of principal CO_2-fixing carboxylation reactions utilized in common autotrophic pathways (abridged from Schidlowski et al. 1983). Note that reduction of CO_2 primarily gives rise to C3 compounds (with 3-carbon skeletons such as phosphoglycerate and pyruvate), C4 compounds (oxaloacetate) and C2 compounds (acetate, acetyl coenzyme A). The quantitatively most important enzymatic carboxylation is the RuBP carboxylase reaction of the Calvin cycle (No. 1) that forms the initial carbon-fixing reaction in C3 photosynthesis and is to be credited for the generation of the bulk of the Earth's biomass (extant and fossil). Groups of organisms for which carbon isotope fractionations are known are identified by asterisks (see also Fig. 3)

1. CO_2 + ribulose-1,5-bisphosphate → phosphoglycerate
   Operated by: *C3 plants, *algae, *cyanobacteria, *purple photosynthetic bacteria (Chromatiaceae), chemoautotrophic bacteria
2. CO_2/HCO_3^- + phosphoenolpyruvate/pyruvate → oxaloacetate
   Operated by: *C4 plants[a], *CAM plants[a], anaerobic and facultatively anaerobic bacteria
3. CO_2 + CO_2 → acetate/acetyl coenzyme A
   Operated by: *Green photosynthetic bacteria (Chlorobiaceae)[b], anaerobic bacteria, *methanogenic bacteria[c]
4. CO_2 + acetyl coenzyme A → phosphoenolpyruvate/pyruvate
   Operated by: *Green photosynthetic bacteria (Chlorobiaceae)[d], autotrophic sulfate-reducing bacteria, *methanogenic bacteria[c]

[a] C4 and CAM plants combine this carboxylation with reaction 1.
[b] Primary CO_2 fixation via succinyl coenzyme A and α-ketoglutarate.
[c] Primary CO_2 fixation probably via C1 acceptors. [Details of the assimilatory pathway of methanogens are as yet poorly known, but the presence of both α-ketoglutarate and pyruvate synthases suggests the involvement of reactions 3 and 4]
[d] Combined with reactions 3 and 2.

thermodynamically controlled *equilibrium fractionations* determine the isotope exchange between CO_2 and HCO_3^- in bicarbonate-utilizing pathways such as C4 and CAM photosynthesis where bicarbonate ion serves as the "active" feeder species. Moreover, equilibrium fractionations have been proposed to also govern the inter- and intramolecular isotope exchange among different classes of biosynthesized metabolites such as proteins, carbohydrates, and lipids (Galimov 1985).

Figure 2 gives a graphic synopsis of the salient isotope-discriminating steps in autotrophic carbon fixation. In concert, the individual fractionations commonly bring about a sizeable shift of the $\delta^{13}C$ values of biosynthesized matter towards negative readings relative to the feeder pool of inorganic carbon (mostly CO_2). This overall shift can be expressed as the difference between the isotopic composition of the cells and that of the parent substrate, i.e.,

$$\Delta = \delta^{13}C_{cells} - \delta^{13}C_{CO_2}. \quad (2)$$

The kinetic isotope effect imposed on the initial diffusion step (Fig. 2, k_1 and k_2) discriminates only slightly against ^{13}C, attaining as a maximum the value for CO_2 diffusion in air (− 4.4‰). Since the "pure" effect for gaseous diffusion is commonly modulated in the natural environment by dissolution, hydration, and liquid transport of CO_2, actually observed fractionations usually stay well below this maximum. With liquid diffusion considerably retarded as compared to the gaseous process, the concomitant isotope effects are commonly small (− 1.6 to − 3.2‰; Vogel 1980), or may even approach unity within a few tenths of a permil (O'Leary 1981). Such

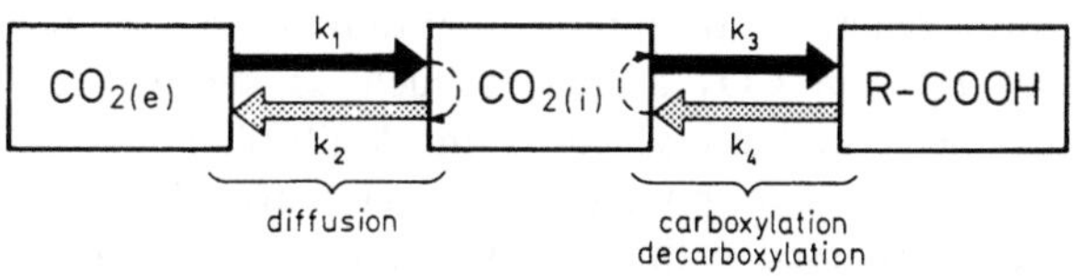

Fig. 2. Synopsis of the principal isotope-discriminating steps in the primary metabolism of CO_2-fixing ("autotrophic") plants and microorganisms (*black* assimilatory reactions; *stippled* dissimilatory and other reverse processes; $k_1 - k_4$ are corresponding rate constants). $CO_{2(e)}$ and $CO_{2(i)}$ stand for environmental and internal (leaf-hosted) CO_2, respectively, and R−COOH represents the product of the first CO_2-fixing enzymatic carboxylation. In concert, these processes lead to a preferential enrichment of light carbon (^{12}C) in the synthesized biomass of the right-hand box; the largest single fractionation effect is commonly associated with the enzymatic carboxylation step

minor fractionations have been specifically observed in aquatic plants and microorganisms whose carbon-fixing pathways are largely diffusion-limited (cf. Benedict et al. 1980; Schidlowski et al. 1984).

Fractionations in the subsequent enzymatic carboxylation step (Fig. 2, k_3) are, on the other hand, significantly larger but highly variable in detail. In the case of the quantitatively dominant ribulose-1,5-bisphosphate (RuBP) carboxylase reaction (Table 1, No. 1), the magnitude of the isotope effect has been shown to mostly range between -20 and $-40‰$. This extended range obviously derives from the fact that fractionations in enzymatic reactions vary widely as a function of pH, metal cofactor, temperature, and a number of other variables (Winkler et al. 1982). Since the carboxylation product emerging from this reaction is a compound with a 3-carbon skeleton (phosphoglycerate) that immediately enters the reductive pentose phosphate (or Calvin) cycle, the corresponding pathway has been termed C3 (or Calvin cycle) photosynthesis. The Calvin cycle constitutes the principal contrivance for the biologically mediated reduction of CO_2 to the carbohydrate level that channels most of the carbon transfer from the nonliving to the living world. It is utilized by all green plants (those relying on it exclusively are called C3 plants), eukaryotic algae, and the bulk of photoautotrophic and chemoautotrophic bacteria.

A quantitatively less important carboxylation reaction that fixes CO_2 as a 4-carbon compound (oxaloacetate) is catalyzed by the enzyme phosphoenolpyruvate (PEP) carboxylase (Table 1, No. 2). This carboxylation figures as the initial carbon-fixing reaction in the C4 dicarboxylic acid (or Hatch-Slack) pathway and entails a minor discrimination against ^{13}C on the order of -2 to $-3‰$ relative to bicarbonate ion that serves as the "active" feeder species in this particular case. The isotope effects brought about by some ferredoxin-linked carboxylation reactions of minor quantitative importance that rely on CO_2 acceptors such as succinyl coenzyme A and acetyl coenzyme A (Table 1, Nos. 3 and 4) are as yet poorly known. However, fractionations observed with organisms that have these reactions integrated in their assimilatory pathways are mostly smaller than of those utilizing the common enzymatic carboxylations.

Apart from fractionations inherent in the assimilatory pathways, enzymatic decarboxylations and related dissimilatory processes (Fig. 2, k_4) may also contribute to the overall isotopic composition, notably of higher plants. Although conflicting results have been reported with regard to the magnitude and even the direction of the isotope effects inherent in these processes, *in vivo* decarboxylations appear to release CO_2 that is isotopically lighter than the parent organic material (O'Leary 1981), with isotope fractionations thus running counter to those of the assimilatory pathway. Most probably, discrepancies between fractionations predicted by the isotope discriminating properties of specific enzymes and those actually observed can be ascribed to the antagonistic effect of respiratory decarboxylations. While, for instance, fractionations caused by RuBP carboxylase mostly average between -25 and $-35‰$, gross fractionations displayed by C3 plants lie close to a mean of $-20‰$ [i.e., at -27 on the δ-scale when atmospheric carbon dioxide ($\delta^{13}C = -7‰$) serves as feeder substrate]. It is well known that photorespiration is particularly pronounced in C3 species due to the "oxygenase" activity of RuBP carboxylase, this also placing limits on the productivity of C3 plants (Walker 1979). Although such processes seem potentially capable of making substantial contributions to the isotope budget of plants, the small number of detailed investigations so far available leaves important aspects of the issue open.

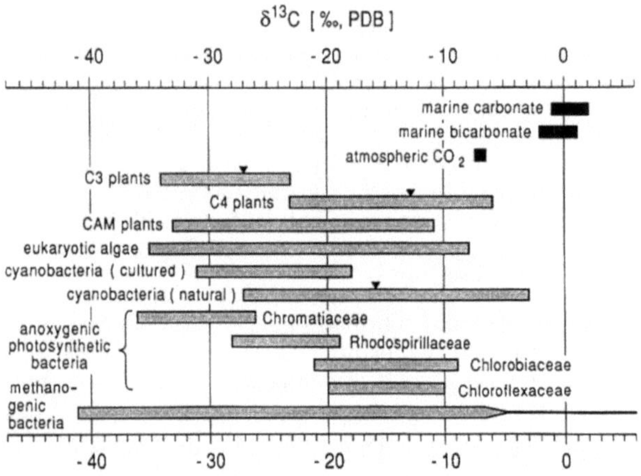

Fig. 3. Carbon isotope spreads of major groups of higher plants and autotrophic microorganisms that figure as principal primary producers in the present biosphere (*triangles* denote approximate means). Respective spreads of oxidized carbon compounds in the surficial environment (CO_2, HCO_3^-, CO_3^{2-}) are shown in *black*. The consistently negative $\delta^{13}C$ ranges of biogenic materials imply an enrichment in light carbon (^{12}C) relative to the inorganic feeder species (mostly CO_2). Note that the positive extremes shown for methanogens were obtained in culture experiments under specific conditions that are irrelevant for natural communities

Figure 3 summarizes the net results of the interplay of these various fractionation processes in the form of observed $\delta^{13}C_{org}$ spreads for the principal groups of extant higher plants, eukaryotic algae, and autotrophic (photosynthetic and chemosynthetic) bacteria [fractionations expressed by these ranges are gross fractionations as defined by Eq. (1)]. It can be inferred from this compilation that, depending on which of the isotope-discriminating steps shown in Fig. 2 and/or which of the carbon-fixing reactions listed in Table 1 become rate-controlling in the specific instance, the isotopic composition of plants and autotrophic microorganisms may vary over an extended range. A detailed and exhaustively referenced discussion of the individual isotope distribution patterns represented in Fig. 3 has been given elsewhere (Schidlowski 1983, 1987; Schidlowski et al. 1983) and will not be repeated here. Parenthetically, mention should be made that the C4 dicarboxylic acid pathway is a relatively late achievement in the evolution of flowering plants (angiosperms) not predating the onset of the Cretaceous (~ 130 million years ago). Hence, the impact of C4 photosynthesis on the geochemical carbon cycle has been as yet negligible. Occurrences in the present world of both C4 and CAM (Crassulacean Acid Metabolism) species are limited or at best moderate, with the former primarily represented by tropical grasses and the latter by succulents.

If weighted for their relative contributions to the contemporary standing biomass, the isotope spreads of the principal groups of primary producers presented in Fig. 3 would indicate that the $\delta^{13}C$ values of the global biomass are, on average, between 20 and 30‰ more negative than those of oceanic bicarbonate, the most abundant inorganic carbon species in our environment (cf. Fig. 1). It should be noted that this conspicuous enrichment of ^{12}C in organic matter derives, for the most part, from the isotope discriminating properties of the key enzyme of the Calvin cycle, RuBP carboxylase. Channeling the bulk of the carbon transfer from the inorganic to the living realm, the RuBP carboxylase reaction was necessarily bound to impose its isotopic signature on the terrestrial biosphere as a whole, imparting to it a principal $\delta^{13}C_{org}$ range of about -26 ± 7‰.

3.2 Isotope Fractionations in Carbonate Formation

Compared to the large fractionations in the biological carbon cycle, isotope discriminations in the operation of the carbon dioxide-bicarbonate-carbonate system [Eq. (1), above] that provides the link between atmospheric CO_2 and sedimentary carbonates are rather moderate. The isotopic compositions of the various oxidized carbon species figuring as intermediates in this system are determined by thermodynamically controlled exchange equilibria that have been investigated by several workers (see, inter alia, Vogel 1961; Deuser and Degens 1967; Wendt 1968; Mook et al. 1974). In the temperature range 0–30 °C, carbon dioxide dissolved in water [$CO_{2(aq)}$] was found to be lower in $\delta^{13}C$ by 1.0–1.2‰ relative to gaseous CO_2, while corresponding fractionations between $CO_{2(aq)}$ and HCO_3^- lead to an increase in $\delta^{13}C$ between 7.4 and 10.8‰ in dissolved bicarbonate ion (Mook et al. 1974). Fractionations between bicarbonate and precipitated carbonate are again small, raising $\delta^{13}C$ in the solid phase just by about 1‰ above the level of its HCO_3^- parent. In biologically mediated carbonate formation, both kinetic and metabolic effects may interfere with inorganic bicarbonate-carbonate equilibria, but the majority of biological calcification processes appear to have occurred under near-equilibrium conditions.

The above equilibria are broadly reflected in the present environment where $\delta^{13}C$ of atmospheric CO_2 lies at -7‰, with the average for marine bicarbonate consequently fixed between -1 and 0‰. The mean for carbonate is then shifted by roughly one more permil in positive direction (see Fig. 3). Hence, marine carbonate rocks closely approach the isotopic composition of their bicarbonate precursors, thus monitoring the $^{13}C/^{12}C$ ratios of the Earth's oceanic bicarbonate pool through geologic time (Keith and Weber 1964; Schidlowski et al. 1975; Veizer and Hoefs 1976).

3.3 Postdepositional Isotope Fractionations in Sedimentary Carbonate and Organic Carbon

Both carbonate carbon (C_{carb}) and organic carbon (C_{org}) are continuously transferred from the surficial environment to the Earth's sedimentary shell where they have formed, over the ages, formidable reservoirs on the order of 10^{22} g each (Fig. 1). During diagenesis and metamorphism, the primary isotopic signatures of both carbon species are apt to suffer secondary overprints as part of the overall changes experienced by their host rocks during a long and annealing geological history. To retain validity as a carrier of biogeochemical information, the isotopic evidence encoded in the record ought to be screened for secondary (postdepositional) alterations that might have affected the primary $^{13}C/^{12}C$ signal. Specifically, a conclusive interpretation of the carbon isotope age curve (see Fig. 5, below) is conditional on an understanding of diagenetic and metamorphic processes capable of causing postdepositional $^{13}C/^{12}C$ changes in surficially formed carbonate and organic matter.

Diagenetically stabilized *sedimentary carbonates* have been shown to preserve the isotopic composition of original carbonate muds within $\pm 1‰$ of the primary value which, in turn, is inherited from a bicarbonate precursor with a shift of about $+1‰$ due to the equilibrium fractionation between HCO_3^- and CO_3^{2-}. Hence, carbonates closely reflect the $^{13}C/^{12}C$ ratio of the parent bicarbonate pool which makes the marine $\delta^{13}C_{carb}$ record a fairly reliable sensor for the isotopic evolution of oceanic HCO_3^-.

During metamorphism, carbonates may suffer more pronounced changes in their isotopic composition. With the exception of pure marbles which retain their $^{13}C/^{12}C$ ratios virtually unchanged through successive stages of metamorphism, common impure (notably siliceous) carbonates undergo decarbonation reactions in which the primary carbonate phases (mostly calcite and dolomite) react with silica to form Ca-Mg-silicates (such as tremolite and diopside) with a concomitant release of CO_2, e.g., 5 dolomite + 8 SiO_2 + 1 $H_2O \rightarrow$ 1 tremolite + 3 calcite + 7 CO_2. Decarbonation processes of this type start already in the greenschist facies (between 300 and 450 °C) and increase with increasing metamorphic grade. They give rise to residual carbonates enriched in ^{12}C since isotopic equilibria between CO_3^{2-} and gaseous CO_2 favor the release of ^{13}C along with the volatile CO_2 phase in this temperature range (Bottinga 1969). As a result, the $\delta^{13}C$ values of metamorphosed carbonates are commonly shifted by 2–5‰ in negative direction (Deines and Gold 1969; Shieh and Taylor 1969; Schidlowski et al. 1979) relative to their primary values around zero permil. Moreover, the magnitude of the observed isotope shift has been found to often correlate with the modal abundances of newly formed calc-silicates (Sheppard and Schwarcz 1970).

Isotope changes during diagenetic reconstitution of *organic matter* are larger than those of carbonates, with shifts in the range from a few to several permil over the maturation pathway of the organic substances. The formation of kerogen, the polycondensed end-product of the diagenetic alteration of biogenic substances (Durand 1980), entails a moderate increase of ^{13}C in high-rank kerogens and derivative graphites, primarily because the isotopically light complement is preferentially relegated to the mobile hydrocarbon fraction that tends to separate itself from the parent kerogenous material. The total isotope change over the maturation pathway sums up the fractionations inherent in a number of discrete steps in the diagenetic process, specifically (1) selective loss of ^{12}C-enriched lipids and hydrocarbons during dehydrogenation of the maturing kerogens, (2) selective loss of isotopically distinctive functional groups during microbial and thermal degradation, (3) preferential splitting of $^{12}C-{}^{12}C$ bonds in response to increasing thermocatalytic stress (with attendant removal of the isotopically light crack products), and (4) high-temperature exposure of kerogenous materials to oxidizing conditions with ensuing preferential mobilization of isotopically light carbon (e.g., $2\,C + 2\,H_2O \rightarrow CH_4 + CO_2$, or $C + 6\,Fe_2O_3 \rightarrow 4\,Fe_3O_4 + CO_2$).

As has been demonstrated by several workers (Chung and Sackett 1979; Peters et al. 1981; Hayes et al. 1983; Galimov 1985; and others), these and related processes bring about a marked enrichment of ^{13}C in high-rank (mature) kerogens, while at the same time assembling the light fraction in the form of oil and gaseous hydrocarbons (preferentially methane). A calculation of the ^{12}C depletion potentially linked to a near-complete dehydrogenation of immature kerogenous materials, as well as $\delta^{13}C_{org}$ shifts recorded from actually observed kerogen maturation series, suggest an isotope effect on the order of $+3‰$ as a most probable upper limit for the total maturation pathway. Process (4) may be capable of inducing larger changes, but the scarcity of oxidizing environments in both the diagenetic and metamorphic realms (which are inherently reducing) places limits on the quantitative role of oxidation processes for promoting isotope changes of the above type. It is in keeping with such inferences that $\delta^{13}C_{org}$ values of graphitized kerogens from carbonate-free sequences stay broadly within the scatter attributable to their biological precursor materials even in high-grade metamorphic terranes (cf. Gavelin 1957).

On the other hand, the processes described cannot account for the extreme depletion in ^{12}C reported for largely graphitized kerogens from selected metamorphic suites where $\delta^{13}C_{org}$ values have apparently been shifted from original means close to $-25‰$ to about $-10‰$ and more (cf. McKirdy and Powell 1974; Eichmann and Schidlowski 1975; Hoefs and Frey 1976; Schidlowski et al. 1979; Hayes et al. 1983). As it turns out, the host sequences of these highly altered organic substances always contain carbonates. Accordingly, there is little doubt that high-temperature isotope exchange between coexisting C_{org} and C_{carb} must be ultimately responsible for the formation of unusually heavy kerogens and derivative graphites from many metamorphic

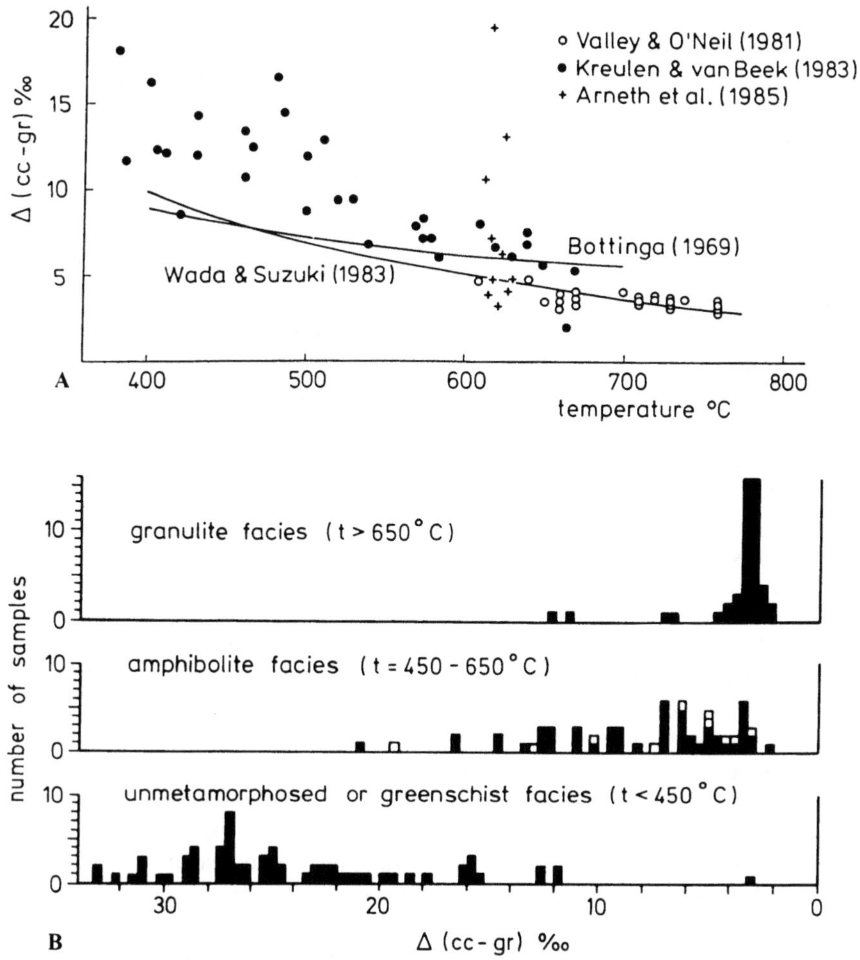

Fig. 4A, B. Carbon isotope fractionation $\Delta(cc-gr)$ between coexisting calcite (*cc*) and graphite (*gr*) as a function of **A** increasing temperature and **B** metamorphic facies (note that C_{org} occurs preferably as graphite at higher metamorphic grades). In **A**, $\Delta(cc-gr)$ values reported by several authors from different metamorphic terranes are shown to scatter around both Bottinga's (1969) function of thermodynamically calculated isotope equilibria and an empirical fractionation curve by Wada and Suzuki (1983) calibrated by dolomite-calcite solvus temperatures (note persistence of single discordant values up to 600–650 °C). **B** Distribution patterns of $\Delta(cc-gr)$ for different metamorphic facies indicating a systematic decrease of average fractionations with increasing metamorphic grade (after Valley and O'Neil 1981; additional data for the amphibolite facies, *open squares*, from Arneth et al. 1985). The reduction of scatter along with the emergence of a slender peak in the granulite facies suggests that isotopic reequilibration between the two carbon species is virtually attained at $t \geqq 650$ °C

terranes. As has been shown by Valley and O'Neil (1981), isotopic reequilibration between kerogen and isotopically heavy CO_2 released by metamorphic decarbonation reactions starts in the lower greenschist facies (300–400 °C), with fractionations between carbonate and reduced carbon steadily decreasing with increasing metamorphic temperatures (Fig. 4A) as dictated by thermodynamically controlled isotope exchange equilibria (Bottinga 1969). As a result, recorded differences in the isotopic compositions of sedimentary carbonates and kerogenous materials demonstrably shrink through metamorphic grades, with thermodynamic equilibrium closely approached in the high-temperature range of the granulite facies (> 650 °C; cf. Fig. 4B).

Summing up the impact of postdepositional processes on the isotopic composition of carbonate and organic carbon, it is obvious that diagenesis and metamorphism are capable of imposing moderate to sizeable overprints on the original $\delta^{13}C$ values of both sedimentary carbon species. However, these secondary changes commonly just blur, and never seriously obliterate, the primary isotopic signatures of the precursor substances. This is even true for the excessively ^{13}C-enriched graphitic constituents of selected high-grade metasediments whose pedigree can be readily traced back to kerogens of common isotopic composition in view of unavoidable high-temperature $^{13}C/^{12}C$ exchange with coexisting carbonate. The potential of H/C ratios for assessing the postdepositional $^{13}C/^{12}C$ exchange in kerogens and derivative graphites is being discussed by Strauss et al. (this Vol.).

4 Record of Organic Carbon and Carbonate Through Geologic Time

As has been definitively confirmed by a fair number of investigations (Schidlowski et al. 1975, 1979, 1983; Veizer and Hoefs 1976; McKirdy and Hahn 1982; Hayes et al. 1983; and others), there exists a continuous record of organic carbon (kerogen) and carbonate in sedimentary rocks that can be traced back to the 3.8-Ga-old Isua Suite of West Greenland (Allaart 1976; Schidlowski et al. 1979). Moreover, systematic assays carried out on the organic carbon fraction of Phanerozoic sediments from both North America (Trask and Patnode 1942) and the Russian Platform (Ronov 1958, 1980) have revealed that the C_{org} content of the average sediment oscillates moderately around a mean of perhaps 0.5% over this time interval. Although both the preserved rock record and the available data base are becoming progressively attenuated when followed up into the more remote geological past, evidence has been presented by many workers that the average content of organic carbon in Precambrian (inclusive of Archaean) sediments is virtually the same as in Phanerozoic rocks (see, *inter alia*, Reimer et al. 1979; Cameron and Garrels 1980; Schidlowski 1982; Hayes et al. 1983). It seems worth noting that the largely graphitized kerogen burden of the 3.8-Ga-old Isua metasediments that mark the beginning of the record amounts to $> 0.6\%$ C_{org} in the case of carbon-rich members of the suite (Schidlowski 1982).

Though at first sight astounding, the surprisingly modern C_{org} content of Archaean sediments may not be unusual at all. It is by now firmly established that microbial communities figure among the biosphere's most productive ecosystems, with specifically benthic prokaryotes (such as cyanobacteria) capable of sustaining prodigious rates of primary productivity between 8–12 g C_{org}/m^2 day (Krumbein and Cohen 1977; Cohen et al. 1980). There is, moreover, little doubt that the early Precambrian was the "Golden Age" of prokaryotic ecosystems (Cloud 1976; Schopf 1983). If such high rates of primary production can be sustained by microbial photoautotrophs operating at the prokaryotic level, a reasonable case can be made that global production of biomass per time (and the pro-rata concomitant C_{org}-flux into sediments) cannot have increased too much during subsequent evolution of photosynthesis. As any generation of biomass is ultimately nutrient-limited, it seems almost cogent to envisage for the Earth's earliest biosphere a scenario characterized by a state of global plenitude or "biotic saturation" (Stanley 1981) in which microbial ecosystems had proliferated in existing aquatic habitats to limits set by the availability of critical resources, notably phosphorus and nitrogen.

The record of sedimentary carbonate appears so well documented as to require no detailed exposition here. Suffice is to say that the very existence of carbonate beds in the Isua metasediments gives eloquent testimony that the solubility product $[Ca^{2+}][CO_3^{2-}] = k$ had been reached already in the Early Archaean ocean. Ever since then, there is evidence of a quasi-continuous carbonate record suggesting that the equilibrium system $CO_2 - HCO_3^- - CO_3^{2-}$ [see Eq. (1)] had governed the low-temperature geochemistry of inorganic carbon at or near the Earth's surface. From the environmental point of view the operation of this system is of crucial importance in that it (1) acts as an effective buffer for both H^+ and OH^- ions in the aqueous (specifically marine) realm and (2) constitutes a "pump" continuously removing CO_2 from the atmosphere with ensuing fixation in the crust. According to Veizer et al. (1989), the geologically oldest carbonates bear a distinct "mantle" signature (notably in their Mn, Fe and $^{87}Sr/^{86}Sr$ characteristics) imparted to them by the proximity to their parent greenstone belt environments of ubiquitous oceanic crust and the associated volcanism. It was only after the large-scale establishment of stable shelf platforms in the Lower Proterozoic ($t < 2.5$ Ga) that the shelf carbonate facies came to dominate the process of marine carbonate formation.

Summing up the evidence, it may be stated with fair confidence that the generation at the Earth's surface of organic carbon and carbonate had commenced at least with the onset of the record 3.8 Ga ago. Since then, both carbon species have continuously entered newly-formed sediments, figuring as common constituents of crustal rocks over the whole of subsequent Earth history. There is reason to believe that the C_{org} content of the average sediment had stayed fairly uniform from Archaean to Recent, oscillating within a few tenths of a percent around a long-term mean of 0.5–0.6%. Accepting the lower bound of this mean and 2.4×10^{24} g as reasonable estimate for the mass of the Earth's sedimentary shell (Garrels and Mackenzie 1971), we would arrive at a figure of 1.2×10^{22} g for the reservoir of sedimentary organic carbon. With various lines of evidence indicating that the reservoir of carbonate-bound carbon is about $4\times$ larger, the mass of C_{carb} fixed in sedimentary rocks would fall just short of 5×10^{22} g. Together, C_{org} plus C_{carb} amount to about 6×10^{22} g equivalent to $\sim 3\%$ of the total sedimentary mass.

5 Operation of the Carbon Cycle on Variable Time Scales

The turnover of carbon in the exogenic cycle as illustrated by the box model of Fig. 1 involves a complex interaction of participating sources, sinks, and reservoirs. A crucial question in this context is whether this cycling has proceeded with a certain degree of uniformity through time allowing the principal reservoirs to approximately persist as steady-state quantities, or whether the operation of the cycle was subject to pulsations with correspondingly changing fluxes and reservoirs. Imbalances in the carbon cycle were apt to have major repercussions on the interlocking sulfur, iron, and oxygen cycles that are coupled with the carbon cycle by a number of geochemically important redox reactions. With photosynthetic oxygen figuring as the principal vehicle of oxidation power in these interactions, imbalances specifically of the C_{org} branch of the cycle have almost certainly affected the redox regime of terrestrial near-surface environments. Likewise, concentration changes during perturbations of the cycle of a radiatively active atmospheric gas like CO_2 would also be of immediate environmental significance.

Any monitoring of changes in the source and sink functions and the concomitant redistribution of carbon among the reservoirs will necessarily depend on the availability of geochemical recorders that reflect these changes with an adequate degree of time resolution. Because of its small quantity and pivotal role in the surficial exchange reservoir, the CO_2-content of the atmosphere would provide an ideal sensor since perturbations in the system would primarily manifest themselves as changes in atmospheric P_{CO_2}. Unfortunately, this index parameter can only be retrieved for the most recent geological past (cf. Sect. 5.3, below). Over Earth history as a whole, carbon isotopes have proved to be the most powerful single sensors of the mode of operation of the cycle, giving specifically information on the emergence and subsequent fluctuations of the sink function of organic carbon.

The sensor properties of $^{13}C/^{12}C$ ratios primarily derive from the fact that the isotopic compositions of C_{org} and C_{carb} are coupled with their relative proportions in the sedimentary reservoir as a result of the contraints of an isotope mass balance. If a primary carbon flux into the atmosphere-ocean system with the isotopic composition of primordial mantle carbon ($\delta^{13}C_{prim} \cong -5‰$; cf. Deines, this Vol.) were partitioned between C_{org} and C_{carb} with a fractionation of $-25‰$ imposed on the formation of biogenic carbon, resulting $\delta^{13}C_{carb}$ values around zero permil (as actually encountered in the bulk of marine carbonates) would indicate a ratio $C_{org}/C_{carb} = 1/4$ in the surficial exchange reservoir. This follows from the relationship

$$\delta^{13}C_{prim} = R\delta^{13}C_{org} + (1-R)\,\delta^{13}C_{carb}, \tag{3}$$

that would render $R = C_{org}/(C_{org} + C_{carb}) = 0.2$ and, consequently, $C_{org}/C_{carb} = 0.2/0.8 = 1/4$ for the above numerical parameters (implicitly, $\delta^{13}C_{org} = \delta^{13}C_{carb} - 25‰$, which shift reflects the average isotope fractionation inherent in biological carbon fixation). If the surficial carbon compartment were flushed at a rapid rate as it is today, with a residence time of oceanic bicarbonate on the order of 10^5 a (see Holland 1978, p. 156), a quasi-steady state in compliance with Eq. (3) would be established within a geologically short time interval.

Hence, any biologically induced isotopic fractionation of the Earth's original endowment of primordial carbon is due to be monitored by the $\delta^{13}C$ values of both exogenic carbon species, i.e., C_{org} and C_{carb}. Since biological fractionations produce a wide scatter of $\delta^{13}C_{org}$ values (cf. Fig. 3), the state of the carbon system in the surficial exchange reservoir is best reflected by the narrow mean of oceanic bicarbonate that is subsequently encoded in carbonate with a minor additional fractionation slightly exceeding $+1‰$. Because of the rapid mixing of the oceans which store $> 80\%$ of total carbon residing in the surficial compartment in the form of HCO_3^-, the global oceanic bicarbonate mean is a most reliable integrator of the surficial carbon reservoir and likely to convey a faithful signal of the isotopic state of the system at any given time. Accordingly, the reflection of this mean in the $^{13}C/^{12}C$ ratios of sedimentary carbonates renders the $\delta^{13}C_{carb}$ record extremely suitable for tracking fluctuations in the relative proportions of organic and inorganic carbon over the history of the carbon cycle.

Given the constraints of Eq. (3), any shift of $\delta^{13}C_{carb}$ from the primordial value close to $-5‰$ in positive direction can be taken as evidence that part of the carbon flux from the mantle to the surficial exchange compartment had been converted to organic carbon since preferential incorporation of ^{12}C in biogenic materials was apt to be monitored by rising $\delta^{13}C_{carb}$ values. Moreover, the magnitude of the observed $\delta^{13}C_{carb}$ shift is a measure of the fraction of the total carbon flux that ends up as organic carbon, with increasing $\delta^{13}C_{carb}$ values reflecting correspondingly increased percentages of C_{org} in the system. In the following, this approach will be applied for elucidating the mode of operation of the carbon cycle on time scales ranging from the largest observable ($10^8 - 10^9$ a) to the interannual (1 – 10 a) and intraannual (seasonal).

5.1 Initiation and Long-Term ($\sim 10^9$ a) Operation of the Biogeochemical Carbon Cycle

As already pointed out above, we may reasonably expect that the equilibrium system $CO_2-HCO_3^- - CO_3^{2-}$ of Eq. (1) has acted as a first sink for atmospheric CO_2 after the formation of the oldest terrestrial hydrosphere on the surface of the juvenile planet. With the residence time of bicarbonate in the ocean short on geological time scales ($\sim 10^5$ a), a steady state should have been established fairly rapidly for the atmosphere-ocean system in which both the quantity and isotopic composition of the carbon input must have equalled that of the output. Hence, if primordial mantle carbon with $\delta^{13}C_{prim} \approx -5‰$ had entered this compartment in the form of CO_2, it was due to leave as carbonate with the same isotopic composition, provided the system had been given adequate time to adapt itself to the prevailing equilibria. Broecker (1970) had been the first to point out explicitly that sedimentary carbonates formed in the absence of a coexisting sink for organic carbon should exhibit $\delta^{13}C_{carb}$ values matching those of the primordial carbon supplied to the surficial reservoir.

The advent of life on the ancient Earth, and notably the first large-scale proliferation of microbial (prokaryotic) ecosystems, were apt to profoundly upset this scenario. With carbon constituting the key element of life, the emergence of CO_2-fixing organisms must have diverted a substantial fraction of the previous carbon flux to the newly established sink sequestering the element with a pronounced bias in favor of the light isotope. Due to the preferential concentration of ^{12}C in organic substances, the residual pool of inorganic carbon became progressively enriched in ^{13}C which increase was, in turn, monitored by rising $\delta^{13}C_{carb}$ values in marine bicarbonate and subsequently encoded in sedimentary carbonates. As follows from the relationship given in Eq. (3), a rise in $\delta^{13}C_{carb}$ from $-5‰$ to zero permil would imply that 20% ($R = 0.2$) of total carbon residing in the surficial exchange reservoir had come to be transformed to C_{org}. Hence, the isotopic signal of the marine bicarbonate-carbonate system as preserved in sedimentary carbonates constitutes a most sensible recorder of changes in the relative proportions of organic and inorganic carbon through time that ultimately reflect pulsations of biospheric activity in the surficial compartment of the global carbon cycle.

This background information, and other relevant facts on carbon isotope fractionations in the geochemical cycle set out in Sect. 3, provide us with the tools for an interpretation of the isotope age functions of organic and carbonate carbon that cover almost four billion years of recorded Earth history (Fig. 5). The envelopes shown for these functions together lump a data base approaching 10000 individual measurements retrieved from a galaxy of published work most of which dating back to the last two decades.

As can be readily inferred from Fig. 5, both the carbon isotope spreads of recent marine bicarbonate and carbonate and those of extant primary producers are transcribed into the record with minor oscillations back to 3.5, if not 3.8 Ga. This constitutes *prima facie* evidence that C_{carb} and C_{org} had been always transferred from the surficial environment to sedimentary rocks with indeed little change in their isotopic compositions that, specifically in the case of organic carbon, stayed well within the scatter of primary values. An immediate conclusion forced upon the investigator confronted with this formidable data collection is that biologically mediated isotope fractionations between inorganic and organic carbon clearly persist to at least 3.5 Ga ago. Accepting an early proposal that the markedly reduced fractionations encountered in the 3.8-Ga-old Isua Suite are due to a metamorphic overprint (Schidlowski et al. 1979), the isotopic evidence would even date back to the very beginning of the sedimentary record. As has been borne out, meanwhile, by a wealth of corroborative data (summarized in Fig. 4), the observed Isua anomaly is fully consistent with the predictable effects of a high-temperature isotopic reequlibration between coexisting sedimentary carbonate and kerogenous materials in response to the amphibolite-grade reconstitution of the host sequence. Such and related findings, along with observational evidence from younger metamorphic terranes have, of late, permitted a fairly conclusive

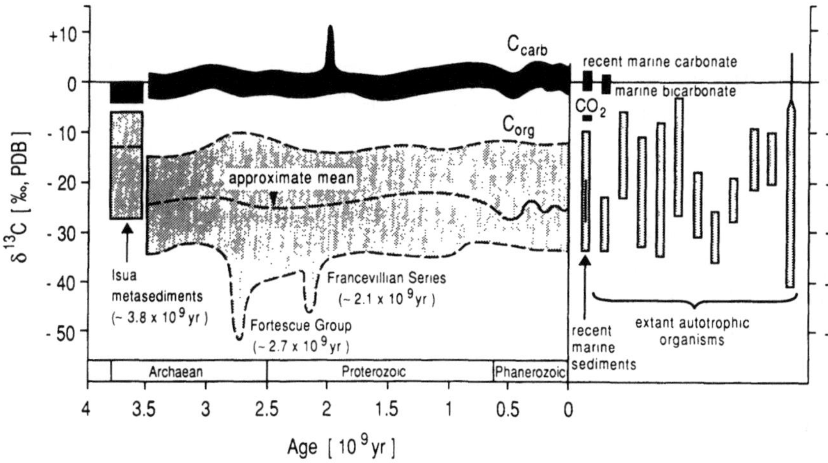

Fig. 5. Isotope age functions of sedimentary carbonate (C_{carb}) and organic carbon (C_{org}) over 3.8 Ga (1 Ga = 10^9 a) of recorded Earth history (right part of graph summarizes spreads of their progenitor materials as shown in Fig. 3). Note that the $\delta^{13}C$ ranges of both marine bicarbonate and the extant biomass are basically transcribed into recent sediments and the subsequent record back to its onset (with the Isua values reset by amphibolite-grade metamorphism, cf. Fig. 4). The envelope shown for fossil organic carbon is an update of the data base presented by Schidlowski et al. (1983), comprising the means of > 150 Precambrian kerogen provinces as well as the currently available information on the Phanerozoic record. Negative excursions of $\delta^{13}C_{org}$ at 2.7 and 2.1 Ga indicate an involvement of methane in the formation of the respective kerogen precursors on a local scale, whereas a prominent positive spike imposed on the $\delta^{13}C_{carb}$ record around 2.0 Ga ago might reflect a major disturbance in the operation of the carbon cycle (for details see text). The *black insert* in the $\delta^{13}C_{org}$ spread of recent marine sediments covers > 90% of some 1600 data points (from Deines 1980) summarized in the *bar*

interpretation of those segments of the early record that bear a metamorphic overprint. Hence, a convincing case can be made that the "normal" sedimentary isotope record had originally extended back to Isua times, allowing little doubt that autotrophic carbon fixation and concomitant biological modulation of the global carbon cycle had already commenced by the time of formation of the Earth's oldest sedimentary rocks.

As for the full sweep over 3.8 Ga of both isotope age curves, students of the sedimentary carbon record had been intrigued since long by the relative constancy of the long-term functions of both $\delta^{13}C_{carb}$ and $\delta^{13}C_{org}$ (cf. Fig. 5). The narrow spread of $\delta^{13}C_{carb}$ through time suggests that the $\delta^{13}C$ values of the bicarbonate precursor of ancient marine carbonates had been closely tethered to the zero permil line, with just occasional departures notably in positive direction. As for the corresponding $\delta^{13}C_{org}$ function, it is evident that the $\delta^{13}C_{org}$ range observed in recent marine sediments faithfully integrates over respective spreads of the principal contributors to the contemporary biomass with only the statistically irrelevant extremes eliminated (Fig. 5, right box). This shows that the impact of a later diagenetic overprint on the primary values is rather limited and usually gets lost within the scatter of primary values that has consequently come to be projected far back into the geologic past (see corresponding envelope in Fig. 5). Over the 3.5 Ga of the unmetamorphosed rock record, isotope fractionations between organic carbon and carbonate thus appear to be demonstrably the same as in the present world, with a characteristic difference in $\delta^{13}C$ between 20 and 30‰ and a corresponding enrichment of isotopically light carbon (^{12}C) in the biogenic phase. Evidently, the kinetic isotope effect inherent in autotrophic (photosynthetic) carbon fixation has been propagated from the living realm into the rock section of the carbon cycle with little alteration, allowing the isotopic signature of the process to be traced far back into the geologic past. Occasional negative offshoots from the data envelope (cf. Fig. 5) seem to be restricted to the Precambrian (Schoell and Wellmer 1981; Hayes et al. 1983; Schidlowski et al. 1983; Weber et al. 1983) and testify to the involvement of methanotrophic pathways in the formation of the respective kerogen precursors. Basically, these excursions are oddities confined to side stages of the carbon cycle.

Summing up, the implications of both long-term isotope age functions depicted in Fig. 5 seem straightforward. There is little doubt that the glaring ^{12}C-enrichment displayed by the data envelope for fossil organic carbon constitutes a coherent signal of autotrophic carbon fixation over 3.8 Ga of Earth history as it ultimately rests with the process that gave rise to the biogenic precursor materials. Moreover, the basic uniformity of this isotopic signal over the same time span attests to an extreme degree of conservatism in the principal biochemical pathway of carbon fixation. In fact, the mainstream of the envelope for $\delta^{13}C_{org}$ depicted in Fig. 5 can be most readily explained as the geochemical manifestation of the isotope-discriminating properties of one single enzyme, namely, ribulose-1,5-bisphosphate (RuBP) carboxylase, the key enzyme of the Calvin cycle. Hence, the above $\delta^{13}C_{org}$ age function may be aptly termed an *index line of autotrophic carbon fixation* over 3.8 Ga of recorded Earth history.

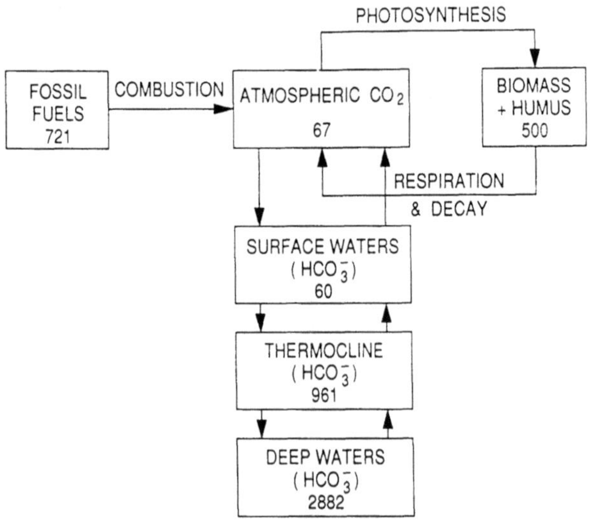

Fig. 6. Box model of the surficial carbon cycle showing sources, sinks, and exchange fluxes relevant at time scales $\leqq 10^4$ a. Reservoir sizes in 10^{16} g C (fossil fuel emissions are cumulative since 1860). For other details see text. (Data from Sawyer 1972; Broecker 1974; Walker 1977; Bolin et al. 1979; and MacIntyre 1980)

5.2 Operation of the Carbon Cycle on Mid-Term Time Scales ($10^6 - 10^8$ a)

It goes without saying that the envelopes of the two carbon isotope age functions shown in Fig. 5 lump a wealth of detailed information that clearly includes distinct time trends. As is particularly evident in the case of $\delta^{13}C_{carb}$, the near constancy of the mainstream of the function over almost 4 Ga gives way, on smaller time scales, to a sequence of moderate oscillations. This is by no means surprising since the isotopic signal ultimately results from a complex interplay of a multitude of factors summarized in Figs. 1 and 6 [with the essentials contained in Eq. (4)] that was unlikely to have proceeded in a time-invariant mode. With oceanic bicarbonate constituting the most faithful integrator of the isotopic state of the surficial exchange reservoir, operational imbalances of the carbon cycle through time are best reflected by the marine $\delta^{13}C_{carb}$ record. It is also this record that gives convincing testimony that both positive and negative excursions are clearly tethered to a time-invariant mean serving as base level for the observed modulations.

Fluctuations in the sedimentary $\delta^{13}C_{carb}$ record indicating changes in the exchange reservoir over periods much in excess of the residence time of oceanic bicarbonate ($\sim 2 \times 10^5$ a) result from an interaction of factors set out in a correspondingly adapted version of Eq. (3), i.e.,

$$\delta^{13}C_{input} = R(\delta^{13}C_{org})_{output} + (1 - R)(\delta^{13}C_{carb})_{output}, \quad (4)$$

where C_{input} is the inorganic carbon entering the ocean as river load or from the atmosphere. The near constancy of the isotopic fractionation between C_{org} and C_{carb} through time with $\delta^{13}C_{org} = \delta^{13}C_{carb} - 25‰$ (cf. Fig. 5) eliminates changes in this parameter as a cause of the observed shifts. Although some workers (e.g., Miller and Fairbanks 1985) have suggested that the river-borne carbon input to the ocean may have varied through time, no independent evidence has as yet been provided to substantiate this inference. Instead, it is common practice to model observed $\delta^{13}C_{carb}$ changes on the assumption that the total carbon input into the ocean and its isotopic composition have remained constant through time (Garrels and Perry 1974; Schidlowski et al. 1975; Holland 1984; Shackleton 1987). If this flux of carbon is partitioned at the outset between two sinks, one the carbonate and the other organic matter with a $\delta^{13}C$ composition that is about 25‰ lighter, then knowledge of the $\delta^{13}C$ content of the carbonates permits the fraction entering the organic carbon pool to be estimated.

Considering the geochemical index function of $\delta^{13}C_{carb}$, the most glaring single excursion of the record appears around 2 Ga ago, with values soaring up to +13‰ (Fig. 5). Originally described as "Lomagundi event" (Schidlowski et al. 1976) and assigned a merely local significance for an extended (300 km) semi-closed basin characterized by excessive deposition of organic carbon, the anomaly has been recently reported also from roughly coeval settings in Scotland and Norway (Baker and Fallick 1989a, b) and from the Baltic shield (Yudovich et al.

Table 2. Excursions of the marine $\delta^{13}C_{carb}$ signal during the Phanerozoic with durations exceeding the residence time of carbon in the modern ocean ($\sim 2 \times 10^5$ a)

Direction of excursion	$\delta^{13}C$ shift (‰)	Stratigraphic age	Duration (10^6 a)	[a]TCM	Reference[b]
Negative	1.5	Pliocene to Recent	7	− 107	1
Negative	0.8	Late Miocene	0.2	− 3	2
Positive	1.0	Mid-Miocene	3	+ 31	2
Negative	1.2	Paleocene/Eocene	3	− 37	3
Positive	2	Late Paleocene	5.5	+ 112	1
Negative	1.5	Cretaceous/Tertiary	0.2	− 5	3, 4
Positive	2	Santonian/Campanian	5.7	+ 114	5
Positive	3	Cenomanian/Turonian	3	+ 94	5
Positive	3	Aptian/Albian	9.7	+ 295	5
Positive	3	Toarcian	0.5	+ 19	6, 7
Negative	4.5	Permo-Triassic	1	− 5	8, 9, 10
Positive	3.5	Late Permian	1	+ 39	10
Positive	3	Mid-Carboniferous	?	?	11
Negative	8(?)	Devonian/Carboniferous	0.5(?)	?	12
Positive	2	Devonian/Carboniferous	7 to 15	?	13
Negative	1.5	Frasnian/Famennian	1	+ 17	14, 15
Negative	1	Ordovician/Silurian	?	?	16
Negative	4	Tommotian (PC/C)	?	?	17
Positive	4 to 6	Latest Vendian	3(?)	+ 187	18

[a] Total Carbon Mass buried (+) or destroyed (−), reported in units of the present atmospheric carbon pool (1 ACP = 7×10^{17} g C), that are required to account for the corresponding $\delta^{13}C_{carb}$ excursions. Calculations were performed using the stepwise integration model of Berger and Vincent (1986), and assuming an oceanic carbon pool of 60 ACP and a carbon flux of 0.5 ACP/10^3 a.

[b] 1. Shackleton (1987); 2. Berger and Vincent (1986); 3. Shackleton (1986); 4. Zachos et al. (1989); 5. Arthur et al. (1985); 6. Jenkyns and Clayton (1986); 7. Jenkyns (1988); 8. Holser et al. (1989); 9. Magaritz et al. (1988); 10. Gruszczynski et al. (1989); 11. Popp et al. (1986); 12. Xu et al. (1986); 13. Lohmann and Walker (1989); 14. Playford et al. (1984); 15. McGhee et al. (1986); 16. Orth et al. (1986); 17. Magaritz (1989); 18. Aharon and Liew (this vol.).

1990). Accordingly, the question of a possible global character of this isotope excursion might still appear open. In any case, the concomitant imbalance of the carbon cycle involving the deposition of prodigious quantities of organic carbon and a corresponding release of oxidation equivalents must have had extraordinary repercussions on the interlocking sulfur, iron and oxygen cycles. Detailed modeling of the event might allow a first feasibility assessment of whether the geochemical consequences of a global anomaly of such magnitude would be compatible with, or rather disruptive to, a smooth operation of parts of the exogenic cycle. Unfortunately, currently available time constraints for this isotope excursion are poor on a high-resolution time scale, obscuring both the exact synchroneity of the anomaly in different parts of the world and the duration of the event as a whole.

Not surprisingly, the quality and the resolving power of reported $\delta^{13}C_{carb}$ records improve as we proceed to the Upper Proterozoic (cf. Knoll et al. 1986; Aharon et al. 1987) and further to the Phanerozoic which primarily reflects the superior quality of stratigraphic and geochronological controls as well as the higher sampling densities available for younger formations. This is particularly true for the Phanerozoic record, and notably the Mesozoic and Cenozoic. Table 2 lists the most prominent Phanerozoic $\delta^{13}C_{carb}$ excursions with durations exceeding the residence time of oceanic bicarbonate, along with their stratigraphic ages and the associated changes in the size of the sedimentary C_{org} reservoir (expressed as multiples of the present atmospheric carbon reservoir) required to produce these excursions. For comparison, the latest positive Precambrian anomaly that occurred in the late Vendian has been included in the tabulation. The following discussion will address a number of points pertinent to the interpretation of these isotope events.

As may be expected, the confidence level of the data generally decreases with increasing geological age. A case in point is offered by the uncertainty concerning the polarity and duration of the $\delta^{13}C_{carb}$ excursion at the Devonian-Carboniferous transition. Whereas Xu et al. (1986) reported a staggering negative $\delta^{13}C_{carb}$ excursion of 8‰ lasting for about 0.5 Ma on the basis of whole rock deep water limestones from one stratigraphic section in China, Lohmann and Walker (1989) have found a 2‰ positive $\delta^{13}C_{carb}$ recorded by carefully selected marine cements bracketed within Famennian to early Visean strata. It is unclear whether these discrepancies stem from poor stratigraphic correlations, diagenetic effects, differences in resolution, or a combination of such factors.

The isotope events listed in Table 2 represent both negative and positive $\delta^{13}C_{carb}$ excursions relative to the background record. In principle, ^{13}C-depletion events can be interpreted as indicators of massive episodes of selective oxidation of organic matter or, alternatively, relatively abrupt decreases in total biomass; conversely, ^{13}C-enrichment events can be interpreted as indicators of an increase in stationary biomass, enhanced deposition of organic matter, or enhanced preservation of the organic fraction in the sediments. Whatever reason applies in the specific case has to be sorted out by a detailed investigation of the single event.

The largest Phanerozoic positive anomaly (3.5‰) has been found in sediments of Late Permian age (Gruszczynski et al. 1989), whereas the largest negative anomaly of 4.5‰ (ignoring the Devonian-Carboniferous excursion because of the uncertainties involved) has been recorded in sediments spanning the Permo-Triassic boundary (Magaritz et al. 1988; Holser et al. 1989). Sediments of Aptian-Albian age have sequestered the largest amount of organic matter equivalent to 295 ACP (atmospheric carbon pool) units, followed by Santonian-Campanian and Late Paleocene sediments with 114 and 112 ACP units, respectively. All changes of the sedimentary C_{org} reservoir tied up to the isotope events listed in Table 2 are brought about by an interaction of many factors including specifically changes in (1) oceanic nutrient inventory, (2) primary productivity, (3) redox state of the ocean, and (4) ocean paleocirculation patterns. However, $\delta^{13}C_{carb}$ excursions alone cannot identify with any certainty which of these factors was predominantly responsible for the inferred time-bound changes of the organic carbon pool. Therefore, independent lines of evidence are necessary to single out the specific causes of the recorded isotope excursions. This evidence may include (1) widespread presence of organic-rich sediments (e.g., black shales) that are contemporaneous with positive $\delta^{13}C_{carb}$ excursions, (2) geochemical tracers indicative of nutrient concentrations and productivity variations (e.g., Cd and Ba), (3) mass extinction events corroborating negative $\delta^{13}C$ excursions, (4) changing $\delta^{13}C_{DIC}$ gradients between surface and deep ocean, and (5) changes indicated by isotope systems which respond to the redox state of the ocean, such as $\delta^{34}S$ of sulfates and sulfides.

Scholle and Arthur (1980) and Jenkyns (1988) have indeed shown that an intimate relation seems to exist between the isotopically positive $\delta^{13}C_{carb}$ excursions recorded by Mid-Cretaceous and Early Jurassic limestones, respectively, and episodes of black shale deposition. This relationship supports the concept of widespread and narrowly synchronous oceanic anoxic events (OAE's of Holser 1977) that have occurred during the Toarcian, Aptian, Cenomanian-Turonian, and Santonian-Campanian. Incidentally, the assumed linkage between positive offshoots of $\delta^{13}C_{carb}$ and the deposition of organic-rich sediments in geographically restricted areas stands in the continuity of respective inferences invoked already by Schidlowski et al. (1976) to account for the Lomagundi event. Also, the Mid-Miocene positive $\delta^{13}C_{carb}$ anomaly ("Monterey Excursion") has been attributed by Berger and Vincent (1986) to deposition of organic-rich sediments in the upwelling regions of the Pacific rim.

In the absence of black shales, other geochemical sensors may indicate the former presence of temporal anoxia in the depositional environment of isotopically distinctive carbonate series. For instance, the positive $\delta^{13}C_{carb}$ anomaly in the Late Paleocene is similar in scale to the Mid-Cretaceous isotope excursions although black shales are not known from the former epoch (Shackleton 1987). However, an increase in the $\delta^{13}C_{carb}$ contrast between surface waters (documented by planktonic foraminifera) and deep ocean (benthic foraminifera) from low values in the early to high values in the Late Paleocene suggest a relationship between increasing oxygen depletion in deep waters and a high rate of organic carbon burial during the Late Paleocene (Shackleton 1987). In the absence of micropaleontological evidence, episodes of ocean stratification may also be recognized by concurrent shifts in the $\delta^{13}C_{carb}$ values of carbonates and $\delta^{34}S$ of sulfates and sulfides because of the coupling between organic matter oxidation and bacterial sulfate reduction in anoxic environments (Goodfellow 1986). In this way, positive $\delta^{34}S$ excursions recorded by both sulfides and sulfates of Frasnian age (Goodfellow and Jonasson 1984) were linked to the observed contemporaneous negative excursion of $\delta^{13}C_{carb}$ (Table 2) to indicate an oceanic anoxic event in the

Late Devonian. Further detailed modeling of the response of marine carbon isotopes to ocean stratification and ventilation events can be found in the paper by Aharon and Liew (this Vol.).

As for the negative anomalies of the record, probably the best case study of a direct link between an abrupt mass extinction event and a negative spike in $\delta^{13}C_{carb}$ has been documented by Zachos et al. (1989) at the Cretaceous-Tertiary (K/T) boundary. This boundary is marked by an extremely rapid (on the order of 10000 a) cessation of oceanic primary productivity, a decrease of 1.5‰ in $\delta^{13}C_{carb}$ of the planktonic microfauna relative to values just below the boundary, a decrease in the flux of barium (a geochemical proxy for organic carbon), and an almost complete obliteration of the carbon isotope gradient between the surface and deep ocean as recorded by planktonic and benthic calcareous tests, respectively ("Strangelove ocean" of Hsu and McKenzie 1985). The presence of an iridium anomaly at the K/T boundary, in conjunction with the indicators above, has led to the proposal of the well-known bolide impact theory of mass extinctions (Alvarez 1986; Hut et al. 1987). Recently, Magaritz (1989) has attempted to generalize the link between mass extinction events and negative $\delta^{13}C_{carb}$ excursions, and to apply it to geologic boundaries other than the K/T (e.g., Precambrian-Cambrian, Fresnian-Famennian, Permo-Triassic, etc.). Such generalization, however, may be rooted in an overly simplistic view of the complex carbon cycle. It is, for instance, known that at some boundaries the extinct fauna has left no traces in the coeval sediment (e.g., Precambrian-Cambrian) and that, in other cases, the negative $\delta^{13}C_{carb}$ excursion can be most readily (and much better) explained by a terrestrial cause (e.g., terminal Ordovician extinction; cf. Orth et al. 1986).

5.3 Operation of the Carbon Cycle on Short-Term Time Scales ($< 10^5$ a)

5.3.1 Reservoirs and Processes

While for intervals exceeding the residence time of carbon in the surficial compartment ($\sim 10^5$ a) the atmosphere-ocean-biosphere reservoirs (cf. Fig. 1) remain in quasi-equilibrium and can be viewed as a unified system, carbon is differentially cycled among this triad at smaller time scales ($< 10^4$ a), with CO_2 as the common currency of exchange. As background for a detailed treatment of the surficial carbon cycle, an elaborated version of the upper part of the box model of Fig. 1 is presented in Fig. 6 that shows the principal reservoirs with subcompartments and respective sizes as well as the interconnecting flux pattern.

As illustrated by Fig. 6, CO_2 is taken up by the vegetation via photosynthesis and stored in plants and soil organic matter. With global net primary productivity just short of 10^{17} g C_{org}/a, the atmospheric CO_2-reservoir (equivalent to 6.7×10^{17} g C) would be consumed by this process within 7 a only. After a residence time that counts in decades, the biomass thus generated is returned to the inorganic carbon reservoir mostly by respiration and other forms of oxidative decay. The ocean provides a much larger carbon reservoir, storing some 60 times more carbon than the atmosphere, but mostly in the form of dissolved bicarbonate ions. The upper layer of the ocean which can store about as much carbon as the atmosphere must be distinguished from the deep ocean on time scales $< 10^4$ a. This is a relatively thin (70 to 100 m) warm layer of water lying above colder and denser water beneath, separated by the transitional thermocline layer. The thermocline is an efficient barrier to the mixing of the lower and upper layers of the ocean between which the CO_2 exchange proceeds mainly via dissolved bicarbonate convected by upwelling and downwelling waters. Studies of ^{14}C concentrations suggest that a complete exchange between the atmosphere and the well-mixed upper layer of the ocean requires some 5 to 10 a, whereas transfer of carbon from the upper layer to the deep ocean and back is a slow process of the order of 1600 a (Broecker 1974). The standing biomass of the oceans is very small ($\sim 0.3 \times 10^{16}$ g C_{org}) and equivalent to less than 0.5% of the atmospheric carbon reservoir, its size being solely controlled by the availability of nutrients such as phosphorus and nitrogen. The settling of organic debris constitutes a drain on the CO_2 of surface waters, and a source of CO_2 in the deep waters. A small fraction of this organic matter is incorporated into sea floor sediments. The dominant processes that exercise a kinetic control of the carbon content of the ocean are the supply of carbon in river water, the return of CO_2 from the surface ocean to the atmosphere, and the accumulation of calcium carbonate in sediments.

There are several natural processes that could disturb the present equilibrium between the carbon stored in the atmosphere and the ocean, such as changes in (1) ocean temperature, (2) ocean volume, (3) oceanic biota, and (4) land biota. A change in ocean temperature or volume would alter the balance because of shifts in the values of the carbonate equilibrium constants and the concentration of the dissolved carbon species in seawater, respectively. A change in ocean fertility would lead to a shift in the carbon concentration in surface waters and thus to a change in the partial pressure of CO_2. In addition, changes in atmospheric CO_2 can result from imbalances in the productivity rates of land biomass, and of respiration and decay. Man's activities can additionally alter the carbon budget. Combustion of fossil fuels since 1860 has released the equivalent of almost 20% of the carbon currently stored as CO_2 and bicarbonate in the atmosphere-ocean system (cf. Fig. 6; see also below).

5.3.2 Recorders of Imbalances in the Short-Term Carbon Cycle

Apart from the carbon isotope approach outlined above, direct monitoring of atmospheric carbon dioxide concentrations provides another powerful tool for retracing the redistribution of carbon among the compartments of the surficial exchange reservoir. Constituting the principal vehicle of carbon transfer, atmospheric CO_2 holds a key position in the surficial carbon cycle, with its reservoir particularly sensitive to change because of its small size and quick interaction with the other reservoirs (cf. Fig. 6). It follows that records of atmospheric CO_2 would constitute ideal sensors of perturbations in the global carbon cycle if they can be accurately retrieved.

Retrieval of atmospheric CO_2 records can be achieved either directly by monitoring the atmospheric CO_2 levels through time, or indirectly by extraction of CO_2 occluded in materials that show continuous deposition with time. Of all natural materials which occlude atmospheric CO_2 only polar ice has so far yielded reliable CO_2 records. Although the concentration of CO_2 in the atmosphere had been measured sporadically ever since the advent of analytical chemistry, systematic monitoring has begun only in 1958 with the establishment of the Mauna Loa baseline station in Hawaii (Keeling et al. 1976). Other monitoring stations, particularly on the southern hemisphere, followed. CO_2 concentration changes measured by the northern and southern hemisphere monitoring stations, respectively, are indicated in Fig. 7A and B. The records show evidence of a quasi-exponential rise in atmospheric CO_2 at a rate of about 1.3 ppm/a over the period 1958–1976. Whereas the overall increase in CO_2 is identical for the antipodian stations, the amplitudes of the respective seasonal variations are dissimilar, with the Hawaiian station showing the greater amplitude. This difference in seasonal amplitudes between the two contemporaneous CO_2 records (cf. Fig. 7A and B) can be explained on the basis of an imbalance in the seasonal growth of land plants, the northern hemisphere containing a much larger proportion of land relative to the southern one (Sawyer 1972; Mook et al. 1983).

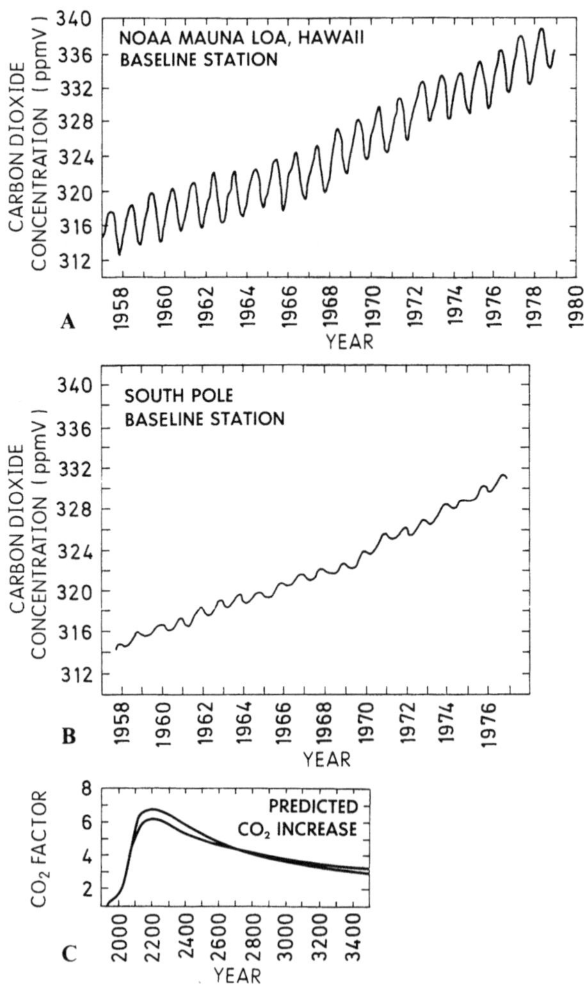

Fig. 7A–C. Changes in atmospheric CO_2 concentrations resulting from anthropogenic emissions. **A** CO_2 levels recorded over a period of 22 a at Mauna Loa, Hawaii (after Rycroft 1982, using data of C. D. Keeling). Note seasonal variations superimposed on the steady CO_2 rise at a rate of 1.3 ppm/a. **B** CO_2 levels recorded over a period of 18 a at the South Pole, Antarctica (after Mook 1986, based on data of C. D. Keeling). Note diminished seasonal CO_2 amplitude relative to the Hawaiian station. **C** Model prediction of atmospheric CO_2 increase based on the premise that industrial CO_2 emissions will continue at present rates. (Keeling and Bacastow 1977)

The continuous increase in atmospheric CO_2 during this century, from an estimated pre-industrial level of 290 ppm to present values exceeding 340 ppm (Rycroft 1982), is generally attributed to the addition of anthropogenic CO_2 from combustion of fossil fuels at an average rate of 4.5% a^{-1} (Broecker et al. 1979) and, to a lesser extent, from deforestation and agricultural practices (Woodwell 1978; Fearnside 1985). Revelle and Suess (1957) have called the observed increase in atmospheric CO_2 an "unprecedented geophysical experiment" because it provides some unexpected glimpses on the short-term *modus operandi* of the carbon cycle. According to model calculations (e.g., MacIntyre 1980), about 5% of the excess CO_2 has been exchanged with the surface ocean and about 34% has already been taken up by the thermocline and deep ocean combined. The remaining 61% of the anthropogenic CO_2 still resides in the atmosphere as confirmed by the observations in Fig. 7A and B.

Presently, the excess CO_2 in the ocean is confined to layers which are also supersaturated with respect to $CaCO_3$. Dissolution of sedimentary carbonate in the ocean would increase the capacity of the oceans to absorb more CO_2 and slow down the increase in the atmospheric CO_2 level (Broecker et al. 1979; MacIntyre 1980). However, according to the predicted industrial CO_2 emissions, excess atmospheric CO_2 will have to increase by a factor of 6 before the buffering capacity of the ocean could play a significant role in the reduction of atmospheric CO_2 (Fig. 7C). The climatic consequences of the current increase in "greenhouse gases", such as CO_2, are beyond the scope of this review.

The question of whether natural changes in atmospheric CO_2 have controlled the waning and waxing of the Pleistocene ice sheets (cf. Sect. 5.3.3) has excited persistent interest ever since it had been demonstrated that ice cores hold clues to past atmospheric CO_2 variations (Berner et al. 1978; Delmas et al. 1980; Neftel et al. 1982). Atmospheric gases, including CO_2, are occluded during transformation of firn to ice; therefore, ice cores offer continuous paleochemical records of CO_2 changes over the duration of ice deposition. Figure 8 shows a compilation of CO_2 data from ice cores taken from Greenland and Antarctica, both spanning the last 50000 a. The record indicates that atmospheric CO_2 was declining during the late last ice age stages, and this depletion culminated around 20000 a ago when atmospheric CO_2 reached about 50% of the pre-industrial CO_2 level. The CO_2 trend reversed rapidly at the onset of deglaciation and reached pre-industrial levels during Mid-Holocene. The chronological order of these CO_2 events is confirmed by the contemporaneous $\delta^{18}O$ record from ice (Fig. 8), with more negative $\delta^{18}O$ values representing lower ambient temperatures corresponding to depressed

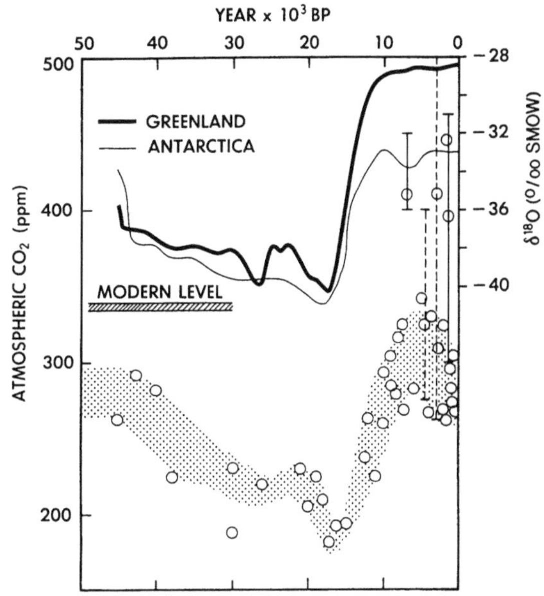

Fig. 8. Atmospheric CO_2 content in ice cores from Camp Century, Greenland, and from Byrd Station, Antarctica, over the past 50000 a. CO_2 reached about 50% of the pre-industrial level during the last glaciation maximum about 20000 a ago. The conspicuous correlation observed between the $\delta^{18}O$ record of the hosting ice (a proxy climatic indicator) and the CO_2 content occluded suggests a cause-and-effect relationship between climatic change during the last ice age and atmospheric CO_2 concentration. (Data from Neftel et al. 1982)

CO_2 levels, whereas the return to more positive $\delta^{18}O$ values (reflecting a climatic warming) corresponds to a return of pre-industrial CO_2 levels.

Apart from the CO_2 ice core record, the carbon isotope approach as outlined above offers additional insight in the operation of the surficial carbon cycle. Pertinent background information is summarized in Fig. 9, which gives a graphic synopsis of the isotopic compositions of the principal reservoirs and subcompartments that are involved in the short-term cycling of surficial carbon.

The atmospheric CO_2 reservoir covers a $\delta^{13}C$ range from − 6.7 to − 7.4‰ [PDB] (Keeling 1961). The $\delta^{13}C$ of dissolved inorganic carbon (DIC, mostly in the form of bicarbonate) in the contemporary oceans varies from about 2‰ at the surface to below − 0.5‰ in the oxygen minimum zone, with the bottom waters generally between − 0.2 and 0.2‰ (Kroopnick et al. 1977). Surface water $\delta^{13}C_{DIC}$ values are controlled by the CO_2 exchange with the atmosphere, and by the organic productivity of the water mass. Deep water $\delta^{13}C_{DIC}$ values are determined by in situ respiration, oxidation, and dissolution of organic and carbonate settling par-

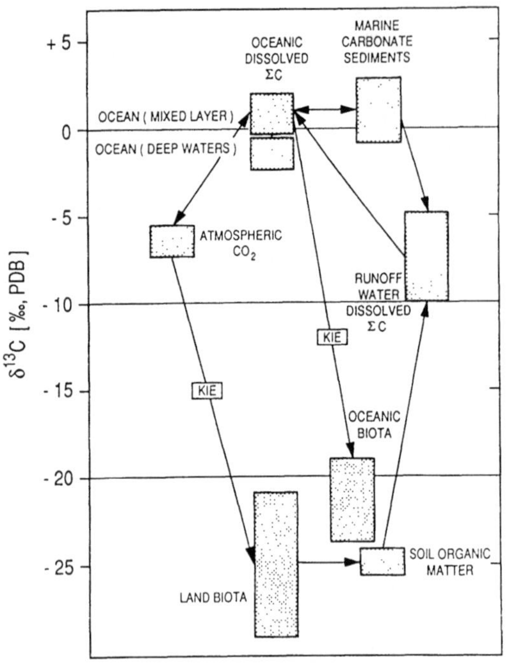

Fig. 9. Principal carbon isotope compositions of the individual subcompartments of the surficial exchange reservoir along with the interconnecting flux pattern (after Aharon 1982). *Tie lines with arrows on both sides* denote pathways governed by isotope exchange equilibria; carbon fluxes into the biosphere are beset with a kinetic isotope effect (*KIE*) responsible for a preferential accumulation of ^{12}C in biogenic matter. Hence, the observed variability in $\delta^{13}C$ is the result of both equilibrium and kinetic fractionations

ticles, respectively, and by the relative rates of upwelling of deep water (Kroopnick et al. 1977). ^{13}C-depleted CO_2 that derives from decomposition of plants and oxidation of soil organic matter, mixed with dissolved bicarbonate stemming from dissolution of carbonates on the continents, constitutes an important source of carbon in the rivers (Fig. 9). At steady state, the flux of riverine carbon into the oceans should balance the removal of oceanic carbon by calcium carbonate precipitation and deposition of organic carbon. Assuming that about 20% of sedimentary carbon exposed in the continental denudation planes is organic carbon (with $\delta^{13}C_{org} \approx -25‰$) and 80% is carbonate carbon (with $\delta^{13}C_{carb} \approx 0‰$), the $\delta^{13}C$ average of riverine carbon should be close to $-5‰$ according to a correspondingly adapted version of the isotope mass balance of Eq. (4). For details of the isotope spreads of the organic components of the surficial carbon system shown in Fig. 9 the reader is referred to Sect. 3.1.

The response of carbon isotopes to imbalances in the carbon cycle is also vividly illustrated by the current addition to the atmosphere of anthropogenic CO_2, and the exchange of the latter with the ocean and biosphere. $\delta^{13}C$ determinations of atmospheric CO_2 over a 22-a-period (1956 to 1978) indicate a decrease in $\delta^{13}C$ of 0.55‰ corresponding to a CO_2 increase of 20.1 ppm (Keeling et al. 1979). The estimated rate of $\delta^{13}C$ change of 0.027‰/ppm is a complex function of $\delta^{13}C$ averages of the burnt fossil fuels ($-24.5‰$ for coal, $-28.5‰$ for oil and $-41.5‰$ for natural gas), and the isotopic fractionations occurring during air-sea and air-biota exchanges (Mook 1986).

Attempts to unravel the relative contributions of fossil fuels and land biota to the excess CO_2 by measuring $\delta^{13}C$ changes in atmospheric samples proved unsuccessful mainly because of insufficient length of time since the beginning of the determinations. Alternatively, extended $\delta^{13}C$ records can be obtained from tree rings on land, or from annual growth increments of reef skeletal material (from corals, giant clams, sclerosponges) in the ocean, on the basis of the predictive relationship between $\delta^{13}C$ of cellulose and atmospheric CO_2, and between $\delta^{13}C$ of skeletal aragonite and dissolved bicarbonate in the surface ocean, respectively. Freyer and Belacy (1983) have demonstrated that $\delta^{13}C$ records from European oak trees show a systematic decrease of about 2‰ over the last 140 years, documenting accurately the $\delta^{13}C$ trend indicated by direct determinations of atmospheric CO_2 (Keeling et al. 1979). For the same time span (i.e., since the start of the industrial revolution), reef skeletal material has yielded a composite $\delta^{13}C$ record suggesting a 0.7‰ depletion in the surface ocean (Fig. 10). The time-dependent $\delta^{13}C$ changes recorded in the surface ocean are in agreement with model predictions that the response of the world ocean to excess CO_2 should amount to about 40% of that of the atmosphere (Oeschger et al. 1975). However, caution must be exercised when interpreting $\delta^{13}C$ records from both tree rings and reef skeletal aragonite because of potential "vital effects" affecting $^{13}C/^{12}C$ fractionations and masking the actual $\delta^{13}C$ shift in either the atmosphere (Francey 1981) or seawater (Aharon 1991).

5.3.3 A Case Study of a Short-Term (10^4-10^5 a) Imbalance: The Pleistocene Example

As pointed out above, variations in the carbon cycle on 10^4-10^5 a time scales primarily involve sources,

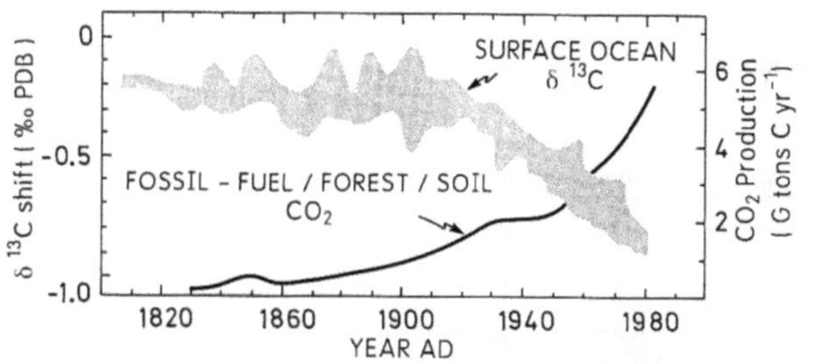

Fig. 10. Composite $\delta^{13}C_{carb}$ record from a Bermuda coral (Nozaki et al. 1978), a Jamaican sclerosponge (Druffel and Benavides 1986), and a giant clam from the Great Barrier Reef of Australia (Aharon 1991) reflecting $\delta^{13}C_{DIC}$ changes in oceanic surface waters over the last two centuries. The inverse trend observed between the increasing anthropogenic CO_2 production and the decreasing $\delta^{13}C_{carb}$ of reef skeletal aragonite suggests a cause-and-effect relationship between the two functions

sinks and reservoirs of the surficial exchange compartment since impacts on the larger crustal (sedimentary) cycle become effective only within a temporal framework exceeding the residence time of HCO_3^- in the world ocean ($\approx 2 \times 10^5$ a). This formidable residence time constrains the response period of the oceanic carbon reservoir to intervals of at least this magnitude under steady-state equilibrium conditions (Broecker 1974). On the other hand, the CO_2 in the atmosphere and ocean is controlled by biological (kinetic) processes rather than by thermodynamic equilibria that just impose broad limits within which the system can vary with a certain degree of freedom. Unlike the thermodynamically fixed state, that controlled by kinetics is apt to undergo temporary changes on the time scale considered here. In the following, we shall summarize the evidence indicative of natural changes in the surficial carbon cycle during the most recent geological past. Also, we shall discuss the possible causes of these changes within the limits set by the confidence levels of the geochemical indicators utilized. Among these, carbon isotopes again appear to be particularly significant (cf. Kroopnick et al. 1977; Shackleton 1977; Broecker 1982).

Recorders of Past Changes in Dissolved Inorganic Carbon (DIC) Chemistry. There are two indicators of changes in the inorganic carbon chemistry of the ocean that have left imprints in the sedimentary record over the last 10^5 a. Variations in the level of dissolved inorganic carbon (DIC) leave a quantitative mark by changing the depth of the lysocline and carbonate compensation depth, and by inducing time-dependent pulses of dissolution-preservation events which can be observed in the carbonate record of deep sea cores (Berger 1968; Crowley 1983). The measurable parameters are the ratios of calcitic planktonic foraminifera susceptible to dissolution to resistant forms such as siliceous radiolaria or benthic foraminifera (Weyl 1970).

The other paleoindicator leaving a signal in the record is the $\delta^{13}C$ of DIC. The $\delta^{13}C_{DIC}$ record of ancient deep ocean water is preserved in benthic foraminifera if allowance is made for the isotopic fractionation between dissolved bicarbonate and the carbonate bodies of the foraminifera (Curry and Lohmann 1982; Zahn et al. 1986). Similarly, the $\delta^{13}C_{DIC}$ record of the upper mixed ocean layer is preserved in the shells of planktonic foraminifera (Crowley and Matthews 1983; Oppo and Fairbanks 1989) and in skeletal carbonate material of emerged coral reefs (Aharon 1982, 1985). Since the interpretation of the $\delta^{13}C_{carb}$ records of planktonic foraminifera was shown to be complicated by factors unrelated to the primary $\delta^{13}C_{DIC}$ values (Bender and Keigwin 1979; Oppo and Fairbanks 1989), we limit our discussion to the contemporaneous $\delta^{13}C_{carb}$ records of benthic foraminifera (deep sea reservoir) and uplifted coral reefs (warm surface water reservoir).

A comparison between the carbon isotope record based on the giant clam *Tridacna gigas* from uplifted coral reefs in New Guinea (Aharon 1982, 1985), and the deep sea benthic record based on *Uvigerina senticosa* from the East Pacific (Shackleton et al. 1983a, b) is illustrated in Fig. 11. The two sequences are plotted to the same scale of isotopic change from present levels. The time scale was defined on the basis of the chronology and $\delta^{18}O_{carb}$ variations shown in Fig. 12, with $\delta^{18}O_{carb}$ changes driven by ice volume and temperature fluctuations that have occurred during the last ice age cycle (Aharon 1983; Shackleton et al. 1983a). The two records reached a minimum contrast (i.e., the present-day difference in $\delta^{13}C_{DIC}$ between surficial and deep ocean of about 2.2‰) during the last high sea level stand (Isotope Stage 5e) and an amplified contrast of 1.5‰ during the last low sea level event (Isotope Stage 2). In spite of their inverse relationship, it is evident that $\delta^{13}C_{carb}$ changes in the two records have a similar periodicity and are synchronous during most of the $\delta^{18}O_{carb}$ stages. The covariance of the long term $\delta^{13}C_{carb}$ variations from the two sites in the Pacific strengthen the view that these variations must reflect seawater chemistry changes on a global scale rather than local or secondary effects.

Causes of $\delta^{13}C$ Changes Between Glacial and Interglacial Periods. The large differences in the sizes

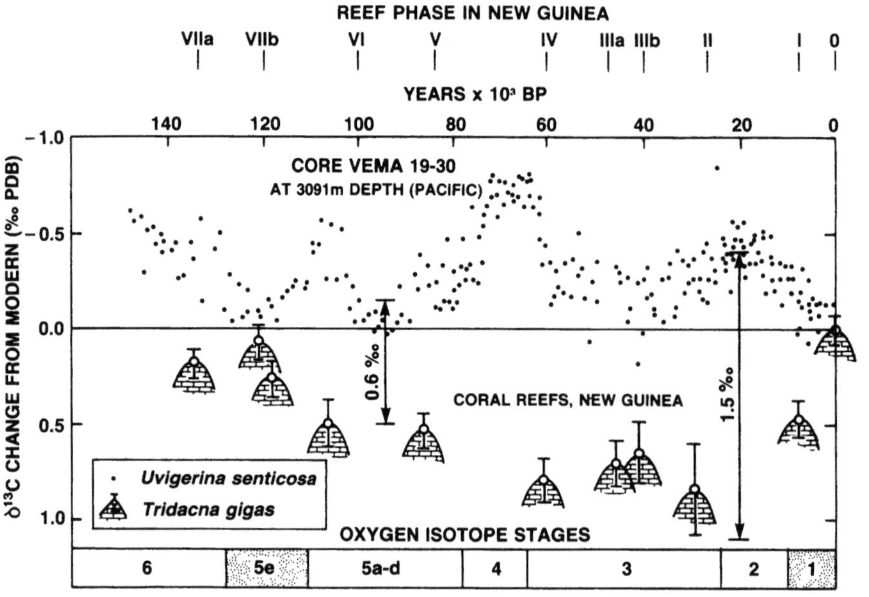

Fig. 11. $\delta^{13}C_{carb}$ contrast separating the shallow water marine carbonates from the deep water marine carbonates over the last 10^5 a, implied by the comparison between the $\delta^{13}C_{carb}$ record from uplifted coral reefs in New Guinea based on the giant clam *Tridacna gigas* (Aharon 1982, 1985), and the deep sea benthic $\delta^{13}C_{carb}$ record based on *Uvigerina senticosa* from the East Pacific (Shackleton et al. 1983a, b). The two sequences are plotted to the same scale of isotope change from present levels, with δ-values adjusted to yield best agreement for the modern samples. The chronology of the isotope records is matched on the basis of the oxygen isotope stages. The two records reached a minimum contrast during the last high sea level stand (Isotope Stage *5e*), and an amplified contrast of 1.5‰ during the low sea level stands of Isotope Stages *3* and *2*

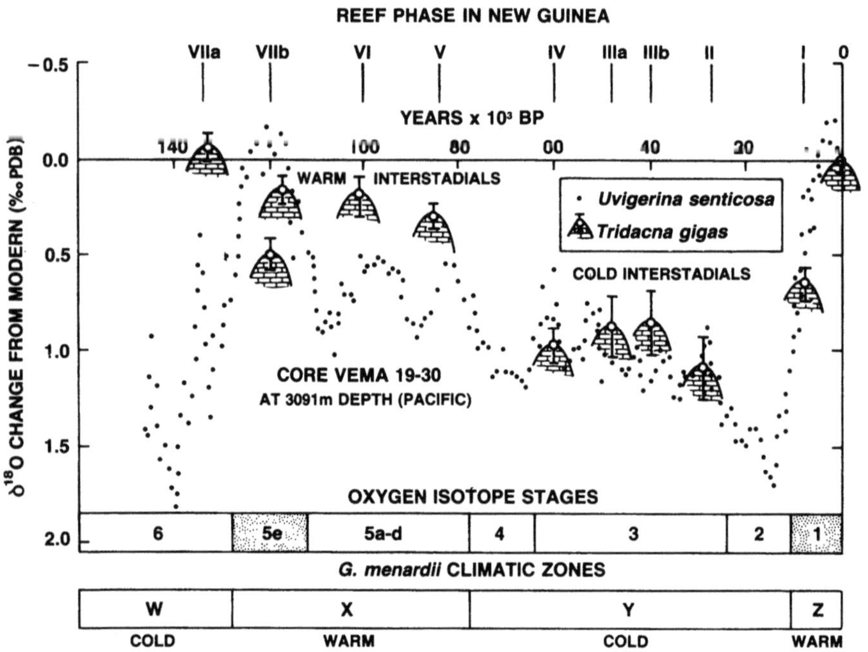

Fig. 12. $\delta^{18}O_{carb}$ records over the last 10^5 a from shallow water marine carbonates (giant clams from uplifted coral reefs in New Guinea – Aharon and Chapell 1986), and from deep water marine carbonates (benthic foraminifera from the East Pacific – Shackleton et al. 1983a). The two sequences are plotted to the same scale of $\delta^{18}O_{carb}$ change from present level. *G. menardii* climatic zones are those of Ericson et al. (1964) with the boundaries of Broecker and Van Donk (1970). Note the excellent agreement between the shallow and deep water climatic records based on the two $\delta^{18}O_{carb}$ age functions that are driven primarily by ice volume changes during the last ice-age cycle

of the carbon reservoirs shown in Fig. 6 and of their respective exchange times (cf. Sect. 5.3.1) impose limits on the possible causes of the recorded $\delta^{13}C_{carb}$ changes. First, as the deep water reservoir is roughly 60 times larger than the surface reservoir, its properties must be close to the mean of the entire ocean system. Second, as about 90% of the ocean-atmosphere carbon resides in the deep sea, the isotopic record for this reservoir should have the leading role in any interpretation. Third, the isotopic indicators of Fig. 11 suggest that the transition from one state of the carbon system to another has occurred within a few thousand years only. Therefore, because of the much longer residence time of carbon in the ocean, nonsteady-state processes must be definitely considered.

Two hypotheses have been offered to explain the evidence from deep sea cores that the $\delta^{13}C_{carb}$ values of glacial benthic foraminifera are, on average, lower by 0.5‰ relative to the interglacial ones (Fig. 11). Although both hypotheses do account for the observed isotope shift, they differ markedly in concept and in the carbon reservoirs believed to be involved in the change. Whereas Shackleton (1977) invokes a land-based cause for the $\delta^{13}C_{carb}$ shift, Broecker (1982) constructs a case in support of a shelf marine origin for the same isotopic difference.

Figure 13 illustrates the carbon transfer between the reservoirs required to account for the glacial-interglacial isotopic shift in the benthic foraminifera. According to Shackleton (1977), a 50% reduction during glacial times in the carbon stored in the land biomass and soil (assumed here to make up a reservoir of 204×10^{16} g C_{org}) and its transfer into the ocean via the atmosphere would lower the oceanic $\delta^{13}C_{DIC}$ average by about 0.5‰. With the onset of the interglacial, the land biomass would expand again to the present size and sequester CO_2 from the ocean via the atmosphere. Contrary to such continental scenario, Broecker's (1982) hypothesis just requires the formation and destruction of a small reservoir of organic-rich sediment (ca. 3% organic carbon by weight) caused by sea level excursions. The area of the present shelves being 3×10^{13} m^2, about 4×10^{16} kg of sediment would have formed a layer 1.5 m thick if spread uniformly.

The point of divergence between the two models is related to the predicted concentrations of oceanic nutrients (e.g., phosphate), because nutrient fluctuations would accompany shelf sediment deposition and removal, but not necessarily expansion or shrinkage of land biota. Therefore, the importance of the nutrients to the carbon cycle must be identified. The nutrient content of seawater is an excellent index

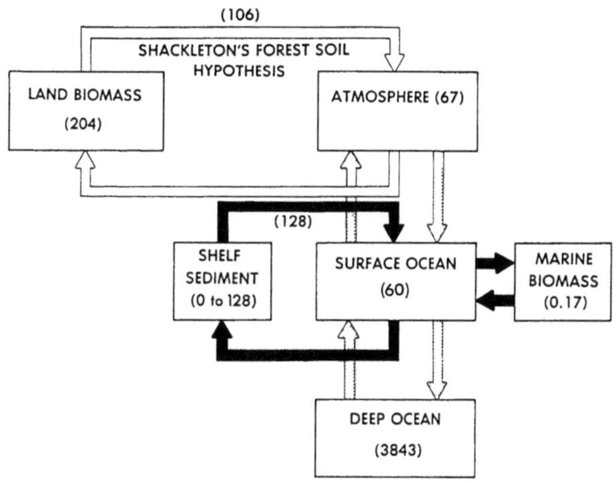

Fig. 13. Carbon transfer between principal reservoirs required to account for the glacial-interglacial $\delta^{13}C_{carb}$ shifts shown in Fig. 11. Whereas Shackleton's (1977) model invokes a land biota origin for the observed $\delta^{13}C_{carb}$ shift, Broecker's (1982) model invokes a shelf marine origin. Carbon reservoir sizes and fluxes are in 10^{16} g C. *Empty arrows* indicate the carbon transfer pathways in Shackleton's model; *full arrows* indicate the carbon transfer pathways in Broecker's model; *stippled arrows* indicate carbon transfer pathways common to both models. A reduction of 50% in the carbon stored in the land biomass during glacial time, and its transfer into the ocean via atmosphere, would lower the deep ocean $\delta^{13}C_{DIC}$ by about 0.5‰. Alternatively, erosion of organic-rich shelf-sediments (128×10^{16} g C) having $C/P \approx 105$ (Redfield ratio) during glacial low sea level stands would (1) increase the nutrient content of the glacial ocean; (2) enhance biomass productivity of the warm surface ocean; (3) lower the surface ocean-atmospheric CO_2 by about 50% through the "biological pump"; (4) lower the $\delta^{13}C_{DIC}$ of the deep ocean by about 0.5‰; (5) enhance the $\delta^{13}C_{DIC}$ contrast between the warm surface ocean and cold deep ocean by 1.5‰ (see also text)

of the extent to which carbon is gained during photosynthesis and lost by respiration. If the nutrient concentration in the glacial ocean was higher than at present (implicit in Broecker's hypothesis), then the biological consumption of the oceanic CO_2 would have been enhanced, thus lowering the pCO_2 of surface waters. In addition, because the mean nutrient content of the ocean generally controls the level of organic productivity of the upper ocean, it also defines the $\delta^{13}C_{DIC}$ contrast between the warm surface water (wsw) and the deep ocean water (dow). The relationship between the shift in the $\delta^{13}C_{DIC}$ contrast of the ocean's two carbon pools and the shift in the nutrient ratio (e.g., $PO_4/\Sigma CO_2$) is described by an equation modified from Broecker (1982),

$$\Delta [(\delta^{13}C_{DIC})_{wsw} - (\delta^{13}C_{DIC})_{dow}] = \varepsilon (C/P)\, \Delta (PO_4/\Sigma CO_2)_{dow}, \qquad (5)$$

where C/P is 105 in marine organic matter ("Redfield ratio") and the isotope enrichment factor $\varepsilon = 25‰$ during aquatic photosynthesis. In the modern ocean, the isotope contrast between the surface and deep ocean reservoirs is 2.05‰ and the $PO_4/\Sigma CO_2$ ratio is 0.00098 (Broecker 1982).

The observed changes through time in the $\delta^{13}C_{carb}$ contrast between the coral reefs and benthic foraminifera (Fig. 11) are likely to reflect temporal changes in the oceanic $PO_4/\Sigma CO_2$ ratio. In terms of Eq. (5), a 1.5‰ shift in the glacial to interglacial $\delta^{13}C_{DIC}$ contrast between the surface ocean and the deep ocean (i.e., from 2.2 to 3.7‰) would correspond to an increase in the $PO_4/\Sigma CO_2$ ratio by a factor of 1.6 relative to present levels. A transfer of 12.8×10^{17} g C from the shelf to the ocean during the onset of glaciation (required to explain the 0.5‰ shift indicated by the benthic $\delta^{13}C_{carb}$ record) would increase the PO_4 content of the glacial ocean by a factor of 1.8 (assuming organic matter with $C/P \approx 105$) which, in turn, would reduce the surface ocean-atmosphere pCO_2 by about 50% relative to present. Unlike Broecker's marine shelf hypothesis, Shackleton's land biota hypothesis predicts no change (or a slight decrease) in the $PO_4/\Sigma CO_2$ ratio of the glacial ocean, and a negative $\delta^{13}C_{DIC}$ shift in the surface glacial ocean comparable to the one in the deep ocean. Thus, in the absence of a nutrient change in Shackleton's hypothesis, the glacial surface ocean-atmosphere pCO_2 must have been slightly higher than that of the interglacial level.

The glacial changes in ocean-atmosphere chemistry predicted by the two models can now be tested against the actually observed variations. First, studies of dissolved gases trapped in Antarctic and Greenland ice cores (Neftel et al. 1982; Delmas et al. 1980; Barnola et al. 1987) indicate a 50% reduction of atmospheric pCO_2 during glacial events (low sea level stands) relative to interglacials (high sea level stands, see Fig. 8). Second, variations of Cd/Ca ratios in the shells of benthic foraminifera (Hester and Boyle 1982; Boyle and Keigwin 1982) suggest an increase by a factor of 2 during severe glaciations. As the oceanic distribution of cadmium is similar to that of the nutrients, Cd/Ca ratios in shells record temporal variations in nutrient distribution. Third, the $\delta^{13}C_{carb}$ records obtained from coral reefs and benthic foraminifera (Fig. 11) indicate an increase in the $\delta^{13}C$ contrast between the surface and deep ocean during low sea level stands relative to high sea level stands. Therefore, the paleoindicators seem to support a mechanism for the observed perturbations of the Pleistocene carbon cycle within the ocean as against an extramarine cause.

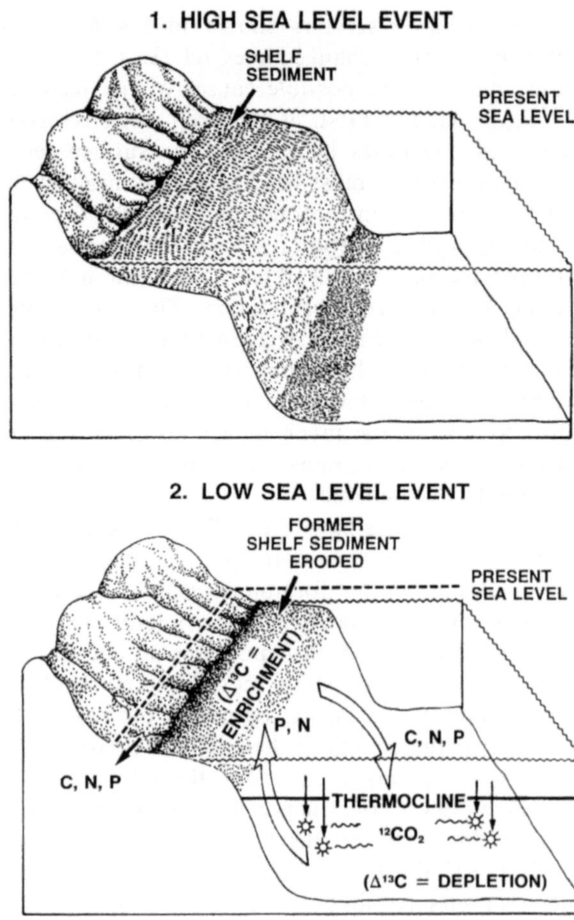

Fig. 14. Model of formation (*1*) and destruction (*2*) of organic-rich shelf sediments caused by sea level excursions, and invoked to account for changes in ocean and atmosphere chemistry (after Broecker 1983). *1 High sea level stand:* River-borne and autochthonous organic matter is trapped and buried on the shelves, thus reducing the dissolved nutrient content of the ocean. *2 Low sea level stand:* Shelf sediments undergo exposure and erosion, releasing the nutrient content of organic matter into the ocean. The dissolved nutrients (e.g., phosphate) trigger a bloom of marine biomass with consequent ^{13}C-enrichment in shallow water marine carbonates and ^{13}C-depletion in deep water marine carbonates

Results obtained from work on uplifted coral reefs in New Guinea have, in fact, demonstrated an inverse correlation between former sea level stands and the $\delta^{13}C_{carb}$ of contemporaneous reef units (Aharon 1985) which can be attributed to the formation and destruction of organic-rich sediments caused by sea level excursions (Fig. 14). The lower the sea level, the more shelf sediment and associated organic matter is transferred to the ocean, thereby increasing the oceanic ΣCO_2 and nutrient concentration, and subsequently the extent to which carbon is gained during photosynthesis at the surface and lost through respiration and chemical oxidation in the deep ocean. This process, in turn, would amplify the $\delta^{13}C_{DIC}$ contrast between the surface ocean and deep ocean, enhancing ^{13}C-enrichment at the surface and ^{13}C-depletion at depth. A rise in sea level would have the opposite effect. Aharon (1985) has estimated an increase by a factor of 2.3 of the net productivity of the marginal ocean coral reefs during the low sea level stand as compared to the high sea level stand.

6 Summary: A Synoptic View of the Carbon Cycle Through Time

With the currently available evidence at hand, there is little doubt that both organic carbon (kerogen) and carbonate had figured as common constituents of sedimentary rocks since the onset of the record $\approx$ 3.8 Ga ago, indicating that biological modulation of the terrestrial carbon cycle had been established early enough in the history of the planet as to be fully operative by the time of formation of the oldest known sediments. Allowing for a metamorphic overprint of the earliest (> 3.5 Ga) record, the persistence over the same time interval of biologically mediated carbon isotope fractionations would be decidely consistent with such conclusion. In concert, these findings support the existence of autotrophy – and specifically photoautotrophy as the quantitatively most important process of biological carbon fixation – as both a biochemical process and a geochemical agent over almost 4 billion years of Earth history. Hence, we may reasonably conclude that ever since its first documentation in the record, the global carbon cycle had been a biogeochemical one characterized by the interplay of an organic and an inorganic branch (Fig. 1). A corollary of these findings is that the planet had been inhabited by life as from about the same time which inference is also supported by the available paleontological evidence (or moderate extrapolations thereof).

As is specifically documented by the carbon isotope age function of marine carbonates (which latter encode the $\delta^{13}C$ signature of their parent bicarbonate precursor), there have occurred marked changes in the intensities of the source and sink functions of both sedimentary carbon species through time, with concomitant imbalances and subsequent repercussions on the sulfur, iron, and oxygen cycles that are coupled with the carbon cycle by geochemically important redox reactions. However, on a long-term (10^9 a) scale, both the $\delta^{13}C_{carb}$ and the $\delta^{13}C_{org}$ time functions display an astounding degree of uniformity (Fig. 5), the consistent enrichment of light carbon in fossil organic matter reflecting the bias in favor of ^{12}C during biological carbon fixation. Accordingly, the mainstream of the envelope for $\delta^{13}C_{org}$ shown in Fig. 5 can be best explained as the geochemical manifestation of the isotope-discriminating properties of ribulose-1,5-bisphosphate (RuBP) carboxylase, the principal CO_2-fixing enzyme of the Calvin cycle that channels most of the carbon transfer from the nonliving to the living realm. This, incidentally, would suggest an extreme degree of evolutionary conservatism in the biochemistry of autotrophic carbon fixation.

Imbalances of the cycle on smaller time scales as evidenced by the oscillations of the carbon isotope functions are due to a variety of reasons, ranging from an inherent inertia in the operation of the exogenic system to possible extraterrestrial causes such as impairments of biospheric activity by bolide impacts. Relying on $\delta^{13}C_{carb}$ as the most faithful sensor of the state of the exogenic carbon system, a glaring operational imbalance of the cycle might be indicated by a positive excursion of the isotope record around 2 Ga ago ("Lomagundi event"), provided the latter had occurred on a global (rather than on a local) scale. Generally, however, the carbon system displays a remarkable degree of resiliency, with both positive and negative excursions limited in magnitude and tethered to an apparently time-invariant mean suggesting the influence of a fairly strong steering agent (such as a possible phosphate control for the proportion of organically fixed carbon). It should be noted that for time scales considerably in excess of the residence time of bicarbonate in seawater ($\approx 10^5$ a), the atmosphere-ocean-biosphere system may be viewed as a unified system in quasi-equilibrium, while at smaller time scales carbon may be differentially redistributed among the three surficial reservoirs (mainly as a result of fluctuations of the standing biomass in response to changes in nutrient supply and other bio-limiting factors). Embedded in the cycle at large, and notably its crustal (sedimentary) compartment, are processes responsible for the formation of fossil fuel (coal, lignite) and hydrocarbon (oil, gas) deposits.

Altogether, the geological record gives eloquent testimony that both the face of the Earth and the geochemical cycle of carbon had been under the continuous influence of life processes as from at least 3.8 Ga ago, implying that the juvenile planet had carried a veneer of prolific microbial life not long after its formation.

References

Aharon P (1982) $^{13}C/^{12}C$ isotope ratio variations over the last 10^5 a in a New Guinea coral reef environment: implications for the fertility shifts of the tropical ocean. In: Galbally IE, Freney JR (eds) The cycling of carbon, nitrogen, sulfur and phosphorus in terrestrial and aquatic ecosystems. Aust Acad Sci, Canberra, pp 119–132

Aharon P (1983) 140000 year isotope climatic record from raised coral reefs in New Guinea. Nature (London) 304: 720–723

Aharon P (1985) Carbon isotope record of late Quaternary coral reefs: possible index of sea surface paleoproductivity. Geophys Monogr 32:343–355

Aharon P (1991) Records of reef environment histories: stable isotopes in corals, giant clams and calcareous algae. Coral Reefs 10:71–90

Aharon P, Chappell J (1986) Oxygen isotopes, sea level changes and the temperature history of a coral reef environment in New Guinea over the last 10^5 yr. Paleogeogr Paleoclimatol Paleoecol 56:337–379

Aharon P, Schidlowski M, Singh IB (1987) Chronostratigraphic markers in the end-Precambrian carbon isotope record of the Lesser Himalaya. Nature (London) 327:699–702

Allaart JH (1976) The pre-3760-Myr-old supracrustal rocks of the Isua area, central West Greenland, and the associated occurrence of quartz-banded ironstone. In: Windley BF (ed) The early history of the earth. John Wiley & Sons, New York London, pp 177–189

Alvarez W (1986) Toward a theory of impact crises. EOS Am Geophys Un 67:649–658

Arneth JD, Schidlowski M, Sarbas B, Goerg U, Amstutz GC (1985) Graphite content and isotopic fractionation between calcite-graphite pairs in metasediments from the Mgama Hills, Southern Kenya. Geochim Cosmochim Acta 49:1553–1560

Arthur MA, Dean WE, Schlanger SO (1985) Variations in the global carbon cycle during the Cretaceous related to climate, volcanism, and changes in atmospheric CO_2. Geophys Monogr 32:504–529

Awramik SM, Schopf JW, Walter MR (1983) Filamentous fossil bacteria from the Archean of Western Australia. In: Nagy B, Weber R, Guerrero JC, Schidlowski M (eds) Developments and interactions of the Precambrian atmosphere, lithosphere and biosphere. Developments in Precambrian Geology 1. Elsevier, Amsterdam, pp 249–266

Baker AJ, Fallick AE (1989a) Evidence from Lewisian limestones for isotopically heavy carbon in two-thousand-million-year-old sea water. Nature (London) 337: 352–354

Baker AJ, Fallick AE (1989b) Heavy carbon in two-billion-year-old marbles from Lofoten-Vesteralen, Norway: implications for the Precambrian carbon cycle. Geochim Cosmochim Acta 53:1111–1115

Barnola JM, Raynaud D, Korotkevich YS, Lorius C (1987) Vostok ice core provides 160000-year record of atmospheric CO_2. Nature (London) 329:408–414

Bender ML, Keigwin LD (1979) Speculation about the upper Miocene change in abyssal Pacific dissolved bicarbonate $\delta^{13}C$. Earth Planet Sci Lett 45:383–393

Benedict CR, Wong WWL, Wong JHH (1980) Fractionation of the stable isotopes of inorganic carbon by seagrasses. Plant Physiol 65:512–517

Berger WH (1968) Planktonic foraminifera: selective solution and paleoclimatic interpretation. Deep Sea Res 15:31–43

Berger WH, Vincent E (1986) Deep sea carbonates: reading the carbon isotope signal. Geol Rundsch 75:249–269

Berner W, Stauffer B, Oeschger H (1978) Past atmospheric composition and climate: gas parameters measured on ice cores. Nature (London) 276:53–55

Bolin B, Degens ET, Kempe S, Ketner R (eds) (1979) The global carbon cycle. John Wiley & Sons, New York, xxxv + 491 pp

Bottinga Y (1969) Calculated fractionation factors for carbon and hydrogen isotope exchange in the system calcite-carbon dioxide-graphite-methane-hydrogen-water vapor. Geochim Cosmochim Acta 33:49–64

Boyle EA, Keigwin LD (1982) Deep circulation of the North Atlantic over the last 200000 years: geochemical evidence. Science 218:784–787

Broecker WS (1970) A boundary condition on the evolution of atmospheric oxygen. J Geophys Res 75:3553–3557

Broecker WS (1974) Chemical oceanography. Harcourt Brace Jovanovich, New York, 214 pp

Broecker WS (1982) Ocean chemistry during glacial time. Geochim Cosmochim Acta 46:1689–1705

Broecker WS (1983) The ocean. Sci Am 249:146–160

Broecker WS, Van Donk J (1970) Insolation changes, ice volumes and the ^{18}O record in deep sea cores. Rev Geophys Space Phys 8:169–198

Broecker WS, Takahashi T, Simpson HJ, Peng TH (1979) Fate of fossil fuel carbon dioxide and the global carbon budget. Science 206:409–418

Buick R (1984) Carbonaceous filaments from North Pole, Western Australia: are they fossil bacteria in Archean stromatolites? Precambrian Res 24:157–172

Cameron EM, Garrels RM (1980) Geochemical compositions of some Precambrian shales from the Canadian Shield. Chem Geol 28:181–197

Chung HM, Sackett WM (1979) Use of stable carbon isotope compositions of pyrolytically derived methane as maturity indices for carbonaceous materials. Geochim Cosmochim Acta 43:1979–1988

Cloud PE (1976) Beginnings of biospheric evolution and their biogeochemical consequences. Paleobiology 2: 351–387

Cohen Y, Aizenshtat Z, Stoler A, Jørgensen BB (1980) The microbial geochemistry of Solar Lake, Sinai. In: Ralph JB, Trudinger PA, Walter MR (eds) Biogeochemistry of ancient and modern environments. Springer, Berlin Heidelberg New York, pp 167–172

Craig H (1953) The geochemistry of stable carbon isotopes. Geochim Cosmochim Acta 3:53–92

Crowley TJ (1983) Calcium carbonate preservation patterns in the central North Atlantic during the last 150000 years. Mar Geol 51:1–14

Crowley TJ, Matthews RK (1983) Isotope-plankton comparisons in a late Quaternary core with a stable temperature history. Geology 11:275–278

Curry WB, Lohmann GP (1982) Carbon isotope changes in benthic foraminifera from the western South Atlan-

tic: reconstruction of glacial abyssal circulation patterns. Quat Res 18:218–235
Deines P (1980) The isotopic composition of reduced organic carbon. In: Fritz P, Fontes JC (eds) Handbook of environmental isotope geochemistry 1. Elsevier, Amsterdam, pp 329–406
Deines P, Gold DP (1969) The change in carbon and oxygen isotopic composition during contact metamorphism of Trenton limestone by the Mt. Royal pluton. Geochim Cosmochim Acta 33:421–424
Delmas RJ, Ascencio JM, Legrand M (1980) Polar ice evidence that atmospheric CO_2 20000 year BP was 50% of present. Nature (London) 284:155–157
Deuser WG, Degens ET (1967) Carbon isotope fractionation in the system CO_2 (gas) – CO_2 (aqueous) – HCO_3^- (aqueous). Nature (London) 215:1033–1035
Druffel ERM, Benavides LM (1986) Input of excess CO_2 to the surface ocean based on $^{13}C/^{12}C$ ratios in a banded Jamaican sclerosponge. Nature (London) 321:58–61
Durand B (ed) (1980) Kerogen – insoluble organic matter from sedimentary rocks. Technip, Paris, 519 pp
Eichmann R, Schidlowski M (1975) Isotopic fractionation between coexisting organic carbon-carbonate pairs in Precambrian sediments. Geochim Cosmochim Acta 39:585–595
Ericson DB, Ewing M, Wollin G (1964) The Pleistocene epoch in deep sea sediments. Science 146:723–732
Fearnside PM (1985) Brazil's Amazon forest and the global carbon problem. Interciencia 10:179–186
Francey RJ (1981) Tasmanian tree rings belie suggested anthropogenic $^{13}C/^{12}C$ trends. Nature (London) 290: 232–235
Freyer HD, Belacy N (1983) $^{13}C/^{12}C$ records in northern hemispheric trees during the past 500 years: anthropogenic impact and climatic superpositions. J Geophys Res 88:6844–6852
Galimov EM (ed) (1985) The biological fractionation of isotopes. Academic Press, Orlando, Fla, 261 pp
Garrels RM, Mackenzie FT (1971) Evolution of sedimentary rocks. Norton, New York, 397 pp
Garrels RM, Perry EA (1974) Cycling of carbon, sulfur, and oxygen through geologic time. In: Goldberg ED (ed) Marine chemistry 5. John Wiley & Sons, New York, pp 303–336
Gavelin S (1957) Variations in isotopic composition of carbon from metamorphic rocks in northern Sweden and their geological significance. Geochim Cosmochim Acta 12:297–314
Goodfellow WD (1986) Anoxic oceans and short-term isotope trends. Nature (London) 322:116–117
Goodfellow WD, Jonasson IR (1984) Ocean stagnation and ventilation defined by $\delta^{34}S$ secular trends in pyrite and barite, Selwyn Basin, Yukon. Geology 12:583–586
Greenberg JM (1984) The structure and evolution of interstellar grains. Sci Am 250:124–135
Grossman L, Larimer JW (1974) Early chemical history of the solar system. Rev Geophys Space Phys 12:71–101
Gruszczynski M, Halas S, Hoffman A, Malkowski K (1989) A brachiopod calcite record of the oceanic carbon and oxygen isotope shifts at the Permian/Triassic transition. Nature (London) 337:64–68
Hayes JM, Kaplan IR, Wedeking KW (1983) Precambrian organic geochemistry: preservation of the record. In: Schopf JW (ed) Earth's earliest biosphere: its origin and evolution. University Press, Princeton NJ, pp 93–134
Hester K, Boyle E (1982) Water chemistry control of cadmium content in recent benthic foraminifera. Nature (London) 298:260–262
Hoefs J, Frey M (1976) Isotopic composition of carbonaceous matter in a metamorphic profile from the Swiss Alps. Geochim Cosmochim Acta 40:945–951
Holland HD (1978) The chemistry of the atmosphere and oceans. John Wiley & Sons, New York, 351 pp
Holland HD (1984) The chemical evolution of the atmosphere and oceans. University Press, Princeton NJ, 582 pp
Holser WT (1977) Catastrophic chemical events in the history of the ocean. Nature (London) 267:403–408
Holser WT, and 14 authors (1989) A unique geochemical record at the Permian/Triassic boundary. Nature (London) 337:39–44
Hsu KJ, McKenzie JA (1985) A "Strangelove" ocean in the earliest Tertiary. Geophys Monogr 32:487–492
Hut P, Alvarez W, Elder WP, Hansen T, Kauffman EG, Keller G, Shoemaker M, Weissman PR (1987) Comet showers as a cause of mass extinctions. Nature (London) 329:118–126
Irvine WM, Knacke RF (1989) The chemistry of interstellar gas and grains. In: Atreya SK, Pollack JB, Matthews MS (eds) Origin and evolution of planetary and satellite atmospheres. Univ Arizona Press, Tucson, pp 3–34
Jenkyns HC (1988) The early Toarcian (Jurassic) anoxic event: stratigraphic, sedimentary, and geochemical evidence. Am J Sci 288:101–151
Jenkyns HC, Clayton CJ (1986) Black shales and carbon isotopes in pelagic sediments from the Tethyan Lower Jurassic. Sedimentology 33:87–106
Junge CE, Schidlowski M, Eichmann R, Pietrek H (1975) Model calculations for the terrestrial carbon cycle: carbon isotope geochemistry and evolution of photosynthetic oxygen. J Geophys Res 80:4542–4552
Keeling CD (1961) The concentration and isotopic abundances of carbon dioxide in rural and marine air. Geochim Cosmochim Acta 24:277–298
Keeling CD, Bacastow RB (1977) Impact of industrial gases on climate. In: Energy and Climate. National Academy of Sciences, Washington DC, 88:72–95
Keeling CD, Bacastow RB, Bainbridge AE, Ekdahl CA, Guenther PR, Waterman LS, Chin JFS (1976) Atmospheric carbon dioxide variations at Mauna Loa Observatory, Hawaii. Tellus 28:538–551
Keeling CD, Mook WG, Tans PP (1979) Recent trends in the $^{13}C/^{12}C$ ratio of atmospheric carbon dioxide. Nature (London) 277:121–122
Keith ML, Weber JM (1964) Carbon and oxygen isotopic composition of selected limestones and fossils. Geochim Cosmochim Acta 28:1787–1816
Knoll AH, Hayes JM, Kaufman AJ, Swett K, Lambert IB (1986) Secular variation in carbon isotope ratios from Upper Proterozoic successions of Svalbard and East Greenland. Nature (London) 321:832–838
Kreulen R, Van Beek PCJM (1983) The calcite-graphite isotope thermometer; data on graphite bearing marbles from Naxos, Greece. Geochim Cosmochim Acta 47:1527–1530
Kroopnick PM, Margolis SV, Wong CS (1977) $\delta^{13}C$ variations in marine carbonate sediments as indicators of the CO_2 balance between the atmosphere and oceans. In: Andersen NR, Malahoff A (eds) The fate of fossil CO_2 in the oceans. Plenum, New York, pp 295–321

Krumbein WE, Cohen Y (1977) Primary production, mat formation and lithification chances of oxygenic and facultative anoxygenic cyanophytes (cyanobacteria). In: Flügel E (ed) Fossil algae. Springer, Berlin Heidelberg New York, pp 37–56

Lohmann KC, Walker JCG (1989) The $\delta^{18}O$ record of Phanerozoic abiotic marine calcite cements. Geophys Res Lett 16:319–322

MacIntyre F (1980) The absorption of fossil fuel CO_2 by the ocean. Oceanol Acta 3:505–516

Magaritz M (1989) ^{13}C minima follow extinction events: a clue to faunal radiation. Geology 17:337–340

Magaritz M, Bar R, Baud A, Holser WT (1988) The carbon isotope shift at the Permian/Triassic boundary in the southern Alps is gradual. Nature (London) 331: 337–339

McGhee GR, Orth CJ, Quintana LR, Gilmore JS, Olsen EJ (1986) Late Devonian "Kellwasser Event" mass extinction horizon in Germany: no geochemical evidence for a large body impact. Geology 14:776–779

McKirdy DM, Hahn J (1982) The composition of kerogen and hydrocarbons in Precambrian rocks. In: Holland HD, Schidlowski M (eds) Mineral deposits and the evolution of the biosphere. Springer, Berlin Heidelberg New York, pp 123–154

McKirdy DM, Powell TG (1974) Metamorphic alteration of carbon isotopic composition in ancient sedimentary organic matter: new evidence from Australia and South Africa. Geology 2:591–595

Miller KG, Fairbanks RG (1985) Oligocene to Miocene carbon isotope cycles and abyssal circulation changes. Geophys Monogr 32:469–486

Mook WG (1986) ^{13}C in atmospheric CO_2. Neth J Sea Res 14:211–223

Mook WG, Bommerson JC, Staverman WH (1974) Carbon isotope fractionation between dissolved bicarbonate and gaseous carbon dioxide. Earth Planet Sci Lett 22:169–176

Mook WG, Koopmans M, Carter AF, Keeling CD (1983) Seasonal, latitudinal, and secular variations in the abundance and isotopic ratios of atmospheric carbon dioxide. J Geophys Res 88:10915–10933

Murphey BF, Nier AO (1941) Variations in the relative abundance of the carbon isotopes. Phys Rev 59:771–772

Neftel A, Oeschger H, Schwander J, Stauffer B, Zumbrunn R (1982) Ice core sample measurements give atmospheric CO_2 content during the past 40000 year. Nature (London) 295:220–223

Nier AO, Gulbransen EA (1939) Variations in the relative abundance of the carbon isotopes. J Am Chem Soc 61:697–698

Nozaki Y, Rye DM, Turekian KK, Dodge RE (1978) A 200 year record of carbon-13 and carbon-14 variations in a Bermuda coral. Geophys Res Lett 5:825–828

Oeschger H, Siegenthaler U, Schotterer U, Gugelmann A (1975) A box diffusion model to study the carbon dioxide exchange in nature. Tellus 27:168–192

O'Leary MH (1981) Carbon isotope fractionation in plants. Phytochemistry 20:553–567

Oppo DW, Fairbanks RG (1989) Carbon isotope composition of tropical surface water during the past 22000 years. Paleooceanography 4:333–351

Orth CJ, Gilmore JS, Quintana LR, Sheehan PM (1986) Terminal Ordovician extinction: geochemical analysis of the Ordovician-Silurian boundary, Anticosti Island, Quebec. Geology 14:433–436

Park R, Epstein S (1960) Carbon isotope fractionation during photosynthesis. Geochim Cosmochim Acta 21: 110–126

Peters KE, Rohrback BG, Kaplan IR (1981) Carbon and hydrogen stable isotope variations in kerogen during laboratory-simulated thermal maturation. Am Assoc Petrol Geol Bull 65:501–508

Pflug HD, Jaeschke-Boyer H (1979) Combined structural and chemical analysis of 3800-Myr-old microfossils. Nature (London) 280:483–486

Playford PE, McLaren DJ, Orth CJ, Gilmore JS, Goodfellow WD (1984) Iridium anomaly in the upper Devonian of the Canning basin, Western Australia. Science 226:437–439

Popp BN, Anderson TF, Sandberg PA (1986) Brachiopods as indicators of original isotopic compositions in some Paleozoic limestones. Geol Soc Am Bull 97:1262–1269

Rankama K (1948) New evidence of the origin of Precambrian carbon. Geol Soc Am Bull 59:389–416

Reimer TO, Barghoorn ES, Margulis L (1979) Primary productivity in an Early Archean microbial ecosystem. Precambrian Res 9:93–104

Revelle R, Suess HE (1957) Carbon dioxide exchange between atmosphere and ocean and the questions of an increase of atmospheric CO_2 during the past decades. Tellus 9:18–27

Ronov AB (1958) Organic carbon in sedimentary rocks (in relation to the presence of petroleum). Geochemistry 1958:510–536

Ronov AB (1968) Probable change in the composition of sea water during the course of geological time. Sedimentology 10:25–43

Ronov AB (1980) Osadochnaya Obolochka Zemli (Earth's sedimentary shell). 20th Vernadsky Lect. Nauka, Moscow, 97 pp (in Russian)

Rycroft MJ (1982) Analysing atmospheric carbon dioxide levels. Nature (London) 295:190–191

Sawyer JS (1972) Man-made carbon dioxide and the "greenhouse" effect. Nature (London) 239:23–26

Schidlowski M (1982) Content and isotopic composition of reduced carbon in sediments. In: Holland HD, Schidlowski M (eds) Mineral deposits and the evolution of the biosphere. Springer, Berlin Heidelberg New York, pp 103–122

Schidlowski M (1983) Biologically mediated isotope fractionations: biochemistry, geochemical significance, and preservation in the Earth's oldest sediments. In: Ponnamperuma C (ed) Cosmochemistry and the origin of life. Reidel, Dordrecht, pp 277–322

Schidlowski M (1987) Application of stable carbon isotopes to early biochemical evolution on Earth. Annu Rev Earth Planet Sci 15:47–72

Schidlowski M (1988) A 3800-million-year isotopic record of life from carbon in sedimentary rocks. Nature (London) 333:313–318

Schidlowski M (1990) Life on the early Earth: Bridgehead from Cosmos or autochthonous phenomenon? In: Gopalan K, Gaur VK, Somayajulu BLK, Macdougall JD (eds) From mantle to meteorites (Festschrift for Devendra Lal). Indian Acad Sci, Bangalore, pp 189–199

Schidlowski M, Eichmann R, Junge CE (1975) Precambrian sedimentary carbonates: Carbon and oxygen isotope geochemistry and implications for the terrestrial oxygen budget. Precambrian Res 2:1–69

Schidlowski M, Eichmann R, Junge CE (1976) Carbon isotope geochemistry of the Precambrian Lomagundi carbonate province, Rhodesia. Geochim Cosmochim Acta 40:449–455

Schidlowski M, Appel PWU, Eichmann R, Junge CE (1979) Carbon isotope geochemistry of the 3.7×10^9-yr-old Isua sediments, West Greenland: implications for the Archaean carbon and oxygen cycles. Geochim Cosmochim Acta 43:189–199

Schidlowski M, Hayes JM, Kaplan IR (1983) Isotopic inferences of ancient biochemistries: carbon, sulfur, hydrogen and nitrogen. In: Schopf JW (ed) Earth's earliest biosphere: its origin and evolution. University Press, Princeton NJ, pp 149–186

Schidlowski M, Matzigkeit U, Krumbein WE (1984) Superheavy organic carbon from hypersaline microbial mats: assimilatory pathway and geochemical implications. Naturwissenschaften 71:303–308

Schoell M, Wellmer FW (1981) Anomalous ^{13}C depletion in Early Precambrian graphites from Superior Province, Canada. Nature (London) 290:696–699

Scholle PA, Arthur MA (1980) Carbon isotope fluctuations in Cretaceous pelagic limestones: potential stratigraphic and petroleum exploration tool. Am Assoc Petrol Geol Bull 64:67–87

Schopf JW (ed) (1983) Earth's earliest biosphere: its origin and evolution. Univ Press, Princeton NJ, xxv + 543 pp

Schopf JW, Packer BM (1987) Early Archaean (3.3-billion to 3.5-billion-year-old) microfossils from Warrawoona Group, Australia. Science 237:70–73

Shackleton NJ (1987) The carbon isotope record of the Cenozoic: history of organic carbon burial and of dissolution cycles. In: Andersen NR, Malahoff A (eds) The fate of fossil fuel CO_2 in the oceans. Plenum, New York, pp 401–427

Shackleton NJ (1986) Paleogene stable isotope events. Palaeogeogr Palaeoclim Palaeoecol 57:91–102

Shackleton NJ (1987) The carbon isotope record of the Cenozoic: history of organic carbon burial and of oxygen in the ocean and atmosphere. In: Brooks J, Fleet AJ (eds) Marine petroleum source rocks. Geol Soc Spec Publ 26:423–434

Shackleton NJ, Imbrie J, Hall MA (1983a) Oxygen and carbon isotope record of East Pacific core V19-30: implications for the formation of deep water in the late Pleistocene North Atlantic. Earth Planet Sci Lett 65:233–244

Shackleton NJ, Hall MA, Line J, Shuxi C (1983b) Carbon isotope data in core V19-30 confirm reduced carbon dioxide concentration in the ice age atmosphere. Nature (London) 306:319–322

Sheppard SMF, Schwarcz HP (1970) Fractionation of carbon and oxygen isotopes and magnesium between coexisting metamorphic calcite and dolomite. Contrib Mineral Petrol 26:161–198

Shieh YH, Taylor HP (1969) Oxygen and carbon isotope studies of contact metamorphism of carbonate rocks. J Petrol 10:307–331

Stanley SM (1981) The new evolutionary timetable. Basic Books, New York, 222 pp

Trask PD, Patnode HW (1942) Source beds of petroleum. Am Assoc Petrol Geol, Washington, 566 pp

Valley JW, O'Neil JR (1981) $^{13}C/^{12}C$ exchange between calcite and graphite: a possible thermometer in Grenville marbles. Geochim Cosmochim Acta 45:411–419

Veizer J, Hoefs J (1976) The nature of $^{18}O/^{16}O$ and $^{13}C/^{12}C$ secular trends in sedimentary carbonate rocks. Geochim Cosmochim Acta 40:1387–1395

Veizer J, Hoefs J, Lowe DR, Thurston PC (1989) Geochemistry of Precambrian carbonates. II. Archean greenstone belts and Archean sea water. Geochim Cosmochim Acta 53:859–871

Vogel JC (1961) Isotope separation factors of carbon in the equilibrium system $CO_2-HCO_3^--CO_3^{2-}$. In: Houtermans FG, Picciotto EE, Tongiori E (eds) Summer course on nuclear geology, Varenna 1960, Lischi, Pisa, pp 216–221

Vogel JC (1980) Fractionation of the carbon isotopes during photosynthesis. Sitzungsber Heidelberger Akad Wiss Math-Nat Kl 3:111–135

Wada H, Suzuki K (1983) Carbon isotopic thermometry calibrated by dolomite-calcite solvus temperatures. Geochim Cosmochim Acta 47:697–706

Walker D (1979) Energy, plants and man. Packard, Chichester, 31 pp

Walker JCG (1977) Evolution of the atmosphere. Macmillan, New York, 318 pp

Walker JCG, Klein C, Schidlowski M, Schopf JW, Stevenson DJ, Walter MR (1983) Environmental evolution of the Archean-Early Proterozoic Earth. In: Schopf JW (ed) Earth's earliest biosphere: its origin and evolution. Univ Press, Princeton NJ, pp 260–290

Weber F, Schidlowski M, Arneth JD, Gauthier-Lafaye F (1983) Carbon isotope geochemistry of the Lower Proterozoic Francevillian Series of Gabon (Africa). Terra Cognita 3:220

Wendt I (1968) Fractionation of carbon isotopes and its temperature dependence in the system CO_2 (gas)–CO_2 (in solution) and HCO_3-CO_2 in solution. Earth Planet Sci Lett 4:64–68

Weyl PK (1970) Oceanography – an introduction to the marine environment. John Wiley & Sons, New York, 535 pp

Winkler FJ, Kexel H, Kranz C, Schmidt HL (1982) Parameters affecting the $^{13}CO_2/^{12}CO_2$ isotope discrimination of the ribulose-1,5-bisphosphate carboxylase reaction. In: Schmidt HL, Förstel H, Heinzinger K (eds) Stable isotopes. Anal Chem Symp Ser 11. Elsevier, Amsterdam, pp 83–89

Woodwell GM (1978) The carbon dioxide question. Sci Am 238:34–43

Woodwell GM, Whittaker RH, Reiners WA, Likens GE, Delwiche CC, Botkin DD (1978) The biota and the world carbon budget. Science 199:141–146

Xu DY, Yan Z, Zahng QW, Shen ZD, Sun YY, Ye LF (1986) Significance of a $\delta^{13}C$ anomaly near the Devonian/Carboniferous boundary at the Muhua section, South China. Nature (London) 321:854–855

Yudovich YE, Makarikhin VV, Medvediev PV, Sukhanov NV (1990) Izotopnye anomalii ugleroda v karbonatakh Karel'skogo Kompleksa. Geokhimiya 1990(7):972–978

Zachos JC, Arthur MA, Dean WE (1989) Geochemical evidence for suppression of pelagic marine productivity at the Cretaceous/Tertiary boundary. Nature (London) 337:61–64

Zahn R, Winn K, Sarnthein M (1986) Benthic foraminiferal $\delta^{13}C$ and accumulation rates of organic carbon: *Uvigerina peregrina* group and *Cibicidoides wuellerstorfi*. Paleoceanography 1:27–42

Precambrian Petroleum: A Decade of Changing Perceptions *

David M. McKirdy [1] and Scott W. Imbus [2,3]

[1] Department of Geology and Geophysics, University of Adelaide, Adelaide, SA 5001, Australia
[2] School of Geology and Geophysics, University of Oklahoma, Norman, OK 73019, USA
[3] Present address: Texaco Inc., E & P Technology Division, 3901 Briarpark, Houston, TX 77042, USA

Abstract

Multidisciplinary studies during the last decade have identified hitherto unrecognized controls on the distribution and composition of organic matter preserved in the Earth's Precambrian sedimentary rock record (> 0.54 Ga). Carbonaceous shales and cherts containing isotopically inhomogeneous kerogen of prokaryotic origin occur sporadically in Archean metasedimentary sequences. The oldest recognizable petroleum source rocks were deposited in anoxic microbasins where methylotrophic bacteria played a key role in the production of oil-prone organic matter highly enriched in ^{12}C. These late Archean sulfide-rich, black shales (TOC up to 15%) are now thermally overmature. During Early Proterozoic time (2.5–1.6 Ga) epeiric seas developed, allowing widespread epicratonic sedimentation and burial of organic matter from an expanding planktonic biomass. The molecular fossil (or biomarker) record indicates that eukaryotic algae became significant kerogen precursors at least 1.7 Ga ago. Along the southeastern margin of the Canadian Shield Aphebian black argillites (or, at Elliot Lake and Blind River, stratiform uraniferous kerogen layers in the Matinenda Formation) were probable sources for widespread vein pyrobitumens found in ~ 2 Ga-old rocks of the Great Lakes region, USA and Canada. Shungite, another epigenetic pyrobitumen in the Jatulian Series of Karelia, USSR, originated from interbedded carbonaceous, pyritic shales. Indigenous Early Proterozoic dry gas was encountered by mineral drillholes in the Pechenga Series, Kola Peninsula, USSR, and McArthur Group, Northern Territory, Australia. Optimum conditions for the formation of thick, extensive petroleum source beds were first attained in Early Proterozoic intracratonic or rift-related tectonic settings, as exemplified by the organic-rich mudstones of the Franceville Basin, Gabon (2 Ga), and the Pine Creek Geosyncline (1.8–2.2 Ga) and McArthur Basin (1.6–1.8 Ga), northern Australia. These ancient source rocks (or carbonized residues of their expelled crude oil) are variously associated with U, Au, Pb, Zn, Ag and Cu mineralization. Among the oldest sediments currently being explored for hydrocarbon resources are those of the 1.4-Ga-old Roper Group, McArthur Basin, and the 1.1-Ga-old Oronto Group, Mid-Continent Rift, USA. Here, extensive organic-rich shales and siltstones were deposited in marine and lacustrine environments, respectively; aliphatic Type I–II kerogen is of appropriate thermal maturity for oil generation; and oil shows are common. The Late Proterozoic (1.0–0.54 Ga) was a time of prolonged oceanic anoxia and high rates of accumulation of organic matter in marine sediments (TOC contents locally as high as 20–30%). Rich source rocks and the world's oldest commercial oil and gas reserves, much of them in giant fields, occur on the Siberian Platform (Lena Tunguska region) and Arabian Shield (Oman) where late Riphean and Vendian siliciclastics and carbonates pass upwards into thick evaporite deposits which acted as regional seals. Peritidal dolomitic carbonates of Sinian age in the Sichuan Basin, southeastern China, include basal argillaceous source beds and the main reservoir of the giant Weiyuan gas field. Proterozoic oils and pyrobitumens have light carbon isotopic signatures ($\delta^{13}C = -30$ to -45‰). The unusual biomarker distributions of the oils and associated organic-rich sediments (where they contain well-preserved kerogen) cast new light on the membrane and cellular lipids of Proterozoic microbial ecosystems.

1 Introduction

The inaugural meeting of IGCP Project 157 (Early Organic Evolution and Mineral and Energy Re-

* This work is dedicated to Thomas C. Hoering, a pioneer investigator of Precambrian organic matter, from whom the senior author as an undergraduate heard his first lecture on organic geochemistry.

M. Schidlowski et al. (Eds.)
Early Organic Evolution: Implications for Mineral and Energy Resources

sources) was held in Jerusalem in July 1978. At that time scientific interest in Precambrian organic-rich sediments capable of generating hydrocarbons in commercial quantities was muted to non-existent. To many explorationists, the idea of "Precambrian petroleum" was a contradiction in terms. This is perhaps surprising, given that the prospectivity of Proterozoic sedimentary basins for petroleum had long been a matter of serious debate (Murray 1965; Vassoyevich et al. 1971), fueled by a burgeoning literature on Precambrian microfossils (mostly marine photoautotrophs), and reports of minor oil and gas shows and various pyrobitumens in rocks as old as 2.7 Ga. Indeed, in retrospect, it is salutary to realize just how much Precambrian petroleum had already been found. More than 20 oil and gas fields (about half deemed commercial) with Late Proterozoic and Early Cambrian reservoirs were discovered in the Lena-Tunguska region of East Siberia between 1962 and 1978 (Meyerhoff 1980; Murray et al. 1980). Heavy, sulfur-rich crude oil, subsequently recognized as originating from Late Proterozoic source rocks, was discovered in Paleozoic glacial clastics at Marmul on the southeastern flank of the South Oman salt basin in 1956 and eventually brought into production in 1980 (de la Grandville 1988). Commercial production of indigenous Precambrian dry gas from the Weiyuan field of the Sichuan Basin, southeastern China, began in 1964. By 1980 exploration in this basin had delineated other medium to large gas accumulations in Late Proterozoic dolomites (Fu et al. 1984).

And what of the chemical and isotopic composition of these ancient hydrocarbons? Although organic matter preserved in Precambrian carbonaceous and bituminous sediments had been the subject of intensive study since the early 1960s (for reviews see Hoering 1967; McKirdy 1974; McKirdy and Hahn 1982; Hayes et al. 1983), organic geochemists by and large were slow to appreciate that "indigenous Proterozoic and Cambrian crude oils (and their source rocks) constitute a vast store of geochemical data on the nature of the early marine and lacustrine biomass, in the form of biological marker compounds e.g. branched, isoprenoid, steroid, and polyterpenoid hydrocarbons" (McKirdy et al. 1983, p. 99). At the 1978 Gordon Conference on Organic Geochemistry in Plymouth, New Hampshire, the late Wolfgang Seifert, a pioneer of modern petroleum biomarker studies, conceded to the senior author that the oldest crude oil he had examined was of Jurassic age. Later, he was to elucidate the unusual sterane distribution of oil from the Siva field in the Ural-Volga region of the USSR (Seifert 1980), one of several Proterozoic crudes in which Russian geochemists had previously identified both steranes and triterpanes (Petrov et al. 1976; Makushina et al. 1978; Aref'ev et al. 1980).

One scientist acutely aware of the potential riches of the geochemical record of ancient life sequestered in well-preserved Precambrian sedimentary rocks was Tom Hoering from the Geophysical Laboratory, Carnegie Institution of Washington. Continuing work begun in concert with other research groups at Harvard University and the University of California, Berkeley (Eglinton et al. 1964, 1966; Barghoorn et al. 1965; Johns et al. 1966), he was then engaged in a long and painstaking search for isotopic signatures and molecular fossils of Middle Proterozoic biota in shale and an associated oil seep from the 1.1-Ga-old Nonesuch Formation at White Pine Mine, Michigan (Hoering and Abelson 1964; Hoering 1965, 1967, 1977, 1978, 1981; Hoering and Navale 1987). Likewise, by the late 1970s, a comprehensive geochemical study of the Proterozoic oils and source rocks of Oman was well underway in the Shell Exploration and Production Research Laboratory, Rijswijk (Grantham 1986; Klomp 1986; Grantham et al. 1988). The tide of awareness was about to turn, ushering in a new era of enquiry into not only the unique biomarker characteristics of Precambrian crude oils, but also the microbiotas, paleoenvironments, stratigraphic distribution, and thermal histories of their source rocks.

This work presents a brief overview of the major advances in our understanding of the world's oldest petroleum arising from geological and geochemical investigations undertaken during the lifetime of IGCP Project 157.

2 Setting the Scene: Organic Matter in Archean Sediments

The first prerequisite for petroleum occurrence in any sedimentary basin is the existence of organic-rich strata, usually fine-grained siliciclastics or carbonates, which have been exposed to sufficient thermal stress to generate and expel hydrocarbons. Sediments that are destined to become effective petroleum source rocks need to contain a minimum concentration of dispersed organic matter (conveniently measured as total organic carbon, TOC). According to Hunt (1979) and Tissot and Welte (1984), these threshold TOC values are 0.3% for carbonates and 0.5% for shales and siltstones. Jones (1984) disagreed, citing a minimum TOC content of 1.5% as a shared characteristic of excellent oil-prone source rocks (regardless of lithofacies), along with deposition in an anoxic environment, a finely laminated texture, and high-quality aliphatic kerogen (initial atomic H/C ratio $\geqq 1.2$). Were these criteria ever satisfied by sediments older than 2.5 Ga?

For the duration of its Archean history the Earth was in a greenhouse state, with elevated atmospheric pCO_2 (at least 10^2–10^3 PAL: Kasting 1987) and surface temperatures in the range 22–40 °C (Worsley et al. 1986). It is also worth noting that, in the absence of extensive higher plant vegetation, nutrient supply is unlikely to have limited the size of the Precambrian aquatic biosphere, although Archean prokaryotes may have required more phosphorus than do contemporary biota (cf. Hayes 1983). Accordingly, primary productivity appears to have been moderately high during the Archean (at least half the present level). This does not necessarily guarantee enhanced organic preservation, even under conditions of globally-low oxygen tension in the atmosphere (10^{-2} to 10^{-6} PAL) and hydrosphere. Nevertheless, above average organic carbon contents are not uncommon in the Archean metasedimentary record. Of the 178 samples from 17 different sequences assembled by Hayes et al. (1983) for organic geochemical analysis, 36 (20%) have TOC values > 1%. Likewise, a weighted average TOC content of 0.7% (n = 406) for late Archean shales from the Canadian Shield (Cameron and Garrels 1980) indicates widespread local environments conducive to the accumulation of organic matter; mean TOC contents at three of the nine localities sampled are in the range 1.4–5.0%. Shales and, to a lesser extent, cherts are the most common Archean carbonaceous lithofacies. Original organic carbon levels in these ancient sediments were almost certainly higher than indicated by their present TOC values because of significant carbon loss during metamorphism. Correlations between carbon abundance and isotopic composition suggest that this inference is valid for Archean (and Proterozoic) shales and carbonates (Hayes et al. 1983).

Secular trends in the stable isotopic composition of carbonate (oxidized) carbon and coexisting organic (reduced) carbon in Precambrian and Phanerozoic sediments provide evidence that photoautotrophy dates back to at least 3.5 Ga (Schidlowski 1988), an interpretation supported by the recognition of stromatolites and associated organic-walled microfossils of prokaryotic affinity in rocks of similar antiquity (Walter 1983; Lowe 1983; Schopf and Packer 1987). However, a feature of the Archean sedimentary carbon record is that its organic fraction is isotopically quite inhomogeneous, with $\delta^{13}C$ values (relative to PDB standard) mostly in the range −45 to −15‰ (McKirdy and Hahn 1982; Hayes et al. 1983; Strauss 1986). This wide spread of $\delta^{13}C_{org}$ values highlights two pertinent aspects of the burial and preservation of organic matter (kerogen) in Archean sediments.

Firstly, it is clear that in many cases the inferred primary isotopic signature of autotrophic carbon fixation (−26 ± 7‰: Schidlowski 1988) has been shifted to heavier values by metamorphic alteration of the kerogen. The enrichment of ^{13}C in highly altered kerogens (atomic H/C < 0.2: McKirdy and Hahn 1982) is due to the preferential rupture of ^{12}C–^{12}C bonds during metagenetic dehydrogenation reactions, with consequent loss of isotopically light methane and other volatiles; and, if exposed to temperatures of 300–350 °C (lower greenschist facies) or higher in a mineral matrix comprising both carbonate and quartz, to isotopic exchange between the residual kerogen and heavy CO_2 released during the formation of metamorphic calc-silicates (Schidlowski et al. 1983). Examples of extreme ^{13}C-enrichment in graphitic kerogen ($\delta^{13}C$ = −18 to −8‰) occur in the lower Onverwacht Group, Swaziland Sequence, South Africa (McKirdy and Powell 1974) and in the amphibolite-grade metasediments of the Isua Supracrustal Belt, West Greenland (Schidlowski et al. 1979). Notwithstanding the difficulty of reliably estimating the initial carbon isotopic composition of the Isua carbonate and organic matter (kerogen H/C = 0.1–0.01: Schopf et al. 1983), isotopic mass balance calculations based on these materials have been used to support the hypothesis that the sedimentary organic carbon reservoir was already 50–60% its present size, and accounted for about 20% of total carbon buried in the crust, by the time the Isua Sequence was deposited ~ 3.8 Ga ago (Schidlowski et al. 1979). A more circumspect view of the evidence from less metamorphosed, 3.4 to 3.5-Ga-old sediments assumes modal $\delta^{13}C_{carb}$ and $\delta^{13}C_{org}$ values of −1 and −35‰, respectively, and yields a somewhat lower estimate for the fraction of early Archean carbon buried as organic matter (C_{org}:C_{carb} = 12:88) (Hayes 1983).

Secondly, it is equally clear that certain late Archean sequences contain reduced carbon which is isotopically very light ($\delta^{13}C$ = −35 to −50‰). When due allowance is made for metamorphic overprint (using $\delta^{13}C_{org}$-atomic H/C curves for reliably analyzed kerogens: Hayes et al. 1983), original signatures highly depleted in ^{13}C are seen to be characteristic of kerogens in many of the Archean sedimentary facies documented by the first Precambrian Paleobiology Research Group at UCLA (Schopf 1983). Where these enigmatic kerogens occur in high concentrations in late Archean (2.6–2.8 Ga) black shales, including those of the Hamersley Formation (TOC = 1.4–6.5%) and Fortescue Group (TOC = 1.2–16.5%) in

Table 1. Carbon isotopic composition of Precambrian pyrobitumens, oils and source rock kerogens

Age (Ga)	Basin/region	Type[a]	Reservoir	Petroleum $\delta^{13}C$ (‰, PDB)	Likely Source(s)	Kerogen $\delta^{13}C$ (‰, PDB)	Ref.[b]
2.7	Witwatersrand Basin, S Africa	PB	Main Fm.	– 38.8 to – 37.0	Carbon Leader Mbr.	– 38.2 to – 28.8	8, 13
2.2	Great Lakes, USA and Canada	PB	Matinenda Fm.	– 33.8 to – 31.4	Matinenda/ McKim Fms.	nd[d]	12
2.1		PB	Biwabik Fm.	– 35.1 to – 34.1	Virginia Fm. (shale)	– 32.7 to – 31.5	4, 12
2.1		PB	Gunflint Fm.	– 39.2 to – 33.0	Gunflint Fm. (shale)	– 34.0 to – 31.9	2, 12, 13
1.9					Rove Fm. (shale)	– 33.6 to – 27.6	7, 14
1.9		PB	Michigamme Fm.	– 31.5 to – 30.5	Michigamme Fm. (shale)	nd	12
1.9		PB	Onwatin Fm.	– 32.0 to – 31.0	Onwatin Fm. (shale)	– 30.7 to – 30.5	7, 12
2.2	N Cape Province, S Africa	PB	Transvaal Dolomite	– 41.2 to – 37.1	Transvaal Dolomite	– 36.7 to – 26.5	5
2.0	Franceville Basin, Gabon	PB	FA (uraniferous)	– 29 to – 21	Lower FB 1c (shale)	– 30 to – 22	3
		PB	FB 2a, FC	– 46 to – 38	Upper FB 1c, FC (shale)	– 46 to – 33	3
1.8	Baltic Shield, USSR	PB	Jatulian Series	– 26.8	Jatulian shales	nd	8
1.4	McArthur Basin, Australia	O	Bessie Creek Sst	– 33.3 to – 32.6[c]	Velkerri Fm. (shale)	– 33.3 to – 32.2	14, 16
1.1	Mid-Continent Rift, USA	O	Nonesuch Fm.	– 31.7 to – 30.8	Nonesuch Fm. (shale)	– 34.7 to – 33.4	1, 9, 10, 11
< 0.7	E Siberian Platform, USSR	O	Not specified	– 34.5 to – 34.1	Unspecified carbonates	nd	15
< 0.7	South Oman Salt Basin	O	Al Khlata/ Gharif Fms.	– 36.0	Ara/Shuram Fms. (carb)	– 35.4 to – 34.4	6
		O	Gharif Fm.	– 30.5	?Huqf Group (shale)	nd	6

[a] PB = pyrobitumen; O = oil.
[b] References: 1, 2, Barghoorn et al. (1965, 1977); 3, Cortial et al. (1990); 4, Douthitt (1982); 5, Eichmann and Schidlowski (1975); 6, Grantham et al. (1988); 7, Hayes et al. (1983); 8, Hoering (1967); 9, Hieshima et al. (1990); 10, Imbus et al. (1988); 11, S.W. Imbus (unpubl. data); 12, Mancuso et al. (1989); 13, 14, Strauss et al. (1991 and this Vol.); 15, Summons and Powell (this Vol.); 16, Zaunbrecher (1988).
[c] Measured on C_{12+} saturates fraction.
[d] nd = No data available.

Western Australia (Hayes et al. 1983) and the Keewatin Group (TOC = 1–15%) within the Superior Province of the Canadian Shield (Strauss 1986), we have possible examples of spent Archean petroleum source beds. Thicknesses up to 7 m have been reported for the Keewatin shales, which have TOC and $\delta^{13}C_{org}$ profiles suggestive of preferential carbon loss (via generation and expulsion of hydrocarbons?) from the upper and lower edges of individual beds (Strauss 1986, fig. 3.). These graphitic, sulfide-bearing shales occur over strike lengths of 100–1400 m within tuffaceous intervals of volcanic piles (Schoell and Wellmer 1981). Their limited lateral extent indicates deposition in small basins (or local depressions) into which hydrothermal brines discharged. The anoxic bottom layer of such density-stratified water bodies was host to methanogenic bacteria. It also trapped and preserved organic debris from photoautotrophs growing nearer the surface, probably in the lower levels of the photic zone to escape lethal UV-radiation. According to the model proposed by Hayes (1983), the carbon source of the photoautotrophs comprised a mixture of atmospheric CO_2 (– 5‰), biogenic CO_2 from the aerobic (– 50‰) and anaerobic (– 10‰) decay of organic matter, and markedly ^{13}C-depleted CO_2 (– 90‰) produced by CH_4-oxidizing bacteria which occupied an intermediate trophic level in the water column. By this time (~ 2.9 Ga) the photosynthesizers included oxygen-producing protocyanobacteria as well as anaerobic bacteria (Schopf et al. 1983). Thus, these late Archean microbasins satisfy all the conditions recognized by Demaison and Moore (1980) as necessary for the deposition of oil-prone source rocks in younger (Phanerozoic) environments. That similar anoxic paleoenvironments may have developed (albeit less commonly) in larger Archean sedimentary basins, such as those with extensive banded iron formations, is attested to by the ^{13}C-depleted kerogen in the aforementioned carbonaceous shales of the Pilbara Block, Western Australia. Hayes (1983) explains how ^{13}C-poor and ^{13}C-rich organic matter can be simultaneously preserved in coeval, but separate, anoxic and oxic sedimentary basins, depending on the amount of reduced mineral matter (e.g., Fe^{2+}) delivered to the basin.

Microbial mats preserved as thin seams and reworked granules of kerogen occur in association with placer concentrates of gold, uraninite and pyrite at several levels within the 2.5 to 2.7-Ga-old Upper Witwatersrand Group. The seams are in places up to 2–5 cm thick and, although patchily developed on fluvial scour surfaces, they are widespread in fan deltas which extend over an area of 9750 km^2 around the margin of the Witwatersrand Basin on the Kaapvaal Craton (Minter 1978). The original aliphatic structure of this kerogen has been severely modified, largely as a result of its prolonged exposure to ionizing radiation emanating from detrital uraninite and zircon trapped in the mat (Zumberge et al. 1978). Its seam-like morphology, protocyanobacterial precursors (Dyer et al. 1988), and primary oil-prone character make it similar in some respects to the so-called kerogenite or kukersite source beds in the Ordovician Red River and Yeoman Formations of the Williston Basin (Kohm and Louden 1979; Fowler, this Vol.). Evidence of hydrocarbon generation from this stratiform kerogen is provided by the isotopically light anthraxolite (Strauss et al. 1991) associated with the Carbon Leader Member ($\delta^{13}C = -36$ to -39‰: Table 1).

Although organic-rich sediments generally may not have been as thick or widespread as in the Proterozoic and Phanerozoic records (e.g., because of the smaller areal extent of stable continental shelves), significant production and preservation of Archean organic matter did occur. Indeed, it is inevitable that petroleum was generated in Archean sedimentary basins. High heat flow in greenstone belts would have countered the requirement for burial of source rocks to depths typical of Phanerozoic basins. However, just as pervasive metamorphism of source beds has destroyed the chemical fossils which may have furnished clues to the lipid and membrane biochemistry of precursor microbiota, so too all evidence of petroleum has been erased from Archean reservoir rocks (save, perhaps, certain thucholites, graphitized pyrobitumens, and thermogenic methane in fluid inclusions). It is tempting to speculate that the otherwise unprecedentedly high organic carbon contents (up to 2.12%, mean = 0.76%) reported by Reimer et al. (1979) for litharenites from the Sheba Formation in the 3.4-Ga-old Fig Tree Group, Swaziland Sequence, may be partly attributable to carbonized reservoir bitumen.

3 Early Proterozoic (2.5–1.6 Ga): Pyrobitumens as Oil Shows

Abundant, well-preserved microbiotas date from about 2.1 Ga (Schopf et al. 1983), providing an indication of the nature of the contemporary biomass. The fidelity of preservation displayed by microbial fossils from Early Proterozoic units in localities as widely scattered as northern Minnesota, USA; southern Ontario and the Belcher Islands, Canada; Ukraine, Karelia and Kola Peninsula, USSR; Pilbara Block, Western Australia; and the McArthur Basin, Northern Territory, Australia (Hofmann and Schopf 1983, and references cited therein), also heralds the concomitant potential for survival of ancient petroleum. Indeed, many of these terrains host tangible evidence of migrated petroleum in the form of pyrobitumen (hitherto variously described as anthraxolite, thucholite, shungite or anthracitic coal) and natural gas.

It was not until the dawn of Proterozoic time that epeiric seas developed and pericratonic sedimentation became widespread following the assembly of the first granite-surfaced supercontinent at ~ 2.7 Ga (Windley 1984) and its fragmentation some 200 Ma later (Worsley et al. 1986). The Early Proterozoic also saw the establishment of a stable aerobic atmosphere and with it an ozone layer increasingly opaque to solar UV radiation (Schopf et al. 1983). As a result the prokaryotic biomass increased and protocyanobacteria occupied new habitats; microplankton gradually expanded into the upper photic levels of the oceans; and the oceanic oxygen minimum layer intensified as dense, oxygenated, surface water at mid-latitudes (where it was more saline) or high latitudes (where it was colder) penetrated deeper ocean levels. The latter cause of oceanic circulation possible became operative between 2.5 and 2 Ga, the age of the oldest recognizable glacial deposits (Walker et al. 1983). Optimum conditons for the accumulation and preservation of sedimentary organic matter prevailed where oxygen minima impinged on the newly created continental shelf and slope in areas of high, near-surface biological productivity.

Such marine environments were widespread at this time on the Canadian Shield. Along the southern and eastern flanks of the Superior Province, shales of Aphebian age (1.6–2.5 Ga) are commonly enriched in organic carbon (mean TOC values at six different localities are in the range 1.1–5.1%, n = 189: Cameron and Garrels 1980), and therefore were potential petroleum source rocks. The widespread occurrence of solid pyrobitumens in Early Proterozoic (1.8–2.4 Ga) sequences throughout the Great Lakes region of Canada and the USA attests to local oil generation. The pyrobitumen occurs in veins and fractures, or less commonly as conformable lenses up to 2 m long and 1 m thick (Mancuso et al. 1989). Host rocks include the Matinenda, Onwatin and Gunflint Formations, Ontario; Biwabik Formation, Minnesota; and Michigamme Formation, Michigan. The last example was long considered to be a Precambrian anthracitic coal (e.g., Tyler et al. 1957; Hoering 1967). Without exception the bitumens are isotopically lighter ($\delta^{13}C = -30$ to -39‰) than the kerogen of their inferred source beds, usually contiguous black shale, argillite or slate ($\delta^{13}C = -28$ to -33‰: Table 1). The highly altered nature of these residual crude oils is shown by their low H/C atomic ratios (0.09–0.57) and high optical reflectance (R_0max = 0.9–9.5%). The pyrobitumen in the Matinenda Formation is the so-called globular organic matter from Elliot Lake and Blind River described by Nagy and Mossman (this Vol.). According to these authors, the globular bitumen originated from a Witwatersrand-style, intraformational, richly uraniferous, stratiform kerogen (Willingham et al. 1985). An alternative and more conventional petroleum source rock is the black shale of the overlying McKim Formation which also may have acted as seal to the Matinenda reservoir (quartz pebble conglomerate).

Isotopically heavy dolomitic carbonates ($\delta^{13}C = +8.2 \pm 2.6$‰) of the lower Logamundi Group (1.95–2.65 Ga) crop out over a distance of 300 km in northeastern Zimbabwe where they provide indirect evidence of massive burial of organic matter in coeval argillaceous sediments in a large closed basin (Schidlowski et al. 1976). However, what is arguably the most impressive accumulation of Early Proterozoic organic matter, and one in which there is a demonstrable genetic relationship between relict petroleum (pyrobitumen) and uranium mineralization, occurs in the 2-Ga-old, essentially unmetamorphosed siliciclastics and subordinate volcanics of the intracratonic Franceville Basin, Gabon (Gauthier-Lafaye 1986; Cortial et al. 1990). Here, shales with mean TOC values in the range 2–11% (and individual beds commonly containing 10–20% TOC) are prominent in the Francevillian Series. The major pelitic formation (FB) is 600–1000 m thick, of which almost 80% is carbonaceous shale. A strong stratigraphic control on the carbon isotopic composition of the organic matter in the Francevillian shales ($\delta^{13}C$ values decrease up section from -21‰ at the base of FB to -37 to -46‰ in formations FC and FD) suggests progressive restriction and stagnation of the basin as biogenic methane became an increasingly important carbon source for the kerogen precursors (Weber et al. 1983).

Originally oil-prone, sapropelic kerogen in the principal Francevillian source rock (FB) has attained the meta-anthracite stage of thermal evolution (mean R_0max = 5.9–6.9%; atomic H/C = 0.1–0.5). Thermal alteration of crude oil in intercalated sandstones and dolomites has produced two varieties of solid bitumen (Table 1). One, in reservoir units FB_2 and FC, is devoid of U mineralization, highly depleted in ^{13}C ($\delta^{13}C = -38$ to -46‰), and apparently derived from isotopically light kerogen ($\delta^{13}C < -33$‰) within stratigraphically adjacent shales of units FB_1c (upper part) and FC. The other type of bitumen is highly uraniferous and oxidized. If occurs in the basal part of the sequence, in sandstones of unit FA and breccias of unit FB_1b, where its redox reaction with dissolved uranyl (IV) species in migrating brines resulted in the precipitation of uraninite and coffinite (Cortial et al. 1990). The distinctly heavier isotopic composition of the mineralized bitumen ($\delta^{13}C = -21$ to -29‰) reflects its derivation from shales of the lower FB_1c unit (Table 1) and subsequent prolonged α-irradiation, as described for Mesozoic high-grade uranium ores by Leventhal and Threlkeld (1978).

In the Pine Creek Geosyncline of northern Australia, 1.85 to 2.2-Ga old sediments of the Mt. Partridge Group and overlying South Alligator Group include extensive carbonaceous shales (grading to graphitic argillites) which host economic U, Pb, Zn, Ag and Au mineral deposits. In the Rum Jungle uranium field, the Whites Formation (formerly Golden Dyke Formation) has organic carbon contents as high as 10% (mean TOC = 3.8%, n = 11: McKirdy 1974; Foster et al. 1990; D. H. Tucker unpubl. data). The organic-rich shales of this formation, together with the magnesitic carbonates of the immediately underlying Coomalie Dolomite (TOC < 1%), also are of interest because they were deposited under lacustrine conditions in a rift basin (Bone 1988). Carbonaceous shales (and metamorphic equivalents) of the younger Koolpin Formation, which are host to major uranium deposits further east in the Alligator Rivers area, typically contain 2–3% TOC but in places are much richer (up to 30%: Berkman 1968).

Kerogen in the Whites Formation around Rum Jungle is extensively dehydrogenated (e.g., at Woodcutters Mine atomic H/C = 0.04, $\delta^{13}C = -15.8‰$: McKirdy and Powell 1974) and clearly overmature. Nevertheless, several lines of evidence point to petroleum genesis and migration within this Early Proterozoic lacustrine sequence (Foster et al. 1990). These include carbon coatings on mineral grains, interpreted as maturation residues of a former mobile hydrocarbon phase; and the presence of aromatic hydrocarbon-rich fluid inclusions in carbonate grains of both the lower Whites Formation and Coomalie Dolomite. Such carbonate-hosted, hydrocarbon-bearing fluid inclusions no doubt account for the otherwise inexplicably high yield (240 ppm) of solvent-extractable organic matter reported by McKirdy (1974) for a core sample of dolomitic carbonaceous shale from the Whites Formation at the Woodcutters locality. However, the survival of biomarker hydrocarbons in high relative abundance in the fluid inclusions (notably pristane and phytane; and n-alkanes and n-alkylcyclohexanes with marked even-carbon-number predominance in the range $C_{14}-C_{24}$ and $C_{16}-C_{20}$, respectively) is inconsistent with a syngenetic origin from the organic-rich, Whites Formation shale. Although the exact source of their hydrocarbons is unclear, the fluid inclusions appear contemporaneous with U ore emplacement (Foster et al. 1990).

Shungite from the Jatulian Series (1.63–1.81 Ga) of Karelia, USSR, is another example of carbonized pyrobitumen commonly mistaken for Early Proterozoic coal (cf. Metzger 1924, as cited by Rankama 1948; Mycke et al. 1988; Glaessner and Foster, this Vol.). At Shunga, the type locality, shungite occurs as bright, jet-black, carbonaceous intercalations in dolomitic limestone and, according to Rankama (1948), was produced by heat from a dolerite intrusion in Jatulian carbonaceous shales (now black, sulfide-bearing slates). Other varieties of shungite (usually with more mineral matter and hence a duller appearance) exist elsewhere in Karelia and eastern Finland. For example, near Suojarvi, shungite is present as discrete layers within its shale source beds. Detailed optical and X-ray diffraction studies by Khavari-Khorasani and Murchison (1979) confirmed that shungite is an epigenetic, non-graphitic pyrobitumen with a degree of carbonization equivalent to meta-anthracite rank. This possibly explains why shungite, of all the non-uraniferous pyrobitumens listed in Table 1, is the least depleted in ^{13}C ($\delta^{13}C = -26.8‰$).

Bright shungite with carbon contents of ~95% (dry, ash-free) and atomic H/C ratios of 0.15–0.45 (Khavari-Khorasani and Murchison 1978) would not be expected to contain intact biomarker hydrocarbons. Nevertheless, using a novel analytical approach, Mycke et al. (1988) discovered that finely disseminated framboidal pyrite in dull shungite has entrapped and preserved a complex array of saturated hydrocarbons (total yield = 39 ppm) including n-alkanes, monomethyl (notably 3-Me and 5-Me) and unassigned branched alkanes, acyclic isoprenoid alkanes, and n-alkylcyclohexanes. Steranes (C_{29} dominant) and hopanes were also detected but only in trace amounts. Collectively, these alkanes were interpreted as molecular fossils of thermoacidophilic bacteria, cyanobacteria and primitive (?Prymnesiophycaean) algae growing in an Early Proterozoic, sulfidic, coal-forming environment. However, the identification of shungite as a pyrobitumen, rather than a coal; the heavily metamorphosed condition of both the pyrobitumen and its host rocks; and the thermal lability of some of the biomarkers (particularly the polycyclic alkanes), make the ultimate source and significance of the hydrocarbons difficult to decipher.

During 1969 and the early 1970s, mineral exploration drilling on the nearby Kola Peninsula, USSR, encountered one of the oldest known accumulations of petroleum-related natural gas within fractured volcanics and metasediments of the 1.71 to 1.89-Ga-old Pechenga Series (Petersil'ye et al. 1974). The gas is very dry ($C_1 = 52-75\%$, $C_2-C_4 = 0.02-0.48\%$; $\delta^{13}C_1 = -46$ to $-60‰$) and rich in non-hydrocarbons ($N_2 = 14-34\%$, $H_2 = 1-23\%$, $He = 1.3-3.7\%$). Detailed logging of various drillholes indicates that the gas, except perhaps for its helium component, originated in marine, sulfide-bearing, carbonaceous shale (mean TOC = 2.35%, n = 11) which, in the vicinity of the Pechenga mine, is up to 600 m thick.

It was in 1979 that a gas blowout from the Coxco Dolomite Member in a base-metal exploration drillhole alerted Amoco and other companies to the petroleum potential of the 1.6 to 1.8-Ga-old McArthur Group in the McArthur Basin of northern Australia. In addition to this uneconomic gas accumulation, there is abundant evidence of migrated hydrocarbons throughout the sequence, including oil bleeding from core of the Barney Creek Formation (also host to the vast HYC stratiform lead-zinc deposit); and reservoir bitumen and dead oil in porous and vuggy facies of the Lynott, Yalco and Looking Glass Formations and Stretton Sandstone (Muir et al. 1980; Womer 1986; Jackson et al. 1988). Features of this Early Proterozoic carbonate sequence are the thickness and quality of its potential source rocks (dolomitic mudstones of the Barney Creek, Lynott and Yalco Formations: Table 2) which were deposited in lacustrine or evaporitic lagoonal environments.

Table 2. Characteristics of selected, presently mature, Proterozoic source rocks

Age (Ga)	Basin	Unit	Maturity	TOC %	HC Potential kg/tonne	Hydrogen index	Kerogen H/C	Organic type	Ref.[a]
1.7	McArthur Basin, Australia	Barney Creek Fm.	Marginally mature[b]	1–10	2–71	150–950	0.70–1.66	I–II	1, 6
		Yalco Fm.	Marginally mature	1–5	4–33	150–590	0.92–1.20	II	1, 6
1.4	McArthur Basin, Australia	Velkerri Fm.	Mature	1–7	8–33	200–640	0.65–1.01	II	1, 6
		Kyalla Mbr., McMinn Fm.	Marginally mature	0.7–4	2–20	100–570	0.86–1.24	II–II/III	1, 6, 7, 10
1.1	Mid-Continent Rift, USA	Nonesuch Fm.	Mature[b]	0.5–3	Up to 8 (1.3)	20–450 (173)	nd[c]	I–II	3, 4, 5
0.85	Grand Canyon, USA	Walcott Mbr., Kwagunt Fm.	Marginally mature to mature	1–7 (~ 3)[d]	Up to 16 (~ 6)	100–200 (135)	0.77–0.93	II III	8, 9
< 0.7	South Oman Salt Basin	Huqf Group	Mature	2–6	[e]	nd	nd	II	2

[a] References: 1, Crick et al. (1988); 2, Grantham et al. (1988); 3, Hieshima et al. (1989); 4, 5, Imbus et al. (1988, 1990); 6, Jackson et al. (1988); 7, McKirdy (1977); 8, Palacas and Reynolds (1989); 9, Summons (1991); 10, Zaunbrecher (1988).
[b] In places overmature.
[c] nd = No data available.
[d] ( ) = Mean value.
[e] Extract yield = 2000–30000 ppm.

Table 3. Biomarker hydrocarbons identified in Proterozoic source rocks and crude oils

Hydrocarbon class	Biomarkers	Microbial precursors and environments	Ref.[a]
Monomethyl alkanes	$C_{12}-C_{22}$ 2-methyl and 3-methyl	Eubacteria	4, 7, 8, 9, 15, 16, 17
	$C_{21}-C_{30}$ mid-chain methyl	Cyanobacteria	
Acyclic isoprenoid alkanes	$C_{15}-C_{20}$ regular	Photoautotrophs	2, 3, 4, 6, 7, 15, 16, 17
	$C_{15}-C_{25}$ regular and $C_{29}-C_{30}$ irregular	Archaebacteria (methanogens, extreme halophiles)	
Aryl isoprenoids	$C_{14}-C_{20}$ irregular	Green sulfur bacteria (meta- to hypersaline environments)	15
Cyclohexyl alkanes	$C_{12}-C_{26}$ n-alkyl $C_{14}-C_{28}$ methyl-n-alkyl	Eubacterial and cyanobacterial fatty acids	7, 8, 16, 17
Tricyclic terpanes	$C_{19}-C_{31}$	Unspecified prokaryotes	13, 16, 17
Pentacyclic triterpanes	$C_{27}-C_{35}$ 17α(H)-hopanes	Eubacteria, cyanobacteria	1, 4, 5, 10, 11, 13, 15, 16, 17
	$C_{29}-C_{35}$ 18α(H)-neohopanes[b]	Eubacteria	
	28,30-bisnorhopane	Eubacteria (?anaerobic)	
	gammacerane	Protozoa (?hypersaline environments)	
	2α-methyl hopanes	Eubacteria (e.g., methylotrophs), cyanobacteria	
Steranes	$C_{26}-C_{30}$ desmethyl	Eukaryotes (algae, heterotrophs)	1, 5, 6, 11, 12, 13, 15, 16, 17
	C_{29} dominant	?Green algae, ?prymnesiophytes	
	C_{27} dominant	?Red algae	
	$C_{28}-C_{30}$ 3β- and 2α-methyl	Unknown	

[a] References: 1, Aref'ev et al. (1980); 2, Barghoorn et al. (1965); 3, Eglinton et al. (1964); 4, Fowler and Douglas (1987); 5, Grantham (1986); 6, Grantham et al. (1988); 7, 8, Hoering (1978, 1981); 9, Klomp (1986); 10, Moldowan et al. (1985); 11, Petrov et al. (1980); 12, Seifert (1980); 13, Summons (1991); 14, Summons and Walter (1990); 15, Summons and Powell (this Vol.); 16, 17, Summons et al. (1988a, b).
[b] Subsequently identified as 17α(H)-diahopanes (R. E. Summons, pers. comm. to D. M. McKirdy 1991).

Remarkably, these source beds are in places still only marginally mature, allowing detailed optical and pyrolytic characterization of their kerogen (Crick et al. 1988: Table 2) and elucidation of their biomarker geochemistry (Summons et al. 1988a; Summons 1991; Table 3). The Type I–II kerogen is essentially monomaceralic (lamalginite) and markedly oil-prone (hydrogen index = 200–950; genetic potential up to 70 kg hydrocarbons/tonne). Its loss of fluorescence (under UV light) and concomitant increase of optical reflectance with increasing thermal alteration affords a useful method of assessing maturation levels in these and other lamalginite-containing pre-Devonian sediments. The major hydrocarbon classes identified were n-alkanes, monomethylalkanes, acyclic isoprenoid alkanes (including the regular C_{21}–C_{25} and irregular C_{29} and C_{30} compounds), n-alkyl and methyl-n-alkylcyclohexanes and, in lesser abundance, hopanes and methylhopanes. They reflect the predominantly prokaryotic (and in part archaebacterial) affinity of this Early Proterozoic organic matter. Of particular note is the detection of steranes ($C_{27} > C_{29} > C_{28} > C_{26} > C_{30}$) in the least thermally altered samples of the Barney Creek Formation. This important finding signifies the otherwise unrecognizable contribution of eukaryotic organisms to these 1.69-Ga-old lacustrine sediments.

4 Middle Proterozoic (1.6–1.0 Ga): The Oldest Petroleum Exploration Frontier

Arguably the oldest sediments currently under active exploration for hydrocarbon resources are the marine siliciclastics of the ~1.4-Ga-old Roper Group, McArthur Basin (Jackson et al. 1988). Early studies had described well-preserved organic-walled microfossils and Type II kerogen in moderately organic-rich, silty shales of the McMinn Formation (McKirdy 1977; Peat et al. 1978). Stratigraphic coreholes drilled during the early 1980s revealed the widespread occurrence of hydrocarbon shows (incuding live oil: Jackson et al. 1986), thick source rocks of excellent quality (Table 2), and extensive sandstones with variable reservoir potential. Since 1984, 11 wildcat wells, many encountering additional oil shows, have been drilled on the northern Bauhinia Shelf. The latest and most encouraging well, Jamison-1, recovered gas (at a rate of 453 m^3/day), 2 m of oil, and 448 m of oil- and gas-cut formation water. The richest, regionally mature source rocks (1–7% TOC; 8–33 kg hydrocarbons/tonne) are the outer shelf marine shales of the Velkerri Formation (and stratigraphically equivalent Lansen Creek Shale) which contain oil-prone Type II kerogen (hydrogen index = 200–640). The most prospective reservoir units, the Bessie Creek Sandstone and Moroak Sandstone Member (McMinn Formation), lie immediately below and above the Velkerri Formation. The upper reservoir is overlain by inner-shelf clastics of the Kyalla Member (McMinn Formation) in which argillaceous interbeds comprise a less prolific source facies (0.7–4% TOC; 5–20 kg hydrocarbons/tonne) containing poorer quality Type II–II/III kerogen (hydrogen index = 100–570). Apart from their very low concentrations of C_{15}, C_{16} and C_{18}–C_{20} isoprenoids and lack of extended isoprenoids, the biomarker geochemistry of the Velkerri and McMinn Formations closely resembles that of the Barney Creek Formation (Summons et al. 1988a). Preliminary oil-to-source correlations identify the lower Velkerri Formation as the source of the second of two distinct oil charges received by the Bessie Creek reservoir; whereas oil shows in the Moroak Sandstone appear to have originated in the Kyalla Member and upper Velkerri Formation (Zaunbrecher 1988). Zircon-cored thucholites are common in the Kyalla Member where they provide evidence of primary oil migration along lamalginite layers (D. M. McKirdy, unpubl. observ.).

Another Middle Proterozoic target which has attracted considerable exploration interest is the alluvial, lacustrine and fluvial sequence of the ~1.1-Ga-old Oronto Group (and stratigraphic equivalents) in the North American Mid-Continent Rift System (Dickas 1984, 1986; Imbus et al. 1988; Elmore et al. 1989). Here, the lacustrine Nonesuch Formation is recognized as a potential source rock. However, reported occurrences of indigenous oil in the northern sector of the rift (Keweenawan Trough, Lake Superior region) are limited to seeps at the White Pine copper mine (Ontonagan County, Michigan: Kelly and Nishioka 1985) and migrated hydrocarbons in mining cores from northern Wisconsin. Gas and condensate accumulations in the Prairie du Chien Formation (Ordovician) of the lower Michigan Peninsula may have a Nonesuch or equivalent source (Seglund 1989). In southeastern Minnesota, organic-rich Middle Proterozoic units are overmature (Hatch and Morey 1985). Drilling in northeastern Kansas has yielded neither gas or oil, nor revealed possible source rocks. Nevertheless, the rift is still considered prospective, as recently released results of drilling in Iowa indicate gas shows and the presence of organic-rich sediments (albeit overmature) similar to those in the Lake Superior region (R. M. McKay, pers. comm. to S. W. Imbus 1990).

The organic-rich lithofacies of the Nonesuch Formation comprises finely laminated, partly calcareous, silty shale with TOC contents up to 3% (Imbus et al. 1988). As much as 77% of the 20–33-m-thick black shale and siltstone unit which represents the perennial lake phase of the formation (cf. Elmore et al. 1989) has TOC values > 0.5% (Imbus et al. 1991). The saturated hydrocarbon distributions of the Nonesuch shales have unusually high pristane/phytane ratios (range = 2.5–6, mean = 4.1) (Hieshima et al. 1989). Such high ratios (pr/ph > 3), commonly associated with organic matter exposed to oxic conditions during deposition and early diagenesis, have not previously been reported from organic-rich Precambrian sediments. The best preserved kerogen (organic petrography 1 of Imbus et al. 1988) resembles lamalginite and has an aliphatic Type I–II composition. The thermal history of the formation in northern Wisconsin and Michigan has been mild enough for it to presently lie at or near the middle of the oil-generation window. The Nonesuch seep oil and shale extracts alike are characterized by high concentrations of C_{12+} n-alkanes, 2- through 6-methyl substituted alkanes, acyclic isoprenoids (up to C_{20}), and alkylcyclohexanes (Barghoorn et al. 1965; Eglinton et al. 1966; Hoering 1978, 1981); and recent advances in GC-MS techniques have at last made possible confirmation of the presence of steranes ($C_{27} > C_{28} > C_{29}$) and hopanoid triterpanes (Summons 1991). Again, a lacustrine biota comprising prokaryotes and subordinate eukaryotes (but lacking the pronounced archaebacterial input evident in the older Barney Creek geolipids) is implicated as the precursor of the Nonesuch organic matter.

The only known commercial production of possible Middle Proterozoic hydrocarbons [other than the (?)Nonesuch-sourced gas and condensate in the Michigan Basin] is from sandstones within the Upper Riphean section in the Larionova and Siva fields near Perm in the Volga-Urals Basin, USSR (Vassoyevich et al. 1971; Kutukov et al. 1977). Here, siltstones at the top of the Kaltasa Formation, one of the producing units, are considered to be the likely source beds. Elsewhere in the basin oil and gas shows occur in both Riphean and Vendian sands. A partially biodegraded, low-sulfur, crude oil from the Siva field contains isoprenoid alkanes (pr/ph = 1.1), hopanes and, in lesser abundance, steranes (C_{27}–C_{29}) and gammacerane (Moldowan et al. 1985). The presence of the latter compound, a C_{30} pentacyclic triterpane of probable protozoan origin, and the dominance of ethylcholestane among the steranes ($C_{29} \gg C_{27}, C_{28}$: Petrov et al. 1976; Seifert 1980) are also features of some Late Proterozoic organic-rich sediments and crude oils (see below). Hence, the source rock of this particular Siva oil may be somewhat younger than Middle Proterozoic, in agreement with the 0.7 Ga age cited by Seifert (1980).

5 Late Proterozoic (1.0–0.54 Ga): The Oldest Giant Oil and Gas Fields

Various strands of sedimentological and geochemical evidence combine to suggest that the world's Late Proterozoic oceans were, for much of their history, characterized by high rates of production, accumulation and burial of organic matter. Prolonged periods of anoxia ensured the preservation of both organic matter and pyrite over large areas of the continental margins and deep ocean (Donnelly et al. 1990). These so-called oceanic anoxic events (OAEs) were punctuated by shorter glacial intervals when increased turnover and mixing led to the reventilation of the sea bottom. Massive sequestration of ^{12}C-enriched, autochthonous organic carbon into marine sediments will necessarily enhance the ^{13}C content of the oceanic bicarbonate reservoir. Hence, pronounced positive excursions in the $\delta^{13}C$ record of Late Proterozoic marine carbonates, as documented in a spate of recent studies (e.g., Knoll et al. 1986; Magaritz et al. 1986; Aharon et al. 1987; Lambert et al. 1987; Pell 1989; Brasier et al. 1990), attest to the widespread deposition of coeval, organic-rich sediments that are potential source rocks for hydrocarbons. These isotopically heavy carbonates span the entire Late Proterozoic from ~0.9 Ga to the base of the Cambrian. Examples of conspicuous organic enrichment in Late Proterozoic sequences include black bituminous shales, up to 20 m thick and with TOC contents up to 30%, in the Upper Vendian Redkino Formation of the Russian Platform (Vassoyevich et al. 1971); and the 150-m-thick, black shale member of the lower Soltanieh Formation in northern Iran, correlatives of which extend through central Iran into Arabia and Oman (Brasier et al. 1990).

The oldest organic-rich, Late Proterozoic sediments with proven oil-source potential are those of the 0.85–0.95-Ga-old Chuar Group in Arizona and Utah, USA (Palacas and Reynolds 1989). More than half the Kwagunt and underlying Galeros Formations in Grand Canyon comprise organic-rich, gray to black mudstone and siltstone. TOC contents as high as 7% (mean ~3%) and pyrolysis yields of up to 16 kg hydrocarbons/tonne (mean ~6 kg/tonne) characterize the lower half of the 281-m-thick Walcott Member, the uppermost unit of the Kwagunt Formation (Table

2). Reported thermal maturities in the oil generation window accord with the presence of well-preserved kerogen (atomic H/C = 0.77–0.92; $\delta^{13}C = -25.7$ to -27.8‰) in argillaceous dolostones of the Walcott Member (Summons 1991). Alkane biomarkers in extracts of the dolostone reflect its deposition in an intermittently hypersaline environment to which both prokaryotes (including eubacteria and halophilic archaebacteria) and eukaryotic protists (including algae and protozoa) contributed organic matter (Summons et al. 1988 b). To date, this is the oldest sediment in which gammacerane has been identified. Several unusual features of the Walcott Member biomarker assemblage, namely abundant desmethyl steranes ($C_{27} \gg C_{26} > C_{28} > C_{29}$), 3β-methyl and 2α-methyl steranes, and putative 18α(H)-neohopanes (Table 3), appear to represent evolutionary innovations in the membrane lipid biochemistry of Late Proterozoic microbiota. Such novel molecular signatures (and those in oils from Siberia and Oman, discussed below) enhance our ability to recognize indigenous Proterozoic crude oils.

Large commercial oil and gas fields with known or inferred Late Proterozoic source beds have been found in the remote Lena-Tunguska region of East Siberia (Meyerhoff 1980), South Oman (Hughes-Clark 1988), and the Sichuan Basin, southern China (Fu et al. 1984). Organic geochemical studies have provided definitive evidence for the intra-Proterozoic origin of hydrocarbons in all three provinces. The famous Markovo field in Siberia, discovered in 1962 and located at the southern end of the NNE-trending Angara-Lena Trough (part of the Baikalian foredeep), produces gas and light oil from Vendian sandstones and Lower Cambrian carbonates. Total proven and probable reserves are 0.62 Tcf gas and 26 MMbbl oil, whereas Vendian sandstones in the adjacent Yaraktin field may contain up to 210 MMbbl oil (Meyerhoff 1980). Some 600 km away on the northern flank of the same trough, there are two giant fields in which production is mostly from Vendian siliciclastic reservoirs: Sredne-Botuobin (149 MMbbl oil and 17.3 Tcf gas recoverable) and Verkhne-Vilyuchanskoye (260 MMbbl oil and 10.3 Tcf gas recoverable) (Carmalt and St. John 1986). The total resource potential of the Lena-Tunguska petroleum province is estimated by Meyerhoff (1982) to be 2000 MMbbl oil and 83 Tcf gas. The oils and condensates are typically light (34–50° API gravity), paraffinic crudes with sulfur contents up to 1.12% in those from Vendian-Cambrian carbonate reservoirs, whereas the gas ranges from wet to relatively dry (C_1 = 72–92%, $C_1/C_1{-}C_4$ = 0.77–0.94, N_2 = 3–9%) (Meyerhoff 1980). Although no specific source horizon has yet been identified for these Siberian hydrocarbons, they could be derived from deeply buried, upper Riphean black shales (TOC = 3–5%) or organic-rich marls (TOC = 1–10%) in the Vendian to Early Cambrian carbonate sequence (Summons and Powell this Vol.; T. G. Powell, pers. comm. 1988). The latter, exemplified by a 10-m-thick bed of argillite (mean TOC = 5.6%) near the top of the Nepa Formation in the Nepa-Botuobin Anticline (Lapichev and Chekanov 1987), are the most likely source rocks of the oils. This inference is strongly supported by significant aspects of the biomarker geochemistry of the Siberian oils (viz., pristane < phytane; virtual absence of diasteranes; high abundance of 28,30-bisnorhopane and norhopane relative to hopane; and presence of aryl isoprenoids attributable to green sulfur bacteria) which all indicate a carbonate-evaporite source affinity (Douglas and Fowler 1987; Summons and Powell, this Vol.). These oils are, moreover, isotopically very light ($\delta^{13}C \sim -34$‰: Table 1). Similar biomarker and isotopic characteristics have been documented in silty carbonates (TOC up to 21%) from the 0.62-Ga-old Khatyspyt Formation in the northern part of the Lena-Tunguska petroleum province (Summons 1991). The gas, on the other hand, may be largely of Riphean origin.

Since 1960 petroleum exploration in Oman has been spectacularly successsful, leading to the discovery and development of some 50 fields with total daily production currently in excess of 500 000 bbl oil, mostly from Cretaceous, Jurassic and Permian reservoirs. A detailed geochemical study by Grantham et al. (1988) identified potential source rocks ranging in age from Late Proterozoic (Huqf Group) to mid-Tertiary and five geochemically distinct end-member oil types (two of Proterozoic origin). The Huqf Group (Vendian-Cambrian) comprises a sequence of shallow marine clastics and carbonates which pass upwards into a major evaporite unit (Ara Formation). Within and beneath these thick, basin-wide salt deposits, rich source beds occur as micritic carbonates and claystones in the Ara and Shuram Formations, typically with TOC contents of 2–6% and extract yields of 2 000–30 000 ppm, and containing amorphous Type II kerogen (Grantham et al. 1988). These oil-prone Proterozoic source rocks are richer and more extensive than the younger marine Paleozoic and Mesozoic source beds, and therefore could have generated the bulk of Oman's identified oil and gas reserves. Using biomarker and carbon isotopic evidence, Grantham and coworkers distinguished two families of Proterozoic-sourced oils, one from Carboniferous, Permian and Cretaceous reservoirs in South Oman (the heavy, sulfur-rich Huqf oils: 10–32° API: 1.3–2.6% S: pr/ph < 1, $\delta^{13}C$

$\sim -30.5‰$) and the other from a Permian reservoir in central Oman (the light, low-sulfur "Q" oils: 42–50° API; <0.5% S; pr/ph ≧ 1; $\delta^{13}C \sim -36‰$). Definitive oil-source correlations were established only in the case of the Huqf crudes. The "Q" oils are believed to have originated from a second source interval (as yet unsampled) deeper in the Huqf Group of the Ara Salt Basin. These two oil families also have contrasting sterane distributions (Huqf: C_{29} dominant, diasteranes <5%; "Q": C_{27} dominant, diasteranes ~20%), but both contain unusually abundant isomeric mid-chain methyl-substituted alkanes (the so-called X-compounds of Klomp 1986), a feature they share with the Siberian and other oils of Late Proterozoic age (Fowler and Douglas 1987; Summons and Powell, this Vol.).

The giant Weiyuan gas field, the largest commercial gas accumulation in China, is one of several medium to large gas fields the southern Sichuan Basin where total annual production from Triassic, Permian, Cambrian and Sinian (Late Proterozoic) carbonate reservoirs is 205 Bcf (Fu et al. 1984). The main natural gas reserves of the Weiyuan field (comprising 80–87% CH_4, 7% N_2, 5% CO_2) are located in Upper Sinian (0.7–0.8 Ga) dolomites and, on the basis of geochemical criteria outlined by Xu Yongchang et al. (this Vol.), are considered to have been derived from thermally overmature source beds within the same sequence (cf. presence of anthraxolitic pyrobitumens). Dark gray to black, argillaceous, micritic carbonates at the base of the Dengying Formation and in the underlying Doushantuo Formation contain >1% TOC (Lambert et al. 1987).

Large areas of unmetamorphosed and structurally little disturbed Late Proterozoic strata in Australia continue to attract exploration. A non-commercial gas flow from the Areyonga Formation (~0.76 Ga) at Ooraminna-1 in 1963 first drew attention to the Proterozoic petroleum potential of central Australia. Activity has centered on two areas: the Amadeus and Officer Basins. Geochemical and organic petrographic studies of Late Proterozoic units from the Amadeus Basin (central Australia) indicate that the Gillen Member of the Bitter Springs Formation (~0.85 Ga) and the Pertatataka Formation (~0.60 Ga) in places have moderate source potential (Jackson et al. 1984). However, organic carbon contents are low (<1% TOC) and the available geochemical data indicate a prevalence of gas-prone Type III kerogen. The contrasting carbonate-evaporite and shale facies of these two units are reflected in differences in their biomarker alkane distributions (Summons and Walter 1990; Summons 1991). The small Dingo gas field, in which the Ediacaran-Early Cambrian Arumbera Formation is the productive unit, is an example of recent exploration success in the Amadeus Basin.

Recent drilling in the Officer Basin has failed to locate economic occurrences of hydrocarbons in either the Adelaidean (Late Proterozoic) or overlying Cambrian sequences (e.g., Townson 1985; Gravestock 1990). Nevertheless, thin potential source beds containing mature, oil-prone lamalginite (TOC = 0.4–1.5%: Summons 1991; D. M. McKirdy, S. D. Pell and M. Smyth unpubl. data) have been identified in the Rodda Beds, a Proterozoic marine unit >1.2 km thick. Moreover, oil shows have been recorded in this unit, in the underlying Murnaroo Formation, and in the overlying Relief Sandstone (basal Cambrian). Preliminary biomarker data on these oils (McKirdy 1989) reveal differences between their isoprenoid alkane, sterane and triterpane distributions and those of the previously analyzed Cambrian oils and extracts (McKirdy et al. 1983, 1984) from the eastern Officer Basin. The new oil shows are possibly of Proterozoic origin. Also of interest is the reported similarity between the alkane patterns of Rodda Beds extracts and those of Late Proterozoic oils from Oman and Siberia (notably a dominance of C_{29} steranes: Summons 1991).

6 Summary and Conclusions

The oldest recognizable sediments with all the necessary attributes of petroleum source rocks are late Archean black shales (or argillites), exemplified by those from the Superior Province of the Canadian Shield. These ~2.7-Ga-old, carbonaceous, pyritic shales (TOC = 1–15%) were deposited in small, anoxic basins scattered across volcanic terrains which became greenstone belts. Although now thermally overmature with respect to hydrocarbon generation, their graphitic kerogens have isotopic signatures ($\delta^{13}C = -35$ to $-50‰$) which reflect the involvement of methylotrophic bacteria in the carbon cycle of the original depositional environment.

Tangible evidence of oil generation from Early Proterozoic source rocks is provided by various pyrobitumens (including thucholite, anthraxolite, shungite and "anthracitic coal") occupying fractures and intergranular pore spaces in reservoir rocks of the same age from the Great Lakes region of USA and Canada; the northern Cape Province, South Africa; the Franceville Basin, Gabon; the eastern Baltic Shield of Karelia and Finland; and the McArthur Basin and Pine Creek Geosyncline, northern Australia. Remarkably, the 1.7-Ga-old lacustrine source beds of the McArthur Basin (Barney Creek, Lynott and Yalco

Formations) are in places still only marginally mature and contain well-preserved Type I–II kerogen. Gas flows from mineral drillholes which penetrated fractured metasediments of the Pechenga Series on the northwestern Kola Peninsula, USSR, and the Coxco Dolomite Member in the Batten Trough, southeastern McArthur Basin, can in each case be linked with adjacent, post-mature source beds of Early Proterozoic age.

The Middle Proterozoic ($\sim$1.4 Ga) marine siliciclastics of the Roper Group, McArthur Basin, are currently the oldest sediments being actively explored for petroleum hydrocarbons. Here, as in the similarly prospective but somewhat younger ($\sim$1.1 Ga) fluviolacustrine Oronto Group of the Keweenawan Trough, Wisconsin and northern Michigan, regionally extensive organic-rich shales and siltstones contain oil and gas-prone Type I–II kerogen with thermal maturities in the oil-generation window.

Although both Early and Middle Proterozoic sediments in the McArthur Basin appear to have generated significant volumes of oil and gas, variable development and/or preservation of reservoir porosity, early hydrothermal fluid movement associated with the emplacement of base metal mineralization along the Emu Fault Zone of the Batten Trough, and lack of structure at the time of peak hydrocarbon generation are important factors which may have limited their ability to trap and retain petroleum in commercial quantities (Crick et al. 1988; Jackson et al. 1988).

Prolonged periods of intense global anoxia in Late Proterozoic oceans ensured the burial of vast quantities of oil-prone kerogen in marine sediments 0.9–0.54 Ga old, most notably just prior to the Precambrian-Cambrian transition (Donnelly et al. 1990). Much of this burial of organic matter occurred within extensive Vendian evaporite basins on the eastern Siberian Platform (Meyerhoff 1980) and the Arabian Shield (Brasier et al. 1990); and during the initial transgressive phase of Sinian peritidal carbonate deposition on the South China Platform (Lambert et al. 1987). The resulting organic-rich sediments were destined to become source rocks for the world's oldest, presently commercial oil and gas fields, some of them in the giant category (Carmalt and St. John 1986).

Proterozoic crude oils, source rocks and other organic-rich sediments containing well-preserved kerogen (atomic $H/C > 0.7-0.8$) have proved to be "a happy hunting ground" for molecular fossils of ancient prokaryotes and the first, single-celled eukaryotic organisms (cf. Summons and Walter 1990). The initial impetus for recent spectacular advances in our knowledge of Proterozoic biomarker hydrocarbons was provided by earlier contributions to IGCP Project 157 (McKirdy et al. 1983, 1984; Fowler 1984) and work on the Omani oils undertaken by Shell geochemists (Grantham 1986; Klomp 1986). However, it was the involvement of Roger Summons and coworkers in the Australian Bureau of Mineral Resources study of the McArthur Basin and in the Precambrian Paleobiology Research Group (Proterozoic) at the University of California, Los Angeles, which was responsible for many exciting new discoveries of microbial biomarkers and insights into the sources of Precambrian organic matter (Jackson et al. 1986; Summons et al. 1988a, b; Summons and Walter 1990; Summons and Powell this Vol.; Summons 1991). The biologically significant hydrocarbons so far detected in Precambrian rock extracts and crude oils are summarized in Table 3. Syngenetic sulfides (e.g., pyrite: Mycke et al. 1988) and carbonate-hosted fluid inclusions (Foster et al. 1990) offer considerable potential, as yet largely untapped, for the preservation of thermally labile biomarker hydrocarbons in Precambrian source and reservoir rocks. Among those biomarkers, which appear to be diagnostic of Proterozoic (or Early Paleozoic) organic matter, are distinctive $C_{21}-C_{30}$ monomethyl alkanes with a preference for mid-chain branching, putative 18α(H)-neohopanes, and 3β- and 2α-methyl steranes. Another unusual feature of certain Proterozoic oils is the dominance of the C_{29} pseudohomologue (24-ethylcholestane) among the $C_{26}-C_{30}$ desmethyl steranes. The presence of significant quantities of extended ($C_{21}-C_{30}$) acyclic isoprenoid alkanes, hopanes and steranes in source rock extracts as old as 1.69 Ga suggests the contribution of membrane lipids from archaebacteria, eubacteria and primitive eukaryotes, respectively, to the kerogen. However, sterane/hopane ratios remain low until the Late Proterozoic ($\sim$0.85 Ga) when there was a major radiation of planktonic algae (Vidal and Knoll 1983). Using biomarker distributions, in conjunction with exceptionally light carbon isotopic values ($\delta^{13}C < -30‰$), it is now possible to "fingerprint" crude oils of Proterozoic origin.

Acknowledgements. This work would not have been completed without the constant encouragement of Manfred Schidlowski. We wish to thank Roger Summons for preprints of several of his recent papers; and Mike Engel, David Gravestock, Nick Lemon and Marjorie Muir for helpful reviews of the manuscript.

References

Aharon P, Schidlowski M, Singh IB (1987) Chronostratigraphic markers in the end-Precambrian carbon isotope record of the Lesser Himalaya. Nature (London) 327:699–702

Aref'ev OA, Zabrodina MN, Makushina VM, Petrov AA (1980) Relic tetra- and pentacyclic hydrocarbons in ancient oils of the Siberian Platform. Izv Akad Nauk SSR Ser Geol 3:135–140 (in Russian)

Barghoorn ES, Meinschein WG, Schopf JW (1965) Paleobiology of a Precambrian shale. Science 148:461–472

Barghoorn ES, Knoll AH, Dembicki H Jr, Meinschein WG (1977) Variation in stable carbon isotopes in organic matter from the Gunflint Iron Formation. Geochim Cosmochim Acta 41:425–430

Berkman DA (1968) The geology of the Rum Jungle uranium deposits. In: Proc Symp Uranium in Australia. Australas Inst Min Metall, Melbourne, pp 12–31

Bone Y (1988) The geological setting of tourmalinite at Rum Jungle, N.T., Australia – genetic and economic implications. Mineral Depos 23:34–41

Brasier MD, Magaritz M, Corfield R, Luo Huilin, Wu Xiche, Ouyang Lin, Jiang Zhiwen, Hamdi B, He Tinggui, Fraser AG (1990) The carbon- and oxygen-isotope record of the Precambrian-Cambrian boundary interval in China and Iran and their correlation. Geol Mag 127:319–332

Cameron EM, Garrels RM (1980) Geochemical compositions of some Precambrian shales from the Canadian Shield. Chem Geol 28:181–197

Carmalt SW, St. John B (1986) Giant oil and gas fields. In: Halbouty MT (ed) Future petroleum provinces of the world. Am Assoc Petrol Geol Mem 40:11–53

Cortial F, Gauthier-Lafaye F, Lacrampe-Couloume G, Oberlin A, Weber F (1990) Characterization of organic matter associated with uranium deposits in the Francevillian Formation of Gabon (Lower Proterozoic). Org Geochem 15:73–85

Crick IH, Boreham CJ, Cook AC, Powell TG (1988) Petroleum geology and geochemistry of Middle Proterozoic McArthur Basin, northern Australia II: assessment of source rock potential. Am Assoc Petrol Geol Bull 72:1495–1514

De la Grandville BF (1982) Appraisal and development of a structural and stratigraphic trap oil field with reservoirs in glacial to periglacial clastics. In: Halbouty MT (ed) The deliberate search for the subtle trap. Am Assoc Petrol Geol Mem 32:267–286

Demaison GJ, Moore GT (1980) Anoxic environments and oil source bed genesis. Am Assoc Petrol Geol Bull 64:1179–1209

Dickas AB (1984) Mid-continent rift system: Precambrian hydrocarbon target. Oil Gas J 82(42):151–159

Dickas AB (1986) Comparative Precambrian stratigraphy and structure along the mid-continent rift. Am Assoc Petrol Geol Bull 70:225–238

Donnelly TH, Shergold JH, Southgate PN, Barnes CJ (1990) Events leading to global phosphogenesis around the Proterozoic/Cambrian boundary. In: Notholt AJG, Jarvis I (eds) Phosphorite research and development. Geol Soc Spec Publ 52. Blackwell, Oxford London, pp 273–287

Douthitt CB (1982) Precambrian coal or anthraxolite: a source for graphite in high-grade schists and gneisses – a discussion. Econ Geol 77:1247–1249

Dyer BD, Krumbein WG, Mossman DJ (1988) Nature and origin of stratiform kerogen seams in Lower Proterozoic Witwatersrand-type paleoplacers – the case for biogenicity. Geomicrobiol J 6:33–47

Eglinton G, Scott PM, Belsky T, Burlingame AL, Calvin M (1964) Hydrocarbons of biological origin from a one-billion-year-old sediment. Science 145:263–264

Eglinton G, Scott PM, Belsky T, Burlingame AL, Richter W, Calvin M (1966) Occurrence of isoprenoid alkanes in a Precambrian sediment. In: Hobson GD, Louis MC (eds) Advances in organic geochemistry 1964. Pergamon, Oxford, pp 41–74

Eichmann R, Schidlowski M (1975) Isotopic fractionation between coexisting organic carbon-carbonate pairs in Precambrian sediments. Geochim Cosmochim Acta 39:585–595

Elmore RD, Milavec GJ, Imbus SW, Engel MH (1989) The Precambrian Nonesuch Formation of the North American Mid-Continent Rift, sedimentology and organic geochemical aspects of lacustrine deposition. Precambrian Res 43:191–213

Foster CB, Robbins EI, Bone Y (1990) Organic tissues, graphite, and hydrocarbons in host rocks of the Rum Jungle Uranium Field, northern Australia. Ore Geol Rev 5:509–523

Fowler MG (1984) Organic geochemistry of pre-Carboniferous sedimentary organic matter. PhD thesis, Univ Newcastle-upon-Tyne

Fowler MG, Douglas AG (1987) Saturated hydrocarbon biomarkers in oils of Late Precambrian age from eastern Siberia. Geochim Cosmochim Acta 11:201–213

Fu Jia Mo, Liu De Han, Jia Rong Fen, Dai Yong Ding (1984) Distribution and origin of hydrocarbons in carbonate rocks (Precambrian to Triassic) in China. In: Palacas JG (ed) Petroleum geochemistry and source rock potential of carbonate rocks. Am Assoc Petrol Geol Stud Geol 18:1–12

Gauthier-Lafaye F (1986) Les gisements d'uranium du Gabon et les réacteurs d'Oklo. Modèle métallogénique de gîtes a fortes teneurs du Protérozoïque inférieur. Mém Sci Géol 78 (Chap 6: Les matières organiques et le potential pétroligène du bassin):109–118

Grantham PJ (1986) The occurrence of unusual C_{27} and C_{29} sterane predominances in two types of Oman crude oil. Org Geochem 9:1–10

Grantham PJ, Lijmbach GWM, Posthuma J, Hughes Clarke MW, Willink RJ (1988) Origin of crude oils in Oman. J Petrol Geol 11:61–80

Gravestock DI (1990) Officer Basin petroleum potential. In: Parker AJ (compiler) South Australia – exploration towards 2000. S Aust Dep Mines Energ, Adelaide, pp 29–32

Hatch JR, Morey GB (1985) Hydrocarbon source rock evaluation of Middle Proterozoic Solor Church Formation, North American Mid-Continent Rift System, Rice County, Minnesota. Am Assoc Petrol Geol Bull 69:1208–1216

Hayes JM (1983) Geochemical evidence bearing on the origin of aerobiosis, a speculative hypothesis. In: Schopf JW (ed) Earth's earliest biosphere: its origin and evolution. University Press, Princeton, pp 291–301

Hayes JM, Kaplan IR, Wedeking KW (1983) Precambrian organic geochemistry, preservation of the record. In: Schopf JW (ed) Earth's earliest biosphere: its origin and evolution. University Press, Princeton, pp 93–134

Hieshima GB, Zaback DA, Pratt LM (1989) Petroleum potential of Precambrian Nonesuch Formation, Mid-Continent Rift System. Am Assoc Petrol Geol Bull 73:363

Hieshima GB, Pratt LM, Hayes JM (1990) Novel isotopic compositions of total organic carbon and bitumen in the Nonesuch Formation, North American Mid-Continent Rift. Am Assoc Petrol Geol Bull 74:674

Hoering TC (1965) The extractable organic matter in Precambrian rocks and the problem of contamination. Carnegie Inst Washington Yearb 64:215–218

Hoering TC (1967) The organic geochemistry of Precambrian rocks. In: Abelson PH (ed) Researches in geochemistry vol 2. John Wiley & Sons, New York, pp 87–111

Hoering TC (1977) The stable isotopes of hydrogen in Precambrian organic matter. In: Ponnamperuma C (ed) Chemical evolution of the early Precambrian. Academic Press, New York London, pp 81–86

Hoering TC (1978) Molecular fossils from the Precambrian Nonesuch Shale. In: Ponnamperuma C (ed) Comparative planetology. Academic Press, New York London, pp 243–255

Hoering TC (1981) Monomethyl acyclic hydrocarbons in petroleum and rock extracts. Carnegie Inst Washington Yearb 80:389–393

Hoering TC, Abelson PH (1964) Chemicals from Nonesuch Shale. Carnegie Inst Washington Yearb 63:262–264

Hoering TC, Navale V (1987) A search for molecular fossils in the kerogen of Precambrian sedimentary rocks. Precambrian Res 34:247–267

Hofmann HJ, Schopf JW (1983) Early Proterozoic microfossils. In: Schopf JW (ed) Earth's earliest biosphere: its origin and evolution. University Press, Princeton, pp 321–360

Hughes-Clarke MW (1988) Stratigraphy and rock unit nomenclature in the oil producing area of interior Oman. J Petrol Geol 11:5–60

Hunt JM (1979) Petroleum geochemistry and geology. Freeman, San Francisco, 595 pp

Imbus SW, Engel MH, Elmore RD, Zumberge JE (1988) The origin, distribution and hydrocarbon generation potential of organic-rich facies in the Nonesuch Formation, central North American Rift System: a regional study. In: Mattavelli L, Novelli L (eds) Advances in organic geochemistry 1987. Pergamon, Oxford, pp 207–219

Imbus SW, Engel MH, Elmore RD (1990) Organic geochemistry and sedimentology of the Middle Proterozoic Nonesuch Formation: hydrocarbon source rock assessment of a lacustrine rift deposit. In: Katz BJ (ed) Lacustrine exploration: case studies and modern analogs. Am Assoc Petrol Geol Mem 50:197–208

Jackson KS, McKirdy DM, Deckelman JA (1984) Hydrocarbon generation in the Amadeus Basin, central Australia. Aust Petrol Explor Assoc J 24(1):42–65

Jackson MJ, Powell TG, Summons RE, Sweet IP (1986) Hydrocarbon shows and petroleum source rocks in sediments as old as 1.7×10^9 years. Nature (London) 322:727–729

Jackson MJ, Sweet IP, Powell TG (1988) Studies on petroleum geology and geochemistry, Middle Proterozoic, McArthur Basin northern Australia I: petroleum potential. Aust Petrol Explor Assoc J 28(1):283–302

Johns RB, Belsky T, McCarthy ED, Burlingame AL, Haug P, Schnoes HK, Richter W, Calvin M (1966) The organic geochemistry of ancient sediments, pt 2. Geochim Cosmochim Acta 30:1191–1222

Jones RW (1984) Comparison of carbonate and shale source rocks. In: Palacas JG (ed) Petroleum geochemistry and source rock potential of carbonate rocks. Am Assoc Petrol Geol Stud Geol 18:163–180

Kasting JF (1987) Theoretical constraints on oxygen and carbon dioxide concentrations in the Precambrian atmosphere. Precambrian Res 34:205–229

Kelly WC, Nishioka GK (1985) Precambrian oil inclusions in late veins and the role of hydrocarbons in copper mineralization at White Pine, Michigan. Geology 13:334–337

Khavari-Khorasani G, Murchison DG (1979) The nature of Karelian shungite. Chem Geol 26:165–182

Klomp UC (1986) The chemical structure of a pronounced series of iso-alkanes in South Oman crudes. In: Leythaeuser D, Rullkötter J (eds) Advances in organic geochemistry 1985. Pergamon, Oxford, pp 807–814

Knoll AH, Hayes JM, Kaufmann AJ, Swett K, Lambert IB (1986) Secular variation in carbon isotope ratios from Upper Proterozoic successions of Svalbard and East Greenland. Nature (London) 321:832–838

Kohm JA, Louden RO (1979) Exploration success depends upon finding Ordovician Red River structures. Oil Gas J 77(29):89–94

Kutukov AV, Vinnikovskiy SA, Shershnev KS (1977) Perspectives on oil and gas in the Vendian section of the Precambrian in the Perm region. Geol Neft Gaza 11:37–43 (in Russian)

Lambert IB, Walter MR, Zang Wenlong, Lu Songnian, Ma Guogan (1987) Palaeoenvironment and carbon isotope stratigraphy of Upper Proterozoic carbonates of the Yangtze Platform. Nature (London) 325:140–142

Lapichev AI, Chekanov VI (1987) Petroleum generating potential of Vendian-Cambrian reservoirs of the Nepa-Botuobinskii anticline and adjacent areas. Geol Geofiz 28(10):26–33 (in Russian)

Leventhal JS, Threlkeld CN (1978) Carbon-13/carbon-12 isotope fractionation of organic matter associated with uranium ores induced by alpha radiation. Science 202:430–432

Lowe DR (1983) Restricted shallow-water sedimentation of Early Archean stromatolitic and evaporitic strata of the Strelley Pool Chert, Pilbara Block, Western Australia. Precambrian Res 19:239–283

Magaritz M, Holser WT, Kirschvink J (1986) Carbon-isotope events across the Precambrian/Cambrian boundary on the Siberian Platform. Nature (London) 320:258–259

Makushina VM, Aref'ev OA, Zabrodina MN, Petrov AA (1978) New relic alkanes of petroleums. Neftekhimiya 18:847–854 (in Russian)

Mancuso JJ, Kneller WA, Quick JC (1989) Precambrian vein pyrobitumen: evidence for petroleum generation and migration 2 Ga ago. Precambrian Res 44:137–146

McKirdy DM (1974) Organic geochemistry in Precambrian research. Precambrian Res 1:75–137

McKirdy DM (1977) The diagenesis of microbial organic matter: a geochemical classification and its use in evaluating the hydrocarbon-generating potential of Proterozoic and Lower Palaeozoic sediments, Amadeus Basin, central Australia. PhD Thesis, Aust Nat Univ, Canberra

McKirdy DM (1989) Biomarker geochemistry of oil shows in the Late Precambrian Murnaroo Formation, Lake Maurice West-1, Officer Basin, S.A. AMDEL Rep F7224/89 S Aust Dep Mines Energ

McKirdy DM, Hahn JH (1982) The composition of kerogen and hydrocarbons in Precambrian rocks. In: Holland HD, Schidlowski M (eds) Mineral deposits and

the evolution of the biosphere. Dahlem Konferenzen. Springer, Berlin Heidelberg New York, pp 123–154
McKirdy DM, Powell TG (1974) Metamorphic alteration of carbon isotopic composition in ancient sedimentary organic matter: new evidence from Australia and South Africa. Geology 2:591–595
McKirdy DM, Aldridge AK, Ypma PJM (1983) A geochemical comparison of some oils from pre-Ordovician carbonate rocks. In: Bjoroy M, Albrecht P, Cornford C, de Groot K, Eglinton G, Galimov E, Leythaeuser D, Pelet R, Rullkötter J, Speers G (eds) Advances in organic geochemistry 1981. Wiley, Chichester, pp 99–107
McKirdy DM, Kantsler AJ, Emmett JK, Aldridge AK (1984) Hydrocarbon genesis and organic facies in Cambrian carbonates of the eastern Officer Basin, South Australia. In: Palacas JG (ed) Petroleum geochemistry and source rock potential of carbonate rocks. Am Assoc Petrol Geol Stud Geol 18:13–31
Meyerhoff AA (1980) Geology and petroleum fields in Proterozoic and Lower Cambrian strata, Lena-Tunguska petroleum province, eastern Siberia. In: Halbouty MT (ed) Giant oil and gas fields of the decade 1968–1978. Am Assoc Petrol Geol Mem 30:225–252
Meyerhoff AA (1982) Hydrocarbon resources in arctic and subarctic regions. In: Embry AF, Balkwill HR (eds) Arctic geology and geophysics. Can Soc Petrol Geol Mem 8:451–552
Minter WEL (1978) A sedimentological synthesis of placer gold, uranium and pyrite concentrations in Proterozoic Witwatersrand sediments. In: Miall AD (ed) Fluvial sedimentology. Can Soc Petrol Geol Mem 5:801–829
Moldowan JM, Seifert WK, Gallegos EJ (1985) Relationship between petroleum composition and depositional environment of petroleum source rocks. Am Assoc Petrol Geol Bull 69:1255–1268
Muir MD, Armstrong KJ, Jackson MJ (1980) Precambrian hydrocarbons in the McArthur Basin, NT. BMR J Aust Geol Geophys. 5:301–304
Murray GE (1965) Indigenous Precambrian petroleum. Am Assoc Petrol Geol Bull 49:3–21
Murray GE, Kaczor MJ, McArthur RE (1980) Indigenous Precambrian petroleum revisited. Am Assoc Petrol Geol Bull 64:1681–1700
Mycke B, Michaelis W, Degens ET (1988) Biomarkers in sedimentary sulfides of Precambrian age. In: Mattavelli L, Novelli L (eds) Advances in organic geochemistry 1987. Pergamon, Oxford, pp 619–625
Palacas JG, Reynolds MW (1989) Preliminary petroleum source rock assessment of Upper Proterozoic Chuar Group, Grand Canyon, Arizona. Am Assoc Petrol Geol Bull 73:397
Peat CJ, Muir MD, Plumb KA, McKirdy DM, Norvick MS (1978) Proterozoic microfossils from the Roper Group, Northern Territory. BMR J Aust Geol Geophys 3:1–17
Pell SD (1989) Stable isotope composition of organic matter and co-existing carbonate in the Late Precambrian of the Officer Basin: stratigraphic relationships with neighbouring basins and environmental significance. BSc Honours Thesis, Univ Adelaide
Petersil'ye IA, Akhmedov AM, Uvad'yev LI (1974) Hydrocarbon gases and organic carbon in Proterozoic rocks of the Kola Peninsula. Dokl Akad Nauk SSSR 215:201–204
Petrov AA, Pustil'nikova SD, Abriutina NN, Kagramanova GR (1976) Petroleum steranes and triterpanes. Neftekhimiya 16:411–427 (in Russian)
Rankama K (1948) New evidence of the origin of Pre-Cambrian carbon. Bull Geol Soc Am 59:389–416
Reimer TO, Barghoorn ES, Margulis L (1979) Primary productivity in an Early Archaean microbial ecosystem. Precambrian Res 9:93–104
Schidlowski M (1988) A 3,800-million-year isotopic record of life from carbon in sedimentary rocks. Nature (London) 333:313–318
Schidlowski M, Eichmann R, Junge CE (1976) Carbon isotope geochemistry of the Precambrian Logamundi carbonate province, Rhodesia. Geochim Cosmochim Acta 40:449–455
Schidlowski M, Appel PWU, Eichmann R, Junge CE (1979) Carbon isotope geochemistry of the 3.7×10^9-yr-old Isua sediments, West Greenland: implications for the Archaean carbon and oxygen cycles. Geochim Cosmochim Acta 43:189–199
Schidlowski M, Hayes JM, Kaplan IR (1983) Isotopic inferences of ancient biochemistries: carbon, sulfur, hydrogen, and nitrogen. In: Schopf JW (ed) Earth's earliest biosphere: its origin and evolution. University Press, Princeton, pp 149–186
Schoell M, Wellmer F-W (1981) Anomalous ^{13}C depletion in Early Precambrian graphites from Superior Province, Canada. Nature (London) 290:696–699
Schopf JW (ed) (1983) Earth's earliest biosphere: its origin and evolution. University Press, Princeton, 543 pp
Schopf JW, Packer BM (1987) Early Archean (3.3-billion to 3.5-billion-year-old) microfossils from Warrawoona Group, Australia. Science 237:70–73
Schopf JW, Hayes JM, Walter MR (1983) Evolution of Earth's earliest ecosystems: recent progress and unsolved problems. In: Schopf JW (ed) Earth's earliest biosphere: its origin and evolution. University Press, Princeton, pp 361–384
Seglund JA (1989) Midcontinent rift contiinues to show promise as petroleum prospect. Oil Gas J 87(20):55–58
Seifert WK (1980) Impact of Treibs' discovery of porphyrins on present-day biological marker organic geochemistry. In: Prashnowsky AA (ed) Proc Treibs Int Symp. Halbigdruck, Würzburg, pp 13–35
Strauss H (1986) Carbon and sulfur isotopes in Precambrian sediments from the Canadian Shield. Geochim Cosmochim Acta 50:2653–2662
Strauss H, Des Marais DJ, Summons RE, Hayes JM (1991) Proterozoic biogeochemistry: concentrations of organic carbon and maturities and elemental compositions of kerogens. In: Schopf JW, Klein C (eds) The Proterozoic biosphere: a multidisciplinary study. Univ Press, Cambridge (in press)
Summons RE (1991) Proterozoic biogeochemistry: abundance and composition of extractable organic matter. In: Schopf JW, Klein C (eds) The Proterozoic biosphere: a multidisciplinary study. Univ Press, Cambridge (in press)
Summons RE, Walter MR (1990) Molecular fossils and microfossils of prokaryotes and protists from Proterozoic sediments. Am J Sci 290A:212–244
Summons RE, Powell TG, Boreham CJ (1988a) Petroleum geology and geochemistry of Middle Proterozoic McArthur Basin, northern Australia: III. Composition of extractable hydrocarbons. Geochim Cosmochim Acta 52:1747–1763
Summons RE, Brassell SC, Eglinton G, Evans E, Horodyski RJ, Robinson N, Ward DM (1988b) Distinctive hydrocarbon biomarkers from fossiliferous sediment of the Late Proterozoic Walcott Member, Chuar Group,

Grand Canyon, Arizona. Geochim Cosmochim Acta 52:2625–2637
Tissot BP, Welte DH (1984) Petroleum formation and occurrence, 2nd edn. Springer, Berlin Heidelberg New York, 699 pp
Townson WG (1985) The subsurface geology of the western Officer Basin – results of Shell's 1980–1984 petroleum exploration campaign. Aust Petrol Explor Assoc J 25(1):34–51
Tyler SA, Barghoorn ES, Barrett LP (1957) Anthracitic coal from Precambrian upper Huronian black shale of the Iron River District, northern Michigan. Bull Geol Soc Am 68:1293–1304
Vassoyevich NB, Vysotskiy IV, Sokolov BA, Tatarenko YI (1971) Oil-gas potential of Late Precambrian deposits. Int Geol Rev 13:407–418
Vidal G, Knoll AH (1983) Proterozoic plankton. Geol Soc Am Mem 161:265–277
Walker JCG, Klein C, Schidlowski M, Schopf JW, Stevenson DJ, Walter MR (1983) Environmental evolution of the Archean-Early Proterozoic Earth. In: Schopf JW (ed) Earth's earliest biosphere: its origin and evolution. Univ Press, Princeton, pp 260–290
Walter MR (1983) Archean stromatolites: evidence of the Earth's earliest benthos. In: Schopf JW (ed) Earth's earliest biosphere: its origin and evolution. University Press, Princeton, pp 187–213
Weber F, Schidlowski M, Arneth JD, Gauthier-Lafaye F (1983) Carbon isotope geochemistry of the Lower Proterozoic Francevillian Series of Gabon (Africa). Terra Cognita 3:22
Windley BF (1984) The evolving continents, 2nd edn. John Wiley & Sons, New York, 399 pp
Willingham TO, Nagy B, Nagy LA, Krinsley DH, Mossman DJ (1985) Uranium-bearing stratiform organic matter in paleoplacers of the lower Huronian Supergroup, Elliot Lake-Blind River region, Canada. Can J Earth Sci 22:1930–1944
Womer MB (1986) Hydrocarbon occurrence and diagenetic history within Proterozoic sediments, McArthur River area, Northern Territorry, Australia. Aust Petrol Explor Assoc J 26(1):363–374
Worsley TR, Nance RD, Moody JB (1986) Tectonic cycles and the history of the Earth's biogeochemical and paleoceanographic record. Paleoceanography 1:233–263
Zaunbrecher ML (1988) Hydrocarbon source potential of the upper Roper Group, McArthur Basin, Northern Territory. BSc Honours Thesis, Univ Adelaide
Zumberge JE, Sigelo AC, Nagy B (1978) Molecular and elemental analyses of the carbonaceous matter in the gold and uranium bearing Vaal Reef carbon seams, Witwatersrand Sequence. Min Sci Eng 10:223–246

Paleontology and Biogeochemical Research: A Powerful Synergy

M. F. Glaessner[1] and C. B. Foster[2,3]

[1] Emeritus Professor M. F. Glaessner, Department of Geology and Geophysics, The University of Adelaide, Adelaide, S.A. 5000. Deceased.
[2] C. B. Foster, Western Mining Corporation Limited, Petroleum Division, 28 Ventnor Avenue, West Perth, W.A. 6005; and WMC-UWA Palynological Laboratory, Department of Geology, University of Western Australia, Nedlands, W.A. 6009
[3] Present address: Bureau of Mineral Resources, GPO Box 378 Canberra, ACT 2601 Australia

Abstract

Traditionally, the systematic affinities of Precambrian and early Paleozoic fossils are determined on the basis of morphologic comparisons and assumed evolutionary theory. Their assignment to either the prokaryotes or eukaryotes on this basis may be equivocal and controversial. Such systematic data are used, however, to predict the physical environment during these periods, which subsequently impacts on the theory of the evolution of the Earth. In contrast to trace and body fossils, molecular fossils, detected by sophisticated gas chromatography and mass spectroscopic techniques (GCMS), can provide unequivocal information about natural affinities. Evidence of eukaryote derived steranes in rocks as old as 1.6 Ga is now available and the marriage between biogeochemistry and paleontology results in a firmer basis for interpretation and prediction. The trend to chemically fingerprint fossil isolates, and their host rocks, is now gaining momentum and must continue. We illustrate the successful application of this technique to determine the probable paleobiology of the microfossil *Gloeocapsomorpha prisca* Zalessky 1917, and show it is the major source for certain Ordovician oils throughout many parts of the world.

1 Introduction

Biogeochemical research is aimed at understanding the quantities and qualities of organic matter preserved in bulk in the Earth's crust. In the last 20 or 30 years improved methods of analysis with newly invented instruments have made it possible to establish links between specific chemical data and morphological characteristics of preserved organic remains of identifiable fossils. Niklas (1982) has given a comprehensive review of what he designates (p. 31) as "the new discipline of paleobiochemistry". He places abundant data on Phanerozoic fossil plants "within a biosynthetic context, as inferred from modern organisms", expecting that thereby "major evolutionary events at the molecular level can be determined" (p. 29). We add to this our expectation that old and new questions raised by paleontology of the Precambrian and early Paleozoic, where taxonomic and evolutionary links of fossil biota with living organisms are difficult to establish, can be brought nearer to solution by new inter-disciplinary methodologies. We conclude with an example of paleobiogeochemical methods leading to successful correlation of early Paleozoic (Ordovician) oils and their source rocks. These new researches are producing important practical as well as theoretical results.

2 The Changing Biosphere

The growth of reliable information on the previously unknown or misunderstood Precambrian fossil record leads to constraints on our perception of the history of the early biosphere from which organic matter preserved in sediments older than 550 million years (Ma) originated (Schopf 1983). Organic matter is present in significant, and therefore potentially economically exploitable, quantities in Early Precambrian and younger Precambrian rocks. Oil and gas has been discovered and hydrocarbon resources are being explored in Late Precambrian and Cambrian strata in Australia, China, Oman, the U.S.A. and U.S.S.R. In the Soviet Union the late Academician A. V. Sidorenko, with his daughter S. A. Sidorenko and other collaborators (Sidorenko and Sidorenko 1971), published much information on the abundance of organic carbon in ancient rocks where earlier geologists had not expected it. About that time analytical methods had sufficiently improved to allow progress to be made from quantitative estimation of organic carbon to critical assessment of its biological origins. Black accumulations of carbon in approximately 1900 Ma old rocks in Finland and elsewhere,

M. Schidlowski et al. (Eds.)
Early Organic Evolution: Implications for Mineral and Energy Resources

named *Corycium enigmaticum* Sederholm 1911, were recognized as rolled-up fragments of microbial mats (Matista 1963, 1969) and compared with the living cyanobacterium *Stigonema collatum* by Ohlson (1962). These conclusions were reached after considerable discussion in the literature of the 1960s and 1970s. The possibly slightly younger fossil coal from Karelia known as shungite, currently exploited as a lightweight building material, may contain significant chemical and morphological fossils. Paleontological evidence currently supports the view that we can expect only bacteria and cyanobacteria in Archean and Early Proterozoic rocks (although future biogeochemical studies may challenge this). This tends to constrain speculation on the early biosphere, atmosphere, and hydrosphere (see Nagy 1982). There is general agreement that ancient sediments contain indications of photosynthetic activity by prokaryotic organisms. Oxygen accession to the atmosphere started in the Archaean but its rate and global yield are still disputed. There is as much uncertainty about configurations of possible environments of life during that time. Future discoveries and interpretations of chemical fossils may lead to answers about the hypothetical differentiation of the early Prokaryota ("bacteria" in the widest sense) into their descendants with their distinctive life styles: plants, fungi and true animals. These are now known in some abundance as fossils in Neoproterozoic sediments, suggesting because of the level of their diversity an earlier origin, or (to more speculative minds) an extraterrestrial origin. Summons and Walter (1990) have given a comprehensive review of molecular fossils, and their probable systematic affinities, recovered from Proterozoic sediments. They suggest, for example, that the consistent presence of steranes in hydrocarbons from Early and Middle Proterozoic sequences of the McArthur Basin, northern Australia (e.g., Barney Creek Formation ca. 1690 Ma) implies a definite presence of eukaryotes. Other groups represented by geochemical signatures in this basinal sequence are the eubacteria and archaebacteria.

3 Carbonaceous and "Chitinous" Integuments – Plants or Animals?

In rocks 1000–700 Ma old there are abundant carbonaceous fossil remains in northern China, India, Iran, Russia, Romania, North America, and elsewhere. They include the abundant disc-shaped fossil *Chuaria*, worm-shaped straight or curved forms (*Tawuia*), and others. The interpretation and systematics of these fossil organisms is still in a confused state (Sun et al. 1986; Sun 1987). *Chuaria* was recently interpreted as possibly related to the cyanobacteria (Nostocales) as a colonial form. Similar, somewhat younger Precambrian fossils have been described as, inter alia: brachiopods, medusae, and green algae. Genera have been delimited by arbitrary divisions in a continuous size range. With very few exceptions (e.g., Summons et al. 1988), the supposedly "carbonaceous" matter of these fossils has not been studied with modern analytical methods. Where found on the surface, the fossils are weathered and iron-stained, while others are only impressions on bedding planes. Fresh specimens of the "carbonaceous" envelopes of these organic forms could be obtained from shallow bores similar to those used by paleomagnetists to obtain fresh rock specimens. If this was tried, perhaps we could learn from their chemical composition whether they are plants or animals. This is important, as no other body fossils of animals are known from rocks 1000 Ma old or older. The recent detailed work by Summons et al. (1988) on *Chuaria*-bearing sediments of the Walcott Member, Chuar Group (Arizona), recognized high abundances of steranes, consistent with input from eukaryotes. They also recorded the oldest occurrence of gammacerane, for which a protozoan origin seems most likely.

Confusion is growing around another fossil ranging through most of the fossiliferous Vendian (Late Precambrian) formations of southern and northern parts of the East European Platform and now reported also from North America. These fossils are often found closely associated in large numbers on bedding planes, mostly as impressions or casts, occasionally described as carbonaceous or chitinous, and cited under different generic and specific names of controversial validity (see Sokolov 1973; Keller et al. 1974; Gureev 1985; Narbonne and Hofmann 1987; Hahn and Pflug 1988). None of these publications indicates clearly the organic substance that caused the impression. Narbonne and Hofmann described them as *Beltanelliformis brunsae* Menner under "Phylum uncertain" while almost identical illustrations are given by Hahn and Pflug (1988, Pl. 2) as *Hagenetta aarensis* n.g., n.sp. (Phylum, Class and Order *indet.*). These fossils are described as having two valves which are mainly organic and subject to plastic deformation. Layering is clearly shown in sections through the "valves" but largely denied in the text (p. 10). Although Narbonne and Hofmann asserted (1987, p. 668) that "neither *Beltanelliformis* nor *Beltanelloides* appear in the Treatise (Glaessner 1979)", it was stated there (p. A 109), after a reference to various larger acritarchs, that "a similar form, ranging up to

44 mm in diameter, *Beltanelloides* Sokolov 1965... = Beltanelliformis Menner 1963 (a *nomen nudum*, figured but not described until 1968)... Some of these Upper Precambrian fossils have been mistakenly considered as Metazoa or Protozoa". That the largest of these fossils are larger than contemporary acritarchs from coeval strata may not be significant since their precise biological nature is also unknown. Clearly, the organic integuments of the originators of these fossil impressions should be searched for and their chemistry examined in order to determine their biohistorical significance. Worm-shaped fossils from India (Lower Vindhyan Rohtas Formation, 900–1000 Ma old) have been described as *Katnia singhi* Tandon and Kumar (see Glaessner 1987). Their impressions, without any traces of organic matter, resemble illustrations of fossil oscillatorian cyanobacteria but are some 20–25 times larger. Some biologists consider such comparison as irrelevant. Apparently, prokaryote cells of similar configuration and size are unknown.

4 Chemical Products of Life Activities

One obvious distinction between plants and animals is the fact that animals produce fecal pellets but plants do not. Robbins et al. (1985) collected samples of dark, laminated shales and slates from anoxic, nonbioturbated environments of Cambrian and older age, where such microfossils could be expected. Treatment with acids yielded pellets of 10–500 µm size. Some were considered as probably fecal pellets on shape evidence. Only organic carbon, nitrogen and phosphorus have been determined (in weight percent). Further studies of similar materials may lead to more definite results. There are other kinds of samples and other methods available. Some Proterozoic sediments contain abundant ellipsoidal bodies, not entirely carbonate, not wholly concentrically laminated, often with peculiar cross-sections. They are known as microphytolites or catagraphia, and used in the Soviet Union for stratigraphic correlation (Zhuravleva 1964). Some of them, particularly from fossiliferous phosphatic rocks, may be of fecal rather than plant or mechanical origin. A sigmoidal string of pellets about 0.37 mm in diameter was discovered in a thin section of Upper Riphean rock (about 950–1000 Ma old). It was figured (Sabrodin 1971, 1972; see also Glaessner 1984, p. 24–5) but not described or even examined in detail. It cannot be the only configuration of this kind. A pioneer study of experimental feeding and excretion by marine crustaceans in the laboratory was undertaken to determine the fate of dietary lipids and their quantities obtained from fecal pellets (Harvey et al. 1987). Although there are obvious limiting factors in this approach, its results will give welcome guidance to biochemical research on the results of life activities of the marine biota.

Stromatolites are the most abundant and widespread configurations in ancient rocks caused by life activities. They are formed either by precipitation of mineral matter from seawater, as the result of local physicochemical changes through life activities of prokaryote organisms, or through trapping and binding of sedimentary particles by them. The resulting flat or variously elevated sedimentary structures are often diagenetically altered. Silicification commonly preserves remains of microorganisms. The ecology of mat- or column-forming marine, hypersaline, or lacustrine stromatolite builders which survived the drastic reduction of their world-wide distribution since the Precambrian-Cambrian transition is now the subject of intensive interdisciplinary studies (Walter 1976; see also Kennard, since 1972). Stromatolites are neither colonial organisms nor true paleo-communities but sedimentary results of life activities, modified by erosion, hydrodynamics, chemical and trophic factors. The stratigraphic significance of some morphological characters of these organo-sedimentary structures is now generally recognized, together with their paleoecological importance in otherwise poorly fossiliferous or unfossiliferous Precambrian and early Paleozoic strata. Extensive changes in the occurrence of stromatolites make paleoecological conclusions drawn from ecological investigations of life activities of survivors somewhat hazardous. Continuing input from current microbiological and biogeochemical studies of Recent stromatolite builders is a most essential requirement (see Subacius and Fairchild 1986).

5 Trace Fossils

Living animals can leave traces of their activities in sediments. Trace fossils occur in many sedimentary sequences and, from morphological and biostratigraphic viewpoints, are closely studied by paleontologists; few have attracted attention from biogeochemists. The geochemical effects on sediments mediated through trace fossils are not necessarily quantitatively significant but the following arguments show that they are not necessarily negligible. Some trace fossils result from movement of animals (e.g., worms, molluscs, and echinoids) that lay down mucus trails to facilitate their own locomotion or lay mucus traps for their prey. Because of some cementation by excreted mucoproteins or mucopolysaccharides, sedimentary particles often remain in characteristic positions in which the animals' movements have placed them. The influence of the original organic composition of the cementing material on the process of diagenesis determines the composition of the resulting structure and in some instances might be worth considering. The great efforts by oil exploration laboratories to study the development of porosity in drill cores in relation to bioturbation and cementation may provide indications of unexpected contributions from the biochemical activities of fossil animals. Trace fossils are often marked by oxidation/reduction processes and not only characteristic colour changes but also mineral formation may result (for a description of trace fossils with "iron walls" see Glaessner and Pledge 1985).

The dwelling tubes made by sedentary worms should be mentioned here. Most of them are of little interest to marine biologists. Some more or less solid tubes can be left (and re-occupied?) by their builders at will. In many instances they are the only fossilizable part of the animal. The tubes consist of mud, selected mineral grains, shell fragments, or secreted calcium carbonate. The particles may be cemented in more or less elaborate patterns by mucus; some tubes consist entirely of it. Could some of the many fossil worm tubes in rocks still contain chemical fossils? A parallel to the tube-making activities of worms can be found in certain foraminifera. They agglutinate their microscopic tests from mineral grains, often specially selected by their pseudopodia. They may incorporate shells or sponge spicules, cemented with mucoproteins or mucopolysaccharides. These structures are either quite firm but brittle, or, as the fossils show, they may be compressed or otherwise deformed without fracture. The organic cement keeps them flexible during their life, and plastic during embedding in the sedimet. Complex calcareous-shelled foraminiferal tests may show internal, organic-rich, dark layers. On recrystallization of the embedding rock matrix this layer can become detached and rolled up into a tangled mass, still showing its chambered structure. That means that the organic layer must have been perfectly flexible at some stage of diagenesis. Its composition in fossils, which remains unknown, would be worth investigating.

6 Biomineralization

The consideration of organisms using organic cement and mineral matter for building their integument leads to the fundamental question of biomineralization, its evolutionary origins and adaptational effects. The physiological and biochemical aspects of this complex problem are being discussed at regular meetings of specialist societies (e.g., Westbroek and de Jong 1983). Their discussion started some years ago after research on pearls and teeth in living organisms.

Equal attention is being paid today to paleontological aspects of the question. The first appearance of shell-bearing and skeletonized organisms in the geological record is taken by some as a marker of the end of the Precambrian Eon. Pearls and teeth and shells are, of course, still important from the experimental viewpoint in the "here and now". The variety of form and function, the variety of minerals employed or produced in these processes, their success, and the evolutionary origination and adoption, by many different organisms, of shells and skeletons, at almost one moment of geological time – these are now exciting problems. The cooperation of biogeochemists in the study of what remains of the organic matrix of hard parts of early Cambrian calcareous algae and animals is required if progress is to be made from current speculations. The cooperation of geochronologists should also be sought, with some urgency. Present views on the age of sediments of the critical period vary by up to 100 million years. What can we know about the rates of processes if that is the best available time framework? The "sudden appearance" of mineralized skeletons may have occurred suddenly or, alternatively, over millions of years. It should also be remembered that perhaps the majority of invertebrate animals living in the present seas can do very well without adopting the wonderful evolutionary novelties of biomineralization (see, e.g., Lowenstam and Margulis 1980).

7 Changes in Biomass?

The question of the biomass existing at different stages in the evolution of the biosphere has great relevance to biogeochemistry as well as to paleontology and, of course, to mineral resources. Hard quantitative data on the biomass at a given time are difficult to obtain and to ascertain. There are indications that the total biomass at the end of the Precambrian was quantitatively inferior to that of today. Marine faunas and acritarch floras from that period are now reasonably well known. They are, however, strikingly limited in their occurrence. There are vast sequences of Precambrian sedimentary rocks, unmetamorphosed and essentially identical with younger, fossiliferous strata, but they do not appear to contain any traces of life;

perhaps, however, this is a function of the state of investigative studies. But the known occurrences of larger fossils show peculiarities in their taxonomic composition and distribution that point towards the same conclusion of quantitatively poorer biota in those early times. On the other hand, students of microfossils and their ecology, e.g., Andrew Knoll, and biogeochemists, among them Manfred Schidlowski, have cautiously pointed out that there is no acceptable evidence of significant quantitative increases in the biomass or standing crop of organisms since early Precambrian times. These divergent hypotheses require further testing. Advances on a firmer, more experimental, and less speculative basis will follow when the quantitative approaches of biogeochemistry are added to the necessarily more qualitative and comparative morphological methods of Precambrian paleontology.

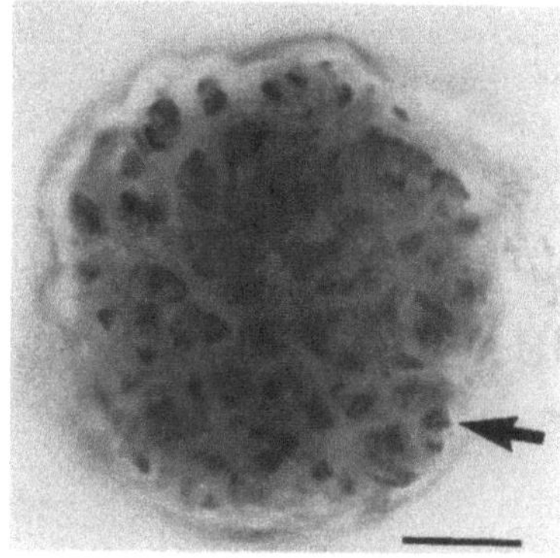

Fig. 1. LM. *Gloeocapsomorpha prisca* Zalessky 1917 from an Estonian kukersite sample. *Arrow* indicates cell voids. An example of a super colony made up of many multilamellate-walled colonies, each of two or more cells. *Bar* 20 μm

8 Middle Ordovician Oil-Source Rock Correlation: Case History of Paleontological-Biogeochemical Research

Heating of organic matter of either modern or fossil origin and from different biological precursors produces distinctly different hydrocarbon products. Lipid-rich taxa, including certain algae, are oil-prone; whereas oxygen-rich material such as woody tissue is gas-prone. This is the basis of a kerogen classification scheme developed by Espitalié and others, and routinely used in presentation and interpretation of results of Rock-Eval pyrolysis (see Tissot and Welte 1978).

The production of hydrocarbons from organic matter is demonstrated unequivocally by retorting organic-rich sediments or oil shales. The Middle Ordovician, marine oil shales, or kukersite deposits, of Estonia in the Baltic Basin have been used as fuel for heating since 1789 (see Bekker 1921), and on retorting have an oil yield of approximately 200–400 litres per tonne. Kukersite is and has been retorted to produce liquid and gaseous hydrocarbons (see Luts 1938), but more importantly it is a fuel to generate electricity and to produce various chemical products. Commercial kukersite deposits in the Baltic Basin are in excess of four billion tonnes and cover an area of more than 3000 km^2; current annual production of kukersite is approximately 25 million tons (Kattai and Puura 1988). Organic productivity and preservation during the Middle Ordovician at this locality was abnormally high.

The organic component of kukersite consists almost entirely of the microfossil *Gloeocapsomorpha prisca* (Fig. 1). Since its description in 1917 by Zalessky, paleontologists have argued its systematic affinities; it has been considered either a cyanobacterium, or a green alga; with initial evidence based mainly on morphology (see Foster et al. 1989 for detailed review). Some early workers, such as Rüger (1926), considered both morphological criteria and evidence from Krogerman's 1922 study of the chemistry of oils produced from kukersite to determine the systematic position of *G. prisca*. He concurred with Lindenbein (1921) that *G. prisca* belonged neither to the cyanobacteria nor to the green algae but to a new class, the Protophyceae, which shared certain characteristics with both the cyanobacteria and Rhodophyta. Vitorović (1980) has given a succinct summary on the various chemical analyses of kukersite determined using oxidative, wet-chemical methods. These studies are significant, but there was no immediately obvious link between the chemistry of the retorted oils, with their highly unsaturated aromatic, aliphatic, and fatty acid components, and the naturally occurring, paraffinic oils derived from other Ordovician sequences, particularly in North America and Australia (Fig. 6). As discussed below, that connection became apparent with hydrous pyrolysis of *G. prisca* under laboratory conditions, and the application of gas chromatographic techniques to analyze products; and increasing knowledge of the fatty acid composition of extant cyanobacteria.

Martin et al. (1963) first commented on the unusual chemistry of oils from three North American Ordovician sequences. They noted the predominance of odd-numbered n-alkanes in the C_{11} to C_{19} range and suggested that they were probably derived from source materials rich in saturated fatty acids (C_{12}–C_{22}), and that simple decarboxylation would yield alkanes of the corrected chain length. Martin et al. further suggested that these unusual oils might result from relatively simple plants or animals growing in a uniformly warm climate. Kenyon (1974), for example, showed that extant cyanobacteria, amongst the "simplest" organisms, yielded fatty acids within the C_{14}–C_{18} range, compatible with the earlier suggestions of Martin et al. As reviewed by Reed et al. (1986) and Longman and Palmer (1987), subsequent authors (e.g., Powell and McKirdy 1972; Williams 1974; Illich and Grizzle 1983; Alexander et al. 1984; Fowler and Douglas 1984) also noted the widespread occurrence of oils having this distinctive n-alkane signature and that such oils are commonly associated with sediments of Ordovician age.

Klesment (1974) and Klesment and Nappa (1980) pyrolyzed kukersite, consisting principally of *G. prisca*, using a Fischer assay technique, and showed that the n-alkanes produced had similarities with those of the Ordovician oils. Their research also concluded that kukersite "is the fossilization product of the fatty acids C_{14}, C_{16} and C_{18}" (Klesment 1974), confirming the views of Martin et al. (1963), and linking these structures with *G. prisca*. The link between *G. prisca* and the recovered oils was further forged by Reed et al. (1986), and Foster et al. (1989), who studied inter alia, palynological content and chromatographic analyses of either the extracted soluble organic fraction from Estonian kukersite or hydrous pyrolysis of those samples. Hydrous pyrolysis converts the unsaturated, less stable olefins to saturates (Lewan et al. 1979) so that the laboratory product is more comparable with naturally occurring oils and similarities are more obvious (Figs. 2, 6).

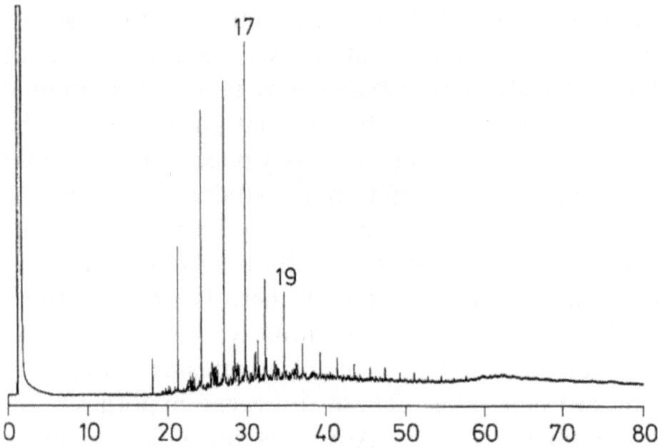

Fig. 2. Chromatogram of alkanes obtained from hydrous pyrolysis of Estonian kukersite. *Bold numbers* n-alkanes

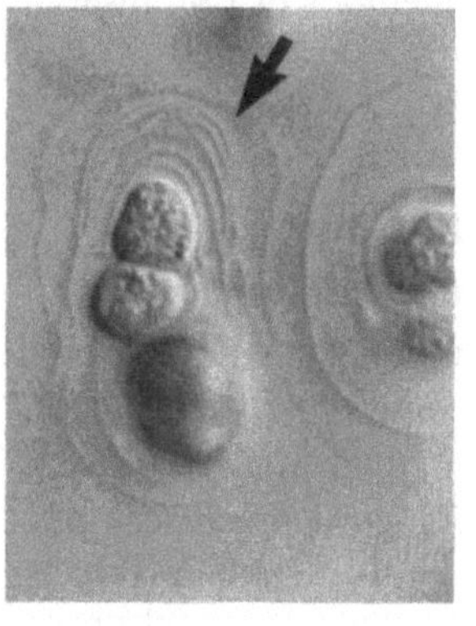

Fig. 3. LM. *Gloeocapsa* sp., University of Texas (UTEX) Culture LB 1938. Living material; note thin lamellae of sheath (*arrow*) surrounding cells. Bar 12 µm

Away from the Baltic Basin, and where characteristic Ordovician n-alkane oils have been reported, palynologists have sought to identify the source rock by analysing for *G. prisca* (e.g., McKirdy et al. 1986; Hoffmann et al. 1987; Jacobson et al. 1988). A recent, detailed morphological study of the type material of *G. prisca* has shown it is represented by at least three distinct morphotypes reflecting growth stages of a colonial organism and/or overprinting effects due to fossilization (Foster et al. 1989). As a consequence, morphologic comparisons need to account for all variations, and not just comparisons with archetype examples as in Fig. 1; this approach might extend the known records of *G. prisca*.

Since Zalessky's (1917) study, a modern analog has been sought for *G. prisca*. As the name implies, Zalessky considered that the fossil was an ancient form of the modern cyanobacterium *Gloeocapsa* Kützing 1843 (Fig. 3). This view was strongly contested by many authors on the grounds of comparative morphology and ecophysiology. Even accounting for modern nomenclatural changes, members of *Gloeocapsa* are predominantly of freshwater or terrestrial origin, whereas *G. prisca* is always intimately associated with marine megafossils and microfossils. Adamczak (1963) noted that from its structure and colonial habit, *G. prisca* could be referred to the Entophysalidaceae whose extant members include the mat and stromatolite-forming *Entophysalis major*. Foster et al. (1989) concluded that on morphological and biochemical grounds *E. major* is closely allied to *G. prisca* and suggest it as a modern analog (cf. Figs. 1, 5). The similarity between the saturated hydrocarbon fractions from the fossil and extant species is striking and although higher molecular weight n-alkanes are also present in the *E. major* sample (Figs. 2, 4), these are derived from higher plant debris (R. E. Summons pers. comm. 1991).

Dombrowski's (1988, see also Friedman 1980) identification of a stromatolite in a core from the

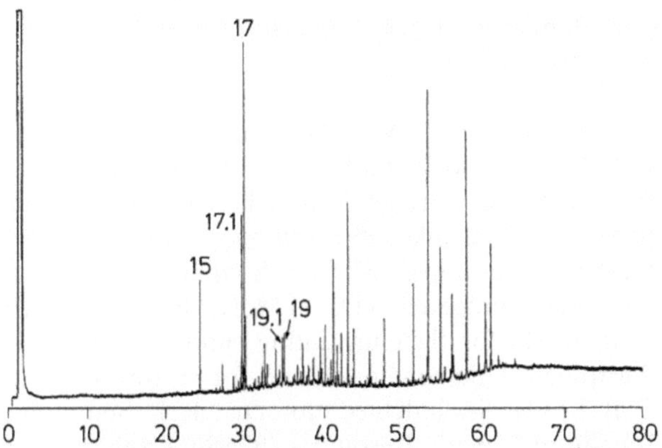

Fig. 4. Chromatogram of aliphatic hydrocarbons extracted from living mat of *Entophysalis major*, from Shark Bay, Western Australia. (Collected by Dr. John Bauld, BMR). *Bold numbers* n-alkanes; *17.1*, *19.1* n-alkenes

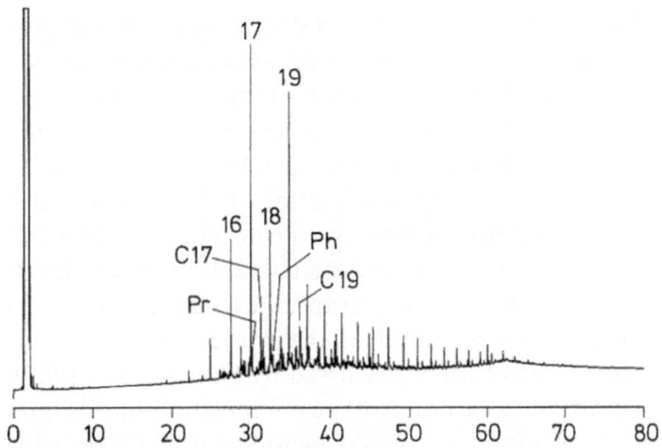

Fig. 6. Chromatogram of alkane fraction of oil extracted from a core of the Red River Formation, Williston Basin, U.S.A. *Bold numbers* n-alkanes; *C19*, *C17* cycloalkanes; *Pr* pristane; *Ph* phytane

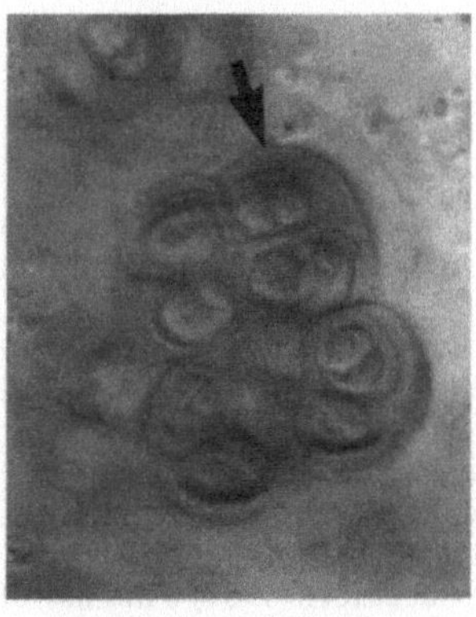

Fig. 5. LM. *Entophysalis major*, modern material collected by Dr John Bauld, BMR. Individual cells are surrounded by multiple polysaccharide layers (*arrowed*) which delineate colony outline. Bar 12 μm

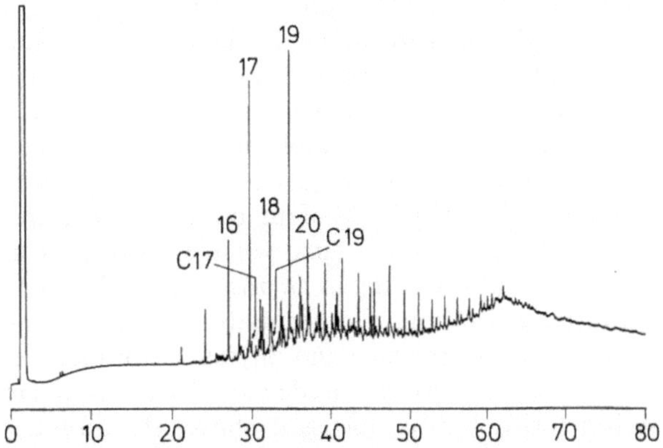

Fig. 7. Chromatogram of alkanes extracted from algal mat material in a core from the Red River Formation, Williston Basin, U.S.A. *Arrowed* peaks are pristane (*left*) and phytane (*right*). *Bold numbers* n-alkanes- *C19*, *C17* cycloalkanes

Middle Ordovician, Red River Formation of the Williston Basin, the same stratigraphic unit from which Martin et al. (1963) had reported oils with the characteristic $C_{11}-C_{19}$ n-alkane signature, was consistent with the proposed paleogeographic setting for *G. prisca* using the *E. major* model. It provided an opportunity to examine for microfossils, and to extract for chromatography and pyrolysis any of the associated organic matter. The core is from a reservoir section and is stained with oil; on extraction, the oil has the chromatographic signature of *G. prisca* (Fig. 6). More important are the alkanes extracted from layered, presumed microbial mat remains in another Red River Formation core (unstained); they too show the n-alkane signature of *G. prisca* (Fig. 7). Microfossils from the last sample can not be identified with certainty because their morphology appears altered by stylolitic-pressure effects. Organic matter associated with the stromatolite has yet to be examined in detail but preliminary analysis shows no recognizable microfossils, only concentrations of amorphous organic material. Nevertheless, the occurrence of a stromatolite, and the record of *G. prisca*-like extracts from oils and associated organic matter lends credence to the proposal of *E. major* as a modern analog.

Reed et al. (1986) noted that the isoprenoid alkanes, phytane and pristane, assumed to be derived primarily from chlorophyll, were essentially absent from a pyrolysate of kukersite. These results were consistent with their analyses of other Ordovician *G. prisca*-bearing rocks and derived oils. Based on these data and sedimentological considerations, Reed et al. suggested that *G. prisca* was a nonphotosynthetic prokaryotic organism. By contrast, Hoffmann et al. (1987) found both pristane and phytane in pyrolysates of *G. prisca* and concluded that it was originally photosynthetic, and based on sterane abundance, possibly a eukaryote. The hydrocarbon composition, however, shows an admixture of bacterial and eukaryotic signals which requires cautious interpretation. Foster et al. (1986) and McKirdy et al. (1986) noted that a mixture of kerogen types occurring with *G. prisca* and including extinct marine microfossils such as chitinozoans, acritarchs, and graptolites, could change the hydrocarbon potential of the source material. This problem was further discussed by Hatch et al. (1987), in relation to carbon isotope composition of oils, and by Jacobson et al. (1988), who suggested that variations in chromatographic characters of oils such as those reported by Reed et al. (1986), Longman and Palmer (1987), and other workers might be explained by admixtures of kerogen types. Brooks et al. (1987), Jacobson et al. (1988), and Foster et al. (1989) reviewed the interpretations of these analyses and concluded that the observed variation in relative abundance of isoprenoids may have multiple causes. These include: primary biological controls, (i.e., chlorophyll present or absent) or evolutionary controls (cf. Teeri 1982); levels of maturation; admixtures of kerogen types; and original volumes of photosynthetic pigment occurring in *G. prisca*, such that with maturity, generation of other n-alkanes will swamp contributions of pristane and phytane (Hoffmann et al. 1987). Further work is needed to solve this problem satisfactorily, but as discussed above, the affinities between *G. prisca* and modern Cyanophyta seem clear.

This study represents a successful union between paleontology and biogeochemistry: it has far-reaching applications in oil exploration, including identification of hydrocarbon source beds, and migration pathways; and more "standard" applications in paleoenvironmental and paleogeographic reconstructions. The wide geographic spread of oils with a *G. prisca* signature in North America (Williston, Michigan, Illinois, Anadarko Basins; Hatch et al. 1987), Australia (Canning and Amadeus Basins) and the U.S.S.R. (Baltic Basin) lends support to a paleoequatorial to mid-latitude setting suggested for these areas (see Foster et al. 1986). Relationships between specific fossils and their biogeochemical signature have been used by Niklas and co-workers in the study of early Paleozoic to Tertiary plants (see Niklas 1982 and references there cited), and examined in various studies by McKirdy and Hahn (1982), Glikson et al. (1985), Glikson and Taylor (1986), Subacius and Fairchild (1986), and applied most recently by Goth et al. (1988). Further work uniting the disciplines of paleontology and biogeochemistry will have equally important economic, paleogeographic, and evolutionary consequences.

Acknowledgments. M. F. Glaessner, informed of C. B. Foster's work and his inability to be present at the Symposium, invited him to be co-author of this paper. He is thanked for willingly contributing an illustrative example of interdisciplinary research. Glaessner is indebted to David McKirdy and Manfred Schidlowski for their unstinting support and to the Australian Unesco Committee for a travel grant. C. B. Foster is indebted to Dr. R. E. Summons, Bureau of Mineral Resources (BMR), Canberra, for chromatograms of extracts and pyrolysates from Red River Formation samples and from *E. major*. Dr. Anna Dombrowski, Shell, Houston generously made a available parts of her study of Red River Formation and provided samples of cores for analysis.

References

Adamczak E (1963) *Gloeocapsomorpha prisca* Zalessky (sinice) z Ordowickich glazów narzutowych Polski. Acta Palaeont Pol 8:465–472

Alexander RA, Cumbers M, Kagi RI (1984) Geochemistry of some Canning Basin crude oils. In: Purcell PG (ed) The Canning Basin. Geol Soc Aust, Petrol Explor Soc Aust Symp Proc Perth, Western Australia, pp 353–358

Bekker H (1921) The Kuckers Stage of the Ordovician rocks of NE Estonia. Acta Comment Univ Dorpat A II 1:1–91

Brooks PW, Snowdon LR, Osadetz KG (1987) Families of oils in southeastern Saskatchewan. In: Carlson CG, Christopher JE (eds) 5th Int Williston Basin Symp, Saskatchewan Geol Soc, pp 253–264

Dombrowski A (1988) Generation of calcrete with chalky porosity in Upper Ordovician Red River grainstone reservoirs, Cedar Creek anticline, southwestern Williston Basin. Am Assoc Petrol Geol Bull 72:179

Foster CB, O'Brien GW, Watson ST (1986) Hydrocarbon source potential of the Goldwyer Formation, Barbwire Terrace, Canning Basin, Western Australia. Aust Petrol Explor Assoc J 26(1):142–155

Foster CB, Reed JD, Wicander R (1989) *Gloeocapsomorpha prisca* Zalessky, 1917: a restudy, part 1; Taxonomy, geochemistry, and paleoecology. Geobios 22(6):735–759

Fowler MG, Douglas AG (1984) Distribution and structure of hydrocarbons in four organic-rich Ordovician rocks. Org Geochem 6:105–114

Friedman GM (1980) Review of depositional environments in evaporite deposits and the role of evaporites in hydrocarbon accumulation. Bull Cent Rech Explor Prod Elf-Aquitaine 4:589–608

Glaessner MF (1979) Precambrian. In: Robison RA, Teichert C (eds) Treatise on invertebrate paleontology, pt A. Geol Soc Am; Univ of Kansas, pp A 79–A 118

Glaessner MF (1984) The dawn of animal life. Univ Press, Cambridge, 264 pp

Glaessner MF (1987) Discussion about some worm-like fossils. Precambrian Res 36:353–355

Glaessner MF, Pledge NS (1985) New trace fossils from the Eocene of Maslin Bay, South Australia. S Aust Dep Mines Energ Spec Publ 5:107–116
Glikson M, Taylor GH (1986) Cyanobacterial mats: major contributors to the organic matter in Toolebuc Formation oil shales. Geol Soc Aust Spec Publ 12:273–286
Glikson M, Gibson DL, Philp RP (1985) Organic matter in Australian Cambrian oil shales and other Lower Palaeozoic shales. Chem Geol 51:175–191
Goth K, de Leeuw JW, Püttmann W, Tegelaar EW (1988) Origin of Messel oil shale kerogen. Nature (London) 336:759–761
Gureev YuA (1985) Vendiata – primitive Precambrian Radialia. In: Problematica of the Late Precambrian and Palaeozoic, pp 92–103 (in Russian)
Hahn G, Pflug HD (1988) Zweischalige Organismen aus dem Jung-Praekambrium (Vendium) von Namibia (SW-Afrika). Geol Palaeontol 22:1–19
Harvey HR, Eglinton G, O'Hara SCM, Corner EDS (1987) Biotransformation and assimilation of dietary lipids by *Calanus* feeding on a dinoflagellate. Geochim Cosmochim Acta 51:3031–3040
Hatch JR, Jacobson SR, Witzke BJ, Risatti JB, Anders DE, Watney WL, Newell KD, Vuletich AK (1987) Possible late Middle Ordovician organic carbon isotope excursion: Evidence from Ordovician oils and hydrocarbon source rocks, mid-continent and east-central United States. Am Assoc Petrol Geol Bull 71:1342–1354
Hoffmann CF, Foster CB, Powell TG, Summons RE (1987) Hydrocarbon biomarkers from Ordovician sediments and the fossil alga *Gloeocapsomorpha prisca* Zalessky 1917. Geochim Cosmochim Acta 51:2681–2697
Illich HA, Grizzle PL (1983) Comment on "comparison of Michigan Basin crude oils" by Vogler et al. Geochim Cosmochim Acta 47:1151–1156
Jacobson SR, Hatch JR, Teerman SC, Askin RA (1988) Middle Ordovician organic matter assemblages and their effect on Ordovician-derived oils. Am Assoc Petrol Geol Bull 72:1090–1100
Kattai V, Puura V (1988) Commercial zonation of the Estonia oil shale deposit. Proc Int Conf Oil shale and shale oil, Beijing. Chemical Industry Press, Beijing, pp 51–58
Keller BM, Menner VV, Stepanov VA, Chumakov NM (1974) New finds of Metazoa in the Vendian of the Russian Platform. Izv Akad Nauk SSSR Ser Geol 12:130–134 (in Russian)
Kennard JM (ed) (from 1972, continuing) Stromatolite newsletters. Bur Mineral Resourc, Canberra, Aust
Kenyon CN (1974) Fatty acid composition of unicellular strains of blue-green algae. In: Blue-green algae, III: current research. MSS Information Corporation, New York, pp 10–27
Klesment I (1974) Application of chromatographic methods in biogeochemical investigations: determination of the structures of sapropelites by thermal decomposition. J Chromatogr 91:705–713
Klesment I, Nappa L (1980) Investigation of the structure of Estonian oil shale kukersite by conversion in aqueous suspension. Fuel 59:117–122
Lewan MD, Winters JC, McDonald JH (1979) Generation of oil-like pyrolysates from organic-rich shales. Science 203:897–899
Lindenbein HAR (1921) Une flore marine sapropélitique de l'Ordovicien moyen de la Baltique. C R Seanc Soc Phys Hist Nat Geneve 38:60–63
Longman MW, Palmer SE (1987) Organic geochemistry of mid-continent Middle and Late Ordovician oils. Am Assoc Petrol Geol Bull 71:938–950
Lowenstam HA, Margulis L (1980) Evolutionary prerequisites for Early Phanerozoic calcareous skeletons. Precambrian Res 12:27–41
Luts K (1938) The occurrence and application of kukersite in Estonia, with particular reference to the plant of the A/G "Esimente eesti Polevkivitootus". (First Estonian oil shale industry). In: Oil shale and cannel coal. Proc Conf, June 1938, Scotland. Inst Petrol, London 1:124–142
McKirdy DM, Hahn JH (1982) The composition of kerogen and hydrocarbons in Precambrian rocks. In: Holland HD, Schidlowski M (eds) Mineral deposits and the evolution of the biosphere. Springer, Berlin Heidelberg New York, pp 123–154
McKirdy DM, Watson BL, Gorter JD, Foster CB (1986) Horn Valley Siltstone, Amadeus Basin: sedimentology and organic facies of an Ordovician petroleum source rock. In: 12th Int Sedimentological Congr, Canberra, pp 208–209
Martin RL, Winters JC, Williams JA (1963) Distribution of n-paraffins in crude oils and their implications to origin of petroleum. Nature (London) 199:110–113
Matista A (1963) Über den Ursprung der Kohlenstoffe in *Corycium*. N Jahrb Geol Palaeontol Monatsh 8:433–441
Matista A (1969) On the microfossils of *Corycium enigmaticum*. Bull Geol Soc Finland 41:199–202
Nagy B (1982) Search for biochemical fossils on Earth and non-biological organic molecules on Jupiter, Saturn and Titan. Naturwissenschaften 69:301–310
Narbonne GM, Hofmann JH (1987) Ediacaran biota of the Wernecke Mountains Yukon, Canada. Palaeontology 30:647–676
Niklas KJ (1982) Chemical diversification and evolution of plants as inferred from paleobiochemical studies. In: Nitecki MH (ed) Biochemical aspects of evolutionary biology. Univ Press, Chicago, pp 29–91
Ohlson B (1962) Observations on Recent lake balls and ancient *Corycium* inclusions in Finland. CR Soc Geol Finlande 33:377–390
Powell TG, McKirdy DM (1972) The geochemical characterisation of Australian crude oils. Aust Petrol Explor Assoc J 12(1):125–131
Reed JD, Illich HA, Horsfield B (1986) Biochemical evolutionary significance of Ordovician oils and their sources. Org Geochem 10:347–358
Robbins EI,Porter GP, Haberyan KA (1985) Pellet microfossils: possible evidence for metazoan life in Early Proterozoic time. Proc Natl Acad Sci USA 82:5809–5813
Rüger L (1926) Die Ergebnisse neuerer Untersuchungen über den estnischen Kuckersit und einige sich daran knüpfende Fragen. Geol Rundsch 17:23–25
Sabrodin W (1971) Leben im Präkambrium. Ideen Exakten Wissens 12/71:835–842 (see also article under that title in Bild Wissensch 1972:586–591)
Schopf JW (1983) Earth's earliest biosphere. Univ Press, Princeton, 543 pp
Sidorenko AV, Sidorenko SA (1971) Organic matter in Precambrian metamorphic sedimentary rocks and certain geological problems. Sov Geol 5:3–20 (in Russian)
Sokolov BS (1973) Vendian of Northern Eurasia. In: Arctic geology, Proc 2nd Symp Arctic geology. Am Assoc Petrol Geol, Tulsa, Oklahoma, pp 204–218

Subacius SMR, Fairchild TR (1986) Allochthonous organic matter in Late Proterozoic stromatolites (*Stratifera* f.) from Brazil. An Acad Brasil Cien 58:243–248

Summons RE, Walter MR (1990) Molecular fossils and microfossils of prokaryotes and lower eukaryotes from Proterozoic sediments. Am Science 290 A:212–244

Summons RE, Brassell SC, Eglinton G, Evans EJ, Horodyski RJ, Robinson N, Ward DM (1988) Distinctive hydrocarbon biomarkers from fossiliferous sediment of the Late Proterozoic Walcott Member, Chuar Group, Grand Canyon, Arizona. Geochim Cosmochim Acta 52:2625–2637

Sun Weiguo (1987) Palaeontology and biostratigraphy of Late Precambrian macroscopic algae: *Chuaria* Walcott and *Tawuia* Hofmann. Palaeontogr Ser B 203:109–134

Sun Weiguo, Wang Guixiang, Zhou Benhe (1986) Macroscopic worm-like body fossils from the Upper Precambrian (900–700 Ma) Huainan District, Anhui, China and their stratigraphic and evolutionary significance. Precambrian Res 31:377–403

Teeri J (1982) Carbon isotopes and the evolution of C_4 photosynthesis and crassulacean acid metabolism. In: Nitecki MH (ed) Biochemical aspects of evolutionary biology. Univ Press, Chicago, pp 93–130

Tissot BP, Welte DH (1978) Petroleum formation and occurrence. Springer, Berlin Heidelberg New York, 538 pp

Vitorović D (1980) Structure elucidation of kerogen by chemical methods. In: Durand B (ed) Kerogen. Editions Technip, Paris, pp 301–338

Walter MR (1976) Stromatolites. Elsevier, Amsterdam, 790 pp

Westbroek P, de Jong EW (eds) (1983) Biomineralization and biological metal accumulation. Reidel, Dordrecht, 533 pp

Williams JA (1974) Characterization of oil types in Williston Basin. Am Assoc Petrol Geol Bull 58:1243–1252

Zalessky MD (1917) On marine sapropelite of Silurian age, formed by blue-green alga. Izv Imp Akad Nauk (IV) 1:3–18 (in Russian)

Zhuravleva ZA (1964) Riphean and Lower Cambrian oncolites and catagraphites of Siberia and their stratigraphic importance. Trans Geol Inst Akad Sci USSR 114:1–73 (in Russian)

Proterozoic Organic Carbon – Its Preservation and Isotopic Record

Harald Strauss [1], David J. Des Marais [2], J. M. Hayes [3] and Roger E. Summons [4]

Precambrian Paleobiology Research Group, Department of Geology and Geophysics, University of California, Los Angeles, USA

Present addresses:
[1] Ruhr-Universität Bochum, Institut für Geologie, W-4630 Bochum 1, FRG
[2] NASA Ames Research Center, Moffett Field, California, 94035, USA
[3] Biogeochemical Laboratories, Departments of Chemistry and Geology, Indiana University, Bloomington, Indiana, 47405, USA
[4] Bureau of Mineral Resources, Geology and Geophysics, GPO Box 378, Canberra City, ACT 2601, Australia

Abstract

The carbon isotopic composition of Proterozoic organic matter has been studied as a function of elemental H/C ratio. Kerogen samples which have experienced various degrees of postdepositional thermal alteration show a correspondingly wide spread of $\delta^{13}C$ values. Setting a minimum H/C threshold of 0.2 to eliminate the kerogens of highest rank immediately results in a marked contraction in the range of $\delta^{13}C$. Analysis of the time-dependence of these data reveals a smooth trend of increasing $\delta^{13}C$ from an Early Proterozoic average of − 34‰ PDB to about − 30.5‰ for the Middle Proterozoic. This contrasts markedly with the record of Proterozoic carbonates, which typically average near 0‰ PDB. Several explanations are offered for this trend. Firstly, a long-term decline occurred in the partial pressure of atmospheric CO_2 during the Proterozoic. An accompanying development of biochemical pumps, which raise the internal CO_2 concentrations of photoautotrophs and allow the maintenance of high levels of carbon fixation, would cause a partially closed system with respect to external CO_2 and a reduction in ^{13}C fractionation. Alternatively, or perhaps additionally, evolutionary pressures toward optimizing the kinetic parameters of the major carboxylating enzyme RUBISCO may have been accompanied by a reduction in the enzymic discrimination against ^{13}C. Late Proterozoic kerogens display substantial isotopic variability. This most likely reflects secular changes in the burial rates of organic versus carbonate carbon.

1 Introduction

It now appears that the biogeochemical carbon cycle has sustained dramatic changes during the evolution of the planet. Evidence for this stems from studies of the abundance and isotopic composition of carbon in its two major crustal reservoirs, namely carbonate and organic matter in shales and carbonates. Our understanding of the landmarks of early biological evolution, such as the onset of distinctive metabolic conversions and the global geochemical cycling of carbon, stems in part from studies of the isotopic record of organic and inorganic carbon in sedimentary systems. Similarly, the isotopic record is informative of the evolution of the atmosphere and the degree to which the biogeochemical cycles of major elements are interconnected and "fine-tuned". Although some authors have emphasized aspects of uniformity in the Archean and Proterozoic carbon isotopic records (Schidlowski 1982, 1987, 1988; Schidlowski et al. 1983), others have noted and discussed variations that can be regarded as significant (Schoell and Wellmer 1981; Hayes et al. 1983). The latter approach follows that which has proven useful in studies of the Phanerozoic carbon-isotopic record (Veizer and Hoefs 1976; Veizer et al. 1980; Holser et al. 1989). Early indications of an oceanic chemical event in the Late Proterozoic stemmed from the work of Holser (1977), who demonstrated secular changes in marine sulfur isotopes (Yudomski event), and Eichmann and Schidlowski (1975) and Veizer et al. (1980), who showed a rapid shift in ^{13}C of marine carbonates near the Cambrian boundary. More recent studies, however, have utilized detailed sampling of continuous sections and improved chronology to reveal dramatic global secular variations in the Late Proterozoic and across the Proterozoic-Cambrian boundary (Hsü et al. 1985; Magaritz et al. 1986; Tucker 1986; Knoll et al. 1986; Lambert et al. 1987; Aharon et al. 1987).

These isotopic excursions are evidently associated with major ecological developments, but the precise mechanistic relationships between the biological and chemical changes remain to be elucidated (e.g.,

M. Schidlowski et al. (Eds.)
Early Organic Evolution: Implications for Mineral and Energy Resources

Awramik et al. 1986). Isotopic changes are also recognized at major faunal boundaries of the Phanerozoic (e.g., Popp et al. 1986; Holser and Magaritz 1987; Magaritz 1989; Zachos et al. 1989). Evidence has recently been presented for a major Early Proterozoic ^{13}C excursion in carbonate carbon (Baker and Fallick 1989).

In this chapter we present further evidence against the inferred uniformity of the organic and inorganic carbon isotope versus age curves in the Precambrian. The data arise from a large number of new samples obtained and analyzed as part of an interdisciplinary study by the Precambrian Paleobiology Research Group-Proterozoic (PPRG-P), detailed in Hayes et al. (in press) and presented in a preliminary report at the final meeting of IGCP Project 157 at Maria Laach, Germany, in September 1988. The new data, combined with available reports from literature, have been examined with particular attention to evidence for the degree of preservation of primary signatures in organic carbon and to detecting related changes in both organic and inorganic carbon phases.

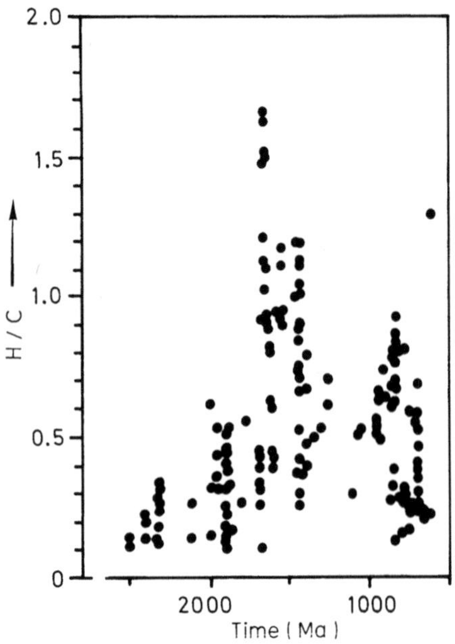

Fig. 1. Plot of H/C ratio vs time for Proterozoic kerogens. (After Hayes et al. in press)

2 Preservation of Proterozoic Organic Matter

Most early studies of organic matter in Precambrian sediments either measured the abundance and isotopic composition of carbon or searched for molecular fossils. Systematic studies on preservation of this organic carbon are rare and the number of analytical results is limited (e.g., Leventhal et al. 1975; Wedeking and Hayes 1983; Hayes et al. 1983; Hoering and Navale 1987). They reveal a consistent picture of severe alteration and a generally poor prognosis for finding well-preserved organic matter, particularly in the Middle Proterozoic and earlier. There are, however, a few notable exceptions (e.g., Jackson et al. 1986).

A reliable and conveniently applied criterion to assess the preservation of kerogen is its elemental composition and particularly the H/C ratio (Durand 1973). Recently deposited organic matter, depending on its source, has an H/C ratio approaching 2.0. Ancient, well-preserved sediments such as oil shales may have values in the range 1.5–1.2 which decline to below 1.0 as the sediments liberate hydrocarbons during thermal maturation. The H/C ratio of organic matter which has sustained extensive thermal alteration will approach zero, the value for amorphous carbon and graphite (e.g., McKirdy et al. 1975). The predominant feature observed in Proterozoic kerogens is an advanced degree of dehydrogenation ($H/C < 0.5$) independent of lithology, facies, or age (Fig. 1). There are a few sediments which have abundant and exceptionally well-preserved kerogen ($H/C > 0.5$) including the Early and Middle Proterozoic McArthur and Roper Groups in Australia (1690–1400 Ma), the Nonesuch Formation (1000 Ma) of the North American midcontinental rift system, and the Late Proterozoic Chuar Group (950–840 Ma) of southern U.S.A. As well, there is the organic matter associated with some latest Proterozoic oil fields in Siberia and Oman (for a review see Hayes et al. in press). These examples provide the paleoenvironmental variety and chronological detail necessary to detect global secular trends over time scales of the order of 100 Ma or less, but not for the whole of the Proterozoic.

The chemical and structural changes to kerogen which take place as a result of postdepositional thermal processes cause progressive loss of biomarker information (e.g., Leventhal et al. 1975; Wedeking and Hayes 1983). This loss of structural detail precedes substantial alteration to the original isotopic composition. Isotopic shifts in kerogen during diagenesis and catagenesis (Tissot and Welte 1984) depend on its homogeneity but are commonly minor (Degens 1969; Redding et al. 1980; Burwood et al. 1988). Once metagenesis and metamorphism commence, substantial displacements in the original isotopic composition are possible. Analyses of sediments and petroleums of various ages (Degens 1969; McKirdy and Powell 1974;

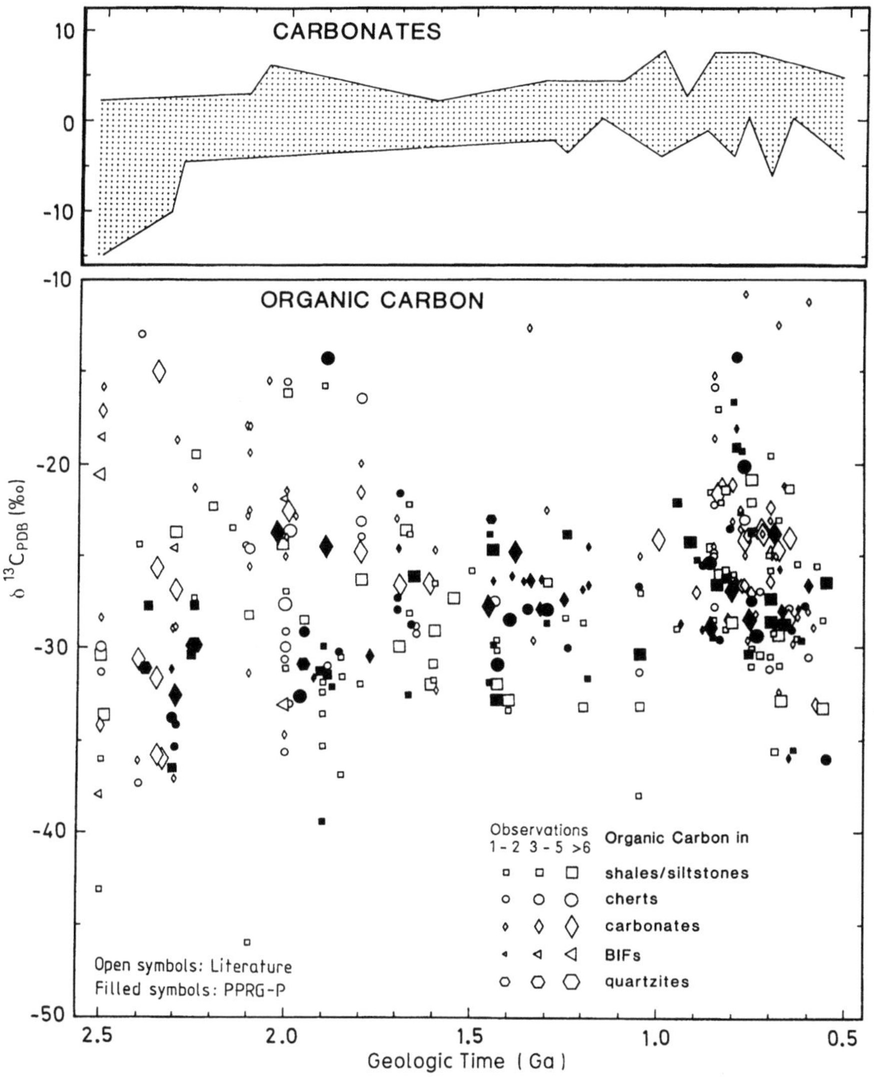

Fig. 2. Carbon isotopic composition of Proterozoic organic carbon irrespective of elemental composition. The corresponding range of values displayed by carbonate carbon is shown for comparison. (After Hayes et al. in press)

Hoefs and Frey 1976; Barghoorn et al. 1977; Stahl 1978; Redding et al. 1980; Peters et al. 1981; Hayes et al. 1983; Burwood et al. 1988) suggest that petroleum generation results in a slight ^{13}C enrichment of the residual organic matter due to the preferential loss of ^{13}C-depleted hydrocarbons. This isotopic shift may be of the order of 2–3‰ (e.g., Burwood et al. 1988) but is often less (e.g., Degens 1969; Peters et al. 1981). McKirdy et al. (1975) demonstrated potential for enrichments of a much greater magnitude by correlating an increase in $\delta^{13}C$ of the order of 10‰ with mineralogic indicators of metamorphism. Data summarised by Hayes et al. (1983) indicate that such isotopic shifts become large in comparison to other uncertainties for $H/C < 0.2$ (see Fig. 5-5 in Hayes et al. 1983). Finally, under conditions encountered during moderate to high-grade metamorphism, exchange of carbon between organic and inorganic phases can lead to very large shifts in δ values as isotopic differences approach equilibrium values (e.g., Valley and O'Neil 1981; Schidlowski 1987). Accurate interpretation of any correlation between the isotopic composition of kerogen and its geologic age therefore requires that its degree of preservation be assessed.

3 Isotopic Composition of Proterozoic Organic Matter

Figure 2 depicts a comprehensive compilation of all Proterozoic organic carbon isotope values gleaned from the literature and determined in the present study. The values for all analyses scatter between − 11‰ and − 47‰ relative to PDB and indicate no discernable time trends. This wide range contrasts markedly to the narrower range of the isotopic composition of Proterozoic carbonates also shown in Fig. 2. It has been suggested that the δ values for organic carbon reflect primary isotopic compositions altered only by minor shifts resulting from postdepositional thermal alteration (e.g., Schidlowski 1982). It has been proposed that values in the range $-26 \pm 7‰$ are broadly representative and convey "a remarkably consistent isotopic signal of autotrophic carbon fixation reflective of an extreme degree of evolutionary conservatism in the biochemistry of photoautotrophy" (Schidlowski 1988).

4 A Revised Approach

In order to avoid the effects of isotopic shifts due either to extreme thermal alteration or to analytical artifacts, we accepted only those kerogens having less than 25% ash and having elemental H/C ratios exceeding 0.2. The primary $\delta^{13}C$ of kerogens having H/C above 0.2 has probably been altered (increased) by less than 5‰ (Hayes et al. 1983). Figure 3A shows the age curve plotted for minimally altered kerogens with the envelope for "total organic carbon" superimposed for comparison. There are often significant uncertainties in the dating of Proterozoic sediments and, thus, Fig. 3B is included to reflect these. Table 1 lists details of the specific minimally altered kerogens used in construction of Fig. 3.

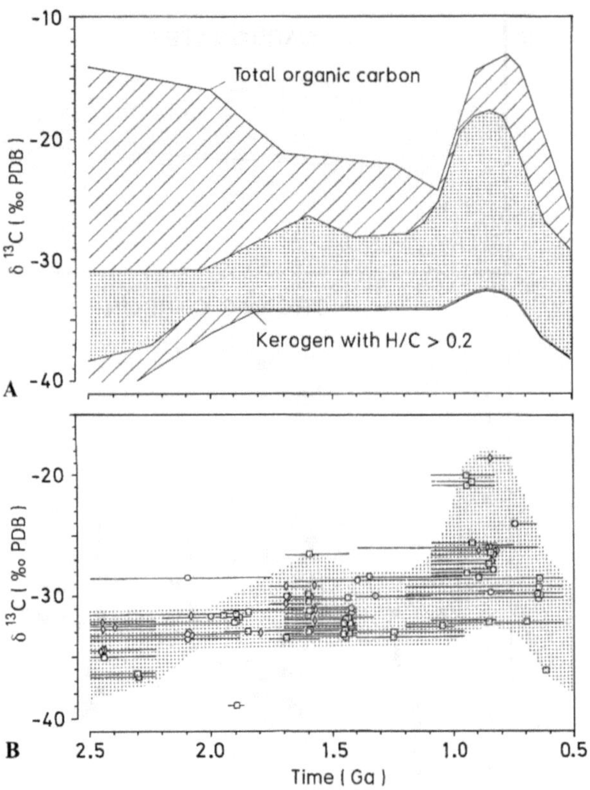

Fig. 3A, B. **A** Carbon isotopic composition of minimally altered Proterozoic kerogens. The range for all samples, irrespective of elemental H/C ratio is shown for comparison. **B** The same data for minimally altered kerogens showing the uncertainties arising from age dating (see also Table 1)

Several aspects of the screened data (H/C > 0.2) in Fig. 3 are significant. Firstly, isotopic variations for individual time intervals (say 200 Ma) rarely exceed 5‰ for the Early and Middle Proterozoic, but larger variations occur in the late Proterozoic. The observation of a significant narrowing in the range of $\delta^{13}C$ for the Early and Middle Proterozoic and the restriction of those values to isotopically lighter values suggests that the larger spread in the "total organic carbon" curve stems predominantly from the effects of thermal alteration rather than from variations in primary signatures, and that the more severely altered kerogens (H/C < 0.2) are frequently more ^{13}C-enriched. Even so, the 5‰ range is remarkably small in comparison to the range of δ values encountered in the modern biosphere. Several forms of selection could be responsible for this. The kerogens analyzed do not represent "all carbon from all conceivable types of environment", but instead "residual organic carbon from environments favored for preservation". A time traveler analyzing Proterozoic mats and plankton, particularly in extreme environments, would be likely to find isotopic compositions ranging over 20‰. The probable occurrence of such extremes no doubt explains some of the divergent δ values in the "total organic carbon" envelope. But the importance of exotic primary δ values within the global carbon budget is uncertain at best. Accurate reconstructions of the carbon cycle must focus on globally averaged δ values for organic carbon immobilized in sediments. Those organic materials are the homogenized residues of biological communities as opposed to collections of primary producers which may be individually distinctive. The isotopic compositions of such residues must record the effects of photoautotrophic carbon fixation, but smoothed by numerous secondary processes. The range of δ values of total organic carbon in organic carbon-rich-sediments of all geological ages is narrow in comparison to materials from the modern biosphere. This reflects the relative homogeneity of environments and communities associated with the

Table 1. H/C-data and carbon isotopic compositions for minimally altered Proterozoic kerogens

Unit	AMIN (Ma)	AEST	AMAX[a]	H/C Range	H/C Avg.	$\delta^{13}C$(‰) Range	$\delta^{13}C$(‰) Avg.	N	Ref.[b]
Khatyspyt Fm		620			1.30		– 36.2	1	1
Sheepbed Fm.	550	650	1200	0.22–0.25	0.23	– 29.9 to – 28.5	– 29.5	5	1
Twitya Fm.	550	700	1200		0.41		– 32.0	1	1
Tapley Hill Fm.	670	750	760		0.25		– 24.2	1	2
Kingston Peak Fm.	650	850	1400		0.27		– 26.0	1	1
Bitter Springs Fm.	760	850	897		0.73		– 18.6	1	3
				0.35–0.39	0.37	– 29.7 to – 26.1	– 27.9	2	1
Walcott Mbr.	830	850	1090	0.69–0.87	0.79	– 27.8 to – 26.4	– 26.9	5	1
Awatubi Mbr.	830	860	1090	0.60–0.79	0.70	– 32.2 to – 27.4	– 29.8	2	1
Carbon Canyon Mbr.	830	900	1090	0.64–0.74	0.69	– 28.4 to – 26.3	– 27.4	2	1
Jupiter Mbr.	830	930	1090	0.50–0.66	0.58	– 25.7 to – 20.6	– 23.2	2	1
Tanner Mbr.	830	950	1090	0.51–0.55	0.53	– 28.1 to – 20.1	– 23.0	3	1
Allamoore Fm.	1000	1050	1200		0.31		– 32.4	1	1
Hongshuizhuang Fm.	958	1250	1400	0.62–0.71	0.67	– 33.4 to – 33.1	– 33.3	2	1
Wumishan Fm.	958	1325	1400		0.54		– 30.1	1	1
Yangzhuang Fm.	958	1350	1400		0.49		– 28.4	1	1
Newland Fm.	1325	1400	1700		0.42		– 31.8	1	1
Dismal Lakes Grp.	1200	1400	1663		0.67		– 28.7	1	1
Gaoyuzhuang Fm.	1400	1425	1950	0.38–0.52	0.45	– 31.4 to – 31.0	– 31.2	2	1
McMinn Fm.	1429	1430	1690	0.40–0.79	0.56	– 32.4 to – 32.0	– 32.2	3	4
					1.17		– 30.2	1	3
				0.30–1.19	0.76	– 33.2 to – 32.1	– 32.6	5	1
Lansen Creek Shale Mbr.	1429	1430	1690		1.00		– 33.4	1	5
Velkerri Fm.	1429	1430	1690	0.65–1.01	0.81	– 33.3 to – 32.2	– 32.9	7	5
Yalco Fm.	1429	1580	1690	0.89–1.12	0.97	– 32.0 to – 29.2	– 30.9	4	5
H.Y.C. Pyr. Shale Mbr.	1429	1600	1690		0.64		– 26.6	1	3
Barney Creek Fm.	1429	1690	1690	0.87–1.66	1.40	– 32.9 to – 29.9	– 31.3	7	5
Bungle Bungle Dolomite	1560	1690	1765	0.31–0.45	0.39	– 33.5 to – 29.3	– 30.7	5	4
Earaheedy Grp.		1800			0.27		– 33.2	1	4
Chuanglinggou Fm.	1400	1850	1950	0.33–0.38	0.36	– 33.0 to – 31.3	– 32.2	2	1
Rove Fm.	1850	1900	2500		0.41		– 32.1	1	4
				0.42–0.51	0.47	– 32.1 to – 31.6	– 31.8	2	1
Union Island Fm.[c]	1865	1900	1928		0.23		– 39.0	1	1
Rocknest Fm.[c]	1760	1925	2300	0.12–0.45	0.22		– 24.7	4	1
Tyler Fm.	1850	1950	2500		0.26		– 31.7	1	1
Duck Creek Dolomite	1850	2000	2200		0.32		– 31.7	1	4
Gunflint Fm.	1880	2090	2500		0.61		– 33.5	1	4
				0.32–0.54	0.42	– 33.3 to – 31.7	– 32.5	5	1
McLeary Fm.	1760	2100	2500		0.26		– 28.4	1	4
Reivilo Fm.	2224	2310	2643	0.31–0.35	0.33	– 36.7 to – 36.4	– 36.6	2	1
Malmani Fm.	2224	2400	2643		0.21		– 32.5	1	4
Gamohaan Fm.	2224	2450	2643	0.25–0.28	0.26	– 34.6 to – 32.1	– 33.7	5	1

[a] AMIN = minimum age, AEST = estimated age, AMAX = maximum age; AMIN and AMAX represent radiometric ages, whereas AEST represents an estimated age for the individual stratigraphic units.

[b] 1 Strauss et al. (in press); 2 McKirdy et al. (1975); 3 McKirdy and Powell (1974); 4 Hayes et al. (1983); 5 Powell et al. (1987).

[c] Samples treated as outliers; Union Is sample is anthraxolite; three out of four Rocknest samples fail H/C > 0.

burial of the bulk of organic carbon. Nevertheless, the very narrow range evident in Fig. 3 is remarkable, particularly considering that the samples represent many different rock sections and localities, have not been screened for lithological effects, and carry unavoidable inaccuracies in time resolution. This latter problem arises because a published estimated age for a thick sedimentary sequence, particularly in the oldest rocks, is usually derived from dates for events which long precede or postdate sedimentation. Often, these estimates cannot be readily adjusted to random samples from individual beds.

A second feature of Fig. 3 is that the minimally altered kerogens essentially represent the base of the "total organic carbon" curve and there is a smooth trend of increasing $\delta^{13}C$ from an average of – 34‰ in the Early Proterozoic to about − 30.5‰ in the Middle Proterozoic. This clearly developed trend to heavier organic carbon is concealed in the "total organic carbon" data and, combined with an "apparent" average of 0‰ for inorganic carbon, suggests a long-term decrease in the fractionation ($\Delta\delta$) between the two carbon phases over this time interval.

A third feature of Fig. 3 is the presence of large fluctuations in the organic carbon record of the Late Proterozoic. Recognition of these variations undoubtedly stems from the improved record of sediments in this interval combined with their tendency to have organic matter with a lower degree of thermal alteration. Present difficulties in temporally correlating the isotope anomalies from the various Late Proterozoic sections are concealing fine detail in the organic carbon isotope curve at this time. Similarly, wider fluctuations about average values may have taken place earlier in the Proterozoic but their existence could be obscured by the absence of a continuous record. Some recent data on carbonate isotope anomalies supports this (e.g., Baker and Fallick 1989). After the dramatic swings in both organic and inorganic carbon in the Riphean and Vendian occurred, the $\delta^{13}C$ for organic carbon returned to a value near − 30‰ in the early Cambrian. This corresponds to a Cambrian marine carbonate composition of −0.57‰ (Veizer et al. 1980; Berner 1989) and a separation ($\Delta\delta$) between "average" organic matter and "average" carbonate of 29.4‰. In their study of isotopic abundances of inorganic and organic carbon from the Upper Proterozoic sections of Svalbard and East Greenland, Knoll et al. (1986) established a consistent separation between coexisting carbon phases of 28.5‰ at this time. Donnelly et al. (1988) studied the primary isotopic separation between coexisting organic and inorganic carbon phases in a Middle Cambrian sediment and reported a value of 30.1‰.

It is now widely accepted that the isotopic compositions of sedimentary organic and inorganic carbon are dependent on the relative rates of their burial. Isotopic mass-balance calculations, therefore, can be used to identify and quantify episodes of enhanced organic carbon burial (e.g., Garrels and Lerman 1984), particularly when there is a global record of sediments with long continuous sections (e.g., Berner 1989). Isotopic fluctuations to ^{13}C-enriched (heavy) carbon generally reflect high rates of burial of organic carbon while movement to ^{13}C depleted (light) values reflects weathering and reoxidation of previously buried organic carbon. Accurately defined periods of ^{13}C enrichment, such as revealed by primary skeletal carbonates in the Permian (e.g., Popp et al. 1986) or whole-rock carbonates and organic-inorganic carbon pairs in the Late Proterozoic (e.g., Knoll et al. 1986; Magaritz et al. 1986; Tucker 1986) have been interpreted to reflect enhanced carbon burial. This is a reliable interpretation when it can be demonstrated that there is a roughly constant $\Delta\delta$ (Berner 1989), as was shown in the study of Late Proterozoic Greenland and Svalbard sediments (Knoll et al. 1986).

Factors other than relative burial rates of organic and inorganic carbon which might lead to major, long-term changes in $\delta^{13}C_{org}$ and $\Delta\delta$ include the availability of CO_2 for photosynthetic carbon fixation and variation in the biological isotope fractionation associated with photosynthesis (Deines 1980). The effects of CO_2 availability are well known through studies of laboratory cultures (e.g., Calder and Parker 1973) and natural systems (e.g., Schidlowski et al. 1984; Estep 1984; Haddad et al. 1988; Des Marais et al. in press). For example, cyanobacterial mat organic matter may be abnormally ^{13}C-enriched (− 6 to − 14‰ PDB) due to reduced availability of inorganic carbon during times of maximal photosynthesis. Haddad et al. (1988) have now demonstrated that the rapid rates of carbon fixation associated with peak photosynthesis in cyanobacterial mat communities is accompanied by zero discrimination against ^{13}C. Ribulose bisphosphate carboxylase oxygenase (RUBISCO), the most significant CO_2-fixing enzyme, shows a maximum fractionation ($\Delta\delta$) of ~ 29‰ for carboxylation of the pure plant enzyme (Roeske and O'Leary 1984). Farquhar et al. (1989) quote a slightly more conservative figure of 27‰ for net C_3 fixation. In combination with the chemical equilibria mediating isotopic exchange between carbonates and dissolved CO_2 (11‰ at 25°), we estimate that this would lead to a maximum fractionation of − 38‰. If, during the Proterozoic, similar maximum fractionations were operating, it would be possible to produce primary photosynthetic organic matter with an extreme isotopic composition of −38‰ from "normal" marine bicarbonate or carbonate near 0‰. However, decreasing CO_2 (or bicarbonate) availability, associated with declining atmospheric CO_2 concentrations, in combination with the effects of reworking prior to burial, would probably result in somewhat less than maximal fractionations taking place (e.g., Hayes et al. 1989; Popp et al. 1989). This could account for the observed continuous decrease in $\Delta\delta$ in the Proterozoic, as has been suggested by Mitzutani and Wada (1982) and the anomalous ^{13}C enrichment observed in Holocene marine organic matter reported and discussed by Arthur et al. (1985). Independent evidence for such a decline in pCO_2 exists (Walker et al. 1983; Des Marais 1985; Kasting 1987).

While a reduction in atmospheric CO_2 partial pressure is proposed as the cause of long-term reduction in ^{13}C fractionation (e.g., Arthur et al. 1985), the precise mechanisms which could produce this effect have not been extensively or effectively analyzed. One phenomenon likely to contribute significantly is the development of biochemical bicarbonate and carbon dioxide pumps (Badger 1987). These are CO_2-concentrating mechanisms which raise the internal CO_2 levels compared to the external environment and allow photoautotrophs to maintain high levels of carbon fixation. Cyanobacteria appear to be particularly efficient in this process, having the capacity to increase their internal pCO_2 by up to three orders of magnitude (Badger 1987). The isotopic consequences of induction of the pump in a green alga have been demonstrated by Sharkey and Berry (1985) and there is, as expected, a significant reduction in ^{13}C fractionation. The pump rapidly depletes the already low inorganic carbon concentration in the immediate vicinity of the photosynthesizing organisms, effectively separating the internal and external inorganic carbon pools.

An evolutionary change to ^{13}C discrimination by RUBISCO might have occurred, together with changes in the predominant types of phototrophy which existed in times past (i.e., oxygenic vs anoxygenic vs partial photoheterotrophy). It is known that RUBISCOs from various plant, algal, and bacterial species differ in the substrate specificities for carboxylase and oxygenase functions (Jordan and Ogren 1981), indicating the possibility of increased carboxylation efficiency during the course of the evolution of photosynthesis. There would have been strong evolutionary pressure for this because increasing atmospheric oxygen and decreasing atmospheric CO_2 concentrations would both be detrimental to efficient operation of RUBISCO. Badger and Andrews (1987) have presented convincing arguments supporting the likelihood of a co-evolution of CO_2-concentrating mechanisms and changes to the kinetic properties of RUBISCO, both adapations being in response to a declining CO_2/O_2 ratio in the external environment. The data of Jordan and Ogren (1981) indicate that the RUBISCOs of eukaryotic algae have, in general, higher specificity factors than do the RUBISCOS of cyanobacteria or photosynthetic bacteria. Hence, the emergence of eukaryotic algae during the Early to Middle Proterozoic may also have contributed to a reduction in discrimination against $^{13}CO_2$ during this time (A. H. Knoll pers. commun. 1989) as their contribution to buried organic carbon increased. The direction of the isotopic consequences of these various biochemical and environmental innovations would be consistent with our observations. However, their actual magnitudes are difficult to predict, due to a shortage of isotopic experimental data from a range of photosynthetic bacteria, cyanobacteria, and algae.

5 Conclusions

Minimally altered ($H/C \geqq 0.2$) Proterozoic kerogens show a narrower and lighter range of $\delta^{13}C$ values than does total organic carbon of unspecified kerogen H/C ratio. They also show a smooth trend from values near − 34‰ at the beginning of the Proterozoic to average values of − 30.5‰ in the middle. The Late Proterozoic is characterised by significant excursions towards "heavy" organic carbon but average values return to about − 30‰ near the Cambrian boundary. The gradual change in average $\delta^{13}C$ values of organic matter during the Proterozoic could indicate a progressive reduction in discrimination against ^{13}C accompanying a reduction in atmospheric CO_2 concentration. This would have caused evolutionary pressure for an increased efficiency of CO_2-fixing enzymes and the development of inducible CO_2-concentrating mechanisms. The shorter-term isotope excursions in the Late Proterozoic, which involve both reduced and oxidized carbon reservoirs, are related to changes in the burial rates of organic carbon at that time and corroborates earlier inferences drawn from the study of carbonate-organic carbon pairs in individual Late Proterozoic basins (e.g., Knoll et al. 1986).

Acknowledgments. We wish to thank J W (Bill) Schopf, Ken Towe, Bruce Runnegar, Malcolm Walter, and the other members of PPRG for stimulating discourse. Andrew Knoll, David McKirdy, Trevor Powell, Manfred Schidlowski, Baruch Spiro and Malcolm Walter provided many suggestions which improved the manuscript. Roger E Summons publishes with permission of the Director, Bureau of Mineral Resources.

References

Aharon P, Schidlowski M, Singh IB (1987) Chronostratigraphic markers in the end-Precambrian carbon isotope record of the Lesser Himalaya. Nature (London) 327:699–701

Arthur MA, Dean WE, Claypool GE (1985) Anomalous ^{13}C enrichment in modern marine organic carbon. Nature (London) 315:216–218

Awramik SM, Conway Morris S, Bengtson S (1986) The Precambrian-Cambrian boundary and geochemical perturbations. Nature (London) 319:696–697

Badger MR (1987) The CO_2-concentrating mechanism in aquatic phototrophs. In: Hatch MD, Boardman NK (eds) The biochemistry of plants, vol 10. Academic Press, New York London, pp 219–274

Badger MR, Andrews TJ (1987) Co-evolution of Rubisco and CO_2-concentrating mechanisms. In: Biggins J (ed) Prog Photosynth Res 3:601–609

Baker AJ, Fallick AE (1989) Evidence from Lewisian limestones for isotopically heavy carbon in two-thousand-million-year-old sea water. Nature (London) 337:352–354

Barghoorn ES, Knoll AH, Dembicki H, Meinschein WG (1977) Variation in stable carbon isotopes in organic matter from the Gunflint Iron Formation. Geochim Cosmochim Acta 41:425–430

Berner RA (1989) Biogeochemical cycles of carbon and sulfur and their effect on atmospheric oxygen over Phanerozoic time. Palaeogeogr Palaeoclimatol Palaeoecol Global Planet Change 75:97–122

Burwood R, Drodz RJ, Halpern HI, Sedivy RA (1988) Carbon isotopic variations of kerogen pyrolysates. Org Geochem 12:195–205

Calder JA, Parker PL (1973) Geochemical implications of induced changes in ^{13}C fractionation by blue-green algae. Geochim Cosmochim Acta 37:133–140

Degens ET (1969) Biogeochemistry of stable carbon isotopes. In: Eglinton G, Murphy MTJ (eds) Organic geochemistry. Springer, Berlin Heidelberg New York, pp 304–329

Deines P (1980) Isotopic composition of reduced organic carbon. In: Fritz P, Fontes JC (eds) Handbook of environmental isotope geochemistry, vol. 1. Elsevier, Amsterdam, pp 329–406

Des Marais DJ (1985) Carbon exchange between the mantle and crust and its effect on the atmosphere. In: Sundquist ET, Broecker WS (eds) The carbon cycle and atmospheric CO_2; natural variation Archean to Present. Geophysical monograph 32. Am Geophys Un, pp 602–611

Des Marais DJ, Bauld J, Risatti JB, Summons RE, Ward DM (in press) The biogeochemistry of modern microbial mats. In: Schopf JW, Klein C (eds) The Proterozoic biosphere: a multidisciplinary study. Univ Press, Cambridge

Donnelly TH, Shergold JH, Southgate PN (1988) Anomalous geochemical signals from phosphatic Middle Cambrian rocks in the southern Georgina Basin, Australia. Sedimentology 35:549–570

Durand B (ed) (1973) Kerogen: insoluble organic matter from sedimentary rocks. Technip, Paris, 519 pp

Eichmann R, Schidlowski M (1975) Isotopic fractionation between coexisting organic carbon-carbonate pairs in Precambrian sediments. Geochim Cosmochim Acta 39:585–595

Estep MLF (1984) Carbon and hydrogen isotopic compositions of algae and bacteria from hydrothermal environments, Yellowstone National Park. Geochim Cosmochim Acta 48:591–599

Farquhar GD, Ehleringer JR, Hubick KT (1989) Carbon isotope discrimination and photosynthesis. Annu Rev Plant Physiol Mol Biol 40:503–537

Garrels RM, Lerman A (1984) Coupling of the sedimentary sulfur and carbon cycles – an improved model. Am J Sci 284:989–1007

Haddad RI, Des Marais DJ, Nguyen H, Bauer J, Nguyen M (1988) Carbon isotope cycling in a recent microbial mat. EOS 69:1083–1084

Hayes JM, Kaplan IR, Wedeking KW (1983) Precambrian organic geochemistry, preservation of the record. In: Schopf JW (ed) Earth's earliest biosphere: its origin and evolution. Univ Press, Princeton, pp 93–134

Hayes JM, Popp BN, Takigiku R, Johnson MW (1989) An isotopic study of biogeochemical relationships between carbonates and organic carbon in the Greenhorn Formation. Geochim Cosmochim Acta 53:2961–2972

Hayes JM, Summons RE, Strauss H, Des Marais DJ (in press) Proterozoic biogeochemistry. In: Schopf JW, Klein C (eds) The Proterozoic biosphere: a multidisciplinary study. Univ Press, Cambridge

Hoefs J, Frey M (1976) Isotopic composition of carbonaceous organic matter in a metamorphic profile from the Swiss Alps. Geochim Cosmochim Acta 40: 945–951

Hoering TC, Navale V (1987) A search for molecular fossils in the kerogen of Precambrian sedimentary rocks. Precambrian Res 34:247–267

Holser WT (1977) Catastrophic chemical events in history of the ocean. Nature (London) 276:403–408

Holser WT, Magaritz M (1987) Events near the Permian-Triassic boundary. Modern Geol 11:155–180

Holser WT, Schönlaub H-P, Attrep M, Boeckelmann K, Klein P, Magaritz M, Orth CJ, Fenninger A, Jenny C, Kralik M, Mauritsch H, Pak E, Schramm J-M, Stattegger K, Schmöller R (1989) A unique geochemical record at the Permian/Triassic boundary. Nature (London) 337:39–44

Hsü KJ, Oberhänsli H, Gao JY, Sun S, Chen H, Krähenbühl U (1985) Strangelove ocean before the Cambrian explosion. Nature (London) 316:809–811

Jackson MJ, Powell TG, Summons RE, Sweet IP (1986) Hydrocarbon shows and petroleum source rocks in sediments as old as 1.7×10^9 years. Nature (London) 322:727–729

Jordan DB, Ogren WL (1981) Species variation in the specificity of ribulose bisphosphate carboxylase/oxygenase. Nature (London) 291:513–515

Kasting J (1987) Theoretical constraints on oxygen and carbon dioxide concentrations in the Precambrian atmosphere. Precambrian Res 34:305–229

Knoll AH, Hayes JM, Kaufman AJ, Swett K, Lambert IB (1986) Secular variation in carbon isotope ratios from Upper Proterozoic successions of Svalbad and East Greenland. Nature (London) 321:832–838

Lambert IB, Walter MR, Zang W, Lu S, Ma G (1987) Paleoenvironment and carbon isotope stratigraphy of Upper Proterozoic carbonates of the Yangtze Platform. Nature (London) 325:140–142

Leventhal J, Suess SE, Cloud P (1975) Nonprevalence of biochemical fossils in kerogen from pre-Phanerozoic sediments. Proc Natl Acad Sci USA 72:4706–4710

Magaritz M (1989) ^{13}C minima follow extinction events: A clue to faunal extinction. Geology 17:337–340

Magaritz M, Holser WT, Kirschvink J (1986) Carbon-isotope events across the Precambrian/Cambrian boundary on the Siberian Platform. Nature (London) 320:258–259

McKirdy DM, Powell TG (1974) Metamorphic alteration of carbon isotopic composition in ancient sedimentary organic matter: new evidence from Australia and South Africa. Geology 2:591–595

McKirdy DM, Sumartojo J, Tucker DH, Gostin V (1975) Organic, mineralogic and magnetic indicators of metamorphism in the Tapley Hill Formation, Adelaide Geosyncline. Precambrian Res 2:345–373

Mitzutani M, Wada E (1982) Effect of high atmospheric CO_2 concentration on $\delta^{13}C$ of algae. A possible cause for the average depletion of ^{13}C in Precambrian reduced carbon. Origins Life 12:377–390

Peters KE, Simoneit BRT, Brenner S, Kaplan IR (1981) Carbon and hydrogen stable isotope variations in kerogen during laboratory-simulated thermal maturation. Bull Am Assoc Petrol Geol 65:501–508

Popp BN, Anderson TF, Sandberg PA (1986) Brachiopods as indicators of original isotopic compositions in some Paleozoic limestones. Geol Soc Am Bull 97:1262–1269

Popp BN, Takigiku R, Hayes JM, Louda JW, Baker EW (1989) The post-Paleozoic chronology and mechanism of ^{13}C depletion in primary organic matter. Am J Sci 289:436–454

Powell TG, Jackson, Sweet IP, Crick IH, Boreham CJ, Summons RE (1987) Petroleum geology and geochemistry, Middle Proterozoic McArthur Basin. Bur Mineral Resour Geol Geophys Aust Record 1987/48, 286 pp

Redding CE, Schoell M, Monin JC, Durand B (1980) Hydrogen and carbon isotopic composition of coals and kerogens. In: Douglas AG, Maxwell JR (eds) Advances in organic geochemistry 1979. Pergamon, Oxford, pp 711–723

Roeske CA, O'Leary MH (1984) Carbon isotope effects on the enzyme catalysed carboxylation of ribulose bisphosphate. Biochemistry 23:6275–6284

Schidlowski M (1982) Content and isotopic composition of reduced carbon in sediments. In: Holland HD, Schidlowski M (eds) Mineral deposits and evolution of the biosphere. Springer, Berlin Heidelberg New York, pp 103–122

Schidlowski M (1987) Application of stable carbon isotopes to early biochemical evolution on Earth. Annu Rev Earth Planet Sci 15:47–72

Schidlowski M (1988) A 3,800-million-year old isotopic record of life from carbon in sedimentary rocks. Nature (London) 333:313–318

Schidlowski M, Hayes JM, Kaplan IR (1983) Isotopic inferences of ancient biochemistries: carbon, sulfur, nitrogen and hydrogen. In: Schopf JW (ed) Earth's earliest biosphere: its origin and evolution. Univ Press, Princeton, pp 149–187

Schidlowski M, Matzigkeit U, Krumbein WE (1984) Superheavy organic carbon from hypersaline microbial mats: assimilatory pathway and geochemical implications. Naturwissenschaften 71:303–308

Schoell M, Wellmer FW (1981) Anomalous ^{13}C depletion in early Precambrian graphites from Superior Province, Canada. Nature (London) 290:696–699

Sharkey TD, Berry JA (1985) Carbon isotope fractionation of algae as influenced by an inducible CO_2 concentrating mechanism. In: Lucas WJ, Berry JA (eds) Inorganic carbon uptake by aquatic photosynthetic organisms. Am Soc Plant Physiol, pp 389–401

Stahl WJ (1978) Source rock-crude oil correlation by isotopic type curves. Geochim Cosmochim Acta 42:1573–1577

Strauss H, Des Marais DJ, Summons RE, Hayes JM (in press) Abundance of carbon, maturity and elemental composition of kerogen. In: Schopf JW, Klein C (eds) The Proterozoic biosphere: a multidisciplinary study. Univ Press, Cambridge

Tissot BP, Welte DH (1984) Petroleum formation and occurrence, 2nd edn. Springer, Berlin Heidelberg New York, 699 p

Tucker ME (1986) Carbon isotope excursions in Precambrian/Cambrian boundary beds, Morocco. Nature (London) 319:48–50

Valley JW, O'Neil JR (1981) $^{12}C/^{13}C$ exchange between calcite and graphite: a possible thermometer in Grenville marbles. Geochim Cosmochim Acta 45:411–419

Veizer J, Hoefs J (1976) The nature of O^{18}/O^{16} and C^{13}/C^{12} secular trends in sedimentary carbonate rocks. Geochim Cosmochim Acta 40:1387–1395

Veizer J, Holser WT, Wilgus CK (1980) Correlation of $^{13}C/^{12}C$ and $^{34}S/^{32}S$ secular variations. Geochim Cosmochim Acta 44:579–587

Walker JCG, Klein C, Schidlowski M, Stevenson DJ, Walter MR (1983) Environmental evolution of the Archean – early Proterozoic earth. In: Schopf JW (ed) Earth's earliest biosphere: its origin and evolution. Univ Press, Princeton, pp 260–290

Wedeking KW, Hayes JM (1983) Carbonization of Precambrian kerogens. In: Bjorøy M (ed) Advances in organic geochemistry 1981. John Wiley & Sons, New York Chichester, pp 546–553

Zachos JC, Arthur MA, Dean WE (1989) Geochemical evidence for suppression of pelagic marine productivity at the Cretaceous/Tertiary boundary. Nature (London) 337:61–64

An Assessment of the Precambrian/Cambrian Transition Events on the Basis of Carbon Isotope Records

Paul Aharon [1,2] and T. C. Liew [1]

[1] Max-Planck Institut für Chemie, Saarstrasse 23, W-6500 Mainz, FRG
[2] On sabbatical leave from the Department of Geology & Geophysics, Louisiana State University, Baton Rouge, LA 70803, USA

Abstract

Renewed interest in the isotope events at the Precambrian/Cambrian (PC/C) transition has led to a recent proliferation of high resolution $\delta^{13}C$ records acquired from sedimentary carbonate sections that are stratigraphically continuous. These $\delta^{13}C$ records generally show a bimodal distribution of values, with a ^{13}C-enriched mode in the end-Precambrian and a sharp transition to a ^{13}C-depleted mode at, or slightly above, the inferred PC/C boundary. The time-bound $\delta^{13}C$ excursions have considerable potential as chronostratigraphic markers which are independent of the controversial biostratigraphic definitions currently used to delineate the PC/C boundary. The relation between $\delta^{13}C$ shifts and oceanic fertility changes is reasonably well known but the circumstances under which these changes have occurred at the PC/C boundary are unclear. Using quantitative arguments derived from a simplified carbon cycle model, an ocean stratification event is proposed to have occurred during the latest Vendian followed by a turnover and a return to a ventilated ocean in the lowest Cambrian. The exact relationship between the changing rates of ocean ventilation and the seemingly contemporaneous faunal turnovers remains to be explored.

An understanding of the process of dolomitization is essential for the correct interpretation of the $\delta^{13}C$ records because of the predominance of dolomites at the end Precambrian. Using the Lesser Himalaya section as a case study, it is shown that samples contain at least two phases of dolomitization and only the early phase contains the seawater imprint. Late dolomite phases are significantly more depleted in ^{18}O and ^{13}C and more enriched in radiogenic ^{87}Sr relative to the early dolomite phases, and preserve the imprint of hot fluids of crustal provenance which advected through the sediments during burial. The linear relationship observed between the $\delta^{18}O$ and $\delta^{13}C$ compositions of all dolomite phases does not offer evidence of primary dolomite precipitation from Precambrian sea water, as recently proposed by Tucker, but attests to the conspicuous absence of soil-derived CO_2 prior to the advent of organic soils.

1 Introduction

Whereas the transition from the Precambrian Eon to the last 500 Ma of Earth history has long attracted the attention of geologists, only recently is its intricate nature becoming unraveled. Profound biotic changes, consisting of the emergence and subsequent extinction of the soft-bodied (Ediacaran) fauna, metazoan explosive radiation, and the emergence of fauna with mineralized skeletons, figure prominently among those events which have occurred during the transition from the Precambrian to the Cambrian over a time interval of ca. 100 Ma (Fig. 1).

Several models have attempted to establish a cause and effect relationship between the documented biotic events at the Precambrian/Cambrian (PC/C) bound-

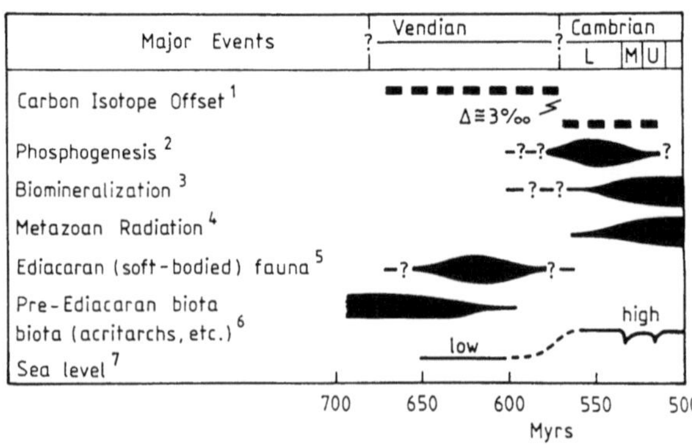

Fig. 1. An "inventory" of major events during the Precambrian-Cambrian transition which left their imprint in the geologic record. Sources for the compilation are: *1* Schidlowski et al. (1975); Veizer and Hoefs (1976); *2* Brasier (1980); Cook and Shergold (1984); Yanshin and Zharkov (1986);*3, 4, 5* Cowie and Glaessner (1975); Brasier (1982); McMenamin (1987); *6* Knoll and Butterfield (1989); *7* Matthews and Cowie (1979); Bjorlykke (1982)

M. Schidlowski et al. (Eds.)
Early Organic Evolution: Implications for Mineral and Energy Resources

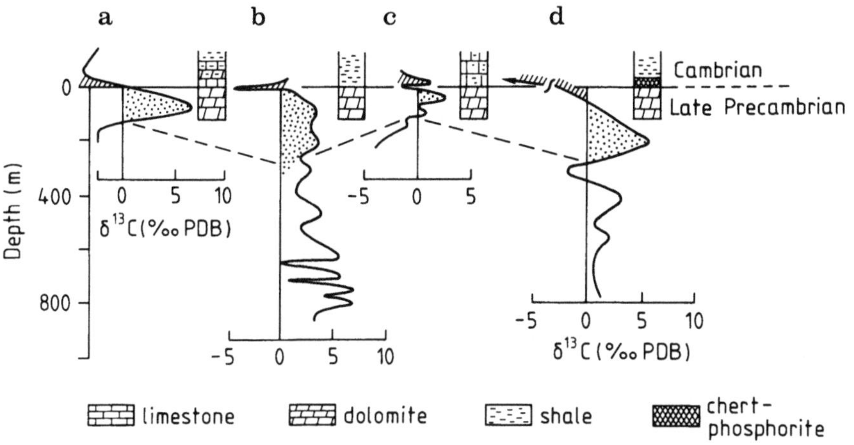

Fig. 2. Carbon isotope records of stratigraphically-continuous sections spanning the PC/C boundary. Correlation is based on the position of the $\delta^{13}C$ excursion relative to the first occurrence of biomineralized skeletons. (After Aharon et al. 1987a; Lambert et al. 1987). Note the lithological changes across the inferred PC/C boundary. *a* Anti-Atlas, Morocco (Tucker 1986; Tucker pers. commun. 1989); *b* Yangtze Platform, China (Hsu et al. 1985; Lambert et al. 1987); *c* Siberian Platform, U.S.S.R. (Magaritz et al. 1986); *d* Lesser Himalaya Aharon et al. 1987a)

ary and the more elusive contemporaneous physical and chemical events. Daly (1909) advanced the hypothesis that an increase in the Ca/Mg ratio in seawater triggered the secretion of hard skeletons by marine organisms. Brasier (1982) suggested that evolutionary explosion was made possible by the opening of vacant habitats resulting from the major sea level rise (Fig. 1). The coincidence of the main phase of phosphogenesis with the faunal diversification and the appearance of the first biomineralized skeletons (Fig. 1) led Cook and Shergold (1984) to propose a link between the chemical and biological events, on the basis of the known key nutrient property of phosphorus in the marine environment. It was a $\delta^{13}C$ negative offset between Precambrian and Cambrian carbonate rocks sampled on a global grid (Fig. 1), first noticed by Schidlowski et al. (1975) and Veizer and Hoefs (1976), which offered the clue that faunal turnovers at the terminal Precambrian were accompanied by a major reorganization of the global carbon cycle.

This isotope displacement has been confirmed recently by $\delta^{13}C$ records acquired from stratigraphically-continuous sections spanning the upper Proterozoic (Knoll et al. 1986) and the PC/C transition (Fig. 2), although its interpretation remains largely controversial. It has been attributed to variations in oceanic biomass fertility and concomitant fluxes of organic burial (Tucker 1986; Magaritz et al. 1986; Lambert et al. 1987; Aharon et al. 1987a), changing rates of ocean ventilation (Goodfellow 1986), onset of biomineralization (Degens et al. 1986), and extraterrestrial causes (Hsu et al. 1985). The transparency of $\delta^{13}C$ records to mass extinction events has been demonstrated at several stratigraphic boundaries (e.g., Cretaceous/Tertiary and Permo/Triassic boundaries: Magaritz 1989) but subscription to such an interpretation for the PC/C $\delta^{13}C$ record (Hsu et al. 1985) has been strongly criticized on the basis of the existing fossil evidence (see Awramik 1986; Morris and Bengtson 1986; Hsu's 1986 reply to criticisms thereof). There is currently a general consensus that changes in oceanic biomass productivity are capable of producing the $\delta^{13}C$ shifts imprinted in marine carbonates, but the circumstances under which these productivity changes have occurred at the PC/C boundary are unclear.

The aim of this chapter is to address several questions regarding the proposed causes of the documented $\delta^{13}C$ offsets at the terminal Precambrian. One such question concerns the consistency of the data with the proposed carbon sources, sinks, and isotope fractionations. Related questions are concerned with the geologic plausibility of the carbon reservoir sizes invoked, and the time scale of the $\delta^{13}C$ excursions. Finally, an appraisal of the quality of the $\delta^{13}C$ records is offered, and the benefits of various strategies devised to minimize the alteration effects on isotope records of Precambrian antiquity are evaluated.

2 Carbon Isotope Records of the Terminal Precambrian

Carbon isotope determinations of carbonates from stratigraphically continuous sedimentary successions have yielded well-defined records of $\delta^{13}C$ variations during the time interval immediately preceding the appearance of shelly metazoans at the base of the Cambrian (Fig. 2). A high quality $\delta^{13}C$ record yielded by samples from the upper Proterozoic successions of Svalbard and East Greenland (Knoll et al. 1986) has been omitted from Fig. 2 because of the occurrence of a hiatus in the sections corresponding to the latest Vendian-earliest Cambrian time. There are remarkable similarities in the patterns of $\delta^{13}C$ variations imprinted in the shallow marine carbonate sequences in spite of their wide geographic distribution in the early Cambrian along the paleotropics (Donovan 1987). In general, the isotope records in Fig. 2 suggest a bimodal distribution of the $\delta^{13}C$ values, with a ^{13}C-enriched mode during the latest Vendian and a sharp transition to a ^{13}C-depleted mode at, or slightly above, the inferred PC/C boundary. Dissimilarities between the $\delta^{13}C$ records are related to the rates of $\delta^{13}C$ shifts, the amplitude of the isotope excursions, and the exact position of the negative $\delta^{13}C$ offset with respect to the PC/C boundary established on the basis of shelly fossil markers.

Assuming continuous sedimentation, the abrupt negative $\delta^{13}C$ excursion of about 5‰ in the Chinese sections (Fig. 2b) has occurred at the staggering rate of about $-0.35‰\,cm^{-1}$ during an estimated time span of 2000 a (Hsu et al. 1985). By contrast, the $\delta^{13}C$ rise of 7.5‰ within the latest Vendian and the subsequent $\delta^{13}C$ decline across the PC/C boundary of the Siberian Platform (Fig. 2c) has occurred at much slower rates of $0.04‰\,m^{-1}$ and $-0.5‰\,m^{-1}$, respectively (Magaritz et al. 1986). We note, however, that presumption of such continuous deposition is controversial and remains a topic of considerable stratigraphic enquiry. In addition, the negative $\delta^{13}C$ anomaly occurring at the top of the isotope record from the Lesser Himalaya (Fig. 2d) corresponds to the China A boundary marker (i.e., first appearance of phosphatic shelly microfauna) according to the correlation of Brasier and Singh (1987), whereas the $\delta^{13}C$ negative excursion at the top of the Chinese record (Fig. 2b) coincides with the China C boundary marker (i.e., first appearance of trilobites) according to Hsu et al. (1985). It is therefore unclear whether the differing rates of the observed $\delta^{13}C$ changes at the PC/C boundary are related to vastly different rates of sedimentation between the individual sections, missing sediment intervals, poor stratigraphic correlations and sampling resolution, or some combination of these factors. Notwithstanding the above uncertainties, it is obvious that time-bound $\delta^{13}C$ excursions have considerable potential as chronostratigraphic tools (Lambert et al. 1987; Aharon et al. 1987a), which may be used in conjunction with the controversial biostratigraphic markers currently providing the criterion for recognition of the PC/C transitional interval (Cowie and Rozanov 1983).

3 Relationship Between $\delta^{13}C$ Excursions and Lithological Changes

The transition between the positive $\delta^{13}C$ excursion at the top of the Vendian and the rapid shift to the negative $\delta^{13}C$ excursion at the base of Cambrian seems to coincide in all cases with lithological changes (Fig. 2). This relationship is further explored here using the Lesser Himalaya section as a case study. The lithology and the fossil succession in the Lesser Himalaya exhibit strong similarities to the Aldan River section (Siberian Platform) which is a candidate stratotype for the PC/C boundary (Cowie and Rozanov 1983). In addition, the massive phosphorite occurrence in the Himalayan locality (Fig. 2d) hosting a Tommotian (lowermost Cambrian) shelly microfauna (Brasier and Singh 1987) correlates with other phosphate deposits that appear to define a world-wide phosphogenesis event at the PC/C transition (Fig. 1). Therefore, the Lesser Himalaya sedimentary succession may hold a key for the study of chemical changes that took place in the marine environment at the terminal Precambrian.

The $\delta^{13}C$ record in Fig. 3 has been reconstructed on the basis of stratigraphically continuous samples from Krol dolomites which went through a careful petrographic and chemical screening (Aharon et al. 1987a), and apatite-calcite pairs from the overlying lower Tal chert-phosphorite member (Banerjee et al. 1986). The lithology of the Krol lower Tal units, their depositional setting and the chronostratigraphy of the sedimentary succession have been documented in recent reports and will not be repeated here (Singh and Rai 1983; Brasier and Singh 1987; Aharon et al. 1987a). Two salient features can be discerned in the $\delta^{13}C$ record shown in Fig. 3.

First, with the exception of one interval, the late Vendian Krol dolomites yield $\delta^{13}C$ values which are generally within the range of late Tertiary dolomites of unequivocal marine origin ($2.0 \pm 0.3‰$, $n = 18$) reported by Aharon et al. (1987b) from mid-oceanic carbonate platforms. The exception to the above can be discerned in the upper Krol D dolomites which contain a distinct positive $\delta^{13}C$ excursion of about 6‰ relative to PDB. Second, the apatites and their paired calcites from the chert-phosphorite lower Tal member yield anomalously negative $\delta^{13}C$ values ranging from

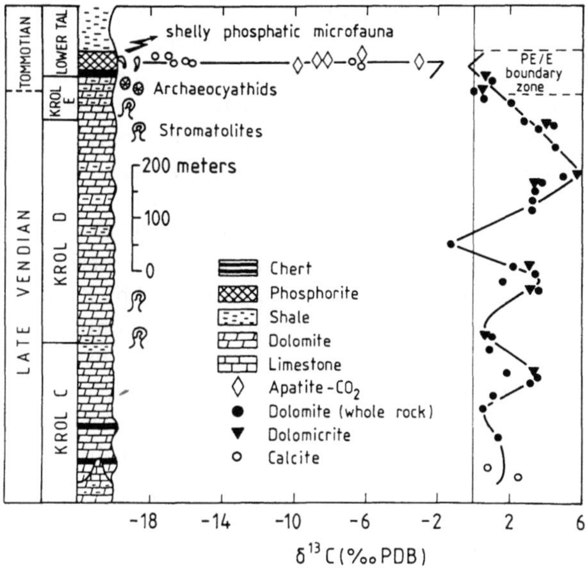

Fig. 3. $\delta^{13}C$ record of upper Krol carbonates (Aharon et al. 1987a) and lower Tal chert-phosphorite member (Banerjee et al. 1986). The line is defined by the isotopic compositions of early cements and homogenous micritic dolomites which best represent the contemporaneous marine $\delta^{13}C$ values (see text). Lithology after Aharon et al. (1987a). Fossil succession after Singh and Rai (1983) and Brasier and Singh (1987)

− 6.5 to − 9.9‰ and − 6.3 to − 17.8‰, respectively (Fig. 3). It is quite clear, however, that the major positive $\delta^{13}C$ excursion and its subsequent decline occurs within the same stromatolitic (algal mat) dolomite facies (upper Krol D/Krol E) before the sharp lithological changes occurring at the Krol E/lower Tal interface.

The large negative shift across the lithological boundary between organic-poor dolomites (C_{org} = 0.02 to 0.03%) and organic-rich phosphorites (C_{org} = 0.4 to 2.2%) is far from being coincidental, as it reflects the imprint of pore fluid evolution under two contrasting boundary conditions. The Krol dolomites are likely to have been formed in well aerated pore waters closely resembling the chemistry of the overlying seawater (Aharon et al. 1987a). The Tal lower member containing Kolodny's (1980) C − Si − P "trinity", bears the fingerprints of oxygen-deficient pore waters with a chemistry which was diagenetically modified by anaerobic bacterial processes (Irwin et al. 1977). The strong ^{13}C-depletions, coupled with the abundance of pyrite within the sediment (Banerjee et al. 1986), suggest that sulfate reduction was the dominant process affecting the chemistry of pore-water carbonate. It is conceivable that the source of P and Si for the phosphorite and chert, respectively, was from overlying sea-water enriched in these nutrient-type elements. This contention is supported by (1) the $\delta^{13}C$ difference of about 10‰ between the paired apatites and calcites (Fig. 3) suggesting the former were deposited before sulfate reduction commenced, and (2) the presence of the phosphatic shelly fauna indicating a readily available source of P in the water overlying the sediment. Having established the effect of lithology on the $\delta^{13}C$ compositions, next we examine the causes of the observed $\delta^{13}C$ shifts at the PC/C boundary.

4 Interpretation of $\delta^{13}C$ Offsets at the Terminal Precambrian

The $\delta^{13}C$ positive shift of about 6‰ relative to PDB in the uppermost Vendian and its subsequent decline toward the PC/C boundary (Figs. 2, 3) raise two intrinsic questions related to (1) the duration of the isotope excursion, and (2) the factor(s) which promoted it. Boundary limits on the time span are discussed first.

4.1 Estimated Duration of the Carbon Isotope Excursion

Since the upper Krol succession in the Lower Himalaya section was deposited on a stable platform dominated by algal laminites facies (Singh and Rai 1983), some indication of the time span represented by the $\delta^{13}C$ positive excursion may be gleaned from the rates of accretion of extant algal mats.

^{14}C determinations on both carbonates and organic phases in core sections taken through algal mats presently forming in restricted pools along the Red Sea indicate a net accretion rate of 0.18 mm a^{-1} (Schidlowski et al. 1985). On the premise that sedimentation rates in the Red Sea pools are typical of both extant and fossil algal mats, and allowing for a 50% compaction due to dewatering and organic matter oxidation, an accretion rate of 90 m/Ma can be estimated. It follows that the rise and subsequent decline of $\delta^{13}C$ in the upper Krol (Fig. 3) lasted for a minimum of 3 Ma (i.e., 300 m of section), on the admittedly tenuous assumption of a uniform sedimentation rate. The order of magnitude of the above estimate can be independently tested by exploiting the $^{87}Sr/^{86}Sr$ seawater evolution at the end Precambrian, shown to increase by 0.00255 units over a duration of about 90 Ma (Keto and Jacobsen 1987). Samples from the base of Krol C and the top of Krol E in the composite section of Fig. 3 yield $^{87}Sr/^{86}Sr$ ratios of 0.70925 and 0.70985, respectively (Aharon et al. 1987a), amounting to a Sr isotope change of 0.0006 units over an 800-m-thick section. It follows from this line of reasoning that the $\delta^{13}C$-positive excursion prior to the PC/C boundary lasted for a time interval not exceeding 7 Ma, in rough agreement with the estimate based on sedimentation rates.

4.2 Factors Controlling Carbon Isotope Compositions in a Ventilated Ocean

On a time scale that exceeds the turnover time of carbon in the exogenic carbon cycle ($\sim 2 \times 10^5$ a), $\delta^{13}C$ composition of a well-mixed ocean is controlled by the carbon fluxes into, and from, the carbonate (C_{carb}) and organic matter (C_{org}) reservoirs Schidlowski et al. 1975). Shifts of carbon between these two major carbon pools would affect the $\delta^{13}C$ balance in the ocean because of the pronounced ^{13}C-deficiency of sedimentary organic matter ($\delta^{13}C \approx -25‰$). The steady-state ratio of the carbon fluxes in the modern ocean is about $C_{org}/C_{carb} = 1/4$ ("Ronov ratio") which maintains an average oceanic carbon isotope composition close to zero permil. In addition, the $\delta^{13}C$ contrast between the surface-ocean and deep-ocean carbon pools, presently maintained at a level of 2.6‰ (Broecker 1982), is also a function of the nutrient fluxes expressed in term of $PO_4/\sum CO_2$ ("Broecker ratio"). This ratio in the modern, well-mixed ocean, is about 10^{-3}.

In any global $\delta^{13}C$ offsets of the marine carbonate record the deep ocean would have a leading role because it contains about 30 times more carbon then the surface-ocean and atmospheric CO_2 pools combined. In favorable circumstances, such as the mid Cretaceous ocean anoxic events (OAEs), the link between the $\delta^{13}C$ positive excursions of the ocean and the enhanced burial of organic matter can be established by the correspondence between the $\delta^{13}C$ shifts in both shallow and deep water marine carbonates, and the increased level of kerogen in the contemporaneous pelagic sediments (Arthur et al. 1985 a, b). Apparent absence of deep-sea sediments of Precambrian/early Cambrian age makes the interpretation of the $\delta^{13}C$ records in Fig. 2, based exclusively on shallow marine carbonates, more tenuous. For example, the 6‰ positive $\delta^{13}C$ shift in the late Vendian (Fig. 3) could be attributed to a change from 1/4 to 1/2.3 in the "Ronov ratio" (i.e., enhanced organic matter burial) or a change by a factor of 3 in the "Broecker ratio" (i.e., enhanced biomass fertility due to excess oceanic nutrients), if a carbon reservoir the size of the modern ocean is considered. This isotopic change could be considerably greater if a smaller ocean carbon mass is involved.

The question of the relative importance of increased marine net productivity versus enhanced marine organic matter preservation with respect to the observed ^{13}C-enrichments in late Vendian carbonates has been explored in a number of studies. Tucker (1986) and Magaritz et al. (1986) have singled out biomass fertility changes whereas Lambert et al. (1987) and Aharon et al. (1987 a) preferred an enhanced burial of organic matter. For the latter case, the episode of ^{13}C-enrichment prior to the PC/C boundary may be viewed as an extension of a period of enhanced organic preservation postulated by Knoll et al. (1986) in the upper Proterozoic, which contributed to the net release of O_2 just before the explosive evolution of the metazoa. However, thick sequences of organic-rich sediments of this age are yet to be found. Under these circumstances, it seems justified to examine the alternative that changes in biomass fertility and/or carbon burial occurred in an end-Precambrian ocean with restricted mixing.

4.3 The Isotope Effect of Biomass Fertility/Organic Preservation Shifts in a Stratified Ocean

Let us consider a simplified global carbon inventory where the size of the ocean-atmosphere carbon reservoir and the annual net flux of C_{org} are the only variables which control positive (or negative) shifts in the $\delta^{13}C$ composition of marine inorganic carbon. Using a simple time-dependent mixing model (see Appendix), it can be shown (Fig. 4) that a net flux of

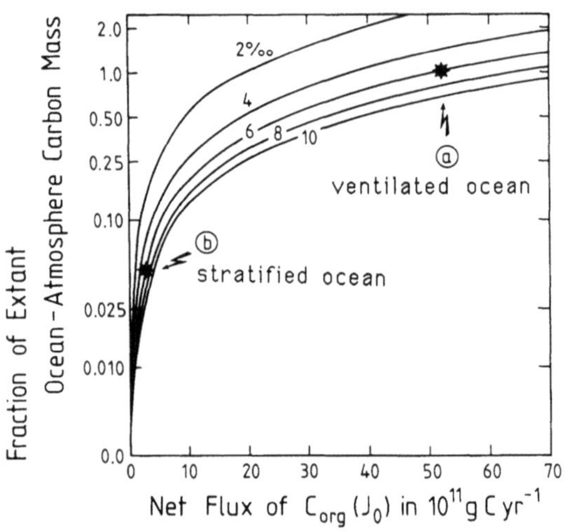

Fig. 4. Model-dependent $\delta^{13}C$ shifts of variable ocean-atmosphere inorganic carbon masses as a function of the net flux of organic carbon (J_o) having a $\delta^{13}C$ composition of −25‰. Duration of event is kept fixed at 1.5 Ma, the estimated half-life of the $\delta^{13}C$ positive excursion in Fig. 3 (see text and Appendix). C_{org} fluxes required to produce a 6‰ shift in a ventilated ocean versus a stratified ocean are represented by points *a* and *b*, respectively

50×10^{11} gC a^{-1} over a time interval of 1.5 Ma (i.e., the estimated half-life of the $\delta^{13}C$ positive excursion in the late Vendian) would be required to shift the $\delta^{13}C$ composition of the extant ocean-atmosphere carbon pool (390×10^{17} gC) by 6‰. Not surprisingly, the annual net flux of C_{org} required to produce the $\delta^{13}C$ shifts would decrease exponentially with the shrinking size of an aerated ocean layer isolated from the deep ocean by stratification (Fig. 4).

Under these circumstances, the question arises whether the late Precambrian biomass was of sufficient size to sustain the flux of C_{org} required to shift the $\delta^{13}C$ composition of the ocean by a large increment. On the basis of constraints imposed by fluvial supply of P on biomass productivity, it an be estimated that at present about 60% of the total organic carbon buried consists of terrestrial organic matter (Arthur et al. 1985a). Evidence seems to suggest that the mid-Cretaceous $\delta^{13}C$ positive excursion of 4‰ resulted from enhanced preservation of kerogen containing in excess of 50% terrestrial C_{org} (op. cit.).

Acritarchs, considered to be cysts of eukaryotic algae, have been reported from Mid to Late Proterozoic sediments (e.g., Vidal and Knoll 1983, Zang and Walter 1989). Their presence in sediments typical of offshore marine environments was taken as evidence for planktic biota in the Proterozoic ocean (op. cit.). However, the absence of "plankton oozes" in the Proterozoic contrasts sharply with the widespread occurrence of rich contemporaneous benthic algal biota evidenced by thick algal laminite deposits. Whereas the Proterozoic history of the plankton is still poorly known (Zang and Walter 1989), and caution is obviously necessary, we suggest on the basis of the existing evidence that plankton was unlikely to be a significant carbon producer in the Proterozoic ocean.

In the absence of late Precambrian terrestrial flora and organic soils (see later), positive $\delta^{13}C$ shifts must therefore be attributed to an unusual proliferation of microbial ecosystems comprising algal mats which quantitatively have constituted the most significant biota prior to the rapid diversification of the metazoa at the PC/C transition (Pratt 1982). However, this apparent proliferation left no particular imprint in the geologic record; and, moreover, it contradicts the prevailing view that the Late Proterozoic was marked by a substantial decline of stromatolite abundance and diversity (Awramik 1971; Walter 1989). Presently, algal mats are restricted to supratidal and hypersaline esoteric environments, and their former basinal to reefal habitats (including algal meadows and coral reefs: Pratt 1982) are now occupied by marine macrophytes which extend over an area of 2×10^{6} km^{2} (Smith 1981). Given an upper limit on primary production of 8.5 gC m^{-2} d^{-1} measured in extant algal mats (Hirsch 1978), the productivity of the dominant biota in the late Precambrian can be estimated at 62×10^{11} gC a^{-1}. It follows that a removal of C_{org} equivalent to 80% of the late Precambrian biota would be required to produce the observed 6‰ shift (point a, Fig. 4) which is quite implausible. By contrast, a more plausible C_{org}-burial rate (e.g., 5% of the biomass) would produce the 6‰ ^{13}C-enrichment in an aerated surface ocean layer containing only 6% of the oceanic carbon mass, and isolated from the deep-ocean carbon pool by stratification (point b, Fig. 4).

Following the above line of reasoning, it can be argued that the size of the oceanic biomass in the late Precambrian was inadequate to produce the observed ^{13}C-enrichments in a ventilated, well mixed ocean; and that shifts in biota fertility and concomitant fluxes of organic matter burial were effective in offsetting the $\delta^{13}C$ of the marine carbon because of episodic ocean stratification. In contrast to a well-mixed ocean, the main feedback mechanisms which tend to maintain the oceanic CO_2 level and its $\delta^{13}C$ composition at a steady state are severely restricted in a stratified ocean. Assuming a stratified ocean, enhanced CO_2 consumption by shallow marine biota, proportional to an increase in the nutrient level, would generate anomalously large $\delta^{13}C$ positive excursions documented in the late Precambrian. These results are consistent with the Ocean Anoxic Event model (OAE) for the terminal Precambrian proposed by Holser (1977) on the basis of the world-wide phosphorite event and the dramatic positive shift in oceanic $\delta^{34}S$ (the Yudomski event).

In summary, the proposed sequence of chemical events at the terminal Precambrian and their effect on the oceanic $\delta^{13}C$ history could have consisted of the following transient steps (Fig. 5):

1. Oceanic stratification during the latest Vendian with an aerated upper layer containing at least 6% of the total ocean-atmosphere carbon mass (i.e., about four times larger than the extant surface-ocean carbon mass) overlying anoxic deep basins scavenging nutrients (including C, P, SiO_2), and ^{32}S as sulfides. At this point, the main feedback mechanisms which tend to maintain the oceanic CO_2 level and its $\delta^{13}C$ composition at steady state are severely curtailed and a small flux of organic matter preservation produced by marine biota of modest size would result in a large $\delta^{13}C$ positive excursion in the aerated ocean layer. A contemporaneous $\delta^{13}C$ negative shift of unknown magnitude, attributed to organic matter degradation by sulfate-reducing bacteria, is postulated for the deep anoxic basins.

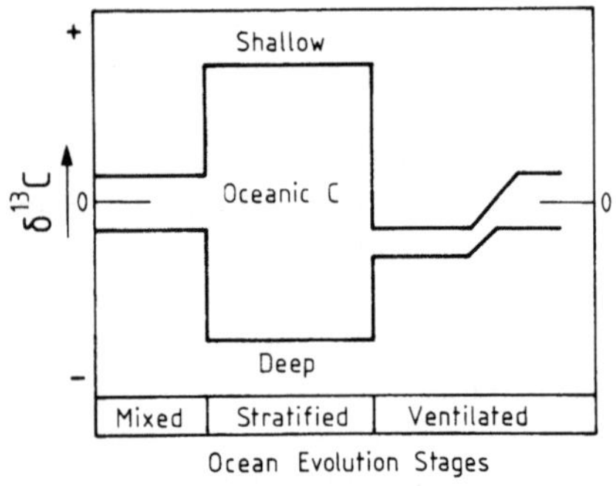

Fig. 5. Conceptual model illustrating the effect of ocean stratification and subsequent turnover on the $\delta^{13}C$ compositions of contemporaneous shallow and deep-ocean inorganic carbon pools. A $\delta^{13}C$ contrast of 2.6‰ between the surface and deep-ocean carbon reservoirs is typical of the "mixed" ocean stage (Broecker 1982). During ocean stratification, the $\delta^{13}C$ contrast is amplified by the concomitant but opposite changes in the two oceanic carbon pools (see text). The "ventilated" ocean stage is represented by the restoration of the "normal" $\delta^{13}C$ contrast between the surface and deep-ocean carbon pools that follows the turnover event

2. The stratification event is followed by progressive ventilation of the deep anoxic basins, and complete turnover at the PC/C boundary. Flooded coastlines with favorable configurations would "trap" the excess phosphate brought to the surface by the turnover (Cook and Shergold 1984), and a transient negative $\delta^{13}C$ excursion would occur on account of the upwelled ^{12}C-enriched CO_2.

3. Establishment of a new steady state with a subsequent restoration of the "normal" $\delta^{13}C$ contrast between the upper ocean layer and the deep ocean ensued.

4. Metazoan "explosive" radiation triggered by the opening of vast vacant habitats (Brasier 1982).

The sequence of chemical events at the PC/C boundary attributed here to "earth-bound" factors thus conflicts with the scenario of Hsu et al. (1985) of an extraterrestrial catastrophic event. Rapid fluctuations in $\delta^{13}C$ marine values accompanied by Ir anomalies at the China C boundary (Fig. 2a) are not unequivocal fingerprints of an impact event but may well represent the effect of turnover of a stratified ocean (Keith 1982) on the latest Vendian depositional environments. The Ir excursion argument is weakened considerably because the "anomalies" tend to "vanish" when normalized against the Al_2O_3 content of the clays, as recently demonstrated by Orth et al. (1986) at the terminal Ordovician (note the sharp lithological change to black shales coinciding with the Ir anomaly at the top of the Chinese sections, Fig. 2b). In this context, the comformable transition from the algal mat dolomites (Krol succession) to the chert-phosphorite-black shale association (Tal lower member) in the Lesser Himalaya section (Fig. 3) may represent distinct lithological expressions of this oceanic anoxic event. However, the link between the chemical events postulated above and the dramatic faunal turnovers at the PC/C boundary remains to be proven.

5 Preservation of Isotopic Memory in Precambrian Carbonates

Stable isotopes tend to preserve an infinite memory of ambient depositional environments for as long as the minerals and rocks in which they are occluded are left undisturbed by postdepositional diagenetic and/or metamorphic processes. With carbonates of Precambrian antiquity, the preservation of pristine isotope compositions may be questionable since surficial meteoric water, deep-burial diagenesis and low-grade metamorphism had ample time and opportunity to alter the original isotope ratios. The "smoothing" of isotope records due to postdepositional effects is particularly potent at times when the alteration transgresses intervals of rapid isotope shifts and/or pronounced high-amplitude fluctuations (Veizer 1989). The quality of isotope records, such as the ones shown in Fig. 2, and the validity of their interpretation are therefore heavily dependent on the postdepositional history of the samples and the degree of preservation of their isotopic compositions.

Several strategies have been exploited with various degrees of success to avoid, or at least to minimize, the inimical effects of postdepositional alteration on stable isotope ratios, most particularly those of oxygen and carbon. Veizer (1983) has devised criteria for the recognition of the "least altered samples" based on elemental chemistry (including Sr/Ca ratios and Mn and Fe levels) of limestones and dolomites. Although Veizer's method offers a relatively rapid screening, the least altered samples would provide only an upper limit for the isotope evolution of the coeval seawater. Another method of sample selection, particularly suited for $\delta^{13}C$ assays, is based on the assumption of an apparent constancy in $\delta^{13}C$ fractionation between coexisting carbonate-kerogen pairs. Knoll et al. (1986) concluded on the basis of 92 upper Proterozoic carbonate-kerogen pairs that samples yielding a $\delta^{13}C$ contrast of 28.5 ± 2‰ would preserve primary carbon isotope compositions. However, documentation of variable isotope fractionations during marine photosynthesis, and carbonate-kerogen $\delta^{13}C$ fractionation

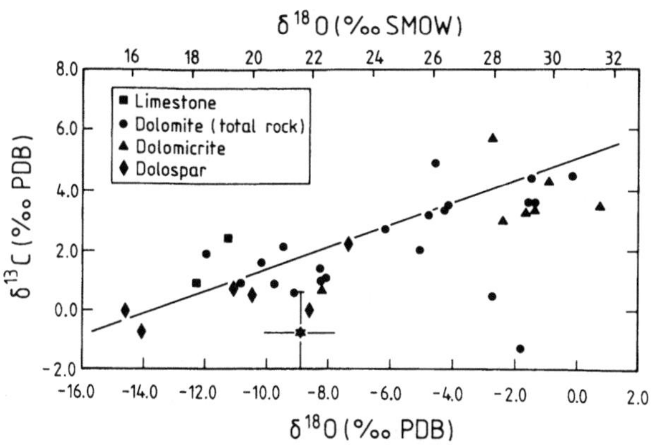

Fig. 6. Oxygen and carbon isotope relationship of early and late dolomite phases coexisting in the Krol dolomites. *Cross* represents the end-Precambrian seawater estimate of Veizer and Hoefs (1976) on the basis of whole-rock carbonates. The *linear array* represents a mixing line between ^{18}0 and ^{13}C-enriched early dolomite phase and a late dolomite phase typified by anomalously ^{18}O and ^{13}C-depleted isotopic compositions (see text)

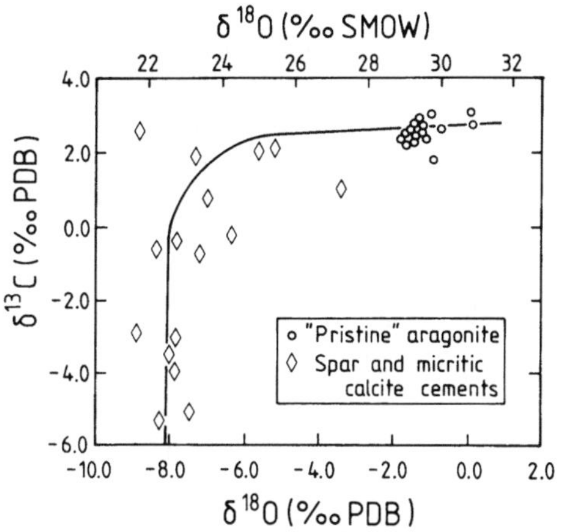

Fig. 7. Oxygen and carbon isotope relationship in late Quaternary reef limestones from Huon Peninsula, New Guinea, showing difference between pristine aragonites and diagenetically-altered late calcite phases. The *inverse "J" shape* is produced by addition of ^{13}C-depleted soil CO_2 during diagenesis coupled with thermodynamic isotope equilibration of the recrystallized calcites with ^{18}O-depleted meteoric water

greater than 28.5‰ in pre-Tertiary times (Arthur et al. 1985b), adds a significant complication to this screening procedure. A third, and perhaps more prudent method of sample selection involves recognition and physical separation of components of the primary rock fabric on the basis of petrography (Tucker 1982; Aharon et al. 1987a). It is our opinion that petrographic "screening" of samples is most rewarding and best suited for Precambrian carbonates with a postdepositional history complicated by successive phases of dolomitization, as illustrated below.

Fabric components coexisting in the Krol dolomites (Fig. 3) form a linear array in coordinates of $\delta^{18}O/\delta^{13}C$ (Fig. 6) that represents a mixing line between two end-members which are vastly different on petrographic and isotopic grounds. At one end of the spectrum there are the "early" dolomicrites which show detailed preservation of the primary fabrics of the sediments (e.g., cryptalgal laminites, fenestrae, ooids, calcareous algae), and yield isotopic compositions ranging from -4 to 1‰ and 2 to 6‰ in $\delta^{13}C$, respectively. At the other end of the spectrum there are "late" diagenetic dolospars which show void-filling and destructive fabrics, and yield anomalous ^{18}O and ^{13}C-depleted values ranging from -8 to -14‰ and -1 to 2‰ in $\delta^{18}O$ and $\delta^{13}C$, respectively. The coexistence of dolomite phases with vastly different isotopic compositions, and the preservation of primary fabrics, rules out postdepositional isotope exchange as a major control on the compositions of the early dolomicrite phases. This result agrees with well-documented Phanerozoic case studies (see papers in Zenger et al. 1980), which indicate that late dolomitization is fabric destructive whereas early dolomitization of metastable carbonate precursors (e.g., high-Mg calcite) tends to preserve the original fabric. In such a case, early dolomites either inherit the $\delta^{13}C$ composition of their marine precursor or, if the fluid flow system which delivers Mg to the sites of dolomitization is sufficiently fast (i.e., high fluid/rock ratio), it might directly represent the $\delta^{13}C$ composition of the overlying seawater (Aharon et al. 1987b). However, the linear relationship observed between $\delta^{18}O$ and $\delta^{13}C$ values in Fig. 6 does not offer evidence of primary dolomite precipitation from late Precambrian seawater (and by implication a late Precambrian seawater chemistry different than present) as proposed by Tucker (1982) in his study of the Precambrian Beck Spring Dolomite. Rather it attests to the conspicuous absence of soil-derived CO_2 prior to the advent of the land plants. This is because CO_2 produced by either aerobic or anaerobic bacteria is conspicuously depleted in ^{13}C (Irwin et al. 1977) and tends to drive the $\delta^{18}O$-$\delta^{13}C$ system into an inverse "J" pattern, as illustrated in Fig. 7 by a well documented late Quaternary reef-carbonate succession from Papua-New Guinea (Aharon and Chappell 1986).

By scaling the mean $\delta^{18}0$ compositions of the early dolomicrite phases in Fig. 6 (-2‰) to the values of Tertiary dolomites of definite seawater provenance ($\delta^{18}O = 2.8$‰, Aharon et al. 1987b) it may be inferred that late Precambrian seawater was more depleted in

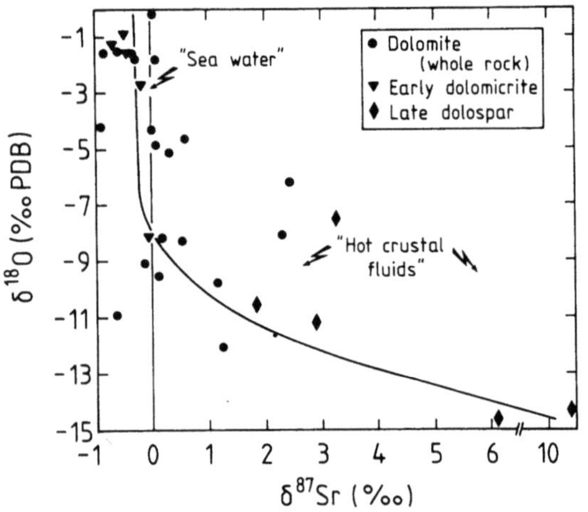

Fig. 8. Cross-plot of $^{18}O/^{16}O$ and $^{87}Sr/^{86}Sr$ compositions of early and late dolomite phases coexisting in the Krol dolomites. Note that whereas early dolomites are scattered around a $\delta^{87}Sr$ value of 0‰ (i.e., seawater value), late dolomites show large positive $\delta^{87}Sr$ excursions due to radiogenic ^{87}Sr enrichments. By analogy to $\delta^{18}O$, $\delta^{87}Sr$ is defined as the departure of $^{87}Sr/^{86}Sr$ in the sample from the modern seawater ratio ($^{87}Sr/^{86}Sr = 0.70920$: Elderfield 1986)

^{18}O by 4.8‰ (or alternatively it was 20 °C warmer) than the Tertiary ocean, and not by 12‰, as indicated by Veizer and Hoefs (1976) on the basis of whole-rock analyses. The late dolomite phases which are significantly more depleted in ^{18}O and ^{13}C relative to the early dolomite phases (Fig. 6), are also substantially enriched in radiogenic ^{87}Sr (Fig. 8) and therefore preserve the imprint of hot fluids (i.e., preferential incorporation of the ^{16}O isotope in the carbonate phase at high temperature) of crustal provenance (i.e., high $^{87}Sr/^{86}Sr$ ratios: Faure 1986) which probably advected through the sediments during burial. On the basis of the above evidence, the prevailing dogma that postdepositional diagenesis would impair the $\delta^{18}O$ systematics but leave the $\delta^{13}C$ values of the whole rock pristine (e.g., Magaritz and Schulze 1980) should be reevaluated.

6 Conclusions

1. Remarkable similarities are observed between the patterns of $\delta^{13}C$ variations imprinted in stratigraphically continuous sedimentary successions of wide geographic distribution that span the Precambrian/Cambrian transition. Therefore, time-bound $\delta^{13}C$ excursions have considerable potential as chronostratigraphic tools which are independent of the controversial biostratigraphic marker definitions currently used to delineate the PC/C boundary.

2. The isotope records generally show a bimodal distribution of $\delta^{13}C$ compositions, with a ^{13}C-enriched mode during the latest Vendian and a sharp transition to a ^{13}C-depleted mode at, or slightly above, the inferred PC/C boundary. The transition between the $\delta^{13}C$ modes seems to coincide with sharp lithological changes in all sections examined here. However, a detailed examination of the $\delta^{13}C$/lithology relationship in the Lesser Himalaya section indicates that the major positive $\delta^{13}C$ shift of 6‰ relative to PDB, and its subsequent decline, occurs before the sharp lithological change from algal dolomite facies to the conformably overlying chert-phosphorite-black shale facies. The anomalously negative $\delta^{13}C$ values (down to -18‰) yielded by apatites and calcites in the phosphorite unit are attributed to deposition in oxygen-deficient environments where bacterial sulfate reduction was intense.

3. In the absence of absolute radiometric ages, the duration of the $\delta^{13}C$ excursion at the PC/C boundary is estimated to have lasted for a minimum of 3 Ma on the basis of rates of accretion of extant algal mats. The order of magnitude of the time estimate is in reasonable agreement with estimates derived from the contemporaneous $^{87}Sr/^{86}Sr$ seawater evolution curve at the end Precambrian.

4. Factors controlling the oceanic $\delta^{13}C$ composition at the PC/C boundary were investigated under two boundary conditions, viz. a ventilated, well-mixed ocean, and a stratified ocean. Using quantitative arguments derived from a simplified model of the carbon cycle, it is concluded that late Precambrian organic biomass was probably of insufficient size to sustain the flux of C_{org} required to shift the $\delta^{13}C$ composition of a ventilated ocean by 6‰. In contrast, biota fertility shifts and concomitant burial fluxes of organic matter amounting to 5% of the late Precambrian biomass were effective in offsetting $\delta^{13}C$ in an aerated surface ocean layer containing 6% of the oceanic carbon mass, and isolated from the deep-ocean carbon pool by stratification. The stratified ocean model proposed here is consistent with the Ocean Anoxic Event (OAE) model for the terminal Precambrian proposed by Holser (1977) on the basis of postulated world-wide phosphorite event and a dramatic positive shift in oceanic $\delta^{34}S$.

5. An understanding of the process of dolomitization is essential for the correct interpretation of the $\delta^{13}C$ records because of a predominance of the dolomite lithology at the terminal Precambrian. Using the Lesser Himalaya section as a case study, it can be shown that samples contain at least two episodes of dolomitization and that only the early phase has the sea-water imprint. Therefore petrographic "screen-

ing" of samples coupled with separation of the early dolomite phases prior to analysis, are prerequisites for avoiding the inimical effects of postdepositional alteration on the stable isotopes, particularly in the case of Precambrian carbonates with histories complicated by successive phases of dolomitization. On this basis, the prevailing view that diagenesis would impair only the $\delta^{18}O$ system, leaving the whole-rock $\delta^{13}C$ system pristine must be reevaluated.

6. The question of biomass fertility in a ventilated versus stratified ocean was brought into focus, but more rigorous tests are necessary to validate the inference of an Ocean Anoxic Event (OAE) at the PC/C boundary. One persisting deficiency is the poor time constraints on the postulated events at the terminal Precambrian. A significant improvement could be achieved by exploiting the chronostratigraphic potential of the proposed monotonic $^{87}Sr/^{86}Sr$ increase in seawater at the end of the Precambrian, as has been done using $^{87}Sr/^{86}Sr$ ratios in Tertiary marine carbonates (Elderfield 1986). Work in progress by Liew and Aharon (in prep.), aiming to produce a high-resolution $^{87}Sr/^{86}Sr$ seawater record for the end Precambrian based on stratigraphically continuous sections, shows considerable promise. The OAE hypothesis at the terminal Precambrian should be tested further using the redox sensitivity of rare earth elements such as Ce which leave a distinct mark in phosphate minerals (Wright 1989).

7. The exact relationship between the postulated ocean-stratification event at the end-Precambrian, the turnover of the ocean at the base of Cambrian, and the documented contemporaneous faunal extinction and radiation events at the PC/C boundary remains to be explored.

Acknowledgments. We thank Manfred Schidlowski for the invitation to attend the Maria Laach meeting, and for his useful comments on the manuscript. The final version of the paper has greatly benefited from the constructive criticism of the Australian "hit squad" of reviewers headed by the Section Editor David McKirdy, and including R. J. F. Jenkins, Terry Donnelly, and Ian Lambert. This study was made possible by a Max-Planck Gesellschaft fellowship to the senior author during 1989.

7 Appendix

Consider an ocean-atmosphere carbon pool in which the inorganic carbon component C_i is the sum of the rare ($^{13}C_i$) and abundant ($^{12}C_i$) isotopes:

$$C_i = {}^{13}C_i + {}^{12}C_i. \tag{1}$$

Given that isotopic ratio $R_i = {}^{13}C_i/{}^{12}C_i$, then

$$^{13}C_i = C_i \frac{R_i}{(1 + R_i)} \tag{2}$$

and

$$^{12}C_i = C_i \frac{1}{(1 + R_i)}. \tag{3}$$

Mixing a carbon source in which $R_0 \neq R_i$ with the ocean-atmosphere carbon pool would result in an isotopic ratio:

$$R_f = \frac{({}^{13}C_i + {}^{13}C_o)}{({}^{12}C_i + {}^{12}C_o)} = \frac{{}^{13}C_f}{{}^{12}C_f}. \tag{4}$$

Replacing Eqs. (2) and (3) into (4):

$$R_f = \frac{C_i \dfrac{R_i}{(1 + R_i)} + C_o \dfrac{R_o}{(1 + R_o)}}{C_i \dfrac{1}{(1 + R_i)} + C_o \dfrac{1}{(1 + R_o)}}. \tag{5}$$

At any time "t", the carbon mass admixed into the ocean would be a function of the flux rate (J_o) in units of mass/time

$$C_o(t) = t J_o. \tag{6}$$

Substituting (6) into (5) and solving for J_o:

$$J_o = C_i \frac{1}{t} \frac{(1 + R_o)(R_i - R_f)}{(1 + R_o)(R_f - R_o)}. \tag{7}$$

Using Craig's (1957) definition of δ:

$$\delta(\text{‰ PDB}) = \left(\frac{R}{R_{std}} - 1\right) 1000, \tag{8}$$

where

$$R = R_{std}\left(1 + \frac{\delta}{1000}\right) \tag{9}$$

and $R_{std} = 0.0112372$. Substituting (9) into (7):

$$J_o = C_i \frac{1}{t} \frac{\left[1 + R_{std}\left(1 + \dfrac{\delta_o}{1000}\right)\right](\delta_i - \delta_f)}{\left[1 + R_{std}\left(1 + \dfrac{\delta_i}{1000}\right)\right](\delta_f - \delta_o)}. \tag{10}$$

Equation (10) allows calculation of the flux rate J_o required to achieve an isotope shift in the ocean-atmosphere carbon mass (C_i) from δ_i to δ_f in time t (see Fig. 4).

The simplified model assumes that isotopic compositions of the resulting mixtures are not modified by fractionation processes after mixing has occurred. This limitation simplifies the equations without compromising the utility of the model.

References

Aharon P, Chappell J (1986) Oxygen isotopes, sea level changes and the temperature history of a coral reef environment in New Guinea over the last 10^5 yrs. Palaeogeogr Palaeoclimatol Palaeoecol 56: 337–379

Aharon P, Schidlowski M, Singh IB (1987a) Chronostratigraphic markers in the end-Precambrian carbon isotope record of the Lesser Himalaya. Nature (London) 327: 699–702

Aharon P, Socki RA, Chan L (1987b) Dolomitization of atolls by sea water convection flow: test of a hypothesis at Niue, South Pacific. J Geol 95: 187–203

Arthur MA, Dean WE, Schlanger SO (1985a) Variations in the global carbon cycle during the Cretaceous related to climate, volcanism, and changes in atmospheric CO_2. In: Sundquist ET, Broecker WS (eds) The carbon cycle and atmospheric CO_2: natural variations Archean to Present. Geophysical monograph 32. Am Geophys Union, Washington DC, pp. 504–529

Arthur MA, Dean WE, Claypool GE (1985b) Anomalous ^{13}C-enrichment in modern marine organic matter. Nature (London) 315: 216–218

Awramik SM (1971) Precambrian columnar stromatolite diversity: reflection of metazoan appearance. Science 174: 825–827

Awramik SM (1986) The Precambrian-Cambrian boundary and geochemical perturbations. Nature (London) 319: 696

Banerjee DM, Schidlowski M, Arneth JD (1986) Genesis of upper Proterozoic-Cambrian phosphorite deposits of India: isotopic inferences from carbonate fluorapatite, carbonate and organic carbon. Precambrian Res 33: 239–253

Bjorlykke K (1982) Correlation of late Precambrian and early Paleozoic sequences by eustatic sea level changes and the selection of the Precambrian-Cambrian boundary. Precambrian Res 17: 99–104

Brasier MD (1980) The lower Cambrian transgression and glauconite-phosphate facies in western Europe. J Geol Soc London 137:695–703

Brasier MD (1982) Sea level changes, facies changes and the late Precambrian-early Cambrian evolutionary explosion. Precambrian Res 17:105–123

Brasier MD, Singh P (1987) Microfossils and Precambrian-Cambrian boundary stratigraphy at Maldeota, Lesser Himalaya. Geol Mag 124:323–345

Broecker WS (1982) Ocean chemistry during glacial time. Geochim Cosmochim Acta 46:1689–1705

Cook PJ, Shergold JH (1984) Phosphorus, phosphorites and skeletal evolution at the Precambrian-Cambrian boundary. Nature (London) 308:231–236

Cowie JW, Glaessner MF (1975) The Precambrian-Cambrian boundary: a symposium. Earth Sci Rev 11:209–251

Cowie JW, Rozanov AY (1983) Precambrian-Cambrian boundary candidate, Aldan River, Yakutia, USSR. Geol Mag 120:129–139

Craig H (1957) Isotopic standards for carbon and oxygen and correction factors for mass spectrometric analysis of carbon dioxide. Geochim Cosmochim Acta 12:133–149

Daly RA (1909) First calcareous fossils and the evolution of the limestones. Am Geol Soc Bull:153–170

Degens ET, Kazmierczak J, Ittekkot V (1986) Biomineralization and the carbon isotope record. Tschermaks Mineral Petrol Mitt 35:117–126

Donovan SK (1987) The fit of the continents in the late Precambrian. Nature (London) 327:139–141

Elderfield H (1986) Strontium isotope stratigraphy. Palaeogeogr Palaeoclimatol Palaeoecol 57:71–90

Faure G (1986) Principles of isotope geology. John Wiley & Sons, New York, 587 pp

Goodfellow WD (1986) Anoxic oceans and short-term carbon isotope trends. Nature (London) 322:116–117

Hirsch P (1978) Microbial mats in a hypersaline Solar Lake: Types, composition and distribution. In: Krumbein WE (ed) Environmental biogeochemistry and geomicrobiology. The aquatic environment 1. Ann Arbor Science, Michigan, pp 189–201

Holser WT (1977) Catastrophic chemical events in the history of the ocean. Nature (London) 267:403–408

Hsu KJ (1986) The Precambrian-Cambrian boundary and geochemical perturbations–replies. Nature (London) 319:697

Hsu KJ, Oberhansli H, Gao JY, Shu S, Haihong C, Krahenbuhl U (1985) "Strangelove" ocean before the Cambrian explosion. Nature (London) 316:809–811

Irwin H, Curtis C, Coleman M (1977) Isotopic evidence for source of diagenetic carbonates formed during burial of organic-rich sediments. Nature (London) 269:209–213

Keith ML (1982) Violent volcanism, stagnant oceans and some inferences regarding petroleum, strata-bound ores and mass extinctions. Geochim Cosmochim Acta 46:2621–2637

Keto LS, Jacobsen SB (1987) Nd and Sr isotopic variations of early Paleozoic oceans. Earth Planet Sci Lett 84:27–41

Kolodny Y (1980) Carbon isotopes and depositional environment of a high productivity sedimentary sequence–the case of the Mishash-Ghareb Formations, Israel. Isr J Earth Sci 29:147–156

Knoll AH, Butterfield NJ (1989) New window on Proterozoic life. Nature (London) 337:602–603

Knoll AH, Hayes JM, Kaufman AJ, Swett K, Lambert IB (1986) Secular variation in carbon isotope ratios from upper Proterozoic successions of Svalbard and East Greenland. Nature (London) 321:832–838

Lambert IB, Walter MR, Wenlong Z, Songnian L, Guogan M (1987) Palaeoenvironment and carbon isotope stratigraphy of upper Proterozoic carbonates of the Yangtze Platform. Nature (London) 325:140–142

Magaritz M (1989) ^{13}C minima follow extinction events: a clue to faunal radiation. Geology 17:337–340

Magaritz M, Schulze KH (1980) Carbon isotope anomaly of the Permian period. Contrib Sedimentol 9:269–277

Magaritz M, Holser WT, Kirschvink JL (1986) Carbon isotope events across the Precambrian/Cambrian boundary on the Siberian Platform. Nature (London) 320:258–259

Matthews SC, Cowie JW (1979) Early Cambrian transgression. J Geol Soc London 136:133–135

McMenamin MAS (1987) The emergence of animals. Sci Am 256:94–102

Morris SC, Bengtson S (1986) The Precambrian-Cambrian boundary and goechemical perturbations. Nature (London) 319:696–697

Orth CJ, Gilmore JS, Quintana LR, Sheehan PM (1986) Terminal Ordovician extinction: geochemical analysis of the Ordovician/Silurian boundary, Anticosti Island, Quebec. Geology 14:433–436

Pratt BR (1982) Stromatolite decline–a reconsideration. Geology 10:512–515

Schidlowski M, Eichmann R, Junge CE (1975) Precambrian sedimentary carbonates: carbon and oxygen isotope geochemistry and implications for the terrestrial oxygen budget. Precambrian Res 2:1–69

Schidlowski M, Matzigkeit U, Mook WG, Krumbein WE (1985) Carbon isotope geochemistry and ^{14}C ages of microbial mats from the Gavish Sabkha and the Solar Lake. In: Friedman GM, Krumbein WE (eds) Hypersaline ecosystems. Ecological studies 53. SpringerBerlin Heidelberg New York, pp 381–401

Singh IB, Rai V (1983) Fauna and biogenic structures in Krol-Tal succession (Vendian-early Cambrian), Lesser Himalaya: their biostratigraphic and palaeoecological significance. J Palaeontol Soc India 28:67–90

Smith SV (1981) Marine macrophytes as a global carbon sink. Science 211:838–840

Tucker ME (1982) Precambrian dolomites: petrographic and isotopic evidence that they differ from Phanerozoic dolomites. Geology 10:7–12

Tucker ME (1986) Carbon isotope excursions in Precambrian/Cambrian boundary beds, Morocco. Nature (London) 319:48–50

Vidal G, Knoll AH (1983) Proterozoic plankton. Mem Geol Soc Am 161:265–277

Veizer J (1983) Trace elements and isotopes in sedimentary carbonates. Rev Mineral 11:265–300

Veizer J (1989) Strontium isotopes in sea water through time. Annu Rev Earth Planet Sci 17:141–167

Veizer J, Hoefs J (1976) The nature of $^{18}O/^{16}O$ and $^{13}C/^{12}C$ secular trends in sedimentary carbonate rocks. Geochim Cosmochim Acta 40:1387–1395

Walter MR (1989) Major features in record of Proterozoic stromatolites. 28th Int Geol Congr Abstr 3:318–319

Wright J (1989) REE in fossil apatite record Phanerozoic OAE. 28th Int Geol Congr Abstr 3:383–384

Yanshin AL, Zharkov MA (1986) Epochs and evolution of phosphate deposition through geologic time. Int Geol Rev 28:390–401

Zang WL, Walter MR (1989) Latest Proterozoic plankton from the Amadeus Basin in central Australia. Nature (London) 337:642–645

Zenger DH, Dunham JB, Ethington RL (eds) (1980) Concepts and models of dolomitization. SEPM Spec Publ 28, 320 pp

Stratiform and Globular Organic Matter in the Lower Proterozoic Metasediments at Elliot Lake, Ontario, Canada

Bartholomew Nagy [1] and David J. Mossman [2]

[1] Laboratory of Organic Geochemistry, Department of Geosciences, The University of Arizona, Tucson, Arizona 85721, USA
[2] Department of Geology, Mount Allison University, Sackville, New Brunswick, Canada EOA 3 CO

Abstract

The lower Huronian strata of the Elliot Lake-Blind River region in Ontario are a good example of the complex evolutionary, diagenetic, chemical, and mobilization processes involved in kerogen paragenesis in ore-bearing metasediments. The present study was undertaken to delineate some of the relationships at this location between uranium-poor kerogen globules and nearby U-rich stratiform kerogen seams. The latter are syngenetic with the host rocks and are probably the progenitors of the younger kerogen globules. Several of the kerogen globules have the morphologies of solidified liquid droplets although some of the other chemically and physically equivalent particles are more angular. Petrological studies suggest that the stratiform kerogens and their host rocks at Elliot Lake underwent diagenetic and mild metamorphic processes, including hydrothermal episode(s) leading to mineral replacement and paragenesis which involved the evolution of the globular kerogens*. This study confirms previous reports which noted that the detrital uraninite grains in the stratiform layers were fractured and invaded by mobilized organic matter. This kerogen at the time of mobilization appears to have been a viscous, liquid substance. Therefore, it is probable that some of the mobilized organic matter also moved away from the stratiform layers under reducing environments to form the globular kerogens, leaving most of the uranium behind. Subsequently, the movement of the globular substance was arrested and it resolidified in situ. The globular kerogens have higher H/C atomic ratios than the stratiform kerogens. The results of vacuum pyrolysis-gas chromatographic-mass spectrometric analyses are compatible with an evolutionary scheme of the globular kerogens from the stratiform layers by thermochemical reactions; one possibility is hydrous pyrolysis.

* Globular kerogen designates solidified bitumen.

1 Introduction

Evolutionary pathways of kerogens in Precambrian sediments have not yet been studied in detail. Sufficient observations are available, however, to consider two Precambrian uraniferous ore deposits associated with organic matter which display certain similarities with the Lower Proterozoic kerogens at Elliot Lake. These examples are useful in attempting to evaluate the nature of kerogens at Elliot Lake. The first example is the thin layers of kerogen in the Witwatersrand Supergroup. These contain more gold but less uranium than the less abundant stratiform kerogens at Elliot Lake (Zumberge et al. 1978; Simpson and Bowles 1981; Willingham et al. 1985). The Witwatersrand kerogens are generally considered to be remnants of Archean cyanobacterial mats (e.g., Pretorius 1974; Zumberge et al. 1978) and show, in several instances, limited remobilization (Willingham et al. 1985). Those at Elliot Lake show suggestive evidence of cyanobacterial mat origin (Mossman and Dyer 1985; Willingham et al. 1985; Dyer et al. 1988) and substantial kerogen remobilization. The depositional environments of the Witwatersrand (Tankard et al. 1982; Smith and Minter 1980) and the lower Huronian sediments (Fralick and Miall 1981; Colvine 1981; Mossman and Harron 1984) were to some extent similar. Both deposits were subjected to diagenesis and, at least at certain locales, to mild metamorphic temperatures (Willingham et al. 1985; Phillips 1987). The second example described by Leventhal et al. (1987) is the uraniferous-organic deposit at Cluff Lake, Canada, where some of the minerals were broken off from the walls of veins and became incorporated into the mobile Proterozoic organic matter as it filled the veins.

The depositional environments and geological history of the Lower Proterozoic Huronian Supergroup (2.25–2.45 Ga old; Roscoe 1973) have been described by Robertson (1978), Colvine (1981) and Mossman and Harron (1984). Detrital uranium minerals, together with some detrital pyrite, were

M. Schidlowski et al. (Eds.)
Early Organic Evolution: Implications for Mineral and Energy Resources

concentrated, mainly by hydraulic processes in the basal Matinenda Formation in the Elliot Lake-Blind River area on the western flank of the Supergroup. Uraniferous kerogens in the Matinenda clastic metasediments at Elliot Lake consist mainly of two types based on their physical and chemical characteristics (Ruzicka and LeCheminant 1984; Mossman 1987). Kerogens of the first group occur in ore as thin discontinuous stratiform seams ($\lesssim 0.5$ cm thick), as matrices or clasts in quartz pebble conglomerates, and as disseminated flakes and subangular grains. Kerogens of these morphological types constitute the stratiform substance, and probably its broken-up fragments. Kerogens of the second group occur as either globules (they may be as large as 5 mm in diameter), cement or fracture/vein fillings. Several of these kerogen grains resemble the shapes of solidified liquid droplets. They generally lack mineral inclusions, except for scattered microlites. The remobilized and globular kerogens in the Denison mine (Denison Mines Limited) and the rehabilitated Stanleigh mine (Rio Algom Limited) are adjacent to or at most only several meters from the stratiform kerogen layers.

2 Samples, Experimental Procedures

The samples consisted of two stratiform kerogen seams (designated RAD 74-11; RAD 83-01) and four globular, or partially globular, kerogens (RAD 84-503; RAD 84-507; No. 38983; No. 95355). All six were obtained from the mines noted above. Samples were removed from behind the walls of the mines with hammer and chisel in order to minimize contamination. The stratiform kerogens were brownish black bands ($\sim$ 4 mm thick) containing detrital quartz, uraninite, some pyrite, clay, probably illite. The kerogen RAD 84-503 consisted of black globular particles (0.5–3 mm diameter). RAD 84-507 comprised moderately rounded, occasionally elongate, brownish black globules, ranging in size from 0.3 to 3.0 mm. The mainly globular kerogens No. 38983 and No. 95355 resembled RAD 84-503 and fell in the same size range.

Analyses were performed for some of the major, minor and trace elements and mineral components of these six Elliot Lake-Blind River kerogens. The characteristic organic moieties of both stratiform samples and of the globular kerogens RAD 84-503 and RAD 84-507 were identified. The analytical procedures were the same as those used by Willingham et al. (1985), except as noted below.

Nonmineral O was determined directly (not by difference) with a TC-136 model elemental analyzer at the Institute for Materials Research at McMaster University. O and H values were, of course, corrected for O and H from minerals determined by X-ray diffraction. Mineral separations were carried out prior to determination of the nonmineral O contents of the kerogens. Sample aliquots were first sieved. The < 60 mesh fraction was washed with a dispersing agent in H_2O, decanted to remove material with a specific gravity < 1.0, and then thoroughly washed with acetone in an ultrasonic cleaner for 3 min to quantitatively remove the dispersing agent. Next, the dried samples were subjected to flotation separation in acetylene tetrabromide diluted with acetone to a specific gravity of 1.6. Again the samples were thoroughly washed to quantitatively remove the organic liquids. Although microscopic examination of the concentrated kerogens showed them to be quite clean, complete removal of mineral matter was probably not accomplished due to the intimate intergrowth of some minerals with the kerogens.

Great care was taken to remove organic contamination before molecular microanalyses. The samples were powdered and then were first extracted with aqueous 0.06 N HCl, followed by extraction with benzene and methanol to remove soluble contaminants. The aqueous HCl extracts were checked for amino acid contamination (amino acids are not sufficiently stable to survive during the long geological history of Precambrian rocks). The benzene and methanol extracts were analyzed for long-chain n-alkanes and isoprenoids, which could indicate contamination by petroleum products in the mines. The sample extracts were analyzed with a Hewlett-Packard 5880 A gas chromatograph equipped either with a SP-2100 (methyl silicone stationary phase, fused silica capillary column, 30 m long, 0.25 mm i.d.) for the detection of hydrocarbons or with a Chirasil-Val III capillary column (chiral silicone stationary phase, 25 m long, 0.25 mm i.d.) for the detection of amino acids. The aliquots used for amino acid analysis were esterified with n-butanol/HCl ($\sim$ 3 N HCl) and acylated with heptafluorobutyric anhydride prior to gas chromatography. Molecular analyses of the kerogen samples were performed with a directly connected high vacuum pyrolysis-gas chromatograph-mass spectrometer system.

3 Experimental Results and Observations

The results of the analytical procedures revealed differences between the stratiform and globular type kerogens. The H, C, and O contents of five of these six

Table 1. Elemental abundances[a] and H/C, O/C atomic ratios of stratiform and globular kerogens at Elliot Lake, Ontario

Abundances expressed in ppm units unless otherwise noted

Element	Stratiform kerogens Sample nos.		Globular kerogens Sample nos.			
	RAD 74-11	RAD 83-01	RAD 84-503	RAD 84-507	95355	38983
U	20%	32%	4.4 ppm	4.9 ppm	3.0 ppm	2.3 ppm
Th	3.6%	4.1%	< 0.2 ppm	1.4 ppm	0.9 ppm	0.6 ppm
Au	1100 ppb	2200 ppb	1400 ppb	40 ppb	13 ppb	25 ppb
Na	n.d.	n.d.	0.015%	0.015%	0.071%	0.045%
Sc	32	32	0.17	0.20	0.04	0.01
Cr	n.d.	n.d.	< 0.5	< 1.0	< 0.6	< 0.5
Fe	2.90%	0.74%	0.061%	0.102%	0.069%	0.051%
Co	75	53	3.0	3.3	2.1	2.2
Ni	<0.1%	1.4%	< 70	< 50	< 50	< 50
Zn	1100	1100	26	47	10	5
As	n.d.	n.d.	180	240	62	62
Se	< 200	< 200	5.7	2.2	< 1.2	7.0
Br	n.d.	n.d.	46	50	33	28
Mo	n.d.	n.d.	< 2	< 2	< 2	< 2
Sb	120	170	0.7	1.0	2.8	1.3
Cs	< 12	21	< 0.5	0.7	< 0.2	< 0.2
Ba	9.5%	0.6%	< 50	< 50	< 20	50
La	3825 ppb (avg)	n.d.	3.3	3.1	1.2	0.8
Hf	150	40	< 0.2	0.4	< 0.2	< 0.2
Ta	77	82	< 0.5	< 0.5	< 0.5	< 0.5
W	n.d.	n.d.	< 1	< 1	< 1	< 1
Ce	110	n.d.	6	6	2	1
Nd	5050 ppb (avg)	n.d.	5	4	< 3	< 3
Sm	2500 ppb (avg)	n.d.	0.14	0.18	0.09	0.07
Eu	123 ppb (avg)	180 ppb	< 0.2	< 0.1	0.06	0.07
Tb	595 ppb (avg)	460 ppb	< 0.1	0.1	< 0.1	< 0.1
Yb	845 ppb (avg)	1100 ppb	< 0.05	0.7	< 0.5	< 0.5
Lu	123 ppb (avg)	107 ppb	0.01	< 0.01	< 0.01	< 0.01
Dy	2100 ppb (avg)	n.d.	n.d.	n.d.	n.d.	n.d.
Sr	< 5000	< 4000	< 100	< 100	< 100	< 100
Rb	< 300	< 400	< 10	< 10	< 10	< 10
C	47.8%	50.7%	80.2%	80.1%	80.3%	79.2%
H	2.4%	1.7%	5.4%	5.3%	4.7%	5.1%
O	5.6%[b]	n.d.	6.9%[b]	6.6%[b]	8.6%[b]	8.2%[b]
S total	4.9%[c]	8.0%[c]	4.2%[c]	3.5%[c]	6.3%[c]	8.1%[c]
H/C	0.60	0.41	0.81	0.79	0.71	0.78
O/C	0.08	n.d.	0.06	0.06	0.08	0.08

[a] Elemental abundances from U to Rb have been determined by neutron activation. Some elements in sample RAD 74-11 have also been analyzed by DCP and ICP spectrometry; these values are expressed as averages of the neutron activation, DCP, and ICP results. The composition of the samples, particularly those of the stratiform kerogens, was not homogenous. Because different portions of the samples had to be used for some of the analyses (e.g., C to S vs. the neutron activation), this undoubtedly introduces a small error range. The high mineral content in the stratiform samples is considered to contribute to this in the C to O and S analytical results although appropriate corrections for minerals have been performed.

[b] Nonmineral O determined directly, and not by difference. The notation n.d. = not detected.

[c] Detrital, authigenic mineral and kerogen S. Detrital pyrite is the most common sulfide mineral phase.

kerogen samples were expressed as H/C and O/C atomic ratios in an attempt to elucidate kerogen maturation and related diagenetic processes (Table 1 and Fig. 1). H/C vs. O/C diagrams have been used repeatedly for characterizing thermally affected organic matter (e.g., Tissot and Welte 1978). The H/C and O/C atomic ratios and the U, Th, and other elemental abundances are shown in Table 1. Based on U content, two types of uranium-kerogen associations appear to be present (stratiform, 20–32% U; globular 4.9–3.0 ppm U). Other elements (e.g., Ba, Au) also show variations which require further consideration.

The results of the analyses of the aqueous HCl, and the benzene and methanol extracts of the kerogen

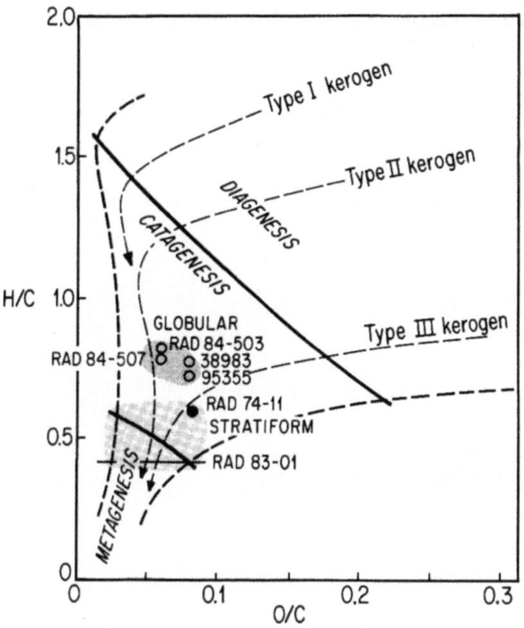

Fig. 1. H/C, O/C atomic ratios of stratiform and globular kerogens plotted on a van Krevelen diagram. O, H, and C values were corrected for O, H, and C from minerals. The oxygen content of stratiform sample RAD 83-01 was not determined. This sample is illustrated by a *horizontal line* corresponding to its H/C ratio. RAD 74-11 is another stratiform sample. The other samples represent globular kerogens

samples showed no contamination by amino acids or soluble hydrocarbons.

Analysis of two globular kerogen samples (RAD 84-503 and RAD 84-507) by high vacuum pyrolysis-gas chromatography-mass spectrometry revealed the presence of organic moieties rather similar to those found in stratiform samples RAD 74-11 and Rad 83-01 (Table 2). The globular kerogens contained slightly longer alkane moieties and more alkyl substitutions (the latter on aromatic hydrocarbons and aromatic sulfur compounds) than did the stratiform kerogens. Also, the stratiform samples contained some condensed aromatic compounds such as acenaphthene and benzothiophenes which the globular kerogens lacked. The globular kerogen RAD 84-503 contained C_4 thiols whereas the stratiform kerogens had only C_2 thiols. The stratiform samples contained only three n-alkanes (C_1-C_3) and the corresponding alkenes (produced during pyrolysis); the globular kerogens contained two more n-alkanes (including n-pentane) and n-alkenes. The stratiform kerogens seem to have yielded more CH_4 than C_2H_6 compared to the globular ones, although the analytical system was not optimized to determine the relative abundances of low molecular weight compounds in the pyrolyzates.

X-ray diffraction analysis showed no graphite in the kerogen samples.

4 Discussion

The stratiform and globular type kerogens in the Elliot Lake-Blind River region have four possible origins. It should be noted that the proposed origins are preliminary because of the limited number of samples. Four possible models are proposed.

First, the stratiform and globular kerogens are not related and their association is fortuitous. Second, the stratiform kerogen evolved from the globular substance. Third, the globular kerogen evolved from the stratiform kerogen. A fourth possibility is that both types of kerogen were derived from a common precursor. However, this is implicit in the second and third genetic models and will not be discussed separately.

The experimental findings and the petrological observations which are noted below, when considered together, imply that the second model is not a viable possibility. Therefore, only the first and the third models will be considered here.

The uraniferous-kerogenous host rocks of sedimentary origin comprising the basal Matinenda Formation display evidence of cyclic deposition, diagenetic to local, mild metamorphic events (Robertson 1981) and hydrothermal effects resulting in the crystallization of some of the authigenic minerals. The cyclic deposition is shown by the well-rounded and well-sorted pebbly layers alternating with, and/or grading into, fine-grained quartzites and subarkosic rocks, some in lenses. Roscoe (1981) noted that deposition was controlled by an efficient hydrodynamic sorting regime. Ruzicka and Steacy (1976) and Ruzicka (1981) pointed out that the uraniferous kerogen layers were deposited in cycles. The deposition started with coarse-grained quartz, and continued with increasingly smaller-sized quartz and pyrite. Higher in this stratigraphic section, even finer grains of pyrite, monazite, and brannerite followed. The sedimentary cycle terminated with the deposition of fine-grained detrital uraninite, in what are now the stratiform kerogen seams. Subsequent cycles repeated this depositional sequence. The presence of the uraniferous kerogen at the top of the finest-grained sediments, where the energy of transport was lowest, is considered to be the strongest suggestive evidence that the progenitors of the stratiform kerogen layers were ancient cyanobacterial mats which trapped the detrital uraninite grains. Such an origin for the detrital uraninite has been noted repeatedly, e.g. by Dyer et al. (1988). Prokaryotic mats would have been torn apart,

Table 2. Characteristic pyrolysis moieties of the stratiform and globular kerogens

Stratiform kerogens	Globular kerogens
Aliphatic hydrocarbons	Aliphatic hydrocarbons
Methane	Methane
Ethane	Ethane
Propane	Propane
n.d.	Butane
n.d.	Pentane
Aromatic hydrocarbons	Aromatic hydrocarbons
Benzene	Benzene
Toluene	Toluene
m, p, o-Xylenes	m, p, o-Xylenes
Ethylbenzene	Ethylbenzene
C_3-Benzene	C_3-Benzene
n.d.	C_4-Benzene (dimethylethylbenzene)
n.d.	C_5-Benzene (dimethylisopropylbenzene)
Naphthalene	n.d.
Methylnaphthalene	Methylnaphthalene
n.d.	Dimethylnaphthalene
Acenaphthene	n.d.
Sulfur compounds	Sulfur compounds
Dithiabutane	Dithiabutane
n.d.	Butanethiol
Thiophene	Thiophene
Methylthiophene	Methylthiophene
Dimethylthiophene	Dimethylthiophene
C_3-Thiophene	C_3-Thiophene
n.d.	C_4-Thiophene
Benzothiophene	n.d.
Methylbenzothiophene	n.d.

n.d. = not detected.

even if they could have grown in environments dominated by the rapid currents which deposited the coarse-grained sediments. Probably the matrix grains, clasts, and disseminated flakes of kerogen are the remnants of stratiform layers disrupted by subsequent rapid currents. Syngenetic organic-rich layers had to evolve at locales of relative depositional quiescence.

After burial, Huronian sediments and the kerogen were affected by diagenesis and by more than one hydrothermal episode. To evaluate possible kerogen paragenesis, hydrothermal and anhydrous thermal events need to be considered. Post-Huronian intrusive rocks, well documented at Elliot Lake (Robertson 1981; Roscoe 1981), may have contributed to the thermal effects. Possibly post-Matinenda intrusions may have been responsible for the hydrothermal event(s) (described and illustrated by Willingham et al. 1985). Based on current information obtained from the Phanerozoic, it is possible that maturation processes analogous to hydrous pyrolysis (Lewan et al. 1979; Lewan 1985) affected the evolution of uraniferous kerogens in the Matinenda Formation. Another alternative would be anhydrous pyrolysis. Temperatures affecting the kerogenous samples probably did not exceed 250 °C, and in places could have been lower. No evidence was found for contact metamorphism from the mineral assemblages, and this is confirmed by the lack of graphite, including disordered graphite (French 1964).

Central to the whole problem of kerogen paragenesis are the relative solubilities of UO_2 and S during later hydrothermal episodes. The stratiform kerogens contain abundant U but the globular matter is U-poor. Garrels and Christ (1965) pointed out that carbonate-bearing aqueous solutions are excellent solvents for uraninite. Probably, if carbonates and authigenic sulfides evolved in the kerogenous host rocks after the syngenetic, detrital uraninite was fully incorporated in hydrophobic organic substances, dissolution of uraninite would have been prevented.

Parks and Pohl (1988) measured the solubility of stoichiometric UO_2 (sintered uraninite) in aqueous solutions between 100 and 300 °C in pressure vessels under helium. They reported that the solubility of uraninite is apparently independent of pH at $4 < pH < 10$ and that the dissolved species is $U(OH_4)(aq)$; $\log S$ [S = solubility, $\Sigma(U)(aq)$, molar] is $\cong 9.47$, a notably low solubility. On the other hand, oxidation increases solubility. It has also been suggested that a change in redox potential may produce a favorable reducing environment for hexavalent uranium as a neutral complex $UO_2(OH)_2^{\circ}$ (Schidlowski 1981).

To further characterize the Elliot Lake kerogens and their host rocks, radiation effects and additional petrologic observations need to be considered. Great age and ionizing radiation would also have affected both the rocks and the organic matter. With increasing age, noted particularly in the Precambrian, kerogens become more and more condensed aromatic substances (e.g., Hayes et al. 1983). Ionizing radiation causes disproportionation-type free radical reactions, which may result in the loss of hydrogen and an increasing abundance of aromatic products. Radical aromatic substitution, a free radical reaction, also leads to the formation of condensed aromatic moieties (Huang et al. 1974). Kerogens containing $\sim 20-32\%$ uranium may contain, under optimum conditions, as much as $\sim 2-3 \times 10^{23}$ free radicals/gC after irradiation for 2 Ga years (Nagy 1975); this, therefore, would cause further aromatization. There is an indication that with increasing U content the H/C ratios decrease.

At Elliot Lake both the stratiform kerogen and the adjacent rocks contain quartz, some fractured uraninite, and highly fractured and shattered authigenic pyrite (Ruzicka and Steacy 1976; Ruzicka 1981; Willingham et al. 1985). Observations with the petrographic microscope revealed that the host rock in the vicinity of the stratiform kerogen consists mainly of quartz with some pyrite, mica and/or fine-grained carbonate. Feldspar, chlorite, palygorskite and other clay minerals (probably illite) occur in lower or trace abundances. Backscattered scanning electron microscopy (BSE) augmented by semi-quantitative energy dispersive spectral (EDS) analysis helped to confirm the mineral composition, textural relationships, the fractured minerals and the hydrothermal episodes. Both the stratiform and the globular kerogens contain mineral and organic bound sulfur (Tables 1 and 2). The minor and trace element abundances are listed in Table 1. They illustrate the complexity of the problem of petrogenesis which will be discussed elsewhere in a separate report.

Although the physicochemical differences between the two groups of Elliot Lake-Blind River kerogens, expressed particularly in terms of their H/C atomic ratios and U contents, could be taken as an indication of unrelated genesis (model 1), it appears that this is not the case. The droplet-like morphology of many of the globular particles resembles that of an organic liquid. Indeed, Nagy and Mossman (1988) suggested that as a result of hydrous pyrolysis the globular kerogen is a migrated derivative of the stratiform kerogen. Also, Mancuso et al. (1989) concluded that the globular organic matter in the Matinenda Formation at Elliot Lake is the result of natural migration and maturation of Precambrian petroleum. Certainly, in the absence of any appropriate conventional petroleum source rocks in the Elliot Lake area, the stratiform kerogen remains the most likely source material. It is instructive to note that the globular kerogens occur in or near the stratiform layers. As is shown in Fig. 2, kerogen of markedly lower reflectance is directly adjacent to some of the main syngenetic, stratiform kerogen layers which contain detrital uraninite, and which have higher reflectances than the adjacent substance. (Quantitative reflectance data will be reported separately). This can be construed as an indication that the low reflectance, uraninite-free organic substance has been mobilized from the stratiform kerogen. This former substance, and by implication the globules, appear to be younger in age than the syngenetic kerogens. If the syngenetic, stratiform kerogens were subjected to thermal episodes involving, for example, exposure to successive hydrothermal solutions (Willingham et al. 1985),

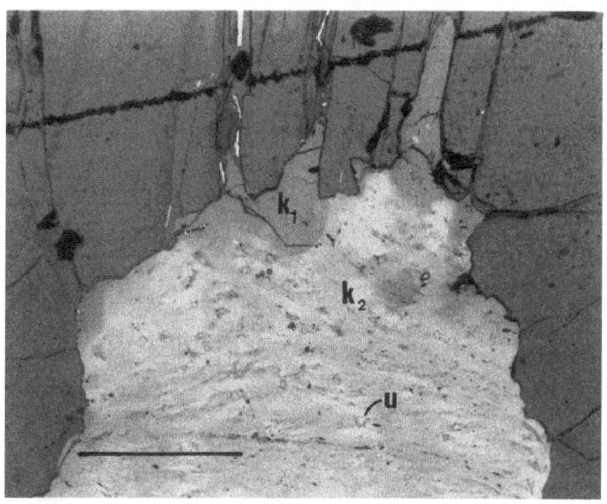

Fig. 2. Uraninite free kerogen (k_1) of low reflectance adjacent to syngenetic kerogen (k_2) of high reflectance and containing small grains of uraninite (u). The kerogen with the low reflectance has been apparently arrested during extraction from its syngenetic parent material. Sample is RAD 85-X4NP, matrix-type kerogen from Quirke mine, Elliot Lake. Surrounding (*dark*) grey material is quartz. Bar 0.1 mm

and the latter caused hydrous pyrolysis, liquids containing hydrocarbons could have been extracted from the parent material. According to current concepts, petroleum-like liquids have been generated in sedimentary rocks up to $\cong 2$ Ga years old (Hoering 1978; Sklarew and Nagy 1981; Kelly and Nishioka 1985; Hoering and Navale 1987; Summons et al. 1988; Mancuso et al. 1989). What is now the solid globular kerogen could have been mobilized during the burial history of the Matinenda Formation, moved as a viscous liquid away from the stratiform substance, and then resolidified. This agrees with the third and preferred model, as outlined by Nagy and Mossman (1988). Under the reducing conditions noted above, dissolution of uraninite would have been curtailed, and the globular kerogen inherited only minimal uranium content.

5 Conclusions

The organic matter at Elliot Lake, Ontario, has two basic morphologies: stratiform and globular. The former kerogen contains abundant U, whereas the globular type is U-poor. It appears that the globular material evolved from the stratiform kerogen. Hydrothermal episode(s), causing, e.g., hydrous pyrolysis, may have been responsible for the evolution of the globular kerogen. Evidence for the hydrothermal events has been reported by us previously. The present investigation showed that contact metamorphism has not affected the samples studied.

Acknowledgments. The authors wish to thank Dr. V. Ruzicka of the Geological Survey of Canada, Ottawa, for providing samples and Mr. J. Prellberg, Mr. J. Reese and Mr. G. Santo of the Laboratory of Organic Geochemistry, Department of Geosciences, The University of Arizona for their assistance with the molecular analyses. This research was supported by NASA Grant NGR-03-002-171 to B.N. and Operating Grant A 8295, Natural Sciences and Engineering Research Council to D.J.M.

References

Colvine AC (1981) Reconnaissance of the Lorrian Formation, Northern Cobalt Embayment. In: Wood J, White OL, Barlow RB, Colvine AC (eds) Summary of fieldwork, 1981. Ontario Geol Surv Misc Publ 100:187–190

Dyer BD, Krumbein WE, Mossman DJ (1988) Nature and origin of stratiform kerogen seams in Lower Proterozoic Witwatersrand-type paleoplacers – the case for biogenicity. Geomicrobiol J 6:33–47

Fralick PW, Miall AD (1981) Sedimentology of the Matinenda Formation. In: Pye EG (ed) Geoscience Research Grant Program, Summary of Research, 1980–81. Ontario Geol Surv Misc Publ 98:80–89

French BM (1964) Graphitization of organic material in a progressively metamorphosed Precambrian iron formation. Science 146:917–918

Garrels RM, Christ CL (1965) Solutions, minerals and equilibria. Harper & Row, New York, 450 pp

Hayes JM, Kaplan IR, Wedeking KW (1983) Precambrian organic geochemistry, preservation of the record. In: Schopf JW (ed) Earth's earliest biosphere; its origin and evolution. Univ Press, Princeton, pp 93–134

Hoering TC (1978) Molecular fossils from the Precambrian Nonesuch shale. In: Ponnamperuma C (ed) Comparative planetology. Academic Press, New York London, pp 243–255

Hoering TC, Navale V (1987) A search for molecular fossils in the kerogen of Precambrian sedimentary rocks. Precambrian Res 34:247–267

Huang RL, Goh SH, Ong SH (1974) The chemistry of free radicals. Arnold, London, 244 pp

Kelly WC, Nishioka GK (1985) Precambrian oil inclusions in late veins and the role of hydrocarbon in copper mineralization at White Pine, Michigan. Geology 13:334–337

Leventhal JS, Grauch RI, Threlkeld CN, Lichte FE, Harper CT (1987) Unusual organic matter associated with uranium from the Claude deposit, Cluff Lake, Canada. Econ Geol 82:1169–1176

Lewan MD (1985) Evaluation of petroleum generation by hydrous pyrolysis experimentation. Philos Trans R Soc of London Ser A 315:123–134

Lewan MD, Winters JC, McDonald JH (1979) Generation of oil-like pyrolyzates from organic-rich shales. Science 203:897–899

Mancuso JJ, Kneller WA, Quick JC (1989) Precambrian vein pyrobitumen: evidence for petroleum generation and migration 2 Ga ago. Precambrian Res 44:137–146

Mossman DJ (1987) Stratiform gold occurrences of the Witwatersrand type in the Huronian Supergroup, Ontario, Canada. S Afr Tydskr Geol 90(2):168–178

Mossman DJ, Dyer BD (1985) The geochemistry of Witwatersrand-type gold deposits and the possible influence of ancient prokaryotic communities on gold dissolution and precipitation. Precambrian Res 30:303–319

Mossman DJ, Harron GA (1984) Witwatersrand-type paleoplacer gold in the Huronian Supergroup of Ontario, Canada. Geosci Can 11:33–40

Nagy B (1975) Carbonaceous meteorites. Elsevier, Amsterdam, p 315

Nagy B, Mossman DJ (1988) The nature and origin of kerogens in the Lower Huronian Supergroup, Elliot Lake region, Ontario, Canada. Terra Cognita 8:219

Parks GA, Pohl DC (1988) Hydrothermal solubility of uraninite. Geochim Cosmochim Acta 52:863–875

Phillips GN (1987) Metamorphism of the Witwatersrand gold fields: conditions during peak metamorphism. J Metam Geol 5:307–322

Pretorius DA (1974) Gold in the Proterozoic sediments of South Africa: systems, paradigms and models. Univ Witwatersrand Econ Geol Res Unit Inf Circ 87:22 pp

Robertson JA (1978) Precambrian stratigraphy and uranium deposits, Elliot Lake area, Ontario. Geol Assoc Can, Mineral Assoc Can Guidebook, pp 192–203

Robertson JA (1981) The Blind River uranium deposits: The ores and their setting. In: Armstrong FC (ed) Genesis of uranium- and gold-bearing Precambrian quartz-pebble conglomerates. US Geol Surv Prof Pap 1161-A-BB: U 1–U 23

Roscoe SM (1973) The Huronian Supergroup, a Paleoaphebian succession showing evidence of atmospheric evolution. Geol Assoc Can Spec Publ 12:31–47

Roscoe SM (1981) Temporal and other factors affecting deposition of uraniferous conglomerates. In: Armstrong FC (ed) Genesis of uranium- and gold-bearing Precambrian quartz-pebble conglomerates. US Geol Surv Prof Pap 1161-A-BB: W 1–W 17

Ruzicka V (1981) Some metallogenic features of the Huronian and post-Huronian uraniferous conglomerates. In: Armstrong FC (ed) Genesis of uranium- and gold-bearing Precambrian quartz-pebble conglomerates. US Geol Surv Prof Pap 1161-A-BB: V 1–V 8

Ruzicka V, LeCheminant GM (1984) Uranium deposit research. In: Current research, pt A, 1983. Geol Surv Can Publ 84-1 A:39–51

Ruzicka V, Steacy HR (1976) Some sedimentary features of conglomeratic uranium ore from Elliot Lake, Ontario. In: Report of activities, pt A. Geol Surv Can Publ 76-1 A:343–346

Schidlowski M (1981) Uraniferous constituents of the Witwatersrand conglomerates: ore-microscopic observations and implications for the Witwatersrand metallogeny. In: Armstrong FC (ed) Genesis of uranium- and gold-bearing Precambrian quartz-pebble conglomerates. US Geol Surv Prof Pap 1161-A-BB: N 1–N 23

Simpson PR, Bowles JFW (1981) Uranium mineralization of the Witwatersrand and Dominion Reef systems. In: Armstrong FC (ed) Genesis of uranium- and gold-bearing Precambrian quartz-pebble conglomerates. US Geol Surv Prof Pap 1161-A-BB: R 1–R 26

Sklarew DS, Nagy B (1981) Pyrolysis of Transvaal kerogens. I. Diagenesis of kerogen in a stromatolite near Zeerust, South Africa: one possible chemical evolution process. Precambrian Res 15:97–111

Smith ND, Minter WEL (1980) Sedimentological controls of gold and uranium in two Witwatersrand paleoplacers. Econ Geol 75:1–14

Summons RE, Powell TG, Boreham CJ (1988) Petroleum geology and geochemistry of the Middle Proterozoic McArthur Basin, northern Australia: III. Composition of extractable hydrocarbons. Geochim Cosmochim Acta 57:1747–1763

Tankard AJ, Jackson MPA, Eriksson KA, Hobday DK, Hunter DR, Minter WEL (1982) Crustal evolution of South Africa. Springer, Berlin Heidelberg New York, pp 115–150

Tissot BP, Welte DH (1978) Petroleum formation and occurrence. Springer, Berlin Heidelberg New York, 538 pp

Willingham TO, Nagy B, Nagy LA, Krinsley DH, Mossman DJ (1985) Uranium-bearing stratiform organic matter in paleoplacers of the lower Huronian Supergroup, Elliot Lake-Blind River region, Canada. Can J Earth Sci 22:1930–1944

Zumberge JE, Sigleo AC, Nagy B (1978) Molecular and elemental analyses of the carbonaceous matter in the gold- and uranium-bearing Vaal reef carbon seams, Witwatersrand sequence. Minerals Sci Eng 10:223–246

Organic Carbon Isotopic Composition of Proterozoic Sedimentary Rocks from India: Preliminary Results

D. M. Banerjee [1], M. Deb [1], and H. Strauss [2]

[1] Department of Geology, University of Delhi, Delhi, India
[2] Institut für Geologie, Ruhr-Universität, W-4630 Bochum, FRG

Abstract

In order to characterize the organic carbon preserved in Proterozoic sedimentary rocks of India, a comprehensive collaborative study has been initiated. The present chapter summarizes the first results from four selected basins: Aravalli-Delhi, Chattisgarh, Vindhyan, and Krol-Tal. Variations in TOC, $\delta^{13}C_{org}$ and H/C ratios from this study, and additional data from earlier published work, are discussed. In general, the carbon isotopic composition falls within the envelope of previously published whole-rock $\delta^{13}C_{org}$ data for the Middle and Late Proterozoic displaying values in the range -33 to -8‰.

1 Introduction

Studies of abundance, preservation, and isotopic composition of organic carbon in Proterozoic rocks may contribute to an understanding of the global carbon cycle which also bears on our interpretation of the evolution of life during this early part of Earth history (e.g., Schidlowski et al. 1983). While extensive carbon isotopic data are available for Proterozoic sedimentary rocks from different parts of the world, little information on the Indian Proterozoic has been recorded. To the best of our knowledge, Banerjee et al. (1986), Deb (1986, 1989), Krishnamurty et al. (1987), and Banerjee (1990) are the only publications wherein isotopic data on organic carbon in the Proterozoic rocks of India are discussed. Carbonate carbon isotopic values, on the other hand, have been reported by Schidlowski et al. (1975), Banerjee et al. (1986), Deb (1986), Sathyanarayan et al. (1987), and Aharon et al. (1987).

In this chapter we report the results of the first part of our investigations concerning organic carbon in sedimentary rocks from selected Proterozoic basins of India.

2 Geological Setting

Proterozoic basins occupy a large area of the Indian subcontinent (Fig. 1). These basins can be subdivided broadly into two major types based on their tectonostratigraphic positions, namely a) those associated with plate margins and affected by postdepositional deformation and/or metamorphism, and b) those formed in intracratonic regimes and which have remained virtually undisturbed and unmetamorphosed.

Basins in the first category include: the North Singhbhum-Gangpur and Kolhan Basins of east-central India; the Sausar, Sakoli and Bijawar Basins of central India; the Aravalli and Delhi Basins of western India; the Cuddapah Basin of south-central India; and the Proterozoic to Lower Paleozoic sedimentary basin of the Lesser Himalaya region (1–6, 15: Fig. 1). The second type of basin is represented by the Chattisgarh, Indravati and Vindhyan Basins of central India; the Pakhal Basin of the Pranhita Godavari valley; the Kaladgi, Bhima and Kurnool Basins of southern India; and the Marwar Basin of western India (7–14: Fig. 1).

The geological information available for each of these basins varies considerably in both quantity and quality. Such variability is governed by the presence or absence of economic mineral deposits, accessibility of the terrain, and the particular interests of the researcher(s) working in a given area. While structural, petrological and field-based sedimentological information is generally available, geochemical and isotopic data for any of these rocks are only poorly recorded. The present study focuses on the organic carbon isotopic record in four of these sedimentary basins, namely (1) the Aravalli-Delhi Basin, (2) the Chattisgarh Basin, (3) the Vindhyan Basin, and (4) the Krol-Tal succession of Garhwal Himalaya. A brief overview of the geology and stratigraphy of these basins, including the absolute ages of their sediments, is given below.

M. Schidlowski et al. (Eds.)
Early Organic Evolution: Implications for Mineral and Energy Resources

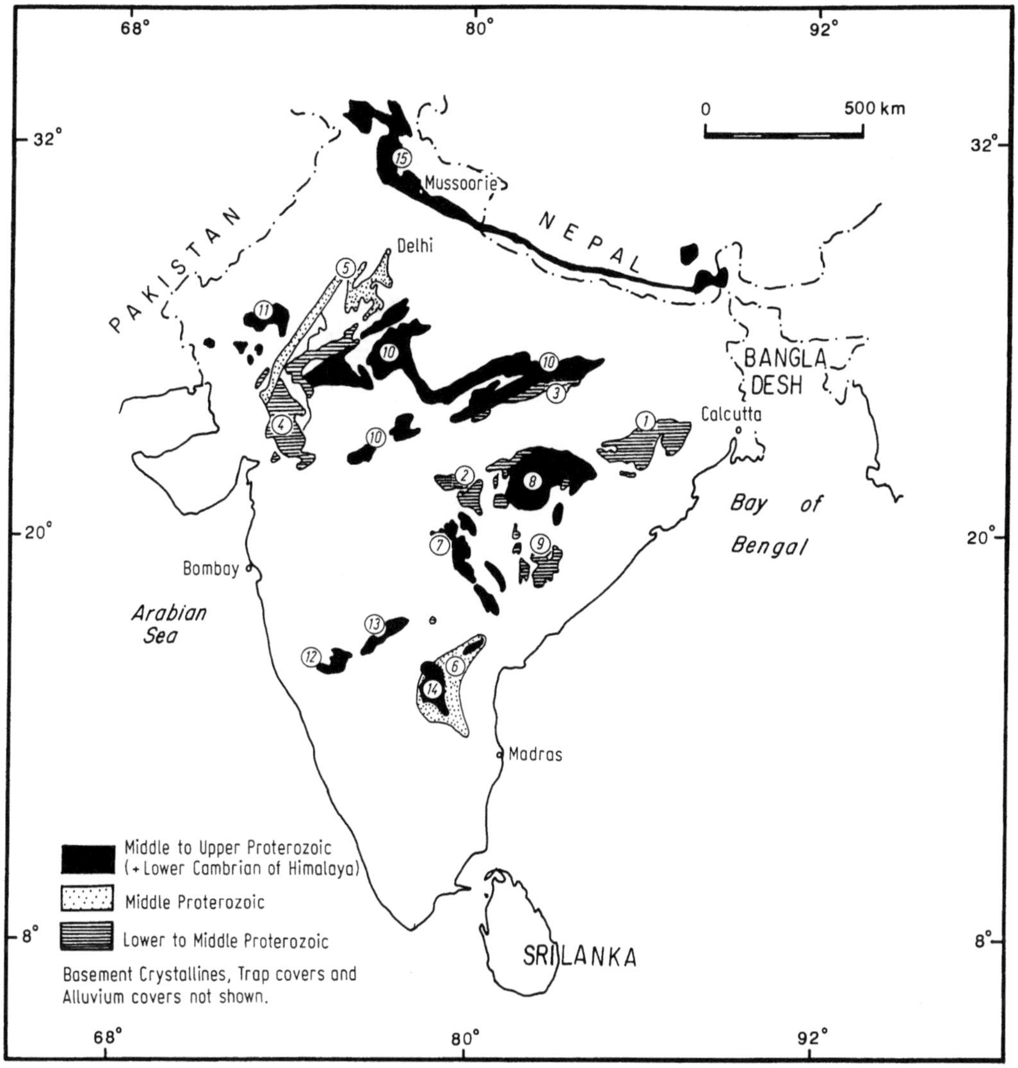

Fig. 1. Distribution of Proterozoic Basins in India: *1* Northern Singhbhum, Gangpur, and Kolhan Basins; *2* Sakoli-Sausar Basins; *3* Bijawar Basin; *4* Aravalli Basin; *5* Delhi Basin; *6* Cuddapah Basin; *7* Pakhal Basin; *8* Chattisgarh Basin; *9* Indravati Basin; *10* Vindhyan Basin; *11* Marwar Basin; *12* Kaladgi Basin; *13* Bhima Basin; *14* Kurnool Basin; *15* Lesser Himalaya Sedimentary Belt

2.1 Aravalli-Delhi Basin

Heron (1953) described the Proterozoic rocks of Aravalli Basin as a predominantly argillaceous sequence. Quartzite, conglomerate, interstratified volcanics, dolomite, phosphorite, greywacke, and several types of phyllite constitute four major lithostratigraphic groups named the Debari, Matoon, Udaipur, and Jharol Groups. These are collectively termed the Aravalli Supergroup (Gupta et al. 1980). The rocks have undergone low-grade metamorphism, mostly greenschist facies, but in places, lower amphibolite facies metasediments are present. The Aravalli terrain in the type area of Udaipur in Rajasthan contains numerous poorly defined enclaves where metamorphic effects are virtually negligible. In such areas, the original sedimentary characteristics of the rocks have been preserved including undeformed columnar stromatolites, cross-bedding, and ripple marks. In contrast, rocks of the Delhi Supergroup (Alwar and Ajabgarh Groups) apparently overlie the Aravalli sequence but their actual relationships are unclear (Naqvi and Rogers 1987). The Delhi Supergroup is represented by a thick succession (some 6000 m) of metaquartzarenite, schist, carbonates (including marble) and metavolcanics, as well as several higher-grade metamorphic units. However, several enclaves show only a low-

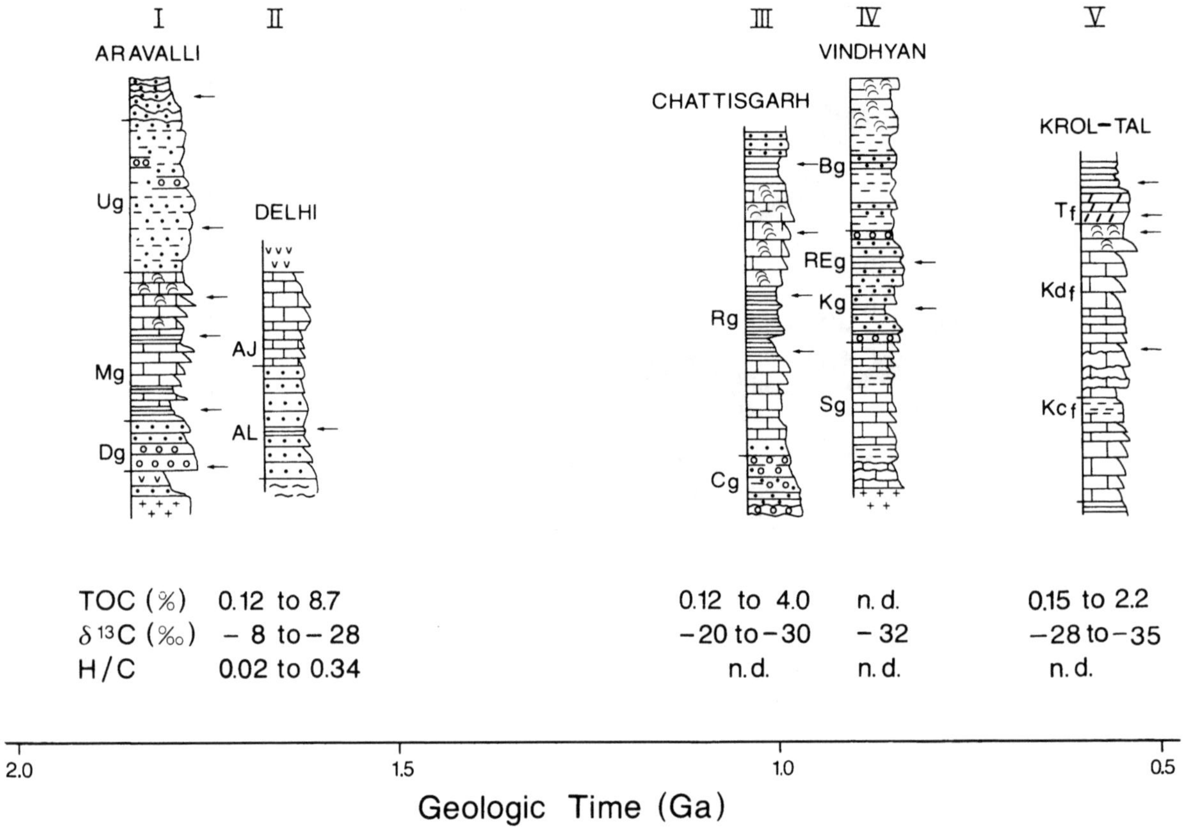

Fig. 2. Generalized lithostratigraphic columns for the geological units sampled in this study. *I* Aravalli Supergroup: *Dg* Debari Group; *Mg* Matoon Group; *Ug* Udaipur Group. *II* Delhi Supergroup: *AL* Alwar Group; *AJ* Ajabgarh Group. *III* Chattisgarh Supergroup: *Rg* Raipur Group; *Cg* Charmuria Group. *IV* Vindhyan Supergroup: *Sg* Semri Group; *Kg* Kaimur Group; *REg* Rewa Group; *Bg* Bhander Group. *V* Krol-Tal succession: *Kcf* Krol C Formation; *Kdf* Krol D Formation; *Tf* Tal Formation

grade metamorphic overprint and invariably retain some of their original depositional features. Generalized lithostratigraphic successions for the Aravalli and Delhi Supergroups are given in Fig. 2.

Carbonates, black and gray shales, phyllites, stromatolitic phosphorites and host rocks for base metal sulfides from Udaipur and environs have been analysed to determine the concentration and isotopic composition of their organic carbon (Table 1).

2.2 Chattisgarh Basin

Exposed in central India (Fig. 1), this unmetamorphosed flat-lying sedimentary sequence was previously studied by Schnitzer (1971) and Murti (1987). It comprises shale, limestone (stromatolitic and nonstromatolitic) and sandstone. A stratigraphic column for rocks of the Chattisgarh Supergroup (Raipur Group) near the town of Raipur is shown in Fig. 2.

Black and gray shales, as well as limestones, from several field exposures around Raipur, have been analyzed (Table 1).

2.3 Vindhyan Basin

Oldham (1856) named the rocks of this intracratonic basin "Vindhyan" and divided them into three "series". Numerous workers have subsequently contributed to our knowledge of the basin (e.g., Auden 1933), and a recent review of its geology has been presented by Soni et al. (1987). The Vindhyan Supergroup is subdivided into the Semri, Kaimur, Rewa, and Bhander Groups, separated from each other by unconformities marked by conglomerates. The sediments have been altered only by burial diagenesis and catagenesis; no signs of a metamorphic overprint are evident. The geographic position and extent of the Vindhyan Basin is shown in Fig. 1. Its generalized lithostratigraphic column

Table 1. Geochemical data for the organic carbon fraction in selected Proterozoic sediments from India

Stratigraphic horizon	Sample no.	Lithology	TOC (%)	$\delta^{13}C$ (‰)	C (%)	H (%)	N (%)	H/C
Aravalli Supergroup								
Debari Group	292	Gray carbon. phyllite	4.00	−26.2	85.0	0.46	0.08	0.08
	447	Gray carbon. phyllite	0.63	−19.6				
	448	Gray carbon. phyllite	1.30	−25.5	67.3	0.74	0.41	0.13
	452	Gray carbon. phyllite	4.72	−23.0	48.0	0.84	0.44	0.21
	453	Carbonaceous limestone	3.39	−28.3				
	457	Gray carbon. phyllite	0.12	−17.2				
	481	Laminated carbon. marble	0.27	−8.1				
	492	Carbon phyllite	1.86	−26.1	63.4	1.04	0.37	0.20
	495	Carbonaceous limestone	1.02	−25.2	40.8	1.15	0.19	0.34
	533	Carbonaceous phyllite	1.54	−19.4				
	537	Carbonaceous phyllite	5.30	−18.2				
	605	Carbonaceous silty shale	0.17	−19.9	68.0	0.13	0.06	0.02
Matoon Group	41	Phosphatic phyllite	2.01	−15.8				
	42	Phosphatic phyllite	1.27	−15.1				
	43	Carbonaceous dolomite	0.43	−9.7				
	46	Carbonaceous phyllite	8.05	−26.6				
	442	Dolomite	0.92	−12.5				
	615	Carbonaceous phyllite	4.49	−24.7				
	DK.1	Stromatolitic phosphorite		−11.7				
	JH.3	Stromatolitic phosphorite		−17.6				
Udaipur Group	388	Carbonaceous phyllite	0.12	−24.5	80.0	1.38	0.20	0.21
	389	Carbonaceous phyllite	0.26	−26.7	76.4	0.56	0.40	0.09
	415	Phyllite	0.29	−24.5				
	417	Carbonaceous slate	0.26	−23.7				
	418	Carbonaceous phyllite	0.70	−24.4				
	579	Carbonaceous phyllite	0.25	−17.7	89.2	0.44	0.09	0.06
	593	Phyllite	0.12	−22.2	92.3	0.21	0.07	0.03
Jharol Group	423	Grayish green phyllite	0.35	−18.7				
	427	Sandy carbon. phyllite	0.27	−24.2				
	428	Carbonaceous phyllite	0.49	−24.1				
	432	Gray phyllite	0.66	−21.6				
	553	Carbonaceous phyllite	8.70	−20.7				
	548	Carbonaceous limestone	1.62	−20.2				
Delhi Supergroup								
Ajabgarh Group	277	Micaceous phyllite	0.14	−19.9				
	213	Carbonaceous limestone	0.29	−20.4				
Chattisgarh Supergroup								
Charmuria Group	32	Phosphatic limestone	0.27	−19.8				
	33	Limestone	0.12	−26.0				
Raipur Group	RP-3	Gray shale	0.21	−25.9				
	RP-4	Carbonaceous limestone	1.51	−24.1				
	34	Carbonaceous shale	3.95	−29.7				
	35	Stromatolitic limestone	0.21	−23.8				
	36	Limestone	0.89	−29.7				
	RP-5	Carbonaceous shale	1.28	−25.9				
Vindhyan Supergroup								
Kaimur Group	I/1	Carbonaceous shale		−23.3				
	I/3	Pyritic shale		−32.8				
	I/3N	Pyritic shale		−33.0				
	I/4	Pyritic shale		−33.2				
	I/4N	Siliceous carbon. shale		−32.9				
	II/8	Siliceous carbon. shale		−31.9				

Table 1 (continued)

Stratigraphic horizon	Sample no.	Lithology	TOC (%)	$\delta^{13}C$ (‰)	C (%)	H (%)	N (%)	H C
Krol-Tal succession								
Krol Formation	KT-1	Stromatolitic carbonate	0.30	– 27.7				
	KT-2	Stromatolitic carbonate	0.31	– 28.3				
	KU-1	Laminated carbonate	0.15	– 30.1				
Tal Formation	KT-3	Cherty phosphorite	0.35	– 33.6				
	KT-5	Shaly phosphorite	0.81	– 33.9				
	KT-6	Carbon. shaly phosphorite	2.19	– 33.6				
	KT-10	Carbon. shaly phosphorite	2.01	– 32.9				
	KT-15	Stromatolitic phosphorite	1.01	– 34.5				
	KT-16	Carbonaceous shale	0.92	– 34.4				
	KT-17	Shaly laminated phosphorite	0.89	– 32.2				

(Fig. 2) provides an overview of the Riphean and Early Vendian sedimentary record in this part of the Indian craton.

Black and gray shales associated with syngenetic pyrite in the Amjhore area of Bihar were analyzed during this investigation (Table 1).

2.4 Krol-Tal Basin

This elongate, northwesterly trending basin is located in the lower parts of the Lesser Himalaya between Solan in the Himachal and Nainital in the Kumaun Himalaya (Fig. 1). Its geology and stratigraphy have been studied by Auden (1934) among others, and Singh (1978). The sedimentary sequence exposed in the Mussoorie hills comprises black shale, phosphorite, and carbonate (Fig. 2).

3 Age Constraints

The absolute time frame of sedimentation, diagenesis, and metamorphism in all these Proterozoic basins is only poorly constrained. An array of age data based on Rb/Sr, K/Ar, U/Th, and Nd/Sm systematics is available. However, these dates are insufficient to assign radiometric ages to all of the units sampled in this study.

For rocks of the Aravalli-Delhi province, the only direct sediment age comes from a recently published Pb-Pb model age for the synsedimentary base metal sulfide deposit of Rajasthan (Deb et al. 1989). This model age constrains the Aravalli base metal formation event between 1.8 and 1.7 Ga. The overlying sediments have ages possible as young as 1.5 Ga (age of metamorphism in the Aravalli sequence). This predates sedimentation of the Delhi Supergroup in the southern province at 1.1 Ga (Deb et al. 1989). All other ages inferred from cross-cutting relationships between magmatic and supracrustal sediments are of dubious value because of uncertainties in the interpreted field relationships of these units.

Sediments of the Chattisgarh Basin are correlated with parts of the Vindhyan sequence of the Son Valley in central India (Dutt 1964). Possibly, these are the result of coeval sedimentation in two separate mid-Proterozoic basins. A K/Ar date from glauconite indicates an age of 0.7 to 0.75 Ga (Kreuzer et al. 1977). In contrast, paleomagnetic studies (Verma et al. 1977) indicate an age of 1.25 to 1.3 Ga for the lowermost part of the sequence (Murti 1987). Stromatolite biostratigraphy suggests a Middle to Upper Riphaen age (1.1 to 0.7 Ga) for the upper part of this sedimentary sequence (Jairaman and Banerjee 1978).

In the Vindhyan Basin sediments were deposited over a large time span starting at around 1.4 Ga in the case of the Semri Group (Vinogradov et al. 1964) and extending up to 0.9 Ga for those of the Kaimur Group (Crawford and Compston 1970). Valdiya (1969) suggest an Upper Riphaen age (0.73 to 0.55 Ga) for the upper part of the sequence. The Vindhyan sediments are flat-lying, undeformed, and unmetamorphosed. Ages based on stromatolite biostratigraphy agree reasonably well with the radiometric ages (Valdiya 1969).

A Late Precambrian to Early Cambrian age for the Krol-Tal succession has recently been established on paleontological evidence, namely stromatolites characteristic of Vendian age for the Krol D Member, archaeocyathids and calcareous algae for the Krol E Member, and phosphatic shelly fossils of unequivocal earliest Cambrian affinity for the lower

Chert Member of the Tal Formation (Brasier and Singh 1987).

4 Analytical Methods

Unweathered samples were collected from field exposures and drill core. Outer surfaces were mechanically removed and rock powder was obtained using tungsten carbide or agate mortars. Organic carbon content was determined via combustion in a Ströhlein Coulomat apparatus. Preparation for organic carbon isotopic analyses involved the initial removal of the carbonate fraction with either dilute (1:4) orthophosphoric acid at 60 °C or with concentrated HCl (37%). Carbonate-free rock powder was mixed with CuO and reacted at 1000 °C using sealed-tube combustion. CO_2 was purified cryogenically and subjected to mass spectrometric analysis. Isotopic measurements were performed on a VG Isogas mass spectrometer at the Max-Planck-Institut für Chemie, Mainz (FRG) and on a Finnigan MAT 250 at the University of California Los Angeles (USA). Results are reported in the usual delta notation as per mil difference from the PDB standard. Reproducibility as determined from duplicate measurements was usually $< \pm 0.2‰$.

For several samples the kerogen fraction was isolated following standard procedures (Durand 1973). Subsequent determination of elemental abundances of carbon, hydrogen and nitrogen was performed using a Hereaus CHN Analyzer at the University of Delhi.

All data for total organic carbon content (TOC), organic carbon isotopic composition ($\delta^{13}C_{org}$), and H/C ratios are reported in Table 1 and presented in Figs. 2 and 3.

5 Results

TOC values for samples from the Aravalli Supergroup range from 0.12 to 8.7% and organic carbon isotopic compositions between −28.3 and −8.1‰. This is in good agreement with previously published results for phosphatic carbonates from the Lower Aravalli Supergroup (Banerjee et al. 1986). H/C ratios vary between 0.08 and 0.34.

Only two samples from the Delhi Supergroup were analyzed in the present study. Their organic carbon contents vary between 0.14 and 0.29% and $\delta^{13}C_{org}$ values lie around −20‰.

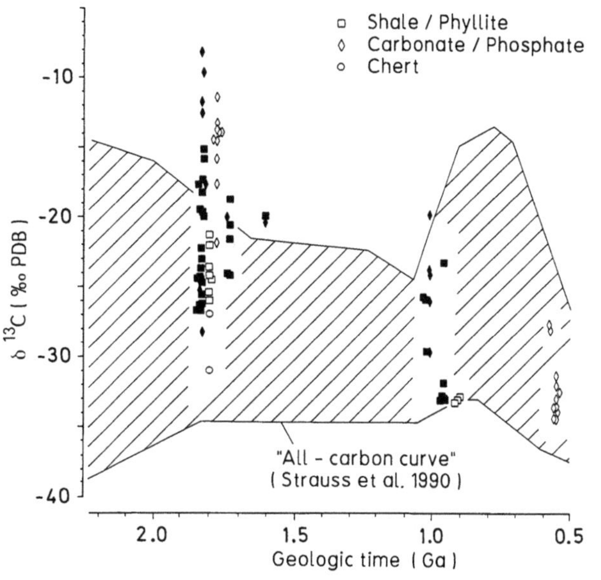

Fig. 3. Carbon isotopic composition for organic matter through Proterozoic time superimposed on the *All-carbon-curve* by Strauss et al. (1992b, and this Vol.). *Filled symbols* are from this study, *open symbols* are literature data

Gray shales and carbonates from the Chattisgarh Basin have TOC values between 0.12 and 3.95% and carbon isotopic compositions between −29.7 and −19.8‰.

Vindhyan black shales associated with syngenetic pyrite show light organic carbon isotopic values mostly around −33‰.

Carbonates and phosphorites from the Krol and lower Tal Formations have TOC values between 0.15 and 2.19%. $\delta^{13}C_{org}$ values range from −34.5 to −27.7‰.

6 Discussion

As anticipated from similar previous studies (for reviews see Hayes et al. 1983; Strauss et al. 1992a, b, and this Vol.), TOC values vary considerably for the Indian Proterozoic sediments so far analyzed. These data, however, neither reflect the initial abundance of organic matter in the sediment nor indicate its possible source. The measured organic carbon content simply represents the fraction of the organic matter deposited with the sediment that has survived the long chain of microbial reworking and postdepositional, thermal alteration during diagenesis and metamorphism.

Of greater significance is the isotopic composition of the organic matter, as this might record changes in the relative burial rates of organic and carbonate carbon or possibly indicate any evolutionary developments (e.g., Strauss et al. 1992b, and this Vol.). The values obtained here show comparatively large variations within a single time period. Much of this variation is likely to be caused by postdepositional alteration and metamorphism accompanied by isotopic shifts from the original to heavier values (Deb 1989; Strauss et al. 1992b, and this Vol.). Alternatively especially where stromatolitic associations are evident, some of these variations may be explained in terms of diffusion-limited ^{13}C-enrichment in cyanobacterially dominated environments (Schidlowski et al. 1985; Banerjee et al. 1986).

Figure 3 displays organic carbon isotopic compositions for individual samples analyzed during this study, as well as previously published measurements on Proterozoic rocks from India. Superimposed on these individual data points is the broad envelope of $\delta^{13}C_{org}$ values for total organic carbon analyses of Proterozoic rocks (hereafter called the "all-carbon curve") compiled and discussed in detail by Strauss et al. (1992b, and this Vol.).

The carbon isotope values plotted in Fig. 3 vary considerably, both with age and within rock sequences of similar age, although the majority fall in the upper part or even above the "all-carbon curve". Strauss et al. (this Vol.) have made a strong plea for the study of preservational aspects of organic matter and indicated a clear-cut relationship between preservation of organic matter, as inferred from the H/C ratio of its kerogen fraction, and its carbon isotopic composition. A limited number of C, H, N analyses have been performed on the present samples (Table 1). Low H/C ratios between 0.08 and 0.3 indicate rather poor preservation of the original organic material. In general, such low values denote an advanced degree of thermal degradation. This is commonly accompanied by a dramatic shift of the initial isotopic signature, most likely towards heavier values (McKirdy and Powell 1974; Hoefs and Frey 1976; Hayes et al. 1983).

Many sample points from the Aravalli and Delhi Supergroups fall above the "all-carbon curve" and isotopic exchange during greenschist facies metamorphic conditions is suggested. In particular, an unusually heavy value of $-8‰$ for a small carbonate enclave at the boundary between the Archean Banded Gneiss Complex and the overlying Aravalli sequence seems to indicate severe alteration of the organic material resulting in a substantial shift of the original isotopic value. A set of data points for little-altered stromatolitic phosphate-rich samples with isotopically heavy organic carbon around $-15 \pm 2.9‰$ (Fig. 3) has previously been suggested to reflect original, diffusion-limited ^{13}C-enrichment (Banerjee et al. 1986). It should be noted, however, that an alternative mechanism has been proposed for cyanobacterial mat organic matter with extremely ^{13}C-enriched isotopic values. Haddad et al. (1988) clearly demonstrated that rapid rates of photosynthetic carbon fixation by cyanobacterial mat communities is accompanied by zero discrimination against ^{13}C.

Results for samples from the Chattisgarh Basin are generally in good agreement with previously published data for this time span. Two isotopically heavier samples come from altered field sections.

Isotopically light $\delta^{13}C_{org}$ values around $-33‰$ characterize black shales of the Vindhyan Supergroup analyzed during this study.

Upper Proterozoic-Lower Cambrian samples taken from the literature (Fig. 3; Banerjee et al. 1986) show $\delta^{13}C_{org}$ values around $-28‰$ (Krol Formation) and from -35 to $-32‰$ (Tal Formation), respectively. This is consistent with the global trend of the marine carbonate carbon isotopic record across the Precambrian/Cambrian boundary (Hsü et al. 1985; Knoll et al. 1986; Awramik et al. 1986; Magaritz et al. 1986; Tucker 1986; Lambert et al. 1987; Aharon et al. 1987; Aharon and Liew, this Vol.).

7 Conclusions

The limited set of organic carbon isotopic data presented here for four selected Proterozoic sedimentary basins of India show a great variability which is consistent with previously published total organic carbon data from Proterozoic sediments on other continents. By inference from independent studies (Strauss et al., this Vol.), part of this variation is suggested to reflect postdepositional thermal alteration resulting in isotopic shifts towards heavier values. A more definite interpretation of the isotopic data hinges on the results of work in progress, particularly on the preservational status of the organic material under study.

Acknowledgments. This paper was presented at the final meeting of the IGCP Project 157 at Maria Laach, FRG and appreciation is expressed to its convenor M. Schidlowski. Mass spectrometric analyses were performed at Mainz and Los Angeles and financial support by the Alexander von Humboldt

Foundation (to D. M. B.) and the Precambrian Paleobiology Research Group – Proterozoic (to H.S.), respectively, is greatly appreciated. Constructive reviews by D. M. McKirdy and M. Schidlowski significantly improved this manuscript. Thanks are also due to P. Bhattacharya for help during preparation of this contribution.

References

Aharon P, Schidlowski M, Singh IB (1987) Chronostratigraphic markers in the end-Precambrian carbon isotope record of the Lesser Himalaya. Nature (London) 327:699–701

Auden JB (1933) Vindhyan sedimentation in Son Valley, Mirzapur district. Mem Geol Surv India 62:141–250

Auden JB (1934) The geology of the Krol Belt. Rec Geol Surv India 67:461–471

Awramik SM, Conway-Morris S, Bengston S (1986) The Precambrian-Cambrian boundary and geochemical perturbations. Nature (London) 319:696–697

Banerjee DM (1990) Carbon isotope variations in Cambrian-Proterozoic rocks – a case for secular global trend. In: Naqvi SM (ed) Developments in Precambrian geology, vol. 8. Elsevier, Amsterdam, pp 453–470

Banerjee DM, Schidlowski M, Arneth JD (1986) Genesis of Upper Proterozoic-Cambrian phosphorite deposits of India: isotopic inferences from carbonate fluorapatite, carbonate and organic carbon. Precambrian Res 33: 239–253

Brasier MD, Singh P (1987) Microfossils and Precambrian-Cambrian boundary stratigraphy at Maldeota Lesser Himalaya. Geol Mag 124:323–345

Crawford AR, Compston W (1970) The age of the Vindhyan System of Peninsular India. Q J Geol Soc London 125:351–371

Deb M (1986) Sulfur and carbon isotope compositions in the stratiform Zn–Pb–Cu sulfide deposits of the Rajpura-Dariba belt, Rajasthan, NW India: a model of ore genesis. Mineral Depos 21:313–321

Deb M (1989) Isotopic composition of carbon in the sulfide ore environments in the Proterozoic Aravalli-Delhi orogenic belt, NW India. Int Geol Congress, Washington (Abstr)

Deb M, Thorpe RI, Cummings GL, Wagner PA (1989) Age, source and stratigraphic implications of the Pb isotopic data for conformable, sediment-hosted base metal deposits in the Proterozoic Aravalli-Delhi orogenic belt, NW India. Precambrian Res 43:1–22

Durand B (ed) (1973) Kerogen: insoluble organic matter from sedimentary rocks. Editions Technip, Paris, 519 pp

Dutt NVBS (1964) A suggested succession of the Purana Formations of the southern part of Chattisgarh M.P. Rec Geol Surv India 93: 143–148

Gupta SN, Arora YK, Mathur RK, Iqbaluddin, Prasad B, Sahai TN, Sharma SB (1980) Lithostratigraphic map of Aravalli region, southern Rajasthan and northeastern Gujarat. Geol Surv India Publ Hyderabad

Haddad RI, Des Marais DJ, Nguyen H, Bauer J, Nguyen M (1988) Carbon isotope cycling in a recent microbial mat. EOS 69:1083–1084

Hayes JM, Kaplan IR, Wedeking KW (1983) Precambrian organic geochemistry, preservation of the record. In: Schopf JW (ed) Earth's earliest biosphere: its origin and evolution. Univ Press, Princeton, pp 93–134

Heron AM (1953) The geology of Central Rajputana. Mem Geol Surv India 79:389 p

Hoefs J, Frey M (1976) Isotopic composition of carbonaceous organic matter in a metamorphic profile from the Swiss Alps. Geochim Cosmochim Acta 40:945–951

Hsü KJ, Oberhänsli H, Gao JY, Sun S, Chen H, Krähenbühl U (1985) Strangelove ocean before the Cambrian explosion. Nature (London) 316:809–811

Jairaman R, Banerjee DM (1978) Preliminary studies on the stromatolites from the Raipur area, Chattisgarh Basin. Geol Surv India Misc Publ 44:57–67

Knoll AH, Hayes JM, Kaufman AJ, Swett K, Lambert IB (1986) Secular variation in carbon isotope ratios from Upper Proterozoic successions of Svalbard and East Greenland. Nature (London) 321:832–838

Kreuzer H, Karre W, Kursten M, Schnitzer WA, Murti KS, Srivastava NK (1977) K/Ar dates of two glauconites from the Chandapur Series (Chattisgarh, India): on the stratigraphic status of the Late Precambrian basins in Central India. Geol Jahrb B 28:23–26

Krishnamurthy RV, Bhattacharya SK, Mathur SM (1987) Carbon isotope ratio of coaly matter from the basal part of the Proterozoic Vindhyan Supergroup. J Geol Soc India 27:119–120

Lambert IB, Walter MR, Zang W, Lu S, Ma G (1987) Paleoenvironment and carbon isotope stratigraphy of Upper Proterozoic carbonates from the Yangtze Platform. Nature (London) 325:140–142

Magaritz M, Holser WT, Kirschvink J (1986) Carbon-isotope events across the Precambrian/Cambrian boundary on the Siberian Platform. Nature (London) 320: 258–259

McKirdy DM, Powell TG (1974) Metamorphic alteration of carbon isotopic composition in ancient organic matter: new evidence from Australia and South Africa. Geology 2:591–595

Murti KS (1987) Stratigraphy and sedimentation in the Chattisgarh Basin. Geol Soc India Mem 6:239–260

Naqvi SM, Rogers JJW (1987) Precambrian geology of India. Univ Press, Oxford, 220 pp

Oldham T (1856) Remarks on the classification of the rocks of Central India resulting from the investigation of the Geological Survey. J Asiat Soc Bengal 25:224–256

Sathyanarayan S, Schidlowski M, Arneth JD (1987) Stable isotope geochemistry of sedimentary carbonates from the Proterozoic Kaladgi, Badami and Bhima Groups, Karnataka, India. Precambrian Res 37:147–156

Schidlowski M, Eichmann R, Junge CE (1975) Precambrian sedimentary carbonates: carbon and oxygen isotope geochemistry and implications for the terrestrial oxygen budget. Precambrian Res 2:1–69

Schidlowski M, Hayes JM, Kaplan IR (1983) Isotopic inferences of ancient biochemistries: carbon, sulfur, hydrogen, and nitrogen. In: Schopf JW (ed) Earth's earliest biosphere: its origin and evolution. Univ Press, Princeton, pp 149–187

Schidlowski M, Matzigkeit U, Mook WG, Krumbein WE (1985) Carbon isotope geochemistry and ^{14}C ages of microbial mats from the Gavish Sabkha and the Solar Lake. In: Friedman GM Krumbein WE (eds) Hypersaline ecosystems: the gavish sabkha 53. Springer, Berlin Heidelberg New York, pp 381–401

Schnitzer WA (1971) Jungpräkambrium Indiens ("Purana System"). Erlanger Geol Abh 85:1–44

Singh IB (1978) Sedimentological evolution of the Krol Belt sediments. Himalayan Geol 8:657–682

Soni MK, Chakraborty S, Jain VK (1987) Vindhyan Supergroup – a review. Geol Soc India Mem 6:87–138

Strauss H, Des Marais DJ, Summons RE, Hayes JM (1992a) Abundance of carbon, maturity and elemental composition of kerogen. In: Schopf JW, Klein C (eds) The Proterozoic biosphere: a multidisciplinary study. Univ Press, Cambridge, (in press)

Strauss H, Des Marais DJ, Hayes JM, Summons RE (1992b) The carbon isotopic record of the Proterozoic. In: Schopf JW, Klein C (eds) The Proterozoic biosphere: a multidisciplinary study. Univ Press, Cambridge, (in press)

Tucker ME (1986) Carbon isotope excursions in Precambrian/Cambrian boundary beds, Morocco. Nature (London) 319:48–50

Valdiya KS (1969) Stromatolites of the Lesser Himalayan carbonate formations and the Vindhyans. J Geol Soc India 10:1–25

Verma RK, Samaddar AK, Murti KS (1977) Paleomagnetic study of Gunderdehi Shale of Raipur Group from Chattisgarh Basin in India. Indian J Earth Sci 4:183–191

Vinogradov AP, Tugarinov AI, Zhykov CI, Stapnikova NI, Bibikova EV, Khorre KG (1964) Geochronology of the Indian Precambrian. Proc 22nd Int Geol Congr 10:553–567

Organic Free Radicals in Precambrian and Paleozoic Rocks: Origin and Significance

Pavle I. Premović [1]

[1] Laboratory for Geochemistry, Department of Chemistry, Faculty of Science, University of Niš, 18000 Niš, Yugoslavia

Abstract

Fourteen Precambrian kerogens including seven isolated from stromatolites were studied by electron spin resonance (ESR). Organic free radicals were detected in only three of these kerogens: those from the Gunflint and Bitter Springs cherts and the Nonesuch shale. All three rocks are known to contain organically preserved microfossils. Comparative studies were conducted on the kerogens of eight fossiliferous Paleozoic rocks and a Jurassic anthracite. Careful measurements were made of g-values, line-widths, line-shapes, and integrated intensities of the observed signals. The kerogen radicals are believed to be polyaromatic structures with unpaired electrons stabilized as π electrons. The marked similarity of the ESR spectral parameters of the free radicals in Precambrian and Paleozoic kerogens and the Jurassic Vrška Čuka anthracite serves to strenghten the view that these radicals are relics of early biochemical processes.

It is suggested that chemical progenitors of Precambrian kerogens and associated free radicals are the corresponding sedimentary humic substances derived from algal and/or microbial sources. Interpretation of significant changes in spin concentration observed during pyrolysis of Precambrian kerogens containing radicals is based on published work on pure bituminous coal macerals (vitrinites and exinites) and anthracites. Experimental pyrolytic data and other evidence suggest that the Gunflint and Bitter Springs rocks have been exposed to temperatures of the order 100–150 °C during their burial histories through proximity to magmatic bodies and subsidence, respectively.

1 Introduction

Following a brief introduction to some of the basic aspects of electron spin resonance (ESR), the geochemical value of organic free radicals in the study of Precambrian kerogens is reviewed. The results of recent ESR studies on organic free radicals in selected Precambrian and Paleozoic kerogens are presented and discussed in terms of their structure, properties, origin, and implications for the origin of organic matter preserved in Precambrian sediments.

2 ESR and Organic Free Radicals

The electron has associated with it a magnetic moment and so it may be treated as a tiny bar magnet. Free electrons (in the absence of a magnetic field) are aligned at random, but in the presence of an external magnetic field, the lower energy state occurs when the electrons are aligned with the magnetic field. A higher energy state exists when the electrons line up in opposition to the magnetic field. The difference in energy between these two states (ΔE) is proportional to the applied field (H_o), thus

$$\Delta E = g \beta_e H_o .$$

The constant of proportionality (g) equals 2.0023 for a free electron and is called the spectroscopic splitting factor, or the g-factor. The value of g varies from 2.0023 when the unpaired electron is no longer free (i.e., when it is in a molecule). β_e is called the Bohr magneton: the magnetic moment of electron.

In quantum mechanical terms, the allowable electron spin states can only take one set of discrete values (i.e. $m_s = +1/2$ and $-1/2$). Electrons in the lower state, $m_s = -1/2$, can be excited to the upper state, $m_s = +1/2$ (Fig. 1 a) and for a given applied field the energy required for the transition will be

$$\Delta E = h\nu = g \beta_e H_o ,$$

where ν is the frequency of the applied electromagnetic radiation. Since the energy gap is very small, the most convenient radiation frequencies occur when the magnetic field is about 0.35 T (3500 gauss) and lie in the microwave region of the electromagnetic spectrum (ca. 9 GHz).

M. Schidlowski et al. (Eds.)
Early Organic Evolution: Implications for Mineral and Energy Resources

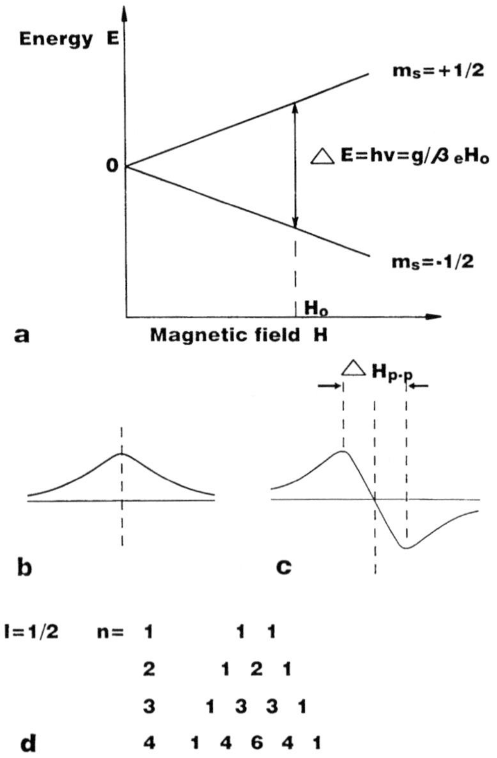

Fig. 1a–d. **a** Effect of an external magnetic field upon the energy levels of an electron and transition induced by an electromagnetic field. **b** Lorentzian zero derivative line-shape. **c** Lorentzian first derivative line-shape. **d** Line intensities for *n* (1–4) equivalent protons ($I = 1/2$)

The intensity, or area, beneath the ESR absorption curve (Fig. 1b) is proportional to the number of unpaired spins in the sample. For instrumental purposes and for improved resolution, an ESR spectrum is usually recorded as the first derivative of the absorption spectrum (Fig. 1c) instead of absorption itself. The linewidth ΔH_{p-p} is the distance, in mT, between the two maxima (peak to peak) of the first derivative curve and thus represents the width of the absorption curve at the level of maximum slope.

The nucleus also has associated with it the property of spin. This is given by the nuclear spin quantum number I which may have values of 0, ½, 1, etc. In an atom subjected to an external magnetic field, the electron and nuclear spins will interact and this interaction yields $(2I + 1)$ ESR lines which are, to a first approximation, equally separated with a spacing *a* (hyperfine splitting constant). If there are n equivalent protons ($I = ½$), the number of lines will be $(n + 1)$. The relative intensities of these lines are given in Fig. 1d.

In normal covalently bonded organic molecules all the electrons occur in pairs. Each pair is contained by the same molecular orbital (i.e., the spins of electrons are opposed); and hence the net magnetic moment of a pair is zero. Organic molecular species are known in which there is one unpaired electron in the molecule, and these are commonly known as "organic free radicals". Such radicals are paramagnetic and can be studied by ESR.

Normally, organic species containing unpaired electrons are unstable, rapidly decomposing, or dimerizing to give a new orbital. Organic free radicals can be relatively stable if the unpaired electrons are delocalized over an extensive conjugated system of double bonds, for example in certain aromatic molecules. These radicals normally are very short-lived but some may be trapped in the solid organic structure, effectively preventing dimerization. Such "trapped" organic free radicals are known to be very stable, although in some cases their ESR signal is broadened by oxygen, which is also paramagnetic.

3 Polyaromatic Paramagnetic Structures in Coals and Kerogens

The discovery of ESR absorption in natural carbons by Uebersfield et al. (1954) and Ingram et al. (1954) prompted researchers to apply the technique to coal and coal-like materials. Marchand et al. (1968) used ESR to demonstrate the presence of stable organic free radicals in kerogens (i.e., insoluble organic matter in carbonaceous sedimentary rocks).

It is generally thought that there are structural similarities between coals and kerogens, especially at higher levels of thermal maturation. Suggested structural models for both materials when relatively immature comprise polyaromatic centers crosslinked mainly by aliphatic bridges (Fig. 2a). With increasing rank the average number of stacked aromatic sheets further increases in coals, as well as in kerogens (Fig. 2b), until (through the catagenetic and metagenetic phases of maturation) a structure resembling graphite is approached.

The ESR spectral parameters of interest are the g-value of the organic free radicals in coals and kerogens, radical concentration, peak-to-peak line-width $\Delta Hp-p$ (of the first derivative signal) and in some cases, the spectral line-shape. The g-value is a parameter which is particulary sensitive to the chemical environment of the unpaired electron. The g-values of organic free radicals in coals and kerogens are greater than that of the free electron (2.0023) and range from 2.0027 to 2.0042 (Yen and Sprang 1977). Spin (radical) concentration is a measure of the number of unpaired

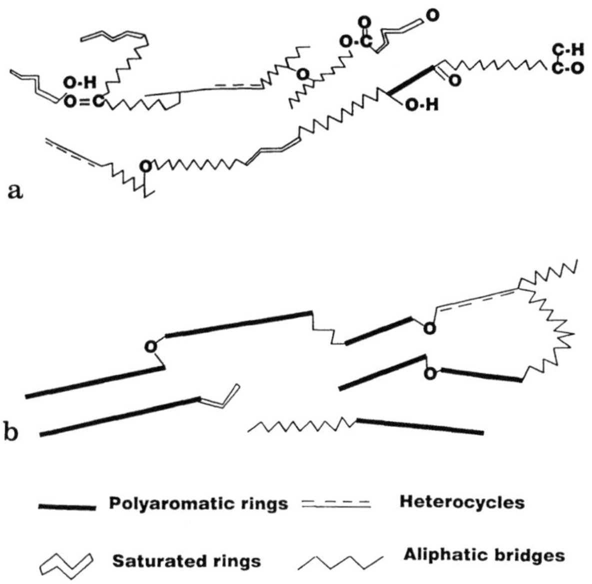

Fig. 2a, b. Simplified structural models for the macromolecular structure of kerogen. **a** Immature and **b** very mature. (Tissot and Welte 1978)

spins (electrons) in a given amount of sample. The spectral line-width, $\Delta Hp-p$, in the case of both coals and kerogens, is probably determined by the proximity of the unpaired electron to nuclei with magnetic moments, such as proton and nitrogen (hyperfine interactions). Typical ESR spectra of both coal and kerogen samples are single symmetric curves without hyperfine structures. Spectral line-shapes in these samples are largely affected by unresolved hyperfine structure and anisotropic effects. Because of the complex mixture of compounds present in both coals and kerogens, the g-value, line-width and line-shape are composite properties of all the organic free radicals present in a sample. The position and characteristics of the ESR signal envelope are greatly influenced by the relative concentrations of the different radicals.

Although the exact nature of the paramagnetic species responsible for the ESR signal has not been established unambigously, various ESR parameters indicate that unpaired electrons, at least in coals and kerogens of high rank, are associated with similar polyaromatic clusters. For these reasons, it seems preferable to refer to free radicals in coals and kerogens as polyaromatic paramagnetic structures (PPS). The term PPS also encompasses polyaromatic charge transfer (CT) complexes (Premović 1982; Pryor et al. 1983). Unpaired electrons of most PPS are stabilized, as π electrons.

Free electron transfer between adjacent aromatic sites is possible and results in exchange effects. Where the exchange frequency is small, compared with the hyperfine splitting constants of protons, the envelope formed by unresolved hyperfine lines adopts a Gaussian shape (Fig. 1 b). On the other hand, if the exchange frequency is much larger than the hyperfine splitting constants and exchange narrowing occurs, the line-shape is Lorentzian (Fig. 1 c). Thus line-shape and line-width together indicate the exchange phenomenon, and the size and compactness of the aromatic clusters. PPS trapped in the matrix of coals and kerogens owe their stability to the immobility of the solid structure of these materials and an inaccessibility to oxygen.

Ignoring natural radiolytic processes there are two possibilities for the formation of PPS in kerogens: (1) PPS formed during diagenesis, or (2) PPS formed during catagenesis (Austen et al. 1966). The second possibility is consistent with the well-documented increase of radical concentration with kerogen rank (Marchand and Conrad 1980) and would also account for the variation of g-values observed by Durand et al. (1977).

The main chemical progenitors of PPS in humic coals (e.g., the Vrška Čuka anthracite) are lignins (Rex 1960; Steelink 1964; Hatcher et al. 1981, 1982). These compounds are phenolic polymers of relatively high chemical stability which occur as major components of the conductive tissues of vascular plants. From an evolutionary viewpoint, lignins first appeared in brown algae (Reznikov et al. 1978).

The composition of a kerogen depends firstly on the nature of its precursor organisms and in part on its progress towards the final stage of maturation (i.e., graphitization). Kerogens may originate from either higher plant debris or microbial remains, or from a mixture of both types of material (Tissot and Welte 1978). At low maturity levels, both bacterial and algal kerogens are complex organic materials containing aliphatic and aromatic moeities (McKirdy et al. 1980), whereas vascular plant kerogens comprise predominantly polyaromatic structures. Thus chemical distinction between higher plant and microbial kerogens is clear at low levels of maturation but is difficult to make in very mature materials. For this reason, it is important to first establish the rank of Precambrian (and Paleozoic) kerogens before interpreting their ESR characteristics.

4 Samples and Methods

The 25 kerogens considered in this paper (Table 1) include seven isolated from stromatolites and range in

Table 1. Evidence of biological activity in selected Precambrian and Paleozoic Rocks

Lithology and rock	Region	Geologic period	Age [Ga]	B	Cb	G	F	V	PPS
Onverwacht Series:	S. Africa	E. Precambrian	3.4						
Kromberg stromatolitic chert									
Hooggenoeg stromatolitic chert									
Fig Tree shale	S. Africa	E. Precambrian	3.1						
Pongola stromatolitic limestone	S. Africa	E. Precambrian	3.0						
Soudan shale	USA	E. Precambrian	2.7						
Bulawayan Group:	Zimbabwe	E. Precambrian	2.7						
Huntsman stromatolitic limestone									
Cheshire stromatolitic limestone									
Gunflint stromatolitic chert	Canada	M. Precambrian	1.9	+	+				+
Michigamme coaly rock	USA	M. Precambrian	1.7						
Sudbury anthraxolite	Canada	M. Precambrian	1.7						
Onwatin slate	Canada	M. Precambrian	1.7						
Paradise Creek stromatolitic chert	Australia	L. Precambrian	1.6	+	+		+		?
Nonesuch shale	USA	L. Precambrian	1.1	+	+	+	+		+
Bitter Springs stromatolitic chert	Australia	L. Precambrian	0.9	+	+	+	+		+
Zvonce black shale	Serbia	M. Silurian	0.4	+	+	+	+		+
Rhynie chert	Scotland	E. Devonian	0.4	+	+	+	+	+	+
Sosnowiec stromatolitic limestone	Poland	L. Devonian	0.4	+	+	+	+		+
Sosnowiec limestone	Poland	L. Devonian	0.4	+	+	+	+		+
Antrim black shale	USA	L. Devonian	0.4	+	+	+	+	+	+
Silicified wood	Croatia	Permian	0.26				+	+	+
Tasmanite	Australia	Permian	0.26	+				?	+
Torbanite	Scotland	Permian	0.26			+		?	+
Vrška Čuka anthracite	Serbia	Jurassic	0.19					+	+

B: bacteria; Cb: cyanobacteria; G: green algae; F: fungi; V: vascular plant.

age from Early Precambrian to Devonian. For comparison, ESR data on the Jurassic Vrška Čuka anthracite and a Permian silicified wood are also included. The microbial origin of the Precambrian samples was established on the basis of their pre Devonian age. In the case of kerogens isolated from younger sedimentary rocks, recognition of their precursor biota depended on paleontological and other organic geochemical evidence.

The basic instrumentation and experimental procedures for the ESR analysis of kerogen have been described by Premović (1986) and Premović et al. (1989). Kerogen was separated from acid-resistant mineral grains by the chloroform sink-float technique of McKirdy and Powell (1974). The limit of PPS detection for the ESR spectrometer employed is about 10^{14} spins g^{-1}.

5 ESR of Precambrian and Paleozoic Kerogens

It can be seen from Table 2 that, unlike the Precambrian kerogens, all the Paleozoic and younger materials contain relatively high concentrations of PPS ($> 10^{18}$ spins g^{-1}). PPS could not be detected in all but three of the Precambrian kerogens, viz. those from the Gunflint and Bitter Springs cherts and the Nonesuch shale. They are, in fact, the most characteristic structural feature of these three kerogens (Jovanović 1989). Their PPS concentrations are comparable with those found in much younger Paleozoic materials Jurassic Vrška Čuka anthracite (Table 2). Figures 3 and 4 show representative ESR spectra of Precambrian and younger materials, respectively, which contain PPS. These spectra are devoid of any hyperfine structure and comprise single, nearly Lorentzian, symmetrical signals (except that of tasmanite which is slightly asymmetric). They have g-values between 2.0027 to 2.0037 and line-widths (ΔH_{p-p}) between 0.32 mT to 0.62 mT. The nearly Lorentzian shape of the signals indicates that exchange and/or delocalization narrowing occur.

The Precambrian kerogens containing PPS in Table 2 have attained very high levels of maturation. This conclusion is based on several lines of geochemical evidence. Firstly, their atomic hydrogen/carbon (H/C) ratios are lower than 0.7 (Jovanović 1989). The various kerogen types (e.g., vascular plant versus algal), seem to be structurally equivalent at H/C ratios less than 0.7 (Hayes et al. 1983). Secondly, their spin concentrations are high (10^{18} spins g^{-1}). It has been shown (Pusey 1973; Hwang and Pusey 1973; Durand

Table 2. ESR parameters of PPS in selected Precambrian and Paleozoic kerogens

Lithology and rock unit	Atomic H/C ratio[a]	$g \pm 0.0003$	ΔHp-p $\pm$ 0.05 mT	Spin concentration $\times 10^{18}$ spins per gram
Gunflint stromatolitic chert	0.50	2.0027	0.32	16
Nonesuch shale	n.a.[b]	2.0037	0.50	4
Bitter Springs stromatolitic chert	0.68	2.0032	0.34	2
Zvonce black shale	n.a.	2.0037	0.55	9
Rhynie chert	0.86	2.0036	0.59	7
Sosnowiec stromatolitic limestone	0.88	2.0030	0.42	4
Sosnowiec limestone	n.a.	2.0030	0.44	11
Antrim black shale	n.a.	2.0034	0.52	8
Silicified wood	n.a.	2.0035	0.47	1
Tasmanite	1.52	2.0036	0.56	0.4
Torbanite	1.56	2.0036	0.62	0.3
Vrška Čuka anthracite	0.51	2.0034	0.49	45

[a] Dry, ash-free corrected kerogen data.
[b] Not analyzed.

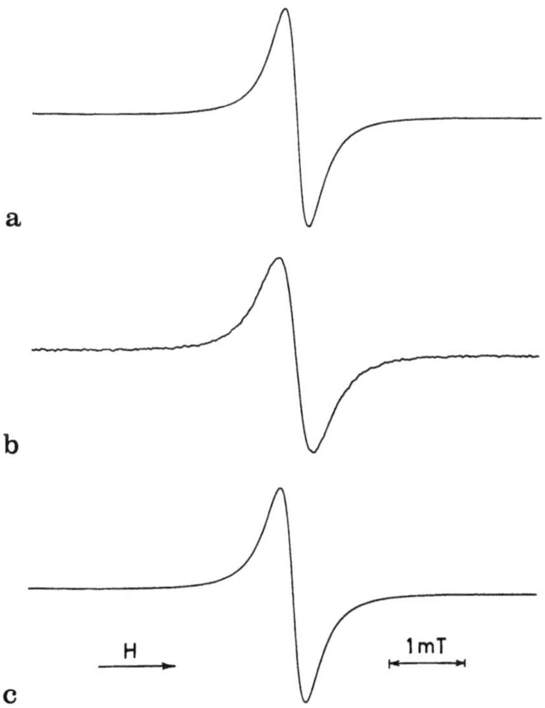

Fig. 3a–c. ESR spectra of Precambrian kerogen PPS. **a** Gunflint chert. **b** Nonesuch shale. **c** Bitter Springs chert

et al. 1977) that spin concentration increases smoothly as the kerogen matures. The very large number of measurements available for correlative studies has made the spin concentration an extremely reliable indicator of kerogen maturity below a rank equivalent to 2% vitrinite reflectance (Durand et al. 1977). Finally, their infrared (IR) spectra are characterized by strong but featureless absorption throughout the range 650–4000 cm^{-1}. Similar spectra were observed for coals of very high rank (e.g., the Vrška Čuka anthracite: Komatinović 1984).

Undoubtedly, the most intriguing result of the present investigation is the striking similarity between the ESR spectral features of PPS in the Precambrian and younger kerogens listed in Table 2. This suggests that the Precambrian kerogens containing PPS, or at least that part of their structure with unpaired spins, may be inherited directly from their microbial precursors.

6 Humic Substances and the Origin of Kerogen-Hosted PPS

Most students of the subject now agree that Devonian and younger kerogens consist of diagenetically altered humic substances (i.e., humic and fulvic acids: Tissot and Welte 1978). Numerous attempts have been made to devise a structural formula representative of humic substances, but none has proved entirely satisfactory (Stevenson 1982). Humic substances cannot be considered true phenolic polymers because other widely diverse structures, such as aliphatic and polyaromatic hydrocarbons, are also present in their macromolecular skeletons.

Four major pathways by which the decomposition products of plants and microorganisms may give rise to humic substances are illustrated in Fig. 5. Briefly, these are as follows:

1. Lignins are degraded by microorganisms and the residues become part of the humic macromolecular structure (Flaig 1972).

2. Phenolic compounds released from lignins during microbial metabolism undergo enzymatic conversion to quinones, which then polymerize in the presence or absence of amino acids to form humic materials (Hurst and Burgess 1967).

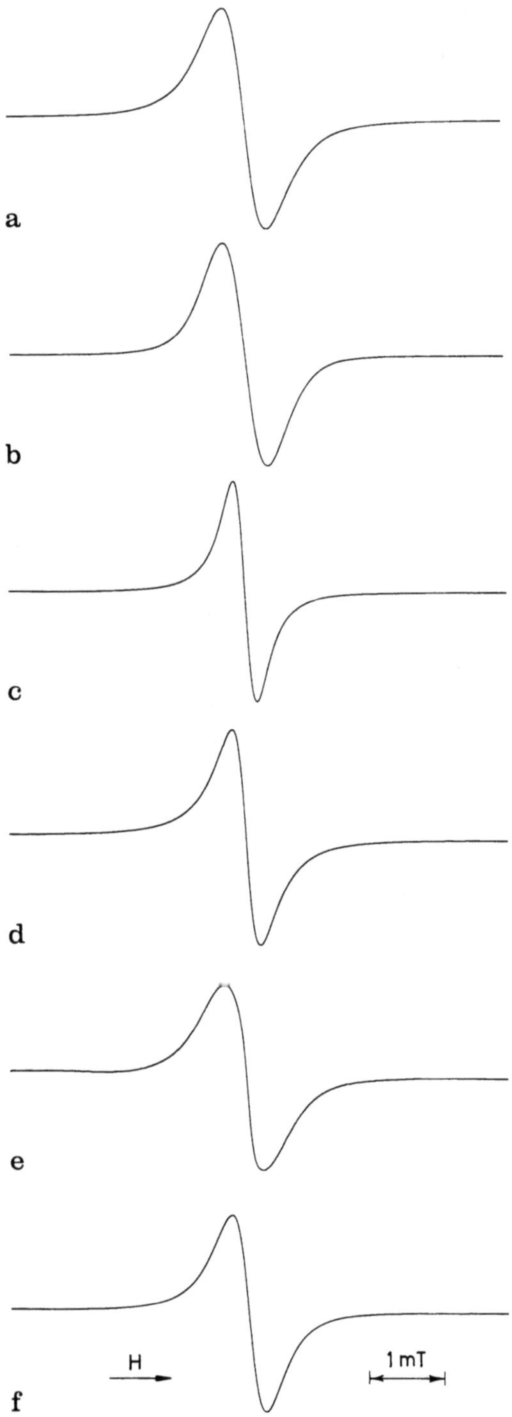

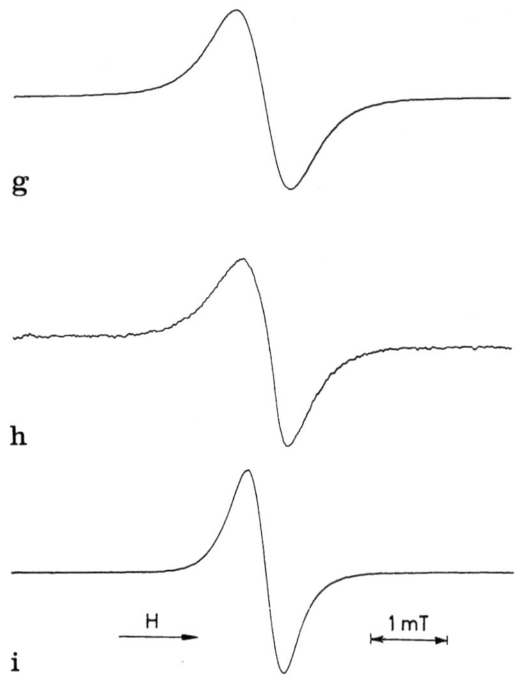

Fig. 4a–i. ESR spectra of PPS in Paleozoic kerogens. **a** Zvonce shale. **b** Rhynie chert. **c** Sosnowiec stromatolitic limestone. **d** Sosnowiec limestone. **e** Antrim shale. **f** Permian silicified wood. **g** Torbanite. **h** Tasmanite. **i** Vrška Čuka anthracite

3. Polyphenols are produced by microorganisms from non-lignin organic sources (e.g., cellulose). The polyphenols are then enzymatically oxidized to quinones and transformed to humic substances (Kononova 1966).

4. Reducing sugars and amino acids, as by-products of microbial metabolism, undergo nonenzymatic polymerization (the Maillard reaction) to form humic-like materials relatively rich in nitrogen (Ikan et al. 1986).

The four pathways suggested in Fig. 5 may operate in a variety of sedimentary environments, but not to the same extent or in the same order of importance. Lignin pathways 1 and 2 may have predominated in Devonian and younger sedimentary environments to which vascular plants contributed detritus (e.g., the Antrim shale), whereas the polyphenol concept (3) or Maillard pathway (4) may have been of considerable importance in Precambrian and early Paleozoic sedimentary environments. Paleontological and other evidence does, indeed, show the absence of vascular plant fossils before the Middle Silurian (Stewart 1983), so it seems plausible to postulate that sedimentary (marine or terrestrial) humic substances during the Precambrian and early Paleozoic were derived from the remnants of nonlignin-bearing (or nonvascular) organisms, i.e., algae and/or bacteria. If this concept is correct then highly branched aliphatic structures would constitute a much larger fraction of pre-Devonian kerogens, as these chemical structures account for a major portion of humic substances derived from algal and/or microbial sources (Hat-

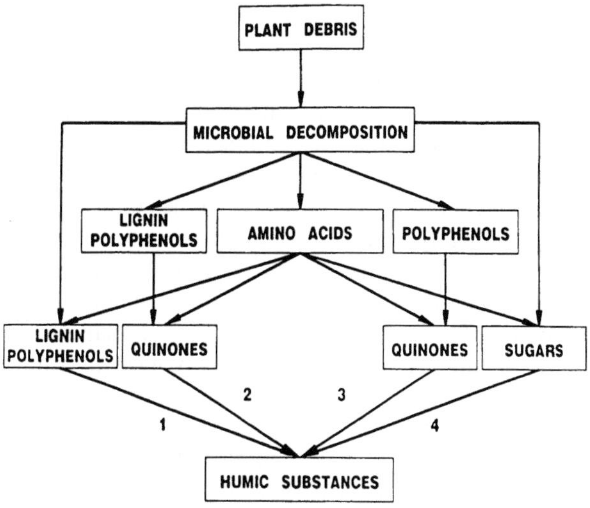

Fig. 5. Mechanisms for the formation of humic substances

cher 1980). Clearly, identification of particular types of microbial progenitor(s) for Precambrian kerogens and their constituent PPS would greatly advance our efforts at reconstruction of the paleoenvironments inhabited by Precambrian microbiota.

Although it is widely believed that there was no land-plant cover prior to the evolution of vascular plants, this conclusion has not been universally accepted. According to Jackson (1973), late Precambrian continents were colonized by green algae. Subsequently, during the early Paleozoic, higher plants may have evolved from these terrestrial green algae, their phenolic secondary metabolites being the evolutionary precursors of lignins. Jackson based his view on quantitative ultraviolet/visible (UV/VIS) and IR spectrophotometric analyses of bitumens extracted from a suite of carbonaceous sedimentary rocks varying in lithology and depositional environment and ranging in age from Early Precambrian to Miocene, with the majority of his samples being of pre-Paleozoic age. Unfortunately our knowledge of paleosols, especially those of Precambrian age, is so meager (Retallack 1981) that it is impossible to say whether or not the geological record bears out Jackson's hypothesis.

Sedimentary humic substances, regardless of chemical composition or place of origin, contain relatively high concentrations of very stable organic free radicals, of the order of 10^{18} spins g^{-1}. An ESR spectrum of sedimentary humic substances consists generally of a single near-Lorentzian profile without hyperfine structure. The g-values and line-widths of these paramagnetic species range from 2.0032 to 2.0035 and 0.36 mT (Ishiwatari 1974), respectively. According to Ishiwatari, organic free radicals in sedimentary humic substances are possibly associated with condensed aromatic ring system which also incorporate some alicyclic rings. The fact that their ESR parameters are similar to those of the kerogen PPS listed in Table 2 suggests that the same structural type may be present in the kerogens. It is clear that these humic PPS (semi-quinone radicals in a polyaromatic ring systems) are much less aromatic (i.e., less mature) than the kerogen PPS (carbon radicals in a polyaromatic ring systems) (Ishiwatari 1974). Nevertheless, it is not unreasonable to suggest that Precambrian kerogens containing PPS (or, at least, those parts of their macromolecular skeletons which are paramagnetic) may be relics of sedimentary humic substances.

7 The Origin of Marine Humic Substances

Lignins have been implicated as a major source material for terrestrial humic substances (Flaig et al. 1975). However, humic materials can accumulate in marine sediments lacking a lignin input (Vasilevskaya et al. 1977; Ertel and Hedges 1984). Elemental and carbon isotopic compositions (Nissenbaum and Kaplan 1972; Stuermer et al. 1978) and spectroscopic properties (Huc and Durand 1977; Hatcher et al. 1981) of marine humic substances differ significantly from those of terrigenous humic materials, indicating that the bulk of the carbon found in many recent marine sediments originates from autochthonous (mainly planktonic and/or bacterial) sources. Moreover, phenolic constituents are less abundant in marine than in terrigenous humic substances (Tissot and Welte 1978).

The amino-carbonyl reaction of sugars with amino acids (Maillard reaction) has been postulated to be of importance in the formation of terrestrial humic substances (Ikan et al. 1986). Most investigators, however, agree that this reaction contributes very little (if any) to the genesis of marine humic material (Stevenson 1975). The evidence supporting this view is diverse, but the most persuasive argument is that the sugar-amino condensation proceeds rather slowly at the temperatures normally found at the seawater/sediment interface or within the sediment. Another unattractive feature of this reaction is that, even at relatively high temperatures (say 90 °C), it requires molar concentrations of reactants (sugars, amino acids etc.) in order to proceed at a moderate rate and yield significant amounts of macromolecular humic-like material (Maillard 1912; Enders and Theis 1938; Hodge 1953; McWeeny et al. 1974; Stevenson 1975).

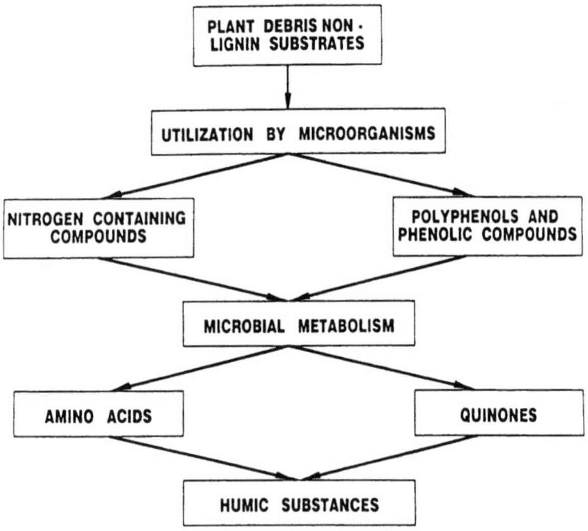

Fig. 6. Schematic representation of the nonlignin polyphenol theory of humic formation

So, there remains to be considered the polyphenol concept (pathway 3 in Fig. 5). A schematic representation of the polyphenolic synthesis of humic substances from non-aromatic substrate by microorganisms is shown in Fig. 6. Numerous phenolic compounds are produced from nonaromatic carbon sources (e.g., polysaccharides) through biosynthesis by microorganisms such as fungi, actinomycetes, and bacteria (Martin and Haider 1971; Steinbrenner and Matschke 1971). The resulting phenolic products (and any uncombined higher plant phenols present) are oxidized to quinones, either by enzymes or autooxidation. The oxidative coupling of phenolic substances and quinones results in the formation of humic-like polymers or melanins (Martin et al. 1972), in which the ESR signal is due to the presence of PPS of the semiquinone type (Schnitzer and Skinner 1969; Rifaldi and Schnitzer 1972; Saiz-Jimenez and Shafizadeh 1985). During diagenesis other materials (including carbohydrates, nitrogenous compounds, and lipids) may be incorporated into the developing polymer (Stevenson 1982). In addition, microorganisms may synthesize intra- and extracellular humic-like materials (Turner 1971). It appears that the PPS content of lignin-derived humic substances is higher than that of the marine humic polymers. In other words, the PPS contents of terrestrial and marine humic substances may reflect their different origins (Steelink 1966; Schnitzer 1971). These immature marine humic PPS are very stable (even in air) and can survive the geochemical processes of humification and diagenesis (immature kerogen PPS) until they are finally converted through catagenesis into mature kerogen PPS.

Although the role of polyphenol mechanism 2 (see Fig. 5) in the synthesis of marine humic substances is unclear, indirect evidence that polyphenols (structurally similar to lignins) contribute to their formation has come from comparative studies of chemical structures of terrestrial and marine humic acids (Hatcher 1980; Hatcher et al. 1981). These studies indicate that complex (highly branched) aliphatic structures constitute a large fraction of both terrestrial and marine humic materials, although overall the former are more aromatic. The authors concluded that similar source materials and/or mechanisms are involved in humic substance genesis in both marine and terrestrial aquatic environments. Recently, Ertel and Hedges (1984) used copper (II) oxide oxidation to examine the distribution of lignin structural units in terrestrial and marine humic substances. No lignin-derived phenols were detected in the marine materials: the sensitivity of the methods employed was about 4 ppm. A similar absence of these compounds in marine humic substances was reported by Vasilevskaya et al. (1977). Thus, one may reasonably conclude that marine humic substances (including those of Precambrian age) could result mainly from polycondensation of the organic residues of microbial (fungal, actinomycete and bacterial) metabolism.

8 Paleontology and Geochemistry of Precambrian Rocks: Origin of Kerogen and Associated PPS

8.1 The Early Precambrian (3.4–2.5 Ga)

Rocks of the Onverwacht and Fig Tree Groups contain possible microfossils resembling modern coccoid or unicellular cyanobacteria (e.g., Muir and Grant 1976; Knoll and Barghoorn 1977). The Pongola and Bulawayan stromatolitic limestones comprise laminae with a microstructure suggestive of construction by finely filamentous organisms of cyanobacterial affinity (Schopf 1972). Similar taxa are suggested for thread-like microfossils in carbonaceous sediment of the Soudan Iron Formation (Cloud and Licari 1968). However, the simple morphology of these microstructures, and their similarity to structured organic matter of nonbiological origins, make their unequivocal identification as fossil entities virtually impossible (Walter 1983; Schopf and Walter 1983).

The origin of Precambrian sedimentary humic substances as outlined in Fig. 6 requires both algal or bacterial residues and heterotrophic microorganisms for the decomposition and conversion of these remnants into the major biochemical components (polyphenols, phenols, sugars, and amino acids). In particular, the polyphenol theory requires appropriate microorganisms for enzymatic conversion of phenolic compounds into quinones.

Lack of detectable PPS in the Early Precambrian kerogens listed in Table 2 suggests either that they are of nonbiological origin or, as is more likely in view of their extreme thermal alteration (atomic H/C <0.2: McKirdy and Hahn 1982; Hayes et al. 1983), that their condensed polyaromatic structures approach that of graphite.

8.2 The Middle Precambrian (2.5–1.7 Ga)

Unmetamorphosed black siliceous stromatolites (cherts) of the Gunflint Formation near Schreiber, Ontario, contain a diverse assemblage of prokaryotic microorganisms including representatives of several extant cyanobacterial families (Chroococcaceae, Oscillatoriaceae, Nostocaceae) and a variety of chemosynthetic bacteria, some of which are morphologically comparable to certain extant actinomycetes and mycobacteria (Barghoorn and Tyler 1965; Cloud 1965). The Gunflint carbonaceous cherts also contain abundant pyrite (Jovanović 1989), suggesting that the bottom waters of the Gunflint sea were anoxic.

According to Schopf (1972), the surface waters of the Gunflint shallow sea were dominated by oxygen-producing photosynthetic microorganisms (planktonic cyanophytes and bacteria). The Gunflint humic substances originated from the debris of these photoautotrophs. Anaerobic bacteria (e.g., sulfate reducers) were the major agents of the decomposition of proteins and polysaccharides in these microbial residues. This is in marked contrast to the formation of humic substances in contemporary aquatic environments, where the principal agents of vascular plant decomposition include fungi which are obligate aerobes (Gäumann 1964).

The presence of a high concentration of PPS in kerogen from the Gunflint carbonaceous chert (Table 2) is interpreted in terms of the initial formation of humic substances (with a significant PPS content) which were subsequently transformed into kerogen by molecular rearrangement and aromatization reactions during the rock's 2-Ga catagenetic history. The low atomic H/C ratio (0.50, Table 2) of the Gunflint kerogen is evidence that it has almost certainly undergone extensive dehydrogenation (Hayes et al. 1983).

Insoluble carbon of coal-like material from the Michigamme Formation (Tyler et al. 1957) was recently investigated using ^{13}C NMR with magic angle spinning (MAS) and cross-polarization (CP) (Premović et al. 1988a). The results of this work indicate that this carbon contains a very low concentration of relatively small polycyclic aromatic structures. The ESR and ^{1}H NMR MAS spectra of the Michigamme carbon reveal the absence of PPS (Table 1) and organic protons, respectively. Similar results were obtained for insoluble organic matter in the Sudbury anthraxolite and the Onwatin slate (Table 1).

8.3 The Late Precambrian (1.6–0.6 Ga)

Sedimentologic and stratigraphic analyses indicate that the Nonesuch Formation is primarily a lacustrine deposit (Elmore 1981; Elmore et al. 1989). It is a heterogenous unit, consisting of claystone, carbonaceous shale, and siltstone (Barghoorn et al. 1965). The Nonesuch carbonaceous shale was deposited under reducing conditions and contains organic matter (up to 2.5% total organic carbon) of algal and bacterial derivation. Algal and possibly fungal or actinomycete remains showing rather poorly preserved cellular details have been isolated from this shale (Meinschein et al. 1965; Barghoorn et al. 1965; Moore et al. 1969). Differential microbial degradation of algal remains was invoked by Imbus et al. (1988) and Elmore et al. (1989) to explain the existence of two distinct types of kerogen, defined using organic petrography in combination with pyrolysis-gas chromatography/mass spectrometry. The first kerogen type fluoresces under incident blue light and is essentially aliphatic in character; whereas the other nonfluorescent variety is either aromatic or aromatic with a significant phenolic component. The latter finding is entirely consistent with the presence of PPS in the Nonesuch kerogen examined in this study.

Carbonaceous cherts from stromatolites of the Bitter Springs Formation formed in a shallow aquatic environment populated by a diverse assemblage of cyanobacteria, bacteria, and green algae (Schopf 1972). Detailed sedimentological work by Southgate (1986) has revealed that these fossiliferous cherts are the results of diagenetic silicification of benthic cyanobacterial communities which grew in shallow saline lakes or ponds. The mean carbon isotopic composition ($\delta^{13}C = -22.18‰$; n = 48) of kerogen from the Bitter Spring chert (Wedeking et al. 1983) is suggestive of intense photosynthetic activity. A sample of this Bitter Springs kerogen is relatively rich in PPS (Table 2). The Bitter Springs microflora is considerably more advanced than that of the Gunflint Iron Formation and includes many species similar to extant fungal organisms (thallophytes). These aerobes may have played a key role in the formation of nonmarine humic substances in the Bitter Springs microbial mat communities.

9 Precursors of PPS in Selected Paleozoic Carbonaceous Sedimentary Rocks

What does the occurrence of PPS in certain Precambrian kerogens tell us about the biogeochemistry of their host sediments? The answer to this question lies in the comparison of the Precambrian PPS with those of demonstrable biological origin found in Paleozoic sediments (Table 2, Fig. 4).

The Zvonce black shale was deposited under reducing conditions in sheltered areas of a Silurian shallow epicontinental sea. This accounts for the observed accumulation of well-preserved graptolite remains and the associated organic matter of bacterial and phytoplanktonic origin (Premović et al. 1986). The depositional environment of the Late Devonian Antrim shale also was highly reducing. Its organic matter is derived mainly from phytoplanktonic and bacterial debris, though fossilized fragments of ancient driftwood (e.g., Callixylon) are also present (Duncan and Swanson 1965). In such anoxic marine depositional environments with relatively high biological (especially bacterial) activity, humic substances form and are subsequently transformed through diagenetic and catagenetic changes into kerogen containing PPS.

The Early Devonian Rhynie chert, one of the earliest records of petrified wood, contains microfossils which may be referred to extant cyanobacteria and green algae. The vascular plants *Rhynia* and *Asteroxylon* growing in a peat bog were killed and fixed, apparently very rapidly, by boiling silica-rich water from a volcanic fumarole (Kidston and Lang 1917–1921). Microscopic examination of the Rhynie chert and the Permian petrified wood indicates that their plant tissues have been extensively attacked by microorganisms, especially fungi, which are likely to have participated in the formation of the humic acid precursors of their kerogen PPS.

The Late Devonian Sosnowiec marine stromatolitic and nonstromatolitic limestones contain abundant bituminous organic matter which is derived from colonial *Eovolvox*-like green algae (Kazmierczak 1975). High contents of PPS ($4-11 \times 10^{18}$ spins g^{-1}, Table 2) of their kerogens are consistent with the marine humic origin of these kerogens.

Both the torbanite (freshwater) and tasmanite (marine) algal kerogens (Table 2) have been examined by ^{13}C NMR CP/MAS, ^{1}H NMR MAS and FTIR spectroscopies (Premović et al. 1986) which showed that these materials have predominantly aliphatic structures with a low total aromatic content. This is consistent with their low PPS concentration ($3-4 \times 10^{17}$ spins g^{-1}, Table 2) and the results of ^{13}C DNP/CP studies (Kitanović 1984).

10 Geothermal History of Precambrian Kerogens Containing PPS

It is of considerable geological interest to know the maximum temperature to which an ancient carbonaceous sedimentary rock may have been exposed during its burial history. This will be particularly so if the kerogen has been thermally altered to such a degree that its carbon isotopic signature is modified (McKirdy and Powell 1974; Schidlowski et al. 1979; Hayes et al. 1983; see also Schidlowski & Aharon and Strauss et al., this Vol.).

Austen et al. (1966) studied the ESR behavior of PPS in a series of bituminous coal macerals (vitrinites, exinites and fusinites) during heating (in nitrogen) from room temperature to about 750 °C. Initial spin concentrations of these macerals were $10-45 \times 10^{18}$ spins g^{-1} for vitrinites and $5-10 \times 10^{18}$ spins g^{-1} for exinites, the ranges reflecting differences in rank. The spin concentrations of vitrinites and exinites were found to increase on laboratory pyrolysis (starting at about 350 °C) to a maximum in the range 500–600 °C, while further increase in the heat-treatment temperature caused an apparent decrease in spin concentration. The changes in spin concentration recorded for vitrinites and exinites are typical of those which have been observed previously with whole coals and other types of organic materials such as cellulose (Milsch et al. 1968). As the pyrolysis temperature increases, the size and number of aromatic ring clusters increase, thus allowing more unpaired electrons to be stabilized. Eventually, a stage is reached where the unpaired electrons become so delocalized that the organic residue (now charred) becomes electrically conducting and spin concentration decreases rapidly. This apparent decrease in unpaired electrons occurs because they become part of the conduction band in a graphite-like structure.

Changes in the ESR characteristics (spin concentration, line-width and g-value) of PPS in bituminous coals and their macerals during laboratory pyrolysis have potential use as a geothermal indicator. Austen et al. (1966) pointed out that if a sedimentary carbonaceous material contains PPS formed at 100 to 200 °C during low-temperature metamorphism, then its ESR characteristics would not be expected to change during laboratory heating until temperatures of 350–450 °C are reached.

The spin concentration of anthracites with the carbon contents > 95% C (high rank) does not conform to the behavior of low-temperature sedimentary carbonaceous materials because metamorphism has made them fairly graphitic (van Krevelen 1981). However, the pyrolytic ESR behavior of low-rank anthracites (e.g., Vrška Čuka anthracite, 92.1% C: Kitanović 1984) does change in the manner described for vitrinites and exinites.

Austen et al. (1966) determined spin concentrations in fusinites (ca. $30-74 \times 10^{18}$ spins g^{-1}) and concluded that an initial high number of spin centers was developed early in the coalification process. They also established that the intrinsic ESR signals of fusinites did not change significantly on laboratory pyrolysis up to 750 °C, and argued that temperatures of 400–600 °C were necessary for fusinite formation (e.g., in forest fires: Cope 1980).

The majority of coal researchers agree that temperature is the most important factor in the coalification process. Temperatures lower than about 200 °C would have been required to form anthracite. On the other hand, bituminous coal formation takes place at temperatures of the order of 100–150 °C (Teichmüller and Teichmüller 1979). Thus, it is likely that the temperatures necessary for the genesis of the Vrška Čuka anthracite were about 150 °C (Pantić and Nikolić 1973) Most coals have at no time been exposed to temperatures above 200 °C. According to Retcofsky et al. (1968), carbon radicals in amounts detectable by ESR can be produced in pure hydrocarbons, as well as in natural cellulose derivatives, by slow laboratory pyrolysis (1–2 years) at this temperature.

The Gunflint and Bitter Springs Formations appear to have had mild thermal histories. The petrology and mineralogy of these cherts indicates general absence of metamorphism. The temperature versus ESR behavior of PPS in the Gunflint (Premović 1982) and Bitter Springs (Premović 1986) cherts during laboratory carbonization is similar to that of vitrinites, exinites (Austen et al. 1966), and low-rank anthracites (van Krevelen 1981) such as the Vrška Čuka coal (Kitanović 1984). The Gunflint chert kerogen is characterized by an atomic H/C ratio of 0.50, and a spin concentration of 16×10^{18} spins g^{-1} (Table 2). These features are comparable to those of low-rank anthracites and imply that the maximum temperature to which the Gunflint chert has been exposed during its geological history was about 150 °C. The Gunflint chert was covered by no more than approximately 150 m of Late Precambrian strata (Goodwin 1956). In this case, the sediment has never been buried deeply enough to attain the minimum temperature (ca. 150 °C) required for the formation of low-rank anthracite-like material. However, there is abundant evidence of intensive magmatic (intrusive and extrusive) activity during and after the Middle Precambrian in the Gunflint locality (Goodwin 1956).

The Bitter Springs kerogen has an atomic H/C ratio (0.68, Table 2) and a spin concentration (2×10^{18} spins g^{-1}, Table 2) that are characteristic of bituminous vitrinites (Austen et al. 1966; Wilson et al. 1984). A normal thermal history for the Late Precambrian Bitter Springs Formation is probable since these sediments were overlain by approximately 2500 m strata by the end of the Middle Cambrian (Schopf 1968). At an average geothermal gradient of 4 °C per 100 m, these rocks would have been exposed to temperature of at least 100 °C over an extensive period of time. It is not unreasonable to conclude that these conditions were sufficient to generate the vitrinite-like properties of its kerogen.

Finally, experimental studies by Oehler (1972) documented the structural and chemical changes in cyanobacteria during simulated fossilization in synthetic chert at 100 °C and at 165 °C. The quality of preservation exhibited by these artificial fossils resembles that of naturally silicified cyanobacterial fossils from the Gunflint and Bitter Springs cherts.

For comparison, the Early Devonian Rhynie chert (Premović 1986) and the Late Devonian Sosnowiec stromatolitic limestone (Jovanović 1989) were pyrolyzed at temperatures ranging from 50 to 800 °C. The resultant ESR changes were typical of what has been reported previously for vitrinites and exinites (Austen et al. 1966). On the other hand, the kerogen H/C ratios (ca. 0.9, Table 2) of these two rocks and in the spin concentrations (2 to 4×10^{18} spins g^{-1}, Table 2) are similar to those of exinites in bituminous coals (Austen et al. 1966; Wilson et al. 1984). Thus, it is likely that the Rhynie and Sosnowiec rocks reached temperatures of the order of 100 °C necessary for the formation of exinite-like material.

The Sosnowiec stromatolitic sample was obtained from a borehole at a present depth of ca. 2400 m (Kazmierczak 1975), leaving no doubt that it has been buried deeply enough to account for the ESR spectrum of its kerogen. In the case of the Rhynie chert, both deep burial (Kidston and Lang 1917–1921) and contact metamorphism contributed to its heating. The silica-rich formation waters of the Old Red Sandstone Formation emanated from hot springs associated with intense volanic activity as indicated by the andesite interbeds in the Rhynie deposit (Gall 1983).

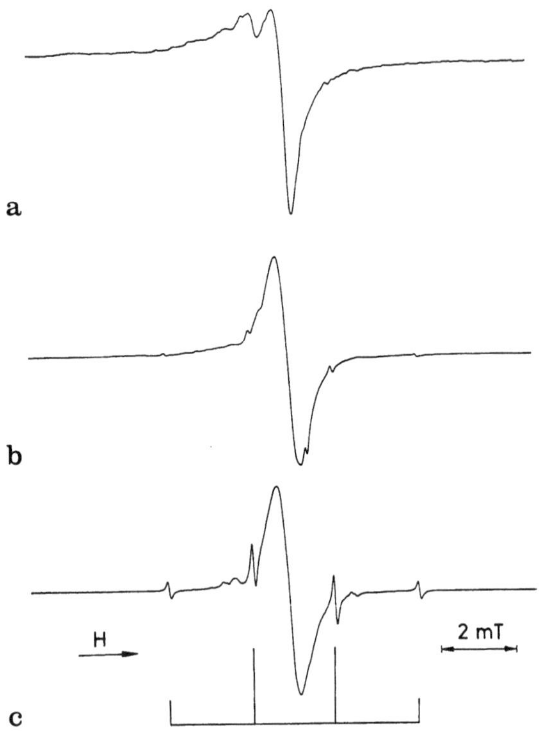

Fig. 7a–c. ESR spectra of CH_3 radicals in: **a** Bitter Springs chert. **b** Rhynie chert. **c** Permian silicified wood

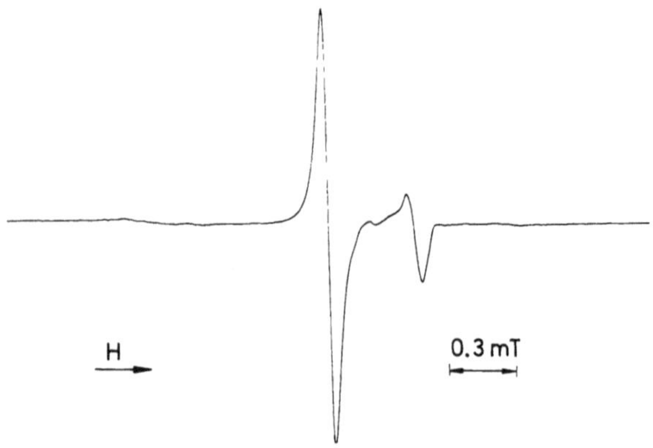

Fig. 8. ESR spectrum of carbon paramagnetic structures in the Paradise Creek chert

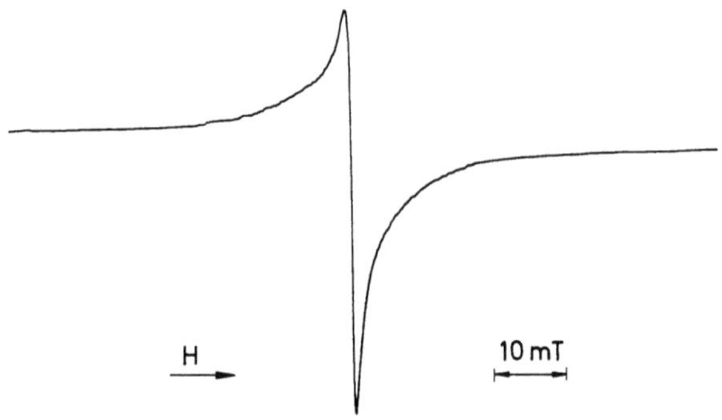

Fig. 9. ESR spectra of insoluble carbonaceous material in the Huntsman stromatolitic limestone

In addition to the ESR signal attributable to kerogen PPS, whole-rock samples of Rhynie chert, Permian petrified wood and, to a lesser degree, the Bitter Springs chert, all show a quartet signal $a_H = 2.22 \pm 0.01$ mT with a relative intensity of nearly 1:3:3:1 (Fig. 7; Premović 1986). A similar quartet spectrum was observed by Ikeya (1982) in several Miocene and Cretaceous petrified woods and attributed to CH_3 radicals among the decomposition products of the woody tissue. Griffiths et al. (1982) detected CH_3 radicals in a Mezozoic chert and suggested that these radicals are trapped in molecular-sized holes in the microcrystalline matrix. In addition, the weak triplet ($g = 2.0029 \pm 0.0001$ and $a_H = 2.17 + 0.01$ mT, with an intensity ratio ca. 1:2:1) characteristic of RCH_2 radicals was observed for different samples of petrified wood. Recently, Ikeya et al. (1986) found that CH_3 radicals appeared after γ-irradiation of granite samples due to the presence of methane trapped in primary or slightly altered quartz. The CH_3 ESR signals is one of the most useful signals in its temperture range for studying the reheating of cherts because of its manifold increase in intensity on heating (Griffiths et al. 1983). Laboratory pyrolysis at 100 °C did not cause a significant change in intensity of the CH_3 signal of unheated Bitter Springs and Rhynie chert samples; and heating at 200 °C produced only a twofold increase in signal intensity. In fact, there was little change in the CH_3 signal until the pyrolysis temperature exceded 200 °C (Jovanović 1989). This finding supports the coalification of the Bitter Springs kerogen at temperatures of the order of 100–150 °C. However, increasing the duration of laboratory heating may shift the maximum signal intensity of the CH_3 radicals towards lower temperatures (Griffiths et al. 1983).

Both the ESR spectrum of kerogen in the Late Precambrian Paradise Creek chert ($g = 2.00263 \pm 0.00005$ and $\Delta Hp-p = 0.12 \pm 0.01$ mT; Fig. 8) and the ESR behavior of its PPS during laboratory heating (Premović et al. 1989) indicate that these structures are similar to those found in fusinites (Austen et al. 1966). Finally, ESR spectra of kerogens from the Huntsman stromatolitic limestone (Fig. 9), the Michigamme anthracite (Fig. 10a) and the Sudbury anthraxolite (Fig. 10b) are characteristic of highly graphitic materials.

The thermal behavior of organic paramagnetic structures in ancient sedimentary rocks is being fur-

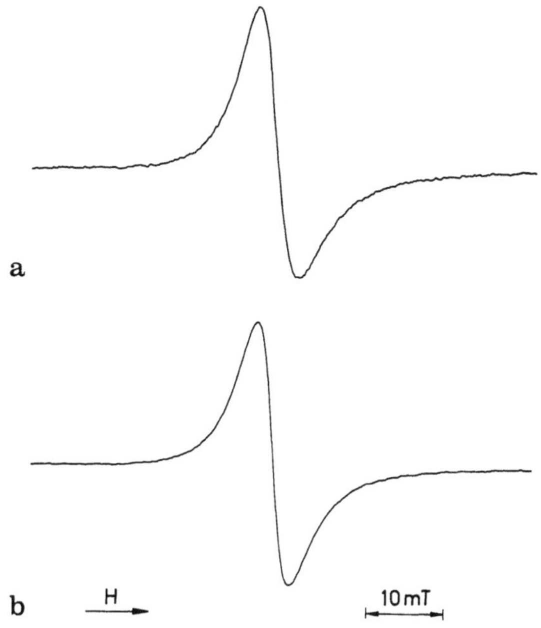

Fig. 10a, b. ESR spectra of insoluble carbonaceous material in: **a** Michigamme coaly sediment. **b** Sudbury anthraxolite

ther studied and is likely to prove useful in determining the maximum temperatures to which these sediments have been exposed.

11 Conclusions

The data presented here address the question: Is it feasible to interpret kerogen-hosted PPS in Precambrian carbonaceous sedimentary rocks as evidence of early biological activity? To answer this question the ESR spectra and other geochemical characteristics of the kerogen preserved in selected unmetamorphosed Precambrian sediments were considered. PPS could be detected in only three of the Precambrian kerogens examined: Those isolated from the Gunflint and Bitter Springs cherts and the Nonesuch shale. They are, in fact, among the most intriguing features of these ancient kerogens. All three rocks contain well-preserved, syngenetically emplaced organic microfossils. For comparison, the kerogens and associated PPS of selected Paleozoic and younger carbonaceous sediments of similar origin, lithology, and thermal history were also investigated. From these studies it is proposed that Precambrian kerogens containing PPS (or, more precisely, that part of kerogen structure with unpaired electrons) may be relics of early biochemical processes. The key to future ESR and other chemical studies of Precambrian biological activity lies not in the analysis of whole (usually highly mature) kerogens, but rather that part comprising intact, organically preserved microfossils. These studies will be particularly effective when properly integrated with existing paleontological and geological information.

Acknowledgments. I am deeply indebted to Prof. Manfred Schidlowski, who made it possible for me to participate in the Maria Laach Conference. Special thanks are extended to Dr. David McKirdy, who painstakingly edited and reviewed earlier drafts of the manuscript. I thank my students for continuous help during much of the research which laid the basis for this paper. The study was supported by RZN of Serbia, project 1.12.

References

Austen DEG, Ingram DJE, Given PH, Binder GR, Hill LW (1966) Electron spin resonance study of pure macerals. Coal Sci Adv Chem Ser 55:334–362

Barghoorn ES, Tyler SA (1965) Microorganisms from the Gunflint chert. Science 147:563–577

Barghoorn ES, Meinschein WG, Schopf JW (1965) Paleobiology of a Precambrian shale. Science 148:461–472

Cloud P (1965) Significance of the Gunflint (Precambrian) microflora. Science 148:27–45

Cloud PE Jr, Licari GR (1968) Microbiotas of the banded iron formations. Proc Natl Acad Sci USA 61:779–786

Cope JM (1980) Physical and chemical properties of coalified and charcoalified phytoclasts from British Mesozoic sediments: an organic geochemical approach to palaeobotany. In: Douglas AG, Maxwell JR (eds) Advances in organic geochemistry 1979. Pergamon, Oxford, pp 663–677

Duncan DC, Swanson VE (1965) Organic-rich shale of the United States and world land areas. US Geol Surv Circ 523:30 pp

Durand B, Marchand A, Combaz A (1977) Etude de kerogènes par résonance paramagnétique électronique. In: Compos R, Goni J (eds) Advances in organic geochemistry, 1975. Enadisma, Madrid, pp 753–780

Elmore RD (1981) The Copper Harbor conglomerate and Nonesuch shale: sedimentation in a Precambrian intracontinental rift, upper Michigan. Thesis, Univ Mich, Lansing 200 pp

Elmore RD, Milavec GJ, Imbus SW, Engel MH (1989) The Precambrian Nonesuch Formation of the North American Mid-Continent Rift, sedimentology and organic geochemical aspects of lacustrine deposition. Precambrian Res 43:191–213

Enders C, Theis K (1938) Die Melanoide und ihre Beziehung zu den Huminsäuren. Brennstoff Chem 19:360–439

Ertel JR, Hedges JI (1984) The lignin component of humic substances: distribution among soil and sedimentary humic, fulvic and base-insoluble fractions. Geochim Cosmochim Acta 48:2065–2074

Flaig W (1972) Some physical and chemical properties of humic substances as a basis of their characterization. In: von Gaertner HR, Wehner H (eds) Advances in organic geochemistry 1971. Pergamon, Oxford, pp 49–67

Flaig WH, Beutelspracher P, Reitz E (1975) Chemical composition and physical properties of humic substances. In: Gieseking JE (ed) Soil components 1. Springer, Berlin Heidelberg New York, pp 1–219

Gall JC (1983) Ancient sedimentary environments and the habitats of living organisms. Springer, Berlin Heidelberg New York, 219 pp

Gäumann EA (1964) Die Pilze. Birkhäuser, Basel, 298 pp

Goodwin AM (1956) Facies relations in the Gunflint Iron Formation. Econ Geol 51:565–595

Griffiths DR, Rabin GV, Seeley NJ, Chandra H, McNeil DAC, Symons MCR (1982) Trapped methyl radicals in chert. Nature (London) 300:435–436

Griffiths DR, Seeley NJ, Symon MCR (1983) Investigation of chert heating conditions using ESR spectroscopy. In: Sieveking G, Hart MB (eds) Proc 4th Int Flint Symp 1983, Brighton. Cambridge Univ Press, London, 262 pp

Hatcher PG (1980) The origin, composition, chemical structure, and diagenesis of humic substances, coals, and kerogens as studied by nuclear magnetic resonance. Thesis, Univ Mar, Baltimore, 283 pp

Hatcher PG, Vanderhart DL, Earl WL (1980) Use of solid-state ^{13}C NMR in structural studies of humic acids and humin from Holocene sediments. Org Geochem 2: 87–92

Hatcher PG, Breger IA, Earl WL (1981) Nuclear magnetic resonance studies of ancient buried wood. I. Observations on the origin of coal to the brown coal stage. Org Geochem 3:49–55

Hatcher PG, Breger IA, Szeverenyi N, Maciel GE (1982) Nuclear magnetic resonance studies of ancient buried wood. II. Observations on the origin of coal from lignite to bituminous coal. Org Geochem 4:9–18

Hayes JM, Kaplan RI, Wedeking KW (1983) Precambrian organic geochemistry, preservation of the record. In: Schopf JM (ed) Earth's earliest biosphere: its origin and evolution. University Press, Princeton, pp 93–134

Hodge JE (1953) Dehydrated foods. Chemistry of browning reactions in model system. J Agric Food Chem 1:928 943

Huc AY, Durand BM (1977) Occurrence and significance of humic acids in ancient sediments. Fuel 56:73–80

Hurst HM, Burgess NA (1967) Lignin and humic acids. In: McLaren AD, Patterson GH (eds) Soil biochemistry. Dekker, New York, pp 297–331

Hwang PTR, Pusey WC (1973) Process for determining hydrocarbon maturity using electron spin resonance. US Pat 3, 740641

Ikan R, Rubinsztain Y, Ioselis P, Aizenshtat Z, Pugmire R, Anderson LL, Woolfenden WR (1986) Carbon-13 cross-polarized magic-angle samples spinning nuclear magnetic resonance of melanoidins. Org Geochem 9:199–212

Ikeya M (1982) Electron spin resonance of petrified woods for geological age assessment. Jpn Appl Phys 22:128–130

Ikeya M, Devine SD, Whitehead NE, Hedenwuist JW (1986) Detection of methane in geothermal quartz by ESR. Chem Geol 56:185–192

Imbus SW, Engel MH, Elmore RD, Zumberge JE (1988) The origin, distribution and hydrocarbon generation potential of organic-rich facies in the Nonesuch Formation, Central North American Rift system: a regional study. In: Mattavelli L, Novelli L (eds) Advances in organic geochemistry 1987. Pergamon, Oxford, pp 207–219

Ingram DJE, Tapley BB, Jackson R, Bond RL, Murnaghan RA (1954) Paramagnetic resonance in carbonaceous solids. Nature (London) 174:797–798

Ishiwatari R (1974) Electron spin resonance of sedimentary humic acids in relation to their aromatic character. Geochem J 8:97–102

Jackson TA (1973) Humic matter in the bitumen of ancient sediments: variations through geologic time. Geology 1:163–166

Jovanović LJS (1989) Chemical structural study of Precambrian kerogens. Thesis, Univ Niš, 172 pp

Kazmierczak J (1975) Colonial Volvocales (Chlorophyta) from the Upper Devonian of Poland and their palaeoenvironmental significance. Acta Palaeontol Pol 20: 73–75

Kidston R, Lang WH (1917–1921) On old red sandstone plants showing structure, from the Rhynie Chert Bed, Aberdeenshire, pt I–V. Trans R Soc Edinburgh 51–52

Kitanović GB (1984) Chemical structure of algal coals: a spectrochemical approach. Thesis, Univ Niš, 182 pp

Knoll AH, Barghoorn ES (1977) Archean microfossils showing all division from the Swaziland system of South Africa. Science 198:396–398

Komatinović BV (1984) Precambrian kerogens: a spectroscopic study. Thesis, Univ Niš, 153 pp

Kononova M (1966) Soil organic matter. Pergamon, London, 554 pp

Maillard LC (1912) Action des acides amines sur les sucres: formation des mélanoidines par votre méthodiques. CR Acad Sci Paris 154:66–68

Marchand A, Conrad J (1980) Electron paramagnetic resonance in kerogen studies. In: Durand B (ed) Kerogen. Editions Technip, Paris, pp 243–270

Marchand A, Libert P, Combaz A (1968) Sur quelques critères physico-chimiques de la diagenèse d'un kérogène. CR Acad Sci Paris Ser D266:2316–2319

Martin JP, Haider K (1971) Microbial activity in relation to soil humus formation. Soil Sci 111:54–63

Martin JP, Haider K, Bondietti E (1972) Properties of model humic acids synthesized by phenol oxidase and autooxidation of phenols and other compounds formed by soil fungi. In: Proc Int Meet Humic substances, Nieuwersluis. Pudoc, Wageningen, pp 171–176

McKirdy DM, Hahn JH (1982) The composition of kerogen and hydrocarbons in Precambrian rocks. In: Holland HD, Schidlowski M (eds) Mineral deposits and the evolution of the biosphere. Springer, Berlin Heidelberg New York, pp 123–154

McKirdy DM, Powell TG (1974) Metamorphic alteration of carbon isotopic composition in ancient sedimentary organic matter: new evidence from Australia and South Africa. Geology 2:591–595

McKirdy DM, McHugh DJ, Tardif JW (1980) Comparative analysis of stromatolitic and other microbial kerogens by pyrolysis-hydrogenation gas chromatography (PHGC). In: Trudinger PA, Walter MR, Ralph BJ (eds) Biogeochemistry of ancient and modern environments. Aust Acad Sci, Canberra, pp 187–200

McWeeny PJ, Knowles ME, Hearne JF (1974) The chemistry of non-enzymic browning in foods and its controls by sulphites. J Sci Food Agric 25:735–746

Meinschein WG, Barghoorn ES, Schopf JW (1964) Biological remnants in a Precambrian sediment. Science 45:262–264

Milsch B, Windsch W, Heinzelmann H (1968) EPR investigations of charred cellulose. Carbon 6:807–812

Moore LR, Moore JRM, Spinner E (1969) A geomicrobiological study of the pre-Cambrian Nonesuch Shale. Yorkshire Geol Soc Proc 37:351–394

Muir MD, Grant PR (1976) Micropalaeontological evidence from the Onverwacht Group, South Africa. In: Windley BF (ed) The early history of the earth. John Wiley & Sons, New York, pp 595–604

Nissenbaum A, Kaplan IR (1972) Chemical and isotopic evidence for the in situ origin of marine humic substances. Limnol Oceanogr 17:570–582

Oehler JH (1972) Experimental studies in Precambrian paleontology: structural and chemical changes in blue-green algae during simulated fossilization in synthetic chert. Geol Soc Am 87:117–129

Pantić N, Nikolić P (1973) Ugalj. Naučna knjiga, Belgrade, 559 pp

Premović PI (1982) Aromatic free radicals in the Gunflint chert. Naturwissenschaften 69:479–482

Premović PI (1986) Electron spin resonance behavior of indigenous organic matter in silicic rocks on laboratory pyrolysis: the Bitter Springs and Rhynie cherts and petrified wood. J Serb Chem 51:63–609

Premović PJ, Stojković SR, Pugmire RJ, Woolfenden WR, Rosenberger H, Scheler G (1986) Spectroscopic evidence for the chemical structure of algal kerogens. In: Rodriguez-Clemente R, Tardy Y (eds) Geochemistry and mineral formation in the earth's surface. CSIC, Madrid, pp 431–440

Premović PI, Komatinović BV, Pugmire RJ, Woolfenden WR (1988a) Solid-state ^{13}C NMR of Middle Precambrian anthracite and related anthraxolite. Naturwissenschaften 75:98–100

Premović PI, Jovanović SLJ, Popović GB, Pavlović NZ, Pavlović MS (1988b) Vanadium in ancient carbonaceous sedimentary rocks of marine origin: the Zvonce black shale. J Serb Chem Soc 53:427–431

Premović PI, Kitanović GZ, Komatinović BV, Stojković SR (1989) ESR study of the Paradise Creek chert. Polyaromatic paramagnetic structures. J Serb Chem 54:83–87

Pryor WA, Hales BJ, Premović PI, Church DF (1983) The radicals in cigarette tar: their nature and suggested physiological implications. Science 220:425–427

Pusey WC (1973) The ESR kerogen method. How to evaluate potential gas and oil source rock. World Oil 176:71–75

Retallack G (1981) Fossil soils: indicators of ancient terrestrial environments. In: Niklas KJ (ed) Paleobotany, paleoecology and evolution, vol 1. Wiley Interscience, New York, pp 17–54

Retcofsky HL, Stark JM, Friedel RA (1968) Electron spin resonance in American coals. Anal Chem 40:1699–1704

Rex RW (1960) Electron paramagnetic resonance studies of stable free radicals in lignins and humic acids. Nature (London) 188:1185–1186

Reznikov VM, Mikhaseva MF, Zil'bergleit MA (1978) The lignin of the alga Fucus vesiculosus. Khim Prirodn Soedin 5:648–650

Rifaldi R, Schnitzer M (1972) Electron spin resonance spectrometry of humic substances. Soil Sci Soc Am Proc 36:301–305

Saiz-Jimenez C, Shafizadeh F (1985) Electron spin resonance spectrometry of fungal melanins. Soil Sci 139:319–325

Schidlowski M, Appel PWU, Eichmann R, Junge CE (1979) Carbon isotype geochemistry of the 3.7 × 10^9-yr-old Isua sediments, West Greenland: implications for the Archaean carbon and oxygen cycles. Geochim Cosmochem Acta 43:189–199.

Schnitzer M (1971) Characterization of humic constituents by spectroscopy. In: McLaren AD, Skujins J (eds) Soil biochemistry, vol 2. Dekker, New York, pp 60–95

Schnitzer M, Skinner SIM (1969) Free radicals in soil humic compounds. Soil Sci 108:383–390

Schopf JW (1968) Microflora of the Bitter Springs Formation, late Precambrian, central Australia. J Paleontol 42:651–688

Schopf JW (1970) Electron microscopy of organically preserved Precambrian microorganisms. J Paleontol 44:1–12

Schopf JW (1972) Precambrian paleobiology. In: Ponnamperuma C (ed) Exobiology. North-Holland, Amsterdam, pp 16–60

Schopf JW, Walter MR (1983) Archean microfossils: new evidence of ancient microbes. In: Schopf JW (ed) Earth's earliest biosphere: its origin and evolution. University Press, Princeton, pp 93–134

Southgate PN (1986) Depositional environment and mechanism of preservation of microfossils, upper Proterozoic Bitter Springs Formation, Australia. Geology 14:683–686

Steelink C (1964) Free radical studies of lignin, lignin degradation products and soil humic acids. Geochim Cosmochim Acta 28:1615–1622

Steelink C (1966) Electron paramagnetic resonance studies of humic acid and related model compounds. Coal Science Adv Chem Ser No 55, 80–90

Steinbrenner K, Matschke J (1971) Synthesis of humic substances by soil fungi. Trans Int Symp Humus Planta 5:111–115

Stevenson FJ (1975) Nonbiological transformations of amino acids in soils and sediments. In: Tissot B, Bienner F (eds) Advances in organic geochemistry 1973. Editions Technip, Paris, pp 701–714

Stevenson FJ (1982) Humus chemistry. Wiley-Interscience, New York, 282 pp

Stewart WN (1983) Paleobotany and the evolution of plants. Cambridge Univ Press, London, 396 pp

Stuermer DH, Peters KE, Kaplan IR (1978) Source indicators of humic substances and protokerogen. Stable isotope ratios, elemental compositions and electron spin resonance spectra. Geochim Cosmochim Acta 42:898–907

Teichmüller M, Teichmüller R (1979) Diagenesis of coal (coalification). In: Larsen G, Chilingar GV (eds) Diagenesis in sediments and sedimentary rocks. Elsevier, Amsterdam, pp 207–246

Tissot BP, Welte DH (1978) Petroleum formation and occurrence. Springer, Berlin Heidelberg New York, 527 pp

Turner WB (1971) Fungal metabolites. Academic Press, New York London, 271 pp

Tyler SA, Barghoorn ES, Barret LP (1957) Anthracitic coal from Precambrian Upper Huronian black shale of the Iron River District, Northern Michigan. Bull Geol Soc Am 68:1293–1304

Uebersfeld J, Etienne A, Combrison J (1954) Paramagnetic resonance, a new property of coal-like materials. Nature (London) 174:614–615

Van Krevelen DW (1981) Coal-typology, chemistry, physics, constitution. Elsevier. Amsterdam, 514 pp

Vasilevskaya NA, Golyashin VN, Denisenko NM, Maximov OB (1977) Chemical study of humic acids in Western Pacific bottom sediments. Oceanology 17:300–307

Walter MR (1983) Archean stromatolities: evidence of the Earth's earliest benthos. In: Schopf JW (ed) Earth's earliest biosphere: its origin and evolution. University Press, Princeton, pp 93–134

Wedeking KW, Hayes JM, Matzigkeit U (1983) Procedures of organic geochemical analysis. In: Schopf JW (ed) Earth's earliest biosphere: its origin and evolution. University Press, Princeton, pp 428–441

Wilson MA, Pugmire RJ, Karas J, Alemany LB, Woolfenden WR, Grant DM (1984) Carbon distribution in coals and coal macerals by cross polarization magic angle spinning carbon-13 nuclear magnetic resonance spectrometry. Anal Chem 56:933–943

Yen TF, Sprang SR (1977) Contribution of ESR analysis toward diagenic mechanisms in bituminous deposits. Geochim Cosmochim Acta 41:1007–1018

On the Atypical Petroleum-Generating Characteristics of Alginite in the Cambrian Alum Shale

B. Horsfield [1], S. Bharati [2,4], S. R. Larter [2,3], F. Leistner [1], R. Littke [1], H. J. Schenk [1] and H. Dypvik [2]

[1] Institute of Petroleum and Organic Geochemistry (ICH-5), Forschungszentrum Jülich (KFA), P.O. Box 1913, W-5170 Jülich, FRG
[2] Geology Department, University of Oslo, Postboks 1047, N-0316 Blindern, Oslo 3, Norway
[3] Present address: NRG in Fossil Fuels and Environmental Geochemistry, The University, Newcastle Upon Tyne, UK
[4] Present address: Geo Lab Nor, P.O. Box 1581 Nidarvoll, 7002 Trondheim, Norway

Abstract

A series of seven Alum Shale samples has been analyzed by pyrolytic, petrologic, and spectroscopic methods. All contain alginitic kerogens at various levels of alteration. Kerogens from four samples appear immature to early mature, whereas three samples are more mature, especially one which is very overmature. Decreasing pyrolysate yields per unit weight of carbon characterize the series. Alum Shale alginite is unlike any other alginite we have studied. Aliphatic carbon is present in all but the highly overmature kerogen, yet only a small proportion of aliphatic moieties is released during high-temperature pyrolysis. Furthermore, all the pyrolysates are very aromatic. The unusual structure of Alum Shale alginite might have been brought about by uranium-related alteration effects. Alternatively, or additionally, its structure might have been directly inherited from biopolymeric or carotenoid-derived precursors in algae and bacteria. Mainly gas and condensate are evolved from Alum Shale kerogen during simulated maturation, and a high proportion (more than 30%) of its originally labile components are progressively and cumulatively converted to "dead carbon". Available data indicate that the Alum Shale behaves similarly during natural maturation.

1 Introduction

The compositional information on kerogen that is obtained from pyrolysis-gas chromatography (Py-GC) appears to be grossly representative of the kerogen structure as a whole, despite the fact that chromatographed products constitute only a rather small mass fraction of the total pyrolysate and that no direct account is taken of the inert kerogen fraction (Horsfield 1989; Larter and Horsfield 1991). Furthermore, it has been shown by correlation studies that certain bulk compositional attributes of kerogen pyrolysates, namely n-alkyl chain-length abundances and the relative abundances of normal versus non-normal hydrocarbons, reflect the composition of related natural petroleums. As an example, Horsfield (1989) showed that alginite derived from *Gloeocapsomorpha* generates a low-wax paraffinic pyrolysate with a notable periodicity in the C_9-C_{18} range, this being closely similar to the composition of naturally generated petroleums from source rocks containing this type of organic matter (Reed et al. 1986 and references therein). Alginite derived from *Tasmanites* generates pyrolysates which display no such periodicity. Nevertheless, they are paraffinic and have a similar restricted chain-length distribution. In contrast, *Botryococcus*-derived alginite yields a pyrolyzate with a high wax content and probably generates petroleum with a similar bulk composition in nature.

It was with these findings, and general petroleum geochemical applications, in mind that a combined study (geochemical, petrological and spectroscopical) of the Alum Shale (U. Cambrian–L. Ordovician) of Sweden and Norway was undertaken. This particular marine shale was chosen because it is exceedingly organic-rich and contains algal liptinite macerals in abundance (Buchardt et al. 1986), and because apparently wide ranges in maturity occur due to strongly differentiated regional subsidence (early Paleozoic) and/or metamorphism by igneous intrusives (Permo-Carboniferous). The Alum Shale is also renowned for its high uranium content, the radioactive decay of which has been noted to have affected the chemical composition of syndeposited organic matter (Leventhal et al. 1986; Dahl et al. 1988; Lewan and Buchardt 1989).

2 Sample Information and Analytical Methods

Seven samples of Alum Shale selected from a larger sample suite (Bharati 1989) were originally obtained

M. Schidlowski et al. (Eds.)
Early Organic Evolution: Implications for Mineral and Energy Resources

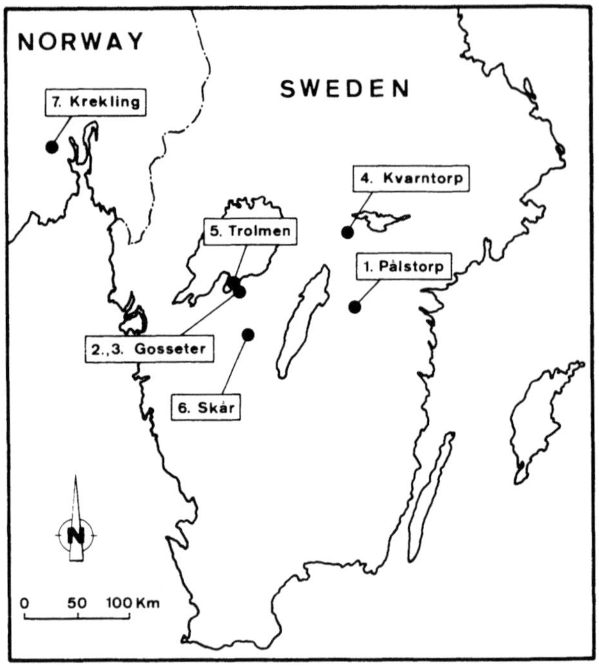

Fig. 1. Alum Shale sample location map

from outcrops at locations in Sweden and Norway (Fig. 1). The presence of fresh pyrite indicated that these rocks were not noticeably weathered. Aliquots of whole rock, as fragments (2–50 mm) and powder (< 200 mesh), and kerogen concentrates were prepared for analysis after the outer rock surfaces had been discarded. Kerogen concentrates were prepared by an acid digestion method (HCl/HF). Total organic carbon (TOC) was determined by the LECO combustion method after carbonates had been removed with hydrochloric acid. Rock-Eval pyrolysis was undertaken in accordance with established procedures using crushed whole rock. The uranium content was determined on crushed whole rocks by neutron activation analysis. Sulfur contents were measured on powdered whole rock by combustion and infrared detection (LECO). Maceral abundances, reflectivities and spectral properties were measured using reflected light and fluorescence-mode illumination of whole rock fragments orientated perpendicular to bedding, as described by Littke et al. (1988). Infra-red spectra were measured on whole rock powder and kerogen concentrates, from which aliphatic carbon (C_{AL}) was quantified, using the method of Schenk et al. (1986).

The concentration of free hydrocarbons existing in freshly powdered whole-rock samples was determined using thermovaporization (300 °C) and on-line GC. Quantitation was achieved using phenylhexane as an internal standard. Programmed Py-GC analysis of kerogen concentrates was performed using the same instrumentation after a thermal extraction step (300–550 °C at 50 °C/min then 10 min at 550 °C). Quantitation again involved the use of phenylhexane, this time as an external standard. Microscale sealed vessel (MSSV) pyrolysis experiments were employed to simulate the maturation of the Alum Shale. As described by Horsfield et al. (1989), weighed aliquots of kerogen were sealed in glass capillary tubes and heated at either 300, 330, or 350 °C for 3 days. Products were then released from the tubes inside the pyrolyzer, with the instrument in thermovaporization mode, using a purpose-built crushing device. Products were quantified using phenylhexane as an internal standard. Residues were then pyrolyzed as outlined above for Py-GC. Quantitation was made with reference to the phenylhexane internal standard from the immediately preceding analysis of MSSV products. Compounds were identified by comparison with authentic standards and by Py-GC-MS.

3 Nature of the Alum Shale

Concentrations of uranium in the Alum Shale (between 24 and 125 ppm; Table 1) are far above the global average for shales, as discussed by Lewan and Buchardt (1989). Similarly, TOC values ranging from 12 to 22%, recorded for the present sample suite, are clearly much higher than average. An appreciable but undetermined proportion of the total sulfur content (1.5–3.7%) exists in the form of pyrite.

The concentration and composition of the organic matter in the Alum Shale are discussed in two ways. First, the results from the various respective analytical approaches are presented independently and numerous possible interpretation scenarios are given (Sect. 3). There then follows (Sect. 4) a discussion of the controls governing the petroleum-generating potential of this rock unit.

3.1 Rock-Eval

The progressive decrease in the Hydrogen Index from 439 (Pälstorp) to less than 1 (Krekling) was used as the basis for ordering the samples into a single suite (Fig. 2) for subsequent and more detailed analysis. A literature review suggests that this compositional change is most likely related to increasing levels of thermal maturity and/or increasing exposure to ionizing radiation (Buchardt et al. 1986; Leventhal et al. 1986; Dahl et al. 1988; Lewan and Buchardt 1989).

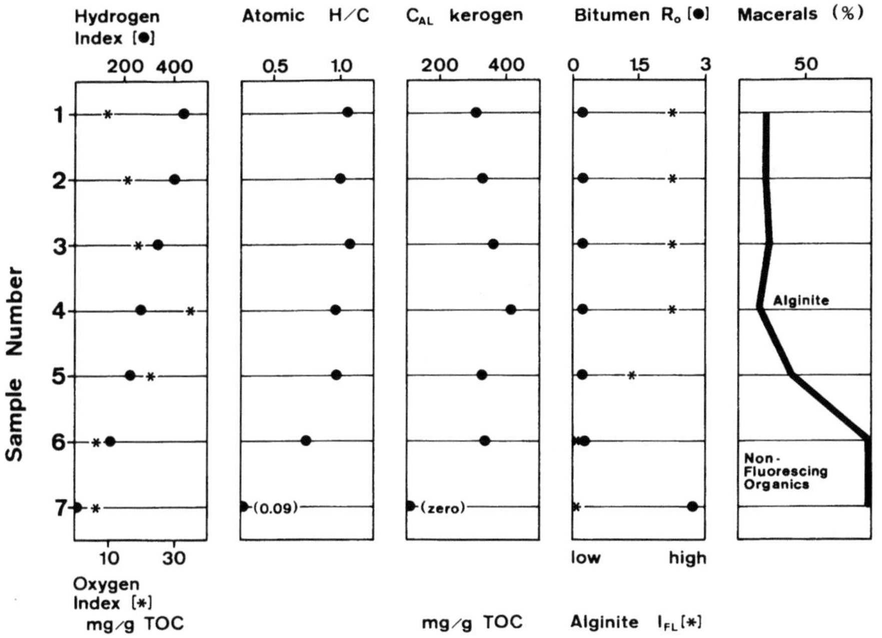

Fig. 2. Compilation of selected analytical data for the Alum Shale

However, other possible influences should not be dismissed, namely variations in precursor biota (despite the absence of land plants), redox potential during diagenesis, the oxidative effects of weathering, especially in producing a pyrolytically more reactive mineralogy (Larter 1984); and clay mineral matrix effects (Espitalié et al. 1980).

Despite the large range in Hydrogen Index values, T_{max} is relatively constant (414–421 °C) and does not change appreciably until sample 7; where a very low T_{max} was recorded. It should be noted that S2 is almost zero (0.02 mg HC/g rock) in the latter case and the T_{max} value is unreliable. Wide variations in maturity need not necessarily be precluded by these T_{max} results as long as the original organic matter had a uniform and homogeneous structure, i.e., provided that the bulk compositions of pyrolysis products and residues are alike (cf. Type I elemental composition in Espitalié et al. 1984). The observed increase, then decrease, in the Oxygen Index could be caused by pyrolysis mineral matrix effects (associated with variable carbonate content or degree of weathering; see Larter 1984), or uranium-organic matter redox processes (Leventhal et al. 1986).

3.2 Organic Petrology

In the first four samples, brightly fluorescing alginite macerals (alginite A, alginite B, liptodetrinite) occur in large amounts, and nonfluorescing oval-shaped "bituminite" (vitrinite-like organic matter) is a subsidiary component. This bituminite has a low reflectance. Sample 5 is similar to the above but appears more mature in that the fluorescence intensity of the alginite is weaker. In contrast to published maturation series (see Bertrand et al. 1986; Lin and Davis 1988 and references therein), this diminished intensity was not accompanied by a "red shift" in the fluorescence spectrum. Fluorescing alginite is absent from sample 6, and the only microscopitcally identifiable component has a low reflectance and does not fluoresce. Sample 7 contains similar material but with a much higher reflectivity, and is unquestionably metamorphosed.

3.3 Elemental Analysis

The relationship between the Hydrogen Index and atomic H/C ratio falls within the broad trend established by Espitalié et al. (1977). Nevertheless, the first five samples have closely similar atomic H/C ratios (0.98–1.07). The sixth has a somewhat lower value (0.78), and the seventh is very low (0.09). There appears, therefore, to be some discrepancy between bulk chemical (H/C) and pyrolytic (Hydrogen Index) estimates of the petroleum-generating potential of the Alum Shale. This might be due to one or more of the following reasons: (1) an appreciable proportion of hydrogen could be in the form of O–H rather than C–H (see I.R. Spectroscopy below), or (2) some of the

Table 1. Selected analytical data for the Alum Shale sample series

Parameter	Sample						
	1	2	3	4	5	6	7
Total organic carbon (%)	21.65	11.51	14.85	11.83	18.52	12.89	12.46
Uranium (ppm)	84	125	100	59	113	57	24
H/C (atomic ratio)	1.05	1.01	1.07	0.98	0.99	0.76	0.08
Rock-Eval							
S1 (mg HC/g rock)	2.3	1.2	1.4	1.6	1.5	1.0	0.1
S2 (mg HC/g rock)	95.0	46.3	48.8	31.8	42.0	19.6	0.0
S3 (mg CO_2/g rock)	2.3	1.9	2.9	4.2	4.2	0.9	0.8
T_{max} (°C)	414	415	415	419	415	421	377
Hydrogen Index/mg HC/g TOC)	439	402	329	269	227	152	0
Oxygen Index (mg CO_2/g TOC)	10	16	19	35	23	7	7
Infrared spectroscopy							
C_{AL} (mg/g rock)	82.9	68.1	60.9	41.2	52.5	36.4	0
Pyrolysis-gas chromatography							
Gas to oil ratio (kg/kg)	0.43	0.39	0.40	0.39	0.47	0.51	0.90

hydrogen which is released as hydrocarbons in nature may be released as water during pyrolysis (Teichmüller and Durand 1983) or (3) the structure of the kerogen changes in such a way as to result in an increasing proportion of the hydrogen being located in moieties which degrade on pyrolysis into involatile polynaphthenic and aromatic ring systems ("dead carbon").

3.4 Infrared Spectroscopy

It has previously been shown that pyrolytic yield and aliphatic carbon concentrations (C_{AL}) for the alginite-rich Posidonia Shale (Jurassic) of Northern Germany are strongly and positively correlated, but that some aliphatic carbon may be retained as involatile residues at high maturities (Schenk et al. 1986). In the case of *kerogens* isolated from the Alum Shale, C_{AL} appears to increase then decrease in parallel with the Oxygen Index. Most importantly, the C_{AL} is greater than the Hydrogen Index in all but the first two samples (Fig. 2). This suggests that a very significant proportion of the total aliphatic carbon is simply not released during Rock-Eval pyrolysis. This major inference is also supported by the comparison of whole-rock IR data (C_{AL}) and S2 values (Table 1) which indicates that between 25 and 50% of C_{AL} is not released as S2 during the pyrolysis of most samples. The same applies if allowance is made for the presence of S1 components, and the possibility that oxidative conversion of aliphatic hydrocarbons to carbon dioxide on weathered mineral surfaces may have contributed to the S3 peak. The extent of conversion to H_2O or H_2S was not assessed. The "least altered" end-member (sample 1) appears to release its aliphatic carbon when pyrolyzed under Rock-Eval conditions. It should also be noted that a large, but as yet unquantified, proportion of aliphatic carbon is contained in methyl, as opposed to methylene, groups. There is no significant hydroxyl group contribution to the Alum Shale kerogens (see Sect. 3.3).

3.5 Thermovaporization and Pyrolysis-Gas Chromatography

The most surprising finding from Py-GC analysis of Alum Shale kerogen is that in all cases the major resolved components are low-molecular-weight hydrocarbons, primarily C_1–C_9 n-alk-1-enes and n-alkanes, C_1–C_3 alkylbenzenes, and C_1–C_2 alkylnaphthalenes (Fig. 3). This result is in contrast with most alginites (those derived from *Botryococcus*, *Tasmanites*, or *Gloeocapsomorpha*) which generate only low proportions of aromatic compounds, and high proportions of n-alk-1-enes and n-alkanes over a wide maturity range. A plot of hydrocarbon-chain-length distributions (Horsfield 1989) shown in Fig. 4, suggests that Alum Shale kerogen is much more gas-prone than are most alginites. Pyrolysates appear to become slightly more aromatic through the series (cf. van Graas et al. 1981), although gas to oil ratios, as determined by the ratio of C_1–C_5/C_{6+} components in the pyrolysates, are rather uniform (0.4–0.5 kg/kg; 1900–2400 SCF/BBL in oilfield units). Thermovaporization of whole rocks results in the release of a relatively simple mixture of low-molecular-weight normal, branched and aromatic hydrocarbons

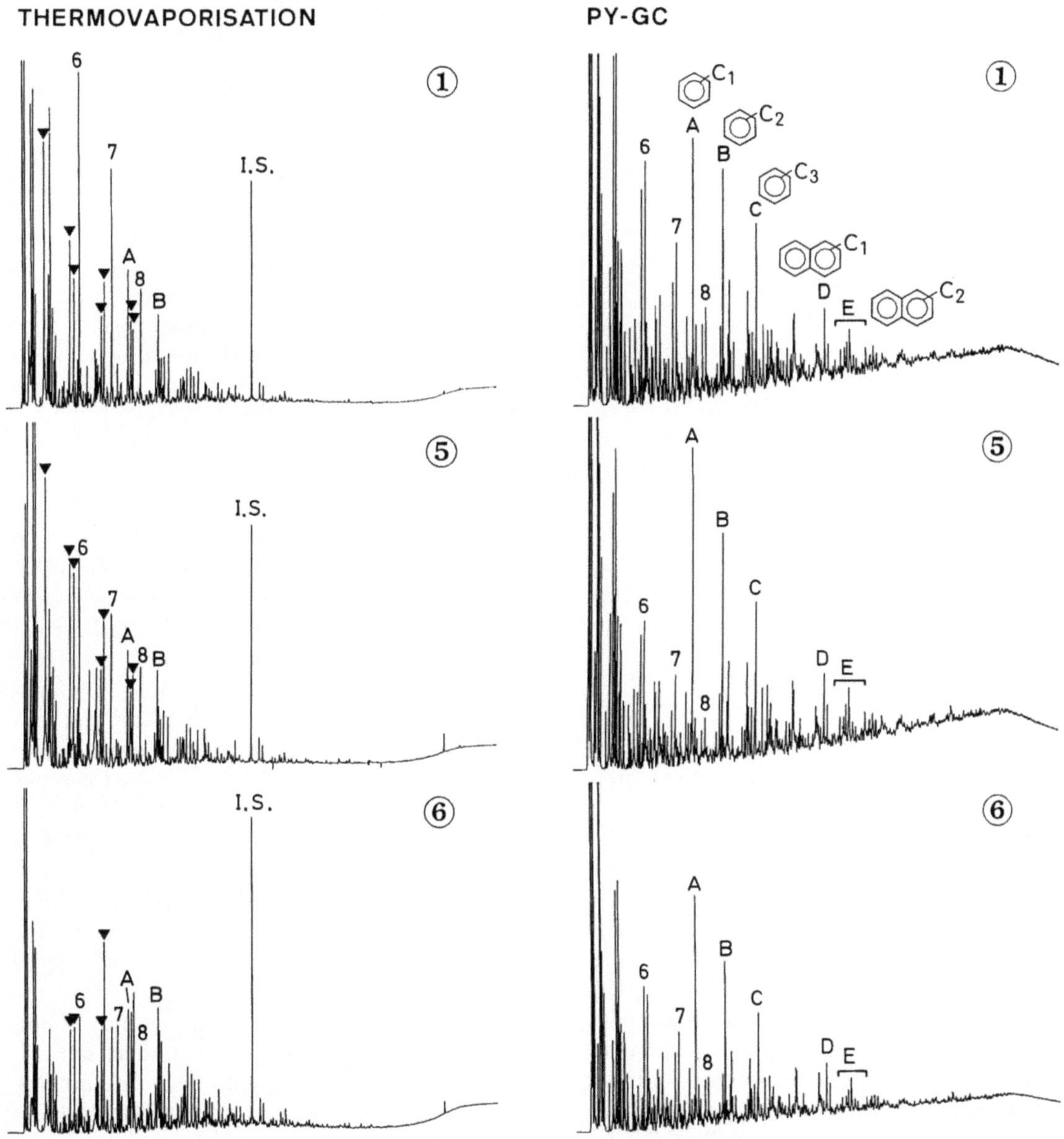

Fig. 3. Pyrolysis-gas chromatography traces for selected isolated kerogens, and thermovaporization traces on whole rocks of the same samples. Normal alkanes are identified according to chain length. Components *A* through *E* are identified in the *right half* of the figure. Peaks labeled by *triangles* are not yet identified, but are considered to represent two homologous series of branched alkanes. *I.S.* is the phenylhexane internal standard peak

(Fig. 3), thus confirming the overwhelmingly gas-prone nature of the Alum Shale organic matter (Fig. 4).

The abnormally high abundance of aromatic compounds in these pyrolysates, despite an abundance of aliphatic structures in the unheated kerogen (IR data), might suggest that aromatization reactions take place during pyrolysis of the Alum-Shale kerogen. If so, aromatized products might be expected to occur in a nonvolatilizable form as "dead carbon". This is a possible cause of the HI-H/C discrepancy discussed earlier. There is a 1:1 relationship between the S2 values (determined on whole rocks by Rock-Eval) and the total Py-GC yield (determined on kerogen concentrates) when both are normalized to organic carbon. This result is rather unusual, because tarry residues are often a major fraction of pyrolysates. The relationship nevertheless serves to demonstrate that mineral matrix effects do not adequately explain the observed Hydrogen Index trend, and that compositional findings from Py-GC data are quantitatively representative of Rock-Eval pyrolysates.

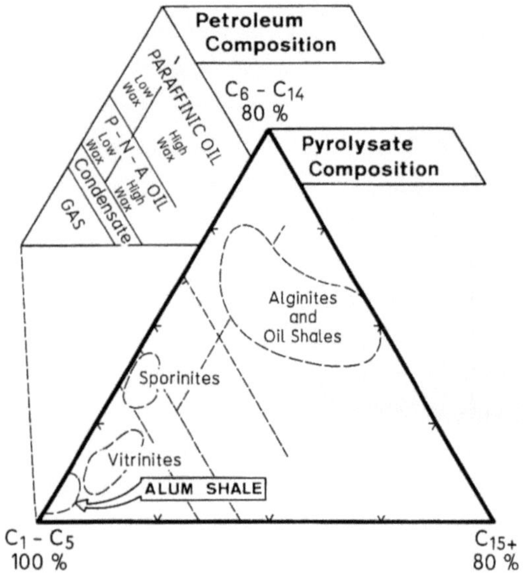

Fig. 4. Comparison of the kerogen quality of the Alum Shale compared with other alginites. (After Horsfield 1989)

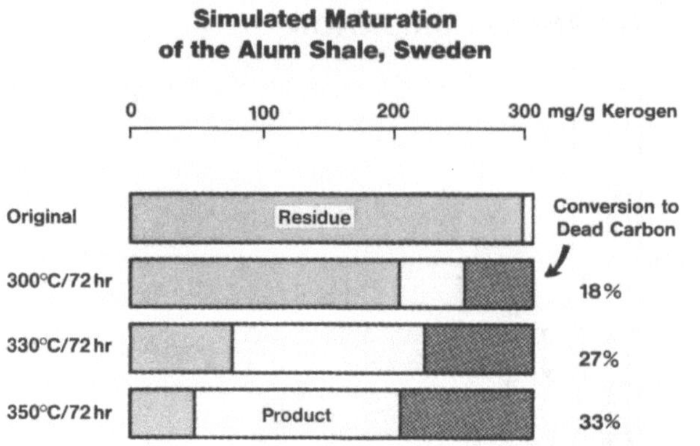

Fig. 5. Mass balance of precursors and products for the simulated maturation of Alum Shale labile kerogen components

3.6 Simulated Maturation Experiments

Simulated maturation experiments were performed on the least mature kerogen (sample 1) by heating at three temperatures (300, 330 and 350 °C) for 3 days (cf. Horsfield et al. 1989). The mildest conditions brought about a significant reduction in the fluorescence intensity of the alginite but no change in its fluorescence spectra. A maturity level was attained which is approximately equivalent to that of sample 5 in the natural series. The higher temperature experiments led to the formation of nonfluorescing blocky materials. The reflectivity of this material was greater than that of sample 6 in the natural series. The amounts of total products and residues formed in these experiments, as determined by thermovaporization and Py-GC (see methods section), are summarised in Table 2 and shown in Fig. 5. A mass balance calculation reveals that 18%, 27% and 33%, respectively, of originally volatilizable (volatile, as in S1, and labile, as in S2) moieties in the unheated kerogen are not detected in the 300, 330 and 350 °C heating experiments. This appears to represent the formation of "dead carbon", and confirms earlier suspicions that the process might be quantitatively very significant not only in the simulation experiments but also in the natural series. Further mass balance calculations reveal that the total Py-GC yield (in mg/g TOC using isolated kerogens) fell from 387 in the unheated sample to 277, 121, and 77 in the sequential heating experiments (TOC associated with S1 was subtracted from the TOC used in the denominator of this calculation). According to these data, the level of alteration induced by these closed-system heating experiments is broadly equivalent to the natural series (whose closed versus open pore system history is albeit unknown) as follows: 300 °C/72 h is approximately the same as sample 4; 330 °C/72 h and 350 °C/72 h lie between samples 6 and 7. These findings agree rather well with the petrological observations.

The composition of the residues from the heating experiments, as determined by Py-GC (Fig. 6), closely resemble those of the kerogens in the natural series. Aromatic hydrocarbons are abundant in all samples

Table 2. Quantitative thermovaporization and pyrolysis data for artificially matured Alum Shale Kerogen (sample 1)

	Thermovaporization of Products (T) in mg/g kerogen	Pyrolysis of Residue (P) in mg/g kerogen	$\frac{(T+P)}{(T+P)_o}$ [a]
Original	8	302	1.00
300°/72 hr	52	203	0.82
330°/72 hr	147	78	0.73
350°/72 hr	156	50	0.67

[a] "o" refers to the values of T and P on the original sample.

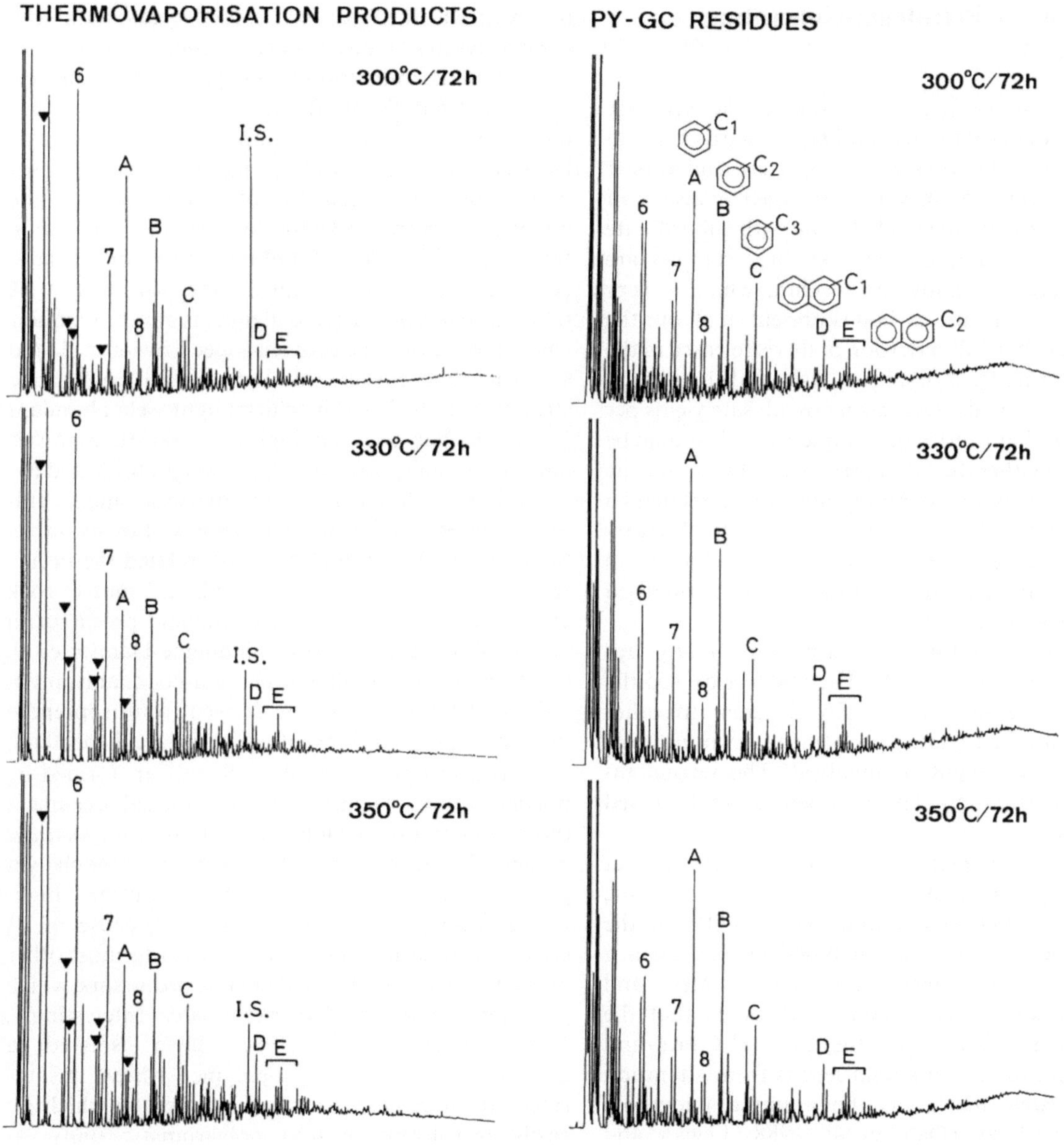

Fig. 6. Pyrolysis-gas chromatography and thermovaporization traces on the residues and products, respectively, from simulated maturation experiments. Key as for Fig. 3

and increase relative to n-alkenes and n-alkanes with increasing rank, though only subtly. The products from the heating experiments, as determined by thermovaporization, also resemble the free hydrocarbons in the natural series in that mainly low-molecular-weight normal, branched and aromatic hydrocarbons are abundant. While there is a small relative increase in branched compounds with increasing temperature (see Fig. 6: cf. Schaefer and Littke 1988), the gross compositions revealed by the gas chromatograms can be considered essentially uniform. The thermovaporization of the least mature sample of the natural series has the same appearance as these GC fingerprints (cf. Figs. 3, 6). In contrast, and despite having similar Hydrogen Indices, sample 6 from the natural series (HI = 152) contains free bitumen that is compositionally different to that generated from the kerogen artificially matured at 330 and 350 °C (HI = 147 and 156, respectively). Namely, there is a significantly greater relative abundance of C_3-C_8 components in the experimental series. This might be related to natural expulsion phenomena, whereby C_3-C_8 components have been preferentially expelled from the Alum Shale in nature.

4 Controls of Petroleum-Generating Potential

Kerogens from the first four samples of the series are essentially identical to one another as regards maceral composition and fluorescence properties, and appear to be immature to early mature. Each consists of mainly the primary maceral alginite with subordinate amounts of bituminite. The next three samples are more mature, especially the seventh, which is very overmature. Aliphatic carbon is present in all but the last kerogen, but only a portion of this is released in the S2 peak during Rock-Eval pyrolysis, and the pyrolysate is very aromatic. Decreasing pyrolysate yields per unit carbon characterize the sample series. It should be noted that neither the falling Hydrogen Index nor the high aromaticity of the pyrolysate can be related to mineral matrix effects (cf. Espitalié et al. 1980; Horsfield et al. 1983), as revealed by the 1:1 relationship between Hydrogen Index and the carbon-normalised pyrolysate yield from Py-GC.

Two types of influences, acting sequentially, appear to explain these results. The first has caused the kerogen to become structurally different to other alginites; uranium-related phenomena or an unusual algal precursor might be involved. The second influence is maturation. These influences are discussed below.

Regarding the presence of uranium, Leventhal et al. (1986) proposed a coeval oxidation of organic matter and reduction of uranium ($U^{6+} \rightarrow U^{4+}$) in the early history of the sediment based in part upon a positive correlation between uranium content and Oxygen Index. Other workers have noted that the negative correlation between the atomic H/C ratio and the concentration of uranium results from alteration of the organic matter as the uranium undergoes radioactive decay (Dahl et al. 1988; Lewan and Buchardt 1989). We find that neither atomic ratios nor Hydrogen Index correlate with uranium concentration, even in the apparently immature part of the section. This might be because weathering has removed uranium in solution from the Alum Shale. The increase in the Oxygen Index that we have observed (see Fig. 2) might be tracking the original uranium concentration of the Alum Shale (cf. Leventhal et al. 1986) prior to the weathering-related solubilization of uranium. In the event that uranium-related phenomena are important, the progressive structural rearrangements of aliphatic moieties in the kerogen (possibly involving oxygen-containing functional groups and cross-linking processes) with increasing exposure to ionizing radiation might explain why 25–50% C_{AL} appears to be retained in an involatile form during pyrolysis, why the Hydrogen Index decreases, and why the pyrolysates themselves are predominantly aromatic, even in the immature-mature part of the sequence.

An alternative (or additional) explanation is that the unusual structure and composition of Alum Shale kerogen is a reflection of the nature of its precursor biota. Thus, the precursors of Alum Shale alginite might have been markedly different in composition from those in other Cambrian seas, or in later geological periods, in much the same way as *Gloeocapsomorpha*-derived alginite is almost uniquely characteristic of the Ordovician (see Fowler et al. and Glaessner and Foster, this Vol.). Suitable precursor structures in the kerogen which might yield abundant C_{1-3} alkylbenzenes on high-temperature pyrolysis and, additionally, low-molecular-weight branched hydrocarbons in low-temperature pyrolysis, might ultimately be derived from carotenoids. For example, Summons and Powell (1987) have related the occurrence of certain aryl isoprenoids in source rock extracts and crude oils to a carotenoid precursor in green sulfur bacteria. Also, Sinninghe-Damste et al. (1988 and references therein) have noted that isoprenoid alkylbenzenes might be derived from isoprenoid quinones or directly biosynthesized by archaebacteria.

It is also noteworthy that Bb rubber, formed by polymerization of complex multi-branched precursors (botryococcenes) in clonal race B of *Botryococcus braunii* (Metzger et al. 1985) is hydrogen-rich, yet generates aromatic hydrocarbons during high-temperature pyrolysis (Larter 1978) because of its cross-linked nature. This is in contrast to most other hydrogen-rich alginites, and alginite precursors, which generate mainly n-alkyl pyrolysate. Interestingly, Chalansonnet et al. (1988) have isolated a cyanobacterial-resistant biopolymer whose unsaturated carbon bonds are located in structures which are easily transformable into polyaromatic units, as evidenced by the considerable amounts of highly aromatic insoluble residue which are formed on pyrolysis. Spectroscopic analyses were noted to have revealed high abundances of methyl groups and tertiary carbons. All these results bear a striking, if superficial, resemblance to those from the Alum Shale reported here.

Kerogen formed in one of the above ways, when exposed to increasing levels of maturation (samples 5–7) appears to result in the generation and migration of low-molecular-weight hydrocarbons whose petroleum product at the surface would be gas plus condensate. Because of its structure, Alum Shale kerogen matures to form "dead carbon" in large amounts, especially under closed-system conditions. Regarding kerogen maturation, it has been noted that the delocalization of electrons in progressively more aromatic macromolecules explains why a shift to longer wavelength in fluorescence spectra occurs during catagenesis (Lin and Davis 1988). The fact that no red-shift was found for the Alum Shale samples suggests that the growth of aromatic ring systems is not an important maturation reaction for this type of kerogen, or that polyaromatic ring systems are isolated rather than ordered. Instead, the recovery of compositionally uniform pyrolysates and the absence of a red-shift are together consistent with the model outlined by Bertrand et al. (1986) whereby fluorescence intensity is progressively quenched through radiationless transfer of energy caused by intermolecular interactions (a dilution-concentration effect).

5 Conclusions

The Cambrian Alum Shale contains alginite with structural pecularities that are due to either an unusual precursor biota or effects related to the presence of uranium (ionizing radiation, oxygen incorporation). Large proportions of aliphatic carbon are released during both pyrolysis and natural maturation as aromatic and branched hydrocarbons, rather than the usual long n-alkyl chains, or retained as involatile "dead carbon". Dead carbon is a major reaction product. Any calculation of petroleum masses generated from the Alum Shale using total organic carbon or Rock-Eval data (cf. Larter 1985; Cooles et al. 1986) might be substantially overestimated because of the tendency of this kerogen to generate diminished amounts of hydrocarbons. Similarly, to use the presence of fluorescing alginite or aliphatic carbon concentration as measures of liquid, or total, petroleum potential would be in gross error. Kerogen aromaticity would also be seriously overestimated using Py-GC data alone. Clearly, a multidisciplinary approach will be required to accurately assess the petroleum-generating potential of the Alum Shale in exploration ventures.

Acknowledgments. BH, FL, RL, and HJS acknowledge the support of Prof. D. H. Welte. BH thanks Dr. U. Mann for his continuing interest and helpful discussions regarding the Alum Shale. SRL acknowledges the financial support given by B. P. Norge and the Nordisk Ministerial Energy Research Program to the Petroleum Geochemistry Program at Oslo. SB is on lave from the Indian Oil and Natural Gas Commission and thanks Dr. J. W. de Leeuw (T. U. Delft) for Py-GC-MS analysis. Fr. B. Schmitz and Frl. S. Bardon typed various drafts of the manuscript. The constructive criticisms and editorial suggestions of Dr. A. G. Douglas, Dr. D. M. McKirdy, and two anonymous reviewers are gratefully acknowledged.

References

Bertrand P, Pittion J-L, Bernard C (1986) Fluorescence of sedimentary organic matter in relation to its chemical composition. Org Geochem 10:641–647

Bharati S (1989) Physical and chemical evolution of source rocks and source rock kerogens of algal origin. Cand Sci Thesis, Univ Oslo

Buchardt B, Clausen J, Thomsen E (1986) Carbon isotopic composition of Lower Palaeozoic kerogen: effects of maturation. Org Geochem 10:127–134

Chalansonnet S, Largeau C, Casadevall E, Berkaloff C, Peniguel G, Couderc R (1988) Cyanobacterial resistant biopolymers. Geochemical implications of the properties of Schizothrix sp. resistant material. In: Mattavelli L, Novelli L (eds) Advances in organic geochemistry, 1987. Pergamon, Oxford, pp 1003–1010

Cooles GP, Mackenzie AS, Quigley TM (1986) Calculation of petroleum masses generated and expelled from source rocks. Org Geochem 10:235–245

Dahl J, Hallberg R, Kaplan IR (1988) The effects of radioactive decay of uranium on elemental and isotopic ratios of Alum Shale kerogen. Appl Geochem 3:583–589

Espitalié J, Laporte JL, Madec M, Marquis F, Leplat P, Paulet J, Boutefeu A (1977) Méthode rapide de charactérisation des roches mères de leur potential pétrolier et de leur degré d'évolution. Rev Inst Fr Petrole 32:23–42

Espitalié J, Madec M, Tissot B (1980) role of mineral matrix in kerogen pyrolysis: influences on petroleum generation and migration. Bull Am Assoc Petrol Geol 64:59–66

Espitalié J, Marquis F, Barsony I (1984) Geochemical logging. In: Voorhees KJ (ed) Analytical pyrolysis techniques and applications. Butterworths, London, pp 276–304

Horsfield B (1989) Practical criteria for classifying kerogens: some observations from pyrolysis-gas chromatography. Geochim Cosmochim Acta 53:891–901

Horsfield B, Dembicki H, Ho TTY (1983) Some potential applications of pyrolysis to basin studies. J Geol Soc London 140:431–443

Horsfield B, Disko U, Leistner F (1989) The microscale simulation of maturation: outline of a new technique and its potential applications. Geol Rundsch 78:361–374

Larter SR (1978) A geochemical study of kerogens and related materials. Ph D Thesis, Univ Newcastle/Upon Tyne

Larter SR (1984) Application of analytical pyrolysis techniques to kerogen characterization and fossil fuel exploration/exploitation. In: Voorhees KJ (ed) Analytical pyrolysis techniques and applications. Butterworths, London, pp 212–275

Larter SR (1985) Integrated kerogen typing and the quantitative evaluation of petroleum source rocks. In: Thomas BM et al. (eds) Petroleum geochemistry in exploration of the Norwegian Shelf. Graham & Trotman, London, pp 269–286

Larter SR, Horsfield B (1991) Determination of structural components of kerogen using analytical pyrolysis methods. In: Engel M, Macko S (eds) Organic geochemistry. Plenum, New York (in press)

Leventhal JS, Daws TA, Frye JS (1986) Organic geochemical analysis of sedimentary organic matter associated with uranium. Appl Geochem 1:241–247

Lewan M, Buchardt B (1989) Irradiation of organic matter by uranium decay in the Alum Shale, Sweden. Geochim Cosmochim Acta 53:1307–1322

Lin R, Davis A (1988) A fluorogeochemical model for coal macerals. Org Geochem 12:363–374

Littke R, Baker DR, Leythaeuser D (1988) Microscopic and sedimentologic evidence for the generation and migration of hydrocarbons in Toarcian source rocks of different maturities. In: Mattavelli L, Novelli L (eds) Advances in organic geochemistry 1987. Pergamon, Oxford, pp 549–559

Metzger P, Berkaloff C, Casadevall E, Coute A (1985) Alkadiene and botryococcene-producing races of mild strains of Botryococcus braunii. Phytochemistry 24:2305–2312

Reed J, Illich HA, Horsfield B (1986) Biochemical Evolutionary significance of Ordovician oils and their source. In: Leythaeuser D, Rullkötter J (eds) Advances in organic geochemistry 1985. Pergamon, Oxford, pp 347–358

Schaefer RG, Littke R (1988) Maturity-related compositional changes in the low-molecular-weight hydrocarbon fraction of Toarcian shales. In: Mattavelli L, Novelli L (eds) Pergamon, Oxford, pp 887–892

Schenk H-J, Witte EG, Müller PJ, Schwochau K (1986) Infrared estimates of aliphatic kerogen carbon in sedimentary rocks. Org Geochem 10:1099–1104

Sinninghe Damsté JS, Kock-van Dalen AC, de Leeuw JW (1988) Identification of long-chain isoprenoid alkylbenzenes in sediments and crude oils. Geochim Cosmochim Acta 52:2671–2677

Summons RE, Powell TG (1987) Identification of aryl isoprenoids in source rocks and crude oils: biological markers for the green sulphur bacteria. Geochim Cosmochim Acta 51:557–566

Teichmüller M, Durand B (1983) Fluorescence microscopical rank studies on liptinites and vitrinites in peat and coals, and comparison with results of the Rock-Eval pyrolysis. Int J Coal Geol 2:197–230

van Graas G, de Leeuw JW, Schenck PA, Haverkamp J (1981) Kerogen of Toarcian shales of the Paris Basin. A study of its maturation by flash pyrolysis techniques. Geochim Cosmochim Acta 45:2465–2474

Distribution and Structure of Hydrocarbons and Heterocyclic Sulfur Compounds Released from Four Kerogens of Ordovician Age by Means of Flash Pyrolysis

A. G. Douglas [1], J. S. Sinninghe Damsté [2], J.W. de Leeuw [2], T. I. Eglinton [2,4] and M. G. Fowler [3]

[1] Newcastle Research Group in Fossil Fuels and Environmental Geochemistry, The University, Newcastle upon Tyne, NE1 7RU, UK
[2] Organic Geochemistry Unit, Technical University of Delft, 2628 RZ Delft, The Netherlands
[3] Institute of Sedimentary and Petroleum Geology, Geological Survey of Canada, Calgary, Canada

Abstract

Kerogen concentrates prepared from four rocks of Ordovician age have been analyzed by pyrolysis-gas chromatography using a sulphur-specific flame photometric detector and by pyrolysis-gas chromatography-mass spectrometry.

All of the kerogen pyrolysates contained homologous series of hydrocarbons including n-alkanes, n-alk-1-enes, alkyl-substituted benzenes and naphthalenes: sulfur compounds which included alkylthiophenes, alkylthiolanes, alkylthianes, and alkylbenzothiophenes were also present.

Fingerprint chromatograms, and mass chromatograms obtained from three of the kerogens were commonly similar to one another, but those from the fourth kerogen, which contained abundant fossilized remains of the alga *Gloeocapsomorpha prisca*, were usually distinctively different. An important aspect of the structures of the homologous series released from this kerogen was the predominance of compounds with linear carbon skeletons.

1 Introduction

The application of flash pyrolysis methods, in combination with high resolution gas chromatography, or gas chromatography-mass spectrometry (GC-MS), to release, separate, and indentify organic compounds that form part of the structure of geopolymers has proved to be remarkably successful. The technique has been applied to the identification of organic molecules released from coal macerals (Larter and Douglas 1978; van Graas et al. 1980; Nip et al. 1991), asphaltenes (Rubinstein and Strausz 1979; Behar and Pelet 1985), and kerogens (Larter et al. 1978; van Graas et al. 1980; van de Meent et al. 1980; Solli et al. 1984; Larter 1984). Included among the series of compounds that have been identified by these methods are homologous series of n-alkanes, n-alk-1-enes, n-alka-α,ω-dienes, cycloalkanes, aromatic hydrocarbons, and heterocyclic sulfur compounds.

More recently, after recognition that petroleums may contain many hundreds of organic sulfur compounds (Sinninghe Damsté et al. 1987), the technique has been applied to kerogens to characterize, in a similar fashion, a variety of distinctive homologies of sulfur-containing moieties released by pyrolysis (Sinninghe Damsté et al. 1988, 1989).

The main aim of the present work was to extend pyrolysis-GC-MS methods to kerogens separated from four North American rocks of Ordovician age with a view to determining whether these kerogens released hydrocarbons and sulfur compounds which differed in any way from each other and from those released from younger kerogens. The four Ordovician rocks chosen had been used in earlier studies, as part of the International Geological Correlation Program's Project 157. Details of the appearance of the kerogens in reflected and ultra-violet light, of their Rock Eval and elemental analyses, and the chemistry of the extracted bitumens have appeared elsewhere (Fowler and Douglas 1984; Fowler et al. 1986), and are again noted by Fowler in this Volume. These investigations showed that the hydrocarbon distributions in the Guttenberg Oil Rock bitumen differed from those in the other three rocks, the Guttenberg being distinguished by n-alkanes with a strong odd/even preference principally at n-C_{17} and n-C_{19}, a rapid "fall off" in n-alkane abundance above n-C_{19}, very little of an unresolved complex mixture, and the presence of n-alkylcyclohexanes and methyl-n-alkylcyclohexanes in the same carbon-number range as the n-alkanes and with similar odd/even distributions.

Microscopically, the Guttenberg rock was also distinguished from the other three rocks. By examina-

[4] Present address: Department of Chemistry, Woods Hole Oceanographic Institution, Woods Hole, Mass. 02543, USA.

M. Schidlowski et al. (Eds.)
Early Organic Evolution: Implications for Mineral and Energy Resources

tion with ultra-violet light, it was seen to contain, in great abundance, and virtually exclusively, remains of the alga *Gloeocapsomorpha prisca*. The nature of this organism was discussed by Zalessky (1917), who proposed, based on morphological criteria, that it had been a planktonic cyanobacterium. More recently, alternative interpretations have been given by Reed et al. (1986), who also noted that the alga was a known major constituent of the Ordovician oil shale kukersite (Estonia) and of a deposit in the Amadeus Basin (Australia) and of several deposits in the United States. Their analyses led them to conclude that *G. prisca* was a chemosynthetic, aerointolerant, benthonic prokaryote. A discussion of existing views and a detailed examination of the chemistry of oils, sediments, and kerogen pyrolysates from *G. prisca*-rich sources subsequently led Hoffmann et al. (1987) to propose that it was "probably planktonic and phototrophic, and possibly an eukaryotic alga, widely distributed in the marine environment during the Ordovician". In more recent work, Foster et al. (1989a) conclude that *G. prisca* and modern *Cyanophyta* have a proven affinity instancing certain species of Entophysalidaceae cyanobacteria, in particular *Entophysalis major*, which is a mat-forming and sometimes stromatolite-forming cyanobacterium (Foster et al. 1989b, 1990); whereas Stasiuk and Osadetz (1991) provide evidence indicating that it had both planktonic and benthic stages during its life cycle.

Based on pyrolysis and spectroscopic studies, Derenne et al. (1988) have proposed that the organic content of kukersites may consist in large part of selectively preserved, resistant aliphatic biopolymers (PRB) similar to those found in the cell walls of *Botryococcus braunii* races A and B, but not race L (Derenne et al. 1989) and in torbanites. These findings are in addition to the claim by Traverse (1955) and Monin et al. (1980) that kukersites are largely composed of fossilized *B. braunii*.

2 Experimental

The four rock samples used in this work are described more fully in Fowler and Douglas (1984). Briefly, the Guttenberg Oil Rock, from a mine in S. W. Wisconsin (Heyl et al. 1959), is a calcareous shale containing large amounts of algal bodies which display a strong yellow/orange fluorescence in ultra-violet light. The Vinini shale from W. Nevada, and the Womble and Big Fork shales from S.E. Oklahoma contain fluorescent amorphous organic matter in which a few discrete algal bodies of the order *Tasmanales* are discernable.

Kerogens were isolated from the powdered shales, after extracting their soluble bitumens, by sequential treatment with hot hydrochloric and hydrofluoric acids followed by further solvent extraction. The four kerogen concentrates were pyrolyzed at 610 °C using a Curie point pyrolyzer attached to a gas chromatograph (Varian 3700) fitted with a flame photometric detector (FPD) or to a gas chromatograph (Hewlett-Packard 5840) connected in tandem with a VG-70S mass spectrometer. Chromatographic separations were achieved using a fused silica capillary column (25 m × 0.32 mm i.d.) coated with CP Sil-5. Full details of the method used have already been published (Sinninghe Damsté et al. 1988). Analyses by GC-MS were conducted by acquiring mass spectra with the instrument operated at a cycle time of 1.8 s

Table 1. Range of occurrence by carbon number of several homologous series present in the pyrolysates[a]

Compound class	Kerogen				
	m/z	Guttenberg	Vinini	Big Fork	Womble
n-Alkanes	57	6–28 (13)	6–36 (10)	6–36 (10)	6–36 (11)
n-Alk-1-enes	55	6–27 (13)	6–36 (10)	6–36 (10)	6–35 (10)
Alkylcyclohexanes	83	8–19 (17)	8–26 (8)	8–26 (8)	8–26 (8)
Alkylbenzenes	91	7–23 (7)	7–32 (7)	7–28 (7)	7–26 (7)
2-Methyl-*n*-alkylbenzenes	105	8–23 (15)	8–33 (9)	8–30 (9)	8–30 (9)
2-Alkylthiophenes	97	5–17 (5)	5–28 (5)	5–10 (5)	5–16 (5)
2-Alkyl-5-methylthiophenes	111	6–19 (6)	6–31 (7)	6–16 (6)	6–23 (6)
2-Alkyl-5-ethylthiophenes	125	7–19 (15)	7–31 (8)	7–23 (9)	7–27 (9)
2-Alkylthiolanes	87	4–19 (6)	4–31 (5)	4–27 (4)	4–30 (5)
2-Alkyl-5-methylthiolanes	101	6–19 (15)	6–27 (7)	n.d.[b]	6–18 (7)
2-Alkylthianes	101	6–17 (15)	6–26 (7)	n.d.	6–18 (7)

[a] Maxima of carbon number distributions are given in parentheses: note that the carbon number includes the carbon atoms in the aliphatic, aromatic, or heterocyclic ring systems.
[b] n.d. = not detected.

(m/z 40–800) and at a resolution of 1000. Mass chromatograms, which illustrated the different fingerprints of homologous series, were reconstructed using specific m/z values for aliphatic and aromatic hydrocarbons, and for sulfur-containing compounds (see Table 1 for appropriate m/z values). In addition, summed mass chromatograms were obtained for alkylbenzenes, alkylnaphthalenes, alkylthiophenes, and alkylbenzo[b]thiophenes in order to illustrate differences (or similarities) in the relative abundances of isomers of increasingly higher molecular weights. Aliphatic and aromatic compounds were identified by comparison of their mass spectra with standard compounds, by comparison of relative retention times (Lee et al. 1979; Schroder 1980), and by linear Kovats plots for homologous series. Sulfur compounds were identified as indicated previously (Sinninghe Damsté et al. 1988). Also, accurate mass chromatograms (mass window 0.05 dalton) were obtained to provide simplified distributions from those using nominal m/z values (for instance mass chromatograms of m/z 97 contain peaks due to both n-alk-1-enes and 2-alkylthiophenes).

3 Results

In this brief overview of the geochemistry of the four chosen Ordovician kerogens, it is not possible to illustrate and discuss all of these results. A more detailed paper has been published elsewhere (Douglas et al. 1991).

All four kerogen pyrolysates were analyzed by gas chromatography (FPD detector) prior to analysis by GC-MS. Figure 1 illustrates differences in the distribution of sulfur-containing compounds in the Guttenberg and Vinini kerogens, differences that were even more clearly seen when selected mass chromatograms were examined. In the chromatograms shown in Fig. 1, the relative abundance of many peaks clearly distinguish these two kerogens: peak identities are recorded in Table 2. More specifically, the Guttenberg pyrolysate contains abundant 2-alkylthiophenes (peak numbers 2, 7, 15, 29, 43, 51, 56, 59), compounds that have a linear carbon skeleton, while the Vinini contains higher relative abundances of thiophenes with branched carbon skeletons (e.g., peaks 19, 23, 32, 37, etc). Chromatograms (FPD) of the Womble and

Table 2. Alkylthiophenes identified in the pyrolysates of Ordovician kerogens

1.	Thiophene	32.	5-Ethyl-2,3-dimethylthiophene
2.	2-Methylthiophene	33.	Ethyldimethylthiophene
3.	3-Methylthiophene	34.	Ethyldimethylthiophene
4.	Thiolane	35.	Ethyldimethylthiophene
5.	Methyldihydrothiophene	36.	Ethyldimethylthiophene
6.	2-Methylthiolane	37.	2,3,4,5-Tetramethylthiophene
7.	2-Ethylthiophene	38.	$C_{4:1}$ Thiophene
8.	2,5-Dimethylthiophene	39.	2-Ethyl-5-propylthiophene
9.	3-Ethylthiophene	40.	3,5-Dimethyl-2-propylthiophene
10.	2,4-Dimethylthiophene	41.	2-Ethyl-4-isopropylthiophene
11.	2-Vinylthiophene	42.	2-Butyl-5-methylthiophene
12.	2,3-Dimethylthiophene	43.	2-Pentylthiophene
13.	3,4-Dimethylthiophene	44.	2,3-Dimethyl-5-propylthiophene
14.	2-Ethylthiolane	45.	Ethyltrimethylthiophene
15.	2-Propylthiophene	46.	Ethyltrimethylthiophene
16.	2-Ethyl-5-methylthiophene	47.	Ethyltrimethylthiophene
17.	2-Ethyl-4-methylthiophene	48.	2-Butyl-5-ethylthiophene
18.	Ethylmethylthiophene	49.	2-Butyl-3,5-dimethylthiophene
19.	2,3,5-Trimethylthiophene	50.	2-Methyl-5-pentylthiophene
20.	2-Methyl-5-vinylthiophene	51.	2-Hexylthiophene
21.	Ethylmethylthiophene	52.	5-Butyl-2,3-dimethylthiophene
22.	$C_{3:1}$ Thiophene	53.	2-Methylbenzo[b]thiophene
23.	2,3,4-Trimethylthiophene	54.	3- and 4-Methylbenzo[b]thiophene
24.	Isopropylmethylthiophene	55.	2-Hexyl-5-methylthiophene
25.	$C_{3:1}$ Thiophene	56.	2-Heptylthiophene
26.	2-Propylthiolane	57.	C_2 Benzo[b]thiophenes
27.	2-Methyl-5-propylthiophene	58.	2-Heptyl-5-methylthiophene
28.	2,5-Diethylthiophene	59.	2-Octylthiophene
29.	2-Butylthiophene	60.	2-Methyl-5-octylthiophene
30.	2-Ethyl-3,5-dimethylthiophene	61.	2-Nonylthiophene
31.	3-Ethyl-2,5-dimethylthiophene		

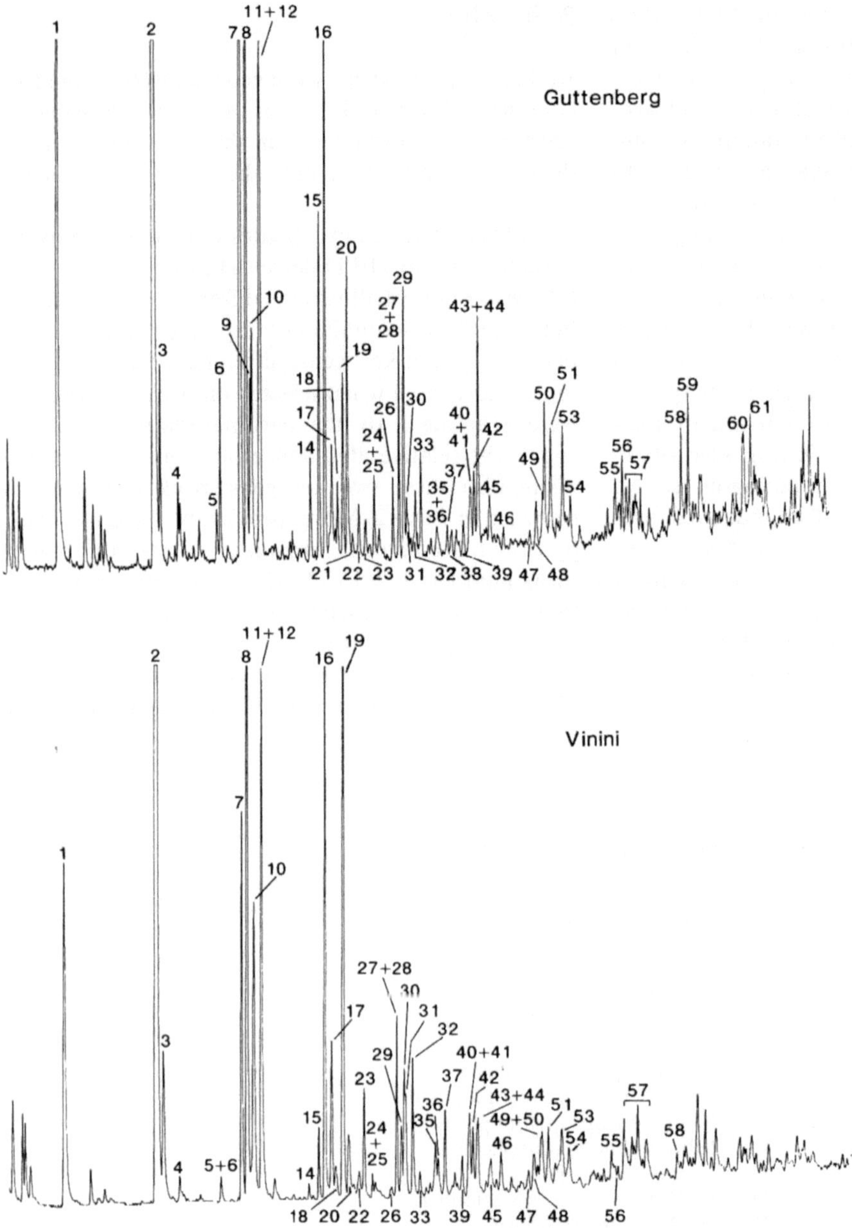

Fig. 1. Pyrolysis-GC of Guttenberg and Vinini kerogens using a flame photometric detector to monitor sulfur compounds. The *numbered peaks* in these partial FPD chromatograms are identified in Table 2

Big Fork pyrolysates (not shown) are similar to that of the Vinini but unlike that of the Guttenberg.

Analyses of all four kerogen pyrolysates by GC-MS indicated that they consisted dominantly of homologous series of n-alkanes and n-alk-1-enes, together with lesser amounts of cycloalkanes, aromatic and heterocyclic sulfur compounds. The Big Fork kerogen pyrolysate contained much smaller quantities of heterocyclic sulfur compounds than did the other three kerogens, whilst in the Womble kerogen the compound with the highest ion intensity was 1,2,3,4-tetramethylbenzene. The presence of extended homologous series of hydrocarbons and sulfur compounds was determined by reconstructing mass chromatograms using specific m/z values as noted previously (Table 1). In all of the kerogens, with the notable exception of the Guttenberg, the distributions, or fingerprints, of these series were in many cases virtually superposable and therefore in some of the figures only one example is used to illustrate their very large differences with fingerprints of the constituents of the Guttenberg pyrolysates. To save repetition the term "grouped kerogens" will be used, when appropriate, to signify the Womble, Vinini, and Big Fork kerogens.

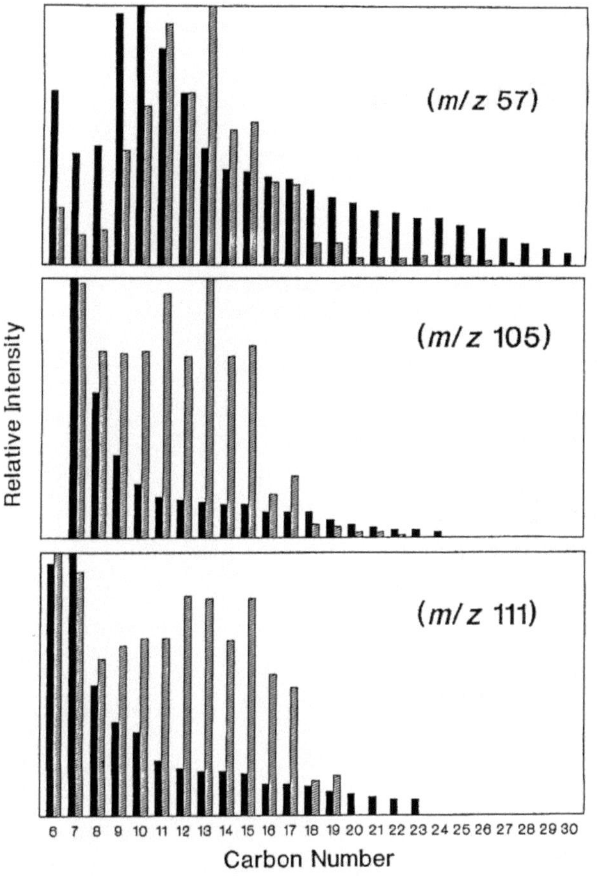

Fig. 2. Histograms prepared from PY-GC-MS data illustrating the distribution of n-alkanes (m/z 57), 2-methyl-n-alkylbenzenes (m/z 105) and 2-n-alkyl-5-methylthiophenes (m/z 111) in Guttenberg (*hatched bars*) and Vinini kerogen (*solid bars*) pyrolysates

3.1 Aliphatic Hydrocarbons

A series of n-alkane/n-alk-1-ene doublets was a major feature of all four pyrograms. In the grouped kerogens, the doublets decreased smoothly in abundance as carbon numbers increased from C_6 to C_{36}. In the Guttenberg kerogen pyrolysate, a quite distinct distribution of doublets was evident and seen to be repeated in most of its other homologous series of hydrocarbons and sulfur compounds. This distribution showed members of the homology in high relative abundance below C_{17} but with a dramatic reduction in abundance of higher homologs. An odd/even predominance was also evident. These differences are well illustrated in Fig. 2, which includes histograms of n-alkane distributions (m/z 57) for the Guttenberg and Vinini kerogens. It is also illustrated for n-alk-1-enes, but less well in the m/z 111 mass chromatograms (Fig. 3).

3.2 Aromatic Hydrocarbons

All four kerogen pyrolysates contained alkyl-substituted benzenes and naphthalenes. The n-alkylbenzene homologies in all four kerogens (not shown) were virtually superposable, a feature not seen in any other mass chromatograms. However, quite distinct and revealing fingerprints were obtained from the summed mass chromatograms of alkylbenzenes containing from one to six carbon-atom substituents. In the Guttenberg trace the major constituent at each molecular weight was the n-alkylbenzene, followed in abundance by the alkyltoluene that could be formed by cyclization of a precursor with a linear carbon skeleton (i.e., *ortho* alkyltoluenes). This feature did not apply to the other kerogens. For instance, the major compound in the Womble and Big Fork kerogens was 1,2,3,4-tetramethylbenzene and in the Vinini kerogen, after toluene, it was *meta* plus *para*-xylene. Mass chromatograms (m/z 105) for n-alkyltoluenes provided virtually superposable traces for the grouped kerogens in which the wide range of 2-methyl, 3-methyl and 4-methyl-n-alkylbenzenes had all approximately equal, low abundances (see, e.g., Vinini kerogen in Fig. 4). However, the Guttenberg mass chromatogram showed a strikingly large relative abundance of the 2-methyl isomers, compounds which, unlike the others, could be formed from a precursor moiety with a linear carbon skeleton. Furthermore, the range of these compounds was restricted, showing much diminished abundance above C_{17} but relatively high amounts of C_7–C_{15} members were observed (Fig. 4).

The ability to distinguish the four kerogens from the summed mass chromatograms of their C_1 to C_3 alkylnaphthalenes was less satisfactory, the main distinguishing feature being the reduced relative abundance of the C_2 and C_3 compounds in the Guttenberg kerogen and the increased abundance of alkylnaphthalenes with linear carbon skeletons.

3.3 Sulfur-Containing Compounds

A large number of sulfur-containing compounds were found in the four pyrolysates, although two series were not present in the sulfur-lean Big Fork kerogen. This may be related to the somewhat higher rank of this kerogen (Eglinton et al. 1988, 1990). The series of heterocyclic sulfur compounds recorded included 2-alkylthiophenes, 2-alkyl-5-methylthiophenes, 2-alkyl-5-ethylthiophenes, 2-alkylthiolanes, 2-alkyl-5-methylthiolanes, and 2-alkylthianes, all of which occurred as homologous series (Table 1).

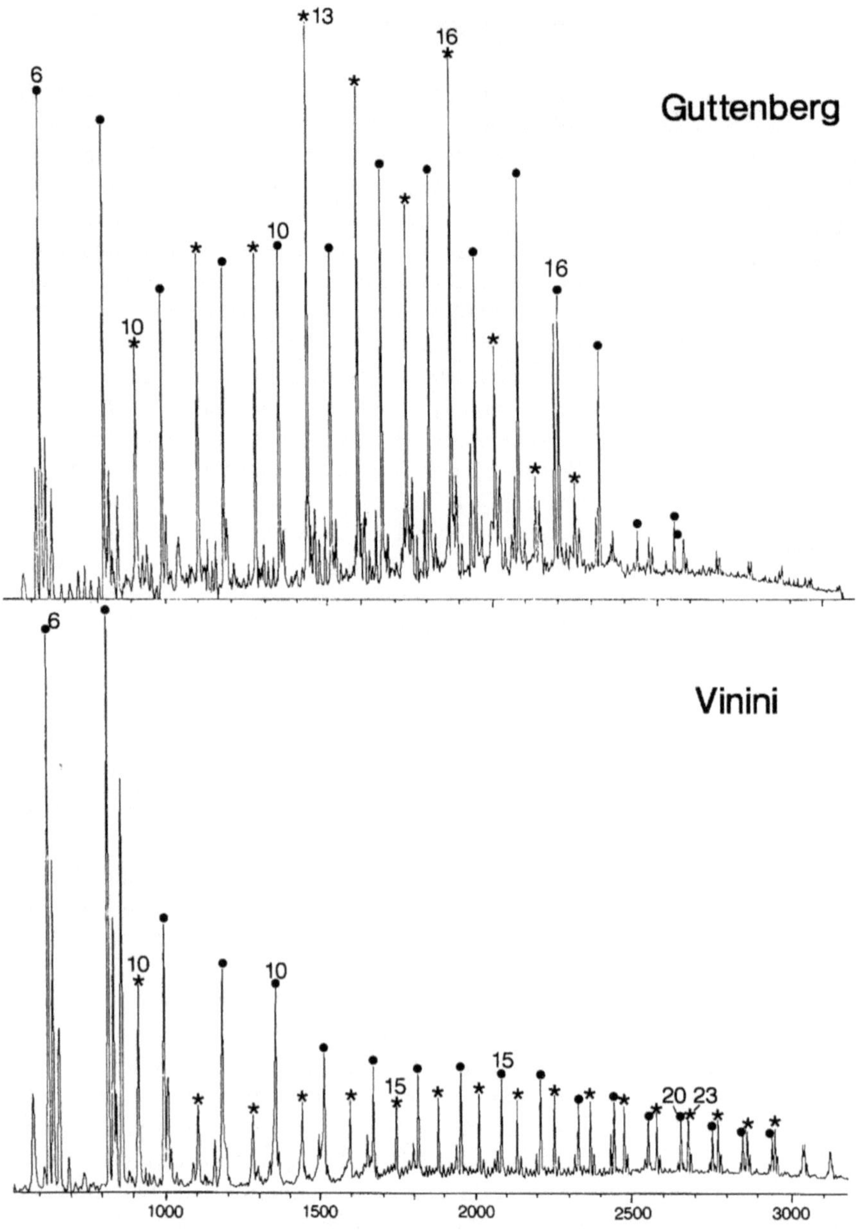

Fig. 3. Mass chromatograms (m/z 111) illustrating the relative distributions of 2-n-alkyl-5-methylthiophenes (*solid circles*) and alk-1-enes (*starred peaks*) in Guttenberg and Vinini kerogen pyrolysates. Peak number refer to the total number of carbon atoms in the molecule. For instance **10* refers to dec-1-ene and • *10* refers to 2-methyl-5-pentylthiophene

The 2-n-alkylthiophenes in all four kerogen pyrolysates had less distinctive distributions than other series, as discussed below, in that there was a decrease in the abundance of homologs as the carbon number increased from C_5 (2-methylthiophene). In the Vinini kerogen, the homology extended to about C_{28} but was below the detection limit beyond C_{10} for the Big Fork kerogen. The 2-n-alkyl-5-methyl and 2-n-alkyl-5-ethylthiophene distributions both showed striking differences between the Guttenberg and grouped kerogens. This is well illustrated in Fig. 3, which shows signals for the 2-n-alkyl-5-methylthiophenes (marked with a solid circle) and the C_8 fragment of n-alk-1-enes (starred peaks). The gradually diminishing abundance of the thiophenes from C_6 to C_{23} shown by the grouped kerogens (Vinini in Figs. 2 and 3) is similar to that reported for a number of other kerogens (Sinninghe Damsté et al. 1989) and contrasts sharply with their high relative abundance up to C_{17} in the Guttenberg pyrolysate. These distributions are much

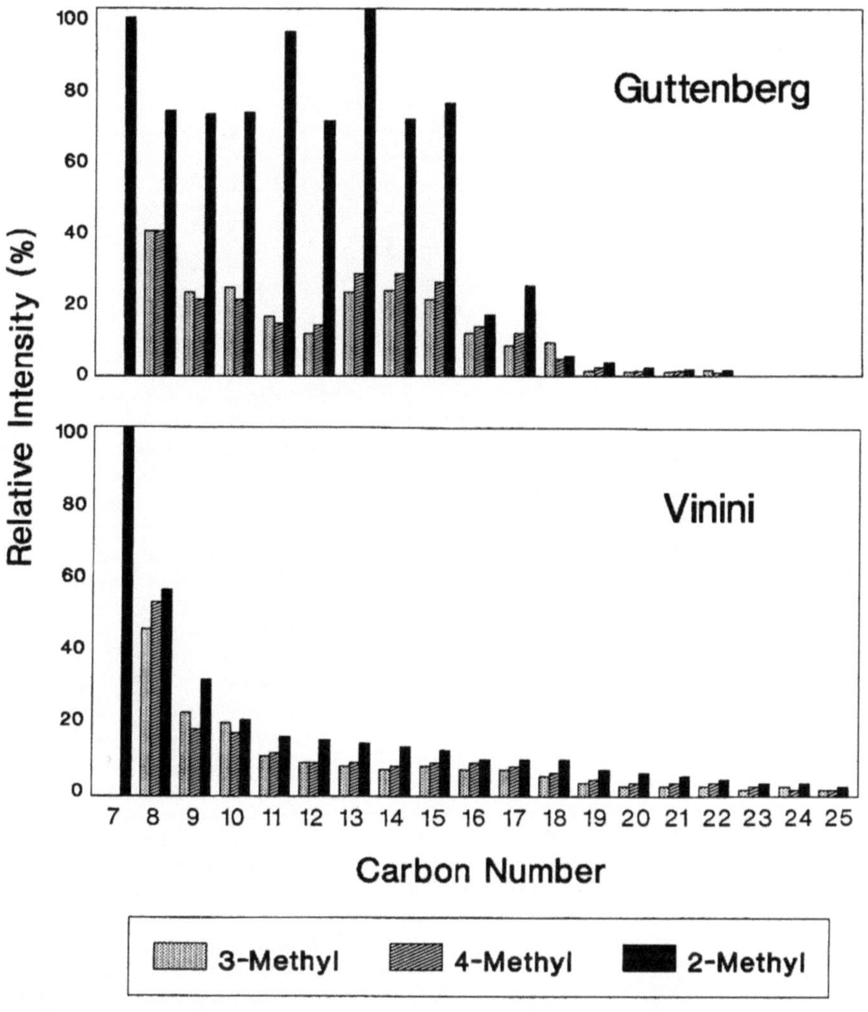

Fig. 4. Histograms prepared from PY-GC-MS data illustrating the distribution of 2-methyl. 3-methyl-, and 4-methyl-n-alkylbenzenes in Guttenberg and Vinini kerogen pyrolysates

like those for the 2-n-alkyl-5-ethylthiophenes (m/z 125) and also for the ortho-substituted n-alkyltoluenes (Figs. 2, 4). Summed mass chromatograms for C_5 (methylthiophenes) to C_{10} compounds also illustrate significant differences in the content of alkylthiophenes in these kerogens. Figure 5 illustrates these differences clearly, indicating that in the Guttenberg kerogen the principal components are 2-n-alkylthiophenes whereas in the Womble (and Vinini), after 2-methylthiophene, the two most abundant compounds are 2,5-dimethylthiophene (peak 8) and 2,3,5-trimethylthiophene (peak 19). Thiophenes identified in the kerogen pyrolysates are listed in Table 2.

Saturated five- and six-membered cyclic sulfides (thiolanes and thianes, respectively) were found in all of the kerogen pyrolysates, but in diminished abundance relative to the other sulfur compounds. Again the Guttenberg fingerprint is quite different from those of the grouped kerogens and displays some of the distinctive distributions shown by the 5-methyl-2-n-alkylthiophenes and the 1-alkyl-2-methylbenzenes.

All of the pyrolysates contained benzo[b]thiophene and a complex mixture of its alkyl derivatives. Summed accurate mass chromatograms for the parent compound and alkyl derivatives containing from one to three carbon atoms for the Guttenberg and Womble samples indicated that all six monomethylbenzo[b]thiophenes were present but in characteristically different relative proportions. In the Guttenberg pyrolysate the most abundant isomer was 2-methylbenzo[b]thiophene, a compound that has a linear carbon skeleton. In the Womble pyrolysate the most abundant benzo[b]thiophene was tentatively identified as the 2,4-dimethyl derivative.

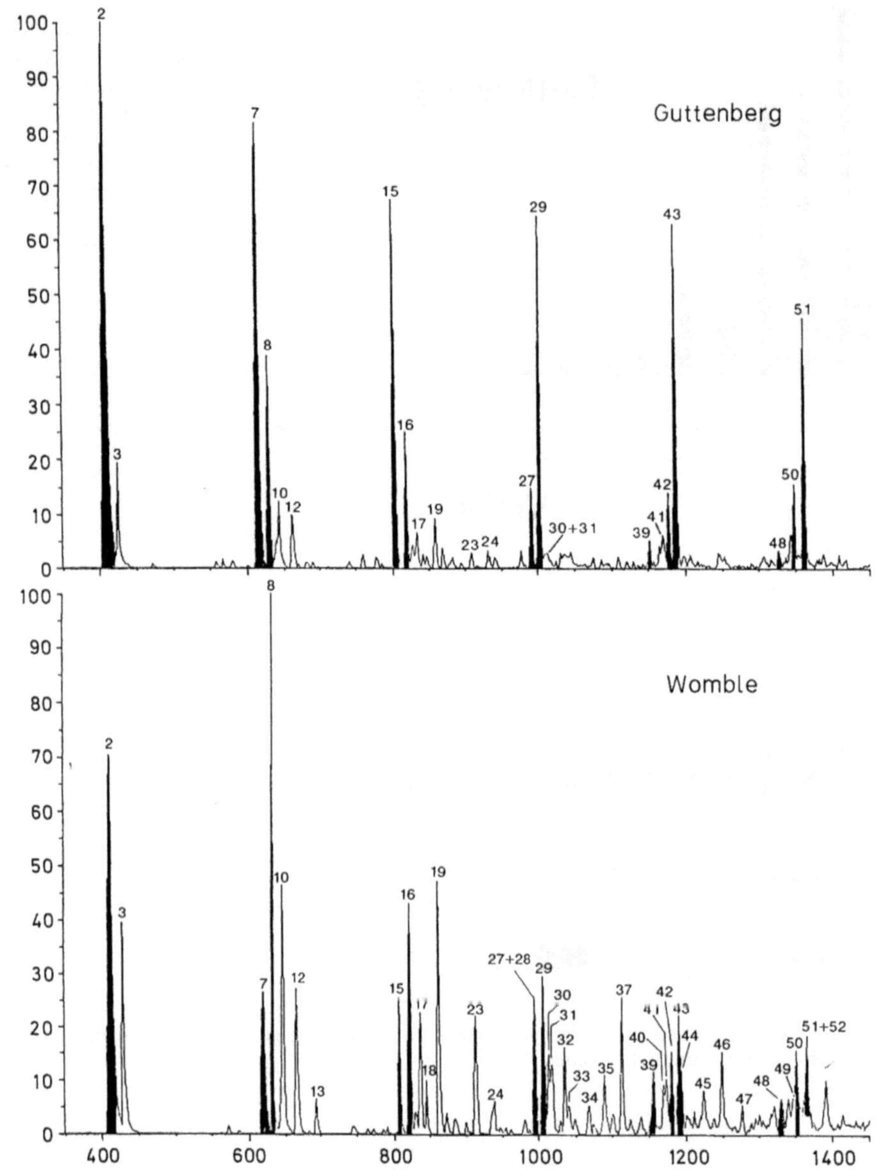

Fig. 5. Summed accurate mass chromatograms (m/z 97.06 + 98.06 + 111.08 + 112.08 + 125.10 + 126.10 + 139.12 + 140.12 + 153.14 + 154.14 + 167.12 + 168.12) for the Guttenberg and Womble kerogen pyrolysates illustrating differences in the distributions of alkylthiophenes. *Solid peaks* indicate alkylthiophenes with linear carbon skeletons. *Peak identities* are listed in Table 2

4 Discussion

Microscopic examination showed that the organic matter in the Guttenberg Oil Rock is composed almost exclusively (99%) of remains of the alga *G. prisca* whereas the other three Ordovician rocks contained mainly amorphous kerogens and occasional intact tasmanitids. In addition to this difference in the morphological appearance of the kerogens, major differences were observed between the carbon number distributions of hydrocarbons, and also sulfur-containing compounds in pyrolysates of the Guttenberg and grouped kerogens. The Guttenberg kerogen pyrolysate contains a dominant suite of n-alkanes and n-alk-1-enes showing high relative abundances and a marked odd/even predominance up to carbon number 17. The three "grouped kerogens" released these hydrocarbons as homologies that gradually decreased in abundance over the carbon number range C_6-C_{30}. These differences are shown for the alkanes (m/z 57) in the Guttenberg and Vinini pyrolysates in Fig. 2. This characteristic distribution seen in the Guttenberg pyrolysate is evident also in the 2-methyl-n-alkylbenzenes and 2-n-alkyl-5-methylthiophenes, also shown in Fig. 2, and in those of the n-alk-1-enes, 2-n-alkyl-5-ethylthiophenes, 2-n-alkylthianes and 2-n-alkyl-5-methylthiolanes which are not shown. The features which distinguish the Guttenberg from the grouped kerogens are therefore an odd-carbon-number predominance with maxima near C_{13} or C_{15}, a restricted carbon-number range with low abundances of compounds above C_{17} and a dominance of isomers possessing a linear carbon skeleton. These distinctive distributions in the Guttenberg pyrolysates have been observed before, principally for hydrocarbons, in other Ordovician oils and bitumen extracts (Fowler, this Vol.) and in pyrolysates of kerogens rich in *G. prisca* (Fowler et al. 1986; Reed et al. 1986; Hoffmann et al. 1987).

Analysis of the cell-wall material of certain extant alga, and of kerogens rich in their fossil counterparts, has shown that the homologous series of n-alkanes and n-alk-1-enes released on pyrolysis is due to their content of highly aliphatic biopolymers. For instance, cell walls of the extant alga *B. braunii* release extended homologies of straight-chain hydrocarbons (Largeau et al. 1984, 1986; Derenne et al. 1988), as do torbanites which consist predominantly of fossilized *B. braunii* (Larter and Douglas 1978, 1980; Allan et al. 1980; Largeau et al. 1986; Derenne et al. 1988). The extant alga *Tetraedron minimum*, and its fossilized remains which occur in the kerogen of Messel Oil Shale behave similarly (Goth et al. 1988). Despite the lack of extant *G. prisca*, which probably became extinct at the end of the Ordovician (Fowler this Vol.), we believe that its cell walls were also rich in an aliphatic biopolymer ("algaenan", Tegelaar et al. 1989b). This would account for not only the aliphatic hydrocarbons discussed herein, but also the homologous aromatic and sulfur compounds that can be formed from moieties with linear carbon skeletons.

Among the aromatic hydrocarbon constituents of the grouped kerogens the most noteworthy finding is the very large relative abundance of 1,2,3,4-tetramethylbenzene in Womble, Big Fork (and Vinini) pyrolysates. Naturally occurring diaromatic carotenoids, as found in photosynthetic bacteria and marine sponges (Liaaen-Jensen 1978a, b) or their symbiotic bacteria (Imhoff and Truper 1976, 1980), and monoaromatic carotenoids from photosynthetic sulfur bacteria (Liaaen-Jensen et al. 1964; Liaaen-Jensen 1978a, b) may become part of these kerogens during diagenesis and be released on pyrolysis as 1,2,3,4-tetramethylbenzene. A wide-ranging series of 1-alkyl-2,3,6-trimethylbenzenes (C_{13} to C_{31}), where the 1-alkyl substituent is an isoprenoid residue, have been reported (Summons and Powell 1987) in reef-hosted oils and source rocks from the Silurian and Devonian in Canada. These compounds have been described as diagenetic products of the diaromatic carotenoid, isorenieratene, which occurs in sulfur bacteria of the family *Chlorobiaceae*. Nicolas and Rozendal's (1975) suggestion that the Womble and Big Fork shales were deposited under shallow marine conditions concurs with the habitat of photosynthetic sulfur bacteria which have a requirement for both light and hydrogen sulfide. The absence of these arylisoprenoid hydrocarbons in our pyrolysates (m/z 133 monitored) can be ascribed to fragmentation of the weak C – C bond (β to the aromatic ring) releasing instead 1,2,3,4-tetramethylbenzene from the kerogen-attached arylisoprenoid moieties.

A most important aspect of the differences in the distributions of the large number of hydrocarbons, and sulfur-containing compounds in the Guttenberg kerogen pyrolysate, as compared with the grouped kerogens, is that compounds deriving from moieties with linear carbon skeletons are always predominant in the former. This is evident in the C_1 to C_6 alkylbenzenes and alkylthiophenes and in the C_1 to C_3 benzo[b]thiophenes and alkylnaphthalenes. Figure 5 nicely illustrates the differences between the Guttenberg and Womble alkylthiophenes, the latter being representative of the grouped kerogens. The formation of the "linear" isomers has been discussed previously (Sinninghe Damsté et al. 1989; Douglas et al. 1991). They are judged to arise by thermal degradation of moieties with a similar structure in the kerogen, the moiety having been formed by the incorporation of sulfur into precursors with linear carbon skeletons that were, or have become, part of a macromolecular structure. In this specific case, we suggest that the precursors of these moieties were already part of the aliphatic cell-wall biopolymer. The source of the alkylcyclohexanes and alkyltoluenes may be from linear moieties that at some stage during diagenesis were liable to cyclize rather than to incorporate sulfur. Elsewhere (Douglas et al. 1991), we have proposed a model which illustrates these suggestions and which also attempts to rationalize the appearance of doublets at C_{16} in some of the homologous series, of which one example can be seen in the mass chromatogram of 2-n-alkyl-5-methylthiophenes (Fig. 3, top). The apparent absence of sulfur compounds of the types discussed in the present biosphere rules out a biological source. Hence, they cannot be considered as biological markers, *sensu strictu*, but they do appear to act as 'pseudo biological markers" for kerogens rich in

G. prisca. This does not hold for the grouped kerogens which contain isomeric sulfur and aromatic compounds with branched substituents. These latter distributions have been found to be comparable in many other Type II kerogens (Sinninghe Damsté et al. 1989), covering a wide spectrum of geological ages and hence their similar appearance in Ordovician kerogens is *prima facie* evidence for at least one origin that does not include higher plants.

Previous reports on the distributions of 2-n-alkylthiophenes, -thiolanes, and -thianes, and also 2,5-di-n-alkylthiophenes and -thiolanes, occurring as wide-ranging homologies in many kerogens, indicate a general pattern of a smooth decrease in the abundance of successively higher homologs (Sinninghe Damsté et al. 1988, 1989). This same feature was found in this work for the grouped kerogens and for the 2-n-alkylthiophenes in the Guttenberg pyrolysate. In contrast, the pyrolysate of the latter kerogen displays a quite distinct distribution for all of the other sulfur compounds, that is a distribution which is similar to that noted earlier for its n-alkanes, n-alk-1-enes and 1-alkyl-2-methylbenzenes. In it, there are no homologies with more than 19 carbon atoms, there is a sharp decrease in their abundance above C_{15} or C_{17} and they exhibit an odd/even carbon-number preference. The distinctiveness of this fingerprint is best conveyed in Figs. 2 and 4 and Table 1, and is supported by additional figures which appear in Douglas et al. (1991).

The results presented here, and in more detail elsewhere (Douglas et al. 1991), together with analyses of the bitumen extracts and hydrous pyrolysates of some of these rocks (Fowler and Douglas 1984; Fowler et al. 1986; Fowler this Vol.) and also work by other authors on sediments rich in *G. prisca*, and oils presumed sourced from *G. prisca*-rich rocks (Summons and Powell 1987; Hoffmann et al. 1987; Alexander et al. 1984; Reed et al. 1986) all attest to the finding of unique distributions of several homologous series of hydrocarbons and sulfur compounds. Evidence is growing that a number of algae contain cell wall biopolymers that are largely aliphatic in nature. The formation of kerogen by the selective preservation of such biopolymers has been proposed by Philp and Calvin (1976), Hatcher et al. (1983), and Tegelaar et al. (1989 a, b, and references therein). How extensive this may prove to be in other extant and fossilized algae is yet to be tested. For *G. prisca* it does appear that the aliphatic cell wall biopolymer provides a characteristic distribution of many compound types in bitumens arising from natural burial of *G. prisca*-rich rocks, or by pyrolysis of their kerogens. One anomaly that has yet to be fully resolved is the absence of phenols in the Guttenberg pyrolysate, but known to be present in pyrolysates of *G. prisca*-rich kukersites (Kogerman 1927): this important difference has been addressed by Derenne et al. (1990). Also remaining to be tested are the effects of differences in the environment of deposition and diagenesis, the chemistry and microbiology of which may modify the findings presented in this paper. The proposal that *G. prisca* flourished in both benthic and planktonic environments (Stasiuk and Osadetz 1991) encourages further investigations into the geochemistry of a wide range of sediments rich in the remains of this organism.

Acknowledgments. We are grateful to Drs A. Gize, B. Horsfield, and J. Winters for rock samples and to Mr. W. Pool for technical assistance.

References

Alexander R, Cumbers M, Kagi RI (1984) Geochemistry of some Canning Basin crude oils. In: Purcell PG (ed) the Canning Basin, W. A. Proc Geol Soc Aust/Petrol Explor Soc Aust Symp, Perth, pp 353–358

Allan J, Bjoroy M, Douglas AG (1980) A geochemical study of the exinite group maceral alginite, selected from three Permo-Carboniferous torbanites. In: Douglas AG, Maxwell JR (eds) Advances in organic geochemistry 1979, Pergamon, Oxford, pp 599–618

Behar F, Pelet R (1985) Pyrolysis gas chromatography applied to organic geochemistry. Structural similarities between kerogens and asphaltenes from related rock extracts and oils. J Anal Appl Pyrol 8:173–187

Derenne S, Largeau C, Casadevall E, Tegelaar EW, de Leeuw JW (1988) Relationships between algal coals and resistant cell wall biopolymers of extant algae as revealed by Py-GC-MS. Fuel Process Technol 20:93–101

Derenne S, Largeau C, Casadevall E, Berkaloff C (1989) Occurrence of a resistant biopolymer in the L race of *Botryococcus braunii*. Phytochemistry 28:1137–1142

Derenne S, Largeau C, Casadevall E, Tegelaar EW, Sinninghe Damsté JS, de Leeuw JW (1990) Characterisation of Estonian kukersite by spectroscopy and pyrolysis: evidence for abundant alkylphenolic moieties in an Ordovician, marine, Type II/I kerogen. Org Geochem 16:873–888

Douglas AG, Sinninghe Damsté JS, Fowler MG, Eglinton T, de Leeuw J (1991) Unique distributions of hydrocarbons and sulphur compounds released by flash pyrolysis from the fossilised alga *Gloeocapsomorpha prisca*, a major constituent in one of four Ordovician kerogens. Geochim Cosmochim Acta 55:275–291

Eglinton TI, Douglas AG, Rowland SJ (1988) Release of aliphatic, aromatic and sulphur compounds from Kimmeridge kerogen by hydrous pyrolysis: a quantitative study. Org Geochem 13:655–663

Eglinton TI, Sinninghe Damsté JS, Kohnen MEL, de Leeuw JW, Larter SR, Patience RL (1990) Analysis of maturity-related changes in the organic sulfur composition of kerogens by flash pyrolysis-gas chromatography. In: Orr WL, White CM (eds) Geochemistry of sulfur in fossil fuels. ACS Symp Ser 429, Dallas, pp 529–565

Foster CB, Reed JD, Wicander R, Summons RE (1989a) Anatomy of an oil-prone maceral: *Gloeocapsomorpha prisca* Zalessky 1917. In: Thomas CG, Strachan MG (eds) Proc Macerals 1989 Symp 16, CSIRO Div Coal Technol, North Ryde, Aust, pp 1–9

Foster CB, Reed JD, Wicander R (1989b) *Gloeocapsomorpha prisca* Zalessky, 1917: a new study part 1: taxonomy, geochemistry and paleoecology. Geobios 22:735–759

Foster CB, Wicander R, Reed JD (1990) *Gloeocapsomorpha prisca* Zalessky, 1917: a new study part II: origin of kukersite, a new interpretation. Geobios 23:133–140

Fowler MG, Douglas AG (1984) Distribution and structure of hydrocarbons in four organic-rich Ordovician rocks. Org Geochem 6:105–114

Fowler MG, Abolins P, Douglas AG (1986) Monocyclic alkanes in Ordovician organic matter. In: Leythaeuser D, Rullkotter J (eds) Advances in organic geochemistry 1985. Pergamon, Oxford, pp 815–823

Goth K, de Leeuw JW, Puttmann W, Tegelaar EW (1988) Origin of messel oil shale kerogen. Nature (London) 336:759–761

Hatcher PG, Spiker EC, Szeverenyi M, Maciel GE (1983) Selective preservation and origin of petroleum-forming aquatic kerogen. Nature (London) 305:498–501

Heyl AV, Agnew AF, Lyons FJ, Belive CH (1959) The geology of the Upper Mississippi Valley zinc-lead district. US Gov Print Off, Washington, US Geol Surv Pap 309, 310 pp

Hoffmann CF, Foster CB, Powell TG, Summons RE (1987) Hydrocarbon biomarkers from Ordovician sediments and the fossil alga *Gloeocapsomorpha prisca* Zalessky 1917. Geochim Cosmochim Acta 51:2681–2697

Imhoff JF, Truper HG (1976) Marine sponges as habitats of anaerobic photosynthetic bacteria. Microbiol Ecol 3:1–9

Imhoff JF, Truper HG (1980) *Chromatium pupuratum* sp. nov., a new species of Chromanaceae. Zentralbl Bakteriol I Abt Orig C1:61–69

Kogerman PN (1927) Estonian shale oil. 1. Isolation and properties of phenols. J Soc Chem Ind 46:138–143

Largeau C, Casadevall E, Kadouri A, Metzger P (1984) Formation of *Botryococcus*-derived kerogens. Comparative study of immature torbanite of the extant alga *Botryococcus braunii*. In: Schenck PA, de Leeuw J, Lijmbach GWM (eds) Advances in organic geochemistry 1983. Pergamon, Oxford, pp 327–332

Largeau C, Derenne S, Casadevall E, Kadouri A, Sellier N (1986) Pyrolysis of immature torbanite and of the resistant biopolymer (PRB A) isolated from the extant alga *Botryococcus braunii*. Org Geochem 10:1023–1032

Larter SR (1984) Application of analytical pyrolysis techniques to kerogen characterisation and fossil fuel exploration/exploitation. In: Voorhees KE (ed) Advances in analytical pyrolysis, vol 2. Butterworths, London, pp 212–272

Larter SR, Douglas AG (1978) Low-molecular-weight aromatic hydrocarbons in coal maceral pyrolysates as indicators of diagenesis and organic matter type. In: Krumbein W (ed) Environmental biogeochemistry and geomicrobiology. Ann Arbor Science, Mich, pp 373–386

Larter SR, Douglas AG (1980) A pyrolysis gas chromatographic method for kerogen typing. In: Douglas AG, Maxwell JR (eds) Advances in organic geochemistry 1979. Pergamon, Oxford, pp 579–583

Larter SR, Solli H, Douglas AG (1978) The analysis of kerogens by pyrolysis-gas chromatography-mass spectrometry using selective ion detection. J Chromatogr 167:421–431

Lee ML, Vassilaros DL, White CM, Novotny M (1979) Retention indices for programmed-temperature capillary-column gas chromatography of polycyclic aromatic hydrocarbons. Anal Chem 51:768–773

Liaaen-Jensen S (1978a) Marine carotenoids. In: Faulkner DJ, Fenical WH (eds) Marine natural products. Academic Press, New York London, pp 1–73

Liaaen-Jensen S (1978b) Chemistry of carotenoid pigments. In: Clayton RK, Sistrom WR (eds) Photosynthetic bacteria. Plenum, New York, pp 233–248

Liaaen-Jensen S, Heggie E, Jackman LM (1964) Bacterial carotenoids XVIII. The carotenoids of photosynthetic bacteria. Acta Chem Scand 18:1703–1718

Monin JC, Durand B, Vandenbroucke M, Huc AY (1980) Experimental simulation of the natural transformation of kerogen. In: Douglas AG, Maxwell JR (eds) Advances in organic geochemistry 1979. Pergamon, Oxford, pp 517–529

Nicolas RL, Rozendal RA (1975) Subsurface positive elements within Ouachita Foldbelt in Texas and their relation to Palaeozoic cratonic margin. Am Assoc Petrol Geol Bull 59:193–216

Nip M, de Leeuw JW, Schenck PA, Crelling JC (1991) The chemical structure of a bituminous coal and its constituent macerals as revealed by flash pyrolysis. Energ Fuels (in press)

Philp RP, Calvin M (1976) Possible origin for insoluble organic (kerogen) debris in sediments from insoluble cell wall materials of algae and bacteria. Nature (London) 262:134–136

Reed JD, Illich HA, Horsfield B (1986) Biochemical evolutionary significance of Ordovician oils and their sources. In: Leythaeuser D, Rullkotter J (eds) Advances in organic geochemistry 1985. Pergamon, Oxford, pp 347–358

Rubinstein I, Strausz OP (1979) Geochemistry of the thiourea adduct fraction from an Alberta petroleum. Geochim Cosmochim Acta 43:1387–1392

Schroder H (1980) Retention indices of hydrocarbons up to C14 for the stationary phase squalane. J High Res Chrom Chrom Commun 3:200–204

Sinninghe Damsté JS, de Leeuw JW, Knock-van Dalen AC, de Zeeuw MA, de Lange F, Rijpstra WI, Schenck P (1987) The occurrence and identification of series of organic sulphur compounds in oils and sediment extracts. I. A study of Rozel point oil (USA). Geochim Cosmochim Acta 51:2369–2391

Sinninghe Damsté JS, Kock-van Dalen AC, de Leeuw JW, Schenck PA (1988) Identification of homologous series of alkylated thiophenes, thiolanes, thianes and benzothiophenes present in pyrolysates of sulphur-rich kerogens. J Chromatogr 435:435–452

Sinninghe Damsté JS, Eglinton TI, de Leeuw JW, Schenck PA (1989) Organic sulphur in macromolecular sedimentary organic matter. I. Structure and origin of sulphur-containing moieties in kerogen, asphaltenes and coal as revealed by flash pyrolysis. Geochim Cosmochim Acta 53:873–889

Solli H, Van Graas G, Leplat P, Krane J (1984) Analysis of kerogens of Miocene shales in a homogeneous sedimentary column. A study of maturation using flash pyrolysis techniques and carbon-13 CP-MAS NMR. In:

Schenck PA, de Leeuw JW, Lijmbach GWM (eds) Advances in organic geochemistry 1983. Pergamon, Oxford, pp 351–358

Stasiuk LD, Osadetz KG (1991) The life cycle and phyletic affinity of *Gloecapsomorpha prisca* Zalessky 1917 from Ordovician rocks in Canadian Williston Basin. Curr Res Geol Surv Pap 90-ID, pp 123–137

Summons RE, Powell TG (1987) Identification of aryl isoprenoids in source rocks and crude oils: biological markers for the green sulphur bacteria. Geochim Cosmochim Acta 51:557–566

Tegelaar EW, de Leeuw JW, Largeau C, Derenne S, Schulten H-R, Muller R, Boon JJ, Nip M, Sprenkels JCM (1989a) Scope and limitations of several pyrolysis methods in the structural elucidation of a macromolecular plant constituent in the leaf cuticle of *Agave americana* L. J Anal Appl Pyrol 15:29–54

Tegelaar EW, de Leeuw JW, Derenne S, Largeau C (1989b) A reappraisal of kerogen formation. Geochim Cosmochim Acta 53:3103–3106

Traverse A (1955) Occurrence of the oil-forming *Botryococcus* in lignites and other tertiary sediments. Micropaleontology 1:343–350

Van Graas G, de Leeuw JW, Schenck PA (1980) Analysis of coals of different rank by Curie-point pyrolysis-mass spectrometry and Curie-point pyrolysis-gas chromatography-mass spectrometry. In: Douglas AG, Maxwell JR (eds) Advances in organic geochemistry 1979. Pergamon, Oxford, pp 485–494

Van de Meent D, Brown SC, Philp RP, Simoneit BRT (1980) Pyrolysis-high resolution gas chromatography and pyrolysis-gas chromatography-mass spectrometry of kerogen and kerogen precursors. Geochim Cosmochim Acta 44:999–1013

Zalessky MD (1917) On marine sapropelite of Silurian age formed by a blue-green alga. Izv Imp Akad Nauk IV Ser:3–18

Thermal Maturity of Early Paleozoic Sediments as Determined By the Optical Properties of Marine-Derived Organic Matter – A Review

F. Goodarzi [1], M. G. Fowler [1], M. Bustin [2] and D. M. McKirdy [3]

[1] Institute of Sedimentary and Petroleum Geology, 3303-33rd St. N.W., Calgary, Alberta, T2L 2A7, Canada
[2] Department of Geological Science, University of Britisch Columbia, Vancouver, British Columbia, V6T 2B4, Canada
[3] Department of Geology and Geophysics, University of Adelaide, P.O. Box 498, Adelaide, S.A. 5001, Australia

Abstract

Higher land plants did not evolve until the Late Silurian. Therefore most pre-Devonian rocks are devoid of vitrinite, making it difficult to assess the thermal maturity of their organic matter using standard optical methods. Pre-Devonian sediments do contain structured marine-derived organic matter including algae, chitinozoa, graptolites and scolecodonts together with amorphous material and bitumen.

In Early Paleozoic rocks (especially Ordovician to Lower Devonian), marine-derived organic matter can be used for maturity determinations, since its optical properties (e.g. reflectance, refractive and absorptive indices) show systematic changes with the maturity of the host rock. Also, the dispersion of optical properties of marine-derived organic matter increases with maturity in a similar manner to that of vitrinite, indicating that similar molecular structural changes occur in these types of organic matter with increasing maturity.

Changes that occur in the optical properties of graptolite skeletal materials with increasing depth, in particular the maximum reflectance in oil, appear to be related to the maximum level of thermal maturity undergone by a rock. Graptolite reflectance increases with increasing conodont alteration index. At present only an indirect correlation exists between graptolite and vitrinite reflectance using conodonts. This indicates that the oil window as defined by vitrinite reflectance (0.5–1.3% Ro) and the conodont alteration index (CAI = 1.5–2.5) coincide with graptolite reflectance values of 0.1–1.7% Ro max. The correlation of graptolite reflectance to the oil window as defined by conodonts is not precise due to the qualitative nature of the conodont alteration index. Values of graptolite reflectance determined in rocks in the vicinity of those generating oil are generally higher by 0.2–0.3% than those expected for vitrinite at the same level of maturity. Future studies of Devonian samples that contain both vitrinite and graptolites should be able to check this relationship. In common with vitrinite, the optical properties of graptolites are also affected by other factors besides temperature, such as the lithology of the host sediment, absorption of bitumen and weathering.

Other fossil remains that can be used for thermal maturity determinations in Early Paleozoic rocks are chitinzoan and scolecodonts. With increasing maturity, both are observed to undergo a series of colour changes under transmitted light, while their reflectance increases under reflected light.

1 Introduction

Petroleum geologists are mostly concerned with the quantity, quality (kerogen type) and maturation level of organic matter in sediments. Probably the most widely used method to determine the maturity of sedimentary organic matter is vitrinite reflectance (Stach et al. 1982; Tissot and Welte 1984). Vitrinite is mostly derived either from humic acids, usually formed through the biodegradation of lignin and cellulose from higher land plants, or from land plant tannins (Stach et al. 1982). As higher land plants did not evolve until the Late Silurian (Banks 1975; Edwards 1980), Early Paleozoic and Precambrian sediments contain effectively no vitrinite. The presence of an optically similar substance to vitrinite has been reported by some workers in rocks of this age (e.g. Sikander and Pittion 1978; Fowler and Douglas 1984; Jackson et al. 1984; Buchardt et al. 1986; McKirdy et al. 1989). This material is possibly derived from algal or bacterial remains such as the prothalluses of algae which contain tannin intercalations (Stach et al. 1982) or asphaltic pyrobitumen (Sikander and Pittion 1978).

Various authors have measured the reflectance of this vitrinite-like material and used it to assess the regional maturity of Early Paleozoic sediments (Sikander and Pittion 1978; Buchardt et al. 1986). A good correlation has been reported between illite crystallinity and "vitrinite" reflectance in highly mature Early

M. Schidlowski et al. (Eds.)
Early Organic Evolution: Implications for Mineral and Energy Resources

Paleozoic rocks (Teichmüller et al. 1979; Kisch 1980). Buchardt et al. (1986) found a positive correlation between their "vitrinite" reflectance measurements in the Late Cambrian Alum Shale and conondont alteration index (CAI) values, and concluded that "vitrinite" reflectance measurements can be used as a rank/maturity indicator in low-rank Early Paleozoic sediments.

Because of the lack of true vitrinite in Early Paleozoic sediments, it has been necessary to develop other optical methods to determine the maturity of their organic matter. Sediments of this age contain structured organic matter derived from marine organisms such as algae, graptolites and scolecodonts, as well as amorphous material and bitumen, and largely inorganic microfossils such as conodonts. In recent years, methods have been developed which make possible the use of all these materials to determine the thermal maturity of their host sediments. This work focuses an structured marine organic matter. The use of conodonts is not discussed although, where possible, maturity parameters based on the other organic entities are compared to CAI determinations. As described elsewhere (e.g. Epstein et al. 1977; Harris 1979; Gorter 1984; Nicoll and Gorter 1984; Nowlan and Barnes 1987) conodont colour alteration has been widely exploited in maturation studies of Paleozoic strata.

2 Technique and Equipment

2.1 Sample Preparation

Two techniques are used in the preparation of rock samples for reflectance measurements on phytoclasts and other organic matter.

2.1.1 Whole Rock

Relief-free polished samples of sedimentary rock are prepared using standard procedures (Mackowsky 1982). However, where the reflectance of graptolites is to be measured, a modification to the normal polishing procedure is required. Attaining a flat, relief-free, polished graptolite surface, parallel to bedding, is essential to ensure accurate reflectance measurements. Following the macroscopic identification of graptolites on a bedding surface, the sample is cut into a 20-mm cube and coated on all sides with a very thin film of epoxy. The sample is then ground on a glass plate covered with a slurry of 600 grit silicon carbide powder and alcohol until a portion (1–5 mm) of the graptolite is exposed. After thorough rinsing of the sample, the specimen is gently polished on water-soaked pieces of soft wood; first in a slurry of 0.3 μm alumina and alcohol, and finally in a slurry of 0.5 μm alumina and alcohol. A detailed description of the preparation procedure is provided by Goodarzi and Stasiuk (1987).

2.1.2 Organic Concentrate

Concentrates are obtained by HCl and HF dissolution of the host rock and prepared as polished thin sections using the method of Bertrand et al. (1985).

Graptolites with a reflectance greater than 10% in oil are in the semi-graphitic to graphitic stage of organic maturation (Teichmüller et al. 1979). Reflectance measurements on such highly metamorphosed graptolites should be determined on their natural surfaces. The reflectance measured on a natural graphite surface is higher than that obtained from a polished surface (Kwiecinska et al. 1977).

2.2 Optical Equipment for Reflected Light Microscopy

A Zeiss MPM II microscope fitted with white (halogen) and fluorescent (HBO high pressure mercury) light sources and attached to a Zonax microcomputer was used for collection of optical data. A motorized monochromator attachment facilitates spectral scanning (400–700 nm).

2.2.1 Reflected White Light

Five optical parameters can be determined for the microscopical study of organic fossils, namely reflectance (in air and in oil), bireflectance (in oil), refractive index and absorptive index.

Reflectance measurement first involves centering upon a polished area of organic fossil, free of relief and scratches. The reflectance of an unknown material can then be measured by comparison with a standard of known reflectance using the following equation:

$$R = Rs \times D/DS,$$

where R is the percentage reflectance of the unknown material and Rs is the percentage reflectance of the standard, and D and DS are the determined values for the unknown (D) and the standard (DS) using microscopic photometry.

The maximum and minimum reflectances are measured at a wavelength of 546 nm using oil ($n = 1.518$) immersion objectives (numerical aperture 0.90; magnification ×40) by rotation of the microscope stage through 360°. The difference between the maximum and the minimum reflectance in oil is defined as the bireflectance (% Ro max minus % Ro min). The visual microscopic observation of bireflectance in a non-quantiative sense is referred to as anisotropy. Reflectance measured on organic particles without any rotation of the microscope stage and with the polarizer out of the light path is referred to as % Ro random.

The refractive and absorptive indices are the two parameters used to estimate the changes in molecular structure of organic materials during metamorphism. To determine these two indices, the reflectance must be measured in two different media of differing refractive index, for example in two of air, oil or water. Modified Beer's equations are used to determine the refractive index (n) and absorptive index (k). For example, with an oil and an air objective, the equation would be:

$$n = \frac{no^2 - na^2/2}{no\,[(1+Ro)/(1-Ro)] - na\,[(1+Ra)/(1-Ra)]};$$

$$k^2 = \frac{Ra\,(n+1)^2 - (n-1)^2}{n^2\,(1+Ra)},$$

where Ra and Ro are the reflectances of the sample in air and oil; n, na and no are respectively the refractive indices of the sample, air and immersion oil; and k is the absorptive index of the sample.

Dispersion curves of the reflectance of graptolites in air and oil are obtained by measuring the maximum percentage reflectance values every 10 nm between the wavelengths of 400 and 700 nm. This procedure is generally referred to as spectral scanning.

2.2.2 Determination of Optical Properties in Fluorescence Mode

Fluorescence spectra are obtained following the standard procedure of Teichmüller and Ottenjann (1977) and using an ultraviolet (365-nm) excitation filter and a 420-nm barrier filter. This combination allowed the determination of fluorescence spectra of hydrogen-rich components in the samples (i.e. algae) with λmax (wavelength of maximum intensity) > 420 nm. The fluorescence properties determined included colour, λmax and red to green quotient (R/G Q). Fluorescence intensities (I) of samples may also be determined at 546 nm using a combination of 450-nm excitation and a 510-nm barrier filter, and a Leitz fluorescence standard, following the recommendations of Jacob (1975). Photomicrographs were taken both under reflected white light and in fluorescence mode.

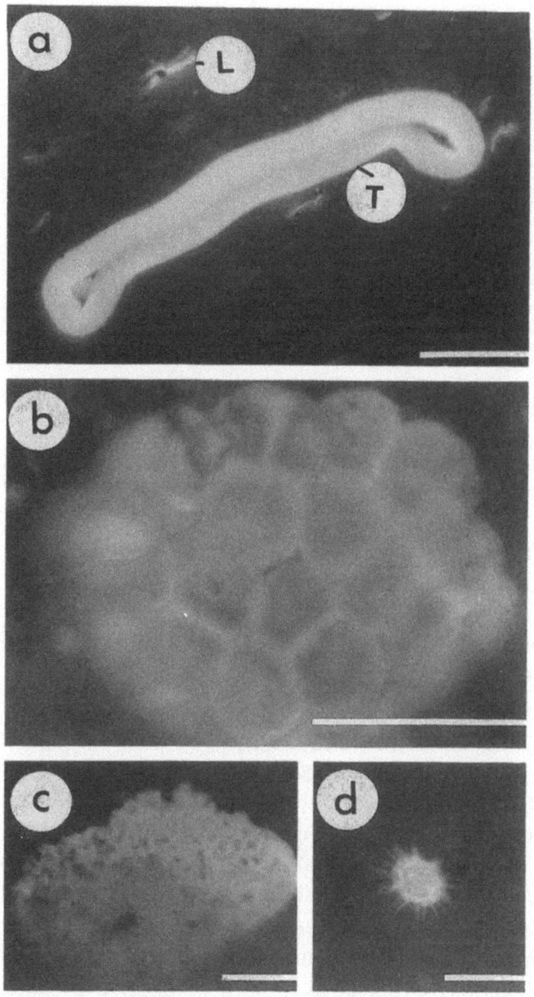

Fig. 1. Algal remains in Early Paleozoic sediments, oil immersion, blue light fluorescence (excitation, 450 nm). *Scale bar* = 50 μm.
a *Tasmanites* algae (*T*) and Liptodetrinite (*L*), Late Silurian shale, southwest Turkey. **b** Algal bodies which display affinities to cup-like structures of colonial alginites, possibly *Gloeocapsomorpha*, Ordovician, Green Point shale, western Newfoundland, Canada. **c** *Gloecapsomorpha*, Ordovician marl, southwest Turkey. **d** Acritarch, Ordovician marl, southwest Turkey.

3 Algae

Acritarchs, *Tasmanites* and *Gloeocapsomorpha prisca* and other algae (telalginite) (Fig. 1) are commonly found in Early Paleozoic sediments (e.g. Alpern 1981; Cook et al. 1981; Hutton 1986). The reflectance and fluorescence properties of telalginite (λmax, R/G quotient, I at 546 nm) can be used in the absence of vitrinite and other organic matter to determine the maturity of a sediment (Teichmüller and Ottenjann

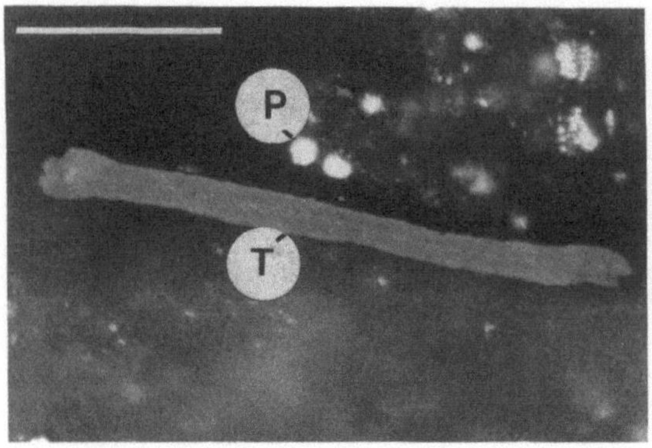

Fig. 2. Matured *Tasmanites* algae (*T*), and pyrite (*P*), Ordovician shale, Grand Banks, Newfoundland, Canada. Scale as in Fig. 1. Reflected white light.

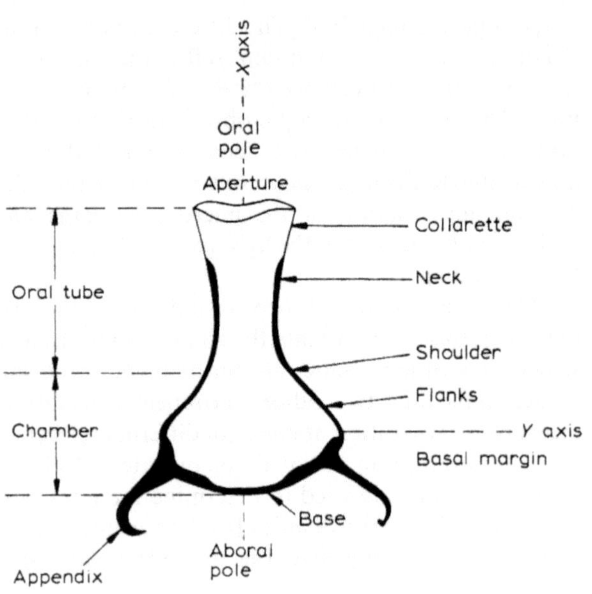

Fig. 3. Generalized lateral view of chitinozoan test (After Jenkins 1970)

1977). Alginite fluoresces throughout diagenesis and through most of the catagenesis stage of oil generation. In reflected light it becomes strongly anisotropic (Fig. 2) during the later stages of catagenesis (Goodarzi et al. 1985).

Telalginite has not been widely used for maturity determination. Relatively few systematic studies have been published, examining the changes of one group of algae with increasing maturity, and none of these was concerned with Early Paleozoic samples. Nevertheless, the results of these studies indicate that the fluorescence properties of alginite have great potential for detailed maturity assessment (e.g. Hufnagel 1977; Teichmüller and Ottenjann 1977; Cook et al. 1981).

Recently, Crick et al. (1988) completed a systematic study of poorly structured algal material (lamalginite) in sediments of Middle Proterozoic age. They found that the reflectance of lamalginite did increase with increasing maturity and compared this to changes in Rock-Eval parameters and the methylphenanthrene index (MPI: Radke and Welte 1983). The reflectance of non-fluorescing lamalginite was observed to be significantly higher than that of the fluorescing form. Moreover, as maturation increased, the non-fluorescing lamalginite became noticeably anisotropic and deformed. The onset of hydrocarbon generation is assumed to be at 0.15% Ro for fluorescent lamalginite, and the base of the oil window at 1.4% Ro for non-fluorescing lamalginite.

4 Chitinozoa

Chitinozoa are vase-like organic fossil remains of an extinct animal group (Fig. 3). They have a worldwide distribution, and stratigraphically range from late Proterozoic to Late Devonian (Bloeser et al. 1977; Jansonius and Jenkins 1978). They have been used in the stratigraphic correlation of isolated outcrops in sedimentary basins and also in the application of time zonation to subsurface stratigraphy (Jenkins 1970).

In whole-rock samples only part of a chitinozoan test is commonly observed (Fig. 4). Chitinozoa are more easily identified under reflected light using polished thin sections of concentrates (Fig. 4), where they range from grey to white in colour and are isotropic (Goodarzi 1985).

Chitinozoa are sensitive to increasing temperature, which make them useful in the evaluation of the maturity of a sediment containing their fragments. Under transmitted light, both a colour change from amber through dark brown to black, and development of lustre have been observed (Correia 1967; Tschudy and Scott 1969; Jenkins 1970; Laufeld 1973). Chitinozoa become increasingly opaque with increasing metamorphism (Jenkins 1970; Laufeld 1973). This increase in opacity is accompanied by increases in both absorption index and reflectance (van Krevelen 1961; Goodarzi 1985). Chitinozoan reflectance was first used by Goodarzi (1985) to evaluate the maturity of Ordovician to Silurian sediments in Turkey. Goodarzi found that chitinozoan reflectance increases with an increase in depth of burial and at a rate higher than that for bitumen reflectance (Fig. 5). This method has also recently been applied to Ordovician and Silurian samples from Anticosti Island, Canada (Bertrand and Heroux 1987).

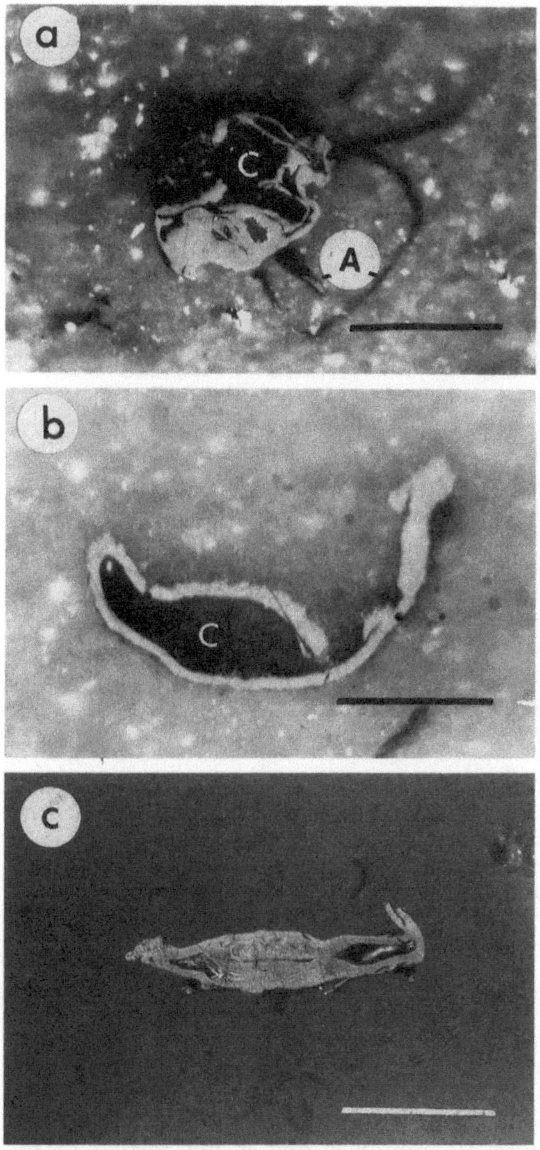

Fig. 4. Chitinozoan fragments, oil immersion, plane polarized; scale bar = 50 µm. **a** Section near base, showing part of chamber (*C*) and appendices (*A*); whole-rock sample, Ordovician marl, southeast Turkey. **b** Broken chamber (*C*), showing internal cavity, whole-rock sample, Silurian marl, southeast Turkey. **c** Basal margin showing internal structure, polished thin section of organic concentrate, Silurian, southeast Turkey

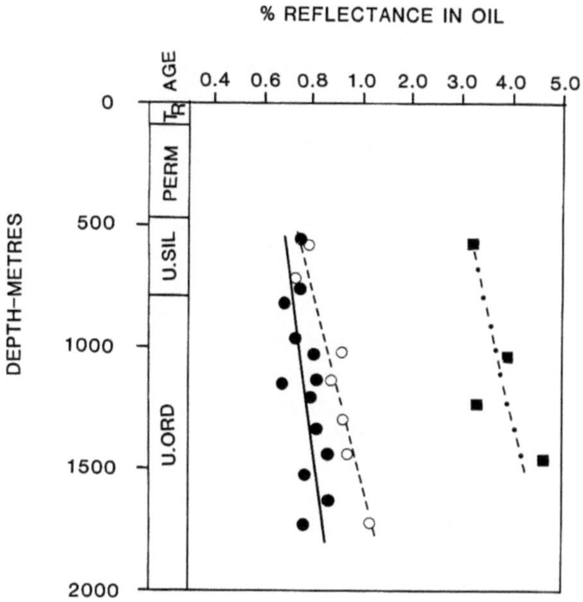

Fig. 5. Reflectance in oil of bitumen (●), chitinozoa (○) and graptolites (■) versus depth for borehole 1, southeast Turkey (After Goodarzi and Norford 1987)

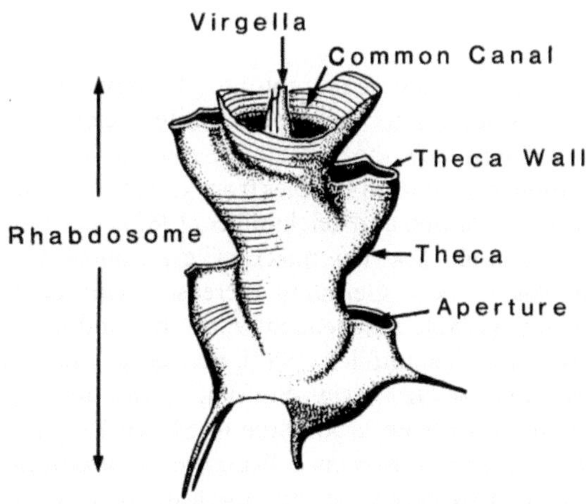

Fig. 6. Simplified morphology of graptolite (Redrawn after Jackson 1973)

5 Graptolites

Graptolites are marine macrofossils that occur mainly in Early Paleozoic rocks (especially Ordovician to Early Devonian). The morphology and terminology of graptolites (figs. 6, 7) have been described by Moore et al. (1952) and Crowther (1981).

The initial studies on the possibility of using the reflectance of the skeletal materials of graptolite as an indicator of sediment maturity were carried out in 1968 and 1970 by A. C. Cook (pers. comm. 1989) and by Watson (1976). However, the first published use of graptolite reflectance was by Kurylowicz et al. (1976) in a study on the maturity of Ordovician sediments in the Amadeus Basin, Australia. They found graptolite reflectance values ranging from 0.80 to 2.37% Ro max, with bireflectance values in the range 0.36 to

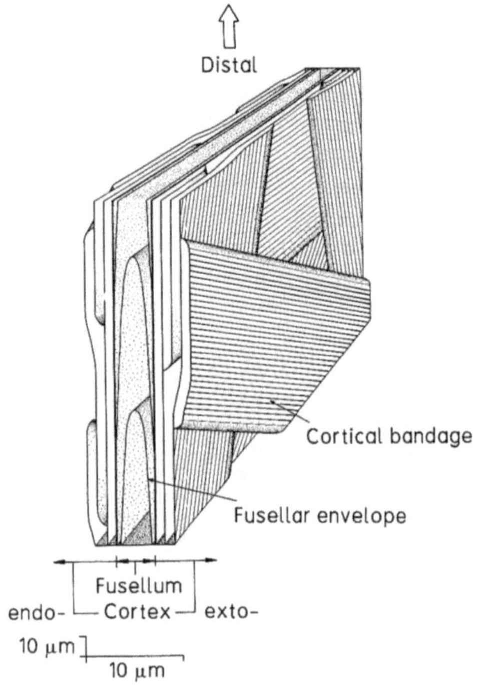

Fig. 7. Simplified graptolite periderm wall (Redrawn after Crowther 1981)

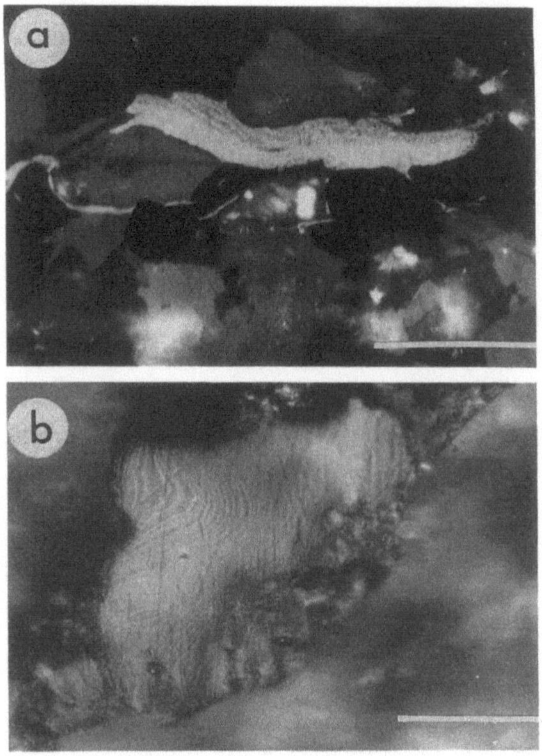

Fig. 8. Graptolite fragments in sediments, oil immersion: scale bar = 50 μm. **a** Granular surface, Early Silurian, argillaceous limestone, Cape Phillips Formation, Canadian Arctic. Plane polarized. **b** Non-granular fragment, showing fusellar layers and complex morphology, Ordovician, limey shale, Road River Group, northwesten Canada. partially crossed polars

1.05%, and considered graptolites to have optical properties similar to those of vitrinite. Teichmüller (1978) determined the reflectance (% Ro max) of graptolite fragments in samples from various locations in Europe. Clausen and Teichmüller (1982) found that the reflectance (% Ro max) of graptolites from boreholes in NW Germany increases with depth. Goodarzi (1982), independently of the findings of Clausen and Teichmüller (1982), arrived at the same conclusion for Ordovician-Silurian graptolite fragments in cores from a borehole in SE Turkey. Both maximum and minimum reflectance of graptolites increased with depth for this borehole in which Ro max values of 3.30 to 4.61% were recorded (Goodarzi 1984). Graptolite reflectance has also been used in the study of Early Paleozoic metamorphic rocks from the British Isles (Kemp et al. 1985; Oliver 1987).

5.1 Morphology

Under reflected white light graptolite fragments may show a segmented structure (Fig. 8). Details of the graptolite exoskeletal material, including cortical bandages and fusellar layers, are often evident (Fig. 8b). Graptolites show two types of surface morphology: granular and nongranular (Goodarzi 1984). The former surface has a fine granular to reticular texture (Fig. 8a) and, depending on the maturity level of the graptolite, may show anisotropy. This anisotropy is superimposed on the fine internal structure of graptolites and is not related to the granular mosaic produced by heating of reactive coal macerals (vitrinite and sporinite) during carbonization or gasification (Goodarzi and Murchison 1972; Goodarzi and Vleeskens 1988). This type of texture is commonly found in a carbonate matrix (Goodarzi and Norford 1985) and, according to Link (1988), may be due to leaching of the graptolite exoskeleton by CO_2. The granular graptolites are softer with regard to polishing, lower reflecting and show weaker anisotropy than do nongranular graptolites (Goodarzi 1984; Goodarzi and Norford 1988). The non-granular graptolite fragments are hard and brittle (Fig. 8b), highly reflecting, strongly anisotropic and are common in shales (Goodarzi and Norford 1985). The relationship between the non-granular and the granular fragments is complicated, particularly when both of these morphologies are found as part of the same graptolite periderm wall. However, the presence of both granular and non-granular morphology may be indicative of different components of the periderm wall.

Riediger et al. (1989) have divided the non-granular graptolites morphologically into two types, blocky and lath-shaped. Lath-shaped graptolites have the higher maximum reflectance and bireflectance (anisotropy) of the two forms.

5.2 Optical Properties of Graptolites

The optical properties of carbonaceous material (coal macerals, bitumens) include reflectance in oil (% Ro) and air (% Ra), bireflectance in oil (B Ro), refractive index (n) and absorptive index (k). It is important to know the behaviour of these parameters with increasing coalfication. The reflectance in oil of organic matter is related to its refractive and absorptive indices (van Krevelen 1961). Therefore, in the case of graptolites, it is necessary to document the variation of these two indices at a wavelength of 546 nm, or over the visible spectrum, before the usefulness of graptolite reflectance as a thermal indicator can be properly assessed. The variation of graptolite reflectance with rank should then be compared to known maturity scales, (e.g., conodont alteration index, and reflectance of bitumen and vitrinite) in order to determine its relationship to the process of hydrocarbon generation.

Whether organic matter is uni- or biaxial is very important in interpreting the significance of its reflectance values. Teichmüller (1978) stated that the graptolites found in highly foliated schists from NW Germany may have a biaxial optical character. Goodarzi (1984) found that graptolite fragments show the true maximum reflectance in sections parallel to bedding and true minimum reflectance in sections perpendicular to bedding (Fig. 9). This is similar to the optical behaviour of vitrinite (Davis 1978). Graptolite fragments in sections perpendicular to bedding have stronger anisotropy (higher bireflectance) than those in sections parallel to the bedding (Goodarzi 1984). This strong anisotropic nature of graptolites suggests that at high maturity they may behave as optically biaxial materials.

5.2.1 Variation of Optical Properties of Graptolites at 546 nm

Goodarzi and Norford (1985) made a detailed study of the variation in the optical properties of graptolites in Canadian sediments with increasing thermal maturation. The maturity level of the samples had previously been established by the conodont alteration index method. All the optical parameters determined (viz. reflectance in air and oil, bireflectance, refractive index and absorptive index) increase with increasing CAI (Fig. 10). The rates of increase in graptolite reflectance and bireflectance follow similar patterns to those known for individual coal macerals and bitumens (Davis 1978; Goodarzi 1984). For vitrinites in coal, reflectance increases slowly with respect to carbon content up to the 90% level, beyond which the rate of increase accelerates (van Krevelen 1961; Davis 1978). Very similar patterns are observed in natural bitumens (Khavari-Khorasani 1975). In the study of Goodarzi and Norford (1985) the reflectance of graptolites ranged from 0.64 to 11.90% which for vitrinite would indicate maturities extending from diagenesis to epimetamorphism (Teichmüller et al. 1979).

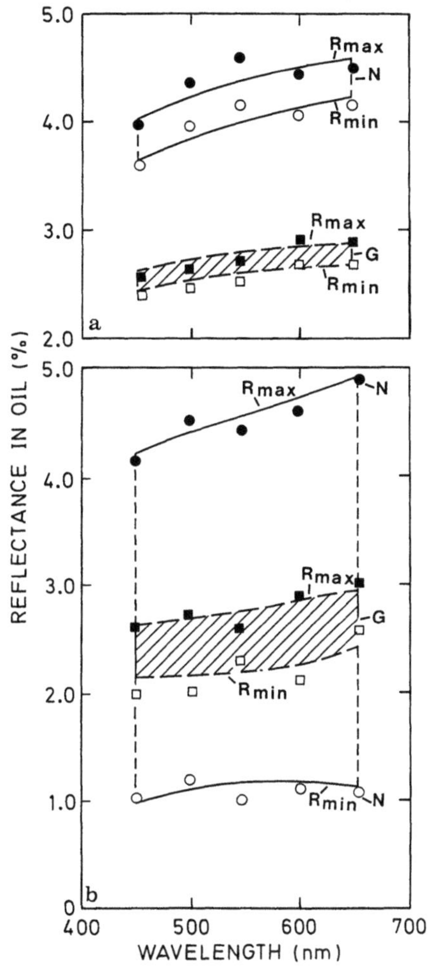

Fig. 9. Dispersion of maximum and minimum oil reflectance for non-granular (*N*) and granular (*G*) graptolites in sections **a** parallel to bedding and **b** perpendicular to bedding (After Goodarzi 1984)

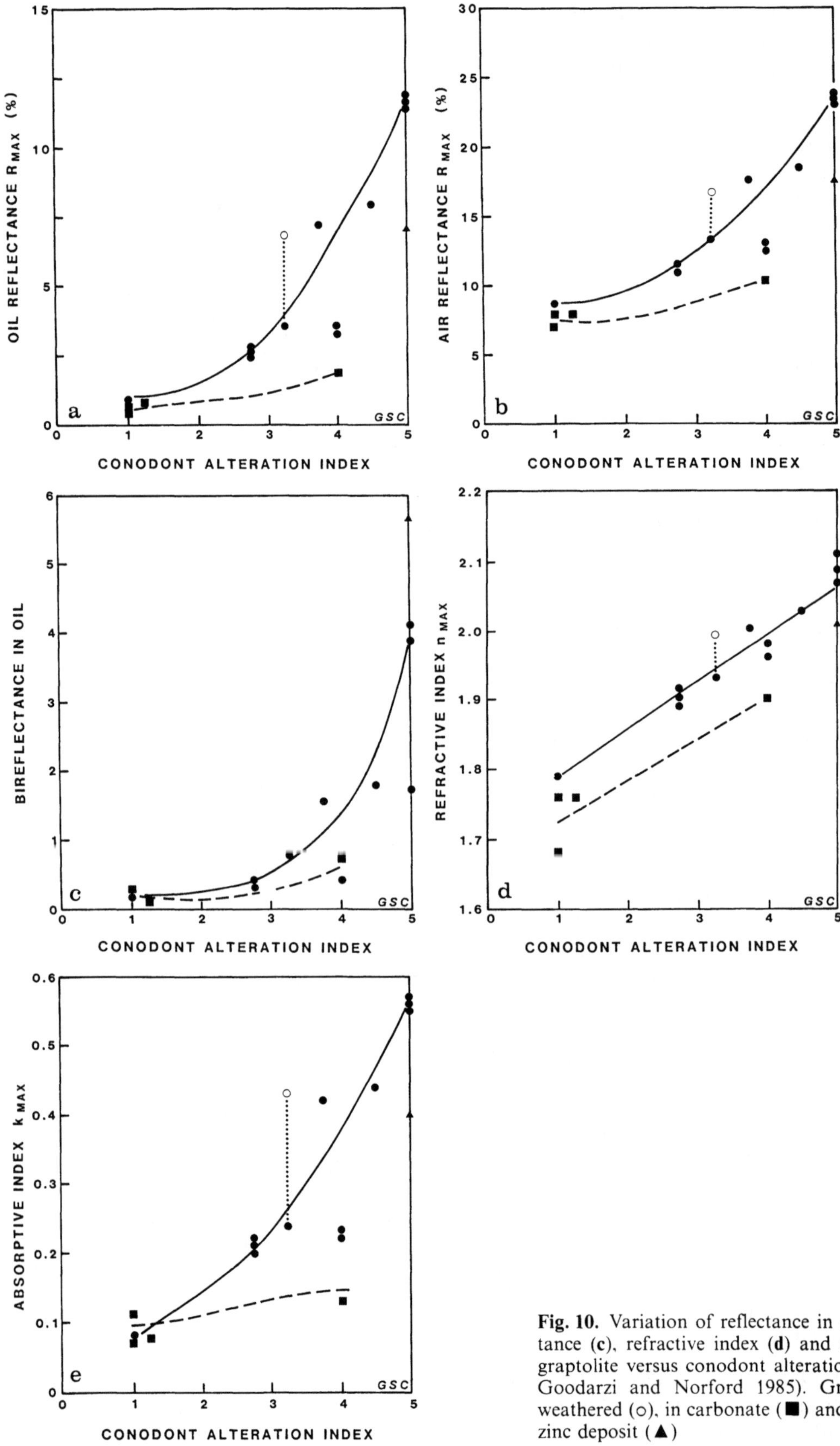

Fig. 10. Variation of reflectance in oil (**a**), air (**b**), bireflectance (**c**), refractive index (**d**) and absorptive index (**e**) of graptolite versus conodont alteration index (redrawn after Goodarzi and Norford 1985). Graptolites in shale (●), weathered (○), in carbonate (■) and in a sedimentary lead-zinc deposit (▲)

The bireflectance of carbonaceous materials is related to the ordering of aromatic carbon lamellae (Goodarzi and Murchison 1972; Khavari-Khorasani 1975; Teichmüller 1982). Vitrinite becomes truly anisotropic at a maximum reflectance of about 0.90% (van Krevelen 1961), whereas graptolites are anisotropic at a maximum reflectance of 0.70% (Goodarzi and Norford 1985). This indicates that during maturation the molecular structure of graptolite develops better ordering at a lower reflectance than is the case for vitrinite. Natural bitumens also develop anisotropy at maximum reflectances lower than the onset of anisotropy in vitrinite (Jacob 1975; Robert 1980). Hence, there may be some similarity between the metamorphism of natural bitumens and graptolites.

5.2.2 Dispersion of Optical Properties of Graptolite at 400 to 700 nm

The molecular structure of coal macerals, kerogens and bitumens consists of ordered carbon (aromatic nuclei) and amorphous carbon (aliphatic, alicyclic side chains, oxygen and other hetero atoms). The ratio of ordered to amorphous carbon increases with increasing maturation (Cartz and Hirsch 1960; van Krevelen 1961). The process of condensation of aromatic carbon in sedimentary organic material is as follows:

1. Molecular stripping and devolatilization of amorphous carbon, which results in development of a polynuclear aromatic structure and an increase in overall aromaticity (Goodarzi and Murchison 1972).
2. Transformation of the polynuclear aromatic structure with two-dimensional (turbostratic) ordering into a mesomorphic structure with three-dimensional (graphitic) ordering (Franklin 1951).

With increasing coalification the slopes of spectral curves of reflectances and refractive indices for both vitrinite and bitumen in visible light (400–700 nm) change gradually from negative to positive (Marshall and Murchison 1971; Khavari-Khorasani and Murchison 1977). This phenomenon is related to the development of absorption bands in the UV part of the spectrum (Friedel and Queiser 1959); the greater the rise of the spectral curves, the greater the aromaticity of the organic materials. The spectral curves for vitrinite and natural bitumen in visible light follow similar patterns to each other with increasing maturation (Gilbert 1960; Hevia Rodriguez 1974; Khavari-Khorasani 1975; Davis 1978). Goodarzi (1986) and Goodarzi et al. (1988a) have determined the dispersion of the optical properties of graptolites in visible light (450–650 nm) with increasing maturity (CAI = 1–5) and found that they follow similar trends to those of vitrinite and natural bitumen. This indicates that the graptolite epiderm is undergoing the same changes in molecular structure that occur in other organic materials (for example, an increase in aromaticity) with increased thermal maturity.

The dispersion of the optical properties of graptolites (Fig. 11a–d) gives three types of spectral patterns that are dependent on maturity (Goodarzi 1986):

1. For low-maturity graptolites, the maximum reflectance and refractive and absorptive indices all decrease with increasing wavelength in the visible spectrum (Fig. 11a–d), suggesting a low proportion of aromatic carbon, as in low-rank vitrinite (Marshall and Murchison 1971; Hevia Rodriguez 1974; Davis 1978) and asphaltic bitumen (Khavari-Khorasani and Murchison 1977).
2. For moderately matured graptolites, the maximum reflectance and refractive index curves stay almost flat, from the blue to red ends of the spectrum (Fig. 11a–d), indicating moderate aromaticity. This is similar to the results obtained from low-volatile bituminous coals and semi-anthracites (Hevia Rodriguez 1974; Davis 1978) and pyrobitumens (i.e. impsonites, Khavari-Khorasani and Murchison 1977).
3. At high maturities, all graptolite optical parameters increase continuously from blue to red wavelengths (Fig. 11a–b) similar to the dispersion of optical properties of highly aromatic organic matter such as anthracite, meta-anthracite and graphite (Marshall and Murchison 1971; Cook et al. 1972; Hevia Rodriguez 1974; Kwiecinska et al. 1977; Davis 1978) and meso- to cata-impsonite (Khavari-Khorasani and Murchison 1977). Goodarzi (1984) determined the dispersion of the maximum and minimum reflectances of high-maturity graptolites (Ro max = 4.61%) and found that both granular and non-granular graptolites in sections parallel and perpendicular to the bedding plane (Fig. 9) are similar to highly aromatic carbon species (e.g. anthracite).

The study of graptolites of increasing maturity confirms that their optical properties and molecular structure undergo systematic changes which are similar to those of vitrinite and bitumen. This work validates the use of graptolite reflectance as a measure of the thermal maturity of the host rock.

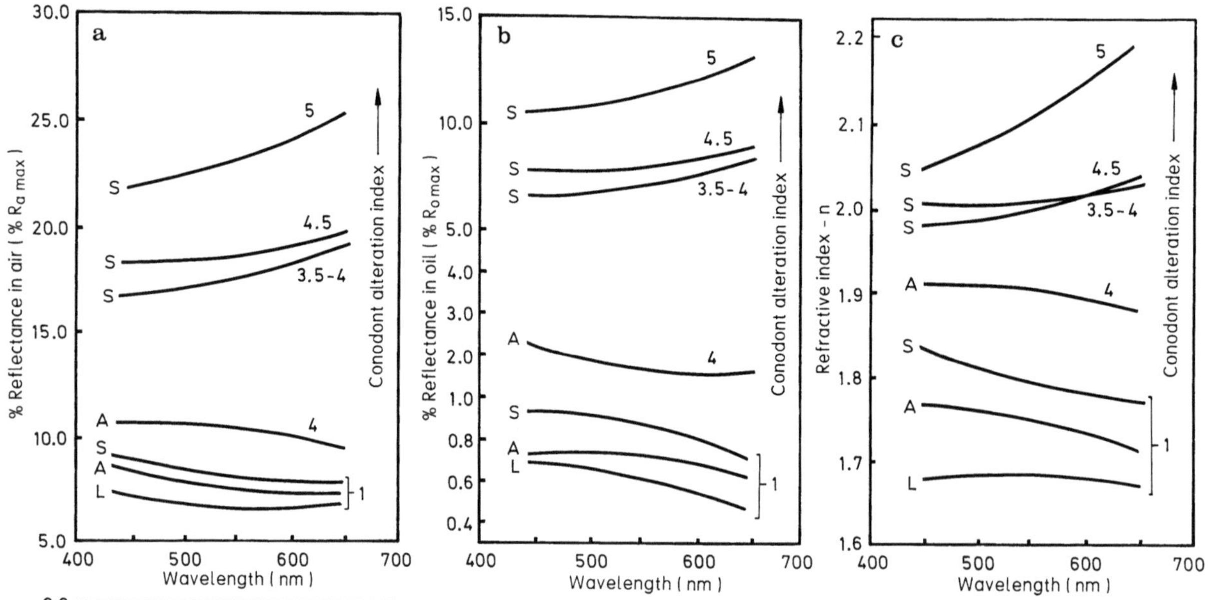

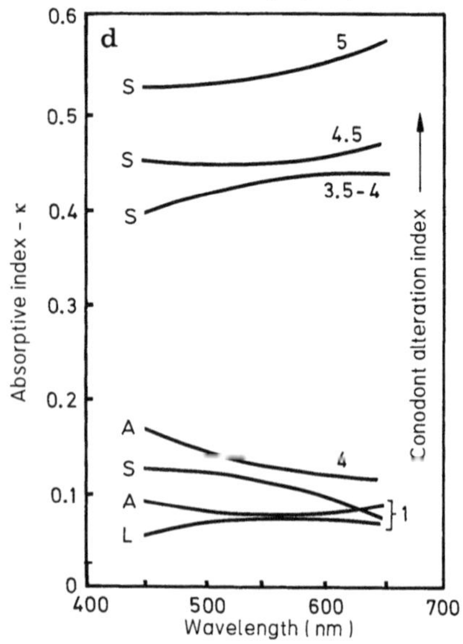

Fig. 11. Variation with wavelength of generalized curves of air (**A**) and oil (**B**) reflectances, refractive index (**C**) and absorptive index (**D**) for graptolites from Canadian Early Paleozoic sediments of increasing maturity (after Goodarzi 1986). *L* Limestone; *A* argillaceous limestone; *S* shale

5.3 Comparison of the Optical Properties of graptolites with Other Maturation Parameters

Goodarzi and Norford (1985) compared the maximum and minimum reflectances of naturally matured graptolites to those of graptolites affected by epithermal fluids (Fig. 12) and found that they followed different paths. Later, Goodarzi and Norford (1987) reported the reflectances of coexisting graptolites, chitinozoans and bitumen in sediments from a borehole in SE Turkey (Fig. 5). The reflectance of all three types of organic matter increased with increasing depth, the graptolites having values higher than those of bitumen and chitinozoans. Bertrand and Heroux (1987) and Bertrand (1990) measured the reflectance of bitumen, chitinozoans and graptolites in a borehole from eastern Canada and found that at low maturities chitinozoans had higher reflectances than those of graptolites.

A relationship between vitrinite reflectance and the conodont alteration index was established by Epstein et al. (1977) and Nowlan and Barnes (1987). Goodarzi and Norford (1988) have used this relationship to correlate the reflectance of graptolites with that of vitrinite (Fig. 13). Graptolites and vitrinite may occur together in late Devonian and younger sediments, making direct comparison of their reflectances possible in such circumstances. The only such comparison to our knowledge has been made by Stasiuk (pers. comm., 1989) who found graptolites with a Ro

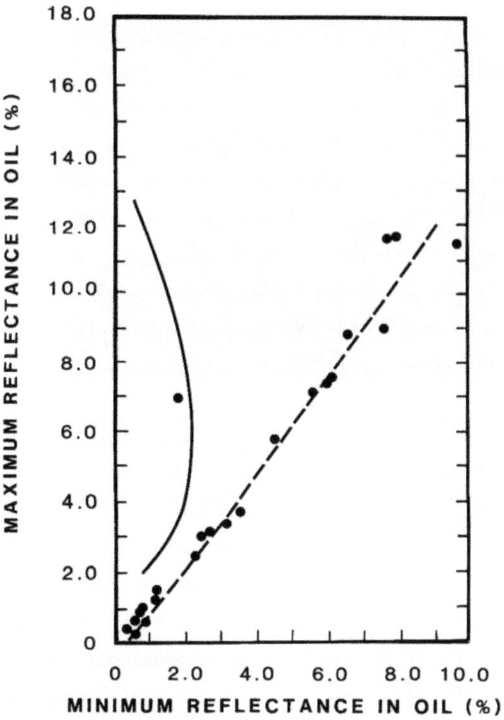

Fig. 12. Maximum and minimum reflectance trends for graptolites (– –) and heat-affected organic materials (—). The latter curve is based on the work of Brown and Taylor (1961), Chandra (1963) and Khavari-Khorasani (1975). The data points (●) are after Goodarzi and Norford (1985), and Goodarzi et al. (1988a). The single point in the "heat-affected" region is the Howard Pass graptolite

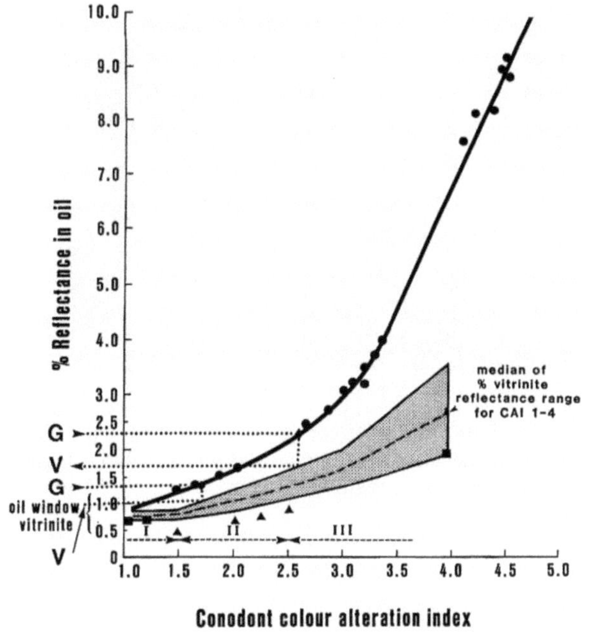

Fig. 13. The reflectance (% Ro max) of non-granular (●) and granular (■) graptolites plotted against conodont colour alteration (Goodarzi and Norford 1985) to define the normal thermal maturation trend for graptolites (—). Also plotted are vitrinite reflectance (Epstein et al. 1977), scolecodont reflectance (▲) (Goodarzi and Higgins 1987); and hydrocarbon generation ranges *I* (immature); *II* (mature); *III* (post-mature) (Nowlan and Barnes 1987). Graptolite reflectance values (*G*) from the Amadeus Basin (after Kurylowicz et al. 1976) are converted equivalent vitrinite reflectance (*V*)

max of 0.52% along with vitrinite (0.62% Ro max) in Carboniferous sediments from Saskatchewan. This observation indicates that graptolites, at the boundary between diagenesis and catagenesis, may have lower reflectance values than vitrinite. A similar relationship exists between the reflectances of bitumen and vitrinite. Bitumen has lower reflectance values than vitrinite during diagenesis and early catagenesis, to almost the peak of oil generation (vitrinite Ro = 0.9–1.0%) (Jacob 1985; Jacob and Hiltman 1988). However, bitumen develops higher anisotropy at a lower reflectance level than does vitrinite. Goodarzi and Norford (1985) stated that graptolites may show similar changes to those of bitumen with increasing maturation. This has recently been supported by the results of Bustin et al. (1989) who found that the residues of vitrinite and graptolites, experimentally carbonized under the same conditions, show a similar level of reflectance at any temperature level. However, the graptolite residue has the higher bireflectance, indicating that graptolites develop a better-ordered molecular structure than do vitrinities.

5.4 Factors Influencing Graptolite Reflectance

The reflectance and bireflectance of graptolite fragments in sediments may be modified by various factors, including

1. Temperature: reflectance of graptolites increases with depth of burial (Goodarzi and Norford 1987) and with rising temperature in laboratory-pyrolyzed graptolitic sediments (Bustin et al. 1989).
2. Rock matrix: graptolite fragments in a carbonate matrix are commonly less reflective than graptolites in shales (Fig. 10; Goodarzi and Norford 1985; Riediger et al. 1989; Link et al. 1989). This is similar to the observed variation of vitrinite reflectance between different host rocks of equivalent maturity (Stach et al. 1982; Goodarzi et al. 1988b). In the case of graptolite fragments, it may be due to the leaching of the graptolite epiderm by carbonate-derived CO_2.
3. Suppression of reflectance: the reflectance of vitrinite, when associated with abundant liptinite (e.g.

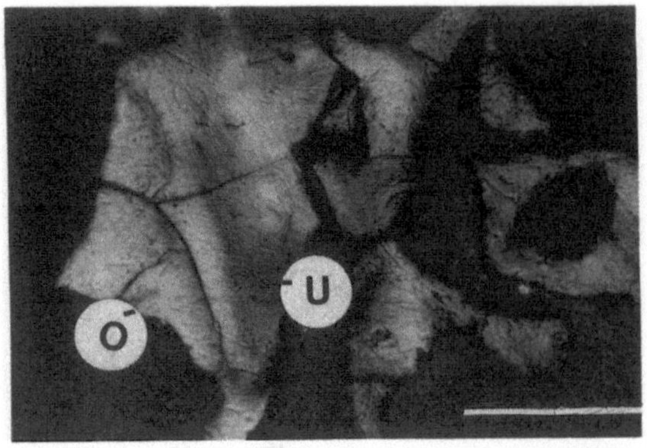

Fig. 14. Weathered fragments of graptolite, Early Silurian, dolomitic silty shale, Road River Group, northwestern Canada. Plane polarized light. Oxidation rim (*O*) has higher relief and reflectance than the unaltered area (*U*). Magnification as in Fig. 1

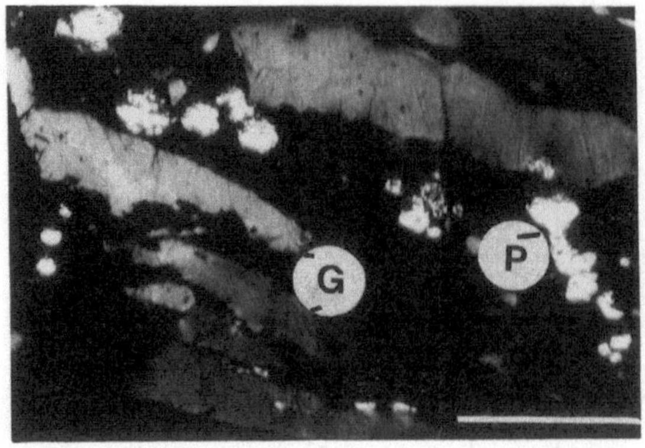

Fig. 15. Thermally metamorphosed non-granular graptolite fragments, showing strong anisotropy, Silurian shale, stratabound lead-zinc deposit, shale, Road River Group, Yukon Territory. Partially crossed polars. Graptolite fragment (*G*), Pyrite (*P*). Magnification as in Fig. 1

in oil shale or oil-prone source rock) is often suppressed due to absorption of oil or bitumen by the vitrinite particles (Hutton and Cook 1980; Price and Barker 1985). Similarily, the reflectance of graptolites in oil shales may be lowered by bitumen absorption. This also enhances the optical differentiation between various parts of the graptolite epiderm.

4. Weathering: organic matter is very sensitive to weathering. Coal macerals particularly the liptinite and vitrinite groups, are especially susceptible to oxidation. Secondary vitrinite (as opposed to primary vitrinite, the variety used for maturation studies) is commonly formed during weathering. The skeletal material of weathered graptolites has highly reflective oxidation rims and cracks (Fig. 14), similar to those of vitrinite in weathered coal (van Krevelen 1961).
5. Thermal perturbations: graptolites are very sensitive to local temperature anomalies, such as those caused by migration of hot mineralizing fluids. Graptolite epiderm collected close to a fault in a sedimentary lead-zinc deposit from Howard Pass, Yukon Territory displays anomalously high anisotropy (Fig. 15) and bireflectance (Fig. 10). This sample plots within the "heat-affected zone" of Fig. 12. A detailed study by Riediger et al. (1989) has indicated that hydrothermal liquids are indeed responsible for the elevated bireflectance of this graptolite. This sensitivity of graptolite reflectance to local thermal perturbations is similar to the elevation of vitrinite reflectance in contact metamorphic aureoles (Bostick 1974; Creaney 1980; Chandra and Taylor 1982).

5.5 Use of Graptolite Reflectance in Regional Maturation Studies

The reflectance (% Ro max) and bireflectance of graptolites are the two most diagnostic parameters available to determine the maturity of Early Paleozoic sediments on a regional scale. Variation of graptolite reflectance versus the conodont alteration index indicates that the former parameter is more sensitive to the maturity changes, particularly at high maturation levels (CAI > 4) (Goodarzi and Norford 1985, 1988; Link et al. 1989; Riediger et al. 1989).

The maturity of graptolite-bearing sediments can be determined indirectly by plotting the measured graptolite reflectance on Fig. 13. In so doing, the corresponding vitrinite reflectance can be assessed and the stage of hydrocarbon generation determined. Using this correlation, the oil window (CAI 1.5–2.5) is defined by the graptolite reflectance range 1.2–2.2% (Ro max). For reasons discussed below, this indirect relationship between vitrinite and graptolite reflectance via the conodont alteration index should be used with caution. One problem is that Fig. 13 is based on that published by Epstein et al. (1977) for the correlation between CAI and vitrinite reflectance. These authors used the average % Ro for vitrinite reflectance while % Ro max is measured for graptolites.

The results of Kurylowicz et al. (1976) on the maturity of Ordovician sediments from the Palm Valley and Mereenie Fields in the Amadeus Basin, Central Australia, can be used as a test of this correlation between graptolite and vitrinite reflectances. The reflectance of graptolites at Mereenie are about 1.27% and at Palm Valley about 2.23%. The

Mereenie field contains light oil and wet gas, whereas Palm Valley is a dry gas field. Figure 13 indicates that these measured reflectivities would be equivalent to vitrinite reflectance values of about 1% at Mereenie and 1.75% at Palm Valley, which is in broad agreement with the maturity of the kerogen in these rocks as expressed by their atomic H/C ratios (0.75–0.78, Mereenie; 0.60–0.65, Palm Valley: Jackson et al. 1984).

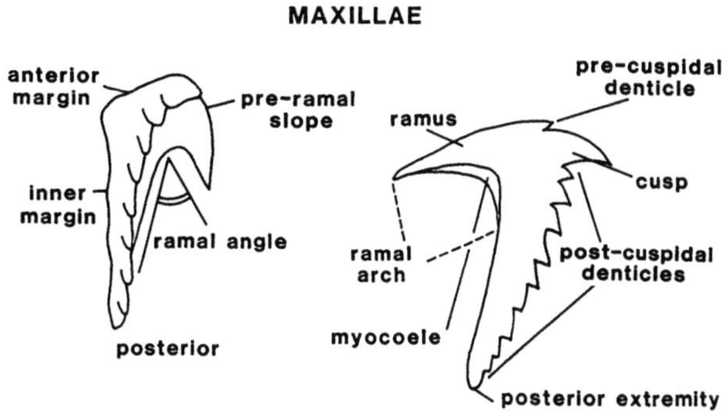

Fig. 16. Simplified morphology of the scolecodonts (After Kielan-Jawaroska 1966)

6 Scolecodonts

Scolecodonts are the partly organic, fossil remains of the jaw apparatus, a tooth-like structure, of eunicid polychaete annelid worms (Fig. 16). They are found in sediments ranging in age from Ordovician to Jurassic. The colours of scolecodants under transmitted light vary from yellow to brown to entirely black, in a fashion similar to the changes observed in conodonts (Epstein et al. 1977) and palynomorphs (Dow 1977) during thermal alteration.

The morphology and optical properties of scolecodonts have been studied under reflected light by Goodarzi et al. (1985) and Goodarzi and Higgins (1987). Scolecodonts are isotropic up to at least a CAI of 3.5 and their colours change from dark to light grey with increasing maturity (CAI = 1.5–3.5). Their surfaces may be smooth or granular and have a porate morphology.

The reflectance and refractive and absorptive indices of scolecodonts increase with increasing maturation as measured by conodont alteration index (Fig. 17). The dispersion of their optical properties within the visible spectrum follows similar trends to those of low-maturity bitumen, vitrinite and graptolites with increasing maturation (Davis 1978; Goodarzi 1986). The negative slopes of the spectral curves of reflectance for low- to medium-maturity scolecodonts (CAI = 1.5–3.5) (Fig. 18) indicate a low ratio of aromatic to amorphous carbon (Khavari-Khorasani 1975; Goodarzi and Higgins 1987).

The reflectances and the refractive and absorptive indices of scolecodonts at 546 nm display a positive correlation with the conodont alteration index, indicating that these optical properties can be used to determine the thermal maturity of sediments in which other rank indicators (e.g. bitumen, graptolites and vitrinite) are absent. The reflectance of scolecodonts is compared with that of graptolites and vitrinite in Fig. 13. Scolecodonts with reflectances of 0.35–0.84% correspond to a vitrinite reflectance range of 0.80–1.1% (CAI = 1.5–2.5) which is the oil-

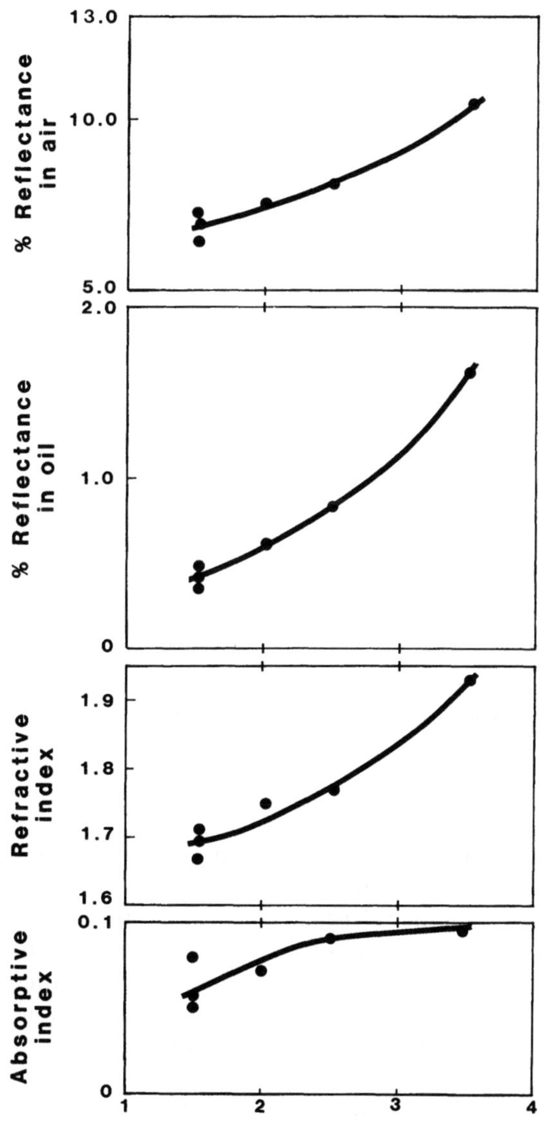

Fig. 17. Variation in optical properties of scolecodonts with conodont alteration index (After Goodarzi and Higgins 1987)

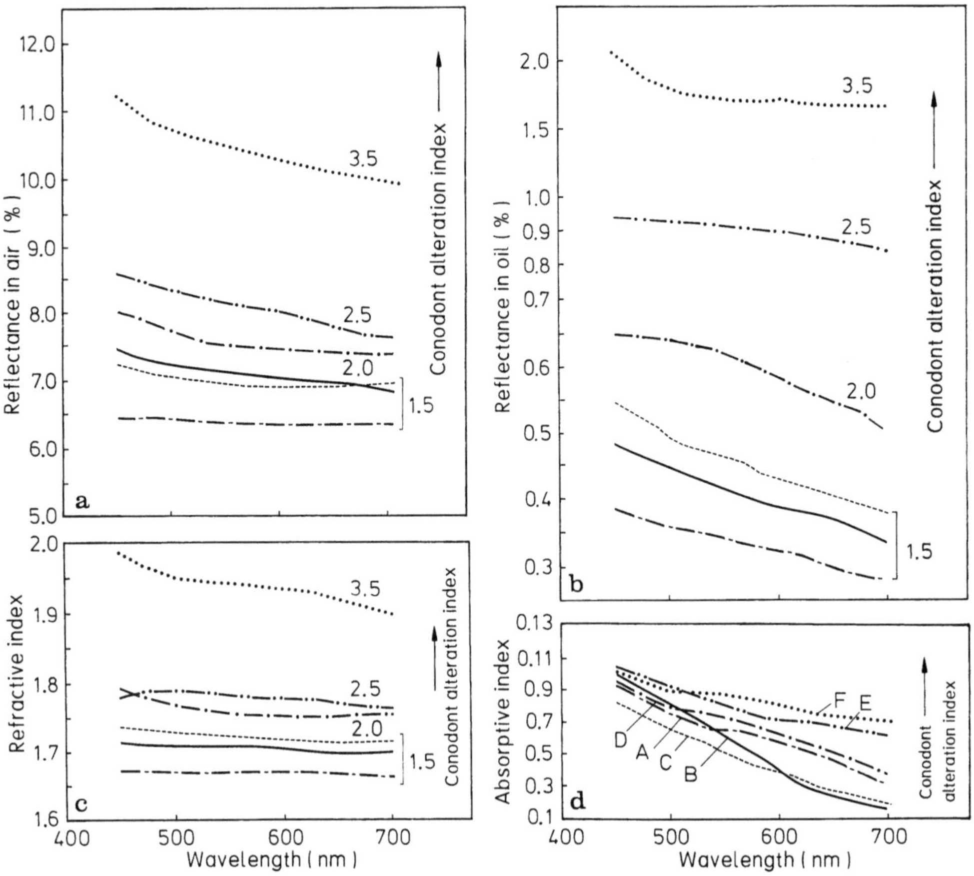

Fig. 18. Variation with wavelength of air (**a**) and oil (**b**) reflectances, refractive index (**c**) and absorptive index (**d**) for scolecodonts of increasing maturity

generation window. Higher reflectance values indicate overmature sediments.

7 Summary and Future Prospects

Determination of the thermal maturity of sediments by measuring the reflectance of marine-derived organic matter, such as algae, chitinozoans, graptolites and scolecodonts, is a relatively recent innovation. Maturation-dependent changes in the optical properties of graptolites are better understood than those of the other types of marine organic matter which are still in the preliminary stages of evaluation.

At present, graptolite reflectance appears to be the most promising optical tool for the evaluation of the maturity of Early Paleozoic sediments. With increasing maturity the optical properties of graptolites at 546 nm, or their dispersion in visible light (450–650 nm), follow a similar trend to those of vitrinite, indicating that similar changes are occurring in the molecular structure of both substances. An indirect correlation between graptolite and vitrinite reflectance has been established using the conodont alteration index. This correlation indicates that the "oil window" as defined by vitrinite reflectance (0.5–1.3% Ro), coincides with a graptolite reflectance of 0.7–1.7% Ro max. More direct correlations between vitrinite (including the "vitrinite-like" material of possible cyanobacterial or algal origin in Pre-Devonian rocks) and graptolites in the same samples are needed to verify this indirect correlation. This should also be done for chitinozoa and scolecodonts.

The use of algal fluorescence, particularly of a single species and in relatively immature sediments also seems promising. For example, initial results of a regional study of kukersites from the Upper Ordovician Yeoman Formation dominated by *Gloeocapsomopha prisca* (Fowler, this Vol.) seem to show a good correlation between fluorescence characteristics and biomarker maturation parameters (Fowler, Stasiuk and Goodarzi, unpubl. results). Reflectance of lamalginite, as recently exploited by Crick et al. (1988), is another method that could be used in maturation studies of pre-Devonian rocks.

Acknowledgments. The authors thank Dr. A.R. Cameron, Institute of Sedimentary and Petroleum Geology, Geological Survey of Canada, Prof. A.C. Cook, Department of Geology, the University of Wollongong and Dr. M. Smyth, CSIRO, North Ryde, NSW, Australia for their valuable comments on earlier versions of this paper: Geological survey of Canada contribution No: 14589.

References

Alpern B (1981) Les Schistes bitumineux: constitution, réserves, Valorisation. Bull Cent Rech Explor Prod Elf-Aquitaine 5:319–352

Banks HP (1975) Early vascular plants: proof and conjecture. Bioscience 25:730–737

Bertrand R, Heroux Y (1987) Chitinozoan, graptolite and scolecodont reflectance as an alternative to vitrinite and pyrobitumen reflectance in Ordovician and Silurian strata, Anticosti Island, Quebec, Canada. Am Assoc Petrol Geol Bull 71:951–957

Bertrand R, Berube J-C, Heroux Y, Achab A (1985) Petrographie du kérogène dans le Paléozoique inférieur: méthode de préparation et exemple d'application. 'Rev Inst Fr Petrol 40:155–167

Bertrand R. (1990) Correlation among the reflectances of vitrinite, Chitinozoans, Graptolites and Scolecodonts. Organic Geochemistry, Vol. 15, No. 6, pp. 565–574

Bloeser B, Schopf JW, Horodyski RJ (1977) Chitinozoans from the Late Precambrian Chuar Group of the Grand Canyon, Arizona. Science 195:676–679

Bostick NH (1974) Phytoclasts as indicators of thermal metamorphism, Franciscan Assemblage and Great Valley Sequence (Upper Mesozoic), California. Geol Soc Am Spec Pap 153:1–17

Brown HR, Taylor GH (1961) Some remarkable Antarctic coals. Fuel 40:211–224

Buchardt B, Clausen J, Thomsen E (1986) Carbon isotope composition of Lower Paleozoic kerogen: effects of maturation. Org Geochem 10:127–134

Bustin RM, Link C, Goodarzi F (1989) Optical properties and chemistry of graptolite periderm following laboratory simulated maturation. Org Geochem 14:355–364

Cartz L, Hirsch PB (1960) A contribution to the structure of coals from X-ray diffraction studies. Philos Trans R Soc London Ser A 252:557–604

Chandra D (1963) Reflectance of thermally metamorphosed coals. Fuel 42:69–74

Chandra D, Taylor GH (1982) Thermally altered coals. In: Stach E, Mackowsky MTh, Teichmüller M, Taylor GH, Chandra D, Teichmüller R (eds) Coal Petrology. Bornträger, Berlin, pp 206–218

Clausen CD, Teichmüller M (1982) Die Bedeutung der Graptolithenfragmente im Paläozoikum von Soest Erwitte für Stratigraphie und Inkohlung. Fortschr Geol Rhein Westf 30:145–167

Cook AC, Murchison DG, Scott E (1972) A British meta-anthracite coal of Devonian age. Geol J 8:83–94

Cook AC, Hutton AC, Sherwood NR (1981) Classification of oil shales. Bull Cent Rech Explor Prod Elf-Aquitaine 5:353–381

Correia M (1967) Relations possibles entre l'étate de conservation des éléments figurés de la matière organique (microfossiles palynoplanctologique) et l'existance de gisements d'hydrocarbures. Rev Inst Fr petrol 22:1285–1306

Creaney S (1980) Petrographic texture and vitrinite reflectance variation on the Alston Block, northeast England. proc Yorkshire Geol Soc 42:553–580

Crick IH, Boreham CJ, Cook AC, Powell TG (1988) Petroleum geology and geochemistry of Middle Proterozoic McArthur Basin, northern Australia II: assessment of source rock potential. Am Assoc Petrol Geol Bull 72:1495–1514

Crowther PR (1981) The fine structure of graptolite periderm. Special papers in Paleontology 26. Palaeontol Assoc, London 119 pp

Davis A (1978) The reflectance of coal. In: Karr C (ed) Analytical methods for coal and coal products, vol. 1 Academic Press, New York, London, pp 27–83

Dow WG (1977) Kerogen studies and geological interpretation. J Geochem Explor 7:77–79

Edwards D (1980) Early land floras. In: Panchen AL (ed) The terrestrial environment and the origin of land vertebrates. Syst Assoc Spec Vol 15. Academic press, New York London, pp 55–85

Epstein AG, Epstein JB, Harris LD (1977) Conodont colour alteration – an index of organic metamorphism. US Geol Surv Prof Pap 995:27 pp

Fowler MG, Douglas AG (1984) Distribution and structure of hydrocarbons in four organic-rich Ordovician rocks. Org Geochem 6:105–114

Franklin RE (1951) Crystallite growth in graphitizing and non-graphitizing carbons. Proc R Soc London 209:196–218

Friedel RA, Queiser JA (1959) Ultra-violet, visible spectrum and the aromaticity of coal. Fuel 39:369–380

Gilbert LA (1960) The reflectivity spectra of coal vitrains in the visible and ultraviolet. Fuel 39:393–400

Goodarzi F (1982) A brief hydrocarbon potential study of southeast Turkey using petrography. Rep Turk Petrol Res Cent, Ankara, 28 pp

Goodarzi F (1984) Organic petrology of graptolite fragments from Turkey. Mar Petrol Geol 1:202–210

Goodarzi F (1985) Reflected light microscopy of chitinozoan fragments. Mar Petrol Geol 2:72–78

Goodarzi (1986) Dispersion of optical properties of graptolite epidermis with increased maturity in Early Paleozoic organic-rich sediments. Fuel 64:1735–1740

Goodarzi F, Stasiuk LD, Lindholm K (1988a) Graptolite reflectance and thermal maturity of Lower and Middle Ordovician shales from Scania, Sweden. Geol Füreing Stockholm Fürhandl 110:225–236

Goodarzi F, Gentzis T, Feinstein S, Snowdon LR (1988b) Effect of maceral subtypes and mineral matrix on measured reflectance of subbituminous coals and dispersed organic matter. Int J Coal Geol 10:383–398

Goodarzi F, Higgins AC (1987) Optical properties of scolecodonts and their use as indicators of thermal maturity. Mar Petrol Geol 4:353–359

Goodarzi F, Murchison DG (1972) Optical Properties of carbonized vitrinite. Fuel 51:322–328

Goodarzi F, Norford BS (1985) Graptolites as indicators of the temperature histories of rocks. J Geol Soc London 142:1089–1099
Goodarzi F, Norford BS (1987) Optical properties of graptolite periderm – a review. Bull Geol Soc Den 35:141–147
Goodarzi F, Norford BS (1988) Variation of graptolite reflectance with depth of burial. Int J Coal Geol 11:127–140
Goodarzi F, Stasiuk LD (1987) Graptolite preparation for reflected light microscope – a technical note. Current Research, PtA. Geol Surv Can Pap 87–1 A:317–322
Goodarzi F, Vleeskens JM (1988) Reactivity of bituminous semianthracitic coals: a reflected light study of their combustion residue (fly ash). J. Coal Qual 7(3):80–85
Goodarzi F, Gunther PR, Jenkins WAW, Snowdon LR (1985) Preliminary organic petrology of Paleozoic rocks from the Grand Banks, Newfoundland. Mar Petrol Geol 2:254–259
Gorter JD (1984) Source potential of the Horn Valley Siltstone, Amadeus Basin. Aust Petrol Explor Assoc J 24(1):66–90
Harris AG (1979) Conodont color alteration and organo-mineral metamorphic index, and its application to Appalachian Basin geology. In: Scholle PA, Schluger PR (eds) Aspects of diagenesis. Soc Econ Paleont Miner Spec Publ 26:3–16
Hevia Rodriguez V (1974) Sur la mesure de la reflectivité des macéraux des charbons en différentes Longueurs d'onde. C R 7th Congr Int Stratigraphie et de geologie du carbonifere, vol 3, Krefeld 1971, pp 189–196
Hufnagel H (1977) Das Fluorezenzvermögen der Dinoflagellaten-Cysten – ein Inkohlungsparameter? Geol Jahrb D 23:59–65
Hutton AC (1986) Classification of oil shales – a petrographic approach. In: Proc 19th Oil Shale Symp, Colorado School of Mines, Golden, CO, pp 27–37
Hutton AC, Cook AC (1980) Influence of alginite on the reflectance of vitrinite from Joadja, NSW and some other coals and oil shales containing alginite. Fuel 59:711–714
Jackson DE (1973) *Amplexograptus* and *Glyptograptus* isolated from Ordovician limestones in Manitoba. Geol Surv Can Bull 2:1–8
Jackson KS, McKirdy DM, Deckelman JA (1984) Hydrocarbon generation in the Amadeus Basin, central Australia. Aust Petrol Explor J 24(1):42–65
Jacob H (1975) Mikroskopphotometrische Analyse natürlicher fester Erdölbitumina. In: Alpern B (ed) Pétrographie de la matière organique des sédiments, relations avec la paléotemperature et le potentiel pétrolier. Ed Cent Natl Rech Sci Paris, pp 103–113
Jacob H (1985) Dispersed solid bitumen as an indicator for migration and maturity in prospecting for oil and gas. Erdöl Kohle 38:365
Jacob H, Hiltman W (1988) Disperse, Feste Erdölbitumina, als Maturitätsindikatoren im Rahmen der Erdöl/Erdgas-Prospektion. Geol Jahrb 89:3–37
Jansonius J, Jenkins WAM (1978) Chitinozoans. In: Haq BV, Boersma A (eds) Introduction to marine micropaleontology. Elsevier, Amsterdam, pp 341–357
Jenkins WAM (1970) Chitinozoa. Geosci Man 1:1–21
Kemp AES, Oliver GHJ, Baldwin JR (1985) Low-grade metamorphism and accretion tectonics: southern uplands terrain, Scotland. Mineral Mag 49:335–344
Khavari-Khorasani G (1975) The properties and structural ordering of some fossil bitumens. PhD Thesis, Univ Newcastle-Upon-Tyne, U.K.
Khavari-Khorasani G, Murchison DG (1977) Application of light microscopy in structural studies of bitumens. In: ICCP Meet, Essen 10 pp
Kielan-Jaworowska A (1966) Polychaete jaw apparatuses from the Ordovician and Silurian of Poland and a comparison with modern forms. Palaeontol Pol 16:1–52
Kirsch HJ (1980) Incipient metamorphism of Cambro-Silurian clastic rocks from the Jüntland Supergroup, central Scandinavian Caledonides, western Sweden: illite crystallinity and "vitrinite" reflectance. J Geol Soc London 137:271–288
Kurylowicz LE, Ozimic S, McKirdy DM, Kantsler AJ, Cook AC (1976) Reservoir and source rock potential of the Larapinta Group, Amadeus Basin, central Australia. Aus Petrol Explor Assoc J 16(1):44–65
Kwiecinska B, Murchison DG, Scott E (1977) Optical properties of graphite. J Microsc 104(3):294–302
Laufeld A (1973) Ordovician chitinozoans from Portixeddu, Sardinia. Bull Soc Paleontol Ital 2:3–7
Link CM (1988) Organic maturation and petroleum source potential of Phanerozoic strata in northern Yukon and northwestern District of Mackenzie. MSc Thesis, Univ British Columbia, Vancouver, 273 pp
Link CM, Bustin MR, Goodarzi F (1990) Petrology of graptolites and their utility as indices of thermal maturity in Lower Paleozoic strata in northern Yukon, Canada. Int J Coal Geol 15:113–115
Mackowsky MTh (1982) Preparation of polished surfaces from particulate samples. In: Stach E, Mackowsky MTh, Teichmüller M, Taylor GH, Chandra D, Teichmüller R (eds) Coal petrology. Borntrӓger, Berlin, pp 296–299
Marshall RJ, Murchison DG (1971) Dispersion of optical properties of carbonized vitrinites. Fuel 50:4–22
McKirdy DM, McHugh DJ, Tardif JW (1989) Comparative analysis of stromatolitic and other microbial kerogens by pyrolysis – hydrogenation – gas chromography (PHGC). In: Trudinger PA, Walter MR, Ralph BJ (eds) Biogeochemistry of ancient and modern environments Aust Acad Sci, Canberra/Springer, Berlin Heidelberg New York, pp 187–200
Moore RC, Lalicker CG, Fisher AG (1952) Invertebrate fossils. McGraw-Hill, New York, pp 715–733
Nicoll RS, Groter JD (1984) Conodont colour alteration, thermal maturation and the geothermal history of the Cumming Basin, Western Australia. Aust Petrol Explor Assoc J 24(1):243–258
Nowlan GS, Barnes CR (1987) Thermal maturation of Paleozoic strata in eastern Canada from conodont colour alteration index (CAI) data with implications for burial history, tectonic evolution, hotspot tracks and mineral and hydrocarbon exploration. Geol Surv Can Bull 367:47 pp
Oliver GJH (1987) Arenig to Wenlock regional metamorphism in the paratectonic Caledonides of the British Isles: a review. In: Harris A L, Fettes D J (eds) The Caledonian-Appalachian orogen. Geol Soc London Blackwell, Oxford, 341–363
Price LC, Barker CE (1985) Suppression of vitrinite reflectance in amorphous rich kerogen – a major unrecognized problem. J Petrol Geol 8:54–84
Radke M, Welte DH (1983) The methylphenanthrene index (MPI): a maturity parameter based on aromatic hydrocarbons. In: Bioroy M, Albercht P, Cornford C, de Groot K, Eglinton G, Galimov E, Leythaeuser, Pelet R Speers G (eds) Advances in Organic Geochemistry 1981. Wiley Heyden, Chichester, pp 504–512

Riediger C, Goodarzi F, Macqueen RW (1989) Graptolites as an indicator of regional maturity in Lower Paleozoic sediments, Selwyn Basin, Yukon and Northwest Territories, Canada. Can J Earth Sci 26:2003–2015

Robert P (1980) The optical evaluation of kerogen and geothermal histories applied to oil and gas exploration. In: Durand B (ed) Kerogen Editions Technip, Paris, pp 385–414

Sikander AH, Pittion JL (1978) Reflectance studies on organic matter in Lower Paleozoic sediments of Quebec. Bull Can Petrol Geol 26:132–151

Stach E, Mackowsky MT, Teichmüller M, Taylor GH, Chandra D, Teichmüller R (1982) Stach's textbook of coal petrology, 3rd edn. Bornträger, Berlin, 535 pp

Teichmüller M (1978) Nachweis von Graptolithen-Periderm in geschieferten Gesteinen mit Hilfe kohlenpetrologischer Methoden. N Jahrb Paläontol Monatsh 7:430–477

Teichmüller M (1982) Origin of the petrographic constituents of coal. In: Stach E, Mackowsky MT, Teichmüller M, Taylor GH, Chandra D, Teichmüller R (eds) Stach's textbook of coal petrology, 3rd edn. Bornträger, Berlin, pp 219–283

Teichmüller M, Ottenjann K (1977) Art und Diagenese von liptiniten und lipoiden Stoffen in einem Erdölmuttergestein auf Grund fluoreszenzmikroskopischer Untersuchungen. Erdöl Kohle 30:387–398

Teichmüller M, Teichmüller R, Weber K (1979) Inkohlung und Illit-Kristallinität. Fortschr. Geol Rheinl West 27:201 276

Tissot B, Welte DH (1984) Petroleum formation and occurrence, 2nd edn. Springer, Berlin Heidelberg New York, 699 pp

Tschudy RN, Scott RA (1969) Aspects of palynology. John Wiley & Sons, New York London, pp 469–474

Van Krevelen DW (1961) Coal. Elsevier, Amsterdam, p 350

Watson SW (1976) The sedimentary geochemistry of the Moffat shales: a carbonaceous sequence in the southern Uplands of Scotland. Ph D Thesis, Univ. St. Andrews, UK

Hydrocarbon Composition of the Late Proterozoic Oils of the Siberian Platform: Implications for the Depositional Environment of Source Rocks

Roger E. Summons and Trevor G. Powell[1]

[1] Bureau of Mineral Resources, GPO Box 378, Canberra, ACT, 2601, Australia

Abstract

Oils reservoired in Cambrian and Vendian (Latest Proterozoic) marine carbonates and terrigenous clastic rocks of the Siberian Platform could have been sourced either by deeply buried basinal shales of Riphean age or, by carbonate/evaporite shelf sediments of Latest Proterozoic to Early Cambrian age. The oils from three fields in the region all have very similar hydrocarbon compositions which include the following distinctive features: high relative abundance of acyclic isoprenoids and Pr/Ph < 1; high relative abundance of $\alpha\alpha\alpha$ and $\alpha\beta\beta$ steranes and virtual absence of rearranged steranes; high abundance of 29,30-bisnorhopane and norhopane relative to hopane; high abundance of 28,30-bisnorhopane, 2α-methyl hopanes and other unusual triterpenoids; and the presence of aryl isoprenoids. These aspects of the biomarker distributions are similar to those found in some carbonates and carbonate/evaporites from the Phanerozoic and evidently correlate with sediment lithology because they relate to biological and chemical conditions prevailing in these depositional environments. Thus, these Siberian Platform oils have likely been sourced within the Vendian/Cambrian carbonate sequence.

Additional aspects of the hydrocarbon assemblage are the very high relative abundances of C_{29} desmethylsteranes, C_{30} 3β-methyl- and 2α-methylsteranes and C_{31} 3β-ethylsteranes, high relative abundance of monomethyl alkanes with a preference for mid-chain isomers and an unusually light carbon isotope signature. These features comprise a pattern seen in other organic matter of Vendian age, including the Huqf Group oils of Oman. They are apparently associated with special features of the biology and palaeoenvironment which existed during the Latest Proterozoic.

1 Introduction

Oils from the Siberian Platform have unusual biomarker compositions (Aref'ev et al. 1980; Fowler and Douglas 1987). They share features, such as the strong predominance of 24-ethylcholestane isomers and high abundances of monomethyl alkanes, with the Late Proterozoic oils from the Huqf Group of Oman (Grantham 1986; Klomp 1986; Grantham et al. 1988). The oils are indigenous to a complex sedimentary sequence of Late Proterozoic and Cambrian rocks (Meyerhoff 1980; Kontorovich et al. 1981). They could be drived from deeply buried, basinal organic-rich shales of Riphean age or the Vendian to Cambrian shelf carbonate and clastic sequence in which they are reservoired. To our knowledge, no specific horizon has ever been identified as the source for the oils.

There is a paucity of data on the characteristics of potential petroleum source rocks in very old sediments. The presence of commercial petroleum deposits in the Proterozoic rocks of Oman (Hughes Clarke 1988) and the Siberian Platform (Meyerhoff 1980) and the existence of oil seeps and rich petroleum source rocks in other areas, notably the McArthur Basin of northern Australia (Jackson et al. 1986, 1988) and the Mid-Continent Rift of the USA (Dickas 1986) provide the impetus to study the nature of the primary organic matter and how it equates with source material from the Phanerozoic. Furthermore, with the availability of improved analytical techniques, there has been renewed interest in whether or not biological information, which may be obtained from the isotopic and biomarker signatures in ancient organic matter, truly reflects the primary signals of primitive biota and the affinities of those biota (for example, see discussions in Hayes et al. 1983; Grantham 1986; Fowler and Douglas 1987; Hoering and Navale 1987; Summons et al. 1988 a, b).

This study was initially undertaken to compare specific details of the hydrocarbons in Siberian Platform oils with the unusual features of organic matter from the McArthur Basin and other Proterozoic

M. Schidlowski et al. (Eds.)
Early Organic Evolution: Implications for Mineral and Energy Resources

deposits. Using these data we sought to determine whether there were any characteristics peculiar to these ancient materials which would be useful for possible oil-source correlations and to delineate possible biological and lithological influences on their composition.

2 Samples and Methods

Three oil samples, Verkhne-Chonskaya (#319), Danilovskaya (#320) and Duliominskaya (#321) from the Lena-Tunguska region (Fig. 1) of the Siberian Platform kindly provided by the East Siberian Research Institute, Geology, Geophysics, and Mineral Resources-Irkutsk of the Ministry of Geology, USSR. Two of the oils came from Vendian clastic reservoirs and the other came from the slightly younger Vendian-Cambrian carbonate-evaporite complex. They were analysed using the separation, GC and GCMS methods detailed by Summons and Powell (1987) and Summons et al. (1988b). The technique of multiple-reaction monitoring (MRM) in the GCMS (Warburton and Zumberge 1983) was used extensively to improve the specificity and reliability of biomarker identifications. Carbon isotope determinations were carried out by Z. Roksandic and R. Drew, BMR.

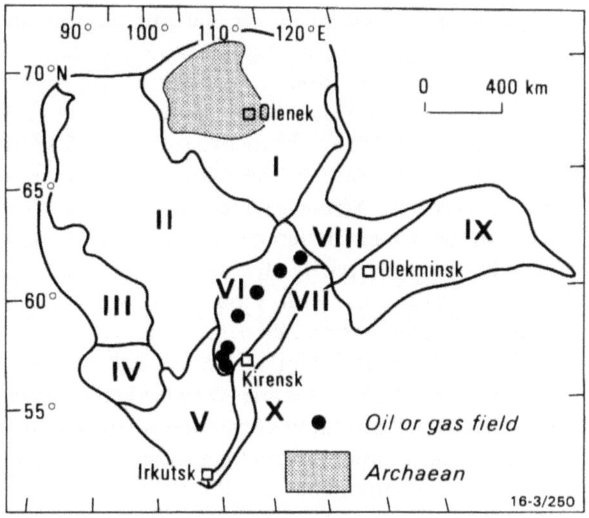

Fig. 1. Map showing location of Lena-Tunguska oil fields of eastern Siberia

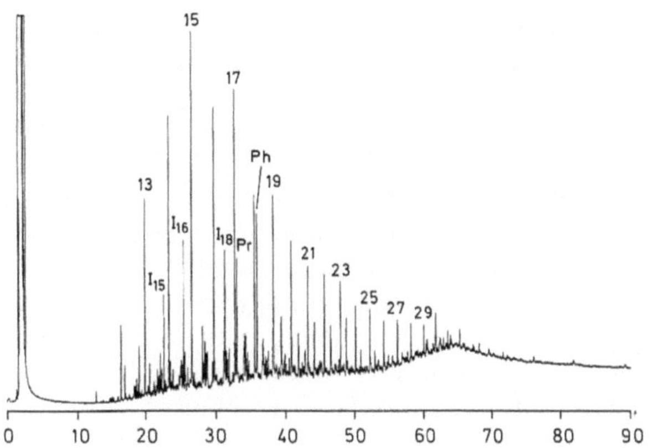

Fig. 2. Gas chromatographic trace (FID response) showing saturated hydrocarbon distribution in the Danilovskaya (BMR #320) oil from the Lena-Tunguska province. *Numbered peaks* are n-alkanes designated according to their carbon-chain length; *Pr* pristane; *Ph* phytane; and I_{15}, I_{16} etc. refer to the carbon numbers of other isoprenoid alkanes. The large peak between each n-alkane in the region $C_{19}-C_{25}$ is a suite of mid-chain, branched monomethyl alkanes

3 Results and Discussion

3.1 Hydrocarbon Assemblage

3.1.1 Acyclic Components

The oils represented in this study and the oils reported by Fowler and Douglas (1987) all show a remarkable similarity in their hydrocarbon compositions. This applies to the distributions of both major and minor components. They have a predominance of low molecular weight alkanes with a slight odd over even carbon number preference. Pristane to phytane ratios are low (< 1). Centrally branched monomethyl alkanes are prominent constituents. All these features are evident on examination of the chromatographic data in Fig. 2. The n-alkane and isoprenoid compositions also resemble those of the upper, and likely hypersaline, carbonate-dominated, section of the Barney Creek Formation, McArthur Basin (Summons et al. 1988b). The presence of small amounts of extended ($> C_{20}$) isoprenoid hydrocarbons and the low Pr/Ph ratio are particularly notable. These features are often reported in organic matter preserved in anoxic marine sediments, carbonate sediments or from hypersaline environments (e.g. Didyk et al. 1978; McKirdy et al. 1983, 1984; ten Haven et al. 1985, 1988; Brassell et al. 1987) and are likely to reflect input of significant amounts of lipid from archaebacteria (Kates 1978; Langworthy 1985).

The chromatographic expansions and GCMS data originally presented by Fowler and Douglas (1987), Makushina et al. (1978) and Aref'ev et al. (1980) and elaborated using MRM by Summons et al. (1988b) demonstrate that the branching position in the monomethyl alkanes changes as the carbon number increases. The $C_{13}-C_{18}$ homologues have terminally branched isomers predominating while the higher homologues ($C_{20}-C_{25}$) have increasing relative abundances of mid-chain, branched isomers. This also appears to be the case in the Oman oils, but is in contrast to the McArthur sediments which show the same kind of isomer preferences at all carbon numbers. This may be partly due to the fact that the McArthur sediments with most abundant methyl alkanes are shales (as opposed to carbonates) resulting in an increased tendency for acid-catalysed diagenetic rearrangements. Methyl alkanes with branching at the centres of the carbon chains are most often reported in petroleum and sediments from the Proterozoic and are less obvious components of Phanerozoic organic matter, at least after the Early Palaeozoic (Hoering 1981; Klomp 1986; Fowler and Douglas 1987; Summons 1987; Summons et al. 1988 a, b). The ubiquitous occurrence of methyl branching in the fatty acids of extant bacteria (Kaneda 1977; Fulco 1983; Dowling et al. 1986; Parkes and Taylor 1986), the isolation of long-chain monomethyl fatty acids from sponges which have bacterial symbionts (Dasgupta et al. 1984), together with the common observation of mid-chain, branched hydrocarbons in cyanobacteria (Gelpi et al. 1970) and in modern environments dominated by cyanobacteria (Boon and de Leeuw 1987; Ward et al. 1989; Robinson and Eglinton 1990; Shiea et al. 1990), suggest that these compounds have derived from prokaryotic organisms which were particularly abundant in the sedimentary environment.

3.1.2 Steroidal Hydrocarbons

A strong preference for the C_{29} 24-ethylcholestane pseudohomologue (Fig. 3) has been observed before in related Siberian Platform oils (Aref'ev et al. 1980; Fowler and Douglas 1987) and in the Huqf oil of Oman (Grantham 1986; Grantham et al. 1988). It is also significant that, in this case, the rearranged sterane isomers are virtually absent and $\alpha\beta\beta$ isomers predominate over $\alpha\alpha\alpha$, features that are often reported from carbonate and/or hypersaline source environments and have been ascribed to the absence of clays and the presence of specific steroidal precursors (e.g. see McKirdy et al. 1983, 1984; Rullkötter et al. 1984; ten Haven et al. 1986). The ratio of 20S to 20R in the $\alpha\alpha\alpha$ steranes [and 22S to 22R in the 17α(H), 21β(H)-homohopanes] corresponds to a "mature" distribution (Mackenzie et al. 1980).

A significant new observation relates to the distinctive isomer and carbon number preferences of the desmethyl and ring-A alkylated steranes (Fig. 3 and 4). There is a significant proportion of a C_{26} desmethyl pseudohomologue and the ratio of 24-ethylcholestane to 24-methylcholestane to cholestane is approximately 10:2:3 (Fig. 3). In Fig. 4, the same ratio is evident in the series of ring-A methylsteranes ($M^{+\cdot} \rightarrow 231$; 24-Et:24-Me:24-H = 10:1.8:3) and in the steranes with two additional carbon atoms attached to ring-A ($M^{+\cdot} \rightarrow 245$; 24-Et:24-Me:24-H = 10:1.8:3.1). Moreover, the predominant C_{30} methylsteranes in the Siberian Platform oils elute earlier than C_{30} 4-methylsterane isomers (Summons et al. 1987; Goodwin et al. 1989) and at retention times consistent with them having the additional methyl group at C-3β and with a minor component with methylation at C-2α (Summons and Capon 1988). These assignments were confirmed by co-injection with authentic synthetic standards. They are inconsistent with the earlier assignment of a major component of 4-methylsteranes by Fowler and Douglas (1987) although these latter biomarkers may still be present as trace components.

The normal biosynthetic pathway to sterols proceeds via 4,4-dimethyl and 4-methyl intermediates but clearly this does not explain the origin of the methylsteranes in the Siberian Platform oils. The monomethyl series could possibly be derived by rearrangement of 4-methyl precursors or by biomethylation of Δ-2 sterenes which are early products in the diagenesis of desmethylsterols (e.g. Mackenzie et al. 1982). The exact structures of apparent "dimethyl" steranes (i.e. steranes with mass spectra showing a base peak at m/z 245, as compared to methylsteranes with m/z 231 and desmethylsteranes with m/z 217) has now been established as 3β-ethylsteranes (Summons et al. unpubl. results), by comparison with synthetic standards. A logical implication of finding a combination of 3β-methyl- and 3β-esthylsteranes is that they have been formed by one- and two-step biomethylation processes acting on Δ^2-sterenes, possibly analogous to the way steroid side chains are alkylated. The discovery of such a ring-alkylation process in fossil steroids is interesting because there appears to be no analogue in the steroid metabolism of extant organisms.

A high abundance of C_{26} desmethylsteranes and the presence of methylsteranes and ethylsteranes which have C-24 alkylation preferences corresponding to desmethylsteranes have been observed in Proterozoic sediments of the Chuar Group (Summons et al. 1988 a) and Palaeozoic oils from Canada (Summons and Powell 1987). In these cases, the most abundant steranes were those bearing no alkyl substituent at C-24. The strong preference for 24-ethylsteranes in Siberian Platform samples, as well as those in the Hugf Group oils of Oman (Grantham 1986) and other

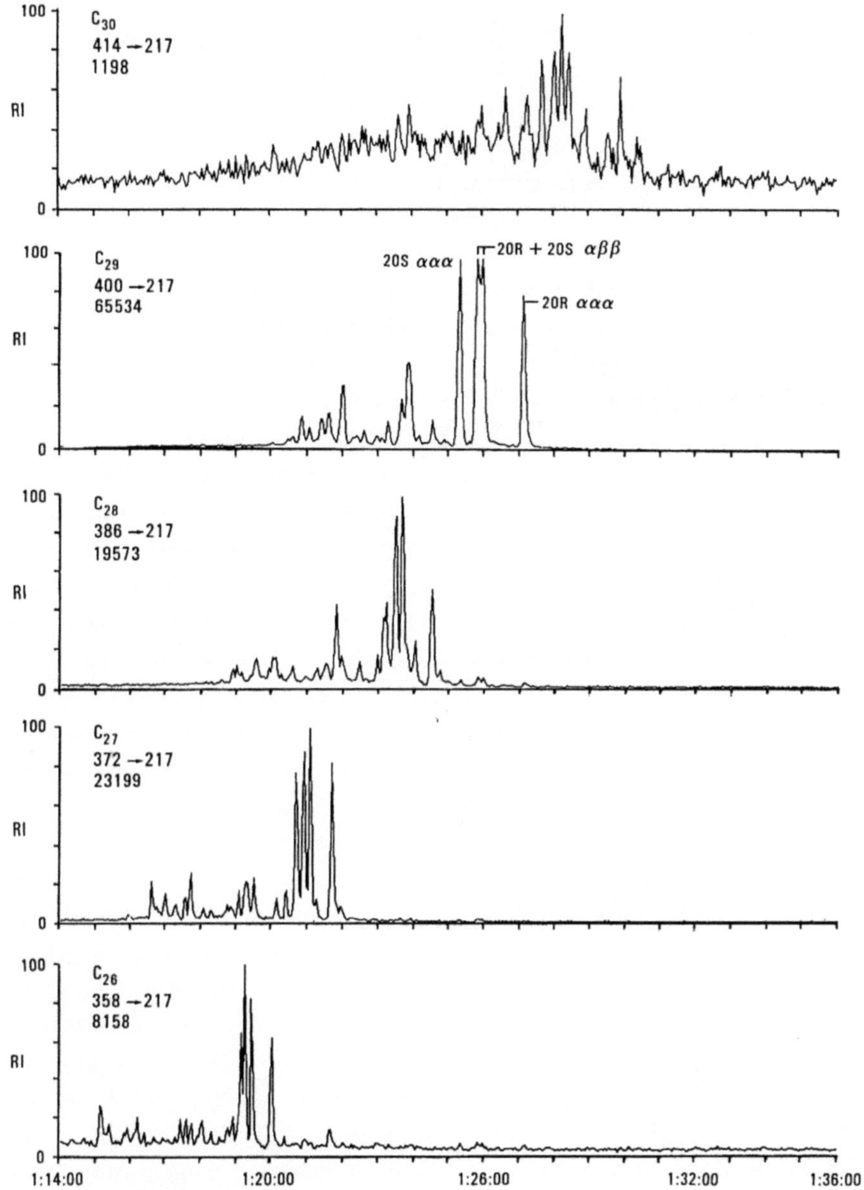

Fig. 3. Sterane distributions from the Danilovskaya oil (BMR #320). The chromatograms were obtained by metastable reaction monitoring (MRM) in GCMS of molecular ions fragmenting to daughter ions at m/z 217. The trace identifiers show relative abundance normalized to the the most intense reaction. GCMS was carried out using an HP Ultra-1 methylsilicone capillary column. Assignments provide details of C-20 stereochemistry (*R* or *S*) and βα, ααα and αββ denote 13β(H),17α(H)-diasteranes, 5α(H),14α(H),17α(H)- and 5α(H),14β(H),17β(H)-steranes, respectively

ancient materials (e.g. Moldowan et al. 1985), and their very high relative abundance, is a feature which could arise from the input of sterols from green algae (e.g. Volkman 1986). Although there have been reports of some cyanobacteria having traces of C_{29} desmethylsterols (e.g. de Souza and Nes 1968) proof of de novo biosynthesis in prokaryotes has not been demonstrated (Ourisson et al. 1987).

3.1.3 Triterpenoid Hydrocarbons

The normal series of 17α(H), 21β(H)-hopanes predominate in the distribution of triterpenoids (Fig. 5). There is a high relative abundance of components with 28 carbon atoms of which the most abundant peak has been assigned as a mixture of isomers of 28,30-bisnorhopane (Moldowan et al. 1984), by its co-

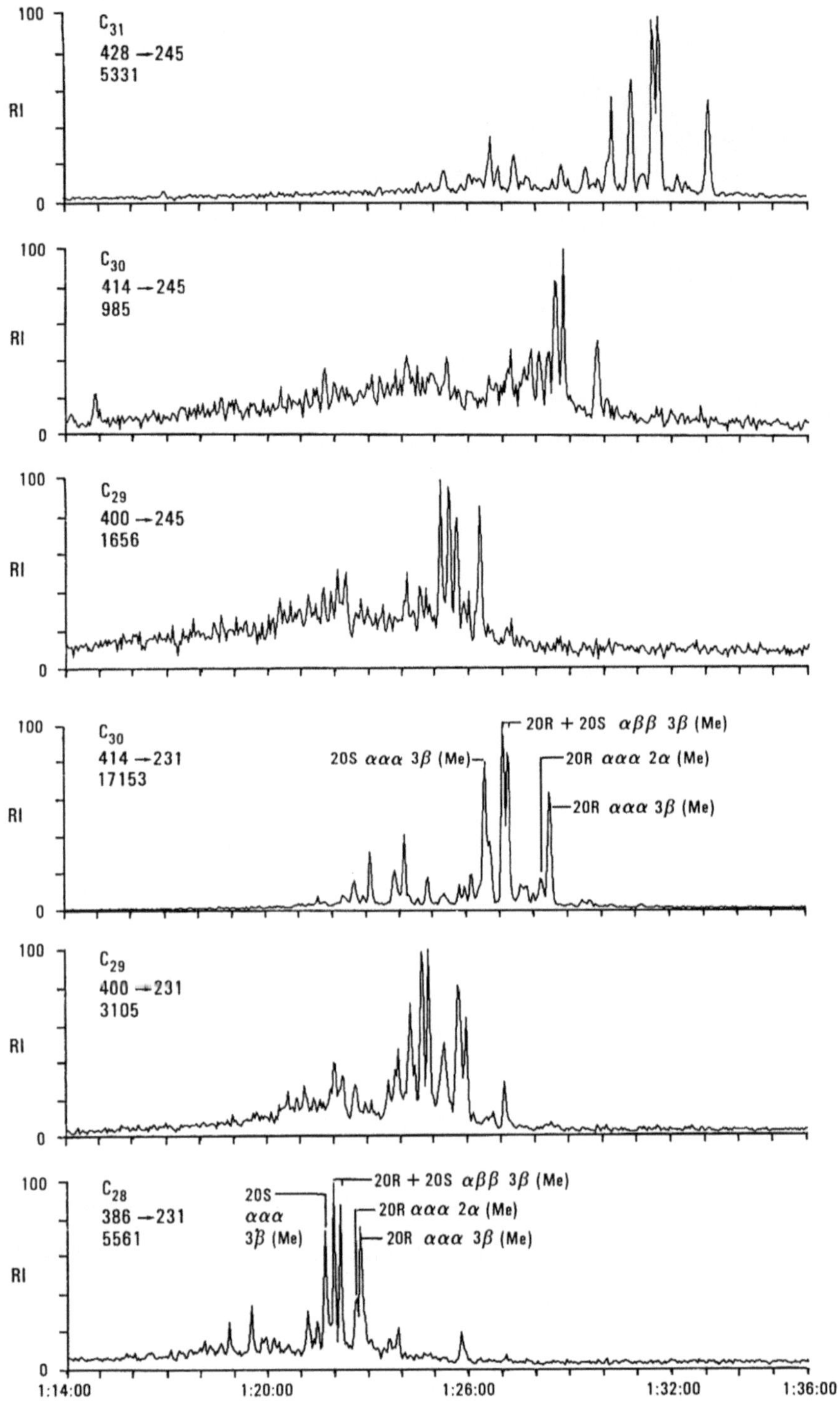

Fig. 4. Methylsterane and ethylsterane distributions from the same GC-MS experiment as shown in Fig. 3. The chromatograms were obtained by metastable reaction monitoring (MRM) of molecular ions fragmenting to daughter ions at m/z 231 and 245. The trace identifiers show relative abundance normalized to the most intense reaction and are proportional to those in Fig. 3. Details of stereochemistry are provided for the major isomers of 3β-methyl-24-ethylcholestane and the minor 2α-methyl analogue

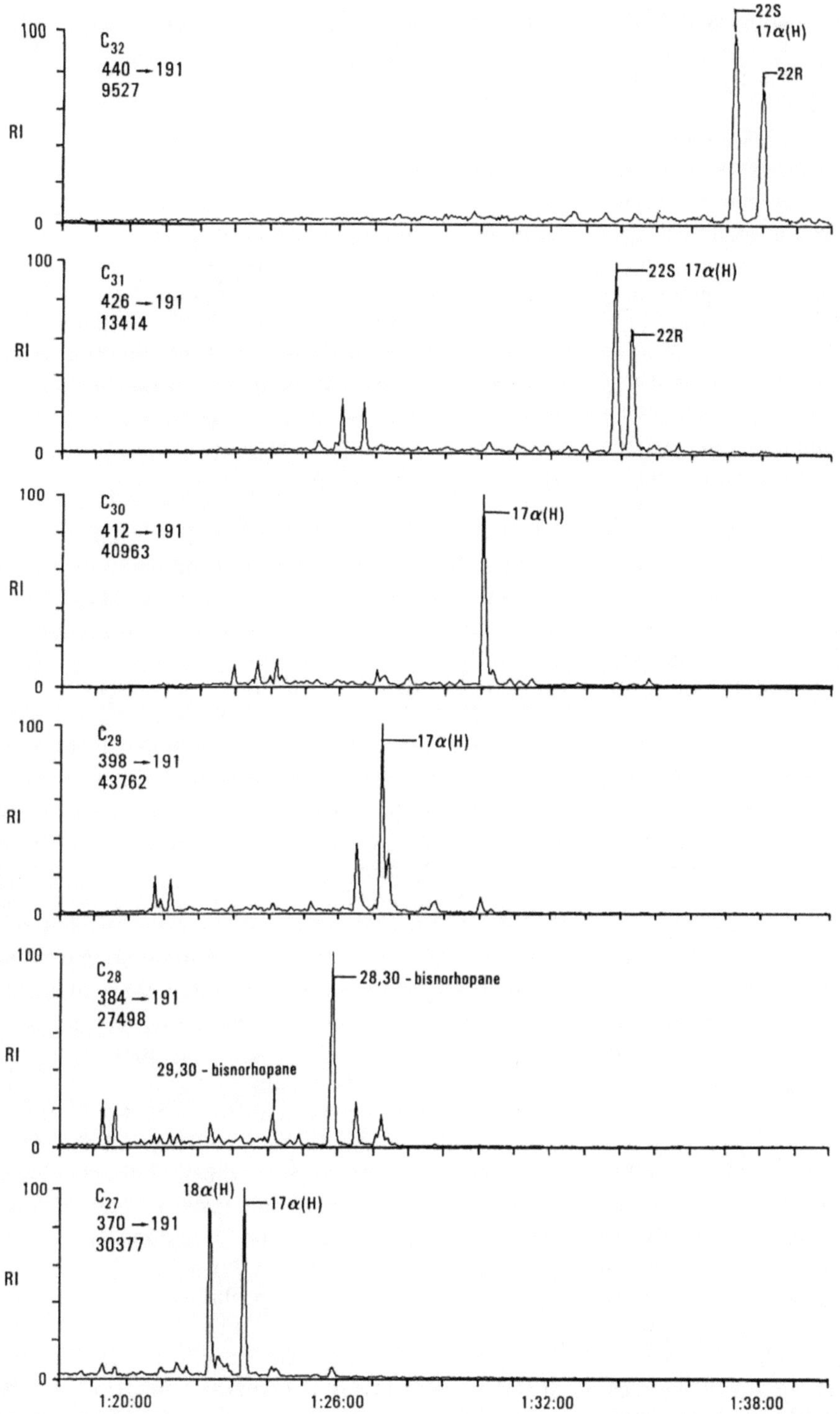

Fig. 5. MRM chromatograms (e.g. $M^{+\cdot} \rightarrow 191$) for $C_{31}-C_{27}$ triterpanes of the Danilovskaya oil (BMR ♯320). The $17\alpha(H)21\beta(H)$-hopanes are denoted $17\alpha(H)$. *R* and *S* refer to the stereochemistry at C-22. 25-norhopane is the major peak eluting before 30-norhopane in the chromatogram of C_{29} components. Gammacerane is the last peak evident in the trace for C_{30} components

elution with the equivalent peak in samples of North Sea oil (Grantham et al. 1980) and the Monterey Formation (Katz and Elrod 1983). Another peak has been assigned as 29,30-bisnorhopane by virtue of its elution position relative to the same compound tentatively identified in Palaeozoic oils from Canada (Summons and Powell 1987). The presence of 29,30-bisnorhopane correlates with the dominance of norhopane over hopane in the normal $\alpha\beta$ hopane series. This pattern of occurrence has been noted previously (Seifert et al. 1984; Price et al. 1988) and may possibly be ascribed to the presence of a series of 30-norhopanes which includes the n-propyl substituted C_{30} 17α(H)-hopane identified by Seifert et al. (1984). The C_{29} component would be common to this and the "normal" 17α(H)-hopane series, thus explaining its high relative abundance compared to C_{30}. Also, in a 30-nor series it is mechanistically feasible to form a C_{28} triterpane (methyl at C-21) by a single bond cleavage of the hopane side chain. This compositional feature is common in triterpane distributions of carbonate petroleum source rocks (cf. McKirdy et al. 1983, 1984; Zumberge 1984; Summons and Powell 1987; Price et al. 1988). The chromatograms presented in Fig. 5 reveal numerous other components. In particular, the homologous series which appear as early eluting doublets are probable tricyclic terpanes inadvertantly detected through reactions of ions in the natural carbon isotope cluster. The C_{29} compound which elutes mid-way between 28,30-bisnorhopane and 17α(H)-norhopane is tentatively identified as 25-norhopane. We have not examined these oils for the other members of this series, although the presence of this compound attests to an episode of biodegradation in the oil reservoir. A trace amount of gammacerane can be observed in the chromatogram of the C_{30} hopanes.

A series of methylhopanes is present with the major member eluting at the trailing edge of, and in about 10% of the abundance of, the corresponding 17α(H), 21β(H)-hopane (Fig. 6). The C_{31} member of the series co-elutes with a synthetic standard of 2α(Me)-hopane (Summons and Jahnke 1990a, b) and thus it is likely that the other members of the series share this same ring-A substitution. The same series of methylhopanes has been reported on numerous occasions and is often associated with sediments from carbonate or hypersaline environments (see e.g. Summons and Powell 1987; Price et al. 1988). Both 2β-methyl and 3β-methyl hopanoids have been isolated from the lipids of extant bacteria (Bisseret et al. 1985; Zundel and Rohmer 1985a, b) and the fossil 2α(Me) series are likely to be derived from biogenic 2β(Me) precursors. Additional unidentified components are evident in the C_{28}, C_{30} and C_{31} methylhopane reaction chromatograms and add to the distinctiveness of these old oils.

3.1.4 Aryl Isoprenoids

An homologous series of aryl isoprenoids is present in the aromatic hydrocarbon fraction of the Siberian Platform oils (Fig. 7). These compounds, which have now been identified in a number of Phanerozoic sediments and oils, are particularly abundant in some unusual Palaeozoic carbonate/evaporite-derived oils from Ontario and Alberta and have been ascribed to an input of carotenoids characteristic of green sulfur bacteria, i.e. *Chlorobium* sp. (Summons and Powell 1986, 1987). Diagenetic hydrogenation of the isoprenoid chain, and cleavages along that chain, could produce the observed mixture from precursor isorenieratene (or possibly chlorobactene). The isomer distributions in the respective sources are, however, quite different. In the Canadian oils, a total of three isomers are detected but there is a very strong preference for a single isomer for each homologue, that is the biological (2,3,6-trimethyl) substitution pattern of the benzene ring and tail-tail isoprenoid branching in the acyclic chain. Many isomers are present in the Siberian Platform oils, as illustrated in Fig. 7. Co-injection experiments (data not shown) with the Devonian Pine Point Zama oil (Summons and Powell 1987) showed that one of the Danilovskaya components co-eluted in the position of the 2,3,6-trimethylalkyl benzene isomers. Reactions of metastable molecular ions of each pseudohomologue (e.g. m/z 190, 204, 218, 232 etc.), fragmenting to the characteristic m/z 134 daughter ions, demonstrate the variety of isomers present in the Siberian Platform oils, although it should be noted that all the compounds are not necessarily trimethylalkylbenzenes. Earlier work, where unambiguously synthesized standards were compared with components in the Canadian oils (Summons and Powell 1987), indicated that minor components probably did not arise through isomerization of aromatic methyl groups. This observation leads to the conclusion that the mixture must involve in-chain isomerization, the variety of which has to be accounted for in mode(s) of diagenetic alteration. Bridging and binding to kerogen through carbon-sulphur bonds has recently been implicated in the preservation and isomerization of other isoprenoid biomarkers, particularly in hypersaline (sulphate) depositional environments (e.g. Sinninghe Damste et al. 1987; Sinninghe Damste 1988). It is possible that similar mechanisms are involved in the

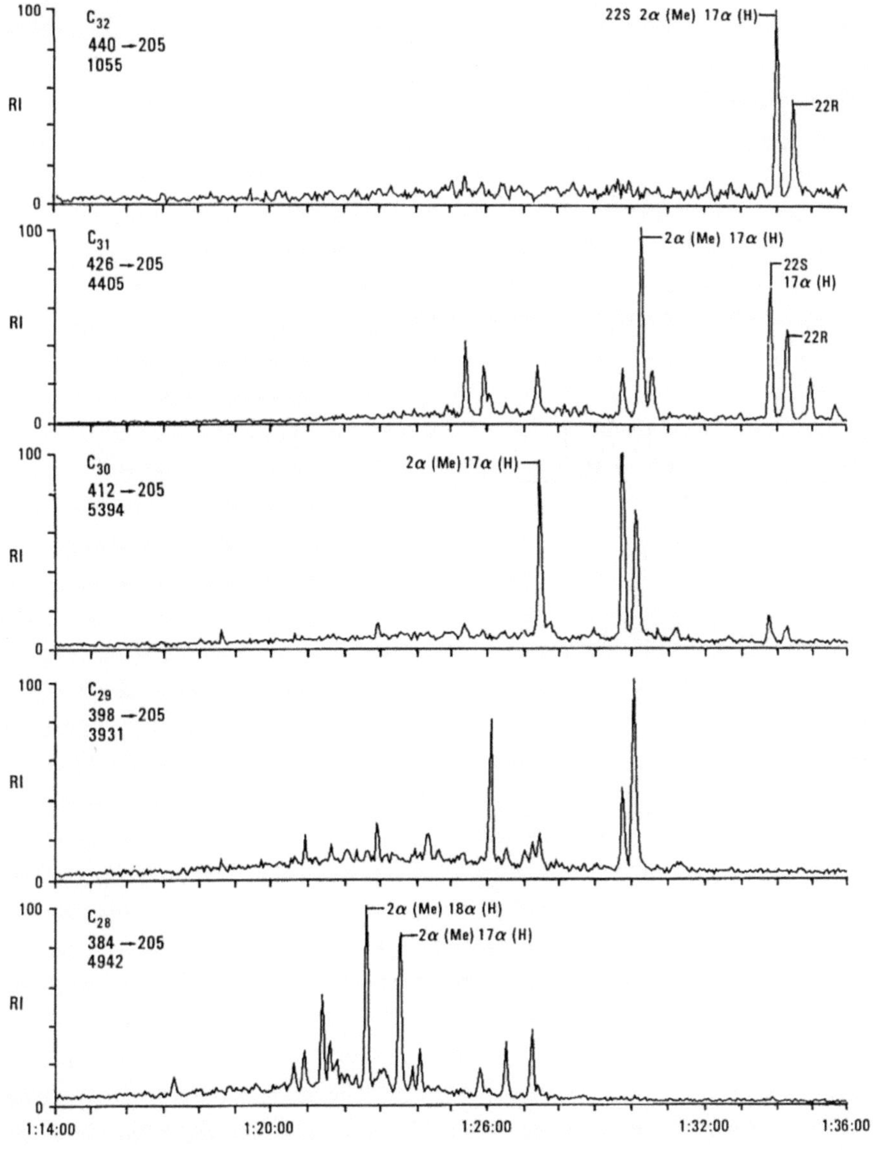

Fig. 6. MRM chromatograms (e.g. $M^{+\cdot} \rightarrow 205$) for $C_{32}-C_{28}$ triterpanes of the Danilovskaya oil (BMR #320). $17\alpha(H)21\beta(H)$-homohopanes appear in the C_{31} trace while major peaks in this and the other traces correspond to 2α-methyl-$17\alpha(H)$, $21\beta(H)$-hopanes. Undesignated peaks are presently unidentified

preservation of carotenoids and the consequent generation of aryl isoprenoids and, under certain conditions, their isomerisation to the variety of structures evident here.

3.2 Carbon Isotope Anomalies

Table 1 shows the carbon isotope signature of whole oils and their subfractions. The values reflect an overall depletion in ^{13}C to an extent unlike anything reported from the Phanerozoic. Although there is a small variation between the samples, the values within each sample closely follow a kerogen evolution trend (Stahl 1978) with the subfractions tending to isotopically heavier values with increasing polarity. There have been numerous recent reports of strong, global fluctuations in the $\delta^{13}C$ values of both carbonate and organic carbon just prior to the Cambrian (e.g. Veizer and Hoefs 1976; Knoll et al. 1986; Magaritz et al. 1986; Tucker 1986; Lambert et al. 1987) whereas earlier in the Proterozoic there had been a slight but smooth trend toward increasing heaviness in organic carbon (Strauss et al. this Vol.).

Moreover, the organic matter in the Late Proterozoic oils of Oman has also been reported as anoma-

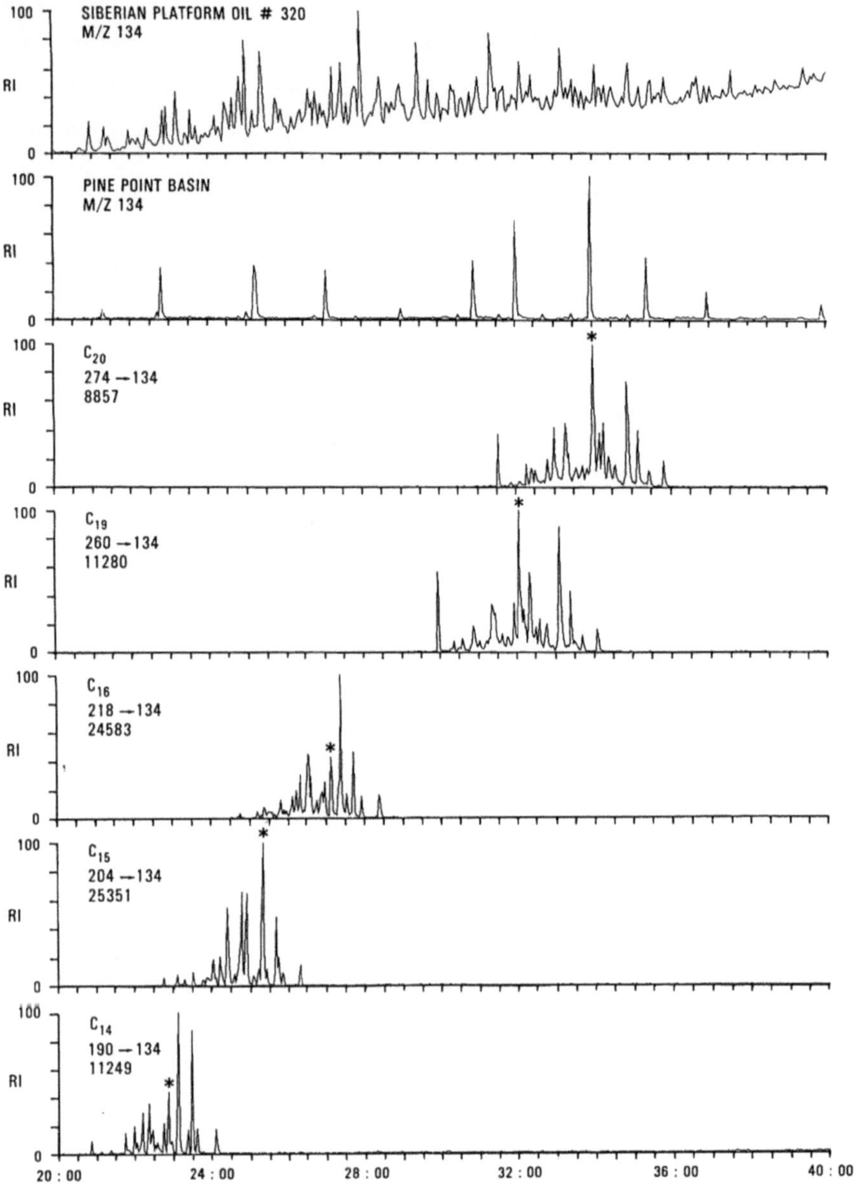

Fig. 7. SIR chromatograms comparing the m/z 134 ion current for oil #320 with the Zama oil from the Pine Point Basin, Canada. The remaining traces are MRM chromatograms showing the distribution of selected aryl isoprenoid homologues.

Table 1. Compositional and carbon isotope data ($\delta^{13}C$ ‰ PDB) for three Siberian Platform oils (fraction BP > 200 °C)

Sample No.	Whole	Saturates		Aromatics		Polars	
	$\delta^{13}C$	%	$\delta^{13}C$	%	$\delta^{13}C$	%	$\delta^{13}C$
319	−34.34	56.7	−34.77	25.4	−34.52	13.9	−33.99
320	−34.53	61.8	−34.66	21.1	−34.03	15.5	−33.03
321	−34.15	65.7	−34.99	16.4	−33.39	15.7	−32.88

lously light (Grantham et al. 1988). It therefore seems likely, from the concurrence of features in the hydrocarbon assemblage, that both deposits were formed in closely similar environments and possibly at about the same time. If the carbon isotope anomaly reflects a global secular trend, it may correlate with a similar trend to very light organic and inorganic carbon signals observed in the Svalbard and East Greenland sequences (Knoll et al. 1986). The possible reasons for such global fluctuations have been discussed at length (see above references). The most plausible explanations for the Siberian Platform carbon anomaly are that some previously formed organic carbon has been oxidized and refixed or that unusually high pCO_2 allowed increased discrimination against ^{13}C. The former may reflect conditions of restricted circulation and poor mixing of carbon pools prevailing locally or globally. Elevated atmospheric CO_2 levels would have been a global phenomenon.

Conclusions

Late Proterozoic oils of the Siberian Platform have distinctive compositional features, several of which suggest that they have been sourced in carbonate sediments deposited in a probable hypersaline environment. This study has revealed an exceptional combination of characteristics which include enigmatic 3β-methyl and 3β-ethylsteranes, several uncommon triterpanes and an ususual depletion in ^{13}C. The branched alkanes and pentacyclic triterpenoids, including methylhopanes, 30-norhopanes and 28,30-bisnorhopane, almost certainly arise from input of bacterial and/or cyanobacterial organic matter while the ring-A alkylated steranes may be the products of bacterial alteration of algal sterols. The very strong predominance of 24-ethylcholestanes in the samples suggest that green algae may have been important phototrophs in the environment where this organic matter was formed. Although the source of the oils is not presently known, their carbon isotope signatures may eventually enable correlation with a probable global carbon isotope anomaly which is evident in accurately dated sediments of comparable age (620–570 Ma) in Oman and in Late Proterozoic sediments from other continents.

References

Aref'ev OA, Zabrodina MN, Makushina VM, Petrov AA (1980) Relic tetra- and pentacyclic hydrocarbons in ancient petroleums of the Siberian Platform. Isv Akad SSR Ser Geol 10:122–129 (in Russian)

Bisseret P, Zundel M, Rohmer M (1985) Prokaryotic triterpenoids 2. 2β-methylhopanoids from *Methylobacterium organophilum* and *Nostoc muscorum*, a new series of prokaryotic triterpenoids. Eur J Biochem 150:29–34

Boon JJ, De Leeuw JW (1987) Organic geochemical aspects of cyanobacterial mats. In: Fay F, von Baalen C (eds) The Cyanobacteria. Elsevier, Amsterdam, pp 471–492

Brassell SC, Eglinton G, Howell VJ (1987) Palaeoenvironmental assessment of marine organic-rich sediments using molecular organic geochemistry. In: Brooks J, Fleet AJ (eds) Marine petroleums source rocks. Geol Soc Spec Publ 26. Blackwell, Oxford, pp 79–98

Dasgupta A, Ayanoglu E, Djerassi C (1984) Phospholipid studies of marine organisms: new branched fatty acids from *Stronylophora durissima*. Lipids 19:768–776

De Souza NJ, Nes WR (1968) Sterols: isolation from a blue-green alga. Science 162:363–365

Didyk BM, Simoneit BRT, Brassell SC, Eglinton G (1978) Organic geochemical indicators of palaeoenvironment and conditions of sedimentation. Nature (London) 272:216–222

Dickas AB (1986) Comparative Precambrian stratigraphy along the Mid-Continent Rift. Am Assoc Petrol Geol Bull 70:225–238

Dowling NJE, Widdel F, White DC (1986) Phospholipid ester-linked fatty acid biomarkers of acetate oxidising sulfate reducers and other sulfide forming bacteria. J Gen Microbiol 132:1815–1825

Fowler MG, Douglas AG (1987) Saturated hydrocarbon biomarkers in oils of Late Precambrian age from eastern Siberia. Org Geochem 11:201–203

Fulco AJ (1983) Fatty acid metabolism in bacteria. Prog Lipid Res 22:133–160

Gelpi E, Schneider H, Mann J, Oro J (1970) Hydrocarbons of geochemical significance in microscopic algae. Phytochemistry 9:603–612

Goodwin NS, Mann AL, Patience RL (1989) Structure and significance of C_{30} 4-methylsteranes in lacustrine shales and oils. Org Geochem 12:495–506

Grantham PJ (1986) The occurrence of unusual C_{27} and C_{29} sterane predominances in two types of Oman crude oil. Org Geochem 9:1–10

Grantham PJ, Posthuma J, De Groot K (1980) Variation and significance of the C_{27} and C_{28} triterpane content of a North Sea core and various North Sea crude oils. In: Maxwell JR, Douglas AG (eds) Advances in organic geochemistry 1979. Pergamon, Oxford, pp 29–38

Grantham PJ, Lijmbach GWM, Posthuma J, Hughes Clarke MW, Willink RJ (1988) Origin of crude oils in Oman. J Petrol Geol 11:61–79

ten Haven HL, de Leeuw JW, Schenck PA (1985) Organic geochemical studies of a Messinian evaporitic basin, northern Appennines (Italy). I. Hydrocarbon biological markers for a hypersaline environment. Geochim Cosmochim Acta 49:2181–2219

ten Haven HL, de Leeuw JW, Peakman TM, Maxwell JR (1986) Anomalies in steroid and hopanoid maturity indices. Geochim Cosmochim Acta 50:853–855

ten Haven HL, de Leeuw JW, Sinninghe Damste JS, Schenck PA, Palmer SE, Zumberge JE (1988) Application of biological markers in the recognition of palaeo-hypersaline environments. In: Kelts K, Fleet A, Talbot M (eds) Lacustrine petroleum source rocks. Geol Soc Spec Publ 40. Blackwell, Oxford, pp123–130

Hayes JM, Kaplan IR, Wedeking KW (1983) Precambrian organic geochemistry, preservation of the record. In: Schopf JW (ed) Earth's earliest biosphere: its origin and evolution. Univ Press, Princeton, pp 93–134

Hoering TC (1981) Monomethyl acyclic hydrocarbons in petroleum and rock extracts. Carnegie Inst Washington Yearb 80:389–393

Hoering TC, Navale V (1987) A search for molecular fossils in the kerogen of Precambrian sedimentary rocks. Precambrian Res 34:247–267

Hughes Clarke MW (1988) Stratigraphy and rock unit nomenclature in the oil producing area of interior Oman. J Petrol Geol 11:5–60

Jackson MJ, Powell TG, Summons RE, Sweet IP (1986) Hydrocarbon shows and petroleum source rocks in sediments as old as 1.7×10^9 years. Nature (London) 322:727–729

Jackson MJ, Sweet IP, Powell TG (1988) Petroleum geology and geochemistry of the Middle Proterozoic, McArthur Basin, northern Australia. Petroleum potential. Aust Petrol Explor Assoc J 28:283–302

Kaneda T (1977) Fatty acids of the genus *Bacillus*: an example of branched chain preference. Bacter Rev 41:391–418

Kates M (1978) The phytanyl ether-linked polar lipids and isoprenoid neutral lipids of extremely halophilic bacteria. Prog Chem Fats Other Lipids 15:301–342

Katz BJ, Elrod LW (1983) Organic geochemistry of DSDP 467, offshore California, Middle Miocene to Lower Pliocene strata. Geochim Cosmochim Acta 47:267–275

Klomp UC (1986) The chemical structure of a pronounced series of iso-alkanes in South Oman crudes. In: Leythaeuser D, Rullkötter J (eds) Advances in organic geochemistry 1985. Pergamon, Oxford, pp 807–814

Knoll AH, Hayes JM, Kaufman AJ, Swett K, Lambert IB (1986) Secular variation in carbon isotope ratios from Upper Proterozoic successions of Svalbard and East Greenland. Nature (London) 321:832–838

Kontorovich AE, Surkov VS, Trofimuk AA (eds) (1981) Geology of petroleum and gas of the Siberian Platform. Nedra, Moscow. 552 pp (in Russian)

Lambert IB, Walter MR, Zang W, Lu S, Ma G (1987) Paleoenvironment and carbon isotope stratigraphy of Upper Proterozoic carbonates of the Yangtze Platform. Nature (London) 325:140–142

Langworthy TA (1985) Lipids of Archaebacteria. In: Woese CR, Wolfe RS (eds) The bacteria. A treatise of structure and function, vol. 8. Academic Press, New York, London, pp 459–497

Mackenzie AS, Patience RL, Maxwell JR, Vandenbroucke M, Durand B (1980) Molecular parameters of maturation in the Toarcian shales, Paris Basin, France. I. Changes in the configuration of acyclic isoprenoid alkanes, steranes and triterpanes. Geochim Cosmochim Acta 44:1709–1721

Mackenzie AS, Brassell SC, Eglinton G, Maxwell JR (1982) Chemical fossils: the geochemical fate of steroids. Science 217:491–504

Magaritz M, Holser WT, Kirschvink J (1986) Carbon-isotope events across the Precambrian/Cambrian boundary on the Siberian Platform. Nature (London) 320:258–259

Makushina VM, Aref'ev OA, Zabrodina MN, Petrov AIA (1978) New relic alkanes of petroleums. Neftekhimiya 18:847–854 (in Russian)

McKirdy DM, Aldridge AK, Ypma PJM (1983) A geochemical comparison of some crude oils from pre-Ordovician carbonate rocks. In: Bjorøy M (ed) Advances in organic geochemistry 1981. John Wiley & Sons, New York Chichester, pp 99–107

McKirdy DM, Kantsler AS, Emmett JK, Aldridge AK (1984) Hydrocarbon genesis and organic facies in Cambrian carbonates of the eastern Officer Basin, South Australia. In: Palacas JG (ed) Petroleum geochemistry and source rock potential of carbonate rocks. Am Assoc Petrol Geol Stud Geol 18. Tulsa, Oklahoma, pp 13–32

Meyerhoff AA (1980) Geology and petroleum fields in Proterozoic and Lower Cambrian strata, Lena-Tunguska Petroleum Province, eastern Siberia. In: Assoc Petrol Geol Mem 30. Tulsa, Oklahoma, pp 225–252

Moldowan JM, Seifert WK, Arnold E, Clardy J (1984) Structure proof and significance of stereoisomeric 28,30-bisnorhopanes in petroleum source rocks. Geochim Cosmochim Acta 48:1651–1661

Moldowan JM, Seifert WK, Gallegos EJ (1985) Relationship between petroleum composition and depositional environment of petroleum source rocks. Am Assoc Petrol Geol Bull 69:1255–1268

Ourisson G, Rohmer M, Poralla K (1987) Prokaryotic hopanoids and other polyterpenoid sterol surrogates. Annu Rev Microbiol 41:301–333

Robinson N, Eglinton G (1990) Lipid chemistry of Icelandic hot spring microbial mats. Org Geochem 15:291–298

Shiea J, Brassell SC, Ward DM (1990) Mid-chain branched mono- and dimethyl alkanes in hot spring cyanobacterial mats: a direct source for branched alkanes in ancient sediments? Org Geochem 15:223–231

Stahl WJ (1978) Source rock-crude oil correlation by isotopic type curves. Geochim Cosmochim Acta 42:1573–1577

Summons RE (1987) Branched alkanes from ancient and modern sediments: isomer discrimination by GC/MS with multiple reaction monitoring. Org Geochem 11:281–290

Summons RE, Capon RJ (1988) Fossil steranes with unprecedented methylation in ring-A. Geochim Cosmochim Acta, 52:2733–2736

Summons RE, Jahnke LL (1990a) Hopenes and hopanes methylated in ring-A: correlation of the hopanoids of extant methylotrophic bacteria with their fossil analogues. In: Moldowan JM, Albrecht P, Philp RP (eds) Biomarkers in sediments and petroleum. (in press)

Summons RE, Jahnke LL (1990b) Identification of the methylhopanes in sediments and petroleum. Geochim Cosmochim Acta, 54:247–251

Summons RE, Powell TG (1986) Chlorobiaceae in Palaeozoic seas revealed by biological markers, isotopes and geology. Nature (London) 319:763–765

Summons RE, Powell TG (1987) Identification of aryl isoprenoids in source rocks and crude oils: biological markers for the green sulphur bacteria. Geochim Cosmochim Acta 51:557–566

Summons RE, Volkman JK, Boreham CJ (1987) Dinosterane and other steroidal hydrocarbons of dinoflagelate origin in sediments and petroleum. Geochim Cosmochim Acta 51:3075–3082

Summons RE, Brassell SC, Eglinton G, Evans EJ, Horodyski RJ, Robinson N, Ward DM (1988a) Distinctive hydrocarbon biomarkers from fossiliferous sediment of the Late Proterozoic Walcott Member, Chuar Group, Grand Canyon, USA. Geochim Cosmochim Acta 52:2625–2637

Summons RE, Powell TG, Boreham CJ (1988 b) Petroleum geology and geochemistry of the Middle Proterozoic McArthur Basin, northern Australia. III. Composition of extractable hydrocarbons. Geochim Cosmochim Acta 52:1747–1763

Tucker ME (1986) Carbon isotope excursions in Precambrian/Cambrian boundary beds, Morocco. Nature (London) 319:48–50

Veizer J, Hoefs J (1976) The nature of $^{18}O/^{16}O$ and $^{13}C/^{12}C$ secular trends in sedimentary carbonate rocks. Geochim Cosmochim Acta 40:1387–1395

Volkman JK (1986) A review of sterol markers for marine and terrigenous organic matter. Org Geochem 9:83–99

Warburton GA, Zumberge JE (1983) Determination of petroleum sterane distributions by mass spectrometry with selective metastable ion monitoring. Anal Chem 55: 123–126

Ward DM, Shiea J, Zeng YB, Dobson G, Brassell SC, Eglinton G (1989) Lipid biochemical markers and the composition of microbial mats. In: Cohen Y, Rosenburg E (eds) Microbial mats: ecological physiology of benthic microbial communities. Am Soc Microbiol, Washington, DC

Zundel M, Rohmer M (1985 a) Prokaryotic triterpenoids. 1. 3β-methylhopanoids from *Acetobacter* sp. and *Methylococcus capsulatus*. Eur J Biochem 150:23–27

Zundel M, Rohmer M (1985 b) Prokaryotic triterpenoids. 3. The biosynthesis of 2β-methylhopanoids and 3β-methylhopanoids of *Methylobacterium organophilum* and *Acetobacter pasteurianus* spp. *pasteurianus*. Eur J Biochem 150:35–39

Zumberge JE (1984) Source rocks of the La Luna Formation (Upper Cretaceous) in the Middle Magdalena Valley, Columbia. In: Palacas JG (eds) Petroleum geochemistry and source rock potenital of carbonate rocks. Am Assoc Petrol Geol Stud Geology 18. Tulsa, Oklahoma, pp 127–134

Biomarkers and Other Hydrocarbons in Upper Sinian Stromatolitic Dolostones from Southwest China

Fan Pu, Li Jinggui, Meng Qianxiang, Yu Xinke and Li Zhenxi [1]

[1] Lanzhou Institute of Geology, Academia Sinica, Lanzhou 730000, China

Abstract

Upper Sinian stromatolitic dolostones from southwest China have been analyzed for biomarkers in their aliphatic and aromatic hydrocarbon fractions. The biomarkers identified are mainly as follows: bicyclic sesquiterpanes, tricyclic terpanes, pentacyclic triterpanes, 8,14-secohopanes, hopane sulfides, alkylcyclohexanes, steranes, triaromatic steroids, aromatized hopanes, and long-chain alkylbenzenes. Other aromatic hydrocarbons which may be used as maturity indicators were also identified, including naphthalenes, phenanthrenes and biphenyls. The relationships between the aforementioned compounds and their microbial precursors, depositional environment and thermal maturity are discussed.

1 Introduction

During field geological investigations of the North China Platform and southwest China, undertaken by the senior author in 1965, 1967 and 1977, potential source rocks for oil and gas in Sinian and Paleozoic strata were sampled for organic geochemical analysis. This work demonstrated the widespread occurrence of source rocks which, in some cases, have generated sufficient hydrocarbons to form commercial oil or gas accumulations. Among the potential source rock units is the Dengying Formation of Sinian (Late Proterozoic) age. This formation comprises stromatolitic dolostones and algal siliceous dolomites. Recently, exploration has addressed the hydrocarbon potential of Sinian strata, resulting in the discovery of a large gas reservoir in carbonates of the Dengying Formation at Weiyuan in the Sichuan Basin, Southwest China (Xu Yongchang et al., this Vol.).

In this study eight core samples of the Dengying Formation from 3109–3662 m depth in the Weiyuan well were collected for solvent extraction and analysis of their saturated and aromatic hydrocarbons by gas chromatography-mass spectrometry (GC–MS). The aim of the investigation was to identify biomarkers and other compounds which provide specific information on (1) the types of precursor microbiota; (2) the nature of the depositional environment; and (3) the thermal maturity of the organic matter preserved in these dark grey Sinian dolostones.

2 Age, Biota and Depositional Environment of the Dengying Formation

Sinian sediments are widespread throughout southwest China where in places they unconformably overlie basement granites with an isotopic age of 840 million years (Ma). These sediments are divided into four stratigraphic units which, in order of decreasing age, are the Chengjiang Formation, the Niantuo Formation (glacial), the Doushantuo Formation and the Dengying Formation. Zircons in tuff from the Niantuo Formation, a correlative of the Chengjiang Formation in the Yangtze Gorges area of Hubei Province (Xing Yusheng et al. 1984), give a U-Pb age of 740 ± 16 Ma (Ma Guogan et al. 1982, cited by Awramik et al. 1985). The Dengying Formation therefore is probably < 700 Ma old.

Dolostones of the Dengying Formation contain abundant fossil remains including filamentous cyanobacteria, calcified cyanobacterial mucilage, stromatolites, vendotaenid algae and, in shaly intervals, sphaeromorphic acritarchs (Cao Ruiji and Zhao Wenjie 1978; Zhao Ziqiang et al. 1980; Cao Ruiji and Xue Yaosong 1983).

The abundance and variety of microfossils, ribbon-like carbonaceous macrofossils and stromatolites suggest a rich and productive biota comprising both cyanobacteria and algae. Sedimentological evidence from the Sichuan-Guizhou Platform (e.g. vadose pisoids, desiccation cracks, pseudomorphs after gypsum) is consistent with the development of hypersaline conditions during deposition of the cryp-

M. Schidlowski et al. (Eds.)
Early Organic Evolution: Implications for Mineral and Energy Resources

tagal dolostones of the lower Dengying Formation (Cao Ruiji and Xue Yaosong 1983).

3 Experimental

All core samples and extracts were well protected from external contamination during grinding and the subsequent analytical procedures. The powdered rock was extracted with chloroform and the extract (chloroform bitumen A) was separated by silica/alumina chromatography into fractions containing aliphatic hydrocarbons, aromatic hydrocarbons and polar compounds. The aliphatics were eluted with hexane, and the aromatics with benzene. The saturated and aromatic hydrocarbon fractions were analyzed by GC-MS using a computerized Hewlett Packard 5890A quadrupole mass spectrometer and data system. The operating conditions were: SE-54 fused silica column (50 m × 0.32 mm i.d.); carrier gas, high-purity He; injector temperature, 300 °C; GC temperature program, 80–200 °C at 6° min^{-1} and 200–300 °C at 3° min^{-1}; ion source temperature, 250 °C; electron energy, 70 eV; and capillary column directly connected to the ion source (EI). Splitless GC sample injection and selected ion monotoring (SIM) were employed because of the low concentrations of extractable organic matter in the samples. Retention times and full-scan mass spectra were used to identify individual compounds.

4 Concentration and General Characteristics of Organic Matter in Stromatolitic Dolostones

Total organic carbon (TOC) values of the core samples are in the range 0.03–0.3%, whereas the concentrations of extractable bitumen are very low (<100 ppm). Saturated hydrocarbons comprise up to 84% of the bitumen which in turn is mostly carbon and hydrogen (90–95%).

These dolomitic carbonates are thermally very mature (kerogen R_0 = 2.6–3.25%). Their indigenous light hydrocarbons have been cracked to natural gas while the heavy components of their bitumen have been almost entirely converted to insoluble pyrobitumen (e.g. anthraxolite, Fu Jiamo et al. 1984). Small vugs and cavities within the dolostones contain carbonized bitumen which is hard and lustrous and has the elemental composition: C = 80.1%, H = 2.93%, S = 5.86%; atomic H/C = 0.44.

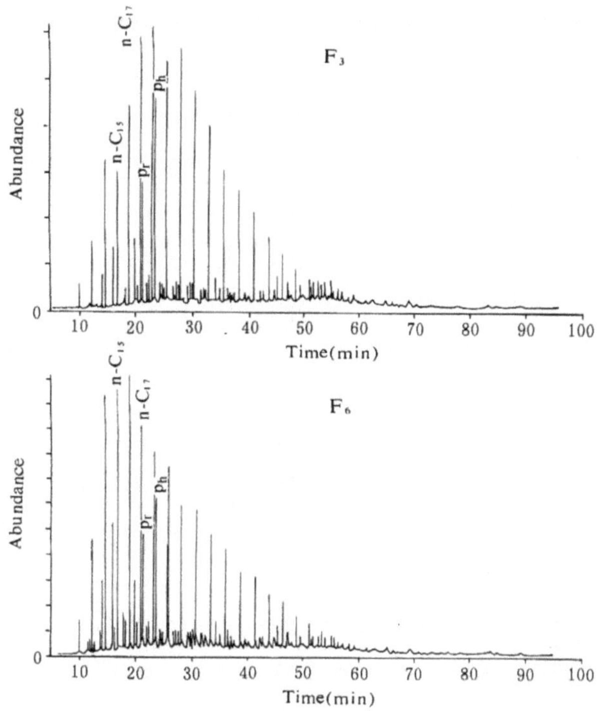

Fig. 1. Mass chromatogram m/z 85 of n-alkanes

The n-alkane distributions of the saturated hydrocarbon fractions range from C_{12} to C_{37}, with a maximum at C_{16} or C_{18}, and display a slight predominance of even-numbered carbon homologues (Fig. 1, Table 1). This even/odd predominance and pristane/phytane ratios less than 1 reflect deposition in a reducing, hypersaline, carbonate-evaporite environment (ten Haven et al. 1985)

5 Biomarkers in the Saturated Hydrocarbons

Representative mass chromatograms showing typical distributions of various classes of saturated biomarker hydrocarbons are presented in Fig. 2–5.

5.1 Bicyclic Sesquiterpanes

Fifteen bicyclic sesquiterpanes ($C_{14}H_{26}$–$C_{16}H_{30}$) were identified in the m/z 123 mass chromatogram (Fig. 2). Among the C_{14} (peaks 1–5), C_{15} (peaks 6–12) and C_{16} (peaks 13–15) compounds, 8α(H)–4,8,9,10-tetramethylperhydronaphthalene (peak 4), 8β(H)-drimane (peak 10), and 8β(H)-homodrimane (peak 15), respectively, are the most abundant. The overall distribution of bicyclic ses-

Table 1. Basic geochemical data on alkanes in the Dengying Formation (Weiyuan well, Sichuan Basin)

Core sample	Depth (m)	Rock type	n-Alkane distribution		
			C_{max}	CPI[a]	Pr/Ph
Zb F_1	3305	grey-black dolostone	C_{16}	1.01	0.07
Zb F_2	3560	grey-black dolostone	C_{18}	0.93	0.59
Zb F_3	3598	grey-black dolostone	C_{18}	1.03	0.57
Zb F_4	3270	grey-black dolostone	C_{17}	1.00	0.84
Zb F_5	3299	grey-black dolostone	C_{17}	1.01	0.66
Zb F_6	3622	grey-green dolostone	C_{16}	0.95	0.64
Zb F_7	3109	grey-white dolostone	C_{18}	0.85	0.49
Zb F_8	3304	dark-grey dolostone	C_{17}	1.01	0.56

[a] CPI = Carbon preference index.

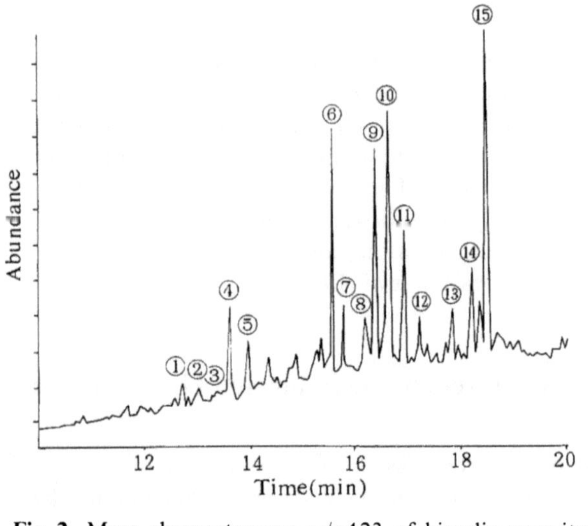

Fig. 2. Mass chromatogram m/z 123 of bicyclic sesquiterpanes (sample F_2)

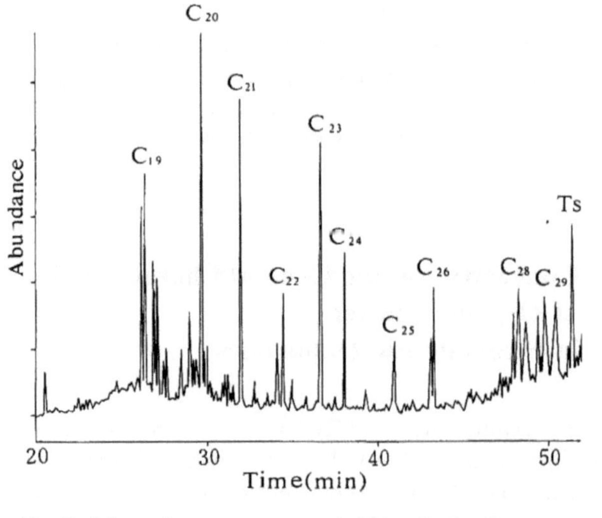

Fig. 3. Mass chromatogram m/z 191 of tricyclic terpanes (sample F_3)

quiterpanes and the relative proportion of 8α(H)-perhydronaphthalene, 8β(H)-drimane and 8β(H)-homodrimane in the dolostones are quite similar to those of crude oils from hypersaline lake facies (Fan Pu et al. 1989). This is further evidence of a hypersaline sedimentary environment for the samples analyzed in the present study.

5.2 Tricyclic Terpanes

Ten compounds ($C_{19}-C_{29}$) of the tricyclic terpane series were detected by the m/z 191 mass chromatogram (Fig. 3). The major peak is a C_{20} tricyclic diterpane. The tricyclic terpanes may have a microbial or algal origin (Aquino et al. 1983).

Table 2. Identification of hopanoids in Fig. 4

No. of peaks	Molecular formula	Molecular weight	Name of compounds
1	$C_{27}H_{46}$	370	18α(H) 22,29,30-trisnorhopane (Ts)
2	$C_{27}H_{46}$	370	17α(H) 22,29,30-trisnorhopane (Tm)
3	$C_{29}H_{50}$	398	17α(H)21β(H) 30-norhopane
Δ	$C_{29}H_{50}$	398	Unknown configuration
X	$C_{30}H_{50}$	412	Unknown configuration, probably γ-lupane
4	$C_{29}H_{50}$	398	17β(H)21α(H) 30-normoretane
5	$C_{30}H_{52}$	412	17α(H)21β(H) hopane
o	$C_{30}H_{52}$	412	
6	$C_{30}H_{52}$	412	17β(H)21α(H) moretane
7	$C_{31}H_{54}$	426	17α(H)21β(H) 31-homohopane-22S
8	$C_{31}H_{54}$	426	17α(H)21β(H) 31-homohopane-22R
9	$C_{30}H_{52}$	412	Gammacerane
10	$C_{31}H_{54}$	426	17β(H)21α(H) 31-homomoretane
11	$C_{32}H_{56}$	440	17α(H)21β(H) 31,32-bishomohopane-22S
12	$C_{32}H_{56}$	440	17α(H)21β(H) 31,32-bishomohopane-22R
13	$C_{32}H_{56}$	440	17β(H)21α(H) 31,32-bishomomoretane-22S
14	$C_{32}H_{56}$	440	17β(H)21α(H) 31,32-bishomomoretane-22R
15	$C_{33}H_{58}$	454	17α(H)21β(H) 31,32,33-trishomohopane-22S
16	$C_{33}H_{58}$	454	17α(H)21β(H) 31,32,33-trishomohopane-22R
17	$C_{33}H_{58}$	454	17β(H)21α(H) 31,32,33-trishomomoretane-22S
18	$C_{33}H_{58}$		17β(H)21α(H) 31,32,33-trishomomoretane-22R
19	$C_{34}H_{60}$	468	17α(H)21β(H) 31,32,33,34-tetrakishomohopane-22S
20	$C_{34}H_{60}$	468	17α(H)21β(H) 31,32,33,34-tetrakishomohopane-22R
21	$C_{34}H_{60}$	468	17β(H)21α(H) 31,32,33,34-tetrakishomomoretane-22S
22	$C_{34}H_{60}$	468	17β(H)21α(H) 31,32,33,34-tetrakishomomoretane-22R
23	$C_{35}H_{62}$	482	17α(H)21β(H) 31,32,33,34,35-pentakishomohopane-22S
24	$C_{35}H_{62}$	482	17α(H)21β(H) 31,32,33,34,35-pentakishomohopane-22R

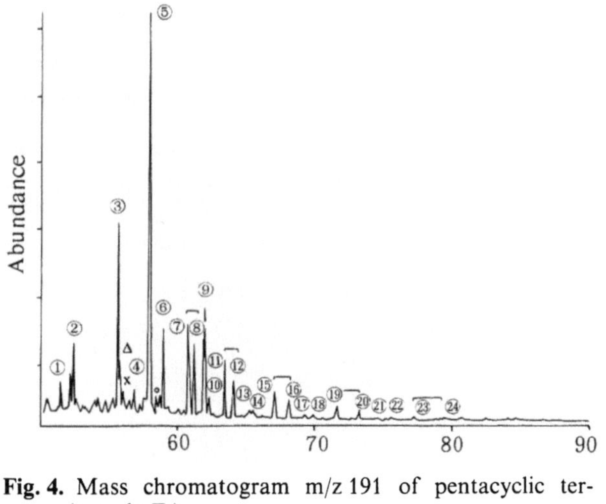

Fig. 4. Mass chromatogram m/z 191 of pentacyclic terpanes (sample F_3)

5.3 Pentacyclic Triterpanes

A complete series of the C_{27} and $C_{29}-C_{35}$ pentacyclic triterpanes is present in the samples (Fig. 4, Table 2), and C_{30} hopane is the major peak. The concentration of the $C_{30}-C_{35}$ hopanes and gammacerane are relatively high. This is similar to the distribution reported for a crude oil derived from source beds deposited in a hypersaline environment (Fan Pu et al. 1984), and constitutes further evidence of a hypersaline paleoenvironment for the Dengying Formation. The hopanes are derived from bacteriohopantetrol (Ourisson et al. 1979), and are found in all sedimentary environments.

5.4 8,14-Secohopanes

A series of $C_{27}-C_{30}$ 8,14-secohopanes is relatively abundant in the m/z 123 mass chromatogram (Fig. 5). These compounds might be either derived directly from algal precursors or formed by the C-ring cleavage of hopanes at a late-mature stage.

5.5 Hopane Sulfides

Two compounds of C_{29} and C_{30} hopane sulfide series (Cyr et al. 1986) were found. Since the same compounds were previously detected in crude oils of hypersaline lacustrine origin (Fan Pu et al. 1989), their presence is consistent with an evaporitic sedimentary environment for the dolostones of the Dengying Formation.

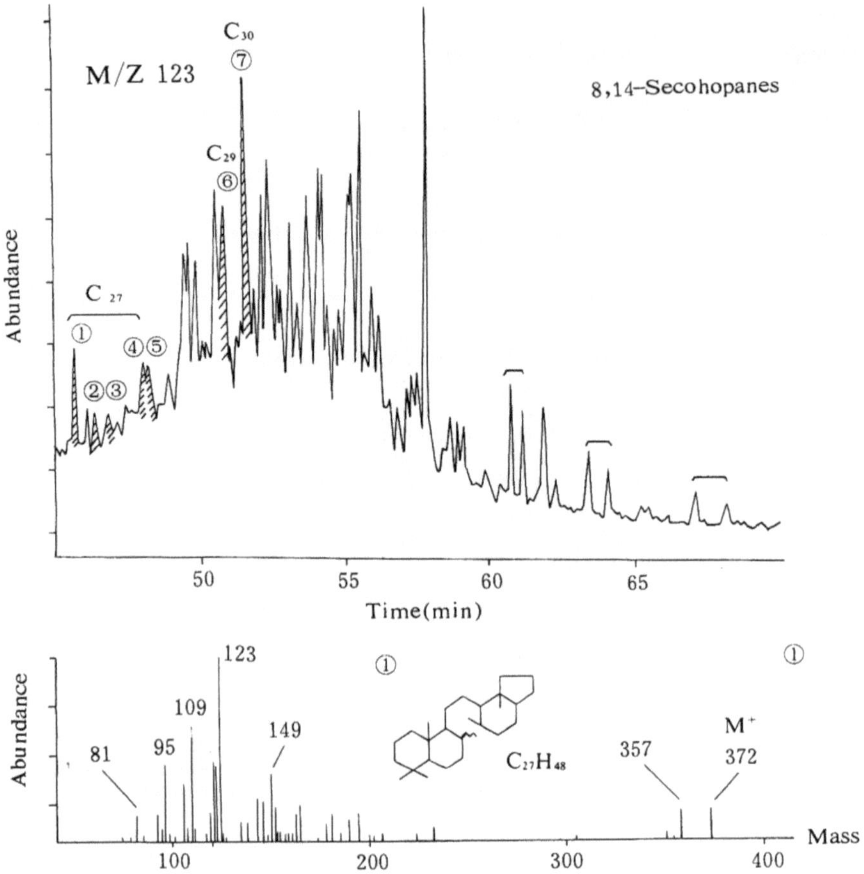

Fig. 5. Mass chromatogram m/z 123 of 8,14-secohopanes and mass spectrum of the main C_{27} compound (peak 1) (sample F_5)

5.6 Steranes

The concentration of $C_{21}-C_{23}$ pregnanes is relatively high. Again, the same feature was noted in samples previously examined from a hypersaline sedimentary environment (Fan Pu et al. 1989). Coeluting with the C_{21} pregnane, there is a prominent component, with a base peak at m/z 219 and a molecular ion of $M^+ = 304$. On the basis of MS analysis, we identify this compound as a $C_{22}H_{40}$ 4,5-secosterane and propose that it is formed by sterane cleavage during late-stage catagenesis.

Among the C_{27}, C_{28} and C_{29} regular steranes, C_{29} sterane has a slight predominance. Recent studies have suggested that C_{29} sterane and its precursor sterol can actually reflect an algal input (Volkman 1986), but its exact source has not been established.

Diasteranes and a total of 13 A-ring methylsteranes were also detected in low concentration. The $C_{28}H_{50}$, $C_{29}H_{52}$ and $C_{30}H_{54}$ A-ring methylsteranes are present as one, two and ten isomers, respectively. These compounds are assigned as A-ring methylsteranes on the basis of their relative retention times in Fig. 6 and their mass spectra.

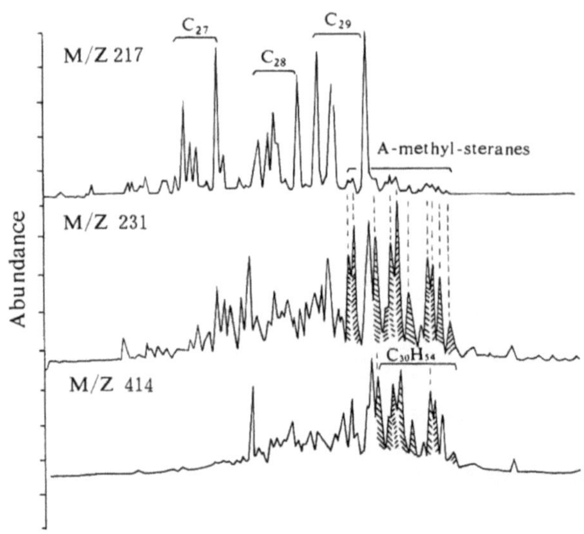

Fig. 6. Mass chromatograms (m/z 217, 231 and 414) of *A*-ring methylsteranes (sample F_7)

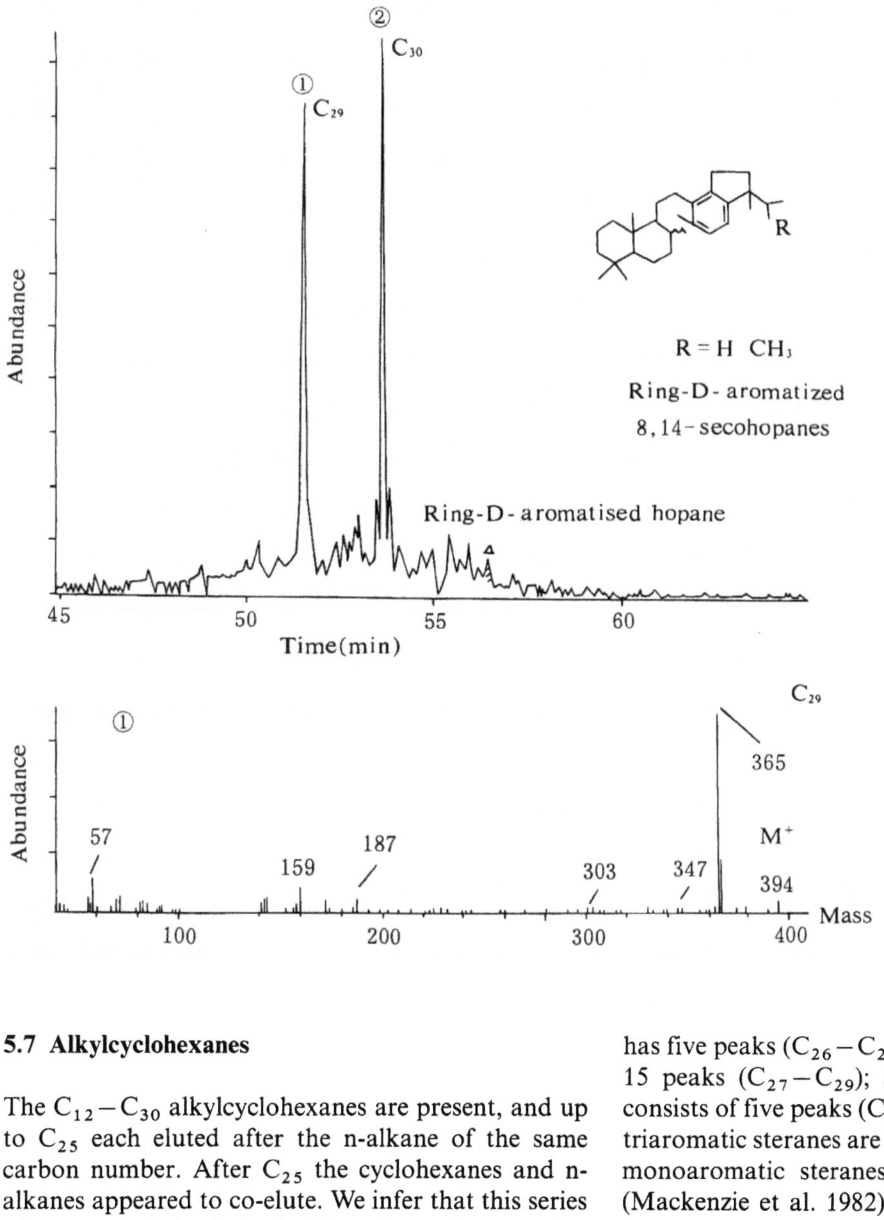

Fig. 7. Mass chromatogram m/z 365 of *D*-ring aromatized 8,14-secohopanes and mass spectrum of the C_{29} compound (peak 1) (sample F_6)

5.7 Alkylcyclohexanes

The $C_{12}-C_{30}$ alkylcyclohexanes are present, and up to C_{25} each eluted after the n-alkane of the same carbon number. After C_{25} the cyclohexanes and n-alkanes appeared to co-elute. We infer that this series of compounds is derived either from algae or, less likely, from bond cleavage of terpenoids at a late stage of maturation.

6 Biomarkers in the Aromatic Hydrocarbons

6.1 Triaromatic Steroids

Triaromatic steranes eluted between 50 and 60 min in the TIC chromatogram. By means of m/z 217, 231, 245 and 259 mass chromatograms, four series of triaromatic steranes (C_nH_{2n-20}) were detected. The m/z 217 series has six peaks comprising C_{25}, C_{26} (3 isomers) and C_{27} (2 isomers) compounds. The m/z 231 series has five peaks ($C_{26}-C_{28}$); the m/z 245 series includes 15 peaks ($C_{27}-C_{29}$); and the m/z 259 series also consists of five peaks ($C_{27}-C_{29}$). These four series of triaromatic steranes are thought to be converted from monoaromatic steranes at high levels of maturity (Mackenzie et al. 1982). No monoaromatic steranes were found in the present samples.

6.2 Aromatized Hopanes

Two D-ring aromatized 8,14-secohopanes, $C_{29}H_{48}$ ($M^+ = 394$) and $C_{30}H_{50}$ ($M^+ = 408$), were detected (Fig. 7). Following the C_{30} D-ring aromatized 8,14-secohopane, there is a compound (labeled Δ in Fig. 7) with a base peak at m/z 365 ($M^+ = 392$), specific fragment ions at m/z 157 and 185 (two H less than the C_{29} D-ring aromatized secohopane), and also having m/z 123 and 201 fragments. On the basis of its mass spectrum (Fig. 8), this peak is thought to be the $C_{29}H_{46}$ D-ring aromatized hopane.

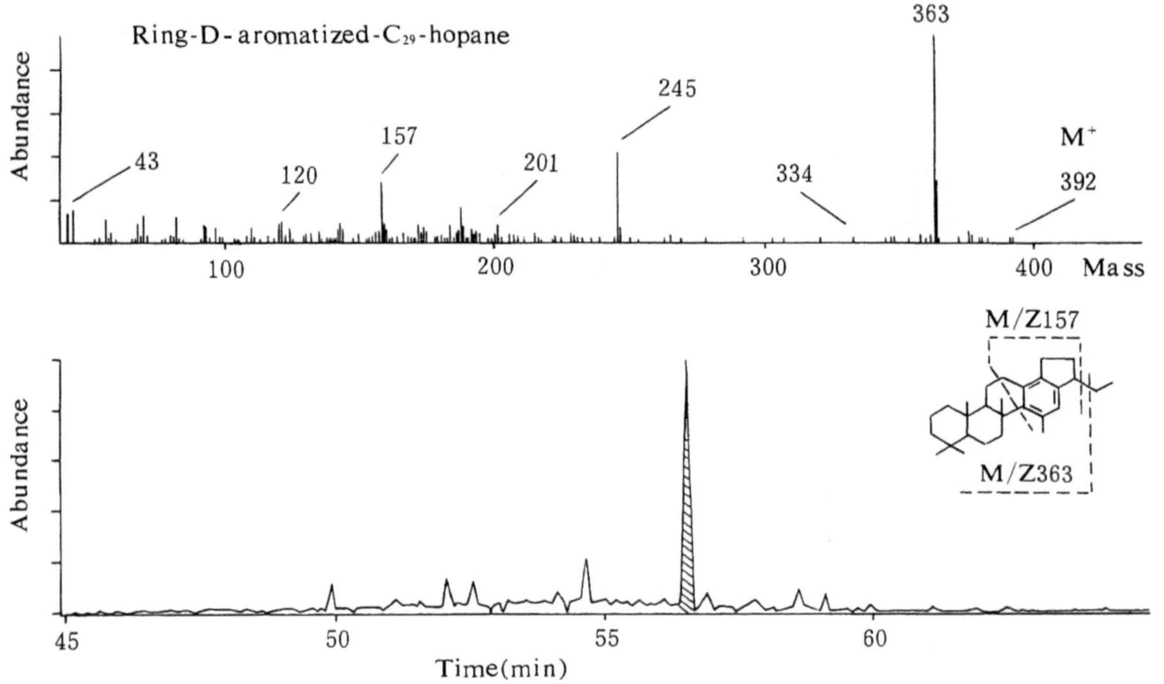

Fig. 8. Mass chromatogram m/z 363 of *D*-ring aromatized hopanes, showing the C_{29} compound and its mass spectrum (sample F_7)

A series of the C_{32}–C_{34} benzohopanes has been detected using the m/z 191 mass chromatogram of the aromatic hydrocarbons. Each of the three aforementioned classes of aromatic hopanoids has been reported to originate from bacteria (Connan et al. 1986).

6.3 Long-Chain Alkylbenzenes

Long-chain alkylbenzenes occur as a small part of the total aromatic hydrocarbons. A complete series of the C_{10}–C_{32} long-chain alkylbenzenes, as well as a series of benzothiophenes in low amount, were revealed by the m/z 105, 119 and 133 mass chromatograms.

7 Other Aromatic Hydrocarbons as Maturity Indicators

Of the aromatic hydrocarbons identified in this study, the naphthalenes and phenanthrenes account for over 70% of the total concentration. Moreover, naphthalenes are more abundant than phenanthrenes (Fig. 9). The content of biphenyls is very low while that of the naphthobenzothiophenes is relatively high.

7.1 Naphthalenes

Naphthalene and the C_1, C_2, C_3 and C_4-naphthalenes were detected by monitoring m/z 128, 142, 156, 170 and 184, respectively. The major compounds are naphthalene and the methyl and dimethylnaphthalene isomers. The methylnaphthalene ratio (MNR = 2-MN/1-MN: Radke et al. 1986) mostly varies between 1.66 and 2.55 (Fig. 9), indicating a high maturity (equivalent vitrinite R_o = 1.0–1.25%).

7.2 Phenanthrenes

Phenanthrene (major peak) and the isomeric methylphenanthrenes (submajor peaks), dimethylphenanthrenes and trimethylphenanthrenes are abundant and were detected by monitoring m/z 178, 192, 206 and 220, respectively. The concentration of tetramethylphenanthrenes (m/z 234), however, is very low. The methylphenanthrene ratio (MPR = 2-MP/1-MP: Radke et al. 1986) ranges from 1.37 to 1.94, and the methylphenanthrene index [MPI-1 = 1.5 (2-MP + 3-MP)/P + 1-MP + 9-MP: Radke et al. 1986] from 0.63 to 1.44, confirming a level of maturity corresponding to the lower oil window.

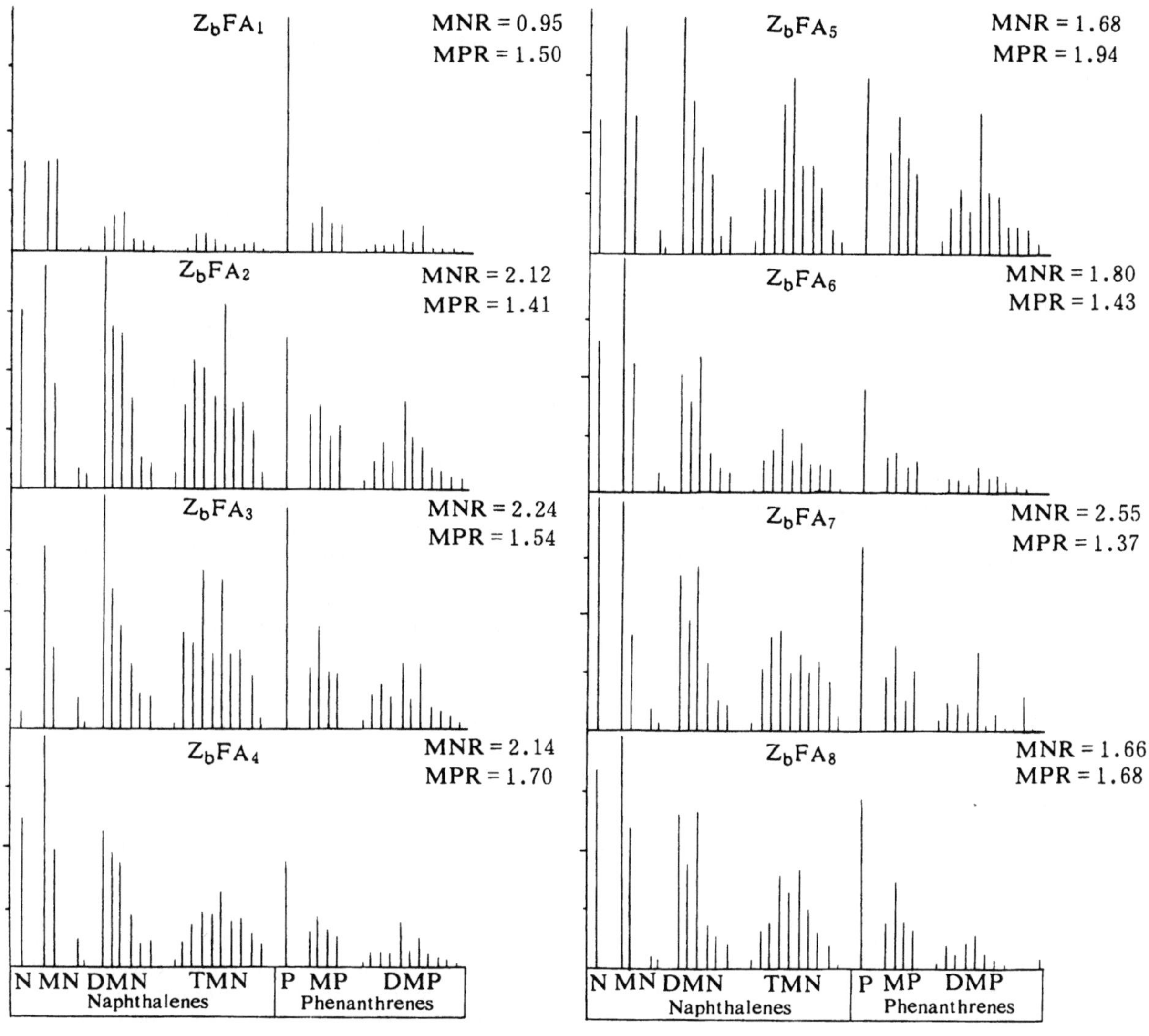

$$MNR = \frac{2\text{-}MN}{1\text{-}MN} \qquad MPR = \frac{2\text{-}MP}{1\text{-}MP}$$

Fig. 9. Relative abundance of naphthalenes and phenanthrenes (samples F_1–F_8)

7.3 Biphenyls

Biphenyls, methylbiphenyls, dimethylbiphenyls and subordinate trimethylbiphenyls were found to be present using the m/z 154, 168, 182 and 196 mass chromatograms. However, the biphenyl series of compounds account for only a small part of the total aromatic fraction.

8 Conclusions

A diverse array of biomarkers is present in the saturated and aromatic hydrocarbons isolated from stromatolitic dolostones of the Sinian Dengying Formation. These biomarkers are derived from various cyanobacteria, algae and bacteria which grew in a hypersaline, carbonate and evaporite-depositing paleoenvironment. Similar Late Proterozoic microorganisms provided the organic matter from which the oldest known commercial accumulations of oil and gas were generated. Assessment of maturity based on diaromatic and triaromatic hydrocarbons are much lower than expected, given the presence of dry gas and carbonized bitumen in the same formation.

Acknowledgements. The authors wish to thank the Geological Exploration and Devolopment Institute, Sichuan Petroleum Administration, for providing samples; and D.M. McKirdy, R.E. Summons and an anonymous reviewer for their helpful comments on the original manuscript.

References

Aquino Neto FR, Trendel JM, Restle A, Connan J, Albrecht PA (1983) Occurrence and formation of tricyclic and tetracyclic terpanes in sediments and petroleum. In: Bjoroy M (ed) Advances in organic geochemistry 1981. John Wiley & Sons, New York Chichester, pp 659–667

Awramik SM, McMenamin DS, Yin Chongyu, Zhao Ziqiang, Ding Qixiu, Zhang Shusen (1985) Prokaryotic and eukaryotic microfossils from a Proterozoic/Phanerozoic transition in China. Nature (London) 315:655–658

Cao Ruiji, Xue Yaosong (1983) Vadose pisolites of the Tongying Formation (Upper Sinian System) in southwest China. In: Peryt TM (ed) Coated grains. Springer, Berlin Heidelberg New York, pp 538–547

Cao Ruiji, Zhao Wenjie (1978) The algal flora of the Dengying Formation (Upper Sinian System) in southwestern China. Mem Nanjing Inst Geol Palaeontol Acad Sin 10:1–28

Connan J, Bouroullec J, Dessort D, Albrecht P (1986) The microbial input in carbonate-anhydrite facies of a sabkha paleoenvironment from Guatemala; a molecular approach. In: Leythaeuser D, Rullkötter J (eds) Advances in organic geochemistry 1985. Pergamon, Oxford, pp 29–50

Cyr TD, Payzant JD, Montgomery DS, Strausz OP (1986) A homologous series of novel hopane sulfide in petroleum. Org Geochem 9, 3:139–143

Fan Pu, King JD, Claypool GE (1984) Characteristics of biomarker compounds (cycloalkanes) in oils formed under various depositional environments. Acta Sedimentol Sin 2, 3:1–7

Fan Pu, Philp RP, Meng Qianxiang, Cheng Xuehui, Li Jinggui, Yu Xinke, Li Zhenxi, Chen Ning (1989) Biomarkers indicating sedimentary paleoenvironments. Sci China Ser 32, 2:251–252

Fu Jiamo, Liu Dehan, Jia Rongfen, Dai Yongding (1984) Distribution and origin of hydrocarbons in carbonate rocks (Precambrian to Triassic) in China. In: Palacas JG (ed) Petroleum geochemistry and source rock potential of carbonate rocks. AAPG Stud 18:11–12

Mackenzie AS, Brassell SC, Eglinton G, Maxwell JR (1982) Chemical fossils: the geological fate of steroids. Science 217:491–504

Ourisson G, Albrecht PA, Rohmer M (1979) The hopanoids. Palaeochemistry and biochemistry of a group of natural products. Pure Appl Chem 51:709–729

Radke M, Welte DH and Willsch H (1986) Maturity parameters based on aromatic hydrocarbons: influence of the organic matter type. In: Leythaeuser D, Rullkötter J (eds) Advances in organic geochemistry 1985. Pergamon, Oxford, pp 51–63

ten Haven HL, De Leeuw JW, Schenck PA 1985) Organic geochemical studies of a Messinian evaporite basin, northern Appenines (Italy) 1. Hydrocarbon biological markers for a hypersaline environment. Geochim Cosmochim Acta 49:2181–2191

Volkman JK (1986) A review of sterol markers for marine and terrigenous organic matter. Org Geochem 9:83–99

Xing Yusheng, Ding Qixiu, Luo Huilin, He Tinggui, Wang Yangeng (1984) The Sinian-Cambrian boundary of China and its related problems. Geol Mag 121:155–170

Zhao Ziqiang, Xing Yusheng, Ma Guogan, Yu Wen, Wang Ziqiang (1980) The Sinian System of eastern Yangtze Gorges, Hubei. In: Wang Rilun (ed) Research on Precambrian geology: Sinian suberathem in China. Tianjin Science and Technology Press, Tianjin, pp 31–55

Natural Gas in Sinian Reservoirs of the Weiyuan Area, Sichuan Province: The Oldest Gas Field in China

Xu Yongchang, Shen Ping and Li Yucheng [1]

[1] Laboratory of Biogeochemistry and Gas Geochemistry, Academia Sinica, Lanzhou 730000, China

Abstract

In the Weiyuan field, Sichuan Province, Sinian carbonates are both source and reservoir for the largest and oldest natural gas accumulation in China. Reserve estimates indicate that there is enough organic matter in these Sinian source beds to account for all the known Sinian-hosted oil and gas. Studies of the molecular and isotopic composition of natural gas in the Sinian reservoirs of the Weiyuan gas field, and of adsorbed C_1-C_3 hydrocarbons in adjacent Sinian source rocks, show that the reservoir gas is authigenic. The isotopic compositions of methane and helium in the natural gas preclude a mantle derivation and, like argon, they appear to have originated entirely from sediments within the crust. Of the inert gases in the Sinian reservoirs only helium may include a minor mantle-derived component. However, much of the Sinian nitrogen in the area studied may be of magmatic origin.

1 Introduction

Most oil reserves in China occur in rocks of Mesozoic and Cenozoic age, but economic gas reserves also are known from Quaternary to Sinian (Late Proterozoic) strata. Older, non-commercial oil and gas reservoirs have also been found in Precambrian metasediments. While Sinian strata form the main reservoir for oil and gas in the Huabei oil field, our research shows that these hydrocarbons are probably sourced from the Tertiary. However, the gas reservoired in the Sinian of the Weiyuan gas field, the largest commercial accumulation of gas found to date in China, may originate from the Precambrian itself. Discussion of the origin of this gas not only has significance for the study of organic evolution during this early stage of Earth history, but also has practical implications for our understanding of the role of microorganisms (algae, bacteria) in the accumulation of economic hydrocarbon reserves.

2 Location and Geological Setting

The Weiyuan gas field lies to the south of Chengdu, the capital of Sichuan Province in southwestern China (Fig. 1). The Weiyuan structure is a large domal feature located on the Weiyuan-Longnusi anticlinal belt (Fig. 2). The main reservoir horizon in the Weiyuan gas field occurs within the 650-m-thick, Late Sinian dolomites which unconformably overlie the Early Sinian Emei Mountain granite (Fig. 3). The upper part of the Sinian Series has been subdivided into four units. The basal unit (Z_1) of dolomitic siltstone (11 – 14 m) is called the Labagang Group. The overlying units (Z_2 to Z_4) comprise the Dengying Group which consists mostly of laminated dolostone containing a large variety of fossil cyanobacteria and algae (Cao Ruiji and Xue Yaosong 1983). The Dengying Group forms the main hydrocarbon reservoir. Porosity is vuggy and interconnected by fracture systems. The seal to the gas reservoir is provided by Early Cambrian mudrocks.

3 Geochemical Characteristics of Gas in the Weiyuan Field

3.1 Molecular Composition

While the Sinian rocks form the main producing reservoirs at Weiyuan, commercial gas is also present in Cambrian, Permian and Triassic strata in the structure. Table 1 contrasts the gas compositions of these younger reservoirs with that of the Sinian accumulation.

The Sinian gas differs in many respects from that in the younger units. It has a high helium content (about 0.2%), making it China's largest single resource of this inert gas. The radiogenic component of its argon is very high (Ar_{ra} generally above 95%). The concentration of nitrogen in the Sinian gas is also very high (4.84 – 15.3%), distinguishing it from the

M. Schidlowski et al. (Eds.)
Early Organic Evolution: Implications for Mineral and Energy Resources

Fig. 1. The geographic location of the Weiyuan gas field, Sichuan Province, China

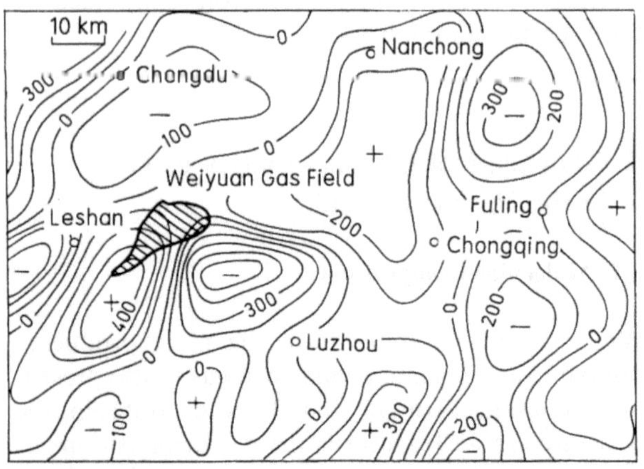

Fig. 2. Structure contour map of the top of the Lower Sinian System in Sichuan Basin (contour interval = 100 m)

Triassic-reservoired gas ($N_2 < 1\%$). Its content of H_2S and CO_2 ($> 5\%$) is significantly greater than that in all the younger reservoirs (1.12–2.80%). The ratio of dry to wet gas for the four Sinian samples analyzed ($C_1/C_{2+} = 373–9080$) has a mean value of 2780, which is higher than the ratios in the younger reservoirs (246–515).

Overall, although the Sinian gas displays certain compositional similarities to the gases in the Cambrian and Permian reservoirs, it is clearly different from the Triassic accumulation.

3.2 Isotopic Composition

The carbon and hydrogen isotope compositions of methane, and the sulphur, argon and helium isotopic compositions of the associated non-hydrocarbon gases from the Weiyuan gas field are shown in Table 2.

The carbon isotope composition ($\delta^{13}C$, ‰) of methane in the Sinian gas ranges from − 31.8 to − 32.2, distinguishing it from methane in the Permian (− 33.6) and Triassic (− 35.7 to − 36.7), but not the Cambrian (− 32.3) gases. The hydrogen isotope composition (δD, ‰) of Sinian methane ranges between − 136 and − 144, which is significantly different from that of the Triassic methane (− 172). However, the Cambrian and Permian hydrogen isotope compositions (− 147 and − 138, respectively) lie close to or within the range of the Sinian gases.

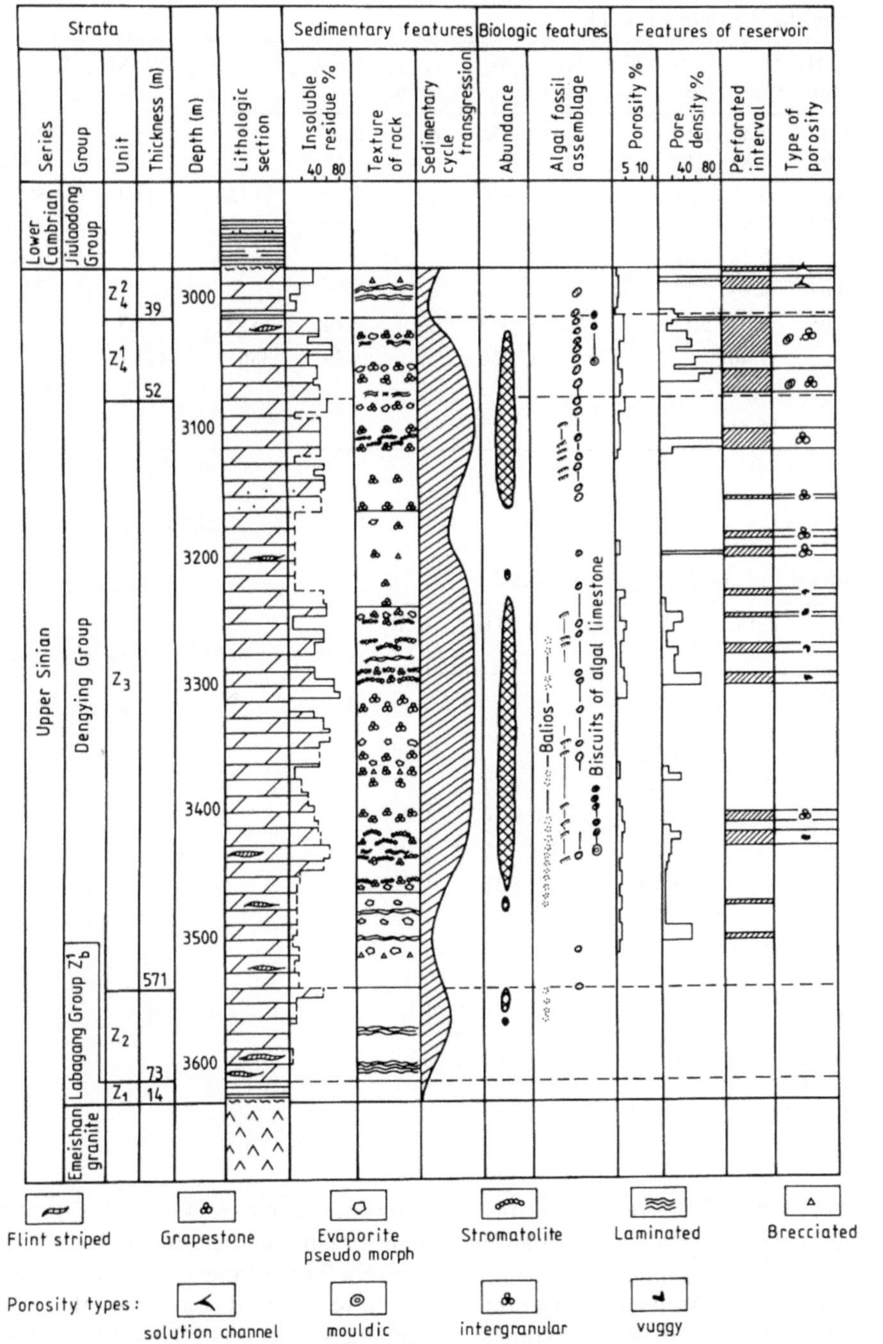

Fig. 3. Stratigraphy and sedimentary features of the Upper Sinian reservoir, Weiyuan gas field

The $^{40}Ar/^{36}Ar$ ratios in the Sinian gases analyzed (7232–9255: Table 2) compare closely with that of the Cambrian gas (7681). Lower values in the Permian gases (2855–5222) and a Triassic sample (561) distinguish them from gases in the older reservoirs. On the other hand, $^{3}He/^{4}He$ ratios (Table 2) suggest that the Sinian ($2.8–2.9 \times 10^{-8}$) and Permian (3.0×10^{-8}) gases are similar and quite distinct from the Cambrian gas (1.8×10^{-8}).

The sulphur isotope composition ($\delta^{34}S$, ‰) of their H_2S component confirms the similarity of the gases in the Sinian (+ 11.5 to + 14.4) and Permian (+ 13.2 to + 13.3) reservoirs. Thus, the gases in these two reservoirs at Weiyuan cannot be distinguished from each other using this parameter. However, elsewhere in the Sichuan Basin, Permian H_2S is isotopically very different from that in Sinian reservoirs (Tables 2 and 4).

Table 1. Components of natural gas in the Weiyuan gas field

Well	Horizon	CH_4 (%)	C_2H_6 (%)	C_3H_8 (%)	N_2 (%)	$H_2S + CO_2$ (%)	He (%)	Ar_{ra} (ppm)	Ar_{atm} (ppm)	He/Ar_{ra}	N_2/Ar_{atm}	He/Ar_{atm}	C_1/C_{2+}
Weiqian 1	$T_{C_1}^4$	98.36	0.33	0.07	0.40	1.12	0.033	54	63	6.11	186	5.24	246
		97.36	0.35	0.04	0.81	1.18							250
Wei 7	P_1^3	94.82	0.22	0.06	5.00	2.31	0.166	320	19	5.19	2632	87.4	388.6
		94.82	0.20	0.04	3.02	2.14							395.0
Wei 26	ε	92.68	0.18		5.12	2.80							514.9
Weiji	Z	89.60	0.17	0.07	4.84	5.27	0.12						373.3
Wei 2	Z	88.54	0.09	0.05	6.67	5.12	0.203	363	12	6.14	5833	185.8	632.4
Wei 23	Z				7.40		0.235	436	19	5.39	3863	123.7	
Wei 28	Z	90.8	0.01		9.17								9080
Wei 29	Z	82.96	0.08		15.31								1036

Ar_{ra} = radiogenic argon.
Ar_{atm} = atmospheric argon.

Table 2. Isotopic composition of natural gas in the Weiyuan gas field[a]

Well	Horizon	$\delta^{13}C_{CH_4}$ (‰)	δD_{CH_4} (‰)	$\delta^{34}S_{H_2S}$ (‰)	$^{40}Ar/^{36}Ar$	$^3He/^4He$	R/R_A[b]
Weiqian 1	$T_{C_1}^4$	− 35.7	− 172		561		
		− 36.7	− 172				
Wei 7	P_1^3			+ 13.2	5222		
				+ 13.3			
Wei 5	P_1	− 33.6	− 138		2855	3.0×10^{-8}	0.022
Wei 26	ε	− 32.2	− 147		7681	1.8×10^{-8}	0.013
Wei 2	Z	− 31.8	− 143	+ 13.7	9255	2.9×10^{-8}	0.021
Wei 23	Z			+ 11.5	7232		
				+ 12.6			
Wei 28	Z	− 32.2	− 144				
Wei 29	Z		− 136			2.8×10^{-8}	0.20

[a] PDB, SMOW, and Canyon Diablo troilite standards were used to calculate the $\delta^{13}C$, δD and $\delta^{34}S$ values, respectively.
[b] $R = R_{sample}$ = sample $^3He/^4He$; $R_A = R_{atmospheric}$ = atmospheric $^3He/^4He$.

Thus isotope measurements on both hydrocarbon and non-hydrocarbon components demonstrate the similarity of natural gas produced from the Sinian, Cambrian and Permian reservoirs of the Weiyuan structure. Gas in the Triassic reservoir appears to have a different origin.

4 Genesis of the Sinian Gas

Much research has been carried out on the origin of the Sinian gas at Weiyuan (Xu Yongchang et al. 1979; Cheng Yaohuang and Cheng Sengji 1982; Shen Ping et al. 1982, 1984; Wang Xianbin 1982; Ji Dehua and Yue Hong 1986). Three different origins have been proposed for the gas: (1) from the mantle; (2) from Cambrian gas-prone source rocks; and (3) from authigenic Sinian organic matter. In the latter case, several authors (Xu Yongchang et al. 1979; Wang Xianbin 1982) have pointed out that the non-hydrocarbon and inert gases might have been derived from the mantle. However, on the basis of our research, we believe that the gas in the Sinian reservoir at Weiyuan, including most of the inert gases, originated from sediments within the Sinian sequence. Our reasons for this conclusion are outlined below.

4.1 Source Rocks

The abundance and composition of organic matter in parts of the Dengying Group (Zhang Shengu and

Tang Chuangji 1986) satisfy the source rock criteria outlined by other authors (e.g. Tissot and Welte 1984). In addition, the common occurrence of bitumen and liquid crude oil in geodes and vugs within Sinian carbonates throughout China (e.g. Fu Jiamo et al. 1982, 1984) provides ample evidence for hydrocarbon generation from intra-Sinian source beds.

4.2 Gas/Source Rock Correlations

The accumulation of helium in sedimentary rocks depends primarily on the radioactive decay of uranium and thorium in the sediments, and the age of the strata. The Dengying Group is mainly a carbonate sequence, and should therefore have low U and Th contents. Our analyses reveal average concentrations of less than 2 ppm for both elements, supporting this conclusion. However, as indicated in Table 1, the Sinian gas at Weiguan has a high He content (about 0.2%). In 1988, we first measured $^3He/^4He$ ratios in gas from the different reservoirs of the Weiyuan field and concluded that the helium content of the Sinian gas was mantle-derived. We now interpret these analytical data to indicate that the helium component of the Sinian gas accumulated over a long period of geologic time from the radioactive breakdown of the autochthonous U and Th, without the introduction of much allochthonous helium.

A similar conclusion is drawn from the high radiogenic argon (Ar_{ra}) content of the Sinian gas. Xu Yongchang et al. (1979) have previously documented the accumulation of argon in sedimentary rocks with increasing age. The mean $^{40}Ar/^{36}Ar$ ratio in the Sinian gas at Weiyuan is 7009, which is the highest value observed for natural gas in any oil or gas field in China. The high value is interpreted to be the result of the accumulation of argon from the long-term decay of ^{40}K in sedimentary strata without the introduction of argon from extraneous sources.

The nitrogen component of natural gas tends to increase markedly when the source rock reaches the metagenesis stage of organic maturation (Tissot and Welte 1984). From Table 1, it can be seen that the nitrogen content of the various gas accumulations at Weiyuan increases with the geologic age of the reservoir. The higher CO_2 plus H_2S component of the Sinian gas (Table 1) may partially reflect the carbonate lithology of the source/reservoir sequence. It is only in the Weiyuan field that Permian and Sinian H_2S are isotopically similar; in other gas fields of the Sichuan Basin they are totally different.

The high C_1/C_{2+} ratio of the Sinian gas is consistent with the advanced thermal maturity of dispersed organic matter in the Dengying Group (Shen Ping et al. 1982). Research elsewhere in the Sichuan Basin (Shen Ping et al. 1982) has shown that the older the reservoir, the higher the ratio of dry to wet gas components. The highest C_1/C_{2+} value is known from the Sinian reservoir at Weiyuan. The carbon and hydrogen isotope compositions of the Sinian methane ($\delta^{13}C = -32‰$; $\delta D = -140‰$) reflect its derivation from overmature marine source rocks. In the Weiyuan area, Sinian carbonates have an estimated equivalent vitrinite reflectance of 4.5% (Shen Ping et al. 1988).

4.3 Comparison of Adsorbed Hydrocarbons with Reservoir Gas

The composition of the reservoir gas at Weiyuan can be compared with that of the light hydrocarbons (C_1–C_3) retained in adjacent fine-grained non-reservoir rocks (Table 3), using a K value. The K value is derived from the following relationship:

$K = (C_2/C_3)$ natural gas $- (C_2/C_3)$ adsorbed hydrocarbons.

Table 3. Analytical data on natural gas and adsorbed hydrocarbons in reservoir formations of the Weiyuan gas field

Horizon		Natural gas (mean)			Adsorbed hydrocarbons	K
		CH_4	C_{2+}	C_2/C_3	C_2/C_3	
Triassic	T_C^5	92.25	0.42	3.7	3.0	0.7
	T_C^4	97.12	0.55	4.9	1.7	3.2
	T_C^3	95.60	1.54	2.5	1.5	1
	T_C^2	95.72	2.58	3.3	1.2	2.1
	T_C^1	95.69	2.59	3.4	1.6	1.8
Permian	P_1^3	96.85	1.10	4.5	3.5	1
	P_1^2	97.17	0.64	2.2	1.2	1
Sinian	Z	89.07	0.23	2.0	1.9	0.1

Independent studies (Shen Ping et al. 1982) have shown that gas in the Permian reservoir originated from Permian sediments which have a K value of 1 (Table 3). From experimental data we conclude that the closer the K value is to zero in the range + 1 to − 1, the closer the genetic relationship between the source rock and the reservoir gas. According to the values in Table 3, the gas from the Sinian reservoir shows the characteristic of an authigenic (i.e. intra-Sinian) origin.

5 Relation of Gas Isotopic Composition to Source Rock

For comparison, Table 4 summarizes the isotope compositions of methane, hydrogen sulphide and inert gases in three Permian and Triassic gas fields located in other areas of the Sichuan Basin (Fig. 1).

As already shown for the Weiyuan field (Table 2), it is not possible to distinguish the Sinian gas from the Permian gases on the basis of the carbon and hydrogen isotopes of methane, as the Permian source beds here and elsewhere are also at the overmature stage of hydrocarbon generation. The Triassic gases, on the other hand, are obviously of lighter isotopic composition.

In general, the Sinian methane is richer in the heavier isotopes of carbon and hydrogen, but is quite different from mantle-derived methane ($\delta^{13}C = -17.9$ to -15‰: Welhan and Craig 1981). On this basis, the characteristics of the Sinian hydrocarbon gas pool at Weiyuan relate more to the thermal evolution of indigenous Sinian organic matter than to mantle degassing.

The average $^{40}Ar/^{36}Ar$ value for the Triassic gas pools of the Sichuan Basin is 875, that of the Permian gas is 925, and the Sinian gases have a mean value of 7009. Since the amount of radiogenic argon increases with the age of the reservoir (Xu Yongchang et al. 1979), the $^{40}Ar/^{36}Ar$ ratio of a gas accumulation may be used to estimate the age of its source rock. In the case of the Sinian gas pool, the source rocks for the argon (and hence also the hydrocarbons) are evidently very old.

6 Source of the Non-Hydrocarbon and Inert Gases

The nitrogen in the Sinian gas at Weiyuan may be related to a magmatic source. In the Wei-28 well, gas recovered from Early Sinian granite has been shown to contain up to 37% nitrogen.

The $^{3}He/^{4}He$ ratio is the main index used to distinguish crustal and mantle-derived helium. Lupton (1983) showed that in helium gas derived from the mantle this ratio is 1×10^{-5}, in air it is 1.4×10^{-6}, and in crustal helium it is 1×10^{-8}. Our measurements of this ratio, the first on natural gas from the Sichuan Basin, give a basin-wide average of 1×10^{-8}, suggesting a mainly crustal origin for its helium.

Since the Dengying Group source beds are much older than those of the Permian and Triassic gas accumulations, the mean $^{3}He/^{4}He$ ratio of the Sinian gas should be appreciably smaller. In fact, it is a little higher than the average Triassic and Permian $^{3}He/^{4}He$ ratios. We suggest that the extra helium in the Sinian gas was contributed from mantle sources. The mantle-derived component is estimated to have been no more than 0.3‰ of the total helium content of the Sinian gas.

7 Conclusions

During the Sinian period, algae and other microflora proliferated. Studies of the abundance and distri-

Table 4. Isotopic composition of natural gas in Tertiary (T) and Permian (P) gas fields of the Sichuan Basin

No. Locality	No. Well	Horizon	$\delta^{13}C_{CH_4}$ (‰)	δD_{CH_4} (‰)	$\delta^{34}S_{H_2S}$ (‰)	$^{40}Ar/^{36}Ar$	$^{3}He/^{4}He$	R/R_A [a]
Zhongba	Zhong 2	T_3X^2	− 36.2	− 175			0.9×10^{-8}	0.006
	Zhong 31	T_3X^2	− 35.9	− 187			2.1×10^{-8}	0.015
Ziliujing	Zi 1	P_1				694		
	Zi 2	P_1^2	− 33.2	− 136	+ 29.1	1069		
	Zi 3	P_{12}^3	− 33.1	− 151	+ 27.8	734		
	Zi 4	P_{12}^3			+ 24.8			
	Zi 6	P_1^3	− 33.4	− 154		899		
Yangaosi	Yang 43	P_1^3	− 33.1	− 140			1.6×10^{-8}	0.011

[a] $R = R_{sample}$ = sample $^{3}He/^{4}He$; $R_A = R_{atmospheric}$ = atmospheric $^{3}He/^{4}He$.

bution of organic matter in Sinian carbonates throughout China reveal the common occurrence of crude oil and bitumen in geodes, and the presence of organic-rich layers which are potential source rocks. Reserve estimates indicate that there is enough organic matter present in these Sinian source beds to account for all the known Sinian oil and gas accumulations.

Studies of the molecular and isotopic composition of natural gas in the Sinian reservoir of the Weiyuan gas field, and of adsorbed C_1-C_3 hydrocarbons in the Sinian source rocks, show that the reservoir gas is authigenic. On the basis of their geological setting and similar geochemical data, it is considered most probable that the gas reserves in the Permian and Cambrian reservoirs of the Weiyuan structure were also sourced from the Sinian.

The isotopic compositions of methane and helium preclude a mantle derivation for these two gases in the Sichuan Basin. Like argon, they appear to be derived almost entirely from sediments within the crust. Of the inert gases, only helium in the Sinian reservoirs may include a minor mantle-derived component. However, at Weiyuan, much of the Sinian nitrogen may be of magmatic origin.

In the Weiyuan field, Sinian carbonates are both source and reservoir for the largest and oldest natural gas accumulation in China. As such they provide a valuable opportunity for the study of hydrocarbon generation from organic matter deposited at this early stage of Earth history.

Acknowledgements. The authors thank Mr. J.D. Gorter (Petroconsultants, Sydney), Dr. G.W.M. Lijmbach (Shell, The Hague) and Dr. D.M. McKirdy (Department of Geology and Geophysics, University of Adelaide) for their assistance in the preparation of this paper.

References

Cao Ruiji, Xue Yaosong (1983) Vadose pisolites of the Tongying Formation (Upper Sinian System) in southwest China. In: Peryt TM (ed) Coated grains. Springer, Berlin Heidelberg New York, pp 538–547

Cheng Yaohuang, Cheng Sengji (1982) Gas sources and exploration in the Sichuan Basin. Nat Gas Ind 2(1):27–34 (in Chinese)

Fu Jiamo, Jia Rong Fen, Liu Dehan (1982) The organic geochemical indicatrix of prospect evaluation for oil and gas reservoirs in carbonate rocks. In: Symp Petroleum geoscience, Academia Sinica. Scientific Press, Beijing, pp 131–138 (in Chinese)

Fu Jiamo, Liu Dehan, Jia Rong Fen, Dai Yong Ding (1984) Distribution and origin of hydrocarbons in carbonate rocks (Precambrian to Triassic) in China. In: Palacas JG (ed) Petroleum geochemistry and source rock potential of carbonate rocks. AAPG Stud Geol 18:1–12

Ji Dehua, Yue Hong (1986) Discussion of oil and natural gas exploration of the Sinian System in southwestern Sichuan. Nat Gas Ind 6(1):10–12 (in Chinese)

Lupton JE (1983) Terrestrial inert gases: isotope tracer studies and clues to primordial components in the mantle. Annu Rev Earth Planet Sci 11:371–414

Shen Ping, Wang Xianbin, Xu Yongchang (1982) Gas source comparison based on isotopic composition of natural gases. Petrol Expl Dev 6:34–38 (in Chinese)

Shen Ping, Wang Xianbin, Shen Qixiang (1984) The isotopic study of adsorbed hydrocarbons for gas-source correlations. In: Symp Sedimentology and Organic Geochemistry. Scientific Press, Beijing, pp 234–249 (in Chinese)

Shen Ping, Shen Qixiang, Wang Xianbin, Xu Yongchang (1988) Characteristics of isotope composition of gaseous hydrocarbons and identification of coal-type gas. Sci Sin Ser B 31(6):734–747

Tissot BP, Welte DH (1984) Petroleum formation and occurrence, 2nd edn. Springer, Berlin Heidelberg New York, 699 pp

Wang Xianbin (1982) Natural gas derived from the deep earth. Sci Rep 17:1069–1071

Welhan JA, Craig H (1981) Hydrocarbons in 21 °N hydrothermal fluids. EOS Trans Am Geophys Un 62(45):913

Xu Yongchang, Wang Xiangbin, Wu Renming, Shen Ping (1979) Noble gas isotope geochemistry in natural gas. Geochemica 4:271–282

Zhang Shenyu, Tang Chuangji (1986) Regional geology and gas content: a review of the Dengying Formation, Sichuan Basin. Nat Gas Ind 6(1):3–9 (in Chinese)

Biological Markers in Paleozoic Sedimentary Rocks and Crude Oils from the Michigan Basin: Reassessment of Sources and Thermal History of Organic Matter

Jürgen Rullkötter [1,3], Roger Marzi [1,4] and Philip A. Meyers [2]

[1] Institut für Erdöl und Organische Geochemie, Forschungszentrum Jülich GmbH, Postfach 1913, W-5170 Jülich, FRG
[2] Department of Geological Sciences, The University of Michigan, Ann Arbor, Michigan 48109-1063, USA
[3] Present address: Institut für Chemie und Biologie des Meeres, Universität Oldenburg, Postfach 2503, W-2900 Oldenburg, FRG
[4] Present address: Amoco Production Co., P.O. Box, Tuka, Oklahoma 74102, USA

Abstract

Paleozoic sedimentary rocks fill the Michigan Basin and provide the sources of the oils produced from this basin. The present geothermal gradient is not high in this shallow sequence, raising questions as to whether Devonian and younger strata are sufficiently thermally mature to have sourced petroleum. The thermal maturities of rock samples from Late Cambrian to Pennsylvanian strata have been determined using the kinetics of apparent transformations of biomarker molecules as an index, augmented by vitrinite reflectance measurements and Rock-Eval pyrolysis data. From the extents of sterane isomerization and the degrees of steroid aromatization in biomarkers extracted from these samples, the geothermal history of the Michigan Basin has been reconstructed. Organic matter in Devonian strata from the center of the basin is thermally mature; it is overmature in Ordovician rocks. The geological history of this basin has evidently been more complex than generally thought, suggesting higher temperatures or lower conduction rates in the past.

1 Introduction

1.1 General Description of the Michigan Basin

The Michigan Basin is composed almost entirely of Paleozoic (Cambrian-Devonian) sediments. The basin is situated within the North American craton and was formed by the gradual subsidence of its central portion throughout Paleozoic time (Dorr and Eschman 1970; Nunn et al. 1984). The result is a nearly circular structure filled with several kilometers of gently dipping strata which were deposited as shallow-water sediments (Fig. 1). The rocks are predominantly limestones and dolomites, with subordinate shales and sandstones. The Silurian sequence includes thick evaporite deposits in which halite is common. Little structural deformation is evident in any of these strata. Rocks of Carboniferous and younger age are a minor part of the stratigraphic succession in the Michigan Basin.

1.2 Michigan Basin During the Paleozoic Era

Many major paleoenvironmental changes occurred during the 200-million-year (Ma) period of Paleozoic time over which the sediments of the Michigan Basin accumulated. Central among these changes are the climatic consequences of plate tectonic shifts in the paleolatitude of the basin, from equatorial in the Late Cambrian to temperate by the Silurian/Devonian boundary and back to equatorial by the Carboniferous (Van der Voo 1988). In the course of these tectonic migrations, this region of what eventually became part of the North American craton twice passed through the arid climate zone between 20° and 30° S, once in the Middle to Late Silurian and once again in the Early to Middle Devonian. The thick sequence of Middle to Late Silurian evaporites records the first stage of this migration. The sedimentary record of the second passage across this climatic belt is abbreviated by unconformities, yet examples of evaporites are again found. As summarized in Table 1, there were periods in the Early Ordovician and in the Late Devonian when the paleolatitudes were the same – tropical southern hemisphere – and indeed the sediments deposited in the basin at these widely separated times display many similarities, indicating that the influence of climate on the paleoenvironments was strong.

Throughout most of its Paleozoic wanderings, the Michigan Basin remained covered by a shallow epicontinental sea. The several gaps in the Paleozoic record, marked by unconformities, probably represent times when the area was emergent. The sea evidently existed in this area during most of the Ordovician and

M. Schidlowski et al. (Eds.)
Early Organic Evolution: Implications for Mineral and Energy Resources

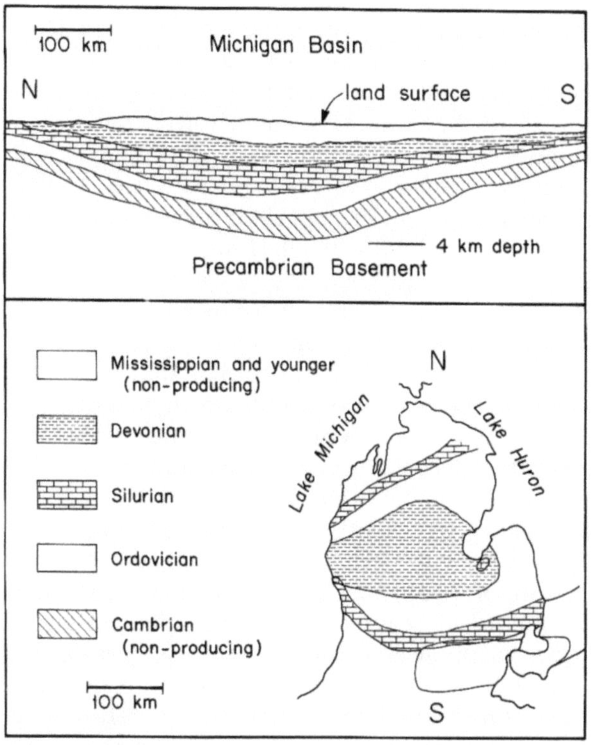

Fig. 1. Shape and cross-section of the Michigan Basin, illustrating the locations of the major Paleozoic rock strata (*above*) and oil-producing reservoir rocks (*below*). Cambrian rocks are potential gas producers. The vertical scale in the cross-section of this shallow basin is greatly exaggerated

gradually deepened, as indicated by a progression from the shallow-water Prairie du Chien sandstones and dolomites to the Black River and Trenton Group limestones, finally capped by the Utica and younger shales from the distant Taconic orogeny at the end of this period. In the Early Silurian, extensive reef banks developed on the margin of the basin, and pinnacle reefs grew on the upper slope, basinward from the margin. In the Late Silurian, shallow marginal banks continued to support reef communities (Niagara Limestones) and created a semi-enclosed evaporitic basin in which the thick Salina Group precipitated. The Early Devonian sequence is attenuated by an unconformity on which the evaporites of the Detroit River Group were deposited. Paleoenvironments apparently alternated between arid land and evaporitic shallow marine during the Middle Devonian. The Late Devonian sequence comprises limestones capped by the Antrim Shale, which represents distal clastics eroded from the young Appalachian Mountains to the east. The deposition of this shale continued into the Early Carboniferous.

Table 1. Paleolatitudes of the center of the Michigan Basin through Paleozoic time (R. Van der Voo, pers. commun., 1989)

Epoch	Ma B.P.	Paleolatitude
Early Permian to Late Carboniferous	282–308	2 °S
Middle Carboniferous to Late Devonian	309–365	9 °S
Late to Middle Devonian	366–378	15 °S
Middle to Early Devonian	379–397	22 °S
Early Devonian to Late Silurian	398–414	43 °S
Late Silurian to Middle Silurian	415–429	23 °S
Early Silurian to Middle Ordovician	430–467	13 °S
Middle Ordovician to Early Ordovician	468–505	9 °S
Late Cambrian	506–542	13 °S
Early Cambrian	543–575	8 °S

1.3 Paleozoic Sources of Organic Matter

Organic matter in sedimentary rocks is the residue of past biota. Over the span of the Paleozoic, several major changes occurred in the evolutionary development of life which have had important consequences for the type of organic matter delivered to sediments of the Michigan Basin.

Foremost among these evolutionary changes was the emergence of vascular plants in the Silurian and their extensive colonization of the continents by the Carboniferous. In the Michigan Basin, which was predominantly a marine setting, fossils of land plants first appear in Devonian strata, and by the Carboniferous coal beds are found. Nonetheless, the fossil record of the Middle Paleozoic is dominated by normal marine fauna with calcareous skeletons such as corals, stromatoporoids, brachiopods, bryozoans, and crinoids.

Geochemical analyses of Ordovician oils from widely separated locations around the world suggest that the organic matter preserved in the sediments from which these oils originated was remarkably uniform, inasmuch as the oils display strong similarities (e.g. Illich and Grizzle 1983; Reed et al. 1986). All of these Ordovician source rocks share the same paleoenvironment – a shallow tropical sea. A lipid-rich alga, *Gloeocapsomorpha prisca*, has been identified as the main source of organic matter in these sediments (Reed et al. 1986; Hoffmann et al. 1987). The life-style of this alga is uncertain; it may or may not have been photosynthetic, and it may or may not have been confined to oxygen-depleted waters. The alga nonetheless seems to have been unique to Ordovician tropical seas (however, see Fowler, this Vol.).

Only specialized organisms could tolerate the hypersaline episodes which led to evaporite formation in the Michigan Basin, particularly in view of the anoxic, sulfide-rich conditions which evidently prevailed. From biomarker and isotopic analyses of oils from the Guelph Formation (equivalent to the Niagara Limestone) in the Ontario portion of the Michigan Basin, Summons and Powell (1987) concluded that sulfur bacteria contributed to the organic matter in their source rocks. They further note that oils from a Devonian hypersaline paleosetting in Alberta also exhibit evidence of sulfur bacteria, showing that similar organic matter can accumulate in equivalent environments of vastly different age.

Consistent with the fossil record, organic matter in the Paleozoic rocks and oils of the Michigan Basin is predominantly marine in origin, yet land plants began to contribute organic matter in the Silurian. Distributions of n-alkanes in Silurian oils commonly contain the long-chain $C_{23}-C_{29}$ components characteristic of the waxes of higher vascular plants. Also, the carbon isotope values of individual longer-chain n-alkanes from a Niagaran oil show the lighter $^{13}C/^{12}C$ ratios typical of land plants (Vogler et al. 1981). Although contributions from land plants are not obvious in the organic contents of most Devonian strata, they become the dominant form of organic matter in the Mississippian and Pennsylvanian coal deposits of the Michigan Basin.

1.4 Tectonic and Thermal History of the Michigan Basin

From its lack of secondary structural deformation, Nunn et al. (1984) have concluded that the Michigan Basin has been tectonically stable since the end of the Paleozoic. Major unconformities exist at the base of Middle Ordovician and at the tops of the Silurian and Devonian sequences, indicating possible uplift and intermittent erosion. With these exceptions, gradual subsidence is believed to have dominated throughout the Paleozoic (Nunn et al. 1984). The almost total absence of rocks younger than Carboniferous obscures the Mesozoic and Cenozoic tectonic history of this basin, yet the little-deformed nature of the underlying strata suggests that nothing dramatic occurred.

The present-day geothermal gradient in the Michigan Basin has been measured in a deep borehole near the center of the basin (Sleep and Sloss 1978), and it is an unexceptional 21 °C/km. If this figure represents the gradient since Paleozoic times, and if the present thickness of sedimentary rocks (ca. 4300 m maximum) represents the total accumulation of sediments, then the amount of heat which could have flowed into the basin constrains petroleum production to the deeper strata. On this basis, only Silurian and older rocks could have produced petroleum (Nunn and Sleep 1978; Vogler et al. 1981; Meyers and Moore 1983; Nunn et al. 1984, 1985). However, if geothermal gradients were once higher, or if a significant thickness of Carboniferous or younger sediments has been lost to erosion, then Devonian strata could have been heated enough to produce petroleum. Cercone (1984) noted higher than expected thermal maturities of organic matter in rocks from various strata and concluded that thermal gradients formerly ranged between 35 and 45 °C/km. Either as much as 1 km of sediments was removed by erosion or, as suggested by Cercone and Pollack (1989), a thinner sequence (500 m) of poorly conducting organic carbon-rich Carboniferous beds has been eroded. The postulated higher heat flow and erosional losses suggest that oil generation may have occurred from Devonian strata. The distinctive molecular compositions of Devonian oils, evident from comparison of oil types in the Michigan Basin, also indicate the existence of Devonian source rocks (Pruitt 1983; Illich and Grizzle 1983, 1985; Powell et al. 1984; Rullkötter et al. 1986).

Contradictions exist between models of the thermal history of the Michigan Basin based on the concept of a stable intracratonic basin and measured geothermal gradients (e.g. Nunn et al. 1984) and models based upon the thermal maturity of organic matter (e.g. Cercone 1984; Rullkötter et al. 1986). Organic matter in this basin exhibits the effects of greater thermal alteration than the present geothermal gradient would predict, but geological evidence for erosion of an adequate thickness of overlying strata is at best circumstantial, and higher heat flows in the past can only be inferred.

1.5 Apparent Biomarker Transformations as Indicators of Geothermal History

There are various chemical reactions, dependent on time and temperature, involving specific transformations of organic molecules with characteristic biochemically controlled structures into well-defined geochemical products (Mackenzie 1984). Those organic compounds, which retain evidence of their biological origin, even though geochemically modified to some degrees, are called biological markers, or biomarkers. The precursor/product relationships of several biomarkers have been sufficiently well established for

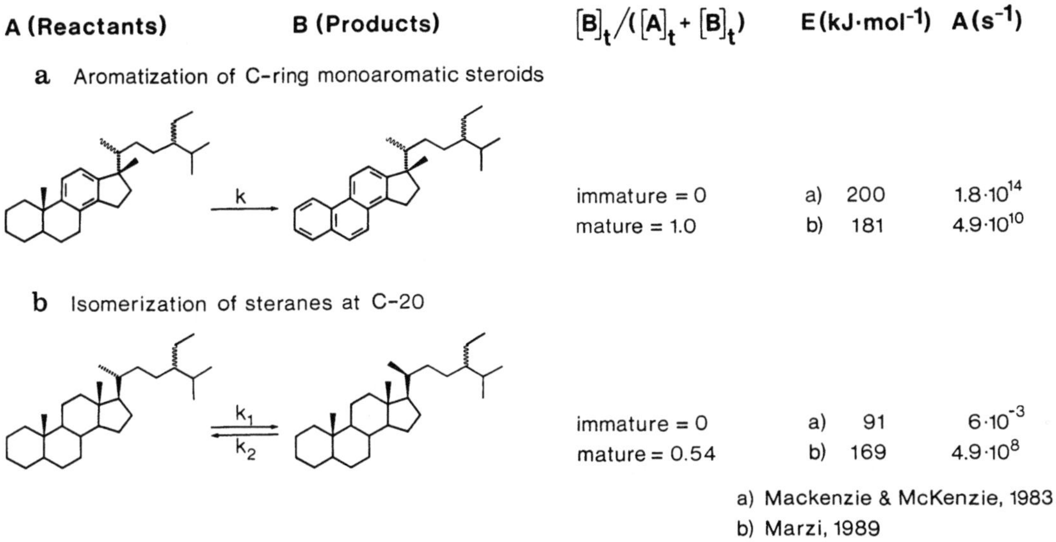

Fig. 2. Schematics of the apparent sterane isomerization and steroid aromatization reactions, with equilibrium ratios, activation energies, and frequency factors for the two processes

their apparent transformation reactions to be used to assess the thermal maturity of crude oils and organic matter in sedimentary rocks. Where these reactions could be simulated by laboratory heating experiments or studied in geological environments with geothermal histories which are reasonably well understood, the kinetics of the different reactions have been determined (e.g. Mackenzie and McKenzie 1983; Marzi 1989). Under the assumption that the reactions follow pseudo-first-order kinetics, the Arrhenius equation can be applied. The derived parameters of activation energy and frequency factor provide information on how the transformations react to the influence of temperature and time during the geological history of a sedimentary layer (Mackenzie and McKenzie 1983).

Two selected biomarker reactions – sterane isomerization and steroid aromatization – for which the kinetic behaviors have been extensively studied (e.g. Mackenzie and McKenzie 1983; Abbott et al. 1985; Marzi 1989) are represented in Fig. 2. In this study of the Michigan Basin, we shall use the kinetic data of Marzi (1989) for sterane isomerization because the kinetics of this reaction could be determined under laboratory conditions, yielding physicochemically useful values of parameters such as the frequency factor.

In principle, the use of data from several different biomarker reactions having different kinetic parameters should provide fairly precise estimates of the geothermal histories of sediments by allowing the separate effects of time and temperature on the maturation of organic matter to be distinguished (Mackenzie and McKenzie 1983). Our laboratory heating experiments by hydrous pyrolysis unfortunately produced fractionation of aromatic steroid hydrocarbons (i.e. triaromatics were preferentially retained in the bitumens relative to monoaromatics: Marzi 1989), so that kinetic data for steroid aromatization could not be used to calculate the geothermal history of sediments in question. Instead, some qualitative implications of this reaction were considered. Determination of the rate and extent of sterane isomerization alone only loosely constrains the different interpretations of the geothermal history of the Michigan Basin.

The biomarker reactions shown in Fig. 2 are useful in the earlier stages of catagenesis only, i.e. up to the peak of oil generation (Mackenzie 1984). As a consequence, we were restricted to investigations of Devonian and younger rocks, from which we extrapolated to estimate the maturity development of older sedimentary rocks in the Michigan Basin. In our application of biomarker transformation reactions, we are aware that there may be no direct epimerization of steranes in the saturated hydrocarbon stage. The "reaction" may be a complex process which combines neoformation from kerogen with destruction at higher thermal stress (Marzi and Rullkötter in press; Marzi et al. 1990; Abbott et al. 1990).

Table 2. Origin and analytical data of rock samples from the Michigan Basin

Sample code	Formation	Age	Depth (m)	County	TOC (%)	Rock-Eval pyrolysis			Extract (mg/g TOC)
						T_{max} (°C)	HI (mg HC/g TOC)	PI	
22084	Coal	Pennsylvanian	Outcrop	Eaton	62.7	426	374	0.16	25.5
25092	Coldwater Shale	Mississippian	364	Ogemaw	0.7	430	187	0.03	15.7
25082	Coldwater Shale	Mississippian	366	Ogemaw	1.0	431	55	0.08	6.3
25563	Antrim Shale	Devonian	1	Alpena	7.8	429	675	0.08	57.1
25564	Antrim Shale	Devonian	1	Alpena	7.9	427	615	0.09	53.9
25565	Antrim Shale	Devonian	18	Alpena	8.1	428	774	0.06	38.6
25091	Antrim Shale	Devonian	745	Missaukee	0.5	434	291	0.25	307.5
22281	Antrim Shale	Devonian	750	Missaukee	0.5	430	325	0.35	285.9
25566	Traverse Lst.	Devonian	33	Alpena	11.7	434	876	0.08	57.5
20826	Dundee Lst.	Devonian	803	Antrim	0.2	n.d.	n.d.	n.d.	103.5
20825	Dundee Lst.	Devonian	994	Gratiot	0.5	n.d.	n.d.	n.d.	70.2
20827	Detroit River Lst.	Devonian	1174	Antrim	0.3	n.d.	n.d.	n.d.	94.4
25090	Bell Shale	Devonian	1167	Missaukee	0.4	446	219	0.06	49.4
22280	Bell Shale	Devonian	1173	Missaukee	0.4	432	74	0.27	56.1
20828	Salina (A-1) Carbonate	Silurian	715	St. Clair	0.0	n.d.	n.d.	n.d.	n.d.
20829	Niagara Lst.	Silurian	795	St. Clair	0.0	n.d.	n.d.	n.d.	n.d.
25089	Queenston Shale	Ordovician	1531	Huron	0.1	432	167	0.10	70.7
22279	Queenston Shale	Ordovician	2417	Missaukee	0.1	n.d.	54	0.34	59.0
25088	Utica Shale	Ordovician	1799	Huron	0.3	457	183	0.06	37.1
25087	Utica Shale	Ordovician	1803	Huron	1.1	445	275	0.11	114.6
25086	Utica Shale	Ordovician	1804	Huron	0.4	439	187	0.08	64.1
22278	Utica Shale	Ordovician	1812	Newaygo	1.0	435	71	0.40	65.2
22277	Utica Shale	Ordovician	1818	Newaygo	0.7	435	74	0.38	64.4
20830	Black River Lst.	Ordovician	1367	St. Clair	0.1	n.d.	n.d.	n.d.	132.4
20824	Eau Claire Sst.	Cambrian	1375	St. clair	0.0	n.d.	n.d.	n.d.	150.0

2 Analytical Procedures

2.1 Samples of Rocks and Oils

Samples of selected Paleozoic strata in the Michigan Basin were obtained from drill cores, except for a coal sample, which was collected from outcrop (Table 2). An earlier study compiled biomarker data on oils from the Michigan Basin (Rullkötter et al. 1986). Some of these data are compared with the organic geochemical characteristics of the rocks examined in the present study.

2.2 Analyses

The rock samples were washed, dried and ground. TOC contents were determined by combustion in a Leco IR-112 apparatus. Rock-Eval pyrolysis was performed as described by Espitalié et al. (1977). Vitrinite reflectance measurements were made on kerogen concentrates embedded in a synthetic resin using a Zeiss Standard Universal reflected light microscope with a 100 × oil immersion objective.

The rock samples were extracted following the procedure of Radke et al. (1978), and the total extracts were fractionated by medium-pressure liquid chromatography (Radke et al. 1980). Instrumental parameters for GC and GC-MS analysis have been described by Rullkötter et al. (1985), except that in this study a VG 7070E mass spectrometer under DS 55 data system control was used.

3 Results and Discussion

3.1 General Properties of Organic Matter in Rock Samples

Results from the analysis of 25 rock samples from the Michigan Basin are compiled in Table 2. These samples were selected from core repositories on the basis of visual evidence (e.g. dark color) of elevated organic matter content. However, the organic carbon concentrations in Table 2 show that the percentages vary considerably, ranging from zero to more than

Table 2 (continued)

Sample code	Liquid Chromatography			Sterane isom	Steroid arom.	Estimated degree of contamination with diesel[a]	Vitrinite reflectance (%)
	Sat. hc. (%)	Arom. hc. (%)	NSO comp. + asphalt. (%)	20 S/(20 S + 20 R)	Tri/(tri + mono)		
22084	12	38	50	0.15	0.35	–	0.43
25092	35	5	60	0.29	0.83	–	n.d.
25082	39	24	37	0.32	0.95	–	1.11
25563	21	23	56	0.31	0.66	–	0.48
25564	25	27	48	0.35	0.72	–	0.46
25565	23	27	50	0.36	0.37	–	0.46
25091	63	8	29	0.44	0.81	–	n.d.
22281	67	8	25	0.49	0.90	–	1.25
25566	31	25	44	0.39	0.80	–	0.47
20826	59	14	27	n.d.	n.d.	n.d.	n.d.
20825	56	20	24	0.54	0.96	–	n.d.
20827	55	18	27	n.d.	n.d.	n.d.	n.d.
25090	49	9	42	0.19	n.d.	+ +	1.47
22280	61	11	28	0.42	n.d.	+ + +	1.20
20828	47	24	29	n.d.	n.d.	–	n.d.
20829	56	27	17	n.d.	n.d.	–	n.d.
25089	22	3	75	0.39	0.49	–	n.d.
22279	55	15	30	0.41	0.28	–	n.d.
25088	42	5	53	0.31	n.d.	+ + +	n.d.
25087	57	8	35	0.42	n.d.	+	n.d.
25086	61	5	34	0.44	n.d.	+ + +	n.d.
22278	78	7	15	0.47	n.d.	+ +	n.d.
22277	75	7	18	0.42	n.d.	+ + +	n.d.
20830	50	21	29	n.d.	n.d.	n.d.	n.d.
20824	34	11	55	n.d.	n.d.	+	n.d.

[a] n.d., Not determined;
+, minor, but detectable, contamination;
+ +, noticeable contamination;
+ + +, minor contamination.

11%, not including the Pennsylvanian coal sample which contains 60% organic carbon. Rock selection was biased towards clastic samples to avoid anomalous biomarker maturation results, especially those based on steroid hydrocarbons, known to occur in carbonates and other clay-free sediments (ten Haven et al. 1986). This explains why some prolific Ordovician and Silurian source rocks may have been missed. In these strata, as also in the Antrim Shale, petroleum source rock properties vary markedly over small intervals. For example, the Ordovician *Gloeocapsomopha prisca* deposits occur in distinct stringers and are not homogeneously distributed in the Trenton/Black River Limestone (e.g. Reed et al. 1986).

Contamination of a number of rock samples with diesel drilling mud additives presented serious analytical problems, particularly in those rocks with low organic carbon contents (Table 2). Parameters affected are the extract values, liquid chromatography data, and Rock-Eval pyrolysis results. Biomarker data should not have been seriously altered by the diesel contamination, although a slight decrease from expected sterane isomerization values appears in the most heavily contaminated rocks.

3.2 Hydrocarbon Contents of Rock Samples

3.2.1 Cambrian and Ordovician Samples

Many of the Cambrian and Ordovician samples contained too little extractable organic matter or were too heavily contaminated to provide reliable hydrocarbon data. The Cambrian Eau Claire Sandstone sample was virtually devoid of organic matter. The same is true for the Black River Limestone sample and the two specimens of Queenston Shale, all of Ordovician age. In the Utica Shale, organic carbon contents

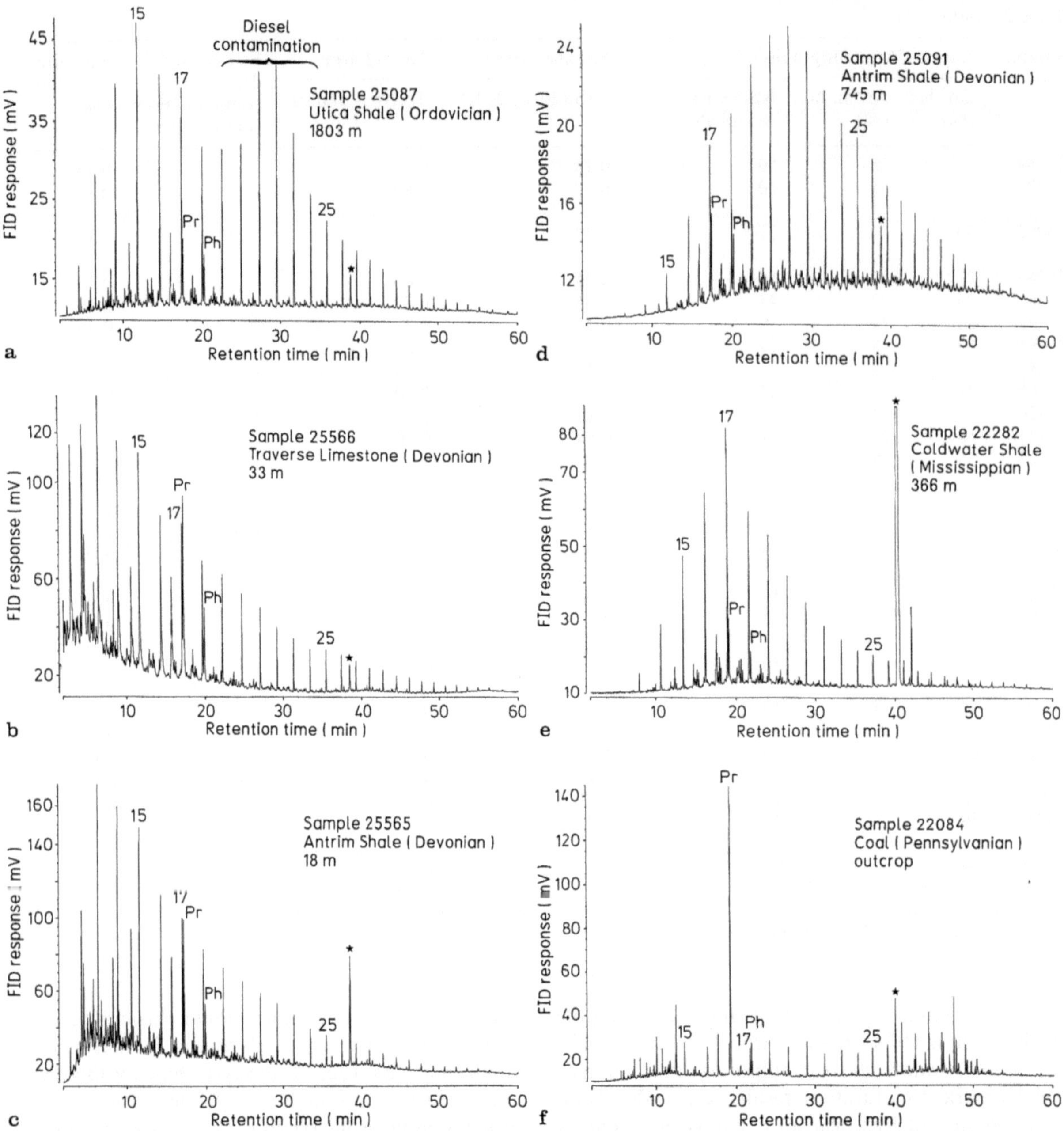

Fig. 3. Representative capillary gas chromatograms of saturated hydrocarbons in extracts of rock samples from the Michigan Basin. **a** Utica Shale, Middle Ordovician; **b** Traverse Limestone, Middle Devonian; **c** shallow Antrim Shale, Upper Devonian; **d** deep Antrim Shale, Upper Devonian; **e** Coldwater Shale, Lower Mississippian; **f** coal outcrop sample, Pennsylvanian. * Squalane internal standard; *Pr* pristane; *Ph* phytane

range up to 1.1%. Hydrogen indices are low to moderate (HI = 70 to 275). However, contamination with diesel is significant in most of these samples. Figure 3a shows the capillary column gas chromatogram of saturated hydrocarbons from the sample with the highest TOC value (sample 25087). Slight diesel contamination is visible in the n-alkane distribution around C_{21}/C_{22}; otherwise the n-alkane pattern shows a uniform decline from the maximum at C_{15} towards the higher molecular-weight range. Pristane and phytane are present in relatively high amounts and indicate that the Utica Shale is not the source of the characteristic Ordovician-type oil occurring worldwide and originating from the Trenton/Black River Limestone in the Michigan Basin (e.g. Reed et al. 1986; Hoffmann et al. 1987).

3.2.2 Silurian Samples

The two Silurian rock samples selected for this study are virtually devoid of organic matter (Table 2) and thus are not representive of the source of Silurian oils in the Michigan Basin (Vogler et al. 1981; Rullkötter et al. 1986). These samples provide no insights into the thermal history of organic matter in the Michigan Basin.

3.2.3 Devonian and Younger Rock Samples

The Bell Shale samples with TOC contents of 0.4% and low to moderate hydrogen indices (HI = 74 and 219) are significantly contaminated with diesel drilling mud additives (Table 2), and thus these data should be regarded with caution. The deeper Bell Shale sample has a very low sterane isomer ratio of 0.19, which contradicts other geochemical data (n-alkane distribution, vitrinite reflectance, T_{max}), indicating high maturity and which cannot be explained as simply diesel contamination. The high relative concentration of steranes in this sample suggests a second source of contamination from immature organic matter of unknown origin. The Dundee and Detroit River Limestones contain 0.3 to 0.5% organic carbon and are not considered good source rocks. The extract yields are not very high (< 100 mg/g TOC), typical of non-prolific source rocks at elevated maturity levels.

The Traverse Limestone with its TOC content of 11.7% and a hydrogen index of 876 mg HC/g TOC exhibits excellent source potential, although the extract yield (57 mg/g TOC) is relatively low. This is probably a consequence of its shallow depth setting and lower maturity compared to most other rocks in this study. The gas chromatogram of the saturated hydrocarbon fraction (Fig. 3b) shows a smooth n-alkane distribution with slight odd-carbon-number predominance in the C_{13}–C_{17} range. Pristane and phytane are abundant and indicate moderate maturity.

As pointed out earlier, the Antrim Shale samples are quite variable in organic matter content and source potential. At the ca. 750-m depth setting in Missaukee County, the Antrim Shale contains only 0.5% TOC and the hydrogen indices are moderate (HI = 300), probably reflecting their elevated maturity rather than an initial poorer quality of organic matter. The three shallow Antrim Shale samples from Alpena County are rich in organic matter and have kerogens with excellent petroleum potential (HI > 600). Relatively low extract yields (38–57 mg/g TOC) indicate moderate maturity, as does their distribution of the saturated hydrocarbons (Fig. 3c) which, although lacking any odd(or even)-carbon-number preference in the n-alkanes, is not very different from that of the Traverse Limestone sample. In contrast to the shallow samples, the deeper Antrim Shale sample has a saturated hydrocarbon distribution which appears more mature (Fig. 3d). In this case, however, there may have been a significant evaporative loss of lower molecular-weight alkanes either during the experimental treatment or during core storage.

The Coldwater Shales have relatively low organic matter contents (TOC = 0.7–1%) and very low extract yields (6–16 mg/g TOC), which are consistent with their poor kerogen quality (HI = 55–187; Table 2). Low saturated hydrocarbon yields, relative to the internal standard, are also obvious from their saturated hydrocarbon gas chromatograms (Fig. 3e).

The Pennsylvanian coal sample has bulk organic geochemical characteristics typical of a lignite. The saturated hydrocarbon distribution is representative of immature terrigenous organic matter (Fig. 3f) with a dominance of pristane over n-heptadecane and abundant polycyclic hydrocarbons in the higher molecular-weight range.

3.3 Thermal Maturity of Organic Matter

3.3.1 Vitrinite Reflectance

Vitrinite reflectance values were determined for a few samples of Devonian and younger age. Low to marginal maturity was observed only for the youngest and shallowest sediments. Below a present depth of a few hundred meters, vitrinite reflectance values for even the relatively young Coldwater and Antrim Shale samples exceed 1%. The organic matter of most samples is well within the window of thermal hydrocarbon generation even though this is not indicated by the Rock-Eval T_{max} values which remain in the range 430 to 435 °C. Although identifiable vitrinite particles are scarce in some of these marine sediments, the vitrinite reflectance data confirm the advanced maturity level of all except the most shallow samples.

3.3.2 Sterane Isomerization and Steroid Aromatization

The conversion of (20R)- to (20S)-24-ethylsteranes (C_{29}) and the transformation of monoaromatic into triaromatic steroid hydrocarbons have been used as molecular maturation parameters (cf. Fig. 2). Results for those rock samples from Table 2 which contained

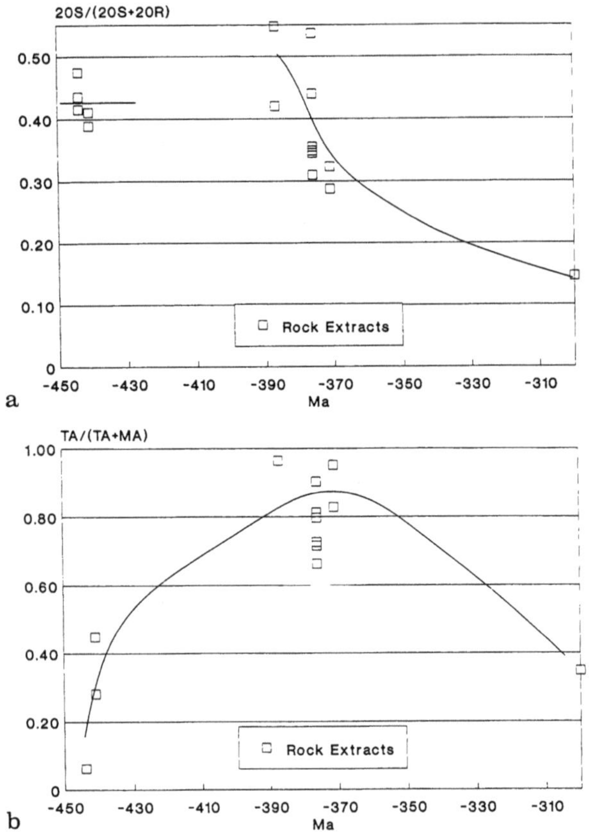

Fig. 4. Measured extents of **a** sterane isomerization and **b** steroid aromatization in Michigan Basin rock samples of various Paleozoic ages listed in Table 2

sufficient quantities of extractable matter are summarized in Fig. 4a, b. The extent of sterane isomerization or steroid aromatization is shown as a function of the age of the respective stratigraphic unit. No allowance was made for different depth settings of sediments of the same age, and this may account for some of the scatter.

Sterane isomerization (Fig. 4a) increases from a low value in the Pennsylvanian coal (0.15) via moderately increased values in the Coldwater Shales and the shallow Antrim Shale samples (around 0.35) to values close to equilibrium (0.54) in the deeper Antrim Shales and in the Dundee Limestone. The Bell Shale sample with its low isomerization value of 0.19 is believed to be heavily affected by diesel drilling mud additive. The extent of sterane isomerization in all the Ordovician samples is lower than in many of the younger and shallower Devonian rocks and significantly different from the equilibrium value. Although some effect has to be attributed to the diesel contamination (particularly the lowest value), recent hydrous pyrolysis experiments have demonstrated that sterane isomerization proceeds through a maximum and decreases when the maturity advances into the overmature range (Marzi 1989; Rullkötter and Marzi 1989). The same inversion effect has also been described for natural samples (Snowdon et al. 1987), but this process is not well understood. The presence of inverted isomerization ratios signifies that Devonian rocks in the Michigan Basin have reached the oil generation zone and that organic matter in the underlying Silurian and Ordovician strata is highly mature. This substantiates the findings of Cercone (1984) regarding thermal maturity of Paleozoic rocks in the Michigan Basin.

Steroid aromatization (Fig. 4b) shows a trend similar to that of sterane isomerization. The increase from the Pennsylvanian coal to the Devonian sediments is greater than that for sterane isomerization. This may indicate a phase of rapid heating in the past which affected steroid aromatization more strongly than sterane isomerization (cf. Rullkötter et al. 1985). Of the Ordovician samples, only three contained measurable amounts of aromatic steroids, which are commonly up to ten times less abundant than the saturated steranes (Rullkötter et al. 1984). The low aromatization values of the overmature Ordovician samples are consistent with those observed in heating experiments at high temperatures (Marzi 1989; Rullkötter and Marzi 1989; Marzi and Rullkötter in press).

3.3.3 Oil Maturities Compared to Rock Organic Matter Maturities

Biomarker maturity data for 27 crude oils from the Michigan Basin have been reported by Rullkötter et al. (1986). Silurian and Ordovician oils have sterane isomerization values between 0.40 and 0.52. Because of the recent finding of decreasing values after the equilibrium maximum (0.54: Marzi 1989; Marzi and Rullkötter in press), these oils are mature to overmature, similar to the rock samples. The same is true for the oils in reservoirs of the Devonian Dundee and Traverse Limestones, which on average have higher sterane isomerization values than the oils in older reservoirs. This, again, is consistent with results obtained from the rock samples (cf. Fig. 4a). In all these oils, the concentration of aromatic steroids is too low for the extent of steroid aromatization to be determined.

Oils produced from reservoirs in the Devonian Berea and Stray Sandstones are postulated to have originated from a Devonian source (Pruitt 1983; Rullkötter et al. 1986). This contrasts with the Devonian Dundee and Traverse oils which are believed to have originated mainly from Ordovician source rocks, with the Traverse-hosted oils in particular containing admixtures of oil from Silurian and possibly Devonian sources (e.g. Vogler et al. 1981; Pruitt 1983; Rullkötter et al. 1986). The Berea and Stray Sandstone oils have sterane isomerization values of about 0.38 which are in the range of those determined for the shallower Upper Devonian rocks in this study, and therefore consistent with a Devonian source for the oils.

3.4 A Biomarker-Derived Model of Geothermal History for the Michigan Basin

The available molecular organic geochemical data now provide a fairly consistent picture of oil sources and present-day maturity of dispersed organic matter in the Michigan Basin. This now needs to be integrated with the geological and geothermal history of the basin. As already pointed out, there is a discrepancy between the more theoretical geophysical model of Nunn et al. (1984) and the concept of Cercone (1984), the latter based on organic petrographic data and a simple kinetic approach (Lopatin/Waples Model). At first glance, our molecular and bulk maturity data agree with the maturity values estimated by Cercone (1984).

Sterane isomerization kinetics allow a tentative reconstruction of the thermal history of the Michigan Basin based on the maturity of its Paleozoic organic matter. We have applied the kinetic data determined by Marzi (1989; cf. Fig. 2) and various constraints imposed by the geologic history of the Michigan Basin. We started with the thermal history proposed by Nunn et al. (1984) and used maximum burial depths and sediment thicknesses in the center of the Michigan Basin. The calculation, explained in more detail by Rullkötter and Marzi (1989) and Marzi and Rullkötter (in press), shows that sterane isomerization in sediments at the base of the Middle Ordovician would have reached equilibrium [$20S/(20S + 20R) = 0.54$] about 200 Ma B.P., whereas the basal Silurian sediments [$20S/(20S + 20R) = 0.50$] would still be approaching it today. In basal Devonian sediments, sterane isomeriation would barely have started (< 0.05).

The assumptions of Cerone (1984) – a geothermal gradient of 45 °C/km during the Carboniferous followed by continuous cooling towards the present gradient of 22 °C/km, plus the erosion of 1 km of Late Carboniferous to Early Pennsylvanian sediments – lead to the result that sterane isomerization reached equilibrium in Ordovician and Silurian sediments more than 350 Ma B.P. and that significantly higher isomerization values are observed in the Devonian strata. This calculation for the sediments in the center of the Michigan Basin leads to slightly higher values than those measured on rock samples in this study. By taking the present depth settings of our samples into account, however, a good match between calculated and measured isomerization values is obtained for the base of the Devonian (Dundee Limestone), the Antrim Shale, and the Coldwater Shale. Nevertheless, a major discrepancy remains between the measured and calculated ratios of the Pennsylvanian coal (0.15 versus 0.01: Rullkötter and Marzi 1989).

The maturity value for the Pennsylvanian coal is best explained by assuming that the geological history was more complex in the latest Paleozoic (and probably also during Early Mesozoic time) than the current literature states (Dorr and Eschman 1970; Nunn et al. 1984). Our final calculation of the development of sterane isomerization values (Fig. 5) starts with the geological and geothermal history proposed by Nunn et al. (1984) for the Ordovician through Devonian. We then assume additional subsidence for the entire Carboniferous, with a maximum amount between 1800 m at the central basin location of the Pennsylvanian coal and about 900 m at the locations of samples from the northern Devonian reef trend (Fig. 1). Erosion was then assumed to have occurred continuously from the end of the Late Carboniferous to the Early Jurassic. During subsidence, the geothermal gradient is presumed to have reached a value of 35 °C/km, remained constant for about 60 Ma into the Permian, and then to have declined through the Permian and the Jurassic to the present value. Under these conditions, all calculated sterane isomerization data match the measured values (Fig. 5), within the limits of experimental error. The values of additional subsidence and heat flow may be slightly overestimated in view of observations that the activation energy of apparent sterane isomerization is lower in coals than in shales (Strachan et al. 1989). At low heating rates and low temperatures, the epimerization of steranes at C-20 is more advanced in coals than in shales. Even so, some amount of deeper burial is required to explain the sterane isomer ratio and the vitrinite reflectance value of the Pennsylvanian coal sample.

The calculations performed herein are a preliminary attempt to define some new constraints on the geological and geothermal history of the Michigan Basin. Despite the limitations of these calculations, the present understanding of the evolution of this Paleozoic basin as it is reflected in the literature has to be modified considerably. The new organic geochemical data presented here indicate that the intracratonic

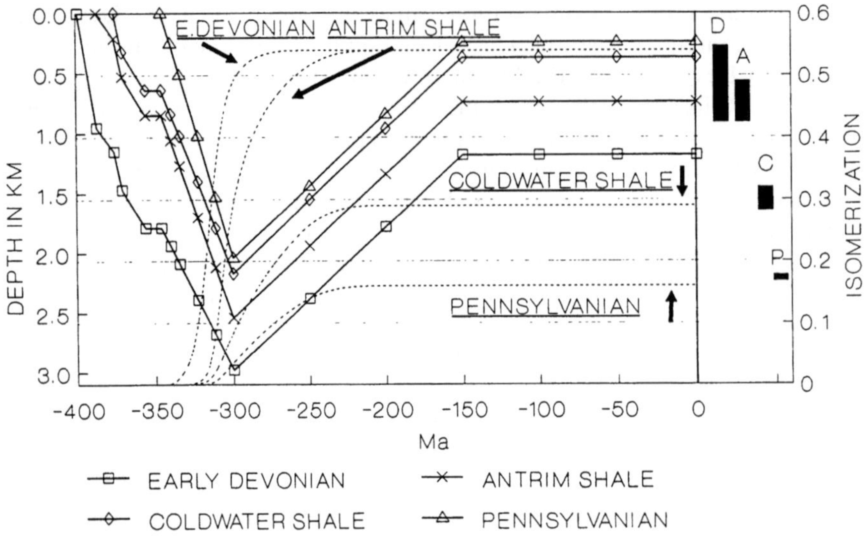

Fig. 5. A model of the subsidence and erosion of the Michigan Basin since the Early Proterozoic, based on the thermal maturation of organic matter as indicated by the isomerization of sterane biomarkers. *Solid lines* are subsidence curves, *dotted lines* represent the progress of sterane isomerization, and the *bars* on the *right-hand side* show the ranges of sterane isomerization measured in Michigan Basin rock samples

Michigan Basin is not as stable as has been previously believed. Rather, it may be a part of, or have a connection to, the mid-American rift system and therefore may have undergone more tectonic movement than heretofore recognized. An investigation of the regional movement of subsurface fluids in the Michigan Basin led Vugrinovitch (1989) to similar conclusions, particularly with respect to higher heat flow in the past.

4 Conclusions

A number of conclusions emerge from this study of the thermal maturation of Paleozoic organic matter in the Michigan Basin:

1. Devonian strata in the central part of the basin are sufficiently mature to have generated petroleum. Sterane isomerization, steroid aromatization, and vitrinite reflectance data show that organic matter is thermally mature.
2. Ordovician strata from the center of the basin are thermally overmature with regard to petroleum generation. The absence of vitrinite in these rocks enhances the importance of the biomarker data. Sterane isomerization and steroid aromatization reactions are past thermal equilibrium.
3. Samples of Devonian Antrim Shale from near the edge of this bowl-shaped basin are significantly less mature than are more deeply buried samples of the same formation. This contrast in maturity with location in the Antrim Shale indicates that it is the depth, not the duration of burial, that has been most important in the thermal maturation of organic matter in the Michigan Basin.
4. The degree of thermal maturation of organic matter in these rocks requires the geothermal history of the Michigan Basin to have been more complex than previously believed, with greater heating having occurred during the Mid- to Late Paleozoic than present heat flow would indicate. Some combination of substantial subsidence, uplift, and erosion (e.g. Cercone 1984) and of higher heat flows (e.g. Vugrinovitch 1989) was probably involved.

Acknowledgements. We thank J. M. Budai for providing access to the rock core repository at The University of Michigan and K. W. Dunham for assisting in sample selection. We are grateful to B. Horsfield, M. Radke, and R. G. Schaefer for allowing use of their laboratory facilities for TOC determination/Rock-Eval pyrolysis, extraction/liquid chromatography, and gas chromatography, respectively. The technical assistance of E. Biermanns, U. Disko, R. Harms, J. Hoeltkemeier, B. Kammer, F. J. Keller, F. Leistner, W. Lüdtke, A. Offergeld, H. Schnitzler, and H. Willsch is much appreciated. Critical comments from J. M. Budai, K. R. Cercone, D. M. McKirdy, and H. N. Pollack improved this manuscript.

References

Abbott GD, Lewis CA, Maxwell JR (1985) The kinetics of specific organic reactions in the zone of catagenesis. Philos Trans R Soc London Ser A 315:107–122

Abbott GD, Wange GY, Eglinton TI, Home AK, Petch GS (1990) The kinetics of sterane biological marker release and degradation processes during the hydrous pyrolysis of vitrinite kerogen. Geochim Cosmochim Acta 54:2451–2461

Cercone KR (1984) Thermal history of the Michigan Basin. Bull Am Assoc Petrol Geol 68:130–136

Cercone KR, Pollack HN (1989) Anomalous thermal maturity of the Michigan Basin: new hypothesis. Bull Am Assoc Petrol Geol 73:342 (abs)

Dorr JA Jr, Eschman DF (1970) Geology of Michigan. Univ Mich Press, Ann Arbor, 476 pp

Espitalié J, Laporte JL, Madec M, Maquis F, Leplat P, Paulet J, Boutefeu A (1977) Méthode rapide de caractérisation des roches mères, de leur potentiel pétrolier et de leur degré d'évolution. Rev Inst Fr Pétrol 32:23–42

ten Haven HL, de Leeuw JW, Peakman TM, Maxwell JR (1986) Anomalies in steroid and hopanoid maturity indices. Geochim Cosmochim Acta 50:853–855

Hoffmann CF, Foster CB, Powell TG, Summons RE (1987) Hydrocarbon biomarkers from Ordovician sediments and the fossil alga *Gloeocapsomorpha prisca* Zalessky 1917. Geochim Cosmochim Acta 51:2681–2697

Illich HA, Grizzle PL (1983) Comment on "Comparison of Michigan Basin crude oils" by Vogler et al. Geochim Cosmochim Acta 47:1157–1159

Illich HA, Grizzle PL (1985) Thermal subsidence and generation of hydrocarbons in Michigan Basin: discussion. Bull Am Assoc Petrol Geol 69:1401–1404

Mackenzie AS (1984) Application of biological markers in petroleum geochemistry. In: Brooks J, Welte DH (eds) Advances in Petroleum Geochemistry, vol 1. Academic Press, New York London, pp 115–214

Mackenzie AS, McKenzie DP (1983) Isomerization and aromatization of hydrocarbons in sedimentary basins formed by extension. Geol Mag 120:417–470

Marzi R (1989) Kinetik und quantitative Analyse der Isomerisierung und Aromatisierung von fossilen Steroidkohlenwasserstoffen im Experiment und in natürlichen Probensequenzen. PhD Thesis, Rheinisch-Westf Tech Hochsch, Aachen

Marzi R, Rullkötter J (in press) Qualitative and quantitative evolution and kinetics of biological marker transformations – laboratory experiments and application to the Michigan Basin. In: Moldowan JM, Albrecht P, Philp RP (eds) Biological Markers in Sediments and Petroleum. Prentice Hall, Englewood Cliffs

Marzi R, Rullkötter J, Perriman WS (1990) Application of the change of sterane isomer ratios to the reconstruction of geothermal histories: implications of the results of hydrous pyrolysis experiments. In: Durand B, Béhar F (eds) Advances in Organic Geochemistry 1989. Pergamon, Oxford, pp 91–102

Meyers PA, Moore WE (1983) Comparison of Michigan Basin crude oils (reply to comments by Illich and Grizzle and by Pruitt). Geochim Cosmochim Acta 47:1161–1162

Nunn JA, Sleep NH (1978) Thermal contraction and petroleum generation in Michigan Basin. Am Assoc Petrol Geol Abstr Pap, pp 99–100

Nunn JA, Sleep NH, Moore WE (1984) Thermal subsidence and generation of hydrocarbons in Michigan Basin. Bull Am Assoc Petrol Geol 68:296–315

Nunn JA, Sleep NH, Moore WE (1985) Thermal subsidence and generation of hydrocarbons in Michigan Basin: reply. Bull Am Assoc Petrol Geol 69:1185–1187

Powell TG, Macqueen RW, Barker JF, Bree DG (1984) Geochemical character and origin of Ontario oils. Bull Can Soc Petrol Geol 32:289–312

Pruitt JD (1983) Comment on "Comparison of Michigan Basin crude oils" by Vogler et al. Geochim Cosmochim Acta 47:1159–1161

Radke M, Sittardt HG, Welte DH (1978) Removal of soluble organic matter from rock samples with a flow-through extraction cell. Anal Chem 50:663–665

Radke M, Willsch H, Welte DH (1980) Preparative hydrocarbon group type determination by automated medium pressure liquid chromatography. Anal Chem 52:406–411

Reed JD, Illich HA, Horsfield B (1986) Biochemical evolutionary significance of Ordovician oils and their sources. Org Geochem 10:347–358

Rullkötter J, Marzi R (1989) New aspects of the application of sterane isomerisation and steroid aromatisation to petroleum exploration and the reconstruction of geothermal histories of sedimentary basins. Preprint Div Petrol Chem Am Chem Soc 34:126–134

Rullkötter J, Mackenzie AS, Welte DH, Leythauser D, Radke M (1984) Quantitative gas chromatography-mass spectrometry analysis of geological samples. Org Geochem 6:817–827

Rullkötter J, Spiro B, Nissenbaum A (1985) Biological marker characteristics of oils and asphalts from carbonate source rocks in a rapidly subsiding graben, Dead Sea, Israel. Geochim Cosmochim Acta 49:1357–1370

Rullkötter J, Meyers PA, Schaefer RG, Dunham KW (1986) Oil generation in the Michigan Basin: a biological marker and carbon isotope approach. Org Geochem 10:359–375

Sleep NH, Sloss LL (1978) A deep borehole in the Michigan Basin. J Geophys Res 83:5815–5819

Snowdon LR, Brooks PW, Williams GK, Goodarzi F (1987) Correlation of the Canol Formation source rock with oil from Norman Wells. Org Geochem 11:529–548

Strachan MG, Alexander R, van Bronswijk W, Kagi RI (1989) Source and heating rate effects upon maturity parameters based on ratio of 24-ethylcholestane diastereomers. J Geochem Explor 31:285–294

Summons RE, Powell TG (1987) Identification of aryl isoprenoids in source rocks and crude oils: biological markers for the green sulphur bacteria. Geochim Cosmochim Acta 51:557–566

Van der Voo R (1988) Paleozoic paleogeography of North America, Gondwana, and intervening displaced terranes: comparisons of paleomagnetism with paleoclimatology and biogeographical patterns. Geol Soc Am Bull 100:325–338

Vogler EA, Meyers PA, Moore WE (1981) Comparison of Michigan Basin crude oils. Geochim Cosmochim Acta 45:2287–2293

Vugrinovitch R (1989) Subsurface temperatures and surface heat flow in the Michigan Basin and the relationship to regional subsurface fluid movement. Mar Petrol Geol 6:60–70

The Influence of *Gloeocapsomorpha prisca* on the Organic Geochemistry of Oils and Organic-Rich Rocks of Late Ordovician Age from Canada

Martin G. Fowler [1]

[1] Geological Survey of Canada, Institute of Sedimentary and Petroleum Geology, 3303-33rd St. N. W., Calgary, Alberta T2L 2A7, Canada
Geological Survey of Canada Contribution No. 22889

Abstract

There has recently been considerable interest in the geochemistry of Ordovician oils and source rocks. Here, organic petrological, Rock-Eval and especially biomarker data are presented for three sets of Late Ordovician organic-rich rocks from Canada that have different geochemical characteristics. The Yeoman kukersites from the Saskatchewan portion of the Williston Basin are representative of those Ordovician samples reported previously to have a distinctive chemical character (e.g. very low concentrations of acyclic isoprenoids and high concentrations of monocyclic alkanes). This distinctive chemistry is principally determined by the main contributor to the organic matter of these samples, *Gloeocapsomorpha prisca*, and the minimal microbial reworking of the primary organic matter during diagenesis. The Collingwood oil shale samples from southern Ontario also show chemical and optical evidence of a significant contribution by *G. prisca* to their organic matter. However, in this case there were also inputs from other algae and especially bacteria (as is evident from the predominantly amorphous appearance of the organic matter) which diluted the "classic" kukersite chemistry. The third group of samples are from Southampton Island in the eastern Canadian Arctic. There is no evidence that these oil shales received any contribution from *G. prisca;* and their biomarker characteristics suggest that they were deposited under hypersaline conditions.

The occurrence of kukersites and "diluted kukersites" are reviewed both in terms of paleogeography and age. Also discussed are geochemical details of a sample from the Middle Cambrian Mount Cap Formation of the Canadian Northwest Territories that shares many of the features of Ordovician kukersites. It is suggested that the organism which contributed to the organic matter of this sample may have been a direct biological precursor of *G. prisca*. Finally, some possible reasons for the absence of *G. prisca* in sediments younger than the Ordovician are suggested.

1 Introduction

The geochemistry of Ordovician oils is of considerable interest because many show distinctive chemical features when compared to oils of other ages. These include a very high concentration of n-alkanes up to C_{19} with a strong odd-carbon-number predominance, very low concentrations of acyclic isoprenoids and polycyclic alkanes, and a relatively high concentration of monocyclic alkanes (Martin et al. 1963; Tissot et al. 1977; Zumberge 1983; Fowler et al. 1986; Reed et al. 1986; Hoffmann et al. 1987). Several authors have discussed the similarities between Ordovician samples showing these characteristics from different parts of the world, including North America (e.g. Fowler et al. 1986; Reed et al. 1986; Longman and Palmer 1987), Australia (Foster et al. 1986; Hoffmann et al. 1987) and Estonia (Reed et al. 1986). The distinctive chemistry of these Ordovician oils is thought to be due to the main organism contributing to the organic matter of their source rocks, which is *Gloeocapsomorpha prisca* Zalessky 1917 (Foster et al. 1986; Reed et al. 1986; Hoffman et al. 1987). In many of these rocks, for example in intervals of the Yeoman Formation of the Williston Basin described later, this organism constitutes greater than 90% of the organic matter. Following Hutton's (1987) classification of oil shales, this type of deposit will be termed a kukersite. They are so named after the Estonian kukersites which were the sediments in which Zalessky (1917) first recognized *G. Prisca*. However, this organism was almost certainly described, but not named, by White (in Bain 1906). *G. prisca* is also the main contributor to oil-prone organic matter "Assemblage A" described by Jacobson et al. (1988) in Middle Ordovician rocks from Iowa.

Understandably, most of the interest in Ordovician organic geochemistry has been concerned with kukersites and, especially, their derived oils. However, it is important to realize that there are Ordovician samples with geochemical characteristics which are less specific to the Ordovician. These include, for

M. Schidlowski et al. (Eds.)
Early Organic Evolution: Implications for Mineral and Energy Resources

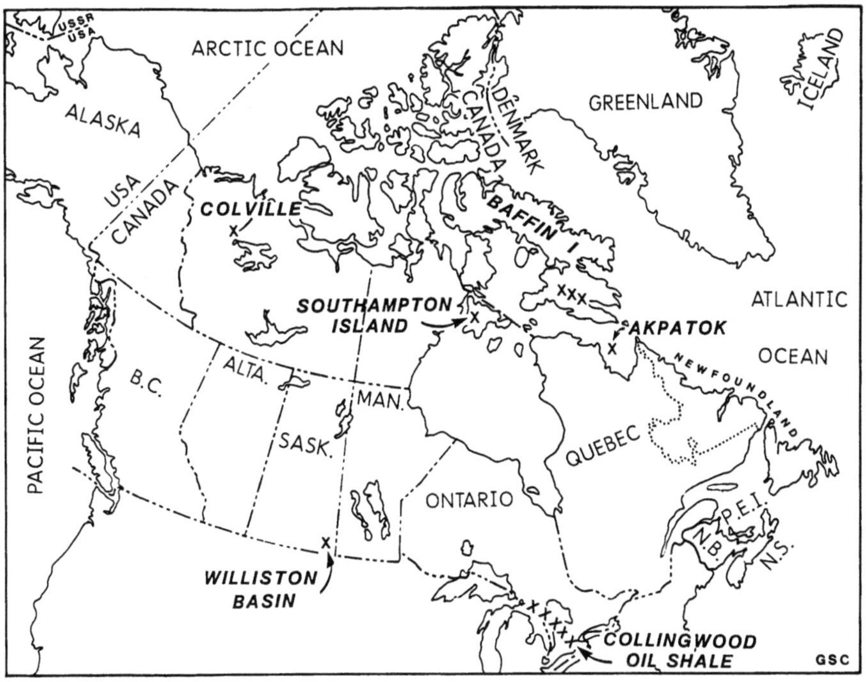

Fig. 1. Map of Canada showing locations of deposits discussed in text

example, the Vinini and Womble shales from the USA (Fowler and Douglas 1984), the Tretaspis Shale from Sweden (Vlierboom et al. 1986) and the Boas oil shale from the Canadian Arctic (Macauley et al. 1990; also discussed below).

Here, data will be presented from three sets of Canadian samples that show different geochemical characteristics representing three types of Ordovician organic-rich deposits. All of these samples were originally thought to have been deposited around the same time during the Late Ordovician (Edenian-Maysvillian) in low-latitude carbonate environments. The first group of samples is from the Yeoman Formation of the Saskatchewan portion of the Williston Basin (Fig. 1). These are examples of kukersite-type deposits as defined above. Hence, their geochemistry is thought to reflect the biochemistry of *G. prisca*. The second group of samples is from the Collingwood oil shales of southern Ontario (Fig. 1). They have geochemical properties that are intermediate between those of kukersites and samples of other ages, and are similar to those described by Jacobson et al. (1988) from their organic matter "Assemblage B". Here, they will be referred to as "diluted kukersites" because, although *G. prisca* makes a significant contribution to their organic matter, it has been diluted by inputs from other organisms. The final set of samples comprises oil shales from Southampton Island in the Canadian Arctic (Fig. 1). Originally, these oil shales were thought to represent a single deposit of Edenian-Maysvillian age (Macauley 1986; Macauley et al. 1990). However, recent work now suggests two separate deposits, both slightly younger than the two other groups of samples in this study (McCracken and Nowlan 1989). The Southampton Island oil shales are examples of deposits that show no evidence of a major contribution from *G. prisca*.

As the kukersites display characteristics that are unique, these rocks will be discussed in most detail. Following the lead of other authors, an attempt will be made to use the organic geochemistry of kukersites to say something about the nature of *G. prisca*. There will also be some discussion on the age range of this organism, with particular reference to a possible related Middle Cambrian deposit and the reasons why kukersites are not found in post-Ordovician sediments.

Yeoman Formation Kukersites

The Yeoman Formation in the Saskatchewan portion of the Williston Basin (Fig. 1) is the basal formation of the Upper Ordovician to lowermost Silurian Big Horn Group, and was probably deposited in Late Edenian-Early Maysvillian time (Macauley et al. 1990). The kukersites occur at several levels within the Yeoman Formation as thin beds of "thinly laminated pale yellowish brown bituminous lime mudstone with a resinous lustre" (Osadetz et al. 1989; see also Kendall

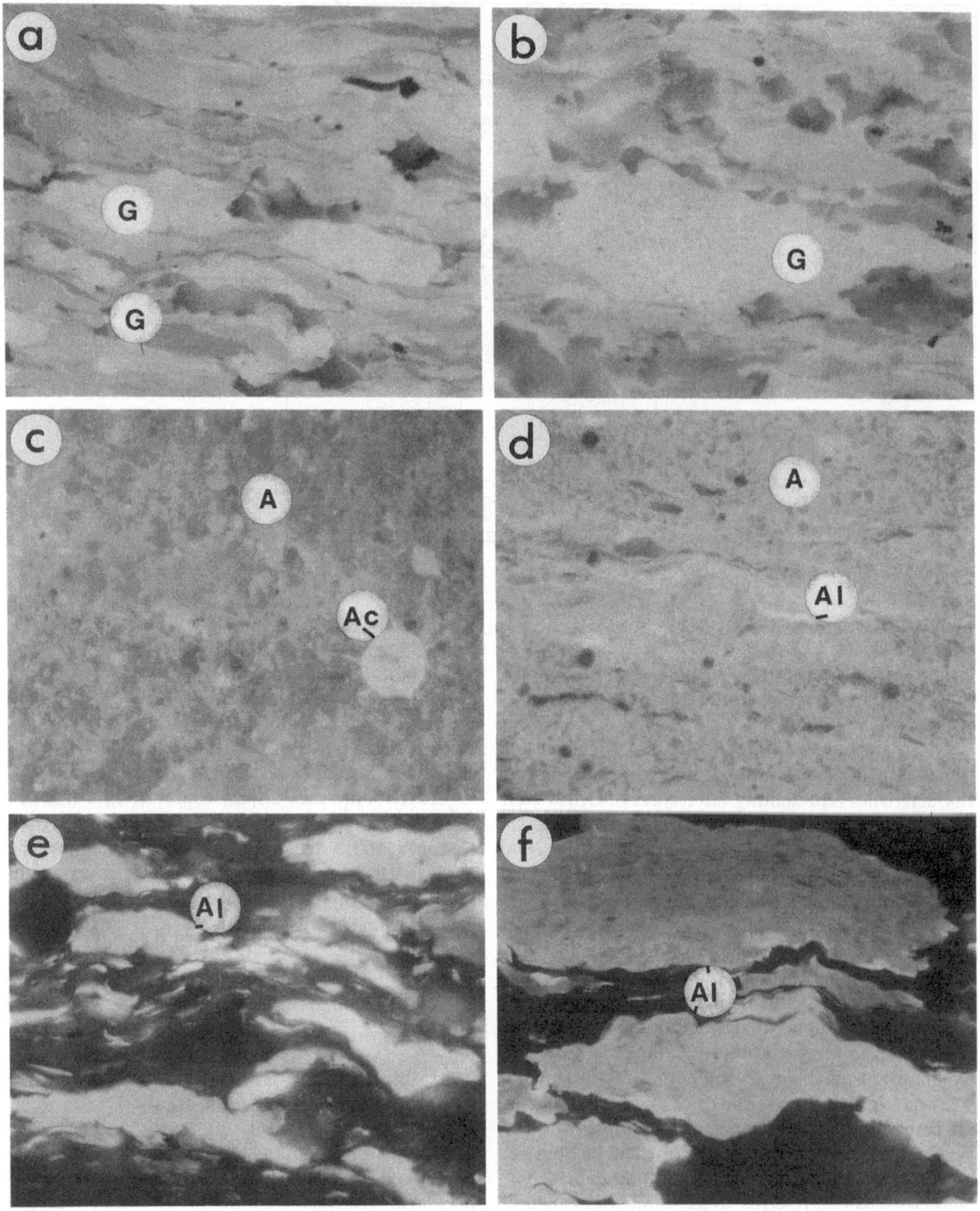

Fig. 2

1976). These beds are equivalent to the "kerogenites" described within the C member of the Red River Formation of the Montana and North Dakota part of the Williston Basin by Kohm and Louden (1978). The Yeoman kukersites are considered by Osadetz et al. (1989) to be the major source of the oils within Ordovician reservoirs of the Williston Basin.

The organic matter of the kukersites is observed to be predominately (> 90%) algal bodies squashed parallel to the bedding plane (Fig. 2a, b). These algae are almost exclusively *G. prisca*. The amount of internal structure shown by the algal bodies varies. Their size can range from 5 up to 100 μ in length and has been related to different growth stages during the life cycle of the organism (Stasiuk and Osadetz 1990). There is very little amorphous organic matter; and only trace amounts of chitinozoa, acritarchs and other algal types are associated with the kukersite laminations. The dominant mineral observed in the organic-rich laminae is dolomite. In some samples the organic matter is intimately associated with the dolomite (e.g. dolomite crystals infilling *G. prisca* bodies: Stasiuk pers. comm. 1988). The appearance of the organic matter of the Yeoman Formation closely resembles that of other kukersites such as the Guttenberg Oil Rock of the Middle Ordovician Decorah Subgroup of southwestern Wisconsin (Fowler 1984; Jacobson et al. 1988).

The results of the Rock-Eval analysis of Ordovician samples within the Saskatchewan portion of the Williston Basin have recently been reported by Osadetz et al. (1989). The kukersite samples from the Yeoman Formation have TOC values up to 34.9%, the average of the 66 samples analyzed being 8.44%. When the hydrogen and oxygen index (HI; OI) values of these samples are plotted on a pseudo-van Krevelen diagram, their organic matter is indicated to be Type I. They show high HI values (commonly above 800) and very low OI values (most below 20). The samples with lower HI and higher OI values tend to be those with lower TOC values (less than 3%). These are also the samples in which the *G. prisca* bodes show less tendency to be laminated, are more dispersed amongst the mineral matter, and hence are not true kukersites.

As the organic matter of the kukersites is almost exclusively telalginite, it is not surprising that it should be defined as Type I kerogen by Rock-Eval analysis. The Yeoman kukersites with TOC values greater than 3% have T_{max} values that range from 446–458 °C. This is well above the T_{max} of 435 °C that is commonly taken as indicating the maturation threshold for oil generation from Type II organic matter. Type I kerogens have high T_{max} values even at relatively low levels of thermal maturity because they are composed almost entirely of long alkyl chains. The amount of energy required to break off volatile components is high (about 80 $k\,cal^{-1}\,mol^{-1}$ or 335 $KJ\,mol^{-1}$ to rupture a carbon-carbon bond) until the chain length is quite small and the kerogen essentially inert. As the OI values indicate, oxygen content is low and hence there is less opportunity for pyrolysis at lower temperatures through the breaking of the more easily ruptured carbon-oxygen bonds. Essentially the same explanation was given by Foster et al. (1986) for the high T_{max} values that they reported for immature kukersite samples from the Canning Basin of Western Australia.

The C_{15+} saturated hydrocarbons of the Yeoman kukersites (Fig. 3a, b) are dominated by C_{15}–C_{19} n-alkanes showing an odd-carbon-number preference. The C_{20} and higher molecular weight n-alkanes are in much lower concentrations and show no odd- or even-carbon-number preference. This is the characteristic n-alkane distribution that has been reported from the extracts of Ordovician rocks from different areas of the world by several authors (Williams 1974; Fowler and Douglas 1984; Foster et al. 1986; Reed et al. 1986; Hoffmann et al. 1987; Jacobson et al. 1988). It shows an obvious relationship to that reported for oils (e.g. Fig. 3c) in Ordovician reservoirs of the Williston Basin (Martin et al. 1963; Williams 1974; Zumberge 1983; Reed et al. 1986; Brooks et al. 1987; Longman and Palmer 1987). Whole-oil gas chromatograms show that the odd-carbon-number preference can extend down to C_9 (Martin et al. 1963; Reed et al. 1986). This n-alkane distribution, dominated by odd-

Fig. 2. Ordovician and Cambrian sediments under blue fluorescent light (filters: excitation 450–490 nm; beam splitter 420 nm; and barrier 510 nm) and oil immersion; lengthwise field of view of photographs is either 300 μ (**a, d, e**) or 120 μ (**b, c** and **f**). **a** Yeoman Formation kukersite (perpendicular section) from the Saskatchewan portion of the Williston Basin. Sample from a depth of 3075.7 m in 15-19-2-14W2. The organic matter of this sample is almost entirely *Gloeocapsomorpha prisca* (*G*) oriented parallel to bedding. **b** Same sample as **a** but algal bodies observed under higher magnification. **c** Collingwood oil shale from the Collingwood area of southern Ontario. Sample collected from a depth of 53.94–54.14 m in Ontario Geological Survey borehole Clgd 01. Organic matter mostly amorphous (*A*) with some acritarch bodies distinguishable (*Ac*). **d** Oil shale collected from an outcrop at Gore Point, Southampton Island. The sample is a laminated marl with mostly amorphous organic matter (*A*) and some rare algal bodies (*Al*). **e** Middle Cambrian Mt. Cap Formation from a depth of 943 m in the Colville D-45 well drilled in the Northwest Territories. Organic matter dominated by deformed algal bodies (*Al*) (oriented parallel to the bedding) that show a resemblance to *Gloeocapsomorpha prisca*. **f** Same sample as **e** but algal bodies (*Al*) observed under higher magnification

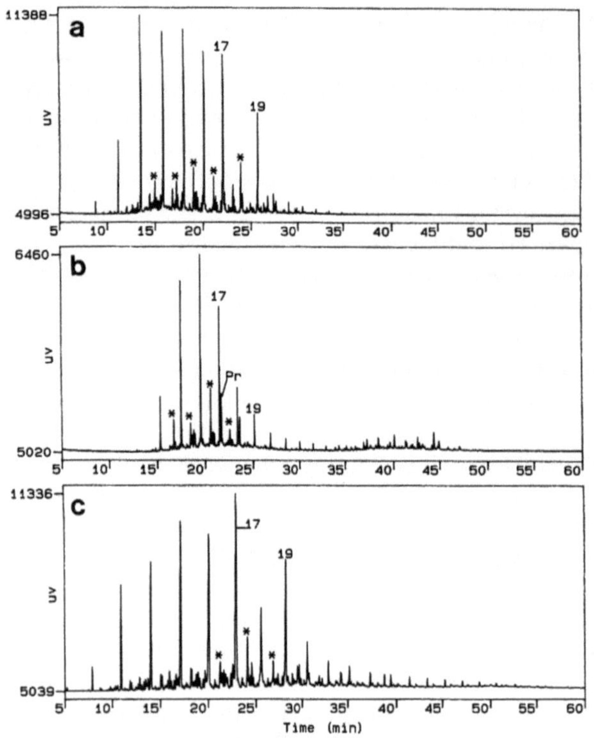

Fig. 3. Gas chromatograms of C_{15+} saturated hydrocarbons in **a** a mature Yeoman kukersite, location 06-32-008-16W2, depth 2459 m; **b** a very immature Yeoman kukersite, location 04-10-033-01 W3, depth 1466 m; **c** Upper Ordovician oil from West Oungre, location 04-22-003-15W2, depth 2929–2936 m. *17* and *19* indicate C_{17} and C_{19} n-alkanes; *Pr* pristane; * indicates n-alkylcyclohexanes

carbon-numbered homologues, most likely results from decarboxylation of the fatty acids of *G. prisca*. The carbon-number range of the n-alkanes favour *G. prisca* being a type of marine eukaryotic algae (Fowler and Douglas 1984).

The composition of the gasoline-range hydrocarbons of the Ordovician oils from Saskatchewan is also unusual. They are dominated by normal and branched alkanes, with only trace amounts of aromatics such as benzene and toluene (Snowdon and Osadetz 1988).

The abundance of acyclic isoprenoids relative to n-alkanes is extremely low in most of the Yeoman kukersite extracts (Fig. 3a) and in the Saskatchewan Ordovician oils (Fig. 3c). This is one of the most frequently noted characteristics of Ordovician samples (e.g. Foster et al. 1986; Reed et al. 1986; Hoffman et al. 1987; Jacobson et al. 1988). The principal source of phytane, pristane and lower-molecular weight isoprenoids in sediments is thought to be the phytyl side chain of the chlorophyll-a molecule (e.g. Volkman and Maxwell 1986), although there are other possible sources such as archaebacterial phospholipids (Nissenbaum et al. 1972; Brassell et al. 1981), algal tocopherols (Goossens et al. 1984) and pristane from zooplankton (Blumer et al. 1963). The very low concentration of acyclic isoprenoids in Ordovician oils was suggested by Reed et al. (1986) to be a consequence of *G. prisca* being a non-photosynthetic organism and hence containing no chlorophyll.

The gas chromatogram of the C_{15+} saturated hydrocarbons in the least mature Yeoman sample analyzed in this study (as indicated by biomarker parameters, notably a high abundance of hopenes: Fig. 7b), is shown in Fig. 3b. This is also the Yeoman sample that shows the greatest concentration of isoprenoids relative to n-alkanes exemplified by its having the highest ratio of pristane to n-heptadecane ($pr/n\text{-}C_{17} = 0.38$). Only two other Yeoman kukersite samples out of more than 60 analyzed have $pr/n\text{-}C_{17}$ values greater than 0.20, and these are also very immature. A similar abundance of pristane relative to n-heptadecane was reported by Hoffmann et al. (1987) for a very immature kukersite sample from the Goldwyer Formation of the Canning Basin, Western Australia. This led these authors to suggest that the low abundance of isoprenoids in Ordovician oils is because *G. prisca* was a very lipid-rich (rather than non-photosynthetic) organism. The ratio of pristane to phytane in the Yeoman samples, where it is possible to measure it, lies between 1.14 and 2.18.

The next most abundant class of saturated hydrocarbons in the Yeoman extracts after the n-alkanes is the n-alkylcyclohexanes. Although peaks due to these compounds can be readily distinguished in the gas chromatograms (Fig. 3), their distribution is more easily observed by using GC-MS and monitoring the m/z 82 fragment ion caused by the loss of the alkyl side chain (Fig. 4a). The distribution of the n-alkylcyclohexanes in the Yeoman samples generally mimics that of the n-alkanes, showing an odd-carbon-number preference over the C_{15}–C_{19} range and much lower concentrations of the C_{19+} members. The distribution differs slightly in that its marked odd/even predominance extends up to C_{23}, beyond which n-alkylcyclohexanes are present in extremely low concentrations. This same n-alkylcyclohexane distribution has been reported for other kukersite-type samples and Ordovician oils (Fowler et al. 1986; Reed et al. 1986; Hoffmann et al. 1987).

Another abundant series of monocyclic alkanes in kukersite samples and their derived oils is the methyl-n-alkylcyclohexanes (Fowler and Douglas 1984; Fowler et al. 1986; Hoffmann et al. 1987). The distribution of these compounds is more difficult to examine because they have similar retention times to the n-alkanes. Moreover, the fragment ion used to

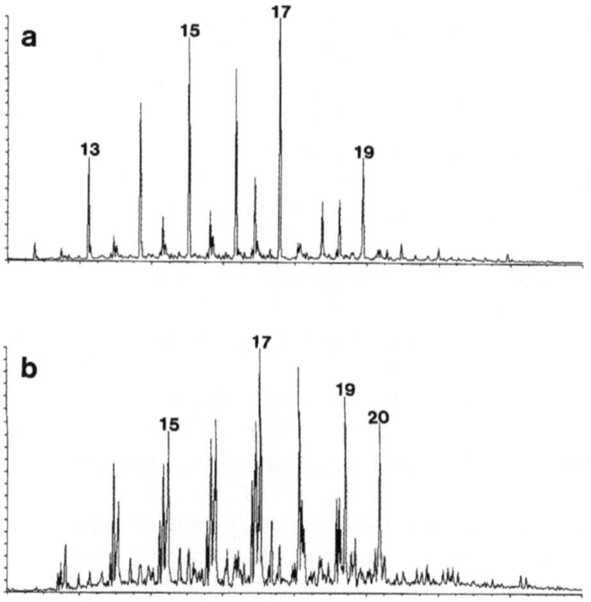

Fig. 4. Mass fragmentograms of saturated hydrocarbons (urea non-adducted fraction) in Yeoman kukersite, location 06-32-008-16W2, depth 2459 m. **a** m/z 82 showing distribution of n-alkylcyclohexanes; **b** m/z 97 showing distribution of methyl-n-alkylcyclohexanes. *Numbers* indicate carbon number of peak

monitor these compounds, m/z 97, is also a significant fragment ion in the mass spectra of n-alkanes. It is therefore necessary to remove the n-alkanes by urea adduction or by using a 5-Å molecular sieve. This extra analytical step is probably the reason why these compounds are rarely reported in geological samples. However, their abundance in Ordovician samples does seem to be considerably greater than in samples of other ages. The identification of these compounds in this work is tentative since no synthetic standards were available. Recently, Hoffmann et al. (1987) have synthesized some methyl-n-alkylcyclohexanes and have confirmed that these are the major series of saturated hydrocarbons with a base fragment ion of m/z 97 in kukersite-type samples.

In the Yeoman samples analyzed for these compounds, the distribution of methyl-n-alkylcyclohexanes is dominated by the C_{15}–C_{20} members (Fig. 4b). Unlike the n-alkanes and n-alkylcyclohexanes they do not show an odd- or even-carbon-number preference. Another significant difference is the higher abundance of the C_{20} methyl-n-alkylcyclohexanes relative to their C_{15-19} homologues; i.e. the sharp decrease in abundance occurs after C_{20}, rather than C_{19} as in the n-alkylcyclohexanes. Fowler et al. (1986) reported this feature in other Ordovician samples, including an oil from the Williston Basin. These authors did find some variation in the distribution of these compounds between different samples. For example, the Williston Basin oil showed an even-carbon-number predominance extending up to C_{24}, whereas a pronounced odd predominance over the C_{15}–C_{19} range was observed for another sample. The distribution of the methyl-n-alkylcyclohexane isomers at C_{18} and C_{20} appears to differ from those at C_{15}–C_{17} and C_{19} in the Yeoman samples when examined in detail. This was also observed by Fowler et al. (1986) in their Ordovician samples and appears to be supported by the published data of Hoffmann et al. (1987).

The similar carbon number ranges of the n-alkylcyclohexanes and the n-alkanes indicate that they may have a common origin probably from cyclization of algal fatty acids (Johns et al. 1966) which are almost entirely even-carbon-numbered (e.g. Weete 1976). Laboratory experiments have shown that this is also a feasible method to obtain methyl-n-alkylcyclohexanes (Rubinstein and Strausz 1979; Hoffmann et al. 1987). The odd-carbon-number predominance of the n-alkylcyclohexanes in Ordovician samples indicates that they are preferentially formed when decarboxylation occurs. The increased amounts of even-carbon-numbered methyl-n-alkylcyclohexanes, as shown by the higher relative abundance of the C_{20} members, suggests that more than one reaction pathway is responsible for these compounds and that at least one mechanism does not involve decarboxylation (Fowler et al. 1986).

Polycyclic alkanes tend to be present in very low concentrations in most of the Yeoman samples, although (like the acyclic isoprenoids) these compounds are commonly more abundant in immature samples. In all the Yeoman samples examined by GC-MS, hopanes are much more abundant than steranes. This feature has also been reported for other Ordovician samples (Reed et al. 1986; Longman and Palmer 1987). As shown by the m/z 217 fragmentograms (Fig. 5) and also by the ternary diagram in Fig. 6, almost all the Yeoman samples show a predominance of C_{29} over C_{27} steranes, with C_{28} steranes in much lower abundance. Figure 6 is very similar to that published by Longman and Palmer (1987, their Fig. 7) which summarized the sterane distributions in Ordovician mid-continent oils. Hoffmann et al. (1987) found a similar sterane-carbon-number distribution in some of their Goldwyer Formation extracts, but in others found a more equal abundance of C_{27}:C_{28}:C_{29} steranes. While some Yeoman kukersite extracts do contain about equal amounts of C_{27} and C_{29} steranes, the C_{28} steranes are always much less abundant (Fig. 6).

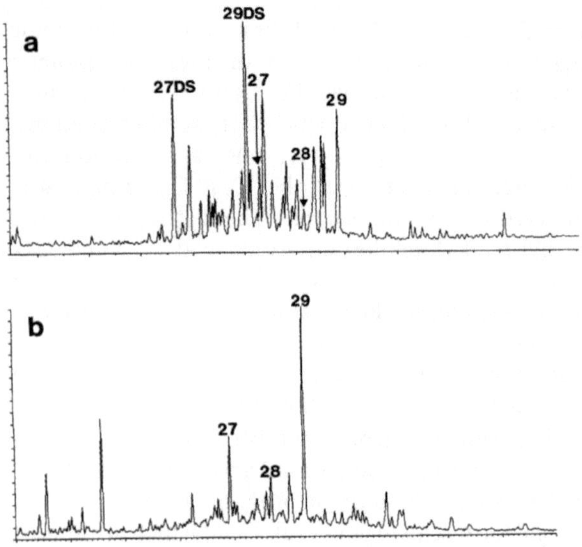

Fig. 5. Mass fragmentograms (m/z 217) of Yeoman kukersite extracts. **a** Sample from location 06-32-008-16W2, depth 2459 m; **b** sample from location 04-10-033-01W3, depth 1466 m. *27–29* are C_{27}–C_{29} 5α(H), 14α(H), 17α(H), 20R steranes. *27DS* and *29DS* are C_{27} and C_{29} 13β(H), 17α(H) 20S diasteranes

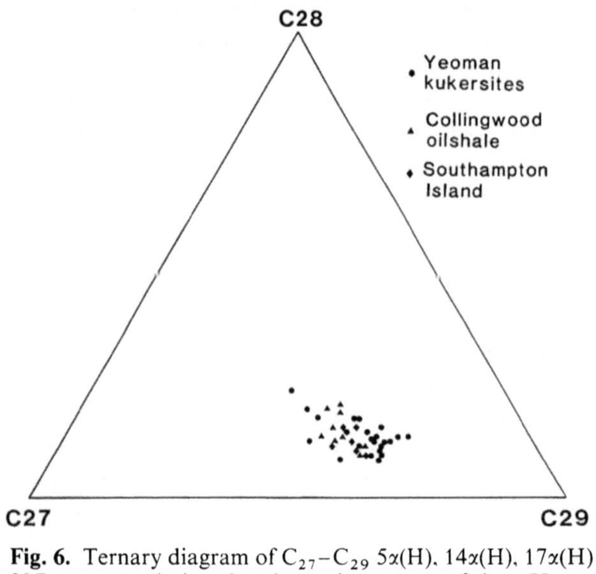

Fig. 6. Ternary diagram of C_{27}–C_{29} 5α(H), 14α(H), 17α(H) 20R sterane relative abundances in extracts of three Upper Ordovician deposits from Canada

Another characteristic of the sterane distributions of all but one of the Yeoman extracts is that they show a predominance of diasteranes over regular steranes. Only the least mature sample analyzed (Fig. 5b) does not show this. The abundance of diasteranes relative to regular steranes in the other samples displays no obvious relationship to their maturity.

The relative abundance of regular steranes and diasteranes in sedimentary rocks is believed to depend on lithology and maturity. Diasteranes are thought to be derived from diasterenes, which are in turn formed from sterenes by acid (clay) catalyzed reactions (Rubinstein et al. 1975; Sieskind et al. 1979). Hence, diasteranes are usually present in lower concentrations than regular steranes in carbonate extracts and carbonate-derived oils (e.g. McKirdy et al. 1983; Rullkötter et al. 1984). The opposite occurs in Yeoman samples where the mineral phase is almost completely dolomite and clays are essentially absent. A predominance of diasteranes over regular steranes was also observed by Longman and Palmer (1987) in Ordovician oils from mid-continent USA, including those from the Williston Basin, which were sourced from so-called kerogenite beds containing "little or no clay". Longman and Palmer (1987) suggested that this may be an indication that clay minerals need not always be present for the occurrence of rearranged steranes.

Hopanes are more abundant than steranes in the Yeoman extracts, as they are in the Ordovician oils from the Saskatchewan part of the Williston Basin. All but one of the m/z 191 fragmentograms are dominated by 17α(H),21β(H)-hopanes (Fig. 7a). One very immature sample is an exception as it contains abundant 17β(H)-hopanes and hopenes (Fig. 7b) besides the 17α(H)-hopanes. The C_{30} 17α(H),21β(H)-hopane is the most prominent peak in all the fragmentograms but the C_{27} and C_{29} hopanes are present in high abundance relative to this compound. The results of a multivariate statistical analysis using terpanes indicated that Ordovician oils may be distinguished from others by their high abundance of C_{27} and C_{29} hopanes (Zumberge 1987). Extended hopanes, especially the C_{33}–C_{35} members, are usually in much lower abundance, as also noted by Longman and Palmer (1987).

The concentration of 17α(H)-22,29,30-trisnorhopane (Tm) exceeds that of 18α(H)-22,29,30-trisnorhopane (Ts) in all samples. The ratio of these two trisnorhopanes is thought to be affected by the same factors that influence the ratio of diasteranes to regular steranes, with the 18α(H)-trisnorhopane favoured by the presence of acid clay catalysts during diagenesis and increasing maturity (Seifert and Moldowan 1978). The predominance of 17α(H)-trisnorhopane over the 18α(H)-trisnorhopane observed in the Yeoman samples contrasts with the aforementioned predominance of diasteranes over regular steranes, and is what would be expected for relatively immature carbonates.

Unusual triterpanes, such as the 28,30-bisnorhopanes, were not detected in any of the Yeoman extracts. Methylhopanes, which have been identified

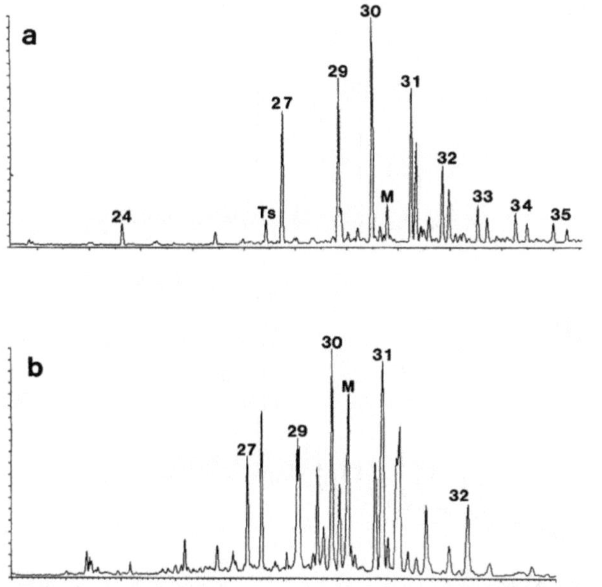

Fig. 7. Mass fragmentograms (m/z 191) of Yeoman kukersite extracts: samples **a** and **b** as in Fig. 5. *27–35* are C_{27}–C_{35} $17\alpha(H)$, $21\beta(H)$, hopanes, *M* is $17\beta(H)$,$21\alpha(H)$ moretane, *Ts* is $18\alpha(H)$-trisnorhopane, *24* is C_{24} tetracyclic terpane

in kukersite-type samples from the Goldwyer Formation (Alexander et al. 1984; Hoffmann et al. 1987), were detected in trace amounts in only one Yeoman sample. Gammacerane is present in some Yeoman samples in very low amounts, but could not be detected in others, even when using GC-MS-MS and monitoring the m/z 412 to 191 transition. In most of the Yeoman samples the C_{24} tetracyclic terpane is present in greater abundance than any of the tricyclic terpanes, and in the remaining samples is of similar abundance to the C_{23} tricyclic terpane (Fig. 7). In the Saskatchewan Ordovician oils, the C_{24} tetracyclic terpane is also present in higher concentrations than any of the tricyclic terpanes. A high amount of the C_{24} tetracyclic terpane relative to the tricyclic terpanes is a common feature in carbonate-derived oils (Aquino Neto et al. 1983).

3 Collingwood Oil Shale

The Collingwood Member is the upper unit of the Middle-Upper Ordovician Lindsay Formation in southern Ontario (Fig. 1). Like the Yeoman Formation, it was deposited in Edenian-Maysvillian time (Macauley et al. 1990) under shallow marine conditions in a low-latitude paleoenvironment. The oil shale beds occur within the lower part of the Collingwood Member. They are characteristically black, organic-rich, laminated and fissile, containing a rich invertebrate fossil fauna including abundant trilobites concentrated in bioclastic beds (Macauley et al. 1985). Calcite is the principal mineral constituent (50%), followed by quartz (35%) and clays (15%) with only minor feldspar (Snowdon 1984).

Oil was retorted sporadically from the Collingwood oil shales between 1859 and 1861 before being replaced by less expensive conventional oil from nearby Oil Springs (Macauley et al. 1990). The Collingwood is thought to be the source of the Cambrian and Ordovician reservoired oils in southern Ontario (Powell et al. 1984).

The organic matter of the Collingwood samples, when examined microscopically, is observed to be almost entirely amorphous material, most ($\sim$ 80%) of which fluoresces in blue light (Fig. 2c). Sparse algal bodies can be distinguished. The most abundant of these are acritarchs but leiosphaeridae-like algae and rare *G. prisca* are also present. Fauna recognized include brachiopods, conodonts and chitinozoa. Compared to the Yeoman Formation, it is evident that a much more diverse assemblage of organisms contributed to the Collingwood organic matter.

The organic carbon content of the Collingwood oil shales can exceed 10% but the average is around 5 to 6% (Macauley et al. 1990). Rock-Eval analysis indicates the organic matter of the less mature Collingwood samples to be Type I–II. Hydrogen indices are commonly in the 500 to 600 range with some samples having values above 650. These are significantly lower than those of the Yeoman kukersites. The T_{max} values of the Collingwood oil shales, unlike those of the Yeoman samples, appear to be a good indication of the maturity of the organic matter and, for example, reflect the regional eastward increase in maturation from western Lake Huron to the Whitby area (Macauley et al. 1990).

Saturate fraction gas chromatograms of the Collingwood oil shales differ from those of the Yeoman Formation by showing higher concentrations of C_{19+} n-alkanes and acyclic isoprenoids (Fig. 8a). The less mature Collingwood samples have a slight odd-carbon-number preference between n-C_{15} and n-C_{19}, but most samples have a smooth distribution of n-alkanes over this range. No sample displays an odd- or even-carbon-number preference in their C_{19+} n-alkanes, which can extend up to at least C_{37}. The pr/n-C_{17} ratio decreases with increasing maturity in the Collingwood samples (which can be seen by plotting pr/n-C_{17} vs T_{max}) as is normally observed in geological samples (e.g. Tissot and Welte 1984, p. 185–186). However, even the more mature Collingwood samples

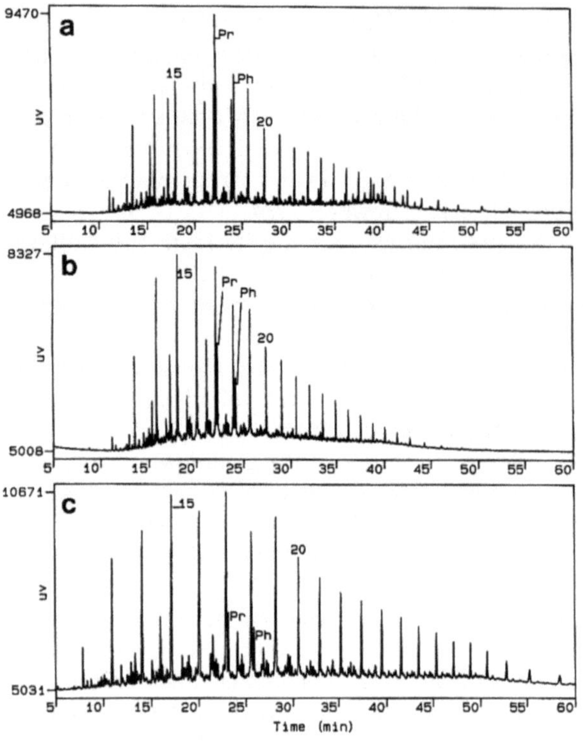

Fig. 8. Gas chromatograms of C_{15+} saturated hydrocarbons in **a** immature Collingwood oil shale from Drummond Island, depth 119.17 m; **b** mature Collingwood oil shale from near Whitby, depth 33.4 m; **c** Ordovician-reservoired oil from southern Ontario. *Numbers* indicate carbon number of n-alkane peaks, *Pr* and *Ph* are pristane and phytane, respectively

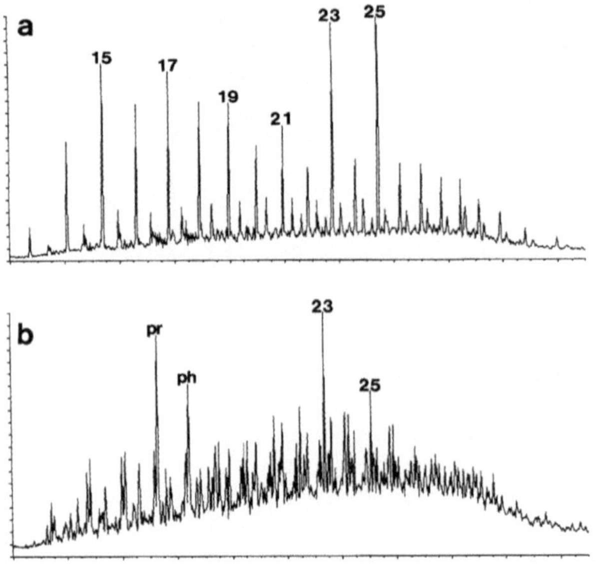

Fig. 9. Mass fragmentograms of saturated hydrocarbons (urea non-adducted fraction) in Collingwood oil shale from the Collingwood area (borehole CLGD 07A, depth 73.1 m): **a** m/z 82 showing distribution of n-alkylcyclohexanes; **b** m/z 97 showing distribution of methyl-n-alkylcyclohexanes. *Numbers* indicate carbon number of peak. *Pr* and *Ph* are pristane and phytane, respectively

(e.g. Fig. 8b), which are within the principal zone of oil generation, have higher amounts of isoprenoids than any of the Yeoman samples. Collingwood samples have pr/ph values of between 1.39 and 1.83. Cambrian and Ordovician reservoired oils (Fig. 8c) from southern Ontario which are believed to be sourced by the Collingwood shales, give similar chromatograms (Fowler 1984; Powell et al. 1984) but have a slightly greater odd-carbon-numbered n-alkane predominance and lower concentrations of isoprenoids relative to the n-alkanes than are shown by the mature Collingwood samples.

After the n-alkanes and acyclic isoprenoids, the next most abundant class of saturated hydrocarbons in the Collingwood extracts is the n-alkylcyclohexanes. The distributions of these compounds in the Collingwood shales are very different from those of the Yeoman kukersites. The immature samples have bimodal distributions with the C_{23} and C_{25} n-alkylcyclohexanes in highest abundance and a pronounced odd-carbon-number preference (Fig. 9a). More mature samples also show a bimodal distribution but have increased amounts of the lower-molecular-weight compounds. An oil from a Cambrian reservoir in southern Ontario did not possess this distribution (Fowler 1984), although an Ordovician-reservoired oil more recently examined does have a distribution which resembles that seen in the mature Collingwood samples. The methyl-n-alkylcyclohexanes are present in much lower concentrations in the Collingwood extracts than they are in the Yeoman samples. This is evident from their m/z 97 fragmentograms where pristane and phytane are major peaks (Fig. 9b). The highest peaks attributable to monocyclic alkanes in the Collingwood m/z 97 fragmentograms are those of C_{23} and C_{25} compounds. The mass spectra of these compounds show base fragment ions of m/z 97 and it is possible that they are methylalkylcyclohexanes.

As the carbon-number distribution of the n-alkylcyclohexanes is different from that of the n-alkanes in the Collingwood extracts, it appears that these two compound classes are at least partially derived from different precursors. Hence, presumably the C_{23} and C_{25} n-alkylcyclohexanes are not derived from algal fatty acids. Although cyclohexyl acids occur in some bacteria (De Rosa et al. 1972; Oshima and Ariga 1975; Suzuki et al. 1981), only C_{17} and C_{19} compounds have been reported. The origin of the C_{23}

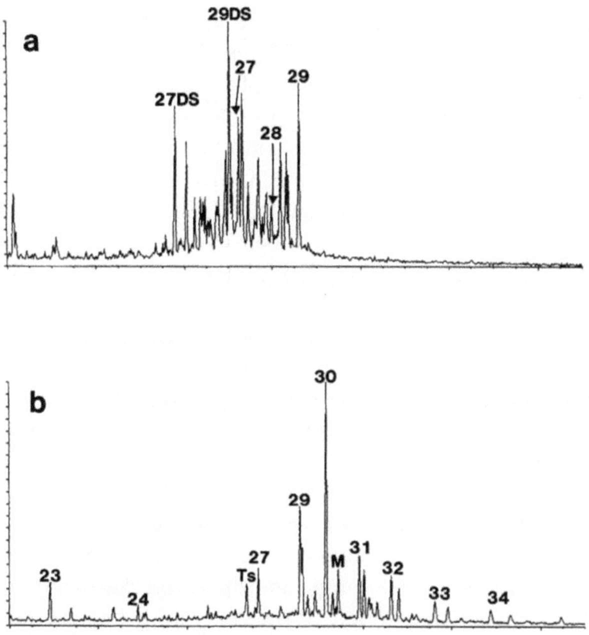

Fig. 10. Mass fragmentograms of a Collingwood oil shale extract from the Collingwood area (borehole CLGD 07A, depth 73.1 m): **a** m/z 217; **b** m/z 191. Annotation as for Figs. 5 and 7, respectively except 23 in **b** is C_{23} tricyclic terpane

and C_{25} n-alkylcyclohexanes and the related C_{23} and C_{25} possible methylalkylcyclohexanes is therefore presently unknown.

Steranes are present in the Collingwood extracts in similar concentrations to the hopanes. The sterane carbon-number distributions of the Collingwood samples are similar to those of the Yeoman samples, showing a predominance of C_{29} over C_{27}, with C_{28} steranes present in much lower abundances (Figs. 6 and 10a). The Collingwood samples exhibit a much greater variation in the relative abundances of diasteranes and regular steranes than do the Yeoman samples, which is probably a function of the wide range in their maturity and their variable clay content.

The m/z 191 fragmentograms of the Collingwood samples (Fig. 10b) differ from those of the Yeoman samples (Fig. 7). These differences include a lower abundance of C_{27} and C_{29} hopanes relative to 17α(H),21β(H)-hopane, greater amounts of gammacerane, a higher abundance of tricyclic terpanes, and the C_{23} tricyclic terpane in greater abundance than the C_{24} tetracyclic terpane in all samples. Although most samples show a predominance of Tm over Ts, with most Tm/Ts values in the 0.90 to 1.80 range, this tends to be considerably less pronounced than in the Yeoman samples.

4 Southampton Island Oil Shales

Organic-rich beds of Ordovician age have been reported over a wide area of the eastern Canadian Arctic, including on Akpatok, Baffin and Southampton Islands (Macauley et al. 1990). Here, the initial results of geochemical studies on samples from Southampton Island (Fig. 1) are briefly discussed. A more detailed study will be published elsewhere. These samples were selected because they were originally thought to have been deposited at the same time as the Collingwood and Yeoman formations. Moreover, there has been some speculation in the literature (Foster et al. 1986) that *G. prisca* was a major contributor to the organic matter of these oil shales. As will be shown below, this is incorrect.

The oil shales on Southampton Island were first described in detail by Nelson and Johnston (1966), and later named the Boas oil shales by Heywood and Sanford (1976). They were originally dated as Edenian-Maysvillian using conodonts (Barnes 1973) and considered to represent a single deposit, although two distinct organic lithotypes were recognized (Macauley et al. 1990). More recent biostratigraphic work suggests the two lithotypes are in fact two units of differing age (McCracken and Nowlan 1989). The upper oil shale is a brownish-black to dark brown, finely laminated, fissile, very organic-rich (10–20% TOC) limestone (Macauley et al. 1989) of Richmondian age (McCracken and Nowlan 1989). The lower unit is a light to medium brown, poorly laminated, hard, organically leaner (2–6% TOC) limestone (Macauley et al. 1990) of Maysvillian age (McCracken and Nowlan 1989). Microscopical examination of the Southampton oil shale samples under ultraviolet light reveals that most of the organic matter is amorphous and concentrated in organic-rich laminae parallel to the bedding plane (Fig. 2d). Only a few fragments of algal bodies and acritarchs can be distinguished. Concentrated along some laminae are occasional grey, rounded bodies with an appearance similar to that of vitrinite in reflected light. These have reflectance values of Ro 0.33 to 0.48% which, if the bodies were vitrinite, would correspond to the level of maturity indicated by other parameters. Also observed in reflected light is framboidal pyrite which has been partially oxidized in some samples. No evidence of *G. prisca* was found.

Rock-Eval analysis of the Southampton Island organic-rich carbonates indicates that they contain immature Type II organic matter (Macauley 1986; Macauley et al. 1990). The upper oil shale samples have slightly higher HI and lower OI values than do the less organic-rich lower unit samples. Their average

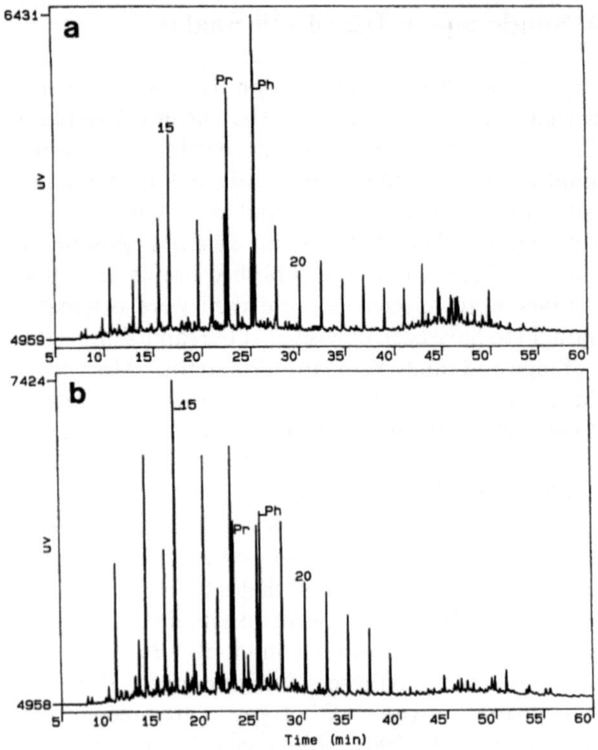

Fig. 11. Gas chromatograms of C_{15+} saturated hydrocarbons in oil shale outcrop samples from Southampton Island: **a** lower oil shale interval from Boas River (6.3% TOC); **b** upper oil shale interval from Sixteen Mile Brook (21.4% TOC). Annotation as in Fig. 8

HI and OI values are 617 and 47, respectively. Hence, compared to the Collingwood and Yeoman samples, these carbonates have lower HI and higher OI values.

The C_{15+} saturate chromatograms of the Southampton Island oil shales (Fig. 11) show n-alkanes up to n-C_{19} in highest abundance, but with a gradual rather than sudden decrease in concentration after n-C_{19}. All samples exhibit an odd-carbon-number preference between n-C_{15} and n-C_{19}. Most also show a predominance of odd-numbered n-alkanes up to C_{23} or C_{25}. This appears to be more pronounced in samples from the lower unit. All the samples possess high concentrations of acyclic isoprenoids relative to n-alkanes although, again, this is more pronounced in those from the lower unit.

The Southampton Island samples differ from the Collingwood and Yeoman samples by having pr/ph values of less than one. This may indicate that the Southampton Island oil shales were deposited under more reducing conditions than were the two other Ordovician units (cf. Didyk et al. 1978). As will be discussed later, this could also be partly due to their deposition under hypersaline conditions and a contribution to the phytane from the lipids of halophilic bacteria (Anderson et al. 1977; ten Haven et al. 1987). Relatively high concentrations of C_{19+} n-alkanes have been reported from other pre-Devonian sediments deposited under hypersaline conditions, such as those from the Cambrian of the Officer Basin, South Australia (McKirdy and Kantsler 1980), and from the Silurian of the Michigan Basin (Vogler et al. 1981; Illich and Grizzle 1983; Rullkötter et al. 1986) and southern Ontario (Powell et al. 1984). It was suggested by McKirdy and Kantsler (1980) that the higher-molecular-weight n-alkanes could be derived from the reworking of algal debris by anaerobic bacteria in a microbial system inhabiting a hypersaline environment.

Monocyclic alkanes are much less abundant in the Southampton Island oil shales than they are in the Collingwood and Yeoman samples. At present, the distribution of these compounds has been examined in detail in only one sample of the lower unit. The m/z 82 fragmentogram shows two classes of compounds, the n-alkylcyclohexanes and an unknown series. The n-alkylcyclohexanes extend up to at least the C_{25} compound and display a slight odd-carbon-number preference between C_{17} and C_{21}. No methyl-n-alkylcyclohexanes were detected. The largest peaks in the m/z 97 fragmentogram are pristane and phytane. Polycyclic alkanes are present in high concentrations. GC-MS analysis indicates that the lower oil shales contain a higher abundance of steranes relative to hopanes than those from the upper unit.

The sterane distributions of the Southampton Island oil shales are dominated by 5α(H),14α(H),17α(H)-20R-steranes indicating that they are very immature (Fig. 12a). There is a predominance of C_{29} over C_{27} steranes, with C_{28} steranes present in much lower abundance (Fig. 6). Diasteranes are much less abundant than regular steranes, probably due to a combination of low maturity and the virtual absence of clays in these carbonates (Macauley 1986). Initial results indicate little difference between the sterane distributions of the upper and lower oil shale intervals.

The two Southampton Island oil shale units do show differences in their hopane distributions. Samples from the lower interval give m/z 191 fragmentograms in which 17α(H),21β(H)-hopane is the highest peak with the C_{27}, C_{29}, C_{31} and C_{32} 17α(H)-hopanes are of similar abundance, and the C_{33}–C_{35} hopanes are present in much lower amounts (Fig. 13a). The upper oil shales differ by having a higher abundance of the C_{34} hopanes (Fig. 13b). As might be expected in limestones of this maturity (Seifert and Moldowan 1978; McKirdy et al. 1983), 17α(H)-trisnorhopane

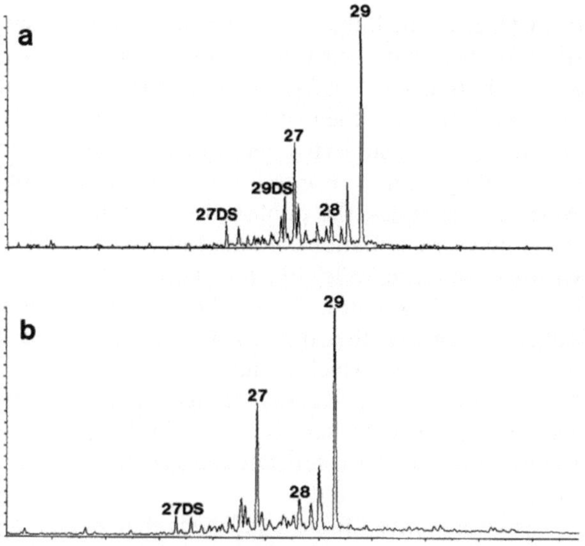

Fig. 12. Mass fragmentograms (m/z 217) of Southampton Island oil shale hydrocarbons: **a** lower unit sample from Boas River; **b** upper unit sample from Sixteen Mile Brook. Annotation as in Fig. 5

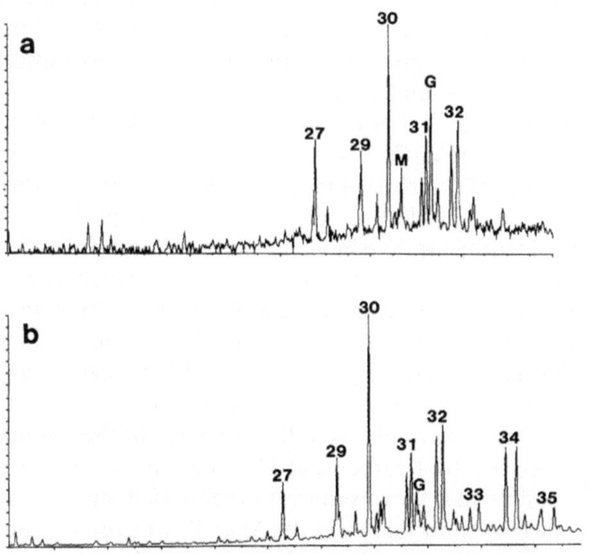

Fig. 13. Mass fragmentograms (m/z 191) of Southampton Island oil shale hydrocarbons: **a** lower unit sample from Boas River; **b** upper unit sample from Sixteen Mile Brook. *G* Gammacerane; other annotations as in Fig. 7

(Tm) is present in much greater abundance than 18α(H)-trisnorhopane (Ts). All samples have a low concentration of tricyclic and tetracyclic terpanes, whereas in samples from the lower unit, gammacerane is particularly abundant relative to the 17α(H)-hopanes (especially in the lower TOC samples).

All the Southampton Island samples contain in their aromatic fractions the 1-alkyl-2,3,6-trimethylbenzenes. These compounds, previously identified in oils and their carbonate source rocks that were deposited under highly reducing and probably hypersaline conditions, are thought to be derived from a carotenoid found in the green sulphur bacteria *Chlorobiaceae* (Summons and Powell 1986, 1987). The concentrations of these compounds relative to other aromatic hydrocarbons are greatest in the lower oil shale unit. The upper oil shale samples contain lesser amounts of these aromatic biomarkers but higher concentrations of sulphur compounds.

5 Comparison of the Organic Geochemistry and Depositional Environments of Ordovician Sediments

The rock samples that show features most specific to the Ordovician are the Yeoman kukersites. These features are thought to be a function of the main organism which contributed to the organic matter of kukersites, *G. prisca*. Other units within the Middle and Late Ordovician of the Williston Basin also contain kukersite intervals. Single core samples from the Icebox Shale Member of the underlying Middle Ordovician Winnipeg Formation and the Gunn Member of the overlying Stoney Mountain Formation have been found to contain organic matter which consists almost entirely of *G. prisca* and has a very similar chemistry to that of the Yeoman kukersites (Fowler, unpubl. results). At present, it is not known how widespread these units are.

Evidence from this and other studies (e.g. Foster et al. 1986) on the occurrence of kukersites and their derived oils indicates that *G. prisca* was a widely distributed organism in warm, low-latitude, epicontinental seas during the Ordovician. Kukersites such as those of the Yeoman Formation were deposited under quite specific conditions that appear either to have optimized primary organic productivity or prevented reworking of the primary organic matter, or both. In a recent detailed sedimentological study, it was concluded that the Yeoman Formation kukersites formed from periodic algal blooms in "approximately 15 to 30 m of water on a low gradient carbonate shelf" in a sediment starved basin (Osadetz, pers. comm. 1989). Additionally, hypersalinity was ruled out as a factor in the formation of kukersites on geological grounds. This is supported by the results of the present study. There is no discernible contribution by *G. prisca* to the Southampton Island oil shales which, from other geochemical evidence, seem to have been deposited under more highly saline conditions than were the

Yeoman and Collingwood sediments. During deposition of the Yeoman kukersites, the bottom waters must have been highly anoxic to prevent bacterial reworking of the algal bodies. Preservation of *G. prisca* would have been helped by the presence in its cell walls of a biopolymer that is relatively resistant to biodegradation, as proposed by Douglas et al. (this Vol.).

The Collingwood oil shales have enough features in common with the Yeoman kukersites to suggest that *G. prisca* was also a significant contributor to their organic matter. In this case, the "classic" kukersite chemistry has been diluted by contributions from other organisms, especially bacteria; and the amorphous appearance of the organic matter under the microscope indicates intensive microbial reworking of the primary organic input during early diagenesis. The Collingwood oil shale and other "diluted kukersites" were probably deposited under normal marine conditions with relatively high biological productivity in the oxic surface waters. The bottom waters were reducing enough to allow reworking of the primary organic matter only by anaerobic bacteria, thus ensuring the preservation of high amounts of organic carbon.

The geochemical characteristics that are least specific to the Ordovician are those shown by the Southampton Island oil shales. These sediments more closely resemble other Canadian samples of Silurian and Devonian age (Summons and Powell 1987), suggesting that they were deposited under similar conditions: that is, in a highly reducing carbonate environment inhabited by the green sulphur-oxidizing bacteria *Chlorobiaceae*. There is no chemical or optical evidence of *G. prisca* in these sediments. This situation is not unique for Ordovician organic-rich sediments. For example, the Vinini and Womble shales from Nevada and Oklahoma, respectively, were deposited in environments different from those of the Southampton Island oil shales and also show no evidence of a *G. prisca* contribution (Fowler and Douglas 1984; Douglas et al., this Vol.).

6 Known Occurrences of *Gloeocapsomorpha prisca*

Kukersite and "diluted kukersite" deposits and their ages are listed in Table 1. This list is certainly not comprehensive but includes those formations for which there is sufficient chemical and/or microscopical evidence to suggest that *G. prisca* was a major contributor to their organic matter. As oils generated from Ordovician kukersites have been more extensively studied than their source rocks by geochemists, a contribution by *G. prisca* to some rocks has been inferred from the chemistry of their derived oils. Obviously, some subjective judgement has been used in compiling Table 1. For example, in the case of the Australian samples, the published data indicate that the Goldwyer Formation contains kukersite intervals, whereas the Horn Valley Siltstone appears to contain only diluted kukersites. Data published on Estonian kukersite suggests that it is a heterogeneous deposit with some intervals which are not kukersites as defined here. Moreover, the variety of ages given for the Estonian "kukersite samples" (Table 1) and their widely differing characteristics suggest that different intervals have been studied.

Examination of Table 1 indicates that almost all the previous reports of sediments containing a significant contribution from *G. prisca* are of Ordovician age. In the older literature (pre-1980) it is possible to find reports of *G. prisca* in sediments ranging in age from Late Precambrian to Tertiary (e.g. see McGregor and Cramer 1971 and references therein). However, almost all of the reports of *G. prisca* outside of the Ordovician appear to be the result of misidentification. For example, many workers erroneously considered *Botryococcus* and *G. prisca* to be the same organism (e.g. Cookson 1953; Traverse 1955). Of the reports of *G. prisca* in rocks younger than the Ordovician, the hardest to eliminate on the grounds of misidentification (since the author had previously examined true Estonian and other kukersites) is a report by Cramer and Diez de Cramer (1972) of what appears to be kukersites in the Lludlovian (Upper Silurian) Read Bay Formation on Somerset and Cornwallis Islands in the Canadian Arctic. However, there is some doubt as to the exact age of the samples described by Cramer and Diez de Cramer and it is possible that they examined organic-rich laminations in what is now known as the Allan Bay Formation and considered to be of Late Ordovician to Early Silurian age (Mayr 1978).

6.1 A Possible Middle Cambrian Kukersite – Mt. Cap Formation

There have been some recent reports of Cambrian deposits in Canada and Australia that are possibly related to kukersites (Monnier et al. 1987; Wielens et al. 1990; McKirdy and Willink 1988, oral presentation). Gas and condensate discoveries have been made in Early Cambrian sandstone reservoirs in the Colville/Tweed Lake area of the Northwest Territories

Table 1. Kukersite and "diluted" kukersite deposits

Formation/group	Basin/area	Country (s)	Age	Ref.
A. Kukersites				
Goldwyer Formation	Canning	Australia	Late Arenigian-Llanvirnian Early-Middle Ordovician	Combaz and Peniguel (1972) Foster et al. (1986) Hoffmann et al. (1987)
Guttenberg oil rock	Wisconsin/Iowa	USA	Mohawkian Middle Ordovician	Fowler and Douglas (1984) Fowler et al. (1986) Hatch et al. (1987) Jacobsen et al. (1988)
Glenwood Formation	Iowa	USA	Mohawkian Middle Ordovician	Hatch et al. (1987) Jacobsen et al. (1988)
Simpson Group	Forest City, Kansas	USA	Mohawkian Middle Ordovician	Longman and Palmer (1987) Hatch et al. (1987)
Trenton Formation	Michigan	USA	Mohawkian Middle Ordovician	Reed et al. (1986) Longman and Palmer (1987)
Yeoman Formation	Williston	Canada-USA	Edenian-Maysvillian Late Ordovician	This paper Osadetz et al. (1989)
Kukersite	Estonia	USSR	?Late Cambrian – Early Silurian	Klesment and Nappa (1980) Vitorovic (1980) Reed et al. (1986)
B. Diluted kukersites				
Horn Valley siltstone	Amadeus	Australia	Arenigian, Early Ordovician	Kurylowicz et al. (1976) Hoffman et al. (1987)
Green Point	Western Newfoundland	Canada	Lower Ordovician	Weaver and Macko (1988) Macauley et al. (1990) Fowler (unpubl. results)
Icebox Mbr, Winnipeg shale	Williston	Canada	Mohawkian Middle Ordovician	This work Osadetz et al. (1989)
Point Pleasant Formation	Ohio	USA	Champlainian-Cincinnatian Middle-Upper Ordovician	Cole et al. (1987)
Maquoteka shale	Illinois	USA	Cincinnatian Upper Ordovician	Longman and Palmer (1987)
Collingwood oilshale	Southern Ontario	Canada	Edenian-Maysvillian Late Ordovician	Powell et al. (1984) Macauley et al. (1990) This work

(Fig. 1). The gasoline fraction of the condensate from the Tweed Lake M-47 well has an unusual composition very similar to that of kukersite-derived oils, such as those from the Williston Basin (Snowdon and Osadetz 1988). Studies of potential source rocks for these hydrocarbons led to the identification of a thin interval of laminated, dark brown, organic-rich shale within the Middle Cambrian Mt. Cap Formation in the Colville D-45 well. The organic geochemistry of this rock has been described in some detail by Wielens et al. (1990). Its organic matter consists predominately of the bodies of a single algal species which closely resembles *G. prisca* (Fig. 2e and f), giving the host sediment a very similar appearance to Ordovician kukersites such as those within the Yeoman Formation. The organic matter is Type I as defined by Rock-Eval analysis and has not generated hydrocarbons. The C_{15+} saturate gas chromatogram (Fig. 14) of the Mt. Cap sample is dominated by C_{15}–C_{17} n-alkanes with a slight odd-carbon-number preference and a sharp drop in abundance after n-C_{17}, rather than n-C_{19} as in the Yeoman samples. The concentration of

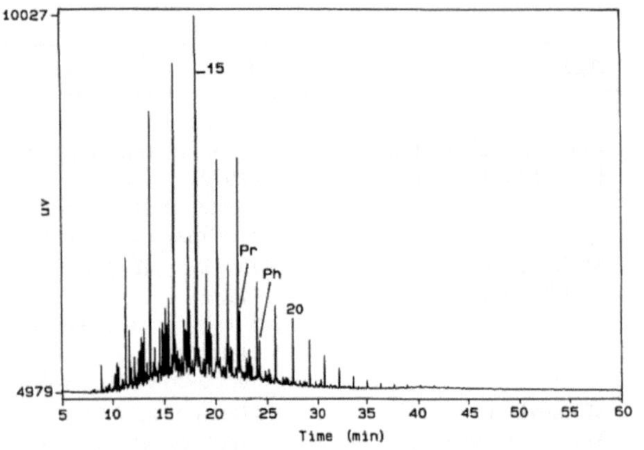

Fig. 14. Gas chromatogram of C_{15+} saturated hydrocarbons extracted from the Middle Cambrian Mount Cap Formation in Colville D-45 well, depth 943 m. Annotation as in Fig. 8

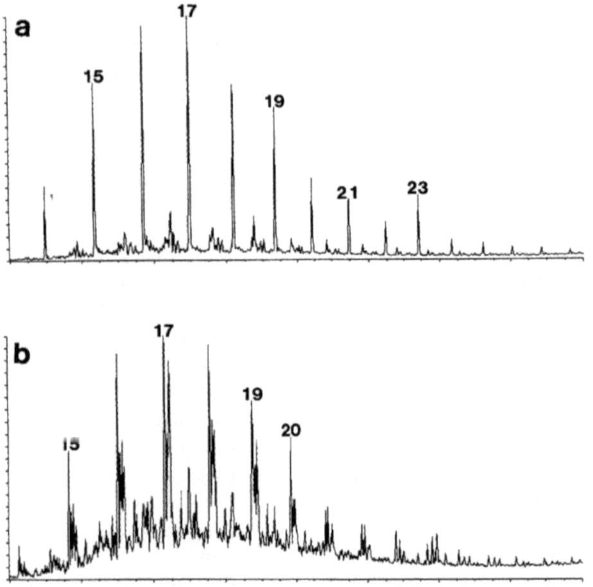

Fig. 15. Mass fragmentograms of the saturated hydrocarbons (urea non-adducted fraction) of Colville D-45, Mount Cap sample extract: **a** m/z 82; **b** m/z 97. Annotation as in Fig. 4

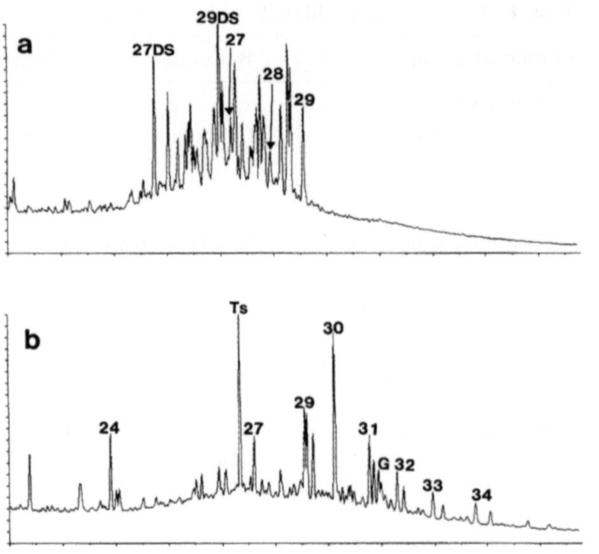

Fig. 16. Mass fragmentograms of Colville D-45 Mount Cap hydrocarbons: **a** m/z 217; **b** m/z 191. Annotation as in Figs. 5 and 7, respectively

acyclic isoprenoids relative to n-alkanes is higher than in Yeoman samples of equivalent maturity.

The Mt. Cap extract has a high concentration of monocyclic alkanes, especially n-alkylcyclohexanes with an odd carbon-number-preference extending up to C_{23} (Fig. 15a). The methyl-n-alkylcyclohexane distribution is smoother than those of the Yeoman samples, but still shows a sharp drop in abundance after the C_{20} rather than the C_{19} compound (Fig. 15b).

Sterane and hopanes are present in the Mt. Cap extract in approximately equal amounts. Once again there is a predominance of C_{29} over C_{27} steranes but the C_{28} steranes occur in higher abundance than in the kukersite samples (Fig. 16a). The predominance of diasteranes over regular steranes in this sample is as would be expected for a shale.

The Mt. Cap sample has a m/z 191 fragmentogram (Fig. 16b) that is very different to those of the kukersite samples. The hopane distribution is unusual because 18α(H)-trisnorhopane is the major component. There are also two unknown C_{30} pentacyclic compounds in relatively high abundance, one of which almost co-elutes with, and one that elutes later than, the 17α(H),21β(H)-30-norhopane. Gammacerane and the tricyclic terpanes are present in higher concentrations than in the Yeoman samples.

In summary, the organic geochemistry of the Mt. Cap sample analyzed by Wielens et al. (1990) shows both similarities and differences to kukersite samples of equivalent maturity. Some of the differences can be attributed to its argillaceous lithology which would favour different diagenetic reactions resulting in, for example, the predominance of Ts over Tm. Many of the similarities are related to the origin of the organic matter, such as the high abundance of n-alkanes up to n-C_{17}, the high concentration of monocyclic alkanes, the C_{29} sterane predominance, and the low abundance of low-molecular-weight aromatics in the Tweed Lake condensate (presumed to be derived from a similar Mt. Cap source rock to that from Colville D-45). These features probably reflect an algal origin but are not attributable to any particular species. However, there are enough similarities to suggest that the organism responsible for the Mt. Cap organic matter may have been a direct precursor of *G. prisca*, albeit with a slightly different biochemistry. The appearance of the algal bodies in the Mt. Cap shale supports this. McKirdy and Willink (oral presentation, 1988) described similar organic matter in sediments of Middle Cambrian age from the Georgina Basin, Australia.

7 The Nature of *Gloeocapsomorpha prisca*

What type of organism was *G. prisca*? The opinions given in the recent literature have been very diverse. This is exemplifed by the very different conclusions reached by two groups (Reed et al. 1986; Hoffmann et al. 1987) who attempted to infer the biochemistry of *G. prisca* on the basis of the unique chemistry of Ordovician kukersites and their derived oils. Reed et al. (1986) thought that *G. prisca* was a non-photosynthetic, prokaryotic, benthonic, mat-forming organism; a "chemosynthetic, aerointolerant, autotroph". According to Hoffmann et al. (1987), it was a widely distributed phototrophic, planktonic, possibly eukaryotic, lipid-rich alga.

7.1 Photosynthetic or Non-Photosynthetic?

The question of whether *G. prisca* was photosynthetic or non-photosynthetic may be addressed on the basis of the abundance of acyclic isoprenoids in Ordovician oils and in the extracts of kukersites, by assuming that these compounds are mostly derived from the phytyl side chain of chlorophyll. Hoffmann et al. (1987) argued that the low abundance of acyclic isoprenoids in Ordovician oils did not necessarily mean *G. prisca* was non-photosynthetic as thought by Reed et al. (1986). Hoffmann et al. (1987) suggested that it could also be due to their dilution by n-alkanes if *G. prisca* was extremely rich in straight-chain lipids. It was also noted by Hoffmann et al. that, in their study, the sample which showed the highest concentration of acyclic isoprenoids relative to the n-alkanes was the most immature of the Goldwyer Formation kukersites. This agrees with the results of the analyses on the Yeoman samples discussed earlier, where again it was the most immature samples that contained the highest relative concentration of acyclic isoprenoids. Hoffmann et al. (1987) also pyrolyzed their immature sample at 250 °C and found relatively high concentrations of isoprenoids in the pyrolysate. Jacobson et al. (1988) repeated these pyrolysis experiments using samples of Guttenberg Oil Rock and obtained similar results. However, Jacobson et al. (1988) pointed out that only a very small part of the kerogen is broken down to hydrocarbons at 250 °C. Hence, the isoprenoids may not originate from *G. prisca* but from the minor part of the organic matter of the sample that is derived from acritarchs and other phytoplankton. Further confirmation that this may be the case is provided by the results of a hydrous pyrolysis experiment on kerogen isolated from the Guttenberg Oil Rock that was heated to 330 °C for 72 h. More than half of the kerogen was converted to organic-soluble products that contained extremely low concentrations of acyclic isoprenoids (Fowler et al. 1986). Similarly, pyrolysis-gas chromatography experiments on a sample of Guttenberg kerogen showed the virtual absence of prist-1-ene in the pyrolysate (Douglas et al, this Vol.). Jacobson et al. (1988) also cited a personal communication from P. Sundararaman reporting that he had found virtually no porphyrins in extracts of *G. prisca*-rich sediments.

Hoffmann et al. (1987, p. 2692) made the following comments: "as far as we are aware, this apparent 'swamping' of biomarkers has not been observed before to the extent seen in these Ordovician sediments. This is also circumstanial evidence for *G. prisca* being a lipid-rich organism". In fact, some previous investigations of other algal-derived kerogens, such as tasmanites (Simoneit and Burlingame 1973; Larter 1978) and torbanites (Allan et al. 1980; Largeau et al. 1986; Saxby et al. 1986) have found them to be also very poor in acyclic isoprenoids. Comparison with some recent work on torbanites and other oil shales (Largeau et al. 1986; Derenne et al. 1988; Goth et al. 1988) has led Douglas et al. (this Vol.) to propose that the Guttenberg Oil Rock kerogen may be made up of selectively preserved cell-wall material of *G. prisca* which, like that of *Botryococcus braunii*, contains a low amount of isoprenoids. Thus, Hoffmann et al. (1987) may be partially correct in suggesting a "swamping effect" as the reason for the low concentration of isoprenoids (and other biomarkers such as hopanes and steranes) in Ordovician oils: it could be due to their dilution by the breakdown products of the cell-wall polymer (e.g. n-alkanes and monocyclic alkanes).

7.2 Prokaryote or Eukaryote?

Ordovician oils from Saskatchewan and the Yeoman kukersite extracts have a considerable predominance of hopanes over steranes. This has been considered by some workers to show that *G. prisca* was a prokaryotic organism (Reed et al. 1986; Longman and Palmer 1987). Despite the finding that hopanes were far more abundant than steranes in their Goldwyer Formation extracts and pyrolysate, Hoffmann et al. (1987) still felt that the "presence of significant amounts of steranes" indicated that *G. prisca* was eukaryotic.

A predominance of C_{29} over C_{27} steranes in pre-Devonian oils and sediments has been noted by many workers (e.g. McKirdy and Hahn 1982; Fowler and Douglas 1984; Rullkötter et al. 1986; Vlierboom et al. 1986; Longman and Palmer 1987; Grantham and Wakefield 1988) and has usually been taken as an indication of a blue-green algal (cyanobacterial) contribution. Certain cyanobacterial species have been reported to contain lipids in which there is a predominance of C_{29} over C_{27} sterols (e.g. De Sousa and Nes 1968; Paoletti et al. 1976b; Volkman 1986). Hence, this sterane distribution could be further evidence of the prokaryotic nature of *G. prisca*. However, this may not be a valid conclusion because there is still considerable doubt as to whether cyanobacteria can synthesize sterols (Ourisson et al. 1987).

Sterane distributions that show a predominance of C_{29} over C_{27} compounds with much lower amounts of C_{28} steranes appear to be very common in pre-Devonian rocks and are not specific to kukersites. Their occurrence has been suggested to be a function of biochemical evolution (Grantham and Wakefield 1988). In the present study, a similar sterane carbon-number distribution was shown by all three sets of Late Ordovician samples (Yeoman, Collingwood, Southampton Island), although they were deposited under apparently very different environmental conditions. This suggests that the sterane distribution has no direct connection with the biochemistry of *G. prisca*. In a review of the occurrence of sterols in unicellular algae, Volkman (1986) pointed out that C_{29} sterols predominate over C_{27} and C_{28} sterols in several algal families including Chlorophyceae (green algae), Chrysophyceae and Xanthophyceae, as well as in cyanobacteria It is possible that the steranes which are present in very low concentrations in kukersites are derived from organisms other than *G. prisca*.

The morphology of *G. prisca* obviously also provides other clues to the type of organism it was. Zalessky (1917) originally gave the name *Gloeocapsomorpha prisca* to the fossil algal bodies in Estonian kukersite because he thought that they resembled the extant blue-green algae, *Gloeocapsa*. Subsequently, other workers have also noted the morphological affinity between *Gloeocapsa* and *G. prisca* (e.g. Adamczak 1963 and references therein; Burns 1982). Two recent studies, which attempted to relate *G. prisca* to living organisms, have arrived at similar conclusions. Foster et al. (1989) suggested that it is closely related to a modern mat-forming, and sometimes stromatolite-forming, cyanobacterium *Entophysalis major*. On the basis of a detailed organic petrographic study of Yeoman kukersite, Stasiuk and Osadetz (1990) infer the life cycle of *G. prisca* to include both a planktonic and benthic life stage. These authors think it likely that *G. prisca* has affinities with chroococcales, prokaryotic photosynthesizers that include the extant genuses *Gloeocapsa* and *Entophysalis*.

7.3 Discussion

Obviously it is not yet possible to say with any degree of certainty what type of organism *G. prisca* was. However, the microscope evidence appears to favour the idea that it was a cyanophyte and hence prokaryotic. The relative abundance of hopanes and steranes and, less reliably, the distribution of steranes could be taken as supporting geochemical evidence. However, the distribution of n-alkanes in kukersites is still not easily explained if these compounds are assumed to be derived from the fatty acids of prokaryotes. This is because cyanobacteria do not contain fatty acids above C_{18} (Paoletti et al. 1976a; Weete 1976). The differing opinions of previous workers (Reed et al. 1986; Hoffman et al. 1987) concerning whether *G. prisca* was mat-forming or planktonic may have been reconciled by the observations of Stasiuk and Osadetz (1990) who believe that, depending on the stage of its life cycle, it could be either.

The suggestion that the organic matter of kukersites is possibly made up of the selectively preserved cell-wall material of *G. prisca*, a highly aliphatic biopolymer (Douglas et al., this Vol.), would explain some of the more unusual features of kukersite-derived oils. For example, the very low concentrations of biomarkers such as acyclic isoprenoids and polycyclic alkanes would be due to dilution by the breakdown products of the biopolymer. Additionally, as the hydrocarbon products need not necessarily reflect the fatty acid composition of the lipid fraction of the precursor organism, arguments for a particular group of organisms based on this fraction could be invalid. The relatively high T_{max} values obtained from Rock-Eval analyses of immature kukersites are an indication of the stability of the biopolymer and the high temperatures needed to convert it to petroleum. The occurrence of such a polymer in kukersites could also be taken as possible evidence that *G. prisca* was a chlorophyte since previous reports of similar polymers have been associated with the remains of green algae (Largeau et al. 1986; Goth et al. 1988). However, at present not enough is known about the occurrence of such polymers to rule out the possibility that they could be derived from cyanobacterial remains.

8 Fate of *Gloeocapsomorpha prisca*

Finally, it is appropriate to consider why there is no evidence that *G. prisca* ever attained the same degree of prominence in any period after the Ordovician. The widespread abundance of *G. prisca* during the Ordovician and its apparent disappearance at the end of the period is possibly a function of the unusual environmental conditions that prevailed during this time and suddenly ended during the latest Ordovician. Especially during Ashgillian times, epicontinental seas were of greater extent than in almost any other period (Jaanusson 1984). At the Ordovician-Silurian boundary one of the five largest mass extinctions of Phanerozoic marine life occurred (Stanley 1988). This extinction has been linked with a major glaciation event (e.g. Brenchley and Newall 1984; Wilde and Berry 1984; Stanley 1988). The groups most affected by this mass extinction appear to have been those living in tropical seas, for example the coral reef communities and certain families of green algae (e.g. Dasycladales) which today are confined to equatorial settings (Stanley 1988). Some authors do not think that a glaciation event on its own could be responsible for the extinctions as no equivalent extinction event appears to be associated with the Pleistocene cooling (e.g. Jaanusson 1984). However, other authors have argued that fauna in the Late Ordovician warm epicontinental seas may have been niche-specific and hence too highly adapted to withstand relatively rapid environmental changes, such as the cooling of surface waters or the reduction of habitable areas following a drop in sea level (Brenchley and Newall 1984). Previous authors have noted that *G. prisca* seems to be confined to sediments that were deposited in tropical epicontinental seas (Cramer and Diez de Cramer 1972; Foster et al. 1986) and therefore it probably lived in the environment most severely affected by the changing conditions. Hence, *G. prisca* could have been one of the species that became extinct at the Ordovician-Silurian boundary.

Since this chapter was originally written, some new observations regarding the nature of *Gloeocapsomorpha prisca* have been made. Foster et al. (1989b, 1990) have examined Middle Ordovician kukersites from Estonia in some detail and have designated a neotype, emending Zalessky's (1917) description. Three morphotypes are recognized which are suggested to arise from growth and life cycle changes that are subsequently overprinted by diagenetic changes. It was suggested that *Entophysalis major*, a modern Entophysalidaceae cyanobacterium is a modern analogue for *G. prisca* (Foster et al. 1989b). Foster et al. (1990) described a possible model for the deposition of the Estonian kukersites presuming that *G. prisca* was a mat-forming benthic cyanobacterium.

Derenne et al. (1990) using a combination of spectroscopic and pyrolytic techniques have shown that the kerogen of an Estonian kukersite contains a substantial contribution of long-chain alkyl chains and phenolic moieties. Douglas et al. (1990) found that the sulphur compound distributions in the flash pyrolysate of a Guttenberg Oil Rock kerogen dominated by *G. prisca* were unique. Derenne et al. (1990) and Douglas et al. (1990) conclude that both of these kerogens are derived from a biopolymer in the cell wall of *G. prisca* that has been selectively preserved during diagenesis. However, their pyrolysates differ slightly in that no phenolic products were detected in the Guttenberg pyrolysate.

Acknowledgements. The writing of this paper would not have been possible without the assistance of the following people who kindly supplied samples, data and advice: P. W. Brooks, F. Goodarzi, G. Macauley, K. G. Osadetz, L. R. Snowdon, L. D. Stasiuk and J. B. W. Wielens. I am especially grateful to my colleagues P. W. Brooks, K. G. Osadetz, L. R. Snowdon and L. D. Stasiuk, and to the reviewers C. B. Foster, D. M. McKirdy and R. E. Summons for making constructive comments on earlier drafts of this paper. I also thank E. M. Northcott, S. Achal and R. J. Fanjoy for their excellent technical assistance; and C. Thompson for word processing.

References

Adamczak A (1963) *G. prisca* Zalessky (sinnice) z Ordowickich glazow narzutwych Poliski. Acta Palaeontol Pol 8:465–472

Alexander R, Cumbers M, Kagi RI (1984) Geochemistry of some Canning Basin crude oils. In: Purcell PG (ed) The Canning Basin. W. A. Proc Geol Soc Aust/Petrol Explor Soc Aust Symp, pp 353–358

Allan J, Bjorøy M, Douglas AG (1980) A geochemical study of the exinite group maceral alginite selected from three Permo-Carboniferous torbanites. In: Douglas AG, Maxwell JR (eds) Advances in Organic Geochemistry 1979. Pergamon, Oxford, pp 599–618

Anderson R, Kates M, Baedecker MJ, Kaplan IR, Ackman RG (1977) The steroisomeric composition of phytanyl chains in lipids of Dead Sea sediments. Geochim Cosmochim Acta 41:1381–1390

Aquino Neto FR, Trendel JM, Restle A, Connan J, Albrecht P (1983) Occurrence and formation of tricyclic and tetracyclic terpanes in sediments and petroleums. In: Bjorøy M (ed) Advances in Organic Geochemistry 1981. Wiley Heyden, Chichester, pp 207–227
Bain HF (1906) Zinc and lead deposits of the upper Mississippi Valley. USGS Bull 294:255 pp
Barnes CR (1973) Ordovician conodont biostratigraphy of the Canadian Arctic. In: Aitken JD, Glass DJ (eds) Canadian arctic geology. Geol Assoc Can/Can Soc Petrol Geol, pp 221–240
Blumer M, Mullin MM, Thomas DW (1963) Pristane in zooplankton. Science 140:974
Brassell SC, Wardroper AMK, Thomson ID, Maxwell JD, Eglinton G (1981) Specific acyclic isoprenoids as biological markers of methanogenic bacteria in marine sediments. Nature (London) 280:693–696
Brenchley PJ, Newall G (1984) Late Ordovician environmental changes and their effect on faunas. In: Bruton DL (ed) Aspects of the Ordovician System. Palaeontol contrib Univ Oslo 295 (Universtetsforlaget):65–79
Brooks PW, Snowdon LR, Osadetz KG (1987) Families of oils in southeastern Saskatchewan. In: Carlson CG, Christopher JE (eds) Proc 5th Int Williston Basin Symp, Bismarck, ND, June 15–17, 1987. Sask Geol Soc Spec Publ 9:253–264
Burns DA (1982) A transmission electron microscope comparison of modern *Botryococcus braunii* with some microfossils previously referred to that species. Rev Esp Micropaleontol 14:165–185
Cole GA, Drozd RJ, Sedivy RA, Halpern HI (1987) Organic geochemistry and oil-source correlations, Paleozoic of Ohio. Am Assoc Petrol Geol Bull 71:788–809
Combaz A, Peniguel G (1972) Etude palynostratigraphique de l'Ordovicien dans quelques sondages du Bassin de Canning (Australie Occidentale). Bull Cent Rech Pau SNPA 6:121–167
Cookson IC (1953) Records of the occurrence of *Botryococcus braunii*, *Pediastrum* and the *Hystrichosphaerideae* in Cainozoic deposits of Australia. Mem Natl Mus Vict Melbourne 18:107–123
Cramer FH, Diez de Cramer MDCR (1972) North American Silurian palynofacies and their spatial arrangement: acritarchs. Palaeontogr Abt B 138:107–180
Derenne S, Largeau C, Casadevall E, Tegelaar EW, de Leeuw JW (1988) Relationships between algal coals and resistant cell wall biopolymers of extant algae as revealed by Py-GC-MS. Fuel Process Technol 20:93–101
Drenne S, Largeau C, Casadevall E, Sinninghe Damsté JS, Tegelaar EW, de Leeuw JW (1990) Characterisation of Estonian kukersite by spectroscopy and pyrolysis: evidence for abundant alkyl phenolic moieties in an Ordovician, marine, type II/I kerogen. Org Geochem 16:873–888
De Rosa M, Gambocorta A, Minale L, Bu'lock JD (1972) The formation of ω-cyclohexyl fatty acids from shikimate in an acidophilic thermophilic *Bacillus*. Biochem J 128:751–754
De Sousa NJ, Nes WR (1968) Sterols: isolation from a blue-green alga. Science 162:363
Didyk BM, Simoneit BRT, Brassell SC, Eglinton G (1978) Organic geochemical indicators of palaeoenvironmental conditions of sedimentation. Nature (London) 272:216–222
Douglas AG, Sinninghe Damsté JS, Fowler MG, Eglinton TI, de Leeuw JW (1990) Unique distributions of hydrocarbons and sulphur compounds released by flash pyrolysis from the fossilised alga *Gloeocapsomorpha prisca*, a major constituent in one of four Ordovician kerogens. Geochim Cosmochim Acta 55:275–291
Foster CB, O'Brien GW, Watson ST (1986) Hydrocarbon source potential of the Goldwyer Formation, Barbwire Terrace, Canning Basin, Western Australia. Aust Petrol Explor Assoc J 26, 1:142–155
Foster CB, Reed JD, Wicander R, Summons RE (1989a) Anatomy of an Ordovician oil-prone maceral: *Gloeocapsomorpha prisca* Zalessky 1917. In: Thomas CG, Strachan MG (eds) Proc Macerals '89 Symp. CSIRO Div Coal Techn, North Ryde, NSW, 16:1–9
Foster CB, Reed JD, Wicander R (1989b) *Gloeocapsomorpha prisca* Zalessky, 1917: a new study part 1: taxonomy, geochemistry and paleoecology. Geobios 22:735–759
Foster CB, Wicander R, Reed JD (1990) *Gloeocapsomorpha prisca* Zalessky, 1917: a new study part II: origin of kukersite, a new interpretation. Geobios 23:133–140
Fowler MG (1984) Organic geochemistry of pre-Carboniferous sedimentary organic matter. PhD Thesis, Univ Newcastle-Upon-Tyne, 218 pp
Fowler MG, Douglas AG (1984) Distribution and structure of hydrocarbons in four organic-rich Ordovician rocks. Org Geochem 6:105–114
Fowler MG, Abolins P, Douglas AG (1986) Monocyclic alkanes in Ordovician organic matter. Org Geochem 10:815–823
Goossens H, de Leeuw JW, Schenck PA, Brassell SC (1984) Tocopherols, likely precursors of pristane in ancient sediments and crude oils. Nature (London) 312:440–442
Goth K, de Leeuw JW, Puttmann W, Tegelaar EW (1988) Origin of Messel oil shale kerogen. Nature (London) 336:759–761
Grantham PJ, Wakefield LL (1988) Variations in the sterane carbon number distribution of marine source rock derived crude oils through geological time. Org Geochem 12:61–73
Hatch JR, Jacobson SR, Witzke BJ, Risatti JB, Anders DE, Watney WL, Newell KD, Vulctich AK (1987) Possible Late Middle Ordovician organic carbon isotope excursion: evidence from Ordovician oils and hydrocarbons source rocks, mid-continent and east-central United States. Am Assoc Petrol Geol Bull 71:1342–1354
Heywood WW, Sanford BV (1976) Geology of Southampton, Coats and Mansell Islands, District of Keewatin, Northwest Territories. Geol Surv Can Mem 382:35 pp
Hoffmann CF, Foster CB, Powell TG, Summons RE (1987) Hydrocarbon biomarkers from Ordovician sediments and fossil alga *G. prisca* Zalessky 1917. Geochim Cosmochim Acta 51:2681–2697
Hutton AC (1987) Petrographic classification of oil shales. Int J Coal Geol 8:203–231
Illich HA, Grizzle PL (1983) Comment on "Comparison of Michigan Basin crude oils" by Vogler et al. Geochim Cosmochim Acta 47:1151–1155
Jaanusson V (1984) What is so special about the Ordovician? In: Bruton DL (ed) Aspects of the Ordovician-System. Palaeontol Contrib Univ Oslo 295 (Universitetsforlaget):1–3
Jacobson SR, Hatch JR, Teerman SC, Askin RA (1988) Middle Ordovician organic matter assemblages and their effect on Ordovician-derived oils. Am Assoc Petrol Geol Bull 58:499–506
Johns RB, Belsky T, McCarthy ED, Burlingame AL, Haug P, Schnoes HK, Richter WJ, Calvin M (1966) The

organic geochemistry of ancient sediments II. Geochim Cosmochim Acta 30:1191–1222

Kendall AC (1976) The Ordovician carbonate succession (Bighorn Group) of southern Saskatchewan. Saskatchewan Dep Mineral Resourc Rep 180:185 p

Klesment I, Nappa L (1980) Investigation of the structure of Estonian oil shale kukersite by conversion in aqueous suspension. Fuel 59:117–122

Kohm JA, Louden RO (1978) Ordovician Red River of eastern Montana and western North America: relationships between lithofacies and production. In: 1979 Williston Basin Symp Economic geology of the Williston Basin. Proc Montana Geol Soc 24th Annu Conf, Billings, Montana, pp 99–117

Kurylowicz LE, Ozimic S, McKirdy DM, Kantsler AJ, Cook AC (1976) Reservoir and source rock potential of the Larapinta Group, Amadeus Basin, central Australia. Aust Petrol Explor Assoc J 16, 1:49–65

Largeau C, Derenne S, Casadevall E, Kadouri A, Sellier N (1986) Pyrolysis of immature torbanite and of the resistant biopolymer (PRB A) isolated from the extant alga *Botryococcus braunii*. Org Geochem 10:1023–1032

Larter SR (1978) A geochemical study of kerogen and related materials. PhD Thesis, Univ Newcastle-Upon-Tyne

Longman MW, Palmer SE (1987) Organic geochemistry of Mid-Continent Middle and Late Ordovician oils. Am Assoc Petrol Geol Bull 71:938–950

Macauley G (1986) Geochemistry of the Ordovician Boas oil shale, Southampton Island, Northwest Territories. Geol Surv Can Open File Rep 1285:15 pp

Macauley G, Snowdon LR, Ball FD (1985) Geochemistry and geological factors governing exploitation of selected Canadian oil shale deposits. Geol Surv Can Pap 85-13:65 pp

Macauley G, Fowler MG, Goodarzi F, Snowdon LR, Stasiuk LD (1990) Ordovician oil shale-source rock sediments in the central and eastern Canada mainland and eastern Arctic areas and their significance for frontier exploration. Geol Surv Can Paper 90-14:51 pp

Martin RL, Winters JC, Williams JA (1963) Distribution of n-paraffins in crude oils and their implications to the origin of petroleum. Nature (London) 199:110–113

Mayr U (1978) Stratigraphy and correlation of Lower Paleozoic formations, subsurface of Cornwallis, Devon, Somerset and Russell Islands, Canadian Arctic Archipelago. Geol Surv Can Bull 276, 55 pp

McCracken AD, Nowlan GS (1989) Conodont paleontology and biostratigraphy of Ordovician carbonates and petroliferous carbonates on Southampton, Baffin and Akpatok Islands in the eastern Canadian Arctic. Can J Earth Sci 26:1880–1903

McGregor DC, Cramer FH (1971) Palynomorphs of the Ordovician Cat Head Member, Lake Winnipeg, Manitoba. Contributions to Canadian paleontology: fossils of the Ordovician Red River Formation (Cat Head Member), Manitoba. Geol Surv Can Bull 20:1–11

McKirdy DM, Hahn JH (1982) The composition of kerogen and hydrocarbons in Precambrian rocks. In: Holland HD, Schidlowski M (eds) Mineral deposits and the evolution of the biosphere. Dahlem Konf. Springer, Berlin Heidelberg New York, pp 123–154

McKirdy DM, Kantsler AJ (1980) Oil geochemistry and potential source rocks of the Officer Basin, South Australia. Aust Petrol Explor Assoc J 20, 1:60–86

McKirdy DM, Aldridge AK, Ypma PJM (1983) A geochemical comparison of some crude oils from pre-Ordovician carbonate rocks. In: Bjorøy M (ed) Advances in Organic Geochemistry 1981, Wiley Heyden, Chichester, pp 99–107

Monnier F, von der Dick H, Fowler MG, Brooks PW (1987) Simultaneous early generation of gaseous and liquid hydrocarbons from a Cambrian lean organic source-rock. In: Abstr 13th Int Meet Org Geochem, Venezia, Sept 21–25, 1987, p 254

Nelson ST, Johnson RD (1966) Geology of Hudson Bay Basin. Bull Can Petrol Geol 14:520–578

Nissenbaum A, Baedecker MG, Kaplan IR (1972) Organic geochemistry of Dead Sea sediments. Geochim Cosmochim Acta 36:709–729

Osadetz KG, Snowdon LR, Stasiuk LD (1989) Association of enhanced hydrocarbon generation and crustal structure in the Canadian Williston Basin. In: Current research, pt D. Geol Surv Can Pap 89-1D:35–47

Oshima M, Ariga T (1975) ω-Cyclohexyl fatty acids in acidophilic thermophilic bacteria. J Biol Chem 250:6963–6968

Ourisson G, Rohmer M, Poralla K (1987) Prokaryotic hopanoids and other polyterpenoid sterol surrogates. Annu Rev Microbiol 41:301–333

Paoletti C, Pushparj B, Florezano G, Capella P, Lercker G (1976a) Unsaponifiable matter of green and blue-green algal lipids as a factor of biochemical differentiation of their biomasses I Total unsaponifiable and hydrocarbon fraction. Lipids 11:258–265

Paoletti C, Pushparj B, Florezano G, Capella P, Lercker G (1976b) Unsaponifiable matter of green and blue-green algal lipids as a factor of biochemical differentiation of their biomasses II. Terpenic alcohol and sterol fractions. Lipids 11:266–271

Powell TG, Macqueen RW, Barker JF, Bree DG (1984) Geochemical character and origin of Ontario oils. Bull Can Petrol Geol 32:289–312

Reed JD, Illich HA, Horsfield B (1986) Biochemical evolutionary significance of Ordovician oils and their sources. Org Geochem 10:347–358

Rubinstein I, Strausz OP (1979) Geochemistry of the thiourea adduct fraction from an Alberta petroleum. Geochim Cosmochim Acta 43:1387–1392

Rubinstein I, Sieskind O, Albrecht P (1975) Rearranged steranes in a shale: occurrence and simulated formation. J Chem Soc Perkin 1:1833–1836

Rullkötter J, Aizenshtat Z, Spiro B (1984) Biological markers in bitumens and pyrolysates of Upper Cretaceous bituminous chalks from the Ghareb Formation (Israel). Geochim Cosmochim Acta 48:151–157

Rullkötter J, Meyers PA, Schaefer RG, Dunham KW (1986) Oil generation in the Michigan Basin: a biological marker and stable isotope approach. Org Geochem 10:359–375

Saxby JD, Bennett AJR, Corcoran JF, Lambert DE, Riley KW (1986) Petroleum generation: simulation over six years of hydrocarbon formation from torbanite and brown coal in a subsiding basin. Org Geochem 9:69–81

Seifert WK, Moldowan JM (1978) Applications of steranes, terpanes and monoaromatics to the maturation, migration and source of crude oils. Geochim Cosmochim Acta 42:77–95

Sieskind O, Joly G, Albrecht (1979) Simulation of the geochemical transformation of sterols: superacid effect of clay minerals. Geochim. Cosmochim Acta 43:1675–1679

Simoneit BRT, Burlingame AL (1973) Carboxylic acids derived from Tasmanian tasmanite by extraction and kerogen oxidations. Geochim Cosmochim Acta 37:595–610

Snowdon LR (1984) A comparison of Rock-Eval pyrolysis and solvent extraction results from the Collingwood and Kettle Point oil shales, Ontario. Bull Can Petrol Geol 32:327–334

Snowdon LR, Osadetz KG (1988) Geological processes interpreted from gasoline range analyses of oils from southeast Saskatchewan and Manitoba. In: Current research, pt D. Geol Surv Can Pap 88-1D:33–40

Stanley SM (1988) Paleozoic mass extinctions: shared patterns suggest global cooling as a common cause. Am J Sci 288

Stasiuk LD, Osadetz K (1990) The life cycle and phyletic affinity of *Gloeocapsomorpha prisca* Zalessky 1917 from Ordovician rocks in Canadian Williston Basin. In: Current research, pt D. Geol Surv Can Pap 90-1D, pp 123–137

Summons RE, Powell TG (1986) *Chlorobiaceae* in Paleozoic sea revealed by biological markers, isotopes and geology. Nature (London) 319:763–765

Summons RE, Powell TG (1987) Identification of aryl isoprenoids in source rocks and crude oils: biological markers for the green sulphur bacteria. Geochim Cosmochim Acta 51:557–566

Suzuki K-I, Saito K, Kawaguchi A, Oduka S, Komagata K (1981) Occurrence of ω-cyclohexyl fatty acids in *Curtobacterium pusillum* stains. J Gen Appl Microbiol 27:261–266

Ten Haven HL, de Leeuw JW, Rullkötter J, Sinninghe Damsté JS (1987) Restricted utility of the pristane/phytane ratio as a palaeoenvironmental indicator. Nature (London) 330:641–643

Tissot BP, Welte DH (1984) Petroleum formation and occurrence, 2nd edn. Springer, Berlin Heidelberg New York, 699 pp

Tissot B, Pelet R, Roucoche J, Combaz J (1977) Utilisation des alcanes comme fossiles géochimques indicateurs des environnements géologiques. In: Campos R, Goni J (eds) Advances in organic Geochemistry 1975. ENADIMSA, Madrid, pp 117–154

Traverse A (1955) Occurrence of the oil-forming alga *Botryococcus* in lignites and other Tertiary sediments. Micropaleontology 1:343–350

Vitorovic D (1980) Structure elucidation of kerogen by chemical methods. In: Durand B (ed) Kerogen. Editions Technip, Paris, pp 301–338

Vlierboom FW, Collini B, Zumberge JE (1986) the occurrence of petroleum in sedimentary rocks of the meteor impact crater at Lake Siljan, Sweden. Org Geochem 10:153–161

Vogler EA, Meyers PA, Moore WA (1981) Comparison of Michigan Basin crude oils. Geochim Cosmochim Acta 45:2287–2293

Volkman JK (1986) A review of sterol markers for marine and terrigenous organic matter. Org Geochem 9:83–99

Volkman JK, Maxwell JR (1986) Acyclic isoprenoids as biological markers. In: Johns RB (ed) Biological markers in the sedimentary record. Elsevier, Amsterdam, pp 1–42

Weaver FJ, Macko SA (1988) Source rocks of western Newfoundland. Org Geochem 13:411–421

Weete JD (1976) Algal and fungal waxes. In: Kolattukudy PE (ed) Chemistry and biochemistry of natural waxes. Elsevier, Amsterdam, pp 350–418

Wielens JBW, von der Dick H, Fowler MG, Brooks PW, Monnier F (1990) Geochemical comparison of a Cambrian potential alginite source rock, and hydrocarbons from the Colville/Tweed Lake area, Northwest Territories, Canada. Bull Can Petrol Geol 38:236–245

Wilde P, Berry WBN (1984) Destabilization of the oceanic density structure and its significance to marine "extinction" events. Palaeogeogr Paleoclimatol Palaeoecol 48:143–162

Williams JA (1974) Characterisation of oil types in Williston Basin. Am Assoc Petrol Geol Bull 58:1243–1252

Zalessky MD (1917) On marine sapropelite of Silurian age formed by a blue-green alga. Izv Imp Akad Nauk IV Ser 1:3–18

Zumberge JE (1983) Tricyclic diterpane distributions in the correlation of Paleozoic crude oils from the Williston Basin. In: Bjorøy M (ed) Advances in organic geochemistry 1981. Wiley Heyden, Chichester, pp 738–745

Zumberge JE (1987) Prediction of source rock characteristics based on terpane biomarkers in crude oils: a multivariate statistical approach. Geochim Cosmochim Acta 51:1625–1637

3. Antiquity and Metallogenic Potential of Bacterial Sulfate Reduction

Sulfur in the Earth's Mantle – A Review

Kurt von Gehlen[1]

[1] Institut für Geochemie, Petrologie und Lagerstättenkunde, Universität Frankfurt a. M., Senckenberg-Anlage 28, W-6000 Frankfurt a. M. 1, FRG

Abstract

The average sulfur content of the Upper Mantle is about 300 to 400 ppm. Basaltic melts contain up to about 1000 ppm sulfur. Sometimes spherules show that an immiscible sulfide melt separated from the silicate melt. Almost all sulfides observed, however, are later reequilibration products formed during cooling or alteration. This also applies to sulfide inclusions in mantle rock xenoliths and even in diamonds. Sulfur isotope ratios in primary mantle material are close to those of meteorites; for the primitive upper mantle, $\delta^{34}S \approx +0.5$ per mill.

1 Introduction

The ultimate source of the sulfur now found in minerals near the Earth's surface is the mantle. Some average sulfur contents given in the literature are compiled in Table 1. Geochemical reasons for the low sulfur content of the upper mantle were discussed by Newsom and Palme (1984) and Schmitt et al. (1989).

Sulfur-bearing phases in samples of mantle origin are mostly sulfides. At least pyrrhotite is stable under p-T conditions of the uppermost mantle (extrapolated from Fig. 1). Pyrrhotite may then contain other metals (Ni, Cu) in solid solution (monosulfide solid solution), Mss, see Sect. 2.3).

Mantle rock fragments are accessible as xenoliths in kimberlites (equilibrated at about 200-km depth) and in alkaline basalts (about 50–100-km depth). The sulfides found in these xenoliths, however, often form associations which are stable only at much lower temperatures (Sect. 2.2). Sulfide inclusions in diamonds should be most reliable as mantle samples (Sect. 2.1). Other sources of information are basaltic melts which rise from the mantle as partial melts derived from pyrolite. They may carry droplets of immiscible sulfide melt (Sect. 3). Finally, volcanic exhalations and fluid inclusions may indicate gaseous phases stable under mantle conditions (Sect. 4).

Sulfur isotope ratios in S-bearing species may give information on their possible primary mantle origin and later changes (Sect. 5).

As not all relevant publications can be discussed in such a short review, only some references have been selected.

Table 1. Sulfur contents of some mantle materials

	S ppm
Average sulfur content of Earth's mantle (Ringwood 1966)	300
Minimum concentration in primitive upper mantle (Schmitt et al. 1989)	390
S content in spinel lherzolites (Basaltic Volcanism Study Group 1981, cited in Schmitt et al. 1989, p. 184)	ca. 100
Garnet pyroxenite xenoliths in Hawaii basalt (recalculated from De Waal and Calk 1975)	300–1400
Eclogite xenoliths from kimberlite "highly altered xenoliths" among them (Robinson 1979)	Up to 900 300–4000
Juvenile sulfur in basaltic magma (submarine glass; Moore and Fabbi 1971)	800 ± 150
Maximum solubility in basaltic magma at 1200 °C (experimental data; Haughton et al. 1974)	500–2000
Maximum solubility in basaltic magma, Hawaii (with globules) at 1065 °C (Skinner and Peck 1969)	380
Average sulfur abundance in Earth's crust (Allègre and Michard 1974)	260
Average sulfur content of continental crust (Sakai et al. 1984; Chaussidon et al. 1989)	780

M. Schidlowski et al. (Eds.)
Early Organic Evolution: Implications for Mineral and Energy Resources

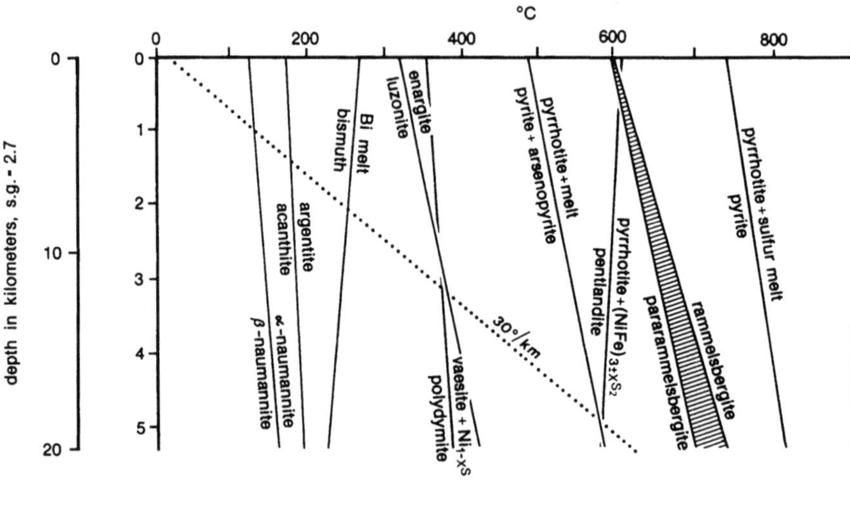

Fig. 1. p (kbar) vs T plot for vapor-free univariant equilibria for ore mineral reactions as summarized by Barton and Skinner 1967 (redrawn from Barton 1970)

2 Sulfides in Mantle Xenoliths from Kimberlites and Basalts and Their Stabilities

2.1 Sulfide Inclusions in Diamonds

Inclusions in diamond "xenoliths" from kimberlites promise to be less affected by later alteration than sulfides in silicate rocks. Pyrite was mentioned as an inclusion already in the last century (Harting 1859). The first to investigate "pyrrhotite" in diamonds was Sharp (1966), who also found minor pentlandite. Meyer and Tsai (1976) gave a review of inclusions in diamond; they distinguished three periods of inclusion formation of which only the first can be called primary (Meyer and Tsai 1979).

Sulfides form the most abundant inclusions in diamonds (Boyd and Gurney 1986), constituting about 40% in diamonds from several African kimberlites (Harris and Gurney 1982). Monosulfide solid solution (see Fig. 2) and its exsolution products – Fe–Ni and Cu–Fe sulfides, with some pyrite and cubanite – were observed. Tsai et al. (1979) also reported pyrrhotite + magnetite, and once pyrite, in diamonds and gave average compositions of the sulfides. Welke et al. (1974) and Kramers (1979) first succeeded in obtaining lead model ages from sulfides enclosed in diamonds.

2.2 Sulfides in Xenoliths

White (1966), Sobolev and Vakhrushev (1967) and Vakhrushev and Kutolin (1970) were among the first to report on this subject. Bishop et al. (1975) gave a general review of sulfides in xenoliths. Sulfide globules mostly consist of intergrowths of pyrrhotite (po) and pentlandite (pn) surrounded by a chalcopyrite (cp) rim. They seem to have developed from sulfide liquids dominated by FeS, containing minor amounts of Cu, Ni and Co plus some oxygen. Lorand and Conquéré (1983) determined equilibration temperatures of 900 to 1200 °C for sulfides in spinel lherzolite xenoliths from basalts.

In massive kimberlites, only minor later alteration of the enclosed xenoliths is likely to occur (Frick 1973). From garnet-peridotite and eclogite (= griquaite) nodules, Frick described primary spherical sulfide globules of 0.001–0.1 mm diameter consisting of nickeliferous pyrrhotite, which seem to constitute immiscible droplets exsolved from the silicate melt. Similar droplets were found by Andersen et al. (1987) in mantle-derived clinopyroxene megacrysts and pyroxenite xenoliths from basalts.

Frick (1973) listed several types of secondary alteration. In garnet-peridotite xenoliths, the most widespread type consisted of increasing amounts of sulfides in a serpentine meshwork. Finally, pyrrhotite was replaced by magnetite and hematite and chalcopyrite by cuprite, even in nonserpentinzed nodules. In "griquaite" xenoliths, small sulfide globules in glassy veins also contained pyrite. Potassium enrichment may lead to the formation of biotite and phlogopite, in addition to talc and calcite. Also, analcime may occur as an alteration product in glassy veins. Pyrite was associated with secondary minerals such as serpentine (Wagner 1914). Lorand (1985) discussed the existence of closed and open-system conditions during incipient alteration of mantle sulfides in peridotites. According to Lorand and Conquéré (1983), present-day sulfide mineral assemblages in xenoliths studied by them result from solid-state reactions down to 250–300 °C.

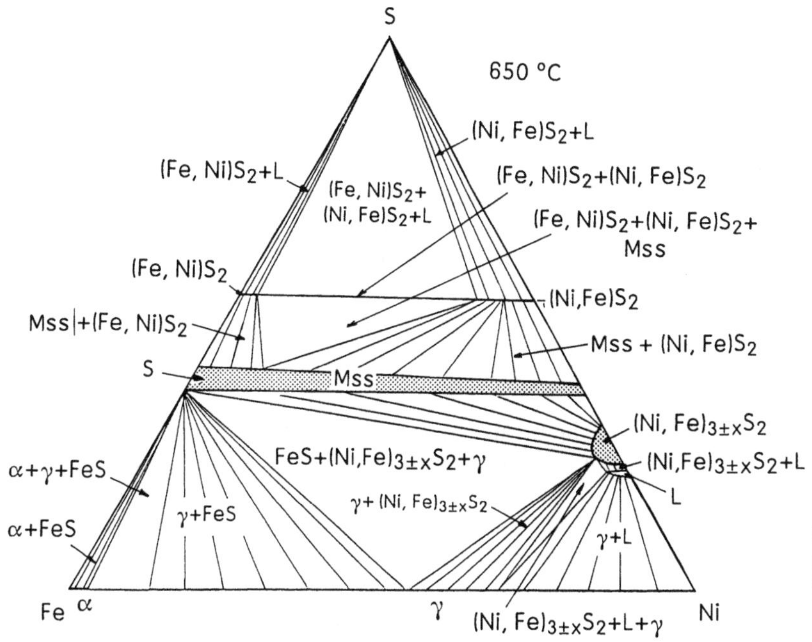

Fig. 2. Phase relations in the Fe–Ni–S system at 650 °C after Kullerud 1963 (Kullerud et al. 1969)

In various types of xenoliths from Yakutian kimberlites, the sulfides partly contain djerfisherite besides pyrrhotite, pentlandite, and chalcopyrite, even in an ilmenite nodule (Ivanov et al. 1977). Djerfisherite, $K_6Fe_{26-x}S_{26}Cl$ is a K-bearing sulfide with Ni and Cu partly replacing Fe (Clarke 1979).

Desborough and Czamanske (1973) observed spherical polymineralic sulfide aggregates in rounded eclogite nodules from the Roberts Victor kimberlite pipe, South Africa. The sulfide assemblage was interpreted to have developed from an immiscible liquid phase and consisted of pyrrhotite, pentlandite, chalcopyrite, $(Fe, Ni)_9S_{11}$, and Ni-bearing pyrite.

In garnet pyroxenite xenoliths in basalt from Oahu, Hawaii, De Waal and Calk (1975) found three chemically distinct sulfide types. "Enclosed" sulfide (0.003–0.029 vol%) formed spherical inclusions, representing Cu–Ni-rich sulfide liquid, coexisting with Cu-rich monosulfide solid solution. "Interstitial" sulfide (0.006–1.117 vol%) in interstices of altered primary silicates was Cu-poor with varying Ni content. "Vein" sulfide (0.005–0.020 vol%) formed fracture fillings. The total sulfide contents of 0.06–0.26 vol% can be recalculated to 0.03–0.14 wt% S (300–1400 ppm S). De Waal and Calk concluded that the mineralogy and chemistry of the enclosed sulfide suggested abrupt cooling of the xenoliths from about 1000 °C upon ejection.

Other sulfides from the Yakutian kimberlite include sphalerite, wurtzite, galena, cinnabar, marcasite and molybdenite (e.g. Bobrievich et al. 1964; see Dawson 1980, p. 72).

In veins and cavities in the kimberlites, also sulfates (barite, anhydrite, gypsum, celestite and epsomite) were found (Wagner 1914; Bobrievich et al. 1964; see Dawson 1980, p. 89). They are definitely not primary.

2.3 Stability Relations of Fe–Ni Sulfides

The dominant system concerning mantle sulfides is Fe–Ni–S. This system was studied experimentally by Kullerud (1963) down to 400 °C. Figure 2 shows that at 650 °C, a complete solid solution of (Fe, Ni) monosulfide exists but no pentlandite is stable. At atmospheric pressure, pentlandite appears below 610 °C, its upper temperature limit decreasing with increasing pressure to about 400 °C at 30 kbar (Fig. 3). Thus, pentlandite is unstable under mantle conditions and only forms upon cooling. For an average mantle sulfide bulk composition (close to the Sudbury ore composition marked S in Fig. 2), pentlandite first forms between 610 and 300 °C when the Mss field narrows so that the bulk composition then falls into the two-phase field between Mss and pentlandite. Figure 4 shows that at atmospheric pressure, Mss (with a narrower range of composition) still persists at 300 °C but at some temperature below 300 °C gives way to other phase combinations, including the coexistence of pyrite and pentlandite. Djerfisherite (K-bearing) has been reported as stable up to 649 °C (Clarke 1979).

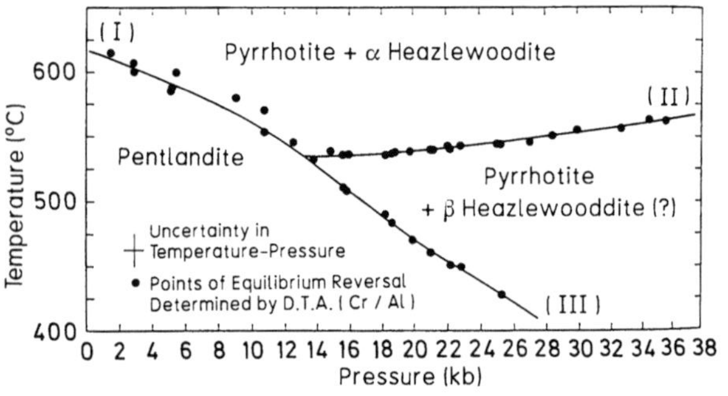

Fig. 3. Upper p, T stability limit of pentlandite for $Fe_{4.5}Ni_{4.5}S_8$ compositions, determined by high-pressure differential thermal analysis (Bell et al. 1964)

Kinetic effects are also important in any discussion of mantle minerals. Sulfides react much faster to changes in thermodynamic stability upon cooling (or decompression) than most silicates and oxides (Fig. 5): pyrrhotite tends to reequilibrate after only 10 years after cooling to 300 °C.

2.4 Reactions Necessary to Explain the Observed Sulfide Assemblages

According to Bishop et al. (1975), the most common assemblage in the sulfide globules is a core of pyrrhotite containing pentlandite inclusions, surrounded by a rim of chalcopyrite. This was interpreted by them to be the result of

1. Crystallization of monosulfide solid solution (Mss) from a sulfide liquid near 1000 °C;
2. Crystallization of chalcopyrite (solid solution) from the remaining Cu-rich liquid near 850 °C;
3. Change of composition of the sulfides with falling temperature;
4. Development of pentlandite from Mss below about 600–300 °C.

The primary liquid probably contained a substantial magnetite component. However, the observed magnetite may also have resulted from late-stage alteration. Euhedral pyrite crystals in globules are not primary because pyrite is stable only below 743 °C (at atmospheric pressure, Kullerud and Yoder 1959) to about 800 °C (at mantle pressures, Fig. 1). The phase coexisting with liquid is Mss. Irregular textures involving magnetite, hematite, mackinawite, millerite and chalcocite must have resulted at least in part from low-temperature reactions (even below 100 °C). The nature and stability of the two M_9S_{11} phases reported by Desborough and Czamanske (1973) are still a problem.

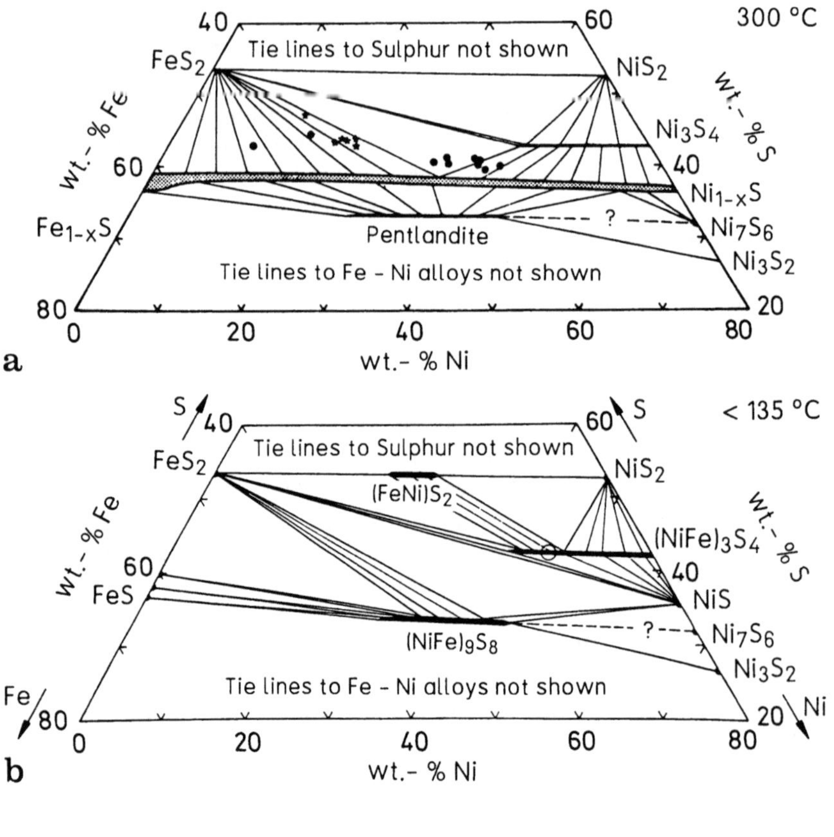

Fig. 4. Central part of Fe–Ni–S system at **a** 300 and **b** <135 °C (After Graterol and Naldrett 1971)

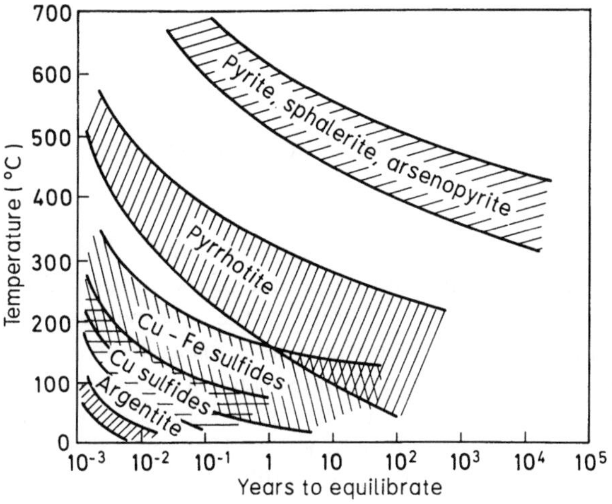

Fig. 5. Time for equilibration for various sulfides as functions of temperature (Barton 1970)

Bishop et al. (1975) listed the following reactions which probably led to a globule of pentlandite-magnetite intergrowths with massive pentlandite found in a spinel lherzolite nodule from the De Beers kimberlite:

1. Mss crystallization starting near 1000 °C from a Fe–Ni–S–O (?) liquid, then cotectic crystallization of Mss + magnetite, finally conversion of Mss to pentlandite below 610 °C with adjustment of bulk composition, or
2. (less probable) formation of a pyrrhotite-pentlandite intergrowth which underwent selective oxidation of the pyrrhotite to magnetite, or
3. (also less probable) formation of a single Fe–Ni–S–O phase which inverted to pentlandite and magnetite.

Bishop et al. (1975, p. 336) concluded: "In general, the present data are consistent with most sulphides developing from a liquid dominated by Fe, Ni, Cu monosulphide of compositions rather similar to those inferred for magmatic sulphide ores, for example those at Sudbury."

3 Sulfides and Sulfate in Basalts

Basalts used as analytical standards contain averages of 500 (BCR-1) and 444 (BR) ppm sulfur (Table 16-E-1 of Schneider 1978).

For basalts, the degassing history plays an important role in limiting the S contents. Thirty-six basalts solidified at the surface averaged 65 ppm S, while 61 submarine basalts gave 782 ppm S (Schneider 1978, p. 16-E-17): "The sulfur content of ascending basaltic magmas may be extrapolated as being 800 to 1000 ppm S", the first figure derived by Moore and Fabbi (1971), the second by McLean (1969). "Accepting the conclusions of Anderson (1974) that sulfur saturation is common in the silicate liquid of ascending basaltic magmas, and considering a certain amount of immiscible sulfide liquid as being associated with these liquids after magma generation in the mantle, basaltic liquids must carry even more than 1000 ppm of sulfur in their initial state" (Schneider 1978, p. 16-E-17).

Skinner and Peck (1969, p. 310 ff) described indications of an immiscible sulfide melt formed during cooling of the Alae (Hawaii) basaltic lava. They found that a differentiated siliceous liquid was saturated with respect to sulfide sulfur with 380 ppm S at 1065 °C. The S-rich liquid was estimated to consist of 61 wt% Fe, 4% Cu, 31% S, and 4% oxygen, while a coexisting Cu-rich solid sulfide consisted of 53 wt% Fe, 9% Cu, 3% Ni, and 35% S. They concluded that immiscible S-rich liquids separating from basic magmas are always FeS-rich and have a significant oxygen content, which upon cooling leads to the formation of magnetite.

The solubility of sulfur in basaltic liquid as related to oxygen and sulfur fugacities and FeO concentration was outlined by Haughton et al. (1974). The maximum solubility of sulfur in a basaltic magma at 1200 °C varies from 500 to 2000 ppm as the FeO content increased from 5 to 20 wt%. A temperature increase of 100 °C would magnify the capacity of a magma to dissolve sulfur by a factor of 5 to 7.

MacRae (1979) in a study of the Newer Basalts of Victoria (Australia) concluded that a CO_2-rich vapor phase must have boiled off during ascent of the basaltic magma and penetrated the xenoliths, resulting in ubiquitous fluid inclusions and partial melting along some fractures and intergranular boundaries. Sulfide globules in the glass and along planes in primary silicates apparently formed during ascent. The sulfur was derived from the invading vapor and not from the ultramafic xenoliths.

The presence of sulfate has been demonstrated in sulfur isotope studies of various basalts (Schneider 1970; Kyser 1986; Fig. 7). Partial oxidation of sulfide to sulfate is possible during alteration of alkaline basaltic rocks (Schneider 1970). Secondary sulfates were even found in kimberlite (Sect. 2.2).

4 Gaseous S-Bearing Species and Fluid Inclusions

The global SO_2 output of volcanoes, which is assumed to be predominantly derived from mantle sources, was estimated by Lambert et al. (1988) at 50×10^6 t/year.

Other sources of information on volatile components in the mantle are fluid inclusions in mantle rocks. Murck et al. (1978) studied more than 200 fluid inclusions from xenoliths associated with basalts and found that most inclusions contained nearly pure CO_2, with in some cases small amounts (about 0.05 to 0.10 mole fraction) of SO_2, H_2S, or COS. Glass lining some of the inclusion walls indicated the coexistence at depth with a melt. Confining pressures during entrapment derived were > 10 kbar, i.e., those of the uppermost mantle.

Sulfur partial pressures were assumed by Mitchell and Keays (1981) to have been $fS_2 < 10^{-5}$ bar at < 1200 °C. Based on his own studies and those of Kogarko (1972), Mitchell (1975) concluded that CH_4 is an important constituent of the fluid phase of alkaline magmas (which are derived from mantle depths) coexisting with high fCO_2, and that H_2S is the dominant S-bearing species (Fig. 6).

Intrinsic oxygen fugacity data on mantle samples (Virgo et al. 1988) indicated two groups of values, one around the quartz-fayalite-magnetite buffer and the other close to the iron-wustite buffer (at 1000 °C, fO_2 ca. 10^{-10} resp. 10^{-14}). This is in agreement with the results of a detailed study by Woermann and Rosenhauer (1985) on fluid phases and the redox state of the Earth's mantle. The CH_4 content of a mantle fluid is strongly affected by the Fe content of the system (Saxena 1989).

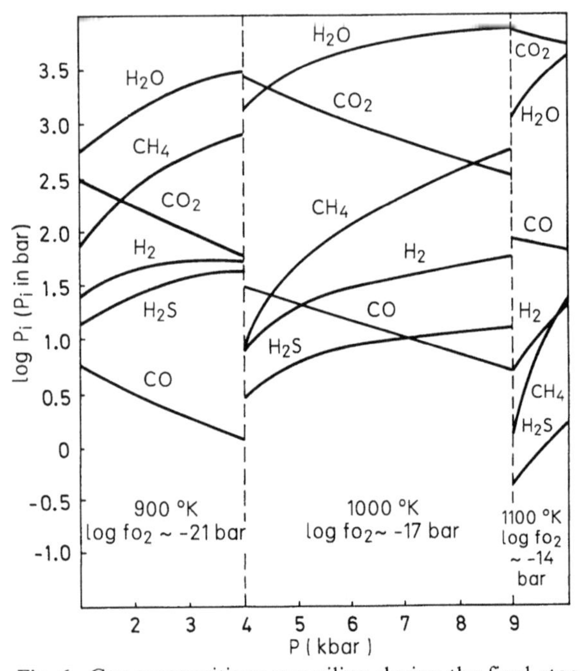

Fig. 6. Gas compositions prevailing during the final stages of kimberlite evolution (Mitchell 1975, Fig. 6)

5 Sulfur Isotopes in Mantle-Derived Minerals

The sulfur isotope ratios of sulfides and sulfates are often used to indicate the source of the sulfur and its geological history. Therefore, the isotope ratio of mantle sulfur is important as a basis from which fractionation by crustal and surface processes started. A condensed review of sulfur isotope ratios in mantle-derived rocks was given by Kyser (1986, pp. 157–158).

Grinenko and Ukhanov (1977) found whole-rock $\delta^{34}S$ values of +0.4 to +2.1‰ in unaltered pyroxenite and garnet peridotite xenoliths from kimberlites. Most recently, Chaussidon et al. (1989) analysed 21 sulfide inclusions in mantle minerals and rocks, including isotope ratios, by ion microprobe. They found a wide range from $\delta^{34}S = -4.9 \pm 1$ to $+8 \pm 1$‰ and postulated a high-temperature S-isotope fractionation of about 3 per mill between liquid sulfide and sulfur dissolved in silicate liquid. Sulfides with high Ni contents (up to 40% pentlandite) in residual peridotite had a mean $\delta^{34}S$ of $+3 \pm 1$‰, while low-Ni sulfides from pyroxenites, ocean island basalts and kimberlites had $+1 \pm 1$‰ (Chaussidon et al. 1989).

Schneider (1970) published $\delta^{34}S$ values for the sulfide and sulfate fractions of continental basalts mostly from the Federal Republic of Germany (Fig. 7). According to Sakai et al. (1984), the SO_4^{2-}/S^{2-} ratio increases with increasing water content of oceanic basalt, with an average $\delta^{34}S$ difference of around +7.5‰.

There are sulfide-bearing deposits such as the Palabora carbonatite which must be assumed to contain pure mantle sulfur; however, only the early sulfides show a narrow, probably primary $\delta^{34}S$ range (Von Gehlen 1967; see Fig. 7).

The $\delta^{34}S$ values shown in Fig. 7 as well as values from mantle-derived "magmatic" sulfide ore deposits also indicate that primary mantle sulfur must have a $\delta^{34}S$ close to the Canyon Diablo meteorite standard (0‰) and a meteorite average (Nielsen 1978) of +0.1‰. Chaussidon et al. (1989) postulated a mean $\delta^{34}S$ of +0.5‰ for the primitive upper mantle. For the subcontinental mantle of Central Europe, Schneider (1970) suggested a range of +1 to +1.5‰. Sakai et al. (1984) concluded that the mantle may not be homogeneous isotopically.

Fractionation and alteration reactions, as well as incorporation of crustal sulfur, lead to larger ranges of

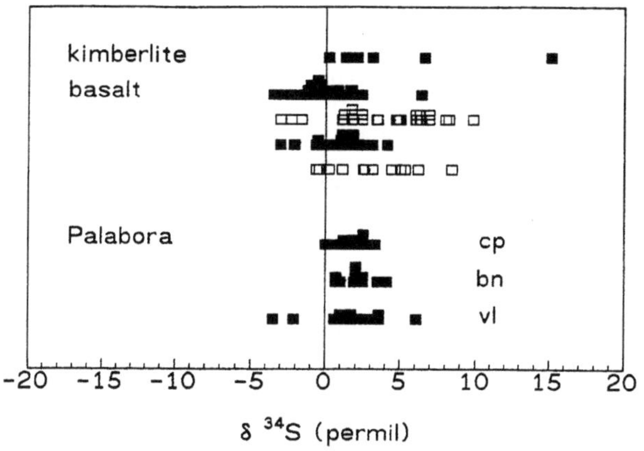

Fig. 7. Sulfur isotope ratios in mantle-derived rocks (*filled squares* sulfide sulfur; *open squares* sulfate sulfur) in kimberlites and basalts (*top five lines*) and one selected mantle-derived ore deposit. *1st line* $\delta^{34}S$ of six sulfides from kimberlites and associated xenoliths (Tsai et al. 1979, Table 6, samples from shale host rocks excluded); *2nd* and *3rd lines* $\delta^{34}S$ of sulfides and sulfates from submarine OIB (ocean island basalts), MORB (mid-ocean ridge basalts), and nodules (sulfides only) (Kyser 1986, Fig. 10) as well as from OIB and MORB (Taylor 1986, Fig. 28); *4th* and *5th lines* $\delta^{34}S$ of sulfide and sulfate sulfur extracted from basalts (tholeiites to nephelinites) from northern Hesse and southern Lower Saxony, FRG (after Schneider 1970, Table 1a, sulfides without values marked b); *last three lines* $\delta^{34}S$ of sulfide minerals from the Palabora (Phalaborwa) carbonatite, South Africa (after Von Gehlen 1967): *cp* chalcopyrite, *bn* bornite, *vl* valleriite

values. One $\delta^{34}S$ value for a shale xenolith from a South African kimberlite of − 37.3‰ clearly indicates "foreign" sulfur. For some sulfides in Yakutian kimberlites, Ilupin (1962) suggested that the sulfur came from evaporite sulfates penetrated by the kimberlite.

As not only the mantle average, but also the crustal average of $\delta^{34}S$ is close to 0‰, it is generally impossible to prove a "mantle" or "igneous" origin solely from sulfur isotope ratios.

References

Allègre C-J, Michard G (1974) Introduction to geochemistry. Reidel, Dordrecht Boston, 142 pp

Andersen T, Griffin WL, O'Reilly SY (1987) Primary sulphide melt inclusions in mantle-derived megacrysts and pyroxenites. Lithos 20:279–294

Anderson AT (1974) Chlorine, sulfur, and water in magmas and oceans. Geol Soc Am Bull 85:1485–1492

Barton PB Jr (1970) Sulfide petrology. Min Soc Am Spec Pap 3:187–198

Barton PB Jr, Skinner BJ (1967) Sulfide mineral stabilities. In: Barnes HL (ed) Geochemistry of hydrothermal ore deposits. Holt Rinehart Winston, New York, pp 236–333

Basaltic Volcanism Study Project (1981) Ultramafic xenoliths in terrestrial volcanisms and mantle magmatic processes. In: Basaltic volcanism on the terrestrial planets. Pergamon, New York, pp 282–310

Bell PM, England JL, Kullerud G (1964) Pentlandite pressure effect on breakdown. Carnegie Inst Washington Yearb 63:206–207

Bishop FC, Smith JV, Dawson JB (1975) Pentlandite-magnetite intergrowth in De Beers spinel lherzolite: review of sulphides in nodules. Phys Chem Earth 9:323–337

Bobrievich AP, Lupin JP, Kozlov IT, Lebedev LI, Pankratov AA, Smirnov GI, Khariv AD (1964) Petrography and mineralogy of the kimberlite rocks in Yakutia. Nedra Izdat, Moscow (in Russian)

Boyd FR, Gurney JJ (1986) Diamonds and the African lithosphere. Science 232:472–477

Chaussidon M, Albarede F, Sheppard SMF (1989) Sulphur isotope variations in the mantle from ion microprobe analyses of micro-sulphide inclusions. Earth Planet Sci Lett 92:144–156

Clarke DB (1979) Synthesis of nickeloan djerfisherites and the origin of potassic sulphides at the Frank Smith Mine. In: Boyd FR, Meyer HOA (eds) Proc 2nd Int Kimberlite Conf vol 2. AGU, Washington, pp 300–308

Dawson JE (1980) Kimberlites and their xenoliths. Springer, Berlin Heidelberg New York, 252 pp

Desborough GA, Czamanske GK (1973) Sulfides in eclogite nodules from a kimberlite pipe, South Africa. Am Min 58:195–202

De Waal SA, Calk LC (1975) The sulfides in the garnet pyroxenite xenoliths from Salt Lake Crater, Oahu. J Petrol 16:134–153

Frick C (1973) The sulphides in griquaite and garnet-peridotite xenoliths in kimberlite. Contrib Mineral Petrol 39:1–16

Graterol M, Naldrett AJ (1971) Mineralogy of the Marbridge No. 3 and No. 4 nickel-iron sulfide deposits, with some comments on low temperature equilibration in the Fe–Ni–S system. Econ Geol 66:886–900

Grinenko VA, Ukhanov AV (1977) Sulfur levels and isotopic compositions in upper-mantle xenoliths from the Obnazhennaya kimberlite pipe. Geochem Int 14(6):169–171 (transl from Geokhimiya 1977:1872–1875)

Harris JW, Gurney JJ (1982) The abundance, mineralogy and chemistry of sulphide inclusions in diamonds. Terra Cognita 2:201

Harting P (1859) Ein Diamant mit eingeschlossenen Krystallen. N Jahrb Mineral Geol Paläontol 1859:192–193

Haughton DR, Roeder PL, Skinner BJ (1974) Solubility of sulfur in mafic magmas. Econ Geol 69:451–467

Ilupin IP (1962) On successive stages in kimberlite intrusion. Geol Geofiz 1972:151–157 (in Russian)

Ivanov VV, Ilupin IP, Starozhitskaya MI, Gross YuI (1977) Distribution of Cu, Ag, and Au in Yakutia plutonic xenoliths and kimberlites. Geoch Int 14(4):48–60 (transl from Geokhimiya 1977:1038–1051)

Kogarko LN (1972) The role of compounds of oxygen, sulphur and carbon in a magmatic gas phase of alkaline rocks. Proc 24th Int Geol Congr 10:20–24

Kramers JD (1979) Lead, uranium, strontium, potassium and rubidium in inclusion-bearing diamonds and mantle-derived xenoliths from southern Africa. Earth Planet Sci Lett 42:58–70

Kullerud G (1963) The Fe–Ni–S system. Carnegie Inst Washington Yearb 62:175–189

Kullerud G, Yoder HS Jr (1959) Pyrite stability relations in the Fe – S system. Econ Geol 54:533 – 572
Kullerud G, Yund RA, Moh G (1969) Phase relations in the Cu – Fe – S, Cu – Ni – S, and Fe – Ni – S systems. Econ Geol Monogr 4:323 – 343
Kyser TK (1986) Stable isotope variations in the mantle. Rev Min 16:141 – 164
Lambert G, Le Cloarec M-F, Pennisi M (1988) Volcanic output of SO_2 and trace metals: a new approach. Geochim Cosmochim Acta 52:39 – 42
Lorand JP (1985) The behaviour of the upper mantle sulfide component during the incipient alteration of "Alpine"-type peridotites as illustrated by the Beni Bousera (northern Morocco) and Ronda (southern Spain) ultramafic bodies. Tschermaks Mineral Petrol Mitt 34:183 – 209
Lorand JP, Conquéré F (1983) Contribution à l'étude des sulfures dans les enclaves de lherzolite à spinelle des basaltes alcalins (Massif Central et Languedoc, France). Bull Min 106:585 – 605
MacLean WH (1969) Liquidus phase relations in the $FeS-FeO-Fe_3O_4-SiO_2$ system, and their application in geology. Econ Geol 64:865 – 884
MacRae ND (1979) Silicate glass and sulfides in ultramafic xenoliths, Newer Basalts, Victoria, Australia. Contrib Mineral Petrol 68:275 – 280
Meyer HOA, Tsai H-M (1976) The nature and significance of mineral inclusions in natural diamond – a review. Min Sci Eng 8:242 – 261
Meyer HOA, Tsai H-M (1979) Inclusions in diamond and the mineral chemistry of the upper mantle. Phys Chem Earth 11:631 – 644
Mitchell RH (1975) Theoretical aspects of gaseous and isotopic equilibria in the system C – H – O – S with application to kimberlite. Phys. Chem Earth 9:903 – 915
Mitchell RH, Keays RR (1981) Abundance and distribution of gold, palladium and iridium in some spinel and garnet lherzolites: implications for the nature and origin of precious metal-rich intergranular components in the upper mantle. Geochim Cosmochim Acta 45:2425 – 2442
Moore JG, Fabbi BP (1971) An estimate of the juvenile sulfur content of basalt. Contrib Mineral Petrol 33:118 – 127
Murck BW, Burruss RC, Hollister LS (1978) Phase equilibria in fluid inclusions in ultramafic xenoliths. Am Min 63:40 – 46
Newson HE, Palme H (1984) The depletion of siderophile elements in the Earth's mantle: new evidence from molybdenum and tungsten. Earth Planet Sci Lett 69:354 – 364
Nielsen H (1978) Sulfur. Isotopes in nature. In: Wedepohl KH (ed) Handbook of geochemistry, vol II-2. Springer, Berlin Heidelberg New York, pp 16-B-1 – 16-B-40
Ringwood AE (1966) The chemical composition and origin of the earth. In: Hurley PM (ed) Advances in earth science. MIT, Cambridge, Mass, pp 287 – 356
Robinson DN (1979) Diamond and graphite in eclogite xenoliths from kimberlite. In: Boyd FR, Meyer HOA (eds) Proc 2nd Int Kimberlite Conf, vol 2. AGU, Washington, pp 50 – 58
Sakai H, Des Marais DJ, Ueda A, Moore JG (1984) Concentrations and isotope ratios of carbon, nitrogen and sulfur in ocean-floor basalts. Geochim Cosmochim Acta 48:2433 – 2441
Saxena SK (1989) Oxidation state of the mantle. Geochim Cosmochim Acta 53:89 – 95
Schmitt W, Palme H, Wänke H (1989) Experimental determination of metal/silicate partition coefficients for P, Co, Ni, Cu, Ga, Ge, Mo, and W and some implications for the early evolution of the Earth. Geochim Cosmochim Acta 53:173 – 185
Schneider A (1970) The sulfur isotope composition of basaltic rocks. Contrib Mineral Petrol 25:95 – 124
Schneider A (1978) Sulfur. Abundances in common igneous rocks. In: Wedepohl KH (ed) Handbook of geochemistry, vol II-2. Springer, Berlin Heidelberg New York, pp 16-E-1 – 16-E-19
Sharp WE (1966) Pyrrhotite: a common inclusion in South African diamonds. Nature (London) 211:402 – 403
Skinner BJ, Peck DL (1969) An immiscible sulfide melt from Hawaii. Econ Geol Monogr 4:310 – 322
Sobolev NV, Vakhrushev VA (1967) Sulfides in pyrope peridotites in kimberlites from Yakutia. Zap Vses Min Obshch 96(4):450 (in Russian)
Taylor BE (1986) Magmatic volatiles: isotopic variation of C, H, and S. Rev Min 16:185 – 225
Tsai H-M, Shieh Y-N, Meyer HOA (1979) Mineralogy and S^{34}/S^{32} ratios of sulfides associated with kimberlite, xenoliths and diamonds. In: Boyd FR, Meyer HOA (eds) Proc 2nd Int Kimberlite Conf, vol 2. AGU, Washington, pp 87 – 100
Vakrushev VA, Kutolin VA (1970) Sulfides in ultramafic inclusions from diatremes of the north Minusinsk depression. Dokl Akad Nauk SSSR 192:157 – 160 (transl from pp 1130 – 1133 in Russian Original of Dokl Akad Nauk SSSR 192)
Virgo D, Luth RW, Moats MA, Ulmer GC (1988) Constraints on the oxidation state of the mantle: an electrochemical and ^{57}Fe Mössbauer study of mantle-derived ilmenites. Geochim Cosmochim Acta 52:1781 – 1794
Von Gehlen K (1967) Sulphur isotopes from the sulphide-bearing carbonatite of Palabora, South Africa. Trans Inst Min Metall 76:B 223
Wagner PA (1914) The diamond fields of southern Africa. Transvaal Leader, Johannesburg, pp 290 – 308
Wedepohl KH (ed) (1978) Handbook of geochemistry, vol II-2: 16 sulfur. Springer, Berlin Heidelberg New York (loose-leaf)
Welke HJ, Allsopp HL, Harris JW (1974) Measurements of K, Rb, U, Sr and Pb in diamonds containing inclusions. Nature (London) 252:35 – 37
White RW (1966) Ultramafic inclusions in basaltic rocks from Hawaii. Contrib Mineral Petrol 12:245 – 314
Woermann E, Rosenhauer M (1985) Fluid phases and the redox state of the Earth's mantle. Extrapolations based on experimental, phase-theoretical and petrological data. Fortschr Mineral 63:263 – 349

Bacterial Sulfate Reduction*: Current Status and Possible Origin

P. A. Trudinger [1]

[1] Division of Continental Geology, Bureau of Mineral Resources, GPO Box 378, Canberra, ACT 2601, Australia
Present address: 144 Dryandra St, O'Connor, ACT 2601, Australia

Abstract

Recent studies on the dissimilatory sulfate-reducing bacteria should help in evaluating their role(s) in ancient geochemical events. As a group, the bacteria are versatile, varied and genetically diverse, and can thrive under a wide range of physico-chemical and nutritional conditions. Discoveries of hydrocarbon-utilizing sulfate reducers, and of organisms that grow above 80 °C, rebut claims that sulfide formation in, for example, oil fields and Mississippi Valley type environments, must of necessity be abiological. In modern anoxic environments, sulfate-reducing bacteria are responsible for a significant, and in some cases the major part of organic remineralization of organic matter.

Other anaerobic bacteria have been studied that require one or more less oxidized sulfur compound (e.g. thiosulfate, sulfite and elemental sulfur) as a terminal electron acceptor. These compounds are also reduced by some sulfate reducers. Based on this information a possible scenario for the evolution of bacterial sulfate reduction is presented which involves the sequential development of sulfur reduction, thiosulfate/sulfite reduction, anaerobic disproportionation of thiosulfate to sulfide and sulfate, and finally sulfate reduction. Comparative biochemical evidence suggests that sulfate reduction may have arisen in the earliest stages of biological evolution.

1 Introduction

Since early this century the notion has been entertained that reduction of sulfate to sulfide by bacteria may have played a part in the formation of economic base metal sulfide deposits (e.g. Siebenthal 1915; Bastin 1926). This idea gained prominence in the 1950s when it was recognized that many stratiform deposits, hitherto thought to have formed by hydrothermal replacement, might in fact be the result of low-temperature processes in sedimentary environments in which bacteria were known to be active (King 1967).

The hypothesis is particularly attractive in the case of those sedimentary deposits where no evidence exists for a hydrothermal or volcanic association and bacterial reduction of sulfate appears to be the only alternative source of sulfide. Despite claims, based on geochemical evidence, that sulfate can be reduced chemically at low temperatures (e.g. Orr 1974; Powell and Macqueen 1984), to date all attempts to demonstrate experimentally abiological reduction below 150 °C under geochemically realistic conditions have proved unsuccessful (Trudinger et al. 1985). On the other hand, rates of biological sulfate reduction in modern sediments often fall within the minimum and maximum values calculated for the deposition of stratiform sulfide deposits (Temple 1964; Trudinger et al. 1972; Rickard 1973).

Stable sulfur isotopic compositions of sulfur-containing minerals are frequently cited as evidence for or against biogenicity (see, for example, Donnelly and Crick, this Vol.; Lambert and Donnelly, this Vol.). Extreme sulfur isotope discrimination is often observed during sulfate reduction by sulfate-reducing bacteria (Kaplan and Rittenberg 1964; Chambers and Trudinger 1979) with fractionation factors well in excess of those found with geochemically relevant chemical processes. Hence, sulfides that are highly enriched in ^{32}S with respect to the presumed values in contemporaneous sulfate are assumed to be of biogenic origin. This assumption in general appears to be valid and accords with isotope fractionation observed in modern euxinic sediments (Chambers and Trudinger 1979; Chambers 1982).

Sulfides with low or moderate ^{32}S enrichment, however, present a problem. Isotope fractionation by

* In this work the term *sulfate (sulfur) reduction* is reserved for *dissimilatory* reduction (sometimes called sulfate/sulfur respiration), where inorganic sulfur is an electron acceptor in energy-yielding reactions, as distinct from *assimilatory* reduction which supplies reduced sulfur for the synthesis of essential sulfur-containing organic molecules.

M. Schidlowski et al. (Eds.)
Early Organic Evolution: Implications for Mineral and Energy Resources

sulfate reducers appears, at least in part, to be inversely related to the rate of reduction per bacterium (Kaplan and Rittenberg 1964; Chambers and Trudinger 1979). Rees (1973) has presented a mathematical model which attempts to explain this phenomenon. Thus, environmental factors affecting reduction rates, such as temperature, organic substrates and sulfate concentration, cause variable isotope fractionation which can result in biogenic sulfides that are similar isotopically to sulfides derived from volcanic and hydrothermal sources (Ohmoto 1972; and references cited in Trudinger 1981). With the Precambrian the problem is more acute since the rarity of sedimentary sulfate minerals prevents an accurate assessment of the isotopic composition of sulfate-sulfur in this period.

Understandably then, there are few instances where we can with any reasonable confidence assign a biogenic origin to sulfides in the ancient geological record. Equally, however, the possibility of biological involvement in the formation of many sedimentary sulfide deposits cannot be excluded. We are essentially concerned with probabilities which must be evaluated in the light of both geochemical and geological data on the deposits themselves and of observations on the biogeochemical processes occurring in present-day sedimentary environments.

In the last 5 to 10 years there have been major advances in our knowledge of the taxonomy, physiology and ecology of sulfate-reducing bacteria which should aid this evaluation. Advances, too, in the whole field of inorganic sulfur microbiology allow for some speculation on the way bacterial sulfate reduction may have evolved.

Table 1. The sulfate-reducing bacteria

Genus	Temperature range [a] (°C)	Organic substrates [b]
Eubacteria [c]		
Desulfotomaculum [d]	20–55	Hexoses, amino acid fatty acids to C18 phenylacetate phenylpropionate benzoate
Desulfovibrio	28–65	Fatty acids to C18 choline
Desulfomonas	37	Lactate, ethanol
Desulfobulbus	28–35	Lactate, ethanol fatty acids to C3
Desulfobacter	28–32	Acetate
Desulfococcus	29–35	Fatty acids to C16 nicotinic acid pimelate, glutarate benzoate
Desulfosarcina	33	Fatty acids to C14 phenylacetate phenylpropionate
Desulfobacterium	20–30	Fatty acids to C20 phenol, *p*-cresol benzoate
Desulfonema	30–32	Fatty acids to C14 benzoate
Thermodesulfobacterium	70	Lactate
Archaebacteria [e]		
"Archaeoglobus"	83	Glucose, peptone yeast extract amino acids cell homogenates

[a] Range of optimum growth temperatures: maximum temperatures are generally from 5 to 20 °C higher.
[b] Eubacterial species will utilize at least some simple organic acids or alcohols.
[c] Abstracted from Widdel (1988).
[d] Spore formers.
[e] Stetter et al. (1987).

2 The Sulfate-Reducing Bacteria

2.1 Speciation and Organic Metabolism

Until 1980 the common conception of the sulfate-reducing bacteria was as a restricted group of microorganisms (genera *Desulfovibrio* and *Desulfotomaculum*) that, under anaerobic conditions, carried out the partial oxidation of a few simple organic acids and alcohols. In that year Widdel (1980) published the results of his studies that characterized several new genera and species of eubacterial sulfate reducers and, in the intervening years, additional genera and species have been described including an extremely thermophilic archaebacterium, provisionally called "*Archaeoglobus fulgidus*", isolated from a submarine hydrothermal source (Achenbach-Richter et al. 1987; Stetter et al. 1987). Table 1 lists most of the known genera and some of their geochemically important characteristics.

As a group the sulfate-reducing bacteria are polymorphic ranging from small spherical forms ca. 2 µm in diameter (*Desulfococcus*) to filaments up to several millimeters in length, formed by 6–8 by 9–13 µm rod-shaped cells (*Desulfonema magnum*; Widdel 1980). Guanine/cytosine ratios of their DNAs ranging from about 34 to 66 are indicative of considerable genetic heterogeny. (For a discussion of genetic relatedness between species of sulfate reducers see Widdel 1988.) Several species are capable of autotrophic growth, synthesizing their organic cell material from carbon dioxide at the expense of hydrogen oxidation (Brysch et al. 1987).

The sulfate-reducing bacteria exhibit considerable versatility in their utilization of organic matter (Table 1). Eubacterial species oxidize higher fatty acids, alcohols, amino acids and several aromatic compounds, while complex organic matter such as peptone and cell homogenates can be metabolized by "*Archaeoglobus*". It should be noted, however, that the metabolic capabilities of individual species vary considerably. For example, *Desulfobacter postgateii* appears to be limited to acetate utilization, whereas *Desulfovibrio saprovorans* oxidizes higher fatty acids up to C18. Some sulfate-reducing bacteria (e.g. *D. saprovorans*) oxidize their organic substrates completely to carbon dioxide and water; others carry out partial oxidation with the production of low molecular weight acids such as acetate and propionate.

Sulfide formation in oil fields and Mississippi Valley type sulfide deposits has been attributed to the reduction of sulfate by hydrocarbons (e.g. Orr 1974; Powell and Macqueen 1984). Two barriers to invoking bacterial sulfate reduction in this process have been the apparently high formation temperatures (80 to ca. 150 °C) and the apparent inability of sulfate reducers to metabolize hydrocarbons (Postgate 1959: it may be noted that there is oblique geochemical evidence for methane-linked sulfate reduction in modern marine sediments; see, for example, Reeburgh 1980; Iversen and Jørgensen 1985; Henrichs and Reeburgh 1987). The first barrier has been at least partly removed with the discovery of extremely thermophilic sulfate reducers. That sulfate reducers in oil-field waters may utilize products of hydrocarbon degradation by aerobic organisms was suggested by Kutznetsova and Gorlenko (1965). Carothers and Kharaka (1980) proposed that aliphatic acids, principally acetate, present in oil-field waters were substrates for bacterial sulfate reduction but considered that the acids were formed by thermal degradation of hydrocarbons (Carothers and Kharaka 1978): thermal alteration of crude-oil/water mixtures with the production of acetate, propionate and oxalate was demonstrated experimentally (Kharaka 1989 pers. comm.). While these are both plausible mechanisms, whereby hydrocarbons may provide energy substrates for sulfate-reducing bacteria, a recent report (Widdel 1988) indicates that saturated aliphatic hydrocarbons per se can be metabolized. From enrichments with hexadecane and oil-field water a rod-shaped organism was separated from acetate- and fatty acid-oxidizing sulfate reducers and shown to reduce sulfate under strictly anaerobic conditions with pentadecane, hexadecane, heptadecane and octadecane as the sole organic substrates. Lower molecular weight hydrocarbons were less effective.

Thus, the sulfate-reducing bacteria now emerge as a varied, genetically diverse group of microorganisms, with an organic metabolic versatility approaching that of aerobic, heterotrophic bacteria. Their sole uniting characteristics are their absolute requirement for anoxic habitats and their ability to use sulfate as an electron acceptor in energy-yielding reactions.

2.2 Sulfate Reduction in the Modern Environment

Sulfate-reducing bacteria appear to be almost ubiquitous in terrestrial environments in which sulfate and organic matter are present and anoxic conditions prevail. The bacteria, or their activities, have been detected in a variety of soils and fresh, marine and brackish waters (Postgate 1984). They are active in Antarctic lakes with water temperatures close to 0 °C (Barghoorn and Nichols 1961; Franzmann et al. 1988) and, as we have seen above, in thermal environments with temperatures in the region of 100 °C. Their activities have been reported in alkaline lakes with pH values approaching 10 (Kharaka et al. 1984) and in acid mine wastes at about pH 2 (Tuttle et al. 1969), although in the latter case the organisms were probably confined to more neutral "microniches" since bacteria isolated from the wastes failed to grow below pH 5. Barophilic sulfate-reducing bacteria have been isolated from oil-bearing waters from a depth of ca. 3300 m (ZoBell 1957; Rozanova and Khudyakova 1974) and from deep-sea sediments at depths of 7000–10 000 m (ZoBell and Morita 1957). Sulfate reduction is adversely affected by hypersalinity (see Skyring 1987) yet sulfate-reducing bacteria are active in Solar Lake, Sinai, with salinities of up to 18% (Jørgensen and Cohen 1977) and their activities have been detected in salt pans which are essentially saturated (Saslawsky and Chait 1929). Skyring (1987) has given a detailed account of sulfate reduction in modern marine environments including estuaries, continental shelf, arid and semi-arid coastal zones, coastal lakes, coral reefs, enclosed seas and fjords.

Generally, sulfate reduction is most active in sediments, although some activity may occur in the water column. In meromictic basins, for example, a zone of sulfate reduction may occur just below the oxic/anoxic interface where anaerobic photosynthetic bacteria produce readily utilized organic matter (Sorokin 1964, 1970). Within sediments the main zone of sulfate reduction is observed in the upper few centimeters (Jørgensen 1982; Skyring 1987) where, presumably,

Table 2. Sulfate reduction rates in modern environments[a]

Ecosystem	Reduction rate[b] (mmol m^{-2} day^{-1})
Salt marsh	0.6–440 (mostly > 20)
Estuary	0–48 (mostly > 5)
Restricted fjord	2.5–130
Tidal/intertidal	0.2–104
Coastal lake	0.02–198
Shallow enclosed sea	12–712
Deep enclosed sea/fjord	0.001–9
Continental shelf	0.02–0.11
Continental slope	0.56

[a] Data abstracted from Skyring (1987).
[b] Rates vary considerably with location, season, etc.

organic matter is most abundant; thereafter sulfate reduction rapidly declines with depth. Nevertheless, in some deep-sea sediments sulfate concentration and $\delta^{34}S_{(sulfate)}$ profiles obey the Rayleigh expression for closed system reduction down to hundreds of meters (Goldhaber and Kaplan 1980), indicating that sulfate reducers continue to be active at these depths. Representative rates of sulfate reduction in modern marine and saline environments are shown in Table 2. Predictably, the highest rates are observed in shallow coastal environments where cyanobacterial and algal productivity, and/or detrital input, is high and decomposition of organic matter by aerobic organisms within the water column is minimal (Skyring 1987; Widdel 1988); average rates per unit area decrease dramatically from the continental shelf through the continental slope to deep-sea basins (Jørgensen 1983; Heggie et al. 1990). Trudinger et al. (1972) estimated that the Upper Proterozoic Copperbelt and Upper Permian Kupferscheifer Mansfeld ore deposits were laid down at an average rate of 0.1 to 1 mmol S/day; somewhat higher rates (up to 4 mmol S/day) were required for the Lower Proterozoic McArthur River and Mount Isa deposits in northern Australia. Comparison of these estimates with the sulfate reduction rates in Table 2 reinforces the notion that there are no inherent kinetic barriers to large-scale biogenic sulfide mineralization.

The importance of sulfate-reducing bacteria in modern marine sedimentary environments is highlighted by the observations that, as terminal oxidizing agents, they can be responsible for the remineralization of between 50 to 100% of sedimentary organic matter (Jørgensen 1977; Skyring et al. 1983). Despite the wide range of organic compounds utilized by sulfate reducers (see above), low molecular weight acids, principally acetate and to a smaller extent propionate and butyrate, appear to be the main substrates in saline sediments (e.g. Sørensen et al. 1981; Skyring 1988). Presumably this is because such acids are the main end products of anearobic fermentation of complex organic matter. Ironically, *Desulfovibrio desulfuricans*, the organism on which the concept of biogenic sulfide ores has been based, appears to be relatively unimportant in marine sediments (see Postgate 1984). Its preferred organic substrate is lactate and it cannot utilize acetate. Moreover, Taylor and Parkes (1985) reported that in marine sediments sulfate reducers were not involved in the initial stages of lactate oxidation. These same authors also showed that stimulation of sulfate reduction in the sediments by hydrogen, which is readily oxidized by *D. desulfuricans*, caused an increase in fatty acid "biomarkers" characteristic of *Desulfobulbus* species and not of those for *Desulfovibrio* spp. Stimulation by acetate caused increases in biomarkers for both *Desulfobacter* and *Desulfobulbus* spp. Earlier, Laanbroek and Pfennig (1981) had reported that *Desulfobulbus* ountnumbered *Desulfovibrio* by ca. 51:22 in a marine sediment and by ca. 24:16 in a fresh-water sediment. Biomarker analysis indicated that *Desulfobacter* species are the predominant sulfate reducers in 400-m-deep sediments of the continental margin of eastern Australia (Heggie et al. 1990).

3 On the Origin of Bacterial Sulfate Reduction

Sulfate reduction is an eight-electron process:

$$SO_4^{2-} + 8\,[H] \rightarrow HS^- + 3\,H_2O + OH^-.$$

(Note: In this and subsequent equations [H] refers to $H^+ + e^-$ which may be derived from hydrogen or organic matter.) In bacteria the reduction proceeds in a series of steps shown in the following pathway (e.g. Postgate 1984):

$$SO_4^{2-} \rightarrow SO_3^{2-} \rightarrow [S_2O_3^{2-},\ S_3O_6^{2-}] \rightarrow H_2S\,.$$

The only well-authenticated inorganic intermediate in this pathway is sulfite but we may assume that sulfur at lower oxidation states is generated during the six-electron reduction of sulfite to sulfide. Under some conditions thiosulfate and trithionate are formed from sulfite in cell-free preparations from sulfate-reducing bacteria. However, whether or not these compounds are true intermediates rather than side products formed from labile, perhaps enzyme-bound, forms of sulfur remains to resolved (Peck and LeGall 1982). Chambers and Trudinger (1975) used ^{35}S as a tracer in experiments designed to detect low levels of intermediates during sulfate reduction by intact cells of *Desulfovibrio* but failed to find any evidence of thiosulfate or trithionate formation. Except for one unconfirmed report of excretion of low concentrations of sulfite and thiosulfate in cultures of *D. desulfuricans* (Vainshtein et al. 1980), sulfide is commonly the only detectable product of sulfate reduction by growing sulfate-reducing bacteria.

The initial stage of sulfate reduction requires an "energy of activation" of about 22 kJ/mol. In sulfate-reducing bacteria this is supplied by adenosine-5'-triphosphate (ATP), a universal energy storage and transfer molecule in the biological world. An enzyme-catalyzed reaction between ATP and sulfate produces adenylyl sulfate (APS) which is the actual substrate for reduction to sulfite (Roy and Trudinger 1970). This energy requirement for sulfate "activation" is reflected in the fact that the growth yields of sulfate-reducing bacteria per unit electron donor (H_2 or acetate) with sulfite or thiosulfate as the electron acceptor are roughly twice those with sulfate (Badziong and Thauer 1978; Cypionka and Pfennig 1986; Widdel and Pfennig 1981).

Bacterial sulfate reduction is thus a complex process that we may confidently assume has evolved from simpler processes involving more reduced sulfur species. There are amongst extant microorganisms bacteria that carry out reactions that repressent possible vestiges of pre-sulfate reduction evolution. Moreover, it is to be expected that volcanic introduction of reduced sulfur species such as H_2S and SO_2 into the early oceans would produce a type of "Wackenroeder" mixture containing a complex mixture of inorganic compounds including elemental sulfur, thiosulfate and polythionates (Lyons and Nickless 1968), all potential pre-sulfate reduction substrates.

3.1 Sulfur-Reducing Bacteria

The simplest reaction producing sulfide from an inorganic source is the two-electron reduction of sulfur at an oxidation level of zero (elemental sulfur or polysulfide):

$$2[H] + S^0 \rightarrow H_2S \qquad \Delta G^{0\prime} = -27.9\ \text{kJ/mol S}.$$

This reaction (with elemental sulfur) supports the growth of several extant non-sulfate-reducing bacteria (Table 3). Most are thermophilic and extremely thermophilic archaebacteria isolated from marine or continental thermal environments which led Belkin et al. (1985) to suggest that sulfur reduction is favored by high temperatures. Nevertheless, the existence of mesophilic sulfur reducers (*Desulfuromonas* and *Campylobacter*) shows that high temperatures are not obligatory, and the prevalence of thermophilic species may merely reflect the fact that elemental sulfur is a common component of thermal environments. Included in the group are *Thermotoga* spp. which currently are the only known eubacterial extreme thermophiles. (It should be noted, however, that these organisms may not be "true" sulfur reducers in the sense of deriving energy from the reaction; Huber et al. 1986.)

Table 3. Sulfur-reducing bacteria [a, b]

Genus	Comments
Eubacteria	
Desulfuromonas	Isolated from anoxic marine and freshwaters.
Campylobacter [c]	Mesophilic (optimum temperature 30 °C). Electron donors limited to simple compounds such as acetate, pyruvate, lactate.
Thermotoga [d]	Extremely thermophilic (optimum temperature 80; max. 90 °C). Utilize sugars, starch and glycogen.
Archaebacteria	
Thermoproteus *Thermofilum* *Desulfurococcus* *Thermococcus* *Thermodiscus* *Pyrodictium* *Acidothermus* *Pyrococcus* *Pyrobaculum*	Diverse group of bacteria isolated from continental and submarine thermal waters. Optimum growth temperatures range from 85 to 105 °C. Electron donors include CO, H_2, low mol. wt. acids and alcohols, polysaccharides, proteins and peptides. *Thermoproteus neutrophilus* and *Pyrodictium occulatum* are obligately autotrophic with H_2 as electron donor. Most grow optimally at a pH of > 5 but *Acidothermus infernus* has an aptimum pH of 2. *Pyrobaculum* also utilizes sulfite and thiosulfate.

[a] Organisms capable of sulfur, but not sulfate reduction.
[b] Data sources: *Thermotoga*, Huber et al. (1986); Jannasch et al. (1988); *Pyrococcus*, Fiala and Stetter (1986); *Pyrobaculum*, Huber et al. (1987); Others, Pfennig and Biebl (1981); Widdel (1988).
[c] Originally called Spirillum 1575.
[d] Not known to derive energy from sulfur reduction.

3.1.1 Syntrophism

Desulfuromonas (Pfennig and Biebl 1976) and *Campylobacter* (Wolfe and Pfennig 1977) readily form syntrophic growth mixtures in which the sulfur reducers utilize the sulfur excreted by photosynthetic green sulfur bacteria during anaerobic, light-catalyzed oxidation of hydrogen sulfide:

$$H_2S \underset{\text{Sulfur reducers}}{\overset{\text{Green sulfur bacteria}}{\rightleftarrows}} S^0$$

Indeed, *Desulfuromonas* was discovered upon examination of a reportedly novel photosynthetic sulfur autotroph, "Chloropseudomonas ethylica" (Shaposhnikov et al. 1960), which proved to be a consortium of a green sulfur bacterium, Chlorobium sp., and *Desulfuromonas acetoxidans* (Pfennig and Biebl 1976). The last authors recognized the analogy between this syntrophic sulfur cycle and the present-day planetary oxygen cycle:

$$H_2O \underset{\text{Aerobic respiration}}{\overset{\text{Green plant oxygenic photosynthesis}}{\rightleftarrows}} O_2$$

The syntrophic cycle is also analogous to the modern "sulfuretum" (Baas Becking 1925):

$$H_2S \underset{\text{Sulfate reducers}}{\overset{\text{Chemical oxidation; Aerobic sulfide-oxidizing organisms; Anaerobic photosynthetic sulfur bacteria}}{\rightleftarrows}} SO_4^{2-}$$

which ensures a constant recycling of sulfur between its most oxidized and reduced forms. It is therefore tempting to speculate that syntrophism between anaerobic photosynthetic sulfide-oxidizing microorganisms and sulfur reducers represents a primitive sulfur cycle that arose before the development of atmospheric oxygen and oceanic sulfate (but see Sect. 4).

3.2 Sulfur Reduction by Sulfate-Reducing Bacteria

Whether biochemical similarities exist between the reduction pathways of the sulfur reducers and sulfate-reducing bacteria is unknown. Certainly there is no evidence that elemental sulfur per se (or polysulfides) is formed by sulfate reducers in either intact cells of cell extracts. Nevertheless, it would appear reasonable that the immediate precursor of sulfide in the sulfate reduction pathway is some form of sulfur at an oxidation state of zero. A few strains of *Desulfovibrio* are able to grow on elemental sulfur (Biebl and Pfennig 1977; Fauque et al. 1980), although the majority of sulfate-reducing bacteria which have been tested, do not. Whether this is due to the lack of the necessary enzymatic machinery for the reduction or to the lack of a mechanism for mobilizing insoluble sulfur is not known.

3.3 Reduction of Sulfite, Thiosulfate and Polythionates

While *Desulfuromonas* spp. and archaebacterial sulfur reducers appear to be limited to reduction of elemental sulfur, *Campylobacter* spp. (Wolfe and Pfennig 1977) and *Pyrobaculum* spp. (Huber et al. 1987) also reduce thiosulfate and sulfite:

$$S_2O_3^{2-} + 2[H] \rightarrow HSO_3^- + HS^- \qquad \Delta G^{0\prime} = -2.1 \text{ kJ/mol } S_2O_3^{2-}$$

$$HSO_3^- + 6[H] \rightarrow HS^- + 3H_2O \qquad \Delta G^{0\prime} = -171.7 \text{ kJ/mol S}.$$

These properties are shared by many sulfate-reducing bacteria and *Campylobacter* therefore may be representative of an intermediate phase in the evolution of bacterial sulfate reduction. Based on this premise it would appear that the critical step is the development of sulfite reduction since thiosulfate reduction per se yields insufficient energy for the generation of ATP. Indeed, Skyring and Donnelly (1982), to account for apparent discrepancies in the Early Proterozoic sulfur isotope record, proposed a scenario in which sulfite reduction was prevalent before sulfate became a significant component of the seas and oceans, an event, they suggested, that coincided with the rise in atmospheric oxygen (see, however, Sect. 5).

Tetrathionate is readily utilized as an electron acceptor by *D. desulfuricans* (Postgate 1951). Nothing is known of its metabolism, although it is likely that the initial step is a reduction to two molecules of thiosulfate:

$$S_4O_6^{2-} + 2[H] \rightarrow 2S_2O_3^{2-} + 2H^+ \qquad \Delta G^{0\prime} = -84.5 \text{ kJ/mol } S_4O_6^{2-},$$

a reaction which is carried out by many facultative anaerobes (see Thauer et al. 1977). The evolutionary significance, if any, of tetrathionate reduction cannot be assessed. Trithionate does not appears to support growth of *Desulfovibrio* (Postgate 1951; Ishimoto et al. 1954).

3.3.1 "Inorganic Fermentation"

The term "inorganic fermentation" has been applied to a recently discovered unique form of inorganic sulfur metabolism in which disproportionation of thiosulfate and sulfite produces both sulfate and sulfide:

$$S_2O_3^{2-} + H_2O \rightarrow SO_4^{2-} + HS^- + H^+ \quad \Delta G^{0\prime} = -21.9\ kJ/mol\ S_2O_3^{2-};$$

$$4SO_3^{2-} + H^+ \rightarrow 3SO_4^{2-} + HS^- \quad \Delta G^{0\prime} = -58.9\ kJ/mol\ SO_3^{2-}.$$

These reactions are carried out, in preference to sulfate reduction, by the sulfate reducer *Desulfovibrio sulfodismutans* and provide the energy for assimilation of organic compounds such as acetate (Bak and Cypionka 1987; Bak and Pfennig 1987). So far, only two other typed organisms, *Desulfobacter curvatus* and *Desulfovibrio vulgaris* (strain Marburg), have been shown to be capable of inorganic fermentation, the latter utilizing only sulfite. However, large numbers of disproportionating bacteria (up to 2×10^6 per ml) have been found in freshwater mud, suggesting that they play a significant role in sulfur cycling (Bak and Pfennig 1987).

4 Possible Evolutionary Status of Bacterial Sulfate Reduction

The archaebacteria are thought to have preceded the eubacteria in the prokaryotic phylogenetic evolutionary tree (Woese 1987) and it as been suggested that sulfur-based metabolism, including sulfate reduction, may have been the earliest biochemistry of the archaebacteria (Achenbach-Richter 1987; Woese 1987). This notion is based, in part, on the observation that "*Archaeoglobus*" contains several cofactors that are characteristic of another major class of archaebacteria, the methanogens (Stetter et al. 1987) and may thus represent a "transition from an ancient thermophilic sulfur-metabolizing phenotype to a methanogenic one" (Woese 1987 p. 258). *Archaeoglobus* lacks coenzyme M, a cofactor in the terminal step of the methanogenic pathway but does in fact produce small amounts of methane. Moreover, several archaebacterial methanogens reduce elemental sulfur (Stetter and Gaag 1983) although whether they derive energy from the reaction is unknown. Enzymatic analysis has shown a strong similarity in the pathways of sulfate reduction in *Archaeoglobus* and sulfate-reducing eubacteria (Speich and Trüper 1988). In 1989 Trüper et al. reported briefly that results of sulfur isotope fractionation experiments with the extreme thermophile, *Archaeoglobus fulgidus* "do not allow geochemical differentiation between archae- and eubacterial dissimilatory sulfate reduction."

APS reductase is the enzyme responsible for converting APS to sulfite in sulfate-reducing bacteria. This enzyme is also present in several anaerobic photosynthetic sulfur bacteria (Trüper and Fischer 1982) and in aerobic autotrophic sulfur bacteria (Roy and Trudinger 1970), where it acts in reverse, oxidizing sulfite to APS:

$$SO_3^{2-} + AMP \rightarrow APS + 2e^-.$$

APS is then converted to sulfate by adenosine-5′-diphosphate (ADP) sulfurylase:

$$APS + P_i \rightarrow ADP + SO_4^{2-}.$$

This enzyme has been detected in *Desulfovibrio* (Peck 1962) and it is an obvious possibility that, in *D. sulfodismutans*, sulfite disproportionation involves a coupling of sulfite reduction with the reverse APS reductase reaction and ADP sulfurylase (Bak and Cypionka 1987). One may speculate then that sulfite disproportionation was the penultimate step in the development of the sulfate reduction pathway, the final stage being the acquisition of a sulfate "activation" mechanism (see Skyring and Donnelly 1982).

Prior to the discovery of archaebacterial species, it had been concluded that the sulfate-reducing bacteria evolved after the anaerobic photosynthetic sulfur bacteria (e.g. Broda 1975) and in effect "inherited" some of the enzymatic machinery for inorganic sulfur metabolism (Trüper 1982). This view was reasonable since anaerobic phototrophic sulfide oxidation and later, with the development of an oxic atmosphere, chemical and microbial oxidation, appeared to be the only means by which sulfate could be formed in significant amounts. As pointed out by Bak and Pfennig (1987), however, the disproportionation reactions described above could have allowed the development of sulfate reduction independent of that of phototrophic sulfide oxidation. Indeed, archaebacterial sulfate reducers might well have preceded the phototrophic bacteria in the evolutionary sequence and have provided the enzymatic framework upon which phototrophic sulfide oxidation was built. Clearly, it will be of great interest to see whether future research uncovers archaebacteria capable of utilizing "inorganic fermentation".

5 Final Remarks

Figure 1, based on the foregoing discussion, outlines a possible scenario for the evolution of bacterial sulfate reduction: thiosulfate and polythionate reductions are assumed to be supplementary processes that developed during or after the evolution of sulfate reduction. Obviously, much more information will be required before the feasibility of this evolutionary sequence can

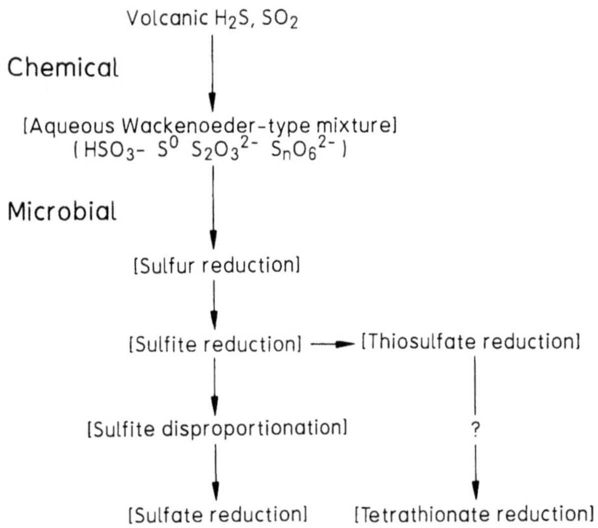

Fig. 1. Possible scenario for the evolution of bacterial sulfate reduction

be fully evaluated. Comparisons of sequences of conservative macromolecules, notably ribosomal oligonucleotides, are providing some clues to the evolutionary relationships between sulfate reducers and other classes of bacteria (Fox et al. 1980; Trüper 1982; Widdel 1988). However, it must be remembered that a metabolic sequence is an expression of but a small fraction of the bacterial genome, and extant organisms are the end products of billions of years of evolutionary change. Therefore, the idea that there should be necessarily any overall genetic similarity between bacteria with similar metabolic capabilities, or between the extant organisms and their primordial counterparts, is clearly untenable. In the words of Postgate (1984, p. 105) "it is probably wisest in considering microbial evolution to disregard existing species and to consider processes, enzymes or even genes." The sulfur-metabolizing anaerobic bacteria provide much scope for this type of investigation.

Disproportionation of sulfite provides an alternative to phototrophic sulfide oxidation as a source of terrestrial sulfate before the rise of atmospheric oxygen and, as we have seen, there are grounds for suspecting that a sulfate reduction pathway developed in the earliest stages of microbial evolution.

Should the ideas on the phylogenic position of the sulfur archaebacteria be correct, it implies that bacterial sulfate reduction predated the emergence of the eubacteria which carbon isotopic and micropaleontological evidence puts at about 3.5 Ga B.P. (see works in Schopf 1983). Why then, we may ask, does the sulfur isotopic record suggest that sulfate reduction developed, or at least became geochemically significant, no earlier than about 2.8 Ga (Schidlowski 1987), and perhaps as late as 2 Ga B.P. (Cameron 1982; Skyring and Donnelly 1982)? One possibility, of course, is than any biogenic sulfide in the Archean/Early Proterozoic was masked by an abundance of "primordial" sulfide from volcanic sources (Skyring and Donnelly 1982). It has also been suggested (Perry et al. 1971) that oceanic sulfate levels in the Archean were too low to permit significant isotopic fractionation by sulfate-reducing bacteria (see Chambers and Trudinger 1978). However, we should also inquire whether our general concepts of the geochemical significance of anaerobic sulfur bacteria need modifying in the light of the proliferation of new genera and species with widely differing properties. For example, do the newly discovered sulfate reducers, in particular perhaps the archaebacteria, exhibit isotopic effects during sulfate reduction similar to those shown by the, possibly geochemically insignificant, *Desulfovibrio*? So far, no species-dependent isotope fractionation has been observed (e.g. Chambers et al. 1975) but the range examined is extremely limited – a few species of *Desulfovibrio* and *Desulfotobaculum*. The Proterozoic sulfur isotopic record discussed by Lambert and Donnelly (in this Vol.) illustrates the fact that our current concepts of isotopic fractionation do not always lend themselves readily to the interpretation of geochemical isotopic data.

Thus, new avenues for exploring the evolution and geochemical impact of anaerobic sulfur biology are emerging at a time, ironically, when IGCP Project 157, whose charter includes these very matters, has come to the end of its term.

Acknowledgements. I thank G. W. Skyring, H. G. Trüper and M. R. Walter for their valuable comments and the Director, Bureau of Mineral Resources, for allowing me access to BMR in my retirement.

References

Achenbach-Richter L, Stetter KO, Woese CR (1987) A possible biochemical missing link among archaebacteria. Nature (London) 327:348–349

Baas Becking LGM (1925) Studies on the sulfur bacteria. Ann Bot (London) 39:613–650

Badziong W, Thauer RK (1978) Growth yields and growth rates of *Desulfovibrio vulgaris* (Marburg) growing on hydrogen plus sulfate and hydrogen plus thiosulfate as sole energy source. Arch Microbiol 117:209–214

Bak F, Cypionka H (1987) A novel type of energy metabolism involving fermentation of inorganic sulphur compounds. Nature (London) 326:891–892

Bak F, Pfennig N (1987) Chemolithotrophic growth of *Desulfovibrio sulfodismutans* sp. nov. by disproportionation of inorganic sulfur compounds. Arch Microbiol 147:184–189

Barghoorn ES, Nichols RI (1961) Sulphate-reducing bacteria and pyritic sediments in Antarctica. Science 134:190

Bastin ES (1926) A hypothesis of bacterial influence in the genesis of certain sulphide ores. J Geol 34:773–792

Belkin S, Wirsen CO, Jannasch HW (1985) Biological and abiological sulfur reduction at high temperatures. Appl Environ Microbiol 49:1057–1061

Biebl H, Pfennig N (1977) Growth of sulfate-reducing bacteria with sulfur as electron acceptor. Arch Microbiol 112:115–117

Broda E (1975) The evolution of bioenergetic processes. Revised reprint. Pergamon, Oxford New York, 231 pp

Brysch K, Schneider C, Fuchs G, Widdel F (1987) Lithoautrophic growth of sulfate-reducing bacteria, and description of *Desulfobacterium autotrophicum* gen. nov., sp. nov. Arch Microbiol 148:264–274

Cameron EM (1982) Sulphate and sulphate reduction in Early Precambrian oceans. Nature (London) 296:145–148

Carothers WW, Kharaka YK (1978) Aliphatic acid anions in oil-field waters and their implications for the origin of natural gas. Bull Am Assoc Petrol Geol 62:2441–2453

Carothers WW, Kharaka YK (1980) Stable carbon isotopes of HCO_3^- in oil-field waters – implications for the origin of CO_2. Geochim Cosmochim Acta 44:323–332

Chambers LA (1982) sulfur isotope study of a modern intertidal environment and the interpretation of ancient sulfides. Geochim Cosmochim Acta 46:721–728

Chambers LA, Trudinger PA (1975) Are thiosulfate and trithionate intermediates in dissimilatory sulfate reduction? J Bacteriol 123:36–40

Chambers LA, Trudinger PA (1979) Microbiological fractionation of stable sulfur isotopes: a review and critique. Geomicrobiol J 1:249–293

Chambers LA, Trudinger PA, Smith JW, Burns MS (1975) Sulfur isotope fractionation during sulfate reduction by dissimilatory sulfate-reducing bacteria. Can J Microbiol 21:1602–1607

Cypionka H, Pfennig N (1986) Growth yields of *Desulfotomaculum orientis* with hydrogen in chemostat culture. Arch Microbiol 143:396–399

Fauque GD, Barton LL, LeGall J (1980) Oxidative phosphorylation linked to dissimilatory reduction of elemental sulfur by *Desulfovibrio*. In: Sulfur in biology. Ciba Found Symp 72. Excerpta Medica, Amsterdam, pp 71–86

Fiala G, Stetter KO (1986) *Pyrococcus furiosus* sp. nov. represents a novel genus of marine heterotrophic archaebacteria growing optimally at 100 °C. Arch Microbiol 145:56–61

Fox GE, Stackebrandt E, Hespell RB, Gibson J, Maniloff J, Dyer TA, Wolfe RS, Balch WE, Tanner RS, Magrum LJ, Zablen LB, Blakemore LB, Gupte R, Bonen L, Lewis BJ, Stahl DA, Luehrsen KR, Chen KN, Woese CR (1980) The phylogeny of the prokaryotes. Science 209:457–463

Franzmann PD, Skyring GW, Burton HR, Deprez PP (1988) Sulfate reduction rates and some aspects of the limnology of four lakes and a fjord of the Vestfold Hills, Antarctica. In: Ferris JM, Burton HR, Johnstone GW, Bayly IA (eds) Biology of the Vestfold Hills, Antarctica, Kluwer, Dordrecht, pp 25–33

Goldhaber MB, Kaplan IR (1980) Mechanisms of sulfur incorporation and isotope fractionation during early diagenesis in sediments of the Gulf of California. Mar Chem 9:95–143

Heggie DT, Skyring GW, O'Brien GW, Reimers C, Herczeg A, Moriarty DJW, Burnett WC, Milnes AR (1990) Organic carbon cycling and modern phosphorite formation on the East Australian continental margin: an overview. In: Notholt AJG, Jarvis I (eds) Phosphorite research and development, Geol Soc Spec Publ 52, pp 87–117

Henrichs SM, Reeburgh WS (1987) Anaerobic mineralization of marine sediment organic matter: rates and the role of anaerobic processes in the oceanic carbon economy. Geomicrobiol J 5:191–237

Huber R, Langworthy TA, König H, Thomm M, Woese CR, Sletyr VB, Stetter KO (1986) *Thermotoga maritima* sp. nov. represents a new genus of unique extremely thermophilic eubacteria growing up to 90 °C. Arch Microbiol 144:324–333

Huber R, Kristjansson JK, Stetter KO (1987) *Pyrobaculum* gen. nov. sp. nov. a new genus of neutrophilic, rod-shaped archaebacteria from continental solfataras growing optimally at 100 °C. Arch Microbiol 149:95–101

Ishimoto M, Koyama J, Omura T, Nagai Y (1954) Biochemical studies on sulfate-reducing bacteria. III. Sulfate reduction by cell suspensions. J Biochem 41:537–546

Iversen M, Jørgensen BB (1985) Anaerobic methane oxidation rates at the sulfate-methane transition in marine sediments from Kattegat and Skagerrak (Denmark). Limnol Oceanogr 30:944–955

Jannasch HW, Huber R, Belin S, Stetter KO (1988a) *Thermotoga neapolitana* sp. nov. of the extremely thermophilic genus *Thermogota*. Arch Microbiol 150:103–104

Jannasch HW, Wirsen CO, Molyneaux SJ, Langworthy TA (1988b) Extremely thermophilic fermentative archaebacteria of the genus *Desulfurococcus* from deep-sea hydrothermal vents. Appl Environ Microbiol 54:1203–1209

Jørgensen BB (1977) The sulfur cycle in a coastal marine sediment (Limfjorden, Denmark). Limnol Oceanogr 22:814–832

Jørgensen BB (1982) Ecology of the bacteria of the sulphur cycle with special reference to anoxic-oxic interface environments. In: Postgate JR, Kelly DP (eds) Sulphur bacteria. R Soc, London, pp 113–131

Jørgensen BB, Cohen Y (1977) Solar Lake (Sinai). 5. The sulfur cycle in benthic cyanobacterial mats. Limnol Oceanogr 22:657–666

Kaplan IR, Rittenberg SC (1964) Microbiological fractionation of sulfur isotopes. J Gen Microbiol 34:195–212

Kharaka YK, Robinson SW, Law LM, Carothers WW (1984) Hydrogeochemistry of Big Soda Lake, Nevada: an alkaline meromictic desert lake. Geochim Cosmochim Acta 48:823–835

King HF (1967) The origins and aims of the Baas Becking Laboratory. Miner Depos 2:142–146

Klemps R, Cypionka H, Widdel F, Pfennig N (1985) Growth with hydrogen and further physiological characteristics of *Desulfotomaculum* species. Arch Microbiol 143:203–208

Kutznetsova VA, Gorlenko VM (1965) Effect of temperature on the development of microorganisms from flooded strata of the Romashkino oil field. Microbiology 34:274–278 (Engl Transl of Microbiologiya 34:329–334)

Laanbroek H, Pfennig N (1981) Oxidation of short-chain fatty acids by sulfate-reducing bacteria in fresh-water and marine sediments. Arch Microbiol 128:330–335

Lyons D, Nickless G (1986) The lower oxy-acids of sulphur. In: Nickless G (ed) Inorganic sulphur chemistry. Elsevier, Amsterdam, pp 509–533

Ohmoto H (1972) Systematics of sulfur and carbon isotopes in hydrothermal ore deposits. Econ Geol 67:551–578

Ollivier B, Cord-Ruwisch R, Hatchikian EC, Garcia JL (1988) Characterization of *Desulfovibrio fructovorans* sp. nov. Arch Microbiol 149:447–450

Orr WL (1974) Changes in sulfur content and isotopic ratios of sulfur during petroleum maturation – study of Big Horn Basin Paleozoic oils. Bull Am Assoc Petrol Geol 58:2295–2318

Peck HD (1962) The role of adenosine-5′-phosphosulfate in the reduction of sulfate to sulfite by *Desulfovibrio desulfuricans*. J. Biol Chem 237:198–203

Peck HD, LeGall J (1982) Biochemistry of dissimilatory sulphate reduction. In: Postgate JR, Kelly DP (eds) Sulphur bacteria. R Soc, London, pp 13–36

Perry EC, Monster J, Reimer T (1971) Sulphur isotopes in Swaziland System barites and the evolution of the Earth's atmosphere. Science 171:1015–1016

Pfennig N, Biebl H (1976) *Desulfuromonas acetoxidans* gen. nov. and sp. nov. a new anaerobic, sulfur-reducing, acetate-oxidising bacterium. Arch Microbiol 110:3–12

Pfennig N, Biebl H (1981) The dissimilatory sulfur-reducing bacteria. In: Starr MP, Stolp H, Trüper HG, Balows A, Schlegel HG (eds) The prokaryotes. Springer, Berlin Heidelberg New York, pp 941–947

Pfennig N, Widdel F, Trüper HG (1981) The dissimilatory sulfate-reducing bacteria. In: Starr MP, Stolp H, Trüper HG, Balows A, Schlegel HG (eds) The prokaryotes. Springer, Berlin Heidelberg New York, pp 926–940

Postgate JR (1951) Reduction of sulfur compounds by *Desulphovibrio desulphuricans*. J Gen Microbiol 5:725–738

Postgate JR (1959) Sulphate reduction by bacteria. Annu Rev Microbiol 13:505–520

Postgate JR (1984) The sulphate-reducing bacteria. Univ Press, Cambridge, 208 pp

Powell TG, Macqueen RW (1984) Precipitation of sulfide ores and organic matter – sulfate reactions at Pine Point, Canada. Science 224:63–66

Reeburgh WS (1980) Anaerobic methane oxidation: rate depth distributions in Skan Bay sediments. Earth Planet Sci Lett 47:345–352

Rees CE (1973) A steady-state model for sulfur isotope fractionation in bacterial reduction reactions. Geochim Cosmochim Acta 37:1141–1162

Rickard DT (1973) Limiting conditions for synsedimentary sulfide ore formation. Econ Geol 68:605–617

Roy AB, Trudinger PA (1970) The biochemistry of inorganic compounds of sulphur. Univ Press, Cambridge, 400 pp

Rozanova EP, Khudyakova AI (1974) A new nonspore-forming thermophilic sulfate-reducing bacterium, *Desulfovibrio thermophilus* nov. spec. Microbiology 43:1908–1912 (Engl Transl of Mikrobiologiya 43:1069–1075)

Saslawsky AS, Chait SS (1929) The influence of the concentration of sodium chloride on several biochemical processes in the limnan. Zentralbl Bakteriol Parasitkd (Abt 2) 77:18–21

Siebenthal CE (1915) Origin of the lead and zinc deposits of the Joplin region. US Geol Surv Bull 606, 283 pp

Schidlowski M (1979) Antiquity and evolutionary status of bacterial sulfate reduction. Orig Life 9:299–311

Schidlowski M (1987) Evolution of the early sulphur cycle. In: Rodriguez-Clemente R, Tardy Y (eds) Geochemistry and Mineral Formation in the Earth Surface. Consejo Superior de Investigacions Cientificas, Madrid, 29–49

Shaposhnikov VV, Kondrat'eva EN, Federov VD (1960) A new species of green sulphur bacteria. Nature (London) 187:167–168

Skyring GW (1987) Sulfate reduction in coastal ecosystems. Geomicrobiol J 5:295–374

Skyring GM (1988) Acetate as the main energy substrate for the sulfate-reducing bacteria in Lake Eliza (South Australia) hypersaline sediments. FEMS Microbiol Ecol 53:87–94

Skyring GW, Donnelly TH (1982) Precambrian sulfur isotopes and a possible role for sulfite in the evolution of biological sulfate reduction. Precambrian Res 17:41–61

Skyring GW, Chambers LA, Bauld J (1983) Sulfate reduction in sediments colonized by cyanobacteria, Spencer Gulf, South Australia. Aust J Mar Freshwater Res 34:359–374

Sørensen J, Jørgensen BB, Revsbech NP (1979) A comparison of oxygen, nitrate and sulfate respiration in coastal marine sediments. Microbial Ecol 5:105–115

Sørensen J, Christensen D, Jørgensen BB (1981) Volatile fatty acids and hydrogen as substrates for sulfate-reducing bacteria in anaerobic marine sediment. Appl Environ Microbiol 42:5–11

Sorokin YuI (1964) On the primary production and bacterial activities in the Black Sea. J Cons Int Explor Mer 29:41–60

Sorokin YuI (1970) Interrelations between sulphur and carbon turnover in meromictic lakes. Arch Hydrobiol 66:391–466

Speich N, Trüper HG (1988) Adenylylsulphate reductase in a dissimilatory sulphate-reducing archaebacterium. J Gen Microbiol 134:1419–1425

Stetter KO, Gaag G (1983) Reduction of molecular sulphur by methanogenic bacteria. Nature (London) 305:309–311

Stetter KO, Lauerer G, Thomm M, Neuner A (1987) Isolation of extremely thermophilic sulfate reducers: evidence for a novel branch of archaebacteria. Science 236:822–824

Taylor J, Parkes RJ (1985) Identifying different populations of sulphate-reducing bacteria within marine sediment systems, using fatty acid biomarkers. J Gen Microbiol 131:631–642

Temple KL (1964) Syngenesis of sulphide ores: an evaluation of biochemical aspects. Econ Geol 59:1473–1491

Thauer RK, Jungermann K, Decker K (1977) Energy conservation in chemotrophic anaerobic bacteria. Bacteriol Rev 41:100–180

Trudinger PA (1981) Origins of sulphide in sediments. BMR J Aust Geol Geophys 6:279–285

Trudinger PA, Lambert IB, Skyring GW (1972) Biogenic sulphide ores: a feasibility study. Econ Geol 67:1114–1127

Trudinger PA, Chambers LA, Smith JW (1985) Low-temperature sulphate reduction: biological versus abiological. Can J Earth Sci 22:1910–1918

Trüper HG (1982) Microbiological processes in the sulfur cycle through time. In: Holland HD and Schidlowski M (eds) Mineral deposits and the evolution of the biosphere. Springer, Berlin Heidelberg New York, pp 5–30

Trüper HG, Fischer U (1982) Anaerobic oxidation of sulphur compounds as electron donors for bacterial photosynthesis. In: Postgate JR, Kelly DP (eds) Sulphur bacteria. R Soc, London, pp 99–112

Trüper HG, Speich N, Leyendecker W, Dahl C, Becker P (1989) Dissimilatory and assimilatory sulfate reduction in archaebacteria. In: 9th Int Symp on environmental biogeochemistry, Moscow, USSR, Abstr, p 157

Tuttle JH, Dugan PR, Macmillan CB, Randles CI (1969) Microbial dissimilatory sulfur cycle in acid mine water. J Bacteriol 97:594–602

Vainshtein MB, Matrosov AG, Baskunov VP, Zyakun AM, Ivanov MV (1980) Thiosulfate as an intermediate product of bacterial sulfate reduction. Microbiology 49:672–675 (Engl Transl of Mikrobiologiya 49:855–858)

Widdel F (1980) Anaerober Abbau von Fettsäuren und Benzoesäure durch neu isolierte Arten Sulfatreduzierender Bakterien. Thesis, Georg-August-Univ, Göttingen, 443 pp

Widdel F (1988) Microbiology and ecology of sulfate- and sulfur-reducing bacteria. In: Zehnder JB (ed) Biology of anaerobic organisms. John Wiley & Sons, New York, pp 469–585

Widdel F, Pfennig N (1981) Studies on dissimilatory sulfate-reducing bacteria that decompose fatty acids. 1. Isolation of new sulfate-reducing bacteria enriched with acetate from saline environments. Description of *Desulfobacter postgateii* gen. nov., sp. nov. Arch Microbiol 129:395–400

Woese CR (1987) Bacterial evolution. Microbiol Rev 51:221–271

Wolfe RS, Pfennig N (1977) Reduction of sulfur by spirillum 5175 and syntrophism with *Chlorobium*. Appl Environ Microbiol 33:427–433

ZoBell CE (1957) Ecology of sulfate-reducing bacteria. In: Sulfate-reducing bacteria, their relation to the secondary recovery of oil. Sci Symp, St Bonaventure Coll, NY, pp 1–24

ZoBell CE, Morita RY (1957) Barophilic bacteria in deep sea sediments. J Bacteriol 73:563–568

Biogeochemistry of Sulfur and the Mechanisms of Sulfide-Sulfate Mineralization in Archean Oceans

Hiroshi Ohmoto [1]

[1] Department of Geosciences, The Pennsylvania State University, University Park, PA 16802, USA and Institute of Petrology, Mineralogy & Economic Geology, Tohoku University, Sendai, 980, Japan

Abstract

The biogeochemistry of sulfur and the mechanisms of sulfide/sulfate mineralization in Archean oceans have been thought by most geoscientists to have been much different from those in younger oceans. Archean oceans contained very little or no SO_4^{2-}, and sulfate-reducing bacteria (SRB) were not active in the oceans. Fe-sulfides in the marine sediments formed directly from magmatic H_2S, rather than by bacterial reduction of seawater sulfate. Bedded sulfate minerals in the marine sediments were formed by local oxidation of magmatic H_2S by sulfide-oxidizing bacteria, rather than by evaporation of sulfate-rich seawater, and the massive Cu−Fe−Zn sulfide deposits associated with submarine volcanic rocks (volcanogenic massive sulfide deposits, VMSDs) were formed by magmatic H_2S, rather than H_2S-bearing fluids that developed from sulfate-bearing seawater through reactions with high temperature country rocks in the hydrothermal plumbing systems.

Critical examinations of the published and unpublished data on the sulfur isotopic characteristics (e.g. the skewness, range and mean values of the $\delta^{34}S$ frequency curves, and the relationships between the rate of sulfate reduction and the kinetic isotopic effects) of marine sediments and VMSDs of various geologic ages and those of various laboratory systems, however, suggest that the biogeochemistry of sulfur and the mechanisms of sulfide/sulfate mineralization in the oceans have basically been the same since at least ~ 3.5 b.y. ago, except for some important differences in the environmental parameters for SRB. Archean oceans, as far back as ~ 3.5 b.y. ago, were probably already sulfate-rich, with a sulfate concentration more than ~ ⅓ of that in modern oceans (i.e. > 10 mM sulfate), and with a $\delta^{34}S$ value of +2‰ (vs. +20‰ for the modern oceans). This model implies that the accumulation of free oxygen in the atmosphere predates 3.5 b.y. ago. SRB were probably more active in Archean oceans than in modern oceans, possibly because the ocean temperatures were about 30–40 °C higher, and more digestible foods were available for SRB. The higher ocean temperature and the higher availability of digestible foods for SRB may be related ultimately to the higher P_{CO_2} ($\sim 10^{-2}$ atm vs. $10^{-3.5}$ atm of the present-day atmosphere) and the lower P_{O_2} ($\sim 10^{-2.5}$ atm vs. the present value of $10^{-0.7}$ atm).

1 Introduction

A theory that the earliest oceans contained virtually no sulfate (< 1/100 of the present level) has been accepted by most geoscientists, because, in the absence of organisms (e.g. sulfide-oxidizing bacteria) and free oxygen, the important mechanism for production of SO_4^{2-} in the oceans was probably photochemical oxidation of volcanic SO_2 (e.g. Walker et al. 1983) or hydration of the SO_2 in submarine environments (see Sect. 4). A rise in the sulfate content of seawater, which was a prerequisite for a rise in the activity of sulfate-reducing bacteria, must have occurred when the atmosphere had already accumulated appreciable amounts of O_2 by photosynthesis, allowing the oxidation of sulfide minerals by O_2 (e.g. $FeS + 2O_2 \Rightarrow Fe^{2+} + SO_4^{2-}$) to become the dominant mechanism for sulfate production. Therefore, questions such as "When did the oceans become sulfate rich and sulfate-reducing bacteria become active?" have usually been related to the more important question of "When did the Earth's atmosphere begin to accumulate free oxygen?".

The sulfate content and the activity of sulfate-reducing bacteria in ancient oceans have also been the concerns of economic geologists, because three major types of sulfide-sulfate mineralization in modern oceans all utilize seawater SO_4^{2-}, but in different ways: (1) formation of gypsum/anhydrite beds by evaporation of seawater; (2) formation of disseminated sulfides (mostly pyrite) by bacterial reduction of SO_4^{2-} in unconsolidated sediments or in euxinic bodies of

M. Schidlowski et al. (Eds.)
Early Organic Evolution: Implications for Mineral and Energy Resources

seawater; and (3) formation of massive sulfide/sulfate ores, commonly referred to as sulfide/sulfate chimneys or volcanogenic massive sulfide deposits (VMSDs), on and/or near the seafloor by the reaction of sulfate-bearing local seawater with high temperature ($T = \sim 100-400\,°C$) fluids. The high temperature fluids appear, from isotopic evidence (see Ohmoto 1986, for a review), to have originated also from the overlying sulfate-bearing seawater, and acquired H_2S during circulation through hot country rocks by two mechanisms: reduction of seawater SO_4^{2-} by ferrous-bearing minerals, and leaching of sulfide minerals in the country rocks. An important question has been whether or not the processes for these three types of mineralization in oceans have always been the same (i.e. nonmagmatic processes) during the entire geologic history.

The dominant view among geoscientists on the Archean (> 2.5 b.y.B.P.) ocean-biosphere system has been that: (1) the oceans were sulfate poor and possibly H_2S rich; (2) sulfate-reducing bacteria were yet to develop in the oceans, at least prior to ~ 2.7 b.y.B.P.; (3) the sulfides in both sediments and VMSDs were formed directly from magmatic H_2S; and (4) the sulfates in sediments and VMSDs were formed by local oxidation of the H_2S by anaerobic bacteria. This view, which has supported a theory that the Archean atmosphere had yet to accumulate free oxygen, has been developed mostly from a simplistic interpretation of the temporal trends in the $\delta^{34}S$ values of sulfides and sulfates in marine sediments and volcanogenic massive sulfide deposits (VMSDs). Previous workers have not critically examined statistical aspects of the sulfur isotope data, such as the relationship between the $\delta^{34}S$ variability and the sample number, or the skewness of $\delta^{34}S$ frequency curves for Archean samples. The skewness of a $\delta^{34}S$ frequency curve provides important information concerning the open vs. closed nature of the system with respect to SO_4^{2-} (or H_2S), and the magnitude of kinetic isotope effects associated with sulfate reduction (or sulfide oxidation).

In the first part of this work, I will suggest that the sulfur isotopic data on Archean marine sediments can be better explained if Archean oceans, as far back as ~ 3.5 b.y. ago, were already sulfate rich, with a sulfate concentration more than ~ ⅓ of that in modern oceans, possessed a $\delta^{34}S$ value around +2‰ (vs. +20‰ for the modern oceans), and if sulfate-reducing bacteria were more active in the Archean oceans than in modern oceans. The average rate of sulfate reduction in the Archean oceans is suggested to have been ~ 100 to ~ 1000 times greater than the average rate of bacterial sulfate reduction in modern oceans. The higher activity of sulfate-reducing bacteria in the Archean oceans, compared to the modern oceans, could have been caused by higher temperature and greater availability of metabolizable organic matter to sulfate-reducing bacteria, which, in turn, could have been caused by the higher P_{CO_2} and the lower P_{O_2} values of the Archean atmosphere. The Archean oceans could have been up to ~ 40 °C warmer than the modern oceans, and the Archean atmosphere probably contained appreciable amounts of free oxygen, although the P_{O_2} value was lower (~ $10^{-2.5}$ atm: Holland 1984) than the present-day value of $10^{-0.7}$ atm. My model contrasts with the one proposed by Cloud (1972) and elaborated on by Walker et al. (1983) and Kasting (1987): in these latter two models, the P_{O_2} value was estimated to be $\leqq 10^{-12}$ atm until around 2.4 b.y. ago.

In the second part, I will examine the sulfur isotopic characteristics of Archean VMSDs, particularly the relationship between the $\delta^{34}S$ variability and the size of orebodies. From these examinations, I will suggest that Archean VMSDs, much like the Phanerozoic ones, were formed by mixing of sulfate-bearing local seawater with high temperature hydrothermal fluids, which were also of sulfate-bearing seawater in origin. It appears, therefore, that the sulfate content in the seawater has been about the same, and that the mechanisms for sulfide-sulfate mineralization in the oceans have been the same (nonmagmatic), since at least about 3.5 billion years ago.

2 Sulfur Isotope Systematics of Archean Marine Sediments

2.1 Previous Models for the Archean Ocean-Biosphere System

The existence of systematic changes with geologic time in the sulfur isotopic composition of marine sediments is well known: the difference in the $\delta^{34}S$ values between contemporary sulfates and sulfides becomes smaller in older sediments (from ~ 50‰ in modern sediments to generally < 10‰ in Archean sediments), and the $\delta^{34}S$ values of both sulfides and sulfates converge toward 0‰ in older sediments (e.g. Franklin et al. 1981; Skyring and Donnelly 1982; Schidlowski et al. 1983; Hattori et al. 1985). Another important characteristic cited by the previous investigators has been the uniformity in the $\delta^{34}S$ values for Archean sediments, less than 10‰ variation in Archean sediments compared with more than 20‰ variation in

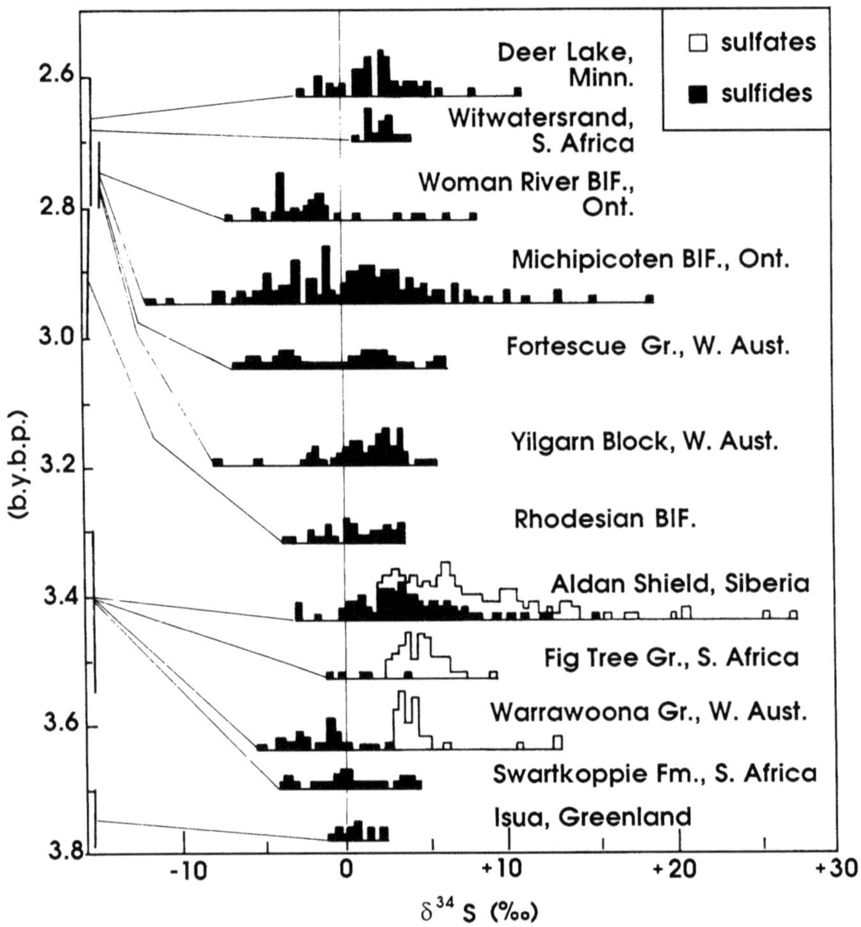

Fig. 1. Histograms of $\delta^{34}S$ values of sulfides and sulfates in Archean marine sediments. *Solid squares* sulfides; *open squares* sulfates. The *smallest square* represents one analysis. See Ohmoto and Felder (1987) for data source

modern sediments. (It will be shown later that this may not be a true characteristic).

Various theories presented by previous researches to explain the above $\delta^{34}S$ trend can be grouped into the following two types. The first theory appeals to sulfide-rich oceans in Archean time, where magmatic H_2S accumulated in oceans to form sedimentary sulfides; sulfate minerals were formed by local oxidation of H_2S (and/or HS^-) by anaerobic phototrophic (green and purple) sulfide-oxidizing bacteria (e.g. Perry et al. 1970; Lambert et al. 1978; Fripp et al. 1979; Franklin et al. 1981; Skyring and Donnelly 1982; Schidlowski et al. 1983; Lambert and Donnelly 1990). The second theory postulates that oceans older than ~ 2.7 b.y.B.P. already contained low to moderate concentrations ($\leqq \sim 1$ mM, compared with 28 mM in modern oceans) of sulfate, and that formation of sedimentary sulfides was occurring by bacterial sulfate reduction (e.g. Perry et al. 1970; Goodwin et al. 1976; Monster et al. 1979; Cameron 1982; Thode and Goodwin 1983; Walker and Brimblecombe 1985). Most investigators, except perhaps the Australian group (Chambers, Donnelly, Fripp, Lambert, Skyring, Trudinger), have interpreted the large $\delta^{34}S$ variation in the ~ 2.7 b.y. old Michipicoten and the Woman River Banded Iron Formations (see Fig. 1) as an indication of a rise in the sulfate concentration and in the activity of sulfate-reducing bacteria (to the present-day levels?) in oceans, and the beginning of accumulation of free oxygen in the atmosphere; the Australian group prefers the ages of 2.4–2.0 b.y.B.P. (Early Proterozoic) for the rise in the activity of sulfate-reducing bacteria.

2.2 Problems with the Previous Models

2.2.1 $\delta^{34}S$ Variation vs. Sample Number

The $\delta^{34}S$ histograms for sulfide and sulfate samples from various Archean formations (Fig. 1) suggest that the observed $\delta^{34}S$ variability may simply be a function

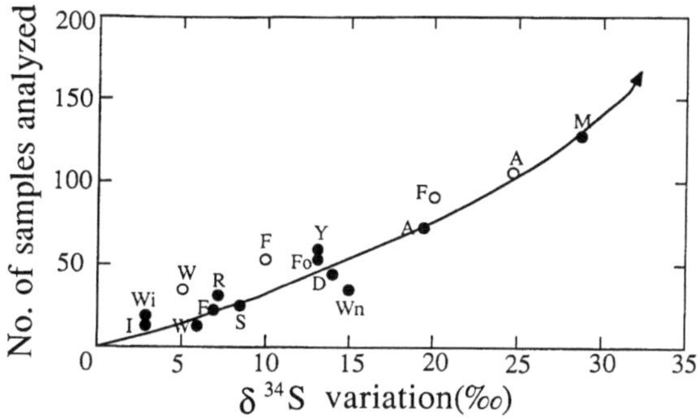

Fig. 2. Relationship between the $\delta^{34}S$ variation and the number of samples analyzed in the studies cited in Fig. 1. *Filled circles* sulfides; *open circles* sulfates. *I* Isua; *S* Swartkoppie; *W* Warrawoona; *F* Fig Tree; *A* Aldan Shield; *R* Rhodesia; *Y* Yilgarn; *Fo* Fortescue; *M* Michipicoten; *Wn* Woman River; *Wi* Witwatersrand; *D* Deer Lake. Two data points for sulfates from the Fig Tree represent two different data sets, one including only the published data and the other including both the published and unpublished data (see text)

of the number of samples analyzed: the more samples analyzed, the more variation found. This suggestion is strongly supported by a comparison of the $\delta^{34}S$ variation vs. the number of samples analyzed (Fig. 2). The $\delta^{34}S$ variation is only 3‰ among 13 sulfide samples of the > 3.8 b.y. old Isua Group and among 17 sulfide samples from the ~ 2.7 b.y. old sediments of Witwatersrand; the $\delta^{34}S$ variation is 19‰ among 72 sulfide samples from the ~ 3.5 b.y. Aldan Shield, and is 28‰ among 128 sulfide samples from the ~ 2.7 b.y. old Michipicoten BIF.

A positive correlation between the $\delta^{34}S$ variation and the number of samples analyzed can be recognized also in modern marine sediments, although the correlation is not as well defined as in the Archean samples (see Ohmoto et al. 1990). The $\delta^{34}S$ variation in modern marine sediments from a given district (e.g. the Santa Barbara Basin, the Black Sea) is typically less than 15‰ when the number of analyses is less than 10, but increases to about 20‰ with about 50 analyses. Therefore, the previously conceived notion that the $\delta^{34}S$ variation is less in Archean marine sediments compared to modern marine sediments may not be a valid one.

Note that the $\delta^{34}S$ values reported in previous literature are almost in all cases those of *bulk* sulfide (or sulfate) minerals that were separated typically from several cm^3 sized rock specimens, rather than the $\delta^{34}S$ values of individual grains of sulfide (or sulfate) crystal, because the conventional mass-spectrometric analyses require typically more than 1 mM of SO_2 (or SF_6) that is converted from ~ 10 mg of sulfur-bearing minerals. The $\delta^{34}S$ variation among *individual* crystals in a rock sample may be much greater than that among *bulk* samples, and is a better indicator of bacterial activity. This is an important aspect which will be discussed further in Sections 2.2.2 and 2.3.2.

2.2.2 Occurrences of Sulfate-Bearing Minerals in Archean Sediments

Sulfate-bearing evaporite bodies are rare in Archean sediments, because the environments suitable for their deposition were rare, or such environments have not been preserved, or they were formed but were dissolved away due to the high solubility of gypsum and anhydrite in low temperature aqueous solutions (e.g. groundwater). Some sulfate beds of Archean age, such as those in the North Pole deposit of the Pilbara Block in Western Australia (3.5–3.4 b.y. old) and those in the Onverwacht and the Fig Tree Groups in the Barberton district of South Africa (3.4–3.2 b.y. old), have been preserved because, during their early diagenetic history, the gypsum was converted to less soluble barite through interaction with Ba-bearing hydrothermal solutions (Heinrichs and Reimer 1977; Lambert et al. 1978; Reimer 1982; Lambert and Donnelly 1990). The discovery of ubiquitous occurrences for sulfate-bearing metamorphic minerals (e.g. scapolite) in > 3.6 b.y. old anhydrite/barite/dolomite-bearing (i.e. evaporite sequence) metasediments of the Aldan Shield, Siberia (Vinogradov et al. 1977) is of special interest; sulfate-rich metamorphic minerals with an anhydrite precursor may be more common than previously recognized in high-grade metamorphic terrains of Archean age, and may be strong evidence of sulfate-rich oceans.

The $\delta^{34}S$ values of sulfate minerals from the three well-studied Archean districts, the Pilbara Block (Australia), the Barberton Mountainland (South Africa), and the Aldan Shield (USSR), all around 3.5 b.y. in ages, are compared in Fig. 3. The Barberton data include the results of our recent analyses (Ohmoto et al. in prep.) of barite samples from the Fig Tree Group.

The barite-bearing units in the Fig Tree Group consist of alternating beds of cherts, dolomites, conglomerates, sandstones, shales and barites, and occur in a well-defined stratigraphic zone which has been traced over a distance of 8 km; they have been identified as sedimentary in origin, although the barite was not a primary sulfate mineral precipitated from the seawater (e.g. Heinrichs and Reimer 1977; Lowe and Knauth 1977; Lowe 1980). All the barite samples studied by previous investigators and by us contained two or three types of barite: rounded detrital grains of barite ("barite sands"), bladed crystals, and "cauliflower" type; the latter two types appear to have

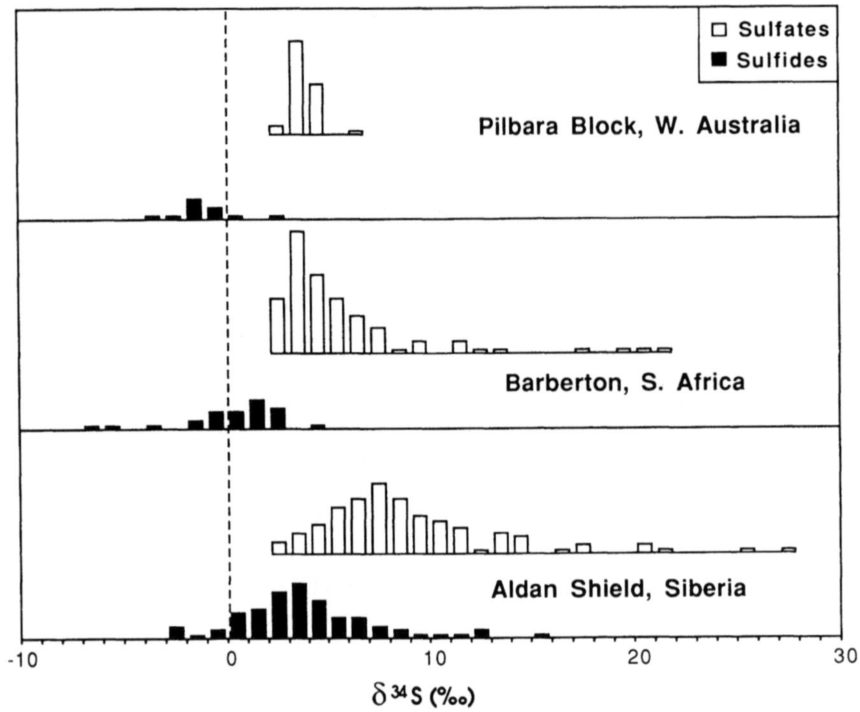

Fig. 3. $\delta^{34}S$ frequency diagrams for evaporite sulfates (mostly barite) and sulfides (pyrite) from sedimentary rocks of ages greater than 3.2 billion years. The Pilbara Block (3.45 b.y.), Australia, data from Lambert et al. (1978) on the North Pole district. The Barberton Mountainland district (3.5–3.2 b.y.), South Africa, data from Perry et al. (1970), Lambert et al. (1978), Vinogradov et al. (1977), Reimer (1982), and Ohmoto et al. (in preparation). The Aldan Shield (> 3.5 b.y.), Siberia, data from Vinogradov et al. (1977)

formed from dissolution/recrystallization of the detrital barite grains during diagenesis of the barite beds. From such textural relationships, the following sequence of events has been suggested by Heinrichs and Reimer (1977) and Ohmoto et al. (in preparation): formation of evaporitic gypsum beds in a shallow sea ⇒ replacement of gypsum by barite (under submarine or subaerial conditions) ⇒ weathering and redeposition of barite beds ("barite sands") ⇒ diagenesis of barite beds ("bladed" and "cauliflower" barites).

A total of 52 analyses has been reported on barite samples (mostly *bulk*, rather than the separated barite types) from the Fig Tree Group by Perry et al. (1970), Vinogradov et al. (1977), Lambert et al. (1978), and Reimer (1982): the $\delta^{34}S$ values range from +2.5 to +13.1‰, with a mean of about +4‰ (see Fig. 1).

The 21 barite-bearing samples studied by us (Ohmoto et al. in prep.) were collected by D.R. Lowe from the same barite beds that had been studied for $\delta^{34}S$ by the previous investigators. We have separated different types of barite crystals from each sample using a dental drill, and analyzed 37 separates (each separate contains 10 to 50 individual grains of crystals) for $\delta^{34}S$. The minimum $\delta^{34}S$ value in our samples (+2.9‰) is very similar to the minimum $\delta^{34}S$ value (+2.5‰) obtained by previous investigators on the *bulk* sulfate samples. The inclusion of our data with the previous data set has not changed the peak $\delta^{34}S$ value (+4.5‰), or the skewness of the frequency curve (strongly skewed towards more positive values) for the Barberton district (cf. Figs. 1 and 3). However, our maximum $\delta^{34}S$ value (+21.4‰) far exceeds the maximum $\delta^{34}S$ value (+13.1‰) for the *bulk* samples.

We have found that the $\delta^{34}S$ values of the "bladed" and "cauliflower" types of barite in some samples differ by more than 10‰, sometimes heavier but other times lighter, than the detrital-type barites; the $\delta^{34}S$ ranges are similar for the three barite types, from the minimum of about +3 to the maximum of about +20‰. Such $\delta^{34}S$ relationships suggest that the $\delta^{34}S$ values of the *detrital* barite grains were highly inhomogeneous, and that the inhomogeneity in these grains was inherited by the bladed and cauliflower barites during recrystallization. Our study further suggests that the maximum $\delta^{34}S$ value for the individual gains of the *detrital* barites (i.e. the maximum $\delta^{34}S$ value for the original gypsum beds) may far exceed +21‰, since each barite separate contained more than ten individual crystals.

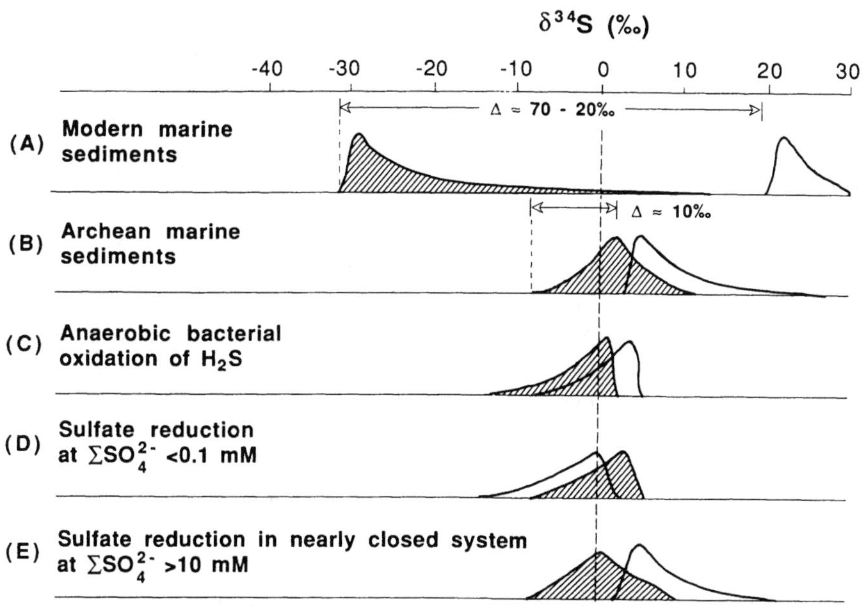

Fig. 4. $\delta^{34}S$ frequency curves for sulfides and sulfates in **A** modern marine sediments (observed relationships); **B** Archean marine sediments (observed relationship); **C** systems where sulfates are produced by oxidation of H_2S-rich seawater by anoxic bacteria (predicted relationship); **D** systems where sulfides are formed by bacterial reduction of low SO_4^{2-} (<0.1 mM) seawater (predicted relationship); **E** systems where sulfides are formed by bacterial reduction of SO_4^{2-}-rich (>1 mM) seawater and are nearly closed to SO_4^{2-} (predicted relationship). The *shaded curves* are for sulfides, and the *open curves* for sulfates

2.2.3 The $\delta^{34}S$ Frequency Curves Expected from the Model of Sulfide-Rich Archean Oceans

Typical $\Delta_{SO_4-H_2S}$ values associated with the oxidation of aqueous sulfides (H_2S and HS^-) by anaerobic bacteria are between 0 and +5‰ (see Chambers et al. 1975; Fry et al. 1984, 1986, for reviews of this subject). Such $\Delta_{SO_4-H_2S}$ values appear, at first glance, to explain the observation that most Archean sulfate minerals are enriched in ^{34}S relative to most Archean sulfides by only 2 to 3‰ (see Fig. 4B), and support the theory of sulfide-rich oceans and formation of sulfate minerals by local oxidation of the sulfide by anaerobic sulfide-oxidizing bacteria.

If the sulfates were produced by oxidation of H_2S which ranged from −7 to +5‰ in $\delta^{34}S$ (i.e. the common range for the Archean sulfides), and if the $\Delta_{SO_4-H_2S}$ values were between 0 and 5‰, the common $\delta^{34}S$ values for Archean sulfates should have been in the range of −7 to +10‰. However, evaporite sulfate samples with $\delta^{34}S < +2$‰ have not yet been reported, while sulfates with $\delta^{34}S > +10$‰ have been found more frequently than previously believed (see Fig. 3).

When comparing the sulfur isotopic characteristics of Archean vs. modern sediments, previous researchers were primarily concerned with the ranges and the means of the $\delta^{34}S$ values of sulfide and sulfate samples. They have paid little attention to the skewness of the $\delta^{34}S$ frequency curves (= histograms), which may provide critical information on the direction of dominant reaction (oxidation of H_2S vs. reduction of SO_4^{2-}), the magnitude of kinetic isotope effect ($\Delta_{SO_4-H_2S}$ value), and the open or closed nature of the system with respect to the reactant (see Schwarcz and Burnie 1973; Ohmoto and Rye 1979). A $\delta^{34}S$ frequency curve is equivalent to a probability curve for finding specific $\delta^{34}S$ values when random sampling is carried out for a particular rock unit.

If the Archean sulfates were formed by oxidation of H_2S in systems that had a limited supply of H_2S, i.e. closed systems with respect to H_2S (parent), the $\delta^{34}S$ frequency curves for *both* sulfates and sulfides should have a skewness toward negative $\delta^{34}S$ values (Fig. 4C), because the removal of isotopically heavier sulfate from a H_2S-bearing system would have caused the residual H_2S (and also sulfates of later generation) to become increasingly more negative in $\delta^{34}S$ and less abundant in quantity. The skewness of the $\delta^{34}S$ frequency curve for Archean *sulfates* is, however, toward more positive $\delta^{34}S$ values (see Figs. 1, 3, 4B, C). Therefore, it is unlikely that the Archean sulfates were produced by oxidation of sulfide-rich seawater. This conclusion is similar to that reached by Holland

(1984) and Walker and Brimblecombe (1985), who have pointed out that oceanic sulfide concentrations should have been very low as a consequence of high Fe^{2+} concentrations in the Archean oceans; H_2S and Fe^{2+} react to form iron sulfides.

It is necessary to correct a serious misconception existing in the literature that relates "reducing", "euxinic" and "anoxic" conditions with "sulfate-poor and H_2S-rich" environments. In fact, anoxic and H_2S-rich bodies of water in modern seas and lakes (e.g. deeper parts of the Black Sea) are created because of the *high* SO_4^{2-} concentrations and the high activity of sulfate-reducing bacteria. The concentration of H_2S in an euxinic body of water is typically less than 3 mM, while that of SO_4^{2-} is similar to normal seawater, i.e. ca. 28 mM (Ohmoto et al. 1990).

2.2.4 $\delta^{34}S$ Frequency Curves Expected from the Model of Bacterial Sulfate Reduction in Low-Sulfate Oceans

The model of Archean oceans containing $\leqq \sim 1$ mM of sulfate has been based largely on the alleged scarcity of Archean evaporites and on the experimental data of Harrison and Thode (1958), who observed a decrease in the $\Delta_{SO_4-H_2S}$ value associated with bacterial sulfate reduction from about 15‰ to about 4 and −3‰, when the SO_4^{2-} content was decreased from > 10 mM to 0.6 and 0.01 mM, respectively, at T = 30 °C.

If the sulfate content of oceans were below $\sim$0.1 mM and the $\Delta_{SO_4-H_2S}$ values associated with the bacterial sulfate reduction were negative, the $\delta^{34}S$ frequency curves for *both* sulfates and sulfides would have had the skewness toward negative $\delta^{34}S$ values, with a tendency of $\delta^{34}S_{sulfides} > \delta^{34}S_{sulfates}$ (see Fig. 4D). This is contrary to the $\delta^{34}S$ frequency curves observed for Archean sediments (cf. Figs. 1, 3, 4B, E).

On the other hand, if the $\Delta_{SO_4-H_2S}$ values associated with the bacterial sulfate reduction were slightly positive (0 to 4‰), and the initial $\delta^{34}S$ value of the SO_4^{2-} was around +2‰ (i.e. the minimum $\delta^{34}S$ value of the Archean sulfates), removal of *more than* 99.8% of the original amount of SO_4^{2-} as sulfide would have been required by the Rayleigh equation to produce sulfate minerals with $\delta^{34}S$ values as high as +27‰ (i.e. the maximum $\delta^{34}S$ value in the Aldan Shield). That is, the probability of discovering sulfate minerals with a $\delta^{34}S$ value of +27‰ in Archean evaporites was *less than* 0.2%; it would have been unlikely to find a sample of $\delta^{34}S = +27$‰ from analyses of $\sim$ 100 samples.

Vinogradov et al. (1977) have suggested that the $\delta^{34}S$ variation in the Aldan Shield samples is an indication that: as far back as 4 billion years ago, (some?) oceans were sulfate-rich, sulfate-reducing bacteria were active, and the atmosphere contained free oxygen. However, their interpretation was neither quantitative nor consistent; they have interpreted that sulfate minerals in the $\sim$ 3.2 b.y. old Fig Tree Group were formed by the local oxidation of sulfides, rather than the sulfides formed by bacterial sulfate reduction.

In order to produce $\delta^{34}S$ probability curves similar to those for the samples from the Barberton district and the Aldan Shield, the kinetic isotope effect (the $\Delta_{SO_4-H_2S}$ value) of about 10‰ or larger is required; more than a few mM of SO_4^{2-} content in Archean oceans is required.

2.3 Proposed Model for the Archean Atmosphere-Ocean-Biosphere System

2.3.1 $\delta^{34}S$ Frequency Curves for the Open and Closed Systems with Respect to SO_4^{2-}

Ohmoto and Felder (1987) have proposed that the differences in the $\delta^{34}S$ frequency curves between the Archean marine sediments and the younger one can be explained by the differences in: (1) the open vs. closed nature of the system with respect to SO_4^{2-} (most natural systems can be considered as open to sulfide); (2) the magnitude of kinetic isotope effect accompanying bacterial sulfate reduction; and (3) the $\delta^{34}S$ value of seawater sulfate. The sulfate content of seawater appears to have remained nearly the same since about $\sim$ 3.5 b.y.B.P.

The open or closed nature of a system with respect to SO_4^{2-} is defined simply by the relative rate of sulfate supply to that of sulfate removal (i.e. reduction; see Table 1). A system is defined as open with respect to SO_4^{2-} when the rate of sulfate supply is much greater than that of sulfate reduction, regardless of the absolute rate of sulfate reduction. A system is considered closed with respect to SO_4^{2-} when the rate of sulfate reduction equals that of sulfate supply, such that all of the SO_4^{2-} is consumed within the system. In nature, a system where sulfate supply rate < sulfate reduction rate will not exist, except for a very short time, because sulfate-reducing bacteria require SO_4^{2-} to survive.

"A sulfur isotope system" refers to a large (> 1 km^3) body of seawater (or lake water) when discussing $\delta^{34}S$ characteristics of *syngenetic* minerals (syngenetic sulfides and evaporite sulfates), but refers to a sediment column (a few milimeters to several meters in thickness, depending on the situation) that underlies the seafloor when discussing $\delta^{34}S$ characteristics of *diagenetic* minerals.

Table 1. Relationships between the environments of modern marine mineralization and open vs. closed nature of systems with respect to SO_4^{2-}

Open with respect to SO_4^{2-}: sulfate reduction rate $\ll$ sulfate supply rate
Closed with respect to SO_4^{2-}: sulfate reduction rate = sulfate supply rate

Environment	Open-bottom vs. closed-bottom with respect to SO_4^{2-}	Type of mineralization
Normal oceans (oxic bottom water)		
Seawater body	Open	No
Sediments, bioturbation zone (< ~20 cm)	Open	Diagenetic sulfides and sulfates
Sediments, below bioturbation zone (> ~20 cm)	Closed	Diagenetic sulfides
Euxinic basins		
Basin as a whole	Open or closed[a]	Syngenetic sulfates
The upper, oxic water body	Open or closed[a]	Syngenetic sulfates
The lower, anoxic water body	Open[a] or closed	Syngenetic sulfides
Sediments beneath anoxic water	Closed	Diagenetic sulfides

[a] More common condition.

It is important to recognize that "coexisting" or "contemporaneous" sulfate and sulfide minerals in marine sediments have formed typically from different subsystems of a large system. For example, sulfate minerals in an evaporite-bearing unit may have formed syngenetically from the upper (oxic) part of a restricted basin which may be closed with respect to SO_4^{2-} because of bacterial sulfate reduction at deeper parts of the basin; sulfides in the same beds may have formed syngenetically from the lower (anoxic) part of the basin which may be open with respect to SO_4^{2-}, or formed diagenetically after the sediments were buried below the seafloor in systems closed with respect to SO_4^{2-}. Formation of sulfides *and* sulfate from the *same* closed system is unlikely, unless the decrease in SO_4^{2-} content due to sulfate reduction is offset by evaporation of water. Therefore, it is necessary to treat the sulfur isotope data of sulfate minerals separately from those of sulfide minerals.

2.3.2 Nature of Systems Indicated by the $\delta^{34}S$ Frequency Curves of Sulfates

The $\delta^{34}S$ frequency curve for bulk sulfate samples from a marine evaporite unit of Phanerozoic-Proterozoic ages (see Fig. 4A) typically exhibits the following characteristics: (1) the minimum and the most frequent (i.e. peak) $\delta^{34}S$ values are similar and about 1–3‰ higher than the $\delta^{34}S$ value of contemporaneous normal seawater, which has fluctuated between +10 and +35‰ during most of the Proterozoic and Phanerozoic time (Claypool et al. 1980; Ohmoto 1986); (2) the width of the curve (i.e., the variation in the $\delta^{34}S$ values) is around 10‰ or more; and (3) the curve is skewed strongly toward more positive values. The first characteristics reflects the equilibrium fractionation factor of around +1.5‰ between gypsum and aqueous sulfate at temperatures around 25 °C (Thode and Monster 1965). The second and third characteristics were produced because: the system for sulfate (e.g. an entire basin) was closed with respect to SO_4^{2-}; both sulfide and sulfate mineralizations were occurring in the same basin (although the sulfide mineralization by bacterial sulfate reduction was taking place mostly in deeper parts of the basins, while the sulfate mineralization in shallower parts of the basins); and the kinetic isotope effect accompanying bacterial sulfate reduction in modern oceans is large ($\Delta_{SO_4-H_2S} > 20‰$). These combined effects caused the sulfate content to decrease and the $\delta^{34}S_{SO_4}$ value to increase continuously with time; sulfate samples with higher $\delta^{34}S$ values become fewer in number.

It is significant that the $\delta^{34}S$ frequency curves for sulfates from three widely separated areas (Western Australia, South Africa, and Siberia) of approximately the same age (~3.5 b.y.B.P.) are very similar in the minimum and the peak $\delta^{34}S$ values, and in the width and skewness of the curve (see Fig. 3). The width and the skewness of their curves are also similar to those for modern sulfates (cf. Figs. 1, 3, 4A, B). Such characteristics suggest that the Archean evaporite sulfates were formed also in systems that were closed with respect to SO_4^{2-} and that the $\Delta_{SO_4-H_2S}$ values were about 10‰ or larger. Such $\Delta_{SO_4-H_2S}$ values require a sulfate concentration greater than a few millimoles, and probably greater than ca. 10 mM (see later

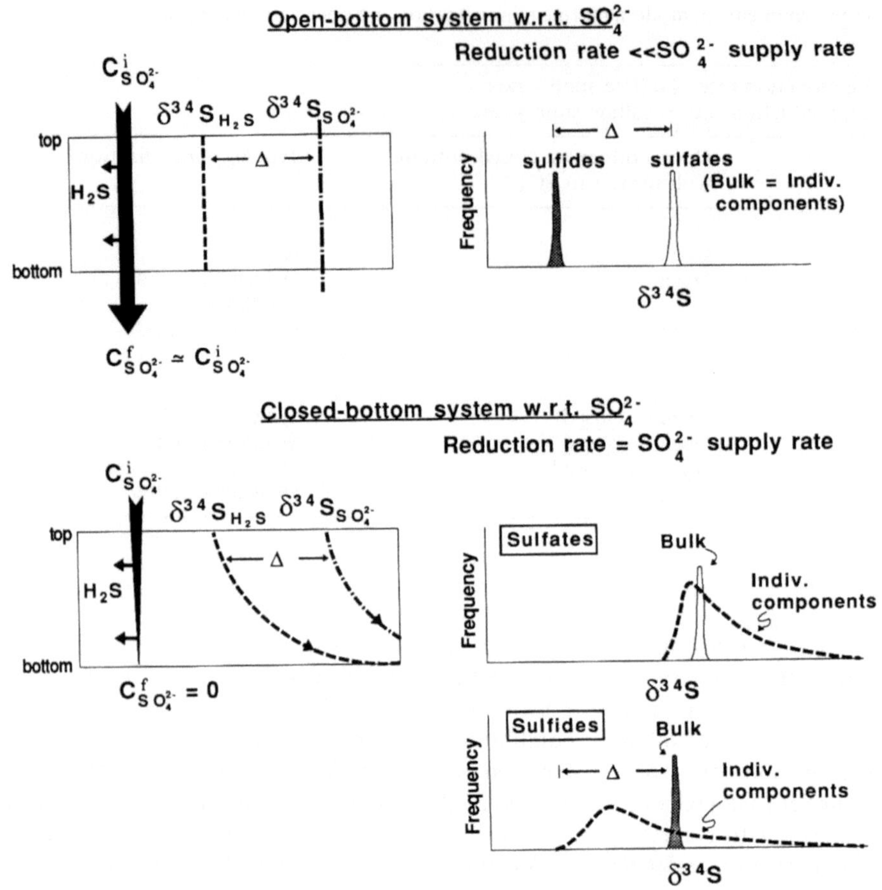

Fig. 5. Schematic representation of chemical and isotopic characteristics of systems that are open-bottom with respect to SO_4^{2-} (*above*), and closed-bottom with respect to SO_4^{2-} (*below*). Δ is the kinetic isotope effect associated with the sulfate reduction. Note that in open-bottom systems with respect to SO_4^{2-}, the $\delta^{34}S$ values of individual sulfide components and the bulk sulfide are identical, which are displaced by Δ from the $\delta^{34}S$ of seawater sulfate. In closed-bottom systems with respect to SO_4^{2-}, the $\delta^{34}S$ value of individual component varies, but the $\delta^{34}S$ of bulk sulfide becomes identical to that of seawater sulfate; the $\delta^{34}S$ of bulk sulfate is shifted slightly toward higher values from that of seawater

section). The only difference between the Archean and the younger evaporite sulfates is that the minimum and the peak $\delta^{34}S$ values lie at about +3‰ for the Archean sulfates, but around +21‰ for the modern ones. This suggests that the $\delta^{34}S$ value of Archean oceans was around +2‰.

2.3.3 Nature of Systems Indicated by the $\delta^{34}S$ Frequency Curves of Sulfides

Normal seawater contains only ~900 ppm S as SO_4^{2-}, but most sediments contain much more than 900 ppm sulfide sulfur (e.g. Ohmoto et al. 1990). This implies that most modern marine systems are continuously replenished with SO_4^{2-}, usually from the top of the systems, as sulfate reduction takes place. Hence, it might be easier to understand the nature of various marine systems, if we consider them to be "open-bottom" or "closed-bottom" with respect to SO_4^{2-}, instead of simply referring to them as "open" or "closed" with respect to SO_4^{2-} (Fig. 5).

The upper ~20–50 cm of sediments under normal, oxic seawater are typically open-bottom systems with respect to SO_4^{2-}, because, despite intensive sulfate reduction to form *diagenetic sulfides*, SO_4^{2-} is supplied at a very fast rate due to the churning of sediments by burrowing organisms (Berner 1980, 1984, 1985; Ohmoto et al. 1990). The anoxic water body at a depth below 200 m in the Black Sea, where *syngenetic sulfide* mineralization is taking place, is another example of an open-bottom system with respect to SO_4^{2-}, because the sulfate content, the $\delta^{34}S_{SO_4}$ and $\delta^{34}S_{H_2S}$ values remain essentially constant throughout the water column (Ohmoto et al. 1990).

In a system that is open-bottom with respect to SO_4^{2-} (see the upper sets of Fig. 5), the concentration and the $\delta^{34}S$ value of SO_4^{2-} in all parts of the system will be the same as that of the incoming seawater; the $\delta^{34}S$ value of H_2S (and sulfide minerals) formed at different parts of the system will also be constant, but displaced from the $\delta^{34}S$ value of seawater sulfate by the kinetic isotope effects associated with the sulfate reduction. In such systems, the $\delta^{34}S$ values of the individual sulfate components (e.g. sulfate-bearing rock samples from a thick evaporite bed) will be the same as the $\delta^{34}S$ value of the bulk sulfate (e.g. the bulk evaporite bed), and the $\delta^{34}S$ values of the individual sulfide components (e.g. individual grains of sulfide minerals in a rock sample) will be the same as the $\delta^{34}S$ value of the bulk sulfide (e.g. bulk rock sample).

Sedimentary sections beneath the bioturbation zone under oxic seawater, and sediments immediately underlying the anoxic water column in euxinic basins are two important environments for *diagenetic sulfides* that are closed-bottom systems with respect to SO_4^{2-}, because the supply of SO_4^{2-} to these systems is mostly due to diffusion.

In a system "closed-bottom with respect to SO_4^{2-}" (see the lower sets of Fig. 5), the SO_4^{2-} content of the water decreases continuously with depth (time). When the $\Delta_{SO_4-H_2S}$ value is positive, the $\delta^{34}S$ values for SO_4^{2-} and for sulfide minerals formed in the system continuously increase downward (later time), following the Rayleigh distillation equation. The $\delta^{34}S$ values of individual components of sulfates *or* sulfides in such systems may exhibit a large variation, and their $\delta^{34}S$ frequency curves are skewed toward more positive values (Schwarcz and Burnie 1973; Ohmoto and Rye 1979). (Remember that formation of sulfides *and* sulfate from the *same* closed system is unlikely, unless the decrease in SO_4^{2-} content due to sulfate reduction is offset by evaporation of water.) The $\delta^{34}S$ values of the bulk sulfide of the system may become identical to the $\delta^{34}S$ of the initial seawater sulfate, because all the sulfate supplied to the system may be quantitatively converted to sulfides.

The $\delta^{34}S$ frequency curve for *bulk sulfide* samples from Phanerozoic-Proterozoic marine sedimentary units typically exhibits the following characteristics (see Fig. 4A): (1) the minimum and the peak $\delta^{34}S$ values that are separated from the $\delta^{34}S$ value of contemporaneous seawater by 20 to 70‰ (average of ~ 50‰); (2) the width of the curve is around 20‰ or more (with more than ~ 50 analyses); and (3) the skewness is strongly toward positive $\delta^{34}S$ values. These characteristics were produced because the kinetic isotope effects accompanying sulfate reduction under modern marine conditions are between 20 and 70‰, and because typical marine sediments acquire about one-half or more of their sulfide contents while they were buried to depths of ~ 20 to ~ 50 cm (= bioturbation zone) which is an open-bottom system with respect to SO_4^{2-}, and the remaining sulfides during burial to depths around 200 cm in a zone that can be considered as an *extended* (i.e. occurs over a wide depth range) closed-bottom system with respect to SO_4^{2-} (Ohmoto and Felder 1987).

In contrast, the $\delta^{34}S$ frequency curves for the *bulk sulfide* samples of Archean sediments typically have skewness near normal or slightly toward negative $\delta^{34}S$ values, and mean $\delta^{34}S$ values within a range of $+1 \pm 2$‰ (see Figs. 1 and 4B). Such features suggest that the systems were nearly closed-bottom with respect to SO_4^{2-}.

In Archean oceans there were no large benthic organisms to stir marine sediments. Therefore, the replenishment of SO_4^{2-} from the overlying seawater to sediment systems would have been carried out by simple diffusion only, creating the condition where the rate of SO_4^{2-} supply equals the rate of sulfate reduction rate. That is, diagenetic sulfides in Archean marine sediments must have been formed in systems closed-bottom with respect to (w.r.t.) SO_4^{2-}. An important question, then, is the size (= depth) of the systems. Remember that the $\delta^{34}S$ values of sediments determined by most investigators are not on individual grains of sulfides, but on *bulk* sulfides typically separated from a several cm^3 of sample. When the size of "a closed-bottom system w.r.t. SO_4^{2-}" is bigger than several centimeters in depth, such as in most recent marine sediments below the bioturbation zone, the individual rock samples for $\delta^{34}S$ analyses can be considered as individual components of the system; the $\delta^{34}S$ values of bulk sulfides may become highly variable. On the other hand, when the size of "a closed-bottom system w.r.t. SO_4^{2-}" is smaller than several centimeters in depth, the rock samples may represent the entire system; the $\delta^{34}S$ values of *bulk* sulfides become the same as the $\delta^{34}S$ of initial SO_4^{2-} (see bottom sets of Fig. 5).

My proposition is that the thickness of the closed-bottom system w.r.t. SO_4^{2-} for Archean marine sediments was generally thinner than a few centimeters. In other words, Archean sediments acquired essentially all their sulfides within several centimeters of burial from the seafloor. This would explain the skewness and the mean $\delta^{34}S$ values of Archean sulfide samples; the $\delta^{34}S$ values of *bulk sulfides* in hand-specimen sized rocks would become essentially identical to the $\delta^{34}S$ value of seawater sulfate when the size of systems is reduced. The requirement for the *thin* system would have been satisfied if the sulfate reduction rate in Archean oceans was, on the average, 100 to 1000 times higher than that in the present oceans.

The difference between the minimum $\delta^{34}S$ value of *bulk* sulfide samples (−12‰ at Michipicoten, and −4‰ at Aldan Shield) and the minimum $\delta^{34}S$ values of Archean sulfate samples (~ +3‰) suggests the kinetic isotope effect accompanying bacterial reduction of sulfate to be ~6 to ~15‰ in Archean oceans, which agrees reasonably well with the $\Delta_{SO_4-H_2S}$ values of about 10‰ or larger, estimated above from the $\delta^{34}S$ frequency curves for the Archean sulfates. In the following section, I will suggest that the smaller kinetic isotope effect (~ 10‰ in Archean oceans vs. ~ 45‰ in modern oceans) may be related to the higher rate of bacterial reduction.

2.4 Relationship Between the Rate of Sulfate Reduction and the Kinetic Isotope Effect

2.4.1 Laboratory Experimental Data

Bacterial sulfate reduction appears to follow zero-order kinetics with respect to sulfate content down to around 1 mM (Goldhaber and Kaplan 1975; Chambers and Trudinger 1979; Berner 1980, 1985; Postgate 1984; Skyring 1987). Various experimental data on bacterial sulfate reduction and measurements on modern marine sediments also indicate that the kinetic isotope effect accompanying bacterial sulfate reduction is independent of the SO_4^{2-} content of solutions down to ~ 1 mM (see reviews by Chambers and Trudinger 1979; Ohmoto et al. 1990).

It has long been recognized that the kinetic isotope effect accompanying bacterial sulfate reduction in laboratory systems generally decreases with increasing reduction rate (e.g. Harrison and Thode 1958; Kaplan and Rittenberg 1964; Chambers et al. 1975; Chambers and Trudinger 1979; Skyring 1987). However, attempts to establish a generalized equation relating the kinetic isotope effect and the sulfate reduction rate (e.g. Chambers and Trudinger 1979) have not been successful, primarily because the biological experiments carried out by different groups of investigators are difficult to compare directly. Different groups of researchers used different strains of bacteria, and normalized their results to different bacterial units, such as rates per organism or per colony of bacteria. It is also difficult to recognize a relationship between the rate of sulfate reduction and the kinetic isotope effect, when all sets of experimental data, regardless of the types of electron donors and SO_4^{2-} contents of the experiments, are treated together.

The rate of bacterial sulfate reduction and the kinetic isotope effect observed in experimental systems that used organic compounds (rather than H_2) as electron donors and solutions with SO_4^{2-} contents in a range of 10 to 100 mM are compared in Fig. 6a–b. Figure 6 shows that the results of experiments obtained different research groups are very similar: the $\Delta_{SO_4-H_2S}$ value decreases by 20 to 30‰ when the reduction rate increases by about 2 orders of magnitude.

2.4.2 Observation in Modern Marine Sediments

Goldhaber and Kaplan (1975) have recognized the existence of excellent correlations among the rate of sulfate reduction, the minimum $\delta^{34}S$ value of sulfides in sediments ($= \delta^{34}S_{seawater\ SO_4} - \Delta_{SO_4-H_2S}$), and the sedimentation rate for modern sediments from about ten different localities, located mostly offshore California (Long Basin, Santa Catalina Basin, Newport Marsh, East Cortez Basin, Santa Barbara Basin, Pascadero Basin, Carmen Basin; see Fig. 7). The kinetic isotope effect decreases from around 60‰ to around 30‰ as the reduction rate increased from $10^{-4.5}$ to $10^{-2.5}$ $mol\,l^{-1}\,yr^{-1}$, which accompanies a sedimentation rate increase from ~50 cm/1000 yr to ~1000 cm/1000 yr.

Sulfate reduction rates and the kinetic isotope effects for sediments in the Black Sea, Spencer Gulf (South Australia), and the Solar Lake (Sinai) are also compared in Fig. 7. The Black Sea data (Vaynshteyn et al. 1986) are for the top 10 cm of sediment cores; the reduction rates, reported by the authors in units of µg S/kg raw ooze/day, are recalculated with a conversion factor of 1 µg S/kg ooze/day = 4×10^{-5} $mol\,l^{-1}\,yr^{-1}$ which assumed a density of 1.8 and water content of 50 vol% for the ooze. Chambers (1982) measured the kinetic isotope effects associated with bacterial sulfate reduction in the top 10 cm of sediment in an intertidal environment at Spencer Gulf over a 13-month period: the kinetic isotope effect varied seasonally from a maximum of 48‰ in winter to a minimum of about 30‰ in summer, exhibiting an inverse correlation with temperature (Fig. 8). The rates of sulfate reduction in these sediments have been reported by Skyring (1987) as 2–104 $mmol\,m^{-2}\,day^{-1}$, correlating positively with a temperature between 16 and 34 °C: these reduction rates are recalculated to be $10^{-1.9}-10^{-0.2}$ $mol\,l^{-1}\,yr^{-1}$ with the conversion factor used by Skyring (1987). The kinetic isotope effects accompanying bacterial sulfate reduction in the algal mats of the Solar Lake are estimated to be about 33‰ about 5 cm below the lake floor, and 43–51‰ at

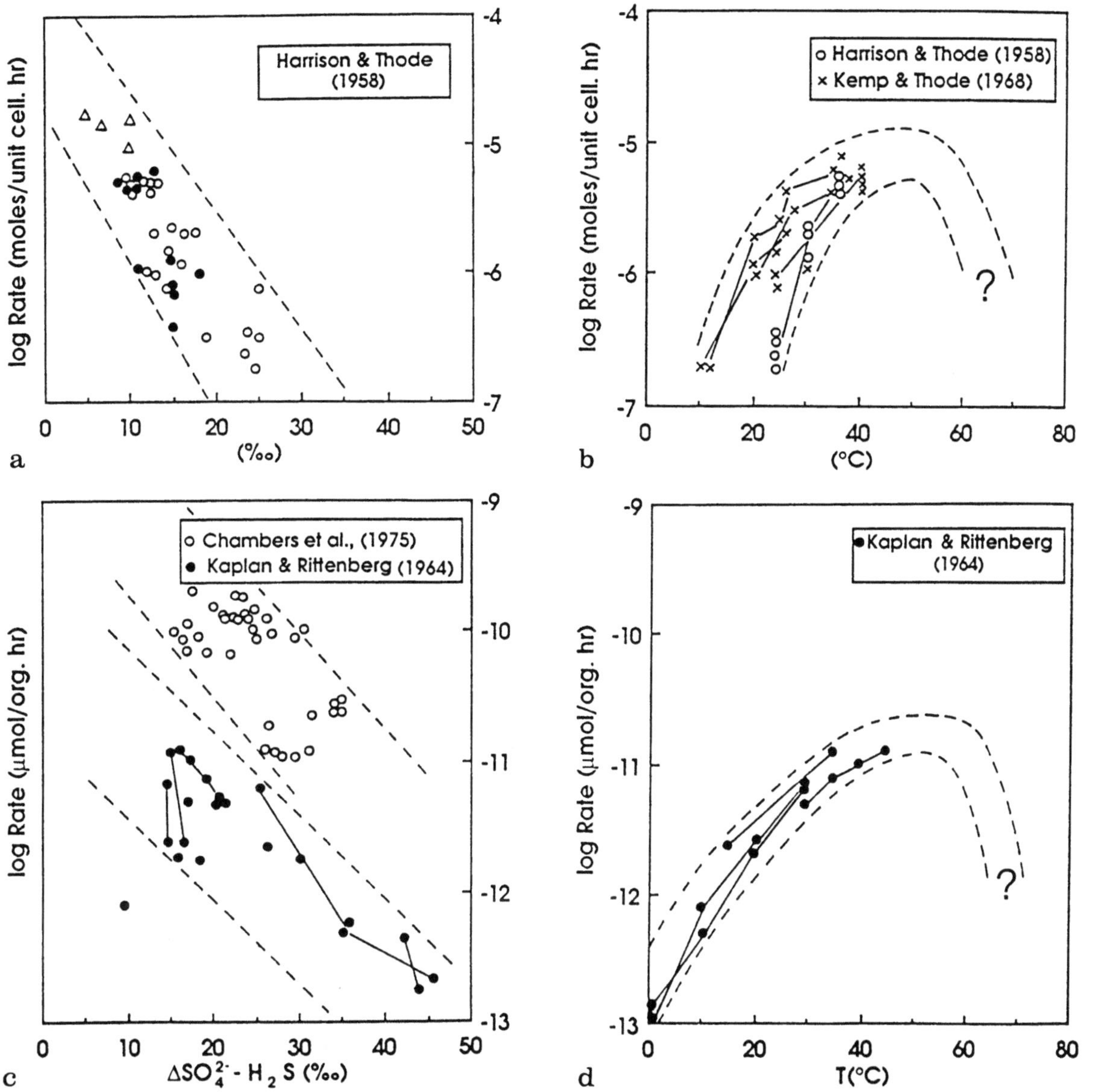

Fig. 6a–d. Relationships among the rates of sulfate reduction, temperatures, and the kinetic isotope effects observed in various laboratory experiments using *Desulfovibrio desulfuricans* (Ohmoto and Felder 1987)

depths of 30–60 cm in the sediment core from the $\delta^{34}S$ profile reported by Aizenshtat et al. (1984). The sulfate reduction rates at the corresponding depths have been measured to be about 500 nmol $cm^{-3} day^{-1}$ ($\approx 0.4\,mol\,l^{-1} yr^{-1}$) and about 3 nmol $cm^{-3} day^{-1}$ ($\approx 0.02\,mol\,l^{-1} yr^{-1}$), respectively (Jørgensen and Cohen 1977).

The rates of sulfate reduction in the sediments of the Black Sea, Spencer Gulf, and the Solar Lake in Figs. 7 and 8 were measured using the radioactive ^{35}S tracer method on sediment cores from the corresponding depths. In contrast, the values of sulfate reduction rates used in Goldhaber and Kaplan (1975) were mostly based on the theoretical calculations using the sulfate-content vs. depth profiles of pore fluids in sediments below the bioturbation zone. This may be the reason why in Fig. 7 the best-fit line for the data set reported in Goldhaber and Kaplan (1975) lies one or two units below that for the new data set (i.e. the Black Sea, Spencer Gulf, and the Solar Lake). Nevertheless, the response of the $\Delta_{SO_4-H_2S}$ value to the reduction rate, i.e. about 30‰ decrease in $\Delta_{SO_4-H_2S}$ with an increase in the reduction rate of 2–3 orders of magnitude, is similar between the two data sets of

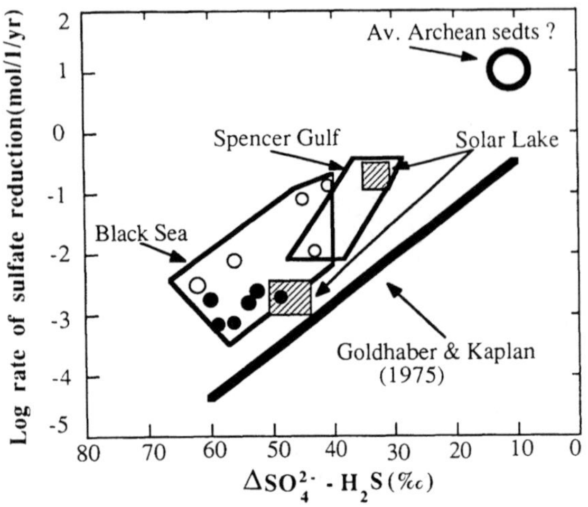

Fig. 7. Relationships between the rates of sulfate reduction and the kinetic isotope effects observed in marine sediments. Spencer Gulf, Australia: the $\delta^{34}S$ data (see Fig. 8) are from Chambers (1982), and the data on reduction rates are from Skyring (1987). Baltic Sea: Vaynshteyn et al. (1986) for both the isotopic and rate data; the *open circles* are for sediments overlain by shallow (< 100 m water depth) and oxic bottom water, the *filled circles* are for sediments overlain by deep (> 300 m) and anoxic water. The Solar Lake, Sinai: isotopic data from Aizenshtat et al. (1984), and the rate data from Jørgensen and Cohen (1977). The *heavy line* represents the best-fit line for a data set presented in Goldhaber and Kaplan (1975)

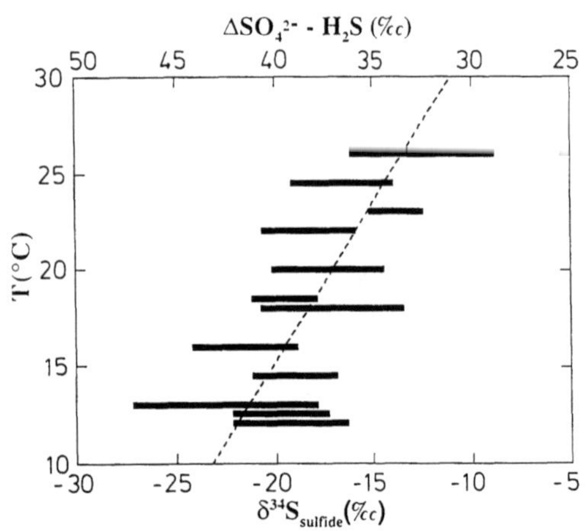

Fig. 8. Relationship between the kinetic isotope effects and the monthly temperatures in the top 10 cm core sediments at Spencer Gulf, Australia: data from Chambers (1982)

natural samples, and is also similar to the experimental data shown in Fig. 6a–b.

An extrapolation of the line connecting the data sets in Fig. 7 suggests that the kinetic isotope effects accompanying bacterial sulfate reduction can be expected to be around 10‰, i.e. the minimum value proposed for an average Archean marine system, when the sulfate reduction rate is around 5–10 $mol\,l^{-1}\,yr^{-1}$. This rate is similar to the one measured by Jørgensen and Cohen (1977) for the upper 0.5 cm of algal mats of the Solar Lake (Sinai), 5.4 $\mu mol\,cm^{-3}\,day^{-1}$ ($\approx 4\,mol\,l^{-1}\,yr^{-1}$), which is about the highest rate of sulfate reduction rate observed on natural systems. (Unfortunately, the kinetic isotope effect associated with the sulfate reduction in this depth zone of the Solar Lake algal mats has not been reported.)

2.5 Possible Causes for Higher Rates of Sulfate Reduction in Archean Oceans

Various parameters that influence the rate of bacterial sulfate reduction, including bacterial species, population, nutrients (such as nitrates and phosphate) and trace metal contents, have been reviewed by many investigators (e.g. Berner 1984, 1985; Postgate 1984; Skyring 1987; Ohmoto et al. 1990). The two most important parameters affecting both directly and indirectly the rate of bacterial sulfate reduction are temperature and the availability of oxidizable organic substrates to sulfate-reducing bacteria; the latter is also linked to temperature (and other parameters) in nature.

2.5.1 Higher Ocean Temperatures

The relationships between the rate of bacterial sulfate reduction and temperature observed in some laboratory experiments using *Desulfovibrio desulfuricans*, the most important sulfate-reducing bacteria in modern oceans, are summarized in Fig. 6c–d. Figure 6 shows that, in solutions containing more than 10 mM SO_4^{2-}, the rate of bacterial sulfate reduction increases by about 2 orders of magnitude as the temperature increases from about 4 °C (i.e. the average ocean temperature of modern oceans) to about 45 °C, accompanied by a decrease in the kinetic isotope effect from about 45‰ (i.e. the average values in modern oceans) to about 10‰ (i.e. the proposed minimum value for the Archean sediments). Excellent correlations among the seasonal temperature, the rate of sulfate reduction, and the magnitude of kinetic isotope effects have also been observed in many modern marine systems (e.g. Chambers 1982; Skyring 1987; Westrich and Berner 1988). These were some of the reasons that led Ohmoto and Felder (1987) to suggest that the temperatures of Archean oceans may have been ~ 30–40 °C.

2.5.2 Higher Availability of Digestible Foods for Sulfate-Reducing Bacteria

The most important parameter affecting the rate of sulfate reduction is the rate of supply of oxidizable organic substrates (i.e. digestable food) available to sulfate-reducing bacteria. Although the rate of sulfate reduction generally correlates positively with the initial organic content, not all organic matter may be oxidizable by sulfate-reducing bacteria (e.g. Skyring 1987). This aspect has been illustrated by Westrich and Berner (1988) using the following reaction for sulfate reduction:

$$2R(CH_2O) + SO_4^{2-} \Rightarrow H_2S + 2HCO_3^- + 2R,$$

where R is the residue of organic matter not metabolized but deposited and preserved in the sediment as "organic carbon". The well-known positive correlation between the amount of pyrite and organic carbon content of normal marine sediments (e.g. Berner 1984, 1985) arises because not all forms of organic matter in normal marine sediments are readily metabolized by sulfate-reducing bacteria. Westrich and Berner (1988) have concluded that about 65% of the plankton from Long Island Sound, and many other nearshore sediments, was rapidly metabolizable by sulfate-reducing bacteria; the nonmetabolized fraction, R, is therefore typically around 35%. However, essentially all the organic matter in cyanobacterial mats is metabolizable by sulfate-reducing bacteria, i.e. $R \approx 0$ (Skyring 1987).

Continuous consumption of metabolizable organic matter by sulfate-reducing bacteria during the burial of sediment results in decreasing the amount of metabolizable organic matter with increasing depth. This is the primary reason for the rapid decrease in the rate of sulfate reduction with depth in most sediments. For example, the rate of sulfate reduction decreases by about one order of magnitude within the first ~ 20 cm of typical muds (e.g. Berner 1984). The rate of sulfate reduction in the algal mat of the Solar Lake decreases by 5 orders of magnitude between the uppermost 0.5 and 20 cm of the sediment section (Jørgensen and Cohen 1977).

Berner (1985) cites a laboratory experiment by Westrich that shows a significant decrease in the rate of sulfate reduction when "aged" dead plankton, rather than "fresh" dead plankton, is used. An implication of this experiment is that a large fraction of metabolizable organic matter will be lost by degradation in environments where it is exposed to oxygenated water for a long duration. This is probably the reason for the well-known positive correlation between the rate of sulfate reduction and the rate of clastic sedimentation (e.g. Goldhaber and Kaplan 1975; Berner 1980, 1984, 1985). Environments with slow rates of clastic sedimentation, such as in deep open oceans, accumulate less metabolizable organic matter, compared to many nearshore environments with rapid rates of sedimentation where less degraded organic matter persists by virtue of rapid burial and insulation from oxidizing seawater.

2.5.3 Lower P_{O_2} and Higher P_{CO_2} in the Atmosphere

In natural systems, the effects of temperature and those of availability of organic matter on the rate of bacterial sulfate reduction may be difficult to separate completely, because temperature could affect the availability of metabolizable organic matter both directly and indirectly many different ways. Ocean temperature and the availability of metabolizable organic matter for sulfate-reducing bacteria may be linked ultimately to the partial pressures of CO_2 and O_2 in the atmosphere (see Fig. 9).

It has been generally accepted among earth scientists that the Archean atmosphere had higher P_{CO_2} and lower P_{O_2} values compared to the Phanerozoic atmosphere (e.g. Holland 1984; Kasting 1987; Kasting and Ackerman 1986). For example, Holland (1984) has suggested that the P_{CO_2} value was $\sim 10^{-2.5}$ atm (vs. the present value of $\sim 10^{-3.7}$ atm) and that the P_{O_2} value was $\sim 10^{-2.5}$ atm (vs. the present value of $10^{-0.7}$ atm) for the atmosphere at ca. 2.5 b.y.B.P.,

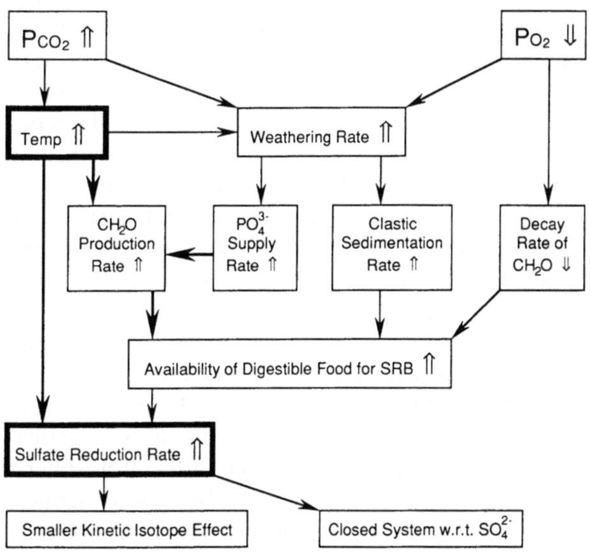

Fig. 9. Relationships among various biochemical parameters affecting the rate of sulfate reduction, and their effects on sulfur isotopic characteristics of sediments

based on the geochemical data from paleosols and the experimental data of Grandstaff (1976) on uraninite dissolution rates. Whether or not the greater greenhouse effect caused by the higher P_{CO_2} value proposed by Holland (1984) could have been sufficient to overcome the sun's lower luminosity and caused higher surface temperatures during Archean time may be debatable (e.g. Owen et al. 1979; Kasting 1987). However, the most plausible mechanism for warmer Archean oceans would be to appeal for higher P_{CO_2} values. The only viable alternative is that the Archean atmosphere was highly reduced and contained appreciable amounts of CH_4 (Kasting pers. comm. 1989), but high CH_4 may not be compatible with sulfate-rich oceans.

The higher P_{CO_2} in the Archean atmosphere would have caused rainwater to be more acidic. The higher surface temperatures could also have enhanced chemical reactions between water and rock. An important consequence of these two effects might have been the higher rate of weathering of continental material, resulting in a higher rate of phosphate supply to the oceans, which is an essential nutrient to photosynthetic organisms, which, in turn, might have increased the production rate of organic matter. The higher rate of sedimentation, resulting from the higher rate of continental weathering, might have contributed to a higher degree of preservation of metabolizable organic matter for sulfate-reducing bacteria. (There is, however, a possibility that the effects on weathering of the higher P_{CO_2} may have been counterbalanced by the smaller continental area).

An effect of the lower P_{O_2} (but with free O_2) in the Archean atmosphere on the activity of sulfate-reducing bacteria is an obvious one. The lesser amounts of oxygen dissolved in oceans would have allowed for plankton and other organic matter to accumulate on the seafloor without much degradation, thus increasing the availability of metabolizable organic matter for sulfate-reducing bacteria.

The above discussion indicates that even if the Archean surface temperatures were similar to the present values, the other effects of higher P_{CO_2} and lower P_{O_2} in the Archean atmosphere would have created marine environments where sulfate-reducing bacteria could have been more active than the present-day settings. The presence of $\leqq \sim 1$ mM sulfate in Archean oceans, as suggested by previous investigators, does not require free oxygen in the atmosphere: this amount can be produced by photochemical oxidation of volcanic SO_2 under anoxic condition (e.g. Walker and Brimblecombe 1985), or by hydration of SO_2 ($4SO_2 + 4H_2O \Rightarrow 6H^+ + 3SO_4^{2-} + H_2S$) from submarine volcanism. However, the sulfate content of > 10 mM in Archean oceans, as suggested in my model, would imply that the accumulation of free oxygen in the atmosphere began more than 3.5 billion years ago. In the next section, I will test the validity of this suggestion (i.e. sulfate-rich Archean oceans) on another important type of sulfide-sulfate mineralization in oceans, the volcanogenic massive sulfide deposits.

3 Sulfur Isotopic Characteristics of Volcanogenic Massive Sulfide Deposits

3.1 $\delta^{34}S$ Characteristics of Phanerozoic and Archean VMSDs (General)

Several papers summarizing isotopic and other geochemical data of submarine volcanogenic massive sulfide deposits (VMSDs) have been published (Ohmoto and Rye 1979; Franklin et al. 1981; Ohmoto and Skinner 1983; Ohmoto 1986). This group of deposits is characterized by massive Fe–Cu–Zn–Pb sulfide ores ($\pm$ massive sulfate ores) associated with submarine volcanic rocks. They appear to have formed on and just below the seafloor by submarine hot springs at temperatures of $\sim 150 - 350\,^{\circ}C$. Modern and recent analogues of these deposits include the sulfide mineralization of the Red Sea and the East Pacific Rise. The δD and $\delta^{18}O$ characteristics of young VMSDs suggest that the hydrothermal fluids were almost entirely seawater that was heated and modified during deep (> 1 km) circulation through hot volcanic rocks (Ohmoto 1986).

Phanerozoic VMSDs exhibit the following sulfur isotopic characteristics (Ohmoto 1986): (1) the maximum and the mean $\delta^{34}S$ values for the sulfide minerals are positive, falling between 0‰ and those of contemporaneous seawater (e.g. $+20$‰ for the modern seawater); (2) the $\delta^{34}S$ values of sulfate minerals (anhydrite and barite), when they occur, are similar to or higher than those of contemporaneous seawater; (3) coexisting sulfate and sulfide minerals exhibit a non-equilibrium isotopic relationship for the depositional temperatures; typically, the observed $\Delta_{sulfate-sulfide}$ values are smaller than the equilibrium $\Delta_{sulfate-sulfide}$ values at the depositional temperatures. For example, in the modern sulfide-sulfate chimneys on the mid-oceanic ridges, such as those on the East Pacific Rise and on Juan de Fuca Ridge, the $\delta^{34}S$ value ranges typically between $+1.5$ and $+7$‰ for sulfide minerals (see Bluth et al. 1988 for a summary) and around $+20.5$‰ for anhydrite (Kusakabe et al. 1982); the observed $\Delta_{sulfate-sulfide}$ values of 13–19‰ are

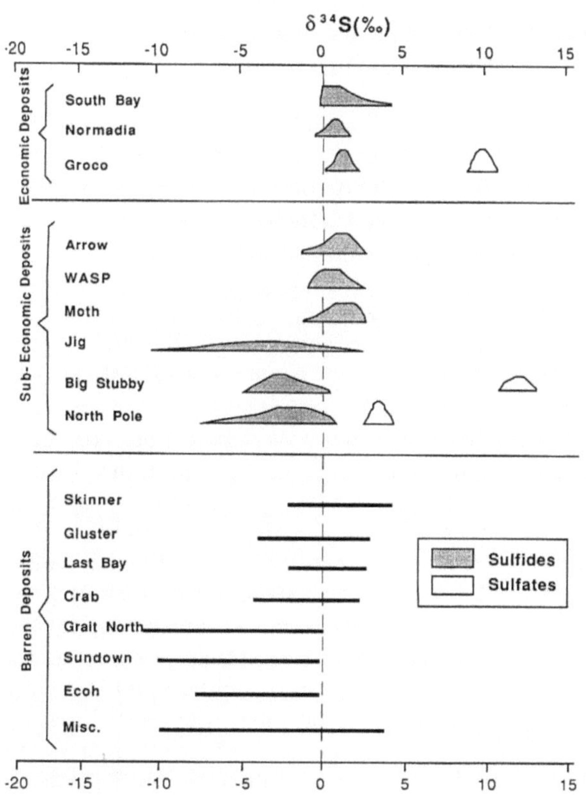

Fig. 10. Comparison of the $\delta^{34}S$ characteristics and the economic importance (i.e. size) of volcanogenic massive sulfide deposits. The *solid bars* for the barren deposits are for sulfides. Data are from Seccombe (1977); Lambert et al. (1978); Franklin et al. (1981)

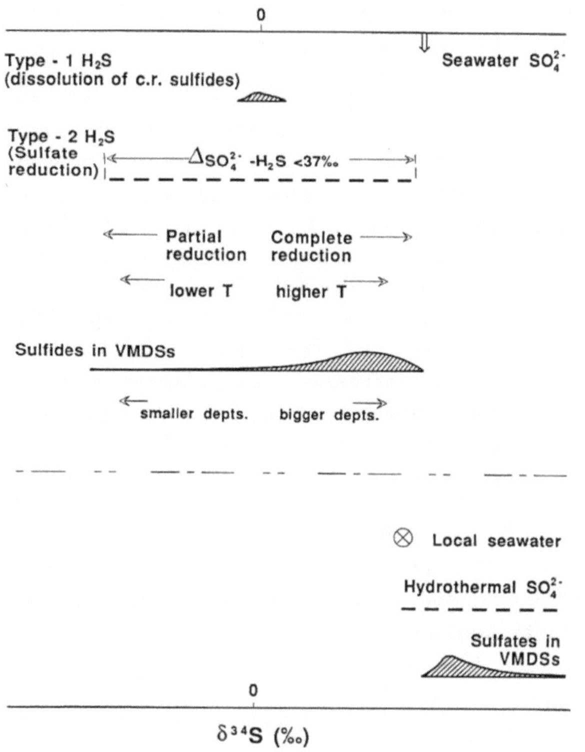

Fig. 11. Schematic representation of the sources of H_2S in submarine hydrothermal fluids responsible for volcanogenic massive sulfide deposits (*above*), and the sources of SO_4^{2-} in sulfate minerals of VMSDs (*below*)

generally smaller than the equilibrium $\Delta_{SO_4-H_2S}$ values of 17‰ (350 °C) to 37‰ (150 °C). In the Miocene Kuroko deposits in northern Japan, the $\delta^{34}S$ values range from −6 to +8‰ with a mean of about +5‰ for the sulfides, and from +19 to +28‰ with a mean of about +22‰; the average $\Delta_{sulfate-sulfide}$ value of 17‰ is much smaller than the equilibrium $\Delta_{SO_4-H_2S}$ value of about 24‰ at the average depositional temperature of 250 °C (Ohmoto et al. 1983).

If the $\delta^{34}S$ value of Archean seawater SO_4^{2-} was around +2‰, as suggested in the preceding section, the Archean VMSDs can be considered to possess the same three types of sulfur isotopic characteristics of the Phanerozoic counterparts (see Figs. 10 and 11). That is, (1) the maximum and the mean $\delta^{34}S$ values for the sulfides fall between 0 and those of contemporaneous seawater (~2‰); (2) the $\delta^{34}S$ values for the sulfate minerals (> 3‰) are similar to or slightly higher than those of contemporaneous seawater; and (3) coexisting sulfate and sulfide minerals exhibit nonequilibrium isotopic relationships with the typical $\Delta_{sulfate-sulfide}$ values of around 5‰. These similarities between VMSDs of Phanerozoic and Archean ages suggest that the formation processes of VMSDs, including the sources for the sulfide and sulfate sulfur in these deposits, have been the same since Archean. The processes that may produce the above three isotopic characteristics of VMSDs of both Phanerozoic and Archean ages are discussed below. The processes discussed below are basically the same as those proposed for the Kuroko deposits and other Phanerozoic VMSDs by Ohmoto et al. (1983).

3.2 The Sources for the Sulfide-Sulfur in Archean VMSDs

The $\delta^{34}S$ characteristics of sulfide minerals in VMSDs suggest the presence of two types of H_2S in the hydrothermal fluids (see Fig. 11). Type-1 H_2S comes from the dissolution of sulfide-sulfur (mostly pyrite and pyrrhotite) in the footwall rocks by the ore-forming fluids (i.e. heated seawater). Type-2 H_2S was generated by high temperature reduction of seawater sulfate in the deeper part of the plumbing system.

Seawater SO_4^{2-} was first fixed in rocks as disseminated anhydrite which, in turn, was reduced by ferrous iron and organic carbon in the rocks.

The $\delta^{34}S$ values of type-1 H_2S generally fall around 0‰, because the $\delta^{34}S$ values of sulfides in submarine volcanic rocks cluster around 0‰, and because the isotopic fractionation during leaching of sulfides is negligible (Ohmoto and Rye 1979). In contrast, the $\delta^{34}S$ values of type-2 H_2S can range from about 37‰ (i.e. equilibrium fractionation factor between SO_4^{2-} and H_2S at 150 °C) to 0‰ lower than the $\delta^{34}S$ values of contemporaneous seawater, depending on the temperature and extent of sulfate reduction at depth. If the sulfate reduction takes place at relatively lower temperatures (ca. 150 °C) and is only partially complete, the $\delta^{34}S$ value of type-2 H_2S will tend to become much less than the $\delta^{34}S$ of the seawater; complete sulfate reduction will result in the $\delta^{34}S$ of type-2 H_2S being the same as that of seawater (see Fig. 11). (An important difference between bacterial and nonbacterial reduction of sulfate is that nonbacterial reduction of sulfate is difficult at temperatures below ~ 70 °C if the reductants are hydrocarbon compounds, and at temperatures below ~ 200 °C if the reductants are ferrous-bearing inorganic compounds. However, when sulfate reduction does occur at temperatures above ~ 150 °C, isotopic equilibrium may be expected between H_2S and SO_4^{2-} in the system; Ohmoto and Lasaga 1982.)

When the submarine hydrothermal processes operate vigorously (i.e. higher temperature and larger scale plumbing system), the hydrothermal fluids will attain a larger quantity of type-1 H_2S ($\delta^{34}S \approx 0$‰) because of increasing solubility of sulfides at higher temperatures; a larger quantity of type-2 H_2S with $\delta^{34}S$ value similar to the $\delta^{34}S$ of seawater will also be produced. A consequence of such processes will be that the $\delta^{34}S$ values of sulfides in bigger VMSDs may tend to fall between 0 and the $\delta^{34}S$ of seawater, with relatively little variation. On the other hand, when the submarine hydrothermal processes operate less intensively, acquisition of type-2 H_2S, which tends to have negative and variable $\delta^{34}S$ value, as well as of type-1 H_2S (~ 0‰), may cause the $\delta^{34}S$ values for the smaller VMSDs to become more variable, with a tendency toward negative values. In fact, such a correlation between the $\delta^{34}S$ variation and the size of Archean VMSDs has been recognized by Seccombe (1977; see Fig. 10), although he did not offer any explanation for the correlation. The existence of such a correlation supports the idea that the Archean VMSDs contain sulfide-sulfur derived from high temperature sulfate reduction, and that the Archean seawater was already sulfate rich.

3.3. The Sources for the Sulfate-Sulfur in Archean VMSDs

The scarcity of sulfate minerals in Archean VMSDs has been used by previous investigators as further evidence for low-sulfate oceans (e.g. Franklin et al. 1981; Lambert and Donnelly 1990). However, the scarcity of sulfate minerals as well as galena in Archean VMSDs can be explained, if the Archean hydrothermal systems were generally more intense than the Phanerozoic ones, due to the higher geothermal gradients in the Archean crust (Ohmoto in prep.).

If Archean seawater was sulfate poor, the only possible sources for H_2S in the submarine hydrothermal fluids would have been sulfides of igneous origin, either type-1 H_2S discussed above or H_2S derived directly from magmas, both of which would have $\delta^{34}S \approx 0$‰. Partial oxidation of such H_2S by Fe^{3+}-bearing minerals, such as magnetite, in the deeper parts of the plumbing systems could produce some SO_4^{2-}, as postulated by Lambert and Donnelly (1990). Such SO_4^{2-} would have equilibrated readily with H_2S; the $\delta^{34}S_{SO_4}$ would have become greater than 15‰ when the $\delta^{34}S_{H_2S}$ was around 0‰ and $T \leqq 400$ °C. However, the $\delta^{34}S$ values of sulfate minerals in Archean VMSDs are all less than 15‰ (see Fig. 10).

It is also unlikely that sulfate in Archean VMSDs was produced by oxidation of magmatic H_2S at the depositional sites, either by sulfide-oxidizing bacteria or by dissolved O_2 in Archean oceans. The O_2 content of Archean oceans would have been too small (< 0.1 mM?) to account for the amount of sulfates in these VMSDs. The kinetic isotope effects of -5.2 ± 1.4‰ accompanying the oxidation of aqueous sulfides by O_2 (Fry et al. 1988) would have resulted in a relation of $\delta^{34}S_{sulfates} < \delta^{34}S_{sulfides}$ instead of the observed relation of $\delta^{34}S_{sulfates} > \delta^{34}S_{sulfides}$.

According to my model of sulfate-rich Archean seawater, the SO_4^{2-} in the hydrothermal fluids (i.e. the residual SO_4^{2-} in the partially reduced high temperature seawater) was probably in equilibrium with H_2S and possessed $\delta^{34}S$ values up to ~ 20‰ higher than that of seawater, before the hydrothermal fluids were discharged to the seafloor. Mixing of such hydrothermal fluids with local sulfate-bearing seawater ($\delta^{34}S$ = ~ 2‰) on or near the seafloor would have produced the observed $\delta^{34}S$ range of sulfate minerals (+2 to 13‰) and the $\Delta_{sulfate-sulfide}$ values (~2 to ~17‰) in the Archean VMSDs.

4 Conclusions

Discussions by all previous investigators on the importance (or unimportance) of sulfate-reducing bacteria (or sulfide-oxidizing bacteria) in ancient oceans have been based on a fundamental assumption that the metabolic response (e.g. the rate of sulfate reduction, the magnitude of the kinetic isotope effect) of sulfate-reducing bacteria (or sulfide-oxidizing bacteria) to various environmental parameters (e.g. temperature, sulfate content, organic content) has been the same during the entire geologic history. In the absence of evidence disproving this assumption, I, too, have adopted the same assumption in this study.

I have suggested in this work that the skewness, range, and mean of $\delta^{34}S$ frequency curves for evaporite sulfates and for marine sulfides of Archean age can be better explained by a model that postulates that Archean oceans, at least 3.5 billion years ago, were already sulfate rich and that sulfides were formed by sulfate-reducing bacteria, rather than by previous models postulating SO_4^{2-} poor oceans and the absence of sulfate-reducing bacteria. Normal seawater probably contained SO_4^{2-} in excess of a few mmol/l (possibly in excess of 10 mmol/l, i.e. about ⅓ of the present value) in quantity with a $\delta^{34}S$ value around +2‰.

The average rate of bacterial sulfate reduction in the Archean oceans could have been around 5–10 mol l^{-1} yr^{-1}, which is about the same as the highest rate of modern sulfate reduction observed today in the cyanobacterial mats of the Solar Lake (Sinai), and is about 100–1000 times higher than the average rate of sulfate reduction in modern oceans. The higher activity of sulfate-reducing bacteria in the Archean ocens, compared to that in modern oceans, could have been caused by higher temperatures and a greater availability of metabolizable organic matter to sulfate-reducing bacteria, which were, in turn, caused by the higher P_{CO_2} and the lower P_{O_2} values of the Archean atmosphere. The Archean oceans could have been up to ~40 °C warmer than the modern oceans, and the Archean atmosphere could have contained appreciable amounts of free oxygen already ~3.5 b.y. ago, although the P_{O_2} value was lower ($\sim 10^{-2.5}$ atm: Holland 1984) than the present-day value of $10^{-0.7}$ atm. My model contrasts with the one proposed by Cloud (1972) and elaborated on by Walker et al. (1983) and Kasting (1987): in these latter two models, the P_{O_2} value was estimated to be $\leqq 10^{-12}$ atm until around 2.4 b.y. ago.

Well-known temporal trends in the $\delta^{34}S$ characteristics of marine sediments, where the $\Delta_{sulfate-sulfide}$ values generally increased, and the skewness for the $\delta^{34}S$ frequency curves for sulfides shifted from a negative to positive direction, in the younger sediments, may indicate a gradual decrease in the rate of sulfate-reducing bacteria in oceans since Archean time. A decrease in the rate of sulfate reduction tends to increase the kinetic isotope effect accompanying bacterial sulfate reduction, and may also change the sulfide-forming systems (e.g. marine sediments) from "the closed-bottom systems with respect to SO_4^{2-}", which produce the negative skewness for the $\delta^{34}S$ frequency curve for the sulfides, to the mixture of "the closed-bottom" and "the open-bottom" systems with respect to SO_4^{2-}, which produce the positive skewness. This transition of the nature of sulfide-forming systems was probably enhanced when large benthic organisms, which may churn the sediments, developed in Phanerozoic time. The decrease in the rate of sulfate reduction in oceans since Archean may have been caused by a gradual decrease in the temperature and the availability of metabolizable organic matter in oceans, which were, in turn, caused by decreasing P_{CO_2} and increasing P_{O_2} in the younger atmosphere.

The sulfur isotope characteristics of Archean volcanogenic massive sulfides can also be better explained if Archean seawater was already sulfate rich, and if the ore-forming fluids developed from sulfate-bearing seawater through high temperature interactions with the underlying country rocks, rather than by magmatic processes. Therefore, the processes of three major types of sulfide-sulfate mineralization in oceans, i.e. formation of sulfate beds by evaporation of sulfate-rich seawater, the formation of disseminated sulfide minerals by sulfate-reducing bacteria, and the formation of massive sulfide ores by high temperature interaction of seawater and rocks, appear to have operated since at least 3.5 billion years ago.

Acknowledgments. Discussions and reviews of the earlier manuscript by S.R. Poulson, J.F. Kasting, L.R. Kump, H.D. Holland, P.A. Trudinger, and E.A. Cameron have contributed greatly to the improvement of this paper. This study was supported by the National Science Foundation under grant Nos. EAR-8508379 and EAR-9003554, and by the Japanese Ministry of Education.

References

Aizenshtat Z, Lipiner G, Cohen Y (1984) Biogeochemistry of carbon and sulfur cycle in the microbial mats of the Solar Lake (Sinai). In: Cohen Y, Castenholz RW, Halvorson HO (eds) Microbial mats: stromatolites. Liss, New York, pp 281–312

Berner RA (1980) Early diagenesis – a theoretical approach. Univ Press, Princeton, 241 pp
Berner RA (1984) Sedimentary pyrite formation: a update. Geochim Cosmochim Acta 48:605–615
Berner RA (1985) Sulfate reduction, organic matter decomposition and pyrite formation. Philos Trans R Soc London Ser A315:605–615
Bluth GJ, Graham UM, Ohmoto H (1988) Sulfide-sulfate chimneys on the East Pacific Rise, 11° and 13° N latitudes, pt 2: sulfur isotopes. Can Mineral 26:505–515
Cameron EM (1982) Sulfate and sulfate reduction in Early Precambrian oceans. Nature (London) 296:145–148
Chambers LA (1982) Sulfur isotope study of a modern intertidal environment, and the interpretation of ancient sulfides. Geochim Cosmochim Acta 46:721–728
Chambers LA, Trudinger PA (1979) Microbiological fractionation of stable sulfur isotopes: a review and critique. Geomicrobiol J 1:249–293
Chambers LA, Trudinger PA, Smith JW, Burns MS (1975) Fractionation of sulfur isotopes by continuous cultures of *Desulfovibrio desulfuricans*. Can J Microbiol 21:1602–1607
Claypool GE, Holser WT, Kaplan IR, Sakai H, Zak I (1980) The age curves of sulfur and oxygen isotopes in marine sulfate and their mutual interpretations. Chem Geol 28:199–260
Cloud PE (1972) A working model of the primitive earth. Am J Sci 272:537–548
Franklin JM, Lydon JW, Sangster DF (1981) Volcanic-associated massive sulfide deposits. Econ Geol 75th Anniv Vol, pp 485–627
Fripp REP, Donelly TH, Lambert IB (1979) Sulphur isotope results for Archaean banded iron formation, Rhodesia. Spec Publ Geol Soc Afr 5:205–298
Fry B, Gest H, Hayes JM (1984) Isotope effects associated with anaerobic oxidation of sulfide by the purple photosynthetic bacterium, *Chromatium vinosum*. FEMS Microbiol Lett 22:282–287
Fry B, Cox J, Gest H, Hayes JM (1986) Discrimination between ^{34}S and ^{32}S during bacterial metabolism of inorganic sulfur compounds. J Bacterial 165:328–330
Fry B, Ruf W, Gest H, Hayes JM (1988) Sulfur isotope effects associated with oxidation of sulfide by O_2 in aqueous solution. Chem Geol 73:205–210
Goldhaber MB, Kaplan IR (1975) Controls and consequences of sulfate reduction rates in recent marine sediments. Soil Sci 119:42–55
Goodwin AM, Monster J, Thode HG (1976) Carbon and sulfur isotope abundances in Archean iron-formations and Early Precambrian life. Econ Geol 71:870–891
Grandstaff DE (1976) A kinetic study of the dissolution of uraninite. Econ Geol 71:1493–1506
Harrison AG, Thode HG (1958) Mechanism of the bacterial reduction of sulphate from isotope fractionation studies. Trans Faraday Soc 54:84–92
Hattori K, Campbell FA, Krouse HR (1985) Sulphur isotope abundances in sedimentary rocks, relevance to the evolution of the Precambrian atmosphere. Geochem Int 22:97–114
Heinrichs TK, Reimer TO (1977) A sedimentary barite deposit from the Archean Fig Tree Group of the Barberton Mountain Land (South Africa). Econ Geol 72:1426–1441
Holland HD (1984) The chemical evolution of the atmosphere and oceans. Univ Press, Princeton, 582 pp
Jørgensen B, Cohen Y (1977) Solar Lake (Sinai). 5. The sulfur cycle of the benthic cyanobacterial mats. Limnol Oceanogr 22:657–666
Kaplan IR, Rittenberg SC (1964) Microbiological fractionation of sulphur isotopes. J Gen Microbiol 34:195–212
Kasting JF (1987) Theoretical constraints on oxygen and carbon dioxide concentrations in the Precambrian atmosphere. J Geophys Res 84:3097–3107
Kasting JF, Ackerman TP (1986) Climatic consequences of very high carbon dioxide levels in the Earth's early atmosphere. Science 234:1383–1385
Kemp AL, Thode HG (1968) The mechanism of the bacterial reduction of sulphate and of sulphite from isotope fractionation studies. Geochim Cosmochim Acta 32:71–91
Kusakabe M, Chiba H, Ohmoto H (1982) Stable isotopes and fluid inclusion study of anhydrite from the East Pacific Rise at 21°N. Geochem J 16:89–95
Lambert IB, Donnelly TH (1990) The paleoenvironmental significance of trends in sulphur isotope compositions in the Precambrian: a critical review. In: Herbert HK, Ho SE (eds) Stable isotopes and fluid processes in mineralisation. Univ West Aust Publ 23:260–268
Lambert IB, Dunlop JSR, Groves DI (1978) Stable isotopic compositions of Early Archaean sulfate deposits of probable evaporitic and volcanogenic origins. Nature (London) 276:806–810
Lowe DR (1980) Archean sedimentation. Annu Rev Earth Planet Sci 8:145–167
Lowe DR, Knauth PR (1977) Sedimentology of the Onverwacht Group (3.4 billion years), Transvaal, South Africa, and its bearing on the characteristics and evolution of the early earth. J Geol 85:699–723
Monster J, Appel PWU, Thode HG, Schidlowski M, Carmichael CM, Bridgewater D (1979) Sulfur isotope studies in Early Archaean sediments from Isua, West Greenland: implications for the antiquaty of bacterial sulfate reduction. Geochim Cosmochim Acta 43:405–413
Ohmoto H (1986) Stable isotope geochemistry of ore deposits. In: Valley JE, Taylor HP Jr, O'Neil JR (eds) Stable isotopes in high temperature geological processes. Rev Mineral 16:491–560
Ohmoto H (in preparation) Formation of volcanogenic massive sulfide deposits
Ohmoto H, Felder RP (1987) Bacterial activity in the warmer, sulphate-bearing, Archaean oceans. Nature (London) 328:244–246
Ohmoto H, Lasaga AC (1982) Kinetics of reactions between aqueous sulfates and sulfides in hydrothermal systems. Geochim Cosmochim Acta 46:1727–1745
Ohmoto H, Rye RO (1979) Isotopes of sulfur and carbon. In: Barnes HL (ed) Geochemistry of hydrothermal ore deposits, 2nd edn. John Wiley & Sons, New York, pp 509–567
Ohmoto H, Skinner BJ (eds) (1983) The Kuroko and related volcanogenic related massive sulfide deposits. Econ Geol Monogr 5, 604 pp
Ohmoto H, Mizukami M, Drummond SE, Eldridge CS, Pisutha-Arnond V, Lenagh TC (1983) Chemical processes of Kuroko formation. Econ Geol Monogr 5:570–604
Ohmoto H, Kaiser CJ, Geer KA (1990) Systematics of sulphur isotopes in recent marine sediments and ancient sediment-hosted basemetal deposits. In: Herbert HK,

Ho SE (eds) Stable isotopes and fluid processes in mineralisation. Univ West Aust Publ 23:70–120

Ohmoto H, Kurai H, Hayashi K, Lowe DR (in preparation) Sulfur isotope study of barite beds in the Fig Tree Formation, Barberton district, South Africa

Owen T, Cess RD, Ramanathan V (1979) Enhanced CO_2 greenhouse to compensate for reduced solar luminosity on early Earth. Nature (London) 277:640–642

Perry EC Jr, Monster J, Reimer T (1970) Sulfur isotopes in Swaziland system barites and the evolution of the Earth's atmosphere. Science 171:1015–1016

Postgate JR (1984) The sulfate-reducing bacteria, 2nd edn Univ Press, Cambridge, 208 pp

Reimer TO (1980) Archean sedimentary baryte deposits of the Swaziland Supergroup (Barberton Mountain Land, South Africa). Precambrian Res 12:393–410

Reimer TO (1982) Sulfur isotopes and the derivation of detrital barytes of the Archean Fig Tree Group (South Africa). In: Sidorenko AV (ed) Sedimentary geology of highly metamorphosed Precambrian complexes. Nauka, Moscow, pp 63–74

Schwarcz HP, Burnie SW (1973) Influence of sedimentary environments on sulfur isotope ratios in clastic rocks: a review. Mineral Depos 8:264–277

Schidlowski M, Hayes JM, Kaplan IR (1983) Isotopic inferences of ancient biochemistries: carbon, sulfur, hydrogen, and nitrogen. In: Schopf JW (ed) Earth's earliest biosphere. Univ Press, Princeton, pp 149–186

Seccombe PK (1977) Sulphur isotope and trace metal composition of straitiform sulphides as an ore guide in the Canadian Shield. J Geochem Explor 8:117–137

Skyring GW (1987) Sulfate reduction in coastal ecosystems. Geomicrobiol J 5:295–374

Skyring GW, Donnelly TH (1982) Precambrian sulfur isotopes and a possible role for sulfide in the evolution of biological sulfate reduction. Precambrian Res 17:41–61

Thode HG, Goodwin AM (1983) Further sulfur and carbon isotope studies of Late Archean iron-formations of the Canadian Shild and the rise of sulfate reducing bacteria. In: Nagy B, Weber R, Guerrero JC, Schidlowski M (eds) Developments and interactions of the Precambrian atmosphere, lithosphere and biosphere. Elsevier, Amsterdam, pp 229–248

Thode HG, Monster J (1965) Sulfur isotope geochemistry of petroleum, evaporites, and ancient seas. Am Assoc Petrol Geol Mem 4:367–377

Vaynshteyn MB, Tokarev VG, Shakola VA, Lein AYu, Ivanov MV (1986) The geochemical activity of sulfate-reducing bacteria in sediments in the western part of the Black Sea. Geochem Int 23:110–112

Vinogradov V, Reimer TO, Leites AM, Smelov SB (1977) The oldest sulfates in the Archean formations of the South African and the Aldan Shields, and the evolution of the Earth's oxygen atmosphere. Lithol Mineral Resourc 11:407–420

Walker JCG, Brimblecombe P (1985) Iron and sulfur in the pre-biologic ocean. Precambrian Res 28:205–222

Walker JCG, Klein C, Schidlowski M, Stevenson DJ, Walter MR (1983) Environmental evolution of the Archean-Early Proterozoic earth. In: Schopf JW (ed) Earth's earliest biosphere. Univ Press, Princeton, pp 260–290

Westrich JT, Berner RA (1988) The effect of temperature on rates of sulfate reduction in marine sediments. Geomicrobiology J 6:99–117

Biological and Abiological Sulfate Reduction in Two Northern Australian Proterozoic Basins

T. H. Donnelly[1] and I. H. Crick[2]

[1] CSIRO Division of Water Resources, PO Box 1666, Canberra ACT, Australia
[2] BMR Division of Continental Geology, PO Box 378, Canberra ACT, Australia

Abstract

Disseminated iron sulfides in carbonaceous fine-grained sediments from the Early Proterozoic Pine Creek Geosyncline and the Middle Proterozoic McArthur Basin of northern Australia have $\delta^{34}S$ values from around − 35 to + 32‰ with most values being positive.

In the Pine Creek Geosyncline, $\delta^{34}S$ values range from around 0 to + 30‰. Pyrite in near-basal sediments have $\delta^{34}S$ values within the magmatic S range of 0 ± 4‰, suggesting that these sulfides did not form as a result of biological sulfate reduction. In formations above these near-basal sediments, $\delta^{34}S$ values range from around 0 to + 30‰. Disseminated and vein sulfides in these overlying formations within contact metamorphic aureoles of granites have magmatic S values. Outside these contact metamorphic aureoles, disseminated sulfides have values up to about + 30‰, indicating that biological reduction took place in an environment which had limited sulfate. However, there are indications that at least some of the pyrite was formed by hydrothermal processes.

In the McArthur Basin, $\delta^{34}S$ values for pyrite range from − 35 to + 32‰ but most values are > + 4‰. In areas of mineralisation, $\delta^{34}S$ values range from around − 5 to + 30‰, and it is considered, because of the large amounts of pyrite in these sediments, that a large proportion of the pyrite formed by hydrothermal processes. In areas away from mineralization, values range from −35 to + 32‰, and it is suggested that whereas biological sulfate reduction played a major role, there are again some indications of addition of hydrothermally formed pyrite.

This study raises the possibility that primary iron sulfides in carbonaceous sediments can be overprinted by hydrothermally formed iron sulfides, particularly in the Early Proterozoic when ocean sulfate levels were probably low, and in mineralized regions where the distribution of geothermal fluids may have been widespread.

1 Introduction

Pyrite in anoxic sediments is formed by the reaction of iron oxides with H_2S produced by sulfate-reducing bacteria using sedimentary organic matter as a carbon source (Berner 1984). In modern marine anoxic terrigenous sediments, which are normally overlain by oxygenated ocean water, the limiting factor in pyrite formation is generally the amount of metabolizable organic matter available to the bacteria. Under these conditions a significant correlation has been found between pyrite S and organic C with a line of best fit through zero and a mean S/C ratio of 0.36. While at times during the Phanerozoic the rate of carbon and sulfur burial in marine anoxic sediments has shown some variation (Berner and Raiswell 1983), S/C ratios in essentially unmetamorphosed ancient sedimentary rocks, deposited under oxic bottom water conditions, are generally similar to those found in the modern ocean (Raiswell and Berner 1986; Donnelly et al. 1988). In contrast, in marine euxinic environments (e.g. the Black Sea) S/C ratios are characteristically high with generally a nonzero intercept on the S axis (Leventhal 1983). This results from the formation of pyrite both within the anoxic sediments and in the anoxic parts of the water column. Pyrite formation in nonmarine anoxic sediments (both modern and ancient) is generally limited by the availability of sulfate and under these conditions S/C ratios are near zero, regardless of organic C content.

In this work we use the above concepts, combined with stable isotope studies, to examine in two northern Australian Proterozoic basins the formation of pyrite in anoxic sedimentary rocks and the environments of their deposition. There is a generalization that when pyrite is found disseminated in anoxic sedimentary rocks, its formation was the result of the activities of sulfate-reducing bacteria. However, we deduce from our examination of the two Australian basins that disseminated pyrite in anoxic sediments was formed by a combination of both biological and abiological processes.

M. Schidlowski et al. (Eds.)
Early Organic Evolution: Implications for Mineral and Energy Resources

2 Geology

The location of the two Proterozoic areas of northern Australia examined, the Early Proterozoic Pine Creek Geosyncline and the Middle Proterozoic McArthur Basin region, are shown in Fig. 1.

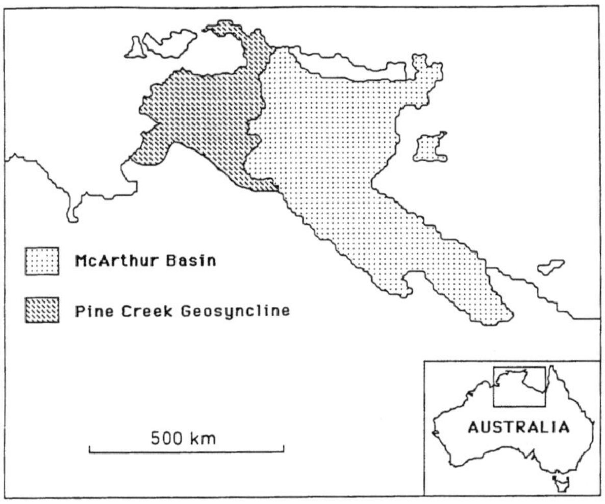

Fig. 1. Location of the Pine Creek Geosyncline and the McArthur Basin in northern Australia

2.1 Pine Creek Geosyncline

The geosyncline is an inlier of Early Proterozoic metasedimentary and volcanic rocks up to 14 km thick, resting on a gneissic and granitic Archean basement and unconformably overlain by relatively undeformed Middle Proterozoic and younger strata. The geology has described been by Needham et al. (1980) and the stratigraphy is summarized in Fig. 2.

The geosynclinal sequence consists mainly of clastic sediments with minor carbonate and felsic to intermediate volcanic rocks. Pelitic rocks predominate and are commonly pyritic and carbonaceous. Basin-wide disconformities to slightly angular unconformities occur beneath the Mount Partridge and South Alligator Groups. Between 1.87 and 1.80 Ga ago the rocks were folded and metamorphosed, mostly to low grade, in places to medium grade, and were intruded by syn- to postorogenic granite plutons. Mafic sills were emplaced prior to regional metamorphism and tectonism (Ferguson and Needham 1978). The age of the Archean granitoid complexes in the western part of the region (Rum Jungle area) is at least 2.4 Ga (Richards et al. 1966, 1977; Page 1976). The age of the South Alligator Group is 1.88 Ga (Page in Bower et al. 1983), whereas the ages of the older Mount Partridge and Namoona Groups are not known but lie between 2.4 and 1.88 Ga.

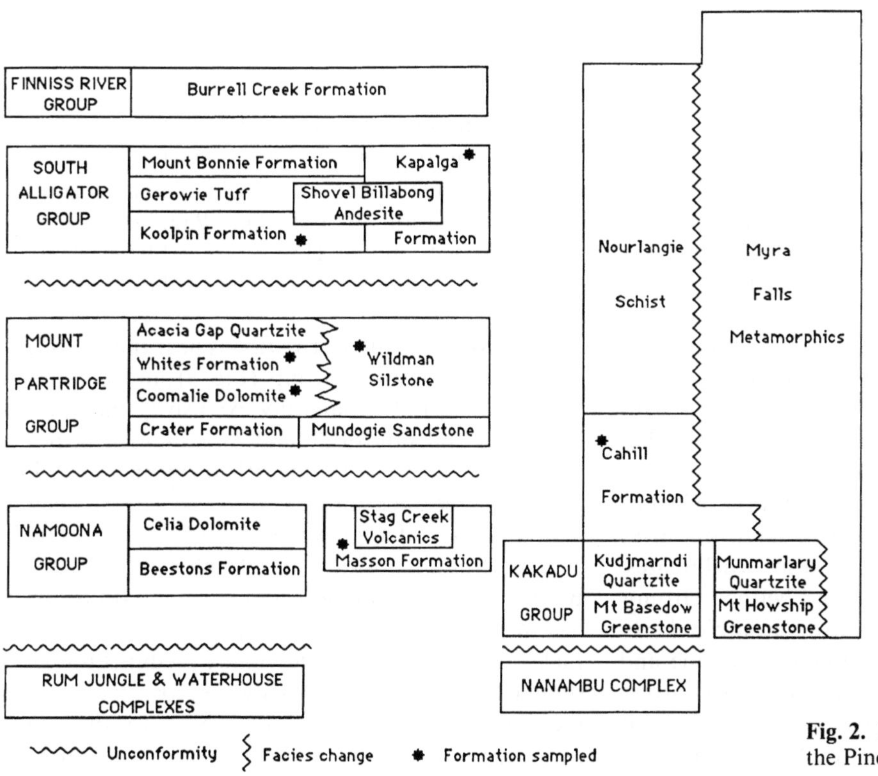

Fig. 2. Early Proterozoic stratigraphy of the Pine Creek Geosyncline

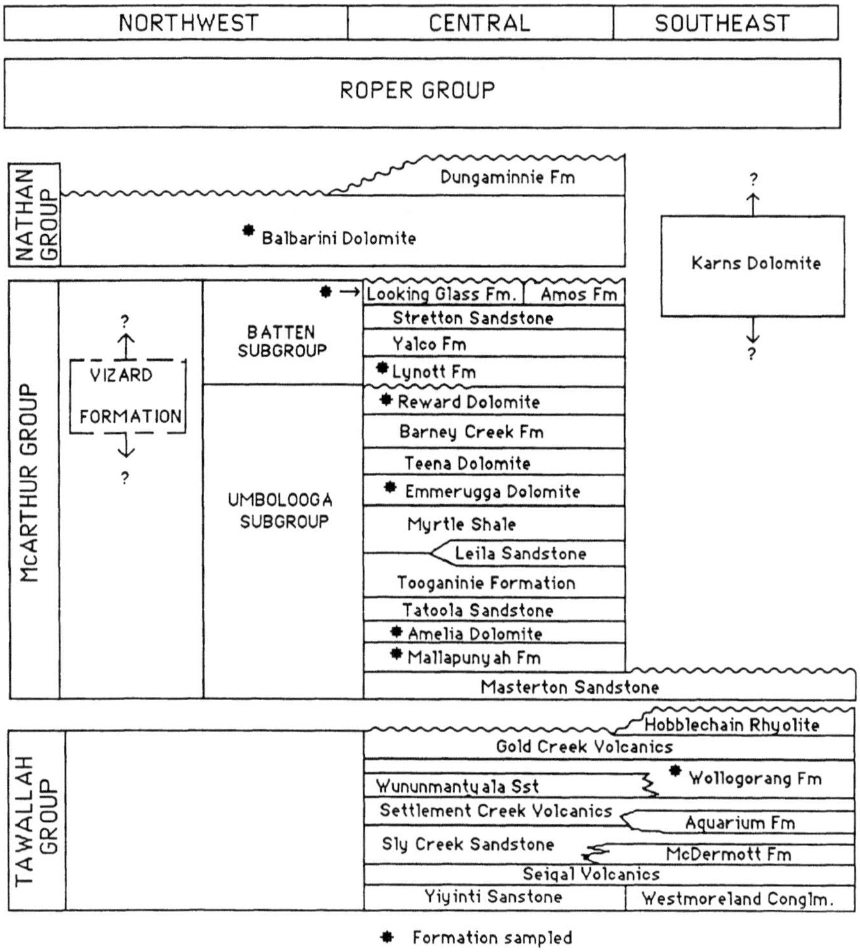

Fig. 3. Middle Proterozoic stratigraphy of the McArthur Basin

2.2 McArthur Basin Region

The geology is described in Jackson et al. (1987a, b) and the stratigraphy is shown in Fig. 3. The basin is located in the northern part of Australia and contains an unmetamorphosed, relatively undeformed, largely sedimentary sequence up to 12 km thick. The older sediments in the basin (Tawallah and McArthur Groups) are significantly different in depositional style and structural setting from the youngest sediments (Nathan and Roper Groups), suggesting that the sequences were deposited in different basins. The Tawallah Group is the oldest group in McArthur Basin and consists largely of quartz sandstone with subordinate volcanics, lutites, and carbonates, whereas the overlying McArthur and Nathan Groups consist mainly of evaporitic and stromatolithic cherty dolostones interbedded with dolomitic siltstone and shale with some thin sandstones. The Roper Group consists of sandstone, siltstone and shale and is intruded by dolerite sills. The McArthur Group is divided into two subgroups, the Umbolooga Subgroup and the overlying Batten Subgroup (Fig. 3), and is approximately 1.6 to 1.7 Ga old. The most precise age determination of 1.690 Ga (Page 1981) comes from tuff beds in the HYC Pyritic Shale Member (Barney Creek Fm.) near the top of the Umbolooga Subgroup.

The Wollogorang Formation is near the top of the underlying Tawallah Group (Fig. 3) and comprises a thick (100–150 m) laterally extensive, fine-grained carbonate and sandstone unit, organic-rich in parts, preserved in an area of some 30 000 km^2 in a sheet-like geometry (see Jackson et al. 1987a). The age of the unit, estimated at approximetaly 1.75 Ga, comes from its stratigraphic position about halfway between the Barney Creek Formation and the McArthur Basin basement, which has been dated at around 1.8 Ga old. Jackson (1981) described many sedimentological similarities between the Wollogorang Formation and the Barney Creek Formation, host to the HYC deposit, the largest of a number of stratiform base metal deposits in this mineralized province. The unit comprises six main lithofacies that were deposited in a continental environment (Donnelly and Jackson 1988). In this study we concentrate on lithofacies 3, a quiet distal (offshore) lacustrine sequence, which consists of black organic-rich pyritic mudstone interlaminated with fine-grained crystalline dololutite. Lithofacies 3 is essentially a carbonate-poor sequence although ovoid dolomite concretions (up to 3 cm diameter) are common in the organic-rich laminates.

3 Geochemistry

3.1 The Pine Creek Geosyncline

A sequence of Early Proterozoic black shales from the middle of the Pine Creek Geosyncline (Fig. 2), dated from > 2.4 to 1.88 Ga age were studied for both their S/C relationships and the sulfur isotope compositions of pyrite (and some pyrrhotite). These sulfides are present mainly in the form of disseminations, but in some cases as coarse-grained and veined iron sulfides. Where possible, extraction of the iron sulfides was carried out using a fine dentist drill. For finely disseminated sulfides, heavy liquids were used to separate a heavy mineral fraction from a crushed rock sample. The heavy mineral fraction was examined under a microscope, and in some cases by X-ray diffraction, to ensure that barite was absent. If more than one sulfide was present, the sulfides were separated under the microscope by hand picking.

3.1.1 Namoona Group

Disseminated pyrite was separated from three black shale samples from three drill cores intersecting the lowermost unit examined in this study, the Masson Formation (Fig. 2). The euhedral pyrite in this unit was fine-grained ($< 10\,\mu$) and intimately associated with organic matter; the black shales are iron-rich (Ewers et al. 1985). If these were Phanerozoic rocks (i.e. deposited in a sulfate-rich ocean), it would be expected that such pyrite would be significantly ^{32}S-rich as a result of the activities of the sulfate-reducing bacteria. The pyrite, however, has $\delta^{34}S$ values (relative to CDT) within the mantle S range of $0 \pm 4‰$. (Fig. 4), similar to that previously found in black shales of the stratigraphically equivalent Cahill Formation (Donnelly and Ferguson 1980) in the eastern Pine Creek Geosyncline. The dominating influence of magmatic S in these sedimentary rocks renders meaningless the interpretation of S/C ratios with respect to the activities of the sulfate-reducing bacteria.

3.1.2 Mount Partridge Group

Disseminated pyrite (some minor pyrrhotite) was separated from black shale samples from three sedimentary units in this Group, the Coomalie Dolomite, Whites Formation and the Wildman Siltstone (Fig. 2). The ^{34}S-enrichment shown by a proportion of the iron sulfides from each of the units examined (up to ca. + 30‰, Fig. 4) indicates the presence of sulfate in these depositional environments. This is because hydrothermal fluids, containing reduced S of mantle origin ($\delta^{34}S_{\Sigma S} = 0 \pm 4‰$) that undergo redox reactions, can form sulfides with significantly negative $\delta^{34}S$ values, but cannot form sulfides with positive $\delta^{34}S$ values > ca. + 4‰ (Ohmoto 1972). Random and positive $\delta^{34}S$ values shown by the iron sulfides from the Mount Partridge Group are typical of those found in low-sulfate lacustrine environments (e.g. the Eocene age Green River Fm., Cole and Picard 1981, and modern lakes, see Ivanov 1983). In such environments the random positive to highly positive $\delta^{34}S$ pyrite values are the result of total sulfate reduction and locally closed system reduction (Rayleigh effect) by sulfate-reducing bacteria. Therefore, while at least some of the iron sulfides in the Mount Partride Group material are biogenic, the sulfides with the more extreme positive $\delta^{34}S$ values are better indicators of the activity of the sulfate-reducing bacteria in this unit. It is important to keep in mind that sulfate reduction by abiological and biological (in low sulfate environments) processes can produce similar positive $\delta^{34}S$ ranges (Ohmoto and Rye 1979), but that abiological processes do not normally produce extreme ^{34}S-enrichments. In contrast, pyrite formed in modern marine black shales and intertidal sediments, where sulfate is usually not a limiting factor, typically has very negative $\delta^{34}S$ values with $\Delta\delta$ values from ca. 30 up to at least 45‰ (Chambers 1982).

The very low S/C ratios from the Whites Formation, which indicate a low-sulfate lacustrine environment (Fig. 5a), are in agreement with the $\delta^{34}S_{pyrite}$ values from this unit. However, S/C ratios from the Coomalie Dolomite and the Wildman Siltstone are very high, suggesting high activity by sulfate-reducing bacteria in environments not limited by either organic matter or sulfate. In contrast the $\delta^{34}S$ results from pyrite in these units suggests that sulfate was also limited in these environments.

3.1.3 South Alligator Group

Disseminated, coarse-grained and veined pyrite and pyrrhotite were separated from black shales of the Koolpin Formation (Fig. 2) that were within contact

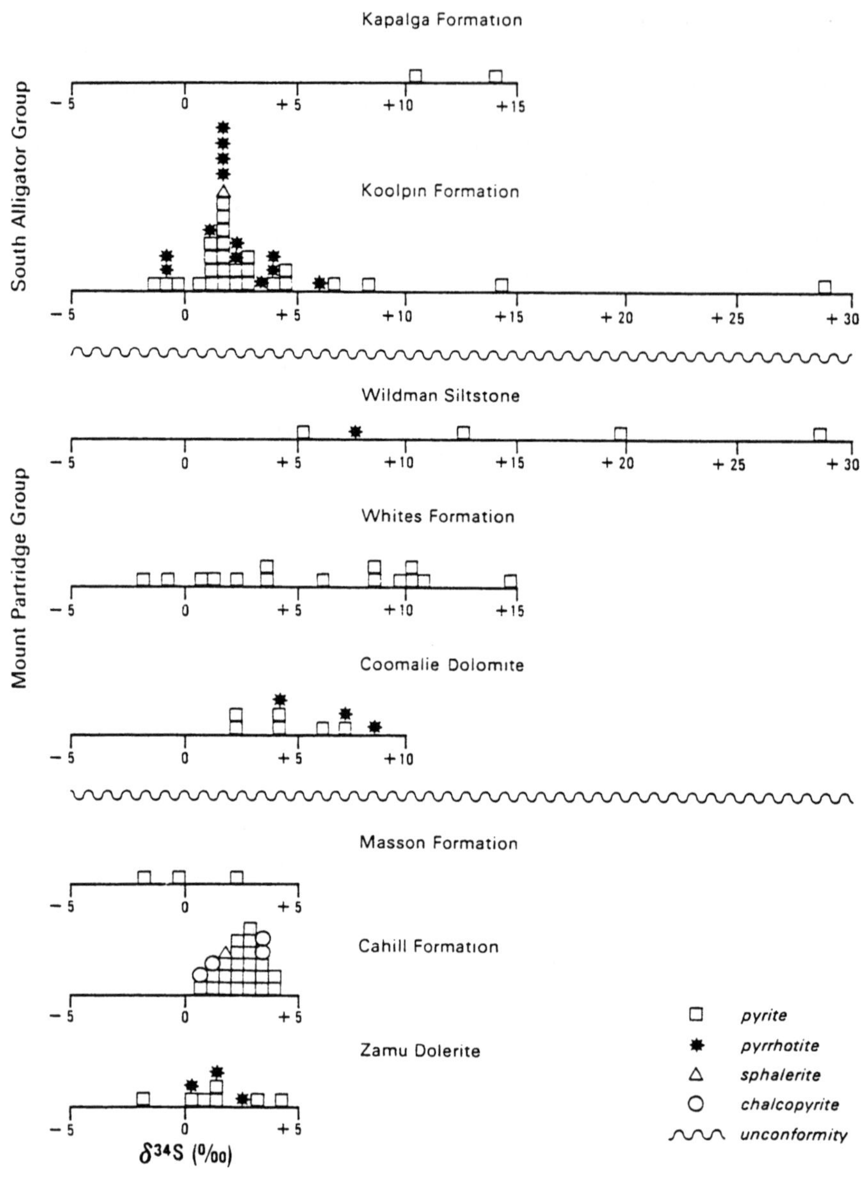

Fig. 4. Sulfur isotopic compositions (relative to CDT) of disseminated iron sulfides in Early Proterozoic carbonaceous sedimentary rocks of the Pine Creek Geosyncline in stratigraphy sequence

metamorphic aureoles of syn- to postorogenic granites. The iron sulfides were present in the black shales in roughly equal proportions. The $\delta^{34}S$ values of these iron sulfides were always within the magmatic S range and averaged ca. + 2‰ (Fig. 4). Disseminated pyrite away from these zones, from the Koolpin and the Kapalga Formations (Fig. 2), has, in general, much higher positive $\delta^{34}S$ values (Fig. 4) similar to those found in the Mount Partridge Group. S/C ratios in these two units are also high, contrasting with their $\delta^{34}S$ values.

In summary, there are indications of the presence of sulfate in the Early Proterozoic depositional environments of the Mount Partridge and South Alligator Groups of the Pine Creek Geosyncline. However, from the black shale sedimentary units in these two groups, sampled away from contact metamorphic zones, only the Whites Formation has S/C ratios in agreement with the $\delta^{34}S$ results, indicating a low-sulfate depositional environment. In a geochemical study of this region, Ewers et al. (1985) commented that the Whites Formation, although locally very enriched in minerals, is, unlike the other sedimentary units they examined, not characterized by regionally high or anomalous metal contents. It is suggested that the abundance of high S/C ratios indicates the widespread distribution of geothermal fluids through this metal-rich region, and the addition to the black shales

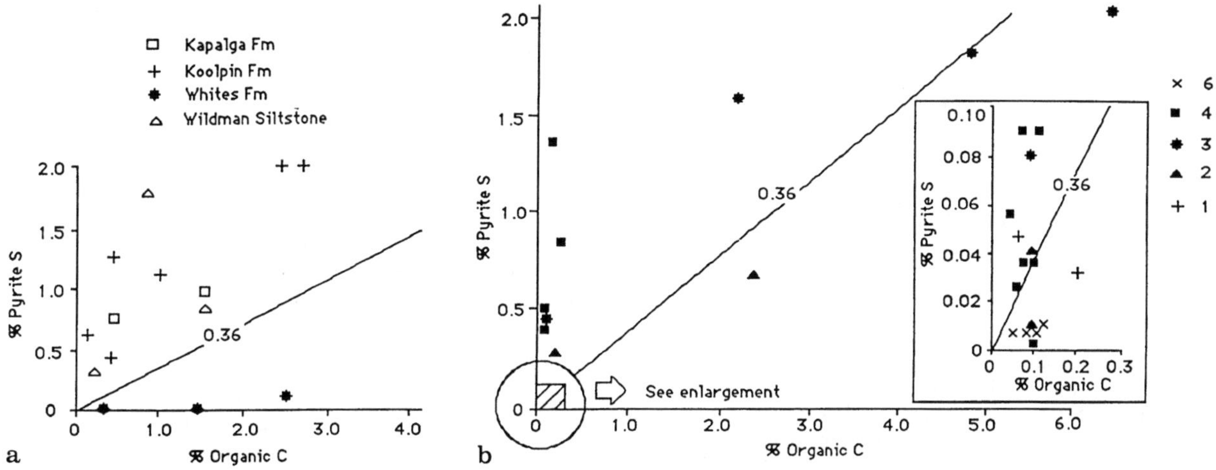

Figs. 5a. Pyrite S/organic C ratios from a sequence of low-grade Early Proterozoic black shales from the Pine Creek Geosyncline. The *0.36 line* is the correlation line for S/C ratios in modern marine anoxic sediments (Berner and Raiswell 1983). **b** Pyrite S/organic C ratios from five of the six main lithofacies (facies sampled shown by different *symbols*) of the essentially unmetamorphosed Wollogorang Formation. Note the unusually high S contents for environments indicated as low-sulfate lacustrine

of hydrothermally formed pyrite with $\delta^{34}S$ values $> +4‰$. In contact metamorphic zones, sulfide from the magmatic intrusion dominates over the country rock sulfide and produces abundant disseminated and vein pyrite and pyrrhotite in the black shales with $\delta^{34}S$ values that average ca. $+2‰$.

3.2 The McArthur Basin Region

A sequence of essentially unaltered Middle Proterozoic organic-rich sedimentary rocks in the McArthur Basin region of northern Australia (Fig. 3), ranging in age from around 1.8 to around 1.6 Ga has been sampled.

3.2.1 Tawallah Group

The Wollogorang Formation in the Tawallah Group (Fig. 3) was the lowermost sedimentary unit examined in this region. In a recent study of this unit, Donnelly and Jackson (1988) demonstrated that the sediments of this formation were deposited in a nonmarine environment, and that one facies (facies 3), consisting of varved carbonate-poor black shales, was deposited in a quiet-water distal (offshore) lake setting. In this study we use data from Donnelly and Jackson (1988) relevant to our discussion of biological versus abiological sulfate reduction and compare the results obtained from the Wollogorang Formation with results from younger sedimentary sequences of the McArthur Basin region and the Pine Creek Geosyncline.

Facies 3 of the Wollogorang Formation contains early diagenetically formed carbonate concretions which are often sites of high pyrite concentrations (Donnelly and Jackson 1988). High pyrite contents can also be found in other parts of the Wollogorang Formation and are visible in outcrop (Jackson 1985). The $\delta^{13}C$ values (relative to PDB) of these early diagenetic carbonate concretions should, in this carbonate-poor environment, be a very sensitive indicator of any extrinsic source of CO_2 in this environment, if the $\delta^{13}C$ value of the introduced CO_2 is significantly different from the value for carbonate in the host sediments. If the activities of sulfate-reducing bacteria were responsible for the relatively high pyrite contents observed, they should provide a CO_2 source with a $\delta^{13}C$ value of around $-28‰$ (Irwin et al. 1977). However, the fact that the concretions have $\delta^{13}C$ values similar to the carbonate of the host sediments ($-0.6‰$) suggests little activity by sulfate-reducing bacteria. Strontium isotope and sedimentological studies of the Wollogorang Formation (Donnelly and Jackson 1988) indicate that this unit was deposited in a continental environment, and evidence for a low-sulfate lacustrine environment comes from the predominantly positive disseminated pyrite $\delta^{34}S$ values (Fig. 6a) from three of the facies (including facies 3). Although such a $\delta^{34}S$ range is typical of pyrite formed in a low-sulfate lacustrine environment, it is also typical of pyrite formed as a result of sulfate reduction in hydrothermal fluids (Ohmoto and Rye 1979). The S/C ratios from five facies (including facies 3) of the Wollogorang Formation (Fig. 5b) examined by Don-

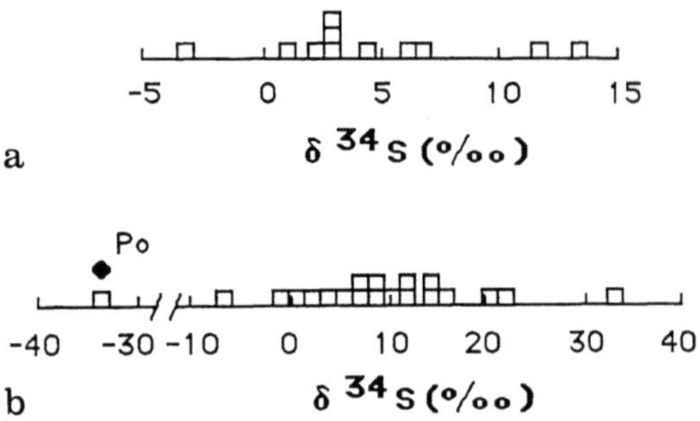

Figs. 6a. $\delta^{34}S$ values of disseminated and coarsely crystalline pyrite from three of the main lithofacies (including facies 3) of the Wollogorang Formation (Donnelly and Jackson 1988). **b** $\delta^{34}S$ values of disseminated pyrite in unmineralized black shales, away from ore deposites, in some McArthur Basin sequences (Muir et al. 1985)

nelly and Jackson (1988) are too high (euxinic values) to be reflecting the low-sulfate lacustrine depositional environment proposed from several strands of evidence for the Wollogorang Formation. We suggest that the high pyrite S values are the result of some addition of hydrothermally formed pyrite S to these units. In this context it is interesting that anomalously high contents of metals in these sediments led Jackson (1981) to suggest that these rocks, like the younger highly mineralized rocks of the McArthur Basin region, may be prospective for base metal mineralization.

3.2.2 The McArthur and Nathan Groups

As part of a study of base metal mineralization in a region which contains a number of giant stratiform lead-zinc deposits of approximately 1.6 Ga age sedimentary rocks (Williams 1980), Muir et al. (1985) examined the S-isotopic compositions of disseminated pyrite in organic-rich sedimentary rocks that were either host to a number of the lead-zinc deposits, or away from mineralized areas.

Early stable isotope studies on some of the very large base metal deposits of this region had assumed that the pyrite in the organic-rich rocks hosting the deposits formed mainly as a result of the activities of sulfate-reducing bacteria (e.g. Smith and Croxford 1973, 1975). These studies ignored (1) the results of an earlier isotope study (Anger et al. 1966) of the somewhat similar Devonian Pd-Zn Rammelsberg ore deposit in Germany, for which it was suggested that a large proportion of the host rock pyrite had a hydrothermal origin; (2) the quantity of pyrite in the host rocks which in the case of the HYC McArthur River deposit vastly exceeds the large amount of lead-zinc ore (Williams 1980); and (3) the similar $\delta^{34}S$ distributions from the pyrite and the lead-zinc sulfide ore.

Middle Proterozoic ocean $\delta^{34}S_{sulfate}$ values have been suggested to be ca. +20 to +25‰ (see Muir et al. 1985), that is, similar to modern ocean values of ca. +21‰. Therefore, if it is assumed that all pyrite in the host rocks of these giant base metal deposits is biogenic, then their predominantly positive $\delta^{34}S$ ranges (up to +45‰; Eldridge et al. 1988) is more indicative of the ranges found for low contents of pyrite in low-sulfate lacustrine environments. Based on sedimentological studies, lacustrine environments have now been proposed for Mt. Isa (Neudert und Russell 1982), the HYC McArthur River deposit (Logan and Williams 1984) and Dugald River (Connor et al. 1982), and a number of authors now suggest that a significant proportion of the pyrite in the host rocks of deposits such as the HYC, Mt. Isa, and the Lady Loretta formed from hydrothermal fluids (Rye and Williams 1981; Muir et al. 1985; Carr and Andrew 1986). In a recent abstract, however, Eldridge et al. (1988) suggest that all pyrite in the host rocks of the HYC deposit formed as a result of the activities of sulfate-reducing bacteria. Although theoretically such enormous accumulations of pyrite as found in the host rocks of the HYC deposit could form from the activities of the sulfate-reducing bacteria (Trudinger et al. 1972), Trudinger (1981) later notes that the relatively low metal sulfide accumulations in modern marine environments, which are analogues of the saline lakes proposed for the ore bodies' host rocks (Williams and Logan 1981), suggest that only low-grade Kupferschiefer-type deposits could be generated biogenically. While the black shale lacustrine sequences of the Barney Creek Formation, host to the HYC deposit, are generally of very low metamorphic grade, levels of maturation of organic matter associated with the lead-zinc deposit are anomalously high (Crick 1989). These maturation levels are in agreement with the inferred involvement of a relatively long-lasting and high temperature hydrothermal fluid in the formation of the base metal sulfides in the HYC deposit (Rye and Williams 1981). Syndepositional formation of abiogenic disseminated pyrite, as a result of some input of hydrothermal fluids into continental depositional environments, would have produced only locally enhanced metamorphic grades close to vent sites. In this regard Muir et al. (1985) found that the relatively low average $\delta^{18}O$ values (+21 to +22‰) for sedimentary carbonates in this region may be due to a slight general increase in water temperature in the depositional environments.

Following their study of mineralization-associated pyrite, Muir et al. (1985) examined disseminated pyrite

$\delta^{34}S$ values in unmineralized black shales (Fig. 6b) from five sedimentary units of the McArthur Basin stratigraphic section (Fig. 3) in an attempt to determine the $\delta^{34}S$ composition of pyrite formed as a result of the activities of sulfate-reducing bacteria in this region. To some degree (depending upon the availability of organic matter and sulfate) pyrite would be expected to have formed in these environments as a result of biological sulfate reduction (Skyring and Donnelly 1982). However, the predominantly positive $\delta^{34}S$ ranges, and often high concentrations of pyrite (and base metals) found in the unmineralized black shales (Lambert 1976; Jackson, pers. comm.) form a pattern similar to that found in mineralized strata in this region, suggesting that a proportion of this pyrite had a hydrothermal origin.

4 Discussion

The examination of disseminated pyrite in organic-rich Middle Proterozoic sediments of the McArthur Basin region has indicated that the formation of some of this pyrite must have involved high temperature reduction of sulfate (e.g. Mottl and Holland 1978) and that abiological sulfate reduction was widespread throughout this region. By the Middle Proterozoic sulfate-reducing bacteria would be expected to be present in anoxic sediments in both marine and nonmarine depositional environments (Skyring and Donnelly 1982). However, the abundance of continental environments now being suggested for the McArthur Group (Jackson et al. 1987a) means that the quantiy of biogenic pyrite formed in these environments (red beds, arenites etc.) would have been generally low. Under these conditions pyrite formed by biological or abiological sulfate reduction will have similar predominantly positive $\delta^{34}S$ ranges (Ohmoto and Rye 1979; Cole and Picard 1981).

The study of disseminated pyrite in Early Proterozoic black shales, through a stratigraphic section in the middle of the Pine Creek Geosyncline, has indicated that pyrite in various parts of the section may have been deposited as a result of four different sets of circumstances and by three different processes. The presence of pyrite in black shales of the Mount Partridge Group with $\delta^{34}S$ values $> +4‰$ indicates that sulfate, and sulfate-reducing bacteria, were present in this Early Proterozoic depositional environment (see Skyring and Donnelly 1982). At least, some of the disseminated pyrite in these Pine Creek sedimentary rocks must have formed as a result of bacterial activities. However, the predominantly positive $\delta^{34}S$ values found for disseminated pyrite in black shales in rocks younger than the Namoona Group (Fig. 2) suggest that most sediments in the Pine Creek Geosyncline were deposited in low-sulfate lacustrine environments. These $\delta^{34}S$ values and evidence of abnormally high pyrite S contents in the black shales (pyrite S/organic C plots; Fig. 5a) suggest there has been some addition of hydrothermally formed pyrite, which was widely dispersed in this mineralized region. A similar result was obtained from our study of the McArthur Basin region. As a result of our examination of these two highly mineralized northern Australian basins, it is suggested that such regions may be characterized by widespread and long-lasting hydrothermal systems and that under occasional favorable circumstances ore deposits will form.

In the Pine Creek Geosyncline, disseminated, coarse-grained and veined pyrite and pyrrhotite in roughly equal amounts were found in black shales in drill core material collected within contact metamorphic aureoles of syn- to postorogenic granites. Sulfide in the fluids associated with the emplacement of the granites dominated over the sedimentary sulfide of the country rock, a process that appears to be typical of these systems (e.g. Shimazaki and Yamamoto 1979; Sun and Kwak 1983). The dominant sulfur source in these Early Proterozoic granites was magmatic, and iron sulfides formed from this source have $\delta^{34}S$ values which averaged ca. $+2‰$.

In the lowermost sedimentary sequences of the Pine Creek Geosyncline, in rocks which could be as old as 2.4 Ga, relatively low concentrations of fine-grained disseminated pyrite were present in black shales with magmatic $\delta^{34}S$ values. This occurred in both areas examined, the middle and eastern areas of the Pine Creek Geosyncline. We are not discussing the presence or absence of ocean sulfate at this time, but we do argue that the dominance of magmatic processes in the Archean (e.g. Veizer et al. 1982) must have provided, under low-sulfate conditions, a dominance of hydrothermally formed pyrite in sedimentary environments with magmatic $\delta^{34}S$ values. Derry and Jacobsen (1988) have shown from Nd and Sr isotope studies that mantle processes dominated seawater chemistry even during the Early Proterozoic and that the Proterozoic was a time of change from older mantle-dominated oceans to oceans significantly influenced, in the Middle Proterozoic, by continental input. As these authors point out, the inferred hot seawater-basalt interactions with vent temperatures only some 10% higher (ca. 385 °C) than present-day values, would have provided sufficient reduced iron to the oceans to form the giant Early Proterozoic banded iron formations. Such a sink of reduced iron would also have had a marked effect on the redox state of the oceans and could have easily taken over the role of the sulfur cycle which, with the carbon cycle buffered atmospheric O_2 levels throughout the Phanerozoic (Veizer et al. 1980). It is, therefore, possible that sulfate contents remained low in Early Proterozoic oceans and that magmatic processes may have continued to dominate pyrite formation in anoxic sedimentary environments, whilst marine evaporites were deposited in whatever restricted near-shore evaporative environments were available at this time.

Acknowledgements. Much of the work on which this paper was based was conducted while one author (THD) was at the Baas Becking Geobiological Laboratory (supported by the Australian Mineral Industry Research Association Ltd., the Bureau of Mineral Resources and the Commonwealth Scientific and Industrial Research Organization). M. D. Muir, T. W. Lyons and P. A. Trudinger are thanked for their very useful comments on a draft manuscript and G. W. Skyring for his helpful discussions on sulfate-reducing bacteria.

References

Anger G, Nielsen H, Puchelt H, Ricke W (1966) Sulfur isotopes in the Rammelsberg ore deposit (Germany). Econ Geol 61:511–536

Berner R A (1984) Sedimentary pyrite formation: an update. Geochim Cosmochim Acta 48:605–615

Berner R A, Raiswell R (1983) Burial of organic and pyrite sulfur in sediments over Phanerozoic time: a new theory. Geochim Cosmochim Acta 47:855–862

Bower M J, Cruikshank B I, Ewers G R, Guy D B, Needham R S, Page R W, Pyke J G, Stuart-Smith PG (1983) Pine Creek Uranium Province. In: BMR Bur Mineral Resourc Aust Yearb 83:51–52

Carr G R, Andrew A S (1986) Preliminary studies of sulfur isotope variations within sedimentary cycles in the Lady Loretta zinc-lead-silver deposit. In: 8th Aust Geol Conv Earth resources in time and space. Geol Soc Aust Abstr 15:41–42

Chambers L A (1982) Sulfur isotope study of a modern intertidal environment, and the interpretation of ancient sulfides. Geochim Cosmochim Acta 46:721–728

Cole R D, Picard M D (1981) Sulfur isotope variations in marginal-lacustrine rocks of Green River Formation, Colorado and Utah. SEPM Spec Publ 31:261–275

Connor A G, Johnson I R, Muir M D (1982) The Dugald River zinc lead deposit, NW, Queensland, Australia. proc Australas Inst Min Metall 283:1–19

Crick I H (1989) Petrological and maturation characteristics of organic matter from the Middle Proterozoic McArthur Basin, Australia. PhD Thesis, Univ, Wollongong

Derry L A, Jacobsen S B (1988) The Nd and Sr isotopic evolution of Proterozoic seawater. Geophys Res Lett 15:397–400

Donnelly T H, Ferguson J (1980) A stable isotope study of three deposits in the Alligator Rivers Uranium Field. In: Ferguson J, Goleby A B (eds) Uranium in the Pine Creek Geosyncline. IAEA, Vienne, Proc Ser, pp 387–406

Donnelly T H, Jackson M J (1988) Sedimentology and geochemistry of a Mid-Proterozoic lacustrine unit from northern Australia. Sediment Geol 58:145–169

Donnelly T H, Shergold J H, Southgate P N (1988) Pyrite and organic matter in normal marine sediments of Middle Cambrian age, southern Georgina Basin, Australia. Geochim Cosmochim Acta 52:259–263

Eldridge C S, Williams N, Compston W, Walshe J L (1988) SHRIMP ion microprobe investigation of the sulfur isotopic composition of the H.Y.C. Pb-Zn deposit, McArthur River, Australia. GSA Abstr 20, 7:A 303

Ewers G R, Needham R S, Stuart-Smith P G, Crick I H (1985) Geochemistry of the low-grade Early Proterozoic sedimentary sequences in the Pine Creek Geosyncline, Northern Territory. Aust J Earth Sci 32:137–154

Ferguson J, Needham R S (1978) The Zamu Dolerite: a Lower proterozoic preorogenic continental tholeitic suite from the Northern Territory, Australia. J Geol Soc Aust 26:309–322

Irwin H, Curtis C D, Coleman M L (1977) Isotopic evidence for source of diagentic carbonates formed during burial of organic rich sediments. Nature (London) 269:209–213

Ivanov M V (1983) The sulfur cycle in lakes and continental reservoirs. In: Ivanov M V, Freeney J R (eds) The global biogeochemical sulfur cycle. Scope 19. John Wiley & Sons, New York London, pp 297–330

Jackson M J (1981) The Wollogorang Formation – a potential host for McArthur River-type base metal deposits? In: Abstr 10th BMR Symp Canberra, ACT. 5–6 May 1981. Bur Mineral Resour. J Aust Geol Geophys 6:279

Jackson M J (1985) Mid-proterozoic dolomitic varves and microcycles from the McArthur Basin, northern Australia. Sediment Geol 44:301–326

Jackson M J, Muir M D, Plumb K A (1987 a) Geology of the southern McArthur Basin. Bur Mineral Resour. Geol Geophys Aust Bull 220:173

Jackson M J, Sweet I P, Powell T G (1987 b) Studies on petroleum geology and geochemistry, Middle Proterozoic McArthur Basin northern Australia I: petroleum potential. Aust petrol Explor Assoc J 1988:283–302

Lambert I B (1976) The McArthur zinc-lead-silver deposit: features, metallogenesis and comparisons with some other stratiform ores. In: Wolf K H (ed) Handbook of strata-bound and stratiform ore deposits 6. Elsevier, Amsterdam, pp 535–585

Leventhal J S (1983) An interpretation of carbon and sulfur relationships in Black Sea sediments as indicators of environments of deposition. Geochim Cosmochim Acta 47:133–137

Logan R G, Williams N (1984) Sedimentary controls on the hydrothermal system that formed the HYC deposit at McArthur River, Northern Territory. In: 7th Aust Geol Conv, Sydney, NSW. Geol Soc Aust Abstr 12:339

Mottl M J, Holland H D (1978) Chemical exchange during hydrothermal alteration of basalt by seawater 1: experimental results for major and minor components of seawater. Geochim Cosmochim Acta 42:1103–1115

Muir M D, Donnelly T H, Wilkins R W T, Armstrong K J (1985) Stable isotope, petrology, and fluid inclusion studies of minor mineral deposits from the McArthur Basin: implications for the genesis of some sediment-hosted base metal mineralization from the Northern Territory. Aust J Earth Sci 32:239–260

Needham R S, Crick I H, Stuart-Smith P G (1980) Regional geology of the Pine Creek Geosyncline. In: Ferguson J, Goleby A B (eds) Uranium in the Pine Creek Geosyncline. IAEA, Vienna, Proc Ser, pp 1–22

Neudert M K, Russell R E (1982) Shallow water and hypersaline features from the Middle Proterozoic Mount Isa sequence, northern Australia. Nature (London) 293:284–288

Ohmoto H (1972) Systematics of sulfur and carbon isotopes in hydrothermal ore deposits. Econ Geol 67:551–579

Ohmoto H, Rye R O (1979) Isotopes of sulfur and carbon. In: Barnes H L (ed) Geochemistry of hydrothermal ore deposits. 2nd edn. John Wiley & Sons, New York, pp 509–567

Page R W (1976) Rum Jungle – Alligator River project. In: Geological branch summary of activities, 1978. Bur Mineral Resourc Austr Rep 196:168

Page R W (1981) Depositional ages of the stratiform base metal deposits at Mount Isa and McArthur River, Australia, based on U-Pb zircon dating of concordant tuff horizons. Econ Geol 76:648–658

Raiswell R, Berner R A (1986) Pyrite and organic matter in Phanerozoic normal marine shales. Geochim Cosmochim Acta 50:1967–1976

Richards J R, Berry H, Rhodes J M (1966) Isotopic and lead-alpha ages of some Australian zircons. J Geol Soc Aust 13:69–96

Richards J R, Ruxton B P, Rhodes J M (1977) isotopic dating of the leucocratic granite, Rum Jungle, Australia. Proc Australas Inst Min Metal 264:33–43

Rye D, Williams N (1981) Studies of base metal sulfide deposits at McArthur River, Northern Territory, Australia, 111. The stable isotope geochemistry of the HYC, Ridge, and Cooley deposits. Econ Geol 76:1–26

Shimazaki H, Yamamoto M (1979) Sulfur isotope ratios of some Japanese skarn deposits. Geochem J 13:261–268

Skyring G W, Donnelly T H (1982) Precambrian sulfur isotopes and a possible role for sulfite in the evolution of biological evolution. Precambrian Res 17:41–61

Smith JW, Croxford NJ (1973) Sulfur isotope ratios in the McArthur lead-zinc-silver deposit. Nature (London) 245:10–12

Smith J W, Croxford N J (1975) An isotopic investigation of the environment of deposition of the McArthur mineralization. Mineral Depos 10:269–276

Sun S-S, Kwak T A (1983) Sulfur isotope study of the Moina skarn deposit. In: Ewers W E (ed) CSIRO Div Mineral Res Rev 1983:123–125

Trudinger P A (1981) Origins of sulfide in sediments. BMR J Aust Geol Geophys 6:279–285

Trudinger P A, Lambert I B L, Skyring G W (1972) Biogenic sulfide ores: a feasibility study, Econ Geol 67:1114–1127

Veizer J, Holser W T, Wilgus C K (1980) Correlation of $^{13}C/^{12}C$ and $^{34}S/^{32}S$ secular variations. Geochim Cosmochim Acta 44:579–587

Veizer J, Compston W, Hoefs J, Nielsen H (1982) Mantle buffering of the early oceans. Naturwissenschaften 69:173–180

Williams N (1980) Precambrian mineralization in the McArthur Cloncurry region with special reference to stratiform lead-zinc deposits. In: Henderson R A, Stephenson P J (eds) The geology and geophysics of northeastern Australia. Geol. Soc. Aust Publ, pp 89–107

Williams N, Logan R G (1981) Depositional environment of the sediments hosting the McArthur River stratiform Pb-Zn deposit. In: Abstr 3, J Geol Soc Aust, 5th Aust Geol Conv, Perth 1981, Sediments through the ages, p 8

Global Oxidation and a Supercontinent in the Proterozoic: Evidence from Stable Isotopic Trends

I. B. Lambert[1] and T. H. Donnelly[2]

[1] Elisian Resources, 2 Bonwick Place, Garran, Canberra, ACT, 2605, Australia
[2] CSIRO, Division of Water Resources, GPO Box 1666, Canberra, ACT, 2601, Australia

Abstract

A significant change in the S-isotope record in the Early Proterozoic is explained by a marked increase in the sulfate content of the hydrosphere around 2.5 to 2.3 b.y. ago. It has been proposed, mainly on the basis of changes in the nature of mineral deposits and sedimentary strata, that an atmosphere with significant levels of oxygen evolved around the same time. These are considered to be related events, which may have resulted from widespread development of environments favorable for the proliferation of photosynthetic organisms. Coupled with declining igneous and hydrothermal activity and consequent decreasing availability of Fe^{2+}, the main oxygen buffer, this would have led to oxygen buildup and surfical oxidation of reduced S species.

Recent results support earlier indications of a predominance of positive $\delta^{34}S$ values for sulfides in Proterozoic carbonaceous strata (and sediment-hosted mineral deposits), although evaporitic sulfate deposits do not provide evidence of unusual ^{34}S enrichment in contemporaneous seawater. These data can be explained in terms of a supercontinent on which there was extensive sedimentation in major intracratonic throughs and platforms. In the many sedimentary basins which had only limited, or no, access to the open ocean, variably ^{34}S-enriched sulfides accumulated by high rates and high degrees of biological sulfate reduction, and also by hydrothermal processes. It is considered that pyrite in Proterozoic open marine strata had predominantly negative $\delta^{34}S$ values, but few such sequences occur in the geological record for this era.

Late Proterozoic carbonates (ca. 900–600 my) from different regions are characterized by mainly positive $\delta^{13}C$ values. These are explained as a result of long-term global oceanic anoxia, which may have been a function of prolonged crustal stability preceding continental dispersion around the Proterozoic/Cambrian boundary. This appears to have been a period of biological diversification leading to the Cambrian "explosion" of organisms, and the overall decrease in oxidized C species would have caused a very significant increase in atmospheric oxygen levels.

The supercontinent model is consistent with the geological evidence for numerous intracratonic troughs and nonmarine basins in the Proterozoic, and with the paleomagnetic data for this era.

1 Introduction

Insights into the evolution of ocean chemistry and sedimentary environments can be obtained from assessments of secular variations in the S-isotopic compositions of sedimentary sulfide and sulfate minerals and the C-isotopic compositions of carbonates.

This work uses trends in the Precambrian S-isotope record to estimate when sulfate became a major part of the exogenic sulfur cycle, and to assess the abundance of sedimentary environments which had limited or no connection with the ocean. The $\delta^{13}C$ trends in Proterozoic carbonates are used to identify periods of abnormally high degrees of preservation or destruction of organic matter. The stable isotope-based conclusions are discussed briefly in the light of geological and paleomagnetic data for the Proterozoic.

These various lines of evidence imply that (1) there was major global oxidation commencing around the Archean-Proterozoic boundary, possibly associated with the widespread development of sedimentary environments favorable for proliferation of photosynthetic organisms; (2) the consequent increase in sulfate concentrations in the hydrosphere resulted in proliferation of sulfate-reducing bacteria; (3) there was a supercontinent from the middle to the end of the Proterozoic on which there was widespread development of restricted, nutrient-rich sedimentary environments; and (4) there were prolonged oceanic anoxic events (OAEs) through the Late Proterozoic.

M. Schidlowski et al. (Eds.)
Early Organic Evolution: Implications for Mineral and Energy Resources

2 Isotopic Background

2.1 S-Isotopes

S-isotope compositions are influenced by the oxidation state because major isotope fractionations characteristically occur between oxidized and reduced S species in biological and hydrothermal systems (see review by Ohmoto and Rye 1979). Therefore, any transition during Earth history from a sulfate-poor hydrosphere to one with significant levels of sulfate should be reflected by a change in the $\delta^{34}S$ ranges of sedimentary and exhalative sulfides from restricted ranges around the mantle [mean Earth] value of approximately 0‰ (relative to CDT) to more variable values (Lambert et al. 1978; Schidlowski et al. 1983; Lambert and Donnelly 1990).

Sulfides formed by bacterial reduction of sulfate in open marine environments typically have wide ranges of negative $\delta^{34}S$ values, reflecting isotopic fractionations ($\Delta\delta$) as large as 65‰. In nearshore environments with high sedimentation rates, abundant nutrients and high rates of sulfate reduction, the $\Delta\delta$ values are commonly in the range of 10 to 25‰. Where sulfate is being used up faster than it is replenished, there is a progressive ^{34}S enrichment in sulfides as biological reduction generates ^{34}S-rich residual sulfate (Rayleigh distillation effect); this can result in sulfides having wide ranges of positive $\delta^{34}S$ values at advanced stages of reduction. Positive $\delta^{34}S$ values are also characteristic of Phanerozoic hydrothermal sulfides formed in volcanic and subvolcanic marine environments by abiological sulfate reduction and rock leaching.

2.2 C-Isotopes

The $\delta^{13}C$ values of marine carbonates (relative to PDB) are commonly close to 0‰ (Veizer and Hoefs 1976; Veizer et al. 1980; Schidlowski et al. 1983). They are considered to reflect photosynthetic activity since the Early Archean (Schidlowski 1987, 1988; Strauss et al. this Vol.), although overall levels of photosynthetic activity may have been lower in the Archean than in younger eras. There are trends away from this mean C-isotopic value at various periods of Earth history, which have been linked to changes in the relative rates of deposition of oxidized and reduced C species. Widespread ^{13}C enrichments in contemporaneous marine carbonates formed in different environments suggest enhanced preservation of (^{12}C-rich) reduced C, as noted for the Cenomanian-Turonian OAE (Schlanger et al. 1987). Conversely, major oxidation of organic carbon generates ^{12}C-enriched bicarbonate which can precipitate as carbonate minerals with distinctly negative $\delta^{13}C$ values.

3 Isotopic Record

3.1 S-Isotopes

Sulfides in Archean sedimentary and igneous rocks exhibit a very marked concentration of $\delta^{34}S$ values in the range of -4 to $+4$‰, implying that the bulk of the hydrosphere did not contain appreciable concentrations of sulfate, and that pyrite formed predominantly from reduced sulfur species of deep-seated origin (Lambert et al. 1978; Lambert and Donnelly 1990). This is consistent with the high levels of igneous and hydrothermal activity and high Fe^{2+} contents in the hydrosphere in this era.

Clearly there were some exceptions to the generally low-sulfate levels through the Archean. As early as 3450 m.y. b.p. sulfate minerals precipitated in certain shallow-water evaporative environments, and this could have been largely a result of local high levels of microbial photosynthesis which oxidized exhalative H_2S (Lambert et al. 1978). In addition, sulfate minerals formed in some Archean hydrothermal mineralization (Lambert et al. 1984; Cameron and Hattori 1985; Lambert and Donnelly 1990), probably as a result of fluid interactions with oxidizing minerals in country rocks.

Wide ranges of $\delta^{34}S$ values in some Archean sulfides, particularly in the Michipicoten and associated iron formations of Canada, have been interpreted as indicating the first appearance of the sulfate-reducing bacteria (Goodwin et al. 1976; Schidlowski et al. 1983). However, there is textural and other evidence that these sulfides, which include abundant vein and massive varieties, are mainly of hydrothermal origin, and it follows that thermochemical redox reactions could have caused the observed isotopic fractionations in these and some other hydrothermal Archean deposits (Lambert et al. 1984; Lambert and Donnelly 1990).

A different model for the Archean (Ohmoto and Felder 1987; Ohmoto et al.) suggests that sulfate levels could have been moderately high throughout the Archean, and that bacterial sulfate reduction was widespread but resulted in only very small isotopic fractionations because of high temperatures in the contemporaneous hydrosphere. This model should be tested by looking for small isotopic fractionations associated with bacterial sulfate reduction in high temperature ($> 50\,°C$) zones of solar ponds.

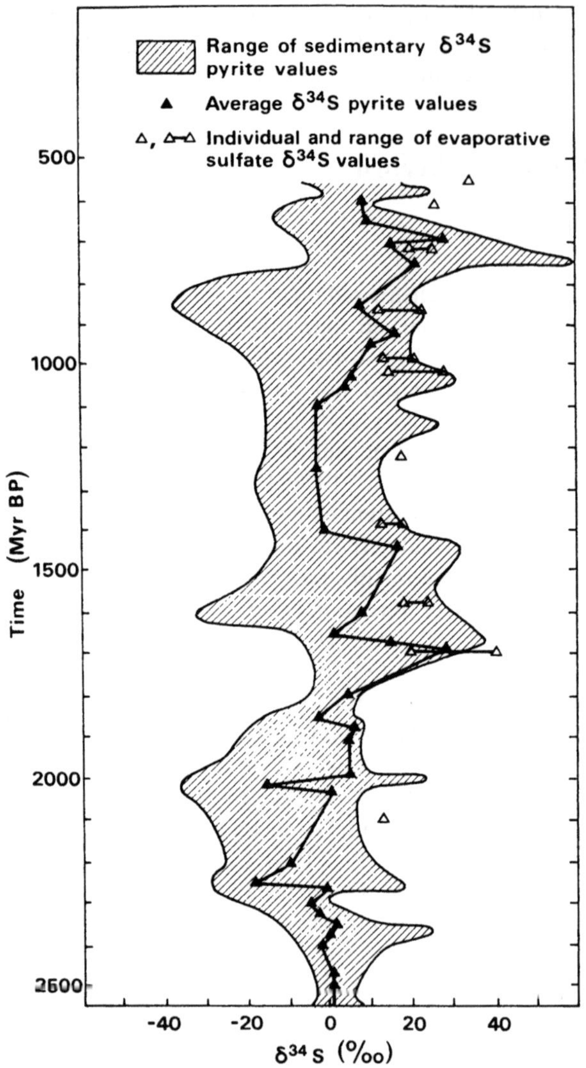

Fig. 1. Sulfur isotopic compositions as ranges and mean values for pyrite and sulfate minerals in Proterozoic sedimentary rocks. Based on data reviewed in Lambert and Donnelly (1990) and Hayes et al. (in press)

The $\delta^{34}S$ values for pyrite from Proterozoic sedimentary strata, and for sulfate from the few available evaporite deposits of Proterozoic age, are summarized in Fig. 1. The sustained change in the S-isotopic compositions in the earliest Proterozoic, to relatively wide $\delta^{34}S$ ranges and mean values commonly well removed from 0‰ (Figs. 1, 2), can be explained by a major increase in the sulfate content of the hydrosphere and the enrichment of this sulfate in ^{34}S as a result of bacterial and hydrothermal reduction processes. Had sulfate concentrations been high in the Archean, as proposed by the Ohmoto and Felder (1987) model, such a marked change in isotopic trends during the Early Proterozoic would have required a more dramatic cooling of the hydrosphere than has been proposed from other lines of evidence (e.g. Knauth and Lowe 1978), or unexpectedly large isotopic fractionations below a threshold temperature reached in the earliest Proterozoic.

Overall, pyrite in Proterozoic black shales exhibits a total $\delta^{34}S$ range of -40 to $+55$‰, but there is a concentration of values between -10 and $+20$‰ (Figs. 1, 2). This contrasts with the strongly negative distribution shown by Chambers (1982) for modern marine sulfides, which have a mean value of ca. -25‰ and a $\Delta\delta$ value of ca. 45‰. There are Early Proterozoic examples of negative $\delta^{34}S$ ranges like those in many younger biogenic sulfides, but negative $\delta^{34}S$ values are subordinate overall, particularly in the Middle and Late Proterozoic record. The sparse $\delta^{34}S$ data from Proterozoic bedded gypsum/anhydrite show values within the common range for younger evaporites, rarely exceeding $+21$‰, and there is a perceptible increase from the Middle through the Late Proterozoic (Figs. 1, 2). However, there are barite occurrences with higher $\delta^{34}S$ values in Middle Proterozoic basins in northern Australia.

The isotopically heavy pyrite may, in part, be reflecting a bias in the sampling. This could result, for example, from the ease of obtaining core from mineralized areas where hydrothermal pyrite may occur for considerable distances from the known ore deposits (see Donnelly and Crick, this Vol.). However, this can only be part of the story as many of the ^{34}S-enriched samples come from unmineralized sequences which do not display any evidence for hydrothermal activity.

3.2 Carbonate-C Isotopes

The broad $\delta^{13}C$ carbonate trends in five time slices from the Archean to the post-Cambrian are related to the $\delta^{34}S$ trends in Fig. 2. With the important exception of the Late Proterozoic, the carbonate $\delta^{13}C$ record for each time slice is dominantly a distribution around zero per mill, as it is for most of the Phanerozoic. Carbonates from differing environments show variably positive $\delta^{13}C$ values, and where measured (e.g. Knoll et al. 1986) constant $\Delta\delta$ values, from just prior to the Proterozoic-Cambrian boundary (Magaritz et al. 1986; Tucker 1986; Aharon et al. 1987; Lambert et al. 1987) to at least 900 m.y. b.p. (Knoll et al. 1986). There is a lack of data from sequences deposited between approximately 900 and 1200 m.y. b.p., when the ^{13}C-enriched Beck Springs Dolomite formed (Zempolich et al. 1988). While ^{13}C-enriched carbo-

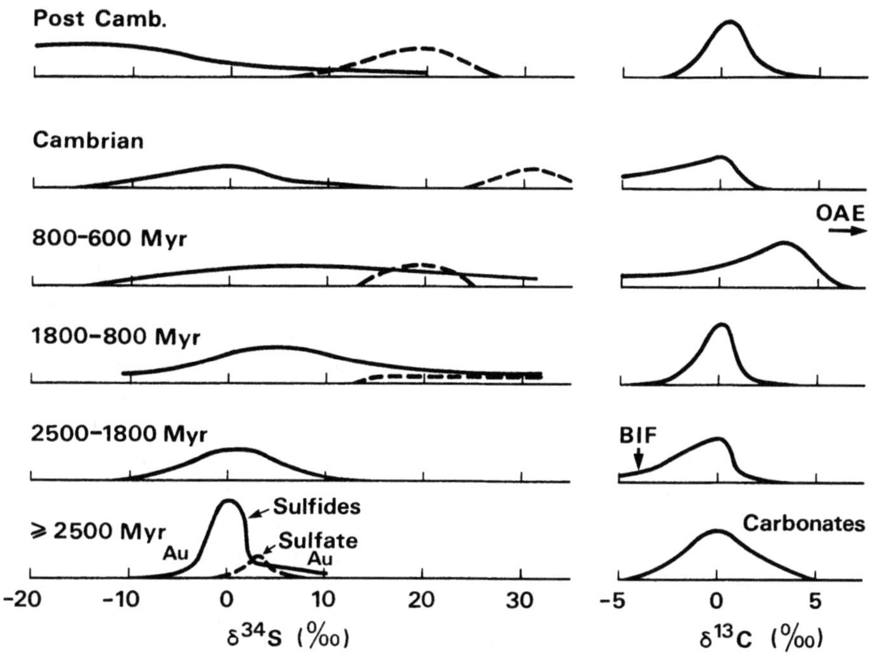

Fig. 2. Secular trends in the isotopic compositions of bedded sulfides, sulfates and carbonates from around > 2500 to < 500 m.y. **B.P.** Proterozoic strata do not display the general inverse correlation between $\delta^{34}S$ values of evaporitic sulfate and $\delta^{13}C$ values of sedimentary carbonate noted by Veizer et al. (1980) in Cambrian and younger strata

nates predominate through the Late Proterozoic, this trend was interrupted by short-lived shifts to normal or negative $\delta^{13}C$ values, at least some of which were associated with glacial events. Also, in the vicinity of the Proterozoic-Cambrian boundary, there is a sharp decrease to negative carbonate $\delta^{13}C$ values, closely followed by values around 0‰ through the Cambrian.

The positive $\delta^{13}C$ values imply that OAEs occurred through much of the Late Proterozoic. The C-isotopic trends for this period are similar to those for OAEs in the Phanerozoic, but persisted for a much longer time than the latter. A much earlier OAE is possible on the basis of the recent documentation of ^{13}C-rich carbonates in an approximately 2000 m.y. old sequence (Baker and Fallick 1989).

A different trend, to quite negative $\delta^{13}C$ values, is seen in the 2500 m.y. old banded iron formations (BIF) of the Hamersley Basin (Baur et al. 1985). This implies the generation of abundant ^{12}C-rich bicarbonate by oxidation of organic matter. Schidlowski (1987) suggested that the occurrence of organic matter with unusually negative $\delta^{13}C$ values in Early Proterozoic BIF was also the result of high levels of oxidation. Again, more data are required from other regions to ascertain whether there is a global scale negative carbonate $\delta^{13}C$ signature resulting from a major period of oxidation in the earliest Proterozoic.

4 Conclusions

The distinct, sustained change from $\delta^{34}S$ values in a narrow range around 0‰ to more variable S-isotopic compositions, which is seen in sedimentary sulfides from approximately 2500–2300 m.y., is most convincingly explained by an increase in the sulfate concentration of the hydrosphere, from generally very low levels. This is considered to have been associated with an increase in atmospheric O_2 levels, as the capacity for oxygen buffering by Fe^{2+} (and, to lesser degrees, organic matter, exhalative H_2S etc.) was exceeded. The fundamental causes may have been the widespread and sustained development of shallow-water, sedimentary environments favorably for proliferation of oxygen-releasing photosynthetic microorganisms, coupled with the generally declining levels of hydrothermal activity which decreased Fe^{2+} availability. It is also concluded that the S-isotopic trends suggest a major increase in the importance of sulfate-reducing bacteria with rising sulfate levels. As discussed by Cloud (1976), Lambert and Groves (1981) and T. C. G. Walker et al. (1983), further support for increased oxidation from around 2500 m.y. b.p. comes from (1) the deposition of numerous BIF in the latest Archean and Early Proterozoic, for which the iron must have been largely introduced in reduced waters; (2) the absence in the Archean of unconformity-related uranium and stratiform copper deposits, which require oxidizing conditions for metal transport to depositional sites; and (3) the widespread deposition of red beds from the Early Proterozoic.

The common mildly to highly positive $\delta^{34}S$ values amongst disseminated sulfides in carbonaceous sedimentary strata, in some cases apparently exceeding the values of contemporaneous seawater sulfate, can be accounted for by the existence of a supercontinent (or a couple of megacontinents) through, at least, the Middle and Late Proterozoic. Numerous troughs and platform environments on the supercontinent would have had limited access to the open ocean, or been completely closed off from it for long periods, and they would have received abundant nutrients, such as phosphate, from weathering of enclosing land. In the early stages of their development, evaporitic sulfates forming in such depositional environments would have had similar isotopic compositions to contemporaneous seawater, and sedimentary pyrite would have been ^{32}S-enriched. It is suggested that in many basins the mean rates of sulfate reduction would have exceeded those of sulfate replenishment, so that the waters of these basins contained sulfate with considerably higher $\delta^{34}S$ values than seawater for geologically long periods. Possible examples are barite in the McArthur Basin sequence (19.5 to 31.5‰; R. N. Walker et al. 1983), and in the Lady Loretta Pb-Zn deposit near Mount Isa (37.4 to 39.4‰; Carr and Smith 1977; Smith et al. 1978); in both of these northern Australian basins, there is widespread evidence for the former existence of calcium sulfate minerals and halite (Muir 1986). Additionally, ^{34}S-rich hydrothermal sulfides could have precipitated at and near the sediment-water interface from fluids which were heated and reduced within the trough sequences (see Donnelly and Crick, this Vol.).

It is considered likely that pyrite in Proterozoic open ocean sediments around the supercontinent had wide ranges of negative $\delta^{34}S$ values, such that the mean $\delta^{34}S_{\Sigma S}$ value (hydrosphere + lithosphere) was maintained close to the mean Earth value of approximately 0‰. Assuming that there is no major sampling bias, the sulfur isotope data imply that open marine strata with 32S-rich pyrite were not commonly preserved or recycled in the Proterozoic. However, the marked lowering of $\delta^{34}S$ values of seawater sulfate in the Ordovician (Claypool et al. 1980) could be reflecting recycling of these Proterozoic marine strata by global tectonic processes.

During prolonged OAEs, deep anoxic waters could have acted as partially closed systems such that the rates of sulfate influx from overlying waters were less than those of sulfate reduction. This would have led to decreasing concentrations and increasing ^{34}S-enrichment of sulfate in the anoxic waters. Abundant continental shelf basins formed as the supercontinent broke up. The trend across the Proterozoic/Cambrian boundary of decreasing $\delta^{34}S$ values for sedimentary pyrite and increasing $\delta^{34}S$ values for seawater sulfate to the highest values recorded in the geological record (Claypool et al. 1980) are considered to be the result of: (1) mixing of ^{34}S-rich waters from marine anoxic and ensialic basin environments, and (2) widespread bacterial reduction under conditions similar to those in modern oceans. The return to $\delta^{13}C$ values of around 0‰ for carbonate strata in the earliest Cambrian are also consistent with homogenization of the hydrosphere during continental dispersion.

The existence of a supercontinent over a long period in the Proterozoic implies global tectonic processes differing in degree, if not style, from those operating in more recent times. Limited rifting preceded the complete breakup of the supercontinent, but a general low level of global tectonic activity is considered to have been a major factor contributing to the existence OAEs through much of the last 300–600 Ma of the Proterozoic, a period some two orders of magnitude longer than for the well-documented younger OAEs such as the Cretaceous Albian-Aptian and the Cenomanian-Turonian events (Schlanger et al. 1987).

The coupling of the exogenic S and C cycles is indicated for the Phanerozoic by the observed inverse relationship between the isotopic compositions of marine sulfates and carbonates (Veizer et al. 1980). The absence of such a relationship in the Proterozoic (Fig. 2) can be accounted for by significant levels of oxygen buffering by Fe^{2+}, together with the restricted ensialic basins and periods of prolonged OAEs. The OAEs in the Late Proterozoic should have resulted in marked increases of oxygen availability (Knoll et al. 1986), largely because less oxygen was being consumed by recycling of organic carbon. This may have been a major factor resulting in the important period of biological evolution culminating with the appearance of numerous macroorganisms around the Proterozoic-Cambrian boundary (Knoll et al. 1986; Brasier in press).

The Proterozoic geological record is in accord with the existence of a supercontinent in that it provides evidence for numerous ensialic troughs and relatively few continental margin environments (e.g. Wyborn et al. 1987; Hoffman 1989). Furthermore, there are many Middle and Late Proterozoic stratiform sediment-hosted and discordant hydrothermal ore deposits which appear to have formed in ensialic rift settings, but very few volcanogenic massive sulfide and porphyry copper deposits (Lambert and Groves 1981; Gustafson and Williams 1983; Meyer 1988; Lambert et al. in press).

Sr- and Nd-isotopic data for chemical sediments can yield information on the relative importance of continental and oceanic (mantle) sources of dissolved constituents in ocean waters, which reflect global scale tectonic processes and/or climates (Veizer et al. 1983; Derry and Jacobsen 1988). Unfortunately, few such data are available for the Proterozoic and some of them are from sequences now regarded as having been deposited in restricted, starved and/or nonmarine basins. The variably low $^{87}Sr/^{86}Sr$ ratios in Late Proterozoic carbonates have been interpreted in terms of major mantle events, as expected in periods of very active ocean-floor spreading, a conclusion at variance with the interpreted global anoxia and relative tectonic stability. Further studies are needed, but it seems feasible that the Sr data are at least partly reflecting marked decreases of continental inputs caused by the major glacial events of this period (see Donnelly et al. 1990), rather than enhanced seawater ocean crust interactions at active mid-ocean ridges. Also, the Sr and Nd isotopic compositions of the Late Proterozoic carbonates which have been studied could be reflecting local factors in closed basin environments.

Finally, paleomagnetic data for the Proterozoic have been interpreted in terms of a supercontinent not greatly different from the Pangaea of the Permian (e.g. Piper 1976; McElhinney and McWilliams 1977). These data appear to preclude tectonic models involving convergence of widely separated (more than roughly 1000 km) continental blocks. Studies of Late Proterozoic glacials indicate that these formed at low latitudes (Embelton and Williams 1986), attesting to the severity of the climate and implying an alignment of continental crust along the equator.

In summary:

1. S-isotopic trends bolster earlier conclusions, based partly on changes in the nature of mineral deposits and sedimentary strata, for a significant oxidation event commencing roughly 2500 m.y. b.p., and
2. stable isotopic trends, geological evidence and paleomagnetic data, combine to provide strong support for the concept of a Proterozoic supercontinent.

Acknowledgements. Much of the research on which this paper is based was conducted while the authors were at the Baas Becking Geobiological Laboratory (supported by the Australian Mineral Industry Research Association Ltd., the Bureau of Mineral Resources and the Commonwealth Scientific and Industrial Research Organization). J. M. Hayes, Shen-Su Sun, P. A. Trudinger and M. R. Walter are thanked for their useful comments on a draft manuscript, as is G. W. Skyring for his advice on the biochemistry of sulfate-reducing bacteria.

References

Aharon P, Schidlowski M, Singh I B (1987) Chronostratigraphic markers in the end-Precambrian carbon isotope record of the Lesser Himalaya. Nature (London) 327:699–701

Baker A J, Fallick A E (1989) Evidence from Lewisian limestones for isotopically heavy carbon in two-thousand-million-year-old sea water. Nature (London) 337:352–354

Baur M E, Hayes J M, Studley S A, Walter M R (1985) Millimeter scale variations of stable isotope abundances in carbonates from banded iron formations in the Hamersley Group of Western Australia. Econ Geol 80:270–282

Brasier M D (in press) Ocean-atmosphere chemistry and evolution across the Precambrian/Cambrian boundary. In: Lipps J H, Signor P W (eds) Origin and early evolution of metazoa. Plenum, New York

Cameron E M, Hattori K (1985) The Hemlo gold deposit, Ontario: a geochemical and isotopic study. Geochim Cosmochim Acta 49:2041–2050

Carr G R, Smith J W (1977) A comparative isotopic study of the Lady Loretta zinc-lead-silver deposit. Mineral Depos 12:105–110

Chambers L A (1982) Sulfur isotope study of a modern intertidal environment and the interpretation of ancient sulfides. Geochim Cosmochim Acta 46:721–728

Claypool G E, Holser W T, Kaplan I R, Sakai H, Zak I (1980) The age curves of sulfur and oxygen isotopes in marine sulfate and their mutual interpretation. Chem Geol 28:199–260

Cloud P E (1976) Beginnings of biospheric evolution and their biogeochemical consequences. Paleobiology 2:351–387

Derry L A, Jacobsen S B (1988) The Nd and Sr isotopic evolution of Proterozoic seawater. Geophys Res Lett 15:397–400

Donnelly T H, Shergold J H, Southgate P N, Barnes C J (1990) Events leading to global phosphogenesis around the Proterozoic/Cambrian boundary. J Geol Soc London Spec Pub 52:273–287

Embelton B, Williams G (1986) Low palaeolatidue of deposition for late Precambrian perriglacial varvites in South Australia: implications for palaeoclimates. Earth Planet Sci Lett 79:419–430

Goodwin A M, Monster J, Thode H G (1976) Carbon and sulphur isotope abundances in Archean iron formations and Early Precambrian life. Econ Geol 71:870–891

Gustafson L B, Williams N (1983) Sediment-hosted stratiform deposits of copper, lead and zinc. In: Skinner B J (ed) Econ Geol 75th Anniv Vol, pp 139–178

Hayes JM, Lambert IB, Strauss H (in press) The sulfur-isotopic record. In: Schopf J W, Klein C (eds) The Proterozoic biosphere: a multidisciplinary study. Univ Press, Cambridge

Hoffman P F (1989) Speculations on Laurentia's first gigayear. Geology 17:135–138

Knauth L P, Lowe D R (1978) Oxygen isotope geochemistry of cherts from the Onverwacht Group (3.4 billion years), Transvaal, South Africa, with implications for secular variations in the isotope compositions of cherts. Earth Planet Sci Lett 41:209–222

Knoll A H, Hayes H M, Kaufman A J, Sweet K, Lambert I B (1986) Secular variation in carbon isotope ratios from Upper Proterozoic successions of Svalbard and East Greenland. Nature (London) 321:832–838

Lambert I B, Donnelly T H (1990) The palaeoenvironmental significance of trends in sulfur isotope compositions in the Precambrian: a critical review. In: Herbert H K, Ho SE (eds) Stable isotopes and fluid processes in mineralisation. Geology Department, Univ of Western Australia, Spec Pub 23:260–268

Lambert I B, Groves D I (1981) Early Earth evolution and metallogeny. In: Wolf K H (ed) Handbook of strata-bound and stratiform ore deposits, vol 8. Elsevier, Amsterdam, pp 339–447

Lambert I B, Donnelly T H, Dunlop J S R, Groves D I (1978) Stable isotopic compositions of Early Archaean sulfate deposits of probable evaporative and volcanogenic origins. Nature (London) 276:808–811

Lambert I B, Phillips G N, Groves D I (1984) Sulfur isotope compositions and genesis of Archaean gold mineralization, Australia and Zimbabwe. In: Foster R P (ed) Gold 82. Balkema, Rotterdam, pp 373–388

Lambert I B, Walter M R, Zang Wenlong Lu Songnian, Ma Guogan (1987) Palaeoenvironment and carbon isotope stratigraphy of Upper Proterozoic carbonates of the Yangtze Platform. Nature (london) 325:140–142

Lambert I B, Beukes N, Klein C, Veizer J (in press) Proterozoic mineral deposits through time. In: Schopf J W, Klein C (eds) The Proterozoic biosphere: a multidisciplinary study. Univ Press, Cambridge

Magaritz M, Holser W T, Kirschvink J (1986) Carbon-isotope events across the Precambrian/Cambrian boundary on the Siberian Platform. Nature (London) 320:258–259

McElhinney M W, McWilliams M O (1977) Precambrian geodynamics; a palaeomagnetic view. Tectonophysics 40:137–15

Meyer C (1988) Ore deposites as guides to geologic history of the Earth. Annu Rev Earth Planet Sci 16:147–171

Muir M D (1986) Facies models for Australian Precambrian evaporites. In: Peryt T M (ed) Lecture notes in earth Science 13. Evaporite basins. Springer, Berlin Heidelberg New York, pp 5–21

Piper J D A (1976) Palaeomagnetic evidence for a Proterozoic super-continent. Philos Trans R Soc London 280:469–490

Ohmoto H, Felder R P (1987) Bacterial activity in the warmer, sulphate-bearing, Archaean oceans. Nature (London) 328:244–246

Ohmoto H, Rye R O (1979) Isotopes of sulfur and carbon. In: Barnes H L (ed) Geochemistry of hydrothermal ore deposits, 2nd edn. Wiley, New York, pp 509–567

Schidlowski M (1987) Application of stable carbon isotopes to early biochemical evolution on earth. Annu Rev Earth Planet Sci 15:47–72

Schidlowski M (1988) A 3,800-million-year old record of life from carbon in sedimentary rocks. Nature (London) 333:313–318

Schidlowski M, Hayes J M, Kaplan I R (1983) Isotopic inferences of ancient biochemistries: carbon, sulfur, hydrogen and nitrogen. In: Schopf J W (ed) The earth's earliest biosphere: its origin and evolution. University Press, Princeton, pp 149–186

Schlanger S O, Arthur M A, Jenkyns H C, Scholle P A (1987) The Cenomanian-Turonian oceanic anoxic event, 1. Stratigraphy and distribution of organic-rich beds and the marine $\delta^{13}C$ excursion. In: Brooks J, Fleet A J (eds) Marine petroleum source rocks. Geol Soc London Spec Publ 26. Blackwell, Oxford, pp 371–399

Smith J W, Burns M S, Croxford N J W (1978) Stable isotope studies of the origins of mineralization at Mount Isa. Mineral Depos 10:369–382

Tucker M M E (1986) Carbon isotope excursions in Precambrian/Cambrian boundary beds. Nature (London) 333:313–318

Veizer J, Hoefs J (1976) The nature of O^{18}/O^{16} and C^{13}/C^{12} secular trends in sedimentary carbonate rocks. Geochim Cosmochim Acta 40:1387–1395

Veizer J, Holser W T, Wilgus C K (1980) Correlations of $^{13}C/^{12}C$ and $^{34}S/^{32}S$ secular variations. Geochim Cosmochim Acta 44:579–587

Veizer J, Compston W, Clauer N, Schidlowski M (1983) $^{87}Sr/^{86}Sr$ in Late Proterozoic carbonates: evidence for a "mantle" event at $\sim$ 900 Ma ago. Geochim Cosmochim Acta 47:295–302

Walker J C G, Klein C, Schidlowski M, Stevenson D J, Walter M R (1983) In: Schopf J W (ed) Earth's earliest biosphere: its origin and evolution. Univ Press, Princeton, pp 260–290

Walker R N, Gulson B, Smith J W (1983) The Coxco deposit – a Proterozoic Mississippi Valley-type deposit in the McArthur River district, Northern Territory, Australia. Econ Geol 78:214–249

Wyborn L A I, Page R W, Parker A J (1987) Geochemical and geochronological signatures in Australian Proterozoic igenous rocks. In: Pharaoh T C, Beckinsale R D, Rickard D (eds) Geochemistry and mineralization of Proterozoic volcanic suites. Geol Soc Spec Publ 33:377–394

Zempolich W G, Wilkinson B H, Lohmann K C (1988) Diagenesis of Late Proterozoic carbonates: the Beck Spring Dolomite of eastern California. J Sediment Petrol 58:656–672

Depositional and Diagenetic C-S-Fe Signatures and the Potential of Shales to Generate Metal-Rich Fluids

R. Raiswell and H. J. Al-Biatty [1]

[1] Department of Earth Sciences, University of Leeds, Leeds LS2 9JT, UK

Abstract

The potential of shales to react with dissolved sulfides introduced during late diagenesis or catagenesis is constrained initially by source area effects (which dictate total iron content) and is subsequently diminished by the removal of iron to form pyrite in early diagenetic sulfate reduction. The efficiency of iron removal during sulfate reduction can be measured by the degree of pyritization (DOP), which represents the ratio of pyrite Fe to pyrite Fe plus HCl-soluble Fe. Acid extraction conditions are chosen such that iron mineral reactivity towards HCl is similar to that towards H_2S during early diagenesis. DOP values so defined demonstrate that the efficiency of reactive iron removal is mainly dependent on depositional environment. Reactive iron removal in marine shales is minimized where deposition occurs from fully oxygenated bottom waters. Here, low DOP values (less than 0.4) result because organic carbon, remaining after benthic activity, is poorly metabolized by sulfate-reducing bacteria, and hence pyrite formation is limited. Still lower DOP values may occur where shales have been rapidly deposited and reworked. In freshwater environments low DOP values also result because the availability of dissolved sulfate limits sulfate reduction and hence also pyrite formation.

The amount of iron reactive towards H_2S during early diagenesis bears a reasonably close relationship to the total iron content, although the latter varies widely depending on source area effects. Acid extraction experiments suggest that a further proportion of the iron which is unreactive towards H_2S during early diagenesis may become so reactive at higher temperatures. However, the suppression of early diagenetic pyrite formation is important because the kinetically most reactive iron phases are consumed, hence the potential for later reaction with dissolved sulfides is significantly modified.

1 Introduction

Recent studies of formation waters (Kharaka et al. 1987) show that these are generally low in metals and high in dissolved sulfide. However, such waters can evolve to become metal-rich and sulfide-poor where there is access to metalliferous sediments such as red beds. Here, organic acids leach out a sufficient volume of metals to precipitate reduced sulfur and hence create metal-rich brines with ore-forming potential. In general, any host rocks which contain Fe minerals capable of sulfidation will also be capable of buffering low levels of dissolved sulfide and hence producing metal-rich fluids. In this contribution we explore the potential of shales to generate fluids with these characteristics, and show how this is modified by depositional and diagenetic processes.

Early diagenetic pyrite is formed as a consequence of microbiological sulfate reduction, by which process H_2S is generated to react with detrital Fe phases in the sediment. The initial products are Fe monosulfides which are then converted to pyrite (Goldhaber and Kaplan 1974). Clearly, Fe which is diagenetically reactive will also be hydrothermally reactive, hence the suppression of early diagnetic pyrite formation will constitute an important ground preparation mechanism by preserving increasing concentrations of Fe which can precipitate dissolved sulfide. Enhanced concentrations of reactive Fe are particularly likely to survive diagenesis in fine-grained Archean sediments because microbially mediated sulfate reduction probably occurred only after 2.7 b.y. (Schidlowski et al. 1983; but see Ohmoto, this Vol.). However, in younger fine-grained sediments the formation and preservation of diagenetic pyrite is also minimized to varying degrees in depositional environments characterized by (1) fully oxygenated bottom waters; (2) low-sulfate concentrations; (3) the presence of little labile organic matter; and (4) extremely rapid sedimentation and/or physical reworking. Furthermore, the initial availability of reactive Fe can itself show considerable intrabasinal or regional variations (Aller et al. 1986;

M. Schidlowski et al. (Eds.)
Early Organic Evolution: Implications for Mineral and Energy Resources

Raiswell and Al-Biatty 1989). The influence of these factors on the potential of ancient sediments to generate metal-rich fluids is evaluated below, following a brief discussion of pyrite formation in modern sediments. There have been important recent reviews of this topic by Berner (1984, 1985), Skyring (1987), and Canfield and Raiswell (1991).

2 Pyrite Formation in Modern Sediments

The process of microbiological sulfate reduction can be represented schematically as

$$2\,CH_2O + SO_4^{2-} \rightarrow 2\,HCO_3^- + H_2S\,. \quad (1)$$

Under anoxic conditions seawater sulfate is reduced to H_2S which reacts with detrital Fe minerals and ultimately forms pyrite (Goldhaber and Kaplan 1974; Berner 1984). The overall process of sedimentary pyrite formation is summarized in Fig. 1. The initial product is not, in fact, pyrite but a series of metastable Fe monosulfides (Rickard 1969; Sweeney and Kaplan 1973) which are readily transformed to pyrite during early diagenesis (by poorly understood processes). Pyrite formation and sulfate reduction are ubiquitous in modern normal marine sediments (those deposited under oxygenated bottom waters), where anoxic conditions are commonly encountered beneath the sediment-water interface. The major factors controlling pyrite formation (Fig. 1) are the amounts of organic matter and reactive Fe minerals deposited in the sediments and the availability of dissolved sulfate.

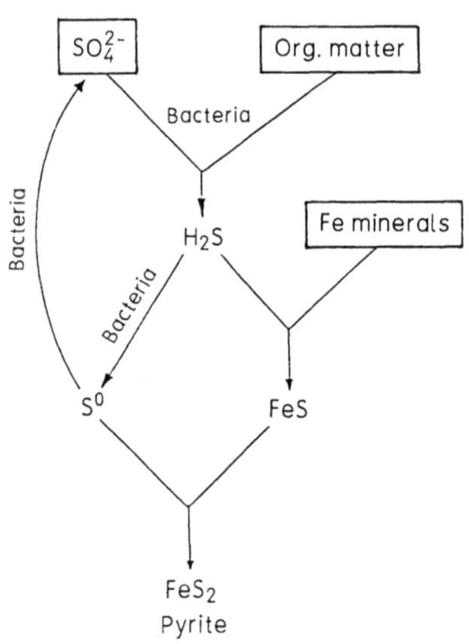

Fig. 1. Mechanisms of pyrite formation

In the following discussion each limiting factor will be examined in turn to show how their relative importance varies from one depositional environment to another.

2.1 Organic Carbon: the Limiting Factor in Normal Marine Sediments

Sulfate reduction occurs only in the absence of oxygen, but such conditions are realized in the majority of shallow-water sediments because sufficient organic C is deposited to act as an oxygen-consuming barrier. Dissolved oxygen is brought into a sediment from the overlying water by molecular diffusion, and by physical stirring of the sediment due to wind, current action and bioturbation (Aller 1980; Berner 1980). However, oxygen is consumed by aerobic bacteria which utilize the oxygen to convert organic matter to CO_2. Clearly, metabolizable organic matter is necessary not only for sulfate reduction but also to create the anoxic conditions under which sulfate reduction is possible. In fact, most organic carbon-bearing sediments contain sufficient metabolizable material to enable oxygen to be rapidly consumed, often within a few millimeters or less of the sediment surface (e.g. Revsbech et al. 1980).

Microbiological sulfate reduction is a complex process which first requires the breakdown of polymeric organic compounds to the simpler molecules which can be utilized by the sulfate-reducing bacteria (Jørgensen 1982; Skyring 1987). Although the overall stoichiometry of Eq. (1) has been proven by Westrich (1983), it should not be taken to indicate that all organic matter is metabolizable. In fact, the organic matter in modern sediments can be divided into a number of groups whose reactivity towards sulfate reduction varies over several orders of magnitude (Westrich and Berner 1984). These groups are used progressively, starting from the most reactive and proceeding to the least reactive but leaving residual comparatively refractory material. Where sedimentation rates are low, there is sufficient time for the destruction of the more metabolizable material by oxic bacteria before the organic matter is buried to a depth where sulfate reduction occurs. Thus, the residual organic matter available for sulfate reduction is both less abundant and less reactive (Westrich and Berner 1984; Canfield 1989a). The converse is true where sedimentation rates are high; more, and more reactive, organic matter survives oxic degradation to reach the zone of sulfate reduction. Once in the zone of sulfate reduction, rates of H_2S generation are directly proportional to the concentrations of reactive organic matter (Westrich and Berner 1984) and higher rates of H_2S

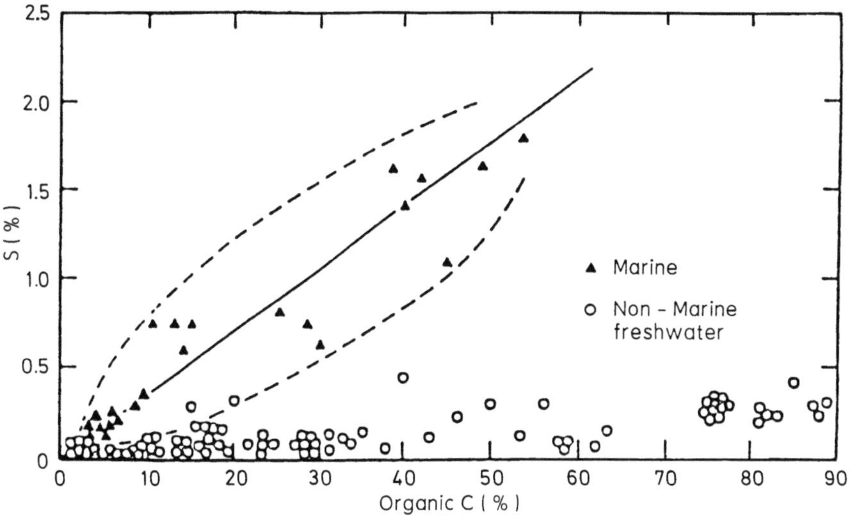

Fig. 2. Organic C and total sulfur relationships in modern normal marine and freshwater sediments (Berner and Raiswell 1983)

generation allow greater proportions of detrital Fe minerals to be transformed to pyrite.

The dominant control of organic matter on sulfate reduction and pyrite formation in most normal marine sediments is demonstrated in Fig. 2 by the good linear relationship between organic C and total (mostly pyrite) sulfur (Berner 1982). The content of organic and sulfate sulfur are usually negligible (see Raiswell and Berner 1986). Neither sulfate concentrations nor reactive Fe contents apparently exert any influence on the amount of pyrite formed in normal marine sediments for the following reasons. Firstly, because most pyrite in normal marine sediments is formed close to the sediment surface, where seawater sulfate is abundant. Therefore, only comparatively small amounts of pyrite form during subsequent burial, even though sulfate is ultimately exhausted (Berner 1984). Secondly, except where biogenic debris is abundant, most fine-grained sediments contain sufficient reactive Fe phases to ensure that their pyrite content is not apparently Fe-limited. Normal marine sediments certainly still contain Fe potentially reactive towards H_2S after pyrite formation has ceased (Berner 1970 and later sections) but these phases are only slowly reactive (e.g. iron silicates react with sulfide with half-lives of 50.000–100.000 years; Canfield and Raiswell, 1991). Little reaction can occur even with slow deposition and large concentrations of H_2S are thus lost from the sediments (Jørgensen 1978; Berner and Westrich 1985). More time, and higher organic carbon contents, are needed to render these phases reactive. However, if reactive Fe is defined with respect to its maximum potential for reaction with H_2S, then the organic matter content emerges as the main limiting factor on pyrite formation. This has important implications as reactive Fe is more completely removed when organic C concentrations are high, hence the potential for subsequent sulfide removal is low.

2.2 Dissolved Sulfate: the Limiting Factor in Freshwater Sediments

Westrich (1983) and Boudreau and Westrich (1984) have shown that dissolved sulfate concentrations exert little influence on rates of sulfate reduction (Fig. 3) at any levels above approximately 5 mM. By the time sulfate has been depleted to this extent in marine sediments the amounts of pyrite which can form are negligible (for a closed system sediment of porosity 85% this would correspond to <0.04% S as pyrite). However, regardless of the rate effect, dissolved sulfate concentrations are sufficiently low in fresh-

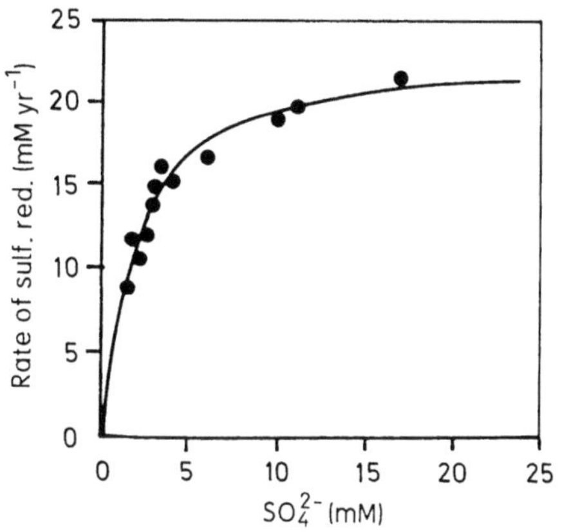

Fig. 3. Variation in rate of sulfate reduction with sulfate concentration (Boudreau and Westrich 1984)

waters (typically several hundred times less than in seawater) for sulfate to be readily depleted from the sediment at depths of only a few centimeters. Thus, little pyrite is formed and a high proportion of organic matter survives sulfate reduction (e.g. Davison et al. 1985). Consequently sulfate, rather than organic matter, is the principal limiting factor on pyrite formation in freshwater sediments and the relationship between organic carbon and total sulfur is very poor (Fig. 2), especially as a substantial proportion of the total sulfur may be organic and not pyritic (Losher and Kelts 1989). Aquatic organisms may contain 0.5–3.0% S by dry weight, so 10% organic C (approx. 25% organic matter) might be accompanied by nearly 1% organic sulfur.

The rapid depletion of sulfate from the porewaters of freshwater sediments leaves a surplus of readily metabolizable organic matter (used for methanogenesis: Rice and Claypool 1981) and reactive Fe. The potential of freshwater sediments to remove dissolved sulfide is accordingly high, unless substantial amounts of reactive Fe have been consumed to form siderite (e.g. Postma 1982).

2.3 Reactive Iron: the Limiting Factor for Sediments with Inhospitable Bottom Waters

Where bottom waters contain little or no oxygen, reactive Fe may often limit pyrite formation (e.g. Raiswell 1982; Raiswell and Berner 1985). Clearly, the reactive Fe content of shales, and hence their metallogenetic potential, is to some extent a function of depositional environment, and degree of bottom water oxygenation (or H_2S concentration). It is therefore necessary *firstly*, to identify in modern sediments the ecological or sedimentological criteria which can be correlated with bottom water oxygen levels and which are applicable to ancient sediments, and *secondly* to demonstrate the links between depositional environment and reactive Fe preservation/consumption.

2.3.1 Sediment Classification by Degree of Bottom Water Oxygenation

All normal marine sediments are capable of sustaining at least some benthic activity. However, the nature and abundance of benthic activity decreases as bottom water dissolved oxygen levels decrease. In modern sediments Rhoads and Morse (1971) have shown that oxygen concentrations ranging from equilibrium with the atmosphere down to 0.7 $ml\,l^{-1}$ have little effect on benthic metazoan activity. These aerobic sediments are bioturbated and contain a diverse calcareous fauna. As oxygen concentrations further decrease from 0.7 to around 0.3 $ml\,l^{-1}$ there is a considerable decrease in benthic activity and these dysaerobic sediments are only weakly bioturbated and mainly contain soft-bodied infaunal deposit feeders. Finally, below 0.1 $ml\,l^{-1}$, even suspension feeders disappear, and these anaerobic sediments are characterized by an absence of benthic fauna and by well-developed laminations.

A rather similar classification has been developed for ancient sediments using a combination of sedimentological criteria and bivalve mode of life (Morris 1979). The divisions recognized in this classification are termed normal, restricted and bituminous, and probably correspond to similar dissolved oxygen levels to those estimated by Rhoads and Morse (1971).

In studies on ancient sediments (e.g. Raiswell and Berner 1985; Raiswell et al. 1988) a modified Morris (1979) terminology was used (i.e. aerobic marine, restricted and inhospitable bottom). Note that the term normal marine now embraces both the aerobic marine and restricted categories and that inhospitable bottom sediments range from those in contact with low concentrations of dissolved oxygen to those in contact with H_2S-bearing water (i.e. euxinic).

An important difference between euxinic and normal marine sediments is that the presence of H_2S in the bottom waters allows the process of pyrite formation to begin in the water column. In fact, formation before burial can occur even in the absence of organic matter because H_2S can be advected in from elsewhere within the basin. For this and the following reasons, the intensity of exposure of Fe minerals to H_2S is particularly high in euxinic sediments. Thus, more H_2S can be fixed per unit of organic carbon buried and euxinic sediments frequently have lower C/S ratios than do normal marine sediments (Fig. 4).

Also important is that the absence of benthic fauna encourages enhanced preservation of more reactive organic matter in the sediment (Demaison and Moore 1980) and hence allows higher concentrations of H_2S to be maintained. Slow rates of deposition are also helpful in enabling detrital Fe minerals to be subjected to higher concentrations of H_2S for longer periods of time. Under these circumstances even comparatively unreactive Fe phases may be converted to pyrite; in extremis, nearly all the Fe in the sediment may be present as pyrite (Berner 1985).

2.3.2 Reactive Iron Measurement

Berner (1970) originally suggested that Fe reactivity towards H_2S could be measured by an extraction with

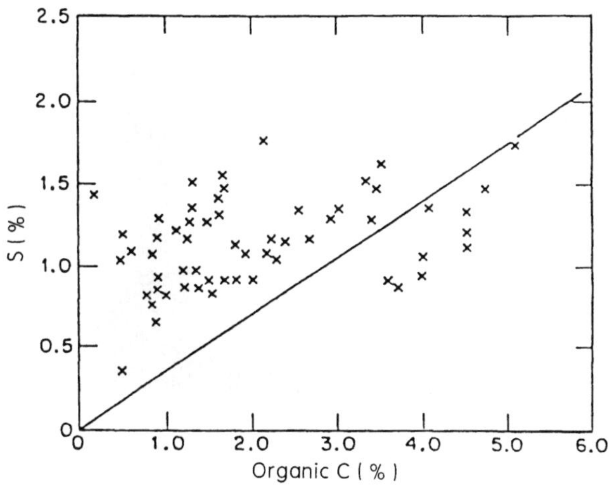

Fig. 4. Organic C and total sulfur relationships in modern euxinic sediments of the Black Sea (data from Rozanov et al. 1974)

boiling concentrated HCl for 1 min. This technique completely dissolved Fe in fine-grained haematite, limonitic goethite and chlorite, together with variable but small amounts from coarse grains of magnetite, haematite, goethite, hornblende, augite, biotite and epidote.

More recently, Canfield (1988, 1989b) suggests that this treatment is too harsh, and that a dithionite extraction is preferable, at least for modern sediments. In both experimental and natural systems the Fe minerals ferrihydrite and lepidocrocite were shown to be more reactive than haematite and goethite and Fe silicates were essentially unreactive. Apparently, only a few fairly reactive Fe phases are capable of conversion to pyrite, and their sulfidation is usually complete shortly after burial in normal marine sediments. This picture is consistent with earlier observations noting that most pyrite is formed quickly, very close to the sediment/water interface (i.e. from highly reactive Fe phases), and that thereafter little extra pyrite is formed (Berner 1984). Furthermore, Berner and Westrich (1985) have shown that large concentrations of H_2S are lost from sediment pore systems (confirming that residual Fe phases are only poorly reactive). Strictly, therefore, the amounts of pyrite formed are always Fe-limited, and the organic carbon control described earlier is observed because it determines the intensity of exposure to H_2S.

In the present context reactive Fe is defined as that extractable by the technique described in Berner (1970) as modified by Raiswell et al. (1988). This is an approach which distinguishes reactive Fe through its potential for sulfidation, given exposure to the fairly extreme conditions encountered in euxinic environments. This takes into account that (1) burial diagenesis of ancient sediments may cause recrystallization of less stable Fe phases and hence render them insoluble in dithionite, but still soluble in HCl, and (2) the harsher HCl extraction may overestimate diagenetically reactive Fe but underestimate hydrothermally reactive Fe (see Sect. 4).

2.3.3 Reactive Iron and Its Sulfidation in Modern Sediments

Berner (1970) presented the first data indicating the extent to which reactive Fe was consumed to form pyrite and described the degree of pyritization (DOP) as:

$$\mathrm{DOP} = \frac{\%\ \text{pyrite Fe}}{\%\ \text{pyrite Fe} + \%\ \text{HCl-soluble Fe}}.$$

Two different sediment depositional environments were studied: (1) normal marine with DOP values ranging from 0.10 to 0.30, and (2) seasonally euxinic (absence of benthic macrofauna) with DOP values ranging from 0.30 to 0.35. Clearly, reactive Fe is more effectively consumed by passing from well-oxygenated (normal marine) to poorly oxygenated/seasonally sulfidic environments with inhospitable bottom conditions.

The influence of increasing quantities of organic C on the degree of pyritization is illustrated (Fig. 5) by the Black Sea data of Rozanov et al. (1974). Although a lack of detail on the HCl extraction technique means the DOP values may not be comparable with other data presented here, these surficial sediments from the deep-water part of the basin show a strong correlation of increasing DOP with increasing organic C content.

Canfield (1988) and Canfield and Raiswell (1991) have also shown that sedimentation rates up to about $2\,g\,cm^{-2}\,yr^{-1}$ (i.e. ca. $2\,cm\,yr^{-1}$) are directly correlated with reduced sulfur burial rates. That is, more rapid burial rates minimize the time that organic matter is exposed to benthic activity, leaving more organic matter to be buried and helping to preserve pyrite against bioturbational re-oxidation. Still faster burial rates could produce an inverse relationship between the sediment and pyrite burial rates, simply because of dilution effects and the decreasing time available for pyrite formation near the sediment-water interface. This effect, however, has not yet been documented (but see Aller et al. 1986). The data of Canfield (1988) show that there are significant differences in pyrite burial rates in different normal marine depositional environments, and hence in the availability of reactive Fe after sulfate reduction has

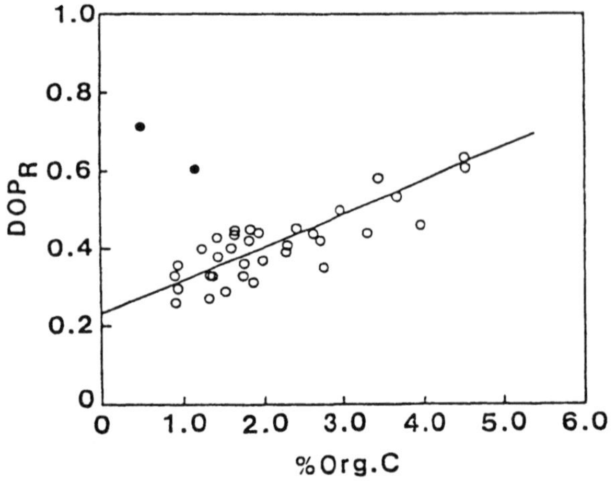

Fig. 5. DOP variation with organic C in euxinic sediments of the Black Sea. Two highly calcareous samples are shown as *filled circles*

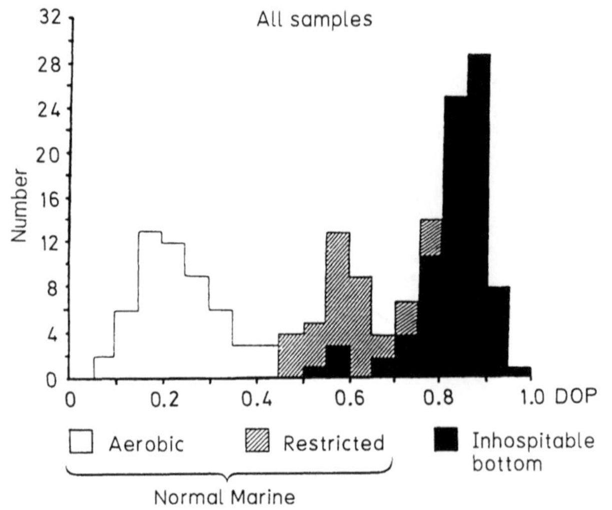

Fig. 6. DOP ranges in aerobic marine, restricted and inhospitable bottom sediments (Raiswell et al. 1988)

been completed. Some of the reasons for these variations are explained below.

3 The Preservation and Abundance of Reactive Iron in Ancient Sediments

The C-S-Fe signatures previously described in modern sediments can also be recognized in ancient sediments, and will be described here initially using the DOP parameter. Through this approach attention is sharply focussed on the preservation/consumption of Fe during diagenesis: subsequently, we will show how the content of reactive Fe varies and how this is linked to the total sediment Fe content.

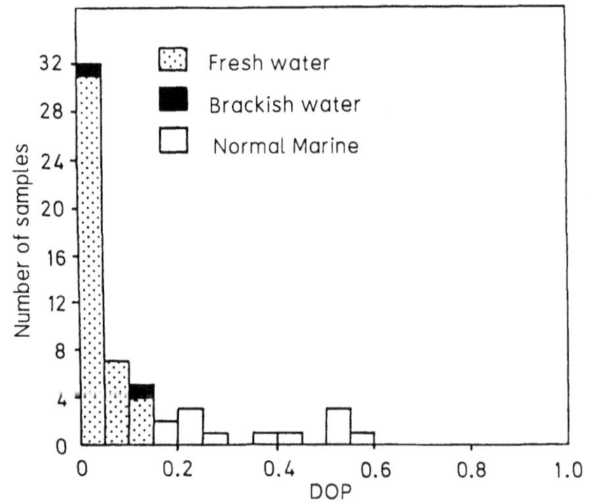

Fig. 7. DOP in Carboniferous freshwater sediments (Leeder et al. 1990)

3.1 Reactive Iron Preservation in Different Paleoenvironments

In general, ancient sediments show the same systematic differences with bottom water oxygenation, salinity and organic C content as have already been documented in modern sediments. Thus, Raiswell et al. (1988) presented data (Fig. 6) showing that sediments (with ages ranging from Devonian to Cretaceous) deposited under different degrees of bottom water oxygenation have systematic differences in DOP. Aerobic marine sediments are confined to DOP values less than 0.42 and are in all cases clearly distinguished from restricted marine sediments. The latter, however, showed some overlap of DOP with sediments deposited under inhospitable bottom water conditions, but a boundary at 0.75 separates more than 90% of the data. A clear distinction between these latter two environments may be impossible for a variety of reasons. For example, the faunal and sedimentological changes are progressive between the two environments: moreover, bottom water conditions may fluctuate between poorly oxygenated and sulfidic. For the present purposes this boundary is treated as transitional, noting as before that most reactive Fe will be preserved where sediments are deposited from well-oxygenated bottom waters.

Leeder et al. (1990) summarize non-marine data from one of the Carboniferous boreholes in the southern North Sea (Fig. 7) which show are no DOP values greater than 0.15, and hence reactive Fe is very effectively preserved in freshwater sediments, whereas Anderson et al. (1987) demonstrate how DOP increases with increasing organic C content in the New Albany Shale of the Illinois Basin (USA). A good linear correlation exists up to DOP = 0.8 for organic C contents of up to approximately 6%. It is apparent that reactive Fe is least well preserved in ancient sediments with high concentrations of organic C.

3.2 Factors Controlling Reactive Iron Abundance

3.2.1 Intrabasinal Variations in Reactive Iron: Depositional or Diagenetic?

It has been assumed that the potential to generate metal-rich fluids is increased as early diagenetic sulfate reduction decreases, producing a low DOP. An additional factor, however, is the actual abundance of reactive Fe in the sediment. Raiswell and Al-Biatty (1989) have measured the reactive Fe contents of more than 300 normal marine sediments. Surprisingly, the data show two distinct features (Fig. 8). One group of sediments (predominantly, but not exclusively, Cambrian-Silurian in age) shows a tendency for reactive Fe to be high and positively correlated with organic C, whilst in younger Devonian-Cretaceous sediments reactive Fe contents are low, fairly uniform and are independent of organic C variations. At low organic C content (<1%) the two groups show considerable overlap. While the reasons for the two

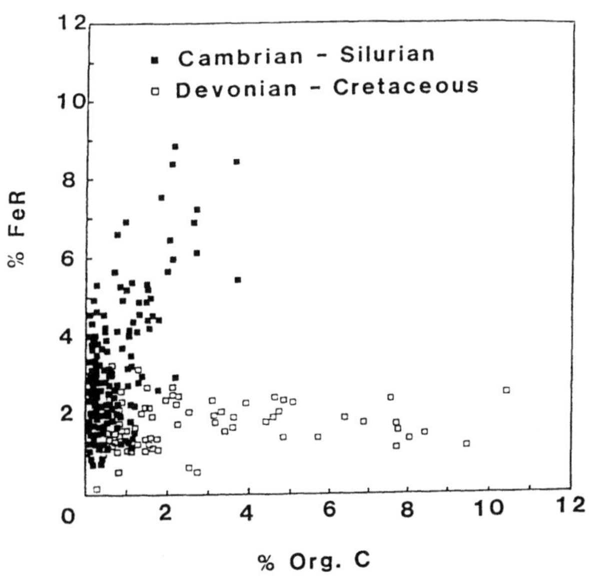

Fig. 8. Variation in reactive Fe with organic C concentration in Cambrian to Silurian and Devonian to Cretaceous normal marine sediments (Raiswell and Al-Biatty 1989)

distributions are not presently known, they are likely the result of regional processes (e.g. depositional and diagenetic processes) rather than any global change in Fe cycling. Regional controls are favoured because;

1. There is as yet no clear evidence for global changes in the Fe cycle, such that more reactive Fe was deposited in Cambrian-Silurian, compared to Devonian-Cretaceous, sediments. Post-Silurian weathering rates are generally believed to have been greatly influenced by the decay of organic matter in soils (which gives enhanced soil-air pCO_2 values and produces organic chelates). However, similar rates could have been maintained in the pre-land plant Cambrian-Silurian periods by a higher atmospheric pCO_2 and a warmer and moister climate (Holland 1984). Donnelly et al. (1990) argue that major weathering events did occur during this time but it is still uncertain whether the production of reactive Fe by weathering would be significantly different in the two time periods in Fig. 8.
2. The two organic C/reactive Fe patterns in Fig. 8 can also be found at different localities where the same horizon is exposed (Raiswell and Berner 1985). Both the Jurassic Jet Rock (UK) and the Posidonienschiefer (Germany) were deposited from inhospitable bottom conditions established during a widespread oceanic anoxic event (Jenkyns 1988). However, the Jet Rock shows a marked correlation between organic C (4–12%) and reactive Fe (4–8%), whereas in the Posidonienschiefer the reactive Fe is uniform ($2.6 \pm 0.4\%$) even though organic C varies from 2–18%.

The large variations in reactive Fe observed in Fig. 8 and in the Jurassic shales discussed above may occur in response to either different styles of diagenesis or different depositional processes/environments.

The initial formation of reactive Fe will be facilitated sometimes by the presence of readily weathered Fe minerals in the sediment source area. For example, in the case of Ordovician ironstones (Van Houten 1985), these have been suggested to be from terrestrial volcanic debris (see also Spirakis 1989). However, a reconnaissance study of 30 terrestrial acid-basic tuffs and ashes indicates that these are not exceptionally high in either their reactive Fe contents (0.3–3.8 Fe%) or in their ratio of reactive to total Fe (0.04–0.55). Because Fe-rich laterites are able to develop from a wide range of rock types, the role of source area geology is probably small compared to deep weathering conditions. Thus, climate and topography are likely to promote reactive Fe formation more than a favourable initial mineralogy.

Whatever the source area geology influence on reactive Fe, an association between Fe and organic C is unexpected, since Fe in modern sediments is usually strongly associated with the siliciclastic fraction (Garrels 1988). However, if the flux of reactive Fe-poor, siliciclastic material can be suppressed, associations between Fe and organic matter (e.g. Baas Becking and Moore 1959) might be able to develop and to dominate observed sedimentary associations, as suggested by the Cambrian-Silurian data of Fig. 8 (Raiswell and Al-Biatty 1989).

Alternatively, a diagenetic explanation for the variations in Fig. 8 would require that the more organic-rich sediments became a sink for Fe mobilized from adjacent organic-poor sediments. Berner (1969) has demonstrated experimentally that the centimeter-scale juxtaposition of organic-rich and organic-poor layers leads to an enrichment in Fe sulfides formed by sulfate reduction at the boundary between the layers. However, the sediments plotted in Fig. 8 are not obviously layered on this scale and the relevance of the experimental system (in which diffusive mass transfer occurs) is uncertain. A better explanation might be to invoke Fe migration from a mobilizing facies to fixing, sulfidic facies, on a considerably larger scale (see below).

3.2.2 Modern Sediments: Reactive Iron Mobilization and Migration

A highly relevant modern case where high sediment reactive Fe contents arise from deeply weathered, lateritic sources are the tropical inner-self muds of the Amazon River mouth (Aller et al. 1986). The total Fe content of these sediments averages approximately 5%, and ranges up to 7%. About half the total Fe can be extracted by dithionite-citrate which, as a milder technique (Canfield 1988, 1989b), underestimates reactive Fe compared to the HCl techniques used here. However, even these reactive Fe contents are reasonably high and are comparable to those found in our Cambrian-Silurian samples plotted in Fig. 8.

In addition to this source area enrichment of reactive Fe, diagenetic effects greatly enhance Fe mobility. The Amazon sediments are frequently and intensely reworked, such that oxidant recharge prevents pyrite preservation and generally maintains a state of suboxic diagenesis (characterized by Fe reduction rather than sulfate reduction). Rapid burial with little labile organic matter subsequently minimizes the consumption of reactive Fe to form pyrite. Hence, the pore waters of these sediments are characterized by high concentrations of dissolved Fe (0.3–0.7 m mol l^{-1}). Here, sulfate reduction is suppressed by reworking and reactive Fe can then be readily reduced and dissolved.

There is an additional consequence of reworking, which Aller et al. (1986) argue greatly assists mobilization. Sediments with Fe-generating capacity can be interbedded with, or lie up-current from, sands which are potential Fe sinks. Mass transfer of Fe from muds to sands could occur by diffusion (where sands and muds are interbedded) or by reworking, which could separate colloidal Fe from the muds and transport it elsewhere for deposition with sands. These mobilization and migration mechanisms are believed to make the Amazon River sediments an incipient or actual analogue for Paleozoic ironstones. It is doubtful whether our Cambrian-Silurian samples could ever have been reworked on this scale, but their high reactive Fe contents may be attributable to similar source area effects, either by drawing on similar continental sources of reactive Fe or by collecting colloidal Fe exported from inner-shelf regions.

Plots of reactive Fe against total Fe also emphasize the similarity between the Amazon inner-shelf sediments and our Cambrian to Silurian data (Fig. 9), both of which are different from the Devonian-Cretaceous shales. The two data sets show that reactive and total Fe are closely related, but the slopes of the best-fit lines are very different (Cambrian-Silurian 0.74 ± 0.04, Devonian-Cretaceous 0.34 ± 0.04). Values of reactive and total Fe are highest in the Amazon inner-shelf sediments and our Cambrian-Silurian samples.

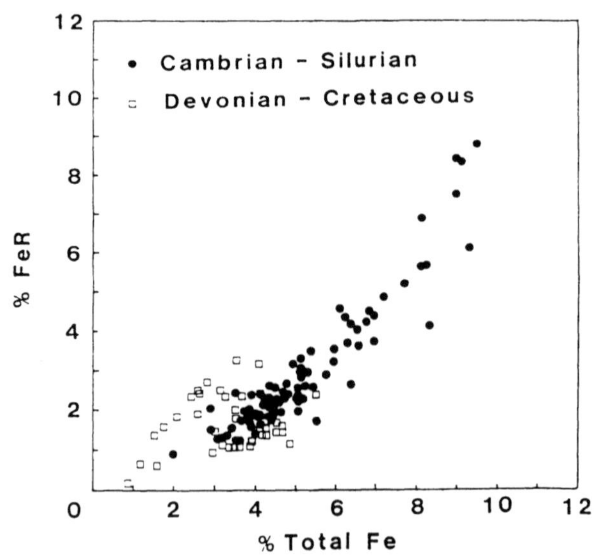

Fig. 9. Relationships between reactive and total Fe for Cambrian to Silurian and Devonian to Cretaceous normal marine sediments

4 Reactive Iron and Hydrothermal Sulfidation

Clearly, Fe which is reactive towards H_2S under low temperature diagenetic conditions will also be reactive under hydrothermal conditions. Unfortunately, there are no experimental data indicating whether, as might be expected, a wider range of Fe phases would become reactive under the more extreme hydrothermal conditions. An obvious stumbling block is that variations in the composition and temperature of the fluids are likely to dictate precisely how much Fe is reactive towards dissolved sulfide, although preliminary data suggest that this factor may not be of great importance.

In a preliminary study of Fe release, we measured the rates of Fe dissolution from an Ordovician shale at 70 °C under reflux with 2 N, 6 N and concentrated HCl. The data (Fig. 10) show an initial stage of parabolic dissolution kinetics (linear plots of %Fe versus the square root of time). These data, from a multiphase system, are not simple to interpret, as there may be a variety of explanations for the occurrence of parabolic kinetics (Velbel 1986). Originally, Holdren and Berner (1979) showed that parabolic dissolution kinetics of feldspars could be attributed to the early and rapid attack on microparticles which were produced when the sample was ground. Dissolution will also be facilitated at defect and dislocation sites, strained surfaces and wherever surface energies are high (Velbel 1986). There is, therefore, a strong probability that the parabolic kinetics of Fe release may also be an artefact of sample preparation.

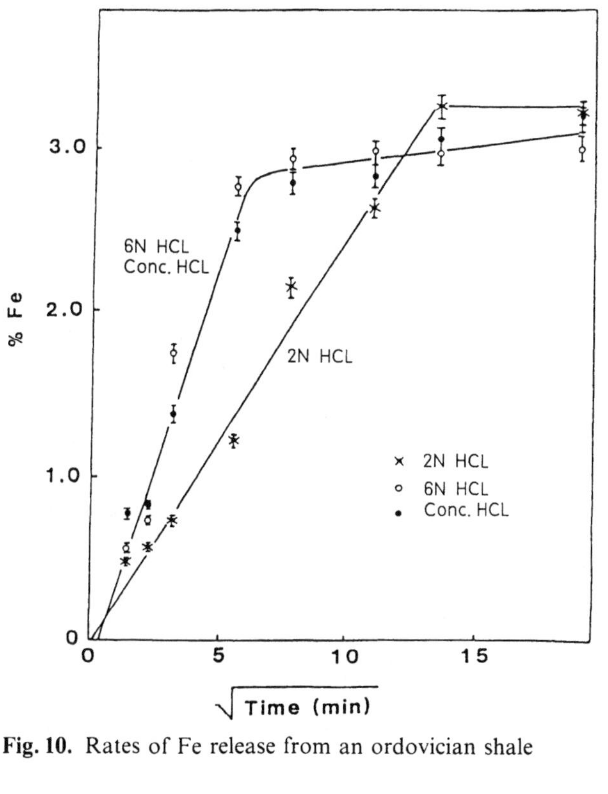

Fig. 10. Rates of Fe release from an ordovician shale

However, the experimental data do show that rates of release were independent of acid strength above 6 N HCl, at these temperatures, and also that all conditions ultimately extracted the same amounts of Fe (approximately 3%, compared to 1.47% by the HCl technique described earlier, and 5.61% non-pyritic Fe). This suggests that not all non-pyrite Fe would be available to remove dissolved sulfide from invading fluids, and also that diagenetic pyrite formation which removed all HCl-soluble Fe would have a significant effect on the potential of a shale to titrate dissolved sulfide and generate metal-rich fluids. These observations suggest that a carefully constructed experimental program might be able to produce a useful estimate of hydrothermally reactive Fe, and of the potential of shales to buffer dissolved sulfide levels.

5 Summary and Conclusions

The process of sedimentary pyrite formation consumes Fe which is potentially capable of hydrothermal sulfidation, and hence changes the capacity of the sediment to destabilize metal-thio complexes. Enough is now known from studies of ancient and modern sediments for the extent of pyrite formation to be at least qualitatively predictable. The variable degree of pyritization (DOP) represents the ratio of pyrite Fe to Fe reactive towards H_2S at deposition, and is a useful parameter for quantifying Fe preservation. Pyrite formation is minimized, and reactive Fe most effectly preserved, where:

1. Sediments are deposited from fully oxygenated bottom waters. In such marine sediments (where high seawater sulfate concentrations favour pyrite formation by sulfate reduction) DOP values of less than 0.4 are common.
2. Sediments are deposited from waters with low concentrations of sulfate. DOP values of less than 0.1 are found in freshwater sediments.
3. Sediments have little labile organic carbon. Even in marine sediments deposited from oxygenated bottom waters, organic C concentrations of less than 1% rarely produce DOP values more than 0.3.
4. Sediments are very rapidly deposited and/or reworked. In the extreme case of the Amazon River sediments, sulfate reduction is almost completely suppressed and the sediments contain large concentrations of reactive Fe (and DOP <0.1).

In addition to these variables (which affect Fe preservation), considerable intrabasinal variations can occur in the amounts of reactive Fe originally contained within the sediments. Lateritic source areas appear to be good suppliers of reactive Fe, which can be augmented by diagenesis. Thus, Fe mobilized within one sediment may migrate into adjacent sediment where there is a fixed capacity (i.e. which are oxidizing, or sulfidic); also sediments can be reworked to separate colloidal Fe for export to nearby down-current depositional areas. There is considerable potential for valuable research into the factors controlling reactive Fe abundances in sediments. Finally, we note that there is also a need for experimental measurements of the reactivity of sediments towards hydrothermal, reduced sulfur-bearing fluids.

Acknowledgements. The authors are pleased to acknowledge the helpful criticism of the Leeds Sulphur Isotope Group, and also the reviewer's comments. HJB received financial support from the Government of Iraq. Permission to reproduce Figs. 2, 3 and 8 was given by Pergamon Press; Fig. 6 by the S.E.P.M.

References

Aller RC (1980) Diagenetic processes near the sediment-water interface of Long Island Sound I: Decomposition and nutrient element geochemistry (S, N, P). In: Saltzman B (ed) Estuarine physics and chemistry: studies in Long Island Sound. Adv Geophys 22:238–348
Aller RC, Mackin JE, Cox RT (1986) Diagenesis of Fe and S in Amazon inner shelf muds: apparent dominace of Fe reduction and implications for the genesis of ironstones. Continent Shelf Res 6:262–289
Anderson TF, Kruger J, Raiswell R (1987) C-S-Fe relationships and the isotopic composition of pyrite in the New Albany Shale of the Illinois Basin, USA. Geochim Cosmochim Acta 51:2795–2805
Baas Becking LGM, Moore D (1959) The relation between iron and organic matter in sediments. J Sediment Petrol 29:454–458
Berner RA (1969) Migration of iron and sulfur within anaerobic sediments during early diagenesis. Am J Sci 267:19–42
Berner RA (1970) Sedimentary pyrite formation. Am J Sci 268:1–23
Berner RA (1980) Early diagenesis: a theoretical approach. Univ Press, Princeton, 241 pp
Berner RA (1982) Burial of organic carbon and pyrite sulfur in the modern ocean: its geochemical and environmental significance. Am J Sci 282:451–473
Berner RA (1984) Sedimentary pyrite formation: an update. Geochim Cosmochim Acta 48:605–615
Berner RA (1985) Sulphate reduction, organic matter decomposition and pyrite formation. Philos Trans R Soc London Ser A 315:25–38
Berner RA, Raiswell R (1983) Burial of organic carbon and pyrite sulfur in sediments over Phanerozoic time: A new theory. Geochim Cosmochim Acta 47:855–862
Berner RA, Westrich JT (1985) Bioturbation and the early diagenesis of carbon and sulfur. Am J Sci 285:193–206
Boudreau BP, Westrich JT (1984) The dependence of bacterial sulfate reduction on sulfate concentration in marine sediments. Geochim Cosmochim Acta 48:2503–2516
Canfield DE (1988) Sulfate reduction and the diagenesis of iron in anoxic marine sediments. PhD Thesis, Yale Univ, 248 pp
Canfield DE (1989a) Sulfate reduction and oxic respiration in marine sediments: implication for organic carbon preservation in euxinic environments. Deep-Sea Res 36:121–138
Canfield DE (1989b) Reactive iron in marine sediments. Geochim Cosmochim Acta 53:619–632
Canfield DE, Raiswell R (1991) Pyrite formation and fossil preservation. In: Allison PA, Briggs DEG (eds) Taphonomy: releasing the data locked in the fossil record. Plenum, New York, Chap 7
Davison W, Lishman JP, Hilton J (1985) Formation of pyrite in freshwater sediments: implications for C/S ratios. Geochim Cosmochim Acta 49:1615–1620
Demaison GJ, Moore GT (1980) Anoxic marine environments and oil source bed genesis. Bull Am Assoc Petrol Geol 64:1179–1209
Donnelly TH, Shergold JH, Southgate PN, Barnes CJ (1990) Events leading to global phosphogenesis around the Precambrian/Cambrian boundary. In: Notholt AJG, Jarvis I (eds) Phosphorite research and development. Geol Soc London Spec Publ 52:273–289
Garrels RM (1988) Precambrian Geochemistry. In: Goldschmidt Conf Abstr. Geochem Soc Baltimore, p 42
Goldhaber MB, Kaplan IR (1974) The sulfur cycle. In: Goldberg ED (ed) Sea 5:569–655
Holden GR Jr, Berner RA (1979) Mechanisms of feldspar weathering. I. Experimental studies. Geochim Cosmochim Acta 43:1161–1171
Holland HD (1984) The chemical evolution of the atmosphere and oceans. Univ Press, Princeton, 582 pp
Jørgensen BB (1978) A comparison of methods for the quantification of bacterial sulfate reduction in coastal marine sediments. I: Measurement with radiotracer techniques. J Geomicrobiol 1:11–27
Jørgensen BB (1982) Ecology of the bacteria of the sulphur cycle with special reference to anoxic-oxic interface environments. Philos Trans R Soc London Ser B 298:543–561
Kharaka YK, Maest AS, Carothers WW, Law LM, Lamothe PJ, Fries TL (1987) Geochemistry of metal-rich brines from central Mississippi Salt Dome basin, USA. Appl Geochem 2:543–561
Leeder MR, Raiswell R, Al-Biatty H, McMahon A, Hardman M (1990) Carboniferous stratigraphy, sedimentation and correlation of Well 48/3-3 in the southern North Sea Gas Basin: integrated use of palynology, wireline logs and carbon/sulphur geochemistry. J Geol Soc London 147:287–300
Losher AJ, Kelts KR (1989) The role of organic matter for the incorporation of sulfur into freshwater lake sediments. Terra Nova, 1:253–261
Morris KA (1979) A classification of Jurassic marine sequences and an example from the Toarcian (Lower Jurassic) of Great Britain. Paleogeogr Paleoclimatol Paleoecol 26:117–126

Postma D (1982) Pyrite and siderite formation in brackish and freshwater swamp sediments. Am J Sci 282:1151–1183

Raiswell R (1982) Pyrite texture, isotopic composition and the availability of iron. Am J Sci 282:1244–1263

Raiswell R, Al-Biatty HJ (1989) Depositional and diagenetic C-S-Fe signatures in Early Paleozoic normal marine shales. Geochim Cosmochim Acta 53:1147–1152

Raiswell R, Berner RA (1985) Pyrite formation in euxinic and semi-euxinic sediments. Am J Sci 285:710–724

Raiswell R, Berner RA (1986) Pyrite and organic matter in Phanerozoic normal marine shales. Geochim Cosmochim Acta 50:1967–1976

Raiswell R, Buckley F, Berner RA, Anderson TF (1988) Degree of pyritisation of iron as a paleoenvironmental indicator of bottom-water oxygenation. J Sediment Petrol 58:812–819

Revsbech NP, Jørgensen BB, Blackburn TH (1980) Oxygen in sea bottom measured with a microelectrode. Science 207:1355–1356

Rhoads DC, Morse JW (1971) Evolutionary and ecologic significance of oxygen-deficient marine basins. Lethaia 4:413–428

Rice DD, Claypool GW (1981) Generation, accumulation and resource potential of biogenic gas. Am Assoc Petrol Geol Bull 65:5–25

Rickard DT (1969) The chemistry of iron-sulphide formation at low temperatures. Stockholm Contrib Geol 20: 67–95

Rozanov AG, Volkov II, Yagodinskaya TA (1974) Forms of iron in surface layer of Black Sea sediments. In: Degens ET, Ross DA (eds) The Black Sea – geology, chemistry and biology. Am Assoc Petrol Geol Mem 20:532–541

Schidlowski M, Hayes JM, Kaplan IR (1983) Isotopic inferences of ancient biochemistries: carbon, sulphur, hydrogen and nitrogen. In: Schopf JW (ed) Earth's earliest biosphere: its origin and evolution. Univ Press, Princeton 146–186 pp

Skyring GW (1987) Sulfate reduction in coastal ecosystems. Geomicrobial J 5:295–374

Spirakis CS (1989) Possible effect of readily available iron in volcanic ash on the carbon to sulfur ratio in Lower Paleozoic normal marine sediments and implications for atmospheric oxygen. Geology 17:599–601

Sweeney RE, Kaplan JR (1973) Pyrite framboid formation. Laboratory synthesis and marine sediments. Econ Geol 68:618–634

Van Houten FB (1985) Oolite ironstones and contrasting Ordovician and Jurassic paleogeography. Geology 13:722–724

Velbel MA (1986) Influence of surface area, surface characteristics, and solution composition on feldspar weathering rates. In: Davis JA, Hayes KF (eds) Geochemical processes at mineral surfaces. ACS Symp Ser 323: 615–634

Westrich JT (1983) The consequences and controls at bacterial sulphate reduction in marine sediments. PhD Thesis, Yale Univ, 530 pp

Westrich JT, Berner RA (1984) The role of sedimentary organic matter in bacterial sulfate reduction: the G model tested. Limol Oceanogr 29:236–249

Genetic Relationship Between Elementary, Organic, and Pyritic Sulfur in Sediments

Alfred V. Hirner [1] and Brian W. Robinson [2]

[1] Institut für Umweltanalytik und Angewandte Geochemie, Universität GH Essen, Universitätsstr. 5–7, W-4300 Essen 1, FRG
[2] Institute of Nuclear Sciences, DSIR, P.O. Box 31312, Lower Hutt, New Zealand

Abstract

The isotopic relationships between cogenetic sulfur species in Mesozoic and Tertiary sediments are examined. Under reducing conditions, inorganic and organic sulfur species interact with each other in the course of the sedimentary sulfur cycle. The latter is discussed in connection with studies of German, New Zealand and Kuwait crude oils.

The intermolecular S-isotopic composition of petroleum may show considerable variations depending on the amount and nature of the organic matter and the amount of pyrite-S present, thus forming a valuable fingerprinting parameter for distinguishing different oil samples. Similar as well as dissimilar $\delta^{34}S$ for pyrite and organic matter have been found, depending on the availability of reactive Fe for the formation of pyrite. Kerogens may be less reactive towards H_2S than sedimentary inorganic Fe phases.

Inorganic and organic forms of sulfur can be used in solving practical problems of exploration and environmental geochemistry: elemental S is incorporated in organic matter without significant isotopic fractionation, and leads to characteristic intermolecular distributions of ^{32}S and ^{34}S in organic fractions. When pyrite sulfur is included, isotopic crude oil/rock extract/rock kerogen correlation studies can be significantly improved.

1 Introduction

Stable sulfur isotope fractionations between sedimentary sulfates and sulfides characteristic of bacterial sulfate reduction can be traced back to about 1.2 Ga (Schidlowski et al. 1983). Archean sulfates, however, are extremely rare and restricted to a few approx. 3.4 Ga old barites, which exhibit $\delta^{34}S$ values nearly identical to those of associated sulfides (Schidlowski et al. 1983; Ohmoto and Felder 1987). While Schidlowski (1987) proposed a large-scale isotopic re-equilibration between marine sulfur and the primordial sulfur reservoir of the early oceanic crust to account for these observations, Ohmoto and Felder (1987; see also Ohmoto, this Vol.) argue for the presence of sulfate-reducing bacteria in sulfate-bearing Archean oceans with temperatures of ca. 30 to 50 °C. However, no measurements have been reported on the stable isotope composition of organic sulfur in Archean sediments.

Much more information is available on the genetic relationships between various sulfur species in sediments of Mesozoic and Tertiary age, in particular, sediments rich in organic carbon, and the subject of geochemical hydrocarbon exploration (e.g. Gransch and Posthuma 1974; Krouse 1977; Orr 1978, 1986; Dinur et al. 1980; Thode 1981; Hirner et al. 1984; Hirner and Robinson 1989).

In this work the isotopic relationships between cogenetic sulfur species are examined, starting with our knowledge of the possible pathways of reduced sulfur in anoxic sediments. These concepts will be applied to oil exploration studies in southern Germany, New Zealand, and Kuwait, and the results compared with pertinent oil-geological models. It is hoped that the results will stimulate geochemists to perform similar studies in Archean systems.

2 The Sedimentary Sulfur Cycle

Sulfur is an essential element for all plants, animals, and microorganisms, and is present in organic combinations at levels up to 1 % of the dry weight biomass (Siegel 1975). In sediments, under anoxic and reducing conditions, inorganic and organic sulfur species interact with each other as indicated schematically in Fig. 1. Plant-S from assimilatory reduction of freshwater and soil sulfate, and H_2S from dissimilatory bacterial sulfate reduction, are the main S-sources in the low temperature geochemical cycle. Usually, dissimilatory sulfate reduction is the dominant S-source, but the

M. Schidlowski et al. (Eds.)
Early Organic Evolution: Implications for Mineral and Energy Resources

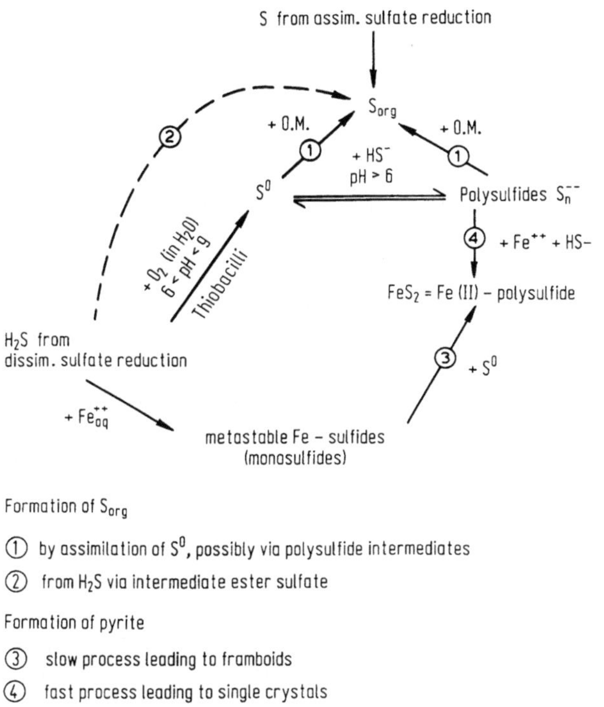

Fig. 1. Pathways of reduced sulfur formation in sediments

assimilatory pathway may be important in certain cases, e.g. for low-S lakes (Cook1981) or low-S coals (Hunt and Smith 1985). Whereas assimilation shows an insignificant isotope fractionation, dissimilatory reduction is accompanied by enrichment of ^{32}S in H_2S of ca. 20 to 70‰ (Rees 1973; Goldhaber and Kaplan 1974; Nielsen 1979). The magnitude of this enrichment depends on environmental conditions, differing between marine and terrigenous environments as well as between systems open and (partly) closed with respect to sulfate supply (Krouse 1977). While in open systems sulfides are up to 70‰ enriched in ^{32}S compared to sulfates, in closed systems both residual sulfates and late-formed sulfides become increasingly enriched in ^{34}S (Rayleigh-type fractionation). Open systems are characterized by a continuous diffusion of sulfate into pore waters together with an unlimited sulfate supply, and are realized in the open ocean or where local sedimentation rates are low (sulfate supply rate $\gg$ sulfate reduction rate). Closed systems prevail in closed basins or where sedimentation rates are fast (sulfate supply rate $\ll$ sulfate reduction rate).

Sulfide is a very reactive ion, and is readily oxidized both chemically and biologically. Pyrite is usually produced from the transformation of (amorphous) iron monosulfides in the presence of elemental S (Berner 1970). A framboidal texture (a spherical aggregate of pyrite microcrystals) develops when an initial intermediate (mackinawite, $FeS_{0.9}$) is transformed into an amorphous sphere of greigite (Fe_3S_4), which in turn adds sulfur during the nucleation and growth of pyrite microcrystals (Sweeney and Kaplan 1973). Alternatively, direct precipitation of non-framboidal pyrite, not involving acid-volatile monosulfides (Rickard 1975), can occur where pore waters are supersaturated with respect to pyrite but undersaturated with respect to monosulfides. Whereas framboidal pyrite formation occurs on a time scale of years (Berner et al. 1979), Howarth (1979) has observed the rapid, direct formation of pyrite crystals within less than 1 day at pH 5 in samples from the Spartina salt marsh.

The ability of humic substances, protokerogen and kerogen to incorporate elemental S has been demonstrated by Prinzler and Pape (1964), Martin and Hodgson (1973), DeRoo and Hodgson (1978), Casagrande and Ng (1979), Dinur et al. (1980) and Idiz (1987). Studies of X-ray absorption near absorption edges (XANES) indicate that the sulfur incorporated in kerogen during hydrothermal activity is probably present as thiophene heterocycles (G. S. Waldo, J. E. Penner-Hahn, pers. comm. 1988). Especially in sediments low in free Fe, polysulfide intermediates may play an important role in the process of S assimilation by organic matter, even at the early stages of diagenesis, probably via free radical reactions (Aizenshtat et al. 1981; Macqueen and Powell 1983; Powell and Macqueen 1984; Nriagu and Soon 1985). On the other hand, Mango (1983) showed that carbohydrates incorporate S from H_2S gas, and François (1987) suggested the possible formation of organic sulfur from H_2S via an intermediate ester sulfate. The feasibility of the latter mechanism was demonstrated by Casagrande et al. (1979) using radioactive tracer techniques. Reinecke et al. (1983) and LaLonde et al. (1985) verified experimentally that different conjugated ene carbonyls, when treated with aqueous alcoholic solutions saturated with alkali polysulfides, produced thiolanes and thiophenes. Low-temperature, carbon-sulfur, bond-forming reactions between nucleophilic S species (e.g. catenated S donors like H_2S_n, HS_n^- or S_n^{2-} whose stabilities are affected by pH) and electrophilic C species (e.g. condensated aldehydes and ketones, coniferyl alcohol units of lignin, carotenoids) are proposed by LaLonde et al. (1987).

Elemental S can be formed by the chemical or biological oxidation of H_2S in the presence or absence of O_2 (Krouse and McCready 1979). Aerobic chemolithothrophs (colorless sulfur bacteria) require the simultaneous presence of H_2S and O_2 established at the oxic-anoxic transition zone (oxycline) of the sediment (Bauld 1986). The role of elemental S as a very dynamic intermediate in the S-cycle is underlined by the very low concentrations (ppm level) normally found in marine sediments and crude oils (Goldhaber and Kaplan 1974).

3 Intermolecular Sulfur Isotopic Distributions

In the geothermal area in New Zealand, Waiotapu crude oil contains high concentrations of sulfur: 9.5 wt% S is organically combined, and more than 2.5 wt% S is dissolved in the oil in elemental form. Hirner and Robinson (1989) assume that the organic S is derived from incorporation of elemental S produced by near-surface H_2S oxidation. Isotope balance calculations show that the transfer reaction from elemental to organic S is accompanied by negligible isotopic fractionation (< 1‰). Of special interest, therefore, is the isotopic fractionation of S during its incorporation in crude oils. While Hirner et al. (1984) observed distinct isotopic ordering, Monster (1972) reported homogeneous intermolecular isotopic distributions. In general, the stable sulfur isotopes in crude oils seem to be distributed between the various organic components in a characteristic manner (Fig. 2). The polarities of the fractions plotted in Fig. 2 increase from top to bottom: from light resins to heavy resins to asphaltenes. In the Waiotapu crude oil of terrestrial origin $\delta^{34}S$ decreases with increasing polarity and molecular weight of the fractions, whereas the opposite is the case for crude oil of marine-brackish origin from the Molasse Basin (southern Germany). A different pattern is shown by samples from the Burgan oil field (Kuwait), where the heavy resin fraction is enriched in ^{34}S compared to the other two fractions. Bacterial degradation of crude oils apparently leads to isotopic homogenization between the fractions (Monster 1972; Orr 1978; Hirner et al. 1984), and has clearly not affected the examples shown in Fig. 2.

The samples represented in Fig. 2 have different geological histories, and the empirical results demonstrate that the intermolecular S-isotopic composition of petroleum may show considerable variations, which are likely to depend on the amount and nature of the accompanying organic matter, the amount of pyrite S present, and the depositional environment (e.g. oxic or anoxic, sedimentation rate). Thus, the intermolecular isotopic distribution may be a valuable fingerprinting parameter for distinguishing different oil samples. Although the results of recent determinations on crude oil and bitumen fractions from various origins show (1) isotopic characteristics similar to those observed for German and Waiotapu oils (Z. Aizenshtat, H. R. Krouse, pers. comm. 1989), or (2) no intermolecular isotopic ordering as with low-S terrestrial crudes from New Zealand (Hirner and Robinson 1989), a scientific rationale for these empirical intermolecular isotopic distributions has still not been found. Progress in the molecular organic geochemistry of sulfur (e.g. Sinninghe Dampste et al. 1987) may help to solve this problem. In this repsect it is of interest that the heterogeneous nature of the organic source material of terrestrial crude oils is generally reflected by heterogeneous intermolecular isotopic distributions like those found for sulfur (Hirner and Robinson 1989), and for carbon within aromatic subfractions in crude oils from New Zealand (Hirner and Lyon 1989).

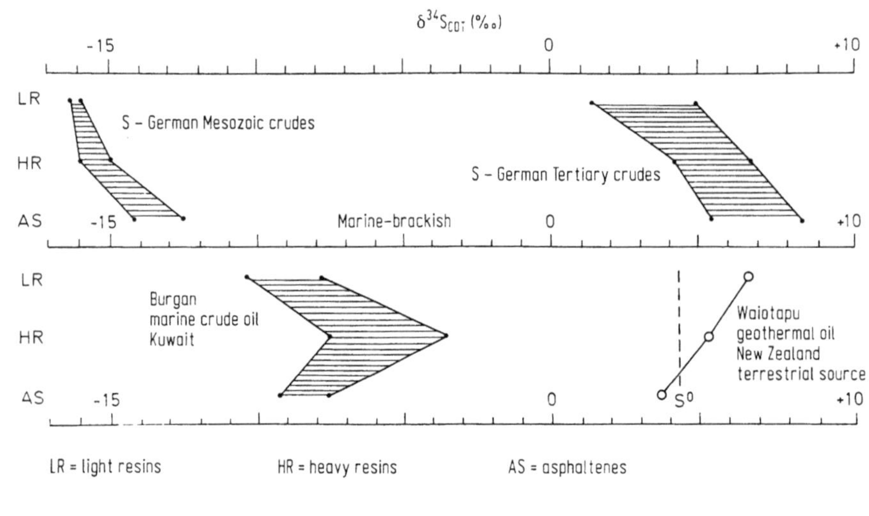

Fig. 2. Intermolecular S-isotopic distributions in selected crude oils

4 Isotopic Relation Between Organic and Pyritic Sulfur

Organic S and pyritic S are the stable end members in the anoxic sedimentary sulfur cycle (Fig. 1). Elemental S, as a dynamic intermediate in their production, is present in negligible concentrations. Therefore, both organic and pyrite sulfur should have nearly identical $\delta^{34}S$ values. Since no major isotope effects are known between reduced sulfur species (Aizenshtat et al. 1981; Fry et al. 1986), no significant isotopic fractionations would be expected if organic and pyrite sulfur were formed at the same time and by the same reduction process from one sulfate source, and the rate of sulfate reduction exceeds the rate of S fixation.

Similar $\delta^{34}S$ values for pyrite and organic matter have often been found in shale and coal samples (Hunt and Smith 1985; Hackley and Anderson 1986; Price and Shieh 1986; Idiz 1987). However, isotopic differences between pyrite and organic matter have also been reported (Dinur et al. 1980; Thode 1981; Orr 1986), generally, because some of the requirements mentioned above are not met, e.g. secondary sulfur from dissimilatory sulfate reduction has been added at a later stage. Furthermore, when the formation of pyrite is severely limited by the availability of reactive Fe (Gibson 1985; Raiswell and Berner 1985), S-rich kerogens may be formed (Gransch and Posthuma 1974) with isotopic compositions different from pyrite S. Data from Monterey shales (Idiz 1987) indicate that a limited supply of reactive Fe led to ^{32}S enrichment of 8 to 12‰ in the early formed pyrite, compared to S subsequently incorporated in kerogen (Orr 1986; Idiz 1987). Because functional reaction sites in immature protokerogen compete with available Fe for the reduced sulfur (Raiswell and Berner 1985), kinetic isotope effects are to be expected in cases when one of the two factors dominates. Presumably, kerogens will be less reactive towards H_2S than are sediment Fe phases, so they would be expected to be isotopically heavier. Gautier (1986) has observed that pyrite $\delta^{34}S$ values depend on the level of oxygenation of the depositional environment, but $\delta^{34}S$ values of coexisting organic-S were not determined.

5 Application to Crude Oil Exploration

The detailed examination of $\delta^{34}S$ distributions in organic fractions of crude oils and source rock extracts may be a promising fingerprinting technique in geochemical oil/oil-correlation and oil/source rock-correlation studies indicating possible genetic relationships between these phases. In addition, the $\delta^{34}S$ values of pyrite associated with source rock kerogen may also be of value. For illustration, some examples selected from measurements of many crude oil and rock samples from New Zealand and Kuwait are presented in Figs. 3 and 4, and are briefly discussed below. Table 1 lists S contents of these samples. While different environments of deposition are reflected in the S contents of oils, the concentrations of inorganic as well as organic sulfur in rocks are very similar.

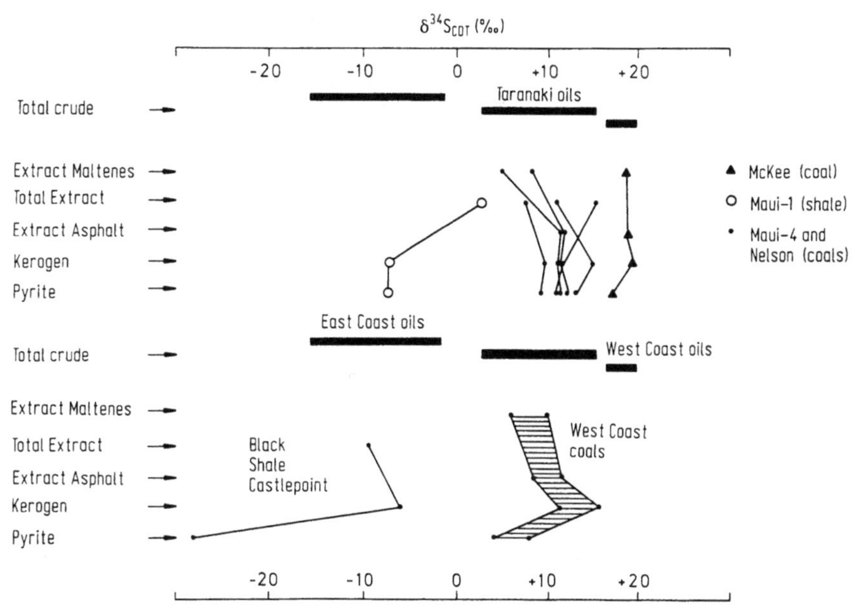

Fig. 3. Selected examples of $\delta^{34}S$ in crude oils, rock extracts, kerogen, and pyrite: New Zealand

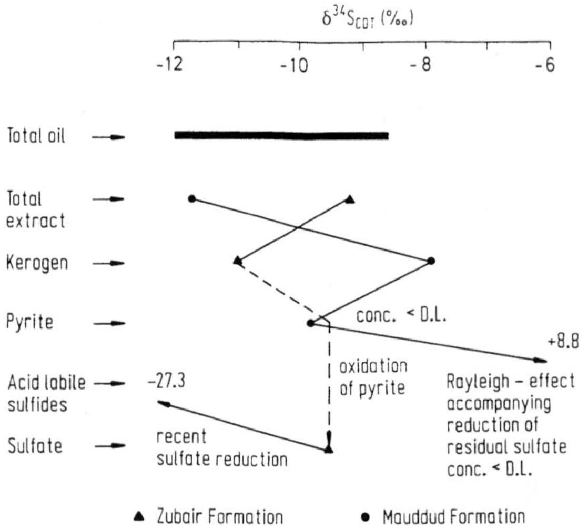

Fig. 4. Selected examples of $\delta^{34}S$ in organic and inorganic oil and rock constituents: Raudhatain oil field (Kuwait)

In the New Zealand samples all the $\delta^{34}S$ values of the Maui-4 and the Nelson coals plot within the range of the corresponding Taranaki oils, whereas those of McKee coal and the West Coast coals plot outside the ranges of the Taranaki and West Coast oils, respectively. The Maui-1 shale appears to have been impregnated by Taranaki oils. In Maui-1 the kerogen and pyrite have similar $\delta^{34}S$ values and were formed from the same source which was different from the Taranaki oil source. This is not the case for the Black Shale Castlepoint where the kerogen and pyrite $\delta^{34}S$ values are quite different. The S in the latter sample, representing basement rocks, should bear no relationship to the S in the oils. In general, the geochemical results reported here are in accord with geological evidence (Pilaar and Wakefield 1984) that Late Cretaceous to Early Tertiary coal measures were the source for the Taranaki oils.

In the case of the two samples from the Raudhatain oil field (Kuwait), rock extract, kerogen and pyrite $\delta^{34}S$ values cover about the same range of the corresponding oils. The $\delta^{34}S$ value of sulfate in the Zubair sample suggests that sulfate arose from complete oxidation of the pyrite (no isotope fractionation); subsequent bacterial reduction yielded acid-labile sulfides enriched in ^{32}S. The isotopically heavy monosulfides of the Mauddud sample are explained by a Rayleigh effect accompanying the reduction of residual sulfates. These examples were selected in order to demonstrate the specificity and sensitivity obtained in S-isotopic speciation. Results of another isotopic study (Robinson et al. 1991) point towards Lower Cretaceous source rocks for Kuwait crude oils in accordance with the oil-geological model presented by Dunnington (1967).

The results described in this chapter illustrate the potential use of relationships between the stable isotope compositions of inorganic and organic sulfur in crude oils, as well as rock extracts and kerogen, in hydrocarbon exploration. Stable sulfur isotope correlation methods may substantially supplement the information gained from other geochemical parameters such as biomarkers or trace element distributions (e.g. Hirner 1987).

6 Summary

Inorganic and organic forms of sulfur are linked to each other in the sedimentary sulfur cycle. Several relationships can be derived, which may be used to solve practical problems of exploration and environmental geochemistry:

1. Elemental S is incorporated in crude oils without significant isotopic fractionation (< 1‰).
2. The incorporation of elemental S in organic components is a specific process, and is reflected by the

Table 1. Concentration of sulfur species in selected crude oils and rocks

Sample locality	Sample type	Elemental S (ppm)	Organic S (wt%)	Pyritic S (wt%)
New Zealand	Terrestrial oils	< 1 to 12	0.06 to 0.38	
	Possible source rocks (coals)	< 0.4 to 54	0.1 to 8	0.01 to 1
S. Germany	Marine-brackish oils	2 to 77	0.23 to 0.91	
Kuwait	Marine oils	32 to 72	2.1 to 4.2	
	Possible source rocks (shales)	< 0.4 to 5	0.1 to 11	0.02 to 0.8

intermolecular distribution of ^{32}S and ^{34}S in organic fractions.

3. When mineral sulfur (in particular, pyrite sulfur) is included, isotopic crude oil/rock extract/rock kerogen correlation studies can be improved significantly.

This work deals with various S-containing phases in sediments of Tertiary and Mesozoic age. Organic S and pyritic S are also found in much older sediments, and it would be valuable to determine whether the relationships and trends described here also apply to more ancient systems, such as those of the Archean, for which models of bacterial sulfate reduction have been proposed (Ohmoto and Felder 1987; Ohmoto, this Vol.).

Acknowledgements. Critical comments by Drs. Rob Raiswell and Phil Trudinger are highly appreciated. This article was written, while the first author was supported by a Heisenberg grant of the Deutsche Forschungsgemeinschaft.

References

Aizenshtat Z, Stoler A, Cohen Y, Nielsen H (1981) The geochemical sulfur enrichment of recent organic matter by polysulfides in the Solar-Lake. In: Bjoroy M (ed) Advances in organic geochemistry 1981. John Wiley & Sons, New York, pp 279–288

Bauld J (1986) Transformation of sulfur species by phototrophic and chemotrophic microbes. In: Bernhard M, Brinckman FE, Sadler PJ (eds) The importance of chemical "speciation" in environmental processes. Dahlem Konferenzen 1986. Springer, Berlin Heidelberg New York, pp 255–273

Berner RA (1970) Sedimentary pyrite formation. Am J Sci 268:1–23

Berner RA, Baldwin T, Holdren GR (1979) Authigenic iron sulfide as paleosalinity indicators. J Sediment Petrol 49:1345–1350

Casagrande DJ, Ng L (1979) Incorporation of elemental sulfur in coal as organic sulfur. Nature (London) 282:598–599

Casagrande DJ, Idowo G, Friedman A, Rickert P, Schlenz D (1979) H_2S incorporation in coal precursors: origin of organic sulfur in coal. Nature (London) 282:599–600

Cook RB (1981) The biogeochemistry of sulfur in two small lakes. Thesis, Columbia Univ, New York, 233 pp

DeRoo J, Hodgson GW (1978) Geochemical origin of organic sulfur compounds: thiophene derivatives from ethylbenzene and sulfur. Chem Geol 22:71–78

Dinur D, Spiro B, Aizenshtat Z (1980) The distribution and isotopic composition of sulfur in organic-rich sedimentary rocks. Chem Geol 31:37–51

Dunnington HV (1967) Stratigraphical distribution of oilfields in the Iraq-Iran-Arabia Basin. J Inst Petrol 53:129–161

François F (1987) A study of sulfur enrichment in the humic fraction of marine sediments during early diagenesis. Geochim Cosmochim Acta 51:17–27

Fry B, Gest H, Hayes JM (1986) Sulfur isotope effects associated with protonation of HS^- and volatilization of H_2S. Chem Geol (Isot Geosci Sec) 58:253–258

Gautier DL (1986) Cretaceous shales from the western interior of North America: sulfur/carbon ratios and sulfur-isotope composition. Geology 14:225–228

Gibson DL (1985) Pyrite-organic matter relationships: Currant Bush Limestone. Georgina Basin, Australia. Geochim Cosmochim Acta 49:989–992

Goldhaber NB, Kaplan IR (1974) The sulfur cycle. In: Goldberg E (ed) The sea, vol 5. Marine chemistry. John Wiley & Sons, New York, pp 569–655

Gransch JA, Posthuma J (1974) On the origin of sulfur in crudes. In: Tissot B, Bienner F (eds) Advances in organic geochemistry 1973. Editions Technip, Paris, pp 727–740

Hackley KC, Anderson TF (1986) Sulfur isotopic variations in low-sulfur coals from the Rocky Mountain region. Geochim Cosmochim Acta 50:1703–1713

Hirner AV (1987) Geochemistry of crude oils of southern Germany. In: Hurst RW, Davis TE, Augustithis SS (eds) The practical applications of trace elements and isotopes to environmental biogeochemistry and mineral resources evaluation. Theophrastus, Athens, pp 217–231

Hirner AV, Lyon GL (1989) Stable isotope geochemistry of crude oils and of possible source rocks from New Zealand 1. Carbon. Appl Geochem 4:109–120

Hirner AV, Robinson BW (1989) Stable isotope geochemistry of crude oils and of possible source rocks from New Zealand 2. Sulfur. Appl Geochem 4:121–130

Hirner AV, Graf W, Treibs R, Melzer AN, Hahn-Weinheimer P (1984) Stable sulfur and nitrogen isotopic compositions of crude oil fractions from southern Germany. Geochim Cosmochim Acta 48:2179–2186

Howarth RW (1979) Pyrite: its rapid formation in a salt marsh and its importance in ecosystem metabolism. Science 203:49–51

Hunt JW, Smith JW (1985) $^{34}S/^{32}S$ ratios of low-sulfur Permian Australian coals in relation to depositional environments. Chem Geol 58:137–144

Idiz EF (1987) Studies on the interactions between organic matter, trace metals and sulfur in recent and ancient sediments. Thesis, Univ Cal, Los Angeles, 301 pp

Krouse HR (1977) Sulfur isotope studies and their role in petroleum exploration. J Geochem Explor 7:189–211

Krouse HR, McCready RGL (1979) Reductive reactions in the sulfur cycle. In: Trudinger PA, Swaine DJ (eds) Biogeochemical cycling of mineral-forming elements. Elsevier, Amsterdam, pp 315–368

LaLonde RT, Florence RA, Horenstein BA, Fritz RC, Silveira L, Clardy J, Krishnan BS (1985) Variable reaction pathways for the action of polysulfide on Michael acceptors. J Org Chem 50:85–91

LaLonde RT, Ferrara LM, Hayes MP (1987) Low-temperature, polysulfide reactions of conjugated ene carbonyls: a reaction model for the geologic origin of S-heterocycles. Org Geochem 11:563–571

Macqueen RW, Powell TG (1983) Organic geochemistry of the Pine Point lead-zinc ore field and region, Northwest Territories, Canada. Econ Geol 78:1–25

Mango FD (1983) The diagenesis of carbohydrates by H_2S. Geochim Cosmochim Acta 47:1433–1441

Martin TH, Hodgson GW (1973) Geochemical origin of organic sulfur compounds: reaction of phenylalanine with elemental sulfur. Chem Geol 12:189–208

Monster J (1972) Homogeneity of sulfur and carbon isotope ratios S^{34}/S^{32} and C^{13}/C^{12} in petroleum. Am Assoc Petrol Geol Bull 56:941–949

Nielsen H (1979) Sulfur isotopes. In: Jaeger E, Hunziker JC (eds) Lectures in isotope geology. Springer, Berlin Heidelberg New York, pp 283–312

Nriagu JO, Soon JK (1985) Distribution and isotopic composition of sulfur in lake sediments of northern Ontario. Geochim Cosmochim Acta 49:823–834

Ohmoto H, Felder RP (1987) Bacterial activity in the warmer, sulphate-bearing, Archaean oceans. Nature (London) 328:244–246

Orr WL (1978) Sulfur in heavy oils, oil sands and oil shales. In: Strausz OP, Loun EM (eds) Oil sand and oil shale chemistry. Chemie, Weinheim, pp 223–243

Orr WL (1986) Kerogen/asphaltene/sulfur relationships in sulfur-rich Monterey oils. Org Geochem 10:499–516

Pilaar WFH, Wakefield LL (1984) Hydrocarbon generation in the Taranaki Basin, New Zealand. In: Petroleum geochemistry and basin evaluation. Am Assoc Petrol Geol Mem 35:405–423

Powell TG, Macqueen RW (1984) Precipitation of sulfide ores and organic matter: sulfate reactions at Pine Point, Canada. Science 224:63–66

Price FT, Shieh YN (1986) Correlation between the $\delta^{34}S$ of pyritic and organic sulfur in coal and oil shale. Chem Geol 58:333–337

Prinzler HW, Pape D (1964) Zur Entstehung und zum postgenetischen Verhalten der organischen Schwefelverbindungen des Erdöls. Erdöl Kohle Erdgas Petrochem 17:539–545

Raiswell R, Berner RA (1985) Pyrite formation in euxinic and semi-euxinic sediments. Am J Sci 285:710–724

Rees CE (1973) A steady-state model for sulfur isotope fractionation in bacterial reduction processes. Geochim Cosmochim Acta 37:1141–1162

Reinecke MG, Morton DW, Del Mazza D (1983) The scope of the synthesis of tetrahydrothiophenes from α,β-unsaturated carbonyl compounds and sodium polysulfide. Synthesis 16:160–161

Rickard DT (1975) Kinetics and mechanism of pyrite formation at low temperatures. Am J Sci 275:636–652

Robinson BW, Hirner AV, Lyon GL (1991) Stable carbon and sulfur isotope distributions of crude oil and source rock constituents from Burgan and Raudhatain oil fields (Kuwait). Chem Geol (Isot Geosci Sec) 86:295–306

Schidlowski M (1987) Evolution of the early sulfur cycle. In: Redriguez-Clemente R, Tardy Y (eds) Geochemistry and mineral formation in the earth surface. Consejo Sup Invest Cient, Madrid, pp 29–49

Schidlowski M, Hayes JM, Kaplan IR (1983) Isotopic interferences of ancient biochemistries: carbon, sulfur, hydrogen, and nitrogen. In: Schopf JW (ed) Earth's earliest biosphere: its origin and evolution. Univ Press, Princeton, pp 149–186

Siegel LM (1975) Biochemistry of the sulfur cycle. In: Greenberg DM (ed) Metabolic pathways. Metabolism of sulfur compounds. Academic Press, New York London, pp 217–286

Sinninghe Dampste JS, De Leeuw JW, Kock-van Dalen AC, De Zeeuw MA, De Lange F, Rijpstra WIC, Schenck PA (1987) The occurrence and identification of series of organic sulfur compounds in oil and sediment extracts. I. A study of Rozel Point oil (USA). Geochim Cosmochim Acta 51:2369–2391

Sweeney RE, Kaplan IR (1973) Pyrite framboid formation: laboratory synthesis and marine sediments. Econ Geol 68:618–634

Thode HG (1981) Sulfur isotope ratios in petroleum research and exploration: Williston Basin. Am Assoc Petrol Geol Bull 65:1527–1535

4. Fossil Microbial Ecosystems of the "Stromatolitic" Type and Their Modern Analogs

The History and Significance of Stromatolites

Stanley M. Awramik [1,2]

[1] Department of Earth and Planetary Sciences Western Australian Museum, Francis Street, Perth, Western Australia 6000, Australia

Abstract

The fossil record of prokaryotes extends back 3500 million years and prokaryotes were only forms of life known for the first 2000 million years of Earth history. Understanding the fossil record of these organisms is important for the comprehension of the interaction of geological factors and the evolution of life. Prokaryotic microbial fossils, however, do not provide sufficient detailed evidence to help a great deal in this. Stromatolites, the biosedimentary products of microbe-sediment interactions, have the potential to provide significant information on the interaction of the biosphere, atmosphere, hydrosphere, and lithosphere throughout the entire history of life on this planet.

1 Introduction

Paleobiologists are identifying patterns in the diversity and distribution of fossil organisms in order to better understand evolutionary processes. This approach has been quite successful for studies dealing with the Phanerozoic Earth and its biota (Sepkoski 1978; Tiffney 1985; Valentine 1986; Allen and Briggs 1989). The Phanerozoic is well suited for such integrated studies because (a) it is reasonably well understood geologically, (b) its chronostratigraphy has high precision in many cases, (c) it contains a rich fossil record in most systems, (d) biostratigraphic correlations can be made with a fair degree of accuracy, (e) many of the fossils studied are well understood paleobiologically, and (f) a great number of geologists and paleontologists have been studying the Phanerozoic for almost two centuries providing a wealth of information. Yet, the Phanerozoic represents only the last 540 Ma of Earth history and, except for the biota and some minor configurational differences, like the positions of the continents, the Phanerozoic Earth was not fundamentally very different than the Earth of today.

Of all the fossils known, only prokaryotes and stromatolites can provide biogeological information on the interaction of biological and geological systems for the full extent of the 3500 million-year history of life known on this planet. Microbial fossils are found preserved in rocks as old as 3500 Ma (Awramik et al. 1983; Walsh and Lowe 1985). Stromatolites, organosedimentary structures produced by the sediment trapping, binding, and/or precipitation activity of microbial communities dominated by photosynthetic bacteria, are also known from 3500-ma-old strata (Walter et al. 1980; Byerly et al. 1986). Other major groups of organisms, such as protoctists or protists (including the algae), animals, and plants, did not appear until much later in geological history, and they provide biogeological information of greater precision and resolution than bacteria and stromatolites. The greatest changes in the history of the Earth, however, occurred before the appearance of animals. Since evolution occurs within the constraints and interactions of a changing Earth and its biota, understanding the fossil record of prokaryotes and stromatolites has a great potential to illuminate these phenomena.

2 Archean and Proterozoic Overview

The Archean and Proterozoic Earth was very different. In terms of time, these two eons (collectively called the Precambrian) span immense amounts of geological time, nearly 75% of Earth's history, from 3960 to 540 Ma ago. The Archean, 3960 to 2500 Ma ago (1460 millions of years in duration or 32% of Earth's history), is unlike any other eon in most of its geological and biological features. Prokaryotes were the only organisms that existed during the Archean. It

Present Address: Department of Geological Sciences, Preston Research Laboratory, University of California, Santa Barbara 106, USA

M. Schidlowski et al. (Eds.)
Early Organic Evolution: Implications for Mineral and Energy Resources

is probable that all extant archaeobacterial and eubacterial phyla had already evolved by Early Archean time. And, if analogy with the metazoan record (see Gould 1989) can be used, it cannot be ruled out that several other phyla evolved during the early diversification of prokaryotes that have since became extinct.

2.1 Archean

The nature of the Earth's crust and lithosphere during the Archean is the subject of some debate (for instance, see Park 1980). It is postulated that the Earth's crust during most of the Archean was probably thinner than younger crust due to higher geothermal gradients (see Ernst 1983). By Late Archean time, the thickness of the Earth's crust may have approached that of the present-day Earth (Newton 1978). Plate tectonics operated in the Archean, but probably involved small "platelets" (Ernst 1983). Sedimentary basins developed at the edges of these platelets and were similar to modern volcanic arc and arc-related marginal basins (DiMarco and Lowe 1989). Growth and stabilization of continental crust occurred by the accretion of island arcs and platelets (Barley et al. 1989).

The Archean Earth also differed from the Earth of younger eons in several ways. (1) The atmosphere lacked O_2 and contained primarily N_2, CO_2, and H_2O (Walker et al. 1983). (2) Ocean chemistry might have been different. Kempe and Degens (1985) suggest high sodium concentrations in the early ocean. Grotzinger (1989) postulates that the Late Archean and Early Proterozoic seas were highly supersaturated with calcium carbonate. Walker et al. (1983), however, suggest that seawater chemistry in the pre-Phanerozoic was not substantially different from younger times. (3) The oceans may have been substantially warmer (Karhu and Epstein 1986). (4) Solar luminosity was lower, possibly by as much as 40% early in Earth's history (Newman and Rood 1977). The lower solar radiation's effects on climate were offset by higher atmospheric CO_2 (Kasting 1989) which produced temperatures that permitted liquid water to exist on the Earth's surface since at least 3800 Ma ago.

2.2 Proterozoic

The Proterozoic eon, 2500 to 540 Ma ago, represents about 43% of Earth's history. It was during this long span of time, some 1960 million years, that the Earth underwent some of its greatest changes. During the Late Archean to Early Proterozoic geothermal gradients decreased, the crust thickened, and seafloor spreading and plate tectonics involving large plates began (Ernst 1983). The atmosphere-hydrosphere changed from essentially anoxic to oxic about 2300 Ma ago (Cloud 1976). Two thousand million years ago, the Sun's luminosity was still probably 15% below present levels (Gilliland 1989).

The Early Proterozoic biosphere was probably prokaryotic through most of its history; however, geochemical evidence (Summons et al. 1988) indicates and paleontological evidence (Hofman and Chen 1981; but also see Zhang 1986) suggests that eukaryotes evolved by 1700 Ma ago. The evolution of eukaryotes was one of the most significant phenomena in the history of life. The possible interplay between prokaryotes in benthic microbial communities and diversifying eukaryotes during the Middle and Late Proterozoic is unknown.

3 Limitations of the Prokaryote and Stromatolite Fossil Records

Although the record of prokaryotes and stromatolites dominate about 85% of the history of life on this planet (up until about 700–800 Ma ago), there are severe drawbacks to using these fossils as the source of detailed biogeological information. Among the most important drawbacks: (1) stromatolites are not that well understood paleobiologically. It is generally accepted that stromatolites are biogenic sedimentary structures produced by the sediment trapping, binding and/or precipitation activity of phototrophic microbes. Cyanobacteria are the predominent organisms responsible for stromatolite construction today and presumably this was true in the past, as far back as the Archean (see Awramik 1990). Numerous examples of Recent stromatolites are known (e.g Logan 1961; Monty 1967; Dill et al. 1986) and a number of occurrences have been studied in some detail (e.g. Walter et al. 1976; Playford 1980). However, the application of the conclusions drawn from studies on the Recent examples has not been very successful in explaining the shape, distribution, and biostratigraphic utility of many ancient stromatolites, in particular those from Proterozoic strata. Most of the stromatolites of Archean, Proterozoic, and Early Paleozoic seas were very different from those found growing today. For instance, two of the most important differences that separate Proterozoic stromatolites from most older and younger stromatolites are (a) the abundance and diversity of columnar stromatolites with elaborate branching patterns, and (b) centimeter- to meter-sized conical stromatolites, some with branches. Empirical data on the distribution in time of different stromatolite shapes indicate that certain definable morphological varieties have restricted time ranges (Keller et al. 1960; Cloud and Semikhatov 1969). Satisfactory solutions are not yet available that explain the morphogenesis of these time-restricted stromatolites. (2) Prokaryotes appear to exhibit limited body plans. Morphological conservatism (sometimes equated with evolutionary conservatism) is a noted feature of prokaryotes when the fossil record is compared with extant forms. Much of the apparent diversity in prokaryotes is manifested in the different metabolic pathways utilized, in other biochemical differences, and in some cases morphology. Most of these metabolic/biochemical and morphological differences which have been the basis of a great deal of bacteriological systematics, are now found to be shared by phylogenetically disparate groups (see Woese 1987). Thus, paleobiology which depends upon the morphology of the fossil studied within the context of populations, paleoecology, and taphonomy, is severely handicapped. The one phylum of Eubacteria that appears to be metabolically unique, phylogenetically circumscribed, and contains a number of distinctive complex morphologies is the Cyanobacteria. Fortunately, representatives of the Cyanobacteria are the most abundant prokaryotes in the fossil record (see Knoll and Golubic, this Vol.).

4 The Stromatolite Record Versus the Prokaryote Microfossil Record

Which fossils, stromatolites or prokaryotic, microfossils, can provide the more useful biogeological information? The strengths and weaknesses of both are reviewed briefly below.

4.1 Prokaryotes

The strengths of the fossil record of prokaryotes rest in its great antiquity, the large number of pre-Phanerozoic localities, the presence in rocks of permineralized three-dimensionally preserved microbes representing the remains of the microbes, the relative abundance of microbial fossils in certain pre-Phanerozoic sedimentary rocks, and the numerous modern analogs available for comparison. There are, however, several important limitations to the prokaryote fossil record.

1. Prokaryotes, although metabolically more diverse than eukaryotes (Carlile 1982), have achieved little visible morphological complexity. Many morphologically similar microbes represent metabolically and phylogenetically diverse kinds of organisms (e.g. see Awramik et al. 1983). Since the paleontology of microbial prokaryotes is based on the morphology of the fossils, the ability to recognize many prokaryotic groups confidently is severely limited. As stressed by Knoll and Golubic (this Vol.), morphological interpretation is maximized when done in the context of environmental and taphonomic data. There is little evidence from the fossil record that prokaryotes have changed systematically with time.

2. The prokaryotic fossil record appears to be strongly biased toward one phylum, the Cyanobacteria. Occasionally other prokaryotes can be recognized (Oehler 1977; Allison and Awramik 1989); however, because of the very small size and simple morphology of many of them, they cannot confidently be assigned to specific prokaryotic phyla. Cyanobacteria, on the other hand, are more easily recognized because they are larger than most other bacteria and they possess greater morphological complexity. Taphonomic factors have greatly influenced the apparent dominance of Cyanobacteria in the fossil record of prokaryotes. The fossilization process appears to favor Cyanobacteria. Innumerable pre-Phanerozoic microbial fossils are found preserved in chert that has permineralized the microbes. Cyanobacteria often produce copious amounts of extracellular mucopolysaccharides in the form of envelopes and sheaths. These outer coverings are commonly more resistant to degradation than cells (Horodyski and Vonder Haar 1975) and may be more readily silicified than other parts of the organism (Golubic and Barghoorn 1977; Knoll and Golubic 1979). Primary and early diagenetic silicification and chert formation, at least in the pre-Phanerozoic, are biased toward very shallow to periodically exposed carbonate environments, environments that are frequently hypersaline (Knoll 1985; Southgate 1986). These environments were, and today still are, common sites for cyanobacteria growth and stromatolite formation.

It is likely that all extant seven major archaebacterial subdivisions and ten eubacterial phyla (Woese 1987) and possibly several extinct phyla had evolved by 3500 million years ago (Awramik in press). Subsequent evolutionary changes were at lower taxonomic levels, involved some metabolic innovations, and morphological change was built upon already established body plans. However, this resulted in little that would lend itself to being preserved in the fossil record. Thus, the ability to recognize changes in the biosphere using prokaryotic fossil taxa is rather limited. Within the Cyanobacteria, which are morphologically larger and more complex than other Eubacteria, changes within this phylum can be more easily recognized in the fossil record. Both classes of Cyanobacteria, the Coccogonae and Hormogonae, might have been present by 3500 million years ago and all orders, except for possibly the Stigonematacea, were present by 2000 million years ago (Awramik, in preparation; see also Knoll and Bauld 1989). Since about 2000 Ma ago, little change at high taxonomic levels has apparently occurred in Cyanobacteria. An increase in size of filaments and coccoid morphs has been observed (Schopf 1977), diversity at the genus and species level has increased (Awramik, unpubl.), and organizational complexity of pluricellular chroococcalean genera has increased during the Middle and Late Proterozoic (Strother 1989).

3. The number of formations with fossilized prokaryotic microbes is not great, probably on the order of a few hundred and this spans 3500 Ma of geological history.

4. The Phanerozoic fossil record of prokaryotes preserved in chert is limited. Except for the Early Cambrian (e.g. Wang et al. 1983) there are probably no more than 20 currently known and most of these are from Devonian strata. However, calcified cyanobacteria appeared at least by the earliest Cambrian (Riding and Voronova 1984) and they are locally abundant in various Paleozoic formations. Except for their remains in oils, the record of prokaryotes in the Mesozoic and Cenozoic is sparse.

4.2 Stromatolites

Stromatolites have an antiquity that matches the record of microfossils. Both types of fossils represent the oldest evidence for life on Earth. Strata as old as 3500 Ma old in Western Australia and South Africa contain stromatolites and microbial fossils.

Ancient stromatolites do not appear to have suffered from many of the same biases that influenced the prokaryote record. Preservation of stromatolites was enhanced by widespread penecontemporaneous and/or early diagenetic precipitation of calcium carbonate and, in some cases chert, producing hard, mineralized structures readily preserved in the rock record. Although not as ecologically widespread as prokaryotes, stromatolites were almost as widespread as the dominant microbes that built them, the Cyanobacteria, and formed in a variety of aqueous environments (Fig. 1) that included: (1) subtidal marine (basin to slope: Hoffman 1974; above fair-weather wave base: Bertrand-Sarfati and Moussine-Pouchkine 1985; lagoons: Fairchild and Herrington 1989); (2) peritidal marine settings (Green et al. 1989); (3) sabkhas or transitional marine-nonmarine settings (Grotzinger 1986); (4) lakes (Karpeta 1989); (5) rivers (Buck 1980); and (6) thermal springs (Walter 1972a). The number of localities and stratigraphic units containing stromatolites is large, probably numbering well into several thousand.

Unlike the majority of prokaryotes found in the fossil record, stromatolites exhibit a wide range in morphology. The more distinctive shapes found include: cone-shaped structures (Komar et al. 1965), laterally-linked cones (Vlasov 1977), conical forms with branches (Shapovalova 1968), unbranched columns (Korolyuk 1960), branching columns (Krylov 1963), domes (Kennard 1981), stratiform stromatolites (or planar-laminated stromatolites or "cryptalgalaminites"; Aitkin 1967), and round structures

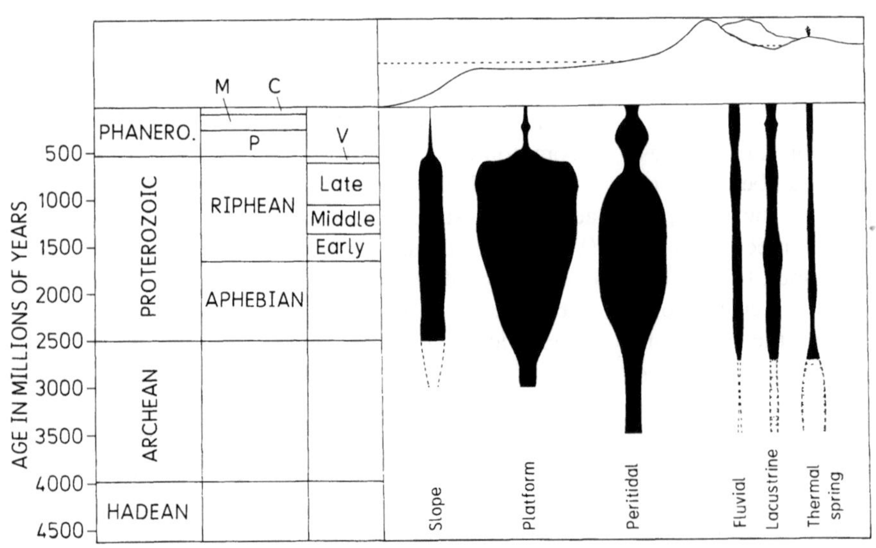

Fig. 1. Hypothetical abundance of stromatolites in major environmental settings.

with encapsulating laminae called oncolites (Zhuravleva 1964). All these can occur in bioherms, biostromes, or they can also occur as isolated structures. On a meso- to microscopic scale, there is also great variability in the nature of the microstructure and laminae.

Size is extremely variable among stromatolites. For instance, conically shaped structures can vary in diameter from less than 1 cm for Recent forms in Yellowstone National Park (Walter et al. 1976) to 10 m in Middle Proterozoic of Canada (Donaldson 1976). However, the size range of any given stromatolite form in outcrop rarely varies by more than a factor of three and in many time-restricted taxa, this variability is often much less.

Stromatolites appear to have the potential to reflect the evolutionary history of Cyanobacteria; not necessarily of individual species but of the interplay of new species, microbial populations, and communities. Today, other microbes besides cyanobacteria, such as numerous different kinds of bacteria, red algae, green algae, and diatoms, belong to stromatolite-building communities. Because of the lack of unequivocal remains of these noncyanobacterial organisms in fossil stromatolites that might have been part of the microbial stromatolite-building community (planktonic microbes are not uncommon in ancient stromatolites), it is difficult to assess their role and impact on ancient stromatolites.

5 Stromatolites as Carriers of Biogeological Information

The fact that particular morphological varieties of stromatolites (in particular among the columnar-branching, conical, and conical-branching types) have limited time ranges in the Proterozoic and Cambrian provides additional empirical evidence for the biogeological importance of stromatolites. Also, the abundance and richness of different morphological varieties of stromatolites provide additional raw data for attempting to understand the interaction of the Earth and biosphere during Earth history before the appearance of animals. The microstructure of stromatolites appears to be under the direct influence of the stromatolite-building microbial community (Gebelein 1974). Communities dominated by filamentous microbes usually produce laminated fabries while coccoid-dominated communities usually produce poorly laminated to unlaminated fabrics. The stomatolites most useful in biostratigraphy are the columnar-branching and conical forms, and these stromatolites are normally laminated. One of the most severe limitations to stromatolite biostratigraphy and the source of strong criticism is that no specific and compelling explanations have been offered that correlate the relationship between microbes and the shape of stromatolites. On the other hand, those researchers who dispute the validity of stromatolite biostratigraphy and contend that environmental rather than biological factors control stromatolite morphology have not satisfactorily explained the fact that stromatolites can be used successfully for age determinations of sedimentary units (see Grey 1984). Until examples of specific biogeological data encoded in stromatolites can be found beyond the generalities mentioned here, we are left to speculate on the relationship between microbes and stromatolite morphology. One such speculation is that stromatolites are sensitive to change at the evolutionary level. The evolution of new microbial taxa within the benthic mat community or taxa that move into the community could bring about changes in the dynamic interactions of the microbial community with sediment and local sedimentary processes. For instance the introduction into the microbial community of a new microbe that produces a sheath or envelope that is stickier than that possessed by the other resident microbes might improve the overall cohesiveness of the microbial mat. This increased cohesiveness of the mat could change the response of the microbial mat to sediment and water movement and result in an overall change in the morphology of the stromatolite produced. The environment, however, must not be overlooked becaused it exerts control on the distribution of the microorganisms. As the environment changes, on an ecological as well as evolutionary time scale, changes will be brought about in the constituent microbes constructing stromatolites. An analysis of the distribution in time and space of stromatolites provides a basis for using stromatolites as a primary source of biogeological data for long-term changes at the surface of the Earth that occurred during the pre-Phanerozoic.

6 Archean Stromatolites

The oldest stromatolites are known from two areas, one in the Pilbara region of Western Australia (Lowe 1980b; Walter et al. 1980; Buick et al. 1981) and the other in the Barberton Mountain Land of South Africa (Byerly et al. 1986). The strata containing these silicified stromatolites are radiometrically dated at about 3500 Ma (Pidgeon 1978; Kronen and Todt 1988). Although not morphologically complex, the stromatolites nevertheless are well laminated and range in shape from undulatory, stratiform structures to domical structures. Oncolites have also been found. Well-defined, columnar or columnar-branching stromatolites have not yet been discovered. A pseudocolumnar geometry in the South African examples is formed by laterally linked domes and short columns (Byerly et al. 1986). Conically shaped, laterally linked stromatolites were reported from some Pilbara stromatolites (Lowe 1983).

Filamentous and coccoidal microbial fossils have been detected in laminated chert near both the Western Australian (Awramik et al. 1983, 1988; Schopf and Packer 1987) and South African (Walsh and Lowe 1985) stromatolite localities. Most of the Western Australian filaments occur within the thin, dark laminae of wavy light-dark lamina couplets; the filaments are commonly oriented parallel to lamination. This arrangement of the microbial fossils is suggestive of stromatolitic laminae (Awramik et al. 1983, 1988). A number of the South African filamentous microbial fossils coat detrital grains, although no well-developed multiple laminae were reported. Microfossils have not yet been discovered in well-defined stromatolites with convex laminae.

Although the Archean represents 1440 million years of geological time and stromatolites appeared within the first 500 million years of the Archean, records of stromatolites are rare during this eon. Fewer than 20 occurrences are currently known, most of these are from the Late Archean, and some are nonmarine (Fortescue stromatolites, see Walter 1983; Ventersdorp stromatolites, see Buck 1980). By the end of Archean time, the majority of the morphological varieties of stromtolites had appeared. These include wavy laminated (stratiform), nodular, domical, pseudocolumnar, columnar, conical, and oncolitic types. Columnar and columnar-branching stromatolites, although present in the Late Archean Fortescue Group (Cloud and Semikhatov 1969; Walter 1972b), do not become abundant and diverse until the Early Proterozoic. The overall shape and microstructure of Archean stromatolities are similar to many Proterozoic and younger examples (Walter 1983).

The presence of stromatolites in 3500-Ma-old Early Archean strata has great paleobiological significance. The construction of a stromatolite requires microorganisms that are metabolically, behaviorally, and morphologically quite sophisticated at a prokaryotic level of evolution (Walter 1983; Awramik 1986, 1990). This indicates that by 3500 Ma ago microbial evolution had already evolved to a rather advanced type of prokaryote. Photoresponsive prokaryotes capable of taxis would be a minimum reasonable level of development required. The overall similarity of the macro- and microstructure of the stromatolites to many younger examples permits one to suggest that cyanobacteria were involved in their construction. The microbial fossils and the stable isotopic geochemistry of the material lend support to these conclusions. Carbon isotope values of kerogen in both the Western Australian and South African cherts yield values around $-30\delta^{13}C$, indicating autotrophic fractionation of carbon, but not necessarily phototrophy (Hayes et al. 1983). The filamentous microfossils from the Western Australian chert present evidence consistent with a cyanobacterial interpretation (presence of sheaths, large size of sheaths, large trichomes, sheaths and trichomes preserved in dark laminae; Awramik et al. 1983). The encapsulated pluricellular coccoids, resembling the chroococcaleans described by Schopf and Packer (1987) from the Western Australian chert, argue well for the existence of cyanobacteria. The cumulative weight of microfossil, stromatolite, and isotopic geochemistry data provides presumptive evidence that Cyanobacteria had evolved by Early Archean time, 3500 Ma ago.

Phylogenetic reconstructions of prokaryotes based on molecular biological techniques indicate that Cyanobacteria are one of the last prokaryotic phyla to evolve (Woese 1987). This supports a similar conclusion based on more traditional lines of research that among prokaryotes, Cyanobacteria are phylogenetically one of the last phyla to appear (e.g. see Margulis and Schwartz 1982, p. 24). Therefore, the presence of Cyanobacteria in 3500-Ma-old rocks is not a trivial issue for it would imply that prokaryotes had undergone a great deal of evolution at the highest taxonomic levels prior to 3500 Ma ago and that most major clades of prokaryotes had already appeared by the end of the Early Archean.

The paucity of Archean stromatolites as a whole and the greater number of 3000-Ma-old and younger Archean stromatolites compared with older Archean examples are likely due to geological rather than biological factors. Most Archean sedimentary basins older than 3000 Ma in age developed within greenstone belts and sedimentation was characterized by the rapid deposition of compositionally immature sediments in subsiding basins or in unstable alluvial settings (Lowe 1980a). Stable, shallow-water environments not flooded with sediments were probably uncommon and/or short-lived in such tectonic settings. Carbonates apparently were scarce (Cameron and Baumann 1972). Such active settings would find it difficult for phototropic microbes to maintain themselves at or near the sediment-fluid interface, influence sedimentation, and produce stromatolites. The development of features approximating continental shelves and platforms during the Late Archean produced sizable, stable, shallow-water environments and resulted in environments conducive to microbial construction of stromatolites, carbonate deposition, and a proliferation of stromatolites (Cloud 1976).

Only a few stromatolites, such as *Alcheringa narrina* (Walter 1972b) and *Hadrophycus immanis* (Fenton and Fenton 1939) have been taxonomically described from Archean strata. The dearth of taxonomically treated stromatolites is due to several factors. (1) Little attention has been paid to Archean stromatolites with regard to their taxonomy and biostratigraphic potential. The emphasis has been on the Proterozoic. (2) Archean stromatolites appear to be quite variable in size and shape in outcrop. (3) Complexly branching columnar morphs, the darlings of stromatolite taxonomists/biostratigraphers, are absent. The lack of complex morphs and the morphological variability may have a paleobiological basis. Although the basic stromatolite-building phenotypes evolved in the Archean, it may not have been until the Early Proterozoic that the diversification of Cyanobacteria at lower taxonomic ranks had their full impact on the expression of stromatolite form.

7 Proterozoic Stromatolites

Stromatolites are more abundant and diverse in the Proterozoic than at any other time in Earth history. Early in the Proterozoic, conical, columnar, and columnar-branching stromatolites became conspicuous components of stromatolite assemblages. Studies on extant stromatolites that elucidated environmental factors important for stromatolite formation (e.g. Black 1933; Logan 1961) and the intensive study of pre-Phanerozoic sedimentary basins during the last four decades, resulted in stromatolites playing a prominent role as important indicators of environmental conditions (Hoffman 1967; Horodyski 1977). Motivated in part by the suggestion of Maslov (1939) that stromatolites might be used for biostratigraphy, scientists in the USSR performed detailed taxonomic and stratigraphic studies on stromatolites (see Semikhatov 1976). These studies resulted in workable schemes for the biostratigraphic correlation of Proterozoic strata the world over (see reviews by Semikhatov 1976; Bertrand-Sarfati and Walter 1981). Over 600 stromatolite taxa have been described from Proterozoic strata (Awramik, unpubl.). This compliments the numerous references to nontaxonomic reports on occurrences of Proterozoic stromatolites (e.g. see Hofmann 1981, for Canadian examples), and provides a robust body of information for the analysis of stromtolites within a biogeological framework.

Stromatolite diversity and abundance have been the subject of discussions in several papers (Fischer 1965; Awramik 1971; Monty 1973; Pratt 1982; Walter and Heys 1985; Zhu et al. 1987; Cao 1988; Awramik, 1991). The database on stromatolite taxa and occurrences by Walter and Heys (1985) contains the time ranges of 485 stromatolite taxa and this greatly expanded the 70 taxa used for the diversity curve published previously by Awramik (1971; see Fig. 2A, B). Examination of the data upon which the Walter and Heys' histograms were based indicated that their database was incomplete. An additional 178 new taxa and an increment of about 300 occurrences/formations were appended to the Walter and Heys' database and used to produce the histogram in Fig. 2C. However, even this histogram represents an incomplete and imperfect representation of the data on the diversity of stromatolite forms. The database is not exhaustive: numerous new taxa remain to be added, the occurrence of many stromatolite taxa in areas outside their type sections have to be included, and the chronology of many stratigraphic units needs to be established with more confidence. Also, the incredibly complex problems of synonymy and validity of described stromatolite taxa need to be resolved. Nevertheless, incorporating all useful data available to date, the frequency distribution of stromatolite taxa is presented in Fig. 2C for discussion. This frequency distribution has important differences when compared to Fig. 2B. An apparent bimodal distribution is evident; the greater mode in the Middle Riphean (1350 to 1050 Ma ago) and the lesser mode in the Late Aphebian (2200 to 1650 Ma ago). At this time, it is uncertain how the frequency distribution of stromatolite taxa will appear after the database is thoroughly revised and updated.

The evolving stromatolite histograms (or "diversity curves"; Fig. 2A–C) illustrate that it may be premature to attempt to decipher stromatolite history from the diversity patterns derived from an incomplete set of stromatolite data. New stromatolite taxa will continue to be described. A great deal of taxonomic revision is required of many of the stromatolite forms described in the literature. In addition, chronostratigraphic data of greater confidence are needed for stromatolite-bearing strata. This should not prevent, however, the analysis of new information every half decade or so in order to determine whether any patterns are apparent, to formulate hypotheses to explain the data analysis, and to test the consequences of the hypotheses.

The following discussion presents the observations and hypotheses that have been formulated based on frequency distributions presented in Fig. 2A, B:

1. The Awramik 1971 "diversity curve" (Fig. 2A). The striking features of this curve were (a) the marked drop in diversity during the Vendian and Cambrian, and (b) the gradual increase in stromatolite diversity from Early Aphebian to Upper Riphean time (Awramik 1971). The increase in diversity was used to support a concept of morphological evolution of stromatolites in the pre-Phanerozoic and the decrease in diversity was hypothesized to be due to the effects of grazing and burrowing animals that appeared in the Vendian (Awramik 1971). Monty (1973) criticized the metazoan disturbance hypothesis as the cause of the diversity decrease, noting that there was no direct evidence of bioturbated Vendian stromatolites. Indeed, even for the Phanerozoic, when compelling evidence for grazing and burrowing metazoans is found associated with stromatolites (Runnegar et al. 1979), direct evidence for burrowed stromatolites is exceedingly rare (e.g. Kepper 1974; Cussey and Friedman 1976). Monty proposed that the decrease was due to competitive exclusion by benthic algae (eukaryotes) and he also suggested that changes in seawater chemistry might have played a role in the stromatolite decrease affecting the ability to form readily calcified structures. Pratt (1982) felt the decline was an artifact of incomplete knowledge of Phanerozoic stromatolite abundance. Monty (1984, p. 428), citing the work of Vidal and Knoll (1982) on acritarch diversity and a Late Proterozoic ecosystem crises due to the effects of glaciation, suggests that glaciation might have contributed to the decline of stromatolites.

2. Walter and Heys' 1985 "diversity curve" (Fig. 2B). The frequency distribution for all stromatolite taxa in Walter and Heys (1985) shows much the same general features as Awramik's 1971 curve: (a) an increase indiversity in the Early and Middle Proterozoic, and (b) a profound decrease in diversity in the Vendian. Walter and Heys' data set was far more robust than Awramik's and utilized the time ranges of 485 taxa, subdividing the stromatolites into stratiform, coniform, and branching-columnar morphs, and also included data on the abundance of stromato-

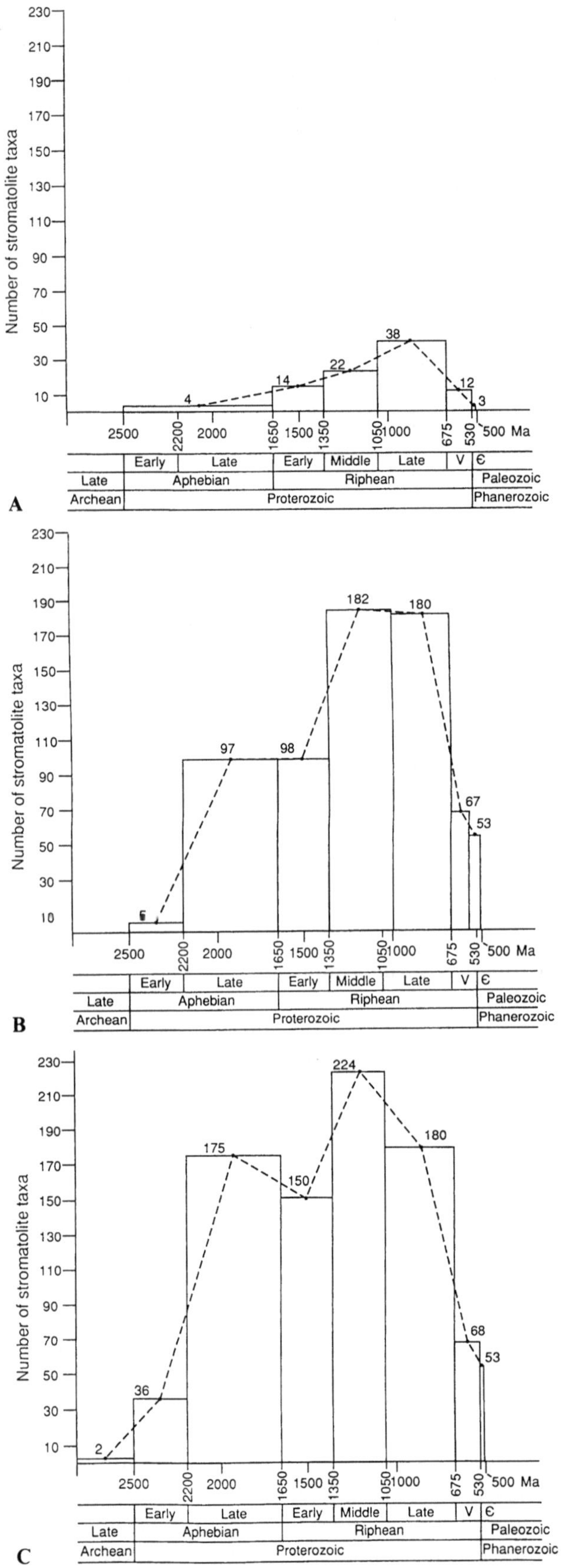

Fig. 2. Frequency distributions of stromatolite taxa through time: the "evolving stromatolite histograms or diversity curves". **A** Data from Awramik (1971). **B** Data from Walter and Heys (1985) to appear in Awramik (1991). **C** The 1990 frequency distribution.

lites. Like Awramik, Walter and Heys concentrated on the profound diversity decline in the Vendian to Cambrian and its causes. They agreed with Awramik that the probable cause for the diversity decline were the effects of metazoans. The data on coniform-shaped stromatolites showed a decline during the Late Riphean, a few hundred million years before the Vendian decline. Walter and Heys attributed this decline of coniform stromatolites, which they noted formed in quiet subtidal environments, to the appearance and evolution of early grazing metazoans. The overall Vendian to Cambrian decrease was due to the appearance of burrowing animals.

Using the data in Walter and Heys (1985), Awramik (1991) noted some additional features in their frequency distribution:

(1) Two diversifications occurred one during the late Early Proterozoic, 2350 to 1925 Ma ago, and the other during the Early to Middle Riphean, 1500 to 1200 Ma ago. (2) Two diversity plateaus occurred, one during the late Early Proterozoic and into the early Early Riphean, around 1925–1500 Ma ago, and the other from Middle to Late Riphean, around 1200–860 Ma ago.

Awramik (1991) went on to offer some possible explanations for these patterns. The first diversification in the Early to Late Aphebian (2350 to 1925 Ma ago) was attributed to adaptive radiation of the microbes that built stromatolites and of the stromatolites themselves (viewing stromatolites as types of microbial ecosystems). Oxygenation of the atmosphere beginning around 2300 Ma ago was postulated to have influenced a cyanobacterial radiation (many extant Cyanobacteria are microaerophilic). In addition, cratons were well established by this time and intercratonic interactions, produced more "basins" (26 noted in Walter and Heys 1985), and thus provided more shallow-water environments for stromatolites.

The second diversification during the Early to Middle Riphean (approximately 1500 to 1200 Ma ago) was more difficult to explain. The diversification followed a diversity plateau. The microbial fossil record during this interval is better than during the first diversification; however, only a slight increase in cyanobacterial taxa from stromatolites is apparent (e.g. Schopf 1977). It was then postulated that the appearance and diversification of eukaryotes influenced stromatolites. Competition between benthic eukaryotic algae and cyanobacteria might have resulted in a further cyanobacterial diversification, giving rise to an increase in stromatolite taxa.

With regard to the diversity plateaus in Fig. 2 B, the overall pattern of diversification, followed by a tapering off (the plateau), resembles the sigmoidal growth-phase curve that increasing populations of various organisms follow and it also resembles the curve of Sepkoski (1979) for the early Phanerozoic diversification of animals, complete with the plateau. Such plateaus or tapering off in diversification are usually explained as a result of filling of available ecological zones by the clades (Bambach 1985). The adaptive ranges of the stromatolite-building microbes, primarily cyanobacteria, might have reached a diversity balance related to the ecological space the organisms were able to utilize.

3. The 1990 stromatolite taxa frequency distribution (Fig. 2C). This frequency distribution is presented for illustration only. It suffers, among other things, from monographic biases of three large monographs. (a) Liang et al. (1984) on the Pseudogymnosolenaceae (a proposed family of predominently columnar-layered ministromatolites with diameters usually less than 2 cm), who describe 50 new taxa from China from 1200 to 1500 Ma old strata (late Early to early Middle Riphean). (b) Zhu et al. (1987) on Early Proterozoic stromatolites from the Mt. Wutai region of China, who described 66 new taxa: 6 from the Early Aphebian and 58 from the Late Aphebian. (c) Makarikhin and Kononova (1983) add 19 taxa to the Aphebian. Also, numerous references describing new stromatolites and noting the new occurrences of stromatolite taxa in formations have yet to be added to the database. Synonymy is a major problem in stromatolite taxonomy and this must be addressed. In addition, chronostratigraphic information on the stromatolites is often too imprecise or vague, and greater resolution is needed. And, as noted above, the taxonomy of stromatolites is far from perfect. A major updating of the database will provide a test for the hypotheses presented. Is there a gradual increase in stromatolite diversity through the Early and Middle Proterozoic? Is the decimation of stromatolite taxa in the latest Proterozoic essentially confined to the Vendian or does it begin well into the Late Riphean? Do different types of stromatolites show different or similar diversity patterns? Are there two diversifications and two diversity plateaus? Are there two diversifications separated by a decrease in diversity (extinction)?

8 Phanerozoic Stromatolites

The 540 million years of the Phanerozoic were an interval of time when stromatolites underwent profound change. The decrease in diversity that began in the Vendian continues into the Cambrian as does the decrease in abundance (Walter and Heys 1985). Two variants of stromatolites become common: coarse-grained stromatolites and thrombolites. Thrombolites, unlaminated, domical to columnar stromatolites with a clotted mesostructure first appear in the Late Riphean (Aitkin and Narbonne 1989) and are most common in Late Cambrian and Early Ordovician strata (Kennard and James 1986). Thrombolites formed bioherms, some in association with metazoans (Pratt and James 1982). Many of Paleozoic stromatolites are coarser grained ($>50\,\mu m$) than the great majority found in the pre-Phanerozoic (Awramik and Riding 1988). These grains include bioclastic debris from skeletonized metazoans which were absent in the pre-Phanerozoic. Unfortunately, comparatively little detailed descriptive/taxonomic work has been done on Phanerozoic stromatolites, therefore, it is difficult to analyze adequately major features on the temporal and spatial distribution of stromatolites. However, it is evident from the fossil record that impressive changes did occur in the Early Paleozoic. Branching columnar stromatolites continued to decline; most branching was bifurcate and branches commonly diverge very little. Domical and nonbranching columnar stromatolites were probably the most common morphs. Although no hard data exist, oncolites may have been more abundant in the Early Paleozoic than at any other time. Since the Middle Ordovician, stromatolites were minor components of marine ecosystems. Occasionally, they did produce bioherms, such as those from the Devonian of the Canning Basin of Western Australia (Playford et al. 1976) and the Permian Zechstein of Central Europe (Paul 1988). Stromatolitic domes and small unbranched columns are not uncommon components within reefs dominated by metazoans (Pratt 1982). There are also hints that stromatolites had a slight resurgence in abundance in the Early Triassic.

Stromatolites are well represented in Phanerozoic nonmarine environments, in particular from lake deposits of the Cenozoic. All varieties of morphology are found in nonmarine settings with domical, columnar, and columnar-branching examples common. The columnar-branching varieties appear to be morphologically more variable than many Proterozoic and Phanerozoic marine examples and the branching is often not complex. The abundance of Cenozoic examples reflects the geologically recent nature of the deposits and tectonic settings favoring the formation of lakes, for instance basin and range with stromatolites (McAllister and Ross 1982) and rift valleys with stromatolites (Casanova 1986).

It appears that the capacity to calcify while accreting is important for marine microbial mats if they are to give rise to stromatolites with a high potential for preservation. The trapping and binding of sedimentary grains rather than the precipitation of calcium carbonate was and is the primary mode of formation for Phanerozoic marine stromatolites. Nonmarine stromatolites did not appear to experience the devastation brought about by animals as did marine stromatolites. There are at least two reasons for this: (1) the relatively reduced numbers of grazers and burrowers in these environments; and (2) precipitation of calcium carbonate often accompanies microbial mat/stromatolite formation. However, the evolution and diversification of lacustrine metaphytes must have had some effect on nonmarine stromatolites. This remains an unexplored area of research.

Today, stromatolites are found forming in a wide range of environments; hypersaline (Logan 1961) and nonhypersaline marine (Dill et al. 1986); saline (Grey et al. 1990) and fresh water (Winsborough and Golubic 1987) lakes, streams (Golubic and Fischer 1975), and thermal springs (Walter et al. 1976). In the marine realm, stromatolites from peritidal environments predominate, few subtidal platform occurrences are known (Dill et al. 1986), and no slope examples have yet been thoroughly documented.

It is difficult to discern meaningful patterns in the spatial and temporal distribution of Phanerozoic stromatolites. Unlike the Proterozoic, few researchers specialize in stromatolites from this eon. The literature on Phanerozoic stromatolites, although recording numerous occurrences of stromatolites, is too dispersed and superficially descriptive when it comes to discussions of stromatolite morphology.

9 Controls on Morphogenesis

One of the great dilemmas in pre-Phanerozoic biogeology is the paleobiological interpretation of stromatolite morphology. Shape is potentially one of the most useful attributes of stromatolites for the interpretation of Earth system changes. The presence of morphologically complex structures with distinctive microstructures restricted to certain intervals of Proterozoic time points to temporally restricted microbes, microbial communities, and/or microbe-sediment interactions, and temporally (and spatially) changing chemical and physical environmental factors.

Studies on extant stromatolites, in particular from Shark Bay, Western Australia, indicate that waves and currents are the major shaping forces on gross morphology (Logan 1961; Logan et al. 1974; Playford

1980). Similar extrabiotic factors have been interpreted as major shaping forces in ancient stromatolites (e.g. Horodyski 1977). If morphology was determined solely by hydrodynamic factors, this could not result in time-restricted stromtolite forms. Other, principally biotic factors, must exert important control on stromatolite morphology. Unfortunately, there is little, if any, direct evidence to support this conclusion. (1) Studies on the Early Proterozoic Gunflint stromatolites (Awramik 1976; Awramik and Semikhatov 1979) show that different stromatolites contain different community organizations of fossil microbes. The species composition among the stromatolites is the same; however, the differences are among the relative abundance of the major microfossil taxa in each of the stromatolites. It must be kept in mind that the interplay between the environment and the microbes will ultimately determine the microbes and resulting benthic microbial community that builds the stromatolite. (2) Microstructure (the fine-scale or microscopic structure of the stromalite) is considered to be under the direct influence of the microbes that build the stromatolites (see Gebelein 1974). Numerous types of microstructure have been described, primarily from Proterozoic stromatolites (Bertrand-Sarfati 1976), and there are several investigators who use microstructure along with gross morphology in their taxonomic treatment of stromatolites (Walter 1972b; Grey 1984). Such taxa that utilize microstructure in their diagnoses are as reliable biostratigraphically as macromorphologically defined taxa. One important caveat to microstructure is its alteration by diagenesis. The effects of diagenesis cannot be dismissed and careful petrography is necessary when analyzing microstructure. However, temporal restriction of certain microstructurally defined stromatolite taxa argues that diagenesis does not completely obliterate or homogenize microbially defined microstructures (temporally restricted geochemical factors, such as seawater chemistry, could be called upon to produce or to contribute to unique microstructures; however, such evolving seawater chemistries might also produce temporally restricted microbial organizations, so we are back to where we started).

Conclusive evidence for microbiological control of stromatolite morphology would be the discovery of a unique or exclusive microfossils in a time-restricted stromatolite taxon. Microbial fossils were found preserved in the silicified Late Proterozoic stromatolite *Conophyton gaubitza* from Chichkan Formation of South Kazakstan (Schopf and Sovietov 1976). The *Conophyton* Group of stromatolites is one of the most distinctive stromatolite taxa known. Schopf and Sovietov found the microbial fossils to be similar to other Late Proterozoic fossil microbes from nonconical and noncolumnar stromatolites; however, the specific community appeared to be distinctive.

Thus, the formation of temporally restricted stromatolites appears to be a function of complex microbial community interaction with the environment within an evolutionary framework. Although Cyanobacteria, the principal organisms responsible for stromatolite formation, are morphologically very conservative, some patterns have been detected. Filamentous stromatolite-building microbes show a general trend toward increasing size through the Proterozoic as do coccoids (Schopf 1977); however, many of the coccoids were planktonic. The number of fossil species of Cyanobacteria and Cyanobacteria-like morphs described is small; estimated at only a few hundred (spread over 3500 million years) and this sample most likely reflects the most abundant taxa that are most readily preserved. Incomplete data on nonacritarch microbial fossils (Awramik, unpubl.) indicate that there is an increases in diversity of fossil filamentous Cyanobacteria and Cyanobacteria-like microfossils throughout the Proterozoic with an apparent maximum diversity within the Middle to Late Proterozoic (late Middle to late Late Riphean; approximately 1000 to 700 Ma ago). Analyses and conclusions drawn from these data require much testing before they are to be taken too literally. The number of reports describing microbiotas from this time interval is greater than older and younger Proterozoic strata. There is also the tendency by many of the practicing taxonomists to erect new species and genera using minor morphological differences without studying species populations and ignoring the effects of degradationally/diagenetically induced modifications to morphology. Looking through the inflated numbers of stromatolite-building taxa, it appears that the number of species described from any given formation of this 1000 to 700 Ma age is usually greater than older and younger Proterozoic formations. Thus, the apparent greater diversity of Middle to Late Proterozoic filamentous microfossils may be real. Nevertheless, Cyanobacteria were probably never very diverse even during the Proterozoic when they were among the dominant organisms on the Earth.

Cyanobacteria are not all that diverse today. The number of species known to build stromatolites may be less than 100, and commonly only one to a few cyanobacteria dominate the stromatolite-building ecosystem. It is not known whether cyanobacterial diversity was greater in the Proterozoic; the fossil record does not appear to indicate that it was. How does one explain the apparently high diversity of stromatolites during the Proterozoic if the species pool of Cyanobacteria is not all that great? Studies on the Gunflint (Awramik 1976; Awramik and Semikhatov 1979) and Chichkan (Schopf and Sovietov 1976) stromatolites would suggest that the microbial community structure is the distinctive feature that produces a singular morphological variety of stromatolite. The shuffling of stromatolite-building taxa by environmental (the sum total of external influences acting on organisms) and ecological (the interrelationships among organisms and their environment) factors determine the microbial community. Within the microbial mat (the surficial layer, a fraction to a few millimeters in thickness, which is the biologically active, accreting portion of the stromatolite) microbes are spatially organized along gradients perpendicular to the mat surface by environmental factors such as light, oxygen, and the ionic composition of the interstitial water (see Golubic 1976). The phenotypic variability of factors other than size within Cyanobacteria (such as rate of taxis, production of mucilage, degrees of pigmentation in sheath and envelope, behavioral interaction with members of its own species and with other species (note that this does not exclude the genetic underpinnings of these factors) may be sufficiently large, coupled with the evolution of new taxa adapted to stromatolite habitats, such that the resultant microbial communities respond in characteristic ways to environmental influence. The environment is primarily responsible for determining the microbes that can become established in any setting, the microbes can produce modified, microbially influenced environments, and it is the dynamic interaction of all these factors with sediments that produce the stromatolites.

10 Summary

It is clear the previous work on stromatolites has just scratched the surface with regard to the potential that stromatolites can offer to better understand microbial evolution, the interaction of the complex microbial communities that commonly build stromatolites, prokaryote-eukaryote interaction, and Earth history. Fossilized prokaryotic microbes, although more satisfying from a paleontological point of view since they constitute the actual remains organisms, have serious limitations with regard to contributing to the understanding of prokaryotic evolution and the interactions between the Earth and its biosphere. Once stromatolite morphogenesis is fully understood, and biotic factors can be separated from other factors influencing shape, stromatolites will become important paleontological entities capable of providing significant paleobiological information on the evolution of microbial ecosystems, and will no longer be viewed as sedimentological phenomena.

Acknowledgments. This paper was written while the author was a Research Associate at the Western Australian Museum. M. R. Walter kindly provided the computer database used in Walter and Heys 1985. A. W. R. Bevan, S. Golubic, K. Grey, and A. H. Knoll critically reviewed the manuscript. Research on stromatolite diversity through time was supported by NSF EAR 87-21192. NASA grant NAGW-1948 also supported this work. This is a contribution of IGCP Project 157 "Early Organic Evolution. Implications for Mineral and Energy Resources" and Project 261 "Stromatolites".

References

Aitkin JD (1967) Classification and environmental significance of cryptalgal limestones and dolomites, with illustrations from the Cambrian and Ordovician of southwestern Alberta. J Sediment Petrol 37:1163–1178

Aitkin JD, Narbonne GM (1989) Two occurrences of Precambrian thrombolites from the Mackenzie Mountains, northwestern Canada. Palaios 4:384–388

Allen KC, Briggs DEG (eds) (1989) Evolution and the fossil record. Belhaven, London, 265 pp

Allison CW, Awramik SM (1989) Organic-walled microfossils from earliest Cambrian or latest Proterozoic Tindir Group rocks, northwest Canada. Precambrian Res 43:253–294

Awramik SM (1971) Precambrian columnar stromatolite diversity: reflection of metazoan appearance. Science 174:825–827

Awramik SM (1976) Gunflint stromatolites: microfossil distribution in relation to stromatolite morphology. In: Walter MR (ed) Stromatolites. Elsevier, Amsterdam, pp 311–320

Awramik SM (1986) New fossil finds in old rocks. Nature (London) 319:446–447

Awramik SM (1990) Stromatolites. In: Briggs DEG, Crowther PR (eds) Palaeobiology. Blackwell, Oxford, pp 336–341

Awramik SM (1991) Archean and Proterozoic stromatolites. In: Riding R (ed) Fossil algae and stromatolites. Springer, Berlin

Awramik SM, Riding R (1988) Role of algal eukaryotes in subtidal columnar stromatolite formation. Proc Natl Acad Sci USA 85:1327–1329

Awramik SM, Semikhatov MA (1979) The relationship between morphology, microstructure, and microbiota in three vertically intergrading stromatolites from the Gunflint Iron Formation. Can J Earth Sci 16:2319–2330

Awramik SM, Schopf JW, Walter MR (1983) Filamentous fossil bacteria from the Archean of Western Australia. Precambrian Res 20:357–374

Awramik SM, Schopf JW, Walter MR (1988) Carbonaceous filaments from North Pole, Western Australia: are there fossil bacteria in Archean stromatolites? A discussion. Precambrian Res 39:303–309

Bambach RK (1985) Classes and adaptive variety: the ecology of diversification in marine faunas through the Phanerozoic. In: Valentine JW (ed) Phanerozoic diversity patterns. Princeton Univ Press, Princeton, pp 191–253

Barley ME, Eisenlohr BN, Groves DI, Perring CS, Vearncombe JR (1989) Late Archean convergent margin tectonics and gold mineralization: a new look at the Norseman-Wiluna Belt, Western Australia. Geology 17:826–829

Betrand-Sarfati J (1976) An attempt to classify Late Precambrian stromatolite microstructure. In: Walter MR (ed) Stromatolites. Elsevier, Amsterdam, pp 251–259

Bertrand-Sarfati J, Moussine-Pouchkine A (1985) Evolution and environmental conditions of *Conophyton – Jacutophyton* associations in the Atar dolomite (Upper Proterozoic, Mauritania). Precambrian Res 29:207–234

Bertrand-Sarfati J, Walter MR (1981) Stromatolite biostratigraphy. Precambrian Res 15:353–371

Black M (1933) The algal sediments of Andros Island, Bahamas. Philos Trans R Soc London Ser B 222:165–192

Buck SG (1980) Stromatolite and ooid deposits within the fluvial and lacustrine sediments of the Precambrian Ventersdorp Supergroup of South Africa. Precambrian Res 12:311–330

Buick R, Dunlop JSR, Groves DI (1981) Stromatolite recognition in ancient rocks: an appraisal of irregularly laminated structures in an Early Archaean chert-barite unit from North Pole, Western Australia. Alcheringa 5:161–181

Byerly GR, Lowe DR, Walsh MM (1986) Stromatolites from the 3,300–3,500-Myr Swaziland Supergroup, Barberton Mountain Land, South Africa. Nature (London) 319:489–491

Cameron EM, Baumann A (1972) Carbonate sedimentation during the Archean. Chem Geol 10:17–30

Cao Rui-ji (1988) Study on stromatolite decline event in terminal Precambrian. Acta Palaeontol Sin 27:737–750 (in Chinese with English summary)

Carlile M (1982) Prokaryotes and eukaryotes: strategies and successes. Trends Biochem Sci 7:128–130

Casanova J (1986) East African Rift stromatolites. In: Fostick LE et al. (eds) Sedimentation in the African rifts. Geol Soc Spec Publ 23:195–204

Cloud P (1976) Major features of crustal evolution. Geol Soc S Afr Annexure 79:1–33

Cloud PE Jr, Semikhatov MA (1969) Proterozoic stromatolite zonation. Am J Sci 267:1017–1061

Cussey R, Friedman GM (1976) Antipathetic relations among algal structures, burrowers, and grazers in Dogger (Jurassic) carbonate rocks, southeast of Paris, France. Am Assoc Petrol Geol Bull 60:612–616

Dill RF, Shinn EA, Jones AT, Kelly K, Steinen RP (1986) Giant subtidal stromatolites forming in normal salinity waters. Nature (London) 324:55–58

DiMarco MJ, Lowe DR (1989) Stratigraphy and sedimentology of an Early Archean felsic volcanic sequence, eastern Pilbara Block, Western Australia, with special reference to the Duffer Formation and implications for crustal evolution. Precambrian Res 44:147–169

Donaldson JA (1976) Paleoecology of *Conophyton* and associated stromatolites in the Precambrian Dismal Lakes and Rae Groups, Canada. In: Walter MR (ed) Stromatolites. Elsevier, Amsterdam, pp 523–534

Ernst WG (1983) The early Earth and the Archean rock record. In: Schopf JW (ed) Earth's earliest biosphere: its origin and evolution. Princeton Univ Press, Princeton, pp 41–52

Fairchild IJ, Herrington PM (1989) A tempestite-stromatolite-evaporite association (Late Vendian, East Greenland): a shore-face-lagoon model. Precambrian Res 43:101–127

Fenton CL, Fenton MA (1939) Pre-Cambrian and Paleozoic algae. Geol Soc Am Bull 50:89–126

Fischer AG (1965) Fossils, early life, and atmosphere history. Proc Natl Acad Sci USA 53:1205–1215

Gebelein CD (1974) Biologic control of stromatolite microstructure: implications for Precambrian time stratigraphy. Am J Sci 274:575–598

Gilliland RL (1989) Solar evolution. Global Planet Change 1:35–55

Golubic S (1976) Organisms that build stromatolites. In: Walter MR (ed) Stromatolites. Elsevier, Amsterdam, pp 113–126

Golubic S, Barghoorn ES (1977) Interpretation of microbial fossils with special reference to the Precambrian. In: Flügel E (ed) Fossil algae. Springer, Berlin, pp 1–14

Golubic S, Fischer AG (1975) Ecology of calcareous nodules forming in Little Connestoga Creek near Lancaster, Pennsylvania. Verh Ver Limnol 19:2315–2323

Gould SJ (1989) Wonderful life. Norton, New York, 347 pp

Green JW, Knoll AH, Swett K (1989) Microfossils from silicified stromatolitic carbonates of the Upper Proterozoic Limestone-Dolomite "Series", central East Greenland. Geol Mag 126:567–585

Grey K (1984) Biostratigraphic studies of stromatolites from the Proterozoic Earaheedy Group, Nabberu Basin, Western Australia. Geol Surv West Aust Bull 130:123 pp

Grey K, Moore LS, Burne RV, Pierson BK, Bauld J (1990) Lake Thetis, Western Australia: an example of saline lake sedimentation dominated by benthic microbial processes. Aust J Mar Freshwater Res 41:275–300

Grotzinger JP (1986) Cyclicity and paleoenvironmental dynamics, Rocknest platform, northwest Canada. Geol Soc Am Bull 97:1208–1231

Grotzinger JP (1989) Facies and evolution of Precambrian carbonate depositional systems: emergence of the modern platform archetype. In: Crevello PD, Wilson JL, Sarg JF, Read JF (eds) Controls on carbonate platform and basin development. Soc Econ Paleontol Min Spec Publ 44, pp 79–106

Hayes JM, Kaplan IR, Wedeking KW (1983) Precambrian organic geochemistry: preservation of the record. In: Schopf JW (ed) Earth's earliest biosphere. Princeton Univ Press, Princeton, pp 93–134

Hoffman P (1967) Algal stromatolites: use in stratigraphic correlation and paleocurrent determination. Science 157:1043–1045

Hoffman PF (1974) Shallow and deep-water stromatolites in Lower Proterozoic platform-to-basin facies change, Great Slave Lake, Canada. Am Assoc Petrol Geol Bul 58:856–867

Hofman HJ (1981) Precambrian fossils in Canada – the 1970's in retrospect. In: Campbell FHA (ed) Proterozoic basins in Canada. Geol Surv Can Pap 81–10:419–443

Hofmann HJ, Chen Jinbiao (1981) Carbonaceous megafossils from the Precambrian (1800 Ma) near Jixian, northern China. Can J Earth Sci 18:443–447

Horodyski RJ (1977) Environmental influences on columnar stromatolite branching patterns: examples from the Middle Proterozoic Belt Supergroup, Glacier National Park, Montana. J Paleontol 51:661–671

Horodyski RJ, Vonder Haar SP (1975) Recent calcareous stromatolites from Laguna Mormona (Baja California) Mexico. J Sediment Petrol 45:894–906

Karhu J, Epstein S (1986) In implication of the oxygen isotope records in coexisting cherts and phosphates. Geochim Cosmochim Acta 50:1745–1756

Karpeta WP (1989) Bedded cherts in the Rietgat Formation, Hartbeesfontein, South Africa: a Late Archaean to Early Proterozoic magadiitic alkaline playa lake deposit? S Afr J Geol 92:29–36

Kasting JF (1989) Long-term stability of the Earth's climate. Global Planet Change 1:83–95

Keller BM, Kazakov GA, Krylov IN, Nuzhnov SV, Semikhatov MA (1960) Novye dannye po stratigrafii rifeiskoi gruppy (verkhnii proterozoi). Izv Akad Nauk SSSR Ser Geol 1960(12):26–41

Kempe S, Degens ET (1985) An early soda ocean? Chem Geol 53:85–108

Kennard J (1981) The Arrinthrunga Formation: Upper Cambrian epeiric carbonates in the Georgina Basin, central Australia. Bur Min Res Bull 211:61 pp

Kennard JM, James NP (1986) Thrombolites and stromatolites: two distinct types of microbial structures. Palaios 1:492–503

Kepper JC (1974) Antipathetic relation between Cambrian trilobites and stromatolites. Am Assoc Petrol Geol Bull 58:141–142

Knoll AH (1985) A paleobiological perspective on sabkhas. In: Friedman GM, Krumbein WE (eds) Hypersaline ecosystems. Springer, Berlin, pp 407–425

Knoll AH, Bauld J (1989) The evolution and ecological tolerance in prokaryotes. Trans Soc Edinburgh Earth Sci 80:209–223

Knoll AH, Golubic S (1979) Anatomy and taphonomy of a Precambrian algal stromatolite. Precambrian Res 10:115–151

Komar VlA, Raaben ME, Semikhatov MA (1965) Conophytons in the Riphean of the USSR and their stratigraphic significance. Tr Geol Inst, Akad Nauk SSSR 131:72 pp (in Russian)

Korolyuk IK (1960) Stromatolites of the Lower Cambrian and Proterozoic of the Irkutsk Amphitheater. Tr Inst Geol Razrab Goryuch Iskop, Akad Nauk SSSR 1:112–161 (in Russian)

Kroner A., Todt W (1988) Single zircon dating constraining the maximum age of the Barberton Greenstone Belt, southern Africa. J Geophys Res 93:15329–15337

Krylov IN (1963) Stolbchaty vetvyashchiesya stromatolity rifeiskikh otlozhenii yuzhnogo urala. Tr Geol Inst Akad Nauk SSSR 69:1–133

Liang Yuzuo, Cao Ruiji, Zhang Luyi et al. (1984) Pseudogymnosolenaceae of Late Precambrian in China. Geol Publ House, Beijing, 200 pp (in Chinese with English abstract)

Logan BW (1961) *Cryptozoon* and associated stromatolites from the Recent of Shark Bay, Western Australia. J Geol 69:517–533

Logan BW, Hoffman P, Gebelein CD (1974) Algal mats, cryptalgal fabrics and structures, Hamelin Pool, Western Australia. Am Assoc Petrol Geol Mem 22:140–194

Lowe DR (1980a) Archean sedimentation. Ann Rev Earth Planet Sci 8:140–194

Lowe DR (1980b) Stromatolites 3,400-Myr old from the Archean of Western Australia. Nature (London) 284:441–443

Lowe DR (1983) Restricted shallow-water sedimentation of Early Archean stomatolitic and evaporitic strata of the Strelley Pool Chert, Pilbara Block, Western Australia. Precambrian Res 19:239–283

Makarikhin VV, Kononova GM (1983) Fitolity nizhnego proterozoya Karelii. Akad Nauk SSSR, Karel Fil, Inst Geol. Nauka, Leningrad, 180 pp

Margulis L, Schwartz KV (1982) Five kingdoms. Freeman, San Francisco, 338 p

Maslov VP (1939) An attempt to determine the age of the Ural's barren strata with the aid of stromatolites. Probl Paleontol 5:277–284 (in Russian with English summary)

McAllister JF, Ross RJ (1982) Steeply inclined algal columns in Late Tertiary Furnace Creek Formation, Death Valley, California. Am Assoc Petrol Geol Bull 62:541 (Abstr)

Monty CLV (1967) Distribution and structure of Recent stromatolitic algal mats, eastern Andros Island, Bahamas. Ann Soc Geol Belg 90:55–100

Monty CLV (1973) Precambrian background and Phanerozoic history of stromatolite communities, an overview. Ann Soc Geol Belg 96:585–624

Monty CLV (1984) Stromatolites and Earth history. Terra Cognita 4:423–430

Newman MJ, Rood RT (1977) Implications of solar evolution for the Earth's early atmosphere. Science 198:1035–1037

Newton RC (1978) Experimental and thermodynamic evidence for the operation of high pressures in Archaen metamorphism. In: Windley BF, Naqvi SM (ed) Archaean geochemistry. Elsevier, Amsterdam, pp 221–240

Oehler JH (1977) Microflora of the H.Y.C. Pyritic Shale Member of the Barney Creek Formation (McArthur Group), Middle Proterozoic of northern Australia. Alcheringa 1:315–349

Park RG (1980) Origin of horizontal structure in high-grade Archaean terrains. In: Glover JE, Groves DI (ed) Archaean geology, 2nd Int Symp, Perth, pp 481–490

Paul J (1988) Stromatolite bioherms and biostroms (Zechstein Basin, central Europe). Terra Cognite 8:220 (Abstr)

Pidgeon RT (1978) 3450 m.y. old volcanics in the Archaean layered greenstone succession of the Pilbara Block, Western Australia. Earth Planet Sci Lett 37:421–428

Playford PE (1980) Environmental controls on the morphology of modern stromatolites at Hamelin Pool, Western Australia. Geol Surv West Aust Annu Rep 1979:73–77

Playford PE, Cockbain AE, Druce EC, Wray JL (1976) Devonian stromatolites from the Canning Basin, Western Australia. In: Walter MR (ed) Stromatolites. Elsevier, Amsterdam, pp 543–563

Pratt BR (1982) Stromatolite decline – a reconsideration. Geology 10:512–515

Pratt BR, James NP (1982) Cryptalgal-metazoan bioherms of Early Ordovician age in the St. George Group, Western Newfoundland. Sedimentology 29:543–570

Riding R, Voronova L (1984) Assemblages of calcareous algae near the Precambrian Cambrian boundary in Siberia and Mongolia. Geol Mag 121:205–210

Runnegar B, Pojeta J, Taylor ME, Collins D (1979) New species of the Cambrian and Ordovician chitons Matthevia and Chelodes from Wisconsin and Queensland: evidence for the early history of polyplacophoran mollusks. J Paleontol 53:1374–1394

Schopf JW (1977) Biostratigraphic usefulness of stromatolitic Precambrian microbiotas. Precambrian Res 5:143–173

Schopf JW, Packer BM (1987) Early Archean (3.3 to 3.5 Ga-old) fossil microorganisms from the Warrawoona Group, Western Australia. Science 237:70–73

Schopf JW, Sovietov YuK (1976) Microfossils in *Conophyton* from the Soviet Union and their bearing on Precambrian biostratigraphy. Science 193:143–146

Semikhatov MA (1976) Experience of stromatolite studies in the U.S.S.R. In: Walter MR (ed) Stromatolites. Elsevier, Amsterdam, pp 337–357

Sepkoski JJ Jr (1978) A kinetic model of Phanerozoic taxonomic diversity. I. Analysis of marine order. Paleobiology 4:223–251

Sepkoski JJ Jr (1979) A kinetic model of Phanerozoic taxonomic diversity. II. Early Phanerozoic families and multiple equilibria. Paleobiology 5:222–251

Shapovalova IG (1968) O novoi gruppe stromatolitov Jacutophyton iz verkhnego proterozoya vostochnogo sklona Aldanskoi anteklizy. In: Tektonika, Stratigrafiya, i Litologiya Osadochnykh Formatsii, Yakutia. Dokl Nauk Sess Yakutsk Fil Sib Otd, Akad Nauk SSSR, 17:97–103

Southgate PN (1986) Depositional environment and mechanism of preservation of microfossils, Upper Proterozoic Bitter Springs Formation, Australia. Geology 14:638–686

Strother PK (1989) Pre-metazoan life. In: Allen KC, Briggs DEG (eds) Evolution and the fossil record. Belhaven, London, pp 51–72

Summons RE, Powell TG, Boreham CJ (1988) Petroleum geology and geochemistry of the Middle Proterozoic McArthur Basin, northern Australia. III. Composition of extractable hydrocarbons. Geochim Cosmochim Acta 52:1747–1763

Tiffney BH (ed) (1985) Geological factors and the evolution of plants. Yale Univ Press, New Haven, 294pp

Valentine JW (ed) (1986) Phanerozoic diversity patterns: profiles in macroevolution. Princeton Univ Press, Princeton, 441 p

Vidal G, Knoll AH (1982) Radiations and extinctions of plankton in the Late Proterozoic and Early Cambrian. Nature (London) 297:57–60

Vlasov FYa (1977) Dokembrickiikie stromatolity iz satkinskoisvity yuzhnogo urala. Akad Nauk SSSR, Ural Nauch Tsentr Materialy Paleontologii Srednego Paleozoya Urala Sibiri, pp 101–128

Walker JCG, Klein C, Schidlowski M, Schopf JW, Stevenson DJ, Walter MR (1983) Environmental evolution of the Archean-Early Proterozoic Earth. In: Schopf JW (ed) The earth's earliest biosphere: its origin and evolution. Princeton Univ Press, Princeton, pp 260–290

Walsh MM, Lowe DM (1985) Filamentous microfossils from the 3,500-Myr-old Onverwacht Group, Barberton Mountain Land, South Africa. Nature (London) 314:530–532

Walter MR (1972a) A hot spring analog for the depositional environment of Precambrian banded iron formations of the Lake Superior region. Econ Geol 67:965–972

Walter MR (1972b) Stromatolites and the biostratigraphy of the Australian Precambrian and Cambrian. Spec Pap Palaeontology 11:190 pp

Walter MR (1983) Archean stromatolites: evidence of the Earth's earliest benthos. In: Schopf JW (ed) Earth's earliest biosphere: its origin and evolution. Princeton Univ press, Princeton, pp 187–213

Walter MR, Heys GR (1985) Links between the rise of the Metazoa and the decline of stromatolites. Precambrian Res 29:149–174

Walter MR, Bauld J, Brock TD (1976) Microbiology and morphogenesis of columnar stromatolites (Conophyton, Vacerrilla) from hot springs in Yellowstone National Park. In: Walter MR (ed) Stromatolites. Elsevier, Amsterdam, pp 273–310

Walter MR, Buick R, Dunlop JSR (1980) Stromatolites 3,400–3,500 Myr old from the North Pole area, Western Australia. Nature (London) 284:443–445

Wang Fuxing, Zhang Xuanyang, Guo Ruihuan (1983) The Sinian microfossils from Jinning, Yunnan, South China. Precambrian Res 23:133–175

Winsborough BM, Golubic S (1987) The role of diatoms in stromatolite growth: two examples from modern freshwater settings. J Phycol 23:195–201

Woese CR (1987) Bacterial evolution. Microbiol Rev 51:221–271

Zhang Zhongying (1986) Clastic facies microfossils from the Chuanlinggou Formation (1800 Ma) near Jixian, North China. J Micropalaeontol 5:9–16

Zhu Shixing, Xu Chaolei, Gao Jianping (1987) Early Proterozoic stromatolites from Wutai Mt. and its adjacent regions. Bull Tianjin Inst Geol Min Res 17:1–221 [in Chinese with English summary]

Zhuravleva ZA (1964) Onkolity i katagrafii rifeya i nizhnego kembriya Sibiri i ikh stratigraficheskoe znachenie. Tr Geol Inst Akad Nauk SSSR 114:73 pp

Proterozoic and Living Cyanobacteria

Andrew H. Knoll[1] and Stejepko Golubic[2]

[1] Botanical Museum Harvard University Cambridge, MA 02138, USA
[2] Department of Biology Boston University Boston, MA 02215, USA

Abstract

The satisfactory systematic and functional interpretation of any fossil requires that its features be comparable to those of living organisms. A conundrum facing paleontologists who study Proterozoic microfossils is that many of the cyanobacteria-like remains that are so abundant in microbial mat assemblages have multiple morphological analogues in the modern biota – organisms that range from oxygenic photoautotrophs (the cyanobacteria) to anaerobic heterotrophs. Interpretation, therefore, requires that careful attention be paid to paleoenvironmental, behavioral, and taphonomic considerations, as well as morphology. Analyses of Proterozoic microfossil populations, including species of *Eoentophysalis*, *Polybessurus*, *Eohyella*, and others, demonstrate how paleobiological interpretation is maximized when careful observations on ancient populations are combined with complementary studies of modern morphological, developmental, and behavioral counterparts living in comparable physical environments.

1 Introduction

It has long been known that many Proterozoic microfossils resemble living cyanobacteria, and in much of the Precambrian paleontological literature, a cyanobacterial origin is presumed for fossils that are morphologically similar to extant blue-greens. On the other hand, it has been known even longer that physiologically diverse Eubacteria exhibit morphologies convergent upon those of common cyanobacteria. Therefore, the question arises of how can one interpret the paleobiology of prokaryotic microfossils with any degree of confidence. This is a difficult issue, and one that may not be resolvable for all fossils – that is, it is probably not possible to provide unambiguous taxonomic or physiological interpretations of all Proterozoic microbial remains. Nonetheless, there are ways in which one can restrict the "degrees of freedom" that surround Proterozoic paleontological interpretation. In this work, we address the specific question of how knowledge of living cyanobacteria can be used as a key to the paleobiological interpretion of Proterozoic cyanobacteria-like microfossils. Our thesis is that the paleobiological potential of Proterozoic microfossils can be realized only when they are studied as populations in a paleoenvironmental context and interpreted in light of systematic, ecological, and taphonomic investigations of living microorganisms found in analogous physical environments.

2 Living Cyanobacteria

The cyanobacteria constitute a eubacterial phylum of unique metabolic potential, morphological diversity, and evolutionary importance. Their metabolic singularity arises from their capacity for oxygenic photosynthesis carried out using chlorophyll a and two photosystems; molecular phylogenetic studies reinforce the idea that this remarkable physiology arose only once within the prokaryotes (Woese 1987). The evolutionary importance of the group also derives from its metabolism. The history of life and environments on Earth is closely tied to the history of atmospheric oxygen, and O_2 is generated largely by the photosynthesis of cyanobacteria and their descendants, the chloroplasts. Further, as noted in the opening paragraph, our sense of the historical role of cyanobacteria is heightened by the preservation of cyanobacteria-like fossils in extremely old rocks. Our ability to recognize possible cyanobacteria in the fossil record, in turn, relates to the third principal characteristic of the group, its morphological variability and complexity.

2.1 Morphology and Development

Living cyanobacteria include simple unicellular rods and cocci (e.g., *Synechococcus*) that closely resemble the cells of other gram-negative Eubacteria. On the

M. Schidlowski et al. (Eds.)
Early Organic Evolution: Implications for Mineral and Energy Resources

other hand, the cyanobacteria also include macroscopic thalloid forms with differentiated cells and a developmental complexity rivaling that of simple bryophytes (e.g., *Stigonema*). Because the phylum is so uniform metabolically, this morphological diversity has tranditionally formed the basis of cyanobacterial taxonomy (e.g., Geitler 1932). More recently, molecular methods have been applied to cyanobacterial classification, with illuminating, but not entirely unexpected results. In a study of DNA base composition in different cyanobacterial taxa, Herdman et al. (1979) found that morphologically simple groups, the unicellular cyanobacteria and – to a lesser degree – nonheterocystous filaments, exhibited a broad range of GC contents, while taxa within the developmentally complex pleurocapsalean, nostocalean, and stigonematalean cyanobacteria showed a restricted range of base compositions. DNA-DNA reassociation experiments supported the phylogenetic unity of heterocystous cyanobacteria, as well as the morphological bases of delimitation in many but not all genera within this group (Lachance 1981). Analysis of 16S rRNA sequences has corroborated and extended earlier results. According to Giovannoni et al. (1988), the three orders that show a restricted DNA base compositional range also form monophyletic clusters in sequence analyses, while unicellular blue-greens and simple, undifferentiated filaments appear to be strongly polyphyletic or paraphyletic. Thus, biochemical studies reinforce the idea that cyanobacteria grouped on the basis of morphological complexity do indeed form monophyletic groups, while simple species without complex (and, therefore, distinctive) developmental cycles encompass a phylogenetically wide spectrum of taxa.

Despite the application of molecular techniques to cyanobacterial systematics (or, perhaps, because of them, insofar as biochemical analyses support traditional classifications), morphology and development remain the principal bases of cyanobacterial taxonomy (Desikachary 1959; Golubic 1976; Waterbury and Stanier 1978; Rippka et al. 1979; Komarek and Anagnostidis 1986). Species are differentiated on the basis of cell size and shape; patterns of cell division; the presence or absence and, if present, type of cellular differentiation; the presence or absence and, if present, characteristics of extracellular envelopes or mucilage; the presence or absence of physiological connections between adjacent daughter cells; and the geometry of pluricellular aggregates or filaments (as determined by division patterns and envelope characteristics). In a few cases, ultrastructure (i.e., the absence of thylakoids in *Gloeobacter*; Rippka et al. 1974) or behavior (see below) contribute to taxonomic distinctions.

In the Geitlerian classification, genera are typically distinguished on the basis of cell or filament shape; cell division and/or differentiation patterns; and envelope characteristics. Species typically reflect quantitative differences among populations sharing qualitative traits. Higher level taxa are delineated according to larger scale developmental features: in coccoidal cyanobacteria, reproductive pattern (binary fission, "exospore" formation, or baeocyte formation), and in filamentous forms, branching geometry and the presence or absence of cellular differentiation. The deepest, or class level, division within the cyanobacteria is between filamentous forms in which adjacent daughter cells retain a physiological connection and coccoidal forms in which cells retain no connections following division (summarized in Golubic and Knoll, in press). Molecular studies question some of the traditionally defined higher taxa – particularly the coccoid-filament dichotomy – but morphological characters remain a reliable means of differentiating taxa at various levels of the taxonomic hierarchy. This has obvious implications for paleontology, where morphological characters are the only ones available for systematic interpretation.

2.2 *Ecology* and Environmental Distribution

The cyanobacteria are fundamentally obligate photoautotrophs, so their distribution is strongly controlled by the availability of light, CO_2, and simple inorganic nutrients. Within the environmental range thereby circumscribed, the cyanobacteria occur in a wide spectrum of habitats, ranging from thermal springs to Antarctic lakes, from fresh water ponds to evaporating playa lakes, from desert crusts to the open ocean plankton (reviewed in Knoll and Bauld 1989). Individual species, however, generally occur in a relatively restricted range of environments. The observation that populations in laboratory culture typically grow under a range of ecological conditions significantly broader than those that they occupy in nature (e.g., Stal and Krumbein 1985) suggests that natural population distributions are determined in part by biological interactions, including competition. Ecological success, in turn, is determined by the physiological as well as morphological properties of cyanobacterial populations. Therefore, environmental distribution provide proxy information on physiological attributes.

Recognition of natural environmental distributions is important in systematic studies because many cyanobacteria exhibit a broad range of morphological variation when grown under different culture conditions. While some authors (notably Drouet and Daily 1956; Drouet 1968, 1978, 1981) have stressed this morphogenetic plasticity and used it to define broadened generic and specific concepts, it is equally important to note that under a given set of environmental conditions, morphogenesis is relatively constant (Wilmotte 1988). Indeed, Anagnostidis and Komarek (1985) have characterized cyanobacterial species as "a complex of populations which occupy [a] particular ecological niche", although in practice, with the exception of endolithic blue-greens, ecological distributions have seldom been incorporated into species diagnoses.

Thus, the paleontologist has two important reasons for conducting paleoenvironmental studies in conjuction with systematic investigations. Environmental distributions may provide insights on the otherwise intractable issue of paleophysiology, and paleoecological control is necessary for the systematic interpretation of morphology.

2.3 Behavior

If behavior can be defined as an organism's physiological response to external stimulus, then cyanobacteria exhibit behavior. Behavioral responses to light are particularly well known. Certain mat-building cyanobacteria grow or move toward light when covered by a thin layer of sediment, a behavior recorded as alternating vertical and horizontal filament orientation in mats (Fig. 1A; Monty 1976; Golubic and Focke 1978). Surficial cells in *Entophysalis* colonies secrete radiation-absorbing pigments in intertidal environments in response to high light intensity (Fig. 3B; Bauld 1981). Most endolithic cyanobacteria are actually negatively phototactic, turning away from the light, at least within a certain range of radiation flux (Fig. 5D). The filamentous endolith *Plectonema terebrans* is particularly distinctive in that it grows down into a substrate until it reaches the light compensation level, at which point it reverses direction and grows back toward the light. The result is a tortuous or knotted growth pattern (LeCampion-Alsumard 1975). Such patterns are useful in taxonomy – the family Hyellaceae is defined on the basis of its endolithic habit – and they provide important ecological information.

2.4 Taphonomy

The role of taphonomy in the interpretation of cyanobacteria-like microfossils has been discussed repeatedly (e.g., Hofmann 1976; Golubic and Barghoorn 1977; Horodyski et al. 1977; Knoll and Golubic 1979) and there is little point in reiterating general arguments here. A point worth pursuing, however, is the observation that rates, patterns, and processes of taphonomic breakdown are not uniform for all microorganisms; neither are they completely random. Patterns of postmortem cellular degradation correlate with environment and taxonomy, so that certain organisms living in particular environments will have

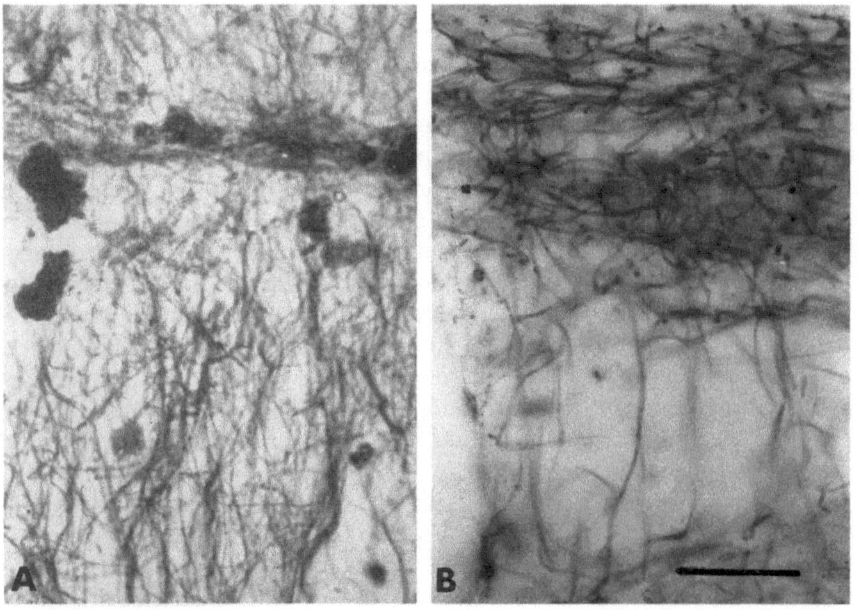

Fig. 1. Microbial mat-building filaments. **A** *Phormidium hendersonii* from Bonaire, showing alternating vertical and horizontal orientation of filaments. **B** *Eomycetopsis robusta*, a Late Proterozoic mat-building population from the Backlundtoppen Formation, Spitsbergen, showing similar filement orientations. *Bar* = 80 μm

a reasonable likelihood of entering the fossil record, while other organisms in other environments will have little chance. Thus, the coupling of actualistic studies of taphonomy with systematic and ecological investigations of living prokaryotes has the potential to place important constraints on the interpretation of the Proterozoic fossil record. Regardless of morphological convergences, the only strong candidates for comparison with specific Proterozoic populations will be those living organisms that both inhabit and may become preserved in environments similar to those inferred for their fossil counterparts (e.g., Knoll 1989).

3 Proterozoic Cyanobacteria and Cyanobacteria-Like Fossils

Of the many observations that can be made on living cyanobacteria, which are potentially recoverable from the fossil record? Certainly ultrastructural, genetic, and biochemical information will be lacking, although in an age of new sophistication in organic geochemistry we might reasonably hope for the recovery of some taxonomically and/or physiologically informative molecular data from rocks bearing Proterozoic cyanobacteria-like fossils (e.g. , Summons et al. 1988). In general, however, what the fossil record presents us for study is information on morphology, orientation, and distribution.

3.1 Morphology

Essentially all of the salient morphological features used in the taxonomic classification of living cyanobacteria can be observed in well-preserved microfossils, despite the pervasive influence of taphonomic alteration on Proterozoic microbial remains. Although beautifully preserved coccoidal cells and trichomes occur in Proterozoic rocks (e.g., Schopf 1968), cellular remains per se typically fare poorly in early diagenesis. Cell walls lyse or shrivel, and cell contents are metabolized by heterotrophic bacteria, leaving the cell represented in the fossil record by a small bleb of organic matter (Figs. 3A, 6C, 6F) or obliterating it completely (Golubic and Barghoorn 1977). In practice, this means that many cyanobacteria have a limited preservation potential. However, some blue-greens form extracellular sheaths or envelopes that are relatively resistant to postmortem alteration. These remains may be preserved three-dimensionally, providing detailed information about cell size, shape, and filament or colony form (Figs. 2–6).

The importance of considering populations rather than individual fossils cannot be overemphasized. Taphonomic effects may be variable even within a single local population, so that full details of morphology may be preserved in only a subset of the remains under study (e.g., Knoll and Golubic 1979). Further, individuals in a population rarely exhibit constant morphology; rather they exhibit a range of variation best communicated in statistical terms.

Development and reproduction cannot be observed in fossils, but under favorable circumstances, cell division patterns can be reconstructed for Proterozoic populations. (The approach is similar to that of a demographic census in which dynamic properties such as growth rate are inferred from age distributions.) When growth and reproduction among members of a population are not synchronized, a death assemblage will include individuals preserved in different stages of their life cycles (Golubic and Hofmann 1976; Golubic and Campbell 1979; Knoll and Golubic 1979; Knoll 1982; Knoll and Calder 1983; Strother et al. 1983; Green et al. 1987, 1988). The geometries and locations of binary fissions can be reconstructed, as can patterns and processes of true and false branching, colony form, or baeocyte formation. In many cases this morphological information will be sufficient to delineate specific and generic taxa using criteria similar to those employed in biological studies. In some instances, morphology will even suffice to identify remains as cyanobacteria, as opposed to some other type of Eubacteria (see below). Often, however, morphological data alone will not permit unambiguous distinction among metabolically disparate Eubacteria. To make further progress toward integretation, one must supplement morphological information with paleoecological and behavioral data.

3.2 Paleoenvironments, Paleoecology, and Behavior

Just as the biological information preserved in fossils is only a subset of the information potentially retrievable from living organisms, so, too, is the environmental information recoverable from sedimentary rocks "degraded" relative to the detailed data that can be collected in present-day environments (e.g., Knoll 1989). Nonetheless, the sedimentological and geochemical record encoded in Proterozoic sediments can constrain environmental interpretations sharply and facilitate comparison with Recent habitats. In such cases, one need not despair when morphological analysis shows that a fossil trichome bears comparison to *Oscillatoria*, *Beggiatoa*, and certain sulfate-reducing bacteria. One can at least ask which of these candidate counterpart

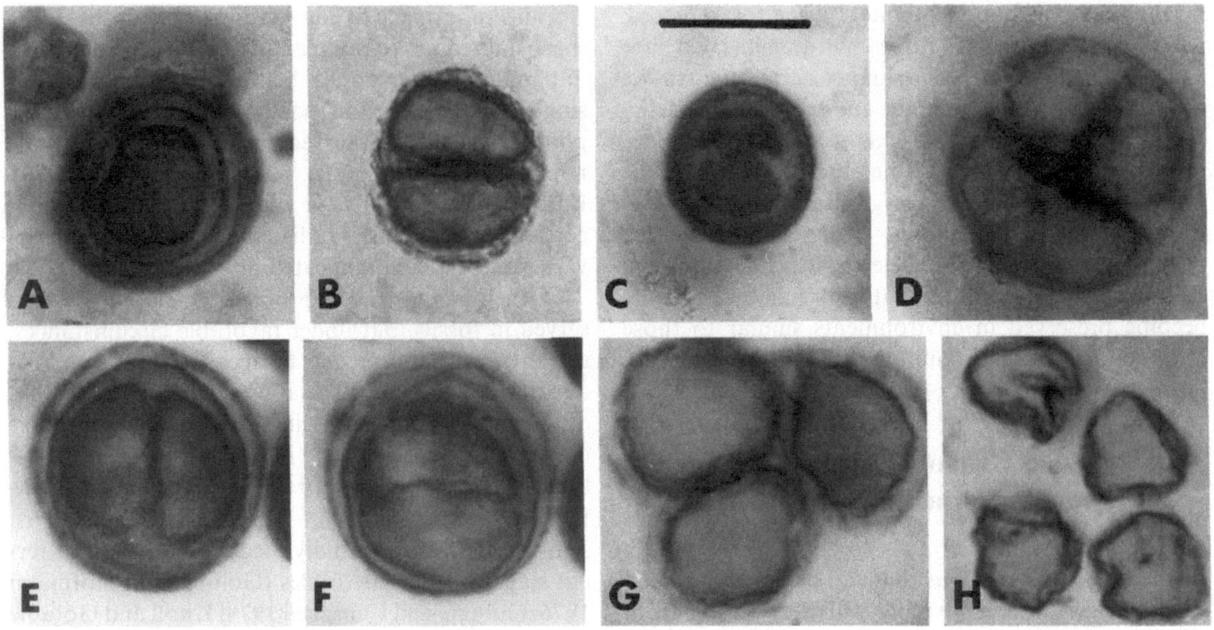

Fig. 2. *Gloeodiniopsis lamellosa*, a coccoidal population from a ponded playa lake environment of the Upper Proterozoic Bitter Springs Formation, Australia. **A** through **H** illustrate different parts of a simple cycle in which two binary divisions in mutually perpendicular planes are followed by envelope rupture and daughter cell release. (In some cases cell slippage gives tetrads a tetrahedral appearance.) This life cycle is comparable to that characterizing living *Chroococcus*. *Bar* = 15 μm

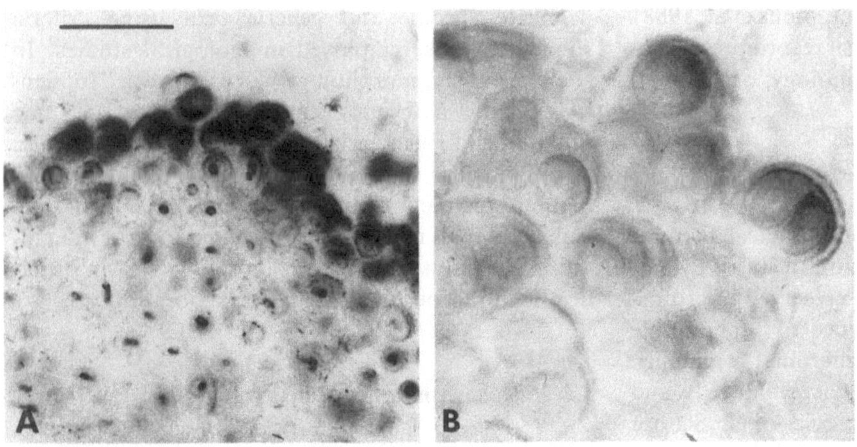

Fig. 3. *Eoentophysalis belcherensis*, from the Lower Proterozoic Belcher Supergroup, Canada. **B** *Entophysalis major* from Shark Bay, Australia. Note packaging of cells in envelopes and pigment-darkening colony surface in both. *Bar* = 40μm for **A** and = 12.5μm for **B**

organisms actually live in habitats similar to the one inferred from the sediments that encompass the fossil population of interest. Equally important, one can examine the taphonomic fate of the various candidate organisms under the postulated environmental conditions. As discussed below, this means that in practice, precision in taxonomic and physiological interpretation will be limited by precision in paleoenvironemental reconstruction.

Within environments, the distributions and orientations of fossil populations record significant features of ecology. In favorable circumstances one can distinguish among mat builder (Figs. 1, 3, 4), mat dweller (Fig. 2), endolithic (Fig. 5), and allochthonous species (e.g., Green et al. 1989). Behavioral traits may also be inferrable from such distributions. The three responses noted in the earlier discussion of behavior have all been documented in the Proterozoic record (see, for example, Golubic and Hofmann 1976; Zhang 1986; Green et al. 1988).

4 Proterozoic Cyanobacteria and Cyanobacteria-Like Fossils: Case Studies

In the following paragraphs, we support the foregoing arguments with specific case studies drawn from the Proterozoic fossil record. The examples have been chosen to illustrate the range of certainties encountered in paleontological investigations, from populations that are unambiguously cyanobacterial to microfossils whose attribution to any suprageneric taxon is a matter of faith, not science.

4.1 Eoentophysalis

One of the first, and still one of the most illuminating comparisons of a Proterozoic microfossil population and a specific recent cyanobacterium is that between the fossil genus *Eoentophysalis* and its modern counterpart, *Entophysalis* (Fig. 3B). As discussed by Golubic and Hofmann (1976), the genus *Entophysalis* is characterized by its coccoidal cells that reproduce by binary fission in three planes. External envelopes are persistent through several cell divisions, giving rise to cell packets that preserve a record of their divisional history. Polarized gel production results in more or less vertical rows of cells within colonies, while crowding of cell packets gives rise to a mamillate mat surface. A brown pigment produced by the cell often characterizes envelopes, especially at colony surfaces subject to intense solar radiation. Species within the genus are distinguished primarily on the basis of differences in cell size.

Modern *Entophysalis* species occur in a variety of peritidal and playa lake environments, but they are particularly abundant and conspicuous along the margins of water bodies characterized by elevated salinity. In such environments, *Entophysalis* populations can form structurally competent and sedimentologically distinctive mats, and participate in the accretion and lithification of stromatolitic structures (Golubic 1983). During times of sediment influx and mat burial, *Entophysalis* populations release gloeocapsoid packets of cells into the environment. These packets, which might easily be mistaken for true *Gloeocapsa*-like remains in the fossil record, recolonize the newly formed sediment surfaces. Patterns of cell degradation during early diagenesis are well understood, with cells shriveling to form dense blebs of organic material and, eventually disappearing, while envelopes sometimes retain much of their three-dimensional arrangement (Fig. 3A; Golubic and Hofmann 1976).

Entophysalis, then, is a cyanobacterium that is well characterized morphologically, ecologically, and taphonomically; significantly, it is also a cyanobacterium whose morphology is diagnostic for the phylum (i.e., the distinctive colony form of *Entophysalis* is not known to be matched in other eubacterial phyla). It is, therefore, an ideal focus for comparison with fossil populations, and research published over the last 15 years indicates that close fossil counterparts to *Entophysalis* are common in Proterozoic strata. The genus *Eoentophysalis* was established for large populations found in the Early Proterozoic Belcher Supergroup, Canada (Hofmann 1976). Belcher *Eoentophysalis* populations conform to living *Entophysalis* species in the details of their cell size and shape, division pattern and life cycle, envelope characters and colony form, taphonomic pattern, and inferred ecology and environmental distribution (Golubic and Hofmann 1976), leaving little doubt about the systematic paleontology and physiological attributes of the nearly 2000-Ma-old fossils. Further *Eoentophysalis* populations, representing several species, have been described from the Middle Proterozoic McArthur (Muir 1976), Nathan (Oehler 1978), and Dismal Lakes Groups (Horodyski and Donaldson 1983), and Gaoyuzhuang Formation (Zhang 1981). Late Proterozoic examples are known from the Narssârssuk (Strother et al. 1983) and Bitter Springs formations (Knoll and Golubic 1979). As discussed by Horodyski and Donaldson (1983), minor differences in environment and taphonomic history complicate species-level comparisions among these biotas, but in all cases, the fossil populations conform morphologically, ecologically, and taphonomically to their living counterpart.

4.2 *Polybessurus*

Morphologically distinctive microfossils assigned to the genus *Polybessurus* were first described from silicified Upper Proterozoic carbonates of South Australia by Fairchild (1975). More recently, Green et al. (1987) have published detailed observations and interpretations of large populations from the Upper Proterozoic Eleonore Bay Group, East Greenland. As in the case of *Eoentophysalis*, *Polybessurus* populations can be matched with living cyanobacteria whose morphology is diagnostic for the phylum (Green et al. 1987).

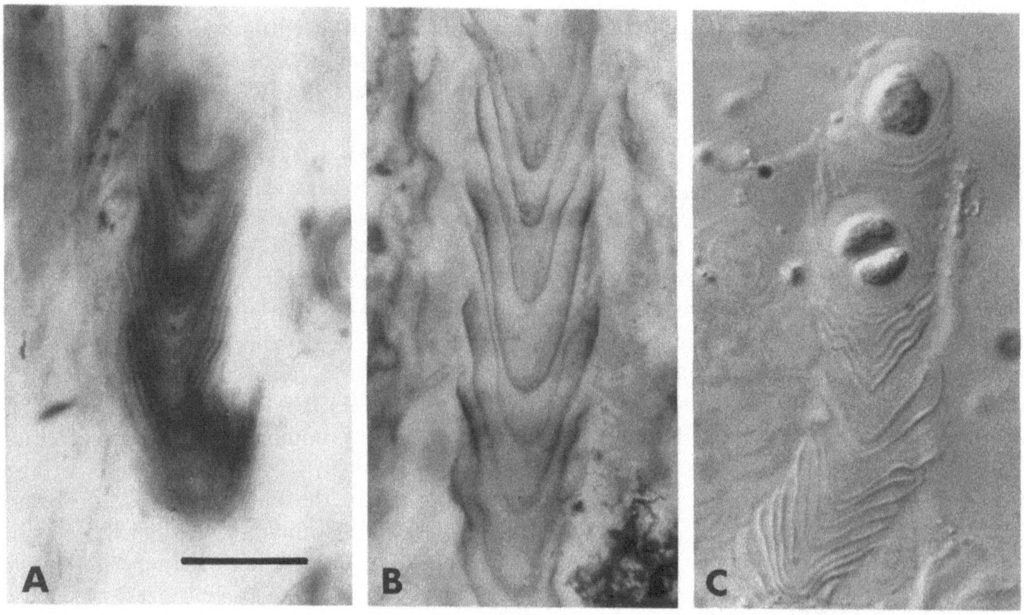

Fig. 4 (A, B). *Polybessurus bipartitus*, from the Upper Proterozoic Eleonore Bay Group, East Greenland, and its modern counterpart **C** from the Bahama Islands. *Bar* = 50 µm for **A** and **B** and = 11 µm for **C**

Polybessurus individuals were spheroidal unicells whose asymmetric envelope production resulted in the formation of a gelatinous stalk that elevated the cell above the sediment-water interface (Figs. 4A, B). Reproduction is inferred to have been by baeocyte formation. *Polybessurus* fossils sometimes occur as isolated individuals within microbial mats, but more conspicuously, large populations formed crusts on fine-grained carbonate substrates in peritidal environments. To date, populations that fit this description are known from Middle and Upper Proterozoic formations in East Greenland (Green et al. 1987), Svalbard (Butterfield and Knoll, unpubl. data), Australia (Fairchild 1975), and Arctic Canada (Butterfield et al., unpubl. data).

Physically analogous environments on the present-day Bahama Banks host cyanobacteria that, except for their smaller size, are comparable to Proterozoic *Polybessurus* populations (Green et al. 1987). This modern counterpart is as yet undescribed, but is similar to the genus *Cyanostylon* (Fig. 4C). It forms gelatinous stalks by asymmetric gel excretion, as sequentially produced envelopes rupture at their upper, thinner side to leave behind the stack of inverted gelatinous cones. The presence of these living cyanobacteria in environments that could be predicted on the basis of the Proterozoic fossil record once again leaves little doubt about the systematic and ecological interpretation of the fossils.

4.3 *Eohyella* and Other Endoliths

Brief mention of a third compelling ancient/modern comparison should serve to cement the point. Ooids found at present in shallow subtidal environments of the Bahama Banks are riddled by endolithic cyanobacteria; up to six species may occur within a single ooid. These cyanobacteria are morphologically, developmentally, and behaviorally distinctive. Late Proterozoic ooids from the Eleonore Bay Group, East Greenland (Green et al. 1988), and the Backlundtoppen Formation, Svalbard (Knoll et al. 1989), host abundant and diverse endolithic microfossils whose morphology, development and reproduction, orientation within the substrate (from which behavior can be inferred), and environmental distribution compare closely with those of the populations found in Recent ooids.

Several species of the cyanobacterial genus *Hyella* occur in Behamian endolithic communities. These populations are united by several characteristics: the presence of pseudofilaments composed of spheroidal to cylindrical coccoidal cells, pseudofilament branching by cell slippage or apical cell dichotomy, the presence of degradation-resistant envelopes that define cell shapes and positions, binary fission accompanied by baeocyte formation, and a distinctive negative phototropism and attendant polarity reversal in development (Fig. 5B, D; LeCampion-Alsumard

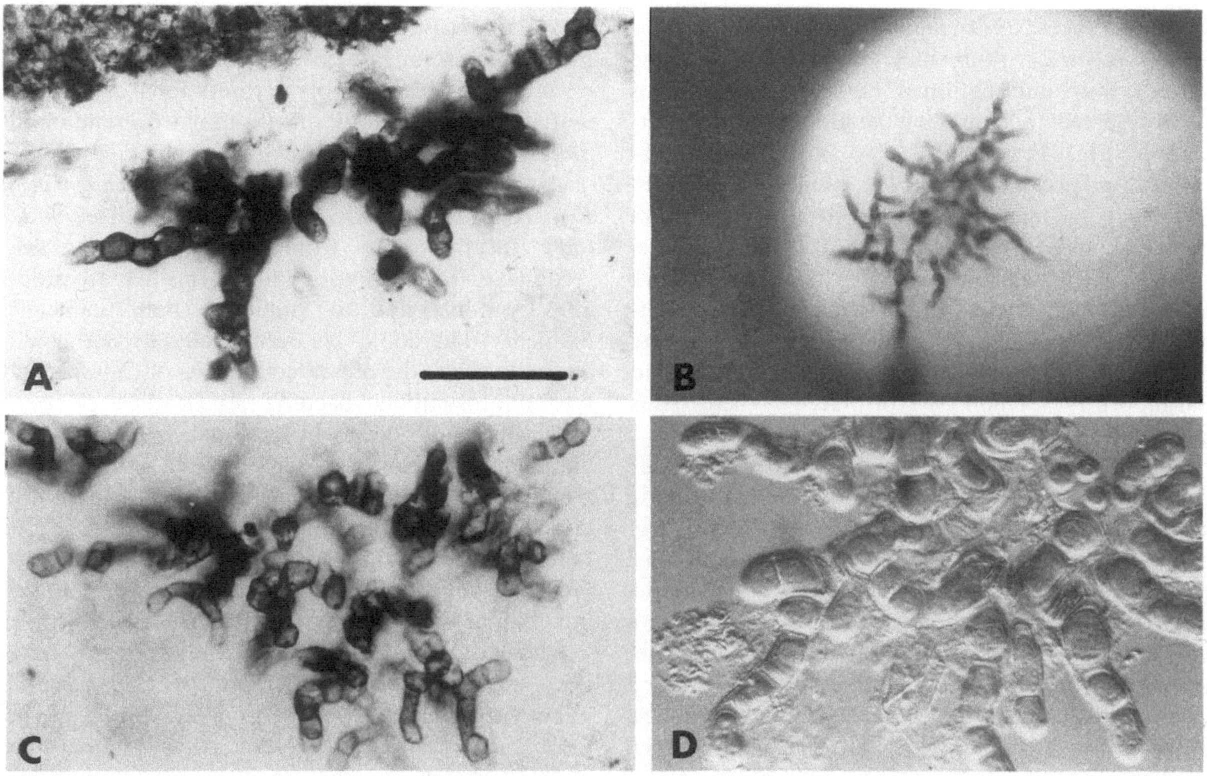

Fig. 5 (A, B). *Eohyella rectoclada*, an endolithic cyanobacterium from oolites of the Upper Proterozoic Eleonore Bay Group, East Greenland, and its modern counterpart **B**, **D** *Hyella gigas* from Bahamian ooids. *Bar* = 80 μm for **A** and **C**, = 400 μm for **B**, and = 40 μm for **D**

1969; Lukas and Golubic 1983; LeCampion-Alsumard and Golubic 1985). Species are distinguished on the basis of size, behavior, and differences in cell-division pattern and pseudofilament characteristics.

Four species attributable to the genus *Eohyella* occur in Late Proterozoic ooids from Svalbard and East Greenland (Fig. 5A, C). Two prominent populations are closely comparable to *Hyella gigas* and *H. caespitosa*, while a third has a living counterpart in an as yet unnamed *Hyella* species. This comparison across 700–800 Ma is much like that cited above, except that behavior joins morphology, life cycle, and environment in interpretation. *Eohyella*, in fact, occurs in 1700-Ma stromatolites from China (Zhang and Golubic 1987), but specific comparisons to modern species cannot be made at present.

4.4 Decreasing Confidence in Interpretation: *Oscillatoriopsis*, Filamentous Sheaths, *Eosynechococcus*, and *Myxococcoides*

The preceding examples provide best-case illustrations of the systematic and paleobiological interpretation of Proterozoic cyanobacteria. The fossils have morphologies not found among noncyanobacterial prokaryotes, and they have close couterparts among living cyanobacteria found in physically analogous environments. Unfortunately, such cases comprise only a small portion of the known Proterozoic fossil record. Many, perhaps most, Proterozoic prokaryotic microfossil populations have morphologies that occur in a taxonomically wide assortment of living bacteria.

The genus *Oscillatoriopsis* provides an illuminating example. The generic name is appropriate for undifferentiated and unbranched, uniseriate trichomes similar in morphology to living *Oscillatoria* (Schopf 1968; Mendelson and Schopf 1982). As in *Oscillatoria*, all cells are inferred to have retained the capacity for continued binary division, a property that can sometimes be demonstrated in fossil populations (e.g., Knoll et al. 1987). Microfossils fitting this description occur throughout the Proterozoic record; they range from forms only a few microns wide to gigantic trichomes more than 100 μm in cross-sectional diameter. With varying justification, species have been established on the basis of cell size and shape.

Most reports of *Oscillatoriopsis* provide little information on environmental setting or taphonomic pattern. Such information is, however, available for well-preserved trichomes of the Bitter Springs Formation (Fig. 6B; Schopf 1968), and these populations, thus, provide a focus for discussion. Silicified Bitter Springs microfossils come from ancient playa lake environments (Southgate 1986). Although no detailed ancient/modern parallel has been drawn between Bitter Springs trichomes and modern oscillatorians from similar environments, the morphology (Schopf 1968) and taphonomic patterns (Golubic and Barghoorn 1977; Knoll 1981) displaced by these fossils make comparisons with *Oscillatoria* plausible.

Unfortunately, what they do not do is exclude comparisons with other trichome-forming bacteria such as the widely distributed sulfur-oxidizer *Beggiatoa* or certain sulfate reducers. Both oxygenic phototrophs such as cyanobacteria and sulfide-oxidizing chemolithotrophs, such as *Beggiatoa*, commonly form mat-like coatings at the sediment-water interface. The former help in creating a steep redox gradient between sulfide-enriched sediments and the oxygenated water column; the latter depend on such gradients for chemotrophic energy generation. Microbiological comparisions with comparable modern playa lake biotas are needed to evaluate the plausibility of various alternative interpretations. One might invoke arguments about the relative likelihood of encountering primary producers versus other types of prokaryotes in the fossil record (Golubic 1980), but this potential argument is compromised by the observation that in the Ross River locality (Schopf 1968), most Bitter Springs trichomes are confined to a single millimeter thick lamina (Knoll 1981). This may suggest that the filamentous forms represent a short-lived bloom that occurred under environmental conditions different from those generally representative of the deposit.

Given all available evidence, the interpretation of Bitter Springs *Oscillatoriopsis* as cyanobacterial is not unreasonable, but it is tinged by uncertainty. A similar assessment can be made for most other Middle to late Proterozoic occurrences, but in the absence of environmental or comparative biological information, uncertainty is magnified. Doubts about the systematic interpretation of *Oscillatoriopsis* probably reach their peak in discussions of populations from Archean and Early Proterozoic formations such as the ca. 2700-Ma-old Fortescue Group (Schopf and Walter 1983) or the ca. 2000-Ma-old Duck Creek Dolomite, Western Australia (Fig. 6E; Knoll et al. 1987). In such ancient deposits, fundamental uncertainties about sedimentary environments or atmosphere/ocean chemistry make comparisons with living organisms in specific modern environments difficult.

A comparable range of interpretational confidence is attached to the empty sheaths that are abundant in most Proterozoic microbial mat assemblages. Variously assigned to the form-genera *Siphonophycus*, *Eomycetopsis*, *Tenuofilum*, *Taeniatum* or *Gunflintia*, these sheaths range from large (>10 μm cross-sectional diameter) mat-building forms that are probably cyanobacterial to minute (<1 μm diameter) filaments that could be cyanobacterial, but could alternatively be the preserved remains of *Cloroflexus*-type photosynthetic bacteria. Ensheathed *Chloroflexus* not only occur in peritidal microbial mat habitats; they can also have a high preservation potential in these environments (Cohen 1984). The large populations of small (1–4 μm diameter) sheaths that occur abundantly in stromatolitic facies of the Gunflint Iron Formation stretch the interpretational possibilities for Proterozoic filaments even further. Classified as *Gunflintia minuta*, these remains have been interpreted as *Leptothrix*-type iron bacteria (Cloud 1965), and given the iron-rich nature of the Gunflint environment, this is a plausible interpretation. Thus, as in the case of *Oscillatoriopsis* trichomes, Proterozoic filamentous sheath populations range from forms whose cyanobacterial interpretation seems likely to those for which confident assignment to the phylum is impossible (Diver and Peat 1979).

Coccoidal microfossils (Fig. 6A, C, F) commonly encountered in Proterozoic microbial mat assemblages include rod-shaped and spheroidal remains of varying colony complexity. The genus *Eosynechococcus* has been established for rod-shaped unicells that divide in one plane (Hofmann 1976). In a few cases colony structure permits the inference of cyanobacterial affinity (Golubic and Campbell 1979), but generally, no such confidence can be placed in interpretation as this general life form is widely distributed within the Eubacteria. Among *Eosynechococcus* populations, there are even a few whose distribution and orientation within the rock *suggest* that they were heterotrophs (Lanier 1986; Butterfield et al. 1988).

Many populations of *Myxococcoides* (Fig. 6C), established for small, unornamented spheroids that are common in many Proterozoic microfossil assemblages (Schopf 1968), defy interpretation even at the kingdom level; the extremely low incidence of dividing cells in many *Myxococcoides* populations suggest that these may well be eukaryotic, but others have a reasonably high probability of being cyanobacterial. In almost all cases, myxococcoid populations provide limited morphological or paleoenvironmental

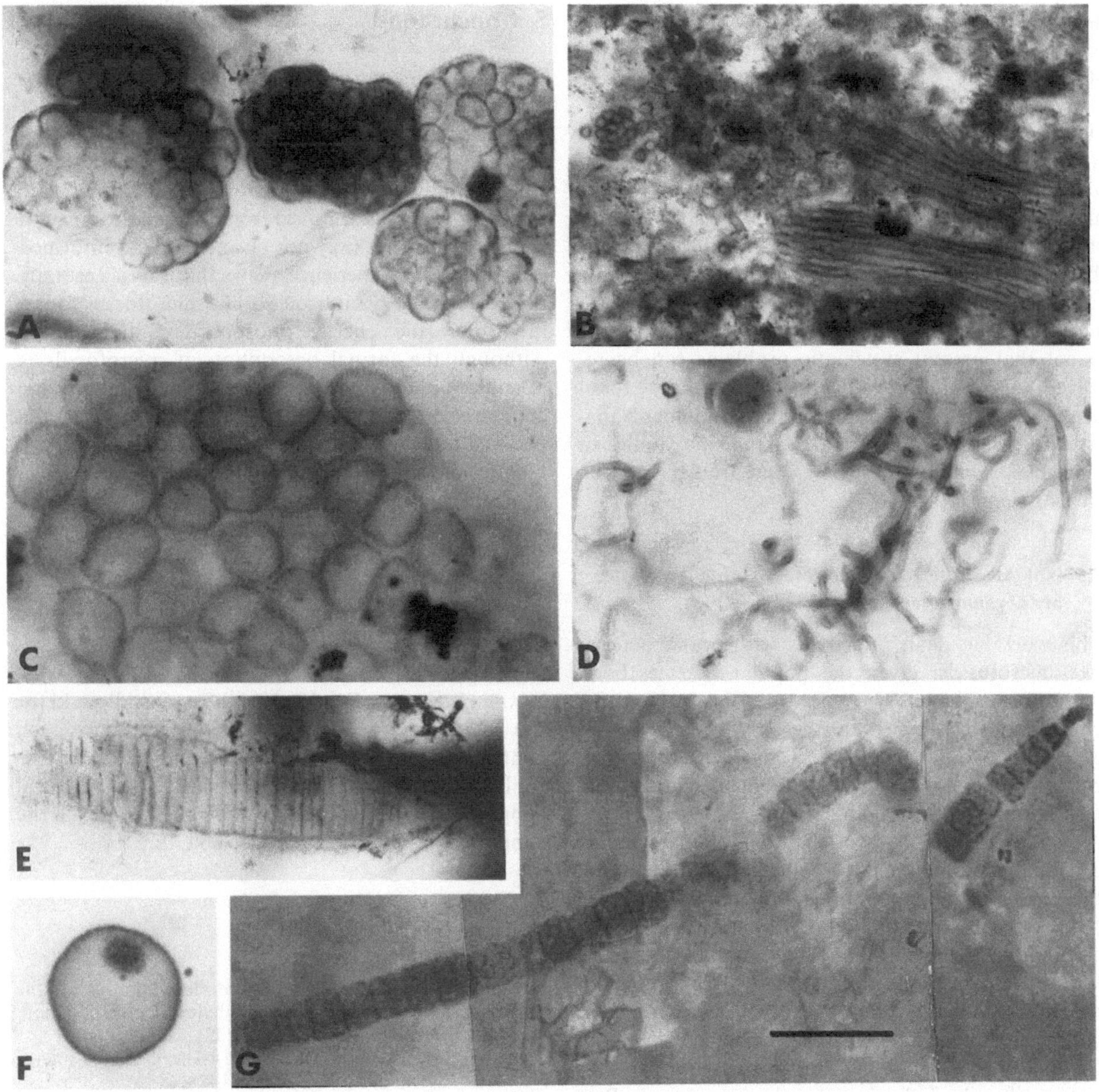

Fig. 6A–G. Miscellaneous Proterozoic silicified microfossils. **A** Pleurocapsoid colonies from tidal flat microbial mats of the upper Proterozoic Draken Conglomerate Formation, Spitsbergen, showing preservation of colony form and division pattern by envelopes. **B** Bundled filamentous envelopes from Upper Proterozoic metasediments, Prins Karls Forland, western Svalbard. **C** *Myxococcoides*, problematic coccoidal microfossils from the Draken Conglomerate. **D** *Tenuofilum septatum*, thin (ca. 1 µm) filamentous sheaths of mat-building prokaryotes from the Bitter Springs Formation; compare to both small oscillatorian cyanobacteria and the anoxygenic phototroph *Chloroflexus*. **E** *Oscillatoriopsis majuscula*, from the Lower Proterozoic Duck Creek Dolomite, Australia. **F** Problematic spheroidal microfossil from the Duck Creek Dolomite. **G** Oscillatorian trichomes preserved in the Bitter Springs Formation. *Bar* = 20 µm in **A, C, D,** and **F**, = 50 µm for **B**, = 100 µm for **E**, and = 15 µm for **G**

information (many appear to be allochthonous; e.g. Knoll 1981), and systematic interpretation is hazardous.

Oscillatoriopsis, *Eosynechococcus*, *Myxococcoides*, and filamentous sheaths are among the most commonly encountered fossils in Proterozoic rocks (Figs. 1 B, 6 B, D). Within each group, there is a broad range of interpretational certainty. Thus, to quote Gerasimerko and Krylov (1983), "The so-called 'natural' or 'biological' classifications adopted in many microfossils are essentially just as formal as the classification of acritarchs" (see also Diver and Peat 1979). To the extent that present systematic problems can be resolved, this resolution will come not from more fossil discoveries, but from careful paleoenvironmental analyses of fossil assemblages and actualistic studies of microbial communities in comparable modern environments.

4.5 Not All Proterozoic Microfossils are Cyanobacteria-Like

This work has dealt principally with cyanobacteria-like microfossils; given the popular and textbook literature on the subject, one would be forgiven for concluding that these were the only, or at least the predominent biological remains found in Proterozoic rocks. We cannot conclude without at least a brief refutation of this mythology.

Cyanobacteria-like fossils do dominate many silicified microbial mat assemblages, but many of these assemblages represent environments where even today cyanobacteria and other prokaryotes form mat communities. One must remember that at least for the Late Proterozoic, silicified mat assemblages are not the most commonly reported microfossils. Far more widely distributed are assemblages of eukaryotic phytoplankton preserved in siliciclastic facies (or, less commonly, silicified subtidal carbonates). Early Proterozoic assemblages contain few unequivocally eukaryotic remains, but they contain quite a few probable noncyanobacterial prokaryotes, especially in iron formations (e.g., Barghoorn and Tyler 1965; Awramik and Barghoorn 1977).

Many of the interpretational problems encountered in studies of cyanobacteria-like remains also plague analyses of other Proterozoic microfossils; thus, some general lessons can be learned from cyanobacterial studies. Nevertheless, typification of the Proterozoic fossil record as the Age of Cyanobacteria is an oversimplication comparable to the characterization of the Paleozoic, Mesozoic, and Cenozoic eras as the "ages" of trilobites, dinosaurs, and mammals, respectively.

5 Conclusions

Systematic paleontology lies at the heart of paleobiology, and systematic interpretation requires that careful attention be paid to population variation, taphonomy, and paleoenvironments of deposition. Isolated individual fossils, fossils in an advanced state of decay, or fossils removed from their environmental frame of reference may not be amenable to informed interpretation. Experience shows that this is certainly the case for the cyanobacteria-like microfossils found in Proterozoic cherts, phosphorites, and shales. Although the formal description of microfossils is possible without paleoecological data or actualistic comparisons, informed intepretation rarely is. At the present time, we know a little about Proterozoic biology and a little more about the organismic biology of microbial mat organisms and other playa lake or coastal marine prokaryotes. A great deal remains to be learned about both, and in particular there is much to be gained from careful comparative studies of Proterozoic and living microorganisms of similar environments.

Acknowledgments. We thank Manfred Schidlowski for the invitation to participate in the Maria Laach conference, and Stanley Awramik and Paul Strother for helpful criticism. AHK acknowledges the financial support of NASA Grant NAGW-893, and SG the support of NSF Grant EAR 8306179.

References

Anagnostidis K, Komarek (1985) Modern approach to the classification system of cyanophytes. 1. Introduction. Arch Hydrobiol Suppl 71(1, 2):291–302

Awramik SM, Barghoorn ES (1977) The Gunflint microbiota. Precambrian Res 5:121–142

Barghoorn ES, Tyler SA (1965) Microorganisms from the Gunflint chert. Science 147:563–577

Bauld J (1981) Geobiological role of cyanobacterial mats in sedimentary environments: production and preservation of organic matter. BMR J Aust Geol Geophys 6:307–317

Butterfield NJ, Knoll AH, Swett K (1988) Exceptional preservation of fossils in an Upper Proterozoic shale. Nature (London) 334:424–427

Cloud PE (1985) Significance of the Gunflint (Precambrian) microflora. Science 148:27–34

Cohen Y (1984) The Solar Lake cyanobacterial mats: strategies of photosynthetic life under sulfide. In: Cohen Y, Castenholz RW, Halvorson H (eds) Microbial mats: stromatolites. Liss, New York, pp 133–148

Desikachary TV (1959) Cyanophyta. Indian Counc Agric Res, New Delhi, 686 pp

Diver WL, Peat CG (1979) On the interpretation and classification of Precambrian organic-walled microfossils. Geology 7:401–404

Drouet F (1968) Revision of the classification of the Oscillatoriaceae. Monogr 15 Acad Nat Sci, Philadelphia, 370 pp

Drouet F (1978) Revision of the Nostocaceae with constricted trichomes. Beih Nova Hedwegia 57, Cramer, Vaduz, 258 pp

Drouet F (1981) Revision of the Stigonemataceae. Beih Nova Hedwigia 66, Cramer, Vaduz, 221 pp

Drouet F, Daily W (1956) Revision of the coccoid Myxophyceae. Butler Univ Bot Stud 12:1–218

Fairchild T (1975) The geologic setting and paleobiology of a Late Precambrian stromatolitic microflora from South Australia. PhD. Thesis, Univ Cal, Los Angeles

Geitler L (1932) Cyanophyceae. Rabenhorsts Kryptogamenflora von Deutschland, Österreich, und der Schweiz, vol. 14. Akademische Verlagsgesellschaft, Leipzig, 1196 pp

Gerasimerko LM, Krylov IN (1985) Postmortem alterations of cyanobacteria in the algal-bacterial films in the hot springs of Kamchatka. Dokl Akad Nauk SSSR 272:201–203 (in Russian)

Giovannoni SJ, Turner S, Olsen GJ, Barns S, Lane DJ, Pace NR (1988) Evolutionary relationships among cyanobacteria and green chloroplasts. J Bacteriol 170:3584–3592

Golubic S (1976) Taxonomy of extant stromatolite-building cyanophytes. In: Walter MR (ed) Stromatolites. Elsevier, Amsterdam, pp 127–140

Golubic S (1980) Early photosynthetic microorganisms and environmental evolution. In: Holmquist R (ed) Life science and space research (COSPAR 18). Pergamon, Oxford, pp 101–107

Golubic S (1983) Stromatolites, fossil and Recent: a case history. In: Westbroek P, Jong EW (eds) Biomineralization and biological metal accumulation. Reidel, Dordrecht, pp 313–326

Golubic S, Barghoorn ES (1977) Interpretation of microbial fossils with special reference to the Precambrian. In: Flügel E (ed) Fossil algae. Springer, Berlin Heidelberg New York, pp 1–14

Golubic S, Campbell S (1979) Analogous microbial forms in Recent subaerial habitats and in Precambrian cherts: *Gloeothece coerulea* Geitler and *Eosynechococcus moorei* Hofmann. Precambrian Res 8:201–217

Golubic S, Focke JW (1978) *Phormidium hendersonii* Howe: identity and significance of a modern stromatolite building organism. J Sediment Petrol 48:761–774

Golubic S, Hofmann HJ (1976) Comparison of modern and Mid-Precambrian Entophysalidaceae (Cyanophyta) in stromatolitic algal mats. J Paleontol 50:1074–1082

Golubic S, Knoll AH (1992) Prokaryotes. In: Lipps J (ed) Fossil prokaryotes and protists. Blackwell, Oxford (in press)

Golubic S, Perkins RD, Lukas KJ (1975) Boring microorganisms and microborings in carbonate substrates. In: Frey RW (ed) The study of trace fossils. Springer, Berlin Heidelberg New York, pp 229–269

Green JW, Knoll AH, Golubic S, Swett K (1987) Paleobiology of distinctive benthic microfossils from the Upper Proterozoic limestone-dolomite "Series", East Greenland. Am J Bot 74:928–940

Green JW, Knoll AH, Swett K (1988) Microfossils from oolites and pisolites of the Upper Proterozoic Eleonore Bay Group, central East Greenland. J Paleontol 62: 835–852

Green JW, Knoll, AH, Swett S (1989) Microfossils from silicified stromatolitic carbonates of the upper Proterozoic Limestone-Dolomite "Series", central East Greenland. Geol Mag 126:567–585

Herdman M, Janvier M, Waterbury JB, Rippka R, Stanier RY, Mandel M (1979) Deoxyribonucleic acid base composition of cyanobacteria. J Gen Microbiol 111:63–71

Hofmann HJ (1976) Precambrian microflora, Belcher Islands, Canada: significance and systematics. J Paleontol 50:1040–1073

Horodyski RJ, Donaldson JA (1983) Distribution and significance of microfossils in cherts of the Middle Proterozoic Dismal Lakes Group, District of Mackenzie, Northwest Territories, Canada. J Paleontol 57:271–288

Horodyski RJ, Bloeser B, vonder Haar S (1977) Laminated algal mats from a coastal lagoon, Laguna Mormona, Baja California, Mexico. J Sediment Petrol 47:680–696

Knoll AH (1981) Paleoecology of Late Precambrian microfossil assemblages. In: Niklas KH (ed) Paleobotany, paleoecology, and evolution, vol 1. Praeger, New York, pp 117–54

Knoll AH (1982) Microfossils from the Late Precambrian Draken Conglomerate, Ny Friesland, Svalbard. J Paleontol 56:755–790

Knoll AH (1989) The paleomicrobiological information in Proterozoic rocks. In: Cohen Y, Rosenberg E (eds) Microbial mats: physiological ecology of benthic microbial communities. Am Soc Microbiol, Washington DC, pp 469–484

Knoll AH, Bauld J (1989) The evolution of ecological tolerance in prokaryotes. Trans R Soc Edinburgh 80: 209–223

Knoll AH, Calder S (1983) Microbiotas of the Late Precambrian Ryssö Formation, Nordaustlandet, Svalbard. Palaeontology 26:467–496

Knoll AH, Golubic S (1979) Anatomy and taphonomy of Precambrian algal stromatolite. Precambrian Res 10:115–151

Knoll AH, Strother PK, Rossi S (1987) Distribution and diagenesis of microfossils from the Lower Proterozoic Duck Creek Dolomite, Western Australia. Precambrian Res 38:257–279

Knoll AH, Swett K, Burkhardt E (1989) Paleoenvironmental distribution of microfossils and stromatolites in the Upper Proterozoic Backlundtoppen Formation, Spitsbergen. J Paleontol 63:129–145

Komarek J, Anagnostidis K (1986) Modern approach to the classification system of cyanophytes. 2. Chroococcales. Arch Hydrobiol Suppl 73,2:157–226

Lachance M-A (1981) Genetic relatedness of heterocystous cyanobacteria by deoxyribonucleic acid-deoxyribonucleic acid reassociation. Int J Syst Bacteriol 31: 139–147

Lanier WP (1986) Approximate growth rates of Early Proterozoic microstromatolites as deduced by biomass productivity. Palaios 1:525–542

LeCampion-Alsumard T (1969) Contribution à l'étude des Cyanophycées lithophytes des étages supralittoral et médiolittoral (Région de Marseille). Tethys 1:119–172

LeCampion-Alsumard T (1975) Etude expérimetale de la colonization d'éclats de calcite par les Cyanophycées endolithes marine. Cah Biol Mar 16:177–185

LeCampion-Alsumard T, Golubic S (1985) *Hyella caespitosa* Bornet et Flahault and *Hyella balani* Lehman (Pleurocapsales, Cyanophyta): a comparative study. Arch Hydrobiol Suppl 71 Algol Stud 38/39:119–148

Lukas KJ, Golubic S (1983) New endolithic cyanophytes from the North Atlantic Ocean: II. *Hyella gigas* sp. nov. J Phycol 19:129–136

Mendelson CV, Schopf JW (1982) Proterozoic microfossils from the Sukhaya Tunguska, Shorikha, and Yudoma formations of the Siberian Platform. J Paleontol 56:42–83

Monty CLV (1976) The origin and development of cryptalgal fabrics. In: Walter MR (ed) Stromatolites. Elsevier, Amsterdam, pp 193–243

Muir MD (1976) Proterozoic microfossils from the Amelia Dolomite, McArthur Basin, Northern Territory. Alcheringa 1:143–158

Oehler DZ (1978) Microflora of the Middle Proterozoic Balbirini Dolomite (McArthur Group) of Australia. Alcheringa 2:269–309

Rippka R, Waterbury J, Cohen-Bazire G (1974) A cyanobacterium which lacks thylakoids. Arch Microbiol 100:419–436

Rippka R, Deruelles J, Waterbury JB, Herdman M, Stanier RY (1979) Generic assignments, strain histories, and properties of pure cultures of cyanobacteria. J Gen Microbiol 111:1–61

Schopf JW (1968) Microflora of the Bitter Springs Formation, Late Precambrian, central Australia. J Paleontol 42:651–688

Schopf JW, Walter MR (1983) Archean microfossils: new evidence of ancient microbes. In: Schopf JW (ed) Earth's earliest biosphere: its origin and evolution. Univ Press, Princeton, pp 214–239

Southgate PN (1986) Depositional environment and mechanism of preservation of microfossils, Upper Proterozoic Bitter Springs Formation, Australia. Geology 14:638–686

Stal LJ, Krumbein WE (1985) Isolation and characterization of cyanobacteria from a marine microbial mat. Bot Mar 28:351–365

Strother PK, Knoll AH, Barghoorn ES (1983) Microorganisms from the Late Precambrian Narrsârssuk Formation, north-western Greenland. Palaeontology 26:1–32

Summons RG, Powell TG, Boreham CJ (1988) Petroleum geology and geochemistry of the Middle Proterozoic McArthur Group, northern Australia. III. Composition of extractable hydrocarbons. Geochim Cosmochim Acta 51:1747–1762

Waterbury JB, Stanier RY (1978) Patterns of growth and development in pleurocapsalean cyanobacteria. Microbiol Rev 42:2–44

Wilmotte A (1988) Growth and morphological variability of six strains of *Phormidium* cf. *ectocarpi* Gomont (Cyanophyceae) cultivated under different temperatures and light conditions. In: Anagnostidis A., Golubic S, Komarek J, Lhodsky O (eds) Cyanphyta. Schweizerbart, Stuttgart, pp 35–46

Woese C (1987) Bacterial evolution. Microbiol Rev 51: 221–271

Zhang Y (1981) Proterozoic stromatolite microfloras of the Gaoyuzhuang Formation (Early Sinian: Riphean), Hebei, China. J Paleontol 55:485–506

Zhang Y, Golubic S (1987) Endolithic microfossils (Cyanophyta) from Early Proterozoic stromatolites, Hebei, China. Acta Micropalaeontol Sin 4:1–12

Zhang Z (1986) Solar cyclicity in the Precambrian rock record. Palaeontology 29:101–111

Lower Precambrian Stromatolite Associations of Karelia

V. V. Makarikhin [1]

[1] Institute of Geology, Karelian Branch, USSR Academy of Sciences, Pushkinskaya Str. 11, Petrozavodsk, 185610, USSR

Abstract

The Archean-Lower Proterozoic strata of the eastern Baltic Shield contain stromatolites, microphytolites, acritarchs, and other problematic fossils which form characteristic assemblages. Each assemblage is characteristic of a certain stratigraphic unit at the regional scale. The distribution of the remains both in time and space is uneven. Stromatolites form the bulk of the fossils found in the Jatulian (2.3–2.2 Ga) carbonate strata which are studied in detail. Stromatolites provide the basis for five biostratigraphic units distinguished upwards as: *Nuclephyton*-, *Sundosia*-, *Omachtenia kintsiensis*-, *Butinella*-, and *Calevia ruokanensis*-bearing layers.

Other strata of different metamorphic grades and isotopic ages, varying from 2.5 to 1.65 Ga, are characterized paleontologically. It has been inferred that the paleontological information obtained can be used for regional biostratigraphy.

1 Introduction

The Precambrian time span under review is represented by metamorphic sedimentary strata in the eastern Baltic Shield which have an isotopic age of 2500 ± 50 to 1650 ± 20 Ma (Bibikova et al. 1989).

Fossil evidence of ancient life, essentially algal in nature, is fairly often found preserved in large quantities in rocks of both pre- and post-2 Ga age (Hofmann 1971; Makarikhin and Kononova 1983; Walter and Heys 1985; Awramik 1986; Semikhatov 1987; etc).

Lower Precambrian organic remains fall into four large groups:

1. Phytogenic structures such as stromatolites, stromatolite-like buildups, and microphytolites (oncolites and catagraphites);
2. organic-walled microfossils, i.e., various small cell-like structures isolated, as a rule, by maceration;
3. mineralized, usually silicified cellular remains which occur as microscopic units studied in thin sections;
4. problematica of obscure, possibly biogenic origin, including trace fossils and relics of organic matter.

In Karelia, the entire Precambrian sequence shows the presence of the above fossil groups, although their distribution is fairly uneven.

2 Stratigraphy

According to the stratigraphic scheme presently adopted for the Karelian region, Lower Precambrian rocks of Karelia are represented by two complexes: The Lopian complex of Late Archean age, and the Karelian complex of Early Proterozoic age. Their upper boundaries have been dated at 2500 ± 50 Ma and 1650 ± 20 Ma, respectively (Fig. 1). The Karelian complex is subdivided into three subcomplexes: Lower, middle, and upper, separated by large stratigraphic unconformities dated 2300 ± 50 Ma and 2100 ± 50 Ma. Each subcomplex comprises two superhorizons separated by minor unconformities; their respective names are based on conventional ethnographic principles: Sumian, Sariolian, Jatulian, Ludicovian, Kalevian, and Vepsian (Kratz 1963; Sokolov 1984). Each superhorizon is characterized by a unique phytolite assemblage in which the leading role is played by stromatolites.

3 Distribution of Phytolites in the Sequence

The problematic alga *Careliella* was first reported from Lopian metamorphic rocks considered to be Archean. Similar rocks were found to contain some microfossils of which only *Symplassosphaeridium* sp. and *Synsphaeridium* sp. are valid (Timofeyev 1982).

M. Schidlowski et al. (Eds.)
Early Organic Evolution: Implications for Mineral and Energy Resources

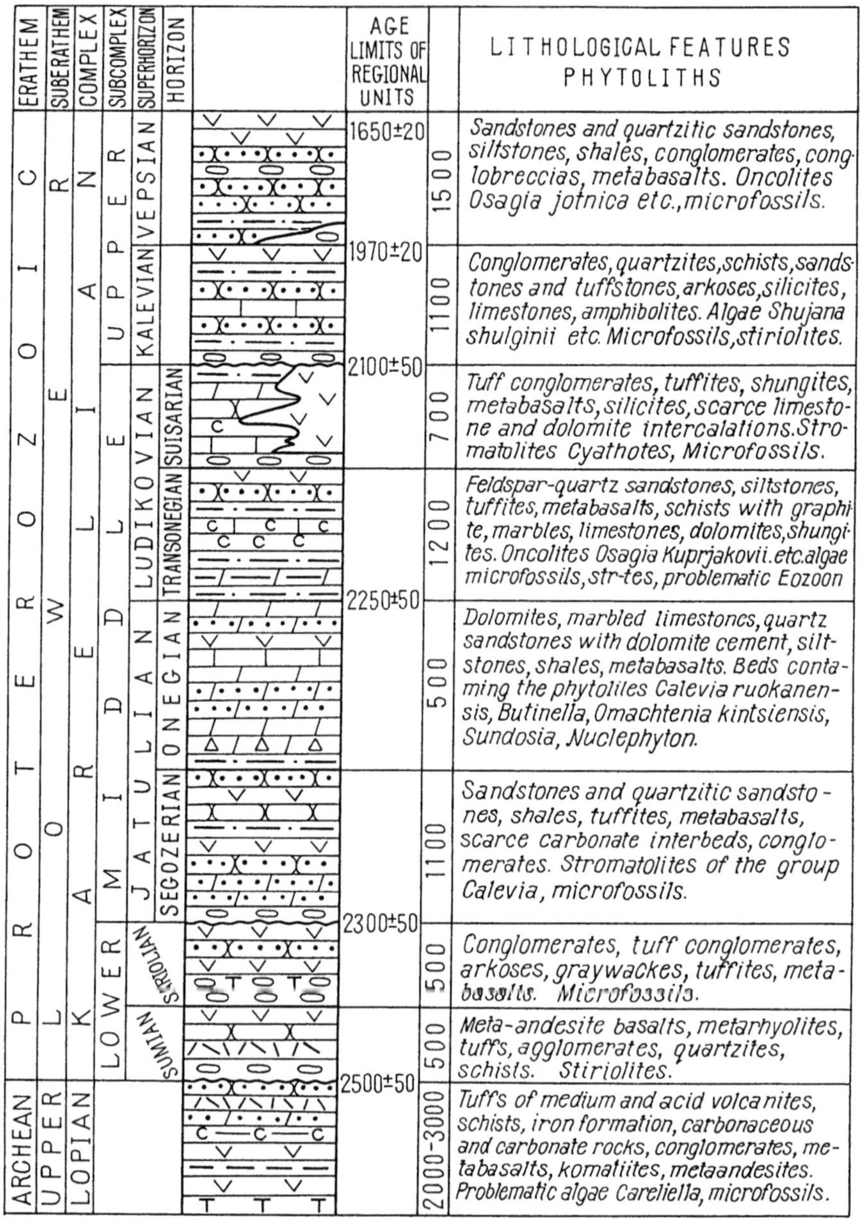

Fig. 1. Stratigraphic column for Lower Precambrian rocks of Karelia

Problematic, essentially siliceous stromatolitic buildups, defined as stiriolites, have been described from the base of the Lower Proterozoic (Sumian superhorizon). The term "stiriolite" was coined by Walter (1976) to denote fossil columnar buildups formed in the areas affected by thermal springs. Such buildups resemble stromatolites, but their primary composition is, as a rule, silicious.

The Sumian unit is overlain by the volcanosedimentary rock of the Sariolian superhorizon which contains *Protosphaeridium densum* Tim., P. *iaccatum* Tim., *P.* sp. and *Trematosphaeridium* sp. No valid stromatolites have been reported from these strata.

Phytolites are most abundant in the Middle Karelian, especially in the carbonate part of the Jatulian referred to as the Onega horizon. The lower part of the Jatulian, represented by essentially terrigenous facies (Segozero horizon), has only been found to contain individual nodular stromatolites of the group *Calevia* and scarce microfossils. Numerous phytolites are present in the upper part of the Jatulian, mainly in dolomites (Onega horizon). The study of the phy-

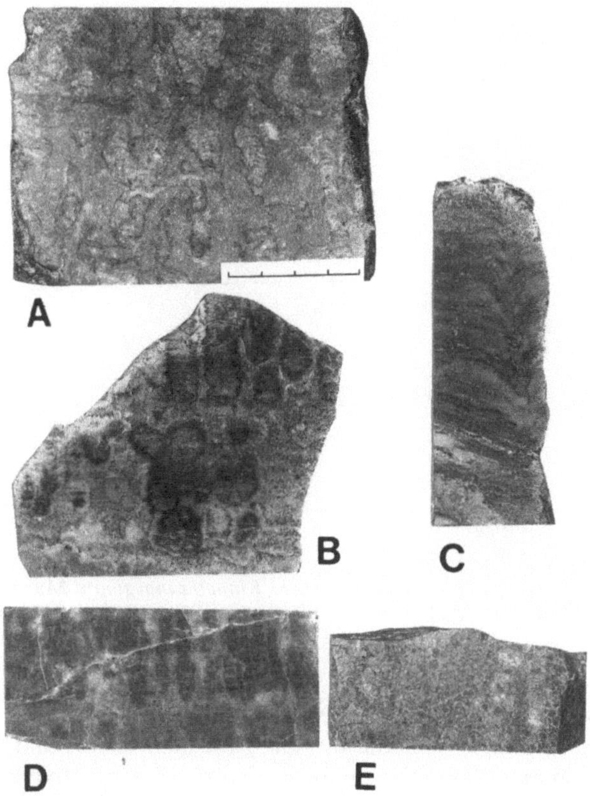

Fig. 2. Stromatolites of the Jatulian superhorizon. Polished samples (bar = 4 cm) of **A** *Sundosia mira* (But.); **B** *Butinella boreale* Mak.; **C** *Omachtenia kintsiensis* Mak.; **D** *Butinella digitus* Mak.; **E** *Klimetia marginata* Mak.

tolites has revealed five nonrecurrent assemblages characteristic of the biostratigraphic units defined as phytolite-bearing layers (Makarikhin 1987):

1. *Nuclephyton*-bearing layers are confined to the lower part of the Jatulian carbonate strata and show the presence of dolomite beds with the pore-forming columnar stromatolites *Nuclephyton confertum* Mak. at the base. For a description of the stromatolites and microphytolites reported here, see Makarikhin and Kononova (1983). The large isolated oncolites *Osagia genulna* G.Kon. have been encountered in the upper part of these layers. *Nuclephyton*-bearing layers are exposed for about 2 km at three localities in the stratotype area. The total thickness of the layers in the stratotype is 17 m.
2. *Sundosia*-bearing layers rest directly on the layers described above. The dolomites contain large elongated biostromes, indistinctly isolated dome-shaped bioherms, or small lens-shaped bodies composed of the rock-forming stromatolites *Carelozoon metzgerii* Mak., *Sundosia mira* (But., Fig. 2A), *Parallelophyton raigubicum* Mak., the algae *Mucostroma carelica* Vol., M. bullata Vol., Leiostroma eleganta Vol., etc. The acritarchs *Protosphaeradium densum* Tim., *P. tuberculiferum* Tim., etc. have been collected from the upper part of the layers which has been traced for about 5 km. The total thickness of the layers in the stratotype section is 12 m.
3. *Omachtenia kintsiensis*-bearing layers occur in the middle portion of the Jatulian carbonate strata. They contain the index species (Fig. 2C), elongated biostromes with the complicated columnar stromatolites *Carelozoon jatulicum* Metz., isolated build-ups comprised of *Colleniella palica* Mak., *Colonnella carelica* Mak., *Parallelophyton strictum* Mak., and the oncolites *Palia septentrionalis* But., *P. bicolor* G.Kon. and *Glebosites palosericus* G.Kon. The layers are exposed at 15 localites over a distance of a few hundred kilometers. Their thickness in the type section is ca. 80 m.
4. *Butinella*-bearing layers are represented by essentially carbonate rocks (limestones and dolomites) which show an abundance of various phytolites such as the stromatolites *Calevia olenica* (Rjab.), *Butinella boreale* Mak. (Fig. 2B), *B. digitus* Mak. (Fig. 2D), *B. ambigua* Mak., *Klimetia marginata* Mak. (Figs. 2E and 3), *K. torosa* Mak., *Stratifera ordinata* Mak., and *Paniscollenia clivosa* Mak.; the oncolites *Asterosphaeroides oleniella* G.Kon., *Radiosus dualis* G.Kon., *R. ignotus* G.Kon., and *R. ignobilis* G.Kon.; the algae *Eoepiphyton jatulicum* But. and *Kareliana zonata* Korde; the acritarchs *Protosphaeridium* sp., *P. flexuosum* Tim., *Stictosphaeridium implexum* Tim., *S. pectinale* Tim., *Gloeocapsomorpha* sp., *Turuchanica ternata* Tim., and *Favosophaeridium* sp. Exposed *Butinella*-bearing layers have been observed at 12 localities and traced for about 200 km. Their thickness in the stratotype is 65 m.
5. *Calevia ruokanensis*-bearing layers make up the upper part of the Jatulian carbonate strata and contain both the individual nodular stromatolites *Calevia ruokanensis* Mak., and small lenticular bodies composed of the rock-forming columnar stromatolites *Djulmekella sundica* Mak. The acritarchs present in the same strata are *Protosphaeridium densum* Tim., *P. laccatum* Tim., *P. planum* tim., *P. flexuosum* Tim., *P. tuberculiferum* Tim., *Trematosphaeridium holtedahlii* Tim., *T.* sp., *Gloeocapsomorpha* sp., *Synsphaeridium* sp. and *Phycomycetes* sp. Usually the thickness of these layers does not exceed 10–15 m. They are exposed at five localities and have been traced for about 50 km.

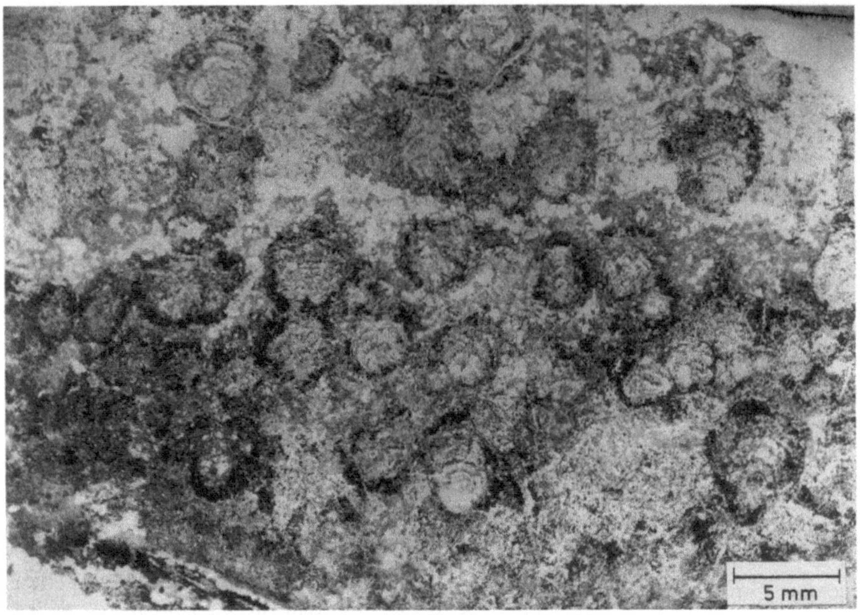

Fig. 3. *Klimetia marginata* Mak. (thin section, magnified)

The shungite-bearing rocks of the Ludicovian superhorizon contain new, poorly studied groups of stromatolites, peculiar low-relief buildups formed by *Cyathotes nigoserica* Mak.; the oncolites *Osagia kuprjakovii* G.Kon., *O. ludmilae* G. Kon., *Volvatella atra* G. Kon., *Asterosphaeroides valericus* G. Kon., and *Volvatella speciosa* G. Kon.; the algae *Agamus shungiticus* Vol., *Gorlovella obvoluta* Vol., and *Grenularia borissovi* Vol.; the acritarchs *Protosphaeridium densum* Tim., *P. acis* Tim., *P. flexuosum* Tim., and some problematica.

The volcanosedimentary rocks of the Upper Karelian contain a few phytolite occurrences which include stiriolites, the problematic alga *Shujana shulginii* Korde, numerous microfossils such as *Protosphaeridium* sp., *P. rigidulum* Tim., *Trachysphaeridium laminaritum* Tim., *Trematosphaeridium holtedahlii* Tim., *Turuchanica ternata* Tim., *Stictosphaeridium implexum* Tim., *Synsphaeridium* sp., and *Careliella lasiantha* Tim., and microphytolites such as *Osagia* sp., *O. jotnica* Mil., and *O. onezhiella* Mil.

3 Conclusion

The distribution of organic remains in the oldest complexes of Karelia is fairly uneven. However, most of the above superhorizons contain specific assemblages with individual characteristics. This allows the use of paleontological information for stratigraphic correlations. Attempts at regional correlations using phytolite assemblages have so far been unsuccessful. According to Semikhatov (1985) and Semikhatov and Komar (1989), the number of formal stromatolite species connecting individual Aphebian assemblages can hardly account for 8–10% of the total number of these taxa described from pre-Riphean strata. The number of connecting generic taxa is higher (25%), but the majority of such genera comprise both individual Aphebian and mixed Aphebian-Riphean types such as *Stratifera*, *Omachtenia*, *Colonnella*, etc. Difficulties in correlating newly found phytolite assemblages, both regionally and globally, are largely due to a variety of phytolite classification systems used by different groups of scholars. This is equally true for both the classification and nomenclature of stromatolites and for the systematics of microphytolites and microfossils. We hope that further elaboration and improvement of relevant classifications and systematics of phytolites will increase their stratigraphic value.

References

Awramik S M (1986) New fossil finds in old rocks. Nature (London) 319:6053:446–447

Bibikova E V, Lobach-Zhuchenko S B, Semikhatov M A, Shcherbak N P (1989) Precambrian geochronological scale for East European Platform and its surroundings. Izv Acad Nauk Ser Geol 4:8–22 (in Russian)

Hofmann H J (1971) Precambrian fossils, pseudofossils and problematica in Canada. Geol Surv Can Bull 189, 146 pp

Kratz K O (1963) Geology of Karelia's Karelides. Acad Nauk Moscow-Leningrad, 210 pp (in Russian)

Makarikhin V V (1987) Results of the paleontological study of Karelia's old sequences. In: Sokolov V A (ed) Geology of Karelia. Nauka, Leningrad, pp 93–100 (in Russian)
Makarikhin V V, Kononova G M (1983) Lower Proterozoic phytolites of Karelia, Nauka, Leningrad, 180 pp (in Russian)
Semikhatov M A (1985) Stromatolites in Precambrian stratigraphy: analysis '84. Izv Acad Nauk Ser Geol 4:3–21 (in Russian)
Semikhatov M A (1987) Taxa of Precambrian stromatolites in space and time. In: 3rd All-Union Symp Precambrian and Early Cambrian paleontology. Petrozavodsk, pp 87–89 (in Russian)
Semikhatov M A, Komar V A (1989) Precambrian stromatolites: biological interpretation, classification and stratigraphic implications. In: Keller B M (ed) Problems in Upper Proterozoic and Phanerozoic stratigraphy. Nauka, Moscow, pp 13–21 (in Russian)
Sokolov V A (ed) (1984) Precambrian stratigraphy of Soviet Karelia. Geol Inst, Petrozavodsk, 116 pp (in Russian)
Timofeyev B V (1982) Early Precambrian microfossils. Nauka, Leningrad, 128 pp (in Russian)
Walter M R, Heys G R (1985) Links between the rise of the Metazoa and the decline of stromatolites. Precambrian Res 29:149–174
Walter M R (1976) Geyserites of Yellowstone National Park: an example of abiogenic "stromatolites". In: Walter M R (ed) Stromatolites. Elsevier, Amsterdam, pp 87–112

Tussocky Microstructure, a Biological Event in Upper Proterozoic Stromatolites; Comparisons with Modern Freshwater Stromatolite Builders

J. Bertrand-Sarfati [1] and A. Pentecost [2]

[1] Centre Géologique et Géophysique, CNRS UP 361, USTL, 34095, Montpellier-Cedex 5, France
[2] Division of Biosphere Sciences, King's College, London, Campden Hill Road, London, W8 7AH, UK

Abstract

An unusual type of microstructure occurs in stromatolites of the Upper Proterozoic cover of the West African Craton. These stromatolites bearing the same microstructure are reported from diverse sections along the northern edge of the Taoudenni basin and from the Pan-African fold belt of the Hoggar (Bertrand-Sarfati 1972). Their microstructure has been defined as "tussocky" (Bertrand-Sarfati 1976).

The stromatolites are entirely built by tussocks, absent from the interspaces between columns. They comprise hemispherical bodies, juxtaposed and overgrowing, roughly superimposed to compose irregular but obvious laminae. The tussocks are made of rods which are interpreted as the remnants of previous filaments. Most of the time they are straight, rarely branched. A common feature of the tussocks is the banding, concentrically arranged bands of brown pigmentation. According to Fairchild et al. (1990), the carbonate of the tussock was originally aragonitic and recrystallization into calcite appeared during burial.

In order to compare the tussocks with modern stromatolite builders, a range of measurements were made on *Rivularia*, *Inzeria*, and *Serizia* to clarify their structural relationships. Among 11 characters listed, 7 are shared by *Rivularia* and the tussocks. The four differences are heliotropism, branching and curvature of the filaments and the interface between adjacent colonies. The observations, however, point to a biogenic origin for the tussocks, but we must accept that they have no modern counterparts. The microorganisms building the tussocks may have been filamentous, weakly heliotropic cyanobacteria with a radiating structure.

1 Introduction

An unusual type of microstructure occurs in stromatolites of the Upper Proterozoic cover of the West African craton. These stromatolites, bearing always the same microstructure, have been reported for the first time, from diverse sections along the northern edge of the Taoudenni Basin (Bertrand-Sarfati 1972). A paleogeographic map shows the probable distribution of the stromatolite biostrome (Fig. 1). This microstructure has been defined as "tussocky" (Bertrand-Sarfati 1976): "stromatolitic laminae defined by the juxtaposition of separate hemispheric tussocks, ranging from 0.1 to 3–4 mm". A few tussock-like structures have been described elsewhere, especially in Australia, where a tussocky microstructure has been recognized by Preiss (1976). However, Grey (1984) considered that this microstructure was slightly different in shape, making more constant laminae, lacking isolated, overgrowing, tussocks. In addition, it never displays any kind of radial fabric. In China, tussock-like microstructures are considered in a paper by Cao and Liang (1974). Recently (pers. comm. 1989), Cao reported hemispheric radiating structures resembling tussocks.

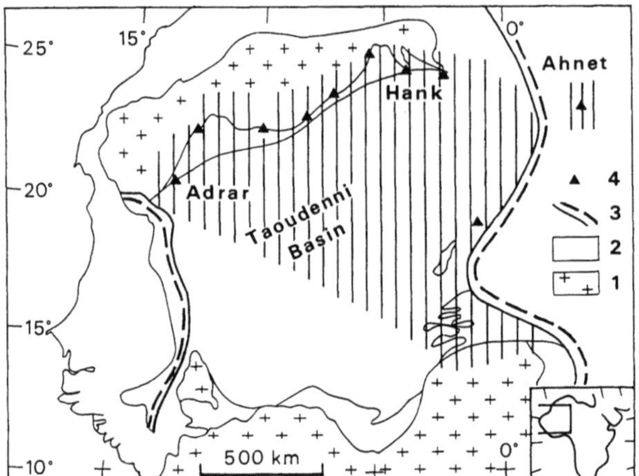

Fig. 1. Paleogeographic map of the West African craton showing the extent of stromatolites vertical lines with a tussocky microstructure. *1* Basement; *2* Proterozoic cover; *3* fold belt limits; *4* sections

M. Schidlowski et al. (Eds.)
Early Organic Evolution: Implications for Mineral and Energy Resources

An organic origin for the tussocks has been proposed (Bertrand-Sarfati 1972, 1976; Bertrand-Sarfati and Moussine-Pouchkine 1988) despite the fact that no true filament molds can be recognized. We propose (1) to discuss the arguments for a biogenic control on the tussocks, inferred from the study of the fossil examples, and (2) to compare tussocks to modern freshwater colonies of *Rivularia* and *Homoeothrix*.

2 Arguments for Biogenic Control of Fossil Tussocks

There are several indirect clues, which are indicative of an organic origin of the tussocks.

1. *Stromatolite. The stromatolites*, entirely built by tussocks *are mainly columnar*, in discrete and juxtaposed clusters, forming biostromes. Inferred environmental conditions are marine, sublittoral. In the Adrar region (Mauritania) the stromatolites belong to the groups *Tarioufetia hemispherica* and *Serizia radians*, in the Hank area (Algeria), about 1000 km toward the east, the stromatolite is mainly *Inzeria globulosa*. Further east, in the Pan-African orogenic belt of the Hoggar, *Serizia radians* and *Inzeria globulosa* are reported from the Adrar Ahnet area (Moussine-Pouchkine et al. 1988).

The columns display a roughly laminated structure, entirely built by tussocks. In some cases, thin micritic films overlie the tussocks (Fig. 3b). Clastic grains, quartz and carbonates, are trapped in between the tussocks (Fig. 3d). The sharp contact between the columns and the sediment filling the interspaces emphasizes the fact that the tussocks are restricted to the stromatolite column (Fig. 3b).

The tussocks are hemispherical bodies (Fig. 3d–e), juxtaposed and overgrowing, roughly superimposed to compose irregular but obvious laminae (Fig. 2a–b). The columns display a laminae convexity which differs from one to the other groups, but is always indicative of a more rapid growth of the tussocks in the center, resulting in a thickening of the laminae. In one of the groups, *Inzeria globulosa*, the central tussocks are much bigger than the others (10 to 20 times), inducing a more convex shape of the laminae (Fig. 3a). In this case, the tussocks have a disymmetric shape that can be interpreted as a possible heliotropic influence. Similar big tussocks are found at random in all stromatolites, they generally present central-basal cracks (Fig. 3a).

Flat-lying stromatolitic carbonates can also display a tussocky microstructure, the organic origin of which may be more difficult to prove (the convex shape of the laminae being a good clue to an organic origin).

2. *Rods making the tussocks* are interpreted as remnants of previous filaments forming the tussocks. They are mostly straight, rarely "branched" and curved, except for a faint divergent habit at the tussock edge. They are present either as lines of small anhedral crystals outlined by dark impurities (Fig. 4a) or as ghosts within anhedral crystals (Fig. 4b–c). Gener-

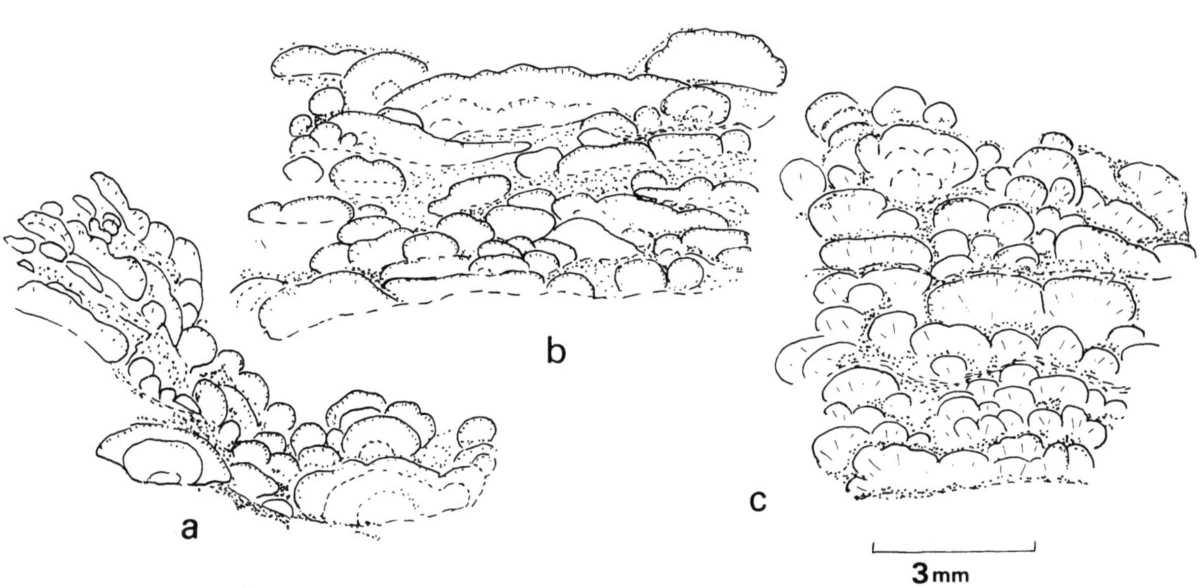

Fig. 2a–c. Superposition of the tussocks forming irregular laminae. **a** and **b** *Tarioufetia hemispherica;* **c** *Serizia radians* (Adrar, Mauritania)

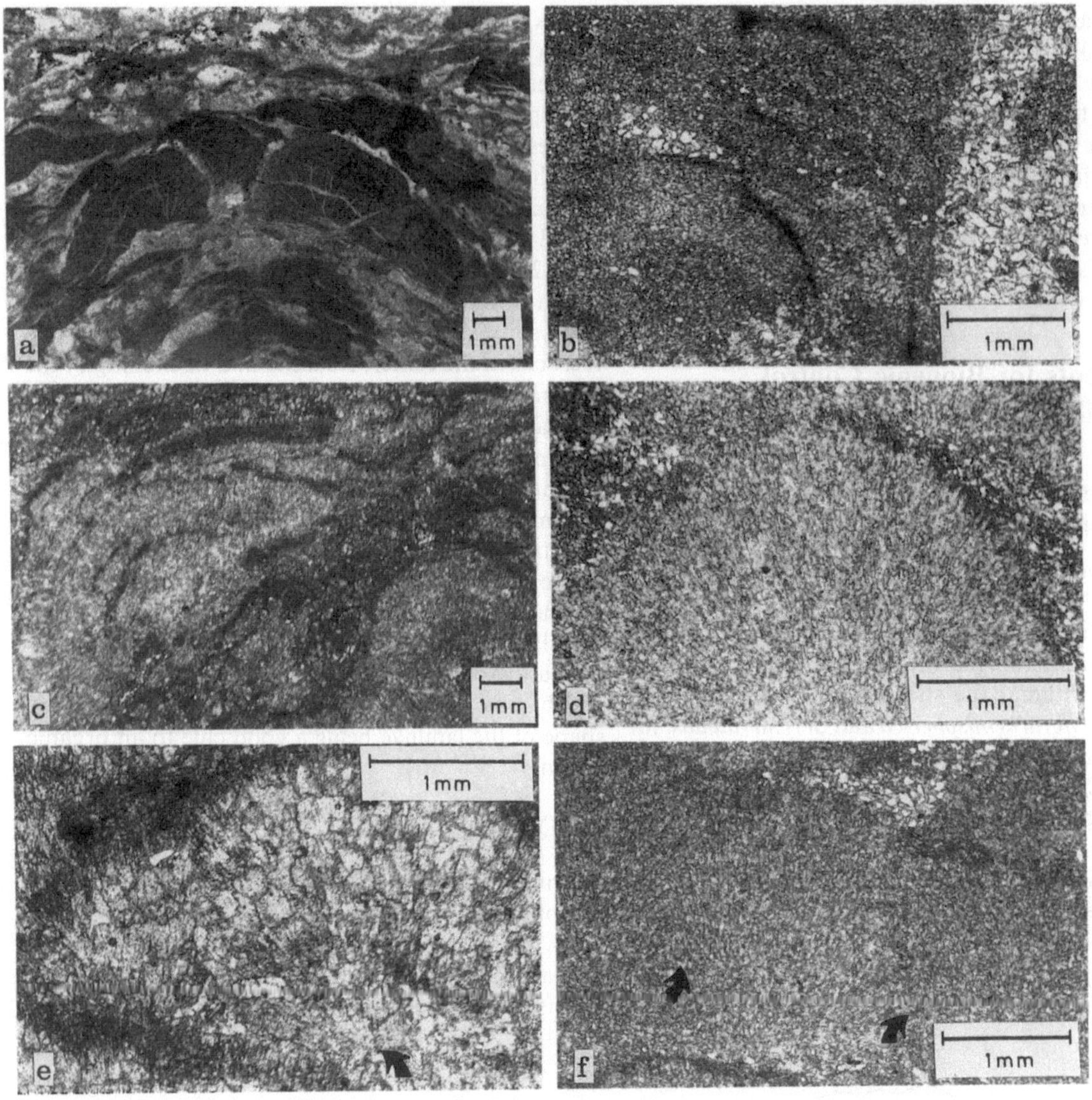

Fig. 3a–f. Tussocks from various stromatolite columns. **a** Very big, overgrowing tussocks, in the central part of *Inzeria globulosa* (Hank, Algeria). Note the cracks in the tussocks. **b** Column-sediment interface in *Tarioufetia hemispherica* (Adrar, Mauritania). **c** Big tussock showing banding in *Serizia radians* (Adrar). **d** High-relief tussock; two generations of tussocks interlayered with clastic (quartz silt) sediment filling up the depression; *Tarioufetia hemispherica* (Adrar). Radial pattern is clearly visible. **e** Radial pattern in *Serizia radians* (Adrar). The "rods" radiate from the center (*arrow*) up to the upper boundary, cross the darker outer rim and stop. Large anhedral crystals are superimposed on the radial fabric of the tussock, except in this outer rim. Rare quartz grains are trapped within the tussocks. **f** Juxtaposed tussocks with a radial pattern of the faint rods, illustrating that the interface between adjacent tussocks (*arrows*) are not sharp when less recrystallized (*Tarioufetia hemispherica*, Adrar). All *bars* = 1 mm

ally, they stop at the tussock boundary, which appears darker. The contacts between adjacent hemispheres are sharp when recrystallized and more diffuse when tussock recrystallization is microsparitic (Fig. 3f). The size of the rods cannot be taken as a diagnostic feature of the previous filaments, but it is meaningful when compared to the crystal size. Measurements show an average size of about 0.025 mm and a range between 0.01 to 0.04.

Detrital quartz grains, mainly in Adrar, within *Tarioufetia* and *Serizia*, are deposited in the depressions between the tussocks (Fig. 3b–d), sometimes in continuous layers. The deposition of quartz grains between the tussocks provides good evidence

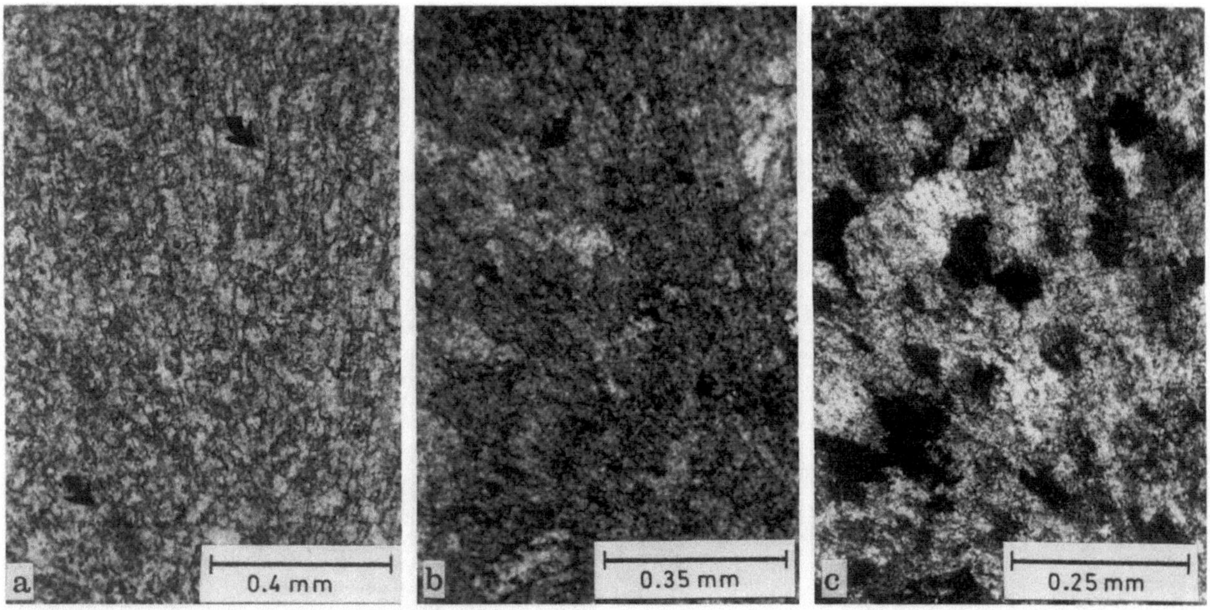

Fig. 4a–c. Radiating "rods" within tussocks (detail). **a** Detail of Fig. 3d; *arrows* point to rods which are made of small, aligned, anhedral crystals with dark limits. **b** and **c** Detail of Fig. 3e; **c** in cross-polarized light; *arrows* point toward rods included within large secondary anhedral crystals. In the *lower part* of **b** a corroded quartz grain is trapped in the tussock

that the tussocks are a primary rather than a diagenetic feature. The fact that tussocks are calcareous, although the sediment supply is dominated by coarse quartz, favours an accretion by precipitation and not by binding. Some microorganisms are known to play an active role in the precipitation of carbonates both in marine and fresh water (Golubic and Campbell 1981). Radial sparitic fabric has been demonstrated to be organic in origin in some *Phormidium* colonies (Monty and Mas 1981).

3. *Banding* is a common feature of the tussocks. Concentrically arranged bands of brown pigmentation occur in the tussocks (Fig. 3c). Such banding is found in tussocks from different stromatolites, both in the Adrar and in the Hank. It may occur in small or big tussocks; in the latter, banding seems to be exclusively of the formation of cracks. The bands are sometimes outlined by the trapping of scarce quartz grains.

4. *Recrystallization* occurs very frequently, if not always. According to Fairchild (1989) and Fairchild et al. (1990), the carbonate of the tussocky stromatolites was probably originally aragonitic, and recrystallization into calcite has been the rule. Thus corresponds in the tussock to the superposition of bushes of acicular crystals restricted to one tussock, or the crystal growth can continue in the space above the tussock, forming a neomorphic rim. Sometimes crystals cover several successive laminae (up to ten). The evidence for a biogenic origin of the tussocks may be obliterated in these examples. When recrystallization occurs in large anhedral crystals (never in the outer micritic rim; Fig. 4b–c), they include rods.

Bushes of crystals (probably aragonitic) occur outside the tussocks, in the column interspace (Fig. 5b–c). Layers of non-stromatolitic carbonates (wackestone to grainstone) with intraclasts, interbedded in siltstones, contain also displacive crystals in bushes (Fig. 5a–d) which can grow downward in the sediment. Differences with the tussock rods are obvious: much bigger crystals, with a larger range between 0.1 to 0.5 mm and a very distinct habit with rectangular or acicular termination. Furthermore, the crystals never display a tussocky shape even if their somewhat hemispherical habit is misleading. These crystals in bushes may have been aragonite. In some recrystallized acicular crystals (Fig. 5e) or cracks in tussocks, the anhedral crystals do not contain rods.

The existence of botryoidal aragonite cement in reef rocks raised the question whether the entire tussocky microstructure is an overprint of aragonite cement on an indiscernible stromatolitic microstructure. The present work justifies, in constrast, a biological control over the tussock shapes.

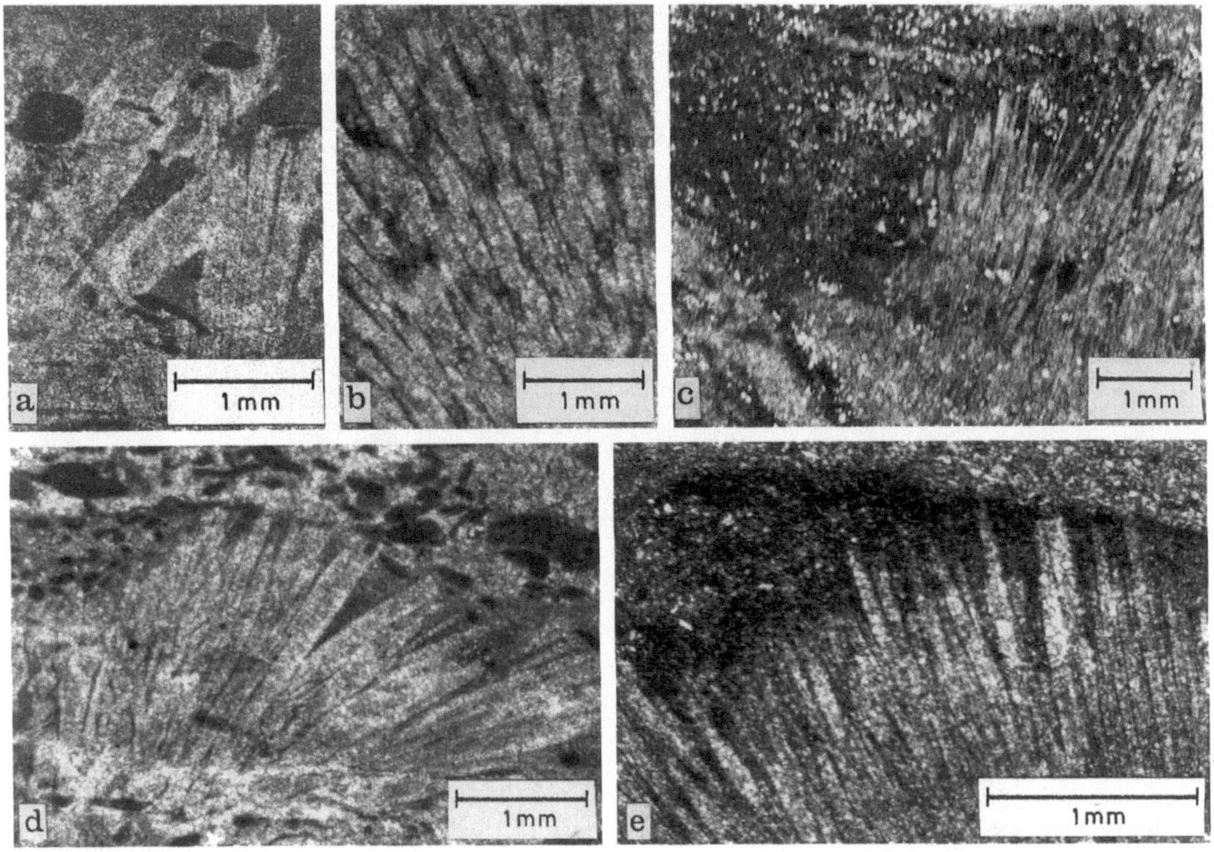

Fig. 5a–e. Bushes of crystals. **a** and **d** Isolated bushes of radiating crystals having a displacive growth in intraclastic carbonates. Crystals are mostly rectangular. Layered, non-stromatolitic carbonates interbedded in siltstones (Adrar). **b** Radiating crystals in the interspaces between columns of *Inzeria globulosa* (Hank-Algeria). **c** Bushes of displacive crystals roughly in rosette, acicular to rectangular in the interspace between columns of *Serizia radians* (Adrar). The tussocks of the stromatolite column are visible in the *lower left corner* of the photograph. **e** Bushes of acicular to rectangular crystals in flat-laminated stromatolitic carbonates, interbedded with the intraclastic carbonates in **a**. All *bars* = 1 mm

3 Comparisons with Modern Stromatolite Builders

A range of measurements were made on *Rivularia*, *Inzeria* and *Serizia* to clarify their structural relationships. The measurements included colony diameter and shape, filament density per unit area, and the pattern of concentric banding.

1. *Colony Shape*. *Rivularia* was chosen for the comparative study because of its close morphological relationship to the tussocks, sharing features of size, shape and internal structure.

Modern stromatolites built of tussocks are unknown. However, a few of the intertidal stromatolites of Shark Bay, Western Australia are formed by *Gardnerula corymbosa*, a close relative of *Rivularia* (Golubic 1976). However, these structures are not directly analogous to tussocks, as they have been formed by a microbe with a distinctly corymbose habit, which is lacking in both *Inzeria* and *Serizia* tussocks. Calcified freshwater microstromatolites of low convexity are also formed by *Homoeothrix*, *Schizothrix* and *Phormidium*, but these are rarely found growing as superposed colonies. This is mainly due to habitat instability which prevents the buildup of large bioherms. They occur in similar habitats, but calcified marine forms are unknown. Bioherms built by *Phormidium* and *Schizothrix* in quieter waters may attain considerable size. These cyanobacteria belong to the Oscillatoriaceae and produce fine filaments of variable length. Occasionally, they form small superposed colonies with a radiating filamentous structure and faint banding (Pentecost and Riding 1986). Micrite, rather than microspar, is precipitated within the colonies, and their structural organization is much lower than that of *Rivularia.*

Although heavily calcified species of *Rivularia* are virtually confined to freshwater today (Schneider et al. 1983; Pentecost 1987), some marine species in the intertidal zone of warm seas precipitate carbonate (Golubic and Campbell 1981). Formerly similar taxa may have grown in shallow seas or lagoons (Elliott 1964; Ott 1972; Richter et al. 1979). In contrast to the freshwater species, which precipitate calcite microspar, the marine species, *R. polyotis* and *R. mesenterica*, contain small aragonite concretions. In the former, the deposits envelope the filaments, forming bush-like structures (Golubic and Campbell 1981).

When colonies of *Rivularia* grow on steep surfaces, they frequently show a phototropic (heliotropic) response, resulting in a ledge-like or bracket-shaped morphology (unpubl. observ.). Heliotropism in modern stromatolites has been studied by Awramik and Vanyo (1986) and is a characteristic feature of many phototrophic microbes. The phenomenon should be a reliable indicator of past biological activity, but it is not always evident, even in modern stromatolites. The apparent absence of heliotropism in *Inzeria* and *Serizia* is therefore insufficient per se to discount a biological origin. The irregular laminae often thicken on the stromatolite tops, which suggests that the tussocks grew more quickly in this position. Modern stromatolite-building cyanobacteria show a similar response, because the total radiation flux is maximal on a horizontal surface, allowing for more photosynthesis and growth.

2. *Filament Density and Branching.* The results of measurements of colony diameter and filament density are shown in Fig. 6. Modern *Rivularia haematites* occur as small, calcified buttons growing on stones in hard water. The colonies often attain diameters exceeding 10 mm, but the average diameter is usually around 5 mm. Filament densities vary from 600–4400 filaments/mm^2 with a tendency for smaller colonies to possess lower numbers of filaments per unit area.

Measurements of the "tubes" or "rods" of *Inzeria* and *Serizia* were more difficult to make because they were not always well delimited. Often, only the outlines, consisting of a reddish-brown amorphous deposit, could be found within crystals of microspar. The continuity of many of the traces through several crystals, and their parallel arrangement, particularly noticeable under high power, gave the impression that they once represented tubular structures, perhaps of organic origin. In thin sections, it was possible to count the tube boundaries within a specified cross-

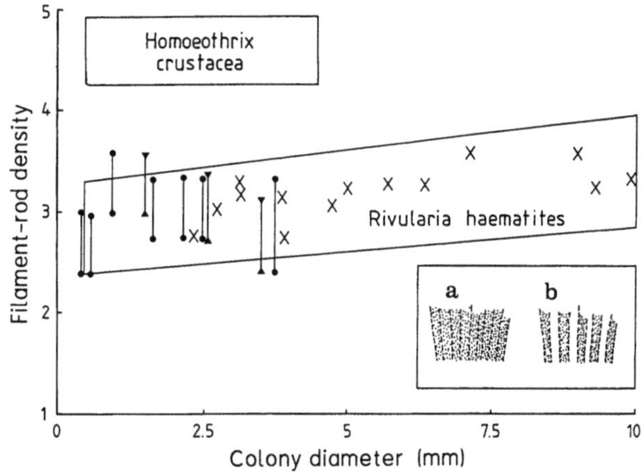

Fig. 6a–b. Measurements in modern and ancient colonies. Relation between filament/"rod" densities and colony diameters for *Homoeothrix crustacea*, *Rivularia haematites*, *Inzeria* (*triangles*) and *Serizia* (*circles*). Densities are expressed as log N where N is the filament density per mm. *Inset* (*bottom right*) shows two interpretations of the "rods" as visualized in thin sections. In **a** the rods are packed closely together, whilst in **b**, each rod (*stippled*) is separated from the others by a space. Estimates of rod numbers based on model **a** would be much higher than those based on **b**. To remedy this, a *vertical line* has been drawn between the estimates in the illustration

section and thus obtain an estimate of their density. This method, however, is prone to the interpretation error illustrated in the lower right inset of Fig. 6. The potential error is illustrated in Fig. 6 as vertical bars which cover the likely range of "tube" densities in these structures. A good correlation is apparent between the tussocks and *Rivularia* values, although the latter are generally greater in diameter. In size, the tussocks correspond more closely to modern *Homoeothrix crustacea* microstromatolites (Pentecost 1988), although in this species, filament densities are much higher (Fig. 6).

The trichomes (cell series) of *Rivularia* undergo a regular type of fragmentation leading to "false" branching of filaments composed of trichomes plus sheath and, although the branch angles are generally smaller than those of "true" branching, it would be difficult to distinguish "true" and "false" branching in fossil material. Fossils of *Rivularia* from the Pleistocene (Richter et al. 1979) and the closely related *Zonotrichites* from the Triassic (Bornemann 1889; Elliott 1964) show clear evidence of branching. Traces of branching in the tussocks are observable but they are not common.

Filaments of *Rivularia* often radiate from a common center but frequently show wave-like curvatures in parts of the colony (Fig. 8a). At colony edges, the

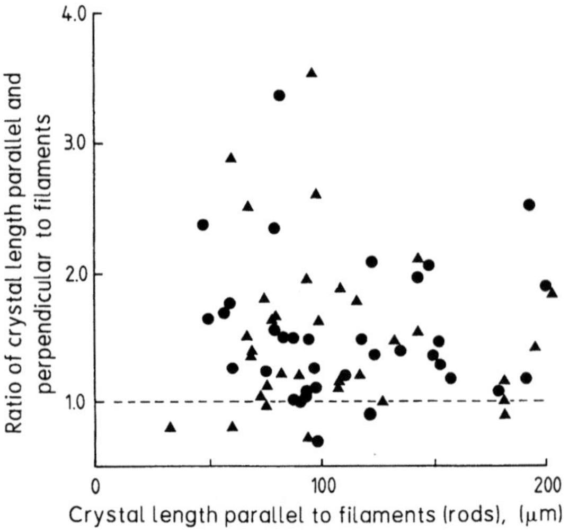

Fig. 7. Crystal matrix orientation. In thin section of *Rivularia* (*circles*), and *Serizia* (*triangles*), the crystal length is parallel to the filaments/rods and is plotted along the abscissa. The ratio of crystal length parallel to the filaments to the length perpendicular to the filaments is plotted along the ordinate. *Broken line* indicates the 1:1 ratio

filaments diverge in a characteristic pattern which is also apparent in fossil material of *Zonotrichites* (Ott 1972; Richter et al. 1979; Richter and Sedat 1983). These features are found only in the tussocks.

Measurements of the matrix crystals in thin sections of the tussocks and *Rivularia* show that in both, crystal size is approximately the same, and the long axes of the crystals are oriented parallel to the direction of the filaments (Fig. 7). This presumably occurs in *Rivularia* because the diffusion of crystal ions is more rapid (less obstruction) parallel to the filaments, leading to more rapid crystal growth in this direction. The degree of emphasis to be placed on the observation of fossil tussocks depends upon the extent to which diagenesis has occurred in the tussocks. It is worth noting that aragonite deposits in marine *Rivularia* can also grow preferentially in the direction of the filaments (Golubic and Campbell 1981).

3. *Concentric Banding Patterns*. Banding is a characteristic feature of *R. haematites*. Bands of calcite crystals are evident in longitudinal section of the colonies and there are at least two types of bands. During winter, when growth is negligible, a broad, dense band is deposited, with a thickness of 0.5–3 mm, consisting of microspar. During the period of growth, from April to November in northern Europe, a series of much narrower bands is deposited, often separated by poorly calcified zones, and occupied by soft mucilage. The thickness of these bands ranges from 0.05–0.4 mm (Fig. 8a). The periodic nature of the calcification suggests a biological influence and it is believed that the narrow bands are formed during periods of low water, when photosynthetic activity by aquatic plants promotes calcite precipitation through a rise in pH. This biological activity, however, is not necessarily due to *Rivularia* alone. In fact, during one investigation (Pentecost and Spiro, unpubl.), it was found to be the result of bryophyte photosynthesis and enhanced degassing caused by a temperature rise in the water. The thick winter bands are believed to be caused by abiogenic precipitation from water supersaturated with calcium carbonate (calcite).

Banding is also found in decalcified colonies. Concentric layers of sheath pigment occur often associated with zones of calcification. Heterocysts are also found in bands and the organized meristematic zones of *Rivularia* might promote calcification (Pentecost 1987).

The band width in *Serizia* was 40–400 μ (mean 90 μ), and these were separated by zones composed of clear sparite. Normally, the tussocks contained six or fewer bands, with a prominent band 50–60 μ wide at the tussock margin. The bands were sometimes crossed completely by crystals of calcite, but crystal size was often reduced in the pigmented zones to 5–25 μ, compared with 25–100 μ outside these areas. Within the bands, dark granules, 1–3 μ across were found, frequently enclosed by calcite. These granules may represent detritus which has accumulated at the tussock surface at regular intervals.

Thin sections of *Inzeria* revealed a similar pattern of banding but the tussocks were often larger and the bands sometimes narrower, ranging from 10 to 260 μ in thickness (mean 90 μ).

Non-carbonate detritus becomes incorporated into colonies of *Rivularia* by entrapment and this can occasionally manifest itself as a series of brownish bands. The reduction in the size of carbonate crystals within the bands may be a reaction to the incorporation of foreign material, or it might represent different conditions of crystallization. Considerable variation in crystal size occurs in *Rivularia* bands (Pentecost 1987), but its causes require more study.

4 Conclusions

The above observations point to a biogenic origin for the tussocks. The most compelling fact is that the tussocks build stromatolites. These columnar stromatolites are characterized by different column "attitudes" and laminae shape and convexity, which are

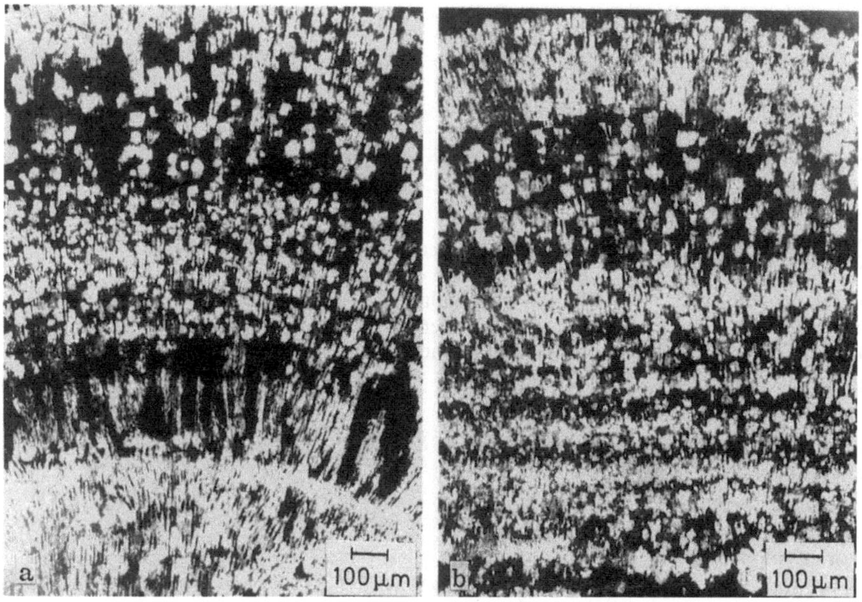

Fig. 8a, b. Longitudinal thin sections of *Rivularia haematites* from northern England. Photographed under crossed polars. **a** Section showing part of a dense winter band (*below*) and a series of narrower bands formed later in the year (*above*). Radiating filaments of *Rivularia* can be seen to "perforate" the calcite crystals in places. Note also the occurrence of "wavy" filaments in the winter band. *Bar* = 100 µm. **b** Selection showing a series of narrow summer bands, about 100 µm wide, from the same locality. *Bar* = 100 µm

primary built by the same kind of hemispheric tussocks. Banding of the tussocks and the existence of radiating rods, visible despite the recrystallization overprint, converge to indicate a biogenic origin of the tussocks.

The comparative study with *Rivularia* is summarized in Table 1. Of the 11 characteristics listed in Table 1, at least 7 are shared by *Rivularia* and the tussocks. The four differences are heliotropism, branching and curvature of filaments and the interface between adjacent colonies. The degree of emphasis to be placed on these differences is to some extent subjective, but the branching and curvature of filaments are biological constraints which should be regarded as more important than colony size and weighted accordingly. Even if this is done, a biological origin remains compelling.

The sharp interface between colonies and the straighter habit of the rods may be due to the complex diagenesis, with aragonite precipitated in the colony, and later neomorphosed into calcite or dolomite. In this context the occurrence of aragonite concretions in marine *Rivularia* as described by Golubic and Campbell (1981) is of interest, although such colonies have only been reported from intertidal sites. Some presumed fossil *Rivularia* colonies have also lost their filaments to rods and the curvature and branching may be obscure (Casanova 1987; Bertrand-Sarfati et al. 1991). Heliotropism is a feature more evident in subaerial or intertidal environments than in sublittoral marine communities, although it could occur in deep, clear water where light scattering is reduced. The fact that heliotropism occurs in modern Rivulariaceae, but not in the tussocks, may indicate that the original organisms were only weakly heliotropic, perhaps belonging to a related but now extinct class of microbes.

We must accept that no modern counterparts of the tussocks are known, but the evidence points toward a biogenic origin. The nature of the organism is uncertain, but it may have been a filamentous, weakly heliotropic cyanobacterium with a radiating structure.

The predominance of a specific type of microstructure raised the question of the possible formation of the stromatolites by a monospecific population of microorganisms. Such accretion has been recorded in modern marine *Rivularia* (Golubic and Campbell 1981) on a smaller scale.

The tussocky microstructure stromatolites are stratigraphically well defined and seem to correspond to a biological event which can be related to a chemical change in the water chemistry (Fairchild 1989; Fairchild et al. 1990).

Table 1. Comparisons of some characters in modern Rivularia and in Precambrian tussocks

Character Colony diameter (mm)	Rivularia 0.5–10, mean 5.0	Tussock 0.05–4
Colony shape	Hemispheres or buttons of low convexity, often marginally fused	The same
Heliotropic modification	Shelf- or bracket-shaped colonies on steep surfaces	No or few modifications
Lamina thickening	Laminae regular and often thicker on horizontal surface	Laminae irregular but often thicker on horizontal surface
Filament/"tube" density mm^2	600–4400	240–3800
Filament branching	Common and often apparent in more recent fossils	Uncommon
Filament curvature	Frequent "waves" of low amplitude and pronounced divergence at colony edge	Curvature rare, no divergence at colony edge
Colony interface	Sharp to diffuse, when sharp, crystals without linear junctions	If recrystallized, crystals with linear junctions
Band width (mm) and compositions	0.05–3 microspar	0.01–0.4, dark granular inclusion and microspar
Band types	Broad "winter" and narrow "summer" bands often distinguishable	Broad and narrow bands occasionally in a regular sequence
Orientation of matrix crystals	Slight elongation parallel to filaments	The same

References

Awramik SM, Vanyo JP (1986) Heliotropism in modern stromatolites. Science 231:1279–1281

Bertrand-Sarfati J (1972) Stromatolites columnaires du Précambrien supérieur, Sahara Nord-occidental. Inventaire, morphologie et microstructure des laminations, corrélations stratigraphiques. These Montpellier CNRS (CRZA) Ser Geol 14:242 pp

Bertrand-Sarfati J (1976) An attempt to classify Late Precambrian stromatolite microstructures. In: Walter MR (ed) Stromatolites. Elsevier, Amsterdam, pp 251–259

Bertrand-Sarfati J, Moussine-Pouchkine A (1988) Is cratonic sedimentation consistent with actualistic models? An example from Upper Proterozoic of the West African Craton. In: Jackson MJ (ed) Aspects of Proterozoic sedimentary geology. Sediment Geol 58:255–276

Bertrand-Sarfati J, Freytet P, Plaziat JC (1991) Microstructures and biogenic remains in nonmarine stromatolites (Tertiary, France). Comparison with some Precambrian microstructures. In: Monty C (ed) Phanerozoic stromatolites II. Springer, Berlin Heidelberg New York (in press)

Bornemann JG (1889) Geologische Algenstudien. Jahrb K. Preuss Geol Landesanst 1886:116–134

Cao R, Liang Y (1974) Fossil-algae and stromatolites from the Sinian System in southern China, description and comparisons. Mem Nanking Inst Geol Palaeontol/Acad Sin 5:16 pp (in Chinese)

Casanova J (1987) Stromatolites et hauts niveaux lacustres pleistocènes du bassin Natron-Magadi (Tanzanie-Kenya). Sci Geol Bull Strasbourg 40:135–153

Elliott GF (1964) *Zonatrichites* (calcareous algae) from the Arabian Triassic. Eclogae Geol Helv 57:567–570

Fairchild IJ (1989) Carbonate mineralization of Proterozoic stromatolites. EUG Strasbourg Abstr SY23c-06:197

Fairchild IJ, Marshall JD, Bertrand-Sarfati J (1990) Stratigraphic shifts in carbon isotopes from Proterozoic stromatolitic carbonates (Mauritania): influence of primary mineralogy and diagenesis. Am J Sci, Spec Issue 290A:46–79

Golubic S (1976) Biology of stromatolites. In: Walter MR (ed) Stromatolites. Elsevier, Amsterdam, pp 113–140

Golubic S, Campbell S (1981) Biogenically formed aragonite concretions in marine *Rivularia*. In: Monty Cl (ed) Phanerozoic stromatolites I. Springer, Berlin Heidelberg New York, pp 209–229

Grey K (1984) Biostratigraphic studies of stromatolites from the Proterozoic Earaheedy Group, Nabberu Basin, Western Australia. Geol Surv W Aust Bull 130:123 pp

Monty Cl, Mas JR (1981) Lower Cretaceous (Wealdian) blue-green algal deposits of the province of Valencia, E. Spain. In: Monty Cl (ed) Phanerozoic stromatolites I. Springer, Berlin Heidelberg New York, pp 85–120

Moussine-Pouchkine A, Bertrand-Sarfati J, Ball E, Caby R (1988) Les séries sédimentaires et volcaniques protérozoïques impliquées dans la chaine pan africaine: la région de l'Adrar Ahnet (N.W. Hoggar, Algérie). J Afr Earth Sci 7:57–75

Ott E (1972) Mitteltriadische Riffe der nördlichen Kalkalpen und altersgleiche Bildungen auf Karaburn und Chios (Agäis). Mitt Ges Geol Bergbau Stud 21:251–276

Pentecost A (1987) Growth and calcification of the cyanobacterium *Rivularia Haematites*. Proc R Soc London Ser B 232:125–136

Pentecost A (1988) Growth and calcification of the cyanobacterium *Homoeothrix crustacea*. J Gen Microbiol 134:2663–2671

Pentecost A, Riding R (1986) Calcification in cyanobacteria. In: Leadbeater BSC, Riding R (ed) Biomineralization of lower plants and animals. Univ Press, Oxford, pp 73–90

Preiss WV (1976) Proterozoic stromatolites from the Nabberu and Officer Basin, Western Australia and their biostratigraphic significance. Geol Surv S Aust Rep Invest 47:51 pp

Richter DK, Sedat R (1983) Brackish-water oncoid composed of blue-green and red algae from a Pleistocene terrace near Corinth, Greece. In: Peryt TM (ed) Coated grains. Springer, Berlin Heidelberg New York, pp 229–307

Richter DK, Herforth A, Ott E (1979) Pleistozäne, brackische Blaugrünalgengriffe mit *Rivularia haematites* auf der Parachorahalbinsel bei Korinth (Griechenland). Jahrb Geol Palaeontol Abh 159:14–40

Schneider J, Schroeder HG, Le Campion-Alsumard T (1983) Algal micro-reefs – coated grains from freshwater environments. In: Peryt TM (ed) Coated grains. Springer, Berlin Heidelberg New York, pp 284–298

On the Association of Ore Deposits with Stromatolites

M. Brongersma-Sanders [1]

[1] National Natuurhistorisch Museum, P.O. Box 9517, 2300 RA Leiden, The Netherlands

Abstract

Several stratiform ore deposits show the following characteristics: (1) close association with stromatolites; (2) occurrence along the borders of a sea with oxygen-poor bottom water; (3) (particularly in the case of Cu deposits) neighborhood of red beds and/or evaporites suggesting a marginal-marine, warm-arid environment. The combination of a warm-arid coast with an oxygen-deficient sea is reminiscent of the rainless deserts along oxygen-deficient seas in areas of strong upwelling.

Seawater may be the source of the metals. The total metal content (dissolved, in living plankton and in particulate organic matter) is relatively high in upwelling systems. Exceptionally high contents occur in slicks (smooth patches on the sea due to capillary wave damping) and foam developing particularly in highly productive, coastal waters. The following hypothesis is presented: metal-containing particulate matter of slicks and foam is trapped by the microbial mats. A high oxygen supply in the surface layer of a mat favors degradation of organic slick complexes, and release of metals as free ions, to be followed by precipitation of metal sulfides in the H_2S-containing bottom layer.

1 Introduction

Many stratiform (or stratabound) ore deposits containing Cu, Pb, Zn and/or precious metals are closely associated with stromatolites. The association has been observed in the Central African Copperbelt (Fleischer 1984); Hank Formation, western Sahara (Cu) (Buffière et al. 1966); Witwatersrand Sequence (Au, U) (Zumberge et al. 1978); Mount Isa Sequence (Cu, Pb, Zn) (Neudert and Russell 1981); Dunphy Formation, Labrador (Cu) (Chevé et al. 1985; Schrijver et al. 1986); North Texas copper deposits (Smith 1976); Mississippi Valley lead-zinc deposits (Mendelsohn 1976).

The genesis of the metal deposits, and a possible role of the stromatolites, are a point of discussion. This work attempts to throw some light on these problems. The habitat of microbial mats, being Recent analogs of stromatolites will be considered as a starting point; the enumeration is limited to mats in the marine environment.

2 Habitat of Microbial Mats

Marine mats occur (1) in intertidal and shallow subtidal areas of hypersaline lagoons in a warm-arid climate (examples summarized in Brongersma-Sanders 1988); (2) on tidal flats in a temperate-humid climate (Gerdes and Krumbein 1987); (3) on the continental shelf and the upper slope: in upwelling systems off Peru-Chile, in the Santa Barbara Basin, and in the Gulf of California in a dysaerobic environment (oxygen content of the water 0.1 – 1.0 ml/l) at the boundary of the oxygen-minimum zone (Gallardo 1977; Soutar and Crill 1977; Reimers 1982; Williams and Reimers 1983; Williams 1984); (4) in the deep sea: near thermal vents (Jannasch and Wirsen 1981).

2.1 Association of Cu Deposits with Red Beds and Evaporites

Cu deposits are frequently associated with red beds and evaporites: often they occur in the vertical sequence: terrestrial red beds → dark shales with Cu → evaporites. This suggests a marginal-marine, warm-arid environment similar to the hypersaline lagoons mentioned in Section 2. Many Pb – Zn deposits show a distant association with red beds and evaporites (Gustafson and Williams 1981).

2.2 Role of Stromatolites and Vicinity of an Anoxic Water Mass

Renfro (1974) suggests that the metals were derived from the hinterland, transported by groundwater in

M. Schidlowski et al. (Eds.)
Early Organic Evolution: Implications for Mineral and Energy Resources

a horizontal direction, carried upwards into a sabkha environment, and precipitated in the H_2S-containing bottom layer of a microbial mat. We assume the open sea to be the source of the metals; metal-containing particulate matter transported by lagoonal water was trapped by the sticky mats; metals becoming free ions in the oxidizing surface layer were precipitated as sulfides in the H_2S-containing bottom layer.

The mats need not be the ultimate repository. Metals released during diagenesis may migrate upwards, the porous materials of overlying mats acting as channelways. Another possibility is that metals released from reefs or mats spread laterally in the lagoonal water; renewed precipitation may occur if anoxic conditions prevail in close proximity. This apparently was the case during the deposition of the Zambian copper ores; the stromatolites themselves are nearly always barren, but high metal contents occur in adjacent deposits. The ores were deposited in anoxic basins adjacent to microbial bioherms (Fleischer 1984). The Mississippi Valley Pb–Zn deposits also developed along the borders of anoxic basins; oil occurs in the basins, the Pb–Zn deposits on regional highs (Dozy 1970). The situation is reminiscent of habitat 3 (Sect. 2), but the border of the anoxic area was probably situated close inshore.

3 Humid-Air Deserts and Upwelling

Where do marginal-marine, warm deserts occur at the present time? Besides normal deserts there are the so-called humid-air deserts (Trewartha 1962); although the humidity of the air is high in the inshore area, they belong to the most rainless regions on earth. They occur along the coasts of Peru-Chile 5°–30° S, of southwestern Africa 15°–32° S, of Mauretania-southern Morocco as far as 30° N, and along California and Baja California, but here the aridity is less intense. In the sea adjacent to these deserts there is upwelling. The combination has a factual basis: the low temperature of the coastal water being one of the controls of the rainless climate (other controls are discussed in Brongersma-Sanders 1971).

Considering the frequent association of stratiform Cu deposits with red beds and evaporites, many authors concluded that the metal enrichment must be related to one of these features. The coupling of rainless coasts to regions of upwelling opens up another possibility: upwelling in the adjacent sea has a bearing on the enrichment.

Lagoons along the arid coasts tend to be hypersaline. A string of hypersaline lagoons containing microbial mats occurs along the Pacific side of Baja California. Similar lagoons with Recent and subfossil mats occur or occurred in the near-past along the Mauretanian coast (Einsele et al. 1974; Schwarz et al. 1975). Properties of these lagoons are related to both the rainless climate and upwelling. The content of clastics in suspension is usually low due to absence of rivers; this implies low dilution of metals in the sediment. The productivity is high in many hypersaline lagoons, but is particularly high if the inflowing water is derived from an area of upwelling.

4 Zoning of Metal Sulfides

Several ore deposits display a zoning of different Cu–Fe sulfides, probably related to differences in sulfide solubilities. If the deposits contain Cu, Pb and Zn, these metals generally occur in the seaward sequence Cu → Pb → Zn; this may also be related to different solubilities, although "there are wide variations in the overlap of different metals, particularly copper and lead, and in the relative abundance of metals" (Gustafson and Williams 1981).

Because the least soluble sulfides occur on the landward side, Renfro (1974) concluded that the zoning rules out the open sea as source of the metals. This conclusion is not wholly warranted. In hypersaline lagoons the circulation is anti-estuarine (Fig. 1): inflow of surface water, sinking in the innermost parts of the lagoon due to increased salinity, and outflow along the bottom. Metals carried into such lagoons by surface inflow may be carried in a seaward direction by the bottom water outflow. This shows that sulfide zoning need not discount the sea as source of the metals.

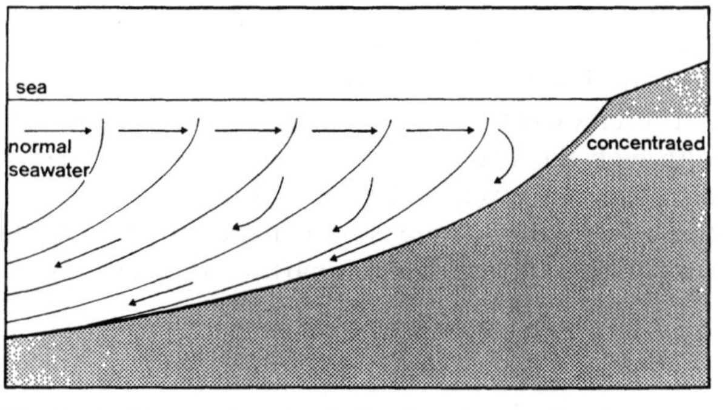

Fig. 1. Anti-estuarine circulation in a hypersaline lagoon

5 Metals with a Nutrient-Type Distribution

Another argument brought forward against the sea as metal source is the low concentration of ocean water. The concentrations of different water masses vary, however, within wide limits. The concentrations of a large number of trace elements vary in the same way as those of the nutrients required for plankton production. Extraction from surface water by living plankton and resolubilization in deeper layers lead to a change in concentration along the vertical profile.

Phosphate regenerates in relatively shallow water, silicate more at depth. The profiles of As and Cd are similar to those of phosphate, the Ba, Cr, Ge and Zn profiles to those of silicate, and the Ni, Pd, Pt and Se profiles to a combination of both. Cu profiles are somewhat different: the concentrations increase linearly with depth, probably governed by surface removal and bottom water injection by regeneration from the sediment (Bruland 1980, 1983; Hodge et al. 1985). The "Zn – Si correlation may be in fact a Zn – Ca correlation" ... "Foraminifera may be important for the marine cycling of zinc" (Boyle 1981). Pb profiles are of no importance, the oceanic lead concentrations being markedly influenced by man's activities.

A combination of biological/chemical processes with oceanic circulation, and particularly with countercurrent systems, leads also to a change in concentration in a horizontal direction, and to an accumulation of nutrients in areas of upwelling. Upwelling systems are large reservoirs of nutrients, and this will also apply to metals with a nutrient-type distribution.

5.1 Metal Concentrations of Sediments in Oceanic Upwelling Areas

A high total metal content (dissolved, in living plankton and particulate organic matter) does not necessarily imply a high concentration in the underlying sediment. In muds deposited under near-anoxic conditions on the Namibian shelf the concentrations of Mo, U, Ba, Cd and Se are locally above normal (Brongersma-Sanders et al. 1980; Baturin 1983; Calvert and Price 1983). Zn concentrations were low in our cores, except for a relatively high value in a sample with high Ba. The high concentration of Ni initially recorded by Calvert and Price was not confirmed in their 1983 paper. There is no marked enrichment of Cu, which may be related in part to complexation.

6 Complexation

Trace metals dissolved in seawater occur as free ions, and as inorganic and organic complexes. Only few metals form organic complexes to a high extent; these are Cu, Fe, and also, but to a lesser extent, Pb and Zn (Kramer 1985). In highly productive waters where there are abundant organic ligands, a high precentage of dissolved Cu is complexed; as a result, the concentration of free ions available for precipitation reactions is small.

6.1 Metal Enrichment at the Sea-Air Interface

Part of the organic ligands providing complexing sites for heavy metals are surface-active. They are carried by rising bubbles and other mechanisms to the sea-air interface. If sufficient materials are available, they form visible slicks, and these are typically marked by foam. Natural slicks are most commonly found in highly productive coastal waters, particularly in bays and lagoons. Release of surfactants is especially abundant during the decline of a phytoplankton bloom (Wallace 1982).

Slicks and foam are remarkably rich in trace metals; the concentrations of Pb, Cd, Cu, etc. may be many times higher than in the directly underlying water mass (Piotrowicz et al. 1972; Lion et al. 1982; Wallace 1982). Slicks are also very rich in both dissolved and particulate organic matter; substrates suitable for adsorption may lead to transformation of metals from a dissolved to a particulate form. A high percentage of the metals in the slicks appears to be adsorbed on particulate matter (Hunter 1980). These materials may float ashore by tide or wind-generated water movements.

Tentatively, we might suggest the following. Particulate metal-containing organic matter of slicks and foam is trapped in hypersaline lagoons by the sticky surface layer of a microbial mat.

A sticky surface may not be a requirement. "Microbial mats are good recipients for slicks of all kinds, including pollution, and I do not think their surfaces need to be particularly sticky to retain them" (S. Golubic, pers. comm.).

Bacterial degradation of organic slick complexes, and a concomitant release of metals as free ions is favored by a high oxygen supply in the phototrophic cyanobacterial surface layer, followed by the precipitation of metal sulfides in the H_2S-containing bottom layer.

7 Anoxic Conditions in Upwelling Systems

The content of dissolved oxygen in the sea depends on demand and supply. Oxygen demand in subsurface waters relates to surface productivity, the supply depends largely on water circulation. High productivity in upwelling systems combined with a limited supply leads to nearly anoxic conditions in the open ocean, and to inflow of oxygen-deficient waters in adjacent basins. Off Peru the oxygen minimum layer impinges on the outer shelf and on the upper slope, near-anoxic conditions prevailing between 100 and 400 m depth from 10° to 25° S (Packard et al. 1983), Off Namibia the minimum layer slopes upward over the outer and inner shelf, the oxygen contents decreasing shorewards. In the Walvis Bay anoxic conditions are observed occasionally (Hart and Currie 1960).

7.1 Anoxic Conditions in Upwelling Systems During Transgressions

Several ore deposits developed during a first transgression over a continental sequence, e.g. Copperbelt, Kupferschiefer, Dunphy Formation (Chevé et al. 1985), and cretaceous deposits in basins bordering the South Atlantic (Caïa 1976). The cretaceous Cu deposits of southern Angola are reminiscent in several aspects of the Zambian Copperbelt (Van Eden 1978), but stromatolites are not recorded; microbial mats might have been present, but were not preserved.

In the case of a first transgression the surface of the inundated land is relatively irregular. An irregular relief favors the development of lagoons in coastal depressions.

"During a transgression of an ocean with a strong and shallow oxygen minimum the area newly covered with low-oxygen water is greatly increased" (Berger 1984). A transgression on the coast of Namibia/Angola would imply extension of anoxic conditions into coastal lagoons.

8 Conclusions

The impact of microbial mats on the development of ore deposits may be the entrapment of metal-containing matter from slicks and foam, followed by sulfide precipitation in the H_2S-containing bottom layer. Mats and/or reefs probably developed in more or less hypersaline lagoons along a sea with strong upwelling.

The sequence of processes is as follows: accumulation of metals in coastal upwelling waters – inflow of fertile, metal-rich surface water into the lagoon – release of surface-active compounds from a declining plankton bloom providing complexing sites for heavy metals – transportation of surfactants with adhering metals to the sea-air interface by rising bubbles – if sufficient materials are available, formation of slicks and foam – trapping of slick material by microbial mats – degradation of organic slick complexes in the upper layer of the mat favored by a high oxygen supply from phototrophic Cyanobacteria – concomitant release of metals as free ions – precipitation of metal sulfides in a H_2S-containing bottom layer.

This need not be the end of the story. In the Zambian Copperbelt the stromatolites themselves are nearly always barren, whereas high metal contents occur in adjacent deposits. Metals released from mats or reefs were precipitated once more along the border of a neighboring, anoxic water mass. Anoxic conditions prevail close inshore in regions of strong upwelling, shifting shorewards during a transgressive period. This might explain the fact that several ore deposits developed during transgressive periods.

References

Baturin GN (1983) Some unique sedimentological and geochemical features of deposits in coastal upwelling regions. In: Suess E, Thiede J (eds) Coastal upwelling. Its sediment record. Nato Conf Ser IV, Mar Sci, 10a, pt B. Plenum, New York London, pp 11–28

Berger WH (1984) Short term changes affecting atmosphere, oceans, and sediments during the Phanerozoic. In: Holland HD, Trendall AF (eds) Dahlem Konferenzen 1–6 May 1983. Springer, Berlin Heidelberg New York, pp 171–205

Boyle EA (1981) Cadmium, zinc, copper, and barium in foraminifera tests. Earth Planet Sci Lett 53:11–35

Brongersma-Sanders M (1971) Origin of major cyclicity, of evaporites and bituminous rocks: an actualistic model. Mar Geol 11:123–144

Brongersma-Sanders M (1971) Origin of major cyclicity of cyanobacterial mats and of Rote Fäule to the metal enrichment of the Kupferschiefer. Proc Kon Ned Akad Wetensch Amsterdam Ser B 91:15–38

Brongersma-Sanders M, Stephan KM, Kwee TG, Bruin M de (1980) Distribution of minor elements in cores from the southwest African shelf with notes on plankton and fish mortality. Mar Geol 37:91–132

Bruland KW (1980) Oceanographic distribution of cadmium, zinc, nickel and copper in the North Pacific. Earth Planet Sci Lett 47:176–198

Bruland KW (1983) Trace elements in sea water. In: Riley JP, Skirrow G (eds) Chemical oceanography, 2nd edn. vol 8. Academic Press, New York London, pp 157–220

Buffière JM, Fahie JC, Ronchin G (1966) Présence d'horizons cuprifères dans la série sédimentaire infratillit-

ique du Hank algérien. Chron Mines Rech Min, Rabat 34, 353:167–173
Caïa J (1976) Paleogeographical and sedimentological controls of copper, lead and zinc mineralisations in the Lower Cretaceous sandstones of Africa. Econ Geol 71:409–422
Calvert SE, Price NB (1983) Geochemistry of Namiban shelf sediments. In: Suess E, Thiede J (eds) Coastal upwelling. Its sediment record. Nato Conf Ser IV, 10a, pt B, Mar Sci. Plenum, New York London, pp 337–375.
Chevé SR, Schrijver K, Tassé N (1985) Cryptalgalaminate dolomite of the Dunphy Formation, Labrador Trough; diagenetic and tectono-metamorphic evolution related to copper mineralisation. Can J Earth Sci 22:1835–1857
Dozy JJ (1970) A geological model for the genesis of the lead-zinc ores of the Mississippi Valley, U.S.A. Trans Inst Min Metall B 79:B163–B170
Einsele G, Herm D, Schwarz HU (1974) Sea level fluctuations during the past 6000 yr on the coast of Mauretania. Quat Res 4:282–289
Fleischer VD (1984) Discovery, geology and genesis of copper-cobalt mineralisation at Chambishi southeast prospect, Zambia. Precambrian Res 25:119–223
Gallardo VA (1977) Large benthic microbial communities in sulphide biota under Peru-Chile subsurface countercurrent. Nature (London) 268:331–332
Gerdes G, Krumbein WE (1987) Biolaminated deposits. Lecture Notes Earth Sciences 9:183 pp
Gustafson LB, Williams N (1981) Sediment-hosted stratiform deposits of copper, lead, and zinc. Econ Geol 75th Anniv Vol, pp 139–178
Hart TJ, Currie RI (1960) The Benguela Current. Discovery report, vol 31. Univ Press, Cambridge, pp 123–198
Hodge WF, Stallard M, Koide M, Goldberg ED (1985) Platinum and the platinum anomaly in the marine environment. Earth Planet Sci Lett 72:158–162
Hunter KA (1980) Processes affecting particulate trace metals in the sea surface microlayer. Mar Chem 9:49–70
Jannasch HW, Wirsen CO (1981) Morphological survey of microbial mats near deep-sea thermal vents. Appl Environ Mikrobiol 41:528–538
Kramer CJM (1985) On the copper complexation capacity in the marine environment. Thesis, Univ Groningen, Neth, 133 pp
Lion LW, Harvey RW, Leckie JD (1982) Mechanisms for trace metal enrichment at the surface microlayer in an estuarine salt marsh. Mar Chem 11:235–244
Mendelsohn F (1976) Mineral deposits associated with stromatolites. In: Walter MR (ed) Stromatolites. Developments in sedimentology, vol 20. Elsevier, Amsterdam, pp 645–662
Neudert MK, Russell RE (1981) Shallow water and hypersaline features from the Middle Proterozoic Mt. Isa Sequence. Nature (London) 293:284–286
Packard TT, Garfield PC, Codispoti LA (1983) Oxygen consumption and denitrification below the Peruvian upwelling. In: Suess E, Thiede J (eds) Coastal upwelling. Its sediment record. Nato Conf Ser IV 10a, pt B, Mar Sci. Plenum, New York London, pp 147–173
Piotrowicz SR, Ray BJ, Hoffman GI, Duce RA (1972) Trace metal enrichment in the sea-surface microlayer. J Geophys Res 77:5243–5254
Reimers CE (1982) Organic matter in anoxic sediments off Central Peru: relations of porosity, microbial decomposition and deformation properties. Mar Geol 46:175–197
Renfro AR (1974) Genesis of evaporite-associated stratiform metalliferous deposits. A sabkha process. Econ Geol 69:33–45
Schrijver K, Bertrand R, Chagnon A, Tassé N, Chevé SR (1986) Fluids in cupriferous dolostones and dolomite veins, Proterozoic Dunphy Formation, Labrador Trough. Can J Earth Sci 23:1709–1733
Schwarz HU, Einsele G, Herm D (1975) Quartz-sandy, grazing-contoured stromatolites from coastal embayments of Mauretania, West Africa. Sedimentology 22:539–561
Smith GE (1976) Sabkha and tidal-flat facies control of stratiform copper deposits in North Texas. In: Wolf KH (ed) Handbook of stratabound and stratiform ore deposits, vol 6. Elsevier, Amsterdam, pp 407–446
Soutar A, Crill PA (1977) Sedimentation and climatic patterns in the Santa Barbara basin during the 19th and 20th centuries. Bull Geol Soc Am 88:1161–1172
Trewartha GT (1962) The earth's problem climates, 2nd edn. Univ Wisconsin Press, Madison, 334 pp
Van Eden JG (1978) Stratiform copper and zinc mineralisations in the Cretaceous of Angola. Econ Geol 73:1154–1161
Wallace GT (1982) The association of copper, mercury and lead with surface-active organic matter in coastal seawater. Mar Chem 11:379–394
Williams LA (1984) Subtidal stromatolites in Monterey Formation and other organic-rich rocks as suggested source contributors to petroleum formation. Bull Am Assoc Petrol Geol 68:1879–1893
Williams LA, Reimers CE (1983) The role of bacterial mats in coastal upwelling regimes: a preliminary report. Geology 11:267–269
Zumberge JE, Sigleo AC, Nagy B (1978) Molecular and elemental analysis of the carbonaceous matter in the gold and uranium bearing Vaal Reef carbon seams, Witwatersrand Sequence. Min Sci Eng 10:222–246

Experimental Hypersaline Ponds as Model Environments for Stromatolite Formation 1. Microbenthos Composition and Biomass Accumulation

I. Dor[1], N. Carl[1], M. Schidlowski[2]

[1] Environmental Sciences Division, School of Applied Science and Technology, The Hebrew University of Jerusalem, Israel
[2] Max-Planck-Institut für Chemie (Otto-Hahn-Institut), W-6500 Mainz, FRG

Abstract

Colonization stages and early development of a mat-forming microbenthic community were studied during a 1-year period in a newly constructed hypersaline solar pond situated in the Dead Sea area. The pond had an inversely stratified water column consisting of a cool convective epilimnion of diluted Dead Sea water, a thermocline, and non-convective hot Dead Sea water at the bottom. The sharp halocline was maintained by adding fresh water to the upper water mass. Unicellular cyanobacteria and diatoms were the pioneer colonizers of the pond slopes. Metazoan grazers were absent, eliminated by the unique chemical composition of the Dead Sea water. Chlorophyll-a concentrations, reflecting photosynthetically active biomass, were highest within the shallow epilimnetic zone of the littoral (reaching 278 mg/m^2), and declined abruptly in the gradient zone. Both temperature and salinity appeared to be major limiting factors in this environment. Cyanobacteria were more resistant to high temperatures, attaining optimal growth up to approximately 48 °C. Within optimal temperature limits, microbenthos development was best in the lower salinities of 50–70 g/l. Salinities of 100–150 g/l seem to be an upper limit for this community. Chlorophyll-a decline at the onset of the gradient zone, where favorable temperatures and salinities prevailed, is attributed to pronounced salinity fluctuations occurring at this level. At the end of the first year, the young stromatolitic mat was several millimeters thick and well consolidated due to gelatinous polysaccharides excreted by cyanobacteria.

1 Introduction

It is well established today that prokaryotic and eukaryotic microbial ecosystems have held dominion over the Earth during the first three billion years of recorded Earth history (cf. Schopf 1983). This ancient microbial world has been primarily preserved in the form of an impressive stromatolite record, i.e., as sequences of fossil microbial carpets originating from the matting behavior primarily of benthic prokaryotes (see also Awramik, this Vol.).

Generally, biosedimentary structures of the "stromatolitic" type represent lithified stacks of superimposed laminated microbial communities, each individual layer standing for a microbial film that once thrived at the sediment-water interface. With lithification resulting from either trapping, binding or biologically mediated precipitation of selected mineral constituents by this primary organic film, such and related structures may also be aptly referred to as "microbialites" (Burne and Moore 1987). Representing the most conspicuous fossil expression of the Earth's microbial benthos, stromatolites have been shown to figure among the most conservative and persistent features of the paleontological record, covering a time span of some 3.5 Ga from their first appearance in the Archaean to their manifestations in the present world (Walter 1983). While their previously postulated biostratigraphic utility appears to be contentious today, they have, nevertheless, proved to be important environmental and facies indicators, and stromatolite-hosting sequences are currently being viewed as potential hydrocarbon source rocks specifically in the Proterozoic during which they had covered vast stretches of the littoral seafloor.

Constituting the oldest indisputable morphological (macroscopic) evidence of life on Earth, stromatolites have attracted increased attention since the revival of interest in Precambrian paleontology in the early 1960s. During these studies, the need was felt for an improved knowledge of present-day benthic microbial communities. With microbial ecosystems on the decline since the emergence of multicellular life at the dawn of the Phanerozoic, some 0.6–0.7 Ga ago (Glaessner 1983), and currently mainly confined to distinctive refugia such as hypersaline and hot spring environments where they are protected from metazoan feeders, the crucial role of microbial mat build-

M. Schidlowski et al. (Eds.)
Early Organic Evolution: Implications for Mineral and Energy Resources

ers in the early history of life had been, in fact, largely obscured by the neobiological perspective. Consequently, for a long time there was little incentive to devote attention to microbial communities thriving in esoteric environments of the present world. Due primarily to the impetus imparted to the field by the study of Precambrian stromatolites, the situation changed radically over the last decades.

Relevant pioneering studies have demonstrated that the ecology and physiology of extant microbial mats may indeed serve as models for a better understanding of their ancient progenitors. Work hitherto performed on recent benthic mats has principally addressed cyanobacterial ecosystems from coastal hypersaline (sabkha-type) habitats of the Sinai Peninsula (Solar Lake, Gavish Sabkha), the Persian Gulf (Abu Dhabi), lower California (Laguna Mormona), and Western Australia (Shark Bay), with representative results summarized, *inter alia*, in several recent publications (cf. Cohen et al. 1984; Friedman and Krumbein 1985; Cohen and Rosenberg 1989; Javor 1989). Though still in their infancy, these investigations have profoundly enhanced our understanding of the nature and geochemical role of stromatolite-forming microbial ecosystems in the present world, and thereby improved our perception of microbially mediated geochemical transformations in the exogenic cycle during the Earth's early history when prokaryotes had dominated the terrestrial biosphere.

As an appropriate complement to the above studies, we here submit the results of an investigation of a benthic microbial community from an artificial environment, namely, an experimental solar pond such as currently tested as a light-harvesting system for collecting solar energy (Tabor 1980; Kirkland et al. 1983). The pond was situated on the premises of Solmat Systems, Ltd. Israel, at the northern shore of the Dead Sea and had been specifically designed to suit a microbiological and biogeochemical investigation program. An intrinsic advantage of the experimental pond approach is that both the microbiology and the biogeochemical data can be related, within fairly narrow limits, to a monitored set of environmental background parameters, specifically temperature, salinity, and light intensity.

2 Objectives

The objectives of the present study were (1) to assess early colonization stages of the shallow pond littoral, leading to the establishment of the benthic microbial mat, (2) to quantify species composition and biomass concentration within the stromatolitic mat *in statu nascendi* in relation to salinity and temperature gradients, and (3) to study stable carbon isotope ($^{13}C/^{12}C$) gradients in the benthic mat along the pond profile as a function of salinity and temperature (see Schidlowski et al., this Vol.).

3 Materials and Methods

3.1 Solar Pond Construction and Operation

The hypersaline solar pond investigated was designed and constructed especially for this project, utilizing the experience of Solmat Systems Ltd., who tested solar ponds as a source of energy for industrial purposes. The pond had a diameter of 10 m, a slope of approximately 20°, and a maximum depth slightly exceeding 1 m. Its floor was lined with a plastic sheath and then covered with local desert soil. The pond had an inversely stratified water column, with a 30–40 cm thick cool and convective epilimnion of diluted Dead Sea water, a gradient layer beneath characterized by a pronounced halocline, and concentrated, hot Dead Sea brine near the bottom. The chemical composition of the epilimnetic layer is given in Table 1.

The pond was protected against excessive wave formation and mixing by plastic nets that were spread on its surface (Fig. 1a). The stratification described above was artificially maintained by the injection of fresh water to the epilimnion, thus compensating for evaporation and preserving the desired salinity. Several times during this study major treatment was needed to readjust the salinity within the gradient zone. The upper convective water layer had a total mean salinity of 54–60 g/l, while concentrations at the bottom were 250–430 g/l.

3.2 Monitoring of Environmental Factors

The pond was monitored on a weekly basis from June 1987 to June 1988. Water density and temperature were measured in the water column every 5–10 cm from the surface to a depth of 100 cm using a density meter model Paar DMA 35 and a YSI field thermo-

Table 1. Chemical composition (mg/l) of pond water at a depth of 10 cm

Na^+	K^+	Ca^{2+}	Mg^{2+}	Cl^-	SO_4^{2-}	HCO_3^-
2500	2100	4020	9533	40353	609	130

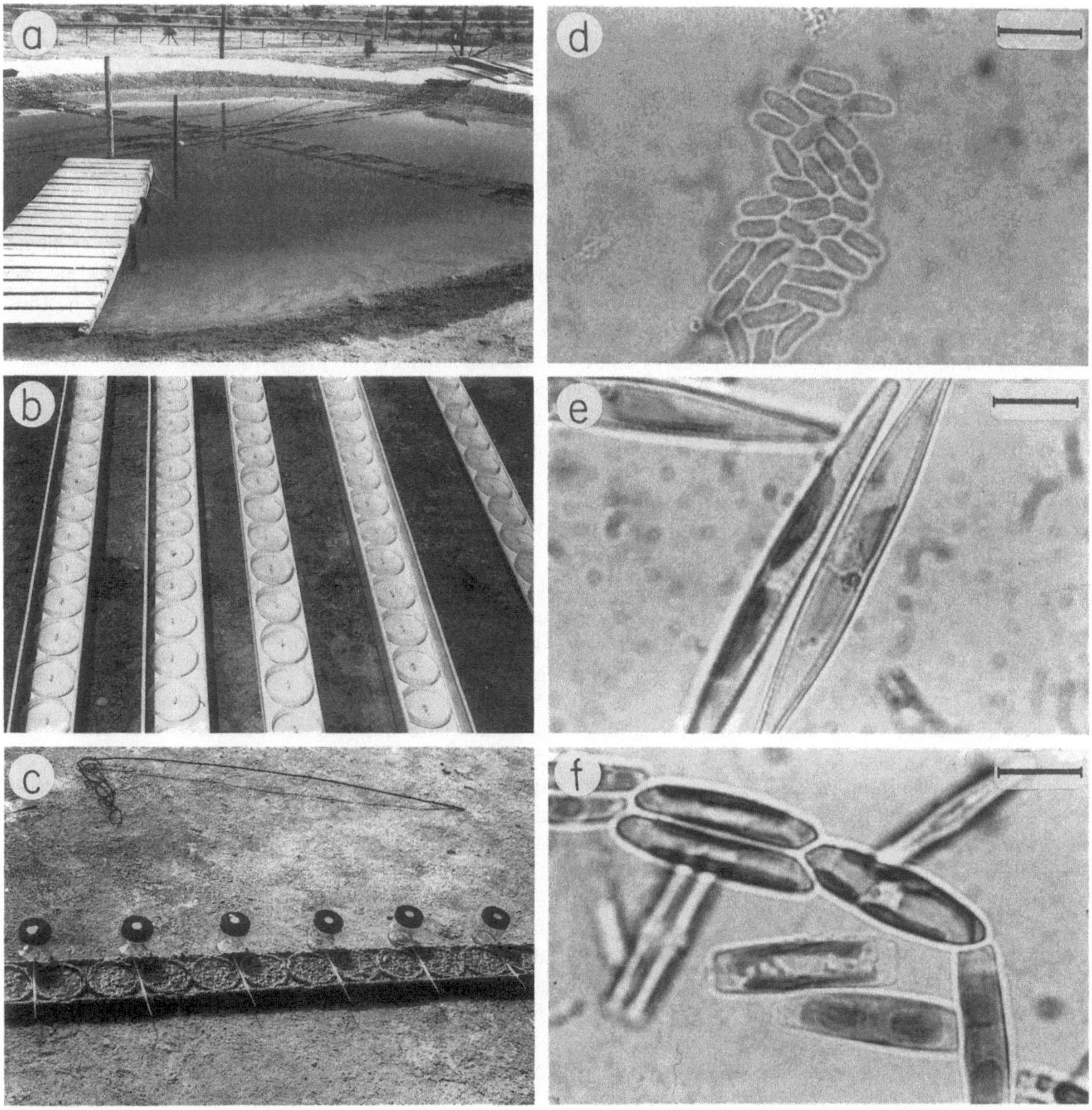

Fig. 1a–f. General view of **a** hypersaline pond investigated. **b** Plastic trays lined with petri dishes before being installed on the pond slope. **c** Trays after an exposure period, ready for mat harvest. Two adjacent petri dishes were transferred to a glass jar and served as a single sample. **d** *Aphanothece halophytica* Frémy, one of the strains. **e** *Nitzschia* sp. aff. *N. rostellata* Hustedt. **f** *Navicula duerrenbergiana* Hustedt

meter, respectively. Density values were converted to approximate salinity values (g/l) using an empirical formula for Dead Sea end brines (Gavrieli 1987).

Light penetration was measured only once on 1.12.1988, with a Lambda Li-Cor 185 Quantum Meter. The pond was sampled for phosphate at two different depths (10 and 100 cm) on 22.2.1988, with analyses performed according to Adin et al. (1981).

A 250-ml brine sample was transferred to the Israel Geological Survey for total chemical analysis. All samples were taken between 8:00–12:00 a.m.

3.3 Sampling Methods

Considering the pond's small diameter, the delicate nature of the mats, and the high temperatures that

Table 2. Sampler exposure periods

In	Out	Harvest season
28.6.87	9.11.87	Summer 1987
28.6.87	10.3.88	Winter 1987/88
28.6.87	13.6.88	Summer 1988

prevailed in the hypolimnion, it was necessary to develop a sampling method that would enable convenient sampling at every level with minimum disturbance of the ecosystem. This was achieved by using samplers especially designed for this study. Each sampler consisted of a 3-m-long plastic tray lined with 30 open petri dishes (Fig. 1b). Such trays were installed on the slopes of the solar pond, traversing salinity and temperature zones characteristic of this environment. Biomass soon started to develop on the dishes, reflecting the natural biotic zonation observed on the pond's slopes. When the assigned exposure period had ended (Table 2), the respective sampling unit was gently pulled out and the total biomass developed on each petri dish was transferred to a separate jar (Fig. 1c).

The samples taken were transported to the laboratory for species identification, cell counts, and chlorophyll-a determination. Each sample was first diluted with pond water from the corresponding depth and stirred vigorously with a magnetic stirrer, until a homogeneous suspension was obtained. Part of the material was transferred to the isotope laboratory for study of $^{13}C/^{12}C$ fractionations.

3.4 Species Identification

Cyanobacteria were identified with the help of Geitlerian taxonomy (Geitler 1932). Diatoms were identified with the kind help of A. Ehrlich from the Geological Survey, Israel.

3.5 Biomass Determination

3.5.1 Cell Counts

Subsamples of 2 ml were preserved in 5% formaldehyde and counted later on a hemocytometer. Since the samples were very turbid, it was necessary to dilute them before counting. In order to decrease turbidity and to minimize the dilution factor, several drops of 10% HCl were added to remove carbonate crystals. Dilution factors ranged between 10–20.

3.5.2 Chlorophyll-a Measurements

Chlorophyll-a was determined according to Talling (1969). One to 5 ml of the above suspension was centrifuged for 30 min and the water was removed. The remaining precipitate was resuspended with 3–5 ml of 90% methanol on a vortex and boiled for 60 s in a test tube immersed in hot water. The sample was then cooled in the dark and recentrifuged. Optical density of the solute was read at 665 nm on a Unicam SP 1800 spectrophotometer. Depending on the sample, one to three extractions with fresh methanol were repeated on the precipitate as needed, until the solvent was colorless. Chlorophyll-a concentration was calculated using a coefficient of 13.9, and normalized to mg chl-a/m^2.

3.6 Primary Productivity

Unpublished results from a former study (Paz 1983) concerning primary productivity in a series of industrial solar ponds operated in the same area are hereby reproduced and discussed. These measurements were carried out seasonally only on the shallow littoral where ambient temperatures prevailed. The ponds were 2 to 4 years old and were operated at different epilimnetic salinities. Samples from the well-consolidated mats were collected quantitatively, using a plastic frame of known dimensions which was laid on the littoral. The mat within the frame was cut into four equal portions and each intact portion was gently slid face up into a glass jar. Duplicate transparent and dark jars were used. These jars (large enough to accommodate an undisturbed sample) were carefully filled with pond water and placed horizontally on the sampled area. Oxygen was measured (YSI 58 Oxygen Meter) before and after an exposure period of about 3 h. From the measurements in the transparent and dark bottles, primary productivity was calculated for the surface unit of the mat.

4 Results

4.1 Environmental Factors

Salinity (cf. Table 1) and temperature were determined for three seasons (Figs. 2–3). A halocline and a corresponding inverse thermocline were maintained in the pond throughout the whole year. The mixed epilimnion was 30–40 cm thick during both winter and summer seasons. A thermal and salinity gradient

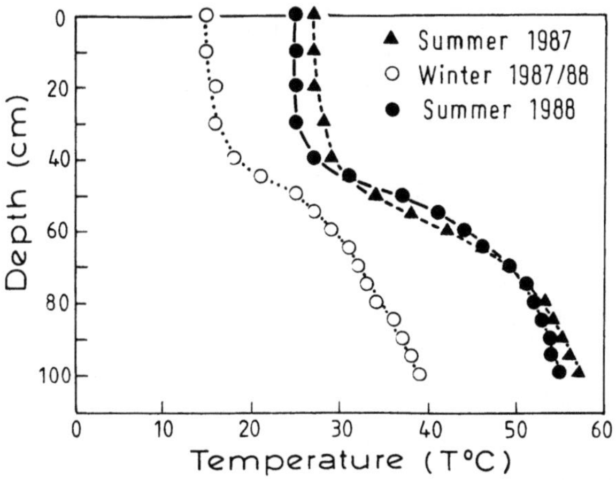

Fig. 2. Seasonal profiles of mean temperature. Maximum standard deviation is 2–5 °C in the epilimnion, and up to 7 °C in the gradient zone and below

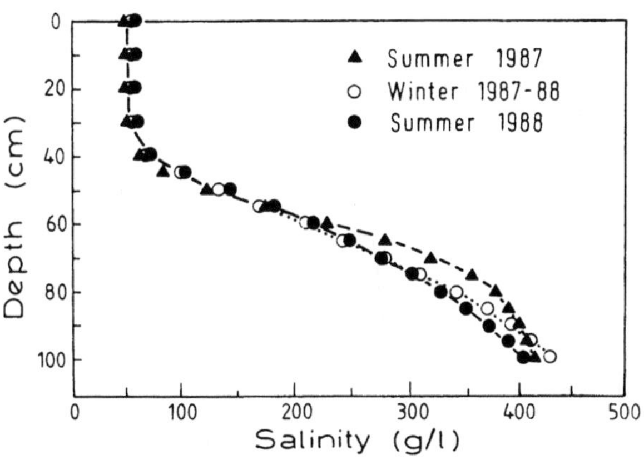

Fig. 3. Seasonal profiles of mean salinity. Maximum standard deviation in the epilimnion down to 30 cm: 11 g/l during the first exposure period and up to 4 g/l during the second and third exposures. Maximum standard deviation in the gradient zone and below up to 73 g/l in the first and up to 37 g/l during the second and third exposure periods

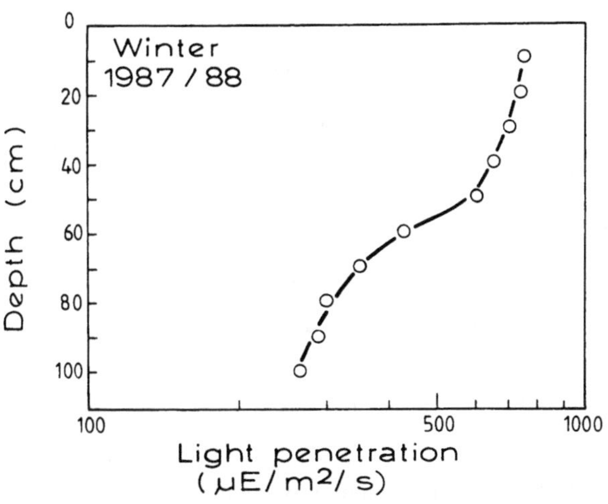

Fig. 4. Light penetration as a function of depth

Table 3. Orthophosphate concentrations

Depth (cm)	PO_4^{3-} (μg at. P/l)
10	9
100	4.8

zone, commencing from a depth of approximately 40–50 cm, extended to the pond's floor. A homogeneous hypolimnetic heat storage layer was not actually established (to achieve this, the pond must be deeper). The pond was thermally quite stable, as seen from the low standard deviation values of about 4 °C in the epilimnion and 7 °C in the gradient zone. However, considerable differences between maximum and minimum temperatures reaching 15–20 °C were exhibited occasionally in both the epilimnion and gradient zone. Mean epilimnetic summer and winter temperatures ranged between 25–30 and 15–20 °C, respectively. Mean bottom summer and winter temperatures ranged between 50–57 and 25–39 °C, respectively. The salinity of the mixed upper water layer was quite stable throughout the study. Mean epilimnetic salinities fluctuated narrowly with a standard deviation of 2–7 g/l, and occasional values of 11 and 15 g/l. However, there was a marked instability in salinity in the gradient zone during the first 14 weeks (summer 1987) compared to the rest of this study. Mean bottom salinities ranged between 250–430 g/l, increasing progressively with depth.

Light measurements in the water column are presented in Fig. 4. Surface light intensity (extrapolated from a subsurface light penetration profile) ranged between 580–780 $\mu E\,m^{-2}\,s^{-1}$. Light transmission at a depth of 60 cm was 55% of the surface intensity corresponding to 430 $\mu E\,m^{-2}\,s^{-1}$ (60 cm was the approximate limit of stromatolitic growth).

Phosphate levels were almost doubled in the epilimnion as compared to the hypolimnion (Table 3), indicating an external input.

The exposure time (Table 2) of sampler 1 was of the shortest duration, representing community build-up during summer 1987. Sampler 2 represents a progressively longer period of exposure, including summer 1987 and winter 1987–1988. Sampler 3, harvested during summer 1988, represents a three-season accumulation of biomass.

4.2 Stromatolitic Mat Development and Community Composition

Continuous observation and sampling of the pond littoral from the very beginning of its construction has demonstrated the sequence of colonization and establishment of the stromatolitic mat community. The pioneer organisms were unicellular cyanobacteria represented by a few strains of *Aphanothece halophytica* Frémy and by several diatoms. The predominant diatom species were *Navicula duerrenbergiana* Hust., *Nitzschia rostellata* Hustedt, and *Amphora coffeaeformis* (Ag.) Kutzing. A few weeks after the pond was filled with water, these unicellular forms appeared, sparsely covering the pond littoral. At this stage, the photosynthetic community developed under aerobic conditions, strong illumination, and an ample supply of nutrients originating from the sediment and from the overlying water. During the next months of summer 1987, the microalgae grew rapidly and at the end of this period formed a continuous green layer about 1 mm thick. However, before the first harvest, an unexpected storm introduced quantities of dust from the surrounding soil into the pond. The resulting turbidity reduced light transmission for several weeks, and the slowly settling particles formed a sediment film which entirely covered the microbenthic community. This incident resulted in a temporary, drastic decline in chlorophyll content, while cell counts remained unchanged. This might explain the discrepancy between the high cell count (10^{11} cells/m^2) and comparatively low chlorophyll concentrations (30 mg/m^2) during the first harvest (Fig. 8). Apart from this effect, biomass distribution, thermal stratification, and the salinity profile remained undamaged.

Soon afterwards, a new layer of microalgae started to develop, creating a new green lamina. After 1 year of growth, the stromatolitic mat was about 5–6 mm thick and well consolidated due to the gelatinous excretions of *Aphanothece*. At this stage, a layer characterized by a distinct H_2S odor appeared under the stromatolitic mat, and single filaments of an anoxygenic, filamentous cyanophyte, *Phormidium hypolimneticum* Campbell, were observed.

During summer 1987, *Aphanothece* attained concentrations of 10^{11} cells/m^2 in the upper littoral (Fig. 5). A steep decline was recorded in the gradient zone, and comparatively low concentrations of 10^9 cells/m^2 were present at the pond bottom. Diatoms reached concentrations of 10^{10} cells/m^2 near the surface, but were completely absent at depths greater than 50 cm. During winter 1987–88, microbial zonation was less prominent (Fig. 6) and became completely blurred during summer 1988 (Fig. 7). Diatoms contributed 6–25% of the total cell count during the first and second harvests. During the third harvest, diatoms contributed a larger fraction to the total cell count, reaching up to 87%, especially from a depth of 50 cm and below.

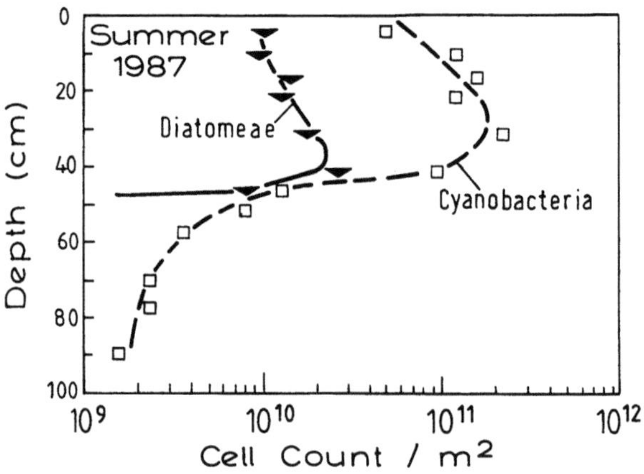

Fig. 5. Microbenthic community composition, first exposure period

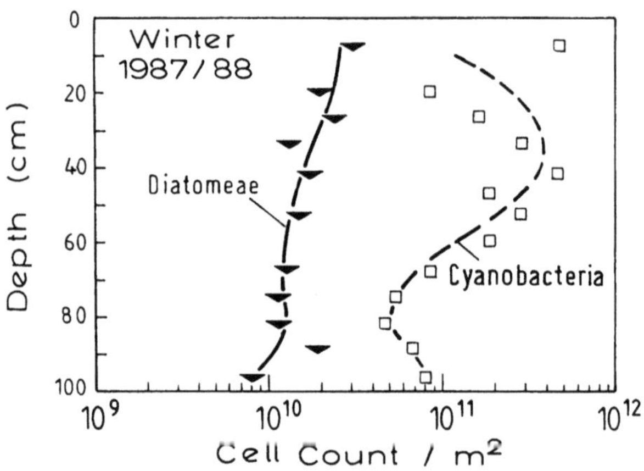

Fig. 6. Microbenthic community composition, second exposure period

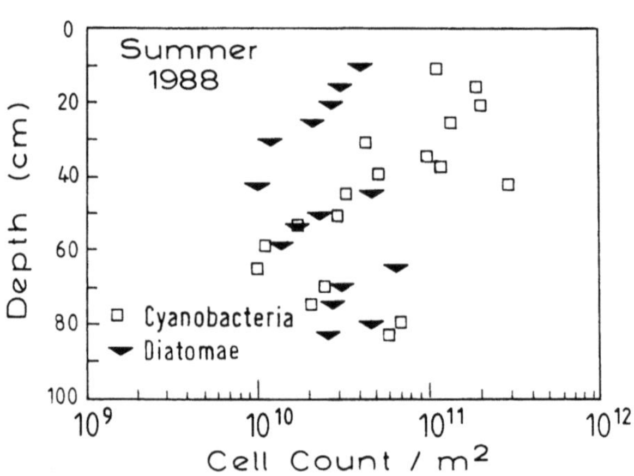

Fig. 7. Microbenthic community composition, third exposure period

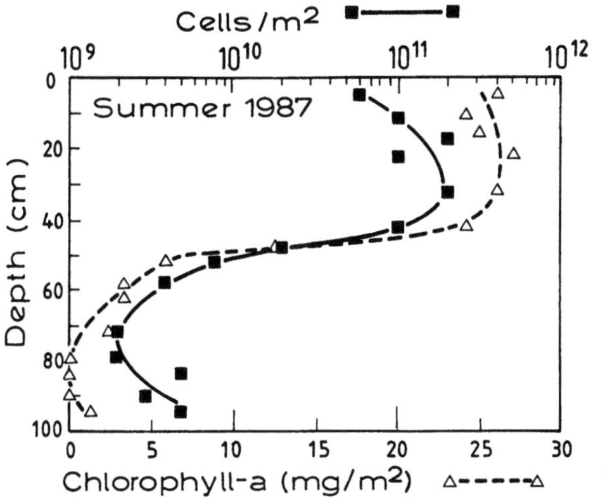

Fig. 8. Chlorophyll-a and total cell count profiles, first exposure period

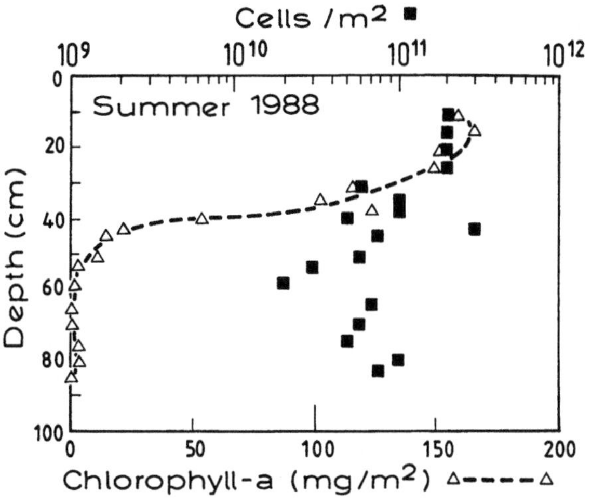

Fig. 10. Chlorophyll-a and total cell count profiles, third exposure period

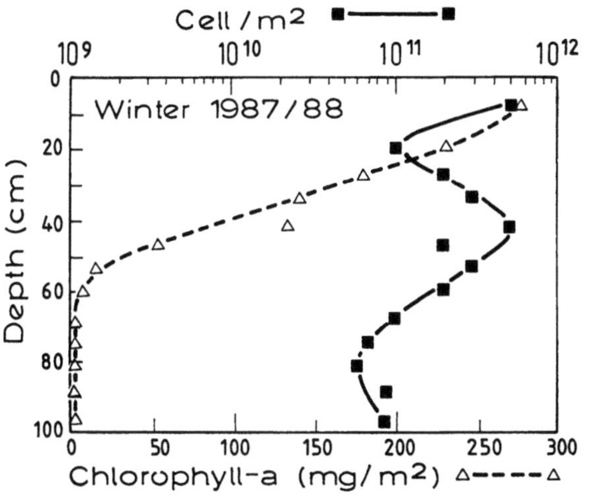

Fig. 9. Chlorophyll-a and total cell count profiles, second exposure period

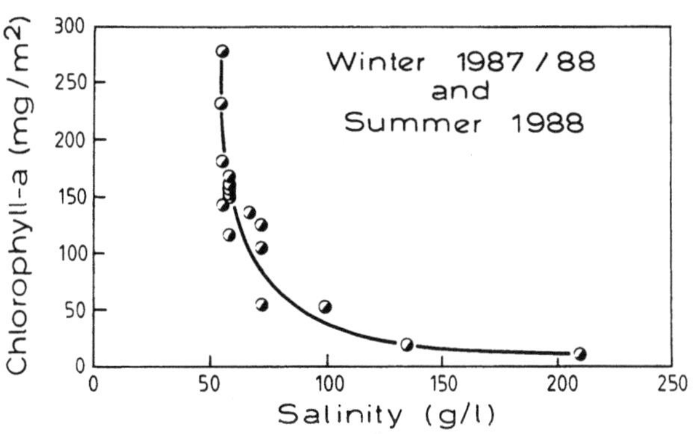

Fig. 11. Chlorophyll-a as a function of salinity. Summary for temperatures not exceeding 48 °C

4.3 Biomass Concentration

Chlorophyll-a and total cell count profiles are presented in Figs. 8–10. In all the seasons the chlorophyll content diminished with depth. Maximum concentrations were recorded in winter in the upper 40 cm of the pond, reaching 278 mg/m^2. A sharp drop appeared at a depth of 40–50 cm, and very low concentrations were recorded down to the bottom. Total cell counts showed the same trend, but only during summer 1987 (Fig. 8), when they dropped from a maximum of 2×10^{11} cells/m^2 in the shallow littoral to only 10^9 at a depth of 60 cm. Cell distribution became more irregular in winter 1987–88 (Fig. 9), while in summer 1988 it was entirely irregular (Fig. 10).

Figure 11 summarizes the effect of increasing salinity on chlorophyll-a concentration. The figure only includes data from areas where maximum temperatures did not exceed 48 °C, i.e., within the optimal growth limits of certain local strains of *Aphanothece halophytica* (Dor and Hornoff 1985b). Chlorophyll-a concentrations declined sharply within the salinity range of 50–150 g/l and were very low at salinities exceeding 150 g/l.

4.4 Primary Productivity

Primary productivity measurements are summarized in relation to salinity (Fig. 12). The highest maximum and mean rates were recorded at 50 g/l which was the lowest salinity tested. A steep decline appeared between 50–150 g/l and very low rates were observed at salinities exceeding 150 g/l.

5 Discussion

5.1 The Solar Pond Ecosystem

The solar pond technology has enabled us to create an artificial hypersaline and thermal ecosystem of controlled salinities and predictable temperatures. The salinity gradient, which coincided with the inverse thermal gradient, produced benthic zones where distinct combinations of salinity, temperature, and light intensities had been attained. Natural heliothermic lakes also show a striking positive correlation between salinity and temperature profiles (Kirkland et al. 1983). The stromatolitic mats typically developing in environments of this kind respond to the combined effects of all these parameters which, however, cannot be clearly separated under natural conditions. The unique composition of Dead Sea brine rich in Mg^{2+} and K^{+} (see Table 1) inhibits the development of metazoan grazers (Nissenbaum 1975; Por 1980). Therefore, biomass growth and accumulation was not affected by grazing activities and responded directly to environmental conditions.

5.2 Sampling

The samplers proved to be very efficient. Negligible disturbance was caused to the pond's littoral and water column during both installation and removal of these units. The biomass that grew on the individual petri dishes corresponded to the natural zones that prevailed on the pond's littoral. This was demonstrated by the color bands which marked zones of algal growth on the littoral slopes of the ponds and which were similar to the bands that appeared on the petri dishes. Upon removal, the delicate structure of the mats that had developed on the petri dishes remained intact. The samplers were easy to construct, simple to operate, and are reusable.

5.3 Salinity Regime

The salinity regime of the pond was successfully maintained by the addition of fresh water (to compensate for evaporation), and by seasonal supplements of brines to the bottom layer. The pond was monitored by density, and it was necessary to convert these measurements to obtain the corresponding salinities. Gavrieli (1987) has confirmed an adequate approximation between calculated and actual salinity values in the Dead Sea brines.

5.4 Thermal Regime

The pond was thermally stratified throughout the study. Summer temperatures in the water column were higher than winter values. This was due to increased solar radiation in summer combined with the fact that the pond was small and shallow, which made it susceptible to heat loss to the surroundings during the winter (cf. Tabor 1981).

5.5 Biomass Distribution

There is probably no single cause responsible for the decline in microbial biomass and activity observed along the salinity-temperature gradient. The highest values may be the result of a diatom bloom, appearing on the shallow littoral; the steep decline at a depth of 30–40 cm corresponds to the gradient zone, and in this case the main limiting factor may be attributed to instability. Finally, salinity and temperature increased above some critical level, obviously suppressing algal activity.

A comparison of chlorophyll-a and cell count profiles shows that both variables had identical trends during the first exposure period, while progressively diverging later. In the second and third harvests, chlorophyll declined consistently with depth, while increasing cell numbers appeared on the deep littoral, where temperature and salinity were above the limits of algal growth and chlorophyll was very low. It seems that our initial assumption, that in the absence of vertical currents in the solar pond there is no vertical biomass transport, could be erroneous.

Evidently, several mechanisms can contribute to biomass dislocation in this environment. First, the pond slope (20°) might have promoted sliding of sediment mixed with algae to the deeper zone; secondly, the plastic nets spread on the pond surface served as a substrate for algal colonization which, when detached, sank to the hot bottom brine. Moreover, part of the littoral mat evolved bubbles of photosynthetic oxygen as a potential transport vehicle, capable of lifting and laterally spreading biomass particles before these finally settled on the bottom. The high bottom salinities preserved the cells for prolonged time intervals. It must be conceded that cell counts may represent an accumulation of dead and living cell communities, thus being a questionable method for evaluating active photosynthetic biomass. Chlorophyll concentrations, on the other hand, provide a more sensitive method and remain in good agreement with primary productivity measurements.

5.6 Predicted Mat Development

It may be reasonably inferred that during further mat maturation the microbenthic community would become more diverse. In similar ponds sampled after 6 years of operation, more than 20 species of cyanobacteria and diatoms were recorded (Dor and Ehrlich 1987). In these mature ponds, *P. hypolimneticum*, an anoxygenic photosynthesizer (Cohen 1984) appeared in large amounts, living beneath the superficial aerobic community and blooming under the lower illumination of fall and winter (Dor and Paz 1989). Further accumulation of decaying organic matter is likely to promote the development of a prominent sulfuretum beneath the oxygenic layer. Such communities are well known from similar environments.

5.7 Limiting Factors

Chlorphyll-a and cell counts served in this study to express concentrations of microbial biomass. Most of the chlorophyll produced was limited to the epilimnetic zone, i.e., from the surface to a depth of 30–40 cm, and winter standing crops were richer than summer ones. The physicochemical factors that might affect chlorophyll-a concentrations are nutrients, light, temperature, salinity and environmental fluctuations.

Nutrients. Extensive research carried out on the Solmat industrial solar ponds from the Dead Sea area consistently demonstrated that the limiting nutrient in the pond waters is inorganic phosphate. Cultures of both diatoms (*Navicula*) and cyanobacteria (*Aphanothece*) were incubated in pond water enriched with either phosphate or nitrogen with vigorous stirring in order to ensure an adequate supply of CO_2 (Adin et al. 1981; Dor 1984b). Increased concentrations of phosphate stimulated growth, while increased concentrations of nitrogen proved to be ineffective, indicating that this nutrient was already present in concentrations adequate to biological demands. Furthermore, algal blooms appeared in the industrial ponds whenever phosphate concentrations reached at least 1–3 μg (at.) P/l. Algae failed to grow in laboratory cultures containing phosphate concentrations lower than 0.5 μg (at.) P/l. It must also be kept in mind that, unlike planktonic algae, which are solely dependent on nutrients supplied by the water column, in a young ecosystem benthic algae are also in direct contact with the bottom soil. Considering the above results, it is evident that phosphate and nitrogen concentrations in the present young pond, similar in every aspect to the above industrial ponds, were clearly in excess (Table 3). The limitation in carbon due to the precipitation of carbonate and the low solubility of CO_2 in concentrated brines might also be factors restricting algal growth. However, since the ponds were very shallow, we have reason to believe that there was a steady influx of CO_2 into the brines from the atmosphere according to need (Lazar et al. 1984). Nevertheless, the heavy demand placed on the pond's inorganic carbon pool by prolific mat development was apt to give rise to some degree of CO_2-limitation coped with by the ecosystem by an isotopically indiscriminate metabolization of the available carbon resources (see Schidlowski et al., this Vol.).

Light. Light intensity (measured on 1.12.88) at a depth of 60 cm (which constituted the lower level of chlorophyll production), was relatively high (55% of surface light intensity). Since these measurements were recorded on a cloudy winter day, they evidently represent a lower limit for light intensity. These values are in good agreement with those of Dor and Paz (1989), who measured light penetration at a depth of 60 cm throughout the year in Dead Sea industrial ponds. The corresponding summer values were always higher and ranged between 700–1200 $\mu E\, m^{-2} s^{-1}$. Compared to light intensities of 40 $\mu E\, m^{-2} s^{-1}$ used for laboratory growth of solar pond cyanobacteria (Dor and Hornoff 1985a, b), these field values are more than adequate and lack of light as a limiting factor can be ruled out. On the other hand, intensive radiation near the surface may have inhibited chlorophyll production (Krumbein et al. 1977; Gerdes et al. 1985). It is most probable that the summer harvests were lower than those of the winter due to the deleterious effect of excess light.

Temperature. As a rule, mean epilimnetic temperatures ranged between 15–18 and 25–28 °C for winter and summer, respectively. Temperatures increased with depth, reaching 66 °C during summer in the pond's hypolimnion. Some of the solar pond microbiota growing in the Dead Sea area are known to thrive in temperatures up to 48 °C (Dor and Hornoff 1985a); therefore, the above high temperatures attained in the gradient and hypolimnetic zones must be a limiting factor, restricting microalgal activity. Temperatures exceeding 48 °C were exclusively recorded at depths > 40 cm. Therefore, we can safely assume that in the littoral zone between 0–40 cm, temperature was within the tolerance limits of the microbial community. Lower winter temperatures may have had a favorable effect on chlorophyll synthesis in the shallow littoral zone. Diatoms usually prefer lower temperatures (Dor and Ehrlich 1987), and the richer winter harvests of chlorophyll can be attributed to diatom blooms in this season.

The major decline in chlorophyll concentrations in both winter and summer coincided with the limit of the gradient layer where summer temperatures exceeded 48 °C. However, winter temperatures, both mean and maximum in the gradient zone, were well below 48 °C and the chlorophyll decline was still observed at the same level. Therefore, temperature does not seem to be the sole limiting factor responsible for the consistent chlorophyll decline observed on the pond littoral.

Salinity. For attempts to attribute the chlorophyll decline to increasing salinity, it is necessary to eliminate the influence of temperature. The following discussion is limited to chlorophyll concentrations recorded from areas that were exposed to temperatures not higher than 48 °C. Increasing total salinity may affect the littoral algal population by either introducing instability in the form of salinity fluctuations or by exerting excessive osmotic pressure.

As far as mean seasonal salinity is concerned, the epilimnetic zone was quite stable throughout this study, and we can rule out seasonal salinity shifts as a factor affecting the seasonal differences observed. Closer scrutiny of each harvest reveals that the major drop in chlorophyll concentrations always occurred at a depth of 30–40 cm. This drop, from 123 to 20 mg/m^2 (summer 1988, Fig. 11) may, at first sight, seem to have been associated with increasing salinity from a mean value of 59 g/l at 40 cm to a mean value of 99 g/l at 45 cm. However, the above steep decline in chlorophyll concentrations vs. depth seems to coincide with the onset of the gradient zone. The gradient zone is an area of instability where large fluctuations in salinity took place as opposed to slight fluctuations observed in the epilimnetic zone. To remain in osmotic equilibrium with its hypertonic surrounding, an organism is forced to make specific adaptations, both physiological and biochemical. Cyanobacteria synthesize organic osmotic solutes, and there is a direct correlation between the type of solute and the salinity in which the specific species resides. The least halotolerant synthesize disaccharides, the intermediate halotolerant glucosylglycerol, and the most halotolerant betaines (Golubic 1980; Reed et al. 1985; Trüper and Galinski 1989). From an energetic point of view, the accumulation of a compatible organic solute is at least ten times more expensive than the utilization of inorganic ions for the same purpose (Reed et al. 1985). Additional stress in the form of an oscillating hypertonic environment may prove to be overtaxing. Miller and Kamykowski (1986) showed in *Nitzschia americana* that a rapid increase in salinity (from 20 to 23.5‰) caused a significant reduction in rates of net carbon fixation, and required a period of at least 24 h and a stable environment to restore previous fixation rates. Dor (1984b) showed that the dilution of a hypersaline culture of *Navicula* with 20–80% distilled water caused a drastic decrease in gross primary productivity which after 8 days was restored only in the least diluted culture (20%), while the other cultures were irreversibly damaged.

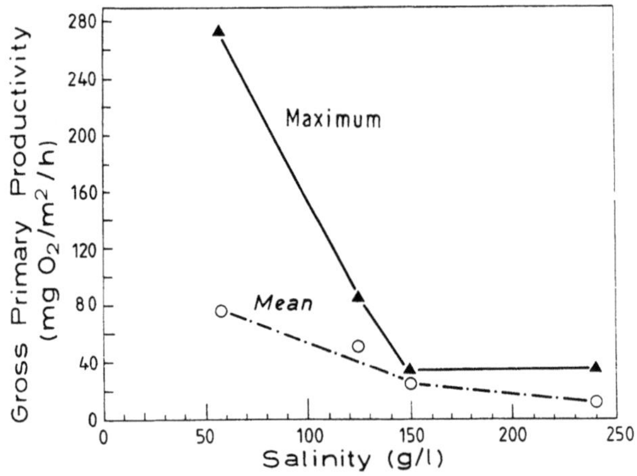

Fig. 12. Primary productivity as a function of salinity. Summary of data from industrial solar ponds operated at different epilimnetic salinities

However, primary productivity measurements performed in the stable epilimnia of a number of solar ponds operated at different salinities (unpublished data of former studies) exhibit a similar decline (Fig. 12). This suggests that the increased salinity itself may be the limiting factor. Indeed, the two independent methods demonstrate a steep drop in algal activity within approximately the same salinity range, thus providing evidence that an increase in salt concentration from 50–70 to 100–150 g/l might be critical for the benthic microalgal community.

The question remains open whether this effect is caused by the specific composition of the Dead Sea water, or whether it is inherently connected with the increase in total salinity and thus of general validity. Future work on chlorophyll contents and primary productivities in marginally marine hypersaline environments will answer this question.

Acknowledgments. We are grateful to B. Buchbinder for the chemical analysis of the pond water, A. Ehrlich for diatom identification, N. Paz for unpublished data on primary productivity, and L. Rubinstein for orthophosphate determination. Excellent technical assistance was rendered by A. Zarczyński, and Solmat

Systems Ltd. are thanked for pond construction and maintenance. We are, furthermore grateful to J. Erez, S. Golubic, and A. Oren for their most valuable critical remarks. Moreover, financial support by the Deutsche Forschungsgemeinschaft is gratefully acknowledged.

References

Adin A, Dor I (1982) Maintaining transparency in electricity generating solar ponds. Scientific report No. 3 to Solmat Systems. Hebrew Univ, Jerusalem, 278 pp (in Hebrew)

Adin A, Dor I, Sela S (1981) Removal of turbidity from electricity generating solar ponds. Scientific report No. 2 to Solmat Systems. Hebrew Univ, Jerusalem, 192 pp (in Hebrew)

Burne RV, Moore LS (1987) Microbialites: organosedimentary deposits of benthic microbial communities. Palaios 2:241–254

Cohen Y (1984) The Solar Lake cyanobacterial mats: strategies of photosynthetic life under sulfide. In: Cohen Y, Castenholz RW, Halvorson HO (eds) Microbial mats: stromatolites (MBL lectures in biology 3). Liss, New York, pp 133–148

Cohen Y, Rosenberg E (eds) (1989) Microbial mats: physiological ecology of benthic microbial communities. Am Microbiol, Washington, 494 pp

Cohen Y, Castenholz RW, Halvorson HO (eds) (1984) Microbial mats: stromatolites (MBL lectures in biology 3). Liss, New York, XVIII + 498 pp

Dor I (1983) Research on algae of solar ponds. Report No. 1 to Solmat Systems. Hebrew Univ, Jerusalem, 46 pp

Dor I (1984a) Epiphytic blue-green algae (cyanobacteria) of the Sinai Mangal: considerations on vertical zonation and morphological adaption. In: Por FD, Dor I (eds) Hydrobiology of the Mangal. Junk, The Hague, pp 35–54

Dor I (1984b) Algal research in solar ponds. Scientific report No. 2 to Solmat Systems. Hebrew Univ, Jerusalem, 46 pp (in Hebrew)

Dor I, Ehrlich A (1987) The effect of salinity and temperature gradients on the distribution of littoral microalgae in experimental solar ponds, Dead Sea area, Israel. P.S.Z.N.I. Mar Ecol 8(3):193–205

Dor I, Hornoff M (1985a) Salinity-temperature relations and morphotypes of a mixed population of coccoid cyanobacteria from a hot, hypersaline pond in Israel. P.S.Z.N.I. Mar Ecol 6(1):13–25

Dor I, Hornoff M (1985b) Studies on *Aphanothece halophytica* Fremy from a solar pond: comparison of two isolates on the basis of cell polymorphism and growth response to salinity, temperature and light conditions. Bot Mar 28:389–398

Dor I, Paz N (1989) Temporal and spatial distribution of mat microalgae in the experimental solar ponds, Dead Sea area, Israel. In: Cohen Y, Rosenberg E (eds) Microbial mats: physiological ecology of benthic microbial communities. Am Soc Microbiol, Washington, pp 114–122

Friedman GM, Krumbein WE (eds) (1985) Hypersaline ecosystems: the Gavish Sabkha (Ecological Studies 53). Springer, Berlin Heidelberg New York, x + 484 pp

Gavrieli I (1987) The source of the halite bodies in the southern Dead Sea. Report No. GSI/11/87. M Sc Thesis, Hebrew Univ, Jerusalem, 123 pp (in Hebrew)

Geitler L (1932) Cyanophyceae. Kryptogamenflora von Deutschland, Österreich und der Schweiz, 14-1. Akademie, Leipzig, 1196 pp

Gerdes G, Krumbein WE, Holtkamp E (1985) Salinity and water activity related zonation of microbial communities and potential stromatolites of the Gavish Sabkha. In: Friedman GM, Krumbein WE (eds) Hypersaline ecosystems, the Gavish Sabkha. Springer, Berlin Heidelberg New York, pp 238–266

Glaessner MF (1983) The emergence of Metazoa in the early history of life. In: Nagy B, Weber R, Guerrero JC, Schidlowski M (eds) Developments and interactions of the Precambrian atmosphere, lithosphere and biosphere. Elsevier, Amsterdam, pp 319–333

Golubic S (1980) Halophily and halotolerance in cyanophytes. Orig Life 10:169–183

Javor B (1989) Hypersaline environments. Springer, Berlin Heidelberg New York, 328 pp

Kirkland DW, Bradbury JP, Dean WE (1983) The heliothermic lake – a direct method of collecting and storing solar energy. Arch Hydrobiol Suppl 65(1):1–60

Krumbein WE, Cohen Y, Shilo M (1977) Solar Lake (Sinai.) 4. Stromatolitic cyanobacterial mats. Limnol Oceanogr 22:635–656

Lazar B, Javor B, Erez J (1984) Total alkalinity in marine-derived brines and pore waters associated with microbial mats. In: Cohen Y, Castenholz RW, Halverson HO (eds) Microbial mats: stromatolites. Liss, New York, pp 84–93

Miller RL, Kamykowski DL (1986) Short-term photosynthetic responses in the diatom *Nitzschia americana* to a simulated salinity environment. J Plankton Res 8(2):305–315

Nissenbaum A (1975) The microbiology and biogeochemistry of the Dead Sea. Microbial Ecol 2:139–161

Paz N (1983) Algal activity in solar ponds. M Sc Thesis, Hebrew Univ, Jerusalem, 149 pp (in Hebrew)

Por FD (1980) A classification of hypersaline waters, based on trophic criteria. P.S.Z.N.I. Mar Ecol 1(2):121–131

Reed RH, Ward SRC, Richardson DI, Moore DJ, Stewart WDP (1985) Multiphasic osmotic adjustment in a euryhaline cyanobacterium. FEMS Microbiol Lett 28:225–229

Schopf JW (ed) (1983) Earth's earliest biosphere: its origin and evolution. Univ Press, Princeton, xxv + 543 pp

Tabor H (1980) Using solar ponds to store power from the sun. Impact Sci Soc 30(4):319–328

Tabor H (1981) Solar ponds. Solar Energy 27(3):181–194

Talling JF (1969) General outline of spectrophotometric methods. In: Vollenweider (ed) A manual on methods for measuring primary productivity in aquatic environments (IBP Handbook 12). Blackwell, Oxford, pp 22–25

Trüper HG, Galinski EA (1989) Compatible solutes in halophilic phototrophic procaryotes. In: Cohen Y, Rosenberg E (eds) Microbial mats: physiological ecology of benthic microbial communities. Am Soc Microbiol, Washington, pp 342–348

Walter MR (1983) Archean stromatolites: evidence of the earth's earliest benthos. In: Schopf (ed) Earth's earliest biosphere: its origin and evolution. Univ Press, Princeton, pp 187–213

Experimental Hypersaline Ponds as Model Environments for Stromatolite Formation 2. Isotopic Biogeochemistry

Manfred Schidlowski[1], Hendrik Gorzawski[1] and Inka Dor[2]

[1] Max-Planck-Institut für Chemie (Otto-Hahn-Institut), W-6500 Mainz, FRG
[2] Environmental Sciences Division, School of Applied Science and Technology, The Hebrew University, Jerusalem 91904, Israel

Abstract

The biogeochemical characteristics of a newly established microbial mat have been traced along the slope of an artificially stratified, eutrophicated hypersaline pond as a function of the principal environmental parameters. The sharp decline in both visual mat development and productivity index functions at the combined thermo- and haloclines shows productivity in the habitat to be primarily temperature- and salinity-dependent. Recordings of $\delta^{13}C_{org}$ over the pond profile have identified high rates of primary productivity as the agent responsible for the generation of isotopically heavy biomass, negating the previously assumed role of salinity as a principal steering variable. Heavy demands placed on the local carbon dioxide pool by high productivity rates sustained by the mat-forming microbenthos obviously give rise to severe CO_2-limitation and, consequently, to a largely indiscriminate metabolization of available CO_2 supplies. The sizeable variations of $\delta^{13}C_{org}$ along the pond slope appear to be smoothed out in the corresponding $\delta^{13}C_{carb}$ functions, with only close scrutiny revealing a sympathetic relationship between the two parameters.

1 Introduction

The last decades have seen a dramatic revival in the study of stromatolites, the fossil relics of laminated microbial mat communities largely dominated by photosynthetic prokaryotes (mostly cyanobacteria). With our view grossly distorted by the neobiological perspective that relegates the modern analogs of such microbial communities to mere niches of the present world, the physiology and ecology of such mat-forming microbenthos and its fossil manifestations had long figured as unimportant side stages of biological and paleontological research. It was only after the profound realization that microbial communities (of both the benthic and planktonic type) range among the most productive ecosystems, and had virtually monopolized the terrestrial biosphere during the first three billion years of recorded Earth history, that their crucial role in both the early evolution of life and the terrestrial environment came to be fully appreciated. There is, for instance, little doubt that microorganisms in the widest sense were the sole producers of biomass in Precambrian times feeding early biogeochemical cycles, and that cyanobacterial photosynthesis is to be credited with the first oxygenation of terrestrial near-surface environments and specifically the atmosphere (which, incidentally, was the most important single event in the evolution of the Earth's gaseous shell). Apart from their utility as environmental and facies indicators notably in Precambrian terranes, stromatolites also hold considerable potential for the solution of a wide spectrum of problems that range from sedimentology to economic geology. A general overview of the field that fully maps out the historical dimension of the topic is given by Awramik (this Vol.).

As part of the first detailed biogeochemical investigations of both fossil and modern stromatolitic ecosystems, isotope studies of their organic carbon content had come progressively into the focus. During these studies, it was noted that the $\delta^{13}C$ values of the kerogen constituents of some of the geologically oldest stromatolites (Hoering 1967; Schopf et al. 1971; Eichmann and Schidlowski 1975) fell right into the isotopic mainstream of sedimentary organic carbon (about $-26 \pm 7‰$ vs. PDB; cf. Hayes et al. 1983; Schidlowski et al. 1983), with perhaps a mild slant toward light values (Fig. 1B). For instance, Eichmann and Schidlowski (1975) had reported an average of $-28.5 \pm 3.6‰$ [PDB] for a total of 22 stromatolite samples covering the whole of the Precambrian.

In contrast, the biomass of extant cyanobacterial mats was found, in the majority of cases, to be distinctly enriched in isotopically heavy carbon (Fig. 1B). Among the first measurements reported

M. Schidlowski et al. (Eds.)
Early Organic Evolution: Implications for Mineral and Energy Resources

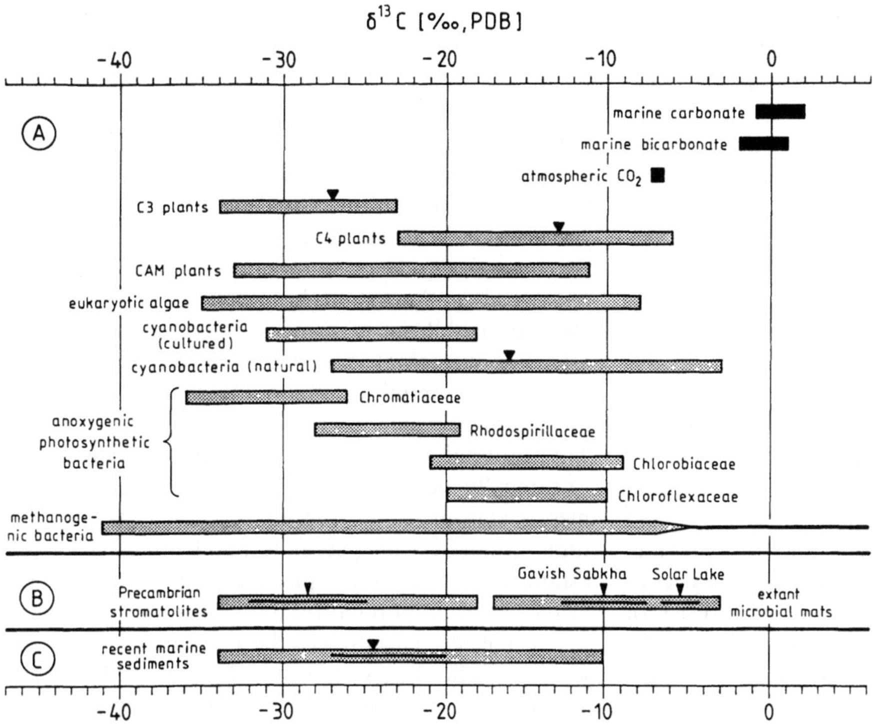

Fig. 1A–C. **A** Isotope spreads of major groups of primary producers (higher plants and autotrophic microorganisms) as compared to the environmental feeder pool of inorganic carbon (mainly CO_2 and HCO_3^-). Note that the isotopic compositions of naturally occurring cyanobacteria and selected groups of anoxygenic photosynthetic bacteria are slanted toward "heavy" carbon, with a $\delta^{13}C_{org}$ mean between − 10 and − 20‰ (similar to that of C4 plants). *Triangles* indicate approximate means. **B** Principal isotope spreads of Precambrian stromatolites (*black line* gives mean and standard deviations of 22 samples reported by Eichmann and Schidlowski 1975) and of their extant counterparts. Inserted in the range of modern microbial mats are the means and standard deviations of the microbial biomass from two classical hypersaline habitats (Gavish Sabka and Solar Lake). **C** Isotope spread for organic carbon constituents of recent marine sediments (Deines 1980). The *black line* covers > 90% of some 1600 data points summarized in the band

were those by Behrens and Frishman (1971) who had obtained $\delta^{13}C$ values between − 14 and − 18‰ [PDB]; other surveys of both naturally occurring and cultured cyanobacteria (e.g., Calder and Parker 1973; Seckbach and Kaplan 1973; Pardue et al. 1976; Barghoorn et al. 1977; Peters et al. 1981; Estep 1982) had likewise recorded a consistent (though variable) slant toward positive readings. As is obvious from a synopsis of the presently known isotope spreads of potentially mat-building cyanobacteria and nonoxygenic photosynthetic bacteria (Fig. 1A), the average isotopic composition of naturally occurring microbial mat communities can be indeed expected to be markedly shifted in a positive direction ($\delta^{13}C$ between − 10 and − 20‰) as compared to respective compositions of C3 plants and eukaryotic algae which make up the bulk of the Earth's standing biomass. The $\delta^{13}C$ mean of the average global biomass (between − 24 and − 28‰) is, in turn, closely approximated by the isotopic composition of organic matter entering present-day marine sediments (Fig. 1C).

Problems related to the isotopic discrepancy between ancient and modern microbial ecosystems were decidedly exacerbated by the recent discovery in cyanobacterially dominated microbial ecosystems of the iosotopically heaviest organic carbon ever encountered in the terrestrial biosphere. The isotopically superheavy microbial biomass occurred in coastal brine pools situated on the eastern shoreline of the Sinai Peninsula, yielding $\delta^{13}C_{org}$-means of − 10.0 ± 2.6‰ ($n = 25$) in the case of the Gavish Sabkha and − 5.4 ± 1.1‰ ($n = 14$) for the Solar Lake south of Eilat, with − 3.5‰ as the single most positive value encountered (Schidlowski et al. 1984, 1985). A tentative explanation of these findings in terms of a diffusion-limited assimilatory pathway imposed by substantially reduced CO_2 solubilities in hypersaline waters in combination with high rates of productivity appeared to be most plausible, but observations from other hypersaline environments (Des Marais et al. 1987) did not clearly support a quantitative relationship between $\delta^{13}C_{org}$ and salinity as the principal predictor variable. Hence, a definitive elucidation of the mechanism responsible for the generation of superheavy microbial biomass seemed to be contingent on the design of an experimental setup allowing one to trace $\delta^{13}C_{org}$ values of the benthic mat as a function of the principal relevant microbiological and environmental background parameters. In this respect, solar ponds of the type operated by Solmat Systems, Ltd., in the Dead Sea area (Israel) promised to be ideal test grounds since they had been shown to harbor benthic communities of microalgae (primarily cyanobacteria) exposed to pronounced gradients of salinity, temperature, and light penetration (Dor and Ehrlich 1987; Dor and Paz 1989). In the following, we present the results of an isotope survey conducted on the benthic mat from one of a series of four experimental ponds that had been specifically designed to trace the isotopic composition of the microbial biomass as a function of varying combinations of the above background parameters. Preliminary results have already been summarized elsewhere (Schidlowski et al. 1988, 1989).

2 Isotopically Heavy Biomass: A Statement of the Problem

It is well established that all pathways of autotrophic carbon fixation discriminate markedly (albeit to varying degrees) against the heavy carbon isotope (^{13}C), the resulting bias in favor of ^{12}C principally deriving from a kinetic isotope effect inherent in the first irreversible enzymatic carboxylation reaction that is responsible for the incorporation of CO_2 into the COOH group of an organic acid (cf. Vogel 1980; O'Leary 1981; Schidlowski et al. 1983). This consequently gives rise to a marked enrichment of ^{12}C in all forms of biogenic matter as compared to the environmental pool of inorganic carbon principally made up of marine bicarbonate, carbonate and atmospheric carbon dioxide (Fig. 1A). As a result, the $\delta^{13}C_{org}$ values of the principal categories of biomass (and notably their averages) are basically locked up within the bounds of the -20 to -30 per mill lines of the δ-scale, with a global mean close to $-25‰$ for the biosphere as a whole. As is obvious from Fig. 1C, this mean has been virtually transcribed into the organic matter from recent marine sediments and subsequently, with a minor diagenetic overprint, into the sedimentary record (see Schidlowski and Aharon, this Vol.).

In contrast to the average global biomass, the ^{13}C-enriched organic matter typically observed in hypersaline microbial habitats such as sabkha environments appears to be remarkably out of pace with the bulk of the Earth's organic carbon pool, with specifically the Solar Lake (Sinai) having furnished the isotopically heaviest organic matter as yet encountered in the extant biosphere (Fig. 1B). These anomalous ^{13}C-enrichments are commonly associated with microbial ecosystems dominated by cyanobacterial communities that form prolific carpets at the sediment-water interface of their brine habitats. The cardinal enigma posed by the enrichment of isotopically heavy carbon in these microbial colonies is epitomized by the question of how this isotopic anomaly could actually build up, as it is well established that the quantitatively predominant cyanobacterial population of such mat communities sequesters carbon via the C3 pathway and the Calvin Cycle known to be characterized by the sizeable isotope fractionations of the ribulose-1,5-bisphosphate (RuBP) carboxylase reaction (cf. Vogel 1980; O'Leary 1981).

We had previously proposed (Schidlowski et al. 1984, 1985; Schidlowski 1985) that the root cause for the blatant reduction of the carbon isotope fractionation in the assimilatory pathway of these microbial photosynthesizers was a dramatic decrease in the solubility of CO_2 in response to increasing salinity and temperature within the aqueous environment of the prokaryotic communities in question. With the currently available data at hand (Fig. 2), a convincing case could be made that the substantially reduced solubility of CO_2 was likely to make this gas a rare commodity in heliothermal brines, bestowing on carbon dioxide the exceptional role of a critical nutrient that is normally reserved for biolimiting elements such as phosphorus and nitrogen. As a consequence, the assimilatory pathway of microbial (prokaryotic) photoautotrophs in such hypersaline habitats should be typically diffusion- (or supply-)limited, with the large fractionations of the RuBP carboxylase reaction apt to be suppressed and therefore remaining cryptic (as they are in the compartmented biochemistry of the C4 pathway). Incidentally, the approximate coincidence in the isotopic compositions of microbial photoautotrophs from extant aquatic and brine environments with C4 plants would, in fact, argue persuasively for a suppression of the isotope-discriminating properties of the carboxylating enzyme by an identical mechanism, notably one involving a transloca-

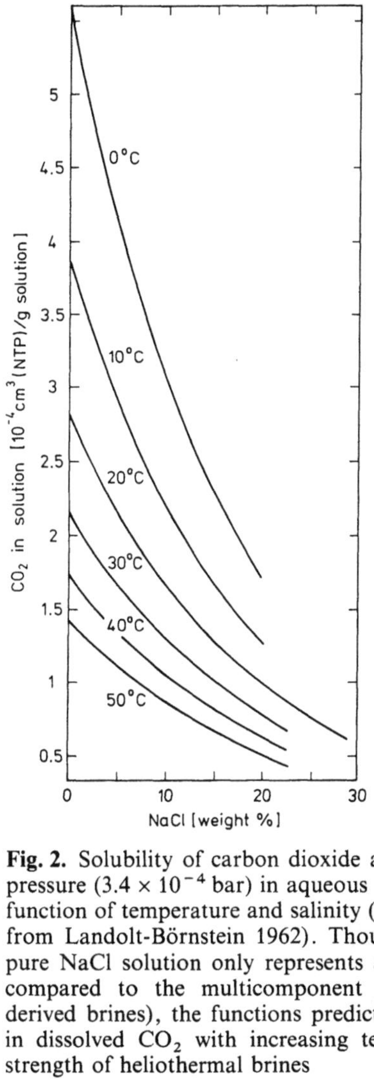

Fig. 2. Solubility of carbon dioxide at atmospheric partial pressure (3.4×10^{-4} bar) in aqueous solution of NaCl as a function of temperature and salinity (functions recalculated from Landolt-Börnstein 1962). Though CO_2 solubility in pure NaCl solution only represents a model situation (as compared to the multicomponent system of seawater-derived brines), the functions predict a dramatic decrease in dissolved CO_2 with increasing temperature and ionic strength of heliothermal brines

tion of CO_2 to the photosynthetically active centers with very little fractionation, and its subsequent processing in the Calvin Cycle under quasi-closed conditions.

Hence, under conditions of an extreme shortage of CO_2 as apparently experienced by hypersaline microbial ecosystems, excessive isotope fractionation is likely to become an intolerable luxury for the microbial photoautotrophs thriving in such habitats, the result being a largely indiscriminate metabolization of all carbon resources available. From such a vantage point, the generation of isotopically heavy biomass could be viewed as a habitat-specific adaptation to extreme degrees of CO_2 depletion in hypersaline environments, permitting the maintenance of sizeable rates of primary productivity only by a large-scale utilization of isotopically heavy carbon.

While very plausible at first sight, there still remained a number of residual questions, notably in view of observations that the postulated predictor function of ionic strength was not an unequivocal one (cf. Des Marais et al. 1987). Therefore, we felt that field experiments utilizing adaptations of commercially operated solar ponds might prove most promising in a further approach to these problems. In particular, such an experimental approach should allow a meticulous quantification of the relationships between the isotopic composition of the microbial biomass and existing gradients of environmental background parameters (notably temperature and salinity). The following report gives a first synopsis of such work carried out in one of these experimental hypersaline ponds.

3 The Experimental Pond as a Microbial Habitat: Physicochemical Environment and Community Composition

The artificial hypersaline pond utilized in this investigation is situated in the Bet Arava area at the northern tip of the Dead Sea (Israel) on the premises of Solmat Systems, Ltd., one of the largest commercial operators of solar ponds. With the results of a detailed investigation of the habitat and its microbial community reported elsewhere (Dor et al., this Vol.), the following account just briefly summarizes the basic background information needed for an evaluation of the biogeochemical and isotopic evidence represented in Figs. 5 and 6.

The pond is circular with a diameter of about 10 m and gently inclined marginal slopes grading into a central bottom area with a depth slightly in excess of 1 m. As may be inferred from the depth functions of the principal hydrological parameters shown in Fig. 3, it represents a meromictic (stratified) hypersaline water body comprising:

1. An upper convective layer (epilimnion) of diluted Dead Sea water of low salinity ($\sim$ 55 g/l) and ambient temperatures (between 15 and 30 °C according to season);
2. a transient nonconvective zone characterized by a pronounced halocline and an *inverted* thermocline (i.e., temperatures are *increasing* with depth); and
3. a nonconvective and largely anoxic bottom stratum (hypolimnion) of highly concentrated or oversaturated brine ($\gtrsim$ 400 g/l) and high temperature (falling just short of 60 °C at the pond bottom).

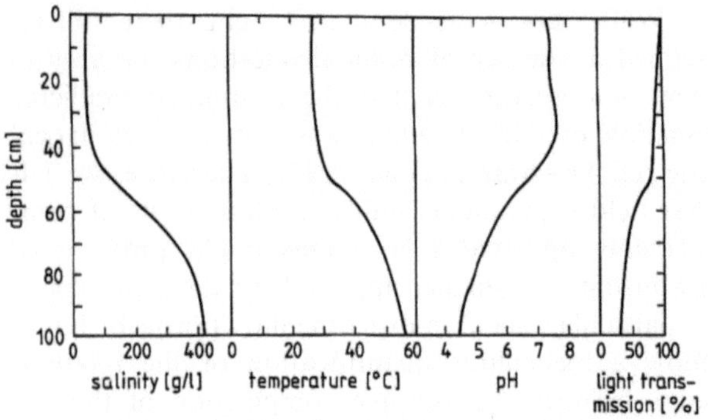

Fig. 3. Principal hydrological background parameters and light transmission (percent of surface radiation) as a function of depth in experimental Pond No. 1 used in this investigation (based on numerical recordings from Carl 1989). The functions show averages for the time period June through October 1987, thus reflecting specifically the environment of the microbial mat community harvested October 30, 1987 (cf. Fig. 5). Note the emergence of a pronounced halocline and thermocline at a depth of about

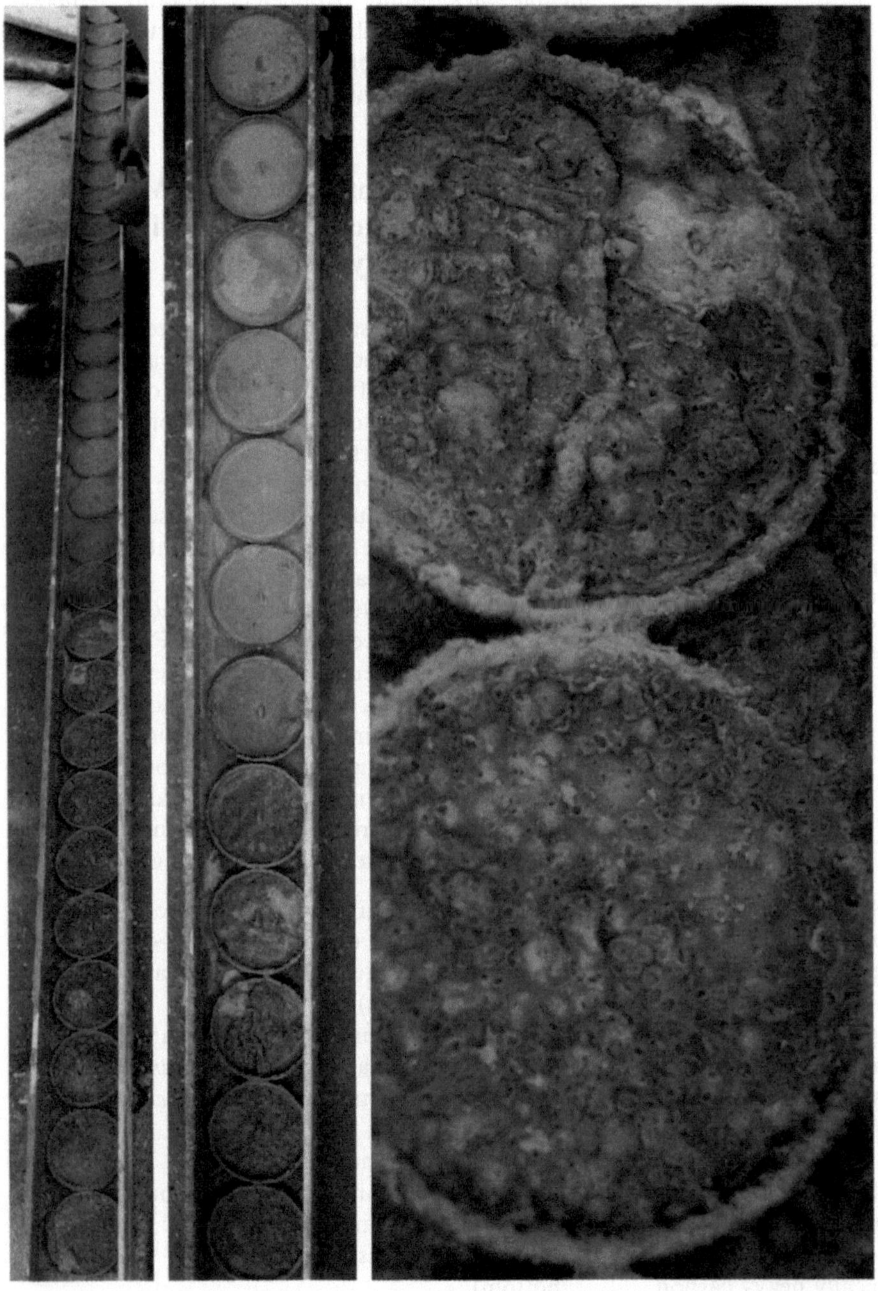

A B C

Fig. 4A–C. A Profile of the microbial mat as preserved in a sampling tray recovered from the slope of the experimental pond (note line-up of petri dishes whose content provided the material for the serial analyses such as represented in Figs. 5 and 6). The fading of the dark-green color marks the end of prolific mat development at the combined thermo- and haloclines. **B** Closeup of the transition zone between epilimnion and hypolimnion. The lush mat (green) typical of the pond littoral peters out upon entry into the gradient layer (as indicated by decreasing shades of brown), the color subsequently changing to dull gray in the zone of retarded biological activity on the central pond bottom. **C** Two adjacent sampling dishes from the littoral zone close to a depth of 30 cm that provides optimum conditions for the growth of the microbial ecosystem. Note pustulous surface and occasional warping of the mat (the pustules resulting from the release of oxygen bubbles evolved during photosynthetic activity of the mat-forming microbenthos)

With these characteristics, the water body as a whole qualifies as a typical solar pond (cf. Tabor 1981; Kirkland et al. 1983) in which specifically the hypolimnic brine underlying the artificially stabilized halocline accumulates a formidable amount of heat. Nonconvecting solar ponds of this kind not only provide an effective means of harvesting solar energy in the multi-kilowatt to multi-megawatt range, but also offer a unique opportunity to study the emplacement and subsequent proliferation of benthic microbial ecosystems in response to gradually changing environmental background parameters that cover extended ranges of salinity, temperature, pH, and light penetration.

After its completion in January 1987, the pond was quickly populated by a community of cyanobacteria and diatoms, with the subsequent establishment on its littoral of a benthic mat (Fig. 4) that was fully developed by the time of the first harvest (October 1987). Colonization of the new habitat had almost certainly proceeded via microorganisms transferred with the brines from neighboring preexisting ponds. As was to be expected, metazoan grazers were virtually kept out of these brines, the habitat being consequently monopolized by primary producers. As is typical for communities of microbial photoautotrophs thriving in extreme environments in the absence of grazing stress, species diversity was significantly reduced. This trend had been shown to be particularly pronounced within Dead Sea solar ponds (Dor and Paz 1989) and ascribed to the increased temperature range of these habitats (as compared to marine littoral and lagoonal environments) as well as to the influence of the specific ionic composition of Dead Sea water characterized, *inter alia*, by unusually high Mg^{2+} and Br^- contents (cf. Nissenbaum 1975; Steinhorn and Gat 1983). The utilization in our pond of Dead Sea end brines (constituting the ultimate residuum of multiple consecutive evaporation processes) turned out to be an inadvertent microbiological experiment in that it demonstrated an absolute climax in species reduction hitherto observed in such environments.

As fully set out in the first part of this study (Dor et al., this Vol.), the pond was populated by just two taxa, namely, cyanobacteria and diatoms. Moreover, the cyanobacterial population proved to be glaringly monospecific, made up solely of different strains of *Aphanothece halophytica*, while the diatom community was more diverse, with *Nitzschia*, *Navicula*, and *Amphora* figuring as main genera. The juvenile mat (as of summer 1987) was shown to be largely dominated by *A. halophytica* accounting for 62–93% of total cell counts in the pond littoral, and for 100% below the halocline (albeit with a reduction in total cell number by two orders of magnitude near the bottom). During the following winter season, both cyanobacteria and diatoms scored increased cell counts in the hypolimnion, i.e., that compartment of the habitat where high temperature and ionic strength heavily retard microbial activity. By the time of the next harvest (summer 1988), cell distributions of cyanobacteria and diatoms appeared to be almost statistical below the halocline, suggesting a certain amount of passive transport from the littoral mat into the underlying hypolimnion.

In summary, it is safe to say that the benthic mat is basically made up of a dominant cyanobacterial species with a minor (but more diverse) diatom component notably in the littoral segment above the halocline. Direct observation (Fig. 4) shows that the visual mat attenuates, and apparently peters out, upon entry into the gradient zone characterized by the combined halo- and thermoclines. This trend is specifically confirmed by the results of chlorophyll-a and organic carbon assays showing consistently declining values from a peak region in the epilimnion to minima in the hypolimnic compartment (cf. Figs. 5A, B and 6A).

4 Sampling and Experimental

Monitoring of the pond's principal environmental parameters (cf. Fig. 3), as well as the recovery of hydrobiological background data such as chlorophyll-a content and cell number, were performed according to the standard techniques currently in use. As these aspects are covered in the first part of this investigation (Dor et al., this Vol.), we shall deal here only with the experimental background of the isotope work.

With the sampling procedure described in detail by Carl (1989) and Dor et al. (this Vol.), it will suffice to draw the attention of the reader to the sampling trays lined with petri dishes as depicted in Fig. 4. Empty trays of this type were immersed along the pond slope and recovered after carefully considered exposure intervals. During their residence time in the pond, the trays were encroached by, and integrated in, the growing microbial carpet, thus providing complete and fairly pristine mat profiles after their recovery. The content of the individual petri dishes furnished the material for the serial analyses graphically summarized in Figs. 5 and 6.

The material recovered was collected in carefully cleaned glass containers and, within 24 h, washed in distilled water to remove the soluble salt fraction. The residue was then dried at low temperature ($t < 60\,^{\circ}C$) and gently powdered in a hand mortar. A first aliquot of the resulting powder was used for a coulometric determination of the C_{org} and C_{carb} contents of every single sample by an automated carbon analyzer (Stroehlein Coulomat 701).

To separate the organic fraction for isotope analysis, the samples were treated with 40% H_3PO_4 at 40 °C for about 24 h. After washing, centrifuging, and drying, the carbonate-free residues were combusted in sealed quartz tubes with an excess of copper oxide at 900 °C for about 15 min. Water released during the combustion process was removed with two cold traps operated at − 90 °C.

For isotope work on the carbonate fraction, the samples were treated with 5% NaOCl for 24 h and afterwards washed several times with distilled water. The residue was collected on carbonate-free glass fiber filters and dried at 50 °C. The finely powdered material was then reacted with anhydrous (100%) phosphoric acid at 25 °C according to the procedures described by McCrea (1950) and Craig (1953).

Carbon and oxygen isotope ratios were determined for the CO_2 obtained from the above procedures. Measurements were performed on a VG PRISM mass spectrometer equipped with a three-collector system permitting the simultaneous collection of the masses 44, 45, and 46. Results are reported as $\delta^{13}C$ and $\delta^{18}O$ values relative to the Peedee Belemnite (PDB) standard, with

$$\delta^{13}C = \left[\frac{(^{13}C/^{12}C)_{sa} - (^{13}C/^{12}C)_{st}}{(^{13}C/^{12}C)_{st}}\right] \times 1000\,[‰]$$

and

$$\delta^{18}O = \left[\frac{(^{18}O/^{16}O)_{sa} - (^{18}O/^{16}O)_{st}}{(^{18}O/^{16}O)_{st}}\right] \times 1000\,[‰],$$

where sa = sample and st = standard. Standard deviations of the measured values were usually below ± 0.1‰.

5 Results

Biogeochemical traverses along the pond profile have been established utilizing material recovered during two consecutive harvests of the microbial ecosystem. The first harvest was performed on a juvenile mat about 4 months old (exposure time of sampling tray July 21–October 30, 1987). The second harvest was done on a more mature mat with an age approximating 1 year (growth period from July 7, 1987 to June 13, 1988). Investigations were carried out on the material that had accumulated during the respective exposure periods in the petri dishes rowed up on the immersed sampling trays, the samples thus comprising both the benthic mat and the associated bottom sediment. Parameters measured were (1) the organic carbon (C_{org}) content, (2) chlorophyll-a content and total cell number, (3) isotopic composition of organic carbon ($\delta^{13}C_{org}$), and (4) isotopic compositions of carbonate carbon ($\delta^{13}C_{carb}$) and carbonate-bound oxygen ($\delta^{18}O_{carb}$). Results are summarized in Fig. 5 (for first harvest) and Fig. 6 (second harvest).

5.1 Quantitative Distribution of Organic Carbon

As can be inferred from the relevant distribution functions shown in Figs. 5A and 6A, the C_{org} content of both profiles sampled during this survey starts with values between 2.3 and 2.5% at the very pond margin and quickly climbs up to maxima between 3.1–3.7% at a depth just below 30 cm. From there, values sharply plummet to a depth of about 40 cm where the function levels off and stabilizes in the range of 1 and 1.5% (in the case of Fig. 6A with the intercalation of a soft slope in the 40–55 cm segment). In the central bottom area, there is a tendency for C_{org} to gently increase again to about 2% (Fig. 5A). Altogether, C_{org} displays a pronounced trend to decrease as a function of depth, with a most conspicuous decline of the values at the onset of the combined halo- and thermoclines. It is worth noting that the sharp drop in C_{org} content at this boundary coincides with the disappearance of the visual mat (cf. Fig. 4).

5.2 Chlorophyll-a Distribution and Cell Counts

These two parameters serve as indicators of the relative abundance of cyanobacterial and microalgal biomass along the pond slopes, with specifically chlorophyll-a concentrations reflecting the presence of a photosynthetically active (i.e., living) microbial film at the sediment-water interface. It is, therefore, heartening to see that both the chl-a and cell number functions (Figs. 5B and 6A) are approximate replicas of C_{org} concentrations measured over the same profiles (Figs. 5A and 6A). Observed resemblances often include details such as the initial increase in the values from the pond margin to a depth short of 30 cm and, in the case of the cell count, also the slight hub in the center of the pond below 80 cm (cf. Fig. 5A, B).

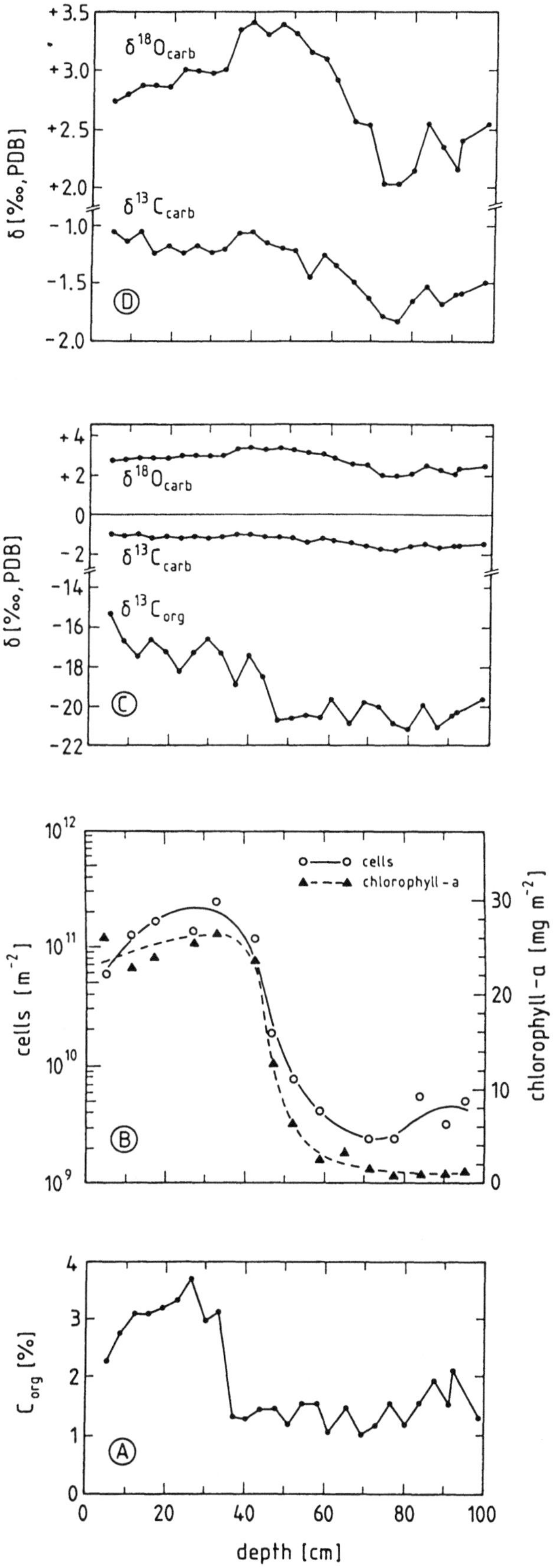

There is, however, a conspicuous difference in absolute chlorophyll-a content between the juvenile mat (Fig. 5B) and its mature counterpart (Fig. 6A), with the maximum in the epilimnic pond littoral soaring up from 26 mg m^{-2} in the case of the former to 166 mg m^{-2} in the case of the latter. The increase of total chlorophyll-a content in the mature mat is not paralleled by a corresponding increase in stationary biomass; in fact, depth-correlated C_{org} concentrations in both mat profiles mostly approach each other within some 0.5%, with those of the mature mat even lying slightly lower (cf. Figs. 5A and 6A). Another difference between the C_{org} and chlorophyll-a functions is that the latter, after plummeting at the halocline, quickly approach minimum values between 0.8 and 10 mg m^{-2} (indicative of a drastic inhibition of photosynthetic activity), while the former tend to stabilize at levels slightly in excess of 1%. The close approach to zero of the chlorophyll-a values at depths $\gtrsim$ 50 cm decidedly supports the conclusion prompted by the visual evidence, namely, that the high temperatures and salinities prevailing in the hypolimnion prevent (maybe in concert with additional inhibitory factors) the establishment of any large-scale benthic microbial ecosystem in this part of the pond.

5.3 Carbon Isotope Composition of the Benthic Mat

As is evident from the relevant pond profiles (Figs. 5C and 6B), the isotopic composition of the microbial ecosystem is generally characterized by decreasing $\delta^{13}C_{org}$ values with depth, i.e., as a function of the principal depth-correlated environmental parameters.

In the case of the juvenile mat (Fig. 5C), values decrease from − 15.3‰ at the pond margin to a minimum of − 21.2‰ near the center, with a soft jump at the halocline. While the function shows a gradual decline over the littoral zone (modulated by second-

Fig. 5A–D. Results of biogeochemical investigations performed on the benthic microbial mat grown on slope of Pond No. 1 during the time span June 21–October 30, 1987 (exposure time of sampling tray). For environmental background parameters of the habitat during this period see Fig. 3. **A** Organic carbon content of the bottom sample (comprising the benthic mat and associated sediment) as a function of depth and covariant hydrological background parameters summarized in Fig. 3. **B** Chlorophyll-a content of benthic mat and associated bottom sediment and results of cell counts over the same depth gradient. **C** Isotopic composition of organic and carbonate carbon and of the oxygen in the carbonate phase over the depth profile. **D** Blowup of $\delta^{13}C_{carb}$ and $\delta^{18}O_{carb}$ from **C** to facilitate identification of small-scale isotopic trends over the depth profile

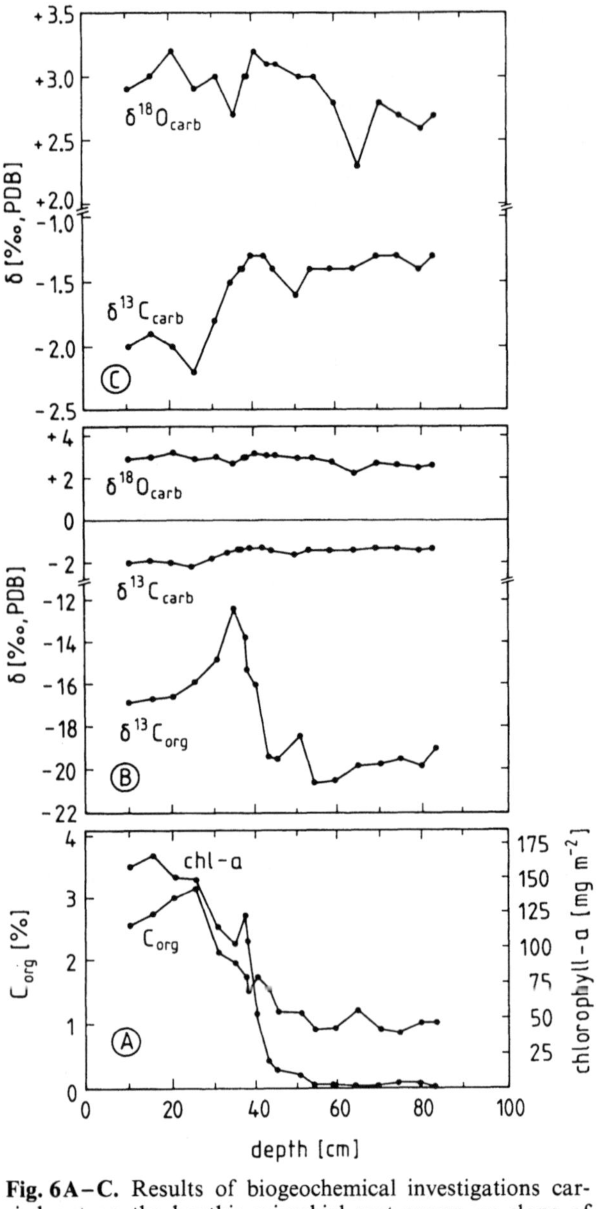

Fig. 6A–C. Results of biogeochemical investigations carried out on the benthic microbial mat grown on slope of Pond No. 1 during the time span July 7, 1987–June 13, 1988. With an exposure time of the sampling tray of almost one year, this harvest was performed on a more mature mat than that dealt with in Fig. 5. With the exception of a temporary drop in temperature during the winter season by 15–20 °C over the whole depth profile, the environmental background parameters during the extended growth period of the microbial community closely approximate those of Fig. 3. **A** Organic carbon and chlorophyll-a content of benthic mat and associated bottom sediment as a function of depth-correlated hydrological parameters (cf. Fig. 3). **B** Isotopic composition of organic and carbonate carbon and of carbonate-bound oxygen over the same depth profile. **C** Blowup of $\delta^{13}C_{carb}$ and $\delta^{18}O_{carb}$ from **B** to show small-scale isotopic trends over the depth profile

order oscillations), it is largely stabilized at a level between −19.6 and −21.1‰ in the hypolimnic segment of the pond spanning the depth range 45–100 cm. This quasi-stable envelope mirrors corresponding stabilizations in the C_{org} content (Fig. 5A) and biomass index functions (Fig. 5B).

The $\delta^{13}C_{org}$ function of the mature mat (Fig. 6B) additionally reflects the initial increase in the stationary microbial biomass (Fig. 5A) over the pond littoral (albeit with a slight displacement of the maxima) before precipitating to the low level characterizing the high-salinity bottom stratum. Along this way, values rise from −16.9‰ at the pond margin to −12.4‰ at a depth of 35 cm, plummeting within the subsequent 8-cm-depth interval to a baseline between −19 and −20‰ (with −20.6‰ as minimum). Closer scrutiny of Fig. 6A, B reveals that the onset of the downward incline in the C_{org} content function slightly precedes that of the $\delta^{13}C_{org}$ curve in the depth profile, while there is a fair coincidence between the slopes of the $\delta^{13}C_{org}$ and the chlorophyll-a functions. A similar trend is apparent also in the juvenile mat (Fig. 5A–C).

5.4 Isotopic Composition of Carbonate

Carbon fixed in the form of carbonate was present over the whole pond profile in amounts varying between 5.7 and 11.1 wt.%, with apparently no correlation to either depth or any other parameter measured during this investigation. These quantities would translate into about 47.5 and 92.5% carbonate in the sampled bottom sediment. Surveys of the carbonate fraction relying on conventional X-ray diffraction techniques showed aragonite to be the prevalent carbonate phase, followed immediately by calcite, while dolomite could be identified only in minor amounts or even as traces. Quartz proved to be the major accessory in the carbonate fraction, while gypsum (commonly present in industrial Dead Sea ponds) was altogether missing in our samples.

In marked contrast to the organic carbon of the benthic mat, the isotopic compositions of carbonate carbon and carbonate-bound oxygen show very little variation over the pond profile, being confined to envelopes bounded by −1.1 and −2.2‰ and +2 and +3.4‰, respectively. If plotted at the same scale, the depth functions of both $\delta^{13}C_{carb}$ and $\delta^{18}O_{carb}$ display a remarkable degree of uniformity (see Figs. 5C and 6B) compared with the pronounced unidirectional trends and intrinsic oscillations of the corresponding $\delta^{13}C_{org}$ curves. The relative constancy of $\delta^{13}C_{carb}$ constitutes proof that the preferential sequestering with depth of isotopically light carbon (^{12}C) by the photoautotrophic microbenthos (manifesting itself by progressively declining $\delta^{13}C_{org}$ values) makes little impact on the isotopic composition of the pond's residual reservoir of dissolved inorganic carbon [a substantial fraction of it being bicarbonate ion (HCO_3^-) that potentially lends itself to precipitation as carbonate (CO_3^{2-}) with a small isotope shift of + 1‰]. This is evident in the scenarios governed by the juvenile as well as by the mature mat; in both instances, the formidable oscillations of $\delta^{13}C_{org}$ over the depth profile are virtually smoothed out in the corresponding $\delta^{13}C_{carb}$ functions due to the dampening effect exercised by the major and better mixed of the two carbon reservoirs.

However, a blowup of $\delta^{13}C_{carb}$ and $\delta^{18}O_{carb}$ at a markedly increased δ-scale (Figs. 5D and 6C) reveals a distinct correlation between the $\delta^{13}C_{carb}$ and $\delta^{13}C_{org}$ functions. In the case of the young pond inhabitated by a correspondingly juvenile microbial ecosystem, both curves are basically covariant, though much detail is demonstrably lost in the $\delta^{13}C_{carb}$ function (cf. Fig. 5C, D). In the mature pond that had served as a microbial habitat for an extended time span, $\delta^{13}C_{carb}$ and $\delta^{13}C_{org}$ could be inversely correlated (Fig. 6B, C). There also exists some correlation between the corresponding $\delta^{13}C_{carb}$ and $\delta^{18}O_{carb}$ functions (cf. Figs. 5D and 6C).

6 Discussion of Results

6.1 Organic Carbon and Biogeochemical Characteristics of the Pond Ecosystem

The interpretation of the various biogeochemical parameters related to the pond's microbial ecosystem seems straightforward and conclusive. The visually evident decline of the benthic mat from a marginal zone of maximum development toward the center (cf. Fig. 4) is matched by the distribution functions of stationary biomass, chlorophyll-a content, cell numbers, and $\delta^{13}C_{org}$ values with an astounding degree of coincidence and precision, a particularly significant feature being the dramatic plummeting of all values at the combined halo- and thermoclines. This latter fact seems to indicate that the ultimate steering function for microbial mat development is exercised by two environmental variables, i.e., temperature and salinity (cf. Fig. 3). The increase in these parameters with depth could have activated, in turn, other and perhaps more specific inhibitors for biomass production in the pond's hypolimnic compartment (see below).

The *organic carbon content* monitored along the depth gradient (Figs. 5A and 6A) reflects, for the most part, stationary biomass densities at the sediment-water interface or mat development, respectively. Apart from being substantiated by the visual evidence (Fig. 4), this conclusion is supported by both the cell count and chlorophyll-a functions (Figs. 5B and 6A) that broadly retrace the principal trends of the C_{org} concentration curve, often with preservation of intricate detail. Cell counts and specifically chlorophyll-a concentrations (providing an index of photosynthetic activity performed by eukaryotic microalgae plus cyanobacteria) have been used in this study as indicators of biomass concentrations (see Dor et al., this Vol.).

Going into the details of the C_{org} concentration curve, we cannot fail to note a zone of instability in mat development at the very pond margin where photic inhibition and fluctuations of the water level usually combine to depress photosynthetic activity and hence biomass production. The situation stabilizes at a depth of about 30 cm where optimum conditions for mat development are indicated by C_{org} maxima between 3 and 4% that are matched by corresponding peaks in cell numbers and chlorophyll-a content. While chlorophyll-a concentrations come close to zero after a sharp drop at the halocline, corresponding C_{org} levels stabilize at 1% or slightly above. Such blatant discrepancy between chlorophyll-a and organic carbon levels would certainly argue for an accumulation of dead biomass at the bottom of the pond's hypolimnic stratum if these C_{org} accumulations were not characterized by a distinctly different isotopic composition. According to the isotopic evidence (Figs. 5C and 6B), the ^{12}C-enriched organic carbon in the hypolimnic compartment cannot be derived from the ^{12}C-depleted "heavy" material generated on the littoral pond slope. Hence, we must necessarily infer that even the harsh environment of these high-temperature hypolimnic brines does not preclude modest rates of primary productivity; the contrast between low chlorophyll-a levels and relatively high C_{org} concentrations could be explained by the retarded decomposition of biomass in this largely anoxic bottom layer. Attention is called to a minor peak of C_{org} in the very center of the pond (Fig. 5A) that is matched by corresponding hubs in the other biogeochemical parameters including a miniscule increase in chlorophyll-a content. The coincident increase also in the cell count function is likely to suggest that the cell numbers obtained in the hypolimnic compartment (ranging two orders of magnitude below those of the littoral zone) do not exclusively record dead cells.

Recorded changes in the *isotopic composition of the benthic microbial mat* over the pond profile appear to have cogent and fairly explicit implications for the key objective of this study, namely, the elucidation of the mechanism responsible for the generation of isotopically heavy biomass. The overall decrease of $\delta^{13}C_{org}$ with depth that is evident in both the juvenile (Fig. 5C) and the mature ecosystem (Fig. 6B), and notably the pronounced plummeting of the values at the halocline, demonstrate conclusively that the microbial biomass tends to become progressively *depleted* in heavy carbon with increasing temperature and salinity of the aquatic habitat. This observation constitutes definitive proof that, contrary to previous inferences (Schidlowski et al. 1984), salinity cannot act as the principal steering variable for the generation of isotopically heavy (^{13}C-enriched) biomass in these and related environments. Although potentially contributing to the overall CO_2 limitation in such habitats (cf. Fig. 2), decreased CO_2 solubility in heliothermal brines demonstrably fails to gain importance as a first-order determinant for the production of ^{13}C-enriched organic matter.

While unequivocally negating the role of ionic strength as the principal predictor variable for the synthesis of heavy biomass, the results summarized in Figs. 5 and 6 offer, at the same time, crucial clues to a definitive solution of this problem. It is obvious from these data that positive extremes in $\delta^{13}C_{org}$ (identifying "heavy" organic matter) appear to be consistently coupled with maxima in biomass density (expressed as C_{org} %), cell counts, and chlorophyll-a content, the latter two representing productivity-related parameters. Conversely, minimum levels in $\delta^{13}C_{org}$ such as characterizing the gradient zone and the hypolimnic compartment of the pond are clearly correlated with minima in these productivity index functions [as pointed out by Dor et al. (this Vol.), specifically chlorophyll-a concentrations appear to be in adequate agreement with the results of primary productivity measurements]. With this evidence at hand, there is little doubt that productivity holds the role of the principal steering variable capable of conditioning microbial photoautotrophs to produce isotopically heavy biomass. We had previously (Schidlowski et al. 1984, p. 306) assigned this factor a merely auxiliary function in this process, while Des Marais et al. (1987) were the first to report evidence suggesting productivity dependence as the main forcing agent. From the data base accrued in this investigation, we may reasonably infer that the excessive CO_2 demand by the resident microbial community due to high productivity rates cannot be satisfied except by a largely indiscriminate drain of CO_2 from the local inorganic carbon pool. With progressive CO_2 exhaustion, assimilatory pathways are bound to become supply- (or diffusion-) limited, with a consequent suppression of carbon isotope discrimination. It is well known that benthic microbial communities figure among the most productive ecosystems (with maximum rates of $8-12$ g C_{org} m^{-2} day^{-1}; cf. Cohen et al. 1980), and that the trend toward metabolization of isotopically heavy carbon is additionally enhanced by high degrees of eutrophication in the environment (O'Leary 1981).

Accepting that the $\delta^{13}C_{org}$ functions shown in Figs. 5C and 6B reflect a productivity gradient, it still remains for us to identify the reason(s) for this gradient. With a detailed discussion presented elsewhere (Dor et al., this Vol.), the following account just briefly summarizes the issue.

Previous work performed on Dead Sea industrial ponds has shown that phosphate (and not nitrogen) is always the limiting nutrient in these habitats, with concentrations around 3 μg (at.) P/l apt to induce microbial blooms. Hence, recorded phosphate levels of about 9 μg (at.) P/l in the epilimnion and 4.8 μg (at.) P/l in the central bottom region characterize the pond throughout as a fairly eutrophicated habitat, with nutrient limitation thus unlikely to be responsible for the observed gradient in mat development. As suggested by its conspicuous covariance with temperature and salinity, this gradient must result from the combined influence exercised by these two key environmental parameters on the benthic microbial ecosystem. The suppression of productivity at the very pond margin is ostensibly due to the deleterious effects of both light inhibition and fluctuation in the water level, which are bound to periodically interfere with mat growth. Optimum conditions for mat development in terms of light, temperature, and salinity obviously prevail at a depth around 30 cm (as testified by coincident maxima in C_{org}, cell number, chlorophyll-a, and $\delta^{13}C_{org}$), whereas changing levels and different combinations of the same variables appear to prevent any large-scale proliferation of the microbial community beyond a depth of some 40 cm. Since lack of light as a limiting factor for photosynthesis is unlikely in this case, the dramatic suppression of microbial activity must be ultimately due to the harsh temperature and salinity regimes established at these depth levels (for details cf. Dor et al., this Vol.). Apart from the inhibitory effect exerted by the sudden increase in both parameters in the zone of instability between 40 and 60 cm, there is empirical evidence that extreme temperatures and salinities as such (single or in concert) are powerful limiting factors for microalgal activity. We would not exclude at this stage, however, that additional inhibitors that might come to bear in these environments have contributed to the observed suppression of the benthic mat (potential candidates being, for instance, increased Mg^{2+} concentrations or the incipient release of H_2S).

6.2 Carbonate in the Pond System

A definitive evaluation of the *inorganic carbon system* of the pond must be postponed until comparative data for sedimentary carbonate and dissolved inorganic carbon have accrued for all four experimental sites investigated. However, with the two high-resolution isotope depth profiles for carbonate currently available (Figs. 5D and 6C), a tentative assessment of the carbonate system might be certainly in order.

It is obvious that the formidable variations in the isotopic composition of the benthic mat as expressed by the decrease in $\delta^{13}C_{org}$ over the pond profile are largely smoothed out in the corresponding $\delta^{13}C_{carb}$ functions which monitor the impact suffered by the pond's feeder pool of inorganic carbon resulting from the preferential fixation in the mat of isotopically light (^{12}C-enriched) carbon. The progressive extraction of light carbon from the reservoir along the depth profile, accordingly, hardly affects the average isotopic composition of inorganic carbon in the overlying water column (let alone the carbon pool as a whole which, by virtue of its over whelming size and short mixing time, substantially dampens the effect of the preferential removal of ^{12}C on the $^{12}C/^{13}C$ ratio of the residual carbon isotope mixture). It is only with a considerable increase in scale (see Figs. 5D and 6C) that factual correlations between $\delta^{13}C_{carb}$ and $\delta^{13}C_{org}$ reveal themselves to the observer.

In the juvenile pond (Fig. 5), a roughly sympathetic relationship seems to exist between the two parameters; specifically, the decline of $\delta^{13}C_{org}$ at the combined halo- and thermoclines is paralleled (albeit considerably dampened) by a corresponding decline of $\delta^{13}C_{carb}$. Since first-order budget considerations would predict that the preferential accumulation of light carbon in the benthic mat should be compensated for by *increasing* $\delta^{13}C_{carb}$ values in the residual reservoir, the actually observed *decreasing* values are likely to argue for an invasion of ^{13}C-depleted CO_2 from the atmosphere into the brine. Such processes have been invoked to explain sharp drops in $\delta^{13}C_{carb}$ within aquatic habitats of limited extension (notably lakes) where excessive consumption of CO_2 during algal blooms (with a concomitant increase in pH) triggers a chemically enhanced rapid diffusion of carbon dioxide across the atmosphere-water interface which is beset with a sizeable negative kinetic fractionation effect of -13 to $-14‰$ (Herczeg and Fairbanks 1987; Lazar and Erez 1990). However, with any potential negative shift in $\delta^{13}C_{carb}$ within our isotope profile demonstrably arrested between the -1 and -2 per mill lines, the quantitative impact of such nonequilibrium (kinetic) fractionations on the carbon isotope budget of the juvenile pond seems to be very small.

In the case of the mature pond, the correlation of the blown-up $\delta^{13}C_{carb}$ curve (Fig. 6C) with the coexisting $\delta^{13}C_{org}$ function seems more equivocal. If the minimum in $\delta^{13}C_{carb}$ would coincide with the maximum in $\delta^{13}C_{org}$ (displaced by about 10 cm, cf. Fig. 6B) as it does with the maximum in C_{org} concentration (Fig. 6A), the scenario would be decidedly consistent with the above-mentioned CO_2-invasion model predicting $\delta^{13}C_{carb}$ minima in the zone of maximum mat development. The ostensible weakness of the carbonate signal would be in good agreement with the fact that the pH in the epilimnion of our pond does not exceed the value of 7.7 (cf. Fig. 3), with the chemical enhancement of CO_2 diffusion (by the reaction $CO_2 + OH^- \rightarrow HCO_3^-$) still being severely retarded. Since, on the other hand, the observed minimum in $\delta^{13}C_{carb}$ is represented by only two aberrant negative values spanning just a few tenths of a per mill, the respective feature of the function might be a fortuitous one. If this were the case, $\delta^{13}C_{carb}$ and $\delta^{13}C_{org}$ would be positively correlated, with their maxima closely coinciding and a subsequent decline of both functions beyond the zone of maximum mat development. We propose that this interpretation is the more probable one.

A problem for a definitive assessment of the pond's inorganic carbon system (and specifically the $\delta^{13}C_{carb}$ profiles obtained) is posed by the probable presence in the sampled bottom sediment of a *detrital* carbonate component likely to be derived, for the major part, from the ubiquitous Judean Limestone of the surrounding countryside (with $\delta^{13}C_{carb}$ close to $-1‰$). However, the systematic changes of both $\delta^{13}C_{carb}$ and $\delta^{18}O_{carb}$ over the depth profile, and their most obvious correlation with the pond's environmental and biogeochemical parameters, give eloquent proof that the isotopic signals recorded are basically those of the *authigenic* carbonate fraction (which latter is often visually evident in the form of solid carbonate coatings on the sampling dishes). A possible way of specifying the parentage of the pond's sedimentary carbonate would be a comparison of its isotopic composition with that of the dissolved inorganic carbon (DIC) in the overlying water column (mostly HCO_3^-) since the $\delta^{13}C$ mean of the latter is transcribed into the precipitated solid carbonate phase with a shift of about $+1‰$ (cf. Mook et al. 1974). With arrangements for DIC analyses having been made, but relevant data not yet available, the potential impact of a detrital carbonate fraction on the $\delta^{13}C_{carb}$ profiles of our pond is still a matter of speculation. In analogy to the situation observed in a neighboring pond, where $\delta^{13}C_{DIC}$ had been determined, we would not exclude, however, that the admixture of a detrital component with the isotopic composition of Judean Limestone might have raised the $\delta^{13}C_{carb}$ level of the pond quasi-uniformly by 1–2‰, albeit with faithful preservation of the modulations imposed by the authigenic carbonate fraction.

7 Summary and Conclusions

From the data base accrued in this investigation (Figs. 5 and 6), it can be inferred beyond reasonable doubt that CO_2 limitation due to high rates of primary productivity sustained by a benthic microbial community under quasi-eutrophicated conditions is the prime cause for the generation of isotopically heavy (^{13}C-enriched) biomass in our hypersaline pond. While markedly decreased CO_2 solubilities in waters of high ionic strength (Fig. 2) can be expected to contribute to the overall CO_2 depletion in such and related habitats, recordings of $\delta^{13}C_{org}$ vs. gradients of temperature and salinity as the salient environmental background parameters show that salinity clearly fails as a dominant steering variable. On the other hand, a straightforward interdependence exists between $\delta^{13}C_{org}$ and productivity-related parameters such as cell numbers and chlorophyll-a levels. Moreover, the observed correlations between $\delta^{13}C_{org}$ and visual mat development over the pond profile (Fig. 4) constitute *prima facie* evidence of a coupling between biomass accumulation and increased $\delta^{13}C_{org}$ levels (see also C_{org} concentration functions, Figs. 5A and 6A). With the habitat eutrophicated throughout with regard to phosphate as potentially limiting nutrient, the observed productivity gradient over the pond slope must have been mainly induced by temperature and salinity as the key environmental variables which assumption is specifically supported by a plummeting of all biological parameters at the combined thermo- and haloclines (Figs. 5 and 6).

With these facts at hand, a reappraisal of the glaring mismatch between the isotope spreads of ancient and modern stromatolitic ecosystems would be in order. Realizing that the bulk of extant microbial mat communities are characterized by isotopically heavy carbon (Fig. 1B), and accepting that the observed ^{13}C enrichment is not intrinsic to the assimilatory pathway of the microbial photoautotrophs, but due to CO_2 limitation resulting from high rates of primary productivity typical of eutrophicated habitats (such as sabkha-type environments), a plausible explanation might be that the increased $\delta^{13}C_{org}$ levels constitute the environmental signature of the ecological refugia to which benthic microbial ecosystems had been forced to retreat after the advent of metazoan life at the dawn of the Phanerozoic (Glaessner 1983). It is, in fact, astounding to note that cultured cyanobacteria (see Fig. 1A) not subjected to the above constraints exhibit $\delta^{13}C_{org}$ spreads that are fairly consonant with average fractionations in C3 photosynthesis dominated by the large $^{13}C/^{12}C$ discriminations of ribulose-1,5-bisphosphate (RuBP) carboxylase as the key enzyme of the C3 pathway. Moreover, it has been demonstrated that an increase in atmospheric CO_2 pressure ($\sim 0.03\%$) by an order of magnitude ($\sim 0.3\%$) may shift the rate limiting step in the assimilatory pathway of aquatic green algae (*Chlorella* sp.) from *diffusion* (entailing a fractionation of $-4‰$ and less) to *carboxylation*, with the concomitant expression of the sizeable isotope discriminations (between -20 and $-30‰$) typical of the enzymatic reaction (Vogel 1980).

Such observations would lend support to the notion that the generation of isotopically heavy biomass by contemporary benthic microbial ecosystems is primarily a response to the severe degree of CO_2 limitation that characterizes their principal host environment in the present world. A reversal of this argument would imply that the same microbial communities were capable of entertaining $\delta^{13}C_{org}$ ranges similar to those of C3 plants and eukaryotic algae if they were not excluded from the normal marine environment by a galaxy of metazoan grazers. Since this restriction did not apply to the Archaean and Proterozoic oceans, the $\delta^{13}C_{org}$ spread recorded for Precambrian stromatolites (cf. Fig. 1B) might not be surprising.

In summary, it seems firmly established that the *ultimate* reason for the observed reduction of biological carbon isotope fractionations in recent microbial mats is CO_2 limitation in their aqueous host habitat enforcing the operation of a diffusion-(supply-)limited assimilatory pathway by the photoautotrophic microorganisms. It is reasonable conjecture that overcoming the diffusion barrier by either a decrease in biological demands placed on the local carbon pool, or by an increase in ambient carbon dioxide pressure, would reinstate the first enzymatic carboxylation reaction of the photosynthetic pathway as the principal isotope-discriminating step, with a consequent rise in carbon isotope fractionations. Both the emplacement of benthic microbial ecosystems in the normal marine realm (as in the shelf regions of the Precambrian seas prior to the advent of heterotrophic grazers), or the possible prevalence of increased CO_2 levels in the Precambrian atmosphere [as previously invoked by Mizutani and Wada (1982)] could bring about such change and might, accordingly, account for the different isotope spreads of organic carbon in ancient stromatolites compared to their modern analogs.

Note Added in Proof

The abstract by Des Marais et al. (1987) cited in the reference list has, meanwhile, matured to the publication "Carbon isotopic trends in the hypersaline ponds and microbial mats at Guerrero Negro, Baja California Sur, Mexico: Implications for Precambrian stromatolites" [in Cohen Y, Rosenberg E (eds), *Microbial Mats*, Am Soc Microbiol Washington, 1989, 191–203].

Acknowledgements. This paper has greatly benefited from our cooperation with N. Carl and G. Josten, from discussions with A. Krause-Mirus, and reviews by P. Aharon and D. J. Des Marais. Financial support by the Deutsche Forschungsgemeinschaft is gratefully acknowledged.

References

Barghoorn ES, Knoll AH, Dembicki H, Meinschein WG (1977) Variation in stable carbon isotopes in organic matter from the Gunflint iron formation. Geochim Cosmochim Acta 41:425–430

Behrens EW, Frishman SA (1971) Stable carbon isotopes in blue-green algal mats. J Geol 79:94–100

Calder JA, Parker PL (1973) Geochemical implications of induced changes in ^{13}C fractionation by blue-green algae. Geochim Cosmochim Acta 37:133–140

Carl N (1989) Microalgae biomass in artificial solar ponds operated at varying salinity regimes. M Sc Thesis, Hebrew Univ, Jerusalem, 56 pp

Cohen Y, Aizenshat Z, Stoler A, Jorgensen BB (1980) The microbial geochemistry of Solar Lake, Sinai. In: Ralph JB, Trudinger PA, Walter MR (eds) Biogeochemistry of ancient and modern environments. Springer, Berlin Heidelberg New York, pp 167–172

Craig H (1953) The geochemistry of stable carbon isotopes. Geochim Cosmochim Acta 3:53–92

Deines P (1980) The isotopic composition of reduced organic carbon. In: Fritz P, Fontes JC (eds) Handbook of environmental isotope geochemistry, vol 1. Elsevier, Amsterdam, pp 329–406

Des Marais DJ, Bui H, Cheatham M (1987) A carbon isotope budget of the hypersaline ponds and microbial mats at Guerrero Negro, Baja California Sur, Mexico. In: Abstr Bat-Sheva de Rothschild Int Sem Microbial mats. Interuniv Inst Eilat, Isr, Sept 13–20, 1987

Dor I, Ehrlich A (1987) The effect of salinity and temperature gradients on the distribution of littoral microalgae in experimental solar ponds, Dead Sea area, Israel. Marine Ecol 8:193–205

Dor I, Paz N (1989) Temporal and spatial distribution of mat microalgae in the experimental solar ponds, Dead Sea area, Israel. In: Cohen Y, Rosenberg E (eds) Microbial mats: physiological ecology of benthic microbial communities. Am Soc Microbiol, Washington, pp 114–122

Eichmann R, Schidlowski M (1975) Isotopic fractionation between coexisting organic carbon-carbonate pairs in Precambrian sediments. Geochim Cosmochim Acta 39: 585–595

Estep MF (1982) Stable isotope composition of algae and bacteria that inhabit hydrothermal environments in Yellowstone National Park. Annu Rep Director Geophys Lab Carnegie Inst Washington 1981–82:403–410

Glaessner MF (1983) The emergence of Metazoa in the early history of life. In: Nagy B, Weber R, Guerrero JC, Schidlowski M (eds) Developments and interactions of the Precambrian atmosphere, lithosphere and biosphere. Elsevier, Amsterdam, pp 319–333

Hayes JM, Kaplan IR, Wedeking KW (1983) Precambrian organic geochemistry: preservation of the record. In: Schopf JW (ed) Earth's earliest biosphere: its origin and evolution. University Press, Princeton, pp 93–134

Herczeg AL, Fairbanks RG (1987) Anomalous carbon isotope fractionation between atmospheric CO_2 and dissolved inorganic carbon induced by intense photosynthesis. Geochim Cosmochim Acta 51:895–899

Hoering TC (1967) The organic geochemistry of Precambrian rocks. In: Abelson PH (ed) Researches in geochemistry. John Wiley & Sons, New York, pp 89–111

Kirkland DW, Bradbury JP, Dean WE (1983) The heliothermic lake – a direct method of collecting and storing solar energy. Arch Hydrobiol Suppl 65(I):1–60

Landolt-Börnstein (1962) Zahlenwerte und Funktionen aus Physik, Chemie, Astronomie, Geophysik und Technik, 6th edn, vol II/2b: Gleichgewichte außer Schmelzgleichgewichten. Springer, Berlin Heidelberg New York, p 1/175

Lazar B, Erez J (1990) Extreme ^{13}C depletions in seawater-derived brines and their implications for the past geochemical carbon cycle. Geology 18:1191–1194

McCrea JM (1950) On the isotopic chemistry of carbonates and a paleotemperature scale. J Chem Phys 18:849–857

Mizutani H, Wada E (1982) Effect of high atmospheric CO_2 concentration on $\delta^{13}C$ of algae. Origins Life 12:377–390

Mook WG, Bommerson JC, Stavermann WH (1974) Carbon isotope fractionation between dissolved bicarbonate and gaseous carbon dioxide. Earth Planet Sci Lett 22:169–176

Nissenbaum A (1975) The microbiology and biogeochemistry of the Dead Sea. Microbiol Ecol 2:139–161

O'Leary MH (1981) Carbon isotope fractionation in plants. Phytochemistry 20:553–567

Pardue JW, Scalan RS, Van Baalen C, Parker PL (1976) Maximum carbon isotope fractionation in photosynthesis by blue-green algae and a green algae. Geochim Cosmochim Acta 40:309–312

Peters KE, Rohrback BG, Kaplan IR (1981) Carbon and hydrogen stable isotope variations in kerogen during laboratory-simulated thermal maturation. Am Assoc Petrol Geol Bull 65:501–508

Schidlowski M (1985) Carbon isotope discrepancy between Precambrian stromatolites and their modern analogs: inferences from hypersaline microbial mats of the Sinai coast. Origins Life 15:263–277

Schidlowski M, Gorzawski H, Dor I (1988) Experimental solar ponds 2. Isotopic composition of microbial biomass as a function of productivity rates and salinity. Terra Cognita 8:229 (Abstr)

Schidlowski M, Gorzawski H, Dor I (1989) Isotopically heavy biomass from microbial mats: predictor variables from experimental hypersaline ponds. Abstr 28 Int Geol Congr Washington 3:45–46

Schidlowski M, Hayes JM, Kaplan IR (1983) Isotopic inferences of ancient biochemistries: carbon, sulfur, hydrogen and nitrogen. In: Schopf JW (ed) Earth's earliest biosphere: its origin and evolution. University Press, Princeton, pp 149–186

Schidlowski M, Matzigkeit U, Krumbein WE (1984) Superheavy organic carbon from hypersaline microbial mats: assimilatory pathway and geochemical implications. Naturwissenschaften 71:303–308

Schidlowski M, Matzigkeit U, Mook WG, Krumbein WE (1985) Carbon isotope geochemistry and ^{14}C ages of microbial mats from the Gavish Sabkha and the Solar Lake. In: Friedman GM, Krumbein WE (eds) Hypersaline ecosystems: the Gavish Sabkha (Ecological Studies 53). Springer, Berlin Heidelberg New York, pp 381–401

Schopf JW, Oehler DZ, Horodyski RJ, Kvenvolden KA (1971) Biogenicity and significance of the oldest known stromatolites. J Paleontol 45:477–485

Seckbach J, Kaplan IR (1973) Growth pattern and $^{13}C/^{12}C$ isotope fractionation of *Cyanidium caldarium* and hot spring algal mats. Chem Geol 12:161–169

Steinhorn I, Gat J (1983) The Dead Sea. Sci Am 249:102–109

Tabor H (1981) Solar ponds. Solar Energy 27:181–194

Vogel JC (1980) Fractionation of the carbon isotopes during photosynthesis. Sitzungsber Heidelberger Akad Wiss Math-Nat Kl 3:111–135

Palynostratigraphy in Phanerozoic and Precambrian Metamorphic Rocks

H. D. Pflug and E. Reitz [1]

[1] Geologisches Institut der Justus-Liebig-Universität, W-6300 Giessen, FRG

Abstract

A review is presented on the current state of palynological research in metamorphic rocks. Recent studies have shown that organic-walled microfossils may be structurally preserved in rocks of the greenschist facies and even those of the lower amphibolite facies from where they can be separated by a special maceration procedure. In transmitted light, these microfossils appear as opaque bodies lacking observable structures, but a combination of transmitted and reflected light usually reveals their morphology. A series of Phanerozoic and Precambrian occurrences have been studied. The results show that palynology can be a useful tool in deciphering the stratigraphic ages of metamorphosed sediments.

1 Problems and Procedures

Rocks of medium to high metamorphic grade are commonly believed to be nonfossiliferous. Consequently, their stratigraphic position remains doubtful in most cases. However, recent studies have shown that organic microfossils such as spores and acritarchs can survive metamorphism under certain conditions. Specimens located in a protected position of the mineral matrix may escape destruction and remain with their morphology intact. On the other hand, the examination of such finds is difficult because their organic wall materials are graphitized and/or pyritized as a result of metamorphic heating. Under transmitted light, the residual particles appear structureless and opaque (Fig. 6, No. 1). We have found that the additional application of reflected light, on the other hand, reveals a wealth of morphological detail (Fig. 6, No. 2). In many cases the preservation is good enough to allow identification of the fossil.

Palynomorphs of subgreenschist alteration, such as those of the Belt and Gunflint formations (Table 1), are sufficiently transparent under transmitted light conditions so as not to require combined illumination. Such conditions can serve as gauges in the comparative study of the different metamorphic ranks. Transparency of palynomorphs generally indicates a H/C ratio greater than 0.25. More exact information is obtained from spectroscopic techniques allowing individual analysis of a microscopic particle. All UV-visible absorption spectra of demineralized palynomorphs have certain features in common (Fig. 1). Absorption increases gradually from the infrared through the visible and ultraviolet. Specimens of low and medium metamorphic rank have a pronounced absorption shoulder in the region 220–240 nm. The feature expands to longer wavelengths with increasing rank. It reaches the visible region if the material is anthracitic. The absorption curve for graphite is parallel and reasonably close to those of highest rank. Raman spectra show which changes occur during organic metamorphism. At first, aliphatic C–H bonds disappear, then aromatic C–H bonds diminish as structures approach extreme levels of graphitization (Fig. 2). In the laser mass spectra, the hydrogen content of the recorded carbon clusters gradually decreases with increasing rank (Figs. 3, 4).

Two criteria may be used to determine whether a find is indigenous to or a later contaminant of the rock. Examination of thin sections must show that the specimen is part of the rock matrix, and not included in rock fissures or cracks. If the fossil is in a graphitized condition, it can hardly be younger than the metamorphic event that has altered the rock.

Thin sections do not clearly show all microscopic details of the finds due to interference by the mineral matrix. It is, therefore, neccessary to isolate the microfossils from the rock by maceration which dissolves the mineral matter, leaving their morphology intact. Transmission electron microscopy is particularly useful because particle in the micron range and smaller stand a better chance to escape the disintegating effect of rock recrystallization than larger ones. The procedures and microscopic techniques are described in previous papers (Pflug 1987; Pflug and Reitz 1987). In

M. Schidlowski et al. (Eds.)
Early Organic Evolution: Implications for Mineral and Energy Resources

Table 1. Microfossil occurrences in metamorphic rocks

Occurence	Metamorphic grade	Typical microfossils	Stratigraphic position	Age (Ma B.P.)
Younger schist cover, Tauern, E-Alps	Greenschist	*Cicatricosisporites* sp., *Batioladinium micropodum*	Lower Cretaceous	140–100
Pulo F., southern Portuguese Zone, SW-Spain	Greenschist	*Retispora lepidophita*, *Grandispora echinata*	Uppermost Devonian	360
Werra anticline Phenish Massif, W-Germany	Low grade	*Retispora lepidophita*, *Hymenozonotriletes explanatus*	Uppermost Devonian	360
Somstal beds, Rhenish Massif, W-Germany	Low grade	*Dictyotriletes gorgoneus*, cf. *Grandispora* sp.	Lower Devonian	390
Taunus phyllites, Rhenish Massif, W-Germany	Lower greenschist	*Emphanisporites schultzii*, *Verruciretusisporites multituberculata*	Lower Devonian	390
Hunsrück phyllites, Rhenish Massif, W-Germany	Lower greenschist	*Verruciretusispora dubia*, cf. *Grandispora* sp.	Lower Devonian	390
Moldanubian Complex, eastern Bavaria	Upper greenschist	*Synorisporites downtonensis*, *Ambitisporites dilutus*	Upper Silurian	410
Spessart Crystalline Complex, NW-Bavaria	Lower amphibolite	*Apiculiretusispora synorea*, *Retusotriletes warringtonii*	Upper Silurian	420
Villé Unit, northern Vosges, eastern France	Low grade	*Timofeevia phosphoritica*, *Cymatiogalea* cf. *cylindrata*	Uppermost Cambrian	500
Moldanubian Complex, eastern Bavaria	Upper greenschist	*Bavlinella faveolata*, Sphaeromorph acritarchs	±Vendian	570
Greenschist sequence of Münchberg Gneiss Massif, E-Bavaria	Greenschist	*Octoedryxium truncatum*, Sphaeromorph acritarchs	Lower Vendian	650
Brévenne Unit, northern Massif Central, France	Greenschist	*Protosphaeridium* sp., *Trachysphaeridium* sp.	Upper Proterozoic	ca. 650
Habach Formation, Tauern, eastern Alps	Greenschist	"Vase-shaped microfossils", *Octoedryxium truncatum*	Upper Riphean/ Lower Vendian	650–700
Belt Supergroup, Idaho, Montana	Low grade	*Nucellosphaeridium* sp., *Trachysphaeridium* sp.	Middle Proterozoic	1100–1300
Gunflint Iron Fm., Ontario	"Subgreen-schist"	*Gunflintia* sp., *Huroniospora* sp., *Eomicrhystridium* sp.	Lower Proterozoic	2000
Campbell Group, Griqualand – West Supergroup, RSA	Lower greenschist	*Stictosphaeridium* sp. *Protosphaeridium* sp.	Lower Proterozoic	2300
Black Reef Shale Transvaal Supergroup, RSA	Lower greenschist	*Trematosphaeridium* sp. *Trachysphaeridium* sp.	Lower Proterozoic	2400
Bulawayan Group, Zimbabwe	Greenschist	Coccoidal cyanobacteria, Filamentous cyanobacteria	Late Archean	2800
Onverwacht Group, Swaziland System, South Africa	Greenschist	Coccoidal microstructures, Filamentous microstructures	Early Archean	3400
Isua Supracrustal Belt, West Greenland	Lower amphibolite	Coccoidal microstructures, Filamentous microstructures	Early Archean	3800

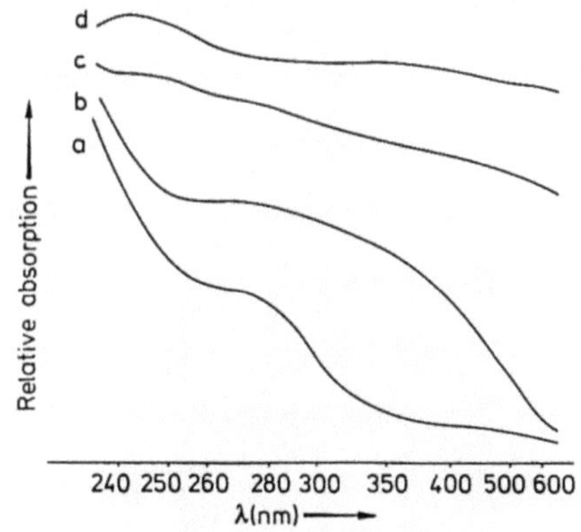

Fig. 1. UV-visible absorption spectra of fossil organic particles after demineralization. **a** Alginite from Permian Boghead coal; **b** *Huroniospora* sp. from Gunflint Banded Iron Formation; **c** spherical microstructure from Isua Banded Iron Formation; **d** graphite. Field of measurements ca. 1 μm diam. (Pflug 1987)

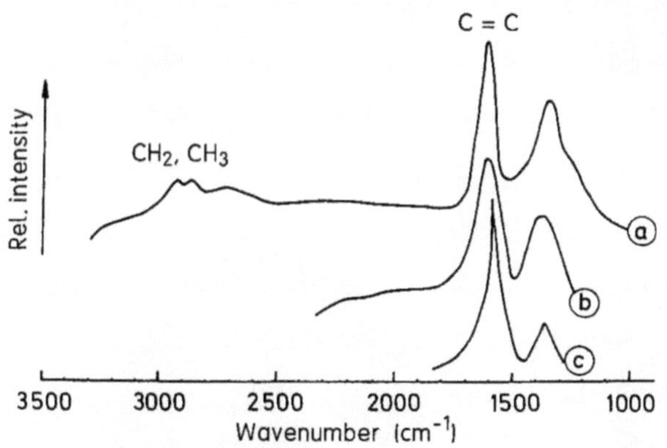

Fig. 2. Laser Raman spectra of **a** *Gunflintia* sp. in thin section of Gunflint chert; **b** spherical microstructure in thin section of Isua graphitic quartzite; **c** graphite. Field of measurements ca. 1 μm diam. *Abscissa*: wavenumber (cm^{-1})

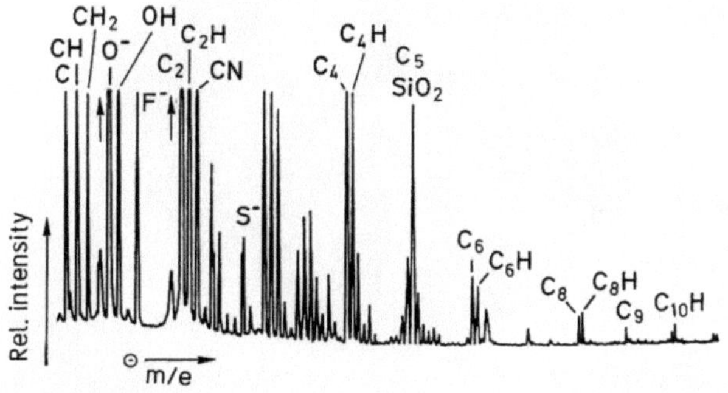

Fig. 3. Laser mass spectrum of *Huroniospora* sp. in thin section of Gunflint chert. Field of measurement ca. μm diam. *Abscissa*: mass/energy (negative ions)

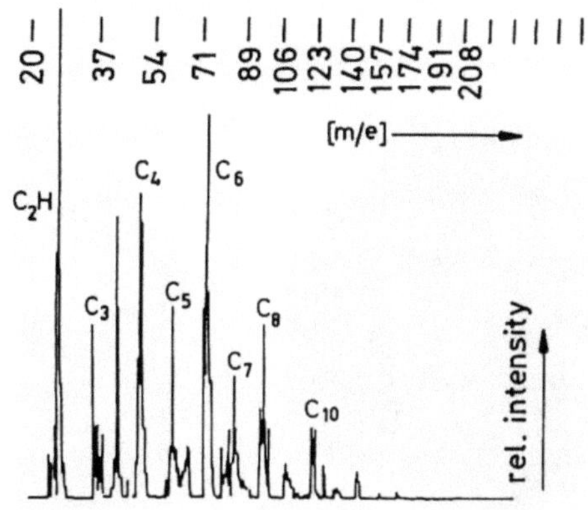

Fig. 4. Laser mass spectrum of Isua spherical microstructure in thin section of graphitic quartzite. Field of measurement ca. 1 μm diam. *Abscissa*: mass/energy (negative ions)

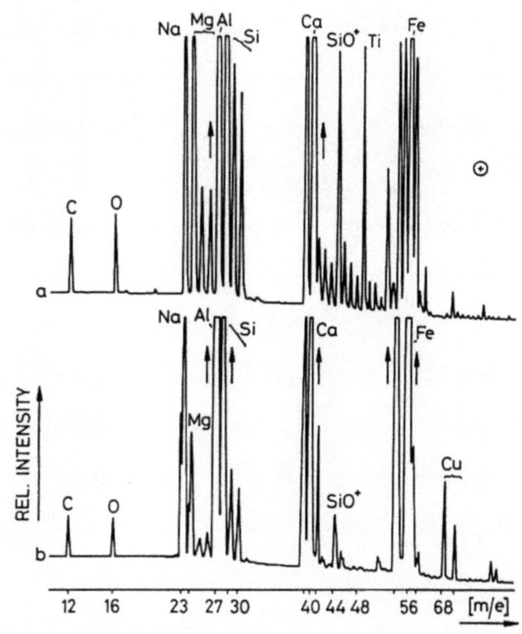

Fig. 5. Laser mass spectrum of encrustations attached to organic walls of microstructures. **a** Gunflint; **b** Isua; from thin sections. Field of measurement ca. 1 μm diam. *Abscissa*: mass/energy (positive ions)

the following, we shall summarize some results of relevant case studies from both Phanerozoic and Precambrian metasediments, including the ~3.8 Ga-old Isua suite.

2 Phanerozoic Occurrences

Greenschist facies metamorphism is widespread in the schist cover of the Tauern Window in the eastern Alps. Spores of Early Cretaceous age were found to

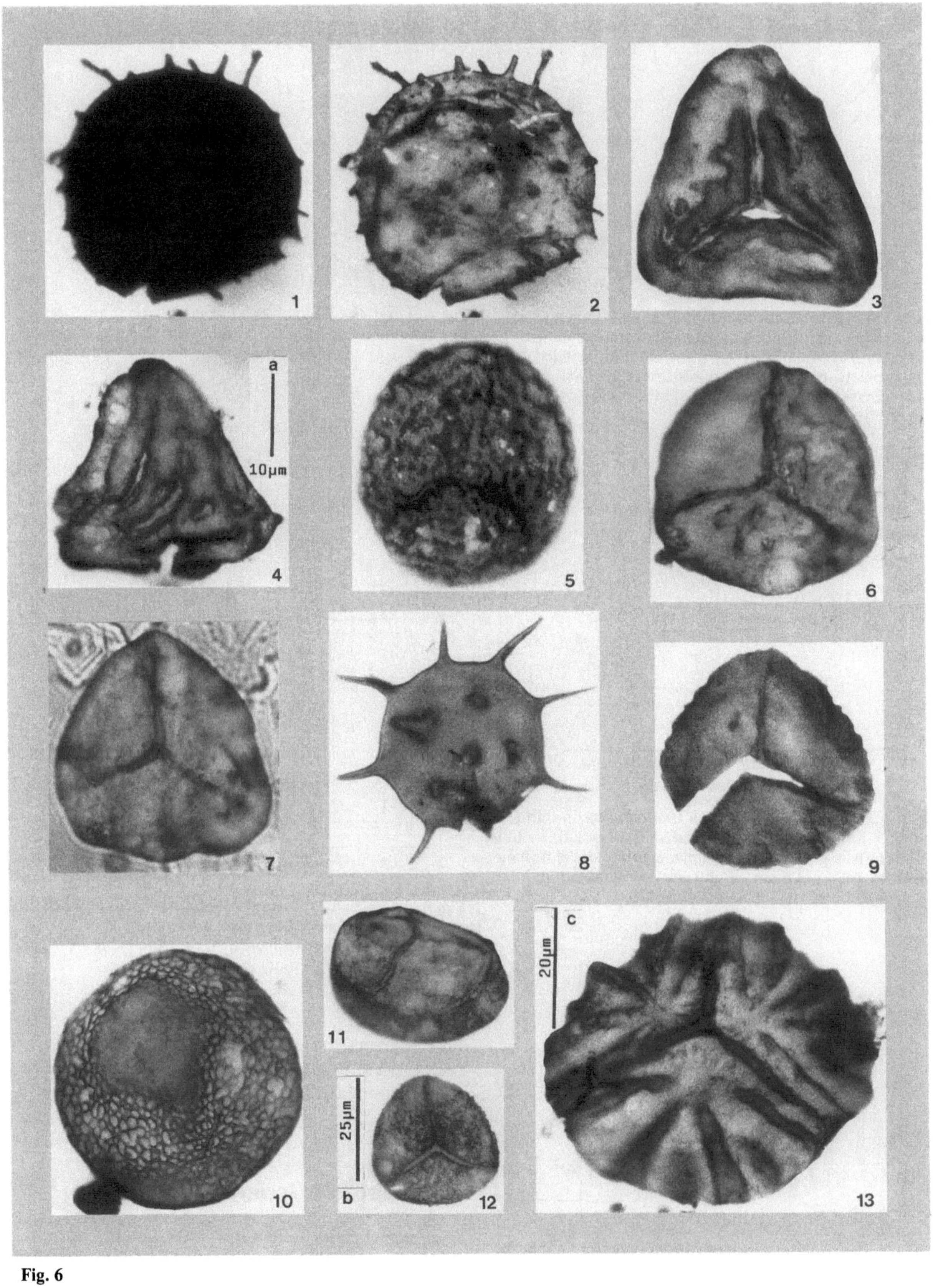

Fig. 6

Fig. 6. 1–2. *Cymatiogalea* cf. *cylindrata* (Rasul) Martin, Villé Unit, northern Vosges Mountains, France (Cambrian). **1** Transmitted light; **2** combined illumination (reflected and transmitted light). **3** *Distaltriangulisporites perplexus* Singh, upper schist cover, Tauern Window, eastern Alps (Upper Proterozoic). **4** *Cicatricosisporites hughesi* Dettmann, upper schist cover, Tauern Window, eastern Alps (Upper Proterozoic). **5** *Synorisporites downtonensis* Richardson and Lister, Moldanubian near Rittsteig, eastern Bavaria (Upper Silurian). **6** *Ambitisporites dilutus* (Hoffmeister) Richardson and Lister, Moldanubian near Rittsteig, eastern Bavaria (Upper Silurian). **7** *Ambitisporites dilutus* (Hoffmeister) Richardson and Lister, Spessart Crystalline Complex, northwestern Bavaria (Upper Silurian). **8** *Micrhystridium vignitispinum* Staplin, northern Phyllite Zone near Kirn, southern Hunsrück, western Germany (Lower Devonian). **9** cf. *Emphanisporites* sp., Moldanubian near Rittsteig, eastern Bavaria (Upper Silurian). **10** *Retispora lepidophyta* (Kedo) Playford 1976; **11** *Aneurospora greggsii* (McGregor) Streel 1974; **12** *Spelaeotriletes* cf. *crustatus* Higgs 1975. **10–12** are from phyllitic slates of the "Albunger Paläozoikum", western Germany (Uppermost Devonian). **13** *Emphanisporites erraticus* (Eisenack) McGregor 1967, Taunus Mountains, western Germany (Lower Devonian). Magnification for **1–9**, see *bar a*; for **10–12**, see *bar b*; for **13**, see *bar c*

be preserved in phyllites of the local Penninic complex. The assemblage is characterized by certain species of the genus *Cicatricosisporites* and other typical representatives of the late Mesozoic (Fig. 6, Nos. 3a–d, 4), (Singh 1971). Dinoflagellates rarely occur. Our findings correct earlier determinations postulating an Early Jurassic age for these members of the Upper Schist Cover.

Phyllitic slates from the "Greywacke-Shale Sequence" of the northern flank of the Pulo do Lobo antiform in southwestern Spain have yielded well-preserved spores and acritarchs. The asemblage is indicative of an Upper Famennian to Lower Tournaisian (Strunian) age. Similar assemblages of uppermost Devonian age were detected in phyllitic slates of the Albunger Paleozoic in western Germany (Fig. 6, Nos. 10–12).

A phyllite zone of greenschist metamorphism forms the southern margin of Taunus and Hunsrück (Rhenish Massif). No fossils had been found there as yet and, accordingly, the stratigraphic position of the suite had remained debatable until recently. We succeeded in isolating a rich and diversified spore assemblage from the phyllites of the Taunus greenschist sequence (Fig. 6, No. 13). The microflora is typical Lower Devonian (Emsian) age.

Phyllites of the Hahnenbach Valley (Hunsrück) yielded similar spores and also contain elements of a marine plankton flora (Fig. 6, No. 8). The Moldanubian complex of the Lam-Bodenmais area in eastern Bavaria contains metamorphic rock series ranging from upper greenschist grade to migmatites. We found spores of early land plants preserved in phyllitic mica schists. The assemblages indicates an Upper Silurian (Pridolian) age (Richardson and McGregor 1986). One of the characteristic forms is *Synorisporites downtonensis* Richardson and Lister 1969 (Fig. 6, No. 5). Our findings correct earlier opinions postulating a Proterozoic age for these parts of the Moldanubian Complex. Trilete spores of early land plants have also been detected in a garnet-bearing micaschist of the Spessart Crystalline Complex of northwestern Bavaria. The characteristic morphologies, such as shapes and sizes of the fossils, as well as the structures and sculptures of their walls are well preserved (Fig. 6, No. 7). The microflora is indicative of Middle to Upper Silurian (mainly Ludlow). *Apiculiretusispora synorea* Richardson and Lister 1969 occurs mainly in the Pridolian, but its first appearance is in the Ludlow. Until now, the age of the Spessart Crystalline Complex was uncertain.

Well-preserved acritarchs have also been detected in phyllites of the Villé Unit in the northern Vosges Mountains of western France (Fig. 6, Nos. 1a–d, 2). The assemblage is indicative of an uppermost Cambrian to earliest Ordovician age (Martin and Dean 1981) and demonstrates, for the first time, a well-defined age for this part of the Villé Unit.

3 Precambrian Occurrences

Proterozoic palynomorphs were detected in the Moldanubian of the Lam-Bodenmais area in eastern Bavaria. Phyllitic micaschists from this area contain characteristic microbiota dominated by *Bavlinella* sp. (Figs. 7, Nos. 3 and 4) and sphaeromorph acritarchs. The assemblage is typical of Late Vendian age. The occurrence is not far from another locality containing Silurian spores which demonstrates the tectonically complex situation in this metamorphic unit.

Phyllites from the Habach Valley, Tauern Window (eastern Alps) have yielded diversified assemblages comprising sphaeromorph and polygonomorph acritarchs as well as "vase-shaped" microfossils (Fig. 7, Nos. 8a–d, 9). The forms are identical to taxa known from the Upper Riphean and Lower Vendian of Scandinavia, the Soviet Union, and North America (Vidal and Knoll 1983). Phyllites from the adjacent Schönbichl Wedge also contain well-preserved acritarchs (Fig. 7, No. 7) and filamentous structures indicative of a Late Proterozoic age.

Low-grade metamorphic phyllites of the Brévenne Unit, northeastern Massif Central, France, host spec-

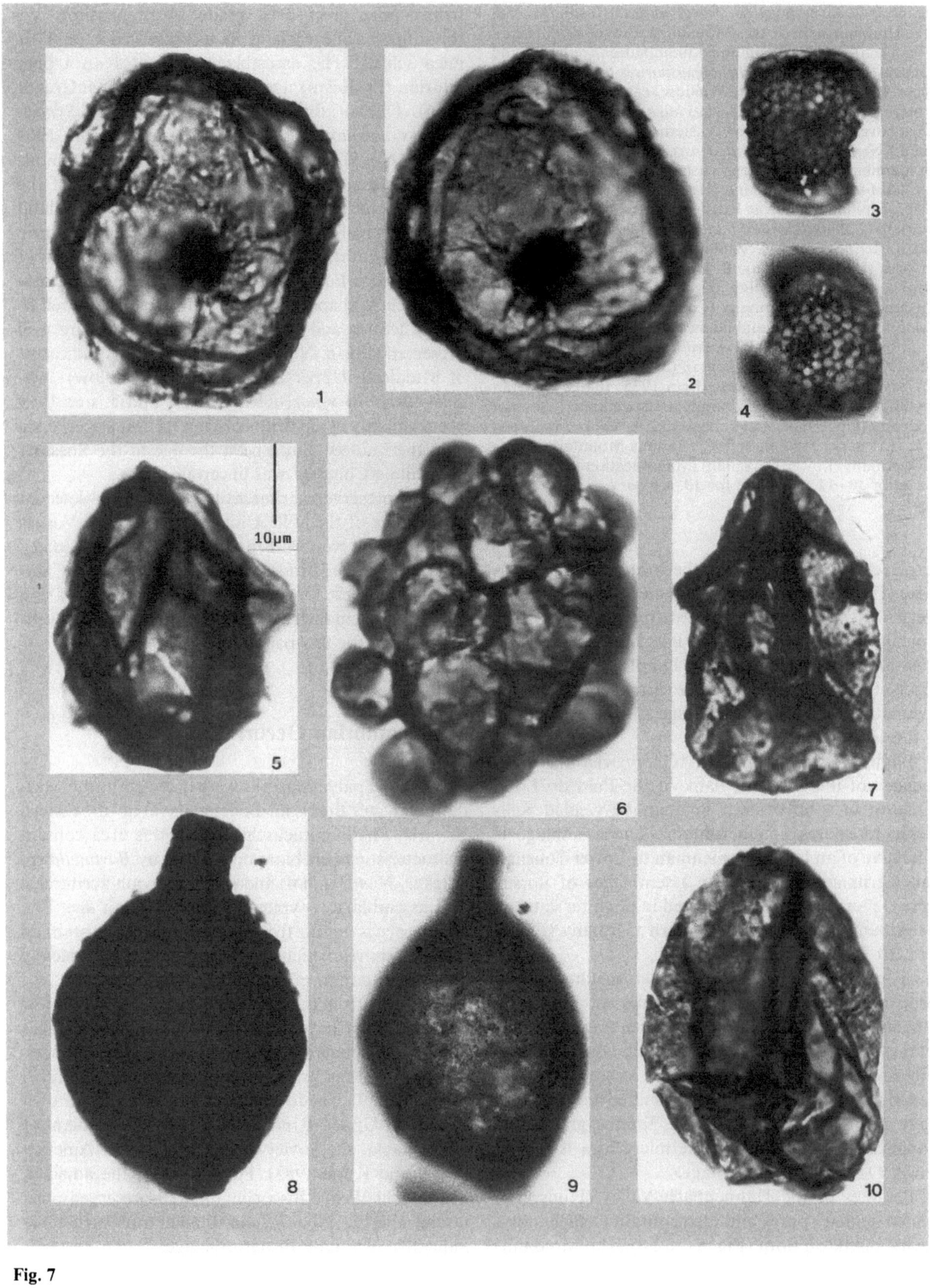

Fig. 7

Figs. 7. 1–2. *Nucellosphaeridium* sp. Timofeev, Striped Peak Formation, Belt Supergroup, NE Idaho (Middle Proterozoic). **3, 4** *Bavlinella faveolata* (Schepeleva) Vidal, Moldanubian near Rittsteig, eastern Bavaria (Vendian). **5, 10** *Leiosphaeridium* sp. Eisenack, Striped Peak Formation, Belt Supergroup, NE Idaho (Middle Proterozoic). **6** Cluster of *Leiosphaeridia* sp., St. Regis Formation, Belt Supergroup, NW Montana (Middle Proterozoic). **7** *Leiosphaeridium* sp. Eisenack, phyllites of the Schönbichl wedge, Tauern Window, eastern Alps (Upper Proterozoic). **8, 9** "Vase-shaped microfossils" *Mellanocyrillium* sp. Bloeser, Habach Formation, eastern Alps, Austria (Upper Proterozoic)

imens of the genus *Protosphaeridium* and other morphologies of Late Proterozoic age. This verifies the existence of Brioverian deposits within a subduction-related tectonic environment in the northern Massif Central.

Microbiota from the ~1.1–1.3 Ga-old Belt Series of Idaho and Montana (USA) are characterized by sphaeromorph acritarchs of simple morphology (Fig. 7, Nos. 1, 2, 5, 6 and 10). The findings suggest that eukaryotic plankton of this time was still in its initial stage of evolution which, according to present evidence, must have started prior to 2 Ga. It was only from about 850 million years ago that the diversification proceeded more rapidly (Sokolov and Ivanovskiy 1985; Hofmann 1987).

Sphaeromorph acritarchs were also found in the 2 Ga-old Gunflint Banded Iron Formation of Ontario (Fig. 8, No. 1). Several species of the acantomorph acritarchs *Eomicrhystridium* are known from there (Fig. 8, Nos. 4 and 7).

The oldest acritarchs hitherto known come from two stratigraphic horizons of the Transvaal system within the time brackets 2.5 and 2.2 Ga ago (Fig. 8, Nos. 6 and 10). Several morphological groups were identified in the assemblages. They correspond to those known from the Krivoi Rog Series in the Ukraine which is approximately the same age or slightly younger (Timofeev 1969, 1973, 1979; Plumb and James 1986).

Microbiota of Archean age are characterized by simple coccoid and filamentous morphologies, most of which are more or less similar to the *Huroniospora* and *Gunflintia* types of the Gunflint Formation (cf. Fig. 8, Nos. 2, 3, 5 and 14–17). The microstructures shown from the Isua Banded Iron Formation (Fig. 8, Nos. 14–16) are from a graphitic quartzite (sample No. GRD 1209 B of the Harvard collection provided by courteousy of H. D. Holland). In thin section under transmitted light, the Isua microstructures appear as black-opaque spheres and filaments (Fig. 9, Nos. 1 and 2). In reflected light, details of preservation become visible (Fig. 9, Nos. 3 and 4). The morphologies closely resemble those known from occurrences of slightly younger age, such as the Warrawoona Group of Australia and the Onverwacht Group of South Africa. The only notable difference is that the Isua sequence has suffered a higher degree of metamorphism that had reached the lower amphibolite facies.

The most abundant micromorphology in the Isua Formation takes the form of sheathed filaments packed together in bundles or layers (Fig. 9, Nos. 1–4). At places, spherical buds originate from individual sheaths (Fig. 9, No. 5). Most cells broken away from the filaments have a cylindrical shape (Fig. 9, Nos. 6 and 9). Laser mass spectra obtained from the specimens show that iron, and some copper, are always associated with the sheaths (Fig. 5b).

Morphologies of similar kind are widely reported from Precambrian and Phanerozoic iron formations (Muir 1978; Robbins 1987). They are generally com-

Fig. 8. 1. *Leiosphaeridia bituminosa* Timofeev, Gunflint Banded Iron Formation, Schreiber locality, Ontario (Lower Proterozoic, ca. 2000 Ma). **2, 3** Spherical microfossils of *Huronispora* type, Gunflint Banded Iron Formation, Schreiber locality, Ontario (Lower Proterozoic, ca. 2000 Ma). **4** *Eomicrhystridium* sp. Deflandre, Gunflint Banded iron Formation, Schreiber locality, Ontario (Lower Proterozoic, ca. 2000 Ma). **5** *Gunflintia minuta* Barghoorn, Gunflint Banded Iron Formation, Schreiber locality, Ontario (Lower Proterozoic, ca. 2000 Ma). **6, 10** *Trachysphaeridium* sp. Timofeev, Black Reef Shale, Hartebeestfontein mine, District Klerksdorp, Transvaal Supergroup (Lower Proterozoic, ca. 2400 Ma). **7** *Eomicrhystridium barghoorni* Deflandre, Gunflint Banded Iron Formation, Schreiber locality, Ontario (Lower Proterozoic, ca. 2000 Ma). **8** Cluster of *Huroniopsora* sp. Barghoorn, Gunflint Banded Iron Formation, Schreiber locality, Ontario (Lower Proterozoic, ca. 2000 Ma). **9** *Huroniospora* sp. Barghoorn, Campbell Group, Ulco locality, Griqualand West Supergroup (Lower Proterozoic, ca. 2300 Ma). **11, 12** *Strictosphaeridium* sp. Timofeev, Campbell Group, Ulco locality, Griqualand West Supergroup (Lower Proterozoic, ca. 2300 Ma). **13** *Protosphaeridium* sp. Timofeev, Black Reef Shale, Hartebeestfontein mine, District Klerksdorp, Transvaal Supergroup (Lower Proterozoic, ca. 2400 Ma). **14–16** Sphaeroidal microstructures, graphitic quartzite, Isua Banded Iron Formation, southwest Greenland (Early Archean, ca. 3800 Ma). **17** Filamentous microstructure, Swartkoppie chert, Onverwacht Group, Barberton Mountain Land (Early Archean, ca. 3400 Ma)

Fig. 9. Microstructures resembling iron bacteria contained in graphitic quartzite of Isua Banded Iron Formation (Early Archean, ca. 3800 Ma). **1–2** Filamentous microstructures from thin sections under transmitted light. **3–4** Filaments in demineralized condition; combined transmitted and reflected light (*g* platelets of graphite). **5, 10, 11** Portions of filaments at higher magnification (TEM); *b* bud. **6–9** Cell-like structures broken away from filaments (TEM). Magnification for **1–4**, see *bar* above No. **1**; for **6**, **7**, **9** see *bar* below No. **7**; for **10, 11**, see *bar* above No. **11**

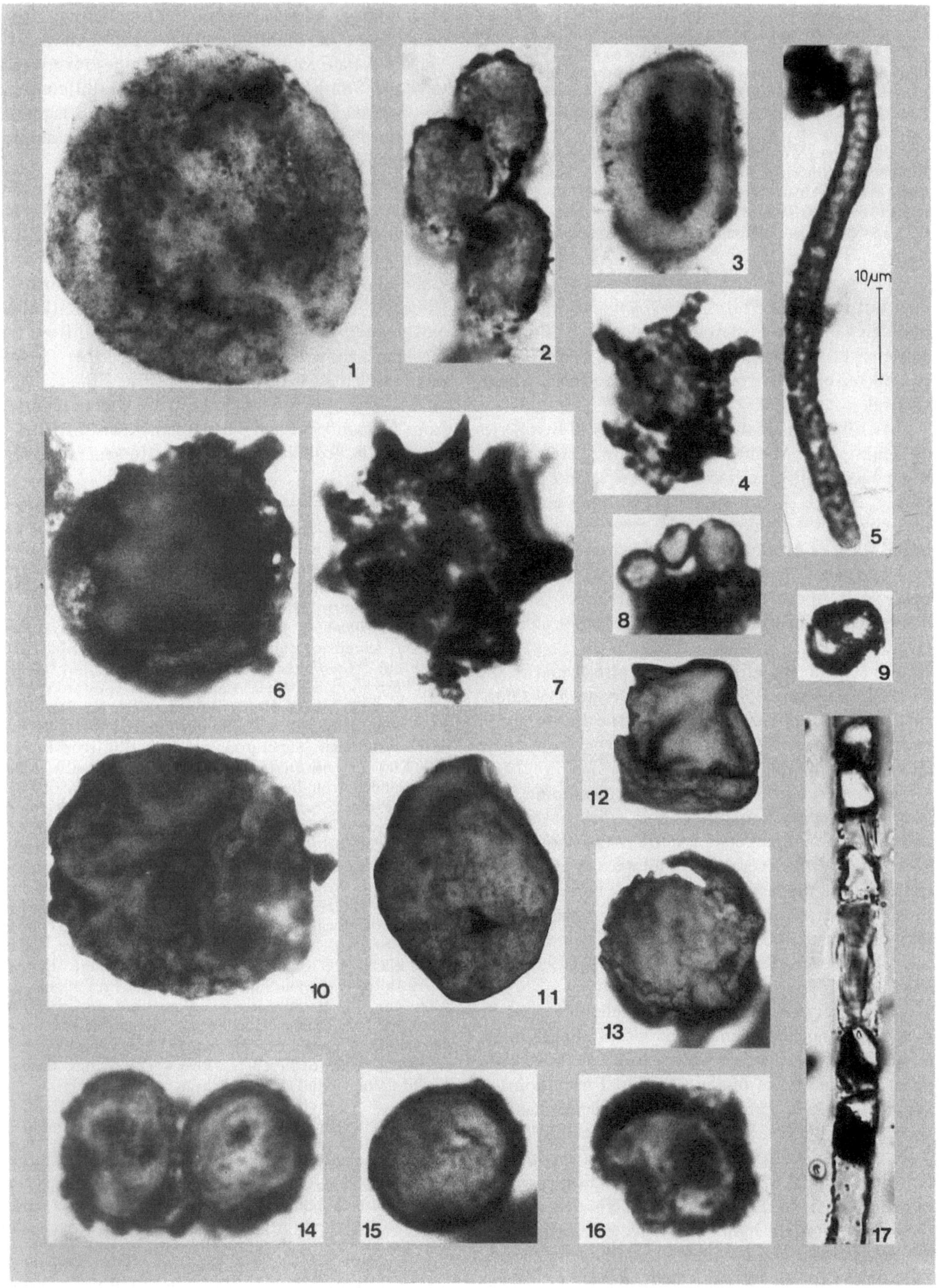

Fig. 8

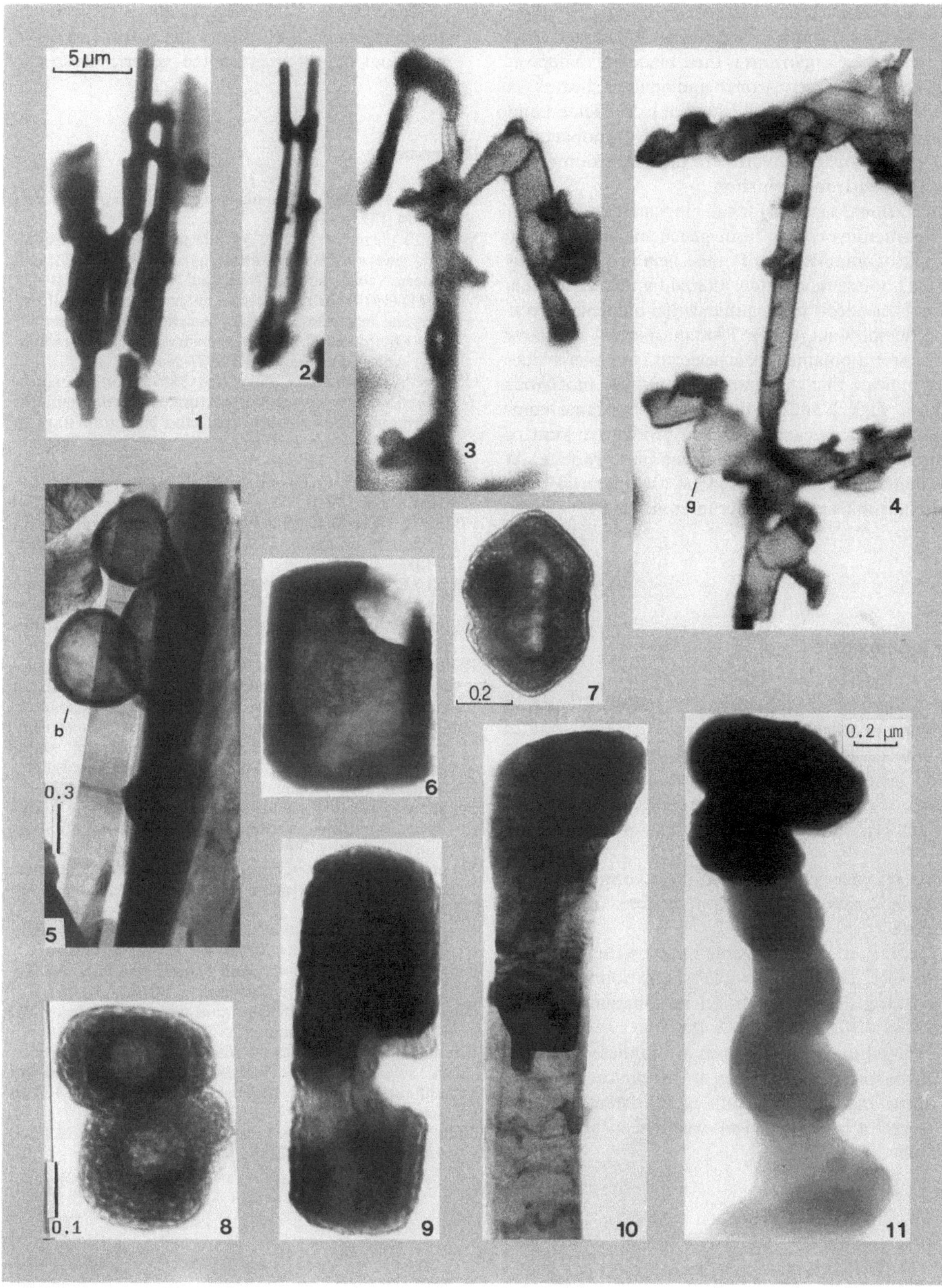

Fig. 9

pared to iron bacteria of the *Sphaerotilus-Leptothrix* or *Siderocapsa* group, respectively. A characteristic feature of these organisms is their tendency to deposit large amounts of ferric iron in and on their sheaths. In summary, the findings indicate that iron bacteria and photosynthetic organisms, especially cyanobacteria, were most probably involved in the deposition of the Isua Banded Iron Formation.

Additional support for the biogenicity of the Isua microstructures comes from geochemical data. The isotopic composition of the Isua carbonaceous matter, though moderately altered by rock metamorphism is decidedly compatible with a biogenic derivation (Schidlowski 1988). Raman spectra and laser mass spectra obtained from selected Isua microstructures indicate that they mostly consist of amorphous carbon (Figs. 2 and 4). Oxygen-rich organic compounds, such as carbohydrates, produce a kind of carbon that cannot be transformed into graphite. At the highest degree of catagenesis, the material turns into amorphic carbon (Oberlin et al. 1980).

4 Conclusion

Our findings show that palynological and biostratigraphic approaches may be applied to the study of metamorphic rocks of greenschist and lower amphibolite facies, if appropriate techniques are applied. These comprise a combined morphological and chemical analysis with respect to the following criteria.

1. Microscopic identification of palynomorphs and related microfossils in thin sections of the host rock.
2. Separation of the biogenic miocrostructures from the rock matrix by suitable maceration techniques allowing their preservation as coherent spherical bodies or filamentous sheaths.
3. Recognition of the presence within these structures of conspicuous quantities of carbonaceous matter whose metamorphic rank is consistent with the degree of metamorphism experienced by the host rock.
4. Recognition of morphological detail in well-preserved specimens that allows the identification of individual microfossils on the generic or species level.

References

Hofmann HJ (1987) Precambrian biostratigraphy. Geosci Can 14:135–154

Martin F, Dean WT (1981) Middle and Upper Cambrian and Lower Ordovician acritarchs from Random Island, eastern Newfoundland. Bull Geol Surv Can 343:1–43

Muir MD (1978) Microenvironments of some modern fossil iron- and manganese-oxidizing bacteria. In: Krumbein WE (ed) Enviromental biogeochemistry and geomicrobiology. Ann Arbor Sci Publ 3:937–944

Oberlin A, Boulmier JL, Villey M (1980) Electron microscopy study of kerogen microtexture. Selected criteria for determining the evolution path and evolution stage of kerogen. In: Durand B (ed) Kerogen. Editions Technip, Paris, pp 191–241

Pflug HD (1987) Chemical fossils in early minerals. Top Curr Chem 139:1–55

Pflug HD, Reitz E (1987) Palynology in metamorphic rocks: indication of early land plants. Naturwissenschaften 74:386–387

Plumb AK, James HL (1986) Subdivision of Precambrian time: recommendations and suggestions by subcommission of Precambrian stratigraphy. Precambrian Res 32:65–92

Richardson JB, McGregor DC (1986) Silurian and Devonian spore zones of the Old Red Sandstone Continent and adjacent regions. Bull Geol Surv Can 364:1–68

Robbins EJ (1987) *Appelella ferrifera*, a possible new iron-coated microfossil in the Isua iron-formation southwestern Greenland. In: Appel PW, LaBerge G (eds) Precambrian iron-formations. Teophrastus, Athens, pp 141–153

Schidlowski M (1988) A 3.800-million-year isotopic record of life from carbon in sedimentary rocks. Nature (London) 333:313–318

Singh C (1971) Lower Cretaceous microfloras of the Peace River area, northwestern Alberta. Bull Res Counc Alb 28:1–542

Sokolov BS, Ivanovskiy AB (eds) (1985) The Vendian system 1: Paleontology. Nauka, Moscow, 212 pp (in Russian)

Timofeev BV (1969) Proterozoic *Sphaeromorphida*. Nauka, Leningrad, 146 pp (in Russian)

Timofeev BV (1973) Precambrian microfossils from the Ukraine. Nauka, Leningrad, 58 pp (in Russian)

Timofeev BV (1979) Microfossils from the Pechenga Series. In: Sokolov BS (ed) Paleontology of the Precambrian and Early Cambrian. Nauka, Leningrad, pp 119–120 (in Russian)

Vidal G, Knoll AH (1983) Proterozoic plankton. Geol Soc Am Mem 161:265–277

5. General Topics

Mineral Deposits and Metallogeny: Indicators of Earth's Evolution

R.W. Hutchinson [1]

[1] Department of Geology and Geological Engineering, Colorado School of Mines, Golden, Colorado 80401, USA

Abstract

Major types of mineral deposits exhibit distinct changes in composition and geological setting from Archean to Recent time. These changes include extinctions of ancient types, appearances of new ones, proliferations, and diversifications, and thus are comparable to those seen in biological evolution. The changes are responses to evolutionary changes in the earth's asthenosphere, lithosphere, hydrosphere, atmosphere and biosphere. All these reflect changing geotectonic environments, and consequently must be considered and explained in conjunction with global tectonic evolution, to which they provide additional, independent guides.

Explanations for the evolutionary changes in the type of deposits are conjectural due to the complexity and imperfect understanding of the processes that form many of them. However, changes in magmatic deposits of Cr, Ni – Cu, and Pt-Group elements may result from changes through time in subcrustal conditions and processes deep within the earth; e.g. changes in core/mantle compositions and boundaries, heat flux, or depth and degree of partial melting. Since about 1 Ga **B.P.**, lithospheric evolution by plate tectonic processes, from initial oceanic crust across continent-margin environments to final, stabilized continental crust, has been paramount and has governed the nature of the hydrothermal systems that form many metalliferous deposits. It has thereby resulted in sequential evolutionary formation of many different types of deposits in these differing supracrustal settings.

An analogous, but temporally much longer and global evolution of lithosphere may be represented by sequential changes from Early to Late Archean greenstone belts, then through latest Archean plutonism and crustal stabilization to the diverse continental crustal tectonic environments of Proterozoic time. This longer term lithospheric evolution also controlled the development of differing types of deposits during the long span of Precambrian time. A remarkable absence of certain earlier types and the appearance and proliferation of certain new ones, or variants, in Mid-Proterozoic time suggests widespread, even global crustal rifting – perhaps due to global expansion.

Atmospheric-hydrospheric oxygenation, probably due to biogenic evolution in Lower Proterozoic time, profoundly altered the nature of ferruginous sedimentary rocks and their contained iron deposits. Primitive algal forms apparently played a role in formation of the great Au deposits of the Witwatersrand, and the appearance of terrestrial organisms in Mid-Paleozoic time created an environment favorable for generation of certain U deposits.

If ore deposits of different types have developed in response to evolutionary changes on the earth, it becomes imperative to explore for them not only in the proper spatial or geological environment but also in the proper time dimension, or age of rock. It would be disastrous to seek trilobites in Cretaceous marine shales or dinosaur bones in Silurian continental clastic sediments. Time, as expressed by geologic age, is of equal importance in mineral exploration.

1 Introduction

Man depends on ore deposits for his supply of essential metals and industrial minerals. Improved knowledge of their descriptive geology and understanding of their origins reveal marked evolutionary changes through time in the unique rocks known as ores, and in the processes that formed them. Ores are products of particular magmatic, sedimentary, and metamorphic processes which are, in turn, controlled by specific tectonic environments. This permits the use of ore deposits and ore-forming processes as additional scientific guides to Earth's tectonic evolution from earliest Archean to Recent times.

M. Schidlowski et al. (Eds.)
Early Organic Evolution: Implications for Mineral and Energy Resources

Understanding of time-dependent changes in ore deposits also adds an important fourth dimension for application in mineral exploration. As stated previously, no paleontologist would seek fossil trilobites in Cretaceous marine shale, nor dinosaur bones in Silurian continental sandstone. The spatial dimensions in both cases are permissive but the temporal ones are prohibitive! Yet analogous examples are revealed by evolutionary changes in mineral deposits. It is equally essential to consider the fourth, or time dimension in exploration for certain types of ores.

As in the evolution of life, that of mineral deposits is unidirectional. Disappearances, i.e., extinctions, of some older types, and sudden appearances of new ones were brought about by poorly understood, but global changes in Earth's geotectonic environments. Rapid proliferation of the new types filled the resulting new environmental niches. Some long-continuing ancient types, and newly appearing ones both underwent diversification and specialization leading toward ever-increasing numbers and varieties of types through time.

This work outlines and discusses evolutionary changes in some important types of ore deposits, and considers the possible significance of these changes to the understanding of Earth's tectonic evolution. Detailed descriptions of the types considered, and discussions of their often controversial genetic processes, however, are beyond the scope of this work and for this information readers are referred to the many cited papers.

2 Mineral Deposits Through Time

2.1 Archean

Archean terranes, spanning time between ca. 3.8 to 2.5 Ga throughout the world, are characterized by four major, and one minor, types of ore deposits. Certain banded iron formations, komatiite-affiliated Ni (and rarely Cr) deposits, greenstone-hosted Au lodes and Cu–Zn-rich massive base metal sulfide ores (Fig. 1) all occur in supracrustal rocks of the greenstone belts which are the focus of mineral exploration in Archean cratons. Although the four major types occur in different lithostratigraphic units within the greenstone successions (Hutchinson 1984), all are fundamental elements of Archean metallogeny (Anhaeusser and Viljoen 1986). All four are commonly, although not always co-regional in important Archean mining districts such as at Timmins, Ontario (Fig. 2). The less important fifth type comprises Li–Cs–Be–Nb–Ta–Sn–Mo-bearing pegmatites which are associated with the Late Archean plutonic granitoids that invade the supracrustal greenstone assemblages.

Algoma-type iron formation (Gross 1966, 1986; Kimberley 1989a, b) is abundant in Archean terranes. It is generally reduced, ferrous predominating over ferric iron, so that siderite is the major mineral in the carbonate facies, pyrite and/or pyrrhotite in the sulfide facies, magnetite rather than hematite in the oxide facies and various amphiboles, chlorites, or serpentines in the silicate facies. All four facies (James 1954) are abundant and rapid facies changes on all scales, both bed by bed across strike and lat-

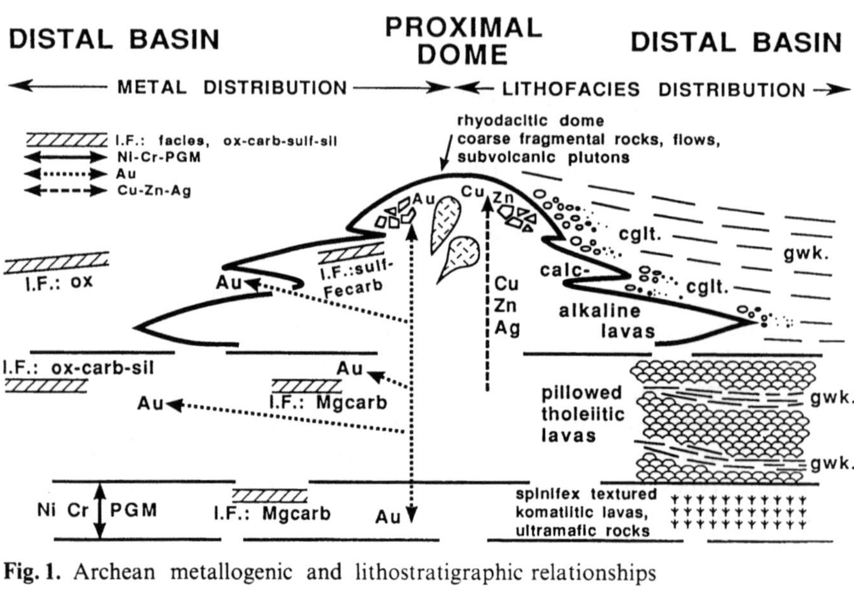

Fig. 1. Archean metallogenic and lithostratigraphic relationships

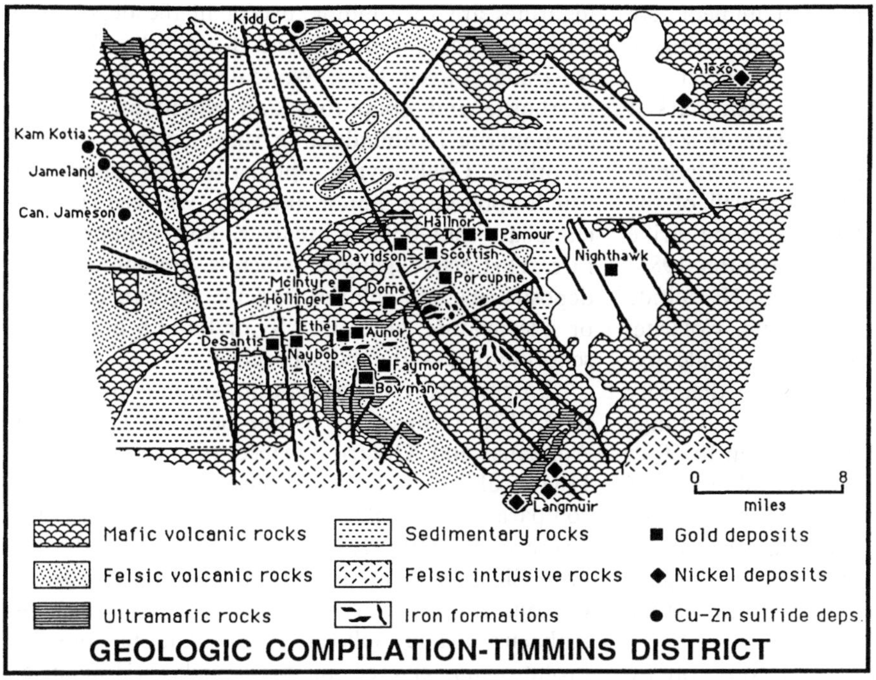

Fig. 2. Geological compilation map showing the distribution within the Timmins District, Ontario, Canada of all four important greenstone-hosted types of Archean ore deposits

erally along strike, are common. Minor Mn substitutes for ferrous iron in minerals of the carbonate and silicate facies, whereas Mn oxides are rare. The primary ores are low in Fe content, containing only 20–30% Fe. Although not widely exploited prior to 1960, they have since been mined and upgraded by benefication to produce premium quality pellets for the steel industry, e.g., at the Adams, Moose Mountain and Sherman Mines in Ontario (Dubuc 1966; Boyum and Hartviksen 1970; Markland 1966). Enrichment by weathering and oxidation produced near-surface ore in these iron formations at Bomvu Ridge in Swaziland (Davies and Urie 1956), and at Koolyanobbing in Western Australia (Broken Hill Proprietary Staff 1975), and primary siderite is sintered to produce high quality blast furnace feed at Wawa Ontario (Goodwin 1962).

These iron formations are closely associated with, and sometimes the host rocks of the Au lodes, massive base metal sulfide, and Kambalda-type Ni deposits (Roberts 1975; Fripp 1976; Friesen et al. 1982). They are up to about 100 m thick, although often thickened in fold hinges, and extend for up to 20 km along strike. They are most widespread, numerous, and largest throughout Archean greenstone belts, with a possible tendency toward Mg-rich ankeritic facies in the stratigraphically lower ultramafic and tholeiitic lavas, Fe-rich sulfidic and sideritic facies in stratigraphically higher felsic volcanic-pyroclastic rocks and magnetite-rich oxide facies in basinal, sediment-hosted examples (Fig. 1; Hutchinson 1984).

Important Ni deposits occur in many Archean greenstone belts, characteristically associated with highly serpentinized, ultramafic, komatiitic lavas which often form their hanging walls, and with tholeiitic basalts on their footwalls. These rocks comprise the stratigraphically lower portions of the belts (Fig. 1). Examples are the many orebodies of the Archean Yilgarn block of Western Australia, particularly those of the Kambalda District, for which this type is named (Gresham and Loftus-Hills 1981; Marston et al. 1981; Gole et al. 1989). Sherlock Bay in the Pilbara Block of Australia, Damba and Shangani in the Zimbabwean Craton and Alexo, Dundonald, Langmuir, Redstone River, and Marbridge in the Superior Structural Province of Canada are all examples in Archean terranes of other continents (Miller and Smith 1975; Williams 1979; Killick 1986; Viljoen et al. 1976; Pyke and Middleton 1971; Coad 1977; Muir and Comba 1979; Buchan and Blowes 1968). The important and unique Cr-rich chromitite lenses of Selukwe, Zimbabwe (Cotterill 1969) are comparable to these Ni ores in their geological setting in an Early Archean greenstone belt.

The Ni orebodies are rich, usually small, but occasionally large lenses of nearly pure pyrrhotite, pyrite, and pentlandite, often with net-textured intercumulate sulfides in the base of an immediately overlying komatiite, and sometimes with stringers of sulfides is underlying basalt. The contact between the komatiite and the basalt is often marked by thin, laminated, pyritic-graphitic slate or schist, albitic chert or cherty iron formation of Algoma-type. In places these form the hanging wall of the massive ore, but elsewhere lie laterally along strike from, and sometimes grade into it (Bavinton 1981; Robinson and Hutchinson 1982). High-Ni sulfide and alloy minerals like millerite, bravoite, violarite, heazlewoodite and awaruite, as well as chalcopyrite, are important accessories. The ores therefore have Ni/Cu ratios that average about 10 and range from about 2 in some massive ores and up to 50 in the net-textured variety. Information about their content of platinum group metals (PGM) is sparse, but the latter are probably present in very minor amounts. These deposits are widespread, largest, most numerous, and thus distinctive, in greenstone belts older than 2.5 Ga.

Greenstone belts in all Archean cratons contain important Au lodes (Hutchinson 1987). These are of widely diverse form on all scales, from district to district, from one orebody to another within districts, from stope to stope in one mine, and even from face to face in one stope. Their many forms, all auriferous, include lenses and layers of stratabound and stratiform pyrite or pyrrhotite, stratiform sericitic-chloritic-pyritic schists, Algoma-type iron formation of all four facies, and discordant veins of several, mutually cross-cutting varieties (Hutchinson and Burlington 1984). Their host rocks and lithostratigraphic settings are equally diverse (Fig. 1). Some lodes occur in komatiitic, basaltic, andesitic and rhyodacitic volcanic rocks, from the stratigraphic base to the top of thick, domical volcanic complexes, some in basinal polymict conglomerate, turbiditic greywacke, and iron formation that flank the complexes, and others near, or rarely in granitoid plutons that invade these rocks (Colvine et al. 1988). However, many important districts and the largest deposits like Hollinger, McIntyre, and Dome in the Timmins District of Ontario (Ferguson et al. 1968) are in the tholeiitic basalts. Other important deposits, like Lupin in the Northwest Territories, occur in metagreywacke and iron formation, whereas smaller deposits occur in the other volcanic and sedimentary rocks, and the least important ones in the plutons.

The lodes have high Au/Ag ratios, averaging about 9, but ranging widely. Electrum, often containing significant Hg, is by far the most common ore mineral, whereas, although widespread, Au–Ag tellurides are very minor, except in the deposits of Kirkland Lake, Ontario and Kalgoorlie, Western Australia (Lovell 1967; Woodall 1975). Pyrite and pyrrhotite and ubiquitous and abundant, whereas arsenopyrite is abundant in some deposits, absent in others. Chalcopyrite, sphalerite, and galena are usually present in minor amounts, whereas stibnite, cinnabar, molybdenite, realgar, and orpiment are rare. Distinctive gangue minerals present in some deposits, include chlorite, sericite, ferroan dolomite or sideroplesite, green Cr-, V- or Ba-rich muscovite, schorlite or dravite and scheelite.

Enigmatically, these deposits are most important in 2.7-Ga Archean greenstone belts, as in Western Australia and Ontario-Quebec (Groves and Batt 1984; Colvine et al. 1988), but less so in otherwise comparable belts older than ca. 3 Ga **B.P.**, as elsewhere in Western Australia, particularly in the Pilbara Block, and in the Kapvaal and Zimbabwean Cratons of southern Africa (Foster and Wilson 1984; Anhaeusser and Viljoen 1986). Another possible evolutionary change in these lodes (Hutchinson 1988; Padgham 1988) is a trend through time toward decreasing amounts of host volcanic rocks, but proportionately increasing turbiditic sedimentary host rocks. Although both are present, volcanic-hosted lodes are more numerous and important in Archean greenstone belts. This may constitute a change through time from predominantly greenstone-hosted to predominantly turbiditic greywacke-hosted varieties of Au lodes (Keppie et al. 1986; Padgham 1986), which are similar in other characteristics.

The largest massive base metal sulfide deposits occur in rhyodacitic rocks that form the stratigraphically higher parts of domical volcanic complexes in Archean greenstone belts (Fig. 1). These orebodies are often overlain by thinly bedded, variably siliceous tuff, or graphitic sedimentary rock that is pyritic near the ore, and sometimes by Algoma-type iron formation. The lens of massive ore is commonly underlain by a pipe-like zone of stringer and disseminated sulfides in highly altered, siliceous-chloritic-sericitic rock. The ores are rich in Zn and Cu, but Pb content is characteristically very low and most deposits contain significant amounts of Ag and Au (Hutchinson 1990). Fe- and Cd-bearing sphalerite, chalcopyrite, and galena are the only important ore minerals, but many minor ones may contain recoverable amounts of Sb, Bi, As, Sn, Co, Ni, Ga, Ge, Te, etc. Pyrite and/or pyrrhotite are always abundant, whereas arsenopyrite is only occasionally present. Their low Pb content identifies these ores as primitive precursors of the Pb-rich, but otherwise similar polymetallic type (Hutchinson 1973, 1980).

Rare element pegmatites accompanied the regional granitic plutonism that affected all Archean greenstone belts during the major orogeny which closed Archean time (Siroonian et al. 1959; Cerny and Trueman 1978; Correia Neves et al. 1983; Meintzer et al. 1984; Robb and Robb 1986; Partington 1990). Pegmatites within the intrusions are usually barren, whereas those nearby in flanking amphibolitized greenstones may contain minerals of Li, Cs, Be, Nb, Ta, Sn, and Mo (Gevers 1936; Hutchinson 1955). Although weakly radioactive, these pegmatites have nowhere been rich enough to be mined for U – Th.

A negative, but important aspect of Archean mineral deposits is the absence of those of certain elements and types. There is no significant production of Pb, U, Th, Sn, W, Nb, rare-earth elements (REE), Mo, or diamonds from Archean terranes throughout the world. Although minor Pb and Sn are recovered from a few massive base metal sulfide deposits, and minor Sn rarely from the pegmatites, the important types of deposits that produce all these elements and appear in younger terranes are absent, or vestigial in Archean ones. Thus, although porphyry-type Cu mineralization is known (Davies and Luhta 1978; Colvine and Sutherland 1979), it is insignificant, both economically and when compared to the great porphyry Cu deposits in much younger rocks. Although known in the Pilbara of Western Australia, the Pongola of the Eastern Transvaal and Hemlo, Ontario, stratiform bedded barite deposits are rare. Also absent are many important types of Fe, Ni, Au, Ag, Cu, Pb, Zn, and Mo deposits that appear only in younger strata. Syenitic-carbonatitic and kimberlitic intrusions are very rare, although the Pooh Bah Lake complex in western Ontario and the Premier kimberlite pipe in the Transvaal are exceptions. The latter may be the source of rare diamonds reported in the Witwatersrand conglomerates. Important types of Archean mineral deposits are therefore relatively few in number.

2.2 Lower Proterozoic

Some important ore deposits of Lower Proterozoic terranes between ca. 2.5 – 1.8 Ga **B.P.** are similar to those of Archean cratons, but others exhibit significant differences. Greenstone belts of this age lack abundant komatiitic lavas, but otherwise exhibit no marked differences from Archean ones (Franklin and Thorpe 1982). Thus, Algoma-type iron formation is present, perhaps more manganiferous than the Archean examples. Also present are volcanogenic massive base metal sulfide deposits that resemble their Archean antecedents but occur in more continental crustal host rocks, contain more Pb and higher proportions of Ag to Au (Hutchinson 1973, 1980, 1990). These represent the first appearance of the typical polymetallic type of these ores, as in the Flin Flon region of northern Manitoba (Byers et al. 1965; Sangster 1972), the Bagdad, Prescott and Jerome districts of Arizona (Baker and Clayton 1968; Gilmour and Still 1968; Anderson and Nash 1972) and the Skellefte and Bergslagen districts of Sweden (Frietsch et al. 1979; Hedstrom et al. 1989). Rare element pegmatites are again related to granitic intrusions of the major orogeny that occurred ca. 1.7 Ga **B.P.** throughout most Precambrian shields and marks the boundary between Lower and Mid-Proterozoic time. Some of these pegmatites and peralkaline intrusions are sufficiently enriched in U – Th to be profitably mineable (Evoy 1986).

Significant changes occur, however, in Lower Proterozoic Fe, Ni, Cr, and Au deposits. The very important Superior type of iron formation, which provides most of the world's direct-shipping iron ore containing 50 – 65% Fe, appears nearly synchronously, ca. 2.3 – 2.1 Ga **B.P.**, on all continents, throughout vast shallow cratonic basins and in great thicknesses (Dorr 1963, 1973; G. A. Gross 1968; Marsden 1968; Tolbert et al. 1971; Bayley and James 1973; Alexandrov 1973; Trendall 1973; Van Deventer et al. 1986; Van Schalkwyk and Beukes 1986). This represents a global metallogenic epoch. These iron formations are much more extensive, thicker, and more oxidized than those of Algoma-type. Hematite, sometimes with manganese oxides, predominates over magnetite in the oxide facies, and reduced carbonate and sulfide facies are less abundant. Facies distribution, particularly oxide, is widespread and persistent; changes are less numerous and abrupt than in the Algoma-type. Unlike the latter, the Superior type is absent from both older and younger successions, thus it is narrowly restricted in its time distribution.

Komatiite-affiliated Ni deposits are rare, perhaps absent in Lower Proterozoic terranes, thereafter disappearing from the geologic record. The large Thompson, and other orebodies of northern Manitoba (Peredery et al. 1982) are of doubtful age because they lie along the boundary between the Superior and Churchill Structural Provinces of the Canadian Shield where high-rank regional metamorphism ca. 1.8 Ga **B.P.** has affected all rocks and ores. Their high Ni/Cu ratios and affiliated komatiitic rocks and iron formations suggest Archean age. Other, apparently Lower Proterozoic examples are the deposits of the Cape Smith belt (Giovenazzo et al. 1989) in northeastern Quebec, which have lower Ni/Cu ratios than most Archean examples.

The rarity of komatiite-affiliated Ni deposits in Lower Proterozoic terranes is compensated by the first appearance and proliferation of important magmatic Ni–Cu-PGM ores in layered, differentiated mafic intrusions. The oldest of these, the Stillwater, Montana and Lac des Isles, western Ontario Complexes are, in fact, Late Archean (Jackson 1969; Sutcliffe et al. 1989). But the important Great Dyke, Bushveld, and Sudbury intrusions are 2.4, 2.1 and 1.8 Ga in age, respectively (Wilson et al. 1989; Worst 1960; Vermaak and von Gruenewaldt 1986; Hulbert and von Gruenewaldt 1985; Viljoen and Hieber 1986; Kinloch 1982; Naldrett 1984). These deposits differ markedly from Archean ones. Unlike the latter, their host rocks are well differentiated from ultramafic to anorthositic/granophyric, but on average are mafic rather than ultramafic, and relatively fresh, not extensively serpentinized or altered. The ores contain more Cu, have lower Ni/Cu ratios which average near unity, and contain significant amounts of PGM (Naldrett 1981). Their host rocks are of continental crustal rather than eugeosynclinal-volcanic affiliation. With the notable exception of Sudbury, these intrusions, especially the Bushveld, contain the world's great layered chromitite deposits (Cameron 1980; Ireland 1986). These, too, differ markedly, particularly in much lower Cr/Fe ratios but also in their geological settings, from the Archean lenticular chromitite bodies at Selukwe, Zimbabwe.

In parts of west Africa and northeastern South America greenstone belts are Lower Proterozoic in absolute age (Black 1980; Gibbs and Barron 1983; Davison et al. 1988) but apparently older than the first major orogeny, and contain important Au lodes as in the Ashanti and Minas Gerais districts of Ghana and Brazil (Kesse 1984; Huot et al. 1986; Ladeira 1988; Leube et al. 1990). But elsewhere in the world greenstone-hosted gold lodes that are so important in Archean belts are present, but much fewer, smaller and rarer, in otherwise comparable greenstone belts of Lower Proterozoic age (Hutchinson 1987). The huge Homestake deposit in South Dakota, variously dated as Late Archean and Early Proterozoic, may be a significant exception to this generalization (Rye and Rye 1974; Rye et al. 1974; DeWitt et al. 1986). However, like the Thompson Ni deposit, its age is obscured because it, too, lies near a major tectonic discontinuity, the Wyoming Lineament (Condie 1982), which separates older Archean rocks to the northwest from rocks that were highly metamorphosed ca. 1.7–1.6 Ga **B.P.**

Au–U–Th paleoplacers (W.H. Gross 1968; Sestini 1973; Robertson 1976; Pretorius 1986) compensate for the rarity of these lodes. Although their appearance is globally diachronous (Fig. 3), taking place following major orogeny ca. 3.1 to 2.8 Ga **B.P.** in India and South Africa but ca. 2.4 to 2 Ga **B.P.** elsewhere (Hutchinson and Viljoen 1988), paleoplacers are characteristic of Lower Proterozoic terranes on most continents. These ores occur distinctively in fluviodeltaic, oligomictic quartz pebble conglomerate with abundant detrital pyrite and many other heavy minerals. They are geochemically similar to the greenstone-hosted lodes and consequently have been attributed to weathering, erosion, and re-

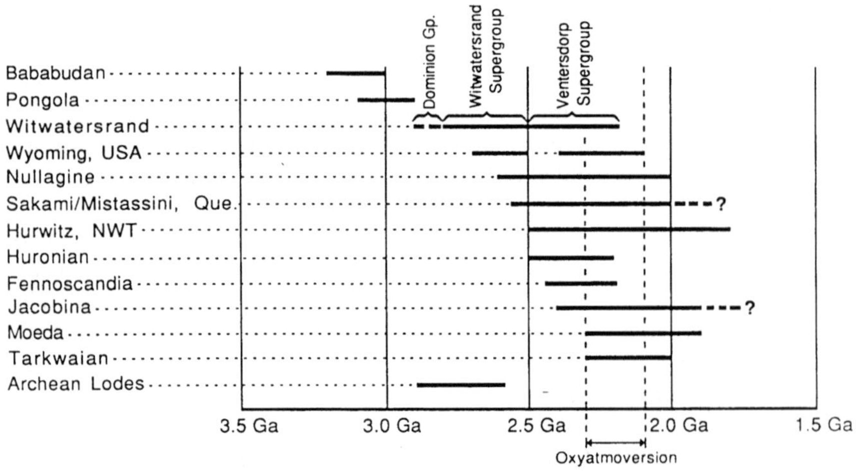

Fig. 3. Diachronous ages of various paleoplacer sequences and deposits around the Earth

deposition of detritus from these lodes, from weathered paleosols or from altered granites in Archean terrane flanking fault-bounded Proterozoic basins (Pretorius 1976; Reimer 1984, 1985; Robb and Meyer 1990). These interpretations have recently been questioned, and an endogenous rather than exogenous source suggested for the immense and unique concentration of Au in the Witwatersrand deposits (Hutchinson and Viljoen 1988).

Rare, Ag–Co–Ni-rich epithermal, polymetallic precious metal veins appear first in Lower Proterozoic terrane, as at Cobalt, Ontario (Berry and Jambor 1971; Andrews 1986; Nichols 1988). But many other types of deposits that are important in younger rocks remain absent, including porphyritic intrusion-related Cu–Mo and sedimentary Cu ores.

Lower Proterozoic mineral deposits thus include an increased diversity of important types. To those of Archean time, some of which continue but others of which become rare, are added new variants or types of Fe, Ni, Cr, Au–U–Th and base metal deposits. Yet evolutionary similarities are recognizable between some of the older and newer types.

2.3 Mid-Proterozoic

Mid-Proterozoic mineral deposits from ca. 1.7 to 1 Ga **B.P.**, are markedly different from those of older time and reflect major evolutionary changes in the types of deposits and their generative environments. Most significant is the disappearance or rarity of all volcanic-hosted Ni, Au–Ag, and massive Zn–Cu sulfide deposits of the older greenstone belts. The last of these gives way to an important sediment-hosted evolutionary variant, and new types of deposits of other metals appear.

Thus, Algoma-type iron formation, although highly metamorphosed, is associated with the ores of Broken Hill, N.S.W. (Johnson and Klingner 1975; Stanton and Vaughan 1979; Stanton 1989) and occurs nearby at The Pinnacles, whereas the Superior type disappears altogether, as do komatiite-affiliated Ni deposits. However, magmatic Ni–Cu-PGM and Cr ores occur in differentiated mafic intrusions like the Duluth Complex and others of western Lake Superior, and the Muskox of the Northwest Territories, which were emplaced late in this span of time ca. 1.2–1 Ga **B.P.** (Faure 1964; Mainwaring and Naldrett 1977; Irvine and Smith 1969). Both paleoplacers and greenstone-greywacke-hosted lodes virtually disappear and Au ores of any kind are rare. The Tennant Creek deposits in Australia (Goulevitch 1975; Large 1975) are a possible rare exception, although they are not geologically comparable, particularly in their high content of magnetite, to the lodes. Epithermal vein deposits remain rare, and like the magmatic ores, are related to the mafic intrusions as at Silver Islet and others near Thunder Bay, Ontario, Bathurst Inlet, Northwest Territories and Königsberg, Norway (Miller 1913; Neumann 1944).

The appearance and proliferation of the important sediment-hosted Pb–Zn–Ag-rich type represent a major evolutionary diversification within the family of massive base metal sulfide deposits. Two varieties apparently represent opposite end members of a spectrum of sedimentary depositional environments; one deep-water, fine clastic, trough or trench, and often turbiditic sedimentation, and the other shallow-water, shelf or platformal, coarse clastic or carbonate sedimentation (Hutchinson 1980). Those in the fine clastic strata predominate in Mid-Proterozoic sequences, and have higher Pb/Zn ratios and Ag contents than the carbonate-hosted ones. With very few exceptions, e.g. Mt. Isa in Queensland – where the Race Course orebodies (Mathias and Clark 1975) are of the clastic sediment-hosted variety but where important Cu orebodies are also present (Swager 1985) – both varieties lack the Cu and Au of older volcanic-hosted types. Mississippi Valley Type (MVT) Pb–Zn deposits, which have many similarities to the carbonate-hosted massive type, also appear first in Mid-Proterozoic carbonates of the Transvaal System in South Africa (Clay 1986), but are not yet exploited.

Another distinctive type to appear in Mid-Proterozoic time was unconformity-related U deposits like those of Australia's Northern Territory and northern Saskatchewan (Eupene et al. 1975; Rowntree and Mosher 1975; Evans 1986). Like their paleoplacer predecessors, they are associated with coarse-clastic sedimentary rocks at a major unconformity. However, they differ from the paleoplacers in their high U/Th ratios, in their time-stratigraphic positions at the next younger (Lower Proterozoic/Mid-Proterozoic) unconformity, in their locations just below, at, or just above the latter, and in their variable host rocks which may belong to the underlying, deformed basement or to the unconformably overlying clastic sedimentary succession.

Alkaline syenite-carbonatite complexes containing important Nb, REE, and phosphate deposits become numerous and important, and in Canada appear to be of earliest and latest Mid-Proterozoic ages; 1.9–1.5 Ga **B.P.** and 1.2–1 Ga **B.P.**, respectively (Deans 1966; Bell and Blenkinsop 1987). Haib, the oldest, although as yet unexploited large porphyry-type Cu deposit (Minnitt 1986), occurs in the Namaqua Metamorphic Belt of southern Namibia in rocks that were metamoprhosed late in Mid-Proterozoic time, and the oldest important granite-affiliated Sn deposits are associated with S-type "tin" granites of this age in northwestern Brazil.

2.4 Upper Proterozoic-Phanerozoic

Many tectonic-orogenic belts of various, Late Precambrian and Phanerozoic ages from ca. 1.1 Ga **B.P.** to Recent time contain similar types of mineral deposits which appear, within each belt, in a similar time sequence (McCartney and Potter 1962; Bilibin 1968). Nevertheless, some marked evolutionary changes in types of deposits also occur within this time span. Despite these changes, their differing individual ages, and multicyclic nature from one belt to another, the important types of Upper Proterozoic-Phanerozoic mineral deposits can be considered collectively to comprise a single group characteristic of this time span. The types are more numerous and diverse than those of earlier time due to reappearances of older types, appearances of new ones, and proliferations or diversifications of still others. The cyclicity results from the many, localized tectonic cycles of differing ages (Taconic-Caledonide, Acadian-Hercynian, Nevadan, Alpine, Laramide, etc.) that characterized this time span at different locations around the earth, each undergoing a similar sequence of evolutionary stages.

Volcanic-hosted massive base metal sulfide deposits, greenstone-hosted Au lodes, and Algoma-type iron formations that were important in Archean and Lower Proterozoic greenstone belts, but absent in Mid-Proterozoic sequences, all reappear with some new evolutionary diversifications. Two new varieties of volcanic-affiliated massive sulfides, the ophiolitic, tholeiitic, basalt-hosted cupreous pyrite, and amphibolite-metagreywacke-hosted Besshi varieties (Hutchinson 1980), appear and represent a further diversification of the older, primitive, and polymetallic types of these deposits. The Au lodes reappear in eugeosynclinal volcanic island arc successions like the Jurassic Foothills-Mother Lode Belt of California and the Alaskan Panhandle. However, they are increasingly hosted in turbiditic greywacke or shale (Keppie et al. 1986), as in the Carolina Slate Belt of the eastern United States, the Meguma rocks of Nova Scotia in Canada, and the Ballarat-Bendigo field of New South Wales-Victoria in Australia, rather than in volcanic rocks. The Algoma-type iron formations are much less common than in the Precambrian greenstone belts, but are again often associated with massive base metal sulfide deposits (Luff 1977). Magmatic Ni–Cu-PGM deposits also reoccur in Phanerozoic mafic intrusions of Noril'sk in the USSR, Insizwa, South Africa, and Skaergaard, Greenland (Tarasov 1970; Naldrett and Macdonald 1980; Maske and Cawthorn 1986; McBirney 1989). Unlike their Precambrian predecessors, however, these intrusions appear to be Cr-deficient, lacking the stratiform chromitite layers of the former.

MVT Pb–Zn ores, porphyry-type Cu–Mo deposits, and epithermal polymetallic Ag–Au–Pb–Zn–Cu veins that were absent, or only vestigial in earlier successions, become important and numerous. And each of these includes diverse varieties which, at least in the porphyry and epithermal deposits, are apparently related to changes through time in tectono-magmatic environments (Lipman 1981).

Thus, MVT deposits contain varying proportions of Pb and Zn, with very little Ag. Moreover, they are hosted in a wide diversity of stratigraphic traps including pinchouts, reefs, karsts, and diapirs. Rare and unimportant in older successions, these ores are present but not mined in Late Precambrian rocks of the Northwest Territories (Hewton 1982). Subsequently, they become very large, widespread, and common in Cambro-Ordovician carbonates of the southeast Missouri, east Tennessee, and Wisconsin Illinois districts, and again in Devono-Mississippian carbonates of the Tri-State, Illinois-Kentucky, and Pine Point districts, but are rare in intervening Silurian strata (Crawford and Hoagland 1968; Sangster 1976; Kyle 1981; Wu and Beales 1981). Very similar and important deposits occur in Cambrian limestone/dolomite of southwestern Sardinia (Zuffardi 1989). Also important in Triassic carbonates in the Silesian and Alpine districts of Europe (Gruszczyk 1967; Holl and Maucher 1976), they virtually disappear again in younger strata.

Porphyry-related Cu deposits, rare with Late Paleozoic intrusions, thereafter proliferate into many, large and widespread deposits, particularly in the Mesozoic-Tertiary circum-Pacific and Tethyan belts. Two distinct varieties are important. A primitive variety, characteristically developed during early stages of continent-margin, island-arc evolution (Gustafson and Titley 1978; Hutchinson 1980), is relatively rich in Au but low in Ag and Mo. It is associated with plutons of intermediate composition that have extensive propylitic rather than potassic alteration and intrude comagmatic volcanosedimentary successions. A more mature, later type (Titley and Hicks 1966), formed within the continent during inward migration of the magmatic arc (Lipman 1981), is richer in Ag and Mo and associated with more silici-alkalic plutons with more extensive potassic alteration that invade virtually all older rocks of the continental crust. A final variant of these are Mo-rich porphyry deposits (Wallace et al. 1968, 1978), with minor W and Nb, but lacking significant Cu. These are farthest from the continental margin and associated with the most silici-alkalic and youngest plutons.

Epithermal veins have widely varying Au/Ag ratios and contents of base metals. Most are very young and associated with intracontinental volcanic or subvolcanic intrusive rocks, particularly felsic ones that exhibit the sequential stages and repetitions of eruptive violence, caldera collapse, and resurgence (Hayba et al. 1985). Also intracontinental, and ranging from Early Paleozoic to Recent in age, extrusive to intrusive alkaline syenite-carbonatite complexes with Nb, REE, and phosphate deposits are present in Canada, but more important in southern and east Africa (Verwoerd 1986). Here also, diamond-rich kimberlites of Mesozoic to Tertiary age are most numerous and important.

Distinctive, sediment-hosted Cu deposits, sometimes with associated basalts and absent from earlier sequences appear first, and almost synchronously at ca. 1 Ga **B.P.**, in the Upper Proterozoic rocks of central Africa, South Australia, northern Michigan, Northwest and Yukon Territories of Canada, eastern Siberia and southern Brazil (Unrug 1988; Fleischer et al. 1976; Lambert et al. 1984; Brown 1971; Brown and Chartrand 1983; Ruelle 1982; Teixeira and Gonzalez 1988; Argall 1976). A second important proliferation of this type occurs in Carboniferous to Cretaceous time in the Bolivian Andes, eastern North America, Texas-Oklahoma, and central Europe (Ljunggren and Meyer 1964; Schenk 1969; Hagni 1988; Vaughan et al. 1989). These important deposits are of two apparently closely associated varieties, one in reduced, black shaley sedimentary strata, the other in oxidized, coarser clastic red beds (Haynes and Bloom 1987a, b; Hoy and Ohmoto 1989).

The Paleozoic, Clinton, or "red ore" (Simpson and Grey 1968) and Mesozoic, Minette, or "yellow ore" varieties of ironstones, make their first appearances, and represent a further diversification of ferruginous sedimentary rocks (Hutchinson 1981). These are smaller, in both strike extent and thickness, than their Archean, and far more so than their Lower Proterozoic iron formation antecedents. They have higher phosphate and alumina contents, and occur in both clastic and carbonate sedimentary rocks of marine littoral or fluviodeltaic facies.

Late Precambrian pegmatites and aplites are differentiates from granites of a third cycle of orogenic partial melting of older sialic crust. Consequently, they contain more U–Th than their older, comparable antecedents and have been more extensively mined (Hewitt 1967; Berning 1986). Early Paleozoic marine black shales are significantly uraniferous (Bates and Strahl 1958; Bell 1978; Andersson et al. 1983) but are mined only in Sweden. Continental sandstone-hosted U deposits (Adler 1974; Rackley 1976) also appear, first in Upper Plaeozoic rocks, and subsequently undergo diversification into organic debris, tabular, roll-front, and stacked subtypes (Hutchinson and Blackwell 1984), many of them related to buried organic matter in Mesozoic-Tertiary sedimentary rocks of continental facies.

As discussed above, time-dependent evolutionary changes are seen in many ore deposits but are particularly well exemplified by those in five major types: ferruginous sedimentary rocks, nickel deposits, gold deposits, massive and related types of base metal sulfides and uranium ores. Evolutionary changes in these are illustrated and summarized diagramatically in Figs. 4 to 8, respectively.

2.5 Ore-Forming Processes

The genetic processes of all these types of deposits are well understood and widely accepted for some, like the massive sulfides (Hutchinson 1973; Sangster and Scott 1976; Franklin et al. 1981; Hannington et al. 1986), but less so and highly controversial for others, like the greenstone-hosted Au lodes (Hutchinson and Burlington 1984; Groves and Phillips 1987). The processes are complex and in some cases multiple, two or more having been superimposed on one another at different times to form some deposits. Their roles and significance in the evolutionary changes are therefor less certain than are the above discriptive metallogenic characteristics. Nevertheless, however well understood, ore-forming processes have, through time, clearly formed very different types of deposits which, despite their differences, are equally clearly related to one another in an evolutionary sequence. Both changing genetic processes and their resulting ores are responses to Earth's evolutionary geotectonics and to resulting changes in deep-plutonic, subcrustal, supracrustal, hydrospheric, atmospheric,

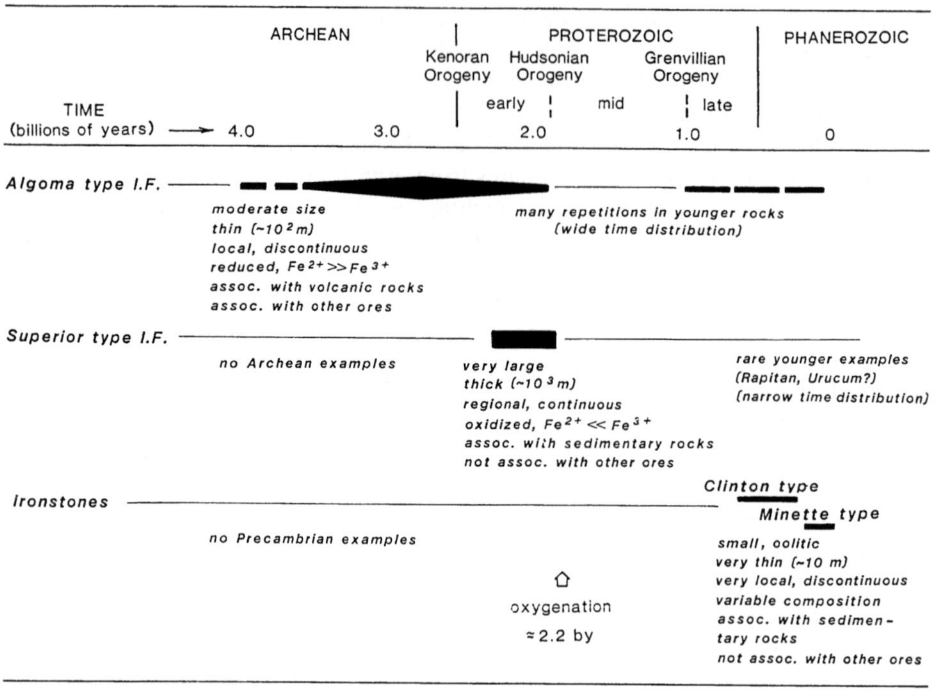

Fig. 4. Evolutionary changes in ferruginous sedimentary rocks

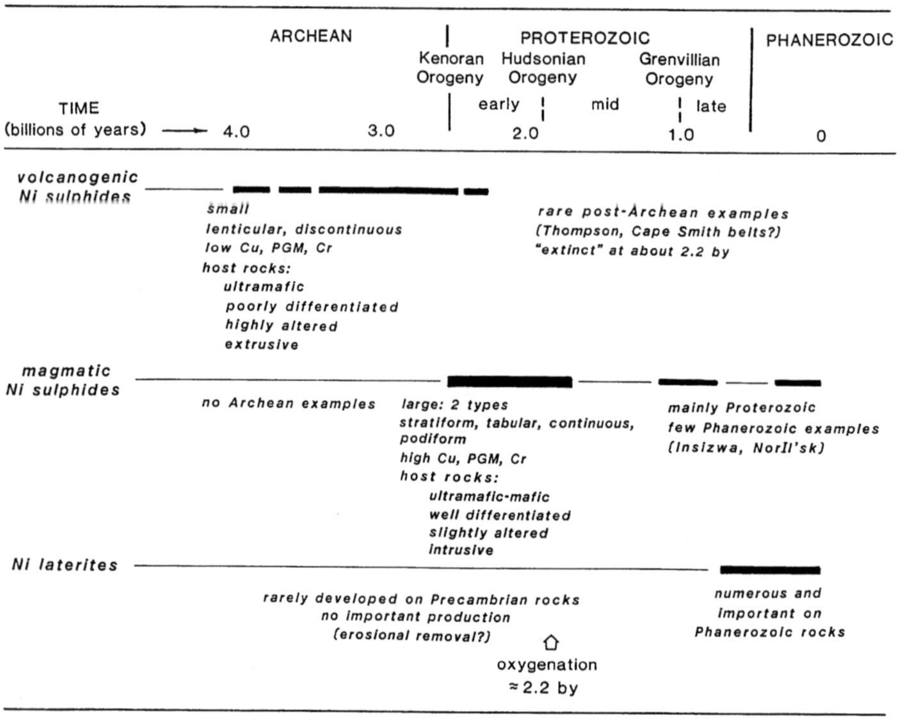

Fig. 5. Evolutionary changes in nickel deposits

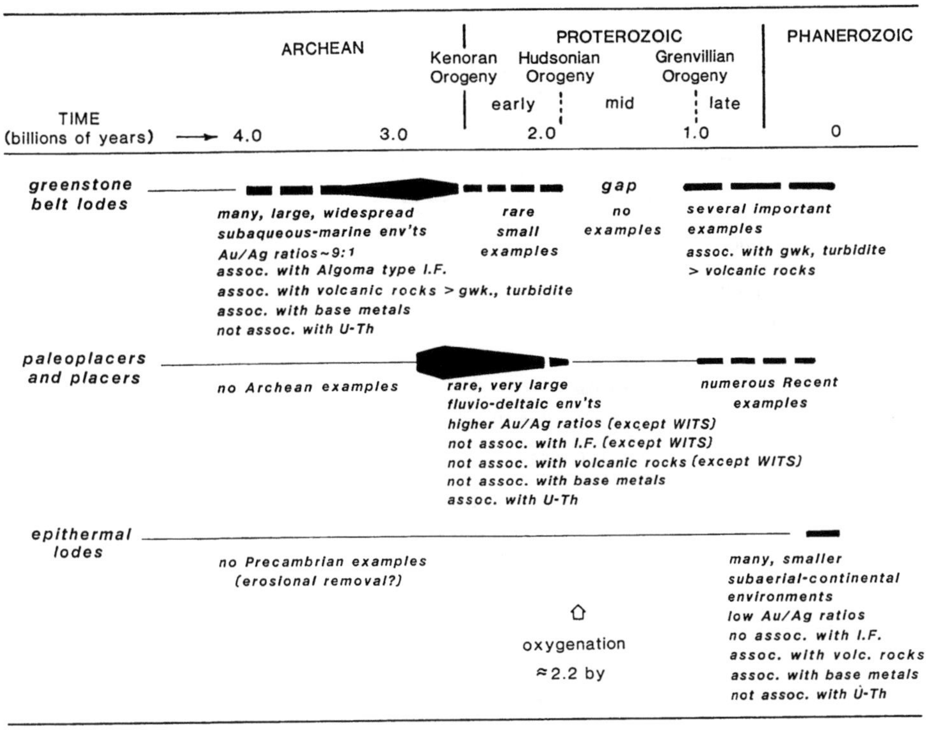

Fig. 6. Evolutionary changes in gold deposits

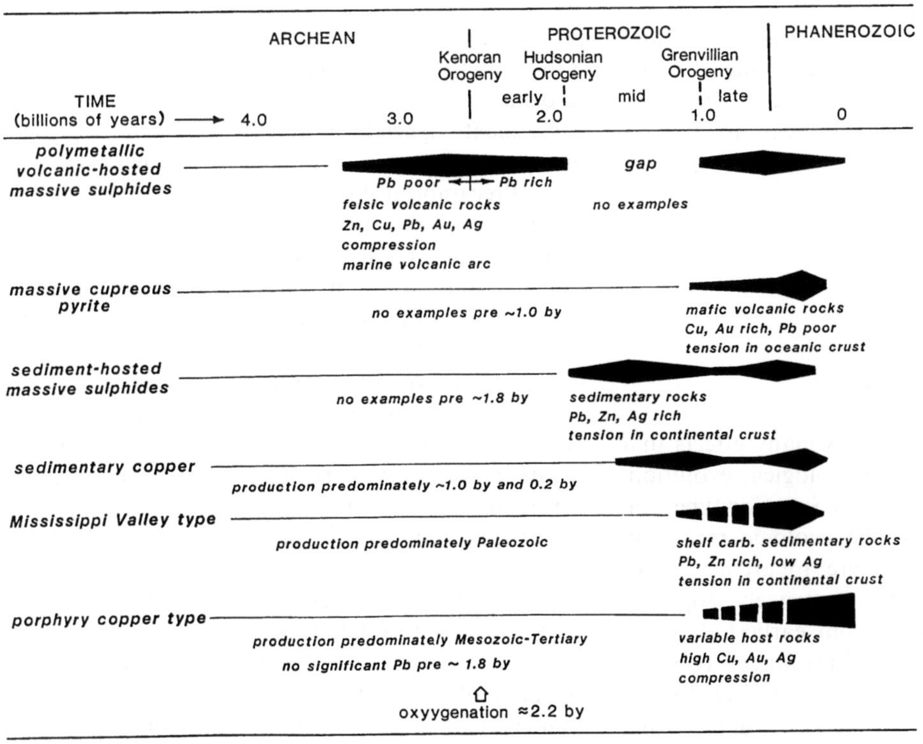

Fig. 7. Evolutionary changes in massive and related types of base metal sulfide deposits

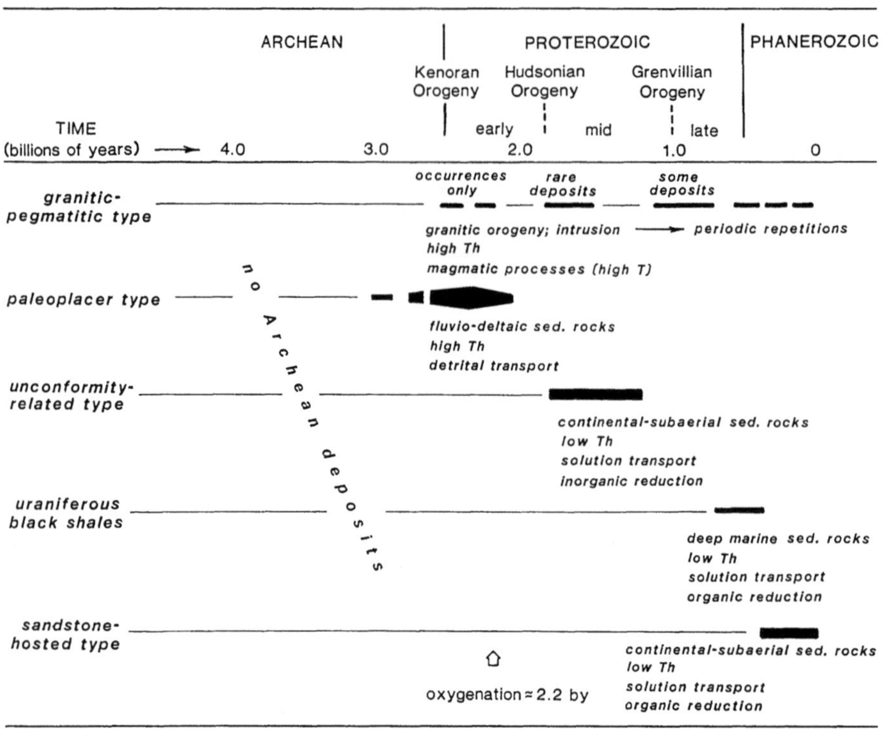

Fig. 8. Evolutionary changes in uranium ores

and biospheric environments. Both processes and products must be considered and explained by any global evolutionary hypothesis. The changes in both thus serve as additional guides to Earth's evolution, against which conclusions from other indicators can be compared independently.

3 Evolutionary Metallogeny

3.1 Fundamental Characteristics

Evolutionary metallogeny exhibits many affects that may be compared to those of biological evolution. Komatiite-hosted Ni deposits and Superior-type iron formations disappear from the geologic record in Lower Proterozoic rocks, and are apparently examples of extinction (Figs. 4, 5). Conversely, Algoma-type iron formation persists from earliest Archean to Recent rocks, an example of prolonged continuation of a primitive type. Superior-type iron formation, clastic sediment-hosted massive sulfide, and sedimentary Cu deposits all appear suddenly in Lower, Mid-, and Upper Proterozoic rocks, respectively (Figs. 4, 7), as if filling some newly evolved environmental niche, and all exhibit a major, global proliferation. A diversification of older volcanic- into sediment-hosted groups occurs in Mid-Proterozoic massive base metal sulfide deposits, and subsequent specialization within each of these groups generates a number of distinct types in Late Precambrian-Phanerozoic sequences. Diversification through time is also distinct in types of U- and porphyry-related deposits (Figs. 7, 8). As in faunal-floral evolution, diversification leads to increasingly varied, numerous, and specialized forms in later time (Fig. 9). Furthermore, these changes are unidirectional; although extinctions remove some older forms, other persist through time, or reappear when their favorable generative environments reoccur, and are joined by newly evolving forms that appear only in later time.

Evolutionary metallogeny is therefore characterized by increasing diversity and complexity of ore deposits and their generative sub- and supracrustal environments through time. Evolutionary metallogeny thus provides an explanation for the absence or rarity of certain types of deposits in rocks of certain ages, and also for diversifications of a single type. The rarity or absence of types formed at shallow depths, and differences within types have commonly been attributed to erosion, especially for the porphyry and epithermal deposits (Lowell and Guil-

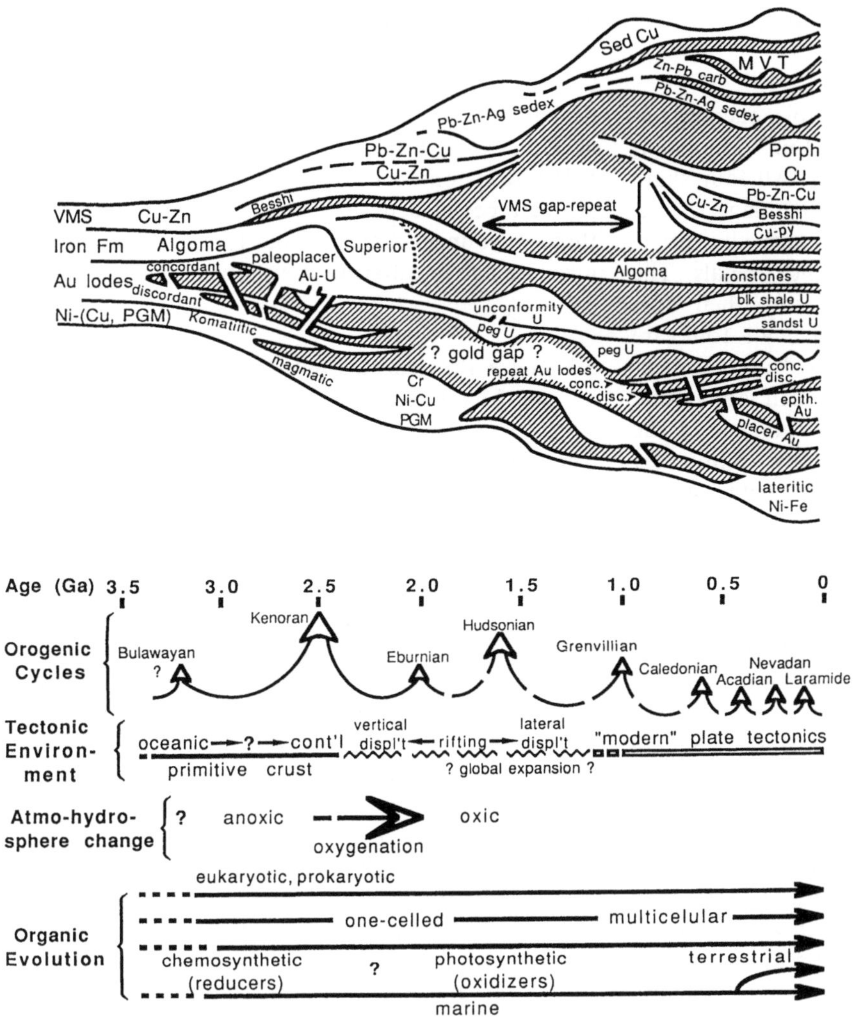

Fig. 9. Evolutionary metallogeny and the development and interrelationships of diverse major types and varieties of important mineral deposits through time

bert 1970; Sillitoe 1973). Obviously, this explanation is partially valid, but has perhaps been overemphasized, and evolutionary metallogeny correspondingly overlooked as the main reason for absences, rarities, or differences.

Many examples contradict the erosional explanation. The central felsic strata of the Noranda volcanic complex are flat lying and in an only moderately tilted, nearly original position at the top of this Archean dome. The Bushveld and Sudbury intrusions, great lopoliths emplaced at very shallow depths in Mid-Proterozoic time, and their granophyric roof rocks, which may have been locally eruptive to surface, are still in place. The epithermal veins of Cobalt and Thunder Bay, Ontario were formed near the surface, the former accompanying intrusion of the huge Nipissing Diabase sill about 2.1 Ga **B.P.** Although some of these were formed in intracratonic settings rather than in orogenic belts, and hence may not have undergone comparable uplift and erosion, none has been deeply denuded despite their great age. Moreover, ophiolite complexes emplaced on to continents by collision-induced obduction along consuming plate margins in northwest Africa, Newfoundland, and Cyprus ca. 0.8, 0.5, and 0.1 Ga **B.P.**, respectively, have also survived later erosional removal. Evolutionary metallogeny controlled by evolving crustal geotectonic processes offers a better explanation than erosional removal for many absences, rarities, and diversifications of types of mineral deposits, and for their spatial and temporal distributions.

3.2 Archean Supracrustal Environments and Metallogeny

Archean metallogeny is relatively simple, encompassing the generation of only a few (four) important types of deposits (Figs. 1, 4, 5, 6, 7). All four share a relatively simple major metal assemblage of Fe, Ni, Cu, Zn, Ag, and Au, albeit in markedly differing proportions and with many minor metals. The latter do not become concentrated into their own deposits until later time. Moreover, all four are narrowly restricted in space within volcanosedimentary greenstone belt successions which apparently represent the only supracrustal environment of the Archean earth. The iron formations (Kimberley 1989b) and primitive massive base metal sulfide deposits formed from seafloor hydrothermal discharge that accompanied the vast Archean submarine volcanism. Although their origins are more complex and controversial, the association of at least some of the Ni and Au deposits with thin, variably cherty-ferruginous-tuffaceous sedimentary strata (Fripp 1976; Gibbons 1979; Bavinton 1981; Robinson and Hutchinson 1982; Bernier and MacLean 1989) suggests a partial role for this same process. Seafloor hydrothermal activity apparently somehow accompanied magmatic processes in forming these Ni deposits and preceded superimposed metamorphic ones in forming the Au lodes (Hutchinson 1987). Whatever their exact roles, these magmatic, exhalative, and metamorphic processes formed and subsequently reshaped deposits throughout the Archean earth, and were the only important ore-forming ones in the widespread greenstone belts.

These belts were ubiquitous and may have been of two differing types, platformal and rift-related (Groves and Batt 1984), but they were dominated by vast, predominantly subaqueous volcanism that progressed cyclically, but generally upward from broad, basal ultramafic-mafic platforms into thick, calc-alkaline to silici-alkalic domical centers (Fig. 1). The latter were flanked by deepening, possibly rift-controlled basins into which coarse volcaniclastic sediments were shed around the flanks of the growing domes, and more distal, finer, turbiditic greywacke and carbonaceous shale (Lowe 1980). Intracontinental, rift-related deposits such as epithermal veins, Nb and REE in alkaline syenite-carbonatite complexes and diamondiferous kimberlites are very rare. These conditions reflect a primitive, thin, unstable, relatively readily defomed or rifted, rapidly subsiding crust.

Greenstone belt generation ceased when supracrustal stabilization was achieved by a major orogeny that formed a thick continental lithosphere for the first time. This event was diachronous around the earth (Hutchinson and Viljoen 1988), occurring in places as early as 3.2 Ga **B.P.**, extensively about 2.5 Ga **B.P.**, and locally as late as 1.8 Ga **B.P.** Greenstone belts and their four distinctive, primitive types of deposits, therefore, range from older through younger Archean, and Lower Proterozoic in absolute age in differing Precambrian terranes.

3.3 Lower Proterozoic Supracrustal Environments and Metallogeny

Three of the four major types of deposits in Lower Proterozoic greenstone belts exhibit evolutionary changes, probably due to this progressive crustal stabilization. The Algoma-type iron formations remain essentially unchanged (Fig. 4). However, primitive komatiite-hosted Ni deposits become rare and disappear in post-Lower Proterozoic rocks (Fig. 5). Enigmatically, greenstone-hosted Au lodes also diminish in number and importance in Lower Proterozoic belts (Fig. 6), whereas Pb, most lithophile of the base metals, becomes increasingly abundant in massive sulfide ores of this age (Fig. 7).

More important than these evolutionary changes in older types are the appearances of new types in response to the evolving stabilized crustal environments. Thickened and stabilized crust underwent rifting with mainly vertical displacement. Upthrown blocks of the older granite-greenstone basement were subaerially weathered, deeply eroded, and the detritus deposited in extensive, shallow but continually subsiding, craton-margin basins like the Witwatersrand of the Transvaal and the Huronian of Ontario. This environment favored proliferation of both paleoplacer Au–U and magmatic Ni–Cu-PGM deposits (Figs. 5, 6), the former in the coarse, fluviodeltaic sediments of the basin edges and the latter in great ultramafic-mafic intrusions that rose through the rifts and intruded the newly deposited supracrustal strata at or near their basal unconformable contacts with the older basement. Examples of the latter are the Bushveld, Sudbury, and Nipissing Diabase intrusions.

Despite their global proliferation during the evolutionary change from unstable granite-greenstone to stable cratonic block-marginal basinal tectonism, the paleoplacers and magmatic sulfide deposits are not synchronous around the earth (Fig. 3) due to the diachroneity of this change. The immense and unique concentration of Au in the Witwatersrand, only one of many paleoplacers districts, is related to the particular absolute age of this basin, which is synchronous not with others like it, but with prolific Au-producing greenstone belts of ca. 2.9–2.6 Ga **B.P.** on other cratons. With other evidence, this temporal relationship suggests that the Witwatersrand Au was perhaps not derived alluvially from an uplifted, older, weathered hinterland. It may, instead, have been of seafloor hydrothermal origin like the Au lodes, deposited originally as auriferous, pyritic exhalite, but subsequently reworked by sedimentary processes along an active, faulted, regressive basin edge (Hutchinson and Viljoen 1988). Whatever the explanation, it is an important and enigmatic aspect of global metallogeny that some 75–80% of all the world's Au has been derived from rocks of this age, much of it from the Witwatersrand conglomerates but significant amounts from the greenstone-hosted lodes (Figs. 3, 6). The U in the paleoplacers is distinctively accompanied by high Th, which is consistent with derivation of accessory U–Th mineral grains from weathering of granitoid rocks and pegmatites in the older, uplifted hinterland of the basins. These were fluvially transported and subsequently redeposited as detrital heavy minerals in the coarse clastic sediments of the basin margin.

The near-synchronous global deposition of huge thicknesses of Superior-type iron formation presents an interesting contrast with the diachronous, but equally global formation of paleoplacers and magmatic sulfides. These ferruginous chemical sedimentary deposits are clearly an evolutionary variant from their Algoma-type antecedents of the greenstone belts to which they show both similarities and differences. Their synchroneity, proliferation, evolutionary characteristics, and disappearance may be explained by Lower Proterozoic changes in the source, transportation, and precipitation of their contained Fe. Prolific subaqueous volcanism and accompanying hydrothermal activity in the greenstone belts flushed Fe into the oceans in large volumes throughout Archean time. Although some was precipitated near the seafloor vents as Algoma-type iron formation, much of this Fe remained in solution as relatively soluble ferrous iron salts in a reduced, oxygen-deficient ocean which became a sink for dissolved Fe. Between about 2.3– 2.1 Ga **B.P.**, oxygen-generating, prokaryotic, photosynthetic organisms apparently proliferated at an enormous rate in offshore, cryptalgal banks. These were associated with platformal carbonate strata in the shallow, craton-margin basins, but remote from the high-energy, nearshore, coarse-clastic repositories. Here, Fe, derived from the oceanic sink, was converted to insoluble ferric oxide and deposited rapidly, over a narrow time span and widespread in great thicknesses on a global scale (Cloud 1972, 1973; Garrels et al. 1973). When the oceanic Fe supply was exhausted, deposition of Superior-type iron formation ceased, never to be repeated in an evolved, now oxygenated ocean. All these rocks were variably deformed, uplifted, and intruded by granitoid batholiths during global Mid-Proterozoic orogeny ca. 1.8–1.6 Ga **B.P.**

3.4 Mid-Proterozoic Supracrustal Environments and Metallogeny

New tectonostratigraphic environments evolved following the Mid-Proterozoic orogeny. Most important were deeply subsiding, probably rift-controlled and aulacogenic, trough-like sedimentary basins like the Belt of western Canada-USA (Kanasewich 1968; Harrison 1972; Harrison et al. 1974), and the Willyamna of western New South Wales, Australia. These may reflect tension-induced separation and lateral displacement along the old Lower Proterozoic rifts. Mafic dikes, sills, and flows of now amphibolitized gabbro or basalt accompanied the rifting, and the basins were infilled with thick sequences of fine terrigenous shale, argillite, and turbiditic greywacke. As subsidence slowed, filling neared completion and shallowing produced coarser clastic strata and flanking, shallow carbonate platforms. Continental conglomerate, sandstone, and shale were deposited in shallow cratonic basins unconformably on upthrown blocks of deformed, crystalline older basement, as in the Athabaska basin of northern Saskatchewan and the Alligator River basin of the Northern Territory, Australia.

These new environments were the generative sites for new types of mineral deposits. Mid-Proterozoic rifting about 1.7 and 1.1 Ga **B.P.** was accompanied by emplacement of many alkaline syenite-carbonatite complexes containing Nb and REE deposits (Bell and Blenkinsop 1989). The greenstone belts of older time disappeared, and with them their four distinctive types of mineral deposits, another enigmatic aspect of evolutionary metallogeny. However, seafloor hydrothermal discharge, perhaps driven by basinal subsidence and expulsion of formational brines rather than by active submarine volcanism as in the greenstone belts, formed many large, sediment-hosted Pb–Zn–Ag-rich massive sulfide deposits (Fig. 7). These are important in the fine, basinal, Mid-Proterozoic clastic sequences, as are higher Zn but lower Ag ones in the somewhat younger or flanking carbonates.

Deep weathering occurred in an oxygenated atmosphere for the first time on the flanking cratons. The Mid-Proterozoic granitoid rocks and their satellitic pegmatites, some of which are the oldest U–Th deposits of this type, are products of a second, global period of orogenic partial melting (Fig. 8) in older, supracrustal rocks. They were thereby twice enriched in U and Th, uplifted during the Mid-Proterozoic orogeny, and weathered. Oxidized, hexavalent, soluble U was separated from its associated but unoxidized and relatively insoluble tetravalent Th, and carried in solution in meteoric waters through the porous, coarse continental clastic sedimentary strata above the basal unconformities. It was redeposited near this horizon as low-Th, unconformity-related U deposits, wherever the transporting waters encountered graphitic, sulfidic, cooling mafic igneous or other rocks capable of reducing soluble hexavalent to insoluble tetravalent U.

3.5 Upper Proterozoic-Phanerozoic Environments and Metallogeny

The common reappearance in Late Proterozoic-Phanerozoic sequences of Algoma-type iron formation, greenstone-hosted Au lodes, and volcanic-hosted massive base metal sulfide deposits (Figs. 4, 6, 7) indicates a repetition of supracrustal environments like those of the ancient greenstone belts. However, in marked contrast to their earlier global distribution throughout large Archean cratons, Phanerozoic deposits of these types are restricted spacially within narrow, linear tectonic belts of accreted, continent-margin, island-arc, and orogenic granitoid rocks. Moreover, the tectonic evolution of these belts progresses from earliest island-arc volcanism through continent-marginal accretion, subsequent cratonization by orogeny and emplacement of granite, to final uplift, weathering, erosion, and denudation, all within a relatively short time span, 0.2–0.4 Ga. Thus, the younger belts and their similar types of deposits were far more restricted, both spacially and temporally than their ancient greenstone belt antecedents. Many Mid- to Late Proterozoic types accompany the more primitive older types in the younger terranes, indicating the presence, in places and at various stages of the tectonic cycle, of the tectonostratigraphic environments in which all were generated. Finally, the proliferation of types that were rare or vestigial in earlier time and the appearance of entirely new types (Fig. 4) further demonstrate the extreme diversity of ore-generating environments in younger time, and their very local spatial and brief temporal distributions.

These metallogenic changes reflect the development, during the last 1 Ga, of new supracrustal environments and processes, the proliferation and maturation of some that were previously rare or immature and poorly developed, and the repetition on reduced spatial and temporal scales, of other very ancient ones. The widespread Phanerozoic sedimentary Cu deposits (Fig. 7) are coeval with initial breakup of Pangeia, and consequently are sometimes associated with mafic igneous rocks and commonly with red, continental, often aeolian sandstone and evaporite. On a global scale they are Carboniferous in the north (Schenk 1969), Permian in central Europe and Oklahoma (Hagni 1988; Vaughan et al. 1989), but Cretaceous-Tertiary, farthest south in Bolivia (Ljunggren and Meyer 1964). This southward younging suggests that the breakup of Pangeia progressed southward through time. The equally widespread Upper Proterozoic ones have similar lithostratigraphic settings and associations. Two periods of widespread, perhaps global tensional rifting of continental crust, accompanied by mafic magmatism, clastic sedimentation in shallow cratonic basins, and arid climatic conditions are thereby indicated, one Late Precambrian and the other Pangeian. Enigmatically, Lower and Mid-Proterozoic continental rifting apparently did not favor formation of similar deposits.

The rarity of Precambrian ophiolite (Glikson 1979) and first appearance, in Late Proterozoic ophiolitic successions, of the massive cupreous pyrite ores suggest that the distinctive oceanic crustal tectonostratigraphic setting in which these ores and rocks were generated is a relatively late development on the earth. Although some older progenitor type of oceanic crust may have floored more ancient oceans, younger oceanic lithosphere, consisting of basal layered ultramafic rock, sheeted diabase, spilitized tholeiitic basalt, and overlying pelagic-siliceous variably metalliferous sediment, may have evolved only in Late Proterozoic time, persisting to the present.

The evolutionary appearance of terrestrial organisms in Devonian time, and their burial preservation thereafter in continental sediments of shallow cratonic basins created the essential new environment for the formation of sandstone-hosted U deposits (Fig. 8). Although comparable in some respects to the Mid-Proterozoic, unconformity-related type, organic debris in these sediments provided a new reductant to precipitate insoluble tetravalent

U from its transporting oxygenated waters (Hutchinson and Blackwell 1984).

Proliferation of MVT Pb–Zn deposits mirrors extensive development of shallow, cratonic marine basins and continent-marginal carbonate platforms in Phanerozoic time (Fig. 7). And proliferation of three varieties of porphyry-type ores, and of continental volcanism-affiliated epithermal veins reflects an evolution of subductive processes and their related magmatism and hydrothermal activity. Relatively steep subduction of oceanic crust occurred at Paleozoic consuming plate boundaries, forming marine volcanic island arcs that contain primitive porphyry Cu deposits, volcanic-hosted Au lodes, and primitive massive base metal sulfides (Hutchinson 1980). This style may have changed in Mesozoic-Tertiary time to lower angles of subduction. In the western United States this carried the magmatic arc increasingly farther into the continent as subduction continued (Christiansen and Lipman 1972; Lipman et al. 1972), resulting in northeastward migration, and later westward retreat of subaerial-continental volcanism and subvolcanic intrusion. Hydrothermal systems accompanying this migratory magmatism (Lipman 1981) changed from marine, and seawater-dominated to continental and meteoric water-dominated with resulting changes in fluid chemistry and composition, types of deposits, their metal assemblages, and rock alteration characteristics. Primitive, island-arc porphyry Cu deposits give way to mature, continental ones, and, farthest "inboard", to porphyry Mo deposits. Primitive volcanic-hosted Au lodes give way to epithermal precious and base metal veins, whereas the massive sulfide ores disappear.

All these types of deposits, and the sequential changes in some, are consequences and evidence of the evolutionary migration of magmatic arcs inward from relatively young, Mesozoic-Tertiary subductive plate margins. The rarity of porphyry deposits in comparable Paleozoic orogenic belts and their virtual absence in greenstone belts strongly suggest that, although subduction may have generated these older belts, it was somehow of a different form or scale, and therefore lacked some of the processes and products of the younger ones.

4 Interpretation of Evolutionary Geotectonics

Just as it reflects evolutionary changes in relatively recent subductive tectonism, evolutionary metallogeny mirrors geotectonic evolution throughout earth history. Tectonostratigraphic processes on the primitive, ancient Archean earth were global, but few and simple compared with those of later time. Tectonic transport was predominantly vertical, involving rapid and deep subsidence, i.e., "sagduction", and reflecting crustal immaturity and instability. Lateral transposition and subduction beneath the margins of primitive sialic crustal plates was minimal, as was inward migration of magmatism, and these processes, with their distinctive types of mineral deposits were therefore vestigial or lacking. Rifting of either primitive sialic or oceanic lithosphere, or perhaps of both in different places, initiated magmatism by extensive partial melting at high temperature of a primitive, poorly differentiated mantle. This formed the earliest platformal, and stratigraphically lowest parts of rift-deposited Archean greenstone belts (Groves and Batt 1984), and provided the generative environment for their distinctive komatiite-hosted Au, Ni, and rare Cr deposits, and Algoma-type iron formations.

The rifting was accompanied by subsidence due to withdrawal of melt from beneath, and loading above. Continued partial melting in the depleted, evolving mantle generated melts that formed the thick, tholeiitic basaltic platforms of the subsiding greenstone belts, the environment for formation of additional iron formations and more important Au lodes. Continuing subsidence caused partial melting of the komatiitic-tholeiitic rocks in a manner comparable to that of lateral subduction of oceanic crust and its partial melting which produces continent-marginal island arcs in modern time. Thus, magmatic evolution progressed to generation of the calc-alkaline and silici-alkalic volcanic rocks that comprise the stratigraphically higher portions of the domical complexes in Late Archean and Lower Proterozoic greenstone belts, and which host the great polymetallic massive base metal sulfide deposits.

In their broad characteristics and products, these processes were akin to, and combined those of modern accreting and consuming plate margins, but in vertically superposed, rather than laterally juxtaposed positions. The felsic volcanic domes of the greenstone belts may be likened to modern continental lithospheric plates, and their flanking sedimentary basins to modern oceans. However, those ancient plates were of primitive, relatively simple island-arc volcanic composition, much smaller and thinner than modern continental lithospheric plates, perhaps reflecting smaller size and greater number of mantle convective cells that drove these processes at that time.

Cratonization due to orogenic plutonism commenced as early as 3.1 Ga **B.P.** and continued at different locations around the globe until as late as 1.8 Ga **B.P.**, producing initial thickened, stable, sialic crust. This represented a different, new supracrustal tectonic environment. Diachronous rifting of this continental lithosphere followed its diachronous cratonization, generating the broad, fault-bounded, deeply subsiding, hinged cratonic basins of Lower Proterozoic age, like the Witwatersrand. Au–U paleoplacers were deposited at the rifted margins of these basins, and farther offshore in shallow algal or carbonate banks, the Superior-type iron formations. Initial displacement on the rifts was mainly vertical, perhaps continuing the dominant subsidence of Archean time. But about 1.8 Ga **B.P.**, roughly coeval with final orogenic cratonization of all greenstone belts, global extension caused widespread lateral separation of the continental lithosphere along the same or younger rifts, perhaps marking the initial appearance of continents and deep ocean basins as now known. Collision between some blocks may account for the global development of high-grade metamorphic terranes of about this age, like the Churchill Structural Province of Canada, and those along the Wyoming lineament in the north-central United States. But subduction of oceanic crust beneath the blocks was minimal or absent, as evidenced by the global lack of obducted ophiolite (Glikson 1979), and also by the rarity of greenstone belt or volcanic island-arc successions with their characteristic mineral deposits in Mid-Proterozoic sequences.

Instead, the opening, deepening ocean basins were apparently infilled with thick terrigenous turbiditic greywacke, argillite, siltstone, and shale accompanied by rift-generated mafic igneous rocks, now amphibolites. These were sites for deposition of the clastic sediment-hosted massive Pb–Zn–Ag sulfide deposits. As the basins filled, the later shallower water strata were sites for deposition of the carbonate-hosted variety of these ores. Both were Pb-rich, reflecting the abundance, by this stage of Earth's evolution, of Pb-bearing, K-feldspar-rich sialic crust. Orogenic granitic plutons generated in the collisional metamorphic terranes from partial remelting of older, Archean supracrust exhibit a second enrichment in U–Th so that their pegmatitic differentiates are sometimes ores of these metals. The first major period of oxygenated, subaerial weathering of these granitoid rocks led to the formation of unconformity-related U deposits.

The geotectonism that might have generated these widespread relationships are speculative, but global expansion (Hutchinson 1981) might account for rifting, and separation, for continent-continent collisions without subduction of oceanic crust and for lack of ophiolite. If so, its cause is even more speculative. Early Proterozoic global rifting might have allowed seawater to first penetrate the upper mantle as in modern oceanic rifts, causing serpentinization and major expansion. Expansion may have been induced by progressive radiogenic heating. An extraterrestrial cause, perhaps involving a slight decrease in the gravitational constant, cannot perhaps be ruled out, just as such a cause remains possible for the origin of life (Schidlowski 1990). Whatever the explanation, the major rifting, fragmentation, separation, and collision of continental blocks, with intervening oceanic basins, might represent the initiation of plate tectonic interactions comparable to those of the modern Earth.

A third major orogeny about 1 Ga **B.P.** affected many parts of the Earth. Like its predecessors, it involved continent-continent collisions that generated granitic magma through partial melting in high-rank metamorphic terranes like the Grenville Structural Province of Canada and the Damaran of Namibia. As predictable in a third differentiate, its granitoid derivatives are still richer in U–Th than the older second, and much more so than the oldest first ones, although perhaps not so much as subsequent ones (Jacob et al. 1986). This episode of cratonization perhaps marks the onset of the plate tectonic processes of later time. These, too, have evolved, becoming increasingly complex, localized, diverse, and cyclic through shorter time spans. The plate tectonic cycle thus evolves from seafloor spreading along oceanic rifts to subduction or transform displacement of oceanic lithosphere at consuming plate margins. Here, volcanic island arcs form, sometimes with back-arc basins and inner, younger arcs, and are ultimately accreted to the continental plates along with obducted slabs of the oceanic crust. Migration of magmatism into the continent leads to emplacement of orogenic granitoid batholiths and cratonization. Accompanying uplift brings denudation and stabilization. Farther inward migration of the magmatic arc generates anorogenic magmatism. And continental rifting with alkaline magmatism closes the cycle, or begins it anew when these rifts open into new ocean basins. However, this entire tectonic cycle duplicates, on more restricted spatial and shorter temporal scales, the same broad sequence of analogous geotectonic environments that has governed metallogenic evolution through all of time. This is analogous to the duplication of phylogeny by ontogeny in biological science.

5 Other Global Evolutionary Changes

Evolutionary changes in many other aspects of the Earth have influenced its changing supracrustal environments and thereby affected metallogenic evolution. These include changes in the asthenosphere, lithosphere, hydrosphere, atmosphere, and biosphere. Earliest Archean mantle-derived partial melts were ultramafic, apparently sulfur-saturated and gave rise to komatiitic Ni deposits, whereas comparable Late Archean-Lower Proterozoic melts formed differentiated, but overall generally mafic, tholeiitic, subvolcanic intrusions with associated magmatic Cr – Ni – Cu-PGM ores. The former reached the surface, penetrating an apparently thin, primitive crust. The latter intruded an evolved, thickened crust. By Mid-Proterozoic time, probable mantle-derived partial melts were extensively alkaline and syenitic-carbonatitic (Bell and Blenkinsop 1989). The changing composition of these melts through time must reflect evolutionary changes in the mantle and its processes, perhaps – through depletion or differentiation – in its composition, or alternatively in size of convection cells, depth or degree of partial melting, heat flow, or others. Analogous evolutionary changes affected the compositions of granitic melts. Three successive major orogenies, each reworking the products of its predecessor, generated orogenic granitoid partial melts that were increasingly enriched in U – Th. And Sn – W-enriched granites did not develop until continental lithosphere had evolved and undergone at least two episodes of partial melting, sometimes in orogenic belts but elsewhere in anorogenic environments. Oxygenation of the hydrosphere and atmosphere (Schidlowski 1976), apparently related to biogenic evolution, profoundly changed the nature of iron formation about 2.3 – 2.1 Ga **B.P.** This also provided an essential new oxygenated weathering environment that allowed the formation of unconformity-related U deposits. Biogenic evolution of terrestrial organisms in Devonian time fostered formation of continental sandstone-hosted U ores. And although uncertain, primitive algae have been implicated in the deposition of gold in the carbon reefs of the Witwatersrand (Hallbauer 1986).

6 Summary

In summary, it appears that most, perhaps all types of mineral deposits have undergone evolutionary changes through time. These changes are primarily responses to Earth's geotectonic evolution, which was foremost in governing the supracrustal lithostratigraphic environments in which the deposits were formed. But complex, interrelated and parallel changes in Earth's subcrustal, hydrospheric, atmospheric, and biospheric environments and processes have also affected metallogenic evolution. Consequently, the deposits and their generative processes serve as useful indicators of evolutionary changes in all of these. Moreover, in terranes where high-rank metamorphism has overprinted all primary geological characteristics, and has perhaps reset many isotopic age relationships, the types of mineral deposits may yet provide a qualitative indication of initial age. Finally, evolutionary metallogeny is useful in mineral exploration because it defines the favorable temporal domaines, or ages, in which particular types of deposits should most profitably be sought.

Acknowledgements. Although they would not agree, either wholly or in part, with this synthesis, many colleagues have contributed to it through discussions and their publications on the various problems and aspects of metallogeny that are considered here. The writer is greatly indebted also to many mining companies throughout the world, and to their geologists for assistance in visiting their mines and for freely provided data on their deposits. Many government agencies, universities, and their geologists have provided similar support and advice. Dr. P. A. Trudinger of the Bureau of Mineral Resources in Canberra, Australia and Prof. E. F. Stumpfl of the Mining University of Leoben in Austria reviewed an early draft of this paper and offered constructive suggestions for its improvement.

References

Adler HH (1974) Concepts of uranium-ore formation in reducing environments in sandstones and other sediments. Formation of uranium deposits. IAEA, Vienna, pp 141 – 168

Alexandrov EA (1973) The Precambrian banded iron formations of the Soviet Union. Econ Geol 68(7):1039 – 1062

Andersson A, Dahlman B, Gee D (1983) Kerogen and uranium resources in the Cambrian Alum Shales of the Billingen-Fahlbygden and Narke areas, Sweden. Geol Foren Stockholm Forh 104:197 – 209

Anderson CA, Nash JT (1972) Geology of the massive sulfide deposits at Jerome, Arizona: a reinterpretation. Econ Geol 67(7):845 – 863

Andrews AJ (1986) Silver vein deposits: summary of recent research. Can J Earth Sci 23:1460–1462
Anhaeusser CR, Viljoen MJ (1986) Archean metallogeny of southern Africa. In: Anhaeusser CR, Maske S (eds) Mineral deposits of southern Africa, I. Geol Soc S Afr, Johannesburg, pp 33–42
Argall GO Jr (ed) (1976) Designing Udokan open pit mine. World Mining, July 1976, Miller Freeman, Brooklyn, New York, pp 51–52
Baker A, Clayton RL (1968) Massive sulfide deposits of Bagdad District, Yavapai County, Arizona. In: Ridge JD (ed) Ore deposits of the United States 1933–1967 (Graton-Sales volume). Am Inst Min Metall Petrol Eng, pp 1311–1327
Bates TF, Strahl EO (1958) Mineralogy and chemistry of uranium-bearing black shales. In: 2nd UN Int Conf, Peaceful uses of atomic energy, Geneva 1958, vol 2, pp 407–411
Bavinton OA (1981) The nature of sulfidic metasediments at Kambalda and their broad relationship with associated ultramafic rocks and nickel ores. Econ Geol 76:1606–1628
Bayley RW, James HL (1973) Precambrian iron formation of the United States. Econ Geol 68(7):934–959
Bell K, Blenkinsop J (1987) Archean depleted mantle: evidence from Nd and Sr initial isotopic ratios of carbonatites. Geochim Cosmochim Acta 51:291–298
Bell RT (1978) Uranium in black shales – a review. In: Kimberley MM (ed) Short course in uranium deposits: their mineralogy and origin. Mineral Assoc Can Short Course Handb 3:307–329
Bernier LR, MacLean WH (1989) Auriferous chert, banded iron formation, and related volcanogenic hydrothermal alteration, Atik Lake, Manitoba. Can J Earth Sci 26 (12):2676–2690
Berning J (1986) The Rossing uranium deposit, South West Africa/Namibia. In: Anhaeusser CR, Maske S (eds) Mineral deposits of southern Africa, II. Geol Soc S Afr, Johannesburg, pp 1819–1832
Berry LG, Jambor JL (1971) The silver-arsenide deposits of the Cobalt-Gowganda region, Ontario. Can Mineral 11:429
Bilibin YA (1968) Metallogenetic provinces and metallogenic epochs. In: Alexandrov EA (ed) Geology bulletin (English Transl of 1955 Pap Geotekhizdat, Moscow). Queens College, Flushing, NY, 35 pp
Black R (1980) The Precambrian of West Africa. Episodes 4:3–8
Boyum BH, Hartviksen RC (1970) General geology and ore grade control at the Sherman Mine, Temagami, Ontario. Can Min Metall Bull 63(701):1059–1068
Broken Hill Proprietary (B.H.P.) Staff (1975) Koolyanobbing iron ore deposits. In: Knight CL (ed) Economic geology of Australia and Papua-New Guinea, I. Metals. Australas Inst Min Metall, pp 940–942
Brown AC (1971) Zoning in the White Pine copper deposit, Ontonagon County, Michigan. Econ Geol 66:543–573
Brown AC, Chartrand FM (1983) Stratiform copper deposits and interaction with coexisting atmosphere, hydrosphere, biosphere and lithosphere. Precambrian Res 20:533–542
Buchan R, Blowes JH (1968) Geology and mineralogy of a millerite nickel ore deposit-Marbridge No. 2 Mine, Malartic, Quebec, Canada. Can Min Metall Bull 61 (672):529–534
Byers AR, Kirkland SJT, Pearson WJ (1965) Geology and mineral deposits of the Flin Flon area, Saskatchewan. Sask Dep Mineral Resourc Rep 62:95 pp
Cameron EN (1980) Evolution of the Lower Critical Zone, central sector, eastern Bushveld Complex and its chromite deposits. Econ Geol 75:845–871
Cerny P, Trueman DL (1978) Distribution and petrogenesis of lithium pegmatites in the Western Superior Province of the Canadian Shield. Energy 3:365–377
Christiansen RL, Lipman PW (1972) Cenozoic volcanism and plate tectonic evolution of the western United States. 2, Late Cenozoic. Philos Trans R Soc London Ser A271, 1213:249–284
Clay AN (1986) The stratigraphy of the Malmani dolomite subgroup in the Carletonville Area, Transvaal: genetic implications for lead-zinc mineralization. In: Anhaeusser CR, Maske S (eds) Mineral deposits of South Africa, I. Geol Soc S Afr, Johannesburg, pp 853–860
Cloud P (1972) A working model of the primitive earth. Am J Sci 272:537–548
Cloud P (1973) Paleoecological significance of the banded iron-formation. Econ Geol 68(7):1135–1143
Coad PR (1977) Nickel sulphide deposits associated with ultramafic rocks of the Abitibi Belt and economic potential of mafic-ultramafic intrusions. Ont Geol Surv Open File Rep 5232:105 pp
Colvine AC, Sutherland IG (1979) Early Precambrian porphyry deposits. Summary of field work, 1979. Ont Geol Surv Misc Pap 90:230–243
Colvine AC, Fyon JA, Heather KB, Marmont S, Smith PM, Troop DG (1988) Archean lode gold deposits in Ontario. Ont Geol Surv Misc Pap 139:136 pp
Condie KC (1982) Plate-tectonics model for Proterozoic continental accretion in the southwestern United States. Geology 10:37–41
Correia Neves JM, Pedrosa AC, Monteiro RLBP, Scliar C (1983) Study of the geology and petrography of the pegmatite region of Virgem da Lapa; Coronel Murta, Minas Gerais. In: Geol Precambrian Bol Soc Brasil Geol, pp 100–114 (in Portuguese with English Summary)
Cotterill P (1969) The chromite deposits of Selukwe, Rhodesia. In: Wilson HDB (ed) Magmatic ore deposits. Econ Geol Monogr 4:154–186
Crawford J, Hoagland AD (1968) The Mascot-Jefferson City zinc district, Tennessee. In: Ridge JD (ed) Ore deposits of the United States 1933–1967 (Graton-Sales volume). Am Inst Min Metall Petrol Eng, pp 242–256
Davies DN, Urie JG (1956) The Bomvu Ridge haematite deposits. Swaziland Geol Surv Spec Rep 3:23 pp
Davies JF, Luhta LE (1978) An Archean "porphyry-type" disseminated copper deposit, Timmins, Ontario. Econ Geol 73:383–396
Davison I, Texeira LBG, Silva MG, Rocha Neto MB, Matos FMV (1988) The Rio Itapicuru greenstone belt, Bahia, Brazil: structure and stratigraphical outline. Precambrian Res 42:1–17
Deans T (1966) Economic mineralogy of African carbonatites. In: Tuttle OF, Gittins J (eds) Carbonatites. John Wiley & Sons, New York, 591 pp
DeWitt E, Redder JA, Wilson AB, Buscher D (1986) Mineral resource potential and geology of the Black Hills National Forest, South Dakota and Wyoming. U S Geol Surv Bull 1580:135 pp

Dorr II JVN (1963) Geology and ore deposits of the Itabira district, Minas Gerais, Brazil. U S Geol Surv Prof Pap 341-C:110
Dorr II JVN (1973) Iron-formation in South America. Econ Geol 68(7):1005–1022
Dubuc F (1966) Geology of the Adams Mine. Can Min Metall Bull 59(646):176–181
Eupene GS, Fee PH, Colville RG (1975) Ranger One uranium deposits. In: Knight CL (ed) Economic geology of Australia and Papua New Guinea, 1. Metals. Australas Inst Min Metall, pp 308–317
Evans EL (1986) Saskatchewan unconformity-associated and sedimentary-hosted deposits of Helikian Age. In: Evans EL (ed) Uranium deposits of Canada. Can Inst Min Metall, pp 123–262
Evoy EF (1986) The Gunnar uranium deposit. In: Evans EL (ed) Uranium deposits of Canada. Can Inst Min Metall, pp 250–262
Faure G (1964) The age of the Duluth Gabbro complex and the Endion sill by whole-rock Rb–Sr method. MIT 12th Annu Prog Rep, pp 255–257
Ferguson SA, Buffam BSW, Carter OF, Griffis AT, Holmes TC, Longley CS (1968) Geology and ore deposits of Tisdale Township, District of Cochrane, Ontario. Ont Dep Mines Geol Rep GR 58:117 pp
Fleischer VD, Garlick WG, Haldane R (1976) Geology of the Zambian Copperbelt. In: Wolf KH (ed) Handbook of stratabound and stratiform ore deposits, vol 6. Elsevier, Amsterdam, pp 223–321
Foster RP, Wilson JF (1984) Geological setting of Archean gold deposits in Zimbabwe. In: Foster RP (ed) Gold '82. The geology, geochemistry and genesis of gold deposits. Balkema, Rotterdam, 753 pp
Franklin JM, Thorpe RI (1982) Comparative metallogeny of the Superior, Slave and Churchill provinces. Geol Assoc Can Spec Pap 25:3–90
Franklin JM, Lydon JW, Sangster DF (1981) Volcanic-associated massive sulfide deposits. Econ Geol 75th Anniversary Vol, pp 485–627
Friesen RG, Pierce GA, Weeks RM (1982) Geology of the Geco base metal deposit at Manitouwadge, Ontario, Canada. In: Hutchinson RW, Spence CD, Franklin JM (eds) Precambrian sulphide deposits (H. S. Robinson memorial volume). Geol Assoc Can Spec Pap 25: 343–364
Frietsch R, Papunen H, Vokes FM (1979) The ore deposits in Finland, Norway, and Sweden – a review. Econ Geol 74:975–1001
Fripp REP (1976) Stratabound gold deposits in Archean banded iron-formation, Rhodesia. Econ Geol 71:58–75
Garrels RM, Perry EA, Mackenzie FT (1973) Genesis of Precambrian iron-formations and the development of atmospheric oxygen. Econ Geol 68:1173–1179
Gevers TW (1936) Phases of mineralization in Namaqualand pegmatites. Geol Soc S Afr Trans 39:331–378
Gibbons W (1979) Gold and Precambrian iron formation in the Northwest Territories. In: Morton RD (ed) Proc Gold Worksh, Yellowknife, NWT. Yellowknife Gold Worksh Commun, pp 258–287
Gibbs AK, Barron CN (1983) The Guiana Shield reviewed. Episodes 1983:7–14
Gilmour P, Still AR (1968) The geology of the Iron King Mine. In: Ridge JD (ed) Ore deposits of the United States 1933–1967 (Graton-Sales volume). Am Inst Min Metall Petrol Eng, New York, pp 1238–1257
Giovenazzo D, Picard C, Guha J (1989) Tectonic setting of Ni–Cu-PGE deposits in the central part of the Cape Smith Belt. Geosci Can 16(3):134–136
Glikson AY (1979) The missing Precambrian crust. Geology 7:449–454
Gole MJ, Barnes SJ, Hill RET (1989) The geology of the Agnew nickel deposit, Western Australia. Can Min Metall Bull 82(929):46–56
Goodwin AM (1962) Structure, stratigraphy and origin of iron formation, Michipocoton area, Algoma District, Ontario, Canada. Geol Soc Am Bull 73:561–586
Goulevitch J (1975) Warrego copper-gold orebody. In: Knight CL (ed) Economic geology of Australia and Papua New Guinea, I. Metals. Australas Inst Min Metall, pp 430–436
Gresham JJ, Loftus-Hills GD (1981) The geology of the Kambalda nickel field, Western Australia. Econ Geol 76:1373–1416
Gross GA (1966) Principle types of iron-formations and derived ores. Can Min Metall Bull 59:150–153
Gross GA (1968) Geology of iron deposits in Canada, 3. Iron ranges of the Labrador geosyncline. Geol Surv Can Econ Geol Rep 22:22 pp
Gross GA (1986) The metallogenic significance of iron formation and related strata. Geol Surv India 28:92–108
Gross WH (1968) Evidence for a modified placer origin for auriferous conglomerates, Canavieiras, Jacobina, Brazil. Econ Geol 63:271–276
Groves DI, Batt WD (1984) Spatial and temporal variations of Archean metallogenic associations in terms of evolution of granitoid-greenstone terranes with particular emphasis on the Western Australian shield. In: Kroner A, Hanson GN, Goodwin AM (eds) Archean geochemistry: the origin and evolution of the Archean continental crust. Springer, Berlin Heidelberg New York, pp 73–98
Groves DI, Phillips GN (1987) The genesis and tectonic control on Archean gold deposits of the Western Australian shield – a metamorphic replacement model. Ore Geol Rev 2:287–322
Gruszczyk H (1967) The genesis of the Silesian-Kracow deposits of lead-zinc ores. In: Brown JS (ed) Genesis of stratiform lead-zinc-barite-fluorite deposits in carbonate rocks. Econ Geol Mono 3, Econ Geol Publ, Lancaster, Penn, pp 169–177
Gustafson LB, Titley SR (1978) Porphyry copper deposits of the southwestern Pacific islands and Australia – preface. Econ Geol 73(5):597–599
Hagni RD (1988) Mineralogy, origin and potential for discovery of Creta, Oklahoma copper shale-type deposits in Oklahoma, Texas and Kansas. In: Kisvarsanyi G, Grant SK (eds) N Am Conf Tectonic control of ore deposits and the vertical and horizontal extent of ore systems. Proc Vol, Univ Missouri, Rolla, pp 455–467
Hallbauer DK (1986) The mineralogy and geochemistry of Witwatersrand pyrite, gold, uranium and carbonaceous matter. In: Anhaeusser CR, Maske S (eds) Mineral deposits of South Africa, I. Geol Soc S Afr, pp 731–752
Hannington MD, Peter JM, Scott SD (1986) Gold in sea floor polymetallic sulfide deposits. Econ Geol 81: 1867–1883
Harrison JE (1972) Precambrian Belt basin of northwestern United States, its geometry, sedimentation, and copper occurrences. Geol Soc Am Bull 83:1215–1240

Harrison JE, Griggs AB, Wells JD (1974) Tectonic features of the Precambrian Belt Basin and their influence on post-Belt structures. U S Geol Surv Prof Pap 866:15 pp
Hayba DO, Bethke PM, Heald P, Foley NK (1985) Geologic, mineralogic and geochemical characteristics of volcanic-hosted epithermal precious-metal deposits. In: Berger BR, Bethke PM (eds) Reviews in economic geology, vol 2: Geology and geochemistry of epithermal systems, vol 2. Econ Geol Publ House, Lancaster, pp 129–167
Haynes DW, Bloom MS (1987a) Stratiform copper deposits hosted by low energy sediments. III. Aspects of metal transport. Econ Geol 82:635–648
Haynes DW, Bloom MS (1987b) Stratiform copper deposits hosted by low energy sediments. IV. Aspects of sulfide precipitation. Econ Geol 82:875–893
Hedstrom P, Simeonov A, Malmstrom L (1989) The Zinkgruvan ore deposit, south-central Sweden: a Proterozoic, proximal Zn–Pb–Ag deposit in distal volcanic facies. Econ Geol 84(5):1235–1261
Hewitt JH (1967) Uranium and thorium deposits of southern Ontario. Ont Dep Mines Mineral Resource Circ 4:76 pp
Hewton RS (1982) Gayna River – a Proterozoic Mississippi Valley-type zinc-lead deposit. In: Hutchinson RW, Spence CD, Franklin JM (eds) Precambrian sulphide deposits (H S. Robinson memorial volume). Geol Assoc Can Spec Pap 25:667–700
Holl R, Maucher A (1976) The stratabound ore deposits in the eastern Alps. In: Wolf KH (ed) Handbook of stratabound and stratiform ore deposits, vol 5. Elsevier, Amsterdam, pp 1–36
Hoy LD, Ohmoto H (1989) Constraints for the genesis of red bed-associated stratiform Cu deposits from sulphur and carbon mass-balance relations. In: Boyle RW, Brown AC, Jefferson CW, Jowett EC, Kirkham RV (eds) Sediment-hosted stratiform copper deposits. Geol Assoc Can Spec Pap 36:135–149
Hulbert LJ, Von Gruenewaldt G (1985) Textural and compositional features of chromite in the Lower and Critical Zones of the Bushveld Complex south of Potgietersrus. Econ Geol 80(4):872–895
Huot D, Sattran V, Zida P (1986) Gold in the Birrimian greenstone belts of Burkino Faso, West Africa (Abstr). Terra Cognita 6:537
Hutchinson RW (1955) Regional zonation of pegmatites at Ross Lake, NWT. Geol Surv Can Bull 34:50 pp
Hutchinson RW (1973) Volcanogenic sulfide deposits and their metallogenic significance. Econ Geol 68(8):1223–1246
Hutchinson RW (1980) Massive base metal sulphide deposits as guides to tectonic evolution. In: Strangway DW (ed) The continental crust and its mineral deposits. Geol Assoc Can Spec Pap 20:659–684
Hutchinson RW (1981) Mineral deposits as guides to supracrustal evolution. In: O'Connell R, Fyfe WS (eds) Evolution of the Earth. AGU-GSA Geodyn Ser 5: 120–140
Hutchinson RW (1984) Archean metallogeny: a synthesis and review. J Geodyn 1:339–358
Hutchinson RW (1987) Metallogeny of Precambrian gold deposits: space and time relationships. Econ Geol 82: 1993–2007
Hutchinson RW (1988) Metallogeny of Precambrian gold deposits: space and time relationships: a reply. Econ Geol 83(6):1287–1288
Hutchinson RW (1990) Precious metals in massive base metal sulfide deposits. Geol Rundsch 79:241–263
Hutchinson RW, Blackwell JD (1984) Time, crustal evolution and generation of uranium deposits. In: Capaldi G, De Vivo B, Ippolito R (eds) Uranium – geochemistry, mineralogy, geology, exploration and resources. Liguori, Naples, It, pp 89–100
Hutchinson RW, Burlington JL (1984) Some broad characteristics of greenstone belt gold lodes. In: Foster RP (ed) Gold '82. The geology, geochemistry and genesis of gold deposits. Balkema, Rotterdam, pp 339–372
Hutchinson RW, Viljoen RP (1988) Re-evaluation of gold source in Witwatersrand ores. J S Afr Geol 91(2): 157–173
Ireland KL (1986) The Winterveld (TCL) Chrome Mine Limited, eastern Bushveld Complex. In: Anhaeusser CR, Maske S (eds) Mineral deposits of southern Africa, II. Geol Soc S Afr, Johannesburg, pp 1183–1188
Irvine TN, Smith CH (1969) Primary oxide minerals in the layered series of the Muskox intrusion. In: Wilson HDB (ed) Magmatic ore deposits. Econ Geol Publ House, Lancaster, pp 76–94
Jackson ED (1969) Chemical variation in co-existing chromite and olivene in chromitite zones of the Stillwater Complex. In: Wilson HDB (ed) Magmatic ore deposits. Econ Geol Publ House, Lancaster, pp 41–75
Jacob RE, Corner B, Brynard HJ (1986) The regional geological and structural setting of the uraniferous granitic provinces of southern Africa. In: Anhaeusser CR, Maske S (eds) Mineral deposits of southern Africa, II. Geol Soc S Afr, Johannesburg, pp 1807–1818
James HL (1954) Sedimentary facies of iron-formation. Econ Geol 49(3):235–294
Johnson IR, Klingner GD (1975) The Broken Hill ore deposit and its environment. In: Knight CL (ed) Economic geology of Australia and Papua New Guinea, I. Metals. Australas Inst Min Metall, pp 476–490
Kanasewich ER (1968) Precambrian rift: genesis of stratabound ore deposits. Science 161:1002–1005
Keppie JD, Boyle RW, Haynes SJ (1986) Turbidite-hosted gold deposits. Geol Assoc Can Spec Pap 32:186 pp
Kesse GO (1984) The occurrence of Gold in Ghana. In: Foster RP (ed) Gold '82. The geology, geochemistry and genesis of gold deposits. Balkema, Rotterdam, pp 645–671
Killick AM (1986) The Damba sulphide nickel deposits, Zimbabwe. In: Anhaeusser CR, Maske S (eds) Mineral deposits of southern Africa, I. Geol Soc S Afr, Johannesburg, pp 263–274
Kimberley MM (1989a) Nomenclature for iron formations. In: Wolf KH (ed) Ore geology reviews, vol 5. Elsevier, Amsterdam, pp 1–12
Kimberley MM (1989b) Exhalative origins of iron formations. In: Wolf KH (ed) Ore geology reviews, vol 5. Elsevier, Amsterdam, pp 13–145
Kinloch ED (1982) Regional trends in platinum group mineralogy of the Critical Zone of the Bushveld Complex. Econ Geol 77:1328–1347
Kyle JR (1981) Geology of the Pine Point lead-zinc district. In: Wolf KH (ed) Handbook of stratabound and stratiform ore deposits, vol 9. Elsevier, Amsterdam, pp 643–741
Ladeira EA (1988) Metalogenia dos depositos de ouro do quadrilatero Ferrifero, Minas Gerais, chap 25. In: Schobbenhaus C, Coelho CES (eds) Principais depositos minerais do Brasil, III. Dep Nacl Prod Mineral, Bras, pp 301–376

Lambert IB, Donnelly TH, Etminan H, Rowlands NJ (1984) Genesis of Late Proterozoic copper mineralization, Copper Claim, South Australia. Econ Geol 79(3):461–475

Large RR (1975) Juno Gold-Bismuth Mine, Tennant Creek. In: Knight CL (ed) Economic geology of Australia and Papua New Guinea, I. Metals. Australas Inst Min Metall, pp 424–430

Leube A, Hirdes W, Mauer R, Kesse GO (1990) The Early Proterozoic Birimian Supergroup of Ghana and some aspects of its associated gold mineralization. Precambrian Res 46:139–165

Lipman PW (1981) Volcano-tectonic setting of Tertiary ore deposits, southern Rocky Mountains. In: Dickinson WR, Payne WD (eds) Relations of tectonics to ore deposits in the southern Cordillera. Ariz Geol Soc Dig, pp 199–214

Lipman PW, Prostka HJ, Christiansen RL (1972) Cenozoic volcanism and plate tectonic evolution of the western United States: I. Early and Middle Cenozoic. Philos Trans R Soc London Ser A 271, 1213:217–248

Ljunggren P, Meyer HC (1964) The copper mineralization in the Corocoro Basin, Bolivia. Econ Geol 59(1): 110–125

Lovell H (1967) Geology and mineral deposits of the Kirkland Lake-Larder Lake mining area of northeastern Ontario. CIM Centennial field excursion guidebook, northwestern Quebec-northern Ontario, pp 72–77

Lowe DR (1980) Archean sedimentation. Annu Rev Earth Planet Sci 8:145–167

Lowell JD, Guilbert JM (1970) Lateral and vertical alteration-mineralization zoning in porphyry ore deposits. Econ Geol 65(4):373–408

Luff WM (1977) Geology of Brunswick No. 12 mine. Can Min Metall Bull 70(782):109–119

Mainwaring PR, Naldrett AJ (1977) Country-rock assimilation and genesis of Cu–Ni sulfides in the Water Hen intrusion, Duluth Complex, Minnesota. Econ Geol 72: 1269–1284

Markland GD (1966) Geology of the Moose Mountain Mine and its application to mining and milling. Can Min Metall Bull 59(646):159–170

Marsden RW (1968) Geology of the iron ores of the Lake Superior region in the United States. In: Ridge JD (ed) Ore deposits of the United States 1933–1967 (Graton-Sales volume). Am Inst Min Metall Petrol Eng, New York, pp 489–507

Marston RJ, Groves DI, Hudson DR, Ross JR (1981) Nickel sulfide deposits in Western Australia: a review. Econ Geol 76:1330–1363

Maske S, Cawthorn RG (1986) The nickel occurrence in the Insizwa Complex, Transkei. In: Anhaeusser CR, Maske S (eds) Mineral deposits of southern Africa, II. Geol Soc S Afr, Johannesburg, pp 2149–2158

Mathias BV, Clark CJ (1975) Mount Isa copper and silver-lead-zinc orebodies – Isa and Hilton Mines. In: Knight CL (ed) Economic geology of Australia and Papua New Guinea, 1. Metals. Australas Inst Min Metall, pp 351–372

McBirney AR (1989) The Skaergaard Layered Series: 1. Structure and average compositions. J Petrol 30(2): 363–396

McCartney WD, Potter RR (1962) Mineralization as related to structural deformation, igneous activity and sedimentation in folded geosynclines. Can Min J 83(4): 83–87

Meintzer RE, Wise MA, Cerny P (1984) Distribution and structural setting of fertile granites and related pegmatites in the Yellowknife pegmatite field, District of Mackenzie. Geol Surv Can Pap 84-1 A: 373–381

Miller LJ, Smith ME (1975) Sherlock Bay nickel-copper. In: Knight CL (ed) Economic geology of Australia and Papua New Guinea, I. Metals. Australas Inst Min Metall, pp 168–174

Miller WG (1913) Lake Superior silver deposits. Rep Ont Bur Mines 19 pt 2, chap 3

Minnitt RCA (1986) Porphyry copper-molybdenum mineralization at Haib River, South West Africa/Namibia. In: Anhaeusser CR, Maske S (eds) Mineral deposits of southern Africa, II. Geol Soc S Afr, Johannesburg, pp 1567–1585

Muir JE, Comba CDA (1979) The Dundonald deposit: an example of volcanic-type nickel-sulphide mineralization. Can Mineral 17 pt 2:351–359

Naldrett AJ (1981) Platinum-group element deposits. In: Cabri LJ (ed) Platinum-group elements: mineralogy, geology, recovery. Can Inst Min Metall, Ste.-Anne-de-Bellevue, Que, Can, pp 197–232

Naldrett AJ (1984) Mineralogy and composition of the Sudbury ores. In: Pye EG, Naldrett AJ, Giblin PE (eds) The geology and ore deposits of the Sudbury structure. Ont Geol Surv, Ont, Can, pp 309–325

Naldrett AJ, Macdonald AJ (1980) Tectonic settings of Ni–Cu sulfide ores: their importance in genesis and exploration. In: Strangway DW (ed) Continental crust and its mineral deposits. Geol Assoc Can Spec Pap 20:633–658

Neumann H (1944) Silver deposits at Königsberg. Norg Geol Unders 162:142 pp

Nichols RS (1988) Archean geology and silver mineralization controls at Cobalt, Ontario. Can Min Metall Bull 81(910):40–48

Padgham WA (1986) Turbidite-hosted gold-quartz veins in the Slave Structural Province, NWT. In: Keppie JD, Boyle RW, Haynes SJ (eds) Turbidite-hosted gold deposits. Geol Assoc Can Spec Pap 32:119–134

Padgham WA (1988) Discussion: of Hutchinson, R.W., 1987. Econ Geol 83(6):1287

Partington GA (1990) Environment and structural controls on the intrusion of the giant rare metal Greenbushes pegmatite, Western Australia. Econ Geol 85(3): 437–456

Peredery WV and Geological Staff (1982) Geology and nickel sulphide deposits of the Thompson belt, Manitoba. In: Hutchinson RW, Spence CD, Franklin JM (eds) Precambrian sulphide deposits (H. S. Robinson memorial volume). Geol Assoc Can Spec Pap 25: 165–209

Pretorius DA (1976) The nature of the Witwatersrand gold-uranium deposits. In: Wolf KH (ed) Handbook of stratabound and stratiform ore deposits, vol 7. Elsevier, Amsterdam, pp 29–88

Pretorius DA (1986) The Goldfields of the Witwatersrand Basin. In: Anhaeusser CR, Maske S (eds) Mineral deposits of southern Africa, II. Geol Soc S Afr, Johannesburg, pp 489–494

Pyke DR, Middleton RS (1971) Distribution and characteristics of the sulphide ores of the Timmins area. Can Min Metall Bull 64(710):55–66

Rackley RI (1976) Origin of western-states type uranium mineralization. In: Wolf KH (ed) Handbook of strata-

bound and stratiform ore deposits. vol 7. Elsevier. Amsterdam. pp 89–152
Reimer TO (1984) Alternative model for the derivation of gold in the Witwatersrand Supergroup. J Geol Soc London 141:263–272
Reimer TO (1985) Volcanic rocks and weathering in the Early Proterozoic Witwatersrand Supergroup. South Africa. Bull Geol Surv Finl 331:33–49
Reimer TO. Mossman DJ (1990) Sulfidization of Witwatersrand black sands: from enigma to myth. Geology 18(5):426–429
Robb LJ. Meyer FM (1990) The nature of the Witwatersrand hinterland: conjectures on the source area problem. Econ Geol 85(3):511–536
Robb LJ. Robb VM (1986) Archean pegmatite deposits in the north-eastern Transvaal. In: Anhaeusser CR. Maske S (eds) Mineral deposits of southern Africa. I. Geol Soc S Afr. Johannesburg. pp 437–450
Roberts JB (1975) Windara nickel deposits. In: Knight CL (ed) Economic geology of Australia and Papua New Guinea. 1. Metals. Australas Inst Min Metall. pp 129–143
Robertson JA (1976) The Blind River uranium deposits: the ores and their setting. Misc Pap Ont Minist Nat Res 65:45 pp
Robinson DJ. Hutchinson RW (1982) Evidence for a volcanogenic-exhalative origin of a massive nickel sulphide deposit at Redstone. Timmins. Ontario. In: Hutchinson RW. Spence CD. Franklin JM (eds) Precambrian sulphide deposits (H. S. Robinson memorial volume). Geol Assoc Can Spec Pap 25:211–254
Rowntree JC. Mosher DV (1975) Jabiluka uranium deposits. In: Knight CL (ed) Economic geology of Australia and Papua New Guinea. 1. Metals. Australas Inst Min Metall. pp 321–326
Ruelle JCL (1982) Depositional environments and genesis of stratiform copper deposits of the Redstone Copper Belt. McKenzie Mountains. NWT. In: Hutchinson RW. Spence CD. Franklin JM (eds) Precambrian sulphide deposits (H. S. Robinson memorial volume). Geol Assoc Can Spec Pap 25:701–737
Rye DM. Rye RO (1974) Homestake gold mine. South Dakota. I. Stable isotope studies. Econ Geol 69(3): 293–317
Rye DM. Doe BR. Delevaux MH (1974) The Homestake gold mine. South Dakota. II. Lead isotopes. mineralization. and source of lead in ores of the northern Black Hills. Econ Geol 69:814–822
Sangster DF (1972) Isotopic studies of ore-leads in the Hanson Lake-Flin Flon-Snow Lake mineral belt. Can J Earth Sci 9(5):500–513
Sangster DF (1976) Carbonate-hosted lead-zinc deposits. In: Wolf KH (ed) Handbook of stratabound and stratiform ore deposits. vol 6. Elsevier. Amsterdam. pp 448–456
Sangster DF. Scott SD (1976) Precambrian stratabound. massive Cu–Zn–Pb sulfide ores of North America. In: Wolf KH (ed) Handbook of stratabound and stratiform ore deposits. vol 6. Elsevier. Amsterdam. pp 129–222
Schenk PE (1969) Carbonate-sulfate-redbed facies and cyclic sedimentation of the Windsorian stage (Middle Carboniferous) Maritime Provinces. Can J Earth Sci 6:1037–1066
Schidlowski M (1976) Archean atmosphere and evolution of terrestrial oxygen budget. In: Windley BF (ed) The early history of the Earth. John Wiley & Sons. New York. pp 525–535
Schidlowski M (1990) Life on the early Earth: bridgehead from cosmos or autochthonous phenomenon. In: Gopalan K. Gaur VK. Somayajulu BLK. Macdougall JD (eds) Mantle to meteorites. Indian Acad Sci. Bangalore. pp 189–199
Sestini G (1973) Sedimentology of a paleoplacer: the gold-bearing Tarkwaian of Ghana. In: Amstutz GC. Bernard AJ (eds) Ores in sediments. Int Un Geol Sci Ser A3: 275–305
Sillitoe RH (1973) The tops and bottoms of porphyry copper deposits. Econ Geol 68:799–815
Simpson TA. Gray TR (1968) The Birmingham red-ore district. Alabama. In: Ridge JD (ed) Ore deposits of the United States 1933–1967 (Graton-Sales volume). Am Inst Min Metall Petrol Eng. New York. pp 187–206
Siroonian HA. Shaw DM. Jones RE (1959) Lithium geochemistry and the source of the spodumene pegmatites of the Preissac-La Motte-La Corne region of western Quebec. Can Mineral 6:320–338
Stanton RL (1989) The precursor principle and the possible significance of stratiform ores and related chemical sediments in the elucidation of processes of regional metamorphic mineral formation. Philos Trans R Soc London Ser A328:529–646
Stanton RL. Vaughan JP (1979) Facies of ore formation: a preliminary account of the Pegmont deposit as an example of potential relations between small "iron-formations" and stratiform sulfide ores. Proc Aust Inst Min Metall 270:25–38
Sutcliffe RH. Sweeny JM. Edgar AD (1989) The Lac des Isles Complex. Ontario: petrology and platinum-group elements mineralization in an Archean mafic intrusion. Can J Earth Sci 26(7):1408–1427
Swager CP (1985) Syndeformational carbonate-replacement model for the copper mineralization at Mt. Isa. northwest Queensland. a microstructural study. Econ Geol 80(1):107–125
Tarasov AV (1970) Structural control of copper and nickel mineralization in Noril'sk I deposit. Int Geol Rev 12: 933–941
Teixeira G. Gonzalez M (1988) Minas do Camaqua. Municipio de Cacapava do Sul. RS. Chapter III. In: Schobbenhaus C. Coelho CES (eds) Principais depositos minerais do Brasil. vol 3. Dep Nac Produaco Mineral. Bras. pp 33–41
Titley SR. Hicks CL. (eds) (1966) Geology of the porphyry copper deposits. southwestern North America. Univ Arizona Press. Tucson. 287 pp
Tolbert GE. Tremaine JW. Melcher GC. Gomez CB (1971) The recently discovered Serra dos Carajas iron deposits. northern Brazil. Econ Geol 66:985–994
Trendall AF (1973) Precambrian iron-formation of Australia. Econ Geol 68:1023–1034
Unrug R (1988) Mineralization controls and source of metals in the Lufilian Fold Belt. Shaba (Zaire). Zambia. and Angola. Econ Geol 83:1247–1258
Van Deventer JL. Eriksson PG. Snyman CP (1986) The Thabazimbi iron ore deposits. north-western Transvaal. In: Anhaeusser CR. Maske S (eds) Mineral deposits of southern Africa. I. Geol Soc S Afr. Johannesburg. pp 923–930
Van Schalkwyk JF. Beukes NJ (1986) The Sishen iron ore deposits. Griqualand West. In: Anhaeusser CR. Maske

S (eds) Mineral deposits of southern Africa, I. Geol Soc S Afr, Johannesburg, pp 931–956
Vaughan DJ, Sweeney M, Friedrich G, Diedel R, Haranczyk C (1989) The Kupferschiefer: an overview with an appraisal of the different types of mineralization. Econ Geol 84(5):1003–1027
Vermaak CF, Von Gruenewaldt G (1986) Introduction to the Bushveld Complex. In: Anhaeusser CR, Maske S (eds) Mineral deposits of southern Africa, II. Geol Soc S Afr, Johannesburg, pp 1021–1029
Verwoerd WJ (1986) Mineral deposits associated with carbonatites and alkaline rocks. In: Anhaeusser CR, Maske S (eds) Mineral deposits of southern Africa, II. Geol Soc S Afr, Johannesburg, pp 2173–2191
Viljoen MJ, Hieber R (1986) The Rustenburg Section of Rustenburg Platinum Mines Limited, with reference to the Merensky Reef. In: Anhaeusser CR, Maske S (eds) Mineral deposits of southern Africa, II. Geol Soc S Afr, Johannesburg, pp 1107–1134
Viljoen MJ, Bernasconi A, van Coller N, Kinloch E, Viljoen RJ (1976) The geology of the Shangani deposit, Rhodesia. Econ Geol 71:76–95
Wallace SR, Muncaster NK, Jonson DC, MacKenzie WB, Bookstrom AA, Surface VE (1968) Multiple intrusion and mineralization at Climax, Colorado. In: Ridge JD (ed) Ore deposits of the United States 1933–1967 (Graton-Sales volume). Am Inst Min Metall Petrol Eng, pp 605–640
Wallace SR, MacKenzie WB, Blair RG, Muncaster NK (1978) Geology of the Urad and Henderson molybdenum deposits, Clear Creek County, Colorado, with a section on a comparison of these deposits with those at Climax, Colorado. Econ Geol 73(3):325–368
Williams DAC (1979) The association of some nickel sulfide deposits with komatiitic volcanism in Rhodesia. Can Mineral 17(2):337–349
Wilson AH, Naldrett AJ, Tredoux M (1989) Distribution and controls of platinum group element and base metal mineralization in the Darwendale subchamber of the Great Dyke, Zimbabwe. Geology 17(7):649–652
Woodall R (1975) Gold in the Precambrian of Western Australia. In: Knight CL (ed) Economic geology of Australia and Papua New Guinea, I. Metals. Australas Inst Min Metall, pp 175–184
Worst BG (1960) The Great Dyke of Southern Rhodesia. S Rhod Geol Surv Bull 47:122–125
Wu Y, Beales F (1981) A reconnaissance study by paleomagnetic measurements of the age of mineralization along the Viburnum trend, southeast Missouri. Econ Geol 76(7):1879–1894
Zuffardi P (1989) Italy. In: Dunning FW, Garrard P, Haslam HW, Ixer RA (eds) Mineral deposits of Europe, vols 4/5. Inst Min Metall Min Soc, Oxford, pp 247–252

Chemical Evolution of Sedimentary and Magmatic Rocks in the Earth's Crust

Alexander B. Ronov [1]

[1] Vernadsky Institute of Geochemistry and Analytical Chemistry, USSR Academy of Sciences, 117975 Moscow V-334, USSR

Abstract

Geochemical analysis of tens of thousands of the three most abundant rock types of the Earth's continental crust (basalts, shales and granitoids), taking into account their volume or areal abundance, indicates that throughout geologic time regular and similar changes occurred in the average chemical composition of these genetically different rock types. With decreasing geologic age for each rock type, Mg, Ni, Co and Cr contents decrease, whereas the contents of K, Rb, REE and various other lithophile elements increase. Compositional changes of these crustal rocks through time result from differentiation of deeply derived material and the products of its transformations at the Earth's surface. This evolution of rock chemistry is compatible with independently assessed changes in the Earth's thermal regime, and with the progressively increased thickness and mass of the continental crust throughout geologic history.

1 Introduction

A systematic study of the abundances, compositions and absolute ages of Phanerozoic and Precambrian sedimentary and magmatic rocks has been carried out. Samples include many thousands of chemical analyses of rocks from various lithologies and geologic ages from type sections of shields, platforms and geosynclines in the Soviet Union, together with tens of thousands from other parts of the world, adopted from the literature. Documented are analyses of over 16000 well-dated specimens of clays and shales, 15000 granitoids and 13000 basalts, the most abundant rocks among sedimentary, intrusive and volcanic formations, respectively, and altogether making up 95% of the Earth's crust. The profusion of data thus available facilitates reconstruction, to a first approximation, of the most common changes through time in the chemistry of the Earth's crust.

2 Procedure

All chemical analyses were performed in the Laboratory of Geochemistry of Sedimentary Rocks at the Vernadsky Institute in Moscow. The validity of the results was confirmed by progressive pooling of the initial data base. Firstly, the influence of provincial compositional variations in rock chemistry was excluded by treating the data on a global basis. Secondly, rock compositions were compared for similar stratigraphic intervals, representing major stages of crustal development. According to Veizer (1984), the duration of these stages must exceed the average life span of sedimentary rocks. Here, the following stages are assumed: Early Archaean (> 3.4 Ga), Middle and Late Archaean (2.5–3.4 Ga), Early Proterozoic (1.6–2.5 Ga), Middle and Late Proterozoic ($\sim$ 0.6–1.6 Ga), and Phanerozoic (Recent to $\sim$ 0.6 Ga). Thirdly, in calculating the average composition of each rock type, their volumetric or areal abundances were taken into account. Lastly, in comparing the chemical compositions of different rock types, the parameters normally used to study magmatic series were employed.

3 Results

The general trends in the chemical evolution of crustal rocks are revealed in an AFM diagram (Fig. 1) where compositional variations through time are indicated for each rock type. With decreasing age for each rock type, the content of alkalis increases, and magnesium decreases. An initial trend towards Fe-enrichment is reversed during the Early Proterozoic (PR_1) stage. Overall, the trend is similar to that of magmatic differentiation.

M. Schidlowski et al. (Eds.)
Early Organic Evolution: Implications for Mineral and Energy Resources

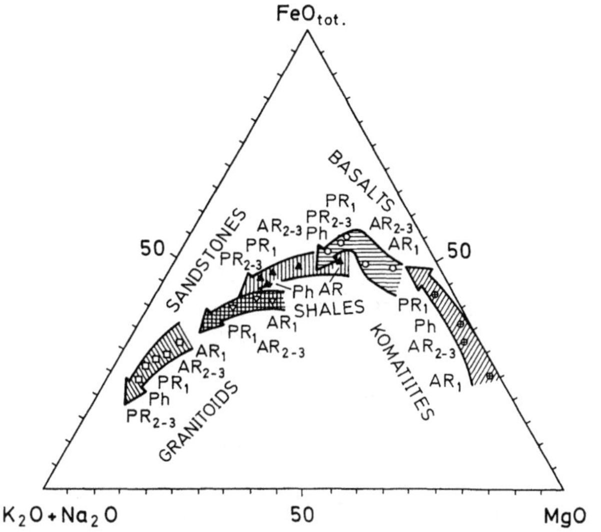

Fig. 1. AFM diagram showing Early Archaean to Phanerozoic trends for the average composition of komatiites, basalts, shales, sandstones and granitoids. AR_1 Early Archaean (> 3.4 Ga); AR_{2-3} Middle and Late Archaean (2.5–3.4 Ga); PR_1 Early Proterozoic (1.6–2.5 Ga); PR_{2-3} Middle and Late Proterozoic (~ 0.6–1.6 Ga); *Ph* Phanerozoic (Recent to ~ 0.6 Ga). *Arrows* show direction of evolution

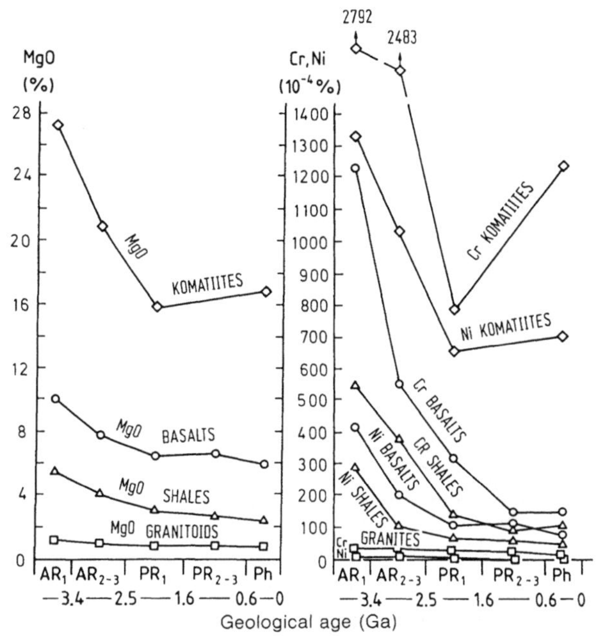

Fig. 2. Secular evolution in the average contents of Mg, Ni and Cr in komatiites, basalts, shales and granitoids. For explanation of time segments, see Fig. 1

Figure 2 illustrates the general depletion of the Mg, Ni and Cr contents of all magmatic rocks with decreasing geologic age. In an intermediate position between basalts and granitoids the trend is faithfully repeated by shales. Conversely, trends towards an increase, with decreasing age, are evident (Fig. 3) for typical lithophile elements (with the exception of sodium) and (Fig. 4) for total rare earth elements (*REE*) as well as for Th, Hf, Nb and Pb (not shown).

Variations through time of the average Fe, V, Ti, Ca and Ba contents of mafic and felsic magmatic rocks have diverse tendencies. Consequently, concentrations of these elements in derivative shales are determined by the proportions of parent rocks (basalt, granite) in the source area. Figure 5 illustrates a trend towards enrichment of shales in barium, and depletion in iron and vanadium, with a decrease in geologic age. This probably results from the relative increase in the area of granite as compared to that of basalt from Archaean through to the present. The trend corresponds well with an earlier scheme advanced by Ronov (1964) showing changes in the ratio of important groups of rocks in continental regions (Fig. 6), and which in general has been substantiated by the results of recent Nd isotopic studies (Allègre and Rousseau 1984; Taylor and McLennan 1985; Ronov et al. 1988).

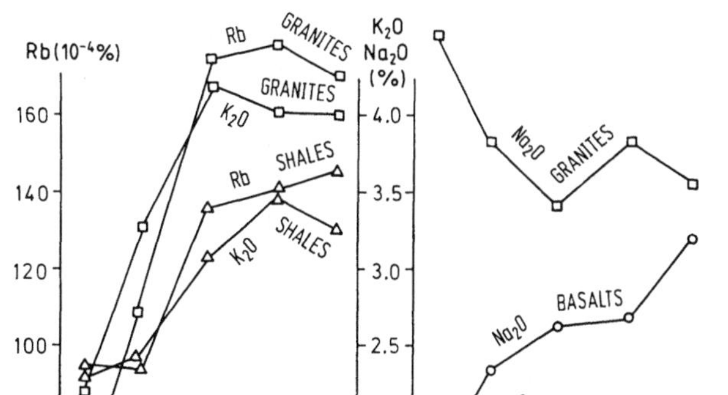

Fig. 3. Secular evolution in the average contents of Rb, K_2O and Na_2O in basalts, shales and granitoids. For explanation of time segments, see Fig. 1

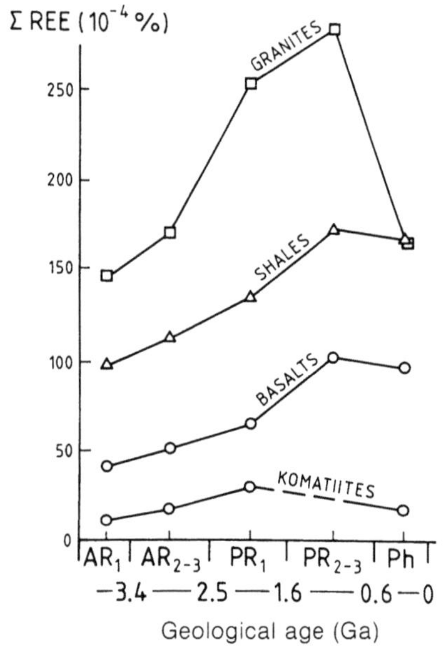

Fig. 4. Secular evolution of rare earth element (*REE*) contents in komatiites, basalts, shales and granitoids. Notation of time segments as in Fig. 1

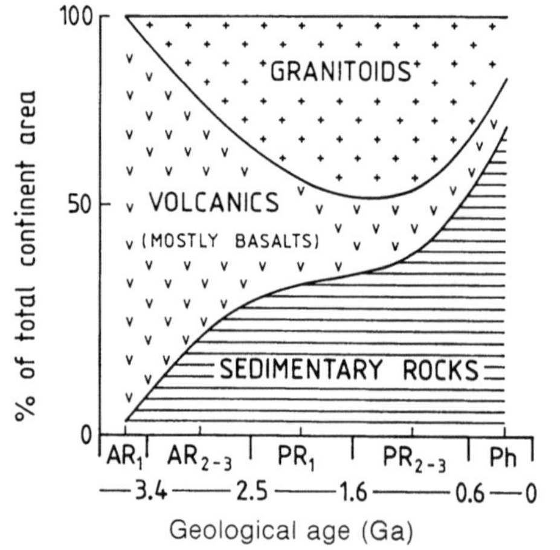

Fig. 6. Secular changes in the relative proportions of the principal groups of rocks that were exposed in areas of continental erosion in the course of geologic time. Time segments as in Fig. 1

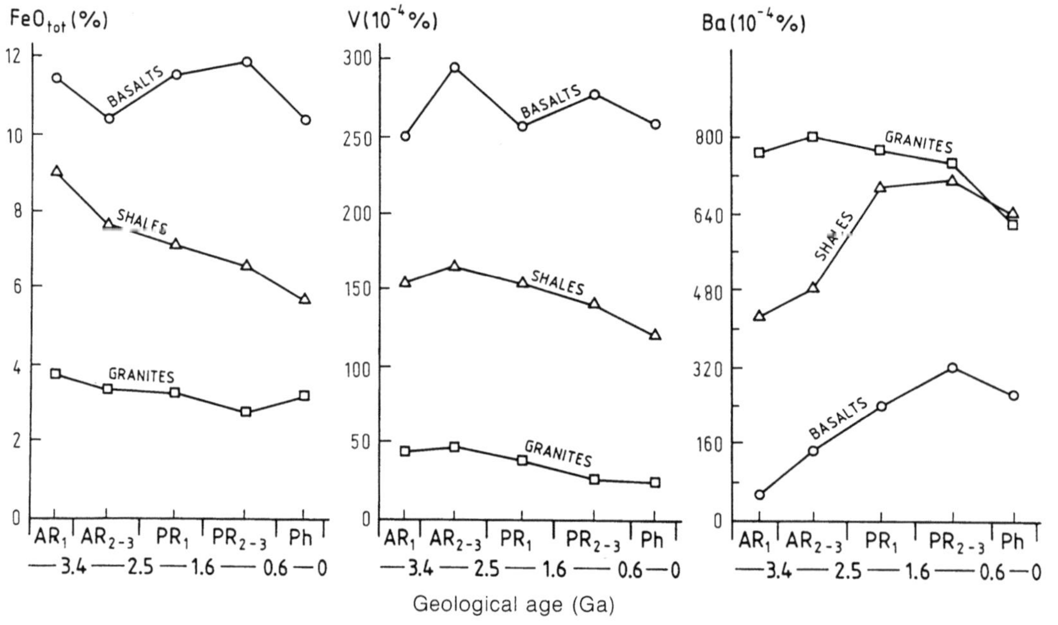

Fig. 5. Secular evolution of the average content of total iron (FeO_{tot}), V, and Ba in basalts, shales and granitoids. For explanation of time segments, see Fig. 1

Lastly, it seems logical that, as a first approximation, changes in the average chemical compositions of magmatic and sedimentary rocks should correlate not only with each other but also with changes in thermal regime and with the growth in mass and thickness of the continental crust throughout the Earth's history. Figure 7 highlights some of these features together with the growth curve of $^{87}Sr/^{86}Sr$ in the ancient ocean, documented by Veizer and Compston (1976) from the investigation of marine carbonates.

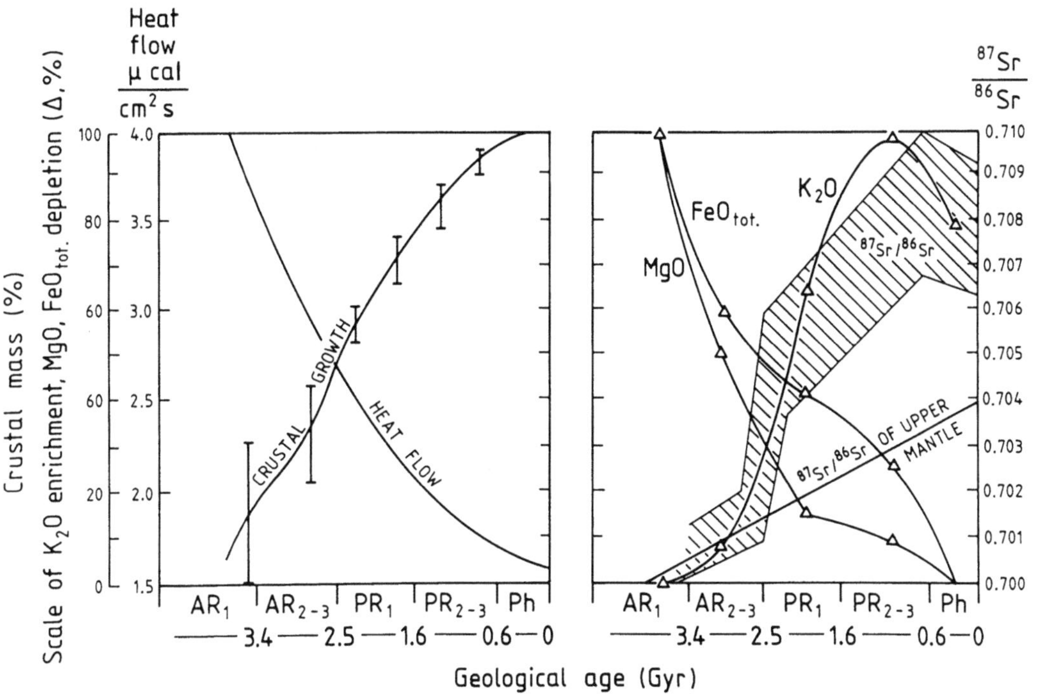

Fig. 7. Secular patterns in mantle thermal flow and in the growth of the mass of continental crust (after Allègre and Rousseau 1984) as well as in the MgO, FeO_{tot} and K_2O contents of shales and the strontium isotopic composition of carbonates (Veizer and Compston 1976). For explanation of time segments, see Fig. 1

3 Conclusions

Concerning the Earth's crustal and exogenic rocks, the following conclusions can be drawn:

1. In shales, basalts and granitoids there are distinct and common changes in chemical composition throughout geologic time.
2. The observed age trend in chemical composition reflects the increasing differentiation of Earth's crust through time.
3. The observed changes in rock composition are, as a whole, compatible with independently determined changes in the thermal regime throughout the Earth's history, and with the increased mass and thickness of the continental crust.

Acknowledgements. Many colleagues have participated in the above study, but among them special acknowledgement is due to A. A. Migdisov and N. V. Bredanova. I am, moreover, grateful to Drs. D. A. Mossman, J. Veizer and M. Schidlowski for their help in transforming the text of my oral presentation into a brief paper. Due to the free winds of "perestroika", I had the fantastic opportunity to present the above results at Maria Laach, FRG, after having been hindered all my life by the Iron Curtain.

References

Allègre CJ, Rousseau D (1984) The growth of the continents through geological time studied by Nd isotopic analysis of shales. Earth Planet Sci Lett 67:19–34

Ronov AB (1964) Common tendencies in the chemical evolution of Earth's crust, ocean and atmosphere. Geochem Int 8:715–743

Ronov AB, Bredanova NV, Migdisov AA (1988) General trends in evolution of chemical composition for sedimentary and magmatic rocks of continental crust. Geokhimiya 2:180–199 (in Russian)

Taylor SR, McLennan SM (1985) The continental crust: its composition and evolution. Blackwell, Oxford, 312 pp

Veizer J (1984) Recycling on the evolving Earth: geochemical record in sediments. In: Proc 27th Int Geol Congr, Moscow, vol 11. Utrecht, pp 325–345

Veizer J, Compston W (1976) $^{87}Sr/^{86}Sr$ in Precambrian carbonates as an index of crustal evolution. Geochim Cosmochim Acta 40:905–914

Subject Index

MIX
Papier aus verantwortungsvollen Quellen
Paper from responsible sources
FSC® C105338

If you have any concerns about our products,
you can contact us on
ProductSafety@springernature.com

In case Publisher is established outside the EU,
the EU authorized representative is:
Springer Nature Customer Service Center GmbH
Europaplatz 3, 69115 Heidelberg, Germany

Printed by Libri Plureos GmbH
in Hamburg, Germany